Neuroradiology

The Requisites Series

SERIES EDITOR

James H. Thrall, MD

Radiologist-in-Chief Emeritus
Massachusetts General Hospital
Distinguished Juan M. Taveras Professor of Radiology
Harvard Medical School
Boston, Massachusetts

THE REQUISITES

Neuroradiology

FOURTH EDITION

Rohini Nadgir, MD
Assistant Professor of Radiology and Radiological Science
Johns Hopkins School of Medicine
Faculty, Division of Neuroradiology
Johns Hopkins Medical Institution
Baltimore, Maryland

David M. Yousem, MD, MBA
Associate Dean for Professional Development
Johns Hopkins School of Medicine
Director of Neuroradiology, Vice Chairman of Program Development
Johns Hopkins Medical Institution
Baltimore, Maryland

ELSEVIER

ELSEVIER

1600 John F. Kennedy Blvd.
Ste 1800
Philadelphia, PA 19103-2899

NEURORADIOLOGY: THE REQUISITES, FOURTH EDITION ISBN: 978-1-4557-7568-2

Library of Congress Cataloging-in-Publication Data

Names: Yousem, David M., author. | Nadgir, Rohini, author.
Title: Neuroradiology : the requisites / Rohini Nadgir, David M. Yousem.
Other titles: Requisites series.
Description: Fourth edition. | Philadelphia, PA : Elsevier, [2017] |
Series:
  The requisites series | Preceded by: Neuroradiology / David M. Yousem,
  Robert I. Grossman. 3rd ed. c2010. | Yousem's name appeared first on the
  previous edition. | Includes bibliographical references and index.
Identifiers: LCCN 2016013344 | ISBN 9781455775682 (hardcover : alk. paper)
Subjects: | MESH: Neuroradiography–methods | Central Nervous System
  Diseases–radiography
Classification: LCC RC349.R3 | NLM WL 141.5.N47 | DDC 616.8/04757–dc23 LC record
available at http://lccn.loc.gov/2016013344

Executive Content Strategist: Robin Carter
Content Development Specialist: Stacy Eastman
Publishing Services Manager: Catherine Jackson
Design Direction: Amy Buxton

Working together
to grow libraries in
developing countries

www.elsevier.com • www.bookaid.org

Printed in China

Last digit is the print number: 9 8 7 6 5 4 3 2 1

To my husband, Kevin, for always helping me see the forest through the trees and my children, Soniya and Dhilan, for showing me that once you become a parent, everything else (outside of parenthood!) becomes much easier.

To my mother, late father, and brothers, for their unending love and support.

To my mentors and trainees past, present, and future, for always keeping me on my toes.

Rohini Nadgir

To my wife Kelly, and my family, friends, and colleagues who sustain me.

To all of the "muses" that have passed through my life.

To Bob Grossman, the best mentor and co-author of all time.

To every person who has walked up to me and said, "Your book(s) helped me get through (a) my neuroradiology rotation, (b) my general radiology boards, (c) my subspecialty certification test, or (d) my MOC recertification test." When I hear that, it makes me contemplate the next edition(s). Thank you for those words of inspiration.

David M. Yousem

Contributors

Rohini Nadgir, MD
Assistant Professor of Radiology and Radiological Science
Johns Hopkins School of Medicine
Faculty, Division of Neuroradiology
Johns Hopkins Medical Institution
Baltimore, Maryland

Aylin Tekes-Brady, MD
Associate Professor of Radiology
Deputy Director, Division of Pediatric Radiology and
    Pediatric Neuroradiology
Charlotte R. Bloomberg Children's Center
Johns Hopkins Hospital
Baltimore, Maryland

David M. Yousem, MD, MBA
Associate Dean for Professional Development
Johns Hopkins School of Medicine
Director of Neuroradiology, Vice Chairman of Program
    Development
Johns Hopkins Medical Institution
Baltimore, Maryland

Foreword

Since the publication of the first edition of *Neuroradiology: The Requisites* in the early 1990s, this book has been one of the most widely read on the subject. Now in its fourth edition, *Neuroradiology: The Requisites* again encompasses its enormous topic in a way that is efficient for the reader while still providing the breadth and depth of material necessary for clinical practice and to meet the expectations for certification by the American Board of Radiology. Drs. Nadgir and Yousem are to be congratulated for this outstanding new addition to the *Requisites in Radiology* series.

After a discussion of cranial anatomy, the rest of *Neuroradiology: The Requisites* is organized by diseases and locations. This organization follows the pattern of the prior editions and based on experience with these prior editions should allow the reader to hone in on a topic of interest very quickly. In addition to the many excellent images used to illustrate key findings, another important strength of this book is the liberal use of drawings, tables, and boxes. Anatomical drawings help orient the reader to radiological features. Tables are used to economically summarize differential diagnoses among other topics, and boxes are used to list and reinforce important characteristics of diseases and conditions.

A challenge held in common in producing each new edition of *Neuroradiology: The Requisites* is the need to address the many new developments that have occurred since publication of the previous edition. These developments include both advances in technology and advances in our understanding of disease. Drs. Nadgir and Yousem have done a masterful job in accomplishing this with many new illustrations and systematically updated material in the text. Methods that were just becoming clinically feasible at the time of the last edition have found important roles in practice. Higher field magnetic resonance imaging has continued to be adopted more widely, and multidetector computed tomography is almost universally available in the United States. To pick just one example, these technologies have helped refine our approaches to the diagnosis and management of stroke.

It is our hope that anyone in radiology, neurology, or neurosurgery with an interest in imaging of the brain or the head and neck will find *Neuroradiology: The Requisites* useful. For residents in radiology, this book will make a challenging subject manageable. In a 1-month rotation, it should be possible to read the entire book. If this is done systematically during each subsequent rotation, the material will be well in hand when the time comes to take the board exams. Fellows in neuroradiology and radiology practitioners will find *Neuroradiology: The Requisites* a useful source for reference and review. Likewise, neurologists and neurosurgeons should find this book useful in better understanding the imaging studies obtained on their patients and the results of those studies.

The original hypothesis behind the *Requisites in Radiology* series was that many textbooks, in trying to be comprehensive, actually make it difficult for the reader to parse out what is truly important to daily practice. One of the guiding principles for the series' authors, captured well by Drs. Nadgir and Yousem, is to only put in the book what you teach to your own residents and fellows. There should be no need to put in obscure things that even the author needs to look up from another source. In this regard, *The Requisites* are not intended to be exhaustive but to provide basic conceptual, factual, and interpretative material required for clinical practice. By eliminating extraneous material using this approach, the reader can focus on what is actually most important in the practice of neuroradiology. Another pitfall in textbooks undergoing revision is to simply "graft on" material without pruning out-of-date material sufficiently. Here again Drs. Nadgir and Yousem have risen to the challenge and done a great job.

I have every confidence that the fourth edition of *Neuroradiology: The Requisites* will join the first three editions as well received and widely read books. I again congratulate Drs. Nadgir and Yousem on their outstanding new contribution. Their book reflects not only their expertise but their willingness to undertake the time and effort to share that expertise with students and seasoned practitioners alike.

James H. Thrall, MD
Radiologist-in-Chief Emeritus
Massachusetts General Hospital
Distinguished Juan M. Taveras Professor of Radiology
Harvard Medical School
Boston, Massachusetts

Preface

When David Yousem initially invited me to work with him to put together the fourth edition of *Neuroradiology: The Requisites,* I was glad the conversation was over the phone so that he couldn't see my bug-eyed expression. After all, this was the book that had guided me throughout my training and had shaped and defined my career thus far. The shoes of the renowned Drs. Yousem and Grossman were enormous to fill, to say the least, and the task of fitting all that's essential to neuroradiology into one place would certainly not be easy. Nevertheless, as a person who derives much of her job satisfaction in molding successful trainees, I was excited about the prospect of making a greater impact on a larger audience.

Fast forward a couple of years, and I'm proud to say this text is just what the doctor (resident-in-training, that is) ordered. We've refreshed and trimmed it down to a digestible 17 chapters of what's most relevant to the neuroimaging trainee and the daily practitioner while preserving the previous editions' buoyant approach to "learning should be fun."

I am so fortunate to be immersed in a circle of inspiration and support from my colleagues, family, and friends in the making of this book. I owe thanks to Dr. Laurie Loevner for years ago taking me under her wing and introducing to me the beauty (and complexity) of neuroimaging. Drs. Osamu Sakai and Glenn Barest have been my personal mentors and close friends since my days in training, exemplifying the best in compassionate patient care and intellectual pursuits. No less importantly, on the home front I would be remiss not to thank my family for their unconditional love, in particular my mother, the kindest person I know, and my brothers, for somehow finding humor in everything. I am grateful to my wonderful husband, Kevin, and spirited children, Soniya and Dhilan, who always help keep things in perspective. Certainly this book would not have been possible without my right-hand woman and nanny to my children, Kayla, who has always been there for me, especially when the going got tough.

Many thanks to Aylin Tekes-Brady for taking on the daunting task of providing a thorough yet concise review of congenital disorders of the brain and spine. And last but certainly not least, I'd like to express my gratitude to Dave for this tremendous opportunity and for having faith in me to step it up and make a difference.

Rohini Nadgir, MD

Neuroradiology: The Requisites was first published in 1994 and was 544 pages in length with 833 illustrations. Bob Grossman was the "brains" behind this neuroradiology textbook and my mentor, guru, division chief, and close friend. Bob insisted that the book have a "style"; it would be the "story" of neuroradiology that a resident could read cover to cover. Its "plotline" would mirror Ben Felson's *Principles of Chest Roentgenology* (the source of the original "Aunt Minnie") with its educational sense of humor.

Over the course of the next two editions, Bob would move quickly to Chairman of Radiology and then Dean of the NYU Medical School. *Neuroradiology: The Requisites* would mature into a larger and larger book, cater to neuroradiology fellows and then academic faculty, and become much more serious. I remained behind, carrying the torch, trying to be a teacher extraordinaire.

With the fourth edition, Bob Grossman no longer graces the title page, but his impact and literally his words (and even some of his old jokes) still populate the tome. I can still hear his cackle from the origin of the longstanding jokes. But, in an effort to return to its roots, we have endeavored to recapture the "essentials only" nature of the first edition.

My colleague and the first author of the fourth edition, Rohini Nadgir, is my former fellow and now a faculty member with me at Johns Hopkins. As Bob did with me, I have ceded the reins of this child of ours to a delightful and outstanding educator. She has delicately massaged the old messages and modernized the text. She has guided the book expertly to this, its 22nd year of adulthood. I look forward to becoming the "silent partner" with Ro, as Bob did with me on the third edition. Thank you to Bob and Rohini and also to Aylin Tekes-Brady, another bright star in the pediatric neuroradiology sphere who gave the congenital lesion chapter its makeover.

We think we hit that sweet spot of enough material to cover the topic thoroughly without overwhelming the reader. Our goal, as it was year one, was to write a book that residents could read in 2 to 4 weeks during their first or second neuroradiology rotation. At the same time, it would be a refresher for all radiologists who read these cases as part of their daily practice.

Thanks to Robin Carter, Rhoda Howell, and Amy Meros on the Elsevier team who assisted in every way and encouraged, rather than pushed.

Enjoy. Live, love, learn, and leave a legacy.

David M. Yousem, MD, MBA

Contents

Chapter 1

Cranial Anatomy

The anatomy and function of the brain is fascinating and complex, and we are still only scratching the surface in terms of our understanding of these structures. Nevertheless, a basic understanding of structure and function is critical in providing meaningful and accurate reporting of the pathology in the brain. Although we will discuss the development of the brain in Chapter 8, we will address the pertinent aspects of normal adult anatomy as it pertains to imaging interpretation in this chapter. Ready? Set? Here we go!

TOPOGRAPHIC ANATOMY

Cerebral Hemispheres

There are four lobes in each cerebral hemisphere: the frontal, parietal, occipital, and temporal lobes. The frontal lobe is separated from the parietal lobe by the central (Rolandic) sulcus, the parietal lobe is separated from the

occipital lobe by the parietooccipital sulcus, and the temporal lobe is separated from the frontal and parietal lobes by the sylvian (lateral) fissure (Fig. 1-1).

The main named areas of the frontal lobe are the precentral gyrus (the primary motor strip of the cerebral cortex) and the three frontal gyri anterior to the motor strip: the superior, middle, and inferior frontal gyri. In front of the motor cortex is, quite naturally, the premotor cortex (Brodmann area 6). The dorsolateral prefrontal cortex (DLPFC or DL-PFC) is a critical area frequently referred to by the functional magnetic resonance imaging (fMRI) gurus who ascribe a great deal of cognition/memory/planning to it. It lies in the middle frontal gyrus of humans. It includes Brodmann area 9 and 46 and lies anterolateral to the premotor cortex. The frontal operculum (superior to the sylvian fissure and in the frontal lobe) contains portions of the Broca motor speech area. On the medial surface of the frontal lobe is the cingulate gyrus just superior to and bounding the corpus callosum, and the gyrus rectus

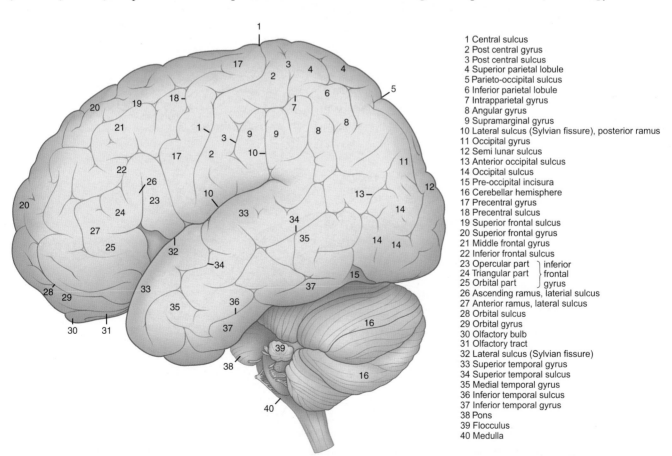

1	Central sulcus
2	Post central gyrus
3	Post central sulcus
4	Superior parietal lobule
5	Parieto-occipital sulcus
6	Inferior parietal lobule
7	Intraparietal gyrus
8	Angular gyrus
9	Supramarginal gyrus
10	Lateral sulcus (Sylvian fissure), posterior ramus
11	Occipital gyrus
12	Semi lunar sulcus
13	Anterior occipital sulcus
14	Occipital sulcus
15	Pre-occipital incisura
16	Cerebellar hemisphere
17	Precentral gyrus
18	Precentral sulcus
19	Superior frontal sulcus
20	Superior frontal gyrus
21	Middle frontal gyrus
22	Inferior frontal sulcus
23	Opercular part ⎤ inferior
24	Triangular part ⎬ frontal
25	Orbital part ⎦ gyrus
26	Ascending ramus, lateral sulcus
27	Anterior ramus, lateral sulcus
28	Orbital sulcus
29	Orbital gyrus
30	Olfactory bulb
31	Olfactory tract
32	Lateral sulcus (Sylvian fissure)
33	Superior temporal gyrus
34	Superior temporal sulcus
35	Medial temporal gyrus
36	Inferior temporal sulcus
37	Inferior temporal gyrus
38	Pons
39	Flocculus
40	Medulla

FIGURE 1-1 Surface anatomy of the brain from a lateral view. Gyri are labeled in this figure. (From Nieuwenhuys R, Voogd J, van Huijen C. *The Human Central Nervous System: A Synopsis and Atlas.* Rev 1st ed. Berlin: Springer-Verlag; 1988.)

extending along the medial basal surface of the anterior cranial fossa (Figs. 1-1, 1-2).

The parietal lobe contains the postcentral gyrus (the center for somatic sensation), the supramarginal gyrus just above the temporal lobe, and the angular gyrus near the apex of the temporal lobe. Two superficial gyri of note are the superior and inferior parietal lobules, which are separated by an interparietal sulcus. On its medial side the precuneate gyrus is present in front of the parietooccipital fissure, with the cuneate gyrus posteriorly in the occipital lobe (see Fig. 1-2).

The temporal lobe contains the brain-functioning elements of speech, memory, emotion and hearing. Superior (auditory), medial, and inferior temporal gyri are seen on the superficial aspect of the brain (see Fig. 1-1). The posterior portion of the superior temporal gyrus subserves language comprehension, the so-called Wernicke area. Deep to the sylvian fissure is the insula, or isle of Reil, which is bounded laterally by the opercular regions and subserves taste function. The inferior part of the insula near the sylvian fissure is called the limen of the insula. The inferior and medial surface of the temporal lobe reveals the parahippocampal gyrus with the hippocampus just superior to it (Fig. 1-3). Anteriorly, the almond-shaped amygdala dominates. In the coronal plane, starting at the right collateral sulcus just inferior to the parahippocampus and traveling northward, you would first hit the entorhinal cortex, then turn at the parasubiculum, pass along the subiculum

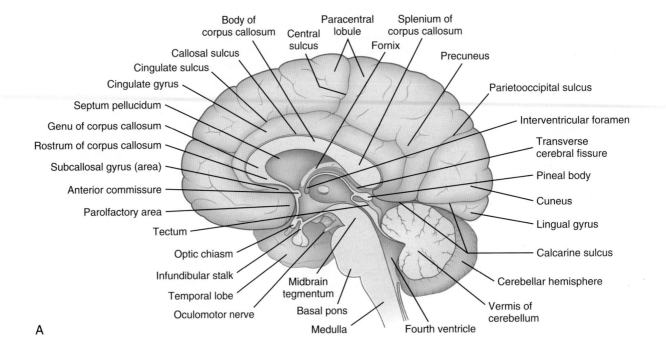

A

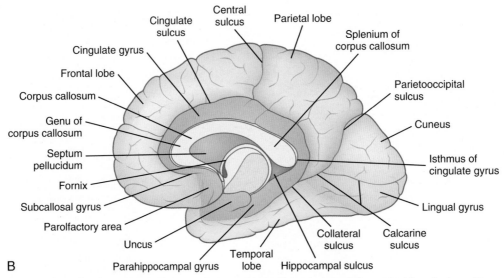

B

FIGURE 1-2 A, Midsagittal view of the brain. **B,** Midsagittal view of the left cerebral hemisphere illustrating the major cortical lobes. Frontal lobe *(blue)*, parietal lobe *(green)*, occipital lobe *(purple)*, temporal lobe *(teal)*, and limbic lobe *(pink)*. (From Burt AM. *Textbook of Neuroanatomy.* Philadelphia: WB Saunders; 1993:159, 160.)

proper, and continue laterally to the presubiculum. All of these represent parahippocampal structures. You would then curl in a spiral into the hippocampus' cornu ammonis and dentate gyrus with the fimbria found superomedially and the alveus on top of the cornu ammonis. Are you dizzy yet? The lateral-most portion of the cornu is particularly sensitive to anoxic injury and is the site where mesial temporal sclerosis occurs.

The occipital lobe is the lobe most commonly associated with visual function. At its apex is the calcarine sulcus, with the cuneate gyrus just above it (posteroinferior to the parietooccipital fissure) and the calcarine gyrus just below it (see Fig. 1-2).

The diencephalon contains the thalamus and hypothalamus. The thalamus has many nuclei, the most important of which (according to your ears and eyes) are the medial and lateral geniculate nuclei associated with auditory and visual functions, respectively. The thalamus is found on either side of the third ventricle and connects across the midline by the massa intermedia. Its other functions include motor relays, limbic outputs, and coordination of movement. Portions of the thalamus also subserve pain, cognition and emotions. The hypothalamus is located at the floor of the third ventricle, above the optic chiasm and suprasellar cistern. The hypothalamus is connected to the posterior pituitary via the infundibulum, or stalk, through which hormonal information to the pituitary gland is transmitted. The hypothalamus is critical to the autonomic functions of the body. Is it getting hot in here?

Brain Stem

Starting superiorly, the brain stem consists of the midbrain, pons, and medulla.

The mesencephalon differentiates into the midbrain. The midbrain is the site of origin of the third and fourth cranial nerves. Additionally, the midbrain contains the red nucleus, substantia nigra, and cerebral aqueduct, or aqueduct of Sylvius (Fig. 1-4). White matter tracts conducting the motor and sensory commands pass through the midbrain. The midbrain is also separated into the tegmentum and tectum, which refer to portions of the midbrain anterior and posterior to the cerebral aqueduct, respectively. The tectum, or roof, consists of the quadrigeminal plate (corpora quadrigemina), which houses the superior and inferior colliculi. The tegmentum contains the fiber tracts, red nuclei, third and fourth cranial nerve nuclei, and periaqueductal gray matter. The substantia nigra is within the anterior border of the tegmentum. Anterior to the tegmentum are the cerebral peduncles, which have somewhat of a "Mickey Mouse ears" configuration. Remember, just as there is only one Mickey, there is only one pair of cerebral peduncles.

The metencephalon develops into the pons and cerebellum. The pons contains the nuclei for cranial nerves V, VI, VII, and VIII (Figs. 1-5, 1-6). Pontine white matter tracts transmit sensory and motor fibers to the face and body. The pons also houses major connections of the reticular activating system for vital functions. One identifies the pons on the sagittal scan by its "pregnant belly."

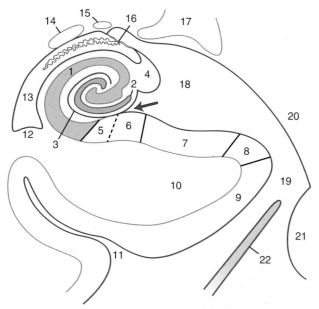

FIGURE 1-3 Hippocampal anatomy, coronal plane. *Arrow* indicates the hippocampal sulcus (superficial part). 1, cornu ammonis (Ammon's horn); 2, gyrus dentatus; 3, hippocampal sulcus (deep or vestigial part); 4, fimbria; 5, prosubiculum; 6, subiculum proper; 7, presubiculum; 8, parasubiculum; 9, entorhinal area; 10, parahippocampal gyrus; 11, collateral sulcus; 12, collateral eminence; 13, temporal (inferior) horn of the lateral ventricle; 14, tail of the caudate nucleus; 15, stria terminalis; 16, choroid fissure and choroid plexuses; 17, lateral geniculate body; 18, lateral part of the transverse fissure (wing of ambient cistern); 19, ambient cistern; 20, mesencephalon; 21, pons; 22, tentorium cerebelli. (Modified after Williams, 1995. From Duvernoy HM. *The Human Hippocampus.* New York: Springer-Verlag; 1998:18. Used with permission.)

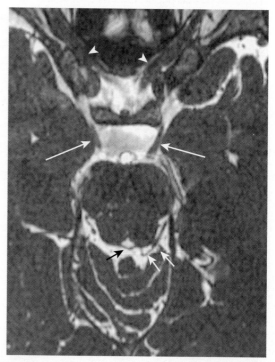

FIGURE 1-4 Midbrain anatomy. This constructive interference steady state (CISS) image shows both oculomotor nerves in their cisternal portions, leading to the cavernous sinus *(long white arrows),* the left trochlear nerve *(double arrows)* emanating from the posterior midbrain and coursing the ambient cistern, and the right trochlear nerve decussating posteriorly in the midline *(small black arrow).* The optic nerves can be seen in the optic canals bilaterally *(arrowheads).*

The myelencephalon becomes the medulla. The medulla contains the nuclei for cranial nerves IX, X, XI, and XII. Again, the sensory and motor tracts to and from the face and brain are transmitted through the medulla. Other named portions of the medulla include the pyramids, an anterior paramedian collection of fibers transmitting motor function, and the olivary nucleus in the mid-medulla (Fig. 1-7).

Cerebellum

The cerebellum is located in the infratentorial compartment posterior to the brain stem. The anatomy of the cerebellum is complex, with many named areas. For simplicity's sake, most people separate the cerebellum into the superior and inferior vermis and reserve the term cerebellar hemispheres for the rest of the lateral and central portions of the cerebellum.

For those interested in details, the superior vermis has a central lobule and lingula visible anteriorly, and the inferior vermis has a nodulus, uvula, pyramid, and tuber on its inferior surface (Fig. 1-8). The superior surface provides a view of the culmen, declive, and folium of the superior vermis. Superolaterally, there is a bump called the flocculus, which may extend toward the cerebellopontine angle cistern. This is a potential "pseudotumor," often misidentified as a vestibular schwannoma. The tonsils are located inferolaterally

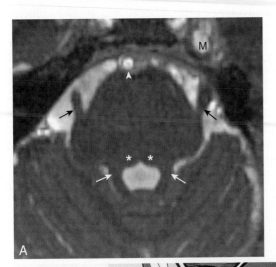

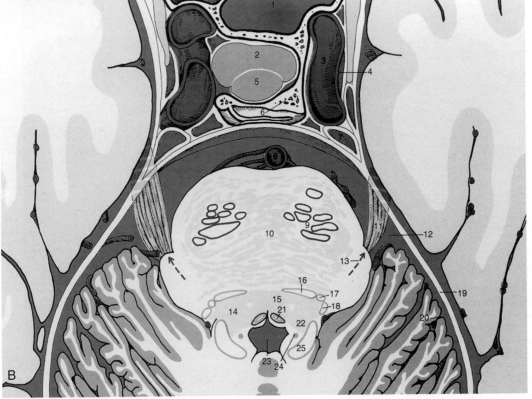

1 Sphenoid sinus
2 Adenohypophysis
3 Internal carotid artery
4 Cavernous sinus
5 Neurohypophysis
6 Dorsum sellae
7 Superior petrosal sinus
8 Basilar artery
9 Corticospinal tract
10 Nuclei pontis
11 Trigeminal nerve
12 Cerebellopontine (angle) cistern
13 Trigeminal nerve (within the slice)
14 Reticular formation (PPRF)
15 Paramedian pontine reticular formation
16 Medial lemniscus
17 Spinothalamic tract
18 Lateral lemniscus
19 Tentorium cerebelli
20 Primary fissure
21 Medial longitudinal fasciculus
22 Locus ceruleus
23 Fourth ventricle
24 Mesencephalic nucleus trigeminal nerve
25 Superior cerebellar peduncle

FIGURE 1-5 Pontine anatomy. **A,** Axial T2 constructive interference steady state (CISS) image shows cranial nerve V exiting the pons *(black arrows).* Note the superior cerebellar peduncles *(white arrows),* the Meckel cave on the left (M), medial longitudinal fasciculus *(asterisks),* and basilar artery *(white arrowhead).* **B,** Pontine anatomy at the level of the superior cerebellar peduncle shows several descending and ascending tracts.

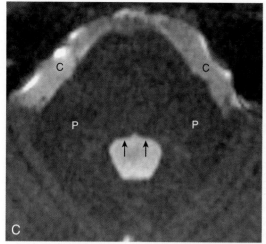

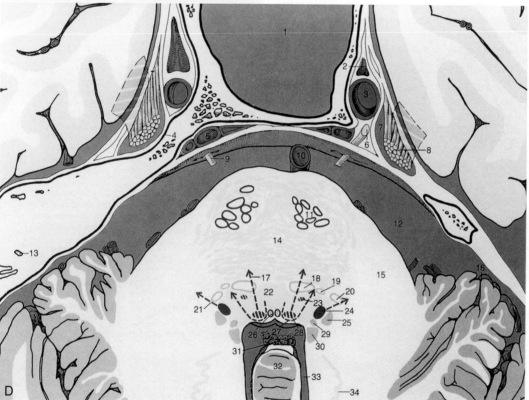

FIGURE 1-5, cont'd C, Facial colliculi *(arrows)* are clearly seen on this axial T2 CISS image. The middle cerebellar peduncle (P) is the dominant structure leading to the cerebellum. Also shown is the cerebellopontine angle cistern (C). **D,** At the facial colliculus one finds numerous cranial nerve nuclei and traversing lemnisci. (**B** and **D** from Kretschmann H-J, Weinrich W. *Cranial Neuroimaging and Clinical Neuroanatomy: Magnetic Resonance Imaging and Computed Tomography.* Rev 2nd ed. New York: Thieme; 1993:139, 137, respectively.)

1 Sphenoid sinus
2 Cavernous sinus
3 Internal carotid artery
4 Trigeminal impression
5 Inferior petrosal sinus
6 Abducens nerve
7 Opening of trigeminal cistern
8 Triangular part of trigeminal nerve
9 Abducens nerve near opening of dura mater
10 Basilar artery
11 Corticospinal tract
12 Cerebellopontine (angle) cistern
13 Anterior semicircular canal
14 Nuclei pontis
15 Middle cerebellar peduncle
16 Primary fissure
17 Abducens nerve (within the slice)
18 Medial lemniscus
19 Spinothalamic tract
20 Lateral lemniscus
21 Portio minor of trigeminal nerve (within the slice)
22 Reticular formation
23 Facial nucleus (in the caudal part of the slice)
24 Motor nucleus of trigeminal nerve
25 Main sensory (pontine) necleus of trigeminal nerve
26 Medial longitudinal fasciculus
27 Facial colliculus
28 Abducens nucleus (within the slice)
29 Mesencephalic nucleus of trigeminal nerve
30 Superior vestibular nucleus
31 Choroid plexus in fourth ventricle
32 Nodule of vermis
33 Posterior recess of fourth ventricle
34 Dentate nucleus

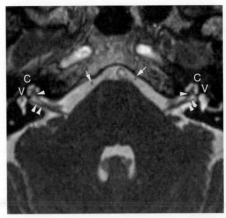

FIGURE 1-6 Lower pontine anatomy. This constructive interference steady state (CISS) image shows the abducens nerve denoted by the *white arrows,* whereas the cochlear (more anterior) and inferior vestibular nerves (more posterior) are seen bilaterally in the cerebellopontine angle cistern (*single and double white arrowheads,* respectively). The fluid-filled cochlea (C) and vestibule (V) are hyperintense on T2.

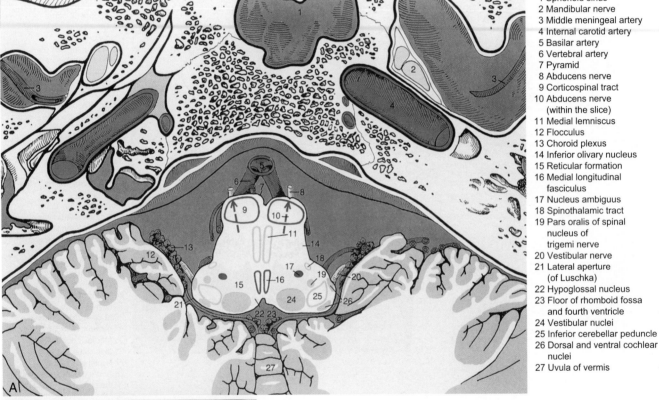

1 Sphenoid sinus
2 Mandibular nerve
3 Middle meningeal artery
4 Internal carotid artery
5 Basilar artery
6 Vertebral artery
7 Pyramid
8 Abducens nerve
9 Corticospinal tract
10 Abducens nerve
   (within the slice)
11 Medial lemniscus
12 Flocculus
13 Choroid plexus
14 Inferior olivary nucleus
15 Reticular formation
16 Medial longitudinal
   fasciculus
17 Nucleus ambiguus
18 Spinothalamic tract
19 Pars oralis of spinal
   nucleus of
   trigemi nerve
20 Vestibular nerve
21 Lateral aperture
   (of Luschka)
22 Hypoglossal nucleus
23 Floor of rhomboid fossa
   and fourth ventricle
24 Vestibular nuclei
25 Inferior cerebellar peduncle
26 Dorsal and ventral cochlear
   nuclei
27 Uvula of vermis

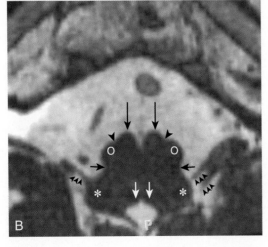

FIGURE 1-7 Medulla anatomy. **A,** This schematic reveals the junction of the vertebral arteries to the basilar artery. The roots of the abducens nerve arise at the border between the medulla oblongata and pons. The upper part of the inferior olivary nucleus is positioned in the medulla oblongata. **B,** Axial T2 constructive interference steady state (CISS) shows the preolivary sulcus (*short black arrows*), the olivary sulcus (*single arrowheads*), pyramidal tract (*large black arrows*) and the inferior cerebellar peduncle (*asterisks*), hypoglossal nuclei (*white arrows*), and nerve complex (cranial nerves IX and X; *small triple arrowheads*). The olive (o) can be seen anteriorly.

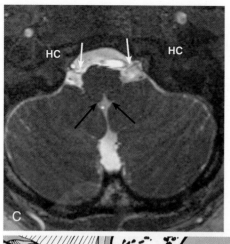

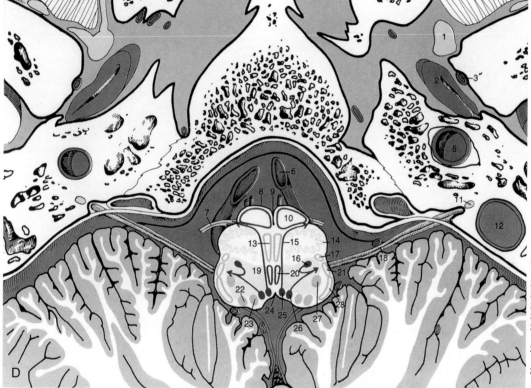

1 Mandibular nerve
2 Auditory (pharyngotympanic) tube
3 Middle meningeal artery
4 Clivus
5 Internal carotid artery
6 Vertebral artery
7 Hypoglossal nerve
8 Pyramid
9 Anterior median fissure
10 Corticospinal tract
11 Glossopharyngeal nerve
12 Bulb of internal jugular vein
13 Medial lemniscus
14 Inferior olivary nucleus
15 Hypoglossal nerve (within the slice)
16 Nucleus ambiguus
17 Spinothalamic tract
18 Vagus nerve
19 Reticular formation
20 Medial longitudinal fasciculus
21 Anterior spinocerebellar tract
22 Cuneate nucleus
23 Solitary nucleus
24 Median sulcus
25 Hypoglossal nucleus
26 Dorsal nucleus of vagus nerve
27 Pars interpolaris of spinal nucleus of trigeminal nerve
28 Inferior cerebellar peduncle

FIGURE 1-7, cont'd C, *White arrows* point out hypoglossal nerves coursing to hypoglossal canals (HC). On either side of the midline posterior cleft are the gracile nuclei *(black arrows).* Lateral to them will be the cuneate nuclei. **D,** This schematic shows the numerous nuclei and tracts that are present at the level of the medulla.

Continued

and are the structures that herniate downward through the foramen magnum in Chiari malformations.

Gray matter masses in the cerebellum include the fastigial, globose, emboliform, and dentate nuclei; the dentate nuclei are seen well on T1-weighted images (T1WI), whereas the fastigial, globose, and emboliform nuclei cannot be discerned. The dentate nuclei are situated laterally in the white matter of the cerebellum, and can be seen on computed tomography (CT) because they may calcify in later life.

Three major white matter tracts connect the cerebellum to the brain stem bilaterally (Fig. 1-9). The superior cerebellar peduncle (brachium conjunctivum) connects midbrain structures to the cerebellum, the middle cerebellar peduncle (brachium pontis) connects the pons to the cerebellum, and the inferior cerebellar peduncle (restiform body) connects the medulla to the cerebellum.

The flocculonodular lobe, fastigial nucleus, and uvula of the inferior vermis receive input from vestibular nerves and are thought to be involved primarily with maintaining equilibrium. Lesions of this part of the cerebellum, the archicerebellum, cause wide-based gait and dysequilibrium.

The superior vermis, most of the inferior vermis, and globose and emboliform nuclei receive spinocerebellar sensory information. Muscle tone information, postural tone,

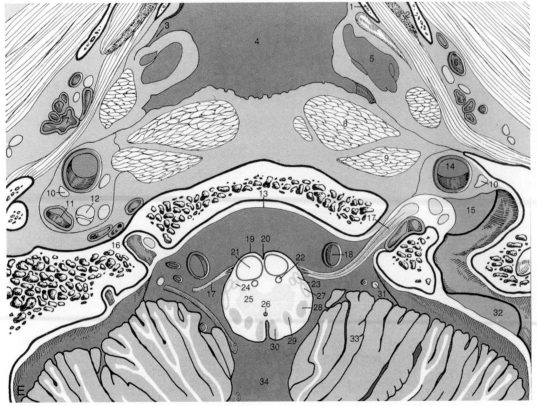

1 Medial pterygoid plate
2 Lateral pterygoid plate
3 Pharyngeal opening of
   auditory tube
4 Nasopharynx
5 Cartilage of auditory tube
6 Maxillary artery
7 Pterygoid venous plexus
8 Longus capitis muscle
9 Rectus capitis muscle
10 Glossopharyngeal nerve
11 Internal jugular vein,
   left-right asymmetry (var.)
12 Vagus nerve
13 Dura mater
14 Internal carotid artery
15 Bulb of internal jugular
   vein
16 Hypoglossal canal
17 Hypoglossal nerve
18 Vertebral artery
19 Pyramid
20 Anterior median fissure
21 Corticospinal tract
22 Medial longitudinal
   fasciculus
23 Anterior spinocerebellar
   tract
24 Spinothalamic tract
25 Reticular formation
26 Central canal
27 Posterior spinocerebellar
   tract
28 Pars caudalis of spinal
   nucleus of trigeminal
   nerve
29 Cuneate nucleus
30 Gracile nucleus
31 Spinal root of spinal
   accessory nerve
32 Sigmoid sinus, left-right
   asymmetry (var.)
33 Tonsil of cerebellum
34 Cisterna magna

FIGURE 1-7, cont'd E, The caudal portion of the medulla oblongata, the rootlets of the hypoglossal nerves, and the hypoglossal canal are included. (**A, D,** and **E** from Kretschmann H-J, Weinrich W. *Cranial Neuroimaging and Clinical Neuroanatomy: Magnetic Resonance Imaging and Computed Tomography.* Rev 2nd ed. New York: Thieme; 1993:133, 131, 127, respectively.)

and coordination of locomotion appear to be influenced by these sites and by their effect on brain stem fibers, the red nuclei, and vestibular nuclei. The hemispheric portions of the cerebellum receive information from the pons and help to control coordination of voluntary movements. Abnormalities within the cerebellar hemispheres result in dysmetria, dysdiadochokinesis (say THAT five times fast!), intention tremors, nystagmus, and ataxia.

Corpus Callosum

The corpus callosum is the large midline white matter tract that spans the two cerebral hemispheres. Its named parts include the rostrum (its tapered anteroinferior portion just above the anterior commissure), the genu (the anterior sweep), the body or trunk (the superiormost aspect), and the splenium (the posteriormost aspect; see Fig. 1-2). Often there may be focal narrowing within the posterior body, the so called "isthmus," which is a normal anatomic variation and should not be confused with focal pathology.

Other white matter tracts that must tread carefully as they cross the midline include the anterior commissure,

located at the inferior aspect of the corpus callosum just above the lamina terminalis, and the posterior commissure, just anterior to the pineal gland near the habenula. The anterior commissure transmits tracts from the amygdala and temporal lobe to the contralateral side. The habenula and hippocampal commissures cross-connect the two hemispheres and thalami.

Deep Gray Nuclei

The basal ganglia are known by a number of names in the neuroanatomic literature. These gray matter structures lie between the insula and midline. The globus pallidus is the medial gray matter structure identified just lateral to the genu of the internal capsule (Fig. 1-10). Lateral to it lies the putamen. The caudate nucleus head indents the frontal horns of the lateral ventricle and is anterior to the globus pallidus; however, the body of the caudate courses over the globus pallidus, paralleling the lateral ventricle and ending in a tail of tissue near the amygdala.

Additional terms used referring to the various portions of the basal ganglia include the striatum (caudate and

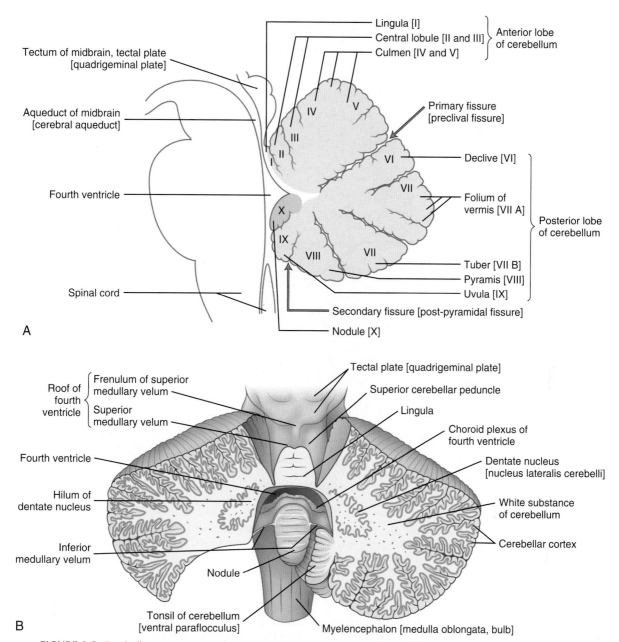

FIGURE 1-8 Cerebellar anatomy. **A,** The parts of the cerebellar vermis. Diagram of a median section. **B,** Coronal diagram of the cerebellar lobes and their lobules. (From Putz R, Pabst R, eds. *Sobotta Atlas of Human Anatomy.* 13th ed. Philadelphia: Lippincott Williams & Wilkins; 1996:292, 293.)

putamen) and the lentiform or lenticular nuclei (the globus pallidus and putamen).

The basal ganglia receive fibers from the sensorimotor cortex, thalamus, and substantia nigra, as well as from each other. Efferents go to the same locations and to the hypothalamus. The main function of the basal ganglia appears to be coordination of smooth movement.

The other deep gray matter structures of interest in the supratentorial space are the thalami, which sit on either side of the third ventricle. The thalamus is subdivided into many different nuclei by white matter striae. The medial and lateral geniculate nuclei, located along the posterior aspect of the thalamus, serve as relay stations for visual and auditory function. The pulvinar is the posterior expansion

of the thalamus. Behind the pulvinar are the wings of the ambient cistern. The massa intermedia connects the thalami across the third ventricle.

In the infratentorial space, the dentate nucleus, the largest deep gray matter structure, has connections to the red nuclei and to the thalami.

Ventricular System, Cerebrospinal Fluid, and Cerebrospinal Fluid Spaces

The normal volume of cerebrospinal fluid (CSF) in the entire central nervous system (CNS) is approximately 150 mL, with 75 mL distributed around the spinal cord, 25 mL within the ventricular system, and 50 mL

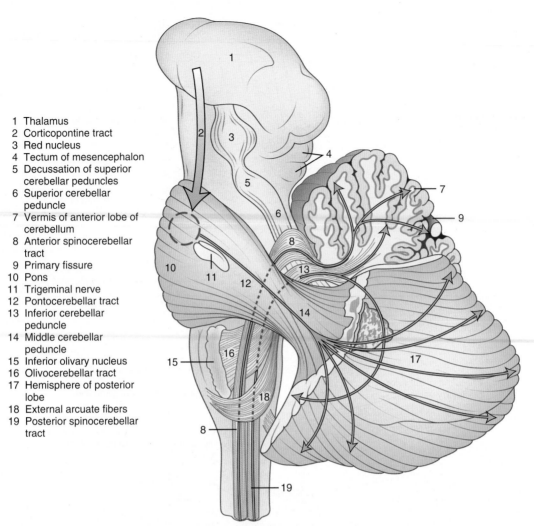

1 Thalamus
2 Corticopontine tract
3 Red nucleus
4 Tectum of mesencephalon
5 Decussation of superior
   cerebellar peduncles
6 Superior cerebellar
   peduncle
7 Vermis of anterior lobe of
   cerebellum
8 Anterior spinocerebellar
   tract
9 Primary fissure
10 Pons
11 Trigeminal nerve
12 Pontocerebellar tract
13 Inferior cerebellar
   peduncle
14 Middle cerebellar
   peduncle
15 Inferior olivary nucleus
16 Olivocerebellar tract
17 Hemisphere of posterior
   lobe
18 External arcuate fibers
19 Posterior spinocerebellar
   tract

FIGURE 1-9 The afferent systems of the cerebellum (lateral view). The left half of the anterior lobe of the cerebellum was removed. The archeocerebellum was separated and removed caudally from the middle cerebellar peduncle. (From Kretschmann H-J, Weinrich W. *Cranial Neuroimaging and Clinical Neuroanatomy: Magnetic Resonance Imaging and Computed Tomography.* Rev 2nd ed. New York: Thieme; 1993:326.)

surrounding the cortical sulci and in the cisterns at the base of the brain. In elderly persons, the intracranial CSF volume increases from 75 mL to a mean of approximately 150 mL in women and 190 mL in men. The normal production of CSF has been estimated to be approximately 450 mL/day, thereby replenishing the amount of CSF two to three times a day. Each ventricle's choroid plexus contributes to CSF production, whereas the reabsorption of CSF occurs at the level of the arachnoid villi into the intravascular system from the extracellular fluid.

The flow of CSF runs from the lateral ventricles via the foramina of Monro, to the third ventricle, out the cerebral aqueduct of Sylvius, and into the fourth ventricle, finally exiting through foramina of Luschka (bilaterally) and Magendie (in the midline; Fig. 1-11). CSF then flows into the cisterns of the brain and the cervical subarachnoid space and then down the intrathecal spinal compartments. The CSF ultimately percolates back up over the convexities of the hemispheres, where it is resorbed by the arachnoid villi into the intravascular space.

There are several named cisterns around the brain stem and midline structures (Fig. 1-12). Contents of these spaces can be compromised depending on the pathology at play, and critical structures coursing through these spaces may be affected and be the source of the patient's presenting complaint. Therefore, awareness of these cisterns and contents is critical in descriptions of different herniation syndromes and other pathologies that can be identified on imaging (Table 1-1).

Meninges and Associated Potential Spaces

The brain is covered in three protective layers of tissue, the meninges (or mater), which consist of pia, arachnoid, and dura. The dura (or pachymeninges) consists of the thickest and toughest layer and is adherent to the inner table of the skull extending to sutural margins. Pathologic conditions may occur in the space between the inner table and the dura, in the so-called epidural space, and result in mass effect on the underlying brain parenchyma. Epidural compartments are separated by sutures and pathologies typically do not cross the sutures when confined to the epidural space. Deep to the dura but superficial to the arachnoid mater is another potential space called the subdural space. When space-occupying pathologies are

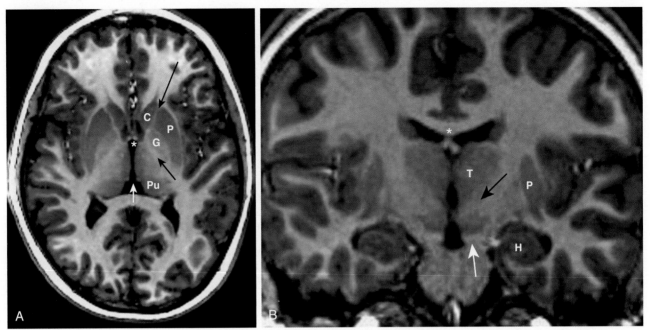

FIGURE 1-10 Deep gray matter anatomy. **A,** Axial T1-weighted image (T1WI) shows the caudate (C), putamen (P), and globus pallidus (G), as well as the anterior limb *(long black arrow)* and posterior limb *(short black arrow)* of the internal capsule. White matter tracts pass between the basal ganglia. The thalamus and periaqueductal gray matter line the third ventricle. The tiny dots of the fornix anteriorly (just ventral to *asterisk*) and the posterior commissure posteriorly *(white arrow),* as well as pulvinar thalamic gray natter (Pu), are also evident. **B,** On this coronal T1WI, the subthalamic nucleus *(black arrow)* and substantia nigra *(white arrow)* can be seen under the thalami (T). The hippocampus (H) is present further laterally. The thalami are joined in the midline at the massa intermedia, and one can also see the forniceal columns (just below *asterisk*) projecting above the thalami.

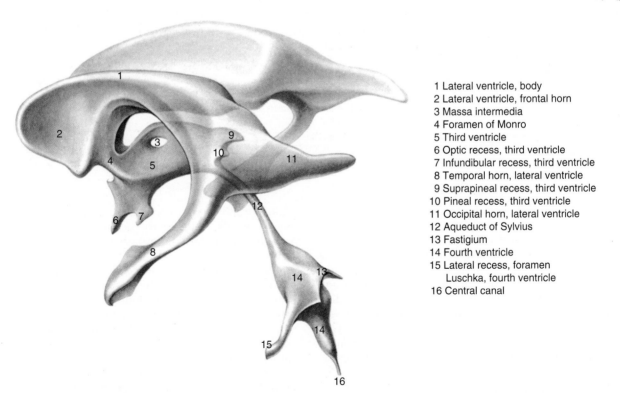

1 Lateral ventricle, body
2 Lateral ventricle, frontal horn
3 Massa intermedia
4 Foramen of Monro
5 Third ventricle
6 Optic recess, third ventricle
7 Infundibular recess, third ventricle
8 Temporal horn, lateral ventricle
9 Suprapineal recess, third ventricle
10 Pineal recess, third ventricle
11 Occipital horn, lateral ventricle
12 Aqueduct of Sylvius
13 Fastigium
14 Fourth ventricle
15 Lateral recess, foramen
   Luschka, fourth ventricle
16 Central canal

FIGURE 1-11 Ventricular system of the brain. Three-dimensional diagram of the ventricular system of the brain is labeled. (From Nieuwenhuys R, Voogd J, van Huijen C. *The Human Central Nervous System: A Synopsis and Atlas.* Rev 3rd ed. Berlin: Springer-Verlag; 1988.)

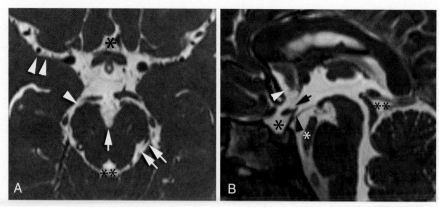

FIGURE 1-12 Cisterns of the brain. **A,** Axial constructive interference steady state (CISS) image shows the interpeduncular cistern *(single arrow)*, ambient cistern *(single arrowhead)*, perimesencephalic cistern *(double arrows)*, sylvian fissure *(double arrowheads)*, and quadrigeminal plate cistern *(double asterisk)*. The cistern of the lamina terminalis is indicated by *single asterisk*. **B,** Sagittal CISS image shows the cistern of the lamina terminalis *(arrowhead)*, suprasellar cistern *(single black asterisk)*, and quadrigeminal plate cistern *(double black asterisk)*. The basilar artery *(white asterisk)* is seen coursing the prepontine cistern. The chiasmatic recess *(black arrow)* and infundibular recess *(black arrowhead)* are also indicated. Note the crowding of structures at the foramen magnum in this patient with borderline Chiari I malformation.

TABLE 1-1   Cisterns of the Brain

Name	Location	Structures Traversing Cistern
Cisterna magna	Posteroinferior to fourth ventricle	None important
Circum-medullary cistern	Around medulla	Posterior inferior cerebellar artery
Superior cerebellar cistern	Above cerebellum	Basal vein of Rosenthal, vein of Galen
Prepontine cistern	Anterior to pons	Basilar artery, cranial nerves V and VI
Cerebellopontine angle cistern	Between pons and porus acusticus	Anterior inferior cerebellar artery, cranial nerves VII and VIII
Interpeduncular cistern	Between cerebral peduncles	Cranial nerve III
Ambient (crural) cistern	Around midbrain	Cranial nerve IV
Quadrigeminal plate cistern	Behind midbrain	None important
Suprasellar cistern	Above pituitary	Optic chiasm, cranial nerves III, IV, carotid arteries, pituitary stalk
Retropulvinar cistern (wings of ambient cistern)	Behind thalamus	Posterolateral choroidal artery
Cistern of lamina terminalis	Anterior to lamina terminalis, anterior commissure	ACA
Cistern of velum interpositum	Above 3rd ventricle	Internal cerebral vein, vein of Galen
Cistern of the ACA	Above corpus callosum	ACA

ACA, Anterior cerebral artery.

present here, cortical vessels that normally traverse the subarachnoid space just deep to the dura become compressed towards the surface of the brain, and the underlying brain parenchyma can also be compressed. Subarachnoid disease processes occur at the surface of the brain, deep to the arachnoid layer but superficial to the pia, and therefore assume a curvilinear configuration, extending along the surface of the sulci and gyri. Subpial processes do occur, however, the pia and arachnoid cannot always be readily distinguished on imaging; the pia and arachnoid layers are collectively referred to as the leptomeninges.

Physiologic Calcifications

The pineal gland calcifies with age. A small percentage (2% of children less than 8 years old and 10% of adolescents) of children show calcification of the pineal gland. By 30 years of age, most people have calcified pineal glands. Anterior to the pineal gland, one often sees the habenular commissure as a calcified curvilinear structure.

The choroid plexus is calcified in about 5% of children by age 15, and most adults by age 40. Such calcifications may be seen in the lateral, third and fourth ventricles, as well as the foramina of Luschka, Magendie, and choroid fissures.

The dura of the falx and/or tentorium is virtually never calcified in children and should be viewed as suspicious for basal cell nevus syndrome in that setting. However, in adults, foci of calcification and even ossification of the dura and falx are not uncommon. The dura shows higher rates of calcification in patients who have had shunts placed or have been irradiated.

Basal ganglia calcification is also rarely observed in individuals less than 30 years of age and should provoke

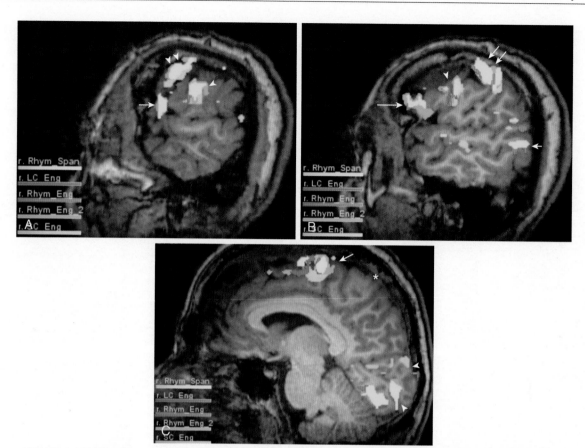

FIGURE 1-13 Functional magnetic resonance imaging (fMRI) with language task. **A,** fMRI blood-oxygen-level dependent (BOLD) activations overlayed onto anatomic T1-weighted images in sagittal plane. The left side of the brain is shown. Convergent activation is seen in the Broca region *(arrow)*. The Broca region typically corresponds to the pars opercularis/pars triangularis of the inferior frontal gyrus (Brodmann area 44 and 45), and in right handed patients is typically lateralized to the left cerebral hemisphere. Even without overt movement, language related motor areas can show concurrent activation, as is seen in the area of convergent activation in the subcentral gyrus *(single arrowhead),* which represents the tongue/facial motor regions. The third convergent area of activation is seen in the ventral premotor cortex, also commonly activated during language tasks *(double arrowheads).* **B,** Convergent activation is seen in the left inferior frontal gyrus along the pars triangularis and pars opercularis corresponding to Brodmann area 44 and 45, compatible with Broca activation *(large single arrow).* Activation is also seen in the left ventral premotor cortex *(arrowhead),* as well as language related motor and sensory areas at the banks of the precentral and postcentral gyri *(double arrows).* Posterior temporal lobe convergent activation represents Wernicke activation *(small single arrow).* There are also smaller foci of language-related convergent activation seen just cranial to this in the supramarginal gyrus. **C,** Convergent activation is seen in the pre-supplemental motor area *(arrow),* which is activated during language tasks. This is anterior to the supplementary motor area, which is in turn anterior to the precentral gyrus. The central sulcus *(asterisk)* is seen as the sulcus immediately anterior to the marginal segment of the cingulate sulcus. Visual areas also demonstrate activation *(arrowheads)* as most of the language tasks used for this patient employed visual language task paradigms. (Courtesy Haris Sair, M.D.)

a search for metabolic disorders or a past history of perinatal infections if seen in youngsters. (See the online Appendix at ExpertConsult.com for causes of basal ganglia calcification.) Over the age of 30, however, basal ganglia calcifications are very common to the point that these do not necessarily need to be mentioned in routine reporting unless true pathology is suspected. Such benign basal ganglia mineralization is typically bilateral, although in some cases it may be more conspicuous on one side compared with the other. Care must be made not to confuse these physiologic calcifications, which are hyperdense on CT, with hemorrhage, which is also hyperdense on CT.

FUNCTIONAL ANATOMY

Understanding the functional anatomy requires a little bit of the cartographer in each of us (or a GPS-enabled smartphone). After having assimilated the destinations and points of departure, one should talk about the entire routes of neuronal travel. For functional anatomy, we can now use fMRI to identify the sites of cortical activation (the points of departure and destinations; Fig. 1-13) and diffusion tensor imaging to perform white matter tracking as the highways between gray matter destinations (Fig. 1-14). The directionality of these white matter tracts can also be inferred now.

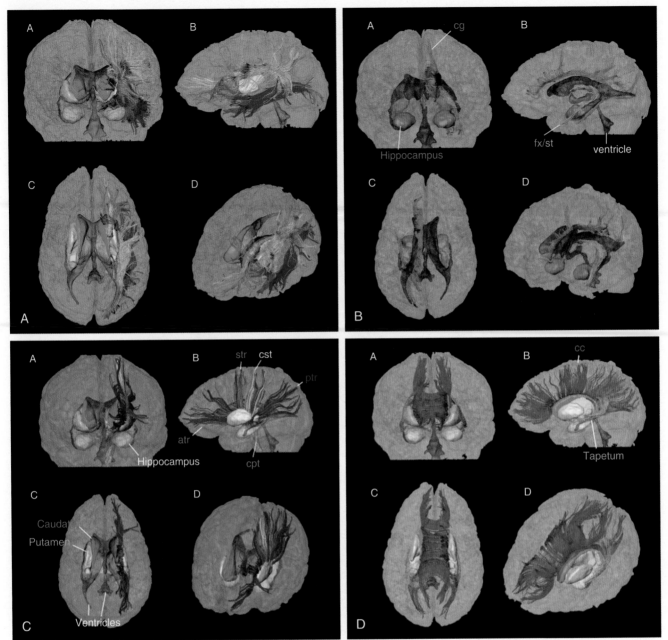

FIGURE 1-14 Diffusion tensor imaging (DTI) tractography. Using DTI data, discrete fiber tracts in the brain can be isolated and color coded for your visual pleasure, demonstrating association fibers (those axon bundles connecting different parts of the brain in the same cerebral hemisphere), projection fibers (those axons connecting the cortex with lower parts of the brain and spinal cord), and commissural fibers (those axons connecting between the two cerebral hemispheres). **A,** Three-dimensional reconstructions of association fibers are depicted, including in the anterior (A), left (B), superior (C), and oblique (left-anterosuperior) (D) orientations. Note the color coded projections of the superior longitudinal fasciculus (yellow), inferior fronto-orbital fasciculus (orange), uncinate fasciculus (red), and inferior longitudinal fasciculus (brown). Thalami are yellow, ventricles are gray, caudate nuclei are green and lentiform nuclei are light green. **B,** Three-dimensional reconstruction results of association fibers in the limbic system viewed from the anterior (A), left (B), superior (C), and oblique (left-anterior) (D) orientations. The hippocampi are depicted in purple. **C,** Three-dimensional reconstruction results of projection fibers viewed from the anterior (A), left (B), superior (C), and oblique (left-superior-anterior) (D) orientations. Depicted are anterior thalamic radiation (atr), corticopontine tract (cpt), corticospinal tract (cst), posterior thalamic radiation (ptr), and superior thalamic radiation (str). **D,** Three-dimensional reconstructions of commissural fibers viewed from the anterior (A), left (B), superior (C), and oblique (left-anterior-superior) (D) orientations. The corpus callosum (cc) is color coded magenta and the tapetum (commissural fibers extending to temporal lobes) is color coded peach. *cg,* Cingulum; *fx,* fornix; *st,* stria terminalis. (Reprinted with permission from Oishi K, Faria AV, van Zijl PCM, Mori S. *MRI Atlas of Human White Matter.* 2nd ed. Philadelphia: Elsevier; 2011:18, 19, 20, 21.)

Brodmann Areas

The functional units of the cerebral hemispheres have been separated into what are called Brodmann areas, and include areas 1 through 47. These numbered areas correspond to different gyri that subserve various functions. The Brodmann areas are the currency with which fMRI scientists transact business and are therefore important to be aware of. In addition, knowing which gyri are responsible for which properties can be critical to identifying lesions and predicting deficits in patients with strokes.

For example, Brodmann area 1 (aka S1) subserves primary somatosensory and position sense, and sits within the postcentral gyrus, in the paracentral lobule of the parietal lobe. Brodmann areas 2 and 3 subserve similar functions and are slightly posterior (2) and slightly anterior (3) to Brodmann area 1. Brodmann area 4 (M1) subserves primary motor function and resides within the precentral gyrus of the frontal lobe. Wernicke areas are comprised of Brodmann areas 21 and 22 (middle and superior temporal gyri respectively), and serve functions of higher order audition and speech reception respectively (can you hear me now?). Also within the superior temporal gyrus are areas 41 and 42 (aka A1-Wernicke and A2-Wernicke), which subserve functions of primary audition and auditory association/speech recognition respectively. Broca areas (Brodmann 44 and 45) located within the inferior frontal gyrus laterally subserve functions of speech expression and motor speech/tongue movement respectively.

Motor System

The primary origin of the stimulus for motor function is the precentral gyrus of the frontal lobe, which receives input from many sensory areas (Fig. 1-15; Table 1-2). Stimulation of the motor area of one precentral gyrus causes contraction of muscles on the opposite side of the body. The motor cortex, like the sensory area, is arranged such that the lower extremity is located superomedially along the paracentral lobule in the midline, whereas the upper extremity is located inferolaterally. The cells innervating the hip are at the top of the precentral sulcus; the leg is draped over medially along the interhemispheric fissure. The face (especially the tongue and mouth) has an inordinately large area of motor and sensory representation along the inferiormost aspect of the precentral motor strip on the surface of the brain, just above the sylvian fissure. The motor contribution to speech is located at the inferior frontal gyrus (frontal operculum regions).

Sometimes finding the central sulcus on imaging can be difficult (Fig. 1-16). This is necessary for discriminating motor from sensory areas particularly when surgery to resect a peri-Rolandic tumor is contemplated. Retaining motor function is desired. Consult Box 1-1 for some clues to identifying the central sulcus.

From the motor cortex of the frontal lobe, the white matter fibers pass into the corticospinal tract, which extends through the white matter of the centrum semiovale to the posterior limb of the internal capsule. From the posterior portion of the posterior limb of the internal capsule, the corticospinal tract continues through the central portion of the cerebral peduncle in the anterior portion of the midbrain. These fibers continue in the anterior portion of the pons to the pyramids of the medulla, where most of them decussate (in the pyramidal decussation) and proceed inferiorly in the lateral corticospinal tract of the spinal cord. Fifteen percent of fibers do not decussate in the medulla. These fibers pass into the anterior funiculus along the anterior median fissure of the spinal cord as the anterior corticospinal tract. The fibers of the pyramidal tract, which include both the lateral and anterior corticospinal tract, synapse with the anterior horn cell spinal cord nuclei.

Motor supply to the face travels from the cortex, through the corona radiata, into the genu of the internal capsule, via the corticobulbar tract. The corticobulbar fibers are located more anteromedially in the cerebral peduncles and have connections to the brain stem nuclei as they descend. Most of the connections to the various cranial nerve nuclei are contralateral to the cortical bulbar tract; however, some ipsilateral fibers are present as well.

The pyramidal tract is responsible for voluntary movement and contains the corticospinal and corticobulbar fibers. The extrapyramidal system includes the corpus striatum, which receives fibers from the cerebral cortex, the thalamus, and the substantia nigra, with connections to the caudate nucleus and putamen. These fibers originate from the cerebral cortex but pass through the internal and external capsule to reach the basal ganglia. The dentate nuclei, found in the cerebellar hemispheres, also send tracts to the thalamus and motor areas of the frontal cortex. The red nucleus of the midbrain receives fibers from the cortical motor area and transmits fibers via the rubrospinal tract to the spinal cord, which also regulates motion.

Abnormalities of the pyramidal system mainly produce weakness, paralysis, or spasticity of voluntary motor function. Extrapyramidal system abnormalities often produce involuntary movement disorders including tremors, choreiform (jerking) movements, athetoid (slow sinuous) movements, hemiballismic (flailing) motions, and muscular rigidity (think: pyramid, paralysis; extrapyramidal, extremity excesses).

Sensory System

Supposedly, humans get the greatest degree of satisfaction from their sense of touch. Certainly, a good back scratch can satiate many a need, but the sense of touch goes beyond merely a light touch on the back. It also includes pain (an inadvertent scratch by the nails), vibration (add a pulsating massager), and position sense (lying on one's stomach).

The sensory system of the CNS is separated into fibers that transmit the sensations of pain and temperature, position, vibration, and general fine touch (Fig. 1-17, *A*; Table 1-3). From the body, pain and temperature primary neuron fibers are transmitted by peripheral nerve fibers. The pain and temperature sensations are transmitted to the dorsal root ganglia of the spinal cord, where fibers may ascend or descend for one or two spinal segments before terminating in the region of the substantia gelatinosa of the dorsal horn. From the secondary neurons of the nucleus proprius of the dorsal horn, the fibers cross the midline in the anterior white commissures of the

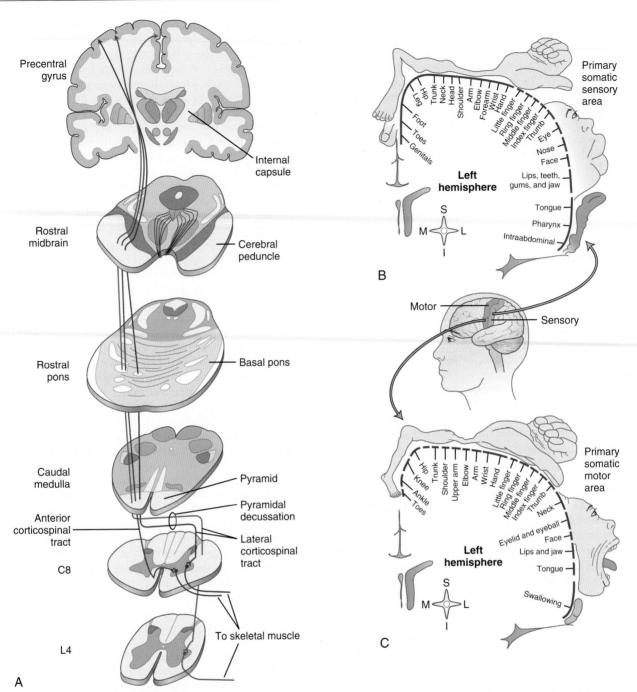

FIGURE 1-15 A, Corticospinal tracts. Fibers from the precentral gyrus and other nearby cortical areas descend through the cerebral peduncles, pons, and medullary pyramids; most cross in the pyramidal decussation to form the lateral corticospinal tract. Those that do not cross in the pyramidal decussation form the anterior corticospinal tract; most of these fibers cross in the anterior white commissure before ending in the spinal gray matter. Most corticospinal fibers do not synapse directly on the motor neurons. They are drawn that way here for simplicity. Primary somatic sensory **(B)** and motor **(C)** areas of the cortex, coronal view. The body parts illustrated here show which parts of the body are "mapped" to correlates in each cortical area. The exaggerated face indicates that more cortical area is devoted to processing information to/from the many receptors and motor units than for the leg or arm, for example. (**A** from Nolte J. *The Human Brain: An Introduction to Its Functional Anatomy.* 4th ed. St. Louis: Mosby; 1999:249. **B** and **C** from Thibodeau GA, Patton KT. *Anatomy and Physiology.* 4th ed. St. Louis: Mosby; 1999:394.)

spinal cord and ascend in the lateral spinothalamic tract. The lateral spinothalamic tract is identified in the lateral midportion of the medulla and centrally in the pons where it is renamed the spinal lemniscus. The spinal lemniscus proceeds through the anterolateral portion of the dorsal pons and along the lateral aspect of the midbrain. From there the fibers synapse with tertiary neurons in the ventral posterolateral thalamic nucleus and then terminate in the somesthetic area of the parietal lobe in the postcentral sulcus region.

TABLE 1-2 Motor Pathways

Pathway	Course	Function
Lateral corticospinal tract	Primary motor cortex to corona radiata to posterior limb of internal capsule to cerebral peduncle to central pontine region to medulla through pyramidal decussation to posterolateral white matter of cord	Motor to contralateral extremities
Anterior corticospinal tract	Primary motor cortex to corona radiata to posterior limb of internal capsule to cerebral peduncle to central pontine region to medulla to anterior funiculus and anterior column of spinal cord	Motor to ipsilateral muscles
Rubrospinal tract	Red nucleus to decussation in ventral tegmentum of the midbrain through the lateral funiculus of the spinal cord to the posterolateral white matter of cord (with lateral corticospinal tract)	Motor control of contralateral limbs
Reticulospinal tract	Pons and medulla to ipsilateral anterior column of cord	Automatic movement of axial and limb muscles (walking, stretching, orienting behaviors)
Vestibulospinal tract	Vestibular nuclei to ipsilateral anterior columns in cord	Balance, postural adjustments, and head and neck coordination

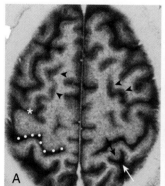 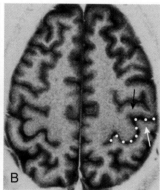

FIGURE 1-16 Central sulcus. **A,** Note the shape of the medial end of the postcentral sulcus, the bifid "y" (between *white* and *black arrows*) and how the superior frontal sulcus *(arrowheads)* terminates in the precentral sulcus *(asterisk).* **B,** The central sulcus is the next sulcus posterior to the precentral sulcus. Note that precentral gyrus' cortical gray matter *(black arrow)* thickness is greater than that of the postcentral gyrus *(white arrow)* cortical thickness. The central sulcus is indicated by dotted line in both images.

BOX 1-1 Localization of the Central Sulcus

1. The central sulcus enters the paracentral lobule anterior to the marginal ramus of the cingulate sulcus.
2. The medial end of the post central sulcus is shaped like a bifid "y" and the bifid ends enclose the marginal ramus of the cingulate sulcus.
3. The superior frontal sulcus terminates in the precentral sulcus and the central sulcus is the next sulcus posterior to the precentral sulcus.
4. The interparietal sulcus intersects the postcentral sulcus.
5. The knob representing the hand motor area is in the precentral gyrus.
6. The precentral gyrus' cortical gray matter thickness is greater than that of the postcentral gyrus thickness. Usually PRE/POST thickness ratio is about 1.5/1.
7. The peri-Rolandic cortex is more hypointense than surrounding cortex on FLAIR.

See Fig. 1-14.

Pain and temperature sensation from the face is transmitted via the primary neuron axons of cranial nerve V, with the nuclei identified in the trigeminal ganglion. The axons from the trigeminal ganglion descend in the spinal trigeminal tract. The fibers terminate in the secondary neuron nucleus of the trigeminal spinal tract, which extends from the lower medulla to the C3 level of the spinal cord. At this point the pain and temperature fibers cross the midline to the contralateral side and ascend as the trigeminothalamic tract, which passes medial to the lateral spinothalamic tract but terminates also in the ventral posterior (lateral) thalamic nucleus. From there tertiary neuron fibers pass to the somesthetic area of the cerebral cortex.

Light touch and pressure from the body are transmitted in the ipsilateral posterior column of the spinal cord and contralateral anterior column (see Fig. 1-17, *B*). The ascending branches may travel up to six to eight segments of the spinal cord before crossing to the contralateral side. Once again, a synapse is present in the nucleus proprius of the dorsal horn. From there, the white matter tracts form the anterior spinothalamic tracts, included as part of the spinal lemniscus. These axons also terminate in the ventral posterior (lateral) thalamic nucleus passing through the anterior portion of the internal capsule and the centrum semiovale to the somesthetic cortex. The spinal lemniscus lies lateral to the medial lemniscus in the posterior pons.

The pathway for light touch of the face is identical to that of the pain and temperature. However, termination of these cranial nerve V fibers occurs in a more superior portion of the nucleus of the trigeminal spinal tract. In addition, these fibers may bifurcate on entering the pons and synapse with the chief sensory nucleus of V within the pons.

The body's sense of proprioception, fine touch, and vibration is transmitted via proprioceptors, which bifurcate in the posterior columns of the spinal cord. A portion of the fibers descend and make up the afferent loop of fiber reflex arcs; however, the ascending portion passes superiorly in the fasciculus gracilis and the fasciculus cuneatus, which terminate in their respective nuclei in the medulla.

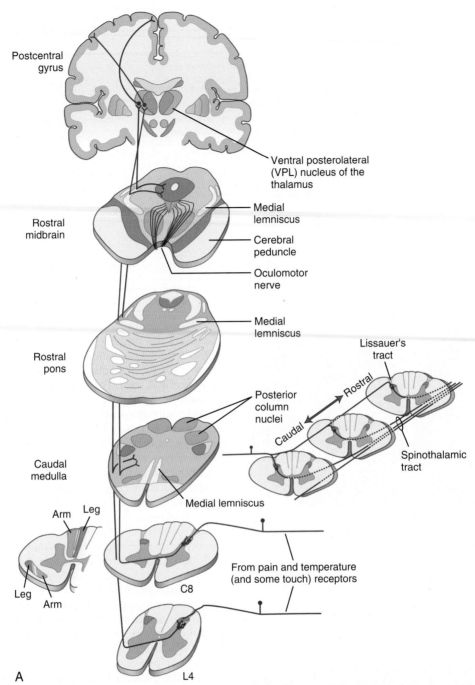

Postcentral gyrus

Ventral posterolateral (VPL) nucleus of the thalamus

Rostral midbrain

Medial lemniscus

Cerebral peduncle

Oculomotor nerve

Rostral pons

Medial lemniscus

Lissauer's tract

Rostral

Posterior column nuclei

Caudal

Spinothalamic tract

Caudal medulla

Medial lemniscus

Arm Leg

Leg Arm

From pain and temperature (and some touch) receptors

C8

A

L4

FIGURE 1-17 The sensory pathways of the body. **A,** Spinothalamic tract. Pain, temperature, and some touch and pressure afferents end in the posterior horn. Second- or higher-order fibers cross the midline, form the spinotha- lamic tract, and ascend to the ventral posterolateral (VPL) nucleus of the thalamus (and also to other thalamic nuclei not indicated in this figure). Thalamic cells then project to the somatosensory cortex of the postcentral gyrus and to other cortical areas (also not indicated in this figure). Along their course through the brain stem, spi- nothalamic fibers give off many collaterals to the reticular formation. The inset to the left shows the lamination of fibers in the posterior columns and the spinothalamic tract, in a leg-lower trunk-upper trunk-arm sequence. The inset to the right shows the longitudinal formation of the spinothalamic tract. Primary afferents ascend several segments in Lissauer's tract before all their branches terminate; fibers crossing to join the spinothalamic tract do so with a rostral inclination. As a result, a cordotomy incision at any given level will spare most of the informa- tion entering the contralateral side of the spinal cord at that level, and to be effective the incision must be made several segments rostral to the highest dermatomal level of pain.

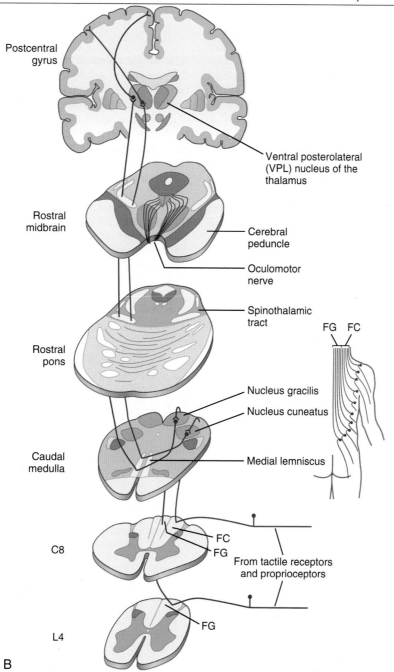

FIGURE 1-17, cont'd B, Posterior column-medial lemniscus pathway. Primary afferents carrying tactile and proprioceptive information synapse in the posterior column nuclei of the ipsilateral medulla. The axons of second-order cells then cross the midline, form the medial lemniscus, and ascend to the ventral posterolateral nucleus of the thalamus. Third-order fibers then project to the somatosensory cortex of the postcentral gyrus. A somatotopic arrangement of fibers is present at all levels. The beginning of this somatotopic arrangement, as a lamination of fibers in the posterior columns, is indicated in the inset to the right. *FC,* Faciculus cuneatus; *FG,* faciculus gracilis. (From Nolte J. *The Human Brain: An Introduction to Its Functional Anatomy.* 4th ed. St. Louis: Mosby; 1999:244, 245.)

TABLE 1-3 Sensory Tracts

Tract	Course	Function
Medial lemniscus	From posterior white matter of cord to dorsal nuclei of medulla, through decussation, to medial lemniscus to thalamus to anterior limb of internal capsule to primary sensory cortex	Touch and limb position sense
Spinothalamic tract	Dorsal horn of cord to spinal decussation to anterolateral spinal tract to reticular formation of pons, medulla, thalamus	Pain and temperature
Lateral lemniscus	From auditory fibers in caudal pons, crossed and uncrossed, to inferior colliculus, to medial geniculate of thalamus to primary auditory cortex	Auditory

These fibers ascend ipsilaterally only. From the nucleus gracilis and nucleus cuneatus the axons cross the midline of the medulla and continue superiorly as the medial lemniscus found in the posterior portion of the medulla and pons before terminating in the ventral posterolateral (VPL) nucleus of the thalamus. From the VPL, the path is through the internal capsule to get to the primary somatosensory cortex.

Fine tactile fibers from the face terminate in the chief sensory nucleus or the mesencephalic nucleus of cranial nerve V. The fibers going to the chief sensory nucleus of cranial nerve V then cross the midline and ascend as the trigeminothalamic tract (ventral trigeminal lemniscus), whereas the fibers going to the mesencephalic nucleus of nerve V ascend ipsilaterally in the dorsal trigeminal lemniscus. From the mesencephalic nucleus in the midbrain, however, these fibers cross the midline at the red nucleus level and ascend to the ventral posteromedial thalamic nucleus. The pathways from both the body and face terminate in the somesthetic area of the cerebral cortex after passing through the posterior limb of the internal capsule.

Visual Pathway

The image received by the rods and cones of the retina is passed to secondary sensory ganglion cells of the retina and is then transmitted along the second cranial nerve—the optic nerve. The optic nerve ascends obliquely through the optic canal to join fibers from the contralateral optic nerve at the optic chiasm (Fig. 1-18). The temporal retina fibers (nasal field) remain uncrossed and pass to the ipsilateral optic tract. The fibers from the nasal retina (temporal fields) decussate to join the nondecussating nasal field fibers from the opposite optic nerve continuing in the postchiasmal optic tract. However, before they cross, some of the inferonasal retinal fibers loop for a short distance up into the contralateral optic nerve in what is termed Wilbrand knee. This accounts for the signs of the "junctional syndrome" found in a lesion that compresses one optic nerve and the looping contralateral Wilbrand fibers. This results in a central scotoma in the ipsilateral eye and a superotemporal visual defect in the contralateral eye. Ninety percent of the fibers in the chiasm are from

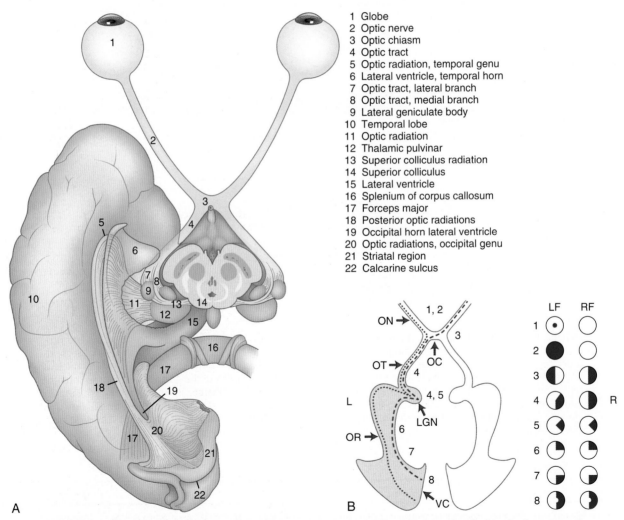

1 Globe
2 Optic nerve
3 Optic chiasm
4 Optic tract
5 Optic radiation, temporal genu
6 Lateral ventricle, temporal horn
7 Optic tract, lateral branch
8 Optic tract, medial branch
9 Lateral geniculate body
10 Temporal lobe
11 Optic radiation
12 Thalamic pulvinar
13 Superior colliculus radiation
14 Superior colliculus
15 Lateral ventricle
16 Splenium of corpus callosum
17 Forceps major
18 Posterior optic radiations
19 Occipital horn lateral ventricle
20 Optic radiations, occipital genu
21 Striatal region
22 Calcarine sulcus

FIGURE 1-18 Visual system anatomy. **A,** The anatomy of the optic nerves, tracts, and radiations is diagrammed. **B,** Optic chiasm: correlation of lesion site and field defect. Note the most ventral nasal fibers (mostly from inferior nasal retina) temporarily travel within the fellow optic nerve in Wilbrand knee. *LGN,* Lateral geniculate nucleus; *OC,* optic chiasm; *ON,* optic nerve; *OR,* optic radiations; *OT,* optic tract; *VC,* visual cortex. (**A** from Nieuwenhuys R, Voogd J, van Huijen C. *The Human Central Nervous System: A Synopsis and Atlas.* Rev 3rd ed. Berlin: Springer-Verlag; 1988.)

the macula of the retina; the crossing fibers lie superiorly and posteriorly in the chiasm. The optic tract encircles the anterior portion of the midbrain before terminating in the lateral geniculate body of the thalamus.

The lateral geniculate body is located in the posterolateral portion of the thalamus within the pulvinar nucleus. A few fibers ascend to the Edinger-Westphal nucleus (cranial nerve III) in the pretectal portion of the midbrain as part of the pupillary reflex, and some connect to the superior colliculus for tracking ability. From the lateral geniculate nucleus, tertiary neuronal fibers pass in the geniculocalcarine tract (optic radiations), course through portions of the posterior limb of the internal capsule and around the lateral ventricle, and terminate in the visual calcarine cortex in the medial occipital lobe. The superior optic radiations (inferior visual fields) pass through the parietal lobe on their way to the superior visual cortex of the occipital lobe. The inferior optic radiations (superior visual fields) are called Meyers loop fibers. They pass over the anteroinferior aspect of the lateral ventricle (around the temporal horn) into the temporal lobe to terminate in the inferior visual cortex just below the calcarine sulcus. An optic radiation lesion is localized depending on whether there are signs of neglect or sensory loss (superior radiation–parietal lobe lesion) or dysphasia and memory loss (Meyers loop- temporal lobe lesion).

Thus, lesions in the optic nerve cause blindness of the ipsilateral eye. Lesions in the midline at the level of the optic chiasm will cause bitemporal defects. Lesions compressing the lateral edge of the optic chiasm cause a nasal hemianopsia of the ipsilateral eye. Lesions of the optic tract extending to the lateral geniculate, and primary lesions of the lateral geniculate nucleus cause a contralateral homonymous hemianopsia. Lesions of the geniculocalcarine tract or visual cortex cause a contralateral homonymous hemianopsia (see Fig. 1-18, *B*). The optic nerve is unique amongst the cranial nerves in that it is consists of an extension of the CNS white matter surrounded by dural sheath, and is therefore not a peripheral nerve like the other cranial nerves. The optic nerve is subject to the pathologies instrinsic to white matter in the brain (such as multiple sclerosis) and its sheath subject to dural based processes.

Sparing of macular vision with cortical strokes is common, and one of four explanations is possible: (1) localization of macular fibers to the watershed area of middle and posterior cerebral artery supply, allowing for dual vascular supply to be present; (2) a very large cortical area devoted to central vision, meaning that small strokes will not affect all fibers; (3) some decussation of macular fibers, so that they are bilaterally represented; and (4) testing artifact because of poor central fixation by the subject.

Auditory System

What goes on in your head when your cell phone chirps? Sound is transmitted through the external ear via vibrations of the tympanic membrane to the middle ear ossicles. It is then transmitted to the hair cells of the organ of Corti in the cochlea. The cochlear division of the eighth cranial nerve runs in the anterior inferior portion of the internal auditory canal. From the canal, the nerve enters the brain stem at the junction between the pons and the medulla. The fibers end in the dorsal and ventral cochlear nuclei, which are identified in the upper part of the medulla along its dorsal surface (Fig. 1-19). After this primary synapse the secondary nerves for hearing may cross the midline at the level of the pons and ascend in the lateral lemniscus in what is termed the trapezoid body. A synapse may occur in the trapezoid body, but other fibers may synapse in the superior olivary nucleus. Some fibers may remain on the ipsilateral side and synapse in the ipsilateral superior olivary nucleus to ascend in the ipsilateral lateral lemniscus. The tertiary neurons of the lateral lemniscus pass through the ventral portion of the pons and midportion of the midbrain before synapsing at the inferior colliculus (thought to be instrumental in frequency discrimination). From the inferior colliculus, the fourth-order fibers pass to the medial geniculate nucleus of the thalamus. It should be noted, however, that fibers may bypass each of these nuclei to get to the next level in the auditory pathway.

Fibers from the medial geniculate course in the posterior limb of the internal capsule as the auditory radiations and terminate in the anterosuperior transverse temporal gyri and superior temporal gyrus. The auditory association cortex is also located in the temporal lobe. Unilateral lesions in the auditory cortex do not induce complete deafness in the contralateral ear, but there is a decrease in auditory acuity because of crossing fibers in the lateral lemniscus and crossed connections between the nuclei of the lateral lemniscus and the inferior colliculi. Most causes of unilateral hearing loss (speak up please!) are at the level of the inner and middle ear.

Limbic System

The main components of the limbic system include the fornix, the mammillary bodies, the hippocampus, the amygdala, and the anterior nucleus of the thalamus. The limbic system controls the emotional responses to visceral stimuli. In addition, portions of memory function are contained within the limbic system.

The olfactory and gustatory systems tie into the limbic system. The olfactory bulb receives nerve fibers located in the upper nasal cavity, the ciliary nerves. The olfactory bulbs feed into the olfactory tracts lying just under the gyrus rectus region in the olfactory sulcus of the frontal lobes. The olfactory tracts penetrate the brain just under the lamina terminalis and send nerve fibers to the septal nuclei, the parahippocampus, the uncus, and the amygdala via medial and lateral striae.

The amygdala is a primary work station for emotions. In addition to input from the olfactory system, it receives fibers from the thalami and the hypothalamus. Efferent fibers are sent from the amygdala to the temporal and frontal lobes, the thalamus, the hypothalamus, and the reticular formation in the brain stem. Lesions of the amygdala and other portions of the limbic system (or reading this chapter without taking a break) may cause anhedonia with a lack of emotional response to what are normally pleasurable stimuli.

The fornices are white matter tracts lying medially beneath the corpus callosum and are the major white matter

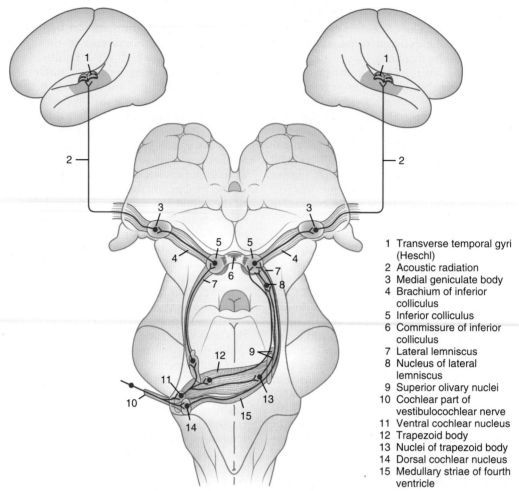

1 Transverse temporal gyri
  (Heschl)
2 Acoustic radiation
3 Medial geniculate body
4 Brachium of inferior
  colliculus
5 Inferior colliculus
6 Commissure of inferior
  colliculus
7 Lateral lemniscus
8 Nucleus of lateral
  lemniscus
9 Superior olivary nuclei
10 Cochlear part of
   vestibulocochlear nerve
11 Ventral cochlear nucleus
12 Trapezoid body
13 Nuclei of trapezoid body
14 Dorsal cochlear nucleus
15 Medullary striae of fourth
   ventricle

FIGURE 1-19 The auditory system in the brain stem and diencephalon (dorsal view) and in the cerebrum (lateral view). (From Kretschmann H-J, Weinrich W. *Cranial Neuroimaging and Clinical Neuroanatomy: Magnetic Resonance Imaging and Computed Tomography.* Rev 2nd ed. New York: Thieme; 1993:284.)

relays from one hippocampus to the other and on to the hypothalamus. The forniceal columns invaginate into the lateral and third ventricles as they sweep anteroinferiorly to end in the mammillary bodies. The anterior portions of the fornices parallel the corpus callosum but are more inferiorly and centrally located.

The hippocampal formation includes the hippocampus (located in the temporal lobe above the parahippocampus), the indusium griseum (a fine gray matter tract situated between the corpus callosum and cingulate gyrus connecting septal nuclei to parahippocampal gyri), and the dentate gyrus (just above the parahippocampal gyrus). The hippocampus is found along the medial temporal lobe adjacent to the inferior temporal horn of the lateral ventricle and the choroidal fissure. The hippocampus is involved primarily with visceral responses to emotions (with the hypothalamus) and memory, with less input into olfaction.

Taste

Taste from the anterior two thirds of the tongue is transmitted by the chorda tympani, a branch of cranial nerve VII that runs with fibers from the third division of cranial nerve V as the lingual nerve. From the chorda tympani, the

fibers run through the otic and geniculate ganglia to end at the cell bodies in the nucleus solitarius. The taste papillae of the posterior third of the tongue are supplied by cranial nerve IX. These fibers course through the petrosal ganglion of cranial nerve IX to reach the cell bodies in the nucleus solitarius (Fig. 1-20). Some bitter taste fibers may be supplied via the vagus nerve's nodose ganglion from the epiglottis.

From the nucleus solitarius, projections are made to the pons, both ventromedial nuclei of the thalamus, hypothalamus, and amygdala. These limbic structures monitor the visceral (nausea, vomiting, sweating, flushing, salivation) and emotional responses (elation, disgust, satiation) to certain foods, like jalapeños. From the thalamus, fibers track up to both sides of the sensory cortex, where the tongue occupies a huge proportion of the homunculus projection of the body on the brain surface.

Speech

Obviously, speech ties into the motor and auditory pathways described previously. Nonetheless, a brief description of the speech pathway is warranted because of its critical role in humans.

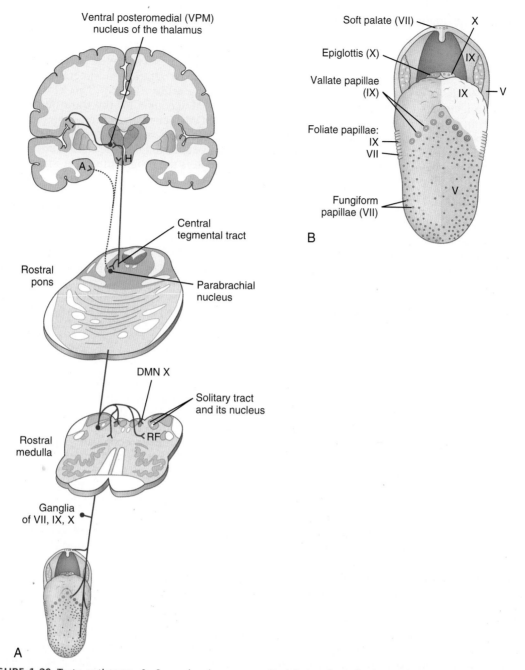

FIGURE 1-20 Taste pathways. **A,** Second-order neurons feed into reflexes both by direct projections (e.g., to the nearby dorsal motor nucleus of the vagus) and by connections with the reticular formation. The projection from the parabrachial nucleus to the hypothalamus and amygdala is dashed because its existence in primates has not been demonstrated conclusively. **B,** Distribution and innervation of taste buds and innervation of the lingual epithelium. The trigeminal nerve (V) subserves general sensation from the anterior two thirds of the tongue, and the glossopharyngeal nerve (IX) has a similar function with taste for the posterior third of the tongue. Taste is controlled anteriorly by the chorda tympani of the facial nerve (VII). *DMN,* Dorsal motor nucleus. (**A** and **B** from Nolte J. *The Human Brain: An Introduction to Its Functional Anatomy.* 4th ed. St. Louis: Mosby; 1999:324, 320, respectively.)

Briefly, speech requires coordination between the left hemisphere's temporoparietal areas assigned to the sensation of speech and the inferior frontal gyrus (Broca area) assigned to motor function. Portions of the arcuate fasciculus connect these two areas. In a few persons (usually left-handed), speech may be localized to the right hemisphere. Portions of the superior and middle temporal gyri and the inferior parietal lobule control ideational language. The auditory association cortex of the superior temporal gyrus (Wernicke area) handles receptive understanding of speech and is deficient in most politicians. If the inferior frontal gyrus is injured, coordination of intelligible expressive speech is lost (motor aphasia). Receptive aphasia develops with lesions in the Wernicke area, and conductive aphasia (disturbance in speech in response to verbal command but not spontaneously [ideomotor dyspraxia]) occurs with arcuate fasciculus lesions.

CRANIAL NERVES

How many medical students over the years have recited one of the classic mnemonics for the cranial nerves: On old Olympus' towering tops, a Finn and German viewed some hops (olfactory, optic, oculomotor, trochlear, trigeminal, abducens, facial, acoustic, glossopharyngeal, vagus, spinal accessory, and hypoglossal nerves)? The phrase remains imprinted forever, but here is a refresher on the anatomy. The cranial nerves can be organized in groups of four. Cranial nerves I through IV arise from the midbrain (oculomotor and trochlear) or above (olfactory and optic nerves). Cranial nerves V through VIII arise from the pons. The last four cranial nerves arise from the medulla. The majority of the cranial nerves are not appreciable by routine noncontrast CT. MRI particularly with thin section images or myelographic CT (performed after intrathecal administration of contrast) are typically performed when cranial nerve pathology is suspected.

Olfactory Nerve

"Hey, what's that smell?" The first cranial nerve, the olfactory nerve, consists of primary afferent neurosensory cells in the roof of both nasal cavities. The cells have efferent axons that pass through the cribriform plate to synapse on secondary sensory neurons in the olfactory bulb situated along the medial and inferior portions of the anterior cranial fossa. The central connections to the brain from the olfactory bulbs and tracts enter the inferior frontal regions via the medial and lateral olfactory striae. Neuronal pathways pass to the subcallosal medial frontal lobe (via medial olfactory stria) and the inferomedial temporal lobes (via lateral olfactory stria). Specifically, the gyrus semilunaris and gyrus ambiens (forming the prepiriform cortex), the entorhinal-parahippocampal cortex (forming the inferomedial margin of the temporal lobe anterior to the amygdala), and the amygdala receive afferents via the lateral stria. The anterior paraterminal gyrus of the frontal lobe and the lamina terminalis appear to receive fibers from the medial stria. The tertiary pathways from these olfactory projections are located in the (1) orbitofrontal cortex, (2) hypothalamus via the septum pellucidum, (3) hippocampus, and (4) limbic system.

The limbic responses to olfaction are derived through connections from the olfactory area to the hypothalamus, habenular nucleus, reticular formation, and cranial nerves controlling salivation, nausea, and gastrointestinal motility.

Optic Nerve (CN II)

The visual pathway has been described previously (see Fig. 1-18).

Oculomotor Nerve (CN III)

The nucleus of the oculomotor nerve is identified in the midbrain just posterior to the red nucleus and anterior to the superior aspect of the cerebral aqueduct. The oculomotor nuclei are paramedian structures that transmit the oculomotor nerve as it courses around and through the red nucleus, with its exit from the midbrain in the interpeduncular cistern. The oculomotor nerve then proceeds anteroinferiorly between the posterior cerebral and superior cerebellar arteries and lateral to the posterior communicating artery (PCOM). This close approximation to the PCOM accounts for the third nerve palsy often seen with PCOM aneurysms. It then enters the cavernous sinus lying in the superolateral portion of the cavernous sinus just above and lateral to the carotid artery. The oculomotor nerve enters the orbit through the superior orbital fissure. The parasympathetic nucleus of the oculomotor nerve, the Edinger-Westphal nucleus, controls the pupillary muscles of the eye for constriction and dilation. The motor portion of the cranial nerve III supplies the extraocular muscles except for the lateral rectus and superior oblique muscles. The levator palpebrae muscle is also supplied by the oculomotor nerve, accounting for the ptosis seen in third nerve palsies. The sympathetic system supplies Müller muscle (superior tarsus portion of levator palpebrae), which is why one gets a ptosis with Horner syndrome (ptosis, anhidrosis, miosis, and enophthalmos).

Trochlear Nerve (CN IV)

Cranial nerve IV, the trochlear nerve, has a nucleus in the midbrain in a location just below the nucleus of cranial nerve III, anterior to the aqueduct. Cranial nerves III, IV, and VI interconnect via the medial longitudinal fasciculus (MLF) to coordinate conjugate extraocular muscle movements. The MLF is located anterior to these three cranial nerve nuclei. The MLF receives connections from many sites, including the nuclei of nerves III, IV, VI, VII, XI, and XII; the vestibular system; and the spine.

Cranial nerve IV is unique in that it is the only one where the nerve exits the posterior portion of the brain stem and it is also the only one that crosses in entirety to the other side (see Fig. 1-4). The fibers of the trochlear nerve decussate just below the inferior colliculi posterior to the cerebral aqueduct. After leaving the brain stem, the cranial nerve IV fibers course anteroinferiorly around the midbrain below the tentorium to enter the cavernous sinus lying just below the cranial nerve III fiber tracts. The trochlear nerve, together with the oculomotor nerve and cranial nerve VI, enters the orbit through the superior orbital fissure. The trochlear nerve supplies the superior oblique muscle.

Trigeminal Nerve (CN V)

The motor and sensory nuclei of cranial nerve V are situated in the posterior aspect of the mid-pons, just ventral to the superior cerebellar peduncle (Fig. 1-21). Cranial nerve V exits the lateral aspect of the pons and courses anteriorly to synapse in the Meckel cave (see Fig. 1-5). The ganglion in the Meckel cave is known as the gasserian or semilunar ganglion. From the gasserian ganglion cranial nerve V trifurcates. Its first division is the ophthalmic nerve, which branches out to the tentorium and then runs anteriorly into the inferior portion of the cavernous sinus, enters the superior orbital fissure, and provides sensory afferents to the upper portion of the face, the eye, the lacrimal gland, and the nose.

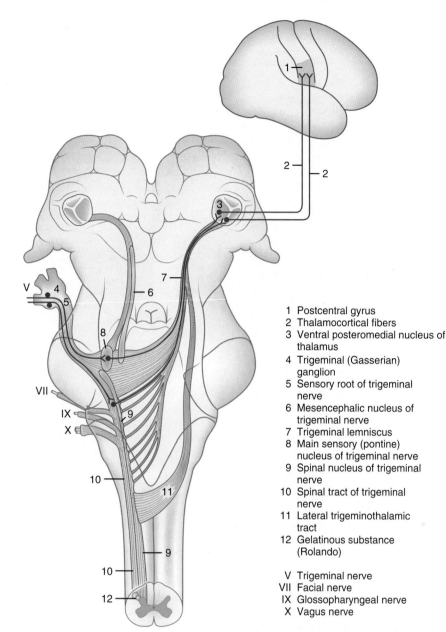

1 Postcentral gyrus
2 Thalamocortical fibers
3 Ventral posteromedial nucleus of thalamus
4 Trigeminal (Gasserian) ganglion
5 Sensory root of trigeminal nerve
6 Mesencephalic nucleus of trigeminal nerve
7 Trigeminal lemniscus
8 Main sensory (pontine) nucleus of trigeminal nerve
9 Spinal nucleus of trigeminal nerve
10 Spinal tract of trigeminal nerve
11 Lateral trigeminothalamic tract
12 Gelatinous substance (Rolando)

V Trigeminal nerve
VII Facial nerve
IX Glossopharyngeal nerve
X Vagus nerve

FIGURE 1-21 The trigeminal system in the spinal cord, brain stem, and diencephalon (dorsal view) and in the cerebrum (lateral view). Roman numerals indicate the other cranial nerves. (From Kretschmann H-J, Weinrich W. *Cranial Neuroimaging and Clinical Neuroanatomy: Magnetic Resonance Imaging and Computed Tomography.* Rev 2nd ed. New York: Thieme; 1993:267.)

The second division of cranial nerve V, the maxillary nerve, also travels the cavernous sinus, but then enters the foramen rotundum and courses anteriorly to the pterygopalatine fossa; from there it supplies sensory innervation throughout the maxillofacial region below the orbits. It has a meningeal branch to the middle cranial fossa. The maxillary nerve enters the inferior orbital fissure, passing through the infraorbital groove to supply the inferior eyelid, upper lip, and nose. The upper teeth and gingiva are supplied by alveolar nerves, which have branches arising from the zygomatic nerve off the maxillary nerve in the pterygopalatine fossa (Fig. 1-22).

The third division of the trigeminal nerve, the mandibular division, supplies motor function to the muscles of

mastication (masseter, temporalis, and medial and lateral pterygoid muscles). Chew on that! The mandibular division also supplies motor function to the tensor tympani, tensor veli palatini, anterior belly of the digastric muscle, and mylohyoid muscle. In addition, sensory fibers contributing to the lower portion of the face, ear, temporomandibular joint, temple, and tympanic membrane are transmitted through the mandibular division of cranial nerve V. From the Meckel cave, the mandibular nerve passes directly through the foramen ovale as it exits through the skull base. A meningeal branch returns through the foramen spinosum with the middle meningeal artery. The posterior branch of the mandibular nerve divides into auriculotemporal, lingual, and inferior alveolar nerves.

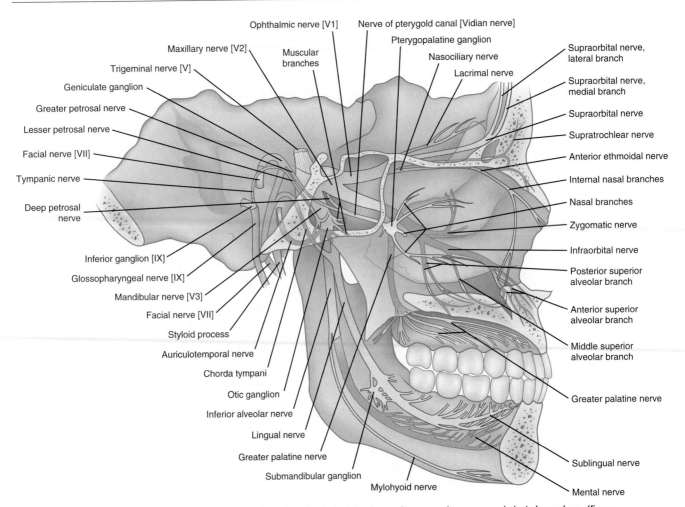

Ophthalmic nerve [V1]
Maxillary nerve [V2]
Trigeminal nerve [V]
Geniculate ganglion
Greater petrosal nerve
Lesser petrosal nerve
Facial nerve [VII]
Tympanic nerve
Deep petrosal nerve
Inferior ganglion [IX]
Glossopharyngeal nerve [IX]
Mandibular nerve [V3]
Facial nerve [VII]
Styloid process
Auriculotemporal nerve
Chorda tympani
Otic ganglion
Inferior alveolar nerve
Lingual nerve
Greater palatine nerve
Submandibular ganglion
Mylohyoid nerve

Muscular branches
Nerve of pterygold canal [Vidian nerve]
Pterygopalatine ganglion
Nasociliary nerve
Lacrimal nerve

Supraorbital nerve, lateral branch
Supraorbital nerve, medial branch
Supraorbital nerve
Supratrochlear nerve
Anterior ethmoidal nerve
Internal nasal branches
Nasal branches
Zygomatic nerve
Infraorbital nerve
Posterior superior alveolar branch
Anterior superior alveolar branch
Middle superior alveolar branch
Greater palatine nerve
Sublingual nerve
Mental nerve

FIGURE 1-22 The nerves of the face: the trigeminal, facial, glossopharyngeal nerves and their branches. (From Putz R, Pabst R, eds. *Sobotta Atlas of Human Anatomy.* 13th ed. Philadelphia: Lippincott Williams & Wilkins; 1996:78.)

The auriculotemporal nerve provides sensory innervation to the skin of the lower face and temporal regions, external auditory canal and tympanic membrane, auricle, and temporomandibular joint. This nerve has two major branches, the larger of which travels lateral to the middle meningeal artery and the smaller of which travels medial to the middle meningeal artery. These branches then coalesce and extend into the parotid gland, where several branches join the facial nerve.

The lingual nerve contains general sensory fibers and taste fibers from the chorda tympani (facial nerve) arising from the anterior two thirds of the tongue. The intimate relationship of branches of V and VII become very important when assessing for perineural tumor spread in the setting of head and neck cancers.

The inferior alveolar nerve enters the mandible through the mandibular (inferior alveolar) canal, giving off motor supply to the mylohyoid muscle and anterior digastric muscle. Its sensory fibers supply the gingiva and teeth of the mandible.

Abducens Nerve (CN VI)

Cranial nerve VI has its nucleus in the middle of the pons near the floor of the fourth ventricle (see Fig. 1-5). The abducens nerve courses from the nucleus anteriorly, leaving the brain stem at the junction of the pons and the pyramids of the medulla. Cranial nerve VI courses through the Dorello canal before entering the cavernous sinus, where it lies lateral to the carotid artery. In this location, it is the nerve closest to the artery and therefore most sensitive to cavernous carotid artery disease. The nerve then enters the superior orbital fissure and innervates the lateral rectus muscle.

Facial Nerve (CN VII)

The facial nerve is a multitasker with numerous and varied functions. It has its motor nucleus in the midpons, situated anterolateral to that of the abducens nucleus (see Fig. 1-5). The facial nerve encircles the nucleus of CN VI and makes a tiny bump on the anterior margin of the fourth ventricle (facial colliculus). It then extends anteriorly to join fibers from the superior salivatory nucleus and the nucleus solitarius of cranial nerve VII, to be joined by the spinal tract of cranial nerve V. The nucleus solitarius receives taste afferents from VII (anterior tongue), IX (posterior tongue), and X (epiglottis). The facial nerve leaves the pons further laterally than the abducens nerve and crosses the cerebellopontine angle cistern to enter the internal auditory canal.

The seventh nerve has a larger motor component and a smaller sensory portion termed the nervus intermedius. The nervus intermedius contains fibers for taste and has parasympathetic fibers as well. The nervus intermedius lies between the motor component of CN VII and CN VIII in the internal auditory canal. The cranial nerve VII fibers run in the anterior superior portion of the internal auditory canal and the sensory fibers have their first synapse at the geniculate ganglion located along the anterior surface of the petrous bone at the level of the ossicles. From the geniculate ganglion, these fibers leave as the greater superficial petrosal nerve, which transmits fibers to the lacrimal apparatus. Cranial nerve VII then turns inferiorly, branching off a nerve to the stapedius muscle and, in the lower mastoid portion of the temporal bone, transmits the chorda tympani branch of the facial nerve. This joins the lingual branch of the mandibular nerve and provides taste sensation to the anterior two thirds of the tongue. On leaving the mastoid portion of the temporal bone through the stylomastoid foramen, the nerve passes laterally around the retromandibular vein in the parotid gland, coursing superficially through the masticator space to innervate the muscles of facial expression.

Vestibulocochlear (Acoustic) Nerve (CN VIII)

The nuclei of cranial nerve VIII are located in the superior aspect of the medulla along the base of the inferior cerebellar peduncle (see Fig. 1-7). The cochlear and vestibular nuclei are situated adjacent to each other, with the vestibular nuclei located more medially. The nerves exit the pontomedullary junction posterior to the inferior olivary nucleus. The cochlear division of the vestibulocochlear nerve runs in the anteroinferior portion of the internal auditory canal, whereas the vestibular branches run in the superior and inferior posterior portions of the internal auditory canal. The nerves end with the cochlear division entering the cochlea through the cochlear aperture and the vestibular components entering the semicircular canals.

Glossopharyngeal Nerve (CN IX)

Cranial nerve IX has its nucleus in the medulla just posterior to the inferior olivary nucleus. The nuclei of the glossopharyngeal nerve include the nucleus ambiguus (for motor function), the inferior salivatory nucleus (for salivation), and nucleus of the tractus solitarius (for taste). The vagus and glossopharyngeal nerves share the nucleus ambiguus and the dorsal vagal nuclei (touch, pain, and temperature to tongue) as sites of origin. The fibers from cranial nerve IX pass through the olivary sulcus to enter the pars nervosa of the jugular foramen. The nerves pass through the jugular foramen, perforate the superior constrictor muscles, and innervate the pharynx and stylopharyngeus muscles. Taste and general somatic sensation to the posterior third of the tongue and part of the soft palate are supplied by nerve IX. Meningeal branches to the posterior fossa, tympanic branches (Jacobson nerve, responsible for glomus tumors), and carotid body fibers are transmitted through cranial nerve IX. The otic ganglion, which receives the lesser petrosal nerve from tympanic branches of the glossopharyngeal nerve, has secretomotor fibers for

the parotid gland from CN VII and may communicate with the chorda tympani and vidian nerve.

Vagus Nerve (CN X)

The vagus nerve (see Fig. 1-7, *D*) has three parent nuclei within the medulla. The nucleus ambiguus gives rise to motor fibers to the larynx and pharynx. The dorsal nucleus of the vagus nerve is identified just anterior to the fourth ventricle in its inferior aspect. It receives sensory information and transmits motor information to and from the cardiovascular, pulmonary, and gastrointestinal tracts. The nucleus solitarius receives CN X taste afferents from the epiglottis and valleculae. The vagus nerve exits the medulla through the olivary sulcus and enters the pars vascularis of the jugular foramen. It receives efferents from the cranial root of the spinal accessory nerve, which provide innervation to the recurrent laryngeal nerves. The main trunks of the vagus nerve then run in the carotid sheath with the internal carotid arteries and internal jugular vein. Recurrent laryngeal nerves loop under the aorta on the left side and subclavian artery on the right, and travel in the tracheoesophageal grooves, before supplying all the laryngeal muscles except the cricothyroids (supplied by the superior laryngeal nerve, also a branch of the vagus).

Spinal Accessory Nerve (CN XI)

The spinal accessory nerve receives fibers from the first three cervical spinal levels as well as the motor cortex. The nerves ascend as the ansa cervicalis (running along the carotid sheath) and course with cranial nerve X out the postolivary sulcus into the jugular foramen's pars vascularis along with the internal jugular vein. The spinal accessory nerve's motor nucleus is in the lower medulla and upper cervical spinal cord and it supplies branches to the sternocleidomastoid muscle and trapezius muscle. The ansa cervicalis (C1-C3) fibers innervate the infrahyoid strap muscles.

Hypoglossal Nerve (CN XII)

The hypoglossal nerve (see Fig. 1-7, *E*) has its nucleus along the paramedian area of the anterior wall of the fourth ventricle in the medulla. The nerve fibers course anteriorly to exit the medulla in the preolivary sulcus. From there, the nerve courses through the hypoglossal canal. It supplies the intrinsic muscles of the tongue as well as the genioglossus, styloglossus, and hyoglossus muscles. Ansa cervicalis branches run with the hypoglossal nerve to supply the anterior strap muscles.

VASCULAR ANATOMY

The advent of cross-sectional imaging modalities has relegated conventional angiography to the role of delineating vascular disease including aneurysms, vasculitis, and arteriovenous malformations, not detection or localization of masses as in yesteryear. Magnetic resonance angiography and CT angiography are largely replacing diagnostic angiography. Catheter based angiography for diagnosis is now mainly a problem solving tool and a prerequisite for the therapeutic intervention for a lesion.

The common carotid artery bifurcates into an external and internal carotid artery approximately at the level of the third or fourth cervical vertebral body in most persons. The angle of the mandible is another good marker for the carotid bifurcation. The external carotid artery typically courses anteromedially to the internal carotid artery from the bifurcation. However, the internal carotid artery typically crosses medial to the external carotid artery at approximately the C1-2 level as it turns to enter the skull.

External Carotid Branches

The branches of the external carotid artery can be remembered by the mnemonic, "She always likes friends over Pop, Mom, and Sis." With this memory aid, you can remember the external carotid artery branches of superior thyroidal, ascending pharyngeal, lingual, facial, occipital, posterior auricular, (internal) maxillary, and superficial temporal arteries (Fig. 1-23). Knowing the branches of the external carotid artery is important for understanding the vascular supply to head and neck and skull base tumors and as an icebreaker at cocktail parties. Anterior and middle cranial fossa meningiomas by and large are supplied by the anterior and posterior divisions of the middle meningeal artery, a branch of the internal maxillary artery. The ascending pharyngeal artery is the artery that is typically implicated in supply of glomus jugulare, glomus tympanicum, and carotid body tumors, while the internal maxillary artery is typically implicated in the supply of juvenile nasopharyngeal angiofibromas.

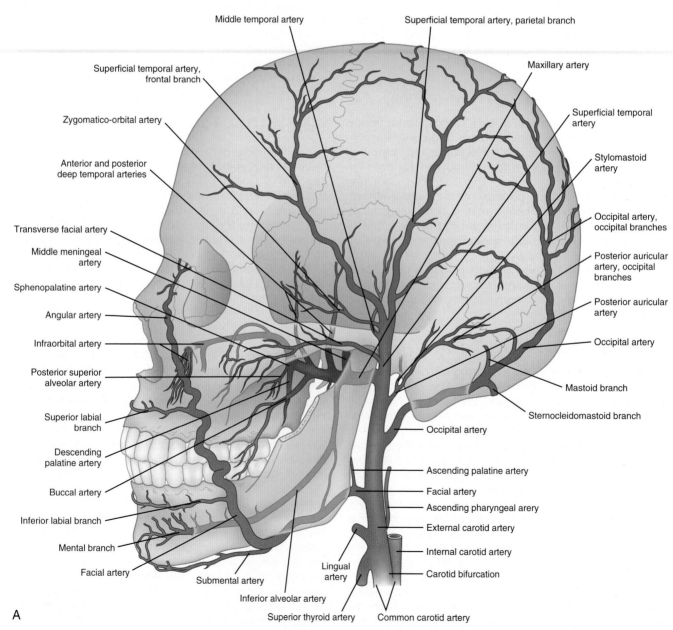

A

FIGURE 1-23 External carotid artery (ECA) anatomy. **A,** The ECA and its branches, lateral aspect.

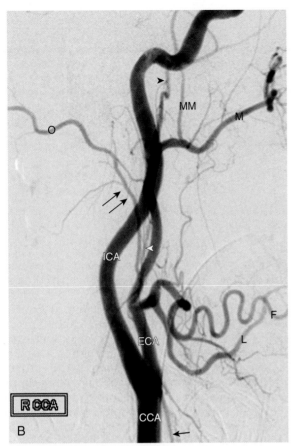

FIGURE 1-23, cont'd B, ECA circulation. Lateral view of a common carotid artery (CCA) injection with opacification of the internal carotid artery (ICA) and ECA. ECA branches can be seen including superior thyroidal *(black arrow),* lingual (L), facial (F), ascending pharyngeal *(white arrowhead),* internal maxillary (M), occipital (O), and superficial temporal *(black arrowhead)* branches. The middle meningeal artery (MM) is seen and the posterior auricular artery *(black double arrows)* is superimposed on the occipital artery. Note common origin of the lingual and facial arteries. (**A** from Putz R, Pabst R, eds: *Sobotta Atlas of Human Anatomy.* 13th ed. Philadelphia: Lippincott Williams & Wilkins; 1996:76.)

The superior thyroidal artery emerges off the anterior aspect of the external carotid artery and supplies the caudal aspect of the thyroid gland. It runs with the external laryngeal nerve (a branch of the vagus' superior laryngeal nerve, which supplies the cricothyroideus muscle) and gives limited blood supply to the larynx. The thyroid gland also receives blood supply from inferior thyroidal vessels from the subclavian artery's thyrocervical trunk.

The ascending pharyngeal artery has three divisions. Whereas its anterior (pharyngeal) branch supplies the posterolateral pharyngeal wall and palatine tonsil, the middle (neuromeningeal) and posterior branches have extensive supply to the jugular foramen region, supplying the vasa nervorum of cranial nerves IX, X, and XI.

The lingual artery arises anteriorly off the external carotid artery and has a characteristic hook in its course. It supplies the floor of mouth, hyoid muscles, tongue, and submandibular and sublingual glands.

The facial artery may arise independently or demonstrate common trunk with the lingual artery. This vessel supplies the region around the mandible, submandibular gland, and maxilla before anastomosing with distal branches of the internal maxillary artery to supply the remainder of the anterior lower facial region. It has labial artery branches for the upper and lower lips, a lateral nasal branch to the nose, and a terminal angular artery at the medial canthus of the eye. Its nasal and orbital ramifications anastomose with ophthalmic branches to the same region.

The occipital artery arises posteriorly off the external carotid artery and branches off to the sternocleidomastoid muscle, posterior belly of the digastric muscle, the back of the head, and the meninges. The stylomastoid branch of the occipital artery may supply skull base lesions such as glomus tumors.

The posterior auricular artery supplies the back of the ear and has meningeal anastomoses to the posterior fossa.

The superficial temporal artery supplies the parotid gland, masseter, buccinator muscles, lateral part of the scalp, and the anterior scalp and cheek. Anterior branches anastomose with ophthalmic artery branches, whereas posterior branches join posterior auricular and occipital artery supply.

Branches of the internal maxillary artery are significant for supplying collateral circulation to the distal internal carotid artery when internal carotid artery occlusion occurs. The middle meningeal artery may anastomose with ethmoidal branches of the ophthalmic artery, or collateral circulation may be achieved by way of the meningolacrimal artery to the ophthalmic artery. On the lateral arteriogram the middle meningeal artery can be distinguished from the superficial temporal artery by (1) its less sinuous appearance, (2) its entrance at the foramen spinosum, (3) its origin from the internal maxillary artery, (4) its persistence within the confines of the skull, and (5) its characteristic course and divisions running anteriorly and posteriorly. Other internal maxillary artery branches besides the middle meningeal artery such as the ethmoidal, temporal, lacrimal, palpebral, and muscular branches may also anastomose to the ophthalmic artery. The artery of the foramen rotundum from the maxillary artery may anastomose with branches of the inferolateral trunk of the cavernous segment of the internal carotid artery. The vidian artery may also anastomose with the cavernous carotid artery. Additional external carotid–internal carotid artery vascular anastomoses include the branches of the ascending pharyngeal artery to the internal carotid artery via petrous and carotid artery branches. The occipital artery may anastomose to the internal carotid via the stylomastoid artery.

External carotid system anastomoses to the vertebral artery may occur via the occipital artery, ascending pharyngeal artery, or posterior auricular artery. The occipital artery sends meningeal branches to the foramen magnum, the jugular foramen, the hypoglossal canal, and the mastoid canal.

Several anastomotic channels that are present in fetal life and connect carotid and vertebral circulations usually regress in utero but can remain patent in rare instances

TABLE 1-4 Persistent Vascular Connections

Persistent Artery	Origin	Feeds	Location	Coexistent Findings
Trigeminal	Cavernous carotid	Top of basilar	Suprasellar cistern	Aneurysms, hypoplastic vertebrobasilar system
Otic	Petrous carotid	Mid basilar	Internal auditory canal	Hypoplastic vertebrobasilar system
Hypoglossal	High cervical internal carotid at skull base	Intracranial vertebrobasilar circulation	Hypoglossal canal	Hypoplastic vertebrobasilar system
Proatlantal type 1	Low internal carotid	Cranial and cervical vertebrobasilar circulation	C2 level	Hypoplastic vertebrobasilar system
Proatlantal type 2	External carotid	Cranial and cervical vertebrobasilar circulation	C2 level	Hypoplastic vertebral arteries

(Table 1-4). The persistent trigeminal artery is the most common persistent anastomosis which connects the cavernous carotid artery to the upper basilar artery in 0.1% to 0.2% of people (Fig. 1-24). This normal variant may be associated with hypoplasia of the vertebrobasilar system below the anastomosis. The persistent otic and hypoglossal arteries are found more inferiorly (and much more rarely) connecting the petrous (otic) and cervical (hypoglossal) internal carotid artery to the lower basilar artery. A persistent proatlantal artery connects the cervical carotid to the vertebral artery at the occipital-atlantal junction (type I) or atlantoaxial junction (type II).

Intracranial Circulation

The intracranial circulation is supplied by paired internal carotid arteries and paired vertebral arteries, the latter of which join to form the basilar artery. Because of the collateral network inherent in the circle of Willis, where the carotid and vertebrobasilar systems are connected, the brain possesses a redundant defense against major-vessel occlusive disease (Fig. 1-25). Such is not the case with the branches distal to the circle of Willis, where collateral circulation is less easily supplied. Therefore, it is important to understand the vascular anatomy and arterial supply to localize vasculopathy.

The internal carotid arteries have cervical (C1), petrous (C2), lacerum (C3), and cavernous portions (C4) before they pierce the dura to enter the intracranial space. The clinoid segment (C5), ophthalmic segment (C6), and communicating segment (C7) follow the cavernous segment. On routine angiography in a normal patient, you do not see branches in the cervical or petrous portions of the internal carotid artery. Occasionally, the ophthalmic artery may arise below the dura from the intracavernous carotid artery, and rarely one may be able to see hypophyseal branches from the cavernous carotid artery. When carotid occlusive disease or vascular masses are present, the small branches from the petrous carotid artery (the tympanic arteries, vidian artery, and caroticotympanic branch) may be enlarged and visible on lateral carotid injections. The cavernous carotid branches include the (1) lateral mainstem artery (inferolateral trunk), (2) meningohypophyseal arteries, and (3) capsular sellar branches of McConnell. The tentorial artery off the meningohypophyseal trunk of the cavernous carotid artery is given the eponym the artery of Bernasconi-Cassinari, which can supply tentorial based lesions such as meningiomas and arteriovenous malformations.

In 80% to 90% of cases, the ophthalmic artery arises intradurally, just below the anterior clinoid process. Branches of the ophthalmic artery anastomose with those from the maxillary artery, providing a rich network for collateral supply in the event of proximal carotid artery occlusion. The ophthalmic artery supplies the orbit, the globe, the frontal scalp region, the frontal and ethmoidal sinuses, and the upper part of the nose. It has anastomoses with facial and internal maxillary artery branches.

The next branch of the internal carotid artery distal to the ophthalmic artery is the posterior communicating artery (PCOM). It connects the anterior carotid circulation with the posterior vertebrobasilar artery circulation. Occasionally, the connection to the basilar artery is absent and the PCOM directly feeds the posterior cerebral artery on that side (a fetal origin posterior cerebral artery [PCA]). When a diminutive proximal PCA is seen before the joining of the PCOM (P1 segment), the PCA is considered to have fetal type origin. The PCOM supplies parts of the thalamus, hypothalamus, optic chiasm, and mammillary bodies. Its named branches are the anterior thalamoperforate vessels. They supply parts of the thalamus, internal capsule, and optic tracts. Posterior thalamoperforate arteries, the artery of Percheron, and thalamogeniculate branches arise from the proximal posterior cerebral artery (the P1 segment), but the PCOM may give minimal supply to these areas. Not uncommonly, the PCOM may be absent or small on one side or both, and this is considered to be normal anatomic variation.

The anterior choroidal artery is the next branch of the supraclinoid intradural carotid artery. The anterior choroidal artery courses through the ambient cistern before entering the choroid plexus of the temporal horn via the choroidal fissure at the plexal point. The anterior choroidal artery supplies (and follows) portions of the optic tract, medial temporal lobe, uncus, amygdala, hippocampus, anterior limb of the internal capsule, choroid plexus of the lateral ventricle, inferior globus pallidus, caudate nucleus, cerebral peduncles, and midbrain. That is a lot of territory for a pretty small artery seen at angiography.

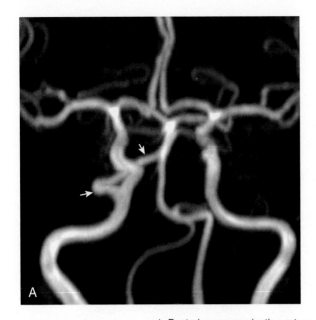

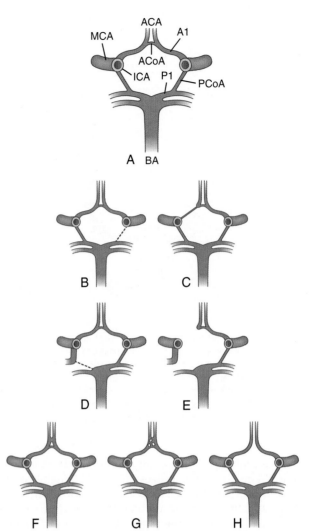

1 Posterior communicating artery
2 Trigeminal artery
3 Otic artery
4 Hypoglossal artery
5 Proatlantal intersegmental artery

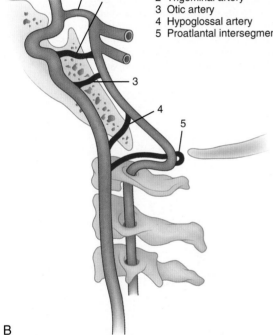

FIGURE 1-24 Persistent fetal connections. **A,** Magnetic resonance angiogram shows the persistent trigeminal artery *(arrows)* connecting the cavernous internal carotid and basilar arteries. The hypoplastic basilar artery below the trigeminal contribution is also evident. **B,** Anatomic diagram depicts the embryonic carotid-basilar and carotid-vertebral anastomoses. The posterior communicating artery is the only vessel that normally persists; the other four, shown in black, usually regress completely. (**B** from Osborne AG. *Diagnostic Cerebral Angiography.* 2nd ed. Philadelphia: Lippincott, Williams & Wilkins; 1998:69, Fig. 3-14.)

FIGURE 1-25 Circle of Willis. Anatomic diagrams depict the circle of Willis **(A)** and its common variations **(B-H). A,** A "complete" circle of Willis is present. Here all components are present and none are hypoplastic. This configuration is seen in less than half of all cases. **B,** Hypoplasia of one or both posterior communicating arteries (PCoAs) is the most common circle of Willis variant, seen in 25% to 33% of cases. **C,** Hypoplasia or absence of the horizontal (A1) anterior cerebral artery segment is seen in approximately 10% to 20% of cases. Here the right A1 segment is hypoplastic. **D,** Fetal type origin of the right posterior cerebral artery (PCA) with hypoplasia *(dotted line)* of the precommunicating (P1) PCA segment, seen in 15% to 25% of cases. **E,** If both a fetal origin PCA and an absent A1 occur together, the internal carotid artery (ICA) is anatomically isolated with severely restricted potential collateral blood flow. **F,** Multichannel (two or more) anterior communicating artery segments, seen in 10% to 15% of cases. Here the duplicated anterior communicating arteries (ACoAs) are complete and both extend across the entire ACoA. **G,** In a fenestrated ACoA, the ACoA has a more plexiform appearance. **H,** Absence of an ACoA is shown. *ACA,* Anterior cerebral artery; *BA,* basilar artery; *MCA,* middle cerebral artery. (From Osborne AG. *Diagnostic Cerebral Angiography.* 2nd ed. Philadelphia: Lippincott Williams & Wilkins; 1999:113.)

Anterior Cerebral Artery System

The internal carotid artery terminates at a bifurcation into the anterior and middle cerebral arteries (Fig. 1-26). The first, horizontal portion of the anterior cerebral artery (ACA) is termed the A-1 segment, and as it turns superiorly along the genu of the corpus callosum it forms the A-2 segment. This then bifurcates into pericallosal and callosomarginal arteries. The ipsilateral A-1 and contralateral A-1 segments are connected via the anterior communicating artery (ACOM). The anterior communicating artery has many anatomic variants, and may be fenestrated or consist

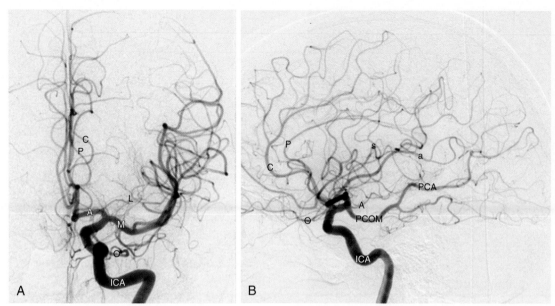

FIGURE 1-26 Internal carotid artery anatomy. **A,** Anteroposterior view of an internal carotid artery injection shows filling of anterior cerebral artery and middle cerebral artery (MCA) branches. The M-1 (M) and A-1 (A) segments are labeled. Note that the pericallosal artery (P) remains closer to the midline than callosal-marginal (C) branches. Lenticulostriate branches (L) and the ophthalmic artery (O) can also be made out. **B,** Lateral view in a different patient with persistent fetal origin of the posterior cerebral artery (PCA) demonstrates filling of the posterior communicating artery (PCOM) and PCA. There is opacification of the pericallosal (P) and callosal-marginal (C) branches. The ophthalmic artery (O), anterior choroidal artery (A), sylvian loops of the insula (s), and angular branch (a) of the MCA can be identified.

of one or multiple arterial connections between right and left ACA. The A-1 segment gives off medial lenticulostriate arteries to the basal ganglia and anterior limb of the internal capsule. Infarctions in the medial lenticulostriate distribution affect motor function of the face and arm. One named branch, the recurrent artery of Heubner, supplies the head of the caudate and anteroinferior internal capsule. The ACOM also gives off small perforating vessels. Not uncommonly, an A-1 segment may be hypoplastic or absent on one side.

Distal branches of the ACA supply the olfactory bulbs and tracts, the corpus callosum, and the medial aspects of the frontal and parietal lobes (Fig. 1-27). Therefore ACA infarctions affect olfaction, thought processes (the medial inferior frontal lobe), motor function of the leg (precentral gyrus medially), sensation to the leg (postcentral gyrus medially), memory, and emotion (the cingulate gyrus).

A series of internal frontal and internal parietal arteries arise from the callosomarginal artery after it gives off anteroinferior orbitofrontal and frontopolar branches. The terminations of the pericallosal artery and the callosomarginal artery are the splenial and cuneal arteries, respectively. These most distal ACA branches anastomose with distal PCA branches, making infarcts of the corpus callosum a rare phenomenon.

Middle Cerebral Artery System

The horizontal portion of the middle cerebral artery (MCA) is termed its M-1 segment, and it too gives off (lateral) lenticulostriate arteries that supply the globus pallidus, putamen, and internal capsule (Fig. 1-28). Lateral lenticulostriate infarctions cause internal capsule damage

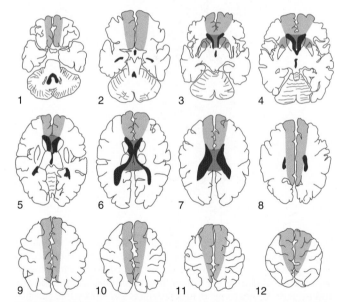

FIGURE 1-27 Anterior cerebral artery (ACA) distribution. Shaded areas of these axial diagrams, arranged in sequence from base to vertex, outline the territory of the ACA including the medial lenticulostriate *(orange)*, callosal *(blue)*, and hemispheric branches *(green)*. (From Latchaw RE. *MR and CT Imaging of the Head, Neck, and Spine.* 2nd ed. St. Louis: Mosby; 1991.)

typically resulting in hemiparesis and sensory deficits on the contralateral side of the body. Speech may be affected because the medial temporal lobes are affected. Vision may be affected because of involvement of the optic radiations as they sweep from the lateral geniculate to the Meyer loop in the temporal lobes.

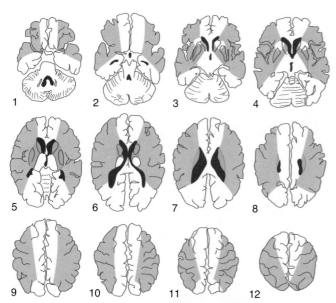

FIGURE 1-28 Middle cerebral artery (MCA) distribution. This diagram of the axial sections, arranged in sequence from base to vertex, outlines the MCA distribution with the lateral lenticulostriate *(orange)* and hemispheric branches *(green)*. (From Latchaw RE. *MR and CT Imaging of the Head, Neck, and Spine.* 2nd ed. St. Louis: Mosby; 1991.)

The M-2 segment refers to the sylvian segment of the MCA after it trifurcates into an anterior division, posterior division, and the anterior temporal artery. The distal branches of the MCA course lateral to the insula, and then loop around the frontal operculum to form the "candelabra" effect of the sylvian triangle on the lateral surface of the cortex (M-3 segment). The last branch of the sylvian vessels of the MCA is the angular artery supplying the angular gyrus just beyond the sylvian fissure. Other branches supply the frontal and parietal lobes. Anterior and inferior temporal branches supply the vast majority of the temporal lobe.

MCA infarctions may affect motor and sensory function of the face, arm, and trunk (lateral precentral and postcentral gyri), speech (inferior lateral frontotemporal gyri), thought processes (anteroinferolateral frontal lobes), hearing (superior temporal gyri), memory and naming of objects (temporal lobe), and taste (insular cortex).

Vertebral Arteries

The vertebral arteries arise from the subclavian arteries and course superiorly between the longus colli and scalene muscles (V-1) before entering the vertebral canal. The artery enters at the C6 transverse foramen in 95% of people and continues upward in the foramina of the vertebrae (V-2) before exiting at the C1-2 level. The left is bigger (dominant) than the right in 75% of people. You know what they say about big verts (just kidding!). Branches of the vertebral artery include muscular arteries to the neck, occipital, segmental spinal arteries, and the anterior spinal artery, which supplies most of the spinal cord. Posterior meningeal branches are found as the vessel travels from the atlas to pierce the dura (V-3), entering the intracranial compartment via the foramen magnum. The first branch of the vertebral artery in the intracranial compartment

(V-4) is the posterior inferior cerebellar artery (PICA). This vessel loops around the medulla and tonsil while supplying the posterolateral medulla, the inferior vermis, the choroid plexus of the fourth ventricle, and the inferior aspect of the cerebellum. The inferior vermian artery and tonsillohemispheric arteries are the terminating branches of the PICA. The size of the PICA is inversely proportional to the size of the anterior inferior cerebellar artery (AICA). Occasionally, one vertebral artery may terminate in PICA, without contributing to the basilar system. Also, common trunk origins of AICA and PICA (AICA-PICA complex) may be seen.

PICA infarcts can induce the lateral medullary (Wallenberg) syndrome. This causes loss of pain and temperature of the body on the contralateral side (lateral spinothalamic tract) and face on the ipsilateral side (descending trigeminothalamic tract), ataxia (cerebellar connections), ipsilateral swallowing and taste disorders (ninth cranial nerve), hoarseness (tenth cranial nerve), vertigo and nystagmus (eighth nerve), and ipsilateral Horner syndrome.

Basilar Artery Branches

The two vertebral arteries join to form the basilar artery at the pontomedullary level (Fig. 1-29). The basilar artery has many tiny branches to the pons and medulla that are never seen at angiography. The first major branch seen from the basilar artery is the AICA. This vessel runs toward the cerebellopontine angle and may loop into the internal auditory canal, where it gives off labyrinthine branches to the inner ear before supplying the anteroinferior part of the cerebellum. Medullary and pontine branches also arise from the AICA.

The superior cerebellar artery is the last infratentorial branch off the basilar artery. The lateral marginal and hemispheric branches supply the upper part of the cerebellum before the vessel terminates in the superior vermian artery. Other fine branches help supply the pons, the superior cerebellar peduncle, and the inferior colliculus. The superior vermian vessel anastomoses with the inferior vermian artery of the PICA.

The basilar artery terminates in the two posterior cerebral arteries (PCAs) and a few small perforating vessels from its vertical dome (Fig. 1-30). Midline perforating arteries (such as the artery of Percheron, which supplies paramedian thalamus and midbrain) arise from the basilar artery and proximal PCAs. Infarctions of these perforating vessels can affect cranial nerves III and IV, causing oculomotor deficits; the cerebral peduncles, affecting motor strength; the medial lemniscus, altering sensation; the red nucleus and substantia nigra, affecting coordination and motor control; and the reticular activating system, affecting the level of consciousness.

The PCAs also have small premammillary, posterior thalamoperforate, and thalamogeniculate branches supplying the hypothalamus, midbrain, and inferior thalami. Infarctions of these vessels may affect memory and emotion (anterior thalamus), endocrine function (hypothalamus), language (pulvinar), pain sensation (thalami), sight (lateral geniculate), and motor control (subthalamic nuclei). The PCAs join the anterior circulation via the PCOMs.

The next branches of the PCA are the medial and lateral posterior choroidal arteries that supply the trigone

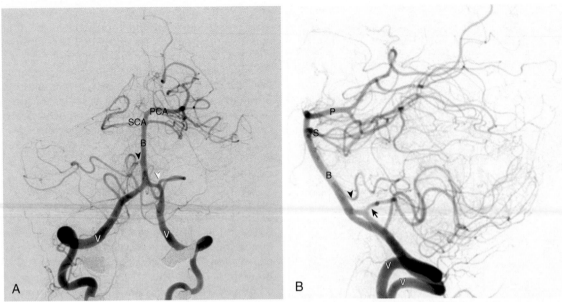

FIGURE 1-29 Vertebrobasilar artery circulation. **A,** Anteroposterior view of vertebral artery arteriogram depicts the vertebral arteries (V), the posterior inferior cerebellar arteries (PICAs; *white arrowhead*), the basilar artery (B), the anterior inferior cerebellar arteries (AICAs; *black arrowhead*), the superior cerebellar arteries (SCAs), and the posterior cerebral arteries (PCAs). Only the left PCA opacifies here because of persistent origin of the PCA on the right. **B,** Lateral view shows the same vascular anatomy. Vertebral arteries (V), basilar artery (B), PICA *(black arrow)*, AICA *(black arrowhead)*, SCA (S) and PCA (P) are shown.

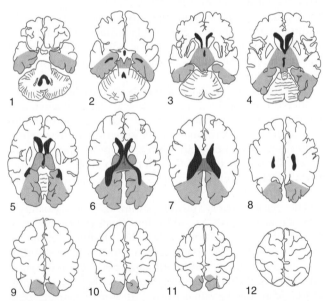

FIGURE 1-30 Posterior cerebral artery (PCA) distribution. Axial diagrams arranged in sequence from base to vertex outline supply from the PCA, the thalamic and midbrain perforators *(orange)*, callosal *(blue)*, and hemispheric branches *(green)*. (From Latchaw RE. *MR and CT Imaging of the Head, Neck, and Spine.* 2nd ed. St. Louis: Mosby; 1991.)

region of the lateral ventricles. Medial posterior choroidal arteries pass around the midbrain and course medially toward the pineal gland. They supply the tectum, the choroid plexus of the third ventricle, and the thalami. The lateral posterior choroidal arteries run behind the pulvinar into the choroidal fissure and supply the choroid of the lateral and third ventricles, the posterior thalamus, and the fornix.

The PCA continues, branching off anterior and posterior temporal arteries and a posterior pericallosal artery before terminating into parietooccipital branches and calcarine arteries to the occipital lobe. PCA infarctions affect vision most commonly (occipital lobes) but also affect memory (posteroinferior temporal lobe), smell (hippocampal region), and emotion (posterior fornix).

Venous Anatomy

Superficial Drainage

Superficial drainage patterns are highly variable from person to person. However, the superficial drainage of the brain is notable for the superficial vein of Labbé (draining from the sylvian fissure laterally into the transverse sinus), and the superior superficial vein of Trolard (draining from the sylvian fissure into the superior sagittal sinus). Superior cerebral veins also empty directly into the superior sagittal sinus (Fig. 1-31). The superficial middle cerebral vein drains from the sylvian fissure into the cavernous sinus. The vein of Labbé may arise from the posterior extent of the superficial middle cerebral vein.

Deep Supratentorial Drainage

The deep venous drainage of the supratentorial space centers on the internal cerebral vein and the vein of Galen. Medullary veins radiate downward from the superficial white matter to drain to the subependymal and thalamostriate veins. Choroidal, caudate, terminal, lateral, atrial, and ventricular veins drain to the medial subependymal septal vein and the lateral subependymal thalamostriate vein. The septal vein courses around the anteromedial aspect of the lateral ventricle before passing behind the foramen of Monro to join the internal cerebral vein at the "true venous angle." If the septal vein joins the internal cerebral vein

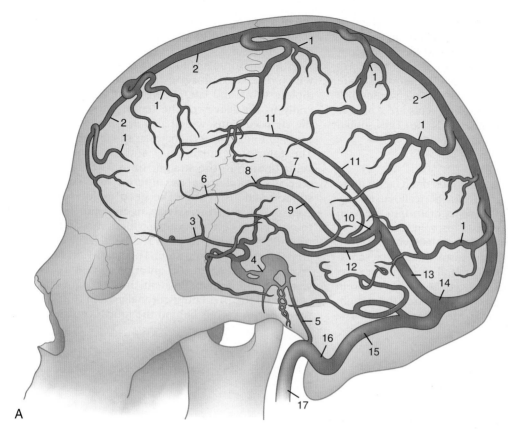

Superficial veins of cortical
regions and their sinuses:
  1 Superior (superficial)
    cerebral veins
  2 Superior sagittal sinus
  3 Superficial middle
    cerebral vein (of Sylvius)
  4 Cavernous sinus
  5 Inferior petrosal sinus

Deep veins of central and
nuclear regions and their
sinuses:
  6 Anterior vein of septum
  7 Superior thalamostriate
  8 Venous angle
  9 Internal cerebral vein
  10 Great cerebral vein (of
    Galen)
  11 Inferior sagittal sinus
  12 Basal vein (of Rosenthal)
  13 Straight sinus
  14 Confluence of sinuses
  15 Transverse sinus
  16 Sigmoid sinus
  17 Internal jugular vein

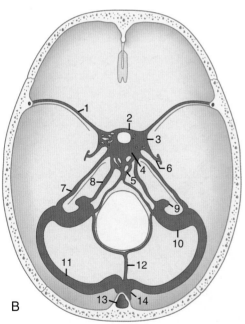

1 Sphenoparietal sinus
2 Anterior intercavernous
  sinus
3 Cavernous sinus
4 Posterior intercavernous
  sinus
5 Basilar plexus
6 Venous plexus of foramen
  ovale
7 Superior petrosal sinus
8 Inferior petrosal sinus
9 Internal jugular vein
  (running caudally)
10 Sigmoid sinus
11 Transverse sinus
12 Occipital sinus
13 Superior sagittal sinus
14 Confluence of sinuses

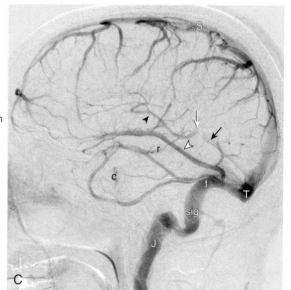

FIGURE 1-31 Venous anatomy. **A,** Lateral view of the head illustrating the cerebral veins and sinuses. The sequence of the numbers takes into account both the areas drained by the veins and the direction of blood flow. **B,** Top down view of the skull base depicts the basal sinuses. **C,** Venous anatomy on angiography. This lateral view shows the internal cerebral vein *(black arrowhead)*, vein of Galen *(white arrow)*, straight sinus *(black arrow)*, torcular (T), and superior sagittal sinus (S). The vein of Labbé *(white arrowhead)* drains into the transverse sinus (t). Sigmoid sinus (sig) and jugular vein (J) are also readily appreciable. Faintly seen are the basal vein of Rosenthal (r) and blush of the cavernous sinus (c). (**A** and **B** redrawn with permission from Kretschmann H-J, Weinrich W. *Cranial Neuroimaging and Clinical Neuroanatomy: Magnetic Resonance Imaging and Computed Tomography.* Rev 2nd ed. New York: Thieme; 1993:214, 215.)

further posteriorly than the demarcation of the foramen of Monro, the junction is called the "false venous angle." Thus, the internal cerebral veins usually begin at the foramina of Monro and run on either side of the roof of the third ventricle (velum interpositum). The internal cerebral veins unite to form the great vein of Galen. The vein of Galen drains into the straight sinus. The internal cerebral veins are a marker for midline shift, behind the foramen of Monro. They should not deviate more than 2 mm from the midline.

Deep Infratentorial Drainage

The anatomy of posterior fossa venous drainage is more complex. Superior vermian, posterior pericallosal, mesencephalic, and internal occipital veins drain into the vein of Galen at the tentorial hiatus. This vein also receives drainage from the basal vein of Rosenthal, which in turn receives venous supply from the insular lateral mesencephalic veins, and deep middle cerebral and anterior cerebral veins in the supratentorial space (Fig. 1-32). The vein of Galen and the inferior sagittal sinus drain to the straight sinus, which in turn drains into the torcular herophili. The

torcular is the common dumping ground (toilet) of the venous system. It also receives the drainage from the superior sagittal sinus.

From the torcular, blood flows to the transverse sinus, which receives drainage from the superior petrosal sinus, diploic veins, and lateral cerebellar veins. It courses laterally in the leaves of the tentorium and continues as the sigmoid sinus, which also drains the occipital sinus. The sigmoid sinus terminates as the internal jugular vein. The inferior vermian veins and superior hemispheric veins drain into the straight sinus. The transverse sinus receives blood from the inferior and superior hemispheric venous system. The deep middle cerebral vein courses deep in the sylvian fissure, and it meets with the anterior cerebral vein (which runs with the corresponding artery) to form the basal vein of Rosenthal arising along the brain stem. The basal vein of Rosenthal receives blood from the anterior pontomesencephalic vein in front of the brain stem, the lateral mesencephalic veins, and the precentral cerebellar vein (just anterior to the superior vermis behind the fourth ventricle).

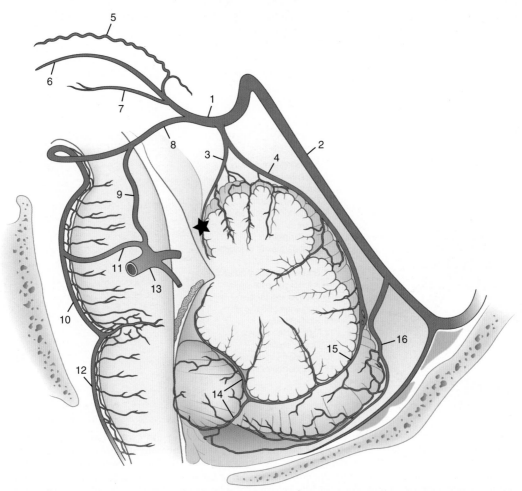

FIGURE 1-32 Anatomic drawing depicts the major posterior fossa veins as seen from the lateral view. The black star represents the colliculocentral point, an angiographic landmark that should be about halfway between the tuberculum sellae and the torcular herophili. 1, vein of Galen; 2, straight sinus; 3, precentral cerebellar vein; 4, superior vermian vein; 5, superior choroid vein (in lateral ventricle); 6, internal cerebral vein; 7, thalamic vein; 8, posterior mesencephalic vein; 9, lateral mesencephalic vein; 10, anterior pontomesencephalic venous plexus; 11, transverse pontine vein; 12, anterior medullary venous plexus; 13, petrosal vein; 14, tonsillar veins; 15, inferior vermian veins; 16, hemispheric vein. (From Osborne AG. *Diagnostic Cerebral Angiography.* 2nd ed. Philadelphia: Lippincott William & Wilkins; 1998:234.)

Anterior venous drainage centers on the cavernous sinuses. These venous channels receive blood from superior and inferior ophthalmic veins, the sphenoparietal sinus, and the superficial middle cerebral vein. The cavernous sinus communicates via (1) an extensive network across midline to the opposite cavernous sinus via the intercavernous sinus, (2) posteriorly via the superior petrosal sinus to the transverse sinuses just before the latter dive inferiorly to the sigmoid sinus, and (3) inferiorly into the inferior petrosal sinuses, which subsequently drain directly to the jugular bulbs. The inferior petrosal sinus also drains the internal auditory canal venous system.

Arachnoid granulations often arise in the transverse and sigmoid venous sinuses and can simulate clots or other lesions. They look like filling defects on angiography, magnetic resonance angiogram (MRA), magnetic resonance venography (MRV), computed tomography angiography (CTA), computed tomography venography (CTV), and CT in or partially outside the sinus with CSF density. They have CSF intensity on magnetic resonance with fibrous septa or vessels appearing as dark signal within them. They occur in approximately 1% of patients. These granulations may also result in bony excavations in the adjacent skull, most commonly at the occipital skull but also along the frontal calvarium and sphenoid bone.

Blood-Brain Barrier

It is probably appropriate after a vascular anatomy section to emphasize the role of the blood-brain barrier (BBB) in neuroimaging and in CNS pathology. The anatomy of the BBB is based on the microanatomy of the capillary endothelial cells. There are tight junctions between normal endothelial cells without gaps or channels. The basement membrane maintains the tubular conformation of the capillary and holds the endothelial cells together. In only a few regions in the brain do channels exist so that direct communication is present between the capillary and extracellular fluid or neurons. Such sites play a role in the feedback mechanism for hormonal homeostasis and as a port of entry into the brain for certain disease processes. They include the choroid plexus of the ventricles, the pineal gland, pituitary gland, median eminence, subcommissural organ, subfornical organ, area postrema, and organum vasculosum of the lamina terminalis.

After injection of contrast, increased attenuation on CT is noted immediately in the vessels of the brain and structures without a BBB. Slightly increased density on CT is also demonstrated in the cerebral parenchyma (gray matter denser than white matter) because of the cerebral blood volume (4% to 5% of total brain volume). The normal cerebral parenchymal enhancement is minimal.

Any alteration of the BBB from factors such as inflammation, infection, neoplasm, infarct, and trauma can produce intraparenchymal enhancement. The alteration is usually in the form of unlocking of the tight junctions, increased pinocytosis of contrast agents, vascular endothelial fenestration (formation of transendothelial channels), or increased permeability of the endothelial membrane. Furthermore, in many neoplastic conditions the BBB is not competent and contrast material is distributed into the extravascular spaces. A minority of enhancement is caused by increased blood volume in certain neoplastic lesions. This has recently been demonstrated by perfusion-weighted magnetic resonance studies of high grade neoplasms. Lack of angiographic vascularity has little bearing on contrast enhancement. Angiogenesis at a microvascular level may also be an important factor in enhancement.

Skull Base Foramina

The skull base foramina and their contents are summarized in Table 1-5 and depicted in Figs. 1-33 and 1-34.

FINAL WORDS

Congratulations—you've made it through Chapter 1! Dense stuff but pretty important. Go celebrate but come back soon. You've only just got your feet wet—now prepare to dive in and get soaked!

TABLE 1-5 Skull Base Foramina and Contents

Skull Base Foramen	Contents	Disease
Cribriform plate	Olfactory nerves, anterior ethmoidal artery	Esthesioneuroblastoma
Optic canal	Optic nerve, ophthalmic artery	Optic nerve gliomas, meningiomas
Superior orbital fissure	Oculomotor, trochlear, abducens, and ophthalmic (V-1) nerves, ophthalmic veins, sympathetic nerve plexus, orbital branch of middle meningeal artery, recurrent branch of lacrimal artery	Schwannomas, meningiomas, PNS
Foramen rotundum	Maxillary (V-2) nerve	Schwannomas, meningiomas, PNS
Foramen ovale	Mandibular (V-3) nerve, accessory meningeal artery, emissary veins	Schwannomas, meningiomas, PNS
Stylomastoid foramen	Facial nerve, stylomastoid artery	PNS, Bell palsy, schwannoma
Internal auditory canal	Facial and vestibulocochlear nerves, labyrinthine artery (branch of AICA)	Schwannomas, meningiomas, epidermoids, arachnoid cysts
Jugular foramen, pars nervosa	Glossopharyngeal nerve, inferior petrosal sinus	Schwannomas, meningiomas, PNS
Jugular foramen, pars vascularis	Vagus, spinal accessory nerves, internal jugular vein, ascending pharyngeal and occipital artery branches	Glomus tumors, schwannomas, meningiomas, PNS
Hypoglossal canal	Hypoglossal nerve, meningeal branch of ascending pharyngeal artery, emissary vein	Schwannomas
Foramen magnum	Spinal cord, vertebral arteries, spinal arteries and nerves	Meningiomas, chordomas, schwannomas

Continued

TABLE 1-5   Skull Base Foramina and Contents—cont'd

Skull Base Foramen	Contents	Disease
Foramen spinosum	Middle meningeal artery, meningeal branch of V-3	
Foramen lacerum	Carotid artery lies on top of it, greater petrosal nerve, vidian nerve pass above it	PNS
Incisive canal/nasopalatine canal	Nasopalatine nerve, palatine arteries	Cysts
Greater palatine canal	Greater palatine nerve, palatine vessels	PNS
Lesser palatine canal	Lesser palatine nerve and artery	PNS
Carotid canal	Internal carotid artery, sympathetic plexus	Aneurysms
Foramen of Vesalius	Emissary veins	
Hiatus for Lesser Petrosal Nerve	Lesser petrosal nerve	
Foramen cecum	Emissary vein	
Vestibular aqueduct	Endolymphatic duct, meningeal branch of occipital artery	Ménière disease, congenital stenosis or enlargement
Condylar canal	Emissary vein, meningeal branch of occipital artery	
Mastoid foramen	Emissary vein, meningeal branch of occipital artery	
Palatovaginal canal	Pharyngeal branches of pterygopalatine ganglion and maxillary artery	PNS
Cochlear aqueduct	Perilymphatic duct, emissary vein	Congenital stenosis or patulousness
Inferior orbital fissure	Maxillary nerve, zygomatic nerve, orbital branches of pterygopalatine ganglion, infraorbital vessels, inferior ophthalmic veins	PNS, schwannomas
Infraorbital foramen	Infraorbital nerve and vessels	Blow-out fractures
Mental foramen	Mental nerves and vessels	Squamous cell carcinoma
Mandibular foramen	Inferior alveolar nerve and vessels	Schwannomas, squamous cell carcinoma
Pterygomaxillary fissure	Maxillary artery, maxillary nerve, sphenopalatine veins	Juvenile angiofibromas
Vidian canal	Vidian nerve and artery	PNS

See Figs. 1-33 and 1-34.
AICA, Anterior inferior cerebellar artery; *PNS,* perineural spread of cancer.

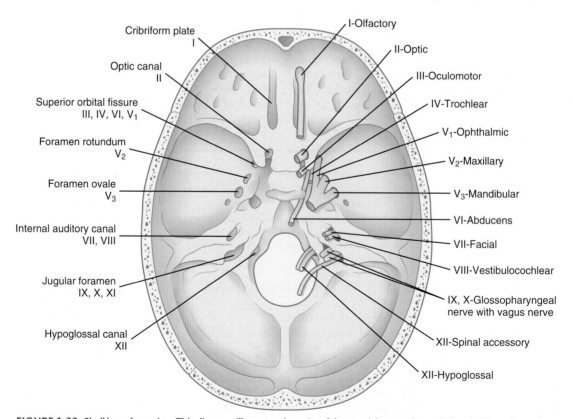

FIGURE 1-33 Skull base foramina. This diagram illustrates the exits of the cranial nerve through the skull base foramina.

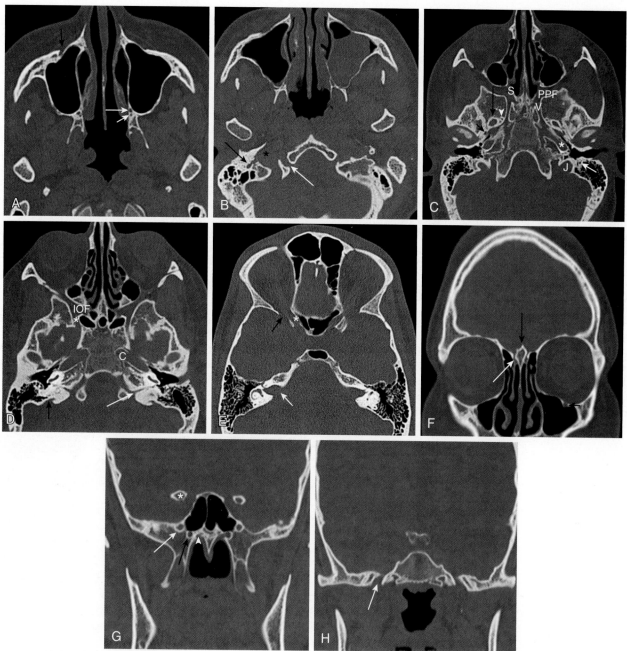

FIGURE 1-34 Skull base foramina on computed tomography (CT). From inferior to superior on axial CT images and anterior to posterior on coronal CT images, skull base foramina are depicted. **A,** At the level of the pterygoid plates, the greater *(long white arrow)* and lesser *(short white arrow)* palatine foramina are clearly seen. The canal of the infraorbital nerve *(black arrow)* can also be seen on this image. **B,** At the posterior cranial fossa, the hypoglossal canal *(white arrow)*, stylomastoid foramen *(black arrow)* and jugular foramen *(asterisk)* are shown. **C,** More superiorly, the pterygopalatine fossa (PPF), sphenopalatine foramen (S), vidian canal (V), and jugular foramen (J) are depicted. The foramen ovale *(large black arrow)* is adjacent to foramen of Vesalius *(white arrowhead)* along its anteromedial margin and the foramen spinosum *(small black arrow)* along its posteromedial margin. The carotid canal is depicted with an *asterisk*. The stylomastoid foramen is shown *(white arrow)*. **D,** Slightly more superiorly, the inferior orbital fissure (IOF), foramen rotundum *(asterisk)*, carotid canal (C), are seen. At the same level within the temporal bone, the vestibular aqueduct *(black arrow)* and cochlear aqueduct *(white arrow)* are indicated. **E,** At the level of the mid orbit, the optic canal is seen *(asterisk)* along with the superior orbital fissure *(black arrow)*. At the same level in the temporal bone, the internal auditory canal is depicted *(white arrow)*. **F,** On this coronal CT, the crista galli is shown at the midline *(long black arrow)* and the olfactory groove at the cribriform plate is seen as well *(white arrow)*. The canal of the inferior orbital nerve is again demonstrated *(short black arrow)* coursing the floor of the orbit. **G,** More posteriorly, the anterior clinoid process is depicted *(asterisk)* and the foramen rotundum *(white arrow)* and vidian canal *(black arrow)* are seen adjacent to the floor of the sphenoid sinus. The palatovaginal canal *(white arrowhead)* is also seen medial to vidian canal. **H,** Even more posteriorly, foramen ovale can be seen *(white arrow)*.

Chapter 2
Neoplasms of the Brain

The World Health Organization (WHO) classification of tumors of the brain remains the worldwide standard. Members of the International Society of Neuropathology, International Academy of Pathology, and Preuss Foundation for Brain Tumor Research met in Lyon France for several bottles of wine and a lengthy discussion of whether to use "tumor" or "tumour" among other topics. Late at night, over brie and baguettes they subsequently produced the most widely used classification of brain tumo(u)rs, one that we will share with you in this book. Vive la France!

EXTRAAXIAL VERSUS INTRAAXIAL TUMORS FOR DUMMIES

One of the essential distinctions that a neuroradiologist must make in describing a lesion is whether it is intraaxial (intraparenchymal) or extraaxial (outside the brain substance; i.e., meningeal, dural, subarachnoid, epidural, or intraventricular). This distinction has been made easier by the multiplanar capabilities of magnetic resonance imaging (MRI) and three-dimensional computed tomography (3D-CT). The quintessential and most common **extraaxial** dural mass is the meningioma, a readily diagnosable and treatable lesion. Extraaxial lesions buckle the white matter, expand the ipsilateral subarachnoid space, and sometimes cause reactive bony changes. On MRI scans you can visualize the dural margin with the cerebrospinal fluid (CSF cleft sign) and determine that the lesion is extraaxial. The prototypical epidural extraaxial mass, a bone metastasis, displaces the dura inward (superficial to the dural coverings), narrows the ipsilateral subarachnoid space, but otherwise may have a similar contour as an intradural extraaxial mass (Fig. 2-1, *A* and *B*).

When you are confronted with a solitary **intraaxial** mass in an adult, the odds are nearly even that the lesion is either a solitary metastasis or a primary brain tumor (and most of these are malignant, so as the fireman says, "You're hosed!"). Fifty percent of metastases to the brain are solitary, so the lack of multiplicity should not dissuade you from considering a metastasis in the setting

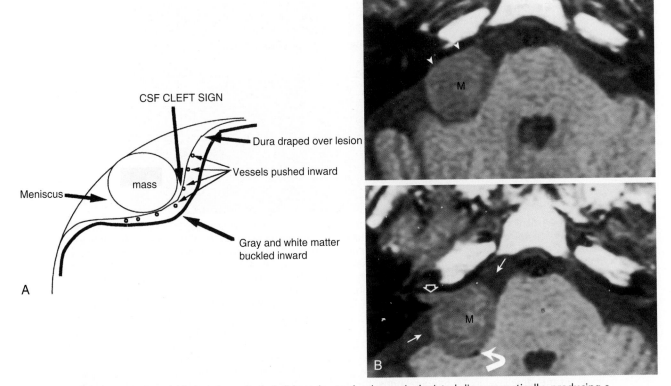

FIGURE 2-1 Extraaxial lesion signs. **A,** A skull-based extradural mass is depicted diagrammatically, producing a meniscus sign, displacement of the subarachnoid veins inward, and buckling the gray-white interface. Dura may be seen stretched over the mass. **B,** Extraaxial mass. The classic extraaxial mass (M) in this patient, Ken L. N. Angle-Masse, has a dural base *(arrowheads)* shown in the upper image, and, on the lower image, is seen to displace blood vessels *(curved arrow)* and widen *(small arrows)* the subarachnoid space. This was a vestibular schwannoma, but on the top image it could just as easily be a meningioma with that dural base. Notice the extension into the internal auditory canal on the lower image *(open arrow),* typical of a vestibular schwannoma. *CSF,* Cerebrospinal fluid.

of a single intraaxial lesion. You can classify a lesion as intraaxial if (1) it expands the cortex of the brain, (2) there is no expansion of the subarachnoid space, (3) the lesion spreads across well-defined boundaries, and (4) the hypointense dura and pial blood vessels are peripheral to the mass.

Occasionally, the distinction between intraaxial and extraaxial lesions may be blurred, because some extraaxial lesions may aggressively invade the underlying brain (aggressive meningiomas, hemangiopericytomas, dural metastases). Conversely, an intraaxial lesion may rarely invade the meninges (glioblastoma, lymphoma, pleomorphic xanthoastrocytomas). Nonetheless, young Jedi, you must localize a lesion as intraaxial or extraaxial so your differential diagnosis will be relevant. Once you have made that decision, you must appreciate its other qualities; its shape (margination), its internal architecture (solid, hemorrhagic, calcified, fatty, cystic), its diffusivity, and its enhancement. These are the secondary features in intraaxial and extraaxial lesions that allow you to arrive at the specific diagnosis. If you get the right diagnosis you are a Jedi neuroradiologist worthy of the "MASTER REQUISITE" distinction!

EXTRAAXIAL TUMORS

Tumors of the Meninges

Meningiomas

Meningiomas constitute the most common extraaxial neoplasms of the brain (Box 2-1). This lesion commonly affects middle-aged women. However, because of its incidence, it represents a significant proportion of the extraaxial neoplasms in men, and in adults of all age groups. The most common locations for meningioma (in descending order) are the parasagittal dura, convexities, sphenoid wing, cerebellopontine angle cistern, olfactory groove, and planum sphenoidale. Ninety percent occur supratentorially. Because meningiomas arise from arachnoid cap cells, they can occur anywhere an arachnoid exists.

On unenhanced CT scans, approximately 60% of meningiomas are slightly hyperdense compared with normal brain tissue (Fig. 2-2). You may see calcification within meningiomas in approximately 20% of cases. Rarely, cystic, osteoblastic, chondromatous, or fatty degeneration of meningiomas occurs. The specific histologic subtypes of

---

BOX 2-1 World Health Organization Classification of Tumors of the Meninges: Tumors of Meningothelial Cells

GRADE I

Meningiothelial, fibrous (fibroblastic), transitional (mixed), psammomatous, angiomatous, microcystic, secretory meningioma, lymphoplasmacyte-rich, metaplastic

GRADE II

Chordoid meningioma, clear cell meningioma, atypical meningioma, hemangiopericytoma

GRADE III

Papillary, rhabdoid, anaplastic
Anaplastic hemangiopericytoma (grade 3)

---

meningioma (e.g., transitional, fibroblastic, and syncytial) cannot be readily distinguished on imaging.

MRI is superior to CT in detecting the full extent of meningiomas, and the possibility of concomitant sinus invasion and/or thrombosis, hypervascularity, intracranial edema, and intraosseous extension. Parasagittal meningiomas may grow into the sagittal sinus and/or through the bone (Fig. 2-3). Classically meningiomas are said to be isointense to gray matter on all MRI pulse sequences, but frankly they can be all over the map, given the internal architecture variability. The "cleft sign" has been described in MRI to identify extraaxial intradural lesions such as meningiomas. The cleft usually contains one or more of the following: (1) CSF between the lesion and the underlying brain parenchyma (Fig. 2-4), (2) hypointense dura (made of fibrous tissue), and (3) marginal blood vessels trapped between the lesion and the brain. One may see vascular flow voids on MRI within and around a meningioma. Avid enhancement after contrast is usually seen, but occasionally, meningiomas may have necrotic centers or calcified portions, which may not enhance. With MRI you may be able to identify the "dural tail"; enhancement of the dura trailing off away from the lesion in crescentic fashion (see Fig. 2-3, *B*), which is typical of meningiomas and has been exhibited in up to 72% of cases (see Fig. 2-4). There has been a debate in the radiologic literature as to whether the dural tail always represents neoplastic infiltration of the meninges or, alternatively, a reactive fibrovascular proliferation of the underlying meninges. In the typical situation where a differential diagnosis of vestibular schwannoma, meningioma, or fifth nerve schwannoma is debated, the dural tail may be a particularly useful sign to suggest meningioma rather than other diagnoses (Table 2-1). Note,

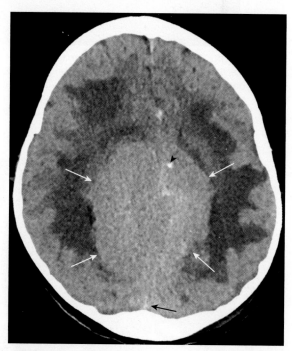

FIGURE 2-2 Meningioma. This patient, Isidora Thicke, shows a hyperdense extraaxial mass arising from the falx and extending into both cerebral hemispheres *(white arrows)*. Note calcification *(arrowhead)*. There is very likely invasion of the superior sagittal sinus *(black arrow)*. Note the extensive vasogenic edema within the surrounding brain tissue on the unenhanced computed tomography scan.

Young Skywalker, that dural metastases may demonstrate a similar finding.

Meningiomas may encase and narrow adjacent vessels. This finding helps in the sella region, because pituitary adenomas, the other culprit in this location, usually does not narrow the cavernous carotid artery.

Although the vast majority of meningiomas do not elicit edema, the larger ones and convexity meningiomas may cause high signal intraparenchymally even without invasion (see Fig. 2-3). This may be secondary to compressive ischemia, venous stasis, aggressive growth, or parasitization of pial vessels. Venous sinus occlusion or venous thrombosis can also cause intraparenchymal edema from a meningioma.

Intraosseous meningiomas may appear as expansions of the inner and outer tables of the calvarium or may even extend into the scalp soft tissues. No dural component may be present at all. This type of meningioma strongly resembles a blastic osseous metastasis and is most commonly found along the sphenoid wing (Fig. 2-5).

Intraventricular meningiomas typically occur around the choroid plexus (80%) in the trigone of the lateral ventricle and have a distinct propensity for the left lateral ventricle (Fig. 2-6). Only 15% of intraventricular meningiomas occur in the third ventricle, and 5% occur in the fourth ventricle. Intraventricular meningiomas calcify in 45% to 68% of cases, and their frequency is higher in children. Multiple meningiomas are associated with neurofibromatosis type 2.

Bony changes associated with meningiomas may be hyperostotic or osteolytic, and occur in 20% to 46% of cases. Hyperostosis is particularly common when the tumor is at the skull base or anterior cranial fossa, and here it may resemble fibrous dysplasia or Paget disease (see Fig. 2-5). The presence of bony reaction may be a helpful feature to distinguish meningiomas from other extraaxial masses, particularly schwannomas, which do not elicit a bony reaction. If the hyperostosis is along the inner table only, you cannot say whether it is because of neoplastic

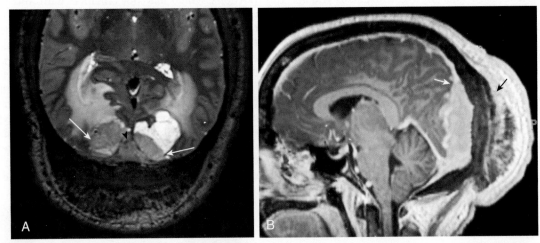

FIGURE 2-3 Parasagittal meningioma. **A,** Patient Doris Tuwyde's meningioma *(white arrows)* is isointense on T2WI but one also has the adjacent bright intraaxial vasogenic edema. Because the lesion crosses the midline outside a normal white matter tract and is invading the dura of the superior sagittal sinus *(black arrowhead)*, it is clearly extraaxial. Oh, and the bone and scalp also are invaded too. **B,** Add a dural tail, tumor growth along the sagittal sinus *(white arrow)*, bone reaction *(black arrow)*, and strong enhancement and you should be dictating this case as a meningioma.

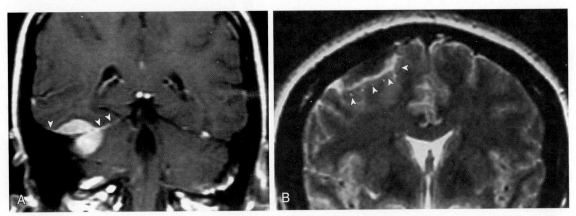

FIGURE 2-4 Another piece of tail in a meningioma. **A,** Coronal enhanced T1-weighted image (T1WI) of this tentorial meningioma in patient X. Axel Tuma shows enhancing tails *(arrowheads)* coursing medially and laterally from the main portion of the tumor. Note that tumor extends to either side of the tentorium. A low signal intensity cerebrospinal fluid (CSF) is seen along the tumor's upper margin on T1WI. **B,** Bright CSF cleft *(arrowheads)* is seen in Perry Filbrayne's convexity meningioma.

TABLE 2-1   Differential Diagnosis of Meningioma versus Schwannoma

Feature	Meningioma	Schwannoma
Dural tail	Frequent	Extremely rare
Bony reaction	Osteolysis or hyperostosis	Rare
Angle made with dura	Obtuse	Acute
Calcification	20%	Rare
Cyst/necrosis formation	Rare	Up to 10%
Enhancement	Uniform	Inhomogeneous in 1/3
Extension into IAC	Rare	80%
MRS	Alanine	Taurine, GABA
Precontrast CT attenuation	Hyperdense	Isodense

CT, Computed tomography; *GABA,* gamma-aminobutyric acid; *IAC,* internal auditory canal; *MRS,* magnetic resonance spectroscopy.

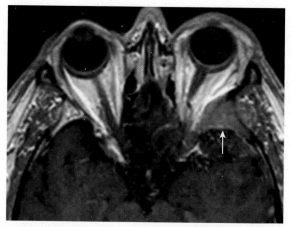

FIGURE 2-5 Intraosseous meningioma. Enhanced T1-weighted image of patient Ina D. Bowen demonstrates evidence of an intraosseous meningioma expanding the left sphenoid wing *(white arrow).* The patient has mild proptosis and a small amount of intraorbital soft tissue. The differential diagnosis would include a metastasis and fibro-osseous primary bone lesions.

invasion or reactive changes. If the outer table is transgressed, tumor is most likely. Pneumosinus dilatans is the expansion of the paranasal sinus from adjacent meningioma (usually anterior cranial fossa meningiomas and large frontal sinuses).

Meningiomas are one of the few tumors where angiography may still play an important role. Meningiomas diagnostically appear as lesions with an angiographic stain (tumor blush) and have both dural and pial blood supply. The characteristics of the stain are classically compared with an unwanted visit from your mother-in-law or Darth Maul—comes early and stays late. Depending on the location of the tumor, you may have to perform internal carotid, external carotid, and vertebral injections. In the typical convexity or sphenoid wing meningioma, the middle meningeal artery is enlarged. At present preoperative embolization of meningiomas is sometimes performed to decrease the vascularity of the tumor.

On magnetic resonance spectroscopy (MRS), meningiomas are characterized by high levels of alanine and absent N-acetyl aspartate. No glutamine is seen. One can use the presence of taurine and/or gamma-aminobutyric acid (GABA) in schwannomas to distinguish meningiomas from schwannomas.

Atypical Meningiomas

Atypical meningiomas are classified as WHO grade II. Histopathologically they have increased mitotic rates, small cells with high nucleus/cytoplasm ratio, prominent nucleoli, sheet-like growth, or foci of necrosis. Accordingly, 2.4% of meningiomas are thereby classified as atypical and these tumors have lower apparent diffusion coefficient (ADC) values on diffusion-weighted images than typical meningiomas. They recur more frequently.

Malignant Meningiomas

Malignant meningiomas (WHO grade III) occur uncommonly and are usually diagnosed when a meningioma exhibits intraparenchymal invasion or markedly rapid growth. Histopathologically they are defined by malignant cytology, very high mitotic rate, anaplasia, or sarcomatous degeneration. They too have restricted diffusion compared with benign meningiomas. They most likely arise from benign tumors gone awry and may

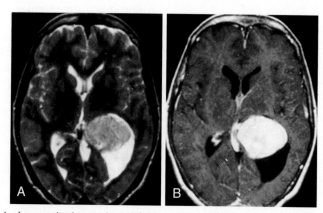

FIGURE 2-6 Intraventricular meningioma. **A,** Axial T2-weighted image shows a well-defined large mass isointense to gray matter in a dilated left lateral ventricular trigone of patient N.E. D. Van-Trickle. **B,** Big time enhancement! Classic I-V meningioma!

aggressively invade the brain. The higher grade is associated with a higher rate of recurrence. Survival varies with the site, size, grade, and extent of surgical removal of the tumor.

Radiation-Induced Meningioma

Radiation therapy induces five times more meningiomas than it does gliomas or sarcomas. The diagnosis of radiation-induced meningioma is made if (1) the meningioma arises in the radiation field, (2) it appears after a latency period (of years), (3) it was not the primary tumor irradiated, and (4) it is not seen in a patient with neurofibromatosis. Multiple meningiomas occur in up to 30% of previously irradiated patients with meningiomas and in 1% to 2% of nonirradiated patients with meningiomas. Recurrence rates are higher in radiation-induced meningiomas than in non–radiation-induced tumors.

The differential diagnosis of primary tumors that mimic meningiomas is broad (Box 2-2).

Mesenchymal Meningeal Tumors

As noted in Box 2-2 there are a number of mesenchymal tumors that can affect the meninges. These are all relatively rare lesions, which may have either osseous (osteoma, osteosarcoma, etc.), chondroid (chondroma, chondrosarcoma, etc.), muscular (leiomyoma, rhabdomyosarcoma, etc.), fatty (lipoma, liposarcoma, etc.), or fibrous (fibroma, malignant fibrous histiocytoma) matrices associated with

---

BOX 2-2 World Health Organization Classification of Tumors of the Meninges: Nonmeningiomatous Extraaxial Masses

MESENCHYMAL, NONMENINGOTHELIAL TUMORS

Lipoma
Angiolipoma
Hibernoma
Liposarcoma
Solitary fibrous tumor
Fibrosarcoma
Malignant fibrous histiocytoma
Leiomyoma
Leiomyosarcoma
Rhabdomyoma
Rhabdomyosarcoma
Chondroma
Chondrosarcoma
Osteoma
Osteosarcoma
Osteochondroma
Hemangioma
Epithelioid hemangioendothelioma
Hemangiopericytoma
Angiosarcoma
Kaposi sarcoma

PRIMARY MELANOCYTIC LESIONS

Diffuse melanocytosis
Melanocytoma
Malignant melanoma
Meningeal melanocytosis

TUMORS OF UNCERTAIN HISTIOGENESIS

Hemangioblastoma

---

them. If you remember that the fibrous falx can be ossified (with bone and marrow fat), you can recall these tumors more readily.

Hemangiopericytoma

These tumors, derived from smooth muscle pericyte cells of Zimmerman around the capillaries of the meninges, are more aggressive than most meningiomas, have a higher rate of recurrence, and can metastasize. Hence some consider them malignant and they are distinct from meningiomas. They tend to be large (over 4 cm in size), lobular, extraaxial supratentorial masses. Hydrocephalus, edema, and mass effect are not uncommon with this entity. The "mushroom sign" refers to a narrow dural attachment despite parenchymal invasion (opposite of what is expected from meningiomas) in some cases. Precontrast CT shows heterogeneous, hyperdense tumors that enhance. Hyperostosis and/or calcification is rare. Hemangiopericytomas affect men more than women, unlike most MENingiomas.

Melanocytic Lesions

Within this category one finds diffuse melanocytosis, melanocytoma, neurocutaneous melanosis, and malignant melanoma. The melanin containing cells are leftovers from neural crest origin. These diagnoses are difficult to distinguish from metastatic melanoma to the dura. The melanocytoma is the most common of the lot and is seen in adults, whereas melanocytosis is more of a pediatric disorder. Although the latter is a diffuse process, the former usually presents as a posterior fossa mass. Hyperintensity on T1-weighted images (T1WI) is the only hope for sealing this diagnosis, but varies with melanin content. Spread of melanocytosis through the Virchow Robin spaces is possible.

Malignant melanoma of the meninges may bleed or spread from the dura to the adjacent nerves, brain, or skull. Neurocutaneous melanosis shows melanocytic nevi of the skin (especially about the face), syringomyelia, and central nervous system (CNS) lipomas. Prognosis is poor as is malignant melanoma of the meninges with distant metastases.

Tumors of Neurogenic Origin

Schwannomas

The intracranial neurogenic tumors (schwannomas, neurofibromas), are similar in appearance. Histologically, schwannomas arise from the perineural Schwann cells. These cells may differentiate into fibroblastic or myelin-producing cells. Two types of tissue may be seen with schwannomas, Antoni A and Antoni B tissue. The Antoni A tissue consists of densely packed palisades of fibrous and neural tissue and typically has a darker signal on T2WI because of the compactness of the fibrils. Antoni B tissue is a looser, myxomatous tissue that is typically brighter on the T2WI. Note that, depending on the degree of Antoni A and B tissue within schwannomas, the signal intensity of these lesions may be brighter than or may directly simulate that of meningiomas. Other terms used for schwannomas include neurilemmomas and neurinomas, but the most accurate term is schwannoma. Malignant schwannomas may also be seen in patients with neurofibromatosis type I.

The distinction between meningiomas and schwannomas is a common one that radiologists must make at the skull base. Table 2-1 should be memorized for the top 10 ways to distinguish the two. Now OB-1! In some instances it is impossible to distinguish between the two (Fig. 2-7). One of the distinguishing features of vestibular schwannomas from meningiomas is the expansion of and growth into the internal auditory canal (IAC) seen in schwannomas. The porus acusticus (the bony opening of the IAC to the cerebellopontine angle cistern) is typically flared and enlarged with vestibular schwannomas, whereas the amount of tissue seen in the IAC with meningiomas is usually small or absent. Vestibular schwannomas account for more than 90% of purely intracanalicular lesions, but only 20% of them are solely intracanalicular (Fig. 2-8, *A*). Approximately 20% of vestibular schwannomas present only in the cerebellopontine angle cistern without an IAC stem (Box 2-3). About 60% of vestibular schwannomas have both a canalicular and cisternal portion (see Fig. 2-8, *B*). Although it is well-known that schwannomas occur most commonly along cranial nerve VIII, the superior and inferior vestibular branches of cranial nerve VIII vie each edition of the *Requisites* as to which is the most common origin of the vestibular schwannoma (NOT "acoustic neuromas" and NOT the cochlear nerve). The Oscar goes to inferior division for the fourth edition. Nonetheless, despite involving the vestibular, not cochlear, nerves, the patients present with hearing loss. Classically, in "neurofibromatosis" type II, bilateral vestibular schwannomas are seen, often of different sizes.

On MRI, schwannomas are usually isointense to slightly hypointense compared with white matter tissue on all pulse sequences. Enhancement is nearly always evident and is homogeneous in around 70% of cases. Peritumoral edema may be seen in one third of cases, usually in the larger schwannomas.

After lesions of cranial nerve VIII, schwannomas of cranial nerves VII (Fig. 2-9) and V are the most common site of intracranial neurogenic tumors. Trigeminal schwannomas may arise anywhere along the pathway from the pons to Meckel cave (Fig. 2-10, *A* and *B*), to the cavernous sinus, and to and beyond the exit foramina (ovale, rotundum, and superior orbital fissure). Outside the brain, the

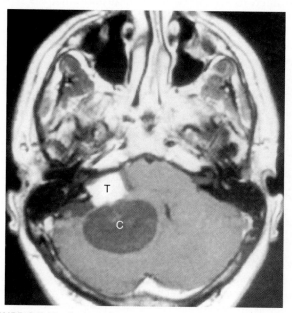

FIGURE 2-7 Vestibular schwannoma. Patient Ike N. Hiryu has a cyst (C) associated with his solidly enhancing (T) tumor nodule on his enhanced T1-weighted image.

BOX 2-3   Cerebellopontine Angle Masses

Vestibular schwannoma (75%)
Meningioma (10%)
Epidermoid (5%)
Facial nerve schwannoma (4%)
Aneurysm (vertebral, basilar, posterior inferior cerebellar artery)
Exophytic brain stem glioma
Arachnoid cyst
Paraganglioma (from temporal bone)
Hematogenous metastasis
Subarachnoid spread of tumors
Lipoma
Exophytic desmoplastic medulloblastoma

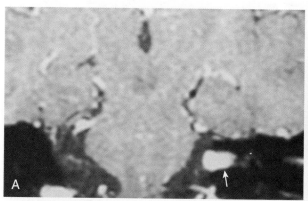
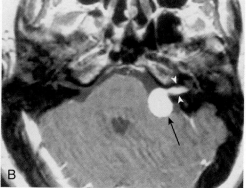

FIGURE 2-8 Intracanalicular vestibular schwannoma. **A,** Left intracanalicular vestibular schwannoma: Post gadolinium enhanced T1-weighted coronal scans on patient Ena Ken Al show an intracanalicular mass *(arrow)* on the left side. Perform a fat saturated scan to ensure that this does not represent fat. **B,** Cerebellopontine angle and intracanalicular vestibular schwannoma. Patient Sister Na Duhl has a brightly enhancing mass with both a cisternal *(black arrow)* and intracanalicular *(arrowheads)* portion. The ipsilateral prepontine cistern is enlarged.

schwannomas of cranial nerve V most commonly occur along the second division. These tumors present with facial pain that is burning in nature. Please note that ANY cranial nerve may be involved by schwannomas (except CN II, which is an extension of the CNS rather than a peripheral nerve).

Postoperatively, it is not unusual to see linear gadolinium enhancement in the internal auditory canal after vestibular schwannoma resection as a dural reaction. This can be followed expectantly. Don't antagonize your insecure neurosurgeon by suggesting residual tumor at the outset. For those cases with progressive, nodular, or masslike enhancement, suggest recurrence.

Jugular schwannomas more commonly grow intracranially than extracranially and typically smoothly erode the jugular foramen. The border of the bone is scalloped as opposed to the paraganglioma that has a much more irregular and nonsclerotic margin. Schwannomas compress the jugular vein, whereas paragangliomas (glomus jugulare tumors) invade the vein. Jugular foramen schwannomas most commonly present with hearing loss and vertigo rather than cranial nerve IX symptoms. Whether the schwannoma arises from one cranial nerve or the next, the imaging appearance is similar.

Neurofibromas

Neurofibromas, strictly speaking, refer to the tumors associated with neurofibromatosis. They are classified as WHO grade I and all ages and sexes are represented. They may occur sporadically as well, though less commonly than schwannomas. The skin and subcutaneous tissues are affected more often than peripheral nerves; spinal nerve neurofibromas are rare and cranial nerve involvement is uncommon. These lesions contain Schwann cells, perineural cells, and fibroblastic cells and may occur in a plexiform aggressive subtype, which appears as a network of diffusely infiltrating masses. Plexiform neurofibromas in the cranial nerves or peripheral nervous system occur in neurofibromatosis type 1 (NF-1) and are one of the seven criteria for the diagnosis

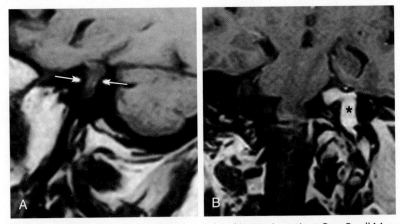

FIGURE 2-9 Facial nerve schwannoma. **A,** Sagittal T1-weighted image in patient Case Ewell Mysse shows a markedly thickened descending portion of the facial nerve *(arrows)*. **B,** The course of this enhancing mass suggests a facial nerve schwannoma by virtue of its tympanic segment and descending portion on this coronal post-contrast T1-weighted image *(asterisk)*.

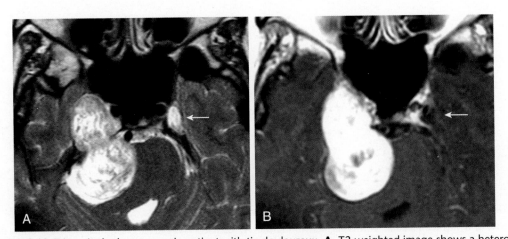

FIGURE 2-10 Trigeminal schwannoma in patient with tic douloureux. **A,** T2-weighted image shows a heterogeneous extraaxial mass in patient Payne N.D. Hedd that courses along the pons to the Meckel cave and the right cavernous sinus (normal Meckel cave on left with *arrow*). **B,** As per most schwannomas, it enhances brightly. With a tumor like this check the ipsilateral pterygopalatine fossa for spread (V-2) and muscles of mastication for atrophy (V-3).

of this disorder. Plexiform neurofibromas, because of their extensiveness, can be distinguished from neuromas and schwannomas. They are generally found in the extremities or in the soft tissues of the head and neck. They have a significant rate of malignant dedifferentiation into a MPNST (malignant peripheral nerve sheath tumor).

Classic descriptions say that neurofibromas can sometimes be distinguished from schwannomas by their eccentricity and more fibrous central low T2 signal on MRI ("target sign"), but in many cases they two look the same.

Neuromas

By strict pathologic definition, neuromas refer to a posttraumatic proliferation of nerve cells rather than to a true neoplasm. The perineural lining and fibroblastic tissue seen in the other lesions just described are not present. These lesions are less common than schwannomas. They are usually seen in the cervical spine when nerves are avulsed, or in an operative bed. Again, one must know where to look for these lesions. Unfortunately, in the common vernacular most people mean "schwannoma" when they say "neuroma" (e.g., "acoustic neuroma," which is a double misnomer as these are "vestibular" not "acoustic" [cochlear] lesions and schwannomas, not neuromas). On scanning, these masses look like small schwannomas.

Metastases

Dural Metastases

Dural metastases usually spread out along the dura as hematogenously disseminated en plaque lesions from extracranial primary tumors (Fig. 2-11). Lung, breast, prostate, and melanoma cancers produce most dural metastases. Some of the dural metastases may arise from spread of adjacent bone metastases. Breast carcinoma is the most common neoplasm to be associated with purely dural metastases. Lymphoma is the next most common but is unique in that the dural lymphoma may sometimes be the primary focus of the neoplasm (Fig. 2-12). Dural plasmacytomas will look identical to dural lymphoma (Fig. 2-13). In children, dural metastases are most commonly associated with neuroblastomas and leukemia. These tumors are also famous for lodging in the cranial sutures, widening them in an infant.

Occasionally, one can identify an adjacent parenchymal metastasis with dural spread (Fig. 2-14), and, alternatively, an osseous-dural metastasis (breast, prostate primaries) occasionally invades the parenchyma (Fig. 2-15). On MRI the T1WI and T2WI characteristics are variable; depending on cellular content (see Fig. 2-11, *A* to *C*). On a CT scan these lesions are identified as isodense thickening of the meninges. Contrast enhancement is prominent. This is a diagnosis where contrast-enhanced T1WI or enhanced fluid-attenuated inversion recovery (FLAIR) scans can readily demonstrate the abnormality.

Inflammatory lesions that may simulate dural metastases include granulomatous infections (mycobacterial, syphilitic, and fungal), Erdheim Chester disease, sarcoidosis, and Langerhans cell histiocytosis.

Subarachnoid Seeding

With subarachnoid seeding (SAS) you see tiny nodules of implanted tumor on the leptomeninges. When the process is diffuse we use the term "sugar-coating" of the subarachnoid space because the whole pial surface is studded with granules. A combination of sugar coating and focal nodules (dare we say chocolate chips?) may also occur. SAS may occur with primary CNS tumors or

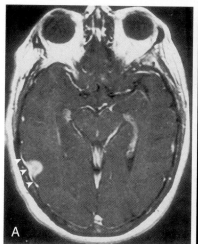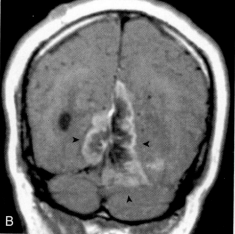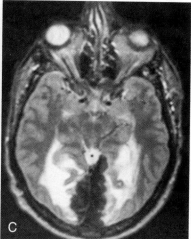

FIGURE 2-11 Metastases with dural tails. **A,** Postgadolinium enhanced T1-weighted image (T1WI) in patient Dora Allcayke with adenocarcinoma of the lung demonstrates a metastasis peripherally in the right temporal region. Although a dural tail *(arrowheads)* is suggestive of a meningioma, it is not pathognomonic for it. **B,** Another patient, Lottie Dima, with mucinous adenocarcinoma metastatic to the dura on coronal unenhanced T1WI shows a dural-based mass with low signal centrally and peripheral high signal intensity *(arrowheads)*. The lesion was calcified on computed tomography. (Observant readers will note the subcutaneous sebaceous cyst in the right upper part of the scalp.) **C,** On axial T2WI the lesion has low intensity but incites high signal intensity edema. The low signal could be from the mucinous content and/or calcification.

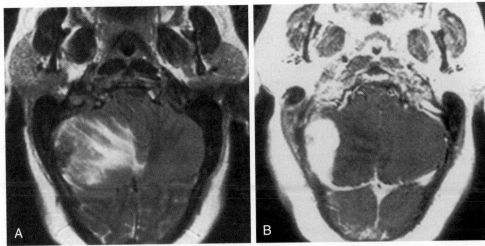

FIGURE 2-12 Dural lymphoma. **A,** Patient Everett N. Hanse has a low signal dural mass on the T2-weighted image that is eliciting the intraparenchymal edema. No? **B,** You cannot miss it (unless cortically blind) on the enhanced scan. Lymphoma will nearly always enhance (unless the neurosurgeons are reluctantly treating it with massive doses of steroids). Include meningioma, sarcoidosis, plasmacytoma, and other dural metastases in the differential diagnosis.

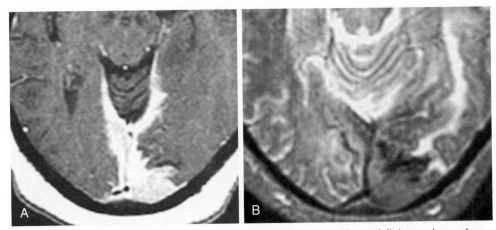

FIGURE 2-13 Plasmacytoma of the dura in patient Denton D'Brayne. **A,** The gadolinium-enhanced scan looks just like one would expect for a meningioma with marked enhancement and a dural-based lesion. The irregular margins (saw-toothed appearance) suggesting pial spread would be funky for a meningioma though. **B,** Even the T2-weighted image shows low signal that simulates a meningioma or lymphoma or sarcoidosis. The way to score on this shot is to have a history of a plasma cell dyscrasia.

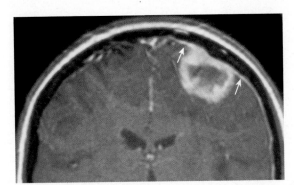

FIGURE 2-14 Leiomyosarcoma with dural-based metastasis. Post-gadolinium coronal scan in patient Hope Anna Prere shows a high left frontal intraparenchymal metastasis, which demonstrates dural invasion *(arrows)* along its superomedial margin. (Ignore the phase-ghosting artifact.)

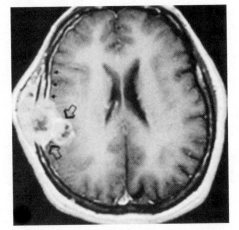

FIGURE 2-15 Intraparenchymal growth of renal cell carcinoma metastasis. Postcontrast scans in patient Vera D. Edley show a lesion centered on the calvarium with extracalvarial, dural, and intraparenchymal invasion *(open arrows)*. Dural enhancement more peripherally is denoted by *small black arrows*.

TABLE 2-2   Sources of Subarachnoid Seeding

CNS Primary	Non-CNS Primary
CHILDREN	
Medulloblastoma (PNET)	Leukemia/Lymphoma
Ependymoma (blastoma5PNET)	Neuroblastoma
Pineal region tumors	
Malignant astrocytomas	
Retinoblastoma	
Choroid plexus papilloma	
ADULTS	
Glioblastoma	Leukemia/Lymphoma
Primary CNS lymphoma	Breast
Oligodendroglioma	Lung
	Melanoma
	Gastrointestinal
	Genitourinary

CNS, Central nervous system; *PNET,* primitive neuroectodermal tumor.

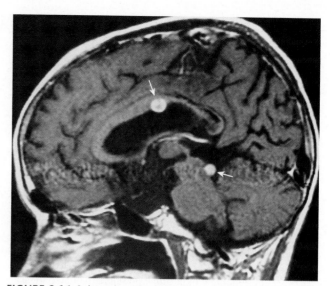

FIGURE 2-16 Subarachnoid seeding from patient Dot O'Whyte's brain stem glioma. Postcontrast T1-weighted image shows contrast-enhancing nodules *(arrows)* in the roof of the lateral ventricle and in the superior vermian cistern. The keen observer will notice the "overly pregnant" belly of the pons.

non-CNS primary tumors (Table 2-2). Lymphoma and leukemia are the most common tumors to seed the CSF. However, because they only rarely invade the meninges and do not incite reactions in the CNS, most leukemic and lymphomatous SAS is invisible on neuroimaging techniques; the diagnosis is usually made by multiple spinal taps for CSF sampling.

When a lesion has spread to the SAS, you may see it most readily on contrast-enhanced studies, and unenhanced and enhanced FLAIR imaging has been shown to be increasingly effective at identifying subarachnoid disease, including metastatic disease. The malignant cells in the CSF and/or the associated elevated protein in the CSF will cause the usually low signal of CSF to be bright on an unenhanced FLAIR scan. Although this may be a difficult diagnosis to make in the basal cisterns where "f-ing" (FLAIR and flow) artifacts abound, the presence of such high signal over the convexities implies subarachnoid seeding, subarachnoid hemorrhage, or meningeal inflammation. FLAIR with contrast enhancement increases the yield even higher!

The typical locations where one identifies SAS are at basal cisterns, in the interpeduncular cistern, at the cerebellopontine angle cistern, along the course of cranial nerves, and over the convexities (Fig. 2-16). Often in non-CNS primary tumors with SAS, one identifies a peripheral intraparenchymal metastasis contiguous with the dural surface of the brain, from which cells are shed into the CSF. With subarachnoid seeding, secondary communicating hydrocephalus may be present.

The other terms you will see for subarachnoid seeding from cancers outside the CNS include meningeal carcinomatosis or carcinomatous meningitis. Clinically, the patients present with multiple cranial neuropathies, radiculopathies, and/or mental status changes secondary to hydrocephalus or meningeal irritation. The cranial neuropathies may be irreversible. Although an initial CSF sample is positive in only 50% to 60% of cases, by performing multiple taps, the positive cytology (and post-LP headache) rate approaches 95%. Patient survival is usually less than 6 months. Breast, lung, and melanoma are the most common non-CNS primaries to seed the CSF.

INTRAVENTRICULAR MASSES (EXCLUDING MENINGIOMA)

Choroid Plexus Masses

Choroid Plexus Papilloma

Choroid plexus papillomas (WHO grade I) comprise 3% of intracranial tumors in children and 10% to 20% of those presenting in the first year of life. Eighty-six percent of these tumors are seen in patients less than 5 years old. In children, 80% occur at the trigone and/or atria of the lateral ventricles; in adults they may also be seen in the fourth ventricle. If you also include adults, 43% are located in the lateral ventricle, 39% in the fourth ventricle, 11% in the third ventricle, and 7% in the cerebellopontine angle cistern. Multiple sites are present in 3.7% of cases.

The tumors present with hydrocephalus and papilledema caused by overproduction of CSF (four to five times more than normal) or obstructive hydrocephalus caused by tumor, hemorrhage, high-protein CSF, or adhesions obstructing the ventricular outlets. Lately, the obstructionists have gained favor over the overproductionists as far as the explanation for hydrocephalus. Calcification occurs in 20% to 25% of choroid plexus papillomas, and hemorrhage in the tumor is seen even more frequently than calcification. Choroid plexus papillomas are typically hyperdense on unenhanced CT, with a mulberry appearance. They are usually of low signal on T1WI and mixed intensity on T2WI from flow voids unless hemorrhage has occurred. These tumors enhance dramatically (Fig. 2-17). Between the calcification, flow voids, and/or hemorrhage, the tumor may have a very heterogeneous appearance, sometimes with a salt-and-pepper appearance from vessel supply.

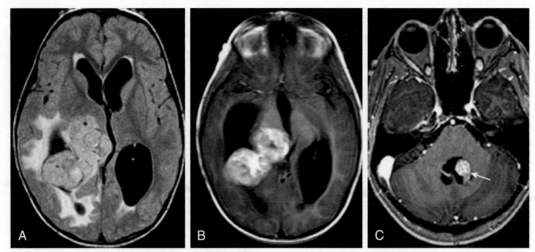

FIGURE 2-17 Choroid plexus papilloma. **A,** This choroid plexus mass in patient Cora P. Lexus-Goombah has a lot of peripheral high signal on fluid-attenuated inversion recovery, signifying edema of the parenchyma. **B,** Note the hydrocephalus and heterogeneous enhancement of the tumor. The voxel placed for magnetic resonance spectroscopy yielded high levels of choline-containing compounds and a complete absence of creatine and the neuronal/axonal marker N-acetyl aspartate. **C,** Fourth ventricular papilloma *(arrow)* in patient C. P. Pee IV shows cauliflower-like shape in a slightly large fourth ventricle in an adult.

Choroid Plexus Carcinomas

Choroid plexus carcinomas (WHO grade III) are much less common than papillomas, with malignant change occurring in fewer than 10% to 20% of papillomas. They are usually seen in the lateral ventricles. CSF dissemination is the rule with choroid plexus carcinomas, occurring in more than 60% of cases, but even benign papillomas may seed the CSF. It is difficult to distinguish a benign papilloma from a malignant one. Parenchymal invasion may suggest carcinoma. The 5-year survival for completely resected choroid plexus papillomas is 100%; for carcinomas it is more like 40%.

Choroid Plexus Hemangiomas

Choroid plexus hemangiomas are benign neoplasms of the choroid plexus usually seen in the lateral ventricle. Although the tumor enhances markedly and may calcify, it is usually seen as an incidental finding in an asymptomatic patient. There is an association with Sturge-Weber syndrome. In this syndrome choroidal hemangiomas ipsilateral to the leptomeningeal vascular malformation may be present (see Chapter 8).

Choroid Plexus Xanthogranuloma

Another benign condition of the choroid plexus is the xanthogranuloma. This incidental lesion may have fat density/intensity within it and is also centered on the glomus of the trigone. Curiously, they are frequently bright on diffusion-weighted imaging (DWI) scans (Fig. 2-18). They are of no clinical import.

Table 2-3 lists common intraventricular neoplasms.

Other Intraventricular Masses (Excluding Meningiomas)

Ependymal Tumors

Ependymomas are one of a variety of ependymal tumors (Table 2-4) that may occur throughout the brain and spinal cord. Although ependymomas usually present before age 10, a second ependymoma peak in the fourth and fifth decade of life is seen.

Posterior fossa ependymomas are usually associated with the fourth ventricle, although lesions arising primarily in the foramina of the fourth ventricle do occur and may present as masses outside the fourth ventricle (Table 2-5). Although it may seem a paradox, 20% of ependymomas arise intraparenchymally, usually in the supratentorial space (thought to be because of ependymal rests left over from radial glial migration). The prognosis with ependymomas varies depending on the site; filum terminale spinal ependymomas (usually WHO grade I) do best, followed by spinal cord, followed by supratentorial space, followed by posterior fossa (usually WHO grade II) ependymomas. There is a 50% 5-year progression-free survival rate with posterior fossa ependymomas.

Ependymomas have a greater incidence of calcification (40% to 50%) than other posterior fossa pediatric neoplasms. The calcification is typically punctate. When cysts are present (15%), they are small. On unenhanced CT scans noncalcified infratentorial ependymomas are typically hypodense to isodense. The lesions demonstrate mild enhancement, to a lesser degree than the medulloblastomas. Hydrocephalus is usually present because of blockage of the fourth ventricular outflow. They often arise as midline lesions that fill the fourth ventricle without displacing it. A classic appearance of a posterior fossa ependymoma is a calcified fourth ventricular mass that extends through and widens the foramina of Luschka and Magendie. When seen in the cerebral hemispheres, the lesions are larger and are cystic 50% of the time.

On MRI the lesions are hypointense on T1WI and tend to be intermediate in intensity on T2WI (Fig. 2-19). Hemorrhage is present in about 10% of cases. Hypointensities on T2WI may be because of calcification.

Ependymomas are another example of brain tumors that have a high incidence of subarachnoid seeding, and the use of contrast material is essential to the detection of subarachnoid spread.

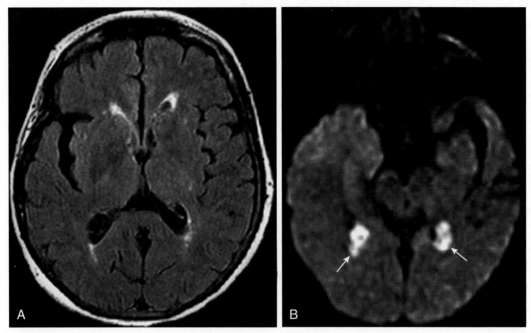

FIGURE 2-18 Xanthogranulomas of the choroid plexus. Although the fluid-attenuated inversion recovery scan **(A)** may be relatively unremarkable, the diffusion-weighted imaging (DWI) scan **(B)** shows the bright signal *(arrows)* in the xanthogranulomas of the choroid plexus. The presence of protein, cholesterol, or other compounds has been cited for why these are bright on DWI.

TABLE 2-3   Intraventricular Masses by Site

Neoplasms	Lateral	Third	Fourth
Choroid plexus papilloma/ carcinoma	Common, pediatric	Common, adult	
Craniopharyngioma	Common from suprasellar growth		
Ependymoma	Common		
Medulloblastoma	Common, growth from vermis		
Meningioma	Common, glomus, atrium	Along choroid plexus	
Metastases	Yes	Yes	Yes
Chordoid glioma	Rare	Typical	
Central neurocytoma	Common, septum	Rare	Never

TABLE 2-4   World Health Organization Classification of Ependymal Tumors

Tumor	Grade	Peak Age (years)
Ependymoma		
Cellular	II	0-9, 2nd peak, 30-50 (spinal)
Papillary	II	0-9
Clear cell	II	0-9
Tanycytic	II	30-50
Anaplastic ependymoma	III	0-9
Myxopapillary ependymoma	I	30-40
Subependymoma	I	40-60

Anaplastic Ependymomas

Anaplastic ependymomas have an unfavorable prognosis with more rapid growth rate and more frequent contrast enhancement. The prognosis is worse with younger age, incomplete resection, subarachnoid dissemination, high cell density, and a higher rate of mitoses histopathologically. Their imaging features are indistinguishable from ependymomas.

Subependymoma

Subependymomas are variants of ependymomas that contain subependymal neuroglia. These tumors are graded WHO grade I and often do not present until late adulthood. They arise intraventricularly or periventricularly and are frequently multiple at autopsy. They most often arise in the lateral recesses of the fourth ventricle (50% to 60%) but can be seen in the lateral ventricles (30% to 40%) attached to the septum pellucidum (Fig. 2-20). Lateral ventricular subependymomas often arise in patients older than 10 to 15 years. The lesion appears isodense on CT scans, isointense on T1WI, and hyperintense on T2WI, although they may be heterogeneous lesions. Most (>60%) subependymomas of the posterior fossa do not enhance, but descriptions of lesions with minimal, moderate, and marked enhancement lead to a nonspecific

TABLE 2-5 Distinctions among Medulloblastoma, Ependymoma, and Astrocytoma in Posterior Fossa

Feature	Medulloblastoma	Ependymoma	Astrocytoma
Unenhanced CT	Hyperdense	Isodense	Hypodense
ADC Value	Low	Intermediate	High
Enhancement	Moderate	Minimal	Nodule enhances, cyst not
Calcification	Uncommon (10%-21%)	Common (40%-50%)	Uncommon (10%)
Origin	Vermis	Ependymal lining	Hemispheric
T2WI	Intermediate	Intermediate	Bright
Site	Midline	Midline*	Eccentric
Subarachnoid seeding	130%	Uncommon	Rare
Age (years)	5-12	2-10	10-20
Cyst formation	10-20%	15%	60-80%
Foraminal spread	No	Yes (Luschka, Magendie)	No
Hemorrhage	Rare	10%	Rare MRS
Metabolite			
NAA	Low	Intermediate	Intermediate
Lactate	Absent	Often present	Often present
Choline	High	Less elevated	High

ADC, Apparent diffusion coefficient; *CT,* computed tomography; *MRS,* magnetic resonance spectroscopy; *NAA,* N-acetyl aspartate; *T2WI,* T2-weighted image.
*Except desmoplastoic variety.

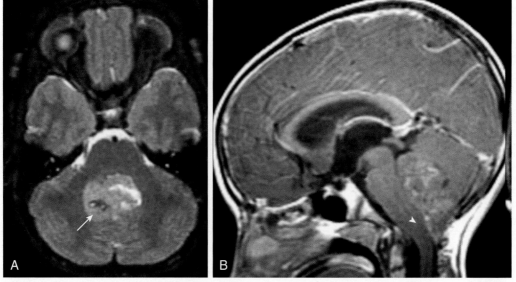

FIGURE 2-19 Fourth ventricular ependymoma. **A,** This mass in patient Dru Batluk fills and expands the fourth ventricle and is heterogeneous on T2-weighted image. The low signal area marked by the *arrow* is a focus of calcification. **B,** Note how the lesion oozes through the foramen of Magendie *(arrowhead)* to squirt into the cervicomedullary junction.

characterization, particularly in the supratentorial space. They are rarely seen in the spinal canal as intramedullary or extramedullary intradural masses. Usually subependymomas are isodense to gray matter on CT and isointense on all MR pulse sequences. They have a benign course with slow growth and a lack of invasiveness. Surgical resection is curative.

Central Neurocytoma

This tumor may grow intraventricularly (WHO grade II tumor of neuronal origin; see Tables 2-3 and 2-6), and because it calcifies, simulates a much rarer intraventricular oligodendroglioma. It may be cystic, and favors the lateral or third ventricle with an attachment to the septum pellucidum (Fig. 2-21). Edema is rare. Central neurocytomas peak in the third decade of life with mean age 29 years. They are isointense to gray matter on all MR pulse sequences and show mild to moderate enhancement. Intraventricular neurocytomas hemorrhage more frequently than oligodendrogliomas, which may suggest that diagnosis. Usually, however, the radiologist cannot differentiate between the two. Immunohistochemical markers for synaptophysin can distinguish intraventricular oligodendrogliomas from neurocytomas. The pathologic distinction is not moot; neurocytomas have a much more benign course than oligodendrogliomas and may not require radiation therapy.

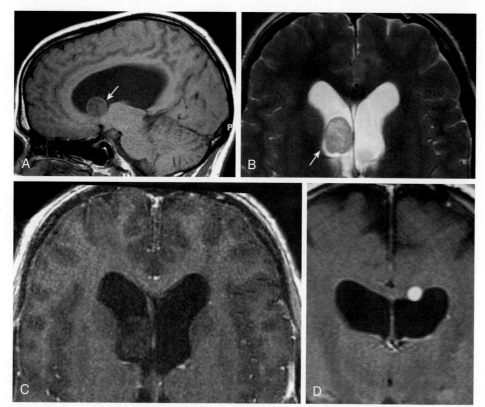

FIGURE 2-20 Supratentorial subependymomas in brothers Lyon D. Penn-Dima and Otto D. Penn-Dima. **A,** An intraventricular mass *(arrow)* is seen on sagittal T1-weighted image (T1WI). **B,** It is well defined on T2WI and does not enhance on post-contrast sequences **(C).** This is best for a subependymoma because a neurocytoma, meningioma, and choroid plexus papilloma will enhance. However, low-grade astrocytoma and ependymoma are in the differential diagnosis. **D,** Subependymomas need not be intraventricular. Here is the brother's lesion, on the outer surface of ventricle, this time enhancing, but the same diagnosis.

TABLE 2-6   World Health Organization Classification of Neuronal and Mixed Neuronal-Glial Tumors

Tumor	Grade	Peak Age (years)
Gangliocytoma	I	0-30
Ganglioglioma	I or II	0-30
Anaplastic ganglioglioma	III	0-30
Desmoplastic infantile astrocytoma/ganglioglioma	I	0-2
Dysembryoplastic neuroepithelial tumor	I	10-20
Central neurocytoma	II	20-30
Cerebellar liponeurocytoma	I or II	45-55
Paraganglioma of filum terminale	I	30-70

Oligodendrogliomas show a predilection for the septum pellucidum as well.

Nonneoplastic "Tumors"

Epidermoids

There is some confusion involved with putting epidermoids and dermoids into a "tumor" chapter because these are congenital epidermal inclusion cysts and dermal inclusion cysts. They really are not neoplastic and merely reflect two entities of ectodermal origin, one with desquamated skin (epidermoid), and one with skin appendages like hair follicles and sebaceous cysts (dermoid). Epidermoids and dermoids grow slowly and are histologically benign. Teratomas, usually lumped in the same category, however, are true neoplasms of multipotential germ cells.

Epidermoids are also called congenital cholesteatomas and, like acquired ones in the temporal bone, have a pearly-white appearance. Men and women are affected equally, with a peak incidence in the 20- to 40-year age range. Epidermoids often occur in the cerebellopontine angle cistern (where they may present with trigeminal neuralgia and facial paralysis), the suprasellar cistern, the prepontine cistern, and the pineal region (Fig. 2-22). Extradural epidermoids are nine times less common than intradural ones and arise intraosseously in the petrous bone and the temporal bone. They appear as well-defined bony lesions with scalloped margins (see Fig. 2-22, *E*).

Epidermoids are typically of low density like CSF on a CT scan. Epidermoids expand to fill the interstices of the CSF space. An epidermoid is a lesion that is quite aggressive in insinuating itself around normal brain structures and often has scalloped borders. CT demonstrates a nonenhancing lobulated lesion. Sometimes epidermoids are hard to distinguish from arachnoid cysts, particularly in the cerebellopontine angle cistern (Table 2-7). Classically, epidermoids do not enhance. A stereotypical appearance on a CT scan is that of displacement of the brain stem posteriorly by what appears to be just a dilated cistern anteriorly. In fact, this cistern is really a CSF-density epidermoid with mass effect.

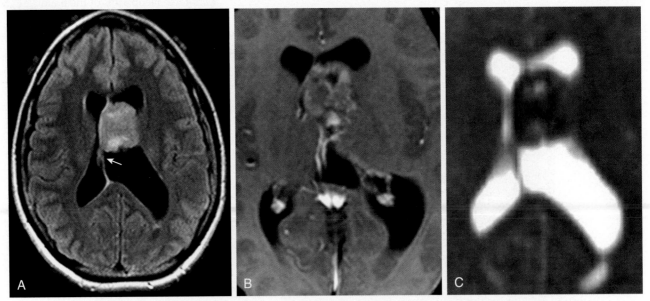

FIGURE 2-21 Central neurocytoma. **A,** Axial enhanced fluid-attenuated inversion recovery scan shows the left lateral ventricular mass in patient Rush Mityo Oh Arr attached to the septum pellucidum *(arrow)*. It enhances **(B)** and also has restricted diffusion on the apparent diffusion coefficient map **(C)** despite being a WHO grade II. Careful that the intrinsic calcification of the lesion does not confabulate the diffusion-weighted image result.

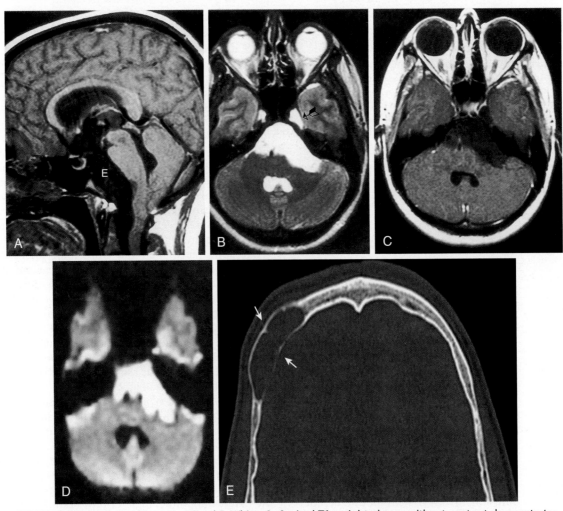

FIGURE 2-22 Epidermoid in patient Pearl E. White. **A,** Sagittal T1-weighted scan without contrast demonstrates a low intensity mass (E) similar to cerebrospinal fluid (CSF) anterior to the brain stem. Note the scalloped margin to the lesion at the pons-medulla-cervicomedullary region. **B,** On the T2-weighted image the lesion again has signal intensity similar to CSF. Note Meckel cave enlargement *(funky arrow)*; is it involved or merely dilated? **C,** No enhancement *(no arrow)* is seen on the postgadolinium T1-weighted scan. **D,** Aha! The diffusion-weighted scan shows that there is restricted diffusion within the mass and it is dissimilar to cerebrospinal fluid based on its very high signal intensity (compare with CSF in fourth ventricle). This is classic for an epidermoid. **E,** Calvarial epidermoid with lytic lesion but sclerotic intact borders *(arrows)* in patient Lucien S. Cull.

TABLE 2-7   **Differentiation of Epidermoid and Arachnoid Cyst**

Characteristic	Arachnoid Cyst	Epidermoid
CT density	CSF	Slightly higher than CSF
Margins	Smooth	Scalloped
Calcification	No	25%
Blood vessel	Displaces	Insinuates between vessels
Intrathecal contrast	Delayed uptake	No uptake, defines borders
Characteristic on MRI sequence sensitive to CSF pulsation (steady-state free precession)	Pulsates	Does not pulsate
Diffusion	Dark	Bright
ADC	Increased	Decreased
FLAIR	Dark	Variable

ADC, Apparent diffusion coefficient; *CSF,* cerebrospinal fluid; *CT,* computed tomography; *FLAIR,* fluid-attenuated inversion recovery; *MRI,* magnetic resonance imaging.

MRI has been very helpful in distinguishing between epidermoids and arachnoid cysts, a distinction that is sometimes blurred on a CT scan. On MRI, these lesions are hypointense on T1WI and hyperintense on T2WI, similar to CSF. Most epidermoids are brighter on FLAIR than CSF, unlike arachnoid cysts. On DWI these lesions are typically very bright, showing restricted diffusion on ADC maps and thus easily distinguishable from arachnoid cysts, which are dark on DWI, and have very high diffusivity. Epidermoids do not demonstrate enhancement unless they have been previously operated or secondarily infected. Rarely, epidermoids may be bright on T1WI; this usually is because of high protein and viscosity in the lesion.

Dermoid

A lesion with the high intensity of fat and signal void of calcification on T1WI suggests a dermoid or teratoma (Fig. 2-23). These lesions more typically occur in the midline as opposed to the epidermoids, which are generally off the midline. Male patients are more commonly affected, and patients are younger than those with epidermoids. The presence of fat may be suggested by an MRI chemical shift artifact seen as a hyperintense and hypointense rim at the borders of the lesion in the frequency-encoding direction. Fat suppression scans decrease the intensity on T1WI. The possibility of a ruptured dermoid should be considered when multiple fat particles are seen scattered on an MRI or when lipid is detected in the CSF (rule out pantopaque myelographic contrast droplets in an old, old patient). Usually the lesions are very well defined. In many cases unruptured dermoids and lipomas are not distinguishable.

Teratoma

Teratomas are congenital neoplasms containing ectodermal (skin, brain), mesodermal (cartilage, bone, fat muscle), and endodermal (cysts with aerodigestive mucosa) elements.

The pineal and suprasellar regions are common sites for teratomas. One will see a lesion of mixed density and intensity. The presence of an enhancing nodule in this neoplasm may help distinguish it from a dermoid.

Lipoma

Lipomas should probably also not be placed in a chapter on neoplasms because they are most frequently developmental or congenital abnormalities associated with abnormal development of the meninx primitiva, a derivative of the neural crest. Lipomas are particularly common in association with agenesis of the corpus callosum, and 60% are associated with some type of congenital anomaly of the associated neural elements (Fig. 2-24). The most common intracranial sites for lipomas are the pericallosal region, the quadrigeminal plate cistern, the suprasellar cistern, and the cerebellopontine angle cistern (unfortunately, overlapping dermoids). Because the tissue is histologically normal but located in an abnormal site, lipomas should best be termed choristomas, not neoplasms. Fat defines the lipoma; look for low density on a CT scan, high intensity on T1WI that darkens when fat suppression techniques are applied.

INTRAAXIAL TUMORS

One thing that gets our goat is when trainees refer to all primary brain tumors as "gliomas." This is not a term befitting readers of our book. You are now an aficionado of neuroradiology; banish "glioma" from your vocabulary and use a more precise term. We will not "friend" anyone saying "glioma." The category of "gliomas" (glial based neoplasms) of the brain includes astrocytomas, glioblastomas, oligodendrogliomas, ependymomas, subependymomas, gangliocytomas, and gangliogliomas. Do not lump all tumors under the umbrella term of "glioma" when you mean to say "astrocytoma."

Of all intraaxial neoplasms, glioblastomas account for roughly 25%, astrocytomas 9%, ependymomas 2.5%, medulloblastomas 2.5%, oligodendrogliomas 2.4%, and gangliogliomas 1%. Metastases account for 35% to 40% of intracranial neoplasms.

We will follow the WHO classification of brain tumors in our description of astrocytomas (Table 2-8), which separates astrocytomas into (benign) circumscribed astrocytomas (grade I), diffuse astrocytomas (grade II), anaplastic astrocytomas (grade III), and glioblastomas (grade IV) on the basis of histologic criteria and gross/imaging appearance. Circumscribed lesions include the pilocytic astrocytomas, subependymal giant cell astrocytomas, and pleomorphic xanthoastrocytomas. Fibrillary, gemistocytic, and protoplasmic astrocytomas are classified as diffuse grade II tumors. Then come the nasties, anaplastic astrocytoma and glioblastoma, that are of the highest grade.

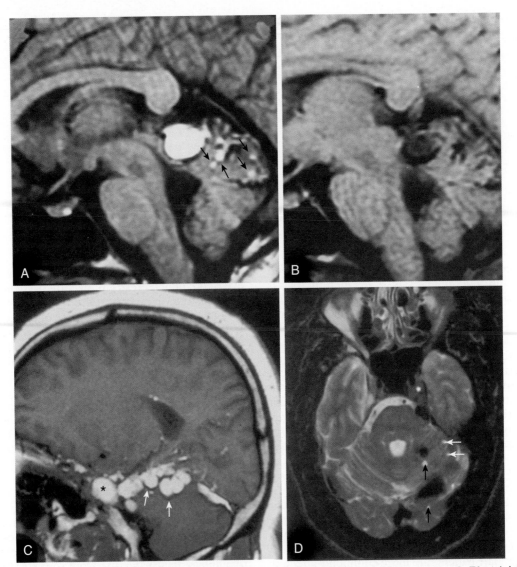

FIGURE 2-23 The many faces of dermoids in twin patients Lee P.D. Lee-Jung and Fah T. Lee-Jung. **A,** T1-weighted image (T1WI) of a ruptured dermoid shows high signal intensity *(arrows)* along the superior surface of the cerebellum. **B,** Fat-suppressed T1WI confirms lesions are fatty. **C,** A larger mass is seen in the Meckel cave region anteriorly *(asterisk).* Chemical shift artifact *(arrows)* is seen along the inferior margin of the fat droplets extending posteriorly. **D,** This T2-weighted fat suppressed scan demonstrates the dark signal of suppressed fat *(star)* in the left Meckel cave region as well as dark signal intensity large *(arrows)* and small fat deposits along the cerebellar folia representing rupture of this dermoid tumor.

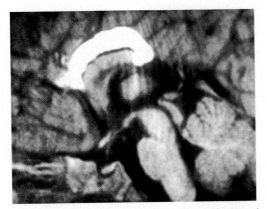

FIGURE 2-24 Lipoma associated with callosal dysgenesis. This bright on T1-weighted image lipoma is associated with an incompletely formed corpus callosum in patient C. C. Fatt.

TABLE 2-8 World Health Organization Classification of Astrocytic Tumors

Tumor	Grade	Peak Age (years)
Pilocytic astrocytoma	I	0-20
Subependymal giant cell	I	10-20 Astrocytoma
Pleomorphic	II	10-20 Xanthoastrocytoma
Diffuse astrocytoma	II	30-40
Fibrillary	II	30-40
Protoplasmic	II	30-40
Gemistocytic	II	30-40
Anaplastic Astrocytoma	III	35-50
Glioblastoma	IV	50-70
Giant cell glioblastoma	IV	40-50
Gliosarcoma	IV	50-70

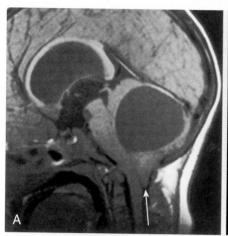
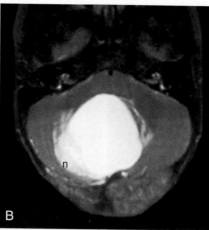
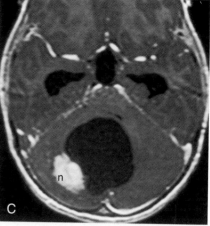

FIGURE 2-25 Benign pilocytic astrocytoma in patient Shlomo Groer. **A,** The large cystic mass in the posterior fossa that causes cerebellar tonsillar herniation *(arrow)* and hydrocephalus in this child with a pilocytic astrocytoma. **B,** T2-weighted image shows the dominant feature of the cyst with a small mural nodule (n), which enhances in **C.**

Grade I: Circumscribed Astrocytomas

Juvenile Pilocytic Astrocytomas

Cerebellar juvenile pilocytic astrocytomas (JPAs) are the most common (infratentorial) neoplasm in the pediatric age group and are classified as WHO grade I astrocytic tumors. Seventy five percent occur before age 20, but they are rare in the first 3 years of life. They account for 70% to 85% of cerebellar astrocytomas. Pilocytic astrocytomas are seen in up to 15% to 20% of all patients with NF-1. They are benign.

Pilocytic astrocytomas are characterized by cysts and mural nodules. In general, pilocytic astrocytomas are well outlined from normal brain, are usually round, and are usually not ominous in appearance (Fig. 2-25). The typical cerebellar astrocytoma in the pediatric age group is cystic (60% to 80%), whereas in older patients cerebellar astrocytomas are more likely to be solid. The cyst and mural nodule in an adult will suggest a differential diagnosis of a hemangioblastomas. On MRI, astrocytomas show hypointensity on T1WI and hyperintensity on T2WI/FLAIR. The cystic portion of the tumor has signal intensity similar to CSF on T1WI and T2WI, but may be hyperintense to CSF on the PDWI or FLAIR because of a larger amount of protein. The lesion is very well defined. The solid portion of the JPA enhances strongly. This is a case where a low-grade tumor shows marked enhancement and high metabolic activity on positron emission tomography (PET) scanning. The adage usually is that enhancement portends a higher-grade tumor.

When removed completely, the lesions are associated with an excellent prognosis (5-year survival rate >90%). Sixty percent of pilocytic astrocytomas occur in the posterior fossa (see Table 2-5) where ependymomas and medulloblastomas also appear in children. Pilocytic astrocytomas also favor the optic pathways and hypothalamus (where they may be associated with neurofibromatosis type 1). The astrocytomas in this region often infiltrate the third ventricle and present with hydrocephalus. Diencephalic syndrome characterized by weight loss despite normal intake, loss of adipose tissue, motor hyperactivity, euphoria, and hyperalertness may occur in cases of chiasmatic/hypothalamic astrocytomas and kids sucking on blow pops with a sugar cream center.

Pleomorphic Xanthoastrocytoma

Pleomorphic xanthoastrocytoma (PXA) tumors are seen in children and young adults, and two thirds occur in patients less than 18 years old. The lesion shows a preference for the periphery of the temporal lobes as they arise from subpial astrocytes. They may show a base at the meninges (which enhances) in more than 70% of cases. Owing to their predilection for the temporal lobes, they usually present with seizures. Nodular enhancement of the cortical mass is common. Cyst formation occurs in about one third to one half of cases, but hemorrhage and calcification are distinctly uncommon. Margins may be well or poorly defined (Fig. 2-26). These tumors may rarely have anaplastic transformation but are considered WHO grade II lesions. Survival is over 80% at 5 years. These tumors are hard to distinguish from dysembryoplastic neuroectodermal tumors (DNETs) but enhance more frequently. PXAs have a meningeal attachment and no cortical dysplasia; these features also help distinguish them from DNETs.

Subependymal Giant Cell Astrocytomas

Another benign variant of "glioma" is the subependymal giant cell astrocytoma (SGCA or SEGA or SEGCA). Classically, this lesion is seen in 2- to 20-year-old patients with tuberous sclerosis. Patients with tuberous sclerosis may have areas of subependymal nodules or tubers (which may be calcified on a CT scan), subcortical tubers, and other hamartomatous lesions. Subependymal giant cell astrocytomas typically occur near the foramina of Monro and in contradistinction with tubers demonstrate moderate to marked enhancement (Fig. 2-27). This tumor may occur in isolation without tuberous sclerosis, but it is an uncommon variant. The tumor is slow growing (WHO grade I) and projects into the ventricular system, where it may appear as a calcified intraventricular mass (see Table 2-3). The outflow of the lateral ventricle may be obstructed, leading to trapping of one or both lateral ventricles with noncommunicating hydrocephalus. Because many subependymal nodules show enhancement on MRI but not on a CT scan, CT can actually be more specific than MRI in

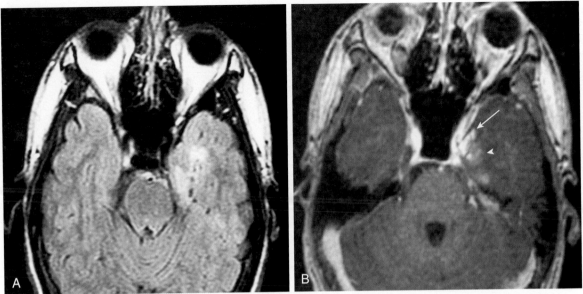

FIGURE 2-26 Pleomorphic xanthoastrocytoma of the temporal lobe **A,** In patient E. Lee Sitseejour the fluid-attenuated inversion recovery (FLAIR) scan shows an ill-defined mass in the left medial temporal lobe. **B,** There is nodular enhancement at the periphery adjacent to the left cavernous sinus *(arrow)* and less well defined in the colocated area of FLAIR signal abnormality *(arrowhead).*

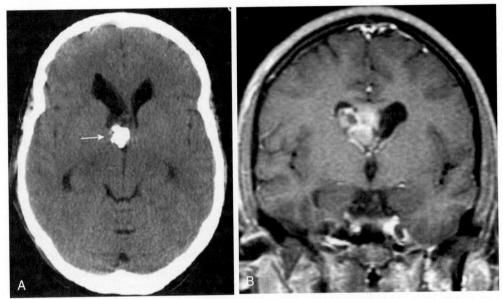

FIGURE 2-27 Grade I benign subependymal giant cell astrocytoma in patient Drew Goode-Locke. **A,** Note the calcified mass *(arrow)* near the right foramen of Monro on unenhanced computed tomography. **B,** On the coronal T1-weighted contrast-enhanced scan one is more impressed with the component of the tumor that extends intraventricularly. The septum also seems to be infiltrated.

distinguishing large subependymal nodules from SGCA by virtue of this feature.

Grade II: Diffuse Astrocytoma

The fibrillary form of astrocytomas is more infiltrative and solid and has a worse prognosis than the pilocytic form. Fortunately, pilocytic astrocytomas constitute 85% of cerebellar astrocytomas and fibrillary the remaining 15%. Fibrillary astrocytomas occur more commonly in older children than the pilocytic type and are the predominant histologic finding in "brain stem gliomas" (OK, 'tis true, brain stem "glioma" is the preferred term over astrocytoma; go ahead and "FB poke" us). Diffuse astrocytomas grade II are more likely to show absence of enhancement. They can occur anywhere (Fig. 2-28), but one third are in the frontal lobes and one third in the temporal lobes. Cystic change and calcification may rarely occur. Gemistocytic astrocytomas are found exclusively in the cerebral hemispheres and are a rare variety of supratentorial astrocytomas that in 80% of cases ultimately convert to glioblastomas. Mean survival time is

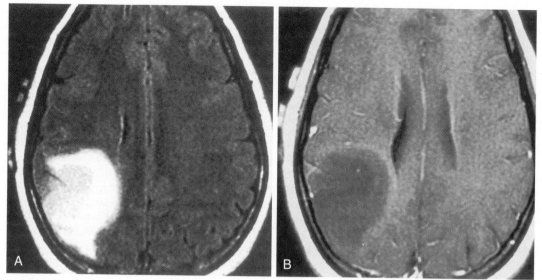

FIGURE 2-28 Grade II astrocytoma in patient Gladys Knott-Wurse. **A,** The mass is somewhat well defined on fluid-attenuated inversion recovery, but still is pretty bulky. **B,** It does not enhance.

more than 3.5 years, slightly better than another variety of astrocytoma, the monstrocellular (or magnocellular) type. The borders are relatively well defined for an astrocytoma in both these varieties, but no other features are distinctive.

Brain Stem "Gliomas"

Brain stem astrocytomas are usually treated with radiation and do not have as high a survival rate as juvenile pilocytic astrocytomas. Nonetheless, they are usually WHO grade II diffuse astrocytic (fibrillary) tumors with a 25% 10-year survival rate. The masses may be isolated to one part of the brain stem and may grow exophytically (20% of cases). Pontine and exophytic brain stem gliomas have a better prognosis than midbrain or medullary ones, and exophytic ones may benefit from surgical resection. Anaplasia develops in 50% to 60% and portends poor outcomes. Pontine brain stem gliomas are most common. These masses may be unapparent on a CT scan because of beam-hardening artifact in the posterior fossa, the decreased soft-tissue resolution of CT, and their subtle expansion of the anatomy. By contrast, MRI provides high sensitivity for the lesion and is well suited to preradiation therapy planning with its multiplanar capability. The lesions are high in intensity on T2WI/FLAIR amid the normal decreased signal intensity of the white matter tracts of the brain stem (Fig. 2-29). They may (33%) or may not (67%) enhance. Cystic degeneration may occur. Subtle enhancement that cannot be seen with CT may be apparent on MRI. Symptoms occur late in the course of the disease because the tumor infiltrates rather than destroys histopathologically. Brain stem astrocytomas also do not produce hydrocephalus until they are far advanced. As they enlarge they often appear to encircle the basilar artery.

Brain stem astrocytomas may occur in adults; however, 80% of the time they are childhood lesions. They comprise 20% of posterior fossa masses in children, less common than cerebellar astrocytomas and medulloblastomas. Other lesions that expand the brain stem in a child include tuberculosis (most common brain stem lesion worldwide), lymphoma, rhombencephalitis (caused by Listeria), and demyelinating disorders (acute disseminated encephalomyelitis and multiple sclerosis).

Grade III: Anaplastic Astrocytoma

Anaplastic astrocytomas (AA) occur most commonly in the fourth and fifth decades of life and usually evolve from lower-grade astrocytomas. They have ill-defined borders with prolific vasogenic edema (Fig. 2-30). They are much more likely to show contrast enhancement, but if necrosis is seen, bump the lesion up to a glioblastoma. When all astrocytomas are considered, anaplasia occurs in 75% to 80% with ultimate dedifferentiation into glioblastoma occurring in 50% of cases. Typical time for progression from AA to glioblastoma is 2 years—short enough to drive the patient to drinking…and the other AA.

Glioblastoma

Of the astrocytomas, glioblastoma (WHO grade IV) is the most common variety, nearly twice as frequent as the anaplastic astrocytomas and the grade I astrocytomas and accounting for 50% to 60% of astrocytic tumors and 15% of intracranial neoplasms. Glioblastoma is the most lethal of the gliomas, having a 10% to 15% 2-year survival rate. These tumors can appear anywhere in the cerebrum but are seen most commonly in the frontal (23%), parietal (24%), and temporal (31%) lobes. The occipital lobes are less commonly affected.

The majority of glioblastomas in the elderly are felt to be primary tumors; that is, they do not evolve from lower-grade astrocytomas. The clinical course is short, a few months. Secondary glioblastomas that arise from dedifferentiation of lower-grade astrocytomas occur in a younger age group and with a more protracted clinical prodrome over a few years in duration.

Glioblastomas are characterized on imaging by the presence of necrosis within the tumor (Fig. 2-31). Histologic grade seems roughly to parallel patients' age in the adults; the older the patient is, the more likely the lesion is to be a higher-grade astrocytoma. Other factors that correlate with higher grade are ring enhancement, enhancement in general, marked mass effect, intratumoral necrosis, and restricted diffusion. Recent reports have found that relative cerebral blood volume (rCBV) and ADC values correlate well with astrocytoma histology and vascularity; the higher the rCBV and the lower the ADC the more likely one is dealing with a glioblastoma. An elevated cerebral blood volume of more than 1.75 is indicative of higher grade. Other groups have shown a 90% sensitivity and specificity for separating low- and high-grade astrocytomas by setting a mean fractional anisotropy (FA) level of 0.141 and maximum FA of 0.244 across all slices through the tumor. FA seems to do better than using ADC values. At the same time, hemorrhage, as depicted by susceptibility sensitive scans, and the presence of high choline/NAA ratios or lactic acid peak on MRS have also been linked to a higher grade (Fig. 2-32). Most people are using a Cho/NAA ratio of

greater than 2.2 to denote a high-grade astrocytoma and the presence of myoinositol to suggest lower-grade lesions. You also have the old fallback option: the older the patient the more likely the tumor is of higher grade.

Glioblastomas infiltrate wildly and are mapped as high signal intensity on long TR images. Enhancement may be solid, ringlike, or occasionally inhomogeneously mild (see Fig. 2-31). Daughter/satellite lesions around the periphery of the mass look like a cluster of grapes. The tumor frequently crosses the corpus callosum, anterior commissure, or posterior commissure to reach the contralateral hemisphere. Of the adult astrocytomas, a glioblastoma is the most common to have intratumoral hemorrhage and subarachnoid seeding (2% to 5% of cases). Occasionally, it coats the ventricles. When you see a lesion involving the corpus callosum (Box 2-4), you should put glioblastoma and lymphoma near the top of the list of neoplasms (Fig. 2-33).

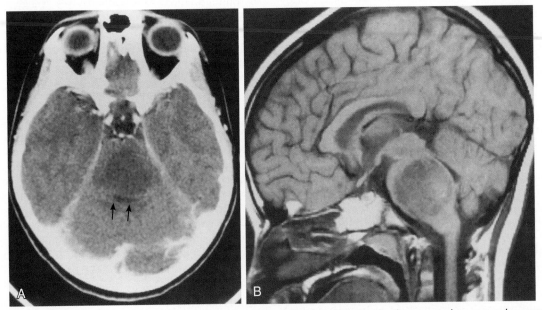

FIGURE 2-29 Brain stem glioma. **A,** Pontine brain stem glioma on this enhanced computed tomography scan in patient Sheila Liv Longue compresses the fourth ventricle *(arrows).* Note lack of enhancement of the tumor. **B,** Sagittal T1-weighted image better defines the extent of the tumor on this midline scan.

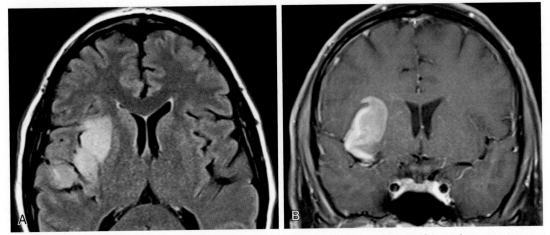

FIGURE 2-30 Patient, Anna P. Layzha, grade III astrocytoma. **A,** The fluid-attenuated inversion recovery scan shows periinsular high signal extending from right temporal lobe to right frontal opercular zone. **B,** After contrast administration enhancement is seen along the perisylvian region on coronal T1-weighted image.

The extent of a neoplasm is not defined by its enhancing rim. In fact, radiation oncologists treat the entire area of abnormal T2WI/FLAIR high intensity with radiation followed by a coned-down portal encompassing the enhancing portion with a 2-cm rim around the enhancing edge. Microscopic infiltration clearly extends beyond the confines of enhancement, and the high signal intensity on T2WI may not represent tumor in all instances.

Recent additions to our armamentarium are the true functional studies. For lesions adjacent to eloquent cortex (by this we mean near the speech, memory, or motor areas) it is helpful to the neurosurgeons that we show them where these critical areas are, so that they can reduce the operative morbidity associated with the resection. Because the surgeons are not going for a complete resection of a glioblastoma, only a gross resection, it would be nice if patients could spend the remaining months of their lives conversant with their family. Functional MRI (fMRI) can be used preoperatively to define the speech areas in relation to the tumor or even intraoperatively to direct the surgical resection to include the maximum amount of tumor and minimal amount of eloquent cortex. A word of caution, the BOLD fMRI contrast effect may be reduced near an astrocytoma secondary to (1) compression of vessels and neurovascular uncoupling, (2) vasoreactive substances (nitric oxide), (3) neurotransmitter substances expressed by tumors, (4) reduced neuronal function, and (5) invasion of vascular structures by the tumor.

Gliosarcomas

Consider this a nasty glioblastoma-mesenchymal tumor. They constitute 2% of all glioblastomas and favor the supratentorial space. Despite their name they may be well defined, superficial, and strongly enhancing.

Multicentric Astrocytomas

Multicentric astrocytomas may be because of true metachronous independent lesions, but more often than not represent contiguous spread of gliomatous tissue in which the connection is unapparent on imaging but present on pathologic study. Multiple independent glioblastomas occur in 2.3% of glioblastomas. NF-1 is associated with multifocal astrocytomas.

Gliomatosis Cerebri

According to the fourth edition of the WHO classification of brain tumors, gliomatosis cerebri is a pattern of disease which involves "at least 3 lobes and is usually bilateral, extending from the cerebral white matter, including the deep and subcortical portions, and often infiltrating the brain stem and the spinal cord" (Fig. 2-34). Peak age is 40 to 50 years, but it occurs in all adult age groups. WHO classification is grade III, analogous to AA. It is not uncommon to have the frontal and temporal lobes involved with the basal ganglia and thalami, and gliomatosis cerebri is bilateral in nearly half the cases. Brain stem involvement is not unusual with the midbrain and pons being affected four times more commonly than the medulla. The WHO classification places the lesion within the astrocytoma and/or oligodendroglial category in its most recent edition.

A CT scan may be interpreted as normal in appearance, but you may see loss of the gray-white differentiation and

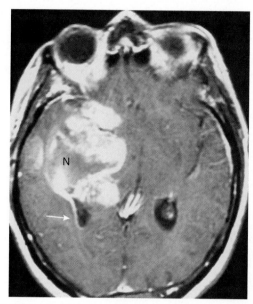

FIGURE 2-31 Glioblastoma. Ms. Sasha Shayme's glioblastoma demonstrates irregular enhancement, necrosis (N) and mass effect with displacement of midline structures to the left side. Bad actor. Note the ependymal spread of tumor *(arrow)*, which also is a bad prognostic sign.

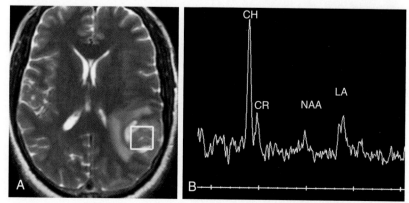

FIGURE 2-32 Glioblastoma. Magnetic resonance spectroscopy in patient Oma Gahd shows voxel placement **(A)** and a lactate peak (LA) at 1.3 PPM and a very high choline (CH) to N-acetyl aspartate (NAA) ratio of peak heights **(B).** Creatine peak (CR) is also indicated.

subtle mass effect. The more sensitive T2W1/FLAIR MRI shows diffuse increased signal intensity throughout. Both gray and white matter may be involved, and the lesion may spread bilaterally. Enhancement, if present at all, is minimal.

The prognosis is equivalent to that of a high-grade astrocytoma and, in fact, gliomatosis cerebri may show an explosive growth rate signifying its transformation to a glioblastoma. Median survival is 15 months.

Embryonal Tumors

Medulloblastomas/Primitive Neuroectodermal Tumors

There has been a recent impetus to rename medulloblastomas and other similar cell line tumors "primitive neuroectodermal tumors (PNETs)." The rationale is that there really is no such cell line as the "medullo cell" (at least that is what the pathologists tell us). In fact, the latest WHO classification lists these as "embryonal tumors" (Table 2-9). In common vernacular, however, the PNET of the posterior fossa is referred to as medulloblastoma.

Medulloblastomas are one of the most common posterior fossa masses in the pediatric population, accounting for more than one third of posterior fossa neoplasms and 50% of cerebellar tumors in children. They are quite malignant and have earned the highest class of aggressiveness (grade IV) by Peter Townsend of The WHO. These tumors are usually seen in the midline arising from the vermis and then growing into the inferior or superior velum of

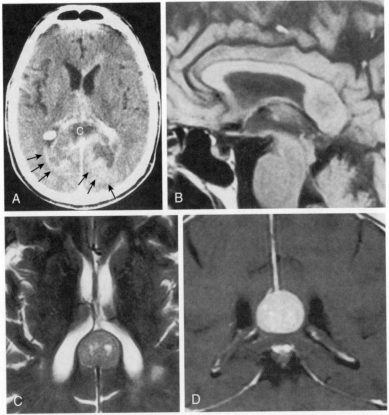

FIGURE 2-33 Lesions invading the corpus callosum in members of the Toomer-Svelling family. **A,** Glioblastoma crossing corpus callosum. Enhanced computed tomography reveals an irregularly enhancing tumor in a garland wreath pattern *(arrows)* crossing the splenium of the corpus callosum. Note that the genu of the corpus callosum also shows subtle enhancement denoting tumor infiltration. C marks the splenium of the corpus callosum, which appears to be somewhat necrotic. **B,** Lymphoma of the splenium: Sagittal T1-weighted image (T1WI) shows expansion of the splenium with abnormal signal intensity. **C,** The T2WI shows a focal mass in the splenium without significant edema. **D,** The lesion enhances strongly on coronal T1WI. Would this favor lymphoma over an intermediate-grade astrocytoma? Try diffusion. (Case courtesy Stuart Bobman.)

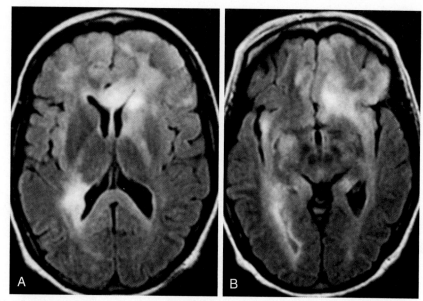

FIGURE 2-34 Gliomatosis cerebri. **A** and **B**, This fluid-attenuated inversion recovery scan in patient Colin D. Undertaker shows multiple lobes, bilateral involvement, and relatively low mass effect for the size of the abnormality, characteristic of gliomatosis. Do not expect avid enhancement.

TABLE 2-9   World Health Organization Classification of Embryonal Tumors

Tumor	Grade	Peak Age (years)
Medulloepithelioma	IV	0-5
Ependymoblastoma	IV	0-2
Medulloblastoma		
Desmoplastic	IV	0-16, 2nd peak 21-40
Large cell	IV	0-16, peak at 7 years
Medullomyoblastoma	IV	2.5-10.5
Melanotic	IV	0-9
Supratentorial PNET		
Neuroblastoma	IV	0-10
Ganglioneuroblastoma	IV	0-10
ATRT	IV	0-5

ATRT, Atypical teratoid/rhabdoid tumors; *PNET,* primitive neuroectodermal tumors.

the fourth ventricle (Fig. 2-35). Medulloblastomas typically occur in the 5- to 12-year age range, boys twice as commonly as girls, and patients usually present with hydrocephalus. Brain stem dysfunction may also be present.

As opposed to the pediatric tumors, the 20% of medulloblastomas that occur in young adults tend to be eccentric in the posterior fossa, residing in the cerebellar hemispheres in more than half the cases. They tend to have a more aggressive course than the pediatric tumors and are less well-defined lesions. A rare variety of medulloblastoma known as the desmoplastic medulloblastoma may grow from the cerebellar hemisphere into an extraaxial location, especially in the cerebellopontine angle.

A CT scan of medulloblastoma typically demonstrates a slightly hyperdense, well-circumscribed mass before enhancement (Fig. 2-36). The hyperdensity on CT is the best finding to suggest medulloblastoma in the posterior fossa (for an intraaxial mass and meningioma for

extraaxial). Occasional calcification and cystic degeneration may occur, more commonly in adults (see Table 2-9). The fourth ventricle, when seen, is displaced anteroinferiorly, and there often is obstructive hydrocephalus.

Sagittal images with MRI are optimal for visualizing the origin of the tumors from the vermis. The masses are usually hypointense on T1WI and isointense on T2WI. The lesions are typically very well defined and do not demonstrate a large amount of edema. A distinguishing feature between medulloblastomas and ependymomas is that the ependymoma classically enlarges the fourth ventricle while maintaining its shape, whereas medulloblastomas distort the fourth ventricle (see Table 2-5). The mass shows moderate contrast uptake. Medulloblastomas have a 10% to 21% incidence of calcification. Cystic change, initially thought to be rare, occurs in 10% to 20% of pediatric cases and 59% to 82% of adult cases. Medulloblastomas have the lowest ADC values of the typical pediatric posterior fossa tumors with ependymomas next and astrocytomas the highest values. Medulloblastoma is one of the tumors that have a special propensity for subarachnoid seeding, seen in one third of cases (see Fig. 2-35, *C* and *D*). For this reason the entire craniospinal axis should be imaged with MRI after contrast.

The spectroscopic signature of a PNET can distinguish it from other posterior fossa masses in children. PNETs show even lower NAA/Ch ratios compared with low-grade astrocytomas and ependymomas. Lactate is usually not present in PNETs but is often seen in pilocytic astrocytomas and ependymomas.

Turcot syndrome (*5q21* gene), hereditary retinoblastoma and nevoid basal cell carcinoma (Gorlin) syndrome (*9q31* gene) are associated with an increased rate of medulloblastomas.

Supratentorial Primitive Neuroectodermal Tumors

Cerebral neuroblastomas are now called supratentorial PNETs. If there are ganglion cells present, they may be termed cerebral ganglioneuroblastomas, but either way

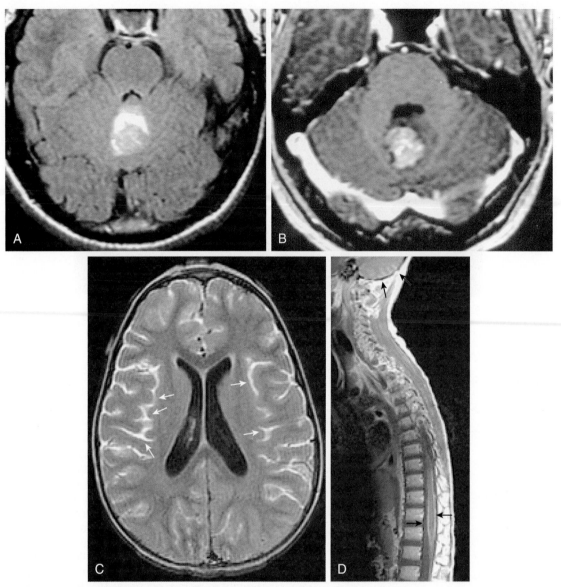

FIGURE 2-35 Medulloblastoma. A, The vermian mass is hyperintense on fluid-attenuated inversion recovery (FLAIR) and in the midline. **B,** The mass enhances. **C,** Seeding by medulloblastoma may be noted by high signal in the subarachnoid space *(arrows)* on FLAIR scanning, coupled with mild hydrocephalus. Postgadolinium FLAIR scans can be exquisitely sensitive to subarachnoid seeding. **D,** The images of the spine confirm subarachnoid dissemination in the posterior fossa *(arrows)* as well as on the thoracic spinal cord and conus medullaris.

they are WHO grade IV, aggressive tumors. The younger the age at diagnosis, the worse the prognosis. These tumors often show cyst formation, calcification (50% to 70%), and hemorrhage and are usually seen in children. They can be seen as hyperdense on unenhanced CT scans. They have a high rate of recurrence and subarachnoid seeding. Usually they are large lesions, 3 to 10 cm in diameter, and inhomogeneous in density and intensity; they enhance in a heterogeneous fashion. Occasionally, they arise in a suprasellar or periventricular location, but cerebral neuroblastomas are most commonly found in frontal, parietal, and occipital lobes. The tumors do not incite much edema and are usually well circumscribed. Eighty percent arise in patients less than 10 years old (25% before 2 years old), but they can be seen in young adults. Three-year survival rate is 60%, 5-year is 34%, and recurrence is common (40%).

Ependymoblastoma

Another in the WHO grade IV pediatric embryonal tumors (PNETs), the ependymoblastoma favors the supratentorial compartment as opposed to ependymomas. The tumors are large at discovery with enhancement, edema, and areas of peripheral necrosis and cyst formation. Nonetheless they are relatively well-circumscribed. They invade the leptomeninges and usually appose the ventricles, having arisen in periventricular neuroepithelial precursor cells. Death because of CSF dissemination usually occurs within 6 to 12 months of presentation and presentation is in young children and neonates.

Atypical Teratoid Rhabdoid Tumors

Atypical teratoid rhabdoid tumors (ATRTs) are WHO grade IV highly aggressive, heterogeneous looking tumors that may have hemorrhage, necrosis, calcification, and cyst

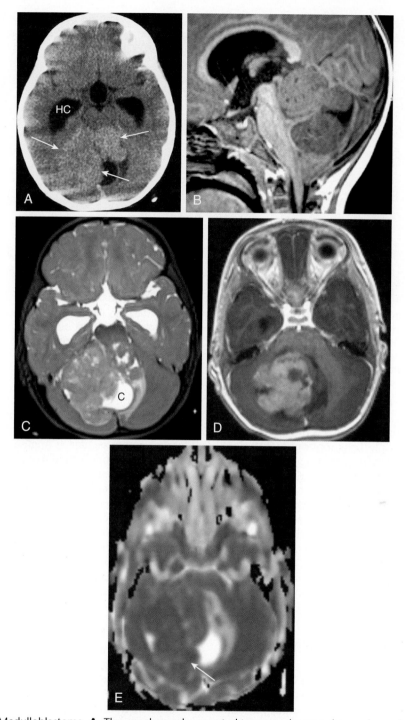

FIGURE 2-36 Medulloblastoma. **A,** The unenhanced computed tomography scan shows a hyperdense posterior fossa mass *(arrows)* causing hydrocephalus (HC) in this child named Xavier Lyfe. **B,** The origin from the superior vermian region is suggested on the sagittal scan. **C,** The associated cystic component is seen on T2-weighted scan (C). **D,** The mass enhances. **E,** One of the features of medulloblastoma is its low apparent diffusion coefficient values *(arrow)*—the lowest of the posterior fossa masses in children.

formation (Fig. 2-37). They are dense precontrast on a CT scan. They occur in the same time frame as primitive neuroectodermal tumors, i.e., in the first years of life. Enhancement is patchy. Subarachnoid dissemination is not uncommon (34%), even at presentation, and death is predictable within 1 year of diagnosis. Fifty-two percent occur in the posterior fossa; the remainder are 39% supratentorial, 5% peripineal, and 2% multifocal or intraspinal in location.

Dysembryoplastic Neuroepithelial Tumor

Dysembryoplastic neuroepithelial (neuroectodermal) tumors (DNET) are most commonly found in the temporal (50% to 62%) and frontal (31%) lobes and usually cause seizures. The lesion is a neuroepithelial tumor and most present by the second and third decade of life in patients with chronic seizures. They are located peripherally in the brain with the cortex involved in nearly all cases and the

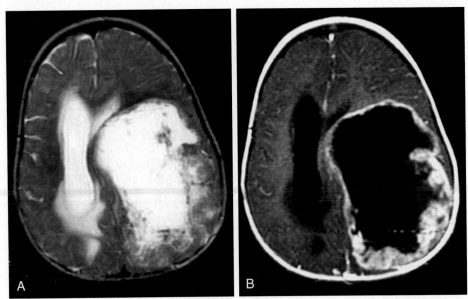

FIGURE 2-37 Rhabdoid tumor. **A,** This is a huge predominantly cystic mass in the left hemisphere of patient Yu Kentmis D'Masse. Note, however, that there is solid tissue posterolaterally. Check out that mass effect. **B,** The rim of the mass enhances, and the periphery belies the "grade IV" nature of this aggressive mass. Hydrocephalus is present with transependymal flow of cerebrospinal fluid evidenced by high signal around the margin of the trapped right lateral ventricle in **A.**

subcortical white matter in most cases leading to a wedge-shaped abnormality (Fig. 2-38). Half have poorly defined contours. On MRI, this tumor is characterized by the presence of cysts, usually multiple. Look for bright T2WI signal. The tumor often has septations and has a triangular pattern of distribution (possibly because of derivation along radial glial fibers). Contrast enhancement is observed in less than a third of the cases and mass effect is variable. The lesion is often hypodense on a CT scan and grows very slowly. DNETs usually do not have edema associated with them and they may remodel the calvarium. They are reported to have coincident focal cortical dysplasias with them in over 50% of cases. They rarely recur after resection, even if partial (WHO grade I). Good actors! If you stop to think about these tumors, remember to start again.

Neuronal, Mixed Neuronal/Glial Tumors

Gangliogliomas

Gangliogliomas are most commonly seen in children and young adults (64% to 80% occur in patients <30 years old). They are the most common mixed glioneural tumors of the CNS (see Table 2-6). They are low-grade tumors with good prognoses. Presentation most commonly is with a seizure. The tumors affect female more than male patients and are characterized by a benign, slow-growing course often associated with bony remodeling that testifies to their indolent growth. The most common sites for these tumors are within the temporal lobes (85%), frontal lobes, anterior third ventricle, and cerebellum. The differential diagnosis of temporal lobe masses is summarized in Table 2-10.

Gangliogliomas of the cerebellum tend to have cystic components but may be solid, mixed, or calcified masses. Over half of them show contrast enhancement. They also frequent the spinal cord and optic nerves. The lesions are cystic in 38% to 50% of cases and are typically hypodense or isodense on CT scans. One third show calcification and half have faint enhancement (Fig. 2-39). MRI features vary

depending on cyst formation. If cysts are present, then the ganglioglioma may be hypointense on T1WI and bright on T2WI. A mural nodule may coexist. The tumor is well defined and avascular at angiography. A cystic ganglioglioma can be distinguished from an arachnoid cyst in that it is clearly intraparenchymal and has higher signal intensity than CSF or arachnoid cysts on PDWI or FLAIR.

The tumor is comprised of neuronal ganglion cells and astrocytic glial cells, hence its name. However, its behavior is defined by the degree of dedifferentiation of the glial portion of the tumor and therefore it may convert into an aggressive lesion, the anaplastic ganglioglioma. The presence of vasogenic edema also is correlated with worse histologic grade, but most gangliogliomas do not produce much edema.

Desmoplastic Infantile Ganglioglioma

Desmoplastic infantile ganglioglioma (DIG) is a variant of the ganglioglioma that is usually seen in the first 2 years of life. The tumors typically occur in the frontal and parietal lobes, are large in size, and have a meningeal base. Cyst formation is the rule and peripheral rim or nodular enhancement usually is present as well. Some may show a calcified rim. Although this lesion may look like a huge necrotic glioblastoma in an infant, it has a good prognosis with benign histology (WHO grade I). The tumor has both glial and ganglionic derivation (no wonder they call it ganglioglioma). The desmoplasia accounts for the rim of low signal intensity tissue on T2W scans. The differential diagnosis will include DNETs and PNETs, but the incidence of calcification and hemorrhage (and age) is higher in cortical PNETs (Fig. 2-40).

Gangliocytoma

As opposed to gangliogliomas, which may undergo malignant degeneration and a more aggressive growth pattern, gangliocytomas have no glial component and no potential for malignant change. Gangliocytomas are usually located

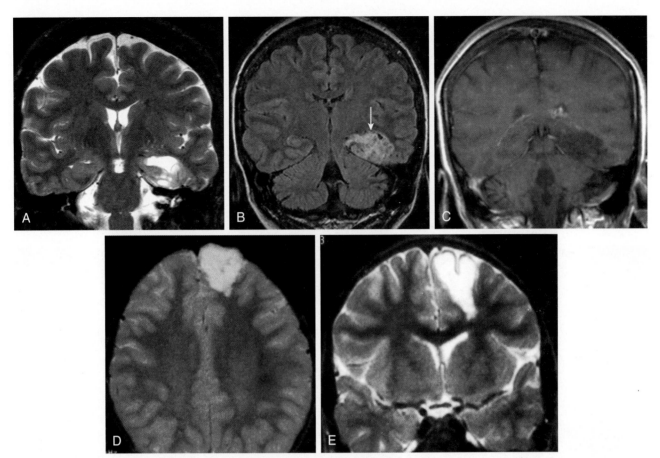

FIGURE 2-38 Dysembryoplastic neuroepithelial tumor (DNET). **A, B, C:** The T2-weighted image (T2WI), fluid-attenuated inversion recovery (FLAIR), and postcontrast T1-weighted magnetic resonance images in patient C. Jur O'Verderr show a peripheral, mainly cortical lesion *(arrow)* that has a Swiss cheese character to it on T2WI and FLAIR scans. No enhancement is seen. **D, E:** The wedge shaped nature of these lesions is better appreciated on the T2WI and FLAIR scan in the patient Bob. L.E. Mosse.

TABLE 2-10   Differentiation of Temporal Lobe Lesions

Tumor	Ganglioma	Low-Grade Astrocytoma	DNET	Oligodendro-glioma	PXA	DIG
Age (years)	0-30	0-30	10-20	30-60	10-35	0-1
Demarcation	Well	Well	Well	Less well	Well, but malignant change in 20%	Well
Edema	Very little	Yes	None	Yes	Uncommon	Occasionally
Percent of tumors causing temporal lobe seizures	40%	26%	18%	6%	4%	<3%
Hemorrhage	Rare	Rare	Common	Variable	Rare	No
Cyst formation	Common	Common	Common, dominant multiple	Variable	Common	Common
Enhancement	Uncommon	Uncommon	Uncommon to variable	Common	Common in mural nodule	Common in nodule desmoplasia
Cortical involvement	Common	Uncommon	Always	Variable	Common, meningeal attachment	Dural attachment
Calcification	Variable	Variable to uncommon	Common	Common	Rare	None

DIG, Desmoplastic infantile ganglioglioma; *DNET,* dysembryoplastic neuroepithelial tumor; *PXA,* pleomorphic xantheastrocytoma.

in the cerebral cortex or the cerebellum. They may be hyperdense on noncontrast CT scans and show little to no enhancement. Gangliocytomas are often isointense on T1WI and T2WI and are best seen by their hyperintensity on FLAIR.

Lhermitte-Duclos disease involves a masslike lesion usually seen in the cerebellum as a diffusely infiltrative process. In the past it was characterized as a hamartoma, or dysplasia, but has been recently reclassified as WHO grade I neoplasm and termed a dysplastic gangliocytoma.

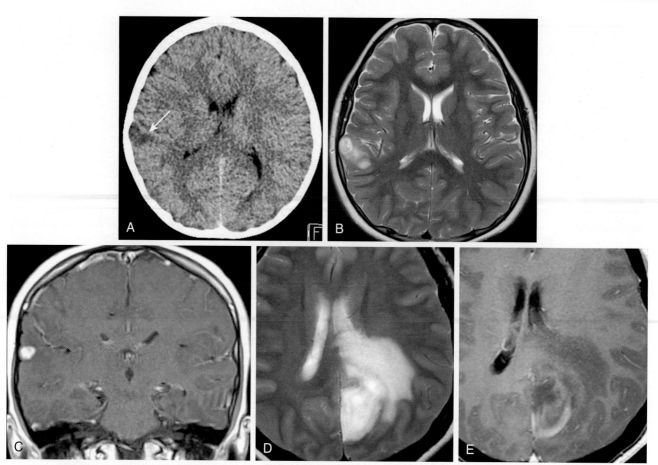

FIGURE 2-39 Multiple faces of ganglioglioma. **A,** Axial computed tomography image in 11-year-old Can U. C. Dis presenting with seizures shows a focal hypoattenuating lesion in the right temporal lobe *(arrow)*. No calcifications in this case. **B,** On T2, the lesion has a T2 hyperintense nodular appearance without much surrounding edema. **C,** Following contrast administration, the nodular component of the tumor shows avid enhancement. **D,** Axial T2-weight imaging (T2WI) in different patient, My Akin Hed, shows a T2 hyperintense lesion in the left parietal lobe with marked surrounding edema. **E,** Enhancement is present on post-contrast T1WI, but less avid than in the first patient.

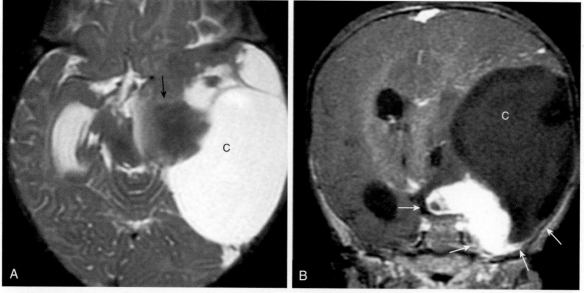

FIGURE 2-40 Desmoplastic infantile ganglioglioma in Dissa D'bikjuan. **A,** This large cystic mass (C) in the temporal region has a solid component *(arrow)* more medially seen as intermediate signal on the T2-weighted image. **B,** Typical of a desmoplastic infantile ganglioglioma, there is a peripheral solidly enhancing component to the mass, which has a dural attachment *(arrows)* and a huge cyst. The differential diagnosis of da desmoplastic infantile ganglioglioma (DIG) is da dysembryoplastic neuroectodermal tumor (DNET) or da pleomorphic xanthoastrocytoma (PXA), but da DIG develops in da infants. Do not delete or defer on da DIG. Dig it: a "BIG-DIG" in Baltimore, not Boston.

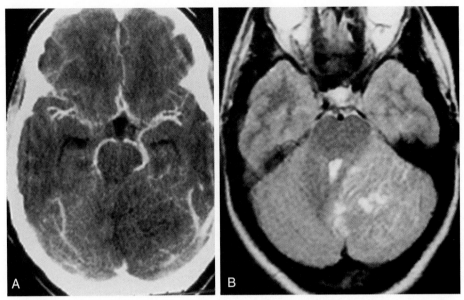

FIGURE 2-41 Patient Dez-Juan Sawryte has Lhermitte-Duclos. **A,** Enhanced computed tomography reveals a nonenhancing mass lesion in the superior left cerebellum. Note the slight low density within the nonenhancing mass. **B,** T2-weighted image of the lesion again demonstrates the obvious mass effect and high intensity striations within the mass. Notice that the intervening parenchyma has similar intensity to normal cerebellum.

The lesion affects cerebellar gray and white matter and is hyperintense on T2WI (Fig. 2-41). Other names for this entity include diffuse ganglioneuroma, Purkinjeoma diffuse hypertrophy, granule cell hypertrophy, and dysplastic gangliocytoma. Lhermitte-Duclos disease usually presents in patients in their early twenties, with symptoms of increased intracranial pressure and/or ataxia. Cowden syndrome, which is associated with multiple hamartomas and neoplasms (especially of the breast) is associated with almost 50% of cases of Lhermitte-Duclos. It is transmitted on chromosome 10q23.

Desmoplastic infantile astrocytoma (desmoplastic cerebral astrocytoma of infancy) probably represents a variant of DIG and has features of glial and mesenchymal histology. Absence of neuronal components histologically distinguishes this tumor from a DIG. It is a benign tumor found in early life. In general, there is a dural-based mass with cystic change. Although the dural-based mass will enhance, the cyst does not, not even on the periphery. Mass effect and vasogenic edema are rare. It also presents in the first 18 months of life and is usually seen supratentorially.

Oligodendroglial Tumors

These tumors are listed in Table 2-11.

Oligodendroglioma

Oligodendrogliomas comprise just 4% to 8% of intracranial gliomas but are typified by their high rate of calcification (40% to 80%; Fig. 2-42). Peak age range is in the fifth and sixth decade, and the tumor favors men by a 2:1 margin. The tumor, when pure, has a benign course (classified as WHO grade II) but beware of anaplasia. Patients with oligodendrogliomas who have loss of heterozygosity on 1p or combined loss of heterozygosity on 1p and 19q are more sensitive to radiation and chemotherapy, especially temozolomide, and therefore survive

TABLE 2-11   World Health Organization Classification of Oligodendroglial Tumors and Mixed Gliomas

Tumor	Grade	Peak Age (years)
OLIGODENDROGLIAL TUMORS		
Oligodendroglioma	II	30-55
Anaplastic oligodendroglioma	III	45-60

substantially longer (mean 10 years) than patients without 1p/19q deletions (mean 2 years). These genetic markers are associated with an indistinct border on T1WI and mixed intensity signal on all pulse sequences and with paramagnetic susceptibility from blood or calcification. The ones with 1p/19q deletions are smaller, more commonly single lobe and temporal lobe tumors, and are less likely to thin overlying bone. This favorable association is not evident in patients with astrocytoma.

Enhancement, when present in oligodendrogliomas, is variable (present in 40%). Hemorrhage occurs in 20% of cases, as does cyst formation. When the tumor encysts, it has a higher rate of malignant astrocytic behavior. On MRI the tumor is hypointense on T1WI and hyperintense on T2WI/FLAIR except in the areas of calcification (Fig. 2-43). Heterogeneity of signal is the watchword. On a CT scan the lesions are hypodense or isodense on unenhanced scans (unless hemorrhage or calcification is present). Edema associated with the mass is typically minimal, a distinguishing point from other more aggressive tumors.

Mixed Tumors

Oligodendrogliomas are often histologically mixed with astrocytic forms (50%); when present, it acts as a medium-grade neoplasm with a high rate of recurrence. The term used for this tumor is oligoastrocytoma and like the oligodendrogliomas and anaplastic oligos populate the frontal and temporal lobes. Calcification in oligoastrocytomas occurs less frequently (14%) but enhancement more frequently (50%).

Median survival for oligoastrocytoma is 6.3 years as opposed to 9.8 years for oligodendroglioma. If anaplasia occurs with oligoastrocytomas, the median survival drops to 2.8 years.

Anaplastic Oligodendroglioma

Anaplastic oligodendrogliomas (WHO grade III) have a worse prognosis than the WHO grade II oligodendrogliomas. They account for one fourth to one half of all oligodendrogliomas with a mean age of 49 years old. Over 90% are found in either the frontal lobe or temporal lobe. Hemorrhage, necrosis, calcification, cystic degeneration, and avid enhancement alone or in combination may occur in these tumors. Five-year survival is approximately 30%.

Neuroepithelial Tumors of Uncertain Origin

Chordoid Glioma

These are tumors of the hypothalamus and anterior third ventricle that are assigned WHO grade II, a glial tumor of

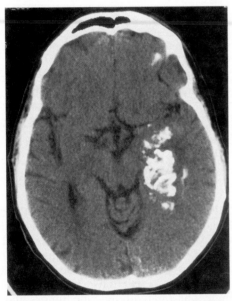

FIGURE 2-42 Calcification in an oligodendroglioma. Although unenhanced computed tomography may seem anachronistic at times, for the evaluation of patient Perry Phil Calque this lesion shows typical calcification of this low-grade tumor. The serpentine nature of the calcification should raise the suspicion of an arteriovenous malformation or even Sturge Weber, so you still have to get the magnetic resonance imaging anyway.

unknown origin. The lesion is slow-growing, solid, well-circumscribed, and avidly enhancing. They look like little olives sitting at the third ventricle—ovoid, sharply delineated, no pimento. They are hyperdense to gray matter on a CT scan and reasonably isointense on standard T1WI and T2WI, but may have central necrosis or cystic regions. They occur in adults over age 30 and can cause acute hydrocephalus because of their obstructive nature.

Astroblastoma

Astroblastomas are rare tumors of variable aggressiveness and of unknown origin. They are tumors of young adulthood affecting the cerebral hemispheres. They are well-circumscribed tumors with peripheral enhancement, central necrosis, and usually large size. Prognosis is good as long as the surgeon does a complete resection.

Hemangioblastomas

The most common primary intraparenchymal tumor in the infratentorial space in adults is a cerebellar hemangioblastoma (HB). This is a benign tumor, WHO grade I, readily curable with surgery. More than 85% of HBs occur in the cerebellum, 9% in the spinal cord, 4% in the medulla, and 2% in the cerebrum. Approximately 10% of posterior fossa masses are HBs. They are far outnumbered by vestibular schwannomas and metastases. Men are more commonly affected than women, and the patients are usually young adults. The common symptoms are headache, ataxia, nausea, vomiting, and vertigo. Polycythemia caused by increased erythropoietin production may be a clinical finding in 40% of cases and is more common with solid HBs. A spinal HB may present with subarachnoid hemorrhage. Twenty-five percent of HBs occur in association with von Hippel-Lindau disease (VHL), which has been mapped to a gene on the third chromosome (3p25-26).

The stereotypical findings of an HB are that of a cystic mass with a solid mural nodule (55% to 60% of cases), which is highly vascular and has signal voids of feeding vessels (Fig. 2-44). However, solid HBs (40%) and, less commonly, purely cystic HBs occur as well. You will see a peripheral cystic mass in the hemisphere or vermis of the cerebellum with a slightly hyperdense mural nodule on the unenhanced CT scan. The mural nodule demonstrates striking enhancement. The cyst and its walls do not enhance. A purely solid HB also demonstrates strong enhancement (see Fig. 2-44).

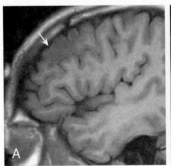
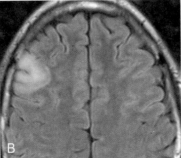
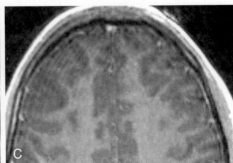

FIGURE 2-43 Typical grade II oligodendroglioma in patient Horace A. Vertted. **A,** Sagittal T1-weighted image shows a far peripheral cortical lesion *(arrow)* with mild mass effect. It lights up on the fluid-attenuated inversion recovery scan **(B)** and has blurred margins. **C,** Only 40% of oligos show contrast enhancement, absent in this case.

On angiographic examination a vascular nodule amidst an avascular mass, usually with serpentine vessels, may be identified with or without draining veins.

The MRI scan demonstrates findings similar to those of the CT scan, with varying components of cystic and solid tissue. However, the advantage of MRI is in showing the large vessels feeding the mural nodule of the HB and additional missed tiny multiple HBs in VHL. The tumors usually reach the pial surface of the brain and therefore may simulate the appearance of a meningioma. Treatment requires removal of the solid nodule of the lesion only, because the cyst is not really neoplastic. Prognosis is very good with a 5-year survival rate of more than 85%.

Because the description above is nearly the same for JPAs you may ask how to distinguish the two. First and foremost is age; JPAs are seen in the 5 to 15 age range versus 30 to 40 for HBs. Given a 23-year-old with a cystic solid mass, go with four findings (1) a pial attachment would suggest HB, (2) a tiny nodule with a huge cyst is more likely

HB, (3) (if pushed) an arteriogram will show the nodule to be hypervascular with HB and hypovascular with JPA, and (4) multiplicity and association with other findings of VHL syndrome suggests HB.

HBs associated with the VHL syndrome (Box 2-5) generally present at an earlier age. VHL syndrome is inherited as an autosomal dominant trait, with nearly 90% penetrance. HBs associated with VHL syndrome may be multiple in the cerebellum, brain stem, and spinal cord. Current criteria for establishing the diagnosis of VHL syndrome are (1) more than one CNS (including retinal) HB, (2) one CNS HB and one visceral lesion (e.g., renal angiomyolipoma, renal cell carcinoma), or (3) one manifestation of VHL syndrome and a positive family history. One difficulty in analyzing patients with VHL disease occurs when the patient has a known metastatic renal cell carcinoma or metastatic pheochromocytoma. Often it is impossible to distinguish hypervascular metastases (which may be single or multiple) from HBs (which also may be single or multiple).

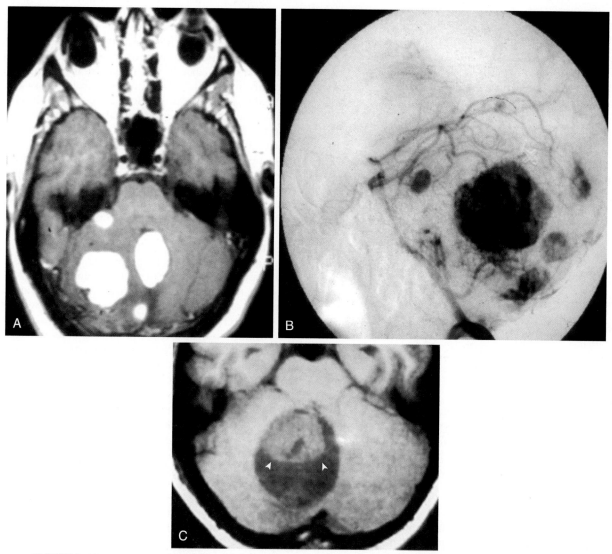

FIGURE 2-44 Hemangioblastomas of the cerebellum. **A,** Note the numerous enhancing masses in the posterior fossa of patient M. I. Cyn Drohmik. There may have been ocular lesions as well judging from the distorted appearance of the globes. **B,** On angiography the hypervascular nature of the masses is evident as well as their multiplicity. The supply is from pial, not dural, branches. **C,** Patient Meral Na Jou has a hemangioblastoma with both a cyst and solid components *(arrowheads)* on this unenhanced axial T1-weighted image.

CENTRAL NERVOUS SYSTEM
Cerebellar HB (66%-80%)
Spinal HB (28%-40%)
Medullary HB (14%-20%)
Extraaxial HB (<5%)
Retinal HB (50%-67%)

RENAL
Cysts (50%-75%)
Hypernephroma (25%-50%)
Hemangioblastoma
Adenoma

PANCREATIC
Cysts (30%)
Adenoma
Adenocarcinoma
Islet cell tumor

LIVER
Hemangioma
Cyst
Adenoma

SPLEEN ANGIOMA

ADRENAL
Pheochromocytoma (10%)
Cyst

LUNG
Cyst

BONE
Endolymphatic sac tumor (10%; temporal bone)
Hemangioma

CARDIAC
Rhabdomyoma

EPIDIDYMIS
Cyst
Cystadenoma

HEMATOLOGIC
Polycythemia (25%-40%)

HB, Hemangioblastoma.

When reviewing images of the posterior fossa in a patient with VHL, don't forget to peruse the temporal bone for endolymphatic sac tumors (ELST), which arise in 11% to 16% of patients with VHL and may be bilateral in 33%. Oh, and the ocular HBs too…nearly missed them.

Metastases

The most common infratentorial and supratentorial malignant neoplasm to occur in the adult population is a metastasis (Fig. 2-45). They are usually seen as well-defined, round masses, which are identified near the gray-white junction. Metastases lodge at the gray-white interface in 80% of cases, basal ganglia in 3%, and cerebellum in 15%. These lesions show contrast enhancement and are one of the lesions of the brain that often causes nodular or ring enhancement. The most common primary extracranial tumors in adults to metastasize to the infratentorial part of the brain are lung and breast carcinomas. Bronchogenic carcinomas spread to the CNS in 30% of cases, although squamous carcinoma is the least frequent histologic type to metastasize to the brain. It is estimated that CNS metastases develop in 18% to 30% of patients with breast cancer. Other neoplasms that have a propensity for metastatic spread to the brain include melanoma (third most common, after lung and breast), renal cell carcinoma, and thyroid carcinoma. Virtually all metastases evoke some vasogenic edema; however, the amount is variable.

Unless the metastatic deposit is hemorrhagic, calcified, hyperproteinaceous, or highly cellular, where it would be hyperdense on a noncontrast CT scan (Fig. 2-46), most metastases are low density on unenhanced CT. Rarely, one may identify calcification in metastatic deposits (Box 2-6). Consider mucinous adenocarcinomas in that scenario. Metastatic deposits may have variable intensity on T2WI. Some lesions are isointense to gray matter on T2WI (probably from hypercellularity) and can be readily distinguished from the high intensity of the edema they elicit. Other metastatic deposits, however, are hyperintense to gray matter on T2WI. Hemorrhagic metastases are usually seen as areas of high signal intensity on T1WI and T2WI with a relative absence of hemosiderin deposition (Box 2-7). Although virtually all metastases enhance to a variable degree, the pattern may be solid, ringlike, regular, irregular, homogeneous, or heterogeneous (Fig. 2-47). As opposed to gliomas, metastases are better defined and have sharper borders. The vasogenic edema is often out of proportion to the size of the metastasis except in cortical metastases, where edema may be minimal or absent and the enhanced studies are essential to their detection.

Hemorrhagic metastases must be differentiated from occult cerebrovascular malformations or nonneoplastic hematomas (Table 2-12). Some primary neoplasms such as melanoma, renal cell, choriocarcinoma, and thyroid (MR, CT!) have a particular propensity to hemorrhage, be it at the primary site or within metastases. However, because lung and breast cancers are so much more common than these primary tumors, a hemorrhagic metastasis is most often from breast or lung. When faced with a solitary hemorrhagic mass in an adult, primary brain tumors such as glioblastoma and oligodendroglioma should be considered along with a solitary hemorrhagic metastasis. First exclude hypertensive bleeds, amyloid angiopathy, and hemorrhagic strokes of course….

In the case of melanoma, one can identify nonhemorrhagic melanotic metastases as lesions that have high intensity on T1WI and isointensity to hypointensity on T2WI caused by intrinsic paramagnetic effects (Fig. 2-48). Although some investigators believe that the paramagnetic effect is a result of paramagnetic cations, others believe it to be caused by free radicals, and still others, an inherent characteristic of melanin. However, an amelanotic melanoma (without melanin) without hemorrhage may have signal intensity characteristics similar to those of other nonhemorrhagic metastases: low signal on T1WI and high signal on T2WI. Furthermore, hemorrhagic melanoma metastases, be they melanotic or amelanotic, have signal intensity similar to other hemorrhagic

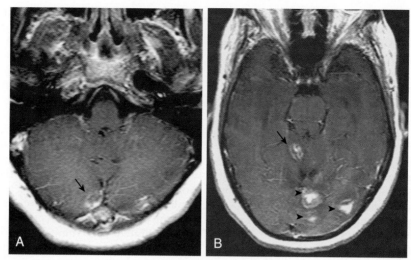

FIGURE 2-45 Multiple posterior fossa metastases in an Cambodian patient Lao Tze Phoo Chorr. **A,** With contrast enhancement, a ring enhancing metastasis *(arrow)* is well seen in the right cerebellum. **B,** Additional metastases are seen in the superior vermis of the cerebellum *(arrow)* and in the left occipital lobe *(arrowheads)* at the gray matter-white matter junction.

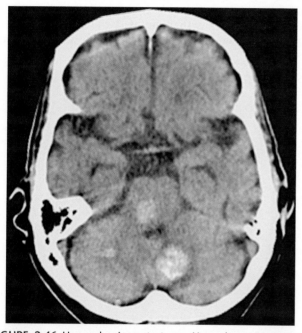

FIGURE 2-46 Hemorrhagic metastases. Hyperdense masses are seen in the brain stem and cerebellum on noncontrast computed tomography in Hedd Lee Ded Lee. Although one might consider cavernomas in the differential diagnosis, this patient had renal cell carcinoma. The edema around the mass in the cerebellum can be seen.

nonmelanotic lesions. Metastases are also the most common masses in the supratentorial space in the adult, making up 40% of intracranial neoplasms. Fifty percent of metastases are solitary. The other 50% are multiple; of these, 20% have two lesions only (Fig. 2-49). The primary tumors that spread to the supratentorial part of the brain are lung (50%) breast (15%), melanoma (11%), kidney, and gastrointestinal primary tumors. Cystic or calcified metastases favor lung, breast, and gastrointestinal primary sites. If the primary site is not clinically or radiographically apparent, the differential diagnosis becomes an astrocytoma. Remember to peruse the calvarium for metastatic

---

> **BOX 2-6** Calcified Central Nervous System Tumors
>
> **METASTASES**
> Mucinous adenocarcinoma (breast, colon, ovary, lung, stomach)
> Osteosarcoma
> Chondrosarcoma
> Chordoma
>
> **PRIMARY CENTRAL NERVOUS SYSTEM TUMORS**
> **INTRAAXIAL**
> Oligodendroglioma
> Ganglioglioma
> Astrocytoma
> Supratentorial ependymoma
>
> **EXTRAAXIAL**
> Meningioma
> Craniopharyngioma
> Pineal region tumors
> Infratentorial ependymoma
> Neurocytoma
> Choroid plexus papilloma
>
> **NEOPLASMS POSTRADIOTHERAPY**

> **BOX 2-7** Hemorrhagic Metastases
>
> Melanoma (most frequent)
> Renal cell carcinoma
> Breast cancer (sheer numbers because of high prevalence)
> Lung cancer (sheer numbers because of high prevalence)
> Thyroid cancer
> Retinoblastoma
> Choriocarcinoma

osseous disease as you look at the scans. In an informal survey of two boneheads, it is our perception that bony metastases of the skull occur more frequently than parenchymal metastases in many primary tumors. This is certainly true for prostate metastases, where parenchymal metastases without bony disease are virtually reportable. Furthermore, a recent study has indicated that breast primary tumors that are estrogen and progesterone receptor positive have a much higher rate of osseous metastases than brain parenchymal ones (and the skeletal ones occur earlier in the course of the breast cancer). Those who are receptor negative develop brain and meningeal but rarely calvarial metastases.

Metastases usually appear as relatively well-defined masses that demonstrate enhancement and moderate edema. The lesions tend to follow vascular flow dynamics, being deposited in the carotid system more commonly than in the vertebrobasilar system (80% to 20%) and favor the middle cerebral artery distribution.

The neurosurgeons will remind you that they are ready and willing to resect some metastases purely because of unresponsive mass effect, impending herniation, and the proximity of their offspring to the tuition of a college education. Recent studies have suggested benefit in resecting (or gamma/cyberknifing radiosurgery) as many as three separate well-circumscribed metastases—especially if Princeton University is on the "short list" for matriculation. One can improve detection and visibility of metastases with the use of double- or triple-dose MRI contrast agents, the use of magnetization transfer suppression techniques,

delayed imaging by 15 to 30 minutes, postcontrast FLAIR scanning, and subtraction techniques.

Lymphomas

The most common type of lymphoma to affect the brain is diffuse histiocytic lymphoma (also known as reticulum cell sarcoma, microglioma, primary cerebral lymphoma). Primary CNS lymphoma is often referred to as PCNSL. CNS lymphoma is often associated with an immunodeficient state including that resulting from acquired immunodeficiency syndrome (AIDS), organ transplantation, Wiskott-Aldrich syndrome, Sjögren syndrome, and prolonged immunosuppressive therapy. Lymphoma can present as multiple masses, coating of the ventricles, clear CSF infiltrated with tumor cells, or one dominant mass (potentially crossing the corpus callosum). CNS lymphoma occurs in 6% of AIDS patients.

The other diagnosis seen in an AIDS population is toxoplasmosis. Toxoplasmosis usually does not abut an ependymal surface as lymphoma does (Table 2-13). Studies have shown that thallium scanning is an excellent means for distinguishing toxoplasmosis from CNS lymphoma. The latter is thallium avid; the former does not show activity with thallium scanning. This is an excellent,

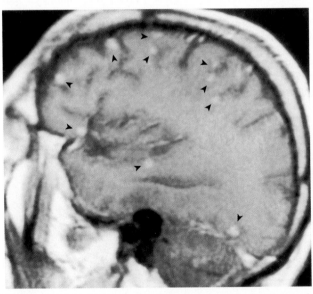

FIGURE 2-47 Multiple small cell carcinoma lung metastases in patient Count D. Spahtz. The preponderance of small enhancing lesions *(arrowheads)* at the gray-white junction on this sagittal postcontrast scan suggests a diagnosis of metastatic disease.

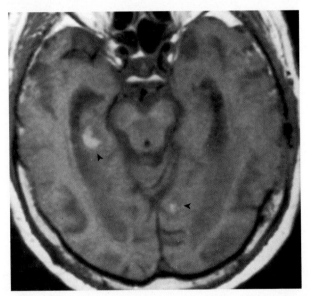

FIGURE 2-48 Paramagnetic high signal in metastases *(arrowheads)*. Noncontrast T1-weighted image in patient Mel E. Nohma. This was a case of metastatic melanoma where the high signal was a result of melanin (or hemorrhage).

TABLE 2-12 **Features of Recently Bled Cavernomas versus Hemorrhagic Metastases**

Feature	Hemorrhagic Cavernomas	Hemorrhagic Metastases
Edema	Only with acute episode, resolves by 8 weeks	Persistent
Mass	Variable but resolves	Moderate to large, persistent
Hemosiderin ring	Complete	Incomplete or absent
Nonhemorrhagic tissue	Absent	Present
Enhancement	Minimal and central	Nodular, ring, or eccentric
Progression of hemorrhagic stages	Orderly	Delayed
Follow-up	Decreases in size with time	Increases in size with time
Calcification	Approximately 20%	Rare

though underused, means to distinguish these two entities and effect an early course of therapy directed at the correct diagnosis. Do it! Another suggested way to differentiate the two, while sticking with your MRI scanner, is to perform perfusion scans. Lymphomas have higher regional cerebral blood volume compared with surrounding tissue whereas toxoplasmosis is hypovascular (see Chapter 5).

Secondary lymphoma most commonly involves the leptomeninges and CSF; but this is rarely detectable on a CT scan or MRI. Hydrocephalus may be the only telltale sign. Dural invasion is a rarity. When one has parenchymal extension by secondary lymphoma, it is usually supratentorial and is more commonly multifocal. Dense enhancement is the norm; however, ring enhancement may also occur. Non-Hodgkin lymphoma is more common than Hodgkin lymphoma, which rarely affects the brain.

MRI findings in CNS lymphoma are varied. Periventricular (40%), subcortical and deep gray matter abnormalities (27%) and mixed patterns (20%) are most common, with masses less than 2 cm in size in patients with AIDS and greater than 2 cm in size in non-AIDS patients. Non-AIDS PCNSL does not calcify and is most commonly seen in the frontal lobes and basal ganglia (Fig. 2-50). Close to 50% abut the ventricular surface from a white matter origin. Multiple lesions occur more commonly in patients with AIDS, making the distinction with toxoplasmosis even more difficult. The signal intensity is variable on T2WI scans with approximately 50% of cases isointense to slightly hypointense. Heterogeneity is the norm. Gadolinium enhancement is marked and homogeneous in over 90% of non-AIDS cases, but beware when steroids are given as treatment because these drugs may suppress the enhancement. Ring enhancement is

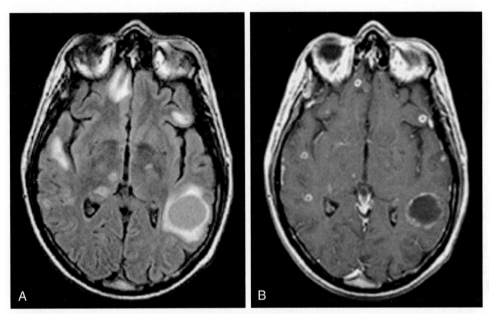

FIGURE 2-49 Supratentorial meatballs. **A,** Cystic and solid metastases are present in this patient Ivana Lyve. The cystic lesion in the left temporal lobe does not have the same intensity as cerebrospinal fluid on fluid-attenuated inversion recovery because of high protein. **B,** The masses enhance along the periphery and the posterior right temporal lesion reveals its nature. The differential diagnosis would include a brain abscess and a diffusion-weighted scan could be useful if bright (suggesting an abscess and differentiating the two).

TABLE 2-13   Lymphoma versus Toxoplasmosis in Acquired Immunodeficiency Syndrome

Feature	Lymphoma	Toxoplasmosis
Restricted diffusion	Solid portion	Cystic portion
Calcification	Rare	Common especially after treatment
Perfusion	Intermediate	Low
Density on NCCT	Hyperdense	Hypodense (unless calcified)
Size of lesions	Can be quite large; >3 cm	Rare over 3 cm
Thallium avidity	Yes	No
FDG PET avidity	Yes	Less
T2 dark	Common	Unusual unless calcified
Subependymal spread	Characteristic	Rare
Response to steroids/XRT	Rapid	Laughable
Hemorrhagic	Rare	More common
MRS	Choline high	Lactate possible

Data from Dina TS. Primary CNS lymphoma versus toxoplasmosis in AIDS. *Radiology.* 1991;179:823-828.
FDG, Fluorodeoxyglucose; *MRS,* magnetic resonance spectroscopy; *NCCT,* noncontrast computed tomography; *PET,* positron emission tomography; *XRT,* radiation therapy.

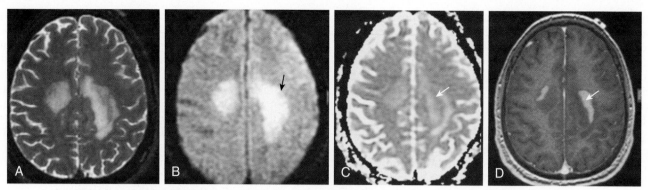

FIGURE 2-50 Low apparent diffusion coefficient (ADC) value in lymphoma. Despite the bright edema on the T2-weighted image **(A)** and diffusion-weighted imaging **(B),** one can make out on the ADC map **(C)** a lower intensity area *(arrows)* indicating restricted diffusion in the tumor amidst the surrounding high intensity vasogenic edema. Hypercellular/small cell tumors can show reduced ADC. **D,** This area also shows enhancement and should be the target for biopsying patient Oliver D. Brayen.

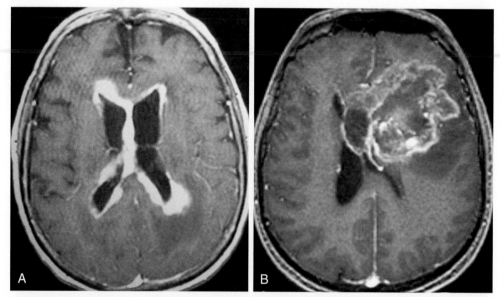

FIGURE 2-51 Fifty shades of lymphoma. Periventricular ependymal enhancement **(A)** is present in patient Dakota Steele Devente, whereas parenchymal necrotic enhancement is seen in another patient, Domans Grey, abutting on the ventricle **(B).**

often seen in immunodeficient (AIDS) patients but is rarely seen in the immunocompetent PCNSL population (Fig. 2-51). Because lymphoma and glioblastoma are often partners in the sentence for differential diagnoses of many malignant looking brain masses, it is useful to note that the FA and ADC values of lymphoma are significantly lower than those of glioblastoma because of its dense cellularity and high nucleus to cytoplasm ratio (see Fig. 2-50, *C*). The choline/creatine ratio is elevated in CNS lymphoma and cerebral blood volume is decreased compared with high-grade astrocytomas (though higher than toxoplasmosis).

The classic teaching used to be that lymphoma was one of the lesions that is typically hyperdense on a noncontrast CT scan and enhances to a moderate degree (Fig. 2-52). Nonetheless, if you see a hypodense infiltrative mass on a noncontrast CT in an adult positive for human immunodeficiency virus, still consider lymphoma. Hemorrhage is distinctly uncommon in lymphomas (<8%).

Sarcoma

The most common forms of sarcomas in the CNS are found along the meninges (meningosarcomas, angiosarcomas, and fibrosarcomas) and have a propensity to invade the brain (see gliosarcoma above).

PINEAL REGION MASSES

The pineal gland grows steadily until age 2 and then the size stabilizes into early adulthood. No difference in size exists between males and females. The normal pineal gland is calcified in 7% to 10% of patients before 10 years of age, 30% of patients in their mid-teens, and peaks by age 20 to 40 at 33% to 40% of individuals. A calcified pineal gland before the age of 6 should be viewed with suspicion for adjacent tumor. African Americans have a lower rate of pineal calcification than Caucasian Americans.

Pineal region masses constitute 1% of all CNS tumors. They are generally separated into two categories: those

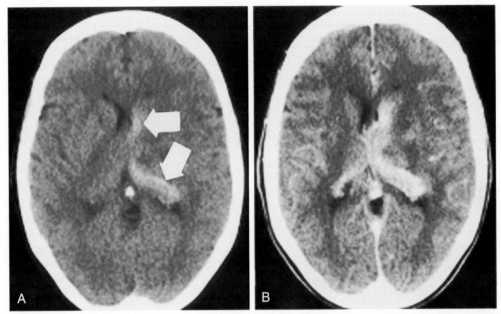

FIGURE 2-52 Periventricular lymphoma in patient Imelda Wistter Royds: Pre- **(A)** and post- **(B)** contrast computed tomography scans demonstrate the classic findings of lymphoma—a hyperdense mass on an unenhanced scan *(arrows)*, which shows enhancement after iodinated contrast administration **(B)** and infiltrates the ependymal surface of the ventricular system.

TABLE 2-14   Pineal Region Masses: Differential Diagnosis of Germ Cell Neoplasms

Characteristic	Germinoma	Choriocarcinoma	Teratoma	Yolk Sac Tumor
Age/sex	Child/M>>F	Child/M>F	Child/M>F	Child/M>F
Density on unenhanced CT	Hyperdense	Variable (hemorrhage predilection)	Variable (fat, calcification, teeth)	Hypodense
Calcification	Accelerates pineal calcification	Rare	Frequent	Absent
Enhancement	Marked	Moderate	Minimal	Rare
Heterogeneity	Homogeneous	Heterogeneous	Heterogeneous	Heterogeneous
Subarachnoid seeding	Frequent	Infrequent	Infrequent	Rare
Serum markers	Placental alkaline phosphatase, sometimes HCG	HCG, human placental lactogen	HCG and alpha-fetoprotein	Alpha-fetoprotein
Hemorrhage	Yes	Yes, yes	Possible	No
Other	Boys, boys, boys	Hemorrhage is the word	Variable density and intensity	

CT, Computed tomography; *HCG,* human chorionic gonadotropin.

of germ cell origin (60%) and those of pineal cell origin (40%; Tables 2-14 and 2-15). The manifestations of pineal region masses are based on their site near many critical structures: the aqueduct, the tectal plate, the midbrain, the vein of Galen. Remember also that the pineal gland may regulate human response to diurnal daylight rhythms. Pineal region masses may cause (a) hydrocephalus through obstruction of the aqueduct of Sylvius, (b) precocious puberty, (c) headache, or (d) paresis of upward gaze (Parinaud syndrome).

Tumors of Germ Cell Origin

Germinoma

The most common pineal tumor of the germ cell line is the germinoma, accounting for 60% of pineal germ cell tumors and 36% of pineal region masses. It has also been termed seminoma and dysgerminoma in the medical literature. This tumor, as in all germ cell tumors, has a distinct male predominance when seen in the pineal region (in some series as high as 33:1) and a slight female predilection when seen suprasellarly. The tumor may be multifocal in suprasellar and pineal locations. It is a tumor of adolescence and young adulthood, rarely seen in patients older than 30 years. Germinomas are far more common in the Asian population. The characteristic appearance of the germinoma is that of a hyperdense mass on unenhanced CT scans, which enhances markedly (Fig. 2-53). The tumor engulfs the pineal gland, and this has led to some confusion regarding whether the tumor calcifies. It is currently believed that there is a high incidence of pineal gland calcification (as a reactive phenomenon) in patients with germinomas but that the tumor itself does not calcify. On MRI, the germinoma

TABLE 2-15   Pineal Region Masses: Differential Diagnosis of Non–Germ Cell Neoplasms

Characteristic	Pineoblastoma	Pineocytoma	Astrocytoma
Age/sex	Child/M = F	Child or adult/M = F	Child or adult/M = F
Density on unenhanced computed tomography	Hyperdense	Hyperdense	Isodense
Calcification	Rare, "exploded" appearance	Present	Absent
Enhancement	Moderate	Moderate	Variable
Heterogeneity	Homogeneous	Homogeneous	Homogeneous

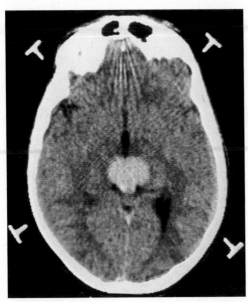

FIGURE 2-53 Germinoma in stereotactic biopsy frame. A hyperdense mass in the pineal region is seen on this enhanced computed tomography of patient Anita Fifty Gray.

has intermediate signal intensity on T1WI and, because of the tumor cells' high nucleus/cytoplasm ratio, a slightly hypointense signal (similar to gray matter) on T2WI. The mass enhances. Germinomas show higher ADC values than the pineal cell tumors, and the patients are younger. Otherwise, there are no definitive imaging characteristics that distinguish pineal cell tumors from germinomas. Germinomas are extremely radiosensitive and also respond well to chemotherapy. CSF seeding is not uncommon. The best imaging study to evaluate for CSF seeding is contrast-enhanced MRI of the entire neuroaxis; nonetheless, repeated CSF cytologic studies are still more sensitive than imaging.

Most of the remaining CNS germinomas occur in the suprasellar cistern (Fig. 2-54), but some have been reported to occur in the basal ganglia and thalami as well. In these locations they still are hyperdense on CT scans but seem to have a higher rate of cystic degeneration and calcification. Ipsilateral cerebral hemiatrophy and brain stem hemiatrophy can occur when the germinomas are located in these sites.

Teratoma

Teratomas may have fat, bone, calcification, cysts, sebaceum, or other dermal appendages associated with them. The lipid and calcification or bone have distinctive

CT and MRI densities or intensities. A chemical shift artifact may signal the presence of fat rather than blood on the T2WI. Enhancement is irregular because of the nonenhancing fatty or calcified component. Teratomas are the second most common pineal region germ cell neoplasm, but they also abound in the suprasellar cistern. In neonates they may diffusely invade a hemisphere or the sacral spine.

Choriocarcinoma and Other Germ Cell Line Tumors

Choriocarcinoma has a distinctive feature: it is commonly hemorrhagic. Yolk sac tumors often show more cystic change than other germ cell line lesions. Embryonal carcinoma is most commonly solid. Teratomas and choriocarcinomas as well as embryonal cell carcinoma and endodermal sinus tumors are more common in male patients and have a worse prognosis. Choriocarcinoma is human chorionic gonadotropin (HCG) and human placental lactogen positive on immunohistochemistry. The others are not. Alpha-fetoprotein titers are negative in embryonal cell tumors but positive in yolk sac tumors. Placental alkaline phosphatase characterizes the embryonal tumor.

Others

The pineal gland is often referred to as the "third testicle" because of the prevalence of the germ cell line of tumors in males. It is also the "third eye"; retinoblastomas occur here as part of the retinoblastoma oncogene complex. The occurrence of bilateral retinoblastomas in association with pineoblastomas (see following section) has been termed trilateral retinoblastoma (though the third primitive neuroectodermal tumor may also occur in a suprasellar location). This occurs in 3% of patients with bilateral retinoblastomas. There is a high rate of subarachnoid seeding.

Tumors of Pineal Cell Origin

Pineoblastoma

The incidence of intrinsic pineal cell tumors, pineocytomas and pineoblastomas, is nearly evenly split between male and female patients and the tumors account for 15% of pineal region neoplasms. The pineoblastoma occurs in a younger age group (peak in first decade of life) than the pineocytoma and is classified as WHO grade IV (despite a 5-year survival of 50% to 60%). Pineoblastomas may occur in association with retinoblastomas and interphotoreceptor retinoid-binding proteins can be seen in pineoblastomas. Their appearance at imaging is nearly identical, but pineoblastomas may be slightly more invasive and larger than pineocytomas and have a higher rate of subarachnoid seeding. Again, because these tumors are of the round cell

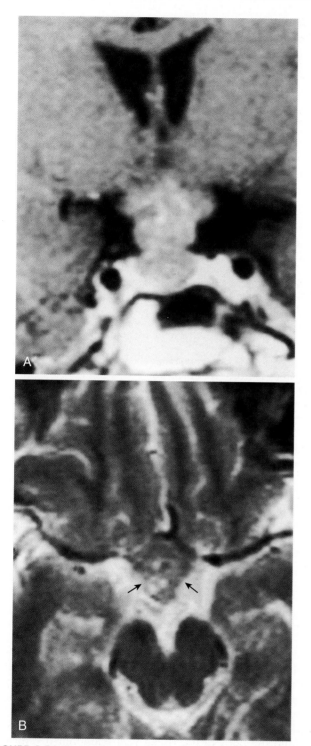

FIGURE 2-54 Suprasellar germinoma in patient I. P. Soh Dyum. **A,** An enhancing suprasellar mass is seen infiltrating the optic chiasm on the coronal scan. The patient had precipitous diabetes insipidus and a sodium of zipidous and a nonserendipidous germinoma. **B,** Axial T2-weighted image shows the low signal you would expect for this tumor *(arrows)*.

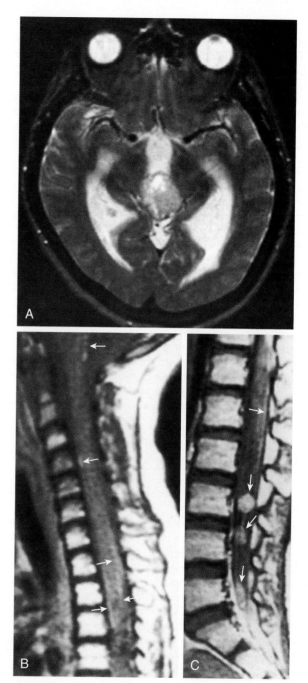

FIGURE 2-55 Pineoblastoma in patient Willie C. DaCienas. **A,** Axial T2-weighted image shows a pineal mass that is intermediate in signal intensity with some heterogeneity to the lesion. Low signal is characteristic of the highly cellular primitive neuroectodermal tumors. Note the dilatation of the third ventricle and occipital horns of the lateral ventricles caused by the compression of the aqueduct, signifying hydrocephalus. Subarachnoid seeding *(arrows)* in the form of sugarcoating **(B)** or gumdrops on the cauda equina **(C)** is not unusual in pineoblastomas.

variety with high nucleus/cytoplasm ratios, they often will be dense on unenhanced CT and intermediate in signal intensity on T2WI with low ADC values (Fig. 2-55). They enhance avidly. Calcification is not common yet may be intrinsic to the tumor rather than within the pineal gland itself. Alternatively, the pineal gland calcification may

appear exploded as it is displaced peripherally by the pineal tumor. We say germinomas "engulf" the pineal gland, pineoblastomas "explode" the gland.

Pineocytoma

Pineocytomas are slower growing pineal parenchymal neoplasms, WHO grade II, and can occur in any age group, with the mean age in the 30s. These tumors are smaller than the pineoblastomas, often less than 3 cm in size, and they may demonstrate a higher rate of calcification or cyst formation than their nastier brother the pineoblastoma. The 5-year prognosis is 86%.

Pineal parenchymal tumors of intermediate differentiation (PPTID) and papillary tumor of the pineal region (PTPR) are newly described WHO grade II and III pineal region masses that look similar to PBs and PCs. PPTIDs constitute 20% to 62% of pineal parenchymal tumors and occur mostly in adults, with a slight female preponderance. Management for PPTIDs includes a combination of surgery, radiation, and chemotherapy. CSF dissemination is more common in high-grade (36%) than in low-grade (7%) PPTID. They enhance more avidly than PTPR. PTPR can occur in both children and adults of a wide age range, from 5 to 66 years, favoring males. Histologically, this tumor is more densely cellular, and darker on T2WI and lower on ADC than PPTIDs but necrotic foci are often seen. A 7% rate of CSF dissemination has been reported.

Because the signal intensity characteristics of pineal parenchymal tumors, malignant germ cell tumors, germinomas, and gliomas may overlap, some investigators have suggested that serum markers may be more specific than imaging features for histology. It is true that some tumors secrete characteristic markers, and these are summarized in Table 2-14.

Some syndromes associated with brain tumors are described in Box 2-8.

NONNEOPLASTIC MASSES

Cysts

Pineal Cyst

Pineal and tectal gliomas, cavernous hemangiomas, meningiomas, and benign cysts also populate this area, but they are peripheral to the pineal gland. Pineal cysts are particularly common, found in 40% of autopsy series, and because some pineal masses (pineocytomas) may be cystic, it is important to attempt to identify a solid portion to the lesion to distinguish the two (Fig. 2-56). Cysts in the pineal region are like those mysterious chocolates in the Godiva box; you never know what's inside them, which ones are good, which ones are bad. Contrary to what has previously been written, pineal cysts may compress or occlude the aqueduct, and may be calcified. They may be round or oblong and can be equal to or greater than 2 cm in size. The key to distinguishing a pineal cyst from a cystic astrocytoma is the lack of growth during long-term follow-up. Nonetheless, it is not necessary to follow every pineal cyst because more than 99% are developmental and nonprogressive: allow the clinicians to use patient symptoms to determine appropriate follow-up. Because pineal cysts are often surrounded by the two limbs of the internal cerebral veins, one must be careful not to misread vascular enhancement as solid mass enhancement (which might suggest a tumor, but be aware that the normal pineal tissue enhances).

Benign pineal region cysts can masquerade as any number of pineal region neoplasms and as such are a pain in the habenula. Another potential pitfall is that, despite their CSF content, pineal cysts do not have the same intensity as CSF on FLAIR, T1WI, or PDW series. This may be because

BOX 2-8 **Syndromes Associated with Brain Tumors**

Basal cell nevus syndrome (chromosome 9q31)/Gorlin syndrome
  Basal cell nevi and carcinomas, odontogenic keratocysts, ribbon ribs, phalangeal deformity and pitting, falcine calcification, craniofacial deformities, scoliosis, and medulloblastomas
Cowden syndrome (chromosome 10q23)
  Multiple hamartomas, mucocutaneous tumors, fibrocystic breast disease, polyps, thyroid adenoma, Lhermitte-Duclos syndrome
von Hippel-Lindau (see Chapter 8)
Li-Fraumeni syndrome (chromosome 17p13)
  Increased rate of breast cancers, soft-tissue sarcomas, osteosarcomas, leukemia
  Autosomal dominant with astrocytomas, primitive neuroectodermal tumors, choroid plexus tumors
  Central nervous system tumors in 13.5%
Maffucci syndrome
  Enchondromas, soft-tissue cavernomas
Neurofibromatosis 1 and 2 (see Chapter 8)
Ollier syndrome
  Multiple enchondromas
Retinoblastoma
  Pineoblastoma (trilateral retinoblastoma)
Sturge Weber
  Choroid plexus hemangioma
Tuberous sclerosis (see Chapter 8)
Turcot syndrome (chromosomes 5q21, 3p21, 7p22)
  Colonic familial polyposis, glioblastoma, rare medulloblastoma

of hemorrhage, hemorrhagic debris, or high protein seen histologically in these cysts.

Colloid Cyst

Colloid cysts (neuroendodermal or paraphyseal cysts) arise in the anterior portion of the third ventricle near the foramen of Monro. They occur with an incidence of three cases per one million individuals per year, usually presenting with positional headaches and/or hydrocephalus in the fourth decade of life. Sudden death because of acute hydrocephalus is one scenario. Usually the lesion is hyperdense on a CT scan because of high protein concentration; the same factor may account for its high signal seen 50% of the time on T1WI (Fig. 2-57). The rim of the cyst may faintly enhance. The lesion is lined by simple to pseudostratified epithelium, and is well circumscribed. MRI is predictive of the ease at which colloid cysts can be aspirated; if the signal intensity is dark on T2WI scans (signifying a high viscosity hyperproteinaceous or cholesterol-laden cyst) it will be a bear to aspirate. Rarely, colloid cysts may occur within the body of the lateral ventricles, fourth ventricle, or outside the ventricular system. Treatment may include biventricular shunting, cyst resection, or endoscopic coagulation.

Neuroepithelial Cyst

Neuroepithelial cysts are most commonly seen as a curiosity in the brain parenchyma (Fig. 2-58) that leads to angst about whether it represents a pilocytic tumor. The fluid in these cysts simulates CSF on CT scans and T2WI MRI but may be bright on T1WI or FLAIR owing to cholesterol debris or high protein concentration. These cysts are lined by epithelium. Differential diagnosis would include epidermoid cysts and traumatized arachnoid cysts.

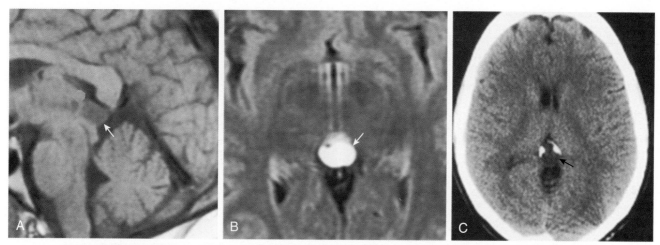

FIGURE 2-56 Cyst or cystic tumor in Otto Liv A. Lowen? Although the signal intensity *(arrow)* on the sagittal T1-weighted image **(A)** and fluid-attenuated inversion recovery image **(B)** is dissimilar to cerebrospinal fluid, this may still represent a benign pineal cyst. **C,** What to do? This cystic lesion in the pineal gland displaces the pineal calcification apart. Is it a benign cyst or a neoplasm? In the absence of an enhancing mass or hydrocephalus, one could probably follow this based on patient symptoms only to assess for growth.

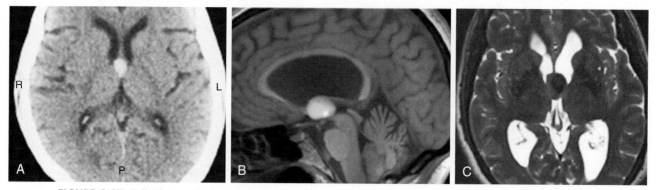

FIGURE 2-57 Colloid cyst. **A,** On computed tomography the colloid cyst is hyperdense from high protein in patient Monroe C. Itz. **B,** On T1-weighted image (T1WI) the colloid cyst is hyperintense due to proteinaceous contents. **C,** Axial T2WI shows the low signal intensity mass in the midline at the foramen of Monro.

When one sees an intraventricular cyst, the differential diagnosis should include a choroid plexus cyst, an ependymal cyst, a colloid cyst, and a cysticercal cyst. Ependymal cysts occur in the frontal horns of the lateral ventricles and are asymptomatic unless they obstruct the foramen of Monro.

Choristomas

Choristomas are masses of normal tissues in aberrant locations, containing smooth muscle and fibrous tissues. They may be hypervascular. Cases have been described in extra-axial locations including the sellar and parasellar regions as well as the internal auditory canals associated with the facial and vestibulocochlear nerves. Choristomas may enhance and hence may simulate schwannomas.

Amyloid

Amyloid may be deposited in the dura and may simulate meningiomas, dural lymphomas, or plasmacytomas, with relatively low signal intensity on T2WI. Intraparenchymal amyloidomas are dense on CT scans and bright on T1-weighted MRI images. Mixed to low intensity on

T2-weighted sequences and contrast enhancement have been reported.

Other Nonneoplastic Masses

Heterotopias and focal areas of abnormal sulcation may also present as masslike lesions. These entities are described more fully in Chapter 8.

Paraneoplastic Syndromes

Paraneoplastic conditions of the brain may occur in association with non-CNS primary tumors. Among these, limbic encephalitis, an abnormality affecting the temporal lobes and causing memory and mental status changes, has been described extensively. Remember the memory deficit. Among these, limbic encephalitis, an abnormality affecting the temporal lobes and causing memory and mental status changes, has been described extensively. (Just checking.) The appearance simulates a herpes simplex encephalitis, usually bilateral (though it may be unilateral in 40%) with extensive disease in the temporal lobes, which is bright on T2WI and may show enhancement (Fig. 2-59). Atrophy of the temporal lobe may coexist, but hemorrhage is exceedingly uncommon. Abnormal signal intensity in the brain

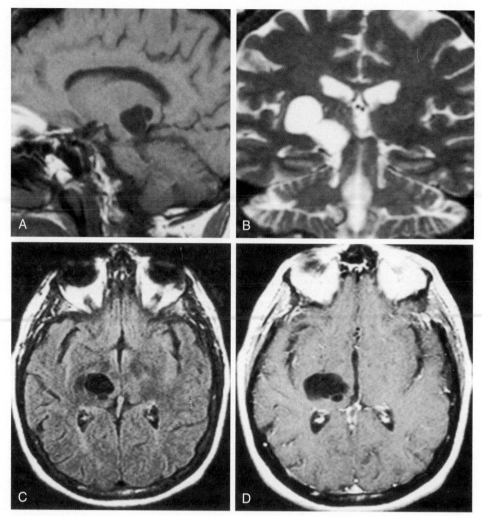

FIGURE 2-58 Neuroepithelial cyst in patient Anne-Udder B. Neinsist. **A,** Unenhanced T1 sagittal scan shows a multiloculated cystic lesion in the right thalamus. **B,** Coronal T2-weighted scan shows no significant mass effect and high signal similar to cerebrospinal fluid. **C,** The fluid-attenuated inversion recovery scan also shows intensity identical to cerebrospinal fluid. **D,** There is no contrast enhancement. This may represent a neuroepithelial cyst, which is a benign lesion that does not require surgical intervention. Differential point: Arachnoid cysts are usually not intraparenchymal, but can exist along perivascular spaces. This could be a huge Virchow Robin space in fact.

stem and/or hypothalamus may be seen in about 10% to 20% of cases of limbic encephalitis. Many different primary tumors have been associated with limbic encephalitis: small cell carcinoma of the lung is classic; testicular germ cell, thymic, ovarian, breast, hematologic, and gastrointestinal malignancies have also been reported. Another paraneoplastic syndrome is that of cerebellar atrophy with clinical manifestations of ataxia. Ovarian carcinoma and lymphoma may cause this finding.

There are a series of antibodies that indicate paraneoplastic syndromes that reads like a Hip Hop lexicon (anti-Yo, anti-Hu, anti-Ma, anti-Ri, anti-CAR, anti-Ta [also called anti-Ma2], anti-What's up?). If you hear these terms, know to look closely at the temporal lobes and cerebellum! Anti-Ri antibodies cause opsoclonus and ataxia and are seen often with breast and lung cancers. Finally, the anti-CAR antibodies attack retinal neurons and cause a retinopathy. They are seen with small cell carcinomas. Antibodies against the voltage gated calcium channels accounts for the Eaton Lambert syndrome with myasthenia gravis like symptoms in patients with thymomas. Treatment of the primary tumor usually results in improvement of the paraneoplastic syndrome.

POSTTREATMENT EVALUATION

Postoperative

Determining whether residual neoplasm is present in the postsurgical tumor bed is one of the most daunting tasks facing a neuroradiologist. What hangs in the balance are prognostic considerations for the patient (Table 2-16 provides survival data for various tumors), potential repeated surgeries, nonsurgical therapeutic decision-making, and the wrath of neurosurgeons regarding your analysis of their fine work. Nothing infuriates the surgeon more than a postoperative scan after a "complete resection" that is interpreted by the radiologist as a small biopsy. Here is how you can avoid this pitfall and why it is such a problem.

Surgical margin contrast enhancement almost always presents after the second postoperative day and is usually thin and linear. The margin may become thicker or more

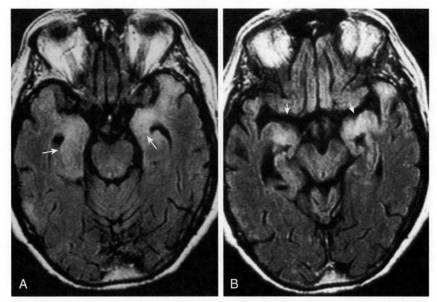

FIGURE 2-59 Limbic encephalitis in patient Assif D'Tumar Wazintinoff. **A,** Bilateral mesial temporal lobe hyperintensity is present on the FLAIR scan *(arrows)* in this patient with ovarian cancer. **B,** Atrophy argues against acute herpes encephalitis. Note prominence of sylvian fissures bilaterally *(arrows)*. In this case, FLAIR hyperintensity is likely from anti-Hu antibodies from a paraneoplastic syndrome.

TABLE 2-16   5-Year Survival Rates by Tumor Type

Neoplasms	Survival (%) at 5 Years
Pilocytic astrocytoma (WHO I)	90%
Ependymoma	64%
Oligodendroglioma	62%
Mixed glioma	59%
Embryonal type	51%
Diffuse astrocytoma (WHO II)	49%
Anaplastic astrocytoma (WHO III)	30%
Glioblastoma (WHO IV)	3%*

WHO, World Health Organization.
*Whoa! That's bad!

TABLE 2-17   Scar versus Residual Tumor

Feature	Scar	Residual Tumor
Enhancement within 1-2 days	No	Yes
Enhancement after 3-4 days	Yes	Yes
Change in size with time	Decreases	Increases
Type of enhancement	Linear, outside preoperative tumor bed	Nodular, solid
Mass effect edema	Decreases	Increases
Perfusion	Low	High (with higher-grade tumor)

nodular after a week. It thus becomes difficult to tell whether enhancing tissue in a surgical bed is because of granulation tissue or marginal tumor enhancement (provided that the tumor enhanced preoperatively). The granulation tissue enhancement may persist for months postoperatively, but intraparenchymal enhancement and mass effect after 1 year should be viewed with suspicion. Dural enhancement is nearly always seen even at 1 year and can persist as long as decades after surgery. Enhancement appears sooner and persists longer on MRI than on CT scans.

This has led neurosurgeons to scan patients soon after an operation, before this scar tissue has time to develop, to identify residual tumor. A scan within 48 hours showing enhancement in the surgical bed should lead one to suggest residual neoplasm (Table 2-17). Unfortunately, the hemorrhagic blood products from the surgery usually have not resolved by 48 hours, and one is forced to interpret enhancement on CT scans and MRI against a bright background of blood products. Herein lies the difficulty and points up the absolute requirement of precontrast scans in the same plane and location as the postcontrast

studies. These scans are viewed side by side to detect the extra thickness of enhancement along a hematoma cavity.

Clearly, the best way to distinguish blood from granulation tissue from recurrent or residual tumor is to scan sequentially. Blood resolves, granulation stays the same or decreases in size, and tumor grows. Unfortunately, neurosurgeons and neurooncologists can have the patience of a 6-month-old child who is hungry and they scream just as loudly at you. Recommending follow-up scans only irritates them, but that is the most reasonable suggestion. If a follow-up scan is not an option, consult Table 2-17, and show it to the surgeon as you lay your head on the block.

Postradiation

Patients placed on the radiation rotisserie to have their brains cooked for primary brain tumors, skull lesions, and/ or intracranial metastasis frequently exhibit central and cortical parenchymal loss. The radiologist's challenge in the postradiotherapy evaluation is to differentiate residual

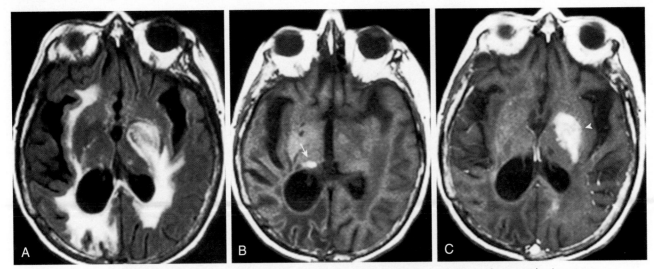

FIGURE 2-60 Radiation necrosis in patient Berne Mai Hedd. This patient had resection of an anaplastic astrocytoma of the right occipital lobe and was treated with radiation therapy. **A,** Fluid-attenuated inversion recovery scan demonstrates high signal intensity without mass effect in the right temporal and occipital lobe with dilatation of the right occipital horn of the lateral ventricle. Note the high signal intensity in the left basal ganglia. **B,** T1-weighted scan shows a high intensity focus in the pulvinar region of the right thalamus secondary to a telangiectasia from radiation therapy *(arrow)*. The right basal ganglia are bright but the left are dark. On the postcontrast scan **(C),** the left basal ganglia enhance *(arrowhead)*. This was an area of radiation injury.

or recurrent tumor from radionecrosis (where the brain has been fried to a crisp). Several factors influence the development of radiation necrosis. These include total dose, overall time of administration, size of each fraction of irradiation, number of fractions per irradiation, patient age, and survival time of patients. As patients survive longer with more effective treatment, the incidence of radiation necrosis will rise, because it is usually a late effect of treatment (Fig. 2-60). The signs and symptoms of radiation necrosis are nonspecific and do not differentiate it from recurrent tumor.

The effects of irradiation have been separated into those occurring early (within weeks) and late (4 months to many years later; Table 2-18). The former is transient, may actually occur during radiotherapy, and is usually manifested by high signal intensity in the white matter caused by increased edema (beyond that associated with the tumor). The delayed effects are separated into early delayed injury (within months after therapy) or late injury (months to years after therapy). Early delayed injury is also a transient effect and is of little consequence other than recognizing it as such (as opposed to tumor growth) directly after therapy. The late effects are usually irreversible, affect white matter to a much greater extent than gray matter, and histologically involve vascular changes that include coagulative necrosis and hyalinization. The late injury to the brain may be focal or diffuse and occurs in approximately 5% to 15% of irradiated patients. Seventy percent of focal late radiation injuries occur within 2 years after therapy. Diffuse late injury to the brain takes the form of severe demyelination, particularly in periventricular and posterior centrum semiovale regions. CT scans demonstrate decreased white matter density, but T2WI/FLAIR is more sensitive and shows high signal intensity in the white matter. Usually the abnormality does not show enhancement.

Disseminated necrotizing leukoencephalopathy is a severe form of radiation-related injury usually seen in conjunction with chemotherapy, whether intrathecal or intravenous. Most patients with disseminated necrotizing leukoencephalopathy do extremely poorly. This entity is described more fully in Chapter 6.

Radiation may induce a mineralizing microangiopathy, causing calcification in the basal ganglia or dentate nuclei, with rare cerebral cortical involvement associated with atrophy of intracranial structures. This generally occurs more than 6 months after radiation and is more common in children than in adults. The cause is thought to be an intimal injury to small vessels with associated tissue hypoxia and dystrophic calcification. Frank radiation vasculitis in large vessels may also be seen as focal narrowed segments on angiograms. Radiation vasculitis of the distal internal carotid arteries (ICAs) may lead to moyamoya phenomenon.

Telangiectasias or cavernomas may occur as a delayed complication to radiation therapy (see Fig. 2-60). This may manifest by hemosiderin laden deposits in the brain most evident as low signal intensity on T2* scans without surrounding edema.

The possibility of focal radiation injury needs to be raised when the lesion is found in the appropriate temporal sequence to treatment. If the lesion is remote from the primary tumor site, then the diagnosis is more easily suggested. Unfortunately, radiation necrosis favors the primary tumor site, probably because of predisposing vascular effects (see Fig. 2-60). The diagnosis of radiation necrosis is made by surgical biopsy but may be suggested by PET scan. Distinguishing radiation necrosis from tumor has been the justification of many PET ventures and has thus generated more radiation than Chernobyl. With residual or recurrent tumor 18–fluorodeoxyglucose PET has increased activity (greater than normal brain tissue), whereas radiation necrosis shows low activity. Overall accuracy of PET is approximately 85% for distinguishing residual or recurrent tumor from radiation necrosis. The accuracy rates are

TABLE 2-18 Types of Radiation Injuries

Feature	Early	Early Delayed	Late Delayed
Time course	During therapy	<3 months after therapy	>3 months after therapy
Manifestation	Transient increase in white matter edema	Transient increase in white matter edema	Focal or diffuse longer-lasting white matter changes
Contrast enhancement	No	Rare	Not uncommon
Long-term sequelae	None	None	Vasculitis, demyelination
Calcification seen	No	No	Yes, in children
Disseminated necrotizing leukoencephalopathy	No	No	Rarely, with chemotherapy
Symptoms reversible	Yes	Yes	No
Telangiectasia	No	No	Yes
Hemorrhage present	No	No	Often
Perfusion	Low	Low	Low

TABLE 2-19 Distinction between Tumor versus Radiation Necrosis

Feature	Residual or Recurrent Tumor	Radiation Necrosis
Timing	Immediate or delayed	Months to years
Mass effect/edema	Present	Present
Enhancement	Yes	Yes, soap bubbles or Swiss cheese
PET (18-FDG)	Positive	Negative
SPECT (Thallium-201, methionine-11C)	Positive	Negative
MRS	Elevated choline	Decreased choline
Perfusion-weighted MR	Elevated rCBV	Decreased rCBV

FDG, Fluorodeoxyglucose; *MR,* magnetic resonance; *MRS,* magnetic resonance spectroscopy; *PET,* positron emission tomography; *rCBV,* relative cerebral blood volume; *SPECT,* single-photon emission computed tomography.

better for high-grade tumors than for low-grade tumors, probably because of inherent differences in tumor growth activity. Removal of the necrotic irradiated nonneoplastic tissue and steroid therapy are the treatments of choice and may be curative. Focal hemorrhage without necrosis also occurs as a result of radiation.

Table 2-19 provides a summary of distinguishing features between residual tumor and radiation necrosis.

Pseudoprogression and Pseudoresponse

As temozolomide (Temodar) and bevacizumab (Avastin) have been used to treat more high-grade astrocytomas, we have experienced the treatment based phenomena of pseudoprogression and pseudoresponse. Essentially, pseudoprogression occurs in 20% to 30% of patients 2 to 6 months after institution of temozolomide and may last 6 to 12 months. What one sees is an exaggeration of the FLAIR signal abnormality and enhancing volume of tumor with increasing necrosis, which looks like tumor progression but is a treatment effect. This can be addressed by noting a decrease in CBF, elevation in ADC, and low uptake on fluorodeoxyglucose (FDG) PET, which would suggest tumor regression rather than progression. *O-6-methylguanine-DNA methyltransferase (MGMT),* a gene located on chromosome 10q26, encodes a DNA-repair enzyme that has been shown to contribute to the resistance of glioblastoma cells to alkylating agents like temozolomide. When the MGMT is methylated, pseudoprogression occurs in more than 90% of glioblastomas,

and the overall survival is also improved, as the tumors are more responsive to chemotherapy.

With respect to pseudoresponse, it makes sense that vascular endothelial growth factor receptor signaling pathway inhibitors like bevacizumab would be associated with diminished gadolinium enhancement of the tumor, thus suggesting tumor regression. Unfortunately this may be a reversible treatment effect. It is a phenomenon seen beginning after 2 weeks of therapy and can last 2 to 3 months after therapy. The findings that suggest a treatment effect and not tumor regression would be the absence of reduction in FLAIR signal abnormality, progression/persistence of ADC reduction and persistent perfusion increases.

Response Assessment in Neuro-oncology

These pseudoresponse and pseudoprogression phenomena have laid waste to the traditional Macdonald criteria for determining tumor response to therapy. These criteria have been used extensively in drug treatment trials. Response is based on the product of two-dimensional tumor measurements of the tumor on the image with the largest contrast-enhancing tumor area. A complete response is determined by resolution of all contrast-enhancing tumor. A partial response is defined by an at least 50% decrease in the product of two orthogonal diameters. Progressive disease is defined by at least a 25% increase in the product of orthogonal diameters.

However, there is a misleading increase in enhanced lesion volume from treatment effect by temozolomide and decreased lesion enhancement volume with anti-VEGF drugs, which can cloud the assessment of treatment response or lack thereof. The response assessment in neuro-oncology (RANO) working group now has advocated new criteria, with the recognition that contrast enhancement is nonspecific during therapy and may not always be an accurate indicator of tumor response. The revision takes into account the increased importance of the nonenhancing component of the tumor, namely the T2/FLAIR signal changes.

Specifically, the RANO response criteria indicate that complete responders (CR) must not be on steroids and be stable or improved clinically and must show no enhancing tissue and stable or decreasing T2/FLAIR signal change. Partial responders (PR) may be on stable or decreasing doses of steroids and be clinically stable or improving and have greater than or equal to a 50% reduction in the volume of enhancing tissue and stable or decreasing T2/FLAIR signal change. Stable disease (SD) shows less than 50% reduction or less than 25% increase in enhancing tumor with stable or decreasing T2/FLAIR signal change. Finally, progressive disease (PD) shows either greater than 25% increase in enhancing tissue or increasing T2/FLAIR signal intensity tissue or declining clinical status.

The tone of this chapter reminds us of a joke. A cannibal goes to the cannibal market because he has a yen for some brains for dinner. He looks at the selection in the frozen meat case and sees "Lawyers' brains—$3.49 a pound," "Artists' brains—$3.49 a pound," "Politicians' brains— $3.49 a pound," and "Neuroradiologists' brains—$24.99 a pound." Now the cannibal shopper calls to the cannibal grocer indignantly, "Hey! What's the big idea? Why is the neuroradiologist brain so damn expensive? You can't be telling me that neuroradiologists are so much smarter than the others. I don't believe it!"

"On the contrary," replies the grocer. "Do you know how hard it is to find enough neuroradiologists to get even one pound of brain? Sheesh."

And so it goes.

Chapter 3

Vascular Diseases of the Brain

Imaging is an essential component in the workup of patients presenting with symptoms of stroke. "Stroke" is a nonspecific clinical term denoting a sudden loss of neurologic function by any cause (e.g., ischemic infarction, spontaneous hemorrhage, post ictal state). A host of new and evolving imaging techniques have been developed in the recent past allowing for ever more accurate and timely detection and characterization of stroke syndromes due to ischemic insult. The development and utilization of these techniques have been spurred by therapeutic advances, most notably the 1996 approval by the U.S. Food and Drug Administration (FDA) of thrombolytic therapy with tissue plasminogen activator (tPA) for intravenous use, as well as the positive results reported from intraarterial thrombolysis and clot retraction in 2015. The imaging techniques available for the assessment of stroke include computed tomography (CT), magnetic resonance imaging (MRI), non-invasive angiography (computed tomographic angiography [CTA] and magnetic resonance angiography [MRA]), catheter angiography, and perfusion imaging (CT and MRI). In this chapter we discuss primary ischemic abnormalities and then turn to the hemorrhagic causes of stroke syndromes. Get a comfy chair and grab a snack—things are about to get pretty intense!

ISCHEMIC CEREBROVASCULAR DISEASE

Clinical Features

Thromboembolic disease consequent to atherosclerosis is the principal cause of ischemic cerebrovascular disease (surprise, surprise). The most common causes of infarction include large-artery atherosclerosis, cardioembolism, and lacunes (small vessel occlusions). Outcomes differ depending on subtype. Large artery lesions have a higher mortality than lacunes. Recurrent strokes are most common in patients with cardioembolic stroke, and have the highest 1-month mortality. Identifying the cause of the infarct has important implications for treatment and prevention of future cerebrovascular events. For example, carotid endarterectomy or stenting might be the more appropriate treatment of choice for large vessel disease whereas anticoagulation therapy is most useful in patients with small vessel disease.

Nonatherosclerotic causes of ischemic stroke include vasculopathies, migraine headache, and systemic/metabolic events (e.g., anoxia/profound hypoxia). They make up a small proportion of strokes in patients over 50 years of age. In younger patients these nonatherosclerotic causes of ischemic stroke are more common in particular in the absence of cardiovascular risk factors (hypertension, diabetes, and hyperlipidemia).

Thromboembolic emboli can arise from arterial stenosis and occlusion in the head and neck arterial vasculature, atherosclerotic debris and ulceration, in the setting of right to left shunts, or cardiac sources (a cardiac source of emboli is responsible for 15% to 20% of ischemic strokes; Box 3-1).

The extent to which narrowing of the arterial lumen contributes to stroke is complex. Even in the absence of severe stenosis, the reduction in flow may decrease the ability to "wash out" distal emboli before they produce ischemia. On the other hand, blood flow may be preserved and infarction may not occur even with complete occlusion of a vessel because of collateral circulation (circle of Willis and leptomeningeal vessels). Patients with complete internal carotid artery (ICA) occlusions in the neck may still have cerebral infarctions from emboli. Emboli may be multiple and simultaneous or a single embolus may break up and produce multiple infarctions.

Atherosclerosis is common and typically affects multiple extracranial and proximal intracranial vessels and/or

---

BOX 3-1 Risk of Cardioembolism

HIGH-RISK SOURCES
Mechanical prosthetic valve
Mitral stenosis with atrial fibrillation
Atrial fibrillation
Left atrial/atrial appendage thrombus
Sick sinus syndrome
Recent myocardial infarction (<4 weeks)
Left ventricular thrombus
Dilated cardiomyopathy
Akinetic left ventricular segment
Atrial myxoma
Infective endocarditis

MEDIUM-RISK SOURCES
Mitral valve prolapse
Mitral annulus calcification
Mitral stenosis without atrial fibrillation
Left atrial turbulence
Atrial septal aneurysm
Patent foramen ovale
Atrial flutter
Lone atrial fibrillation
Bioprosthetic cardiac valve
Nonbacterial thrombotic endocarditis
Congestive heart failure
Hypokinetic left ventricular segment
Myocardial infarction (>4 weeks, <6 months)

multiple regions within the same vessel. Thirty-five percent of patients over 50 years of age have severe stenosing atherosclerotic changes in cervical cerebral arteries but only one third of these individuals have symptoms of vascular disease. Primary stenosis/occlusion most often results in infarction when there is a preexistent stenosis with either new occlusion and/or a period of systemic hypotension. Acute extracranial carotid occlusion may produce large areas of infarction involving the deep (ganglionic) and superficial (cortical) middle cerebral artery (MCA) distribution (Fig. 3-1). In these cases the infarcts are likely the result of large distal emboli associated with the proximal occlusion. The anterior cerebral artery (ACA) territory is typically spared because of collateral supply from the contralateral anterior cerebral artery via anterior communicating artery of the circle of Willis. Combined MCA/ACA ("holo-hemispheric") infarcts are rare and usually fatal. They most often occur in patients with acute myocardial infarction and atrial fibrillation because of the combination of large emboli and poor cardiac output.

Carotid stenosis or occlusion can result in "watershed" or borderzone infarction. Vascular watersheds are the distal arterial territories often at borders between two vascular distributions, which are relatively less well vascularized compared with other territories. Major borderzones are found between the anterior and middle cerebral arteries and the middle and posterior cerebral arteries. Borderzone infarcts occur in the posterior parietal region (middle and posterior cerebral arteries borderzone), the frontal lobes (the anterior and middle cerebral arteries borderzone) and the basal ganglia (Fig. 3-2). These infarcts are often small and may be confused with lacunar infarcts. The key to diagnosis is the presence of multiple infarcts at the interface between different vascular territories and evidence of carotid occlusion, stenosis or slow flow. Other sites in the brain are selectively jeopardized by hypoxia and/or hypotension because of increased susceptibility to ischemia resulting from increased metabolic rate and a lack of redundancy of blood supply. These include the hippocampus (Ammon's horn), globus pallidus, amygdala (anterior choroidal artery–posterior cerebral artery watershed territory), cerebellum and occipital lobes, in that order. In utero and perinatal, watershed zones are much different and center along the deep gray matter interfaces (see Chapter 8).

Interest in the detection and treatment of extracranial carotid artery disease has been heightened by the results of two large trials for the treatment of symptomatic and asymptomatic patients, the North American Symptomatic Carotid Endarterectomy Trial (NASCET) and Asymptomatic Carotid Atherosclerosis Study (ACAS). The NASCET trial demonstrated benefit of endarterectomy in patients with 70% or higher stenosis of the cervical ICA in symptomatic patients, and the ACAS study showed benefit of endarterectomy in patients with more than 60% cervical ICA stenosis in asymptomatic patients. The widespread availability of noninvasive vascular imaging (ultrasound, MRA and CTA) and the introduction of stent devices for the carotid artery have resulted in a dramatic increase in the number of individuals being treated for carotid stenosis without the same strength of evidence that these treatments provide similar significant benefits to patients. The assessment of degree of stenosis is complicated by the existence of various methods for measuring stenosis. NASCET uses the ratio of

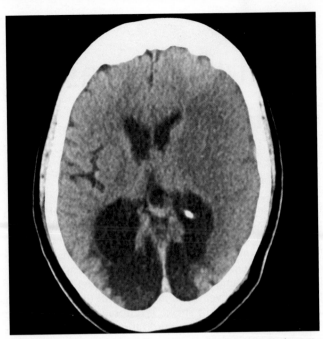

FIGURE 3-1 Thromboembolic infarct. Axial computed tomographic image at the level of the lateral ventricles shows large wedge-shaped region of hypoattenuation in the left middle cerebral artery distribution. There is loss of the gray-white differentiation and sulcal effacement in the affected territory with mild mass effect on the left lateral ventricle. These changes indicate acute infarct. Note encephalomalacia in both occipital lobes related to chronic ischemic insult in the posterior cerebral artery distribution.

the tightest point of carotid stenosis to the "normal lumen" distal to the stenosis while ACAS and the European based studies use the degree of stenosis relative to the *estimated* normal lumen at the same site. Each method has its limitations. The NASCET criteria can lead to underestimation of stenosis when the distal lumen narrows as a result of the severe proximal stenosis "string sign." The ACAS method is problematic because the observer must extrapolate what is thought to be the true lumen (Fig. 3-3).

Intracranial embolic occlusion most commonly produces infarction in the midsection (posterior frontal, anterior parietal and superior temporal) of the MCA distribution. Emboli entering the carotid artery preferentially lodge in these MCA branches (not because these branches are cuter, but because they are along the straightest route an embolus can take). Pure ACA embolic infarcts are rare. Isolated ACA infarcts typically occur as a result of intrinsic arterial disease and occlusion (e.g., diabetes, hypertension, vasospasm, and vasculitis) or from severe subfalcine herniation rather than emboli. The location and extent of the infarct are determined by the site of embolic occlusion and extent and location of collateral supply to the brain distal to the occlusion. Occlusion of the distal carotid bifurcation and proximal MCA and ACA vessels ("T" occlusions) may result in infarction of the cortical (superficial) and ganglionic (deep) portions of the MCA territory. If there is good cortical collateral supply, the infarct may be confined (at least initially) to the basal ganglia and insula in part because of lenticulostriate branch obstructions. Embolic infarcts in the vertebrobasilar system may affect single or multiple vessels. Complete basilar occlusion produces cerebellar and brain stem infarcts and variable bilateral infarction of the inferior medial temporal and occipital lobes and posterior

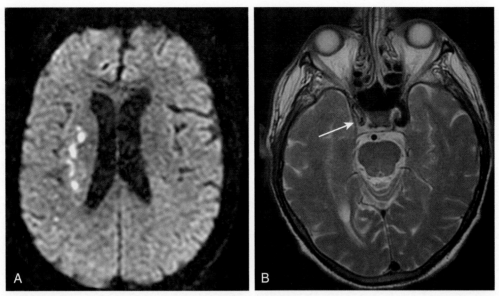

FIGURE 3-2 Borderzone (watershed) infarcts. **A,** Diffusion-weighted image reveals foci of hyperintensity at the right middle cerebral artery-anterior cerebral artery border. **B,** Hyperintensity within the distal right internal carotid artery *(arrow)* indicates slow flow and high-grade stenosis.

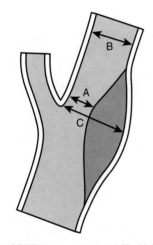

NASCET % stenosis = 100 [B–A] / B
ECST % stenosis = 100 [C–A] / C

FIGURE 3-3 Drawing of North American Symptomatic Carotid Endarterectomy Trial (NASCET) and European Carotid Surgery Trial (ECST) criteria for evaluation of carotid stenosis. **A,** The diameter of the residual lumen at the point of maximal stenosis. **B,** The diameter of the normal artery distal to the stenosis. **C,** The estimated "true" lumen at the point of stenosis. (Courtesy P. Kim Nelson, MD.)

thalami while basilar tip occlusions spare the posterior fossa structures. The extent of posterior cerebral artery involvement is dependent upon the status of the posterior communicating arteries. Focal occlusion of the distal vertebral artery produces infarcts in the distribution of the posterior inferior cerebellar artery (PICA) leading to infarcts in the inferior cerebellum and lateral medulla (Wallenberg syndrome; Fig. 3-4). Multiple infarcts in several vascular distributions in both cerebral hemispheres raise concern for central embolic source, such as the heart (Fig. 3-5).

Lacunar infarcts are small lesions produced by occlusion of deep perforating arteries. "Lacune" is a venerable pathologic term indicating a fluid filled hole in the brain (Fig. 3-6). A lacune by definition is an infarct less than 15 mm in size. Lacunes have a predilection for the basal ganglia, internal and external capsules, pons, and corona radiata. Occlusion of brain stem perforating arteries produces distinctive infarcts that are paramedian, unilateral and tubular in appearance on axial imaging reflecting the location and course of the pontine perforating arteries. Although originally thought to arise from small vessel atherosclerosis and lipohyalinosis associated with hypertension, many other causes of lacunar infarcts have been proposed including emboli, hypercoagulable states, vasospasm, and small intracerebral hemorrhages.

Transient ischemic attack (TIA) is a sudden functional neurologic disturbance limited to a vascular territory that usually persists for less than 15 minutes, with complete resolution by 24 hours. The diagnosis of TIA is difficult because it is by definition retrospective. Although TIAs have a variety of causes, the common pathway is temporarily inadequate blood supply to a focal brain region. TIAs are not benign events. Almost one third of patients will eventually progress to completed cerebral infarction (20% within 1 month of the initial TIA). Quantitative measurement of apparent diffusion coefficients (ADCs) from magnetic resonance (MR) diffusion-weighted images (DWIs) may reveal mild decreased diffusion (<25%) in symptomatic areas without signal abnormality on DWI indicating that while there is no permanent functional deficit, neurons have been lost (25% in some animal studies). Thus proceeding with the workup after the TIA is urgent, with the goal to treat the cause before the development of a completed cerebral infarct. A reversible ischemic neurologic deficit (RIND) lasts less than 7 days and symptoms should resolve. MRI DWI is positive in about 50% of these cases even with symptom resolution.

PATHOLOGY OF ATHEROSCLEROSIS

The process begins with subendothelial fatty deposition (fatty streak) consisting of smooth muscle cells, foam cells, T lymphocytes, and an extracellular matrix of

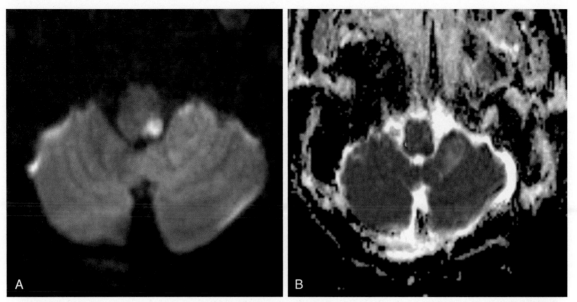

FIGURE 3-4 Wallenberg syndrome. **A,** Diffusion-weighted imaging shows hyperintensity in the posterolateral medulla with restriction seen on **(B)** the apparent diffusion coefficient (ADC) map. These findings indicate acute infarct in the left posterior inferior cerebellar artery territory. There is adjacent gliosis in the left cerebellar hemisphere because of prior surgery best seen on ADC map.

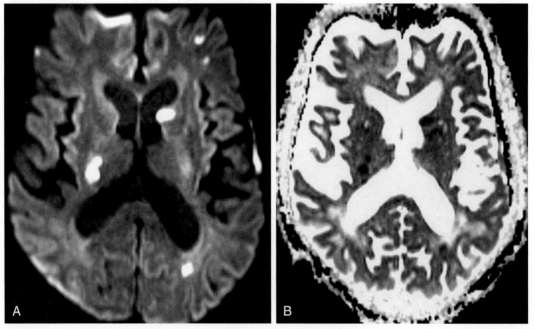

FIGURE 3-5 Embolic infarcts. **A,** Diffusion-weighted imaging shows numerous foci of hyperintensity within both cerebral hemispheres and in multiple vascular territories. Cortical, subcortical and deep structures are affected. **B,** Restriction is confirmed on apparent diffusion coefficient maps, indicating that these represent acute infarcts. This distribution raises concern for central embolic source, such as cardiac source.

lipid and collagen in an arterial vessel (gross, right?). Fat is discharged into the extracellular space precipitating intimal thickening, proliferation of smooth muscle cells and inflammatory changes eventually resulting in fibrosis and scarring. The endothelial surface of the plaque may degenerate with ulceration of the fibrous cap of the plaque, and subsequent discharge of lipid and/or calcified debris into the vessel lumen. Platelets may accumulate on the ulcerated intimal surface and expose thrombogenic collagen or fat, leading to thrombus formation and platelet emboli. Arterial bifurcations are subject to the greatest mechanical stress and are especially prone to atherosclerosis. The composition of plaques is variable, with some becoming large and fibrotic, producing luminal narrowing, whereas others accumulate lipid and cholesterol. The composition of the plaque may have significant prognostic and therapeutic implications. Plaques with thick fibrous caps may be stable and asymptomatic even while producing significant stenosis. These lesions may require no therapeutic intervention. Plaques with

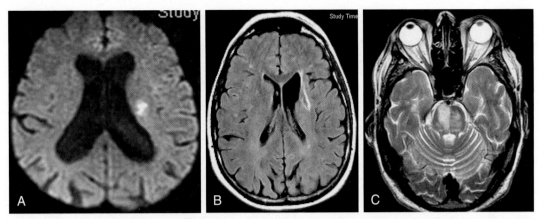

FIGURE 3-6 Lacunar infarction. **A,** Diffusion-weighted imaging (DWI) reveals a round focus of hyperintensity in the left periventricular white matter indicative of acute lacunar infarct. **B,** Fluid-attenuated inversion recovery image reveals central fluid intensity with peripheral T2 hyperintensity indicative of chronic lacunar infarct. **C,** Paramidline T2 hyperintense lesion in the right pons with sharp medial border at the midline indicates acute infarct (confirmed on DWI, not pictured), superimposed on bilateral chronic ischemic change.

a thin or absent cap (unstable plaque) with exposed lipid and or hemorrhage are prone to development of thrombus and embolization. Aggressive therapy may be warranted regardless of the degree of stenosis. High resolution surface coil black blood imaging can reveal (1) fibrous plaques show gadolinium enhancement, (2) plaque hemorrhage (blood intensity), (3) calcification (dark on all sequences), and (4) platelet accumulation at a site of plaque disruption through a thin enhancing fibrous cap.

IMAGING TOOLS

Brain

CT has been the mainstay of stroke imaging since its inception in the mid-1970s. Unenhanced CT scans are fast and readily available. They are excellent for detecting large ischemic infarcts of over 6 to 8 hours' duration. Nonischemic causes of stroke including hemorrhage, infection, and tumor are readily detected although often poorly characterized. There are, however, significant limitations to CT. It does not reliably detect infarcts of less than 4 hours' duration and the extent of the infarct is often difficult to characterize. Acute lacunar infarcts often go undetected and are typically difficult to distinguish from chronic lacunar infarcts. Overall detection rates for acute infarction are approximately 58% in the first 24 hours.

Detection of hyperacute infarction (<6 hours) on unenhanced CT is not for the tame of heart! It is a skill that requires expertise and experience, and shockingly, a good clinical history. Knowing the neurologic deficit and time of onset of symptoms can really help in picking up the subtle changes of infarct that would otherwise be below the threshold for calling abnormal. No, it's not cheating; it's good patient care! Narrow CT window widths and levels should be applied to detect early infarct evidenced by loss of the gray-white matter distinction. Unenhanced CT can sometimes identify a dense vessel sign of acute embolic occlusion (Fig. 3-7) but often provides no information on the status of the brain that surrounds the already infarcted tissue.

MRI is much more sensitive than CT in the detection of hyperacute infarction. Certainly, the advent of DWI has greatly enhanced our ability to detect hyperacute infarction and characterize all infarctions. While "routine" MR has 85% sensitivity for infarction within 24 hours, MR with DWI has a sensitivity of approximately 95% in this period including the first 3 hours after infarction when CT typically does not demonstrate any detectable parenchymal abnormality.

Diffusion imaging is a technique that is sensitive to the movement of water molecules (Brownian motion). In pure water, protons move about and jostle each other, and the extent of water molecule motion (self-diffusion) will be determined by temperature. The higher the temperature the more energy the protons possess and the further they will move. Biologic tissues are more complex. The water molecule encounters various barriers and impediments to motion including cell membranes, intracellular organelles and extracellular proteins. The term "apparent" is applied to modify the word "diffusion" connoting the uncertainty of the water motion in biologic samples caused by these barriers. In gray matter these structures are relatively randomly arrayed so diffusion is the same in all directions (isotropic). In white matter diffusion is constrained by the orientation of the white matter tracts. Water will diffuse preferentially along rather than across these tracts and is therefore anisotropic. The distance traveled by a particular proton will depend on the number of impediments it encounters and period of time during which the molecule is "observed" during the MR sequence. If the observation time is too short, the paths of most molecules will not be differentially affected by cellular barriers (membranes, proteins, etc.); however, when the observation time is long enough encounters with barriers will restrict diffusion. Thus DWI is unique among all imaging techniques in that it is a direct window into the spatial scale of molecules and cells.

You still with us? Good! The ADC can be calculated by using images with varied gradient strengths (different b values). At a minimum, ADC can be calculated if there are at least two b values, one of which must be set to approximately 0 sec/mm^2, that is, with no diffusion weighting. In clinical practice two b values are generally used; however, four or more b values can be measured to improve accuracy of measurement. Commonly used values include a b value of 800 to 1200 sec/mm^2, TE 90 to 120 msec.

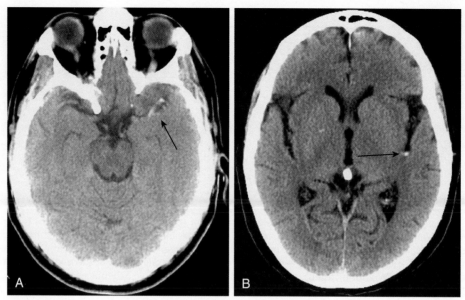

FIGURE 3-7 Hyperdense middle cerebral artery (MCA) sign. **A,** Axial image at the level of the circle of Willis at 3 hours reveals hyperdensity in the proximal left middle cerebral artery, indicating proximal embolic occlusion *(arrow)*. **B,** Focal hyperdensity in the left sylvian fissure is indicative of distal embolus *(arrow)*.

In clinical practice, DWI sequences include approximately 30 slices with individual images obtained in approximately 20 msec. Four acquisitions are obtained at each location (total acquisition time for the brain <1 minute). One acquisition is acquired with no diffusion gradients (the b0 image—a T2 and susceptibility-weighted image) and three sets of orthogonal (anterior-posterior, superior-inferior, and right-left) images are acquired with a b value of around 1000. The three orthogonal images are averaged to produce a "trace" image that is insensitive to the anisotropy created by the orientation of white matter tracts. For instance, on a DWI acquired with the diffusion gradients applied in the anterior-posterior direction, the corpus callosum will appear bright because there is almost no anterior-posterior motion of water molecules in the highly organized right to left oriented callosal fibers. On the other hand, on images where the diffusion gradients are applied in a right-left orientation, the vertically oriented white matter of the corticospinal tract will appear bright. The trace image is the average of these three acquisitions that eliminates the effects of fiber tract orientation on signal intensity. In clinical practice only the trace image is viewed because in processes like infarction and other diseases it is the magnitude not the directionality of diffusion that is important.

However, information on the direction of diffusion and the degree of anisotropy are obtained and can be used to create images that record the direction and integrity of white matter tracts. This technique called diffusion tensor imaging (DTI) requires image acquisition in at least six planes rather than the three planes used for clinical DWI to completely describe the diffusion tensor (a tensor—aside from the radiology boards—is any measurement with at least three components).

The diffusion data can be used to generate ADC maps by performing a voxel by voxel calculation of ADC using the trace diffusion and b0 image. Subtractions of the diffusion and b0 data can also be used to generate "exponential" diffusion images. Generation of these maps is fast and simple. In clinical practice it is common to generate and view DWI, ADC and exponential images. ADC maps and exponential maps eliminate the T2 component of intensity ("T2 shine through") on diffusion sequences (see later).

All DWIs start as T2-weighted images (T2WI) from which signal is subtracted based on the extent of diffusion, and therefore with routine DWI there is always a contribution of T2 to signal intensity. It is also helpful to have the b0 images available for viewing. Because of speed of acquisition, these images are rarely motion degraded, and therefore in uncooperative patients or in patients having very rapid MR studies the b0 can serve as a "poor man's" T2-weighted and/or susceptibility-weighted image (SWI).

Substances that most nearly approximate water will have the highest rates of diffusion (high ADC) and will lose signal more rapidly than those with low ADC. Thus cerebrospinal fluid (CSF) appears dark on diffusion-weighted images as the water molecules can freely diffuse for relatively large distances, while gray matter is light gray and white matter is slightly darker gray in adults. On ADC maps, contrast is reversed. Increased diffusion is bright and therefore CSF is bright while brain tissue is dark. Some clinicians prefer exponential diffusion images to ADC maps because the relative signal intensities are the same as with DWIs (high diffusion such as CSF is bright). In reality the reversal of signal between DWI and ADC maps is not a problem if one simply remembers that CSF has the highest diffusion and that lesions with low diffusion will appear as the opposite of CSF. In tissues where diffusion is more restricted than in normal brain (e.g., hyperacute infarction) there will be less water molecule motion than in normal tissue and therefore less signal loss during the diffusion acquisition. These regions will appear bright on DWI and dark on ADC maps. When water motion is increased in tissue because of vasogenic edema (increased extracellular water) or gliosis (decreased cellularity), tissue will appear isointense on DWI and hyperintense on ADC maps. Tissues with increased diffusion are typically isointense rather than hypointense on DWI because of T2 effects. Increased tissue water (vasogenic edema) increases the T2 of the tissue, and therefore the effects of increased T2 (increased signal) and increased diffusion (decreased signal) tend to cancel

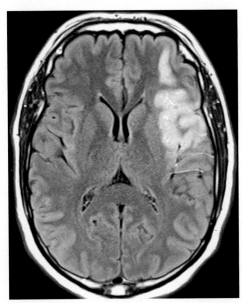

FIGURE 3-8 Slow flow. Axial fluid-attenuated inversion recovery image shows evidence of parenchymal infarct in the anterior division branch of the left middle cerebral artery. Note prominent hyperintense vessels along the surface of the brain in the infarcted territory, reflecting slow flow in distal arterial branches.

each other out. In circumstances when diffusion is equal to normal brain but T2 is increased (subacute infarction) the tissue will look bright on DWI and isointense on ADC maps, a phenomenon known as "T2 shine through." Phew!

T2-weighted fluid-attenuated inversion recovery (FLAIR) scans have a sensitivity of 85% within the first 24 hours. Some have suggested that a DWI positive, FLAIR negative stroke suggests a time frame of less than 6 hours, but this will vary depending on type and location of the stroke and the collateral substrate. MR has been shown to also be more sensitive than CT (even in the hyperacute phase) in the detection of hemorrhage (either within the infarct or as an independent cause of stroke). Detection of hemorrhage has been greatly facilitated by the routine use of gradient echo (GrE) and more recently SWI sequences. Other causes of stroke including venous thrombosis, vascular malformations, infections, and tumors are detected and characterized with greater accuracy than is possible with CT. Arterial and venous occlusion and/or slow flow can be detected on MR, in particular with the use of gradient echo scans and FLAIR (Fig. 3-8). Focal acute embolus in a major vessel (the corollary of the dense vessel sign on CT) is best detected on gradient echo scans (see below), and slow flow can be seen on FLAIR and enhanced T1-weighted images (T1WI). Towel off for the time being, but come back soon—there's more to come.

Vessels

It is obviously important to have knowledge of the arteries and veins in assessing individuals presenting with "stroke." Identification of occlusion and/or stenosis of extracranial and intracranial arteries can confirm the ischemic nature of a lesion and help to determine whether an infarct is because of slow flow, proximal (e.g., MCA) embolic occlusion or small vessel disease. Direct visualization of the dural venous sinuses and

cortical veins is often critical to the correct diagnosis of venous thrombosis in particular given the protean clinical manifestations, etiologies and imaging findings in this disorder. In the past, assessment of vascular structures required invasive catheter angiography, but there are now multiple noninvasive ways of assessing the cervicocerebral vessels, including CTA, MRA, and ultrasound. Each of these techniques has its advantages and limitations, and therefore the choice of the technique or combination of techniques to be utilized will depend on the clinical circumstances, diagnostic questions and treatment options in each case. Catheter angiography is reserved for those cases in which noninvasive studies do not provide a definitive diagnosis and most importantly when endovascular intervention (e.g., angioplasty, stenting, aneurysm coiling) is performed.

Carotid Ultrasound/Transcranial Doppler

Ultrasound uses sound waves to image structures or measure the velocity and direction of blood flow. Color-coded Doppler ultrasound can depict the residual lumen of the extracranial carotid artery more accurately than conventional duplex Doppler. However, the results from color-coded Doppler ultrasound examinations are operator dependent and can be confounded by artifacts related to plaque contents and limited by vessel tortuosity. Problems include distinguishing high-grade stenosis from occlusion, calcified plaques interfering with visualization of the vascular lumen, inability to show lesions of the carotid near the skull base, difficulty with tandem lesions, and inability to image the origins of the carotid or the vertebral arteries. In the NASCET study, Doppler measurements were 59.3% sensitive and 80.4% specific for the detection of stenosis greater than 70%.

Transcranial Doppler ultrasound is a noninvasive means used to evaluate the basal cerebral arteries through the infratemporal fossa. It evaluates the flow velocity spectrum of the cerebral vessels and can provide information regarding the direction of flow, the patency of vessels, focal narrowing from atherosclerotic disease or spasm, and cerebrovascular reactivity. It can determine adequacy of middle cerebral artery flow in patients with carotid stenosis and evidence of embolus within the proximal middle cerebral artery. It is very useful in the detection of cerebrovascular spasm following subarachnoid hemorrhage or after surgery in the intensive care unit setting on site, and can rapidly assess the results of intracranial angioplasty or papaverine infusions to treat vasospasm.

Magnetic Resonance Angiography

MRA is a critical and important tool for assessing the extracranial and intracranial vascular system. The technique is noninvasive and does not involve use of ionizing radiation. MRA can be performed without or with contrast. In some cases, injection of contrast material may be problematic in patients with compromised renal function, given recent concern for development of nephrogenic systemic sclerosis in patients with relatively low glomerular filtration rates (<30 mL/min/1.73 m^2).

There are three different techniques used to generate MRA: time-of-flight (TOF), phase contrast (PC), and contrast-enhanced angiography. Once the imaging data is gathered it may be processed by several display techniques. The one most commonly used is termed maximum intensity projection (MIP), which finds the brightest pixels along a ray and projects them along any viewing angle. MIP is fast and insensitive to low-level variations in background intensity.

In two-dimensional and three-dimensional TOF MRA (the most commonly used technique), protons not immediately exposed to an applied radio frequency (RF) pulse (unsaturated spins) flow into the imaging volume and have higher signal than the partially saturated stationary tissue (which has lost signal secondary to the RF pulse). This is a T1 effect and flowing blood will appear bright. To visualize the arteries without interference from the veins, an initial superiorly positioned nonspatially localized saturation pulse is applied. With TOF magnetic resonance venography (MRV), the saturation pulse is applied inferiorly to saturate the arterial blood.

The two-dimensional TOF techniques are very sensitive to slow or moderate flow (as flow related enhancement is maximized), whereas three-dimensional techniques are better than two-dimensional MRA for rapid flow and have higher resolution. A pitfall in the evaluation of TOF MRA can occur when there are T1 hyperintense lesions or structures within the tissues. These areas of T1 hyperintensity will be visible on the MRA images because the MIP images will include all regions with signal intensity above a predefined threshold. Thus subacute hematomas and fat containing lesions will appear bright and might confound interpretation of the MRA. Subacute intramural clot in dissections and venous sinus thrombosis will also appear bright and may be mistaken for flow.

The advent of 3 Tesla MR scanners has produced a dramatic improvement in TOF MRA. At 1.5T visualization of second order intracranial branches (e.g., intrasylvian MCA branches) is limited, and therefore detection of distal occlusions, vasculopathy, and arterial spasm is not reliable. At 3T these vessels and even smaller arteries (e.g., lenticulostriate arteries) are well visualized in most cases (Fig. 3-9). Therefore it is preferable to perform MRA studies on 3T scanners if available.

In phase contrast (PC) MRA, bipolar flow-sensitizing gradients of opposite polarity are used to "tag" moving spins (protons) that are then identified owing to their position change at the time of each gradient application. The operator chooses the flow velocities that the angiogram will be

sensitive to, termed the VENC, ranging from about 30 cm/sec for arterial flow to 15 cm/sec for venous flow. Complex subtraction of data from the two acquisitions (one of which inverts the polarity of the bipolar gradient) will cancel all phase shifts except those resulting from flow. This technique provides excellent background suppression to differentiate flow from other causes of T1 shortening such as subacute hemorrhage or fat. However, PC MRV is routinely used for suspected venous thrombosis because of its ability to differentiate between flow and subacute (bright) thrombus that obviates the need for TOF MRV.

Contrast enhanced MRA (CEMRA) provides rapidly acquired (<30 seconds) high-resolution images of the extracranial and proximal intracranial vessels with typical coverage from the aortic arch to the circle of Willis (Fig. 3-10). Timing is critical as enhancement of veins confounds the ability to demonstrate arterial anatomy. Because it is not affected by turbulence, it is superior to noncontrast MRA for evaluation of carotid bifurcations and cervical and intracranial vertebrobasilar system. It also can decrease ambiguity in cases with flow reversal such as subclavian steal (Fig. 3-11).

MRA is a good tool for the noninvasive evaluation of the extracranial vasculature for the presence of a hemodynamically significant lesion of the carotid arteries, dissection of the vertebral and carotid arteries, extracranial traumatic fistula, extracranial vasculitis such as giant cell arteritis, or congenital abnormalities of the vessels such as fibromuscular disease. Because it is noninvasive and does not utilize ionizing radiation, it is an excellent screening test for cervical vascular disease. Keep in mind that cervical MRA tends to overestimate moderate stenosis, particularly if only unenhanced two-dimensional TOF methods are used. Thus, an apparent severe stenosis (>85%) may actually be moderate (~50%).

Intracranial MRA can be used to reliably detect proximal stenosis and occlusion as well as vasculopathy (at 3T). MRA has been shown to accurately detect aneurysms (90% accuracy for aneurysms >3 mm). It is therefore a useful tool for screening asymptomatic patients with a risk of intracranial aneurysm (e.g., patients with polycystic kidney disease or individuals with a first degree relative with a history of ruptured aneurysm). It can also be used to follow patients with known nonruptured aneurysms and patients who have undergone endovascular coiling of aneurysms. In the setting of acute subarachnoid hemorrhage, however, detectability of aneurysms by MRA is limited as the flow

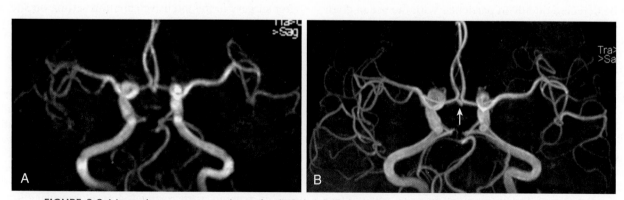

FIGURE 3-9 Magnetic resonance angiography (MRA) 1.5 Tesla (1.5T) versus 3T. **A,** 1.5T and **(B)** 3T maximum intensity projection reconstructions from cranial MRA shows improved visualization of small and peripheral vessels at 3T. The aneurysm at the anterior communicating artery complex is more clearly defined on the 3T image *(arrow).* (Courtesy Kevin Chang, MD.)

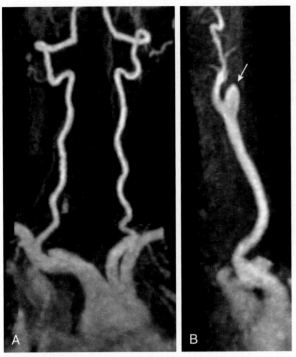

FIGURE 3-10 Contrast-enhanced magnetic resonance angiography (CEMRA). **A,** Maximum intensity projection (MIP) reconstruction from CEMRA with cervical carotid systems excluded shows the great vessels arising from the aortic arch as well as bilateral cervical vertebral arteries to be widely patent. The origins of the great vessels at the arch can be difficult to reliably assess on computed tomography angiography because of streak artifact from bolus injection and shoulders. **B,** MIP reconstruction from CEMRA shows abrupt tapered occlusion of the internal carotid artery just beyond the carotid bifurcation *(arrow),* indication dissection.

related enhancement in vessels may be camouflaged behind the T1 hyperintense signal of acute subarachnoid blood in the cisterns. Although MRA may easily detect arteriovenous malformations (AVM), the superimposition of feeding arteries and draining veins limits the value of this technique in evaluating AVMs. Four-dimensional CTA and MRA in which a time element is superimposed to show inflow and outflow have shown promise with more detailed noninvasive imaging of arteriovenous malformations and fistulas, although conventional catheter angiogram remains the gold standard for imaging of these lesions.

Computed Tomographic Angiography

CTA (Fig. 3-12) has emerged as an alternative to MRA for imaging both the extracranial and intracranial blood vessels with the development of multirow detector scanners. Current 16-128 row scanners can provide excellent visualization of extracranial and intracranial vessels without venous contamination (assuming accurate timing of contrast bolus injection). New 320 row detector scanners can acquire data from the entire brain simultaneously and therefore, with multiple acquisitions, produce time resolved angiographic studies that mimic catheter angiography in their appearance.

Dual-energy CT acquires two image datasets using two different tube energies applied in the form of kVp. Tissue density may be variable between a high-energy spectrum and a low-energy spectrum, and this attenuation difference allows a more nuanced examination of tissue characteristics. Iodine and calcification in vessels may be more easily differentiated on images because they have different responses to high (120 kVp) and low energy (50 kVp) radiation.

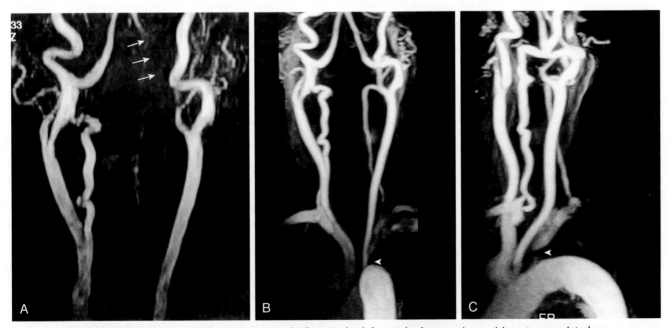

FIGURE 3-11 Subclavian steal. **A,** Anterograde flow in the left vertebral artery *(arrows)* is not appreciated on time-of-flight magnetic resonance angiography (MRA). This is because flow within this vessel is in the opposite direction and has been suppressed along with the venous structures by the intentionally applied saturation pulse. **B,** Gadolinium enhanced MRA now shows the left vertebral artery, which is opacifying in a retrograde fashion. The cause is a stenosis of the proximal left subclavian artery *(arrowhead),* which is hard to believe on this projection, but much more plausible *(arrowhead)* on the oblique view **(C).**

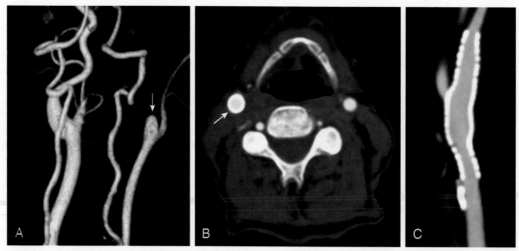

FIGURE 3-12 Computed tomography angiography (CTA) of the neck. **A,** Three-dimensional volume rendered reconstructions of the cervical arterial vasculature with bone subtraction elegantly shows abrupt occlusion of the internal carotid artery (ICA; *arrow*) indicating dissection, with remainder of vessels normal in appearance (same patient as in Fig 3-10, *B*). **B,** Axial CTA source image in a different patient to assess patency of a right ICA following stent placement shows the stent to be intact with normal enhancement in the vessel lumen *(arrow)*. **C,** Sagittal maximum intensity projection reconstruction shows the same stent in the craniocaudal dimension to be widely patent.

CTA requires the placement of a catheter usually in the antecubital vein with rapid injection of approximately 50 to 125 mL of iodinated contrast material. After a short delay following contrast injection, calculated by test bolus run before the formal CTA exam to optimize arterial of venous flow sensitivity, imaging commences and a three-dimensional data set is acquired. CT advances have resulted in thinner images improving resolution. Computer postprocessing is necessary for MIP images and excluding the bony base of the skull structures.

As an imaging technique, CTA has several advantages when compared with MRA at the cost of radiation exposure and iodinated contrast dye. Faster imaging acquisition and higher spatial resolution allow for accurate assessment of vascular morphology, such as in the setting of aneurysms or extracranial stenoses. Calcification does not cause the same artifacts that are seen on MR and extremely slow flow and tandem lesions are more reliably detected on CTA than MRA. Intracranial embolic occlusion is more easily seen and focal thrombus within proximal intracranial vessels may be directly visualized. The superb quality of CTA has prompted many neurosurgeons to operate directly on the basis of CTA findings reserving catheter angiography for those cases where CTA findings are inconclusive or when endovascular treatment is to be performed.

The limitations of CTA include (1) risks of intravenous iodinated contrast injection; (2) exposure to radiation; (3) obscuration of vessels at the base of the skull because of bone; (4) obscuration of aneurysms by extensive subarachnoid hemorrhage; (5) extensive atherosclerotic calcifications in the walls of the vessels which can obscure opacification within the vessel; (6) normal osseous structures such as the anterior clinoid process may obscure the adjacent vessel and less frequently mimic the appearance of an aneurysm on CTA surface rendered reconstructions; and (7) the three-dimensional reconstruction process is still operator dependent. While workstations have improved the ability to detect aneurysms near the skull base, in particular within and adjacent to

the cavernous sinus, skill (of the nunchuck, bow hunting, and computer hacking variety) at image manipulation is often required to make aneurysms in this region visible.

Conventional Catheter Angiography

Arterial catheter angiography is the definitive imaging modality for vascular lesions of the brain and great vessels of the neck but has been relegated to a secondary role in the diagnosis of stroke. In the hyperacute strike setting, catheter angiography is primarily used in stroke treatment for planning and execution of thrombolysis and stenting.

Patients are referred for angiography for the following reasons: (1) if the MRA, CTA, or/and carotid ultrasound are equivocal; (2) if MRA is contraindicated (e.g., in patients with pacemakers); (3) if cardiac output is too low to produce a diagnostic CTA; (4) to evaluate complex aneurysms or vascular malformations responsible for an intracranial hemorrhage; and (5) for the evaluation of vasculitis. The advent of rotational three-dimensional digital subtraction angiography has made it possible to combine the advantages of selective arterial injection of contrast and that of the three-dimensional imaging intrinsic to CTA (Fig. 3-13).

For assessment of arteriovenous malformations and fistulas, selective catheter angiography is necessary to obtain time resolved images that separate arterial and venous components of the malformations. While high field MRA and CTA may suggest the correct diagnosis of vasculitis, the absence of evidence of vasculopathy does not exclude this diagnosis. Because the treatment of this disorder is not without risk, catheter angiography may be performed to confirm or exclude the diagnosis and may be used to determine the best site for biopsy if necessary. Catheter angiography is a safe (but not completely harmless) study and in many situations provides crucial information. The incidence of all complications for femoral artery catheterizations is approximately 8.5% with the range of permanent

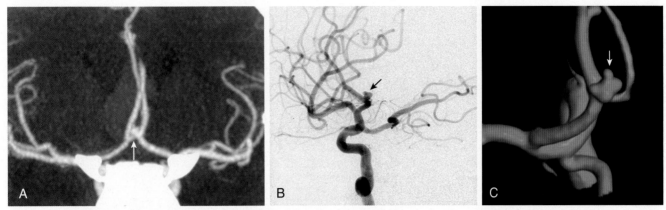

FIGURE 3-13 Aneurysm work-up. **A,** Coronal maximum intensity projection from computed tomography angiography examination in this patient presenting with diffuse subarachnoid hemorrhage shows a lobulated aneurysm arising from the anterior communicating artery complex *(arrow).* **B,** Conventional catheter angiogram redemonstrates the aneurysm to better advantage *(arrow),* with **(C)** rotational three-dimensional reconstructed images better depicting the lobulated morphology of the aneurysm *(arrow).*

complications (the most significant of which is stroke) from 0.1% to 0.33%, a 2.6% incidence of transient complications, and a 4.9% incidence of local complications.

In individuals with acute or chronic ischemic disease, catheter angiography is used in selective cases, in particular if endovascular intervention is contemplated. It is an excellent albeit invasive method for determining whether a lesion is hemodynamically significant in the carotid circulation (Box 3-2). Assessment of collateral circulation distal to a stenosis or occlusion is most easily determined with catheter angiography, where serial images show the presence, source, and extent of collateral supply to the brain.

Detection of ulcerated plaques is more accurate with conventional catheter angiography than noninvasive angiography. However, on all types of angiographic exams, it is difficult to distinguish ulceration from irregularity. The most reliable angiographic sign is the penetrating niche, but depression between adjacent plaques and intraplaque hemorrhage may produce a similar appearance (Fig. 3-14). Luminal bulging secondary to destruction of the media with an intact intima can also appear as an ulcer. One should appreciate that the association of ulcer and stroke is also controversial. Many asymptomatic plaques are ulcerated, and many symptomatic plaques are not. Generally, however, ulceration is frequently found on the symptomatic side in association with significant stenosis. High-resolution surface coil enhanced MR is an excellent way to evaluate ulcerated plaque but requires hands-on study to optimize planes of section and flow suppression. The best approach presently is for the radiologist to describe the plaque as smooth or irregular, and if niche is present, the term ulceration can be used. It is in the province of the physician caring for the patient to base therapy on the severity of findings and on the patient's symptoms.

Perfusion

Perfusion imaging aims to characterize microscopic flow at the capillary level. The key concept to remember in perfusion imaging is the central volume principle:

$$CBF = CBV/MTT$$

> **BOX 3-2** Conventional Angiographic Findings in Hemodynamically Significant Lesions of the Extracranial Carotid Arteries
>
> Lesions with 50%-70% reduction of vessel lumen diameter
> Less than 2 mm residual lumen corrected for angiographic magnification
> External carotid artery opacification contributing to internal carotid artery opacification
> Delayed ocular choroidal blush (>5.6 sec for patients over 30 years of age) with injection of contralateral carotid or vertebral arteries, angiographic filling of ipsilateral carotid circulation

Cerebral blood flow (CBF) is determined by the ratio of cerebral blood volume (CBV) divided by the mean transit time (MTT). The CBF of the normal brain ranges between 45 and 110 mL/min/100 g of tissue. Cerebral oligemia (about 20 to 40 mL/min/100 g) is defined as underperfused asymptomatic region of brain that will recover spontaneously, whereas an ischemic hypoperfused brain is symptomatic and at risk to develop irreversible infarct without revascularization. The ischemic threshold identified in animal experiments when there is cessation of action potential generation occurs around 20 mL/min/100 g and the infarction threshold, associated with irreversible neuronal damage, is at approximately 10 mL/min/100 g. Therefore, ultimately it is CBF that determines whether tissue will live or die, but changes in MTT and CBV reflect the pathophysiologic processes that precede and then determine when CBF decreases to nonviable levels. The initial event is an increase in MTT because of an occlusion or stenosis. MTT will be determined by the site of occlusion or stenosis and the presence and type of collateral supply to the affected brain. The autoregulatory response of the brain is vasodilatation of vascular bed distal to the occlusion or stenosis and increased oxygen extraction from the blood. Vasodilatation increases CBV, and therefore initially CBF is maintained or at least does not decrease to the level where neuronal death occurs. However, once maximal vasodilatation is achieved, any further increases in MTT (because of progressive occlusion, new embolization, or decrease in systemic blood

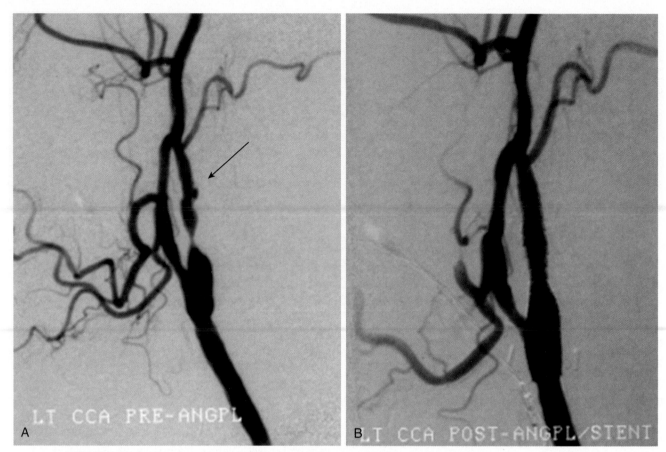

FIGURE 3-14 A, Common carotid angiogram showing high-grade stenosis of the left internal carotid artery. Notice the ulceration in a distal plaque *(arrow).* **B,** The patient underwent an angioplasty and stenting procedure. Observe the improved flow and the obliteration of the ulcer by the stenting. (Courtesy P. Kim Nelson, MD.)

pressure) will result in decrease in central perfusion pressure, collapse of the vascular bed and decrease in CBV and consequent decrease in CBF.

Perfusion imaging can be performed in a number of ways, but by far the most common technique in clinical practice involves an intravenous injection of contrast material on CT or MRI. Rapid sequential imaging of all or part of the brain allows the visualization of the effect of the contrast agent as it traverses the vascular system. This "bolus tracking" technique is used for both MR perfusion (MRP) and CT perfusion (CTP). In CTP the density of the brain increases while the iodinated contrast agent passes the vascular supply, and with MRP the intensity of the brain decreases because the paramagnetic gadolinium agent causes T2 shortening (dynamic susceptibility imaging). In both cases one obtains direct measurement of CBV (it is the area under the curve of the density/intensity change). The time that it takes the contrast to traverse the brain is the MTT, and therefore the CBF can be calculated using the central volume principle. However, to precisely measure CBV and MTT it is necessary to eliminate the contribution of contrast within small arterioles and venules. This requires mathematical "deconvolution" of the data. This is easy with CTA where data from the arterial input and venous output (obtained by measuring the changes in density within large arteries such as the anterior cerebral arteries and large venous channels such as the superior sagittal sinus) can be obtained. With MR this is more difficult because of the contribution of flow effects within

large vessels. Therefore the values obtained from CTA are precise mathematical measures of the three perfusion parameters, whereas those obtained with MRP are relative values (e.g., rCBV, rCBF, rMTT). With both CTA and MRA, parametric "MTT," "CBV," and "CBF" maps are generated and evaluated qualitatively. Advances in software now allow for quantitative measurement of these parameters as well.

The parametric maps provide somewhat different information and each has its advantages and limitations. Because the initial event in an infarct is increase in MTT, the MTT maps are the most sensitive to early ischemic changes, but because not all areas of elevated MTT go on to infarction, MTT maps tend to overestimate final infarct volume. What measure best correlates with the size of the final infarct? It depends upon many factors including what literature you read. CBV maps appear to have the best correlation with the ultimate infarct volume. However, this is controversial, with some reports indicating that relative CBV (rCBV) underestimates final infarct volume, whereas relative CBF (rCBF) can more accurately estimate it or overestimate it. Such differences may, in part, be related to when the measurement is made (12 hours versus 24 hours) and also by the type of postprocessing techniques employed.

Perfusion imaging is critical to determining whether or not there is salvageable brain that can be protected by use of intravenous (IV) or intraarterial thrombolytic (tPA) or clot retraction/removal therapy, or medical therapy (Figs. 3-15, 3-16). All of these treatments are associated with an increased risk of intracranial hemorrhage and therefore

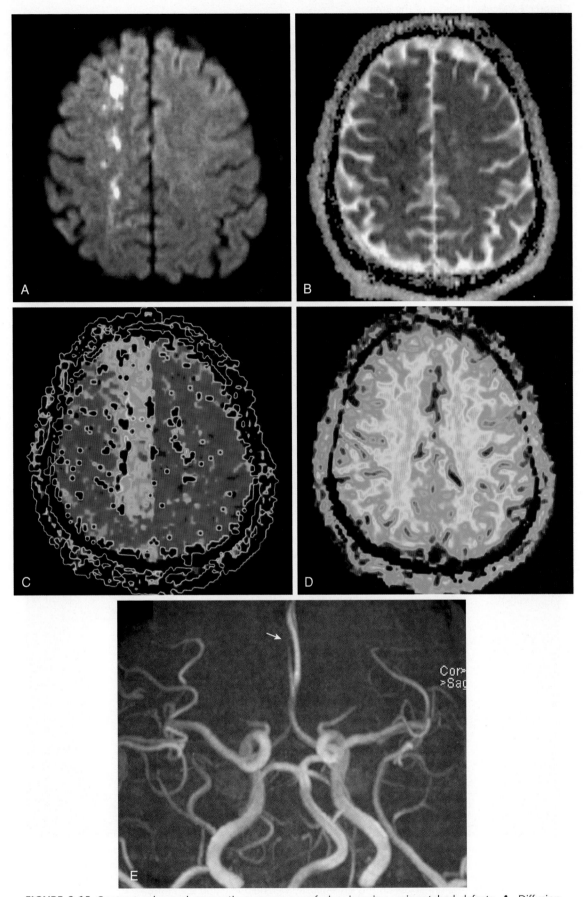

FIGURE 3-15 Contrast-enhanced magnetic resonance perfusion imaging, mismatched defects. **A,** Diffusion-weighted imaging and **(B)** apparent diffusion coefficient maps show multiple acute infarcts in the right anterior cerebral artery (ACA) distribution. Relative mean transit time (rMTT) maps **(C)** show prolonged relative mean transit time in the entire right ACA distribution, while relative cerebral blood flow (rCBF) maps **(D)** show no asymmetry, indicating that the uninfarcted right ACA territory is at risk of infarct. **E,** Three-dimensional time-of-flight magnetic resonance angiography on this patient shows an abrupt occlusion in the right A2 segment *(arrow),* compatible with thrombus.

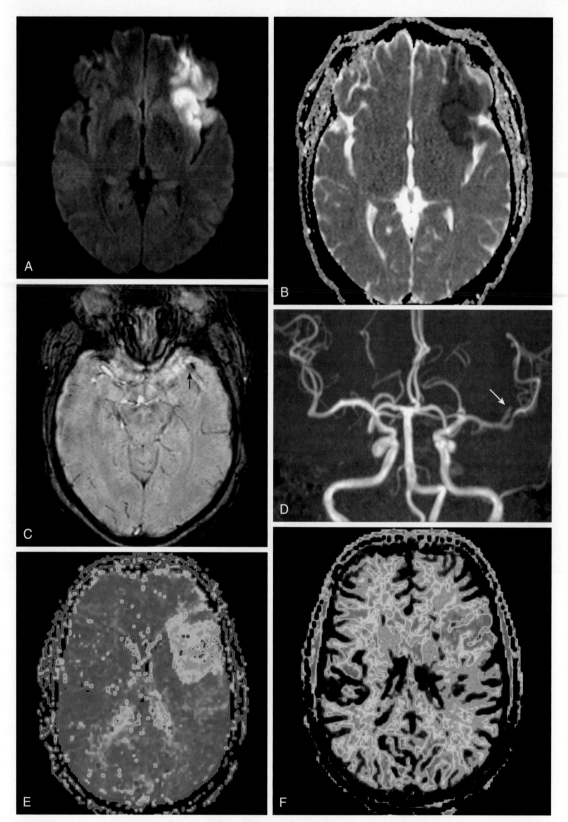

FIGURE 3-16 Contrast-enhanced magnetic resonance perfusion imaging, matched defects. **A,** Diffusion-weighted imaging (DWI) and **(B)** apparent diffusion coefficient maps in this patient show acute infarct in the anterior left middle cerebral artery (MCA) distribution. **C,** Thrombus is evident in the anterior division branch of the left MCA as focal signal loss on this susceptibility-weighted imaging scan *(arrow)*. **D,** Three-dimensional time-of-flight cranial magnetic resonance angiography shows abrupt occlusion of this vessel *(arrow)*. Relative mean transit time (rMTT) maps **(E)** and relative cerebral blood flow (rCBF) maps **(F)** show matched defects corresponding to region of acute infarct on DWI. There is no at-risk brain tissue, and aggressive clot retrieval is therefore not warranted in this case.

treatment should be reserved for individuals who can benefit from recanalization. Individuals in whom the area of infarction matches the area of abnormal perfusion should not be treated regardless of other factors (time from onset of symptoms, extent of infarcted brain) because there is no brain tissue to protect. On the other hand, in patients where volume of brain at risk is greater than the already infarcted brain by more than 20%, treatment may result in improved outcome. More recent data indicate that the extent of collateralization of distal branches beyond the occlusion has a major impact on outcome as well (the better the collaterals, the better the prognosis), and must also be taken into consideration when triaging patients for recanalization. However, large volumes of at-risk brain tissue are also susceptible to complications of reperfusion hemorrhage, and therefore should be considered for potential treatment only with great caution.

The brain at risk is described as the ischemic penumbra. On MR, the penumbra is the brain tissue surrounding the core diffusion "positive" (hyperintense on DWI) infarcted brain that has normal diffusion but abnormal relative perfusion (diffusion-perfusion mismatch). On CT there is no easy direct way to measure the extent of the already infarcted brain, and therefore it is necessary to use quantitative measures of perfusion to define the predicted infarcted brain (<10 mL/min/100 g) and the penumbra (10 to 30 mL/min/100 g; Fig. 3-17).

There are other methods of measuring brain perfusion that deserve brief attention. Nuclear medicine studies including positron emission tomography (PET) and single photon emission computed tomography (SPECT) can be used to generate perfusion maps but have little use in the work-up of acute infarction. Xenon CT involving inhalation of stable xenon gas to act as a contrast agent to estimate cerebral blood flow can be performed quickly but is not readily available in most centers. Arterial spin labeling (ASL) is an MRP technique that requires no exogenous contrast agent (see Fig. 3-40, *E*). In this technique the protons in arteries at the base of the brain are subjected to an MR pulse that inverts their spins. The tagged protons can then be measured as they pass through the brain. In both xenon CT and ASL the perfusion agent (xenon and tagged water molecules, respectively) freely diffuses across the blood-brain barrier and therefore it is possible to directly measure CBF. However, while CBF is the critical determinant of brain tissue viability, knowledge of MTT and CBV allows for understanding of the status of the vascular system, not just the brain. ASL is technologically demanding and not in common clinical use in many imaging centers. One advantage of the technique is that, because no contrast injection is necessary, ASL perfusion studies can be repeated as often as necessary.

Dynamic susceptibility contrast (DSC) enhanced MRP is often measured in CBF and CBV; however, another parameter commonly used in MRP studies is time to peak (TTP) contrast concentration. This refers to the time it takes until the maximum T2 shortening effect of the first-pass gadolinium in the vessel (the lowest signal intensity). When highly concentrated as in an MRP bolus, the primary effect seen is a T2 shortening within the vessel and the subsequent parenchyma from the bolus and perfusion of tissue respectively. TTP maps can be measured directly from the time-intensity curve of an MRP study by comparing the ischemic and the contralateral normal voxels. Using a TTP delay of more than 4 seconds relative to the contralateral hemisphere can correctly identify 84% of hypoperfused and 77% of normoperfused tissue. Other sites are currently using an area under the curve

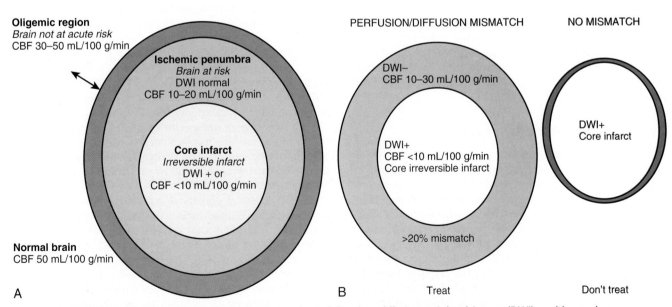

FIGURE 3-17 Ischemic penumbra. **A,** Core infarct defined as diffusion-weighted image (DWI)-positive region (area of irreversible infarction) on magnetic resonance or cerebral blood flow (CBF) <10 mL/100g/min on computed tomography. Ischemic penumbra is the region that is DWI normal with CBF of 10-20 mL/100g/min. There may be a region of relative oligemia (CBF >30 mL/100g/min) that is not at risk for acute infarction but that might be at risk if there is further compromise of the arterial supply. **B,** Thrombolytic therapy is indicated when the mismatch between core infarct and penumbra is more than 20% if all other treatment criteria are met (<4.5 hours' duration, no hemorrhage, infarct <30% of the vascular distribution). If mismatch is less than 20%, thrombolytic therapy should not be undertaken because there is insufficient target brain at risk to warrant treatment.

TABLE 3-1   Magnetic Resonance Imaging Signal Changes in Infarct

	T1	T2	FLAIR	DWI	ADC	GrE/SWI	Postcontrast T1WI
Hyperacute infarct	Isointense	Iso to mildly hyperintense	Iso or mildly hyperintense	Bright	Dark	Signal loss may be seen at site of intraluminal thrombus	Slow flow distal to clot may be seen
Acute infarct	Isointense	Hyperintense	Hyperintense	Bright	Dark	Signal loss indicates parenchymal hemorrhage	Slow flow distal to clot may be seen, leptomeningeal enhancement
Early subacute infarct	Mildly hypointense	Hyperintense		Variable	Variable	Signal loss indicates parenchymal hemorrhage	Parenchymal gyriform enhancement
Late subacute infarct	Hypointense	Hyperintense	Hyperintense	Mildly bright	Bright	Signal loss indicates parenchymal hemorrhage	Parenchymal gyriform enhancement
Chronic infarct	Hypointense (cortical hyperintensity indicates laminar necrosis)	Hyperintense	Hyperintense gliosis, hypointense cystic encephalomalacia	Isointense to bright	Bright	Signal loss indicates chronic hemorrhage	No enhancement

ADC, Apparent diffusion coefficient; *DWI,* diffusion-weighted imaging; *FLAIR,* fluid-attenuated inversion recovery; *GrE,* gradient echo; *SWI,* susceptibility-weighted imaging; *T1WI,* T1-weighted imaging.

Tmax parameter between 5 and 6 seconds to identify the ischemic penumbra. A Tmax bolus delay of 6 seconds has become the accepted threshold for the definition of a relevant hypoperfusion in recent stroke thrombolysis trials (EXTEND, EPITHET).

IMAGING OF INFARCTION

CT and MR findings change rapidly in the initial week after an infarct reflecting underlying relatively stereotypical pathophysiologic changes. In this section we will describe each phase based on time from infarction, predominant underlying event and CT and MR imaging findings (Table 3-1).

Hyperacute Infarction (0 to 6 hours)

The initial event that leads to infarction is vascular insufficiency because of focal proximal or distal occlusion or stenosis. In most instances routine imaging will not demonstrate the occlusion except when there is embolic occlusion of large vessels (e.g., dense MCA or basilar artery sign). Vascular occlusion leads to decreased perfusion, which when sufficiently severe and/or sufficiently prolonged initiates the "ischemic cascade." Within 5 minutes of hypoxia, the membrane pumps that maintain the disparity between the normal high concentration of extracellular sodium and the lower intracellular sodium fail. Sodium enters the cell and the influx of sodium produces an osmotic gradient and water passively enters the cell creating "cytotoxic" edema. In addition, calcium enters the cell, which in turn activates intracellular enzymes that begin to lyse intracellular organelles and precipitates proteins. This produces cell lysis and the release of excitatory amino acids (glutamine and glutamate) and vasoactive substances, which further compromise the metabolic status of adjacent cells. Hence apoptosis.

CT: During the hyperacute phase, CT may be normal or may demonstrate the "dense vessel" sign when there is an embolic occlusion of a proximal vessel (see Figs. 3-7, 3-18). The dense vessel can sometimes be more conspicuous on thinner section CT imaging (i.e., <1 mm thick cuts). The initial parenchymal finding is loss of normal gray matter density without mass effect. The gray matter becomes isodense to adjacent white matter leading to loss of the normal "cortical ribbon" and/or loss of the ability to differentiate the basal ganglia and/or thalamus from the internal capsule (Figs. 3-19, 3-20). Loss of cortical density may occur as early as 3 hours but more typically takes 4 to 6 hours to develop. This finding is subtle and is often missed by inexperienced observers and those without the advantage of the specific clinical history. The advent of PACS reading stations has facilitated detection of hyperacute infarction. One can improve detection of loss of gray matter density by narrowing the window on CT images thus accentuating gray/white density differences (Fig. 3-21).

Loss of cortical density is typically described as cytotoxic edema. Although it is true that cytotoxic edema is occurring, it is likely that the loss of normal gray matter density is not a direct result of this process. We think of edema as hypodense because the most common cause of brain edema is disruption of the blood brain-barrier (vasogenic edema) leading to increased tissue water that in turn produces hypodensity. In cytoxic edema there is shift of water from the extracellular space to the intracellular space without an increase in the total amount of tissue water. In addition, at this stage of infarction there is often little or no hyperintensity on FLAIR and T2WI (see later). Because T2WI is much more sensitive than CT to changes in tissue water, it is unlikely that subtle changes in water would be detected on CT and not MR.

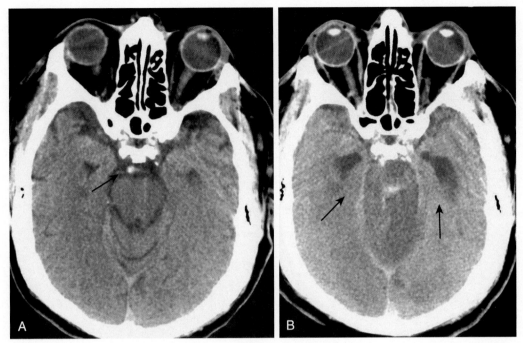

FIGURE 3-18 Hyperacute infarct computed tomography—dense vessel sign. **A,** Hyperdensity in the basilar tip is present at 4 hours without other evidence of infarction *(arrow)*. **B,** Repeat examination at 24 hours reveals persistent basilar hyperdensity with new edema of the brain stem and left superior cerebellum, indicative of acute infarction. Note hydrocephalus with dilated temporal horns *(arrows)* secondary to acute cerebellar infarct.

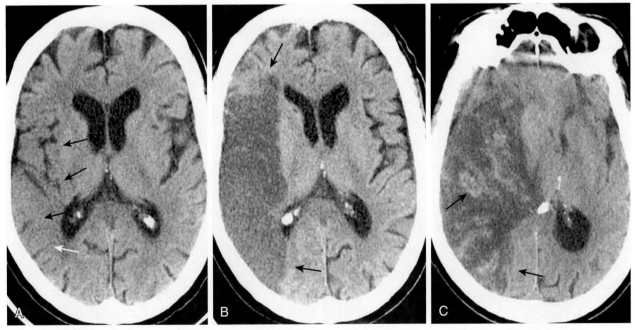

FIGURE 3-19 Hyperacute–subacute infarct computed tomography. **A,** Initial scan at 4 hours reveals subtle loss of normal cortical density along the insula ("insular ribbon sign") and the convexity gyri *(black arrows)*. Note that sulci are visible because there is no mass effect *(white arrow)*. **B,** Repeat examination at 36 hours reveals absolute uniform hypodensity of gray and white matter of the right middle cerebral artery distribution. Mass effect is present with sulcal obliteration. Margins of infarct are more discrete *(arrows)*. **C,** Repeat examination at 4 days reveals marked increase in mass effect with transfalcine herniation. Streaky hyperdensity within the infarct represents reperfusion hemorrhage *(arrows)*.

A more likely cause of the initial changes on CT is decreased cerebral blood volume. Gray matter is denser than white matter in large part because it has a higher blood volume. Decreased blood volume renders gray matter isodense to white matter. This concept helps to explain several observations concerning acute infarction. For instance, it typically takes approximately 24 hours for ganglionic hypodensity to be seen in acute anoxic injury (e.g., smoke inhalation and near drowning). This relative delay in development of hypodensity likely reflects the

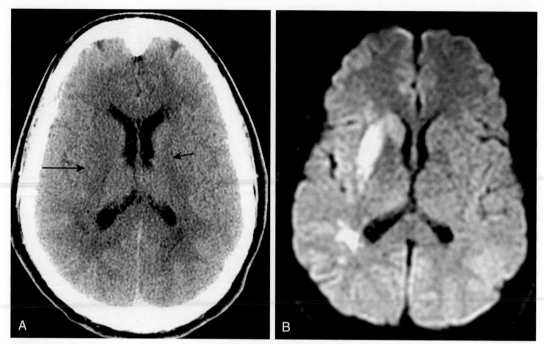

FIGURE 3-20 Acute ganglionic infarct. **A,** Computed tomography (CT) scan at approximately 4 hours reveals relative hypodensity in the right basal ganglia *(long arrow)* compared with left *(short arrow)*. The normally hypodense internal capsule visible on the left *(short arrow)* cannot be differentiated from adjacent basal ganglia because of this hypodensity. **B,** Diffusion-weighted magnetic resonance scan approximately 1 hour after CT reveals obvious hyperintensity.

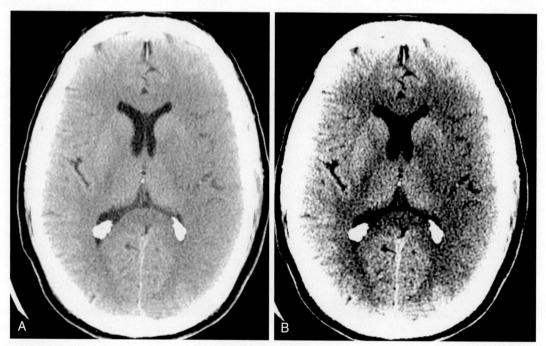

FIGURE 3-21 Use of narrow stroke windows. **A,** A computed tomography scan at approximately 5 hours reveals loss of normal gray matter density in the left insula, left cortical gyri, and lateral basal ganglia (note the inability to identify left internal and external capsule white matter). **B,** The same section with narrow windows improves visualization of loss of normal gray matter density.

fact that in anoxic injury there is no decrease in blood flow, but rather there is a decrease in blood oxygen level. It has been observed that infarcts that become apparent on CT within 4 hours of symptom onset have a worse prognosis than similar sized infarcts that do not become apparent until 6 to 12 hours. This most likely is the result of the more profound perfusion deficit that must be necessary for these infarcts to become apparent in the first few hours. One way to improve infarct detection is to evaluate CTA "source" images. The normally perfused gray matter becomes hyperdense compared with the underperfused infarcted brain (Fig. 3-22).

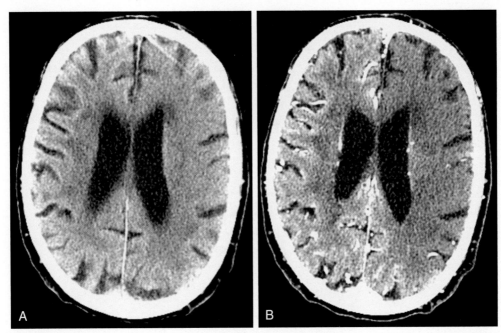

FIGURE 3-22 Computed tomography angiography (CTA) source images in the detection of hyperacute infarction. **A,** Unenhanced scan reveals subtle loss of normal gray matter density in the left middle cerebral artery (MCA) distribution. **B,** Source image from CTA reveals obvious relative hypodensity in the left MCA and anterior cerebral artery distributions. Lesion is more conspicuous and extensive than on unenhanced scan.

Although large infarcts in the middle cerebral artery distribution can be detected within 6 hours in about 75% of cases (at least by experienced readers) the overall sensitivity for detection of all infarcts by CT is only 45% in the first 24 hours. The low rate occurs because of the poor performance of CT in detecting small cortical infarcts, cerebellar infarcts and white matter infarcts. In addition, even when an infarct is detected, its true extent is difficult to determine. One of the major contraindications to the use of IV tPA is large infarct size (infarcts that involve more than one third of the MCA distribution or 70 mL or core infarct). Therefore noncontrast CT is not in and of itself an accurate tool for assessment of infarction, in particular if IV or intraarterial therapy is contemplated. Its major role is to identify hemorrhagic infarctions and hyperdense clots in vessels and to exclude processes such as nonischemic hemorrhage (e.g., hypertensive hemorrhage), masses or infections presenting as stroke. Note that the vast majority of infarcts in the first 24 hours are not hemorrhagic.

Some neurologists and interventionalists find the Alberta stroke program early CT score (ASPECTS) useful in patient selection for thrombolytic therapy in patients presenting with MCA strokes. Starting with an initial score of 10, 1 point is deducted if infarction in the following structures is evident on the initial unenhanced CT: caudate, putamen, internal capsule, insular cortex, frontal operculum (M1), anterior temporal lobe (M2), posterior temporal lobe (M3), anterior MCA territory superior to M1 (M4), lateral MCA territory superior to M2 (M5), posterior MCA territory superior to M3 (M6) (Fig. 3-23). An ASPECTS score of 7 or under predicts worse functional outcome as well as symptomatic hemorrhage.

CT evaluation of infarction in the hyperacute phase should be performed in conjunction with CTA of the head and neck. CTA can demonstrate the presence and location of stenosis, tandem stenoses or occlusion. CTP can also be performed to determine if there is viable brain that can be saved by thrombolytic therapies.

MR: There are several MR findings indicative of flow-limiting stenosis that can be detected even without the benefit of MRA. Remember that the typical intraluminal hypointensity is a result of flow effects rather than the intrinsic signal of blood. Blood is a proteinaceous fluid that is relatively T1 isointense and T2 hypointense. After contrast administration, slow flowing blood becomes T1 hyperintense. The MR correlate of the CT dense vessel sign of acute embolic occlusion is focal "blooming" at the level of intraluminal thrombus (marked hypointensity that often extends beyond the lumen of the vessel) on GrE or SWI. If diffusion images are reviewed carefully, restricted diffusion at the site of the thrombus can often be appreciated as well. Chronic occlusion and/or extremely slow flow in large vessels (e.g., cavernous carotid artery) is manifested by isointensity to hyperintensity on T1WI and hyperintensity on T2WI rather than typical dark flow voids seen on T2WI. In the presence of proximal occlusion or severe stenosis, intraluminal hyperintensity is present on FLAIR images in distal branches because of slow flow (see Fig. 3-8). If contrast is given, intraluminal hyperintensity will be more extensive distal to an occlusion than in normal circulation.

With regard to parenchymal signal changes on MRI, hyperacute infarction is T1 isointense and T2 isointense to mildly hyperintense. T2 hyperintensity is best appreciated on FLAIR (sometimes only in retrospect) and is typically confined to the gray matter in thromboembolic infarction. In the first 24 hours, FLAIR hyperintensity is seen in approximately 80% of cases, but it is seen in less than two thirds of cases studied within 6 hours (Figs. 3-24, 3-25). DWI increases the sensitivity for detection of acute infarction to greater than 90% in the hyperacute (<6 hours) period. DWI hyperintensity with ADC map hypointensity can be seen within minutes of the onset of ischemia in animal models and in clinical cases where patients have the misfortune of developing an infarct during or just preceding the MR exam.

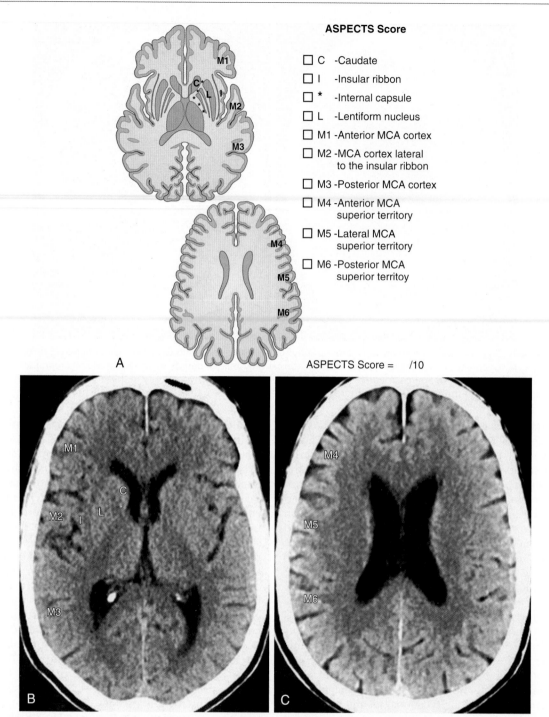

FIGURE 3-23 Alberta stroke program early computed tomography scoring (ASPECTS). Schematic **(A)** and anatomic **(B** and **C)** axial noncontrast computed tomographic images at the level of the basal ganglia and corona radiata show the specific regions to be assessed for involvement by infarct to provide ASPECTS results. Starting with an initial score of 10, 1 point is deducted if infarct is observed in the following structures caudate (C), lentiform nucleus (L), internal capsule, insular cortex (I), frontal operculum (M1), anterior temporal lobe (M2), posterior temporal lobe (M3), anterior middle cerebral artery (MCA) territory superior to M1 (M4), lateral MCA territory superior to M2 (M5), posterior MCA territory superior to M3 (M6). A score of 7 indicates that the patient is not an ideal candidate for thrombolytic therapy.

These early changes are the result of cytotoxic edema. In the vast majority of cases the restricted diffusion is an indicator of irreversible neuronal damage and death.

In less than 5% of cases diffusion changes are reversible (thank goodness for small miracles!). In most of these instances there is early (<3 hours) spontaneous or therapeutic recanalization of the occluded vessels. The initial ADC reduction is often less than that seen in most infarcts (~25% as compared with >75%). It should be noted that in some of these cases subsequent exams reveal return of restricted diffusion and DWI hyperintensity. These unusual cases probably represent examples of borderline ischemia. If flow is rapidly reestablished, damaged cells may return to normal metabolic function at least transiently. These

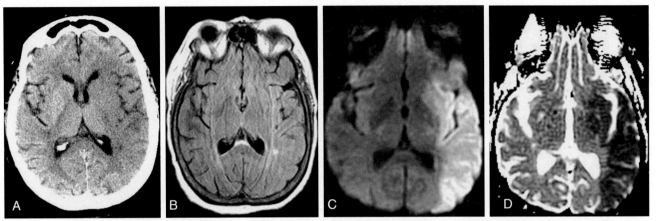

FIGURE 3-24 Hyperacute infarction. Computed tomography (CT) can be better than fluid-attenuated inversion recovery (FLAIR). **A,** CT scan at 3 hours reveals loss of normal gray matter density in the left basal ganglia insula and frontal and parietal cortex without mass effect. **B,** FLAIR image obtained at 4 hours reveals no hyperintensity in the affected area. **C,** Diffusion-weighted image reveals extensive ganglionic and cortical hyperintensity indicative of hyperacute infarction. **D,** Apparent diffusion coefficient map reveals diffuse hypointensity indicative of restricted diffusion.

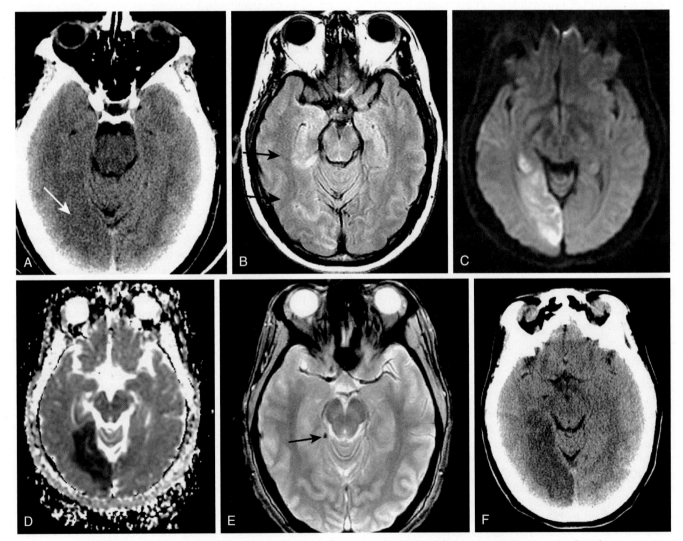

FIGURE 3-25 Hyperacute embolic infarct. Computed tomography (CT) at 3 hours, magnetic resonance imaging at 3 hours 30 minutes. **A,** CT reveals loss of gray-white differentiation in the right occipital lobe *(arrow)*. **B,** Fluid-attenuated inversion recovery image reveals subtle T2 hyperintensity in the temporal and occipital gyri *(arrows)*. **C,** Diffusion-weighted image reveals hyperintensity in the affected gyri. **D,** Apparent diffusion coefficient map reveals obvious hypointensity, indicative of restricted diffusion. **E,** A gradient echo scan reveals focus of susceptibility hypointensity in the region of the posterior cerebral artery branch, indicative of intraluminal clot *(arrow)*. **F,** Follow-up CT scan at 30 hours reveals more pronounced hypodensity and sulcal effacement.

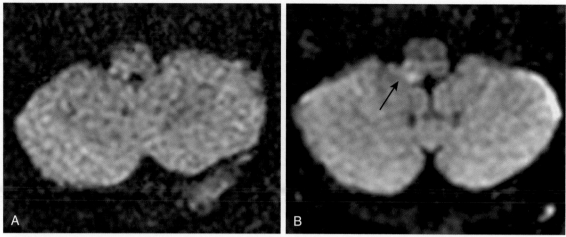

FIGURE 3-26 Normal appearing diffusion-weighted image (DWI) in setting of acute infarct. **A,** Initial DWI shows no acute infarct despite clinical certainty of stroke in a patient with lateral medullary syndrome at 8 hours. **B,** Repeat exam at 24 hours reveals subtle DWI hyperintensity in the right posterolateral medulla *(arrow)*.

cells may fully recover, but many will go on to cell death (with return of DWI hyperintensity) because of continued ischemia or apoptosis (programmed cell death). Venous ischemia is another setting where ADC values may reverse with restoration of normal flow patterns.

In 5% to 10% of cases the initial DWI study is normal when an infarct is present (as confirmed by clinical findings and/or subsequent imaging studies; Fig. 3-26). Most of these cases are small inferior brain stem or cerebellar infarcts that are obscured by susceptibility artifact from the skull base. In some cases the normal initial DWI study may be because of pseudo-normalization. DWI sensitivity seems to drop in the 8- to 16-hour range, a period of time where partial recovery of cellular function may result in transient resolution of DWI and ADC abnormalities.

As opposed to CT, unenhanced MR is sufficient to detect essentially all hyperacute infarcts. Therefore multimodal MR (MRI, MRA and MRP) is of most value when thrombolytic therapy, clot retraction, or other aggressive medical interventions are contemplated. Craniocervical MRA allows for detection of stenosis, occlusion and dissection of the entire cerebrovascular system. MRP allows for identification of areas of hypoperfusion that might be the target of thrombolytic therapy.

Hyperacute Stroke Management

To appreciate how multimodal CT and MR are used in treatment decisions it is necessary to understand the risks and benefits of those options. It has been shown that IV administration of tPA improves outcomes (e.g., residual disability) in stroke patients in the following circumstances: (1) treatment must begin within 4.5 hours of symptom onset; (2) no CT evidence of hemorrhage; and (3) infarct must not be large (exceed one third of the distribution of the MCA territory or 70 mL). With multimodal imaging, additional criteria have been established including (a) no evidence of occlusion of the distal internal carotid artery and proximal MCA and ACA ("T" occlusion) and (b) the presence of a penumbra of salvageable brain that represents at least 20% of the overall region of abnormal perfusion. Application of these criteria eliminates more than 90% of patients presenting to emergency departments with signs of acute infarction. These criteria are strict

because IV tPA treatment is not benign. The risk of intracerebral hemorrhage, often massive, is high in particular in patients treated too late or in whom an infarct is too large. IV tPA is ineffective if there is no salvageable brain or there is a proximal occlusion that is unlikely to lyse with IV treatment.

Intraarterial (IA) treatment can extend the therapeutic window to up to 8 hours in the anterior circulation and up to 24 hours in the vertebrobasilar or retinal circulation. In addition to extending the time window for treatment, IA therapy can be used to eliminate proximal clots that are unlikely to lyse with IV tPA. IA treatments include direct injection of tPA into the clot via super-selective catheter placement, manual clot disruption, and removal of the clot with a mechanical device.

The process of patient selection for IA treatment continues to evolve as we learn more about the hemodynamics of ischemic strokes. Several major recent trials have shown that in patients presenting with proximal vessel occlusion in the anterior circulation up to 8 hours following stroke onset, endovascular thrombectomy showed improvement in patient disability and function at 90 days posttreatment compared with medical management (including IV tPA) alone. The endovascular treatment for small core and proximal occlusion ischemic stroke (ESCAPE) trial showed that in patients with proximal vessel occlusion and small infarct core but good collateral circulation, prompt endovascular treatment resulted in improved functional outcomes and lower mortality. This trial and other investigations have found that identification of good collateral circulation should be a strongly considered criterion in patient selection for endovascular treatment of stroke and may be more important than time between symptom onset and directed treatment. Mechanical thrombectomy and clot retrieval devices also continue to evolve, with new generation devices showing improvement in outcomes compared with previous generation devices just in the past few years.

Based on these evolving concepts, in the clinical setting of hyperacute stroke with potential for IA therapy, many stroke centers now advocate excluding hemorrhage (noncontrast head CT), assessing core infarct size (noncontrast head CT/ASPECTS versus diffusion MRI), identifying vessel occlusion and presence of adequate collateralization (CTA), and assessing clinical penumbra at the bedside (NIH stroke scale/neurologist evaluation). The addition of perfusion imaging by CT or MR remains of questionable value in the hyperacute stroke setting.

Currently, in those centers that have access to interventional neuroradiology, IV tPA is being followed by intraarterial clot retrieval in the hyperacute (<6 hours setting) setting of anterior circulation strokes with proximal vessel occlusions. Such were the recommendations from the MR CLEAN EXTEND-IA, SWIFT-PRIME, and ESCAPE trials. Most of these trials required demonstration of good collateral flow either by CTA or, by inference, CTP (or MRP).

Acute Infarction (6 to 36 hours)—Cytotoxic and Vasogenic Edema

With continued ischemia, neuronal damage and death (cytotoxic edema) increases. Endovascular cells are damaged resulting in opening of the blood-brain barrier and leakage of fluid into the extravascular space. With increased tissue water, local brain swelling occurs. Red cell extravasation may also occur although hemorrhage is usually absent or mild. Clot within proximal vessels may persist or dissolve and wash "downstream" into distal vessels. Leptomeningeal collateral vessels can dilate to provide some perfusion to the affected brain. The extent and rate at which vasogenic edema develops depends on the blood flow to the affected brain. If there is no reperfusion, edema is mild and takes longer to develop. If flow is rapidly reestablished (spontaneously or because of treatment) and the vascular bed is damaged, edema will develop rapidly and hemorrhage may occur. This is known as reperfusion hemorrhage.

CT: Vasogenic edema produces hypodensity in the affected brain. In thromboembolic infarcts the gray matter becomes hypodense and swollen resulting in sulcal effacement. The hypodensity is initially confined to the gray matter but then progresses with loss of gray-white distinction. It is homogeneous and has well-defined smooth to convex borders. Lacunar infarcts are visible as discrete round to oval foci of hypodensity without mass effect. When these areas of hypodensity are subcentimeter in size it can be difficult to distinguish between acute, subacute, and chronic lacunar infarcts. Hemorrhage is typically not detected in the acute stage. Clot within a proximal vessel will still be visible. There is no parenchymal enhancement at this stage of infarct evolution.

MR: T1 isointensity and T2 hyperintensity (best appreciated on FLAIR) are present in the infarcted brain. In thromboembolic infarcts the T2 hyperintensity is confined to the gray matter (Fig. 3-27). Focal swelling and sulcal effacement are present. The infarct is DWI hyperintense and there is ADC hypointensity (indicating reduced ADC) indicative of restricted diffusion. While the extent and degree of T2 hyperintensity increases during the acute phase of infarction, the extent of DWI abnormality remains relatively stable unless there is actual progression of the infarct. DWI lesion volume measured within 48 hours has been suggested to be a reasonable predictor of outcome in stroke. Lacunar infarcts present as foci of T1 isointensity and T2 hyperintensity. As is true with CT, it is difficult to distinguish acute from chronic infarcts on T2WI without DWI, in particular when there is a background of white matter T2 hyperintense foci from small vessel chronic ischemic injuries. DWI makes detection of acute lacunar infarcts simple because the acute lesions are hyperintense while chronic white matter lacunar infarction is DWI isointense. Hypointensity on susceptibility-weighted sequences (GrE and SWI) may be present, indicative of hemorrhage (Fig. 3-28). Several studies have shown that susceptibility-weighted sequences are more sensitive than CT in detecting subtle hemorrhagic transformation of infarction. In addition, these sequences allow for detection of chronic petechial hemorrhages (microhemorrhages) typically the result of hypertension and/or cerebral amyloid angiopathy (CAA). The presence of microhemorrhages implies vascular fragility and carries an increased risk of hemorrhagic transformation of infarction and an increased risk of future infarction. Evidence of proximal intraluminal clot and/or slow flow (see above) can be seen at this stage of infarction although with slightly lower frequency than in hyperacute phase, because of clot resolution. If contrast is given, slow flow within vessels distal to a clot may be present, and sulcal enhancement may be seen as a result of leptomeningeal collaterals.

Early Subacute Infarction (36 hours to 5 days): Reperfusion

Blood flow to the affected portion of the brain is typically reestablished 24 to 72 hours after infarction. Proximal and distal clots are lysed or break up and move downstream. Leptomeningeal collaterals become prominent during this phase. By day 3 or 4, ingrowth of new vessels into the area of infarction commences. These immature vessels have "leaky" blood-brain barriers. As a result of these changes, vasogenic edema increases with progressive mass effect that typically peaks at around 5 days. In large infarcts, mass effect can lead to transfalcine and/or transtentorial herniation. Hemorrhagic transformation most commonly occurs during this phase of infarction. While vasogenic edema increases, cytotoxic edema may actually decrease as neuronal death leads to cell lysis. Of course if there is ongoing ischemia, new areas of infarction with cytotoxic may develop.

CT: The infarcted area is hypodense. In thromboembolic infarcts hypodensity involves both the gray matter and adjacent white matter. Density is more heterogeneous than in the acute phase with streaky mild "gyral" hyperdensity representing either reperfused cortex or hemorrhagic transformation. Frank hyperdense gyriform hemorrhage may occur. The margins of the infarct are indistinct. Mass effect increases. The degree of edema and mass effect is determined by the size of infarction and the extent of arterial recanalization. In severe cases ("malignant infarct") there may be transfalcine and/or transtentorial herniation (Fig. 3-29) for which craniectomy may be required for decompression. Contrast enhanced scans may demonstrate parenchymal enhancement in the infarcted territory. In cortical infarction the enhancement is typically gyriform. In deep gray matter (ganglionic and thalamic), enhancement is often peripheral and may mimic that seen in necrotic masses.

MR: The infarcted brain is mildly T1 hypointense and markedly T2 hyperintense. The T2 hyperintensity involves both gray and white matter and the margins are ill-defined. Differentiation between bland reperfused gray matter and hemorrhagic transformation is straightforward on MR.

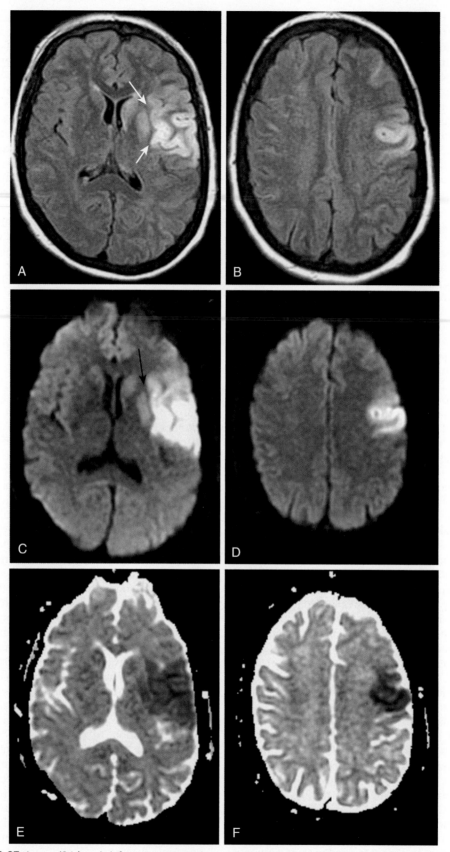

FIGURE 3-27 Acute (24-hour) infarct on magnetic resonance imaging. Fluid-attenuated inversion recovery **(A** and **B)** and diffusion-weighted **(C** and **D)** images reveal hyperintensity in the left basal ganglia and insula with apparent sparing of the subjacent white matter (*arrows* in **A** and **C**). **E** and **F,** Apparent diffusion coefficient maps reveal hypointensity indicative of restricted diffusion.

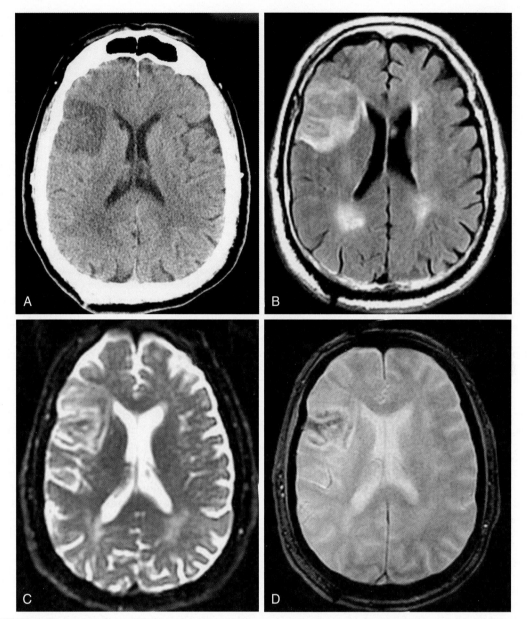

FIGURE 3-28 Acute infarct with hemorrhagic transformation. **A,** Computed tomographic scan at 36 hours reveals a discrete hypodense right frontal middle cerebral artery acute infarct with sulcal effacement. Mild central heterogeneous density is present, but there is no definite evidence of hemorrhage. **B,** Fluid-attenuated inversion recovery reveals heterogeneous hyperintensity with relative isointensity of gyri. **C,** B0 image demonstrates T2 hyperintensity surrounding relatively isointense gyri. **D,** Gradient-echo image demonstrates obvious hypointensity indicative of hemorrhage.

Hemorrhagic transformation produces mild to moderate T2 hypointensity and marked hypointensity on susceptibility-weighted sequences. Intensity on DWI and ADC maps is variable at this stage reflecting the balance of the extent of cytotoxic edema (decreased ADC) and vasogenic edema (increased ADC; Fig. 3-30). In most cases DWI hyperintensity persists, but ADC hypointensity becomes less apparent or resolves if cytotoxic edema decreases and/or there is extensive vasogenic edema. In some cases DWI hyperintensity may decrease or completely resolve during this phase. If contrast is administered, parenchymal gyriform enhancement may be encountered that is similar to that seen on CT. Leptomeningeal enhancement becomes less apparent or resolves.

Late Subacute (5 to 14 days): Resolving Edema and Early Healing

Over time edema is resorbed with resultant decreased mass effect. Macrophages and glial cells enter the area of infarction and begin to remove dead neuronal tissue. Cytotoxic edema resolves. Blood flow is reestablished. Mild reperfusion hemorrhage can occur but symptomatic hemorrhagic transformation is rare.

CT: Density becomes more heterogeneous. The infarct typically remains hypodense. However, as edema resolves and cortical density is at least partially reestablished, there may be transient period when the infarct is isodense to normal brain ("fog effect"; Fig. 3-31). Mass

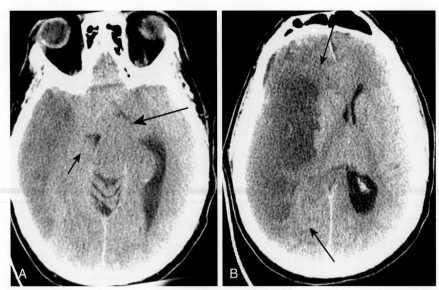

FIGURE 3-29 Subacute "malignant infarct." Computed tomography scan 40 hours after onset of symptoms with progressive obtundation. **A,** Scan at level of suprasellar cistern reveals marked hypodensity in right middle cerebral artery (MCA) distribution. The suprasellar cistern is obliterated *(long arrow)*, the right temporal horn *(short arrow)* is medially displaced, and the left temporal horn is dilated, indicative of transtentorial herniation and trapped ventricle. **B,** Scan at level of lateral ventricles reveals hypodensity throughout the right middle cerebral artery territory with somewhat ill-defined anterior and posterior margins *(arrows)* and marked mass effect with transfalcine herniation.

effect resolves and there may be early evidence of focal atrophy. If significant hemorrhagic transformation has occurred the hemorrhage will undergo typical evolutionary changes. Lacunar infarcts appear as nonspecific foci of hypodensity in the deep gray matter or periventricular white matter. If contrast is administered, parenchymal enhancement often occurs and is increased in extent compared with that seen in the early subacute phase. The presence of enhancement in isodense regions of subacute infarction improves detection but may create a diagnostic dilemma because it may be mistaken for neoplastic or inflammatory disease. As always, clinical information is critical in differentiating between disease processes, in particular if the initial imaging occurs during the late subacute phase of infarction.

MR: T1 hypointensity and T2 hyperintensity persists. There is no MR equivalent with the "fog effect" seen on CT. In thromboembolic infarcts these intensity changes are most marked in the subcortical white matter beneath the infarcted cortex. The overlying infarcted gray matter may be nearly isointense to normal cortex on T1- and T2-weighted sequences. DWI reveals isointensity to mild hyperintensity. ADC maps demonstrate hyperintensity indicative of increased diffusion. Therefore residual DWI hyperintensity is the result of "T2 shine through." Susceptibility-weighted sequences may reveal hypointensity because of subacute to chronic hemorrhage (hemosiderin staining). Since pathologic studies reveal small amounts of hemorrhage in most infarcts, improvements in detection of susceptibility effects (e.g., high-field MR, SWI) will inevitably lead to increased detection of small amounts of hemorrhage that are not clinically significant. Lacunar infarcts are T1 hypointense and T2 hyperintense. DWI hyperintensity has typically resolved although mild residual hyperintensity because of T2 shine through may be present (see Fig. 3-31). ADC maps

reveal increased intensity because of vasogenic, not cytotoxic, edema. Enhancement frequency and pattern are similar to that seen on CT with the same caveats about differentiation between subacute infarction and other disease processes. When the infarct involves the corticospinal tract (e.g., posterior limb of internal capsule), wallerian degeneration occurs resulting in mild T2 hyperintensity and mass effect in the ipsilateral cerebral peduncle and pons that should not be mistaken for an additional area of infarction.

Chronic Infarction (More Than 2 Weeks): Healing

Edema has completely resolved. Dead neuronal tissue is removed and replaced by gliosis and cystic degeneration (cystic encephalomalacia). Infarcted cortex demonstrates pseudo-laminar necrosis (pseudo-laminar because it is not confined to a specific cortical layer). Lacunar infarcts are typically small fluid filled cavities surrounded by zones of gliosis (a "true" pathologic lacune). There is focal volume loss. Depending on the size and location of the infarct, this results in focal cortical atrophy and/or focal dilatation of the adjacent ventricle. If the infarct involves the corticospinal tract there will be wallerian degeneration producing atrophy of the ipsilateral cerebral peduncle and ventral pons.

CT: Hypodensity is present in the infarcted brain. With thromboembolic infarction, this is most marked in the subcortical white matter with portions of the overlying gray matter appearing normal to mildly hyperdense. Note that while the overlying cortex may be normal in density, it is not functional. Cortical CT hypodensity that is present in the hyperacute and acute phase of infarction evolves into subcortical hypodensity with relative cortical hyperdensity. Lacunar infarcts appear as discrete foci of hypodensity

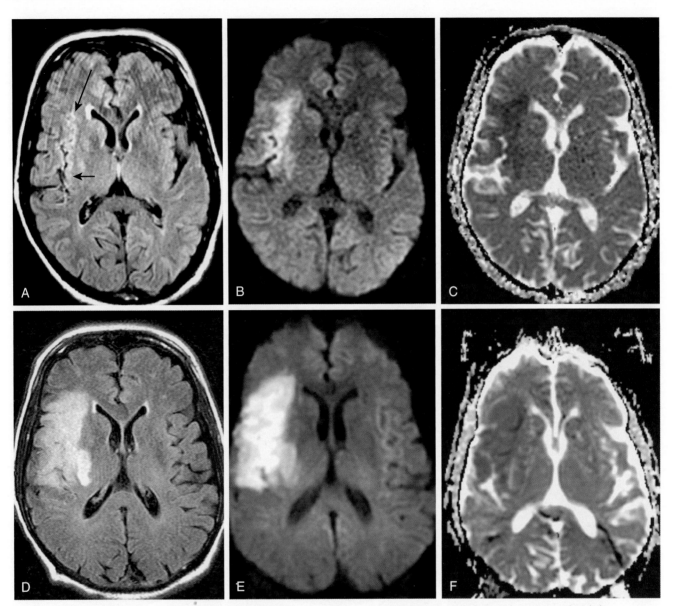

FIGURE 3-30 Early subacute infarct magnetic resonance imaging. **A,** Hyperacute infarct (4 hours) reveals subtle hyperintensity in the insular cortex *(long arrow)* and intravascular hyperintensity *(short arrow)* on fluid-attenuated inversion recovery (FLAIR). Diffusion-weighted image (DWI) hyperintensity **(B)** and apparent diffusion coefficient (ADC) hypointensity **(C)** are present in the insula and right frontal cortex. **D,** At 3 days the infarct is hyperintense on FLAIR with involvement of both gray and white matter. **E,** Infarct is hyperintense on DWI. **F,** ADC map hypointensity is relatively mild compared with prior examination indicative of resolving restricted diffusion.

that are difficult to differentiate from acute lacunar infarcts and chronic ischemic white matter disease. Focal atrophy leads to sulcal enlargement and/or local ventricular dilatation (Fig. 3-32). Mild enhancement may persist for up to 2 months but more often has resolved by the end of 3 weeks. Wallerian degeneration is manifested by focal atrophy of the ipsilateral cerebral peduncle and ventral pons. Laminar necrosis can lead to mild hyperdensity to the cortical margin of the infarcted tissue.

MR: The infarcted brain is T1 hypointense and T2 hyperintense. The affected cortex is often T1 hyperintense secondary to laminar necrosis (not hemorrhage or calcification; see Fig 3-31). Cystic encephalomalacia appears as a central region of cerebrospinal fluid intensity (T1 hypointense, T2 hyperintense and FLAIR hypointense) surrounded by T2 hyperintensity

representing gliosis that is best appreciated on FLAIR. On DWI chronic infarction is isointense to mildly hypointense. On ADC maps the infarct is hyperintense because of increased diffusion in the hypocellular infarcted brain. Lacunar infarcts have the same intensity characteristics as cystic encephalomalacia albeit on a smaller scale. FLAIR is critical for differentiating chronic lacunar infarcts from chronic ischemic change (absence of central hypointensity with chronic ischemic change) and dilated perivascular spaces (absence of peripheral FLAIR hyperintensity). As on CT, mild enhancement can persist for up to 2 months. Wallerian degeneration produces focal atrophy and minimal T2 hyperintensity in the cerebral peduncle and pons (Fig. 3-33). Laminar necrosis is usually seen as hyperintensity on T1WI ± hypointensity on T2WI. Be careful, without looking at

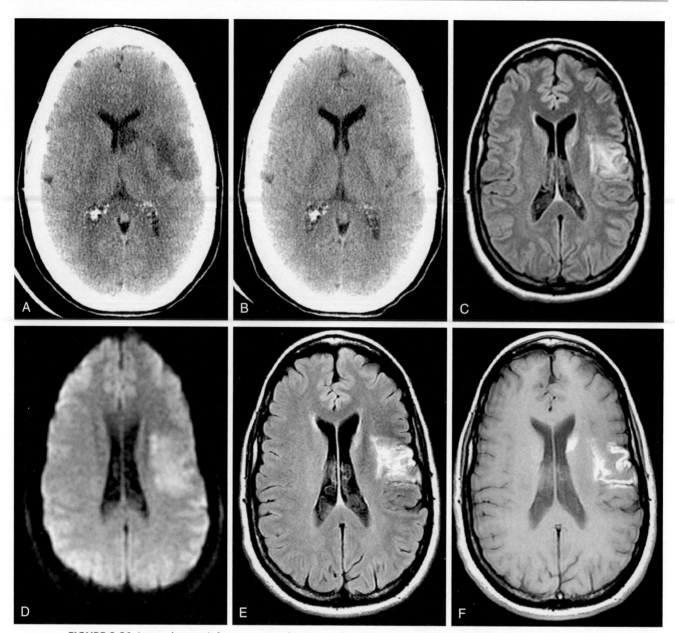

FIGURE 3-31 Late subacute infarct computed tomography (CT) and magnetic resonance imaging. **A,** CT scan 3 days after onset of symptoms reveals focal hypodensity in the left frontal lobe and caudate nucleus with mild mass effect. **B,** Repeat CT at 11 days reveals apparent resolution of hypodensity (fog effect). Fluid-attenuated inversion recovery (FLAIR) **(C)** and diffusion-weighted image (DWI) **(D)** on same day as **B** reveals obvious T2 hyperintensity within the infarct **(C)** with mild residual DWI hyperintensity **(D)** because of T2 shine-through. **E,** Magnetic resonance at 25 days reveals little apparent change on FLAIR. **F,** Note marked T1 cortical hyperintensity within infarct on T1-weighted image because of laminar necrosis, not hemorrhage.

precontrast T1WI the postgadolinium images may look like gyriform enhancement.

Acute Cerebellar Infarcts

Cerebellar infarction is a special case. The frequency is less than 5% and has a male predominance and a mean age of 65 years. The acute clinical findings include the abrupt onset of posteriorly located headaches, severe vertigo, dysarthria, nausea and vomiting, nystagmus, ipsilateral dysmetria, and unsteadiness of gait. Cerebellar infarction can be treacherous, with delayed alteration of consciousness seen in 90% of patients with mass effect. This can occur rapidly

(within a few hours) or up to 10 days after the ictus. The cerebellum swells with (1) an infarction involving more than one third of its volume, (2) a basilar artery occlusion with poor collateral supply, (3) an embolus with reperfusion, and (4) a massive superior cerebellar artery infarction. These infarcts are often difficult to identify on CT because beam-hardening artifact or partial volume averaging masks subtle regions of low density in the posterior fossa. On the other hand, dilatation of the great horizontal fissure and adjacent sulci may produce wedge shaped areas of hypodensity that can be confused with infarction. It is important to visualize the fourth ventricle and quadrigeminal plate cistern because subtle asymmetry may be

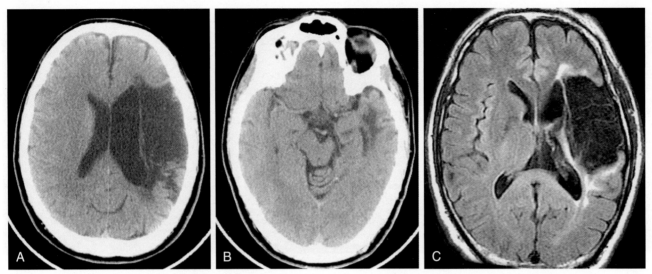

FIGURE 3-32 Chronic infarct computed tomography (CT) and magnetic resonance imaging. **A,** CT scan reveals a large discrete focus of hypodensity in the left frontal lobe. Lesion is more hypodense than acute infarct and it has irregular, somewhat concave margins. There is ex vacuo dilatation of the left lateral ventricle. **B,** CT scan at lower level reveals atrophy of the ipsilateral cerebral peduncle (wallerian degeneration). **C,** Fluid-attenuated inversion recovery scan performed 1 day after CT reveals large fluid collection with T2 hyperintense margins indicative of cystic encephalomalacia.

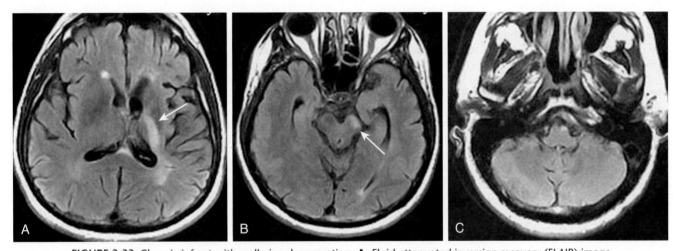

FIGURE 3-33 Chronic infarct with wallerian degeneration. **A,** Fluid-attenuated inversion recovery (FLAIR) image at the level of the basal ganglia reveals T2 hyperintensity in the posterior limb of the left internal capsule *(arrow)* with dilatation of the ipsilateral lateral ventricle. **B** and **C,** Axial FLAIR images at the level of the mesencephalon *(arrow),* and medulla reveal focal atrophy of the cerebral peduncle and T2 hyperintensity of the entire corticospinal tract.

the result of cerebellar edema (Fig. 3-34). Hydrocephalus as manifested by enlargement of the temporal horns (an early sign of obstructive hydrocephalus) may occur and carries a poor prognosis without decompression or shunting. Subtle imaging characteristics (minimal mass effect and slight enlargement of the ventricles) can rapidly evolve to large volume strokes with compression of the brain stem and cerebellar herniation. The superior vermis can herniate upward through the tentorium, whereas the tonsils and inferior vermis may herniate downward into the foramen magnum. Given that death is a highly probable outcome in these types of acute infarcts, there is a bit more leeway in terms of timing of aggressive recanalization; intervention may be performed even up to 24 hours after symptom onset. Treatment of acute cerebellar infarction producing

such mass effect can also involve ventricular drainage and cerebellar/posterior fossa decompression often with bilateral occipital bone craniectomy and/or parenchymal resection.

ANOXIA, HYPOXIA, AND BRAIN DEATH

Anoxic and hypoxic injuries to the brain occur when there is decreased oxygen content of the blood. Anoxic injuries occur when there is near complete absence of oxygen in the blood for more than 5 minutes while hypoxia occurs when there is partial but more prolonged hypoxemia. Clear separation of these entities can be somewhat difficult because a short period of anoxia can give way to hypoxia and both conditions can be complicated by hypotension

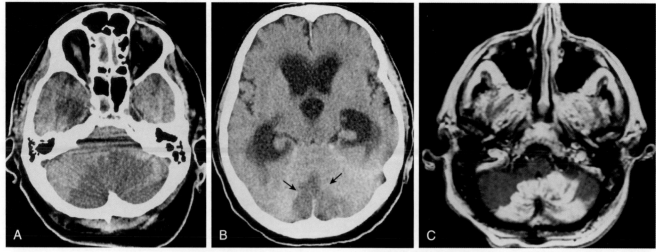

FIGURE 3-34 Acute cerebellar infarction. **A,** Computed tomography scan demonstrates an acute cerebellar infarct having a variegated anterior border, producing significant mass effect with compression of the pons and fourth ventricle. Note complete effacement of the posterior fossa cerebrospinal fluid spaces as well. **B,** Higher section reveals acute hydrocephalus from compression of the fourth ventricle by the cerebellar mass effect. The superior vermis is also involved *(arrows)* and the swollen cerebellum compresses the quadrigeminal plate cistern. **C,** Enhancement on T1 in the infarcted cerebellum is seen.

(either as a cause or effect). Anoxia can be seen in cardiac arrest, prolonged seizures, strangulation/hanging, near drowning and smoke/carbon monoxide inhalation.

In anoxic injuries, it is the metabolically active areas of the brain that are most severely affected, including the basal ganglia and Ammon's horns (dentate nucleus and hippocampus; Fig. 3-35). The earliest CT finding is loss of the ability to differentiate the basal ganglia and thalami from the internal capsules because of subtle loss of normal gray matter hyperdensity. With time, obvious hypodensity and mass effect become apparent. It is important to stress that CT abnormalities take at least 12 hours to become manifest. Thus a normal CT scan obtained after a cardiac arrest does not imply a good prognosis. The delayed appearance (as opposed to that seen in hyperacute infarction) probably results from the fact that cerebral blood volume is maintained (in hyperacute infarction it is diminished), and therefore CT changes are not visible until vasogenic edema has developed. On MR, anoxic injuries are visualized much earlier (3 hours) because of cytotoxic edema and cell death resulting in DWI hyperintensity and restricted diffusion on ADC maps. If the patient survives, chronic anoxic injury results in basal ganglia and hippocampal atrophy with secondary dilatation of the temporal and frontal horns of the lateral ventricles. The frontal horns lose their normally concave contour and become flattened or convex.

Carbon monoxide (CO) toxicity produces anoxic injury by preventing binding of oxygen to hemoglobin. Changes are similar to those seen in anoxia but are most marked in the bilateral globus pallidus. A delayed encephalopathy begins 2 to 3 weeks after recovery and occurs in 3% of patients, resulting in additional findings of high intensity on T2WI in the corpus callosum, subcortical U fibers, and internal and external capsules associated with low intensity on T2WI in the thalamus and putamen. There may be a diffuse anoxic leukoencephalopathy that looks like a dysmyelinating disorder and is associated with mental status deterioration delayed from the CO exposure. Strange as it may seem, this delayed encephalopathy may show restricted diffusion on ADC maps.

With prolonged hypoxia, the basal ganglia and hippocampi are relatively spared. CT will initially be normal, but there is subsequent development of diffuse cerebral edema with loss of definition of gray white interface. MR reveals T2 and DWI hyperintensity with restricted diffusion at the gray white junction bilaterally. Because the process is diffuse, it is possible to overlook the changes on DWI, particularly in infants.

In severe cases, anoxia and/or hypoxia progress to the point where there is diffuse edema with sulcal and cisternal obliteration. The increased intracranial pressure produces transtentorial and tonsillar herniation with complete cessation of cerebral blood flow (brain death). On CT the brain is diffusely hypodense with no gray white matter differentiation and is nearly completely featureless. The ventricles are small and the sulci and cisterns are not visible. Beware! The vessels around the circle of Willis and the falx and tentorium remain relatively hyperdense and may be mistaken for subarachnoid and subdural hemorrhage (pseudosubarachnoid hemorrhage). The relatively more normal attenuation of the cerebellar hemispheres can be strikingly conspicuous (dense cerebellar sign; Fig. 3-36).

VASCULOPATHIES

The vasculopathies are an interesting and heterogeneous group of diseases. Vasculopathy is preferred to the traditional term "vasculitis" because some of these diseases do not have an inflammatory component. Vessel changes may be because of endothelial damage and thrombosis produced by circulating antigen-antibody complexes, mural edema, and/or spasm. Inflammation, when present, may be the cause of the vascular process or a late phenomenon occurring as a result of the vascular insult. Prolonged insults may result in fibrosis and fixed narrowing regardless of the initial insult. Many of these diseases have an

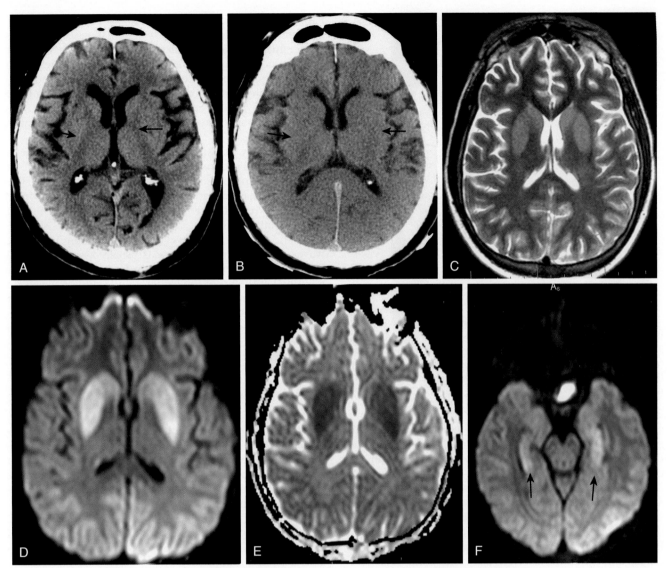

FIGURE 3-35 Anoxic injury. **A,** Computed tomography (CT) scan obtained 8 hours after cardiac arrest reveals normal density of basal ganglia, thalamus, and cortex with good visualization of the normal hypodense internal capsules *(arrows)*. **B,** Repeat examination at 36 hours reveals hypodensity in the basal ganglia and thalami (note inability to identify the internal capsule, *arrows*). Diffuse brain edema is present with early loss of gray matter density and sulcal obliteration. **C-F,** Magnetic resonance (MR) scan at 6 hours in a different patient. CT scan 1 hour before MR was normal. Mild T2 hyperintensity **(C)** is present in the basal and ganglia bilaterally. Diffusion-weighted image (DWI) hyperintensity **(D)** and restricted diffusion **(E)** is present. Hippocampal DWI hyperintensity is also present *(arrows)* **(F).**

immunologic basis resulting in vascular injury. In other cases the inflammation may be extravascular (e.g., brain parenchyma or leptomeninges) with compression of the vessel leading to spasm and then mural inflammation.

Catheter angiography remains the imaging "gold standard" for detection and characterization of vasculopathy. CTA and MRA are capable of documenting proximal and occlusion and stenosis but do not provide sufficient detail for reliable detection or exclusion of vasculopathic changes in secondary or tertiary intracranial vessels (e.g., sylvian and convexity MCA branches). 3T MRA allows for routine visualization of the secondary and tertiary vessels as well as the lenticulostriate arteries and it is therefore the preferred noninvasive angiographic exam for the work-up of vasculopathy. Catheter angiographic studies are often normal (10% of patients undergoing catheter angiography for vasculitis actually have it angiographically

documented) because many of these diseases affect small arteries and arterioles that are too small to be detected even with high resolution catheter angiography. High resolution black blood and postcontrast MRI scans may show thickening and enhancement of vessel walls in some florid cases of vasculitis.

Brain imaging features depend on the location and extent of the vascular pathology as well as systemic abnormalities. "Territorial" (e.g., MCA, ACA) infarcts are seen with proximal vessel involvement while small deep white matter and ganglionic infarcts are encountered when there is involvement of deep perforating arteries. Parenchymal and superficial subarachnoid hemorrhage may occur because of distal arterial disease. Many of the vasculopathies are systemic diseases and therefore laboratory, clinical and imaging evidence of involvement of other organs provide important clues as to correct diagnosis. In cases where the

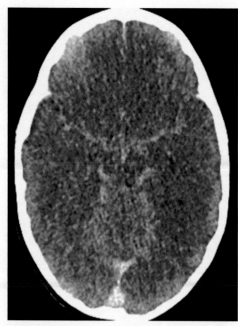

FIGURE 3-36 Pseudosubarachnoid hemorrhage. Diffuse cerebral edema and global parenchymal hypoattenuation along with cisternal effacement results in increased conspicuity of the vessels in the subarachnoid space, mimicking the appearance of subarachnoid hemorrhage.

angiographic studies are normal or nonspecific and the clinical and laboratory findings do not yield a definitive diagnosis, brain biopsy may be necessary to establish the correct diagnosis.

Because of the large number of causes of vasculopathy and the similarity of the appearances of many of these diseases, it is easiest to discuss these processes based on the location of the abnormalities rather than the etiology. Broadly, the vasculopathies can be said to affect (1) extracranial and extradural arteries; (2) arteries at the skull base at or near the circle of Willis; (3) secondary and tertiary branches of the carotid and/or basilar arteries (e.g., sylvian and convexity branches of the MCA); and (4) small perforating arteries (e.g., lenticulostriate arteries). Table 3-2 contains an extensive list of disease processes and potential patterns of involvement for your enjoyment. The following section is a summary of this topic.

Extracranial and Extradural Arteries

Fibromuscular Dysplasia

Fibromuscular dysplasia (FMD) produces nonatheromatous fibrous and muscular thickening alternating with dilatation of the arterial wall, producing an appearance characterized as a "string of beads" (Type 1 FMD; Fig. 3-37). Less common appearances include unifocal or multifocal tubular stenosis (Type 2) or lesions confined to only a portion of the arterial wall (Type 3). Although all layers of the artery may be involved, the media is most commonly affected with hyperplasia producing arterial narrowing and thinning associated with disruption of the internal elastic lamina, producing saccular dilatations. The internal carotid artery, approximately 2 cm from the bifurcation (around C2) is most commonly affected (90% of cases). The vertebral artery is involved in approximately 12% of

FMD cases. Multiple vessel involvement is common (bilateral carotid involvement occurs in 60% of cases), whereas intracranial FMD is rare. Dilated regions are always wider than the normal lumen, and narrowing is usually less than 40% diameter stenosis. Complications of FMD include dissection (which may be difficult to differentiate from FMD alone) and cavernous carotid fistulae. A higher rate of intracranial aneurysms may be due in part to pseudoaneurysm formation.

The etiology of FMD is unknown. The condition has a marked female predominance (4 to 1) with a mean age of 50 years. Symptoms and findings such as headache, TIAs, stroke, vascular dissection, or subarachnoid hemorrhage have been reported. The differential diagnosis of FMD includes atherosclerotic disease, vascular spasm secondary to the catheter, and (on catheter angiography) standing waves. Atherosclerotic disease is usually asymmetric and has a propensity for the bifurcation. Catheter spasm can be identified at the tip of the catheter, and standing waves do not usually have the constrictive picture characteristic of FMD and should resolve on subsequent injections. Dilatation of the vessel is not seen with catheter spasm. When diagnosis is in doubt, evaluation of systemic vessels including the renal arteries may confirm diagnosis.

Skull Base/Circle of Willis

Moyamoya

There is often confusion and inconsistency on the terminology describing the condition of moyamoya vasculopathy (Fig. 3-38). Moyamoya syndrome is an epiphenomenon of numerous vasculopathies that lead to proximal artery stenoses including neurofibromatosis, radiation vasculopathy, severe atherosclerosis, and sickle cell disease. However, in patients with no known associated condition, the imaging findings of bilateral high-grade ICA and proximal branch stenosis with collateral circulation, the term moyamoya disease is applied.

In either case, there is progressive stenosis and then occlusion of the distal internal carotid arteries and their proximal first-order branches (the circle of Willis). Because the process develops over a long period of time and occurs in young patients, extensive collaterals develop to supply the brain distal to the circle of Willis. These collaterals include dural vessels (e.g., external carotid artery to orbital branches of the internal maxillary artery to transethmoid collaterals to the inferior frontal ACA branches), leptomeningeal collaterals from the posterior cerebral arteries (splenial branch to pericallosal artery to distal ACA and MCA territory), and deep perforating lenticulostriate arteries. These collateral vessels produce a classic hazy appearance on angiography termed moyamoya, which in Japanese translates to "puff of smoke."

As cool as the name sounds, the clinical sequelae are not. The disease may be divided into pediatric and adult subgroups on the basis of clinical course and disease features. In children, moyamoya has a more progressive course, presenting with symptoms of cerebral ischemia including TIAs and stroke, whereas in adults intraparenchymal and subarachnoid hemorrhages are the most common presentations. Over time dementia develops because of progressive compromise of the vascular system and chronic hypoxia.

TABLE 3-2 Vasculopathy

Disease	Age (years)	Sex	Etiology	Special Features
EXTRACRANIAL				
Fibromuscular dysplasia	>50	F > M	Unknown	Extracranial internal carotid artery (C2) and vertebral arteries; multiple vessels
Giant cell arteritis	70	F > M	Associated with polymyalgia rheumatica	Extracranial vessels in particular superficial temporal artery
SKULL BASE—CIRCLE OF WILLIS				
Moyamoya disease	10-30	M = F	q25.3, on chromosome 17	Childhood and adult variants
Sickle cell disease	10-20	M = F	Sickle cell	More common in children; transfusions reduce risk; may mimic moyamoya disease
Basal meningitis	5-15	M = F	Tuberculosis and fungal disease	Also affects basilar artery; deep collaterals less common
Cocaine abuse	20-40	M = F	Chronic vasospasm leads to fibrosis	Rare
SECONDARY AND TERTIARY VESSELS				
Inflammatory Granulomatous				
(1) Primary angiitis of the central nervous system (PACNS)	50	F = M	Autoimmune	Middle Eastern decent, brain involvement
(2) Polyarteritis nodosum	30-50	M > F	Autoimmune	
(3) Wegner granulomatosis	20-60	M > F	Unknown	
(4) Sarcoidosis	20-40	F > M	Unknown	
(5) Behçet disease	20-30	M > F	Autoimmune—HLA-B51?	
Infectious				
(1) Herpes zoster	>50	F = M	Spread along fifth nerve from facial zoster infection	Often immune compromised (e.g., HIV)
(2) Tuberculosis and fungal	20-40	M = F	Often in association with basal meningeal disease	
(3) Neurosyphilis	>50	M > F	Late tertiary phase of disease	
Noninflammatory				
(1) Drug related	20-50	M > F	Vasospasm and mural edema; inflammation late, vasospasm edema, eclampsia	Acute hypertension may produce PRES
(2) Pregnancy, puerperium, birth control pills	20-40	F (duh)		Cocaine, amphetamines, sympathomimetic amines (e.g., Ephedrine)
Lymphomatoid Granulomatosis	>50	M > F	Epstein Barr–induced lymphoma	
SMALL VESSEL				
Collagen Vascular Diseases				
(1) Systemic lupus erythematosus (SLE)	20-50	F > M	Autoimmune	Relative sparing of periventricular white matter
(2) Anticardiolipin and antiphospholipid antibody syndrome	20-50	F > M	With or without SLE	Cortical infarcts because of emboli (Libman-Sacks endocarditis)
(3) Sjögren syndrome	40-60	F > M	Autoimmune	
(4) Radiation change	Any	M = F	Fibrinoid necrosis	Confluent white matter disease; months to years after treatment; focal mass like lesions less common
(5) Migraine headache	20-50	F > M	Vasospasm	Few lesions, subcortical frontal lobes
(6) HIV encephalitis	<15	M = F	Inflammatory vasculitis	Deep gray; basal ganglia calcification in children
(7) Susac syndrome	20-40	F > M	Idiopathic	Corpus callosum involvement frequent; lesions smaller than in MS; microinfarcts in cortex
Cerebral Autosomal Dominant Arteriopathy with Subcortical Infarcts and Leukoencephalopathy (CADASIL)	30-50	M = F	NOTCH 3 gene on chromosome 19	Predilection for the anterior frontal and temporal lobes (60%-100%)

HIV, Human immunodeficiency virus; *MS,* multiple sclerosis; *PRES,* posterior reversible encephalopathy syndrome.

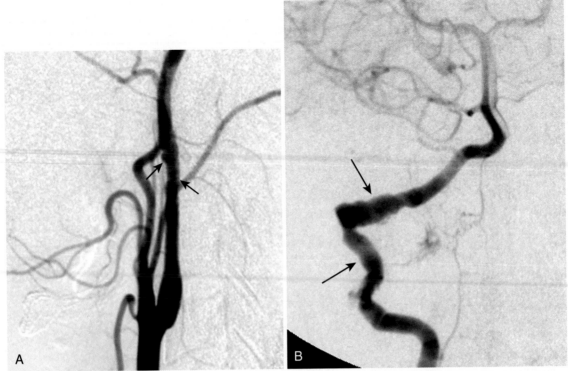

FIGURE 3-37 Fibromuscular disease. "String of beads" appearance of both the internal carotid **(A)** and vertebral **(B)** arteries on conventional catheter angiogram. Abnormal segments are indicated with *arrows*. (Courtesy Philippe Gailloud, MD.)

CTA and MRA are capable of demonstrating the stenotic and/or occluded arteries at the base of the brain and information about collateral supply but these studies cannot provide the necessary detail and time resolved images for full evaluation of the collateral supply to the brain, in particular if revascularization procedures are contemplated. Catheter angiography demonstrates distal internal carotid artery and proximal first-order branch stenoses/occlusions with the extensive collaterals from the vertebral and external carotid arteries. The lenticulostriate and other perforating arteries are dilated and irregular, producing the characteristic moyamoya appearance. On CT, mottled hyperdensity may be visible in the basal ganglia. On MR, foci of T2 hypointensity are seen in the basal ganglia. Sagittal and coronal images reveal these foci to be curvilinear, indicative of the extensive basal ganglionic vascular collateral network that is virtually pathognomonic of this entity. Slow flow luminal hyperintensity may be visible on FLAIR in the affected arteries. Parenchymal changes include deep ganglionic and subcortical infarcts, ganglionic hemorrhages and superficial subarachnoid hemorrhage. Serial studies demonstrate development of atrophy.

Granulomatous Meningitis (e.g., Tuberculosis)

Patients with severe basal cisternal meningeal inflammation may develop proximal stenoses and occlusion because of arterial constriction and spasm. In acute bacterial meningitis this is uncommon and when it occurs it typically leads to massive catastrophic infarction. In chronic granulomatous meningitis, stenosis and occlusion may develop slowly and persist for long periods of time leading to a "moyamoya pattern." However, in these diseases, involvement of the basilar artery and its branches is common and therefore posterior circulation changes occur as well.

In many cases of granulomatous meningitis the disease affects the small perforating arteries that arise within the leptomeninges leading to deep gray matter (basal ganglia and thalamus) infarcts. The combination of communicating hydrocephalus and deep infarction is highly suggestive of granulomatous meningitis.

Secondary and Tertiary Carotid and Basilar Branch Vessels

A large number of disease processes can produce the classic findings of "cerebral vasculitis." Long segments of vessels are involved circumferentially. There is circumferential tapered stenosis in arteries alternating with regions of normal vessel caliber. Multiple short segments of narrowing may produce a "string of beads" appearance. Typically vessels in multiple vascular territories are involved. Focal occlusion may lead to leptomeningeal collateral filling of portions of vascular territory distal to the occlusion. Atherosclerosis (in particular in diabetic and/or hypertensive patients) can also produce narrowing of these arteries, but the stenosis is typically asymmetric and irregular. Vasospasm associated with subarachnoid hemorrhage produces more diffuse narrowing with no intervening regions of normal vessel caliber often centered at arterial bifurcations.

The imaging work-up begins with a CT or MR scan. Presenting symptoms are quite variable and often nonspecific. Patients may present with nonfocal findings such as headache and/or seizures. Focal deficits occur when there is secondary infarction or hemorrhage and depend on the location of these lesions. These studies may reveal one or more cortical infarcts (often hemorrhagic) typically within the distal portions of a vascular territory. Focal parenchymal

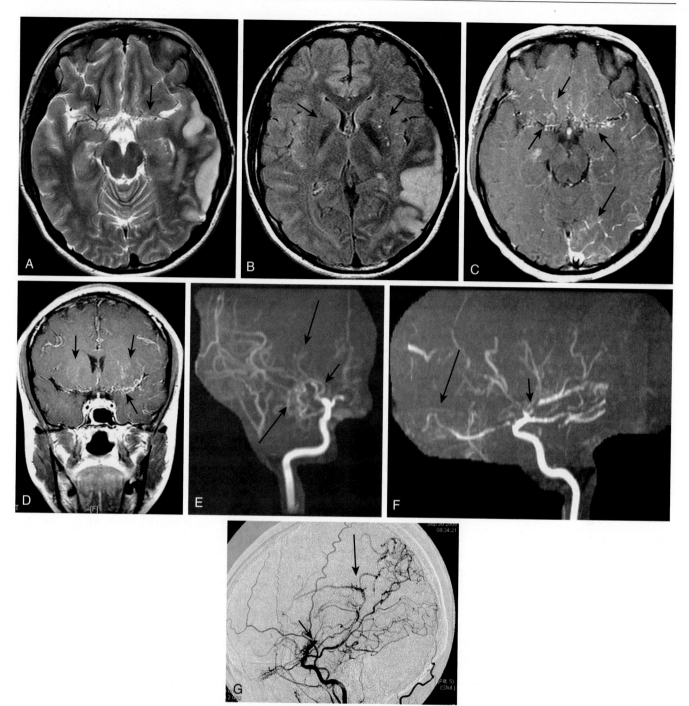

FIGURE 3-38 Moyamoya. **A,** Axial T2-weighted image (T2WI) at the level of the suprasellar cistern reveals absence of normal flow voids of the distal internal carotid arteries. Small arterial branches along the sylvian fissures are present instead of the expected dominant M1 segment flow voids *(arrows).* **B,** Axial fluid-attenuated inversion recovery image at the level of the basal ganglia demonstrates multiple T2 hyperintense foci within the basal ganglia, representing slow flow within dilated lenticulostriate arteries *(arrows).* Axial **(C)** and coronal **(D)** contrast-enhanced T1WI demonstrate multiple small vascular structures along the expected course of the bilateral proximal MCAs. Small serpentine enhancing vessels along the gyrus rectus and dilated lenticular arteries are present *(small arrows* in **C** and *arrows* in **D**). Leptomeningeal enhancement is present in the occipital lobes *(long arrow* in **C**). Frontal **(E)** and lateral **(F)** magnetic resonance angiography images of the right carotid artery demonstrate occlusion of the supraclinoid internal carotid artery *(short arrows)* with numerous small collateral branches in the region of the M1 segment of the MCA *(long arrows* in **E**). Note enlarged ophthalmic artery *(long arrow* in **F**) and poor filling of distal MCA branches. **G,** Catheter angiogram confirms occlusion of the distal internal carotid artery *(short arrow)* with filling of the right posterior cerebral artery and retrograde filling of the pericallosal artery via leptomeningeal collaterals *(long arrow).* Note acute left MCA infarct in **A** and **B**.

hemorrhage and/or subarachnoid hemorrhage may occur. Deep structures (basal ganglia, thalami, brain stem) are less commonly involved. On MR, intraarterial hyperintensity on FLAIR and extensive heterogeneous arterial enhancement on enhanced T1W images may indicate the presence of slow flow. GrE and/or SWI may reveal evidence of microhemorrhage. 3T MRA will typically reveal involvement of secondary arterial vessels (e.g., intrasylvian MCA branches); however, CTA and MRA could be normal too. MR or CT perfusion studies can reveal microcirculatory abnormalities. Catheter angiography remains the gold standard for detecting and characterizing the extent of "vasculitis."

Primary Angiitis of the Central Nervous System

Primary angiitis of the central nervous system (PACNS, aka granulomatous angiitis) affects parenchymal and leptomeningeal arteries with a predilection for small arteries and arterioles (200 to 500 mm in diameter). This can be a rapidly progressive, frequently fatal disease. The sedimentation rate is elevated in more than two thirds of patients, and cerebrospinal fluid (CSF) demonstrates elevated protein and pleocytosis in more than 80% of cases. Often there is an association with varicella zoster virus exposure possibly from the reactivation of the trigeminal ganglion storehouse of viral load.

Wegener Granulomatosis

This is a necrotizing systemic vasculitis which affects the kidneys, upper and lower respiratory tracts. It can affect the brain producing stroke, visual loss, and other cranial nerve problems. The peak incidence is in the fourth to fifth decade with a slight male predominance. History plus positive c-antineutrophil cytoplasmic antibody (c-ANCA) tests help make the diagnosis.

Polyarteritis Nodosa

This is a multisystem disease characterized by necrotizing inflammation of the small and medium size arteries with central nervous system (CNS) involvement occurring late in the disease in more than 45% of cases. It is an immune-mediated disease with about 30% of patients having hepatitis B surface antigen. Polyarteritis is closely related to allergic angiitis and granulomatosis (Churg-Strauss). Aneurysms, which are common in the renal and splanchnic vessels, are unusual in the CNS.

Neurosarcoidosis

This condition can rarely produce a CNS vasculitis characterized by frank granulomatous invasion of the walls of the arteries with or without ischemic changes in the supplied brain parenchyma. These patients usually have a history of systemic sarcoid, although sarcoid can rarely affect only the CNS (see Chapter 5). Sarcoidosis may also cause inflammation of the small veins and lead to microinfarcts from the venous side. The veins are often inflamed in the meningeal form of sarcoidosis.

Infectious Vasculitis

Tuberculosis and haemophilus influenzae infections may on occasion affect secondary and tertiary arteries. Meningovascular syphilis affects both arteries and veins, particularly in the middle cerebral distribution. Herpes zoster infections may spread to the cavernous sinus from the face along the trigeminal nerve branches and then produce an extensive vasculitis with multiple areas of infarction mimicking PACNS. Fungal sinusitis from aggressive mucor or aspergillosis may also affect the cavernous sinuses.

Vasculopathies

The "vasculitis" pattern may also be seen in conditions where inflammation of the vessel wall is not present (based on postmortem studies). Women who are pregnant, peripartum or taking birth control pills can develop changes that have been traditionally thought to be inflammatory but are now believed to be primarily the result of edema and or spasm. The risk of stroke is increased 13 times in pregnancy or puerperium. Conditions responsible for this increase include hypercoagulability, embolism, migraine, vasculitis, and vasospasm, which result in arterial occlusion or venous thrombosis. The same is true for drug related "vasculitis" (e.g., amphetamines and cocaine). Proposed mechanisms for cocaine-induced strokes include increased platelet aggregation with thrombosis, hypertension, direct or indirect arterial constriction, or migraine phenomena induced by the drug. Cocaine induced hypertensive episodes have been thought to be responsible for preexisting aneurysmal rupture and bleeding from arteriovenous malformations. The use of cocaine has been implicated specifically in infarction of the spinal cord and retina, and in intraparenchymal, intraventricular, and subarachnoid hemorrhage. In all of these conditions rapid development of systemic hypertension may lead to the development of posterior reversible encephalopathy syndrome (PRES).

Reversible Cerebral Vasoconstriction Syndrome

Reversible cerebral vasoconstriction syndrome (RCVS) describes vasospasm in cerebral arterial vessels. Numerous predisposing conditions include pregnancy, eclampsia, drug use (both licit and illicit), and recent head trauma/neurointervention, although most cases are idiopathic. About 20% of affected patients have subarachnoid hemorrhage on CSF analysis. Young adult and middle-aged women are most commonly affected. Patients present with recurrent headaches ("thunderclap") and angiography during symptomatic periods can show beaded narrowing of the vessels of the circle of Willis (Fig. 3-39). These areas of narrowing can spontaneously resolve on the order of weeks to months; resolution of findings helps confirm the diagnosis. The brain parenchyma findings may resemble posterior reversible encephalopathy syndrome with confluent white matter lesions predominantly posteriorly that may resolve in a matter of weeks with antihypertensive medications. Subarachnoid blood is also evident on CT and FLAIR.

Lymphomatoid Granulomatosis (Neoplastic Angioendotheliosis)

This very rare malignant lymphoma is restricted to the intracranial vessels. It presents with recurrent strokes or strokelike symptoms, encephalopathy, and seizures. There are multiple high intensity lesions on T2WI in the cerebral white matter often extending along the perivascular spaces with associated enhancement. Angiography shows evidence of medium-sized vascular occlusions. Brain biopsy is needed to establish the diagnosis.

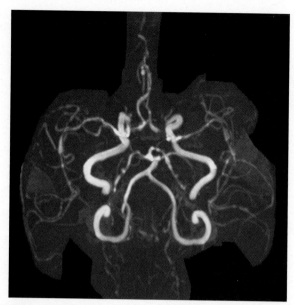

FIGURE 3-39 Reversible cerebral vasoconstrictive syndrome. Three-dimensional time-of-flight magnetic resonance angiography (MRA) of the circle of Willis shows beaded appearance of the proximal and distal anterior cerebral arteries, middle cerebral arteries, and posterior cerebral arteries and distal branches bilaterally. MRA returned to normal several months later, confirming the diagnosis.

Small Vessel Diseases

Involvement of small perforating arteries and arterioles characterizes several disease processes. Because of the small caliber of the vessels involved, angiographic studies are virtually always normal. Imaging features consist of deep gray matter, white matter and subcortical infarcts. Lesions tend to have a different shape (irregular and parallel) along the periventricular white matter helping to differentiate chronic ischemic disease from multiple sclerosis (flame shaped and perpendicular). Parenchymal hemorrhage is less common than in other forms of vasculopathy. Earlier age of onset, rapid progression, and more extensive subcortical involvement are clues to the correct diagnosis in patients with vasculitis compared with chronic small vessel disease.

Collagen Vascular Diseases

These include systemic lupus erythematosis, anticardiolipin, and antiphospholipid antibodies syndromes.

Cerebral vasculitis is rarely associated with collagen vascular disease. The primary CNS lesion seen at autopsy in patients with CNS systemic lupus erythematosis is perivascular inflammation or endothelial cell proliferation (similar to pathologic changes in chronic hypertension). True vasculitis is rarely if ever present (7%) and may be related to infection. The causes of stroke in patients with collagen vascular diseases include cardiac valvular disease (Libman-Sacks aortic and mitral valve vegetations in lupus), an increased tendency toward thrombosis (or reduced thrombolysis) related to antiphospholipid antibodies such as lupus anticoagulant and/or anticardiolipin antibodies, and atherosclerosis accelerated by hypertension or long-term steroid use. Venous thrombosis is also a risk. Atrophy is commonly found in these patients, related either to the encephalopathy itself or to the effect of steroid treatment.

Sjögren Syndrome

This autoimmune disease is characterized by focal or confluent lymphocytic infiltrates in the exocrine glands producing clinical features of dry eyes and dry mouth. However, 25% of these patients have central nervous system complications including infarction. The etiology of the stroke may be a small vessel vasculitis; however, these patients also have antiphospholipid antibodies (another risk factor). Cerebral angiography has been reported as positive in about 20% of cases.

Migraine

White matter T2 hyperintensity can be seen in 10% to 25% of patients with migraine headaches. Lesions are typically few in number and have a predilection for the subcortical white matter of the frontal lobes. These lesions are felt to be the result of spasm in small arteries associated with migraine attacks. There is significant controversy concerning the frequency of this phenomenon. Migraine is often invoked as a cause of white matter lesions when these are encountered in healthy young individuals undergoing MR scans for a variety of clinical findings. Rarely, patients with migraines will present with clinical findings of stroke syndrome (hemiplegic migraine) usually in the distribution of the anterior or middle cerebral arteries (Fig. 3-40). Migraines occur with greater frequency in patients with cerebral autosomal dominant arteriopathy with subcortical infarcts and leukoencephalopathy (CADASIL), and women with migrainous auras are at highest risk for subsequent development of migraine-associated strokes.

Cerebral Autosomal Dominant Arteriopathy with Subcortical Infarcts and Leukoencephalopathy

This is a rare disease that has received much attention because of its direct genetic causation. The disease presents between 30 and 50 years of age with variable findings including multiple infarcts, migraine headaches with aura, depression and dementia. Imaging features, at least early in the disease, are rather characteristic (Fig. 3-41). There is focal and/or confluent T2 hyperintensity and volume loss in the superior frontal lobes, external capsule, and anterior temporal lobes in over 75% of cases. Lacunar infarcts in these regions are common (80% of cases), and microhemorrhages are present in the majority of patients on GrE and SWI MR sequences. As the disease progresses, involvement becomes more widespread and lesions become confluent.

INTRACRANIAL HEMORRHAGE

Now it's time to switch gears and discuss another unwelcome ouchie: the head bleed. Intracranial hemorrhage is the primary event in 15% to 20% of strokes. We now consider CT and MR characteristics of hemorrhage and attempt to explain why we see what we see with these modalities.

Computed Tomography

The x-ray attenuation values of a substance or a structure determine its visibility on CT. With hemorrhage there is a linear relationship between CT attenuation (density) and hematocrit. The attenuation of whole blood with a hematocrit of 45% is approximately 56 Hounsfield units (HU). Normal gray matter ranges from 37 to 41 HU, and normal

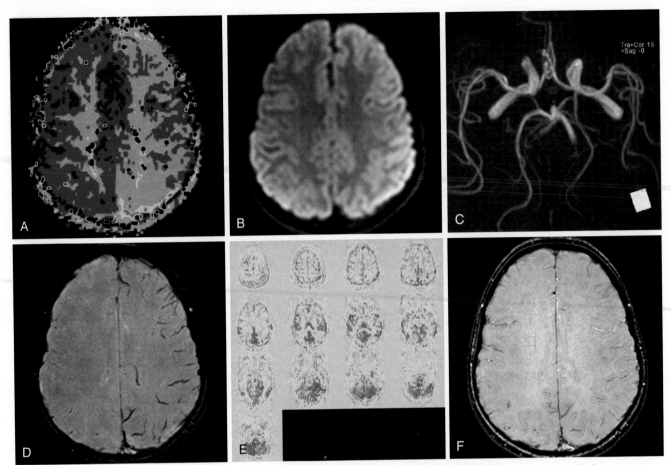

FIGURE 3-40 Migraine-related vasospasm. **A,** Marked asymmetry with prolonged relative mean transit time (rMTT) is seen in the left cerebral hemisphere on this contrast enhanced magnetic resonance perfusion study in a young patient presenting with acute right sided weakness in conjunction with her usual migraine headache. Although no diffusion abnormality is seen **(B)** and three-dimensional time-of-flight magnetic resonance angiography appears normal **(C)**, the susceptibility-weighted imaging (SWI) scan shows increased prominence of cortical vessels in the left cerebral hemisphere, suggesting venous engorgement **(D)**. Her symptoms improved over the next several hours and **(E)** arterial spin labeling perfusion imaging performed the following day shows resolution of the perfusion deficit involving the right cerebral hemisphere. **F,** SWI scan shows symmetric appearance of distal vessels on the follow-up exam as well.

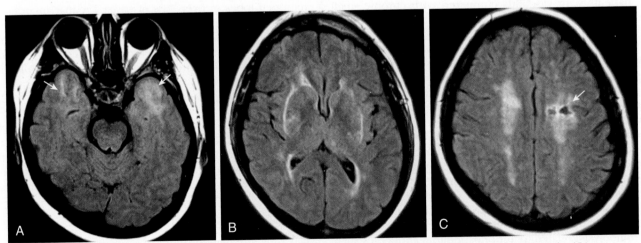

FIGURE 3-41 Cerebral autosomal dominant arteriopathy with subcortical infarcts and leukoencephalopathy (CADASIL). **A-C,** Axial fluid-attenuated inversion recovery images show characteristic features of CADASIL, including subcortical infarct in the left superior frontal lobe (*arrow* in C) as well as confluent T2 hyperintensity in the superior frontal lobes, external capsule, and anterior temporal lobes (*arrows* in A).

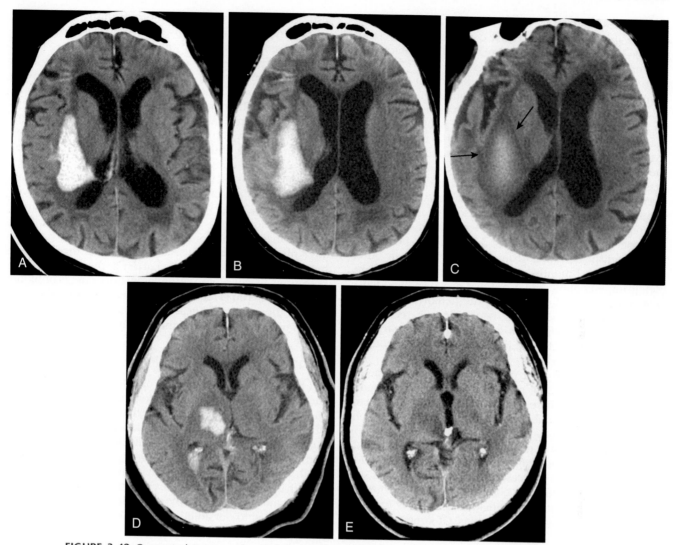

FIGURE 3-42 Computed tomography (CT) scan of hematoma evolution. **A,** CT scan 6 hours after onset of symptoms in a patient with chronic hypertension reveals a large lenticular nucleus homogeneous hyperdense hematoma with mild surrounding edema and relatively little mass effect. **B,** Follow-up exam at 6 days reveals decreased density at the margin of the hematoma. There is increased mass effect on the right lateral ventricle. **C,** Examination at 3 weeks reveals decreased central hyperdensity that fades gradually at the periphery of the hematoma *(arrows),* where hematoma is now hypodense. **D,** A CT scan in a different hypertensive patient at 18 hours reveals discrete hyperdense right thalamic hemorrhage with surrounding edema and mass effect on the third ventricle. **E,** Follow-up exam at 2 weeks reveals complete resolution of hyperdensity. Hypodense hematoma still has mass effect on the third ventricle.

white matter is from 30 to 34 HU. Thus freshly extravasated blood in a patient with a normal hematocrit can readily be demonstrated on CT. In severely anemic patients, there is a small possibility that the acute hemorrhage will be isodense to brain because of low hematocrit (<10%). Conversely, in infants with high hematocrit or patients with polycythemia, the dural sinuses, large veins and proximal arteries may normally appear somewhat dense (mimicking thrombosis).

After the extravasation, density progressively increases for approximately 72 hours and then subsequently decreases (Fig. 3-42). This is caused by increasing hemoglobin concentration attributed to clot formation and retraction. The presence of fluid levels within a hematoma suggests active bleeding and/or absence of coagulation often because of medication effects of heparin or Coumadin. A (vanilla fudge—ok, that's gross) swirled appearance

of hematoma can be seen with actively bleeding hematomas. After the third day, the attenuation values of the clot begin to decrease and during the next 2 weeks, the hemorrhage fades to isodensity with brain. Loss of density is because of red blood cell lysis, with dilution, and subsequent digestion of the blood products by peripheral macrophages. Eventually (rarely more than 1 month), no residual high density can be seen. Factors affecting evolution of density include (1) size of initial hematoma (the larger the hematoma, the longer it takes to become isodense); (2) recurrent hemorrhage (particularly common in subdural hemorrhage); (3) mixture of blood and fluid; (4) initial hematocrit; (5) location of hemorrhage; (6) cause of hemorrhage (tumor, AVM, cavernoma, aneurysm); and (7) rate of resorption and removal of blood (e.g., subarachnoid blood is rapidly removed from the CNS because of resorption with spinal fluid by the pacchionian granulations).

Each of these factors are more or less involved with hemorrhage in different intracranial compartments accounting for differences seen in parenchymal, subarachnoid, subdural, and epidural hemorrhages.

To best understand changes it is helpful to use parenchymal hemorrhage as the "index" condition. Initially on CT there is a thin zone of low attenuation surrounding the high-density hyperacute intraparenchymal hemorrhage that is caused by the serum from the retracted clot. Extensive edema around an acute hematoma (<12 hours) should raise suspicion for an underlying preexistent process (tumor, infection, or prior hemorrhage). The circumferential hypodensity increases and reaches a maximum at approximately 5 days because of vasogenic edema. Focal parenchymal hematomas result in less atrophy than hemorrhagic infarcts or contusions because the hematoma tends to displace rather than destroy brain tissue. After approximately 2 months a small hematoma may be completely imperceptible on CT or there may be a hypodense cleft representing the residua of the hemorrhagic event. Subtle marginal hyperdensity is occasionally seen likely because of hemosiderin deposition or minimal dystrophic calcification. Large chronic hematomas produce nonspecific regions of hypodensity with focal volume loss.

In the evaluation of hyperacute, acute, and subacute hemorrhages, CT is fast, available, easily performed, and accurate. However, there are limitations to CT. Small hemorrhages in the posterior fossa, the anterior inferior frontal and temporal lobes (the most common locations for contusions) and adjacent to the calvarium are difficult to detect because of artifact from adjacent bone and/or partial volume effects. In cases where hemorrhage is suspected but not confirmed, coronal and sagittal reformations of the initial data set can help to confirm or exclude hemorrhage in problematic locations. Of course, MR can also be performed to provide more definitive information on presence, extent, and cause of hemorrhage.

The use of intravenous iodinated contrast in patients with hemorrhage is unnecessary in most situations. Identification of underlying mass lesions is best done with enhanced MR (see below for discussion of enhancement patterns of hemorrhagic masses). However, contrast is routinely used in CTA and CT venography (CTV) to evaluate for vascular anomalies (aneurysms and arteriovenous malformations) and thrombosis (arterial or venous). Focal extravasation of contrast (the "dot sign") into the center of a hematoma on source images is a sign of active hemorrhage and is often associated with continued growth of the hematoma (Fig. 3-43). If contrast is given, intraparenchymal hemorrhage is often associated with a peripheral rim of enhancement from approximately 6 days to 6 weeks after the initial event. This enhancement is the result of breakdown of the blood-brain barrier at the margin of the hematoma because of toxic/inflammatory effects of blood products. The rim is relatively smooth and follows the contour of the original hematoma. A resolving hematoma is one of many causes of peripheral ring enhancing lesions.

Magnetic Resonance Imaging

MR has dramatically improved our diagnostic prowess in the assessment of hemorrhagic conditions. MR is more sensitive than CT in the detection of hemorrhage at all stages of parenchymal hematoma evolution. Detection of associated and underlying lesions is facilitated and the intensity pattern on multiple sequences allows for accurate "aging" of the hematoma. The MR imaging features and their underlying causes are complex when compared with CT. Understanding these complex MR characteristics is, however, essential to our ability to take advantage of MR in the

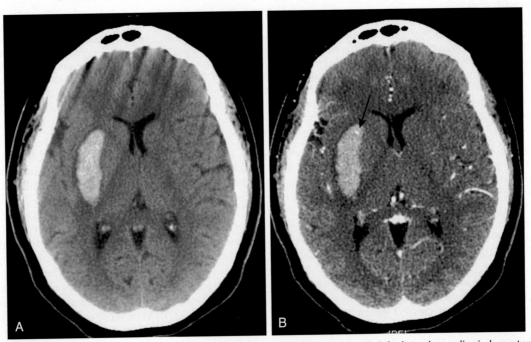

FIGURE 3-43 Hyperacute intracerebral hemorrhage. **A,** Hyperacute (3 hours) right lateral ganglionic hypertensive hemorrhage. Note the lack of surrounding edema and paucity of mass effect. **B,** Computed tomography angiography source image reveals "dot sign" *(arrow)* within the anterior portion of the hematoma, indicative of acute extravasation and active hemorrhage.

evaluation of hemorrhagic lesions. The progression of imaging hallmarks is well understood, although the specific time intervals for these changes are variable. Temporal variation is the result of the same factors described above for CT.

Let us now briefly consider some of the biophysical mechanisms necessary to understand the MR appearance of blood. Warning: This discussion might bring back some painful memories from your organic chemistry days. But you survived that, right? So this should be a piece of cake!

Structure of hemoglobin: Hemoglobin, the primary oxygen carrier in the bloodstream, is composed of four protein subunits. Each subunit contains one heme molecule consisting of a porphyrin ring and an iron atom (Fe^{2+}), which sits near the center of the porphyrin ring and binds to oxygen (O_2). Oxyhemoglobin functionally has no unpaired electrons and is diamagnetic. When a hemoglobin subunit loses its O_2, it forms deoxyhemoglobin, which has four unpaired electrons and can normally be oxidized to methemoglobin in a reversible fashion. In hemorrhage, however, irreversible oxidation to methemoglobin takes place. The iron atom of methemoglobin has five unpaired electrons.

Susceptibility effects: When placed in a magnetic field, certain substances generate an additional smaller magnetic field, which either adds to or subtracts from the externally applied field. Diamagnetic substances have no unpaired electrons and generate very weak fields that subtract from the externally applied fields. Paramagnetic materials such as deoxyhemoglobin and methemoglobin have unpaired electrons that generate much larger local fields surrounding the paramagnetic molecule that add to the externally applied field.

When deoxyhemoglobin or methemoglobin is encapsulated within red blood cells (RBCs), the effective local field is greater within the RBC than outside the cell because of the greater susceptibility of the intracellular paramagnetic hemoglobin solution compared with extracellular plasma. As protons move (diffuse) through these locally varying gradients, they accumulate a different amount of phase change depending on the time spent at different effective field strengths. These phase dispersions produce signal loss on T2WI called proton relaxation enhancement. This phase dispersion leads to loss of signal (T2* effect – hypointensity). Hence, signal loss on T2WI from deoxyhemoglobin, intracellular methemoglobin, and hemosiderin can be attributed to susceptibility effects and proton relaxation enhancement. Gradient-refocused echo and SWI sequences are more sensitive to susceptibility changes.

Proton-electron dipole-dipole interaction: The paramagnetic iron atom in methemoglobin generates a local field approximately 1000 times greater than the local field generated by the proton nucleus. When water binds heme, both T1 and T2 are shortened by proton-electron dipole-dipole interaction. The proton dipole interactions are best observed on T1WI, where T1 shortening produces high signal intensity.

In summary, two key effects are created by the hemoglobin molecule: (1) the paramagnetic effect secondary to the iron within the heme molecule, which can produce susceptibility effects/proton relaxation enhancement in the case of intracellular deoxyhemoglobin, methemoglobin and hemosiderin leading to T2 shortening, and (2) proton-electron dipole-dipole interactions with methemoglobin (both intracellular and extracellular) leading to T1 shortening. Susceptibility effects are field strength dependent but proton-electron dipole-dipole interactions are not.

Temporal Changes in Parenchymal Hemorrhage

Temporal changes in parenchymal attenuation and signal are summarized in Table 3-3.

TABLE 3-3 Stages of Hemorrhage

Stage	CT	T1WI	T2WI	Mass Effect	Time Course	Explanation
Hyperacute	High density	Mild hyper-intensity	High intensity with peripheral low intensity	+++	<6 hours	CT: High protein T1WI, T2WI: Central oxyhemoglobin in with peripheral deoxyhemoglobin (deoxy-Hb)
Acute	High density	Isointense to low intensity	Low intensity	+++	<6 to 72 hours	CT: High protein T1WI: High protein, susceptibility (deoxy-Hb) T2WI: Susceptibility (deoxy-Hb)
Early subacute	High density	High intensity	Low intensity	+++/++	<3 days to 1 week	CT: High protein T1WI: PEDDI (intracellular methemoglobin [met-Hb]), high protein T2WI: Susceptibility (intracellular met-Hb), high protein
Late subacute	Isodense	High intensity	High intensity with rim of low intensity	±	1 to 2 weeks to months	CT: Absorption of high protein T1WI: PEDDI (free met-Hb), absence of susceptibility effects (from intracellular met-Hb), dilution of high protein T2WI: PEDDI (free met-Hb), absence of susceptibility effects, dilution of high protein, susceptibility effects from hemosiderin and ferritin in peripheral rim
Chronic	Low density	Low intensity	Low intensity	–	2 weeks to years	CT: Atrophy T1WI: Susceptibility effects from hemosiderin and ferritin (T2 effect on T1WI) T2WI: Susceptibility effects from hemosiderin and ferritin

CT, Computed tomography; *Hb,* hemoglobin; *PEDDI,* proton-electron dipole-dipole interaction; *T1WI,* T1-weighted image; *T2WI,* T2-weighted image.

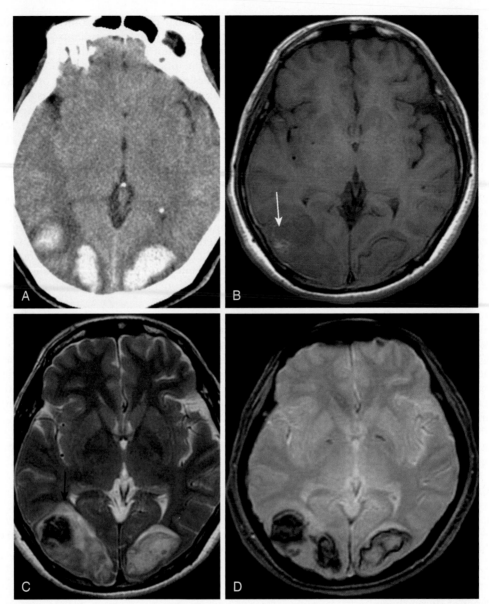

FIGURE 3-44 Imaging of hyperacute and acute intracranial hemorrhage. Patient presented with left field cut 2 days before the examination. Approximately 4 hours before examination, the patient became blind. Both hyperacute left and acute right hematomas are present. **A,** Computed tomography scan several hours after onset of new field cut reveals bilateral occipital hematomas. Right hematoma has more edema than left, suggesting slightly "older" bleed. **B,** T1-weighted image (T1WI) reveals that the left hematoma is relatively isointense with a hypointense margin. The right hematoma has a focus of hyperintensity in its lateral margin *(arrow)* indicative of acute to subacute hematoma. **C,** T2WI reveals that left hematoma is hyperintense while right hematoma has regions of marked hypointensity *(arrow)*. **D,** Gradient-echo scan reveals that left hematoma has a hypointense margin and a relatively hyperintense center. Right hematoma is more diffusely hypointense.

Hyperacute Hemorrhage (0 to 4 Hours)

In the first 3 to 6 hours after extravasation the intact red cells contain mostly diamagnetic oxyhemoglobin. There is early clot formation without clot retraction. Hyperacute hemorrhage (Figs. 3-44, 3-45) is, in essence, a highly protenaceous cellular fluid and it is mildly T1W hyperintense/ hypointense or isointense (because protein decreases T1 but edema/water increases T1) and T2 hyperintense. The periphery of the hematoma contains red blood cells containing hemoglobin that has started to desaturate (deoxyhemoglobin). This produces peripheral hypointensity in particular on susceptibility-weighted (T2*) sequences

and at higher field strength. At this stage T2 hyperintense peripheral vasogenic edema is mild.

Acute Hemorrhage (4 to 72 Hours)

During this period (Fig. 3-46) there is conversion of oxyhemoglobin to deoxyhemoglobin because of local hypoxia and acidosis. Clot formation and retraction leads to decreased water content. These effects produce profound T2 and T2* hypointensity that begins at the periphery and extends to the center of the hematoma. Hypointensity is more marked and appears earlier on T2* sequences. At this stage hematomas are mildly hypointense on T1 images because

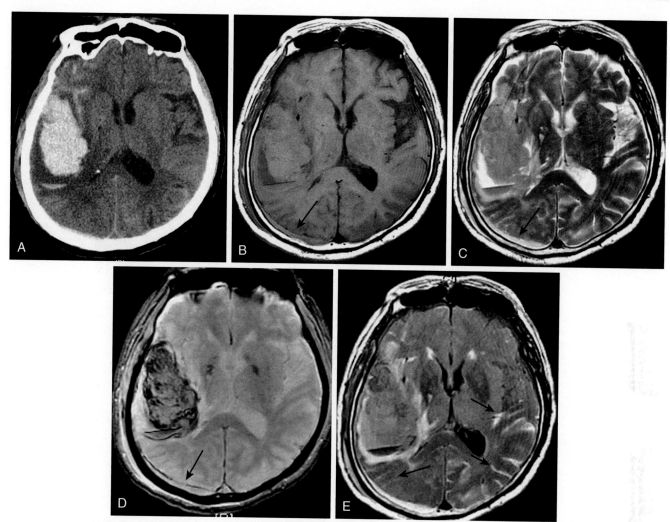

FIGURE 3-45 Imaging of acute hemorrhage. **A,** Computed tomography (CT) scan at the level of the foramen of Monro reveals a large right parasylvian hematoma. **B,** T1-weighted image 1 hour 30 minutes after CT reveals mildly hyperintense mass in the right parasylvian region. Small subdural hematoma is also slightly hyperintense *(arrow)*. **C,** T2-weighted image reveals hematoma is insointense to minimally hyperintense to normal cortex (but overall less intense than would be seen in hyperacute hematoma). Note fluid level in posterior portion of the hematoma. Subdural hematoma is hyperintense *(arrow)*. **D,** Gradient-echo scan reveals hypointensity within the hematoma. Subdural hematoma is hyperintense with a hypointense rim *(arrow)*. **E,** Fluid-attenuated inversion recovery scan reveals hematoma to be minimally hyperintense to gray matter. Note extensive sulcal hyperintensity *(arrows)* indicative of subarachnoid hemorrhage not apparent on CT.

of the susceptibility effects and increased water content of edema. Because water molecules are unable to approach close enough to the iron atom of deoxyhemoglobin; no T1 shortening is caused by proton-electron dipole-dipole interactions. There is an increase in T2 hyperintense peripheral vasogenic edema with a concomitant increase in mass effect.

Early Subacute Hemorrhage (3 to 6 Days)

During this period (Fig. 3-47) there is oxidation of deoxyhemoglobin to methemoglobin inside the RBC. Unlike deoxyhemoglobin, water molecules are able to approach the paramagnetic heme of methemoglobin, permitting proton-electron dipole-dipole interactions that shorten T1. This effect gives subacute hematomas their characteristic hyperintensity on T1WI. Oxidation of deoxyhemoglobin to methemoglobin proceeds from the periphery to the center of the clot during the first week after the ictus, and

therefore T1 hyperintensity is initially seen at the edge of the hematoma (2 to 3 days) with progressive "filling in" of the center of the hematoma (7 to 10 days). Because the paramagnetic methemoglobin remains encapsulated within the RBC, marked hypointensity is present on the T2WI and in particular T2* images because of the susceptibility mechanism described previously for deoxyhemoglobin. Towards the end of the early subacute phase T2 hypointensity decreases because of cell lysis. Once paramagnetic methemoglobin is no longer sequestered within the red cells, local field inhomogeneity begins to decrease and susceptibility induced T2* hypointensity begins to resolve. Peripheral edema and mass effect continue to increase during this period.

Late Subacute Hemorrhage (6 to 14 Days)

During this period (Fig. 3-48), red cell and clot lysis occur. Methemoglobin is less stable than deoxyhemoglobin,

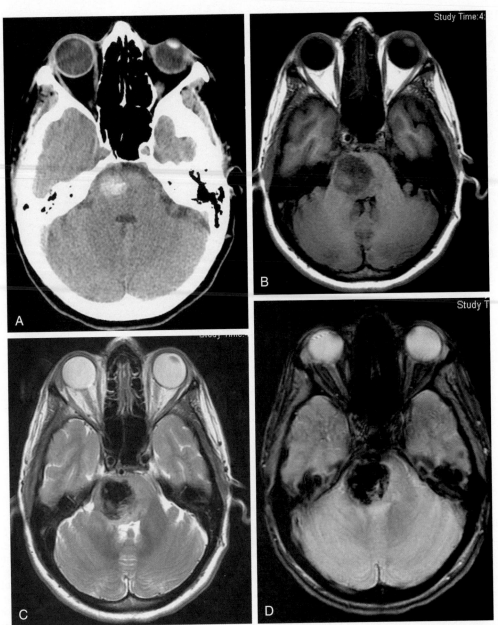

FIGURE 3-46 Imaging of acute intracranial hemorrhage. **A,** Computed tomography scan 4 hours after onset of symptoms in chronic hypertensive patient reveals right pontine hematoma. T1-weighted image (T1WI) **(B)** and T2WI **(C)** at 36 hours reveals large focus of hypointensity. **D,** On gradient-echo scan hypointensity is increased.

and the heme group can spontaneously be lost from the protein molecule. This free heme and/or other exogenous compounds promote RBC lysis. Concomitantly, there is protein breakdown and dilution of the remaining extracellular methemoglobin. Hyperintensity persists on the T1WI because of the T1-shortening effects of intracellular and extracellular methemoglobin even at relatively low concentrations. The hematoma intensity progressively increases on T2WI, approaching that of CSF, because of a loss of local field inhomogeneity and reduction in proton relaxation enhancement that results from RBC lysis and a decrease in the protein concentration. Paralleling the breakdown of the methemoglobin is an accumulation of the iron molecules hemosiderin and ferritin within macrophages at the periphery of the lesion. The result is a hypointense rim at the margin of the lesion, variably visible on T1WI but increasingly prominent on T2WI and SWI. In addition, peripheral edema and mass effect decrease.

Chronic Hemorrhage (Greater Than 2 Weeks)

The hematoma becomes progressively smaller (Fig. 3-49). The central T1 and T2 hyperintensity and peripheral T2/T2* hypointensity persists but peripheral edema and mass effect completely resolve. After months there is nearly complete breakdown and resorption of the fluid and protein within the clot such that the bright signal on T1WI and T2WI resolves. The iron atoms from the metabolized hemoglobin molecules are deposited in hemosiderin and ferritin molecules that are trapped permanently within the brain parenchyma because of restoration of the blood-brain barrier. The susceptibility effects of the superparamagnetic iron

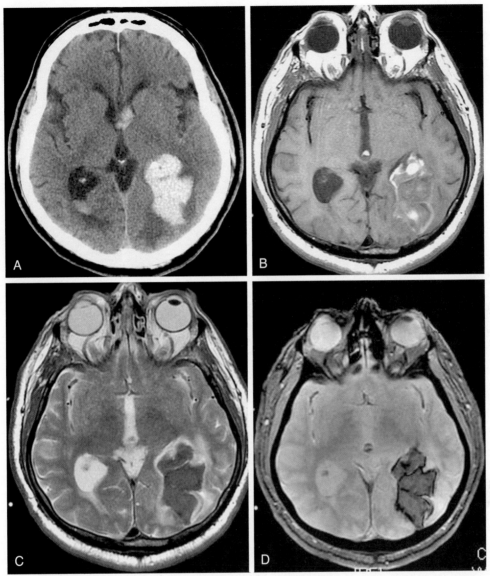

FIGURE 3-47 Imaging of subacute hematoma. **A,** Computed tomography scan at 1 day reveals a large irregular hyperdense left parietooccipital hematoma with intraventricular hemorrhage. Magnetic resonance imaging performed at 3 days. **B,** T1-weighted image (T1WI) reveals hyperintensity at the margin of the lesion and within the anterior and posterior portions of the hematoma. **C,** T2WI reveals homogeneous marked central hypointensity with mild surrounding T2 hyperintense edema. **D,** Gradient-echo scan reveals diffuse hypointensity with a more hypointense rim.

cores of hemosiderin produce permanent hypointensity on all sequences but are most prominent on the gradient echo and SWI sequences where "blooming" of the hypointensity is observed. Therefore chronic hemorrhage is directly visible on MR as opposed to CT where only the sequelae of the hemorrhage (lacunes, hypodensity, volume loss) are apparent. The pattern of hemosiderin "scarring" depends on the size, location, and etiology of the original hemorrhage. Small hematomas produce peripheral hypointense clefts (hemosiderin slit) while large hematomas, hemorrhagic infarcts, and contusions produce areas of encephalomalacia with marginal or gyral hypointensity. Small petechial hemorrhages (microbleeds) produce foci of T2* hypointensity that cannot be detected on CT or other MR sequences. Microbleeds typically occur in hypertensive cerebrovascular disease, amyloid angiopathy, and as result of head trauma

with axonal injury. They may also be seen with multiple cavernous malformations and after radiation therapy.

ETIOLOGIES OF INTRACRANIAL HEMORRHAGE

Intraparenchymal Hemorrhage

Trauma is a common etiology for parenchymal hemorrhage(s), discussed in detail in Chapter 4. Spontaneous (nontraumatic nonischemic) parenchymal hemorrhage accounts for approximately 10% of stroke syndromes (Box 3-3). The most common cause of parenchymal hemorrhage is hypertension and these lesions are most often seen in the deep gray matter (basal ganglia and thalamus), brain stem and cerebellum. Lobar hemorrhages may be because of

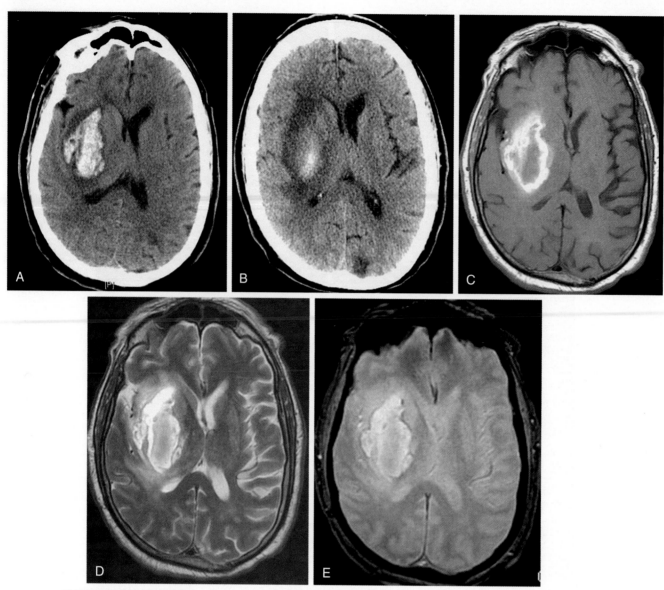

FIGURE 3-48 Late subacute hematoma. Computed tomography scans at 24 hours **(A)** and 14 days **(B)** reveal evolution of right ganglionic hematoma. Magnetic resonance (MR) scan also performed at 14 days reveals peripheral hyperintensity and central isointensity on T1-weighted image (T1WI) **(C)**; marked peripheral hyperintensity, central mild hyperintensity, and a subtle hypointense margin on T2WI **(D)**; and central hyperintensity with a peripheral hypointense margin on gradient-echo image **(E)**. Note that size of hematoma is more easily appreciated on MR than computed tomography (CT) at this stage. Hematoma is approximately the same size on MR as it was at time of initial CT.

infarcts, hypertension, or may be drug induced (i.e. cocaine) or from supratherapeutic anticoagulation (i.e. warfarin) as well. In patients over 50 years of age, lobar hemorrhages can be the result of cerebral amyloid angiopathy (CAA), but these patients also get microhemorrhages, hemosiderosis, and extraaxial bleeds. Venous thrombosis can lead to parenchymal hemorrhages typically in the white matter adjacent to the thrombosed dural sinus or cortical vein. Hemorrhage into underlying neoplastic lesions (primary or metastatic) or from vascular abnormalities (arteriovenous malformations, cavernous angiomas, and aneurysms) can occur in any location and at any age.

Hypertension

Whatever you do, stay calm and keep your blood pressure down in this section. In hypertensive cerebral vascular disease, damage to small perforating arteries arising from proximal vessels (e.g., MCA and basilar arteries) leads to fibrinoid necrosis. Hemorrhages occur most frequently in the basal ganglia followed by the thalamus, brain stem, and cerebellum (see Fig. 3-48). Initially hypertensive hemorrhages may have very little mass effect. Over the first 24 to 36 hours, the hemorrhages often enlarge and always develop vasogenic edema and increasing mass effect. These changes account for the clinical observation that patients with hypertensive hemorrhage typically deteriorate over the first few days.

Imaging studies typically reveal other evidence of hypertensive cerebrovascular disease. Chronic lacunar infarcts or hemorrhages and extensive microvascular white matter disease are typically present. On GrE and SWI MR sequences microbleeds (punctate foci of T2* hypointensity) are

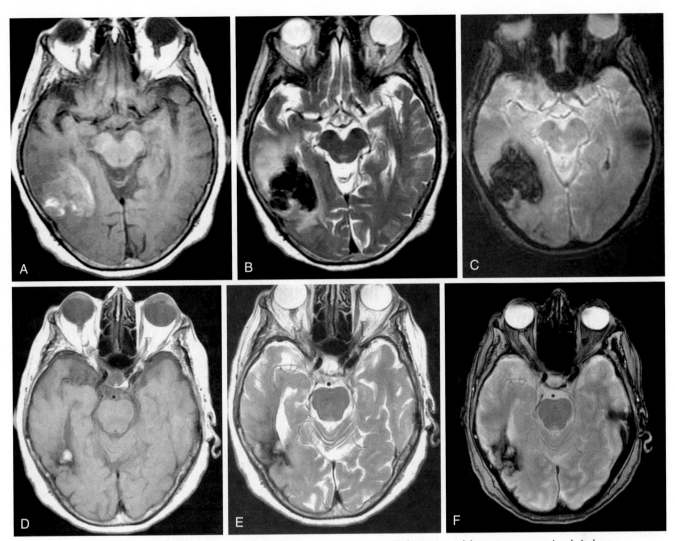

FIGURE 3-49 Magnetic resonance study of chronic hematoma. Right temporal hematoma examined 4 days after ictus reveals typical features of early subacute hematoma with peripheral hyperintensity on T1-weighted image (T1WI) **(A),** diffuse hypointensity on T2WI with surrounding edema and mass effect **(B),** and marked hypointensity on gradient echo **(C). D-F,** Repeat examination at 2 months reveals marked contraction of clot with focal volume loss and no edema. Residual central T1 **(D)** and T2 **(E)** hyperintensity is present. There is extensive T2 and susceptibility hypointensity in the adjacent tissue, and there is mild sulcal hemosiderin deposition in the leptomeninges (siderosis) adjacent on the gradient-echo scan **(F).**

encountered in the deep gray matter, posterior fossa, and subcortical regions. While chronic hypertension alone can lead to parenchymal hemorrhage, rapid episodic increase in blood pressure may occur with cocaine use, dialysis, and fluid overload resulting in parenchymal hemorrhage.

Cerebral Amyloid Angiopathy

Cerebral amyloid angiopathy (CAA) results from deposition of amyloid (an eosinophilic, insoluble extracellular protein) in the media and adventitia of small and medium-sized vessels of the superficial layers of the cerebral cortex and leptomeninges, with sparing of the deep gray nuclei. Amyloid deposition increases with age but does not correlate with hypertension or the presence or absence of systemic amyloidosis. It does correlate with brain parenchymal amyloid deposition so is more common in patients with Alzheimer disease. Amyloid protein replaces the normal constituents of the vessel wall in particular the elastic lamina leading to microaneurysm formation and fibrinoid

degeneration resulting in vascular fragility. Hemorrhages are usually lobar, typically involving the frontal and parietal lobes including the subjacent white matter. Subarachnoid and subdural hemorrhages have also been reported as a result of the superficial vessels involved in CAA. There is a propensity for recurrent hemorrhage in the same location and/or multiple simultaneous hemorrhages.

The presence of multiple simultaneous lobar hemorrhages in an elderly patient should raise suspicion for CAA although multiple hemorrhagic metastases must always be considered. Because CAA may produce recurrent hemorrhage in a single location, hematomas are often complex in appearance on both CT and MR. Variable CT density and MR intensity within the same lesion reflect hemorrhage of different ages. One finding that can help to establish the diagnosis of CAA is presence of cortical and/or subcortical microbleeds on T2*-weighted MR imaging (Fig. 3-50). In virtually all cases CAA is accompanied by extensive microvascular white matter ischemic change with sparing of

BOX 3-3 Hemorrhagic Causes of Stroke

PRIMARY INTRACEREBRAL HEMORRHAGE

Aneurysm (mycotic, congenital)
Hemorrhagic infarction (arterial, venous)
Hypertensive, arteriosclerotic hemorrhage
Neoplasms (primary or metastases)
Trauma
Vasculitis
Vascular malformations
Amyloid angiopathy
Drugs
    Cocaine
    Amphetamine
    Phenylpropanolamine
    L-Asparaginase
Hematologic causes
    Antithrombin III deficiency
    Protein C+S deficiency
    Antiphospholipid antibodies
    Factor VII deficiency
    Factor IX deficiency
    Factor VIII deficiency
    von Willebrand factor deficiency
Acquired coagulopathies
    Thrombocytopenia and platelet dysfunction
    Disseminated intravascular coagulopathy
    Uremia
    Multiple myeloma
    Myeloproliferative disorders
    Lymphoproliferative disorders
    Leukemia

SUBARACHNOID HEMORRHAGE

Aneurysm
Arteriovenous malformation, nonaneurysmal perimesencephalic hemorrhage
Dural malformation
Hemorrhagic tumor
Trauma
Vascular dissection

---

subcortical U fibers. Hemosiderosis and evidence of subarachnoid hemorrhage can be present. Since the advent of SWI and its ability to detect microhemorrhages, the age range has shifted from 70 years of age and older down to 50 year olds for the onset of this disease.

Rarely, there may be an inflammatory component to CAA (CAA-related inflammation or CAA-RI), with patients presenting with marked cognitive decline and seizures, and MRI showing asymmetric unifocal or multifocal T2 and FLAIR hyperintensities extending to the subcortical white matter in a background of parenchymal microbleeds. Patients can respond well to immunosuppressive treatment if the diagnosis is made sooner rather than later.

Venous Thrombosis

In our humble opinion, venous thrombosis (Box 3-4) is among the most challenging diagnoses to be made in neuroradiology. Symptoms and signs are nonspecific, and depend on the location of the clot and the acuity of the process. It may occur secondary to a large number of conditions. Imaging findings on routine CT may be subtle and overlooked if clinical suspicion is low. Fortunately, the advent of MR

and the development of CT angiographic techniques have dramatically improved our ability to make this diagnosis.

Venous thrombosis is the great mimicker of other diseases. It occurs in the setting of dehydration (e.g., infants with systemic infection, severe nausea and vomiting), hypercoagulation (autoimmune diseases, collagen vascular disease, inflammatory bowel disease, pregnancy puerperium, and birth control pills and other medications), acute bacterial infection, and hereditary conditions. Calvarial and/or dural tumors (e.g., meningiomas) may cause venous thrombosis and it may be seen in association with skull fractures that traverse dural sinuses. In a significant number of cases an etiology is not discovered. Hypercoagulability from malignancy (Trousseau phenomenon) is a consideration in hospitals with "thriving" cancer centers. Venous thrombosis has also been implicated as the cause of dural arteriovenous malformations and of pseudotumor cerebri.

Sinus thrombosis is often encountered in young women and children because they are more likely to have the conditions that predispose to venous thrombosis. Clinical findings are often nonspecific and include headache and seizures. Signs on examination may be nonspecific and/or may indicate increased intracranial pressure with papilledema. Localizing neurologic findings result from regional venous infarcts and parenchymal hemorrhage. In severe cases venous thrombosis can lead to intracranial hypertension, coma, and death. The temporal pattern of the disease is also variable and dependent on the cause of the thrombosis, location and extent of initial thrombosis, rate of progression or spontaneous regression, and the pattern of collateral venous drainage. Some patients present with an acute ictus and rapid deterioration suggestive of infarction. In most cases clinical onset is less abrupt and progression occurs over days mimicking the time course of intracranial infection. In rare cases the onset is insidious with patients presenting with slowly progressive symptoms (months to years) mimicking neurodegenerative processes.

Venous thrombosis typically begins in a dural venous sinus (superior sagittal sinus or transverse sinus). The thrombosis may progress to involve other portions of the dural venous system or extend into adjacent cortical veins. Isolated cortical vein thrombosis is rare but does occur. Thrombosis of the deep venous system (e.g., internal cerebral veins and vein of Galen) can occur in children.

Spontaneous resolution of venous thrombosis can lead to rapid resolution of symptoms. Use of heparin to dissolve venous clots and/or endovascular treatment to remove clot often results in dramatic improvement in clinical outcome even in the presence of parenchymal hemorrhage and/or long standing symptoms.

Because treatment can dramatically improve outcome, an understanding of the imaging diagnosis of venous thrombosis is extremely important. Imaging findings can be divided into three categories: (1) direct identification of clot in a dural sinus or vein, (2) identification of collateral venous channels, and (3) identification of the complications of venous thrombosis. Clot identification is complicated by the fact that, like all hematomas, clots within the venous system undergo time dependent CT and MR intensity changes and therefore look different at different times.

On unenhanced CT the acutely thrombosed vein appears hyperdense and is often enlarged (Fig. 3-51, *A*). The diagnosis of transverse sinus thrombosis is complicated by

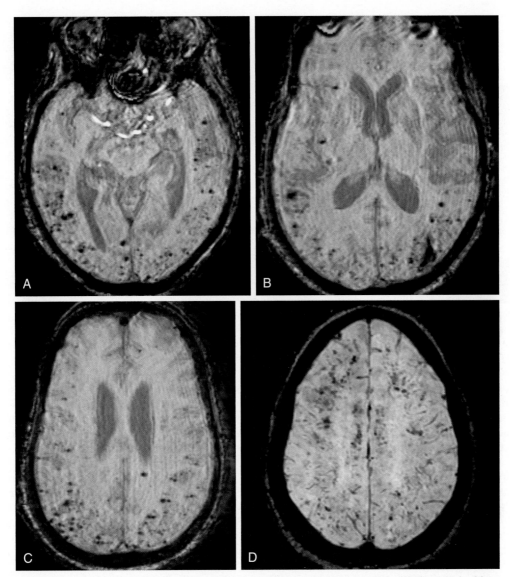

FIGURE 3-50 Cerebral amyloid angiopathy. **A-C,** Susceptibility-weighted imaging (SWI) scans in 89-year-old woman show innumerable foci of signal loss predominantly in cortical/subcortical parenchyma indicating microhemorrhages. Note absence of such signal changes in basal ganglia and brain stem, as might be seen in the setting of chronic hypertension. This distribution of microhemorrhage in a patient of this age is most compatible with cerebral amyloid angiopathy. **D,** Be aware there is a differential for such microhemorrhages based on clinical context and patient demographics. In this young adult patient with sickle cell disease, the innumerable microhemorrhages on SWI *("starfield" pattern)* are a consequence of chronic fat embolic infarcts occurring secondary to avascular necrosis of long bones.

BOX 3-4   Causes of Venous Thrombosis

Acute dehydration (diarrhea)

Chemotherapeutic agents (l-asparaginase)

Cyanotic congenital heart disease

Hypercoagulable states and coagulopathies: sickle cell disease, hemolytic anemia, polycythemia, use of oral contraceptives, inflammatory bowel disease, nephritic syndrome, protein S and protein C deficiencies, antithrombin III deficiency

Iatrogenic (indwelling venous catheters)

Infection involving sinuses, mastoids, and leptomeninges

Malignancy, including leukemia (also associated with coagulopathies)

Malnutrition

Pregnancy

Trauma

---

the fact that the transverse sinuses are often asymmetric with one sinus (more typically the left) smaller or absent. In patients with asymmetric sinuses, the larger sinus may appear abnormally dense and be mistaken for thrombosis. Remember that in newborns with expected normal polycythemia leading to increased vascular density, the relative hypodensity of the brain, and the frequent occurrence of minimal perinatal paratentorial hemorrhage can mimic the appearance of sinus thrombosis.

Cortical vein thrombosis can produce a dense superficial structure that extends along the convexity surface of the brain in the region of the dural venous sinus ("cord sign"). Dilated superficial veins indicative of collateral drainage rather than thrombosis are occasionally visible on unenhanced CT. Within a week, clot in a dural sinus becomes isodense to hypodense and sinus expansion

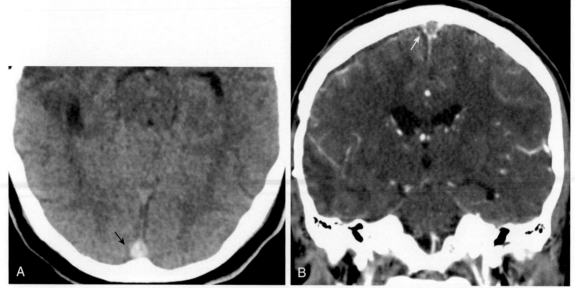

FIGURE 3-51 Venous sinus thrombosis on computed tomography (CT). **A,** Axial unenhanced CT showed hyperattenuation within the expanded sigmoid sinus *(arrow).* **B,** Following contrast administration for venogram study, the same clot appears relatively hypodense *(arrow)* on coronal reformatted CT because of adjacent hyperattenuating contrast material and parenchymal contrast enhancement ("empty delta sign").

resolves. Therefore direct identification of sinus thrombosis on unenhanced CT becomes impossible. Use of dynamic thin section CT venogram protocols with multiplanar reconstruction has made the diagnosis of venous thrombosis more straightforward. On contrast-enhanced CT venograms, nonenhancing clot within the thrombosed vein ("the empty delta sign") is seen (see Fig. 3-51, *B*). In addition, ancillary findings including dilated cortical venous collateral vessels and thick enhancement of the tentorium and/or falx can also be seen.

Additional concomitant findings can include subcortical hemorrhage near the thrombosed sinus as well as subarachnoid hemorrhage and subdural hemorrhage. The brain may appear swollen but this is difficult to differentiate from normal in particular in young patients on CT. Narrowing and/or obliteration of basal cisterns (e.g., suprasellar, prepontine) helps to distinguish normal from abnormal studies.

The diagnosis can also be made with MR. Detection of clot within a dural sinus or vein is dependent upon understanding of two factors: (1) the normal appearance of the dural venous sinuses on different pulse sequences, and (2) knowledge that intraluminal clots have some of the same intensity and temporal changes in intensity as parenchymal hematomas but things evolve more slowly and hemosiderin is absent.

Normal venous flow voids will be hypointense on both T1- and T2-weighted sequences. However, because venous flow is slower than arterial flow, hyperintense flow signal can also normally be seen in the major venous sinuses and cortical veins. On GrE/SWI scans, flow related hypointensity is normally seen within or adjacent to the transverse and superior sagittal sinus. Acute sinus clot is T1 isointense to mildly hyperintense and T2 hypointense and thus similar in appearance to flowing blood. The key finding at this stage is marked hypointensity with blooming on GrE scans within the affected venous sinus or cortical vein. With thrombosis, the affected dural sinuses are typically expanded. Acute clot can also show diffusion restriction. In addition, dilated collateral cortical veins are readily visible on unenhanced T2W images. It is therefore possible to make the diagnosis of venous thrombosis on routine noncontrast MR scans even in the absence of clinical suspicion of the correct diagnosis.

Diagnosis on MR is facilitated by performance of MRV using phase contrast (PC) and/or enhanced or unenhanced TOF techniques (Fig. 3-52). As previously stated PC MRA can be made sensitive to slow flow rates and therefore can be performed such that only the veins are visualized. Two-dimensional PC MRV images in sagittal and coronal plane can be obtained in less than 1 minute and allow for rapid assessment of dural venous sinuses. TOF techniques take longer but provide greater detail about both the dural sinuses and the cortical veins. One potential limitation of TOF MRV is that subacute T1 hyperintense intrasinus clot will appear hyperintense and mimic flow. It is therefore helpful to perform both PC and TOF venograms. Also, by adding contrast-enhanced venogram sequence one can often sort out the slow flow or turbulent flow areas that mimic sinus thrombosis and/or sinus stenosis. Normal pacchionian granulations may appear as filling defects in normal sinuses on enhanced scans. Fortunately, the arachnoid granulations have CSF signal and are typically very bright on T2WI. Chronically occluded sinuses are small in caliber. In many cases there is at least partial recanalization of dural sinuses and development of additional venous collaterals. MRV and CTV studies reveal irregular areas of narrowing and/or focal occlusion in chronic thrombosis.

MR is superior to CT in detecting the complications of sinus thrombosis. Detection of hemorrhage is equal to or superior to that seen with CT. A general rule of thumb is that the parenchymal hematoma clot will look "younger" (more acute) than the intraluminal clot because it develops at some point after the thrombus forms. Detection of focal and diffuse T2 hyperintense

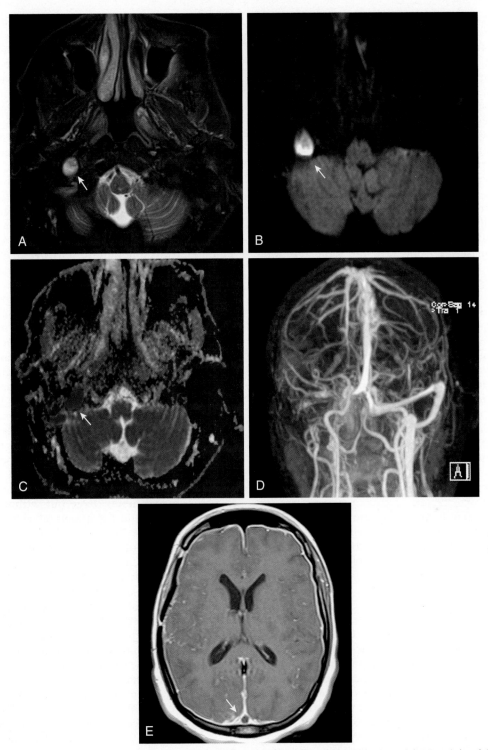

FIGURE 3-52 Venous sinus thrombosis on magnetic resonance imaging (MRI). **A,** Axial T2-weighted image (T2WI) shows abnormally hyperintense T2 signal in the right jugular vein *(arrow)*. **B,** Diffusion-weighted imaging and **(C)** apparent diffusion coefficient maps show diffusion restriction in this same location *(arrows)*. **D,** Contrast-enhanced magnetic resonance venogram shows absence of opacification within the right transverse and sigmoid sinuses, as well as the right jugular vein, indicating occlusive venous sinus thrombosis. **E,** Contrast-enhanced T1WI in a different patient (same patient as in Fig. 3-51) shows the MRI correlate of the "empty delta sign" indicating thrombosis *(arrow)*.

edema is much easier on MR than CT, and use of DWI allows for differentiation between irreversible venous infarction (DWI hyperintense with restricted diffusion) from reversible venous edema (DWI isointense with increased diffusion).

Intratumoral Hemorrhage

Brain tumors associated with hemorrhage are usually malignant primary astrocytoma (World Health Organization [WHO] grade III—anaplastic astrocytoma and grade IV—glioblastoma) and metastases. In the general population the most common hemorrhagic metastases encountered on imaging include in order of decreasing frequency: lung and breast metastases (very common lesions that bleed occasionally; melanoma metastases (rather common lesions that bleed in most cases); renal and thyroid metastases (uncommon lesions that bleed in many cases); and choriocarcinoma metastases (rare lesions that bleed in most cases). Other tumors associated with hemorrhage include pituitary adenomas, hemangioblastomas, dysembryoplastic neuroepithelial tumours, ependymomas, and craniopharyngiomas.

Hemorrhage into neoplasm is often recurrent and takes place over days to weeks. Therefore, hemorrhagic tumors typically produce more heterogeneous CT density and MR intensity than "simple" parenchymal hematomas and they do not have the same temporal progression. With tumoral hemorrhage, the deoxyhemoglobin state may be prolonged with central hypointensity existing for more than a week; a complete rim of hemosiderin induced T2 hypointensity is usually absent. There is generally more edema and mass effect than with a simple hematoma. Edema is present at the time of the initial ictus and does not resolve over time. Enhancement within or adjacent to the hematoma should always raise the suspicion of underlying neoplasm, in particular when encountered within a few days of ictus. In some cases a large acute hematoma will completely obscure an underlying neoplasm making correct diagnosis impossible until the true nature of the lesions becomes apparent after hemorrhage resolves.

ETIOLOGIES OF EXTRAAXIAL HEMORRHAGE

Subdural and epidural hemorrhages are most commonly seen in the trauma setting, and are discussed in detail in Chapter 4. The discussion below will focus on subarachnoid hemorrhage and the various etiologies for hemorrhage in the subarachnoid space.

Subarachnoid Hemorrhage

Subarachnoid hemorrhage (SAH) has different CT and MR imaging features than other types of hemorrhage because the blood mixes with CSF. Spinal fluid dilutes the blood (hematocrit is typically <5%) and antifibrogenic elements in CSF prevent or inhibit clot formation. The unclotted blood is rapidly cleared from the subarachnoid space via the pacchionian granulations. Finally the subarachnoid CSF has a relatively high oxygen tension (43 mm Hg) and therefore the deoxyhemoglobin concentration is low (28%).

On CT, acute SAH produces hyperdensity in the affected sulci, cisterns and fissures. The hyperdensity changes quickly over the first week with complete resolution within 2 days for small amounts of isolated SAH and within 5 to 7 days for most aneurysmal SAH. When hemorrhage is brisk the anticlotting mechanisms of the CSF may be overcome and focal clot will form at the site of bleeding producing an expanded hyperdense subarachnoid space with local mass effect and edema in the adjacent brain. Density will persist more than 1 week in focal subarachnoid clot or large volume bleeds. Depending upon the source and

extent of SAH, dissection of hemorrhage into the adjacent parenchyma as well as ventricular system may coexist.

Historically, MR was felt to be insensitive to SAH. On routine T1WI SAH produces only subtle T1 hyperintensity ("dirty" CSF) and normal intensity on T2WI. Rapid dilution and removal of blood, absence of clot formation and presence of high O_2 concentration (which limits the amount of deoxyhemoglobin) prevents the development of T2 and T2* hypointensity and subacute T1 hyperintensity. The advent of FLAIR imaging has markedly improved the sensitivity of MR to SAH. Blood in the subarachnoid space changes the T1 of the CSF sufficiently to prevent suppression of signal by the 1800 msec inversion pulse. Because CSF signal is not suppressed, T2 effects are visible and the bloody CSF is bright on FLAIR. In fact, for any given concentration of blood FLAIR is much more sensitive than CT (Fig. 3-53). SWI has also been shown to be sensitive to SAH, seen as dark signal in the SAS.

Ironically, while sensitivity is high, specificity is low. Subarachnoid FLAIR hyperintensity may be seen in inflammatory and neoplastic leptomeningeal disease. Artifactual FLAIR hyperintensity occurs when metal is present (e.g., shunt valves, dental appliances) and when patients are breathing pure oxygen (anesthesia) during the exam (O_2 is paramagnetic and its presence within the subarachnoid CSF shortens T1 sufficiently to prevent suppression of signal). The most common artifactual cause of FLAIR CSF hyperintensity is pulsation artifact at the skull base. This limits the utility of FLAIR in the region of the basal cisterns where aneurysmal SAH is most common. When focal clot does form, the MR intensities encountered are similar to those seen in parenchymal hemorrhage but without hemosiderin. Acute clot is T2 and T2* hypointense and subacute clot is T1 hyperintense. Because acute clot is T2 hypointense, it is isointense (not visible) to CSF on FLAIR but readily apparent on T2W images.

Hemosiderin deposition on the surface of the brain (leptomeningeal and subpial) and cranial nerves can occur as the result of chronic recurrent subarachnoid hemorrhage (superficial siderosis). It is typically not seen with a single aneurysmal subarachnoid hemorrhage no matter how severe because the blood is cleared from the spinal fluid before it can be converted to hemosiderin. Recurrent bleeding occurs with conditions such as chronic low-grade neoplasms, vascular malformations, CAA, and recurrent trauma and is more common when patients have bleeding diatheses. Hemosiderin is neurotoxic and therefore when it involves the cranial nerves, patients may develop specific neuropathic symptoms. Siderosis cannot be diagnosed on CT. On MR hypointensity, coating the surface of the brain and cranial nerves is visible on T2WI and more extensively on T2*-weighted images (Fig. 3-54). Hemosiderin deposition has also been noted on the ventricular ependyma after neonatal intraventricular hemorrhage.

Causes of Subarachnoid Hemorrhage

SAH may result from a variety of circumstances including trauma, ruptured aneurysm, arteriovenous malformation, vasculopathy, venous thrombosis and extension of parenchymal hemorrhage into the subarachnoid space. The two

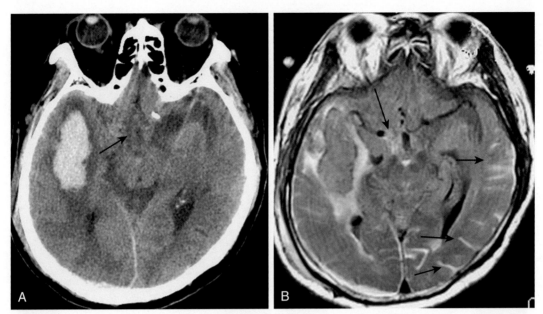

FIGURE 3-53 Subarachnoid hemorrhage (SAH) on computed tomography (CT) and fluid-attenuated inversion recovery (FLAIR). **A,** CT scan at the level of the suprasellar cistern reveals acute right sylvian fissure hematoma secondary to rupture of middle cerebral aneurysm hematoma. No other areas of SAH are similarly hyperattenuating on CT (*arrow,* suprasellar cistern). **B,** FLAIR image on same day reveals hyperintensity in the suprasellar cistern (*long arrow*) and superficial sulci (*short arrows*) indicative of SAH not as readily visible on CT. Hyperacute right sylvian fissure hematoma is isointense.

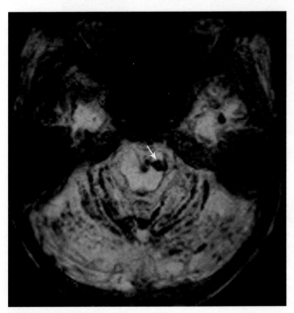

FIGURE 3-54 Superficial siderosis. On this susceptibility-weighted imaging scan there is signal loss along cerebellar folia and along surface of the brain stem, indicating superficial siderosis from chronic recurrent subarachnoid hemorrhage. In this case, the hemorrhage is related to known cerebral neoplasm (not shown). Note signal loss in the brain stem suspected to be related to cavernoma *(arrow).*

most common causes of SAH are trauma and ruptured aneurysm.

Trauma. In most cases of severe head trauma SAH accompanies and is adjacent to parenchymal and/or extraaxial hematomas. Extensive basal SAH without parenchymal or extraaxial hemorrhage is uncommon. Isolated SAH

overlying the cerebral hemispheres is relatively common in mild and moderate head trauma. The small amount of sulcal hemorrhage will often resolve quickly (within 24 hours) and may be seen to migrate towards the vertex on serial exams. It may be difficult on initial exams to differentiate between focal SAH and small cortical contusions and serial exams may reveal evolution of parenchymal contusions at the site of focal SAH. Even when there is a history of trauma it is necessary to exclude the possibility of an underlying aneurysm because spontaneous SAH may be the precipitating event in an episode of head trauma—the chicken and egg theory at it again!

Aneurysmal Subarachnoid Hemorrhage. Aneurismal SAH (aSAH) is typically centered around the base of the brain, in particular within the suprasellar cistern (90% of berry aneurysm occur in the region of the circle of Willis; Fig. 3-55). While hemorrhage is typically diffuse, the region with the most accumulation of blood is likely adjacent to the source of hemorrhage. Sensitivity of CT in the first 48 hours after aneurysmal SAH approaches 95%. Sensitivity decreases to about 50% by postictal day 5 as blood is resorbed. False negative CT scans occur when the initial hemorrhage is mild, when it results from posterior fossa aneurysms or when the initial CT is obtained several days after ictus. Recognition of subtle signs of aneurysmal SAH is of paramount importance. Subtle SAH is relatively isodense to adjacent brain and therefore the basal cisterns are not visible. The hypodense CSF within anterior third ventricle and temporal horns of the lateral ventricles and sometimes the aneurysm itself stands out because normal cisternal CSF density is not present. Careful inspection of sylvian fissures and the anterior interhemispheric fissure will typically reveal mild hyperdensity or at the least no hypodensity. Posterior fossa SAH (e.g., from PICA aneurysms) is also difficult to detect because the cisterns are

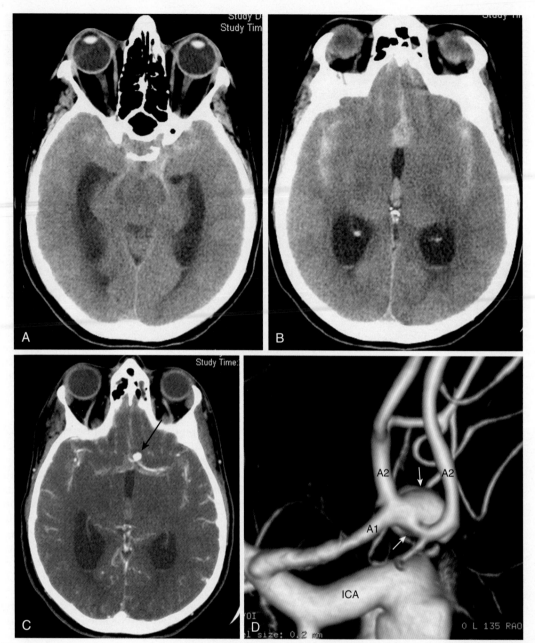

FIGURE 3-55 Aneurysmal subarachnoid hemorrhage. **A** and **B,** Computed tomography (CT) scans at the level of the suprasellar cistern and inferior third ventricle reveal diffuse hyperdensity in the suprasellar cistern, sylvian fissures, and anterior interhemispheric fissure. Intraventricular hemorrhage and hydrocephalus are present. **C,** CT angiography source image reveals left anterior communicating aneurysm *(arrow).* **D,** Three-dimensional reconstruction from catheter angiogram reveals relationship between aneurysm neck *(arrows)* and adjacent A1 and A2 segments of the anterior cerebral artery and anterior communicating artery. *ICA,* Internal carotid artery.

commonly small and beam hardening artifact from the skull base may obscure the CSF spaces. When subtle SAH is suspected, always check the occipital horns of the lateral ventricles and the fourth ventricle for hyperdensity as extension of hemorrhage into the ventricular system is common. The intraventricular hemorrhage (IVH) layers and becomes relatively hemoconcentrated in the dependent portions of the ventricles. Presence of IVH therefore helps to confirm the diagnosis of SAH in cases with subtle findings.

Nonaneurysmal Subarachnoid Hemorrhage. One cause of angiographically negative SAH is benign perimesencephalic SAH (Fig. 3-56). The clinical presentation is similar to that seen in mild aSAH except patients are typically younger and there is a male predominance. Anecdotal evidence suggests that it may be associated with coitus. These cases have an uncomplicated clinical course (hydrocephalus and vasospasm may develop but are typically mild) and rehemorrhage does not occur (hence the name "benign"). Current opinion is that hemorrhage is venous in origin and arises from the rich retroclival venous plexus.

Benign perimesencephalic subarachnoid hemorrhage is a diagnosis of exclusion. Only after a negative angiogram and a characteristic appearance of the site of hemorrhage adjacent to the basilar tip or in the interpeduncular cistern, can one begin to raise the presumption

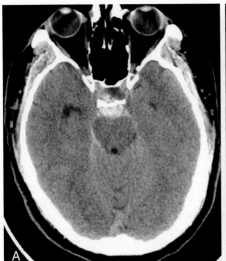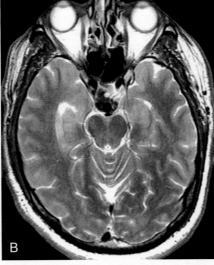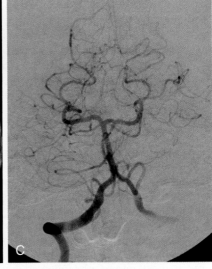

FIGURE 3-56 Perimesencephalic subarachnoid hemorrhage. **A,** Computed tomography scan reveals focal subarachnoid clot posterior to the clivus anterior to the brain stem. **B,** T2-weighted image reveals acute hypointense clot in the interpeduncular cistern. **C,** Right vertebral artery anterior posterior view from catheter angiogram is normal. There is no evidence of posterior circulation aneurysm.

that it is nonaneurysmal. At least one follow-up angiographic study (CTA or catheter angiogram) should be performed to confirm that no aneurysm is present. MR is also a useful adjunct. Focal T1 hyperintense clot surrounding the distal basilar artery is usually seen from 3 to 10 days after SAH.

VASCULAR ABNORMALITIES

Cerebral Aneurysms

An aneurysm is a focal dilatation of an artery that may be associated with numerous clinical syndromes and conditions (Box 3-5). Many different types of aneurysms involve the CNS and are listed in Box 3-6. The most frequent aneurysm encountered in the CNS is the saccular ("berry") aneurysm. A fusiform aneurysm is a diffuse long segment enlargement of a vessel, most commonly the distal vertebral, basilar or proximal middle cerebral artery. Fusiform aneurysms are most often the result of severe atherosclerosis but may also be seen in traumatic and spontaneous arterial dissection, in association with vasculopathies and congenital conditions including collagen disorders and neurofibromatosis I. Septic emboli may lead to development of mycotic aneurysms, which are typically small and arise from distal vessels most frequently in the MCA distribution. Aneurysms may be seen on feeding vessels and or within the nidus of high flow AVMs. Neoplastic aneurysms result from tumor emboli and subsequent growth of the neoplasm through the vessel wall (pseudoaneurysms).

Saccular aneurysms are outpouchings from a parent vessel most often at a point of vessel branching. They form because of damage to the endothelium, thinning of the tunica media, and fragmentation of the internal elastica likely the result of shear forces where a formerly straight vessel curves and gives off branches. The etiology of aneurysms remains controversial. The advent of noninvasive vascular imaging has made it clear that

---

BOX 3-5 Disorders Associated with Intracranial Aneurysms

3M syndrome
Alkaptonuria
Anderson-Fabry disease
Autosomal dominant polycystic kidney disease (10% of asymptomatic patients)
Behçet disease
Coarctation of the aorta
Collagen vascular disease
Ehlers-Danlos syndrome type IV
Familial idiopathic nonarteriosclerotic cerebral calcification syndrome
Fibromuscular dysplasia
Hereditary hemorrhagic telangiectasia
Homocystinuria
Marfan syndrome
Moyamoya disease
Neurofibromatosis type I
Noonan syndrome
Pseudoxanthoma elasticum
Sickle-cell disease
Systemic lupus erythematosus
Takayasu disease
Tuberous sclerosis
Wermer syndrome
α-Glucosidase deficiency
α1-Antitrypsin deficiency

BOX 3-6 Aneurysm Types

Dissecting
Fusiform or atherosclerotic
Mycotic
Neoplastic
Pseudoaneurysm
Saccular or berry
Traumatic

saccular aneurysms are rarely seen in children or young adults unless there is a predisposing factor. Aneurysms are typically acquired lesions that usually develop after the fourth decade of life. Although not congenital, there are clearly genetic forces at work. Multiple aneurysms are present in approximately 20% of patients presenting with aneurysmal SAH (aSAH) and there is a higher than expected incidence of aneurysms (7% to 20%) in individuals with first-degree relatives who have suffered aSAH. Predisposing conditions include fibromuscular dysplasia, polycystic kidney disease, collagen disorders, Marfan syndrome, Loeys-Dietz syndrome etc. Pathologic evaluation of vessel wall histology in patients who have had aSAH reveal no differences from normal vessels. Conditions that are believed to increase the incidence of aSAH include cigarette smoking, rapid increase in blood pressure (e.g., cocaine use), and binge drinking.

Anterior communicating artery complex aneurysms rupture most commonly, followed by distal ICA (posterior communicating, anterior choroidal, ophthalmic segment origins and carotid terminus) and MCA aneurysms. Vertebrobasilar aneurysm rupture is less frequently seen compared with anterior circulation aneurysms. The incidence of unruptured aneurysms based on angiographic and autopsy studies is about 5%.

Incidentally discovered unruptured aneurysms that are smaller than 3 to 5 mm are unlikely to bleed, and the incidence of hemorrhage increases with aneurysm size. This fact leads to an apparent paradox because many ruptured aneurysms are less than 5 mm (some as small as 1 to 2 mm). The paradox is solved if one considers the natural history of aneurysms. An aneurysm forms and grows acutely. During the acute stage, the aneurysm may bleed (even when small) or stabilize. Aneurysms that stabilize at smaller than 5 mm are less likely to grow and unlikely to bleed. Aneurysms that stabilize at more than 5 mm are likely to grow and many may eventually hemorrhage. Overall we say that aneurysms bleed at a rate of 1% to 2% per year (and some say that is an overstatement). So, to treat or not to treat—that is the question. There are no clear guidelines on the appropriate treatment (surgical excision or endovascular ablation or even watchful waiting) for small unruptured aneurysms. Although size matters, it is not the sole determining factor. Aneurysm morphology, patient comorbidity, patient age, skill/experience of particular treating neurosurgeon and/or interventional neuroradiologist, and most importantly the patient's wishes must all be factored into the therapeutic plan. If a decision is made not to treat an aneurysm, serial noninvasive studies (CTA or MRA) should be performed because increase in aneurysm size and/or change in morphology are indications of instability and possible future hemorrhage.

While the treatment of unruptured aneurysms remains controversial, the treatment of ruptured aneurysms is not. Aneurysmal SAH is a devastating and often fatal disease. Approximately 12% of patients with aSAH die before reaching the hospital. Forty percent of hospitalized patients die within 1 month of ictus and one third of the survivors have major neurologic deficits. Rebleeding occurs in 20% of untreated patients who have already bled within 2 weeks of initial hemorrhage, in 30% by 1 month, and in 40% by 6 months. Rebleeding is associated with a mortality rate in excess of 40%. Even after 1 year the rebleeding rate is 2% per year. New aneurysms have been reported to develop in at least 2% of patients with previously ruptured aneurysm with risk of rupture 6 to 10 times higher in patients with previous SAH who have a new aneurysm.

Work-up of Aneurysmal Subarachnoid Hemorrhage

The work-up of aSAH has evolved over the past few years under the influence of rapidly advancing technology (CTA) and new treatment options (endovascular versus open surgical). The initial imaging exam remains a CT scan (see Fig. 3-55). In addition to documenting the presence of SAH, the pattern of hemorrhage will typically be a good indicator of the site of hemorrhage if multiple aneurysms are present. Anterior communicating artery (ACoA) aneurysms produce symmetric SAH centered on the suprasellar cistern, anterior interhemispheric fissure and cistern of the lamina terminalis (the CSF space below the frontal horns of the lateral ventricles). Extension into the lateral ventricles from the cistern of the lamina terminalis is common. When hemorrhage is brisk, focal clot may develop in the inferior medial frontal lobe. Posterior communicating (PCoA)/anterior choroidal and carotid terminus aneurysms produce hemorrhage centered on the suprasellar cistern often with extension into the ipsilateral sylvian fissure and anterior portion of the perimesencephalic cistern. MCA aneurysms typically arise from the lateral portion of the horizontal segment of the MCA where the vessel trifurcates (M1-M2 junction) into its sylvian branches. SAH from MCA aneurysms has a similar pattern to that seen with PCoA aneurysms, but SAH is most marked near the root of the sylvian fissure rather than in the suprasellar cistern. Brisk hemorrhage from an MCA aneurysm may produce thick clot that expands the sylvian fissure and mimics a parenchymal hematoma. Basilar tip aneurysms produce SAH centered in the interpeduncular cistern with extension into the suprasellar cistern and the anterior and posterior perimesencephalic and prepontine cisterns. PICA aneurysms produce SAH that is confined to the posterior fossa and upper cervical region and are therefore often difficult to detect. Presence of apparently isolated IVH in the fourth ventricle should raise suspicion of a ruptured PICA aneurysm.

At the time of the initial exam, some degree of hydrocephalus is virtually always present because of partial obstruction of CSF outflow. The presence of severe hydrocephalus requires emergent placement of a ventricular drain and therefore should be reported to the referring physician. In addition to providing information on the likely site of hemorrhage, the initial CT also provides prognostic information. Poor outcome is associated with "thick" SAH (because of the likely development of vasospasm) and extensive IVH.

The Fisher grade of subarachnoid hemorrhage reflects patient prognosis based on size and extent of intracranial hemorrhage: Grade 1 has no SAH present, Grade 2 is less than 1 mm thick on CT, Grade 3 is more than 1 mm thick on CT, and Grade 4 shows IVH or parenchymal blood. The higher the grade the worse the prognosis. The Hunt and Hess classification scheme is also referred to and estimates long-term prognosis based on clinical parameters at presentation (Table 3-4).

In most institutions the noncontrast CT is followed immediately by a CTA. With modern multidetector CT scanners these exams can be completed in minutes as long as there is good venous access. If the patient is deemed to be a surgical candidate, the anatomic information provided by the CTA is often sufficient to proceed to surgery, eliminating the need for catheter angiography. If the patient is deemed to be a candidate for endovascular treatment, diagnostic angiography and endovascular intervention are performed in the same sitting. In all cases of confirmed or strongly suspected aSAH, a negative CTA should not be considered sufficient for exclusion of aneurysm as the aneurysm may be transiently thrombosed at the time of imaging. Catheter angiography with visualization of all intracranial vessels should be performed in these cases.

The goal of angiography (however performed) is to characterize the aneurysm and its relationship to adjacent vessels so that appropriate therapy can be planned (Fig. 3-57). Key findings that need to be recognized and reported include (1) aneurysm size including maximum diameter and width; (2) aneurysm neck morphology—the neck may be narrow or broad based and may be incorporated into the wall of parent vessel and/or the origin of the branching vessel (e.g., PCoA aneurysms) requiring vascular reconstruction at the time of intervention (e.g., stent placement); (3) shape and contour of the aneurysm—multilobed aneurysms, irregular aneurysms, and aneurysms with focal outpouchings at the dome ("Murphy's tit") are indicators of prior hemorrhage and/or propensity for future hemorrhage; (4) filling defects within aneurysms are indicative of acute clot that carry an increased risk for embolization; (5) wall thickness—thick walled aneurysms are likely chronic with episodes of prior intraluminal or intramural hemorrhage; (6) precise relationship to adjacent vessels; (7) normal anatomic variations of the circle of Willis and vertebral basilar system have a significant effect on treatment choices: during treatment, extra care must be taken when an aneurysm arises near the origin of a "dominant" PCoA that provides all or most of the arterial supply to the posterior cerebral artery, or when there is an ACoA aneurysm in an individual with a hypoplastic A1 segment of anterior cerebral artery; (8) presence of focal or diffuse vasospasm; (9) aneurysm wall and neck calcification, which will make successful clipping more difficult because the "squeeze" cannot be applied with such "hard" calcified walls; and (10) additional aneurysms. When multiple aneurysms are present the likely source of SAH must be determined. Indicators of source of hemorrhage include pattern of SAH on CT, size and shape of aneurysm (the largest and/or most irregular is the most likely culprit) and location of vasospasm if present. If the site of hemorrhage cannot be determined on the basis of CT and angiography, MR can be performed. Focal clot can

TABLE 3-4 Hunt and Hess Grading System

Grade	Signs and Symptoms	Survival
1	Asymptomatic or minimal headache and slight neck stiffness	70%
2	Moderate to severe headache; neck stiffness; no neurologic deficit except cranial nerve palsy	60%
3	Drowsy; minimal neurologic deficit	50%
4	Stuporous; moderate to severe hemiparesis; possibly early decerebrate rigidity and vegetative disturbances	20%
5	Deep coma; decerebrate rigidity; moribund	10%

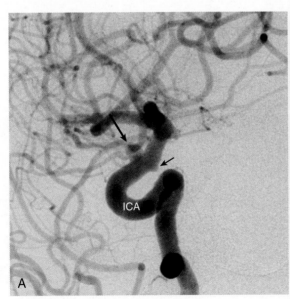
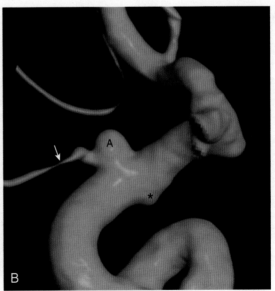

FIGURE 3-57 Conventional angiogram in the work-up of aneurysm. **A,** Following suspicion for aneurysm on time-of-flight magnetic resonance angiography head (not shown), selective internal carotid artery (ICA) injection in the lateral projection shows the ICA, a small outpouching along the posterior aspect of the ICA reflecting infundibular origin of the posterior communicating artery *(small arrow),* and an aneurysm arising from the paraophthalmic segment of the ICA *(large arrow).* **B,** Three-dimensional reconstruction from rotational angiogram beautifully depicts the aneurysm (A), with the ophthalmic artery *(arrow)* arising from the base of the aneurysm, which will add complexity to the patient's treatment options. Infundibular origin of the posterior communicating artery is indicated by an *asterisk.*

be more readily identified on MR than CT and the location of clot is an excellent indicator of site of hemorrhage.

Aneurysms larger than 3 mm rarely present diagnostic difficulty on CTA and MRA; however, small aneurysms may be difficult to diagnose even on catheter angiography. A small looping arterial branch may be mistaken for an aneurysm on a single angiographic view. Three-dimensional CTA and rotational catheter angiography have made it relatively easy to differentiate loops from small aneurysms. Some vessels, in particular the posterior communicating arteries, may have a somewhat dilated origin (infundibulum) that may mimic an aneurysm. Typically the branch vessel is seen to "arise" from the apex of the conical vessel origin. When the branch vessel is small it may be difficult to visualize on CT angiographic images, and therefore careful evaluation of source images is necessary to differentiate between infundibulae and aneurysms. It is of course imperative to evaluate the entire vascular system. Aneurysms in unusual locations can be overlooked unless a systematic search of the angiographic exam is undertaken.

In the week following initial diagnosis and treatment, clinical deterioration may occur for a number of reasons: (1) recurrent hemorrhage in particular if ablation of the aneurysm cannot be achieved. Recurrent hemorrhage will often extend into the brain parenchyma because the aneurysm has become adherent to the adjacent brain; (2) hydrocephalus may progress if a ventricular catheter is not in place or is not functioning; or (3) the most common and serious cause of deterioration is vasospasm as a result of chemical irritation to the arterial wall

(Fig. 3-58). Spasm usually begins approximately 3 days after the initial bleed and can persist or worsen over the next 2 weeks if not treated. It may be focal or diffuse. If unchecked, vasospasm can lead to infarction distal to the narrowing. Medical treatment includes hypertension, hypervolemia, and hemodilution ("Triple H" therapy sometimes supplemented with calcium channel blockers). Endovascular treatments including angioplasty and stenting may be performed when medical therapy is ineffective. The diagnosis of vasospasm (in particular before it leads to permanent infarction) is difficult. On CT, vascular distribution hypodensity is visualized only after infarction has occurred. Transcranial Doppler ultrasound and CTA can demonstrate luminal narrowing and decreased flow before infarction occurs, but many patients have "clinical vasospasm" without angiographic abnormality. CT perfusion can be performed in conjunction with CTA (as it is in acute stroke) to assess blood flow and to guide treatment.

Assuming that the patient survives the initial postictal period, serial CT scans will show resolution of SAH and/or postoperative hemorrhage and edema. In most individuals hydrocephalus will also resolve. Less than 10% of patients with aSAH require permanent shunting. Residual gliosis and volume loss will be present at sites of prior parenchymal hemorrhage and/or severe vasospasm.

Whatever the treatment, it is important to assess for aneurysm regrowth. Even minimal residual filling of an aneurysm can lead to progressive enlargement of the aneurysm with peripheral displacement of the clips and or coils. Currently follow-up aneurysm checks are performed

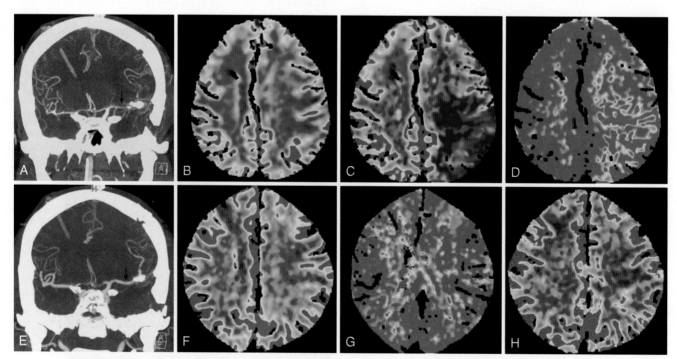

FIGURE 3-58 Postsubarachnoid hemorrhage vasospasm. **A,** Coronal maximum intensity projection (MIP) reconstruction from computed tomography angiography (CTA) performed on patient with treated left middle cerebral artery (MCA) aneurysm shows severe narrowing with beaded appearance of the left M1 segment *(arrow)*, indicating vasospasm. **B,** Symmetric appearance of cerebral blood volume (CBV) map from computed tomography perfusion study suggests that there is no infarcted tissue. However, cerebral blood flow (CBF) map and mean transit time (MTT) maps (**C, D,** respectively) show at-risk brain tissue in the left MCA distribution. **E,** Following medical therapy, coronal MIP reconstruction from repeat CTA shows improved patency of the left M1 segment *(arrow)*. Not surprisingly, the CBV, MTT, and CBF maps (**F, G, H,** respectively) show no asymmetries, indicating restored perfusion to this region.

with MRA for coiled aneurysms and CTA for clipped aneurysms because coils produce more artifact on CT and clips produce more artifacts on MR. Aneurysm clips and coils cause artifacts on both CT and MR that obscure the brain at the levels of the treated aneurysm. Catheter angiography is performed when aneurysm regrowth is suspected, for retreatment planning.

Angiographically, negative spontaneous SAH occurs in approximately 10% of cases. It may be seen when there is severe vasospasm or when the aneurysm has filled with clot, when the source of hemorrhage is within the spinal canal, or when hemorrhage arises from the venous plexus posterior to the clivus (benign perimesencephalic nonaneurysmal subarachnoid hemorrhage). Spontaneous thrombosis of an aneurysm may seem like a good thing but unfortunately these aneurysms virtually always recur and subsequently bleed. Therefore when the initial angiographic studies are negative or reveal only vasospasm but the pattern of SAH on CT or MR (e.g., focal clot at a common location for aSAH) is suggestive of aneurysmal hemorrhage repeat angiography within 1 week of the initial bleed is necessary. This is a truly precarious situation because recanalization can occur at any time and hemorrhage (often massive) often follows shortly thereafter. On rare occasions SAH occurs secondary to a vascular anomaly in the cervical spine such as an AVM or AV fistula. MR studies of the cervical spine can be performed to exclude this possibility in particular if large amounts of SAH are identified around the foramen magnum or in the upper cervical spine on routine head CT scans.

Capillary Telangiectasias

These lesions have a propensity for the pons, although they may be seen in other regions of the brain. They are angiographically occult but can be identified on enhanced MR or, rarely, CT. They vary in size but average about 3 mm in diameter. Cerebral capillary telangiectasias are rarely seen in hereditary hemorrhagic telangiectasia (HHT). Telangiectasias are difficult to diagnose because the vast majority do not hemorrhage and therefore have no imaging features. They are typically isodense on noncontrast CT and isointense on MR and therefore are only appreciated on enhanced scans as small foci of nodular enhancement. SWI or GrE scans may show a focal area of slightly hypointense signal credited to hemosiderin deposition. They are almost always asymptomatic and of concern only in that they may mimic other processes, in particular small metastatic lesions that may also be isodense/isointense on unenhanced images. Lack of progression on serial exams allows for differentiation from more aggressive lesions. Radiation therapy is associated with development of telangiectasias.

Cavernous Angioma (Cavernous Hemangioma, Cavernous Malformation, Cavernoma)

Like telangiectasias these lesions are also angiographically occult. They can be associated with a developmental venous anomaly (DVA, formally called venous angiomas). Cavernomas are more common than telangiectasias and while the majority are asymptomatic, recurrent hemorrhage may occur leading to acute focal deficits and/or seizures. Pathologically a cavernoma consists of dilated thin walled venous channels without intervening normal brain tissue. Areas of hemorrhage of varying ages in and around the cavernoma are present with extensive hemosiderin deposition and in some cases dystrophic calcification. While most lesions are small, some become quite large because of recurrent hemorrhage and can have associated heme-filled cavities ("hemic cysts"). Cavernomas can be congenital and multiple cavernomas (10% to 20% of cases) are typically familial. On the other hand, cavernomas may be acquired lesions occurring after radiation therapy. They have a predilection for the brain stem but may occur anywhere within the brain. In most instances the hemorrhages are small and have a good prognosis but on occasion a cavernoma may bleed repeatedly and require surgical excision or focused radiosurgery.

On noncontrast CT, cavernomas appear as high density regions with or without associated calcification. In the absence of recent hemorrhage they lack edema and mass effect. Acute hemorrhage may obscure the underlying malformation. MR features are highly characteristic (Fig. 3-59). There is a hypointense rim on all sequences that is most dramatic on T2 and susceptibility-weighted sequences (GrE, echo planar imaging [EPI], SWI). The hypointensity extends from the margin of the lesion "staining" the adjacent parenchyma. The center of the lesion is more variable. If the lesion is small and there has been no recent hemorrhage the entire cavernoma may be hypointense. In larger cavernoma heterogeneous T1 and T2 hyper- and hypointensity are often present producing a highly characteristic "popcorn pattern." Mulberry, raspberry, and blackberry patterns have also been ascribed. Take your pick; they are all really good snacks. Enhancement is uncommon. Presence of significant T1 hyperintensity indicates recent hemorrhage. Adjacent parenchymal edema is an indicator of acute hemorrhage.

Developmental Venous Anomaly (Venous Angioma)

A DVA consists of a network of dilated medullary veins converging in a radial fashion to a large vein that drains into either deep or superficial veins. The surrounding brain parenchyma is normal. The incidence of DVAs is 2.6%. It is hypothesized that DVAs arise because of prenatal occlusion of a vein draining into the deep or superficial system. As a consequence, a large collateral vein forms that drains through the white matter to the other venous system. DVAs are usually asymptomatic. Rarely, hemorrhage occurs, in particular in association with a cavernoma.

While DVAs have classic angiographic findings, they are more difficult to appreciate on CT or MR in particular without contrast. On CT, the intrinsic density is similar to brain, and on MR, the flow within the DVA is most often too slow to produce hypointensity. On enhanced scans small veins near either the surface of the brain or the subependymal region coalesce on a linear vascular structure that traverses the white matter. Medusa head DVAs may drain from different angles to a single large vein (see Fig. 3-59, *D*). Vague amorphous enhancement represents the small dilated venules that drain into the DVA. DVAs are not visible on MRA because flow is slow. On occasion, DVAs may be quite large draining most of a cerebral hemisphere. These lesions are visible on unenhanced MR scans.

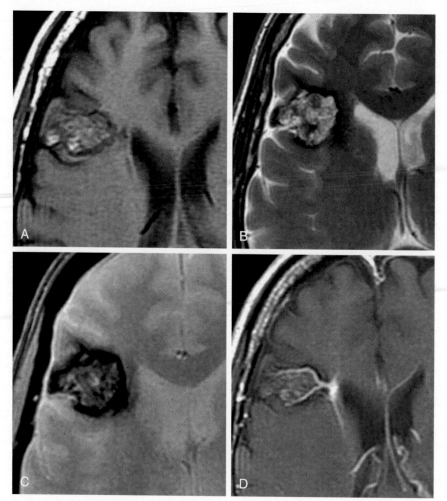

FIGURE 3-59 Cavernous malformation and developmental venous anomaly (DVA). **A,** T1-weighted image (T1WI) reveals a mixed-intensity lesion with small foci of hyperintensity and mild peripheral hypointensity. **B,** T2WI reveals a peripheral hypointense margin with central foci of both hypointensity and hyperintensity. **C,** Gradient-echo scan reveals marked peripheral hypointensity that extends into and "stains" the adjacent parenchyma. **D,** Enhanced T1WI reveals linear enhancing DVA that surrounds the cavernous malformation and drains into the subependymal venous system.

One important aspect of this lesion is that the DVA is a compensatory drainage route for normal brain. Sacrifice of this pathway can produce venous infarction of tissue being drained. Therefore these lesions are not amenable to surgical or endovascular treatment unless hemorrhage is either life threatening or resection of the adjacent brain will not cause major deficit.

Arteriovenous Malformations

True AVMs are considered congenital anomalies of blood vessels that arise in fetal life but usually become symptomatic in the third or fourth decade of life. AVMs are common causes of spontaneous parenchymal hemorrhage in adults (20 to 50 years of age). Hemorrhage may occur in any portion of the brain. They have a tendency to hemorrhage at a rate that has been estimated in a study from Finland to be 4% annually (2% to 3% in USA) with an annual mortality rate of 1% and a mean interval between hemorrhagic events of 7.7 years.

Steal phenomenon is a recognized complication of AVMs in which blood preferentially seeks the AVM and adjacent brain parenchyma is hypoperfused. Steal can produce focal neurologic symptoms, seizures, and ultimately parenchymal loss in the affected part of the brain even without hemorrhage. High flow aneurysms can be detected on feeding arteries and may be a source of hemorrhage rather than the AVM proper.

The AVM consists of a feeding artery or arteries, which are usually dilated, and a cluster of entangled vascular loops (the core or nidus) connected to abundant vascular channels where the arterial blood is shunted, finally terminating in enlarged draining veins. The draining veins are typically much larger than the feeding arteries. Intranidal aneurysms can be detected in over 50% of AVMs. In unruptured AVMs, the presence of nidal aneurysms and relative small size and number of draining veins has been associated with an increased risk of hemorrhage. AVMs can be supplied by a single or multiple vascular systems (e.g., anterior and middle cerebral arteries). Dural arteries can be recruited to supply superficial AVMs, in particular in the posterior fossa. AVMs have been classified according to their size, location, and venous drainage using a grading system proposed by Spetzler and Martin: the higher the cumulative score, the greater the surgical risk for treatment (Table 3-5).

TABLE 3-5   Arteriovenous Malformation Classification

Feature	Score
SIZE	
<3 cm	1
3-6 cm	2
>6 cm	3
ELOQUENT	
No	0
Yes	1
VENOUS DRAINAGE	
Superficial only	0
Any deep	1

From Spetzler RF, Martin NA. A proposed grading system for arteriovenous malformations. *J Neurosurg.* 1986;65:476-483.

The diagnosis of AVM without associated hemorrhage can be made in most cases on routine CT or MR (Fig. 3-60). On CT, the dilated vessels (mostly the draining veins) in or adjacent to the brain parenchyma are mildly hyperdense without contrast (blood pool effect) and have a serpentine, punctate, or an irregular mélange configuration. Curvilinear or speckled calcification may be present. Surrounding hypodensity indicative of gliosis from chronic ischemia or prior hemorrhage may be present. The vascular nature of these lesions is apparent on enhanced scans where the intensely enhancing curvilinear structures are visible. On MR, curvilinear flow voids in dilated veins are observed and modestly dilated proximal arteries can also be noted. AVMs are visible on routine MRA and CTA, but these studies do not yet provide much useful temporal information because the feeding arteries and draining veins are visualized simultaneously. The introduction of time resolved enhanced MRA and CTA has dramatically improved the utility of these techniques in assessment and treatment planning.

In approximately 30% of patients with parenchymal hemorrhage secondary to AVMs the abnormal vessels cannot be detected on routine CT or MR. Be sure to carefully observe the enhanced MR scans because the AVM may be apparent on careful scrutiny of these images as a small tuft where veins drain. Missing the AVM occurs most often with small AVMs and when the AVM is located near a major dural venous sinus. In these cases the draining veins need to travel only a short distance and may not be apparent on routine imaging. Less commonly, the hemorrhage will temporarily obscure the underlying AVM. Therefore a normal CT, MR, CTA and/or MRA does not "rule out" an underlying AVM.

The definitive study for detection and characterization of AVMs is catheter angiography. The diagnosis is made by demonstrating enlarged feeding arteries, the nidus, and early filling of enlarged draining veins. If the initial angiogram is negative but other causes of hemorrhage have been excluded repeat angiography is warranted because small AVMs may be temporarily obscured by mass effect from acute hemorrhage.

Treatment of AVM depends on its size, location, and angioarchitecture. It consists of endovascular therapy, surgery, and radiation therapy, or combinations thereof. In cases of AVMs less than 3 cm in diameter, stereotactic radiotherapy is the treatment of choice (depending on location) with cure rates of 80% to 90% within 2 to 3 years. An immediate complication associated with surgical or endovascular treatment of AVMs with extensive steal is brain swelling (sometimes resulting in rapid herniation and death) and/or hemorrhage. The chronically hypoperfused brain is unable to autoregulate because of long-standing ischemia and the abrupt change in hemodynamics after obliteration of the AVM. Staged treatment is advocated to prevent perfusion breakthrough from occurring. The rate of cure for endovascular treatment alone is 10% to 20%.

Vein of Galen aneurysmal malformations, better termed as median prosencephalic vein malformations, are a special type of vascular malformation presenting most commonly in the pediatric setting. While these malformations typically present in infancy with signs of hydrocephalus (rapidly enlarging head size) or high output cardiac failure, the increasing use of prenatal ultrasound is rendering the diagnosis in the prenatal setting as well. The malformation consists of abnormal arteriovenous fistulous connections between primitive choroidal vessels and the median prosencephalic vein of Markowski, a precursor to the vein of Galen, occurring at about 6 to 11 weeks' gestation. This prosencephalic vein fails to regress and becomes aneurysmal. It drains into the straight sinus or a persistent falcine sinus. A normal vein of Galen does not form. Other venous anomalies can coexist, including anomalous dural sinuses and sinus stenosis. On imaging, an enlarged persistent prosencephalic vein is seen draining superiorly from the tent/falx junction into the straight or persistent falcine sinus. Hydrocephalus is virtually always present.

Common complications of vein of Galen malformations include intraventricular hemorrhage and focal dural sinus thrombosis. The brain may be atrophic or at least delayed in development because of in-utero ischemia as a result of chronic steal phenomena. Prognosis is poor without treatment and as rule of thumb the earlier the presentation the worse the outcome. Treatment is endovascular occlusion. Both diagnostic and therapeutic catheter angiography is challenging because flow is very high and total allowable contrast dose is limited by the small size of the infant and often tenuous cardiac status. In many cases treatment must be staged to slowly reduce flow through the malformation.

Dural Arteriovenous Fistula

Dural arteriovenous fistulae (DAVF) are acquired lesions that are the consequence of dural sinus thrombosis with subsequent often sudden recanalization of the sinus resulting in direct communication between the small dilated arteries in the sinus wall and the sinus lumen. This results in increased pressure within the sinus and impaired outflow from the adjacent brain. Development of "downstream" stenosis because of turbulent high flow may further compromise venous drainage and increase venous back pressure. As a result, cortical veins may dilate, which may in turn result in parenchymal hemorrhage, SAH, venous infarction, and elevated intracranial pressure. Increased risk of neurologic deficits is associated with dural fistulae that (1) drain into

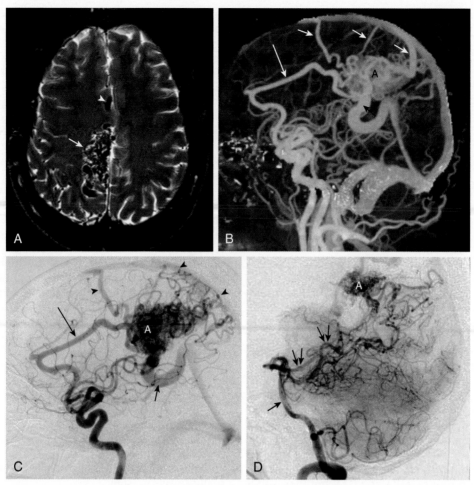

FIGURE 3-60 Arteriovenous malformation (AVM). **A,** Axial T2-weighted image shows large tangle of vessels *(arrow)* representing the nidus of an arteriovenous malformation in the right parietal lobe. Large caliber vessels are seen around the AVM *(arrowhead),* which may reflect arterial supply or venous drainage for this lesion. **B,** Three-dimensional maximum intensity projection reconstruction from subsequently performed computed tomography angiography (CTA) head shows the nidus (A). The major arterial supply from an enlarged pericallosal artery *(white arrow)* and both superficial *(small white arrows)* and deep *(black arrow)* venous drainage are depicted. **C,** Selective internal carotid artery injection from conventional catheter angiogram in the lateral projection again shows arterial supply from an enlarged pericallosal artery *(large arrow)* to the AVM nidus (A), with both early superficial *(arrowheads)* and deep *(small arrow)* venous drainage. **D,** The superiority of conventional angiography over others in diagnosing AVMs is shown with this selective vertebral artery injection in the lateral projection. The posterior aspect of the AVM nidus (A) is also supplied by bilateral posterior cerebral arteries *(double arrows),* which was not appreciated on magnetic resonance or CTA. Basilar artery is indicated by *single arrow.*

the deep venous system, (2) have associated retrograde venous flow, (3) have venous aneurysms, (4) have stenotic channels through which they must drain, and (5) are complex.

Symptoms depend on the location, size of the malformation, and venous drainage pattern. Parasellar malformations draining into the cavernous sinus with retrograde flow in the superior ophthalmic vein (supplied from the meningohypophyseal trunk, accessory meningeal artery, middle meningeal artery, branches of the ascending pharyngeal artery, and other vessels) create a "low flow" nontraumatic carotid-cavernous fistula (CCF; Fig. 3-61). CT and MR reveal enlargement of the cavernous sinus and superior ophthalmic vein as well as stranding of intraorbital fat, enlargement of extraocular muscles and proptosis. Patients present with proptosis and visual loss because of glaucoma

and ophthalmoplegia. Slow flow CCFs can spontaneously resolve (sometimes aided by intermittent ipsilateral carotid artery compression—but please don't try that at home) or can improve after incomplete embolization or even after diagnostic angiography. Endovascular particulate occlusion of feeding arteries can be performed when less invasive measures fail. Fistulae arising from the sigmoid sinus can present with tinnitus and bruit over the temporal bone (what? WHAT?). Transverse sinuses and superior sagittal sinus fistulae may present with parenchymal hemorrhage and increased intracranial pressure associated with sinus stenoses ("pseudotumor cerebri/ idiopathic intracranial hypertension"). Rarely, patients may present with progressive dementia because of chronic hypoperfusion. Even patients who have been symptomatic for years may demonstrate resolution of dementia after treatment of the fistula.

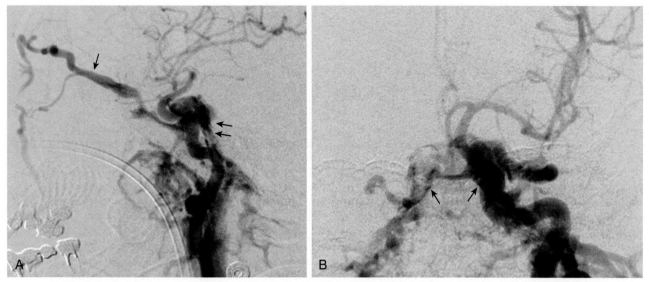

FIGURE 3-61 Cavernous carotid fistula. **A,** Lateral projection from conventional catheter angiogram with left internal carotid artery (ICA) injection shows abnormal opacification and enlargement of the superior ophthalmic vein *(single arrow)* and abnormal enhancement of the cavernous sinus *(double arrows),* petrosal sinuses, and jugular vein in the arterial phase because of the presence of a cavernous-carotid fistula. **B,** Frontal projection from left ICA injection shows similar early enhancement of major inferior venous sinuses, including bilateral cavernous sinuses *(arrows),* which communicate at the midline. This fistulous connection was subsequently coiled.

Dural fistulae associated with the superior sagittal sinus transverse sinuses and sigmoid sinuses are often difficult to diagnose. Routine CT and MR may be normal or demonstrate a "tight" brain with small ventricles and sulci. Subtle enlargement of cortical veins and dural sinus enlargement may be present but is easily overlooked. In other cases, CT and MR may demonstrate prominent vessels associated with a parenchymal hematoma mimicking an AVM. Chronic fistulae may produce dilated vessels throughout the cranial cavity with little hint as to the actual location of the malformation. CTA and MRA may appear normal or may demonstrate dilated middle meningeal and/or posterior fossa arteries. MRV and CTV may show regions of venous sinus stenosis or focal occlusion.

Catheter angiography is the key to the diagnosis of this lesion. The initial internal carotid or vertebral artery injections may reveal little or no abnormality (often a surprise in patients with parenchymal hemorrhages associated with dilated vessels). Injections into the common carotid arteries or selective injection of the external carotid arteries will demonstrate the dilated meningeal arteries, the site of the fistulae and delayed filling of cortical veins.

Dural malformations can be treated with success by embolization (arterial and/or venous), although permanent closure may demand use of tissue adhesives, which are riskier than particulate embolic agents. These increased risks include skin necrosis, cranial nerve palsy, and visual loss. Another approach to reduce venous hypertension is to reopen by stenting the previously thrombosed sinus with restoration of normal flow patterns.

Stupefy! We guess Harry Potter's stunning spell rendering victims unconscious will have little effect on you as you are probably already out like a light after reading this chapter. Let us know when you wake up, because there's still much more left to tell you!

Chapter 4

Head Trauma

Let's start off with some depressing statistics. Traumatic brain injury (TBI) has an incidence of over 500,000 cases per year, and is the leading cause of disability and death in children and young adults in the United States. The annual cost of TBI, including direct costs and lost income, is estimated to be above $25 billion. It has a peak incidence in 15- to 24-year-olds with males being injured two to three times more frequently than females. The simplest form of TBI is concussion, which may be defined as a transient impairment of neurologic function associated with or without loss of consciousness occurring at the time of injury.

Imaging of head trauma has a primary role both in diagnosing the extent of the traumatic injury and in expediently determining the appropriate therapy. The most efficient method of triage for acute trauma remains computed tomography (CT). It is fast and usually very accurate at detecting acute hemorrhage (Fig. 4-1). CT is excellent for assessing facial and skull fractures. Treatable lesions are readily identifiable on CT and include epidural hematomas (EDHs), large subdural hematomas (SDHs), actively bleeding parenchymal hemorrhages, brain herniation, and depressed skull fractures.

CT does have some pitfalls of which the radiologist should be aware. Not all hemorrhage is high density. Isodense to low density hemorrhages are seen in patients who are severely anemic or in those patients with disseminated intravascular coagulopathy or in the hyperacute actively bleeding stage (Fig. 4-2). Fluid-fluid levels imply active bleeding and

sedimentation, which may be prolonged in patients on anticoagulants. Extracerebral intracranial hemorrhages of the infratemporal region, subfrontal region, and posterior fossa can be difficult to detect on CT. CT is also less sensitive than magnetic resonance imaging (MRI) in detecting diffuse axonal injury and vascular injury.

When evaluating trauma patients on CT, review of the images with brain, subdural, and bone algorithms at the workstation is essential. Wide window settings aid in separating high-density blood from the high density of bone and are particularly useful in acute subdural and epidural hematomas, which can be thin and difficult to differentiate from the calvarium. Bone windowing is essential in the search for fractures, as is the combination of coronal, sagittal, and axial images with very thin sections, particularly of the skull base, temporal bone, orbit, or face (Fig. 4-3).

Patients with TBI have been classified by the Head Injury Interdisciplinary Special Interest Group based on (1) a period of loss of consciousness (LOC), (2) loss of memory for events immediately before or after the trauma, (3) change in mental status, and (4) focal neurologic deficits. The Glasgow Coma Scale (GCS) was devised to provide a uniform approach to the clinical assessment of patients with acute head trauma and is the most frequently used means of grading the severity of injury within 48 hours (Table 4-1).

One issue with the GCS is the marked heterogeneity of location and type of brain lesions in moderate to severe TBI patients. Thus individuals with the same GCS could have markedly different outcomes depending upon the nature of the causative lesion.

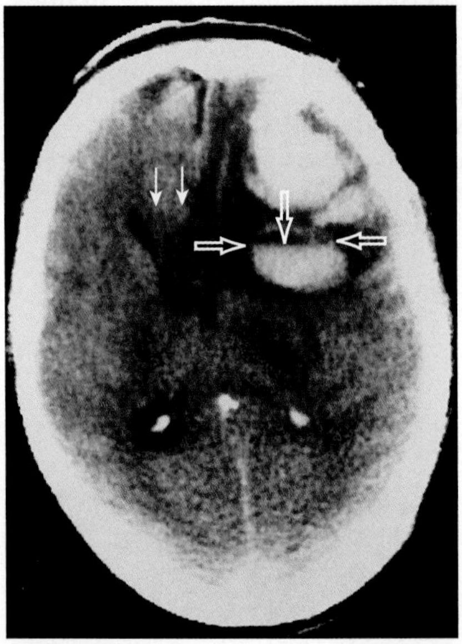

FIGURE 4-1 Computed tomography of acute traumatic intraparenchymal hematoma. Multiple large intraparenchymal hematomas are identified as high-density mass lesions associated with edema in the frontal region. Subfalcine herniation is present *(arrows)*. Also, note hemorrhagic cavity with blood level, which can be observed in traumatic intraparenchymal hemorrhage *(open arrows)*.

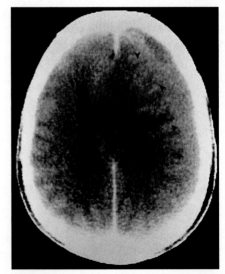

FIGURE 4-2 Computed tomography of isodense subdural hematoma. *Arrowheads* indicate junction of cortex and subdural hematoma. Note that the density of the subdural hematoma is similar to that of cortex.

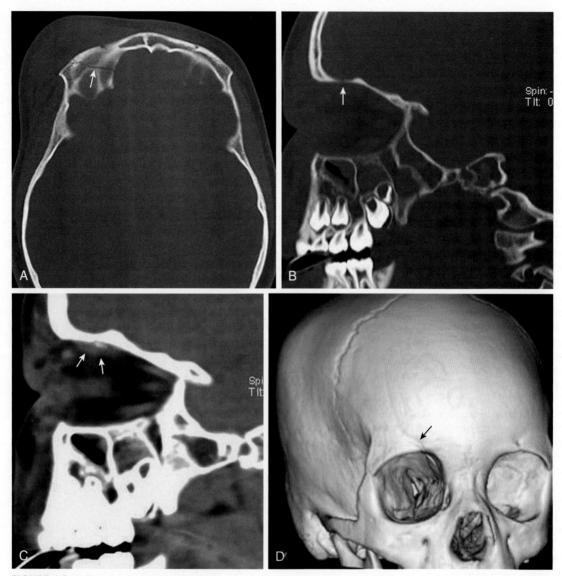

FIGURE 4-3 Right orbital roof fracture. **A,** Axial computed tomographic image in bone windows shows linear nondisplaced fracture coursing the roof of the right orbit *(arrow)*. Note overlying periorbital soft-tissue hematoma. **B,** Sagittal reconstructed image shows the same fracture *(arrow)*. **C,** For those nonbelievers out there, this sagittal soft-tissue reconstruction shows the associated subperiosteal hematoma within the orbit *(arrows)*. **D,** Three-dimensional surface rendered reconstruction of the skull shows the same nondisplaced orbital roof fracture *(arrow)* from a different perspective.

TABLE 4-1   Glasgow Coma Score Grading

Score	Eye Opening	Motor	Verbal
1	No response	No response	No response
2	To painful stimulation	Extensor posturing to pain	Incomprehensible sounds
3	To speech	Flexor posturing to pain	Inappropriate words
4	Spontaneous	Withdraws to pain	Conversant but disoriented
5	Localized movements to pain	Oriented X 3	
6	Follows commands		

Severity definition score: Severe = total score 3–8; moderate = total score 9–12; mild = total score 13–15.

Approximately two thirds of patients with head trauma in the United States are classified as having minor head injury, but less than 10% of these cases have positive CT findings, and less than 1% require neurosurgery. The following clinical parameters in patients with head trauma are associated with positive CT: headache, vomiting, over 60 years of age, drug or alcohol intoxication, deficits in short-term memory, physical evidence of trauma above the clavicles, coagulopathy, and seizure.

MR is more sensitive than CT in detecting all the different stages of hemorrhage (i.e., acute, subacute, and chronic); within the first 4 hours; however, for hyperacute hemorrhage,

CT may be more sensitive than low-field MR. Detecting acute subarachnoid hemorrhage is difficult if fluid-attenuated inversion recovery (FLAIR) imaging is not performed with the MR exam (Fig. 4-4). The direct multiplanar capability of MR and absence of beam hardening artifact enables excellent visualization of the inferior frontal and temporal lobes and the posterior fossa (Fig. 4-5). It is more sensitive than CT in demonstrating diffuse axonal injury, isodense SDHs, minor deposits of hemorrhage with susceptibility-weighted imaging (SWI; particularly at higher field), and coexistent ischemia.

Skull films have little role in the triage or care of the patient with significant head trauma. They are useful only to ascertain whether linear skull fractures are present. CT is excellent at demonstrating depressed or comminuted skull fractures; however, fractures parallel to the plane of the scan slice such as those involving the vertex of the skull may be missed if submillimeter scans from multidetector computed tomography (MDCT) are not reviewed (word to the wise: maintain and view the thin section raw data in trauma cases). Particularly in pediatric patients, three-dimensional surface rendered reconstructions of the skull can improve detectability of skull fractures (Fig. 4-6). Always assess the scout view from the CT for evidence of fracture or high cervical injury.

Computed tomography angiography (CTA) and magnetic resonance angiography (MRA) are both excellent means for detecting vascular dissection. At present, angiography is indicated to define the anatomy of traumatic fistulas, particularly before therapeutic endovascular obliteration and, in some cases, for diagnosis of vascular dissection and pseudoaneurysm. Another indication for angiography is the source and treatment of an expanding hematoma in the neck.

An organized approach to the categorization of head trauma is essential in examining patients undergoing diagnostic imaging. Accurate classification enables prediction of prognosis and institution of appropriate therapeutic management.

CLASSIFICATION OF INJURY

Traumatic brain damage may be separated into two major categories: (1) primary injury and its associated primary complications are directly related to immediate impact damage and (2) secondary complications resulting from the primary injury over time. The radiologist plays an important role in initial assessment but also in anticipating and evaluating for secondary complications which may be preventable or treatable.

Open head injury involves the intracranial contents communicating through the skull and scalp. In closed head injury, there is no communication between the intracranial contents and the extracranial environment.

Primary Injury

Primary injury is the consequence of brain damage occurring at impact. Primary injury can be divided further into focal or diffuse injury. Focal injuries represent contusions and other mass lesions such as epidural or subdural hematomas, which may generate shift of the midline brain structures, which may cause secondary injuries such as herniation with compression of and damage to the brain stem. Contusions may be defined as bruises that are associated with petechial hemorrhages, and are the most common primary intraaxial lesion. Diffuse axonal injury (DAI) is

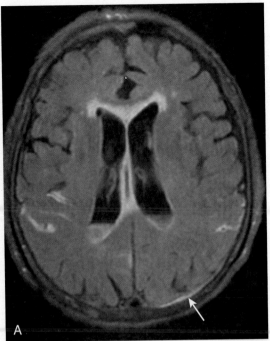

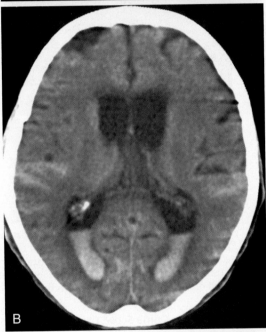

FIGURE 4-4 Subarachnoid hemorrhage on fluid-attenuated inversion recovery (FLAIR). **A,** FLAIR scan shows intraventricular blood as well as high signal in the sulci of the posterior sylvian fissure and inferior frontal sulci bilaterally. Small subdural hematoma on the left is also noted *(arrow)*. **B,** Here is the computed tomography scan showing subarachnoid hemorrhage in similar distribution.

produced by stress-induced movements of the head that severely disrupt axons either completely or incompletely. In its fulminant form, there is diffuse disruption of axons throughout the brain, resulting in permanent coma or death.

Two different mechanisms are principally responsible for primary injury: (1) direct contact between the skull and an object (contact phenomena) and (2) inertial injury resulting from the differential accelerations between white and gray matter. The former is responsible for scalp laceration, skull fractures, intracerebral and extracerebral hematoma, and contusion underlying the immediate contact (coup injuries).

Such contact phenomena generate stress waves, which create skull base fractures, contusions, extraaxial collections, and hemorrhages remote from the site of contact (contra-coup injuries). Severe contact injury is associated with SDH and massive intraparenchymal hemorrhage. Inertial injury results in DAI (usually with a rotational component) and pure SDH after shearing of bridging veins.

Secondary Complications

Secondary complications are temporally removed from the original trauma and result in brain injury including (1) raised intracranial pressure (ICP), leading to increased cerebral edema and hydrocephalus; (2) hypoxia; (3) infection; (4) infarction; and (5) brain herniation. The rapid release of excitatory neurochemicals (glutamate, aspartate)

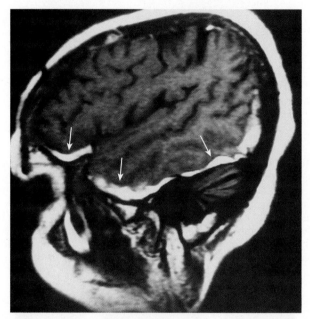

FIGURE 4-5 Sagittal T1-weighted image demonstrates inferior frontal and inferior temporal lobe subdural hematoma extending along the tentorium posteriorly *(white arrows)*.

can cause cellular swelling, unsolicited calcium and magnesium influx into the cells, generation of free radicals, cytokine release, caspase (proteases that cleave proteins that induce apoptosis activation), mitochondrial dysfunction, breakdown of the cytoskeleton, and additional neuronal death. Many of these specific cascades have been targeted with pharmacologic treatments. In the acute period (hours to days), cytotoxic swelling (detectable by diffusion-weighted imaging [DWI]) is the most important factor responsible for increased intracranial volume.

There are factors noted on imaging that may predict TBI outcome: (1) Levels of reduction in NAA (N-acetyl-aspartate) on MR spectroscopy; (2) mesencephalic cistern effacement (the risk of elevated ICP is increased threefold if the cistern is obliterated; (3) subarachnoid hemorrhage (increases the possibility of death twofold); (4) SDH and DAI (two thirds of deaths in head injury are associated with these abnormalities); (5) positive DWI (implies worse outcome and more severe damage); and (6) perfusion deficits (indicating coexistent ischemia). Still, 10% to 42% of patients with severe head injury have normal-appearing initial CT scans, and elevated ICP still develops in 10% to 15% of these patients.

EPIDURAL HEMATOMA

The potential space between the inner table of the skull and the dura is the epidural space. The most common cause of EDH is head trauma with skull fracture of the temporal bone (90% of cases) lacerating the middle meningeal artery or vein. In children, the greater elasticity of the skull permits meningeal vascular injury without coexistent fracture. Tears of the middle meningeal artery (60% to 90%) or venous structures (middle meningeal vein, venous sinus, or diploic veins [10% to 40%]) result in the extravasation of blood and acute EDH. After injury there may be a lucid interval (50% of patients) as the lesion expands before causing midline shift and deterioration (as opposed to DAI, where coma occurs immediately after the injury). Slower bleeding from the meningeal vein or dural sinus temporally delays symptoms. Chronic EDH has also been observed.

EDHs are usually biconvex and are the result of the firm adherence of the dura to the inner table and its attachment

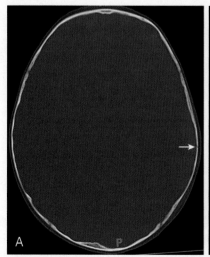
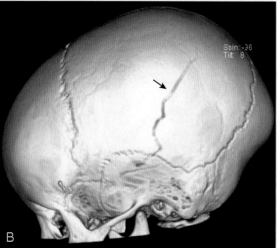

FIGURE 4-6 Utility of three-dimensional surface rendered reconstructions of the skull. **A,** Axial computed tomographic image in bone algorithm shows subtle nondisplaced fracture of the left parietal skull *(arrow).* **B,** This fracture is much more conspicuous on three-dimensional surface rendered reconstruction of the skull *(arrow).*

to the sutures. The collection should not cross a suture (as opposed to SDH). When fractures run through the middle meningeal artery and vein, occasionally a fistula may develop between the two. Pseudoaneurysm of the meningeal artery may also occur. EDHs of the posterior fossa result from tearing of the venous sinus (transverse or sigmoid most commonly) and may be continuous with the supratentorial and infratentorial space. Occasionally the superior sagittal sinus at the vertex may be involved leading to a collection that crosses the midline (in this case across the sagittal suture).

CT reveals a high-density extraaxial mass acutely (Fig. 4-7). There may be some low density in the acute hemorrhagic mass, possibly representing serum extruded from the clot. In these cases, carefully review bone window settings to look for associated fracture (see Fig. 4-6). Chronic EDHs reveal low density and peripheral enhancement, and they may be concave. Chronic EDHs must be distinguished from other epidural masses such as infection, inflammation, and tumor, and from subdural lesions such as empyemas.

On MR, the extraaxial hemorrhage has different appearances depending on the interval between the traumatic event and imaging (Fig. 4-8). If imaged less than 12 hours after trauma, acute hemorrhage generally demonstrates low intensity on T2-weighted images (T2WI) and isointensity on T1-weighted images (T1WI), whereas if the hemorrhage is imaged a few days after the incident (subacute hematoma), it is high intensity on T1WI. One may visualize the medial dural margin as a hypointense rim on all pulse sequences. While skull fractures are elegantly demonstrated on CT, they are poorly visualized on MR.

SUBDURAL HEMATOMA

Hemorrhage into the potential space between the pia-arachnoid and the dura is termed a subdural hematoma.

Acute SDHs result from significant head injury and are caused by the shearing of bridging veins (Fig. 4-9). This occurs because of rotational movement of the brain with respect to the fixed position of these veins at the adjacent venous sinus or dura. The subdural portion of the vein is not ensheathed with arachnoid trabeculae as is the subarachnoid portion and therefore is the weakest segment of the vein. SDHs may also occur in the setting of intracerebral hematoma where the parenchymal clot is in direct continuity with the subdural space. Penetrating injury can also result in acute SDH.

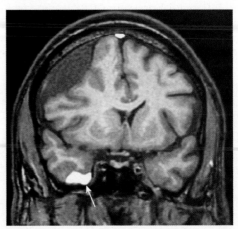

FIGURE 4-8 Extraaxial hemorrhages on magnetic resonance imaging. On this coronal T1-weighted image, there is large hypointense acute epidural hematoma overlying the right frontal lobe and causing mass effect on the adjacent brain parenchyma. There is also a small T1 hyperintense subacute hematoma along the floor of the middle cranial fossa *(arrow)*, which would be difficult to detect on computed tomography because of beam hardening artifact, partial volume averaging, and reduced contrast sensitivity.

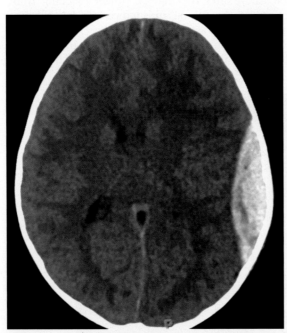

FIGURE 4-7 Computed tomography of acute epidural hematoma (same patient as in Fig. 4-6). Note the convex shape of the epidural hemorrhage.

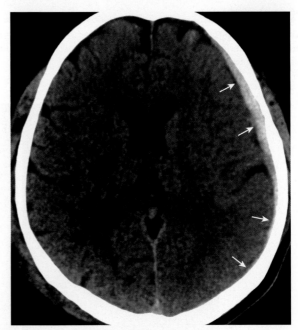

FIGURE 4-9 Subdural hematoma. Axial computed tomographic image demonstrates hyperattenuating left hemispheric acute subdural hematoma *(arrows)*. There is mild effacement of the subjacent sulci but no midline shift is present.

SDHs are seen in approximately 30% of patients with severe closed head trauma. In this circumstance the subdural hemorrhage might be associated with acute parenchymal injuries. Patients with subdural hemorrhages without parenchymal injury have a better prognosis than those patients who also suffer parenchymal injury. Acute SDH is associated with a 35% to 50% mortality rate, and of those who survive, most have functional limitations. Strange though it may seem, SDHs are more common contracoup to the site of the initial contact in head injury than directly at the point of contact (coup injury). This is because of the sloshing around of the brain in the intracranial cavity from the aforementioned contact waves.

The SDHs in the elderly, at discovery, are larger because of the generalized loss of brain parenchyma allowing greater unimpeded growth before mass effect and symptoms develop. So perhaps there is some advantage to a shrunken brain after all.

The temporal nomenclature of the SDH is arbitrary. Lesions occurring at the time of the initial injury are considered acute, although symptoms may take up to a few days to become manifest. Those lesions becoming symptomatic between approximately 3 days to 3 weeks are subacute, whereas those lesions that are diagnosed after 3 weeks are considered chronic. Hemorrhages present on CT acutely but not discovered until weeks later are considered misses!

Acute SDHs on CT are high-density crescentic extracerebral intracranial masses (see Fig. 4-9). SDHs are not confined by the cranial sutures and may extend over an entire cerebral hemisphere but will not cross the midline falx. SDHs generally arise over the convexities but can also occur in the posterior fossa, middle cranial fossa, and/or along the tentorium. The thickness of the hematoma can range from pencil-thin to large and can at times be convex inwardly. Rarely, acute SDHs have been reported to be isodense or hypodense to brain on CT. This may be correlated with

significant anemia (hemoglobin level 8 to 10 g/dL), disseminated intravascular coagulopathy, ongoing bleeding, or tears in the arachnoid membrane, leading to dilution of the red blood cells with cerebrospinal fluid (CSF). This seems to be especially common in children (so called hematohygromas).

Subacute SDHs may be isodense to hypodense; mass effect because of the hematomas can render them conspicuous (see Fig. 4-2). Intermediate window settings on CT are useful in separating thin SDHs from the bony calvarium (Fig. 4-10). Subacute and chronic subdural lesions show dural-based enhancement. This is caused by the vascularization of the subdural membranes formed between 1 to 3 weeks following injury. These vessels are not associated with tight junctions. They leak contrast and may easily be torn resulting in growth of the subdural. Repeated episodic bleeding results in fibrous septations and compartments within the hematoma. Subacute and chronic SDHs may also display layering with dependent cells and cellular debris and an acellular supernatant. The dependent portion is of higher density than the supernatant. Hematomas may also occur between the leaves of the tentorium. Tentorial subdurals are sneaky; they tend to fade out laterally and stop abruptly at the medial margin of the tentorium (Fig. 4-11).

Chronic SDHs are usually low density. High- and low-density levels observed in these lesions may be caused by rebleeding into the chronic subdural collection. There may be no history of trauma or only a history of minor head trauma. Other contributing conditions include coagulopathy, alcoholism, increased age, epilepsy, and surgery for ventricular shunts. Over 75% of chronic subdurals occur in patients over 50 years of age. Chronic subdurals occur in infants as a result of birth injury, vitamin K deficiency, coagulopathy or child abuse. In fact, the presence of multiple-aged SDHs should raise concern for nonaccidental trauma. Rarely, chronic SDHs may ossify or contain fat (Fig. 4-12).

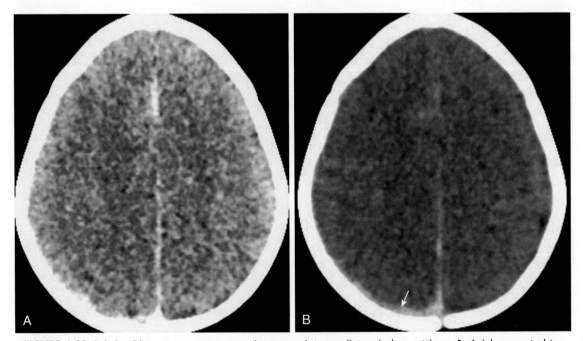

FIGURE 4-10 Subdural hematoma more conspicuous on intermediate window settings. **A,** Axial computed tomographic image in this child in soft-tissue window does not clearly depict the presence of subdural hematoma. **B,** The same image with narrower window settings shows an acute subdural hematoma overlying the right posterior parietal lobe *(arrow)*. Note poor definition of gray-white distinction indicating concurrent ischemic brain injury in this child presenting in the setting of nonaccidental trauma.

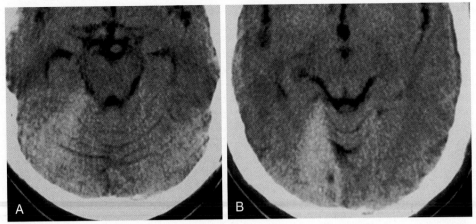

FIGURE 4-11 Computed tomography of tentorial subdural hematoma. Observe that the high density on image **A** is lateral to the high density on image **B.** Image **B** is slightly superior to image **A.** The hemorrhage is contained within the leaves of the tentorium.

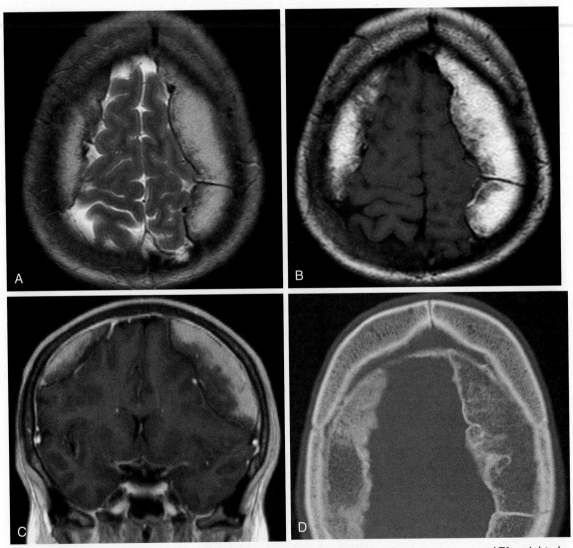

FIGURE 4-12 Extraaxial hematoma ossification. Axial T2 **(A),** axial T1 **(B),** and postcontrast coronal T1-weighted image **(C)** show fatty marrow signal (ossification) of chronic extraaxial hematomas overlying both cerebral hemispheres causing mass effect upon the underlying parenchyma. **D,** Axial computed tomography confirms ossification within the extraaxial spaces. Morphologically, these have the configuration of epidural hematomas, but they are also noted to cross suture lines, suggesting that these lie in the subdural compartment. As a compromise we could say they are extraaxial. Talk about a thick skull—this one's double thick!

It is easy to overlook isodense SDHs, particularly if the lesions are bilateral. In such a situation both lateral ventricles are compressed and may appear symmetric. Evaluation of the sulci can be helpful in making the diagnosis. Large

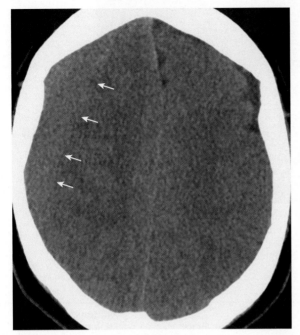

FIGURE 4-13 Computed tomography scan of isodense subdural hematoma. The edge of an isodense subdural hematoma overlying the right frontal lobe is indicated *(arrows)*. Despite its size, the hematoma could be overlooked without utilizing narrow windows. Effacement of right sided cerebral sulci and leftward deviation of the falx should raise suspicion for space occupying process in the right hemisphere. Note the significant amount of white matter buckling with preservation of the gray-white interface.

subdurals usually occur in older patients who should have prominent sulci. Anytime those sulci are poorly visualized, consider the possibility of a subdural hemorrhage. Furthermore, if the gray-white interface is displaced deeper than its normal expected position, then consider the presence of an extraaxial process (Fig. 4-13). Reviewing thin section MDCT images in the coronal plane is often useful for detecting small subdurals, especially isodense ones.

On MR, SDHs follow the intensity of blood over time like EDHs (Fig. 4-14). Chronic SDHs more closely approximate attenuation and signal of CSF on CT and MRI, respectively, but this can vary depending on protein content within the collections (Fig. 4-15). Usually they are bright on FLAIR sequences, your best friend for detecting SDHs. Chronic SDHs rarely have significant amounts of hemosiderin deposition except after recurrent hemorrhages.

Helpful hints in distinguishing epidural and subdural hematomas are outlined in Table 4-2.

SUBDURAL HYGROMA

Collections of fluid within the subdural space that have similar imaging characteristics to CSF, although they may have higher protein, are termed hygromas. They can result from trauma and either can occur acutely as a tear in the arachnoid membrane with CSF collecting in the subdural space and have the potential to grow, or they can result from the chronic degradation of a SDH. Abnormal CSF absorption because of trauma may lead to acute growth and mass effect. Oh, my aching hygroma!

Problems may arise in distinguishing subdural hygromas from atrophy (Fig. 4-16). In subdural hygroma, the collection is between the dura and the arachnoid, forcing the arachnoid inward along with the cortical veins. Atrophy is present if cortical veins on the surface of the brain extend

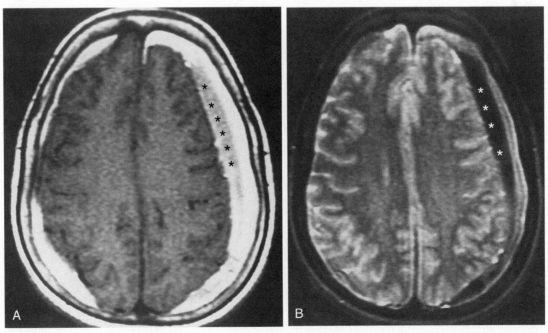

FIGURE 4-14 Subdural hematomas. **A,** T1-weighted image (T1WI) demonstrates acute and subacute hematoma. Acute hematoma is relatively isointense to gray matter *(asterisks)*, whereas the subacute hematoma is high intensity on T1WI and is bilateral. **B,** T2WI shows marked hypointensity of the acute subdural hematoma and the higher intensity of the subacute subdural hematoma. Acute subdural hematoma in both cases is marked by *asterisks*. Subacute subdural hematoma is peripheral superficial to the acute hematoma, and is bilateral.

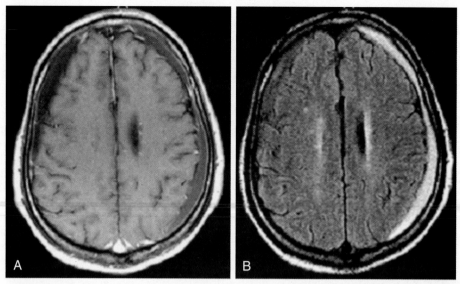

FIGURE 4-15 Magnetic resonance scan of chronic subdural hematomas. **A,** T1-weighted image of bilateral chronic subdural hematomas of different intensities secondary to protein content. **B,** Fluid-attenuated inversion recovery image shows the left-sided subdural hematoma to be high intensity and the right to be isointense to low signal intensity. The intensity differences reflect the protein content of the fluid.

TABLE 4-2   Subdural versus Epidural Hematomas

Type	"Bleeder"	Acute Shape	Chronic Shape	Nontraumatic Etiologies
Subdural hematoma	Bridging veins	Crescentic	Elliptical	Aneurysm Amyloid Menkes disease Postoperative Post-shunt Coagulopathy
Epidural hematoma	Middle meningeal artery, middle meningeal vein, venous sinus	Biconvex	Crescentic	Postoperative

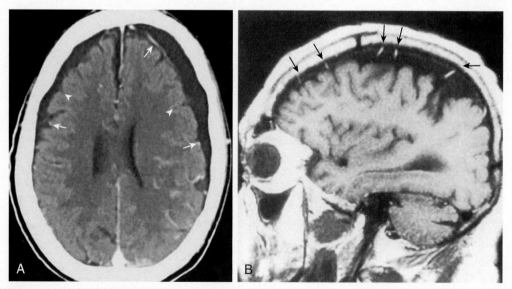

FIGURE 4-16 Subdural hygromas versus atrophy. **A,** Postcontrast axial computed tomographic image shows bilateral subdural collections isoattenuating to cerebrospinal fluid, which displace cortical vessels towards the surface of the brain *(arrows)*. Note shallow appearance of the subjacent cerebral sulci *(arrowheads)*. **B,** Cortical veins in cerebral atrophy. Note the high intensity of the bridging cortical veins on this sagittal T1-weighted image *(arrows)*. The high intensity is secondary to flow-related enhancement. Observe that they cross the extracerebral space of the parietal region. Because these veins are identified coursing through this space, there is no subdural hygroma, and the patient has cortical atrophy.

across the fluid collection towards the inner table of the calvarium. If the cortical veins are not detected crossing the fluid collection, then most likely a subdural hygroma is present. MR makes the diagnosis of hygroma more straightforward because of high protein (bright on FLAIR) and small amounts of residual hemorrhage. Benign macrocephaly (otherwise known as benign external hydrocephalus) of childhood shows dilatation of the subarachnoid space in a manner that may simulate bifrontal subdural hygromas (Fig. 4-17). It is because of immature incomplete resorption of CSF by the arachnoid villi, which consequently leads to increased head circumference in the 1- to 2-year-old range. The condition resolves on its own.

Besides trauma, subdural hygromas may be seen after ventricular decompression by ventriculostomy tubes, in cases of meningitis, after marsupialization of an arachnoid cyst, and in the setting of intracranial hypotension. Spontaneous intracranial hypotension is a source of postural headaches often found in the setting of dural tears and CSF leakage and is likely the source of headache after lumbar punctures. Intracranial pressure readings at lumber puncture (LP) are low and might not even be detectable as the low pressure CSF has no desire to make a presence in the needle hub. The work-up may include CT myelography, radionuclide cisternography, and high-resolution MR to search for a CSF leak site. In the case of post-LP headaches, imaging is not required and an epidural blood patch at the site of initial puncture is usually curative. In addition to subdural collections, cerebellar tonsillar descent, brain stem sagging, enlarged pituitary gland, and pachymeningeal enhancement can be seen intracranially in cases of intracranial hypotension (Fig. 4-18, Box 4-1). Imaging of the spine can show enlarged venous plexus, nerve root sleeve thickening, subdural collections, pachymeningeal enhancement as well as

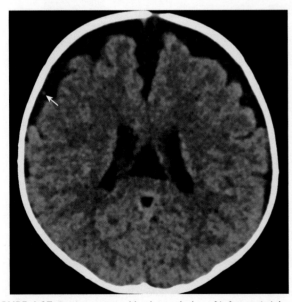

FIGURE 4-17 Benign external hydrocephalus of infancy. Axial computed tomographic image shows prominence of the extraaxial spaces overlying the frontal lobes and prominence of the ventricular system. A cortical vein *(arrow)* is noted to course immediately subjacent to the inner table of the skull, indicating that the subarachnoid space is expanded as it is typically seen in this condition. A subdural hygroma would be expected to displace cortical veins onto the surface of the brain, which is not the case on this image.

possible site of CSF leak if you're really, really lucky. Those benign nerve root sleeve cysts in the neural foramina occasionally leak and can be causative, so don't shrug them off in a patient with suspected intracranial hypotension.

HEMORRHAGIC CONTUSION

Contusions are brain parenchymal bruises where petechial hemorrhage is visualized in the cortex. The hemorrhage may extend into the white matter, and, more rarely, into the subdural and subarachnoid space. Focal edema is associated with the hemorrhage. Two mechanisms of injury include direct trauma (where the head is not in motion) resulting from depressed skull fracture, and acceleration (e.g., boxing injury) or deceleration (e.g., car accident where the head strikes the dashboard) of the skull. In the case of head motion there is an increased propensity for cortical injury adjacent to roughened edges of the inner table of the skull, along the floor of the anterior cranial fossa, the sphenoid wings, and petrous ridges. The inferior, anterior, and lateral surfaces of the frontal and temporal lobes are particularly vulnerable. Contusions occurring at the site of impact are referred to as coup contusions, whereas diametrically opposed contusions are termed contrecoup contusions. These terms are controversial, but they are simple and connote location as opposed to mechanism. Contusion of brain may also occur in the brain parenchyma along the falx and tentorium and may also involve deeper structures of the brain.

Initially, contusions may be difficult to detect on CT with relatively small hypodense or subtle hyperdense cortical/subcortical abnormalities (Fig. 4-19, *A*). These lesions become more prominent and obvious over hours to days. Contusions can range in size from punctate to very large (see Fig. 4-19, *B*). On CT, high density is noted at the site of injury. Focal regions of low density representing edema more conspicuously surround the acute hemorrhage over the first few days. On T2WI, acute hemorrhagic contusions are hypointense from deoxyhemoglobin and surrounded by high signal intensity vasogenic edema. Gradient echo or T2* or SWI scans are more sensitive for detecting these focal hemorrhagic areas as regions of low signal abnormality (see Fig. 4-23).

In the patient with significant head injury, occasional blood-fluid levels may be identified in the brain parenchyma (see Fig. 4-1). Patients with this injury tend to have a poorer outcome. Blood-fluid levels can also be seen early after trauma in patients with coagulopathies. Remember that the contrast-enhanced CT dot sign may predict hematoma growth (see Fig. 3-43).

PENETRATING INJURIES

Penetrating injuries produce lacerations of the brain, occurring as a consequence of bullet wounds, stab wounds, or bone fragments (Fig. 4-20). With any penetrating wound of the head or neck there is the risk of injury to vital structures traversed by the penetrating object. Injury to blood vessels in the brain and neck can lead to dissection, acute hemorrhage, and pseudoaneurysms with delayed hemorrhage. With respect to penetrating wounds of the brain, CT is the most efficient method for triage of these emergency patients. It can determine whether a surgical lesion exists (e.g., acute hematoma) and the amount and location of foreign bodies lodged in the brain. In all cases, infection is an ever-present danger. Intracranial air (pneumocephalus) implies a communication between the intracranial compartment and an extracranial air-filled space such as a sinus or outside environment. Rarely, air can be trapped in the brain with a "ball-valve mechanism" so that the air cavity gradually expands and compresses normal brain. This is termed a pneumatocele (aka airhead). Long-term follow-up of these lesions demonstrates resolution of the hemorrhage, residual hemosiderin, and loss of brain substance (encephalomalacia). The dura and the surface of the brain may be adhered to each other by fibroglial scar, a source of seizures in posttraumatic patients.

ACUTE VASCULAR INJURY

Acute dissection occurs most often as the result of neck injury. The injury is often mild and symptoms may take hours to days to develop so the association between the injury and the dissection may be clinically obscure. Blunt and penetrating traumas are obvious causes of dissection, but in many cases the initial insult is the result of rapid neck turning (e.g., in chiropractic manipulation, motor vehicle accidents, excessive coughing) or hyperextension (nod emphatically if you agree... on second thought, don't!). Intracranial dissections can occur as a result of skull base or penetrating injuries. Dissection can occur spontaneously in association with hypertension,

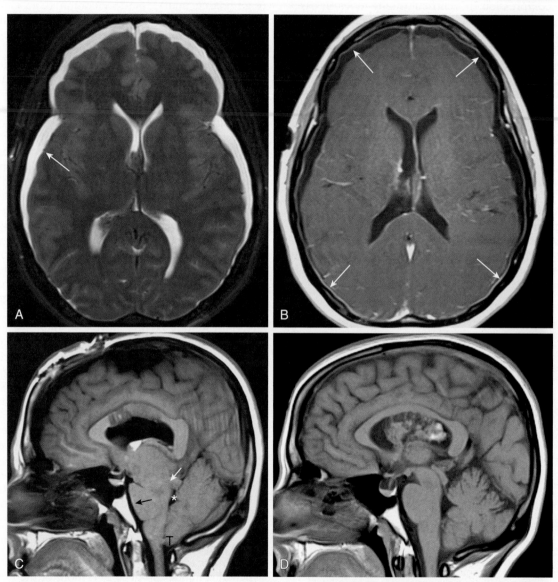

FIGURE 4-18 Intracranial hypotension in a patient after resection of intraventricular neurocytoma. **A,** Axial T2-weighted image (T2WI) shows prominent subdural collections isointense to cerebrospinal fluid (effusions) over both cerebral hemispheres. Cortical vessels are displaced onto the surface of the brain *(arrow)*. There is hemorrhage layering in the posterior horns as a consequence of recent surgical tumor resection. **B,** Postcontrast T1WI shows diffuse pachymeningeal enhancement *(arrows)*. **C,** Sagittal T1WI shows sagging of the brain stem and effacement of the basal cisterns, with flattening of the pons against the clivus *(black arrow)* resulting in effacement of the prepontine cistern. The cerebral aqueduct is effaced *(white arrow)* and the fourth ventricle is small *(asterisk)*. The cerebellar tonsils have descended below the foramen magnum (T). Compare this image to **D,** Sagittal T1WI performed before surgical resection where the pons, cerebral aqueduct, fourth ventricle, and cerebellar tonsils show normal morphology. Note intraventricular mass representing neurocytoma.

fibromuscular dysplasia, and collective tissue diseases (e.g., Ehlers-Danlos IV syndrome, Marfan syndrome). Idiopathic intracranial dissection can produce fusiform (pseudo)aneurysms with heterogeneous intramural clot, usually along the internal carotid artery (ICA) or M1 middle cerebral artery (MCA) branches. It should also be considered when there is isolated stenosis of a proximal intracranial vessel without other clinical or imaging evidence of atherosclerosis.

Dissection occurs when an intimal tear allows blood to enter the arterial wall. The blood divides the layers of the wall producing stenosis, occlusion or pseudoaneurysm formation. The blood can extend within the wall creating a false lumen for several centimeters and then re-enter the true lumen. One or more vessels may be involved.

BOX 4-1 Findings in Spontaneous Intracranial Hypotension

INTRACRANIAL

Bilateral subdural hygromas
Tonsillar descent
Brain stem sagging
Enlarged pituitary gland
Pachymeningeal enhancement
Subcortical white matter decreased signal intensity on FLAIR

SPINAL

Enlarged venous plexus
Subdural collections
Pachymeningeal enhancement
Nerve root sleeve thickening
CSF leakage site

CSF, Cerebrospinal fluid; *FLAIR,* fluid-attenuated inversion recovery.

Dissection of the carotid or vertebral artery should always be considered as a cause of stroke symptoms in young otherwise healthy patients. Symptoms result from the primary cervical vascular injury, thromboembolism, and secondary ischemia but pain in the neck may also accompany the event. Extracranial carotid dissection can present with neck and face pain, headache, ptosis, and miosis (Horner syndrome) because of the expanding intramural hematoma and compression of sympathetic nerves adjacent to the internal carotid artery.

Traditionally, dissections are detected by angiographic studies. In the extracranial internal carotid artery, four patterns of luminal abnormality can be detected with catheter and noninvasive techniques: (1) smooth tapering to a pointed occlusion; (2) long segment asymmetric narrowing (occasionally spiraling around the lumen); (3) double lumen with an intimal flap (the true lumen is typically smaller than the false lumen); and (4) pseudoaneurysm formation. In the vertebral arteries, luminal patterns are less specific. Occlusion or stenosis of the midsection of the vertebral artery within the foramen transversarium (often around C5 where rotation tends to be maximal) and/or presence of a false lumen in the distal vertebral artery are most frequently observed.

Luminal changes are well seen with both catheter angiography and CTA (Fig. 4-21, *A* and *B*). Source images from CTA allow for detection of intimal flaps and double lumens and are particularly valuable in assessing the small vertebral arteries. MRA is an excellent tool for identifying dissection although intraluminal intensity changes secondary to turbulent and in plane flow artifacts may be difficult to differentiate from flaps and false lumens. On MRI, the mural hematoma is commonly T1 hyperintense and T2 hypointense to hyperintense (patients with dissections are most often imaged 3 to 7 days after the initial event; see Fig. 4-21, *C* through *F*). Within the expanded T1 hyperintense artery wall the small hypointense residual lumen is

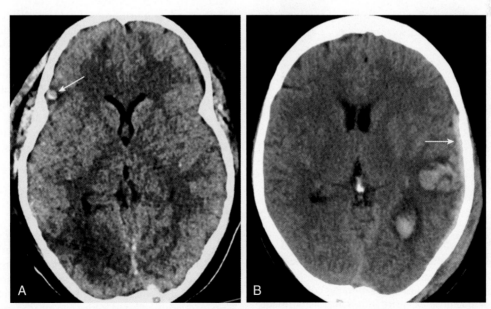

FIGURE 4-19 Contusion on computed tomography (CT). **A,** Axial noncontrast CT shows subtle hyperattenuation at the gray-white interface at the right frontal opercular region with surrounding hypoattenuation *(arrow)* indicating contusion in this patient presenting to the emergency room after motor vehicle collision. Hypoattenuation in the right occipital lobe is artifactual. **B,** Axial noncontrast CT image in a different patient also presenting after motor vehicle collision shows multiple large contusions with surrounding vasogenic edema. Note acute subdural hematoma overlying the left cerebral hemisphere *(arrow)*.

often present (target sign). Fat-suppressed T1WI sequences may be useful to distinguish periarterial fat from intramural subacute hemorrhage, particularly in the neck. Because the mural hematoma is typically hyperintense, care must be taken in evaluating MRA sequences to make sure that hyperintense mural clot is not mistaken for hyperintense normal flow. The mural clot is typically less hyperintense than the lumen and it has an amorphous appearance. Mural clot is less often visualized in vertebral artery dissections. MR is also useful for following the dissection to visualize when the hemorrhage is reabsorbed, when the normal lumen dimensions are reestablished, and if there is progression to pseudoaneurysm or occlusion.

The principal complications of extracranial vascular dissection, infarcts, and transient ischemic attacks result from luminal compromise and, more commonly, embolic phenomena. Ischemic infarction is usually the result of distal embolization from the cervical dissection. Embolic infarction typically occurs several hours to days after the onset of the dissection but may be delayed for several weeks. Treatment (anticoagulation) is directed toward preventing recurrent emboli. Most dissections of the neck heal spontaneously.

Other terrifying sequelae of traumatic vascular injury include laceration, extravasation, occlusion, pseudoaneurysm, and arteriovenous fistula (Fig. 4-22).

Trauma to the cavernous carotid artery can produce direct communication between the carotid artery and cavernous sinus (see Fig. 3-61). The fistula may drain from the cavernous sinus into the orbit through the superior and inferior ophthalmic veins, posteriorly through the petrosal veins, inferiorly through veins around the foramen ovale, or superiorly through middle cerebral veins. Patients may develop

intraparenchymal hemorrhage. Other complications of carotid cavernous fistula include subarachnoid hemorrhage, intraorbital hemorrhage, epistaxis, otorrhagia, glaucoma, dilated conjunctival vessels, progressive proptosis, ophthalmoplegia, and increased ICP. The diagnostic findings and angiographic workup of this condition are discussed in Chapter 9.

DIFFUSE AXONAL INJURY

DAI is the injury responsible for coma and poor outcome in most patients with significant closed head injury resulting from automobile accidents (although it can result from other kinds of trauma as well). At the time of injury the stress induced by rotational acceleration/deceleration movement of the head causes some regions of the brain to accelerate or decelerate faster than other regions resulting in axonal shearing injury. Axons may be completely disrupted together with adjacent capillaries, or some axons may be incompletely disrupted, the so-called injury-in-continuity. Patients with severe DAI are unconscious from the moment of injury and may remain in a persistent vegetative state or be severely impaired.

DAI characteristically involves the corpus callosum, brain stem, internal capsule, and gray-white interface (Box 4-2). Rotationally induced shear-strain may also produce cortical contusion and lesions in the deep gray matter. The unbending (like your mother-in-law) falx, which is broader posteriorly, prevents the cerebral hemispheres from moving across the midline, whereas anteriorly the falx is shorter so that the brain can transiently move across the midline. The fibers of the splenium and posterior corpus callosum are thus under greater risk of shearing than the anterior fibers.

Lesions may be ovoid or elliptic, with the long axis parallel to fiber bundle directions. CT is not nearly as sensitive as MR in detecting DAI in the brain (Fig. 4-23). On CT, one may visualize focal punctate regions of high density that may be surrounded by a collar of low-density edema. These are hemorrhagic shearing injuries most likely associated with complete axonal disruption. On CT, it may be difficult to detect nonhemorrhagic shearing injury early and initial CT may appear normal. When DAI is clinically suspected and CT is unrevealing, MRI should be performed as it has a much higher sensitivity for detecting shear injuries. MRI reveals high intensity on T2WI/FLAIR at injured sites, which, through the magic of SWI sequences, are shown to frequently be associated with hemorrhage. DWI may be positive reflecting the hemorrhagic lesions or nonhemorrhagic shearing injuries. Splenial lesions in particular seem to restrict diffusion.

TRAUMATIC BRAIN INJURY IN THE LONG TERM

Cognitive deficits resulting from TBI (and reading *Neuroradiology Requisites*) include decreased speed in information processing, poor attention, concentration, and memory, and impaired logical reasoning skill, as well as more focal deficits including impairment of language or constructional abilities. The correlation of head injury symptoms and their sequelae is controversial yet there is growing evidence that head injury, even mild in nature, may have greater consequences than previously assumed. Indeed, TBI is now considered a risk factor in the development of Alzheimer disease.

There are many reports describing loss of brain parenchymal volume as the sequelae of TBI in particular

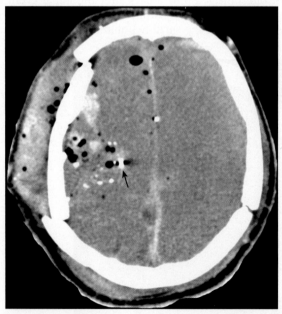

FIGURE 4-20 Bad day at the office. Axial computed tomographic image following gunshot wound to the head shows comminuted and depressed skull fractures, extensive intracranial hemorrhage including parenchymal, subarachnoid, and subdural hemorrhages, poor gray-white matter differentiation indicating diffuse anoxic injury, and sulcal effacement indicating diffuse cerebral edema. Note the hyperattenuating bullet fragments (largest indicated by *arrow*). Should this guy survive the immediate trauma (guess what: he didn't), the presence of foreign body material in the brain renders the patient at risk for infection.

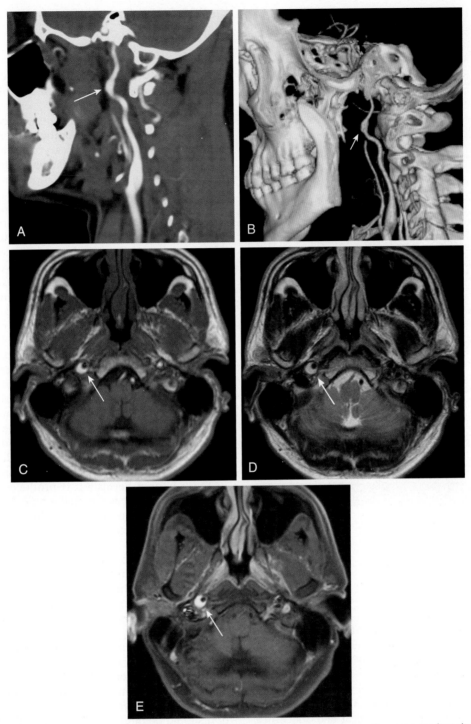

FIGURE 4-21 Dissection on computed tomography angiography (CTA) and magnetic resonance imaging (MRI). **A,** Sagittal maximum intensity projection and **(B)** surface rendered reconstructions of the left cervical carotid artery system from CTA shows abrupt focal enlargement of the caliber of the internal carotid artery (ICA) below the skull base *(arrows),* indicating dissection with associated pseudoaneurysm. The dissection is not flow limiting as contrast opacifies the vessel beyond the level of caliber change. On MRI, in a different patient, axial T1-weighted image (T1WI) **(C)** and axial T2WI **(D)** show crescentic mural hematoma within the wall of the right ICA at the skull base *(arrows).* Fat-suppressed postcontrast T1WI **(E)** confirms presence of T1 hyperintense intramural hematoma *(arrow).* Note preservation of the flow void within the residual ICA lumen on these MRI images (target sign).

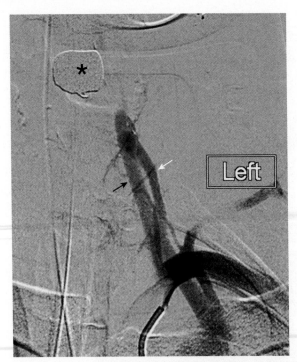

FIGURE 4-22 Posttraumatic arteriovenous (AV) fistula. Conventional catheter angiogram in a patient status post gunshot wound to left neck shows opacification of the proximal left vertebral artery *(white arrow)*. This vessel occludes abruptly and there is direct communication with draining vein *(black arrow)* which opacifies in the arterial phase, indicating AV fistula. *Asterisk* indicates bullet fragment in the neck.

BOX 4-2   Locations of Diffuse Axonal Injury

Gray-white junction
Corpus callosum
Brain stem
Superior cerebellar peduncle
Internal capsule
Basal ganglia

---

structures including the caudate nucleus, corpus callosum, hippocampus, fornix, and thalamus. These injuries result in enlargement of the ventricles (width of the third ventricle, bicaudate span of the lateral ventricles, area of lateral ventricles, total ventricular volume, and temporal horn or ventricular/brain ratio). Posttraumatic thalamic atrophy has been correlated with the presence of cortical and subcortical (but nonthalamic) lesions. This suggests that such traumatic lesions may be responsible for transneuronal degeneration. Hypertrophic olivary degeneration (see Chapter 7) may result from traumatic injuries affecting the red nucleus, dentate nucleus, central tegmental tract or cerebellar peduncles. Wallerian degeneration of the corticospinal tract will often be seen after contusions/shears of the motor area of the frontal cortex or white matter tracts respectively.

The prevalence of cavum septum pellucidum has been observed to be increased in professional boxers and soccer players (and schizophrenics). This has been related to repetitive head trauma with tearing of the septal walls and secondary CSF dissection of the septi forming a cavum.

SECONDARY COMPLICATIONS

Herniation Syndromes

The brain lives in a rigid container (the skull) and is compartmentalized by inelastic dural reflections (tentorium, falx cerebri, falx cerebelli). Much like playground bullies, swelling or mass lesions, when large enough, force the brain from one compartment into another location. Herniations of brain from one region to another can produce both brain and vascular damage. Five basic patterns can be encountered (Fig. 4-24): inferior tonsillar and cerebellar herniation, superior vermian herniation (supratentorial herniation), temporal lobe/uncal herniation, central transtentorial herniation, and subfalcine herniation. It is important to alert the physicians caring for the patient with incipient herniation, as these are potentially life-threatening situations.

When mass effect occurs in the posterior fossa, the fourth ventricle is compressed, producing obstructive supratentorial hydrocephalus (dilated temporal horns are particularly sensitive indicators of such hydrocephalus). The cerebellar tonsils and cerebellum are usually pushed inferiorly through the foramen magnum. On CT or MR, the obvious finding of foramen magnum herniation syndrome is absence of the normal CSF around the foramen magnum and the presence of tonsillar and cerebellar tissue in that site (Fig. 4-25). In circumstances where there is superior cerebellar mass effect, the forces exerted produce upward herniation of the vermis (Fig. 4-26) in what may be termed upward/ascending transtentorial herniation. Cerebellar tissue can be identified obliterating the superior vermian, ambient and quadrigeminal cisterns. Compression of the aqueduct by such mass effect often leads to obstructive hydrocephalus.

Transtentorial herniation is produced by mass lesions whose vector force is directed inferiorly and medially. The temporal lobe uncus shifts over the tentorium, compressing to a variable extent the oculomotor nerve and its parasympathetic fibers (with ipsilateral pupillary dilatation), the posterior cerebral and anterior choroidal arteries, and the midbrain. Compression of the contralateral cerebral peduncle against the edge of the tentorium produces ipsilateral motor weakness and is termed Kernohan-Woltman notch phenomenon (false localizing sign). It is associated with high intensity on T2WI in the midbrain from compression and occlusion of perforating arteries at the tentorial notch. Vascular compression of the posterior cerebral and anterior choroidal arteries by the medial temporal lobe against the tentorium or by the petroclinoid ligament results in infarction in these vascular distributions. Attention should be directed to the cerebral artery distribution and diencephalon, respectively, for evidence of ischemia or infarction in situations where supratentorial mass effect is significant. Direct central caudal transtentorial herniation forces the diencephalon and midbrain down through the tentorial incisura. Herniation can result in small hemorrhages in the brain stem. Hemorrhages in the tegmentum of the pons and the midbrain caused by uncal herniation have been termed Duret hemorrhages.

Ipsilateral ambient cistern widening from the interposition of the temporal lobe, contralateral temporal horn dilatation, rotation and/or contralateral sliding of the brain stem, contralateral cisternal obliteration, and uncal tissue hanging over the incisura confirm descending transtentorial herniation. Sometimes, when things are not so bad

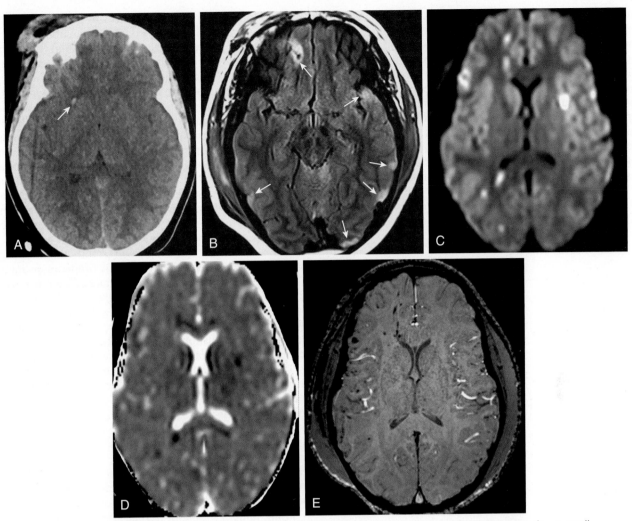

FIGURE 4-23 Diffuse axonal injury (DAI). **A,** Axial noncontrast computed tomography (CT) image shows small focus of hemorrhage with surrounding edema in the right external capsule *(arrow)*. **B,** Axial fluid-attenuated inversion recovery image performed on subsequent magnetic resonance imaging shows numerous additional lesions that are not appreciable by CT which lie at the gray-white interface *(arrows)*. **C,** Diffusion-weighted imaging and **(D)** apparent diffusion coefficient map show numerous additional lesions also at gray-white interface and at/near the splenium of the corpus callosum. **E,** Axial susceptibility-weighted imaging is the most sensitive in picking up tiny hemorrhages and confirming the diagnosis of DAI. Every dark spot on this image indicates a focus of shear injury.

looking and there is less downward vector, we use the term uncal herniation; the uncus/parahippocampus is over, but the same degree of compromise of the brain stem/cisterns is not present. Coronal reconstructions or direct imaging can be quite elegant in showing the effects of the inferomedial forces.

Another herniation pattern is the subfalcine variety. Here, supratentorial mass effect can be directed medially, causing the cingulate gyrus to shift beneath the falx cerebri with possible compression of the anterior cerebral artery (ACA) or internal cerebral veins (Fig. 4-27). The frontal horns are shifted and rotated to one side and the septum pellucidum deviates over at the level of the foramen of Monro. The CSF space may be widened along the contralateral side of the anterior falx. The ACA distribution may be infarcted if an ACA branch is severely compressed along the falx.

In the setting of herniation syndromes, death is a very likely outcome unless urgent surgical decompression is performed. The neurosurgeon's decision to decompress is based on both clinical and imaging parameters (Table 4-3).

BRAIN DEATH

Brain death is defined as the irreversible cessation of all cerebral and brain stem functions. The absence of cerebral blood flow (from increased ICP) is generally accepted as a definite sign of brain death. In the clinical setting of absence of spontaneous respirations, brain stem reflexes, and flat serial electroencephalograms (EEGs), a radionuclide perfusion study (SPECT, 99mTc HMPAO) can confirm the diagnosis of brain death. Absence of the cerebral flow on MRA or CTA has also been reported to be specific in confirming clinical brain death.

CHILD ABUSE/NONACCIDENTAL TRAUMA

Radiologists must display increased sensitivity to the diagnosis of child abuse. Caffey first recognized this in 1946 in children with long-bone fractures and chronic SDHs. Shaken baby syndrome is caused by sudden acceleration-deceleration forces in the process of violent shaking.

There may be little evidence of external injury, but retinal hemorrhages (intraretinal and preretinal) are present in many instances and associated with subdural (particularly interhemispheric) hematomas and/or subarachnoid hemorrhage (Fig. 4-28). Contusions and diffuse cerebral swelling can also be seen. Hematomas in the upper cervical spinal cord have been described. The abused child can have epidural, subdural, subarachnoid, intraparenchymal, and intraventricular hemorrhages as well as DAI. MR is useful in determining the ages of these lesions by noting the various stages of blood products (acute, subacute, and/or chronic or mixed hematohygromas). Other common clinical features include seizures, ecchymoses, vitreous hemorrhage, and hemiparesis. Box 4-3 lists the central nervous system abnormalities associated with the shaken baby syndrome.

Attention should be directed to the presence of skull fractures, particularly depressed fractures with a history of "mild trauma," fractures that cross the midline, and those involving the occiput without a known significant event. In reviewing CT images on pediatric patients, it is imperative to review thin section data in bone algorithm to detect these subtle fractures. Additionally, three-dimensional surface rendered reconstructions are very useful for identifying fractures occurring in the transverse plane that might be overlooked if reviewing axial images only. Surface rendered reconstructions can also be useful in demonstrating wormian bones and sutures as normal anatomic structures that might otherwise be confused for acute fracture. Scalp edema identified on thin sections in soft-tissue algorithm can be very useful in directing attention to the adjacent skull to evaluate for subtle fracture.

Global ischemia from strangulation, smothering, aspiration, or other conditions generating hypoxia produces distinctive CT findings. There is loss of the gray-white differentiation and generalized low density of the cerebral cortex (Fig. 4-29). The basal ganglia may show high or low density. The cerebellum is more resistant to hypoxia and therefore demonstrates normal density in this situation ("white cerebellum sign"). MR can reveal diffuse cerebral swelling with loss of the normal sulci.

The result of significant cerebral insults from repeated traumatic events is the malacic atrophic brain with or without collections of hemorrhages in different compartments of varying ages (see Fig. 4-28).

SKULL/SKULL BASE FRACTURE

Careful review of the skull and skull base should be performed routinely in trauma cases. Presence of fractures can be obvious or very subtle, depending on trauma mechanism. A strong understanding of normal anatomy is necessary to identify bony injuries and to avoid calling normal structures (i.e., cranial sutures) as skull fracture. Identification of soft-tissue swelling can be a helpful indicator to carefully examine the bony structures that are immediately adjacent.

There are several types of common skull fractures: (1) linear, (2) diastatic, (3) comminuted, and (4) depressed. When fractures are identified, careful review for associated lesions, including contusion, shearing injury, or extracerebral collection should follow. Depressed fractures are usually comminuted and can produce underlying brain injury along with extraaxial hemorrhage. Comminuted fractures commonly have a depressed fragment. It is important for the radiologist to comment on the extent of the depression (usually defined in relation to the thickness of the skull table), and associated brain injury. Attention should be focused on open fractures, and fractures through paranasal sinuses along the skull base which increase the risk of intracranial infection. Longitudinal fractures of the temporal bone have a high correlation with temporal lobe injury. Fractures can extend along sutures resulting in an asymmetrically widened (diastatic) appearance.

Skull base injuries, including fractures of the carotid canal, jugular foramen, and skull fractures adjacent to major venous sinuses should prompt angiographic evaluation to assess integrity of housed vascular structures. Clival and central skull base fractures can be associated with extraaxial hematomas along the craniocervical junction, so when such fractures are seen, a careful review of soft tissues

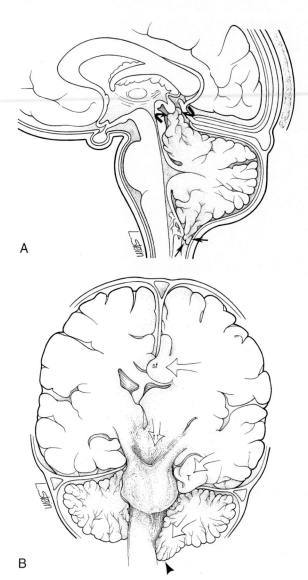

FIGURE 4-24 Herniations of the brain. **A,** Sagittal diagram of cerebellar herniation, with *curved arrows* demonstrating upward herniation of the superior cerebellum and superior vermis and *straight arrows* demonstrating tonsillar and inferior vermian herniation. **B,** Coronal diagram with temporal lobe herniation (T), central transtentorial herniation (tt), tonsillar herniation *(arrowhead)*, and subfalcine herniation (sf). Lines of force are demonstrated by *large open arrows*. Note the pressure on the brain stem from these herniation patterns. Ouch!

to exclude associated hemorrhage should ensue. Identification of temporal bone fractures should prompt a review of middle and inner ear structures addressing integrity of ossicles and otic capsule structures (see Chapter 11).

In children, most linear skull fractures heal in time, whereas in adults evidence of these fractures is present for years after injury, although the margins are less distinct. When the dura is torn with the skull fracture, the arachnoid can insinuate itself into the cleft of the fracture. Rarely, the pulsations of the CSF enlarge the cleft between the fracture fragments, producing either linear widening of the fracture margins or multiloculated cysts with smooth, scalloped margins. This appearance

has been termed a "growing fracture" or "leptomeningeal cyst" and is seen most commonly in children.

FACIAL AND ORBITAL TRAUMA

CT is the most efficient imaging modality for visualizing facial fractures. Axial MDCT thin (less than or equal to 1 mm) sections with coronal and sagittal reconstructions are essential for detecting the full extent of the injury. Certain classic fracture eponyms have attained historical notoriety and at times are useful in describing what tends to be a complex of fractures. What is more important is that these

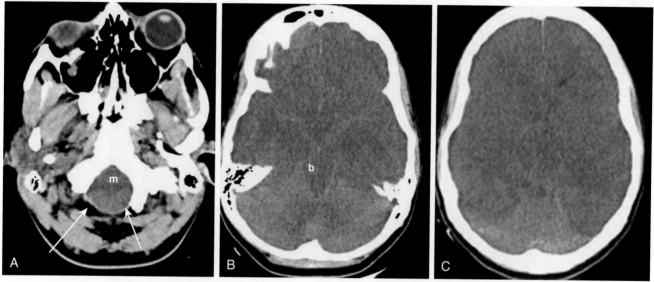

FIGURE 4-25 Cerebral edema. **A,** The cerebellar tonsils *(arrows)* have herniated through the foramen magnum in this patient with rapidly progressive meningoencephalitis. No surrounding cerebrospinal fluid (CSF) is seen and the medulla (m) is displaced anteriorly. **B,** Note the elongated squeezed midbrain (b) and the absence of gray-white differentiation and CSF in the basal cisterns. Often the relatively denser vessels will simulate subarachnoid hemorrhage. **C,** Further superiorly, things are equally tight with compressed lateral ventricles.

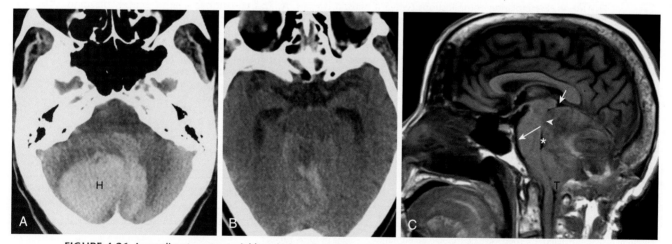

FIGURE 4-26 Ascending transtentorial herniation. **A,** Axial computed tomographic image through the posterior fossa shows a large hematoma (H) in the right cerebellar hemisphere resulting in leftward shift of posterior fossa structures and complete effacement of sulci. **B,** More superiorly, there is effacement of the quadrigeminal plate cistern from superior migration of the vermis because of mass effect in the posterior fossa. Note enlarged temporal horns of the lateral ventricles, indicating early hydrocephalus. **C,** Despite urgent suboccipital craniectomy for decompression, the fourth ventricle *(asterisk)* remains nearly completely effaced, the quadrigeminal plate cistern *(arrowhead)* is effaced, and the superior vermian cistern *(small arrow)* is partially effaced. The pons is flattened against the clivus *(large arrow)* and there is crowding of structures at the craniocervical junction because of downward descent of the cerebellar tonsils (T).

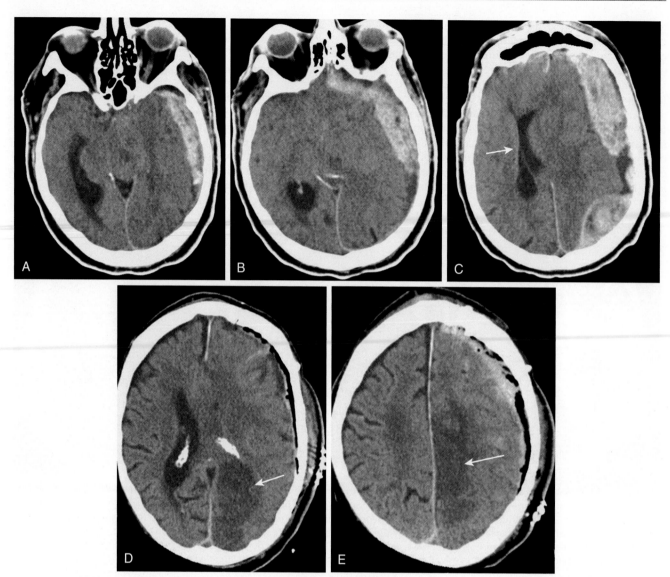

FIGURE 4-27 Herniation complications (when it rains, it pours!). **A,** This patient had a large, actively bleeding subdural hematoma. Note the compression of the uncus from left to right, the contralateral temporal horn dilation of the right lateral ventricle, and rotation of the midbrain indicative of transtentorial herniation. **B,** Basal cisterns are effaced and the brain stem is distorted. **C,** More superiorly, the lateral ventricles *(arrow)* are grossly displaced across the midline with subfalcine herniation. **D,** Two days later, after evacuation of the hematoma, one can see a left occipital lobe infarct *(arrow)* from the compression of the right posterior cerebral artery associated with the downward transtentorial herniation. **E,** A left anterior cerebral artery infarction *(arrow)* is also present, likely from the stretching by the subfalcine herniation.

TABLE 4-3   Clinical and Imaging Indications for Surgical Intervention versus Observation in Patients with Intracranial Hemorrhage

Lesion	Indications for Surgery—Clinical	Indications for Surgery—Imaging	Nonsurgical Management (close clinical monitoring, serial imaging)
Parenchymal hemorrhage	Progressive neurologic deterioration Refractory intracranial hypertension GCS 6-8	Signs of mass effect Midline shift >5 mm Cisternal effacement Hematoma volume >50 mL (Temporal contusion >20 mL)	No neurologic compromise Controlled ICP No mass effect on imaging
Acute EDH	Focal deficit GCS <8	Volume >30 mL Volume <30 mL *and* >15 mm thickness *or* >5 mm midline shift *and* clinical indications for surgery	Volume <30 mL *and* <15 mm thickness *and* <5 mm midline shift *and* GCS >8 without focal deficit

TABLE 4-3 Clinical and Imaging Indications for Surgical Intervention versus Observation in Patients with Intracranial Hemorrhage—cont'd

Lesion	Indications for Surgery—Clinical	Indications for Surgery—Imaging	Nonsurgical Management (close clinical monitoring, serial imaging)
Acute SDH	GCS <9 Decrease in GCS by 2 points Asymmetric, fixed, dilated pupils ICP >20 mm Hg	Thickness >10 mm *or* >5 mm midline shift Thickness <10 mm and <5 mm shift *and* GCS <9 *and* clinical indications for surgery	No neurologic compromise
Posterior fossa lesions	Neurologic deterioration	Signs of mass effect (displacement/effacement of fourth ventricle, effaced basal cisterns, obstructive hydrocephalus)	No mass effect on CT No neurologic compromise
Depressed open (compound) skull fracture	Clinical evidence of dural penetration	Depression > thickness of skull	No clinical/imaging evidence of dural penetration No significant intracranial hemorrhage <1 cm depression No frontal sinus involvement No significant cosmetic deformity No wound infection No pneumocephalus

Adapted from Winn HR. *Youmans Neurological Surgery*. Philadelphia: Elsevier; 2011.
CT, Computed tomography; *EDH,* epidural hematoma; *GCS,* Glasgow coma scale; *ICP,* intracranial pressure; *SDH,* subdural hematoma.

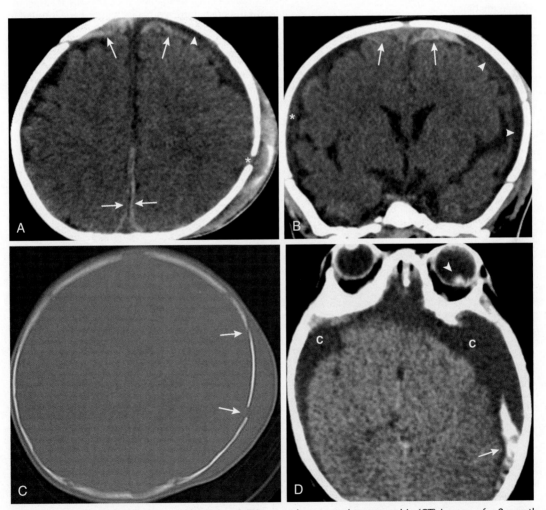

FIGURE 4-28 Nonaccidental trauma. **A,** Axial and **(B)** coronal computed tomographic (CT) images of a 2-month-old child presenting with lethargy shows mixed density subdural hematomas overlying both cerebral hemispheres, indicating blood products of different ages. More acute blood is hyperdense *(arrows)* and more chronic blood is hypodense *(arrowheads)*. Asterisk in **B** indicates a more intermediate age heterogeneous subdural hematoma overlying the right opercular region. In this particularly gruesome case, *asterisk* in **A** indicates a calvarial defect through which brain tissue is herniating with extensive overlying scalp swelling. **C,** Bone algorithm confirms presence of a comminuted parietal skull fracture and overlying soft-tissue swelling *(arrows)*. **D,** Axial CT image from a 1-year-old child presenting with seizures shows mixed attenuation subdural hematomas overlying both cerebral hemispheres (C), indicating blood products of different ages *(arrow indicates more acute hemorrhage)*. Note retinal hemorrhage also *(arrowhead)*. When you see intracranial blood products of different ages in vulnerable populations, suspect nonaccidental trauma.

BOX 4-3   Imaging Findings in Nonaccidental Trauma

SKELETAL

Skull fracture (diastatic, comminuted, linear)
Vertebral compression fracture, fracture-dislocation, disk space narrowing, spinous process fracture
Cervical spinal cord hematoma
Growth plate/avulsion/metaphyseal fractures
Rib fractures

BRAIN INJURY

Contusion
Diffuse cerebral swelling
Subdural (interhemispheric) hematoma
Diffuse axonal injury
Focal, multifocal, or diffuse ischemic injury
Subarachnoid hemorrhage
Epidural hematoma
Intraventricular hemorrhage

ORBIT

Retinal hemorrhage (more readily appreciated on fundoscopic exam rather than imaging)
Vitreous hemorrhage

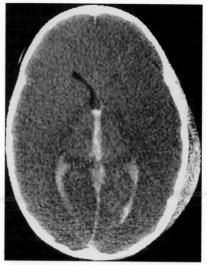

FIGURE 4-29 Computed tomography (CT) scan of diffuse cerebral swelling with intraventricular hemorrhage. Loss of differentiation on CT between gray and white matter and the inability to visualize basal ganglionic structures are features of hypoxic insult to the brain. Intraventricular hemorrhage is noted in the third and lateral ventricles.

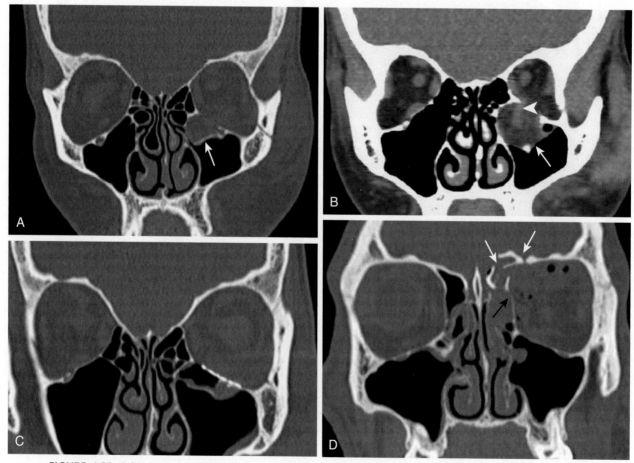

FIGURE 4-30 Orbit fractures. **A,** Coronal computed tomographic image in bone window shows orbital floor fracture fragment *(arrow)* projecting inferiorly (trap door appearance), also taking out the inferomedial orbital wall. **B,** The fat *(arrow)* and muscular *(arrowhead)* herniation through the fracture gap led to restriction of motion. **C,** Subsequent repair with plate and screws of the orbital floor led to anatomic alignment. **D,** This inmate was stabbed in the eye and incurred an unusual combination of orbital roof *(white arrows)* and medial orbital wall *(black arrow)* fractures. Note the air and hematoma and traumatized muscles in the superomedial orbit.

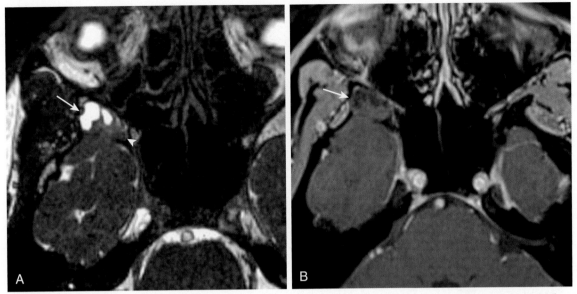

FIGURE 4-31 Posttraumatic meningoencephalocele. In this patient presenting with recurrent headaches following motorcycle crash, axial T2 CISS **(A)** image shows a meningoencephalocele *(arrow)* extending through a defect in the sphenoid wing. Note fluid and brain *(arrowhead)* contents within the herniation. **B,** As expected, the meningoencephalocele does not enhance on postcontrast T1-weighted image *(arrow)*.

mechanistic descriptions allow you to raise your suspicion for associated fractures elsewhere. We will cover just a few of the more common fractures.

Signs of facial and orbital trauma include soft-tissue swelling about the face or orbit, and fluid and blood in the maxillary or paranasal sinuses. In orbital fractures, note the location of the fracture, its extent, whether muscle or fat is entrapped in the fracture site, and whether associated orbital hemorrhage including subperiosteal hematomas is present (see Fig. 4-3). In the orbit, the most common fracture is the so-called blowout fracture caused by a direct (blunt) injury resulting in fracture of the orbital wall and entrapment of the orbital contents. This can involve the orbital floor, with the fracture usually through the orbital plate of the maxilla medial to or involving the infraorbital canal (Fig. 4-30). The fracture is commonly hinged on the medial side, appearing on coronal CT as a "trapdoor." Medial orbital blowout fractures involve the lamina papyracea of the ethmoid bone. Rarely, the blowout can occur upward into the frontal sinus. Besides entrapped tissue, air may be noted in the orbit (orbital emphysema), and air fluid levels can be seen in the involved sinuses. This can be an excellent indicator of a subtle ethmoidal fracture. Blowout floor fractures are associated with enophthalmos, diplopia (on upgaze), and ocular injuries but generally do not involve the orbital rim. Focus on the rectus muscles and orbital fat, which may be displaced through the fracture site. This can result in clinically apparent entrapment syndromes with limitation of movement of the globe. Hematomas involving the orbital muscles can also produce limitations in range of motion. Involvement of the orbital rim indicates a more severe injury. This is particularly true when it is the superior rim which can be associated with intracranial injury. The weakest portion of the roof is near the superior orbital fissure and optic canal. Fracture of the orbital roof can predispose to pseudomeningocele (Fig. 4-31).

Ocular trauma can result in perforation of the globe and ocular hypotony (flat tire sign; see Chapter 9). Orbital hemorrhage can occur in the intraconal space, within the nerve

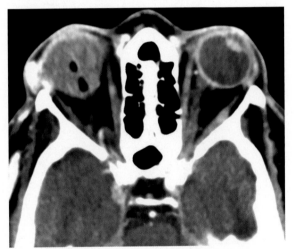

FIGURE 4-32 Ocular injury. Axial unenhanced computed tomographic image through the orbits shows heterogenous attenuation of fluid in the posterior segment of the globe indicating vitreous hemorrhage. Presence of air within the globe as well as loss of normal globe morphology indicates globe rupture.

sheath (subdural), in the extraconal space, subperiosteal space, and sub-Tenon space (between the sclera and Tenon capsule, which is a fibrous membrane adjacent to the orbital fat). Hemorrhage within the globe can be seen in the anterior chamber (anterior hyphema) and posterior segment (separated by the lens). Posterior segments hemorrhages can be centered in the vitreous or can be seen with retinal or choroidal detachments. Lens dislocations or acute traumatic cataracts (low in density) can be seen in the setting of ocular trauma. Foreign bodies may be seen within the orbit and within the globe. Presence of air and/or foreign body material within the globe in the trauma setting implies globe rupture, even if globe morphology appears preserved (Fig. 4-32). Certainly, a collapsed, misshapen globe in the setting implies globe rupture. These findings are critical to report to the clinical service, as such findings may be difficult to appreciate on physical exam when the eye is swollen shut.

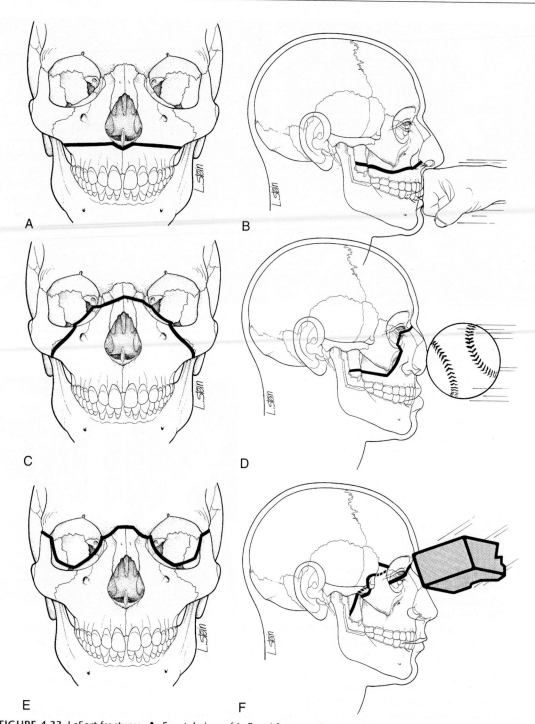

FIGURE 4-33 LeFort fractures. **A,** Frontal view of LeFort I fracture. **B,** Lateral view of LeFort I fracture. **C,** Frontal view of LeFort II fracture. **D,** Lateral view of LeFort II fracture. **E,** Frontal view of LeFort III fracture. **F,** Lateral view of LeFort III fracture.

A direct blow to the maxillary sinus can cause a "blow-in" fracture, with elevation of the orbital floor into the orbit. These fractures cause proptosis and restrict ocular motility. Blow-in fractures with superior orbital rim fractures are associated with frontal lobe contusion and EDH.

MR can have a special role in detecting optic nerve sheath hematomas. These tend to be more obvious on MR than on CT. Diagnosis is important because vision can be rapidly lost and operative nerve sheath decompression can be restorative. On MR, hemorrhage can be observed along the nerve sheath, which may be swollen and irregular. On CT, the nerve sheath complex may be enlarged.

The zygomatic arch can be fractured alone or as part of the zygomaticomaxillary (ZMC) complex (tripod) fracture, which includes (1) fracture of the lateral wall of the orbit, usually as diastasis of the zygomaticofrontal suture, (2) fracture of the inferior orbital rim and floor, at times injuring the infraorbital nerve, (3) fracture of the zygomatic arch, and (4) fracture of the zygomaticomaxillary strut (see Fig. 4-34). The zygoma can be displaced posteriorly and medially, causing difficulty with the normal motion of the jaw. When this occurs, the lateral wall (at times the anterior and posterior walls as well) of the maxillary sinus is involved (the fourth foot in the tripod) in addition to the floor of the orbit.

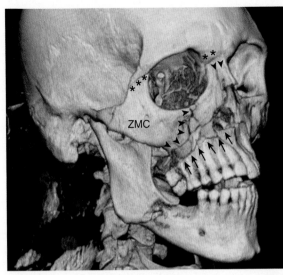

FIGURE 4-34 Facial smash. Three-dimensional surface rendered reconstruction of the face says it all and shows coexistence of multiple facial injuries simultaneously. LeFort I injury is indicated by *black arrows,* LeFort II injury is indicated by *black arrowheads,* and LeFort III injury is indicated by *asterisks.* Zygomaticomaxillary complex (ZMC) fracture is present, with the separated fragment indicated by ZMC. There is a comminuted displaced fracture of the mandible as well. Don't text and drive!

"Simple" fractures are those without displaced fragments, whereas "complex" ones are associated with displaced and rotated fracture fragments that may produce secondary airway compromise, trismus or vascular injury.

Naso-Orbital-Ethmoid Complex Fracture and LeFort Injuries

The naso-orbital-ethmoid (NOE) complex fracture is an injury involving direct trauma to the midface and nasal bones. Consequences of this injury include telecanthus (widening of the interorbital distance caused by rupture of the medial canthal ligament), transection of the naso-lacrimal system associated with epiphora or lacrimal mucocele, and CSF rhinorrhea. The fractures typically involve the nasal bones, medial maxillary buttresses, nasal septum, ethmoid sinuses, and medial orbital walls.

LeFort has classified maxillary fractures into three basic forms based on laboratory experiments on skulls (Fig. 4-33). A LeFort I (transmaxillary fracture) refers to a fracture that extends around both maxillary antra, through the nasal septum and the pterygoid plates. The maxilla is free from the rest of the facial bones (floating palate) and is usually displaced posteriorly. A LeFort II, or pyramidal fracture, starts at the bridge of the nose and extends obliquely lateral to the nasal cavity, traversing the medial wall of the orbit, the floor of the orbit, the inferior orbital rims, the maxillary antra, and the pterygoid plates. It results in disarticulation (usually posteriorly) of the nose and maxilla from the remainder of the face. In the LeFort III fracture (cranial-facial separation) the nose, zygoma, and maxilla are disarticulated from the skull. The fracture lines run from the nasofrontal area across the medial, posterior, and lateral orbital walls, the zygomatic arch, and through the pterygoid plates. Commonly, combinations of the tripod and LeFort I, II, or III fractures can occur (Fig. 4-34).

Mandible Fractures

In the trauma setting, mandible injuries can be unilateral or bilateral with presence of fractures and/or dislocations. Fractures can range from nondisplaced to severely displaced and comminuted. If the mandibular canal, which conveys the mandibular nerve, is involved by fracture, this should be reported. Additionally, injury to adjacent teeth including loosening and/or fracture should be commented upon, because in the obtunded patient, these can be aspirated and result in future complications. The temporomandibular joint should be assessed for displacement, keeping in mind physiologic anterior subluxation of the condylar head relative to the glenoid fossa in the open mouth position. Three-dimensional surface rendered reformations can be very useful to the clinician when surgical reconstruction is being planned.

TRAUMATIC CRANIAL NERVE INJURY

This can occur following head trauma and knowledge of the anatomy is useful in the search for the lesion. The olfactory bulb and tract can be injured in frontal brain trauma or from surgery. This can result in anosmia. Fractures of the optic canal/orbital apex or direct injuries to the optic nerve result in visual loss (injury to the optic nerve). Chiasmal injury has been reported secondary to mechanical, contusive, compressive, or ischemic mechanism. Fractures of the sella, clinoid processes, or facial bones should initiate a careful evaluation of the chiasm. Also, be mindful of associated pituitary stalk or hypothalamic injuries. Third nerve injury can occur in the absence of skull fracture from rootlet avulsion and distal fascicular damage secondary to a shearing type mechanism. It is also stretched during downward transtentorial herniation (DTTH). Hemorrhage at the exit site of the nerve and high intensity in the midbrain can be identified by MR. Horner syndrome from traumatic carotid dissection should also be considered with third nerve symptoms. Isolated fourth nerve palsy is common (43% of trochlear lesions) following traumatic injury. The trochlear nerve may be compressed with DTTH as it runs through the perimesencephalic cistern. The trigeminal nerve can be injured in orbital floor, roof, or apex fractures as well as central skull base injuries. The sixth cranial nerve can be affected from basilar skull fractures (Dorello canal), and injuries to the cavernous sinus/orbital apex or secondary to increased intracranial pressure. It has been acknowledged to be particularly sensitive to injury because of its long intracranial course. The seventh nerve can also be injured from longitudinal or transverse fractures through the petrous bone involving the facial canal (see Chapter 11). Associated injuries will include disruption of the ossicular chain, hematotympanum, otorrhea, and injury to the temporomandibular joint. Mechanisms of posttraumatic peripheral facial nerve palsy include transection, extrinsic compression by bony fragment or hematoma, or intrinsic compression within the facial canal secondary to intraneural hematoma/edema. Enhancement has been identified in the distal intrameatal segment, labyrinthine and proximal tympanic segments and in the geniculate ganglion. Fractures of the jugular foramen and hypoglossal canal, though uncommon, are frequently associated with VII-XI and XII injuries respectively. To add insult to injury (pun intended!), you're not even halfway done with this book.

Chapter 5

Infectious and Noninfectious Inflammatory Diseases of the Brain

A plethora of infectious and noninfectious inflammatory diseases affects the central nervous system (CNS). The normal brain responds to these insults in a rather limited, unimaginative, and stereotypical masculine manner. Initially it gets rubor (increased perfusion), calor (hot), tumor (edematous) but without dolor (pain) unless the meninges (surface) are affected. In most cases, there is a concomitant abnormality of the blood-brain barrier with associated enhancement. Later, if the insult results in neuronal death, the previously swollen organ shrinks and becomes atrophic/flaccid (another masculine trait). That's our "50 Shades of Gray" moment for the gray and white matter!

Imaging techniques are relatively sensitive for detecting an abnormality, localizing it, and in many cases, categorizing the lesion into infectious/inflammatory disease versus neoplastic or vascular disease. Diffusion-weighted imaging (DWI) is particularly useful in this scenario because restricted diffusion is characteristic of some stages of some infections. With the aid of clinical history, physical examination, and the patient's age, the radiologist can more accurately interpret the particular images and make an educated guess at a probable differential diagnosis.

Localization of lesion(s) is the critical first step in the differential diagnosis! Is it epidural, subdural, subarachnoid, intraventricular, or intraparenchymal? Is it confined to a particular region of the brain such as the temporal lobe, which would imply a specific pathogen (Table 5-1)?

ANATOMICALLY BOUNDED INFECTIOUS PROCESSES

Anatomy of the Spaces around the Brain and Its Coverings

Three membranes cover the brain; these layers of connective tissue are collectively called the meninges. They are named, from the outermost layer inward, the dura mater, arachnoid mater, and pia mater. The dura mater (literally, "tough mother") is also referred to as the pachymeninges and is composed of two layers of very tough connective tissue. The outermost layer is tightly adherent to the skull and represents the periosteum of the inner table. The inner layer is covered with mesothelium and lines the outer subdural space. The two layers separate to form the venous sinuses. The inner layer reflects away from the skull to give rise to the tentorium cerebelli, the falx cerebri, the diaphragma sellae, and the falx cerebelli. The space between

the inner table of the skull and the dura mater is the epidural space and it is most closely adherent at the suture lines. The space between the dural covering and the arachnoid is the subdural space. The subdural space is a potential space containing bridging veins, which drain blood from the cortex into the venous sinuses, and outpouchings of the arachnoid (arachnoid villi), which project into the venous sinuses.

Beneath the subdural space are two other layers of connective tissue, the arachnoid mater and pia mater, which together constitute the leptomeninges. The arachnoid is a delicate outer layer that parallels the dura and is separated from the pia by the subarachnoid space, which contains the cerebrospinal fluid (CSF). The pia is closely applied to the brain and spinal cord and carries a vast network of blood vessels. Figure 5-1 illustrates this anatomy.

Epidural Abscess

Epidural abscess is most often the result of infection extending from an operative bed, the mastoids, paranasal sinuses, or infected skull. The imaging findings of epidural abscess are those of a focal epidural mass of low density on computed tomography (CT), low intensity or isointensity on T1-weighted images (T1WI), high signal on T2-weighted images (T2WI/fluid-attenuated inversion imaging [FLAIR]), and restricted diffusion on DWI. Enhancement, particularly on a thickened dural surface or along the margins of

TABLE 5-1 Location and Favored Pathogen

Location	Favored Pathogen	Useful Hint
Cerebellum/brain stem	*Listeria*	Hard to culture
Perivascular spaces	*Cryptococcus*	AIDS host
Deep gray matter	Tick/mosquito borne	Very sick
Basal meninges	TB	Endemic locale
Periventricular	TORCH	Neonate, calcifications
Cavernous sinus	Fungi	Spread from sinusitis
Subdural space	*H. flu*	Effusions with meningitis

AIDS, Acquired immunodeficiency syndrome; *H. flu,* Haemophilus influenzae; *TB,* tuberculosis; *TORCH,* toxoplasmosis, other infections, rubella, cytomegalovirus, herpes simplex.

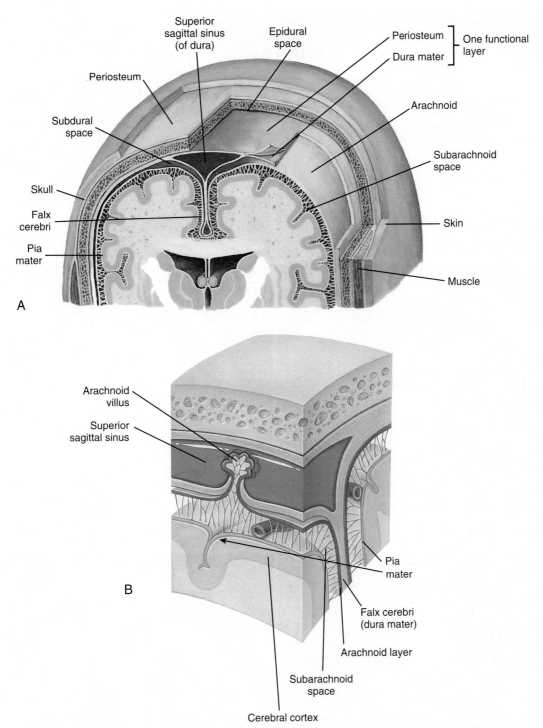

FIGURE 5-1 Coverings of the brain. **A,** Multiple layers of tissue can be seen, covering the brain with fluid in between. Pass from scalp to bone to dura mater, between arachnoid, and pia to find cerebrospinal fluid water. **B,** Arachnoid villi pooch into sinus space, blood vessels in subarachnoid fluid run in place. (From Thibodeau GA, Patton KT. *Anatomy and Physiology.* 5th ed. St. Louis: Mosby; 2003:376, 379.)

the collection, is observed. An epidural abscess can extend into the subgaleal space through emissary veins or intervening osteomyelitis to appear outside the skull, a finding more frequent when the abscess occurs as a postoperative complication. These same veins can lead to thrombophlebitis of draining cerebral veins and sinuses.

Epidural abscesses can dissect across the dural sinuses and thereby cross the midline, distinguishing them from subdural empyemas, which are usually confined by the midline falx. Epidural abscesses, however, like epidural hematomas, are typically confined by sutures (Fig. 5-2).

Subdural Empyema

Disruption of the arachnoid meningeal barrier by infection leads to the formation of CSF collections within the potential compartment of the subdural space. These may present acutely or chronically, and can be sterile or infected at

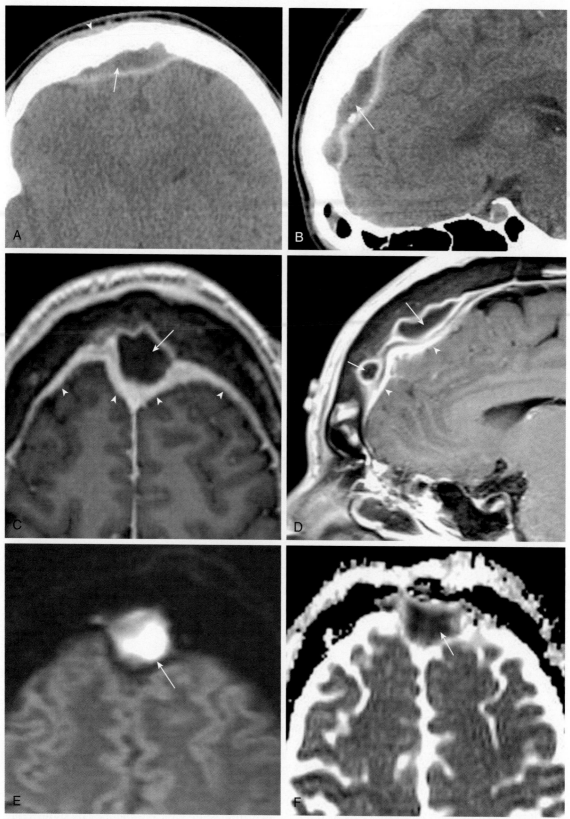

FIGURE 5-2 Epidural abscess resulting from frontal skull osteomyelitis. On **(A)** axial and **(B)** sagittal contrast-enhanced computed tomography, a collection *(arrow)* that crosses the midline is noted, thus the "epidural space" must be quoted. On the other side of the frontal bone, the scalp is thickened and you can hear it groan *(arrowhead)*. **C,** On axial and sagittal **(D)** postcontrast T1, the abscess *(arrows)* is demonstrated. The location above dura is confirmed *(arrowheads)*. **E,** Diffusion-weighted imaging (DWI) and **(F)** apparent diffusion coefficient (ADC) map; restricted diffusion is plain as can be *(arrows)* on images of DWI and ADC.

time of presentation. Empyema rather than abscess is the appropriate term for a purulent infection in this potential space. Box 5-1 lists the causes of subdural empyema. Among the several possible mechanisms by which a subdural empyema is thought to form are (1) a distended arachnoid villus, which could rupture into the subdural space and infect it; (2) phlebitic bridging veins (secondary to meningitis), which may infect the subdural space; (3) the subdural space, which may be infected by direct hematogenous dissemination; and (4) direct extension, which may occur through a necrotic arachnoid membrane from the subarachnoid space or from extracranial infections.

Clinical signs and symptoms in this group of patients include fever, vomiting, meningismus, seizures, and increased intracranial pressure. Venous thrombosis or brain abscess develops in about 10% of patients with subdural empyema. The mortality rate from subdural empyema has been reported to range approximately from 12% to 40%. Prompt treatment with appropriate antibiotics and drainage through an extensive craniotomy can result in a favorable outcome.

Features of subdural empyema are those of extracerebral collections over the convexities and within the interhemispheric fissure, which on magnetic resonance (MR) display isointensity on T1WI and high signal on T2WI/FLAIR, and on CT show an isodense to low density extraaxial mass (Fig. 5-3). Empyema may be distinguished from subdural effusion on DWI scans; that is, empyemas are hyperintense on DWI (low apparent diffusion coefficient [ADC]), whereas sterile effusions are low intensity on DWI and have ADCs similar to CSF. There may be effacement of the underlying cortical sulci and compression of the ventricular system. A rim of enhancement may be observed. This enhancement occurs from granulation tissue that has formed over time in reaction to the adjacent infection.

Leptomeningitis

(Lepto) meningitis is inflammation of the pia and arachnoid mater (Fig. 5-4). Leptomeningeal inflammation most often occurs after direct hematogenous dissemination from a distant infectious focus. Pathogens also gain access by passing through regions that may not have a normal blood-brain barrier, such as the choroid plexus. Direct extension from sinusitis, orbital cellulitis, mastoiditis, or otitis media is also common. After septicemia, bacteria may lodge in venous sinuses or arachnoid villi and precipitate inflammatory changes, which in turn can interfere with CSF drainage leading to hydrocephalus. With stagnation of CSF flow, bacteria are offered the opportunity to invade the meninges and indulge themselves. Early in the course of infection, there is congestion and hyperemia of the pia and arachnoid mater. Later an exudate covers the brain, especially in the dependent sulci and basal cisterns. The leptomeninges become thickened. Clinical features are related to patient age. Infants and particularly neonates may have a perplexing clinical picture, lacking physical signs that directly demonstrate meningeal irritation. Young children and adults often declare symptoms of fever, headache, photophobia, and neck pain. Symptoms in the elderly can be perplexing too; these patients not uncommonly present with confusion, depressed levels of consciousness, and stupor.

There may be no abnormal imaging findings in early and successfully treated meningitis. Findings for acute meningitis are (1) high intensity of the subarachnoid fluid on FLAIR; (2) acute cerebral swelling indicative of encephalitis (often leading to herniation and death); and (3) communicating hydrocephalus, with enlargement of

---

BOX 5-1　Causes of Subdural Empyema

Hematogenous dissemination
Osteomyelitis of calvarium
Otomastoiditis
Paranasal sinusitis
Postcraniotomy infection/calvarial osteomyelitis
Posttraumatic
Purulent bacterial meningitis
Thrombophlebitis

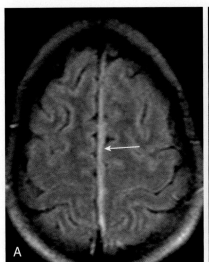
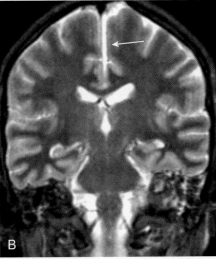
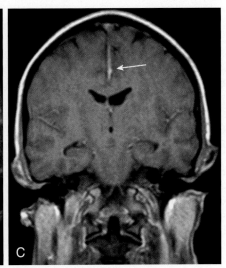

FIGURE 5-3 Subdural empyema. Although they are rare, the subdural empyemas are seen best on fluid-attenuated inversion recovery (**A,** *arrow*). They fade into cerebrospinal fluid on T2 (**B,** *arrow*) and may enhance with gadolinium when thick goo (**C,** *arrow*).

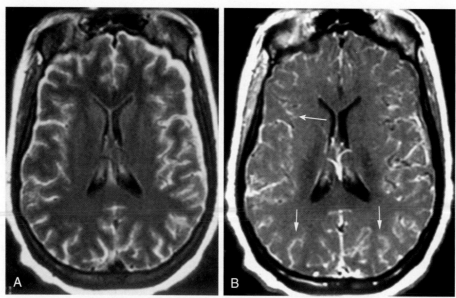

FIGURE 5-4 Leptomeningitis. Fluid-attenuated inversion recovery **(A)** scan shows bright cerebrospinal fluid. **B,** Postcontrast T1 image shows enhancement in sulcal depths *(arrows)*. While one may also see enhancing swollen veins, look hard at the surface for the pial enhancing stains.

the temporal horns and effacement of the basal cisterns. Shortly thereafter, one may identify marked enhancement of the leptomeninges, better visualized on MR than on CT (and much more common with bacterial > viral lesions). Postgadolinium FLAIR images appear to be very sensitive for subarachnoid disease (see Fig. 5-4, *B*). Concurrent parenchymal abnormalities may occur from encephalitis or venous infarction. Vasculitis may involve either arteries or veins; hence, patterns of infarction associated with meningitis differ depending on the location, number, and type of vessels involved.

Many additional complications may occur as a result of inflammation involving the meninges. These sequelae are better imaged and characterized than the manifestations of the meningitis itself. Communicating hydrocephalus can occur as both an early and a late manifestation of leptomeningitis, often becoming symptomatic to the point of requiring ventricular shunting. The subacute imaging findings of complicated leptomeningeal infection are those of atrophy, encephalomalacia (infarction), focal abscess, subdural empyema formation, and basilar loculations of CSF (Box 5-2).

Meningitis from sinusitis can produce septic thrombosis of adjacent venous sinuses and pseudoaneurysm formation if the cavernous sinus is affected (Fig. 5-5). In the temporal bone, labyrinthitis ossificans (see Chapter 11) may occur as a late finding, secondary to infected CSF, communicating with the endolymph and perilymph presumably through the cochlear aqueduct.

Neonates represent a special case with respect to the cerebral sequelae of bacterial leptomeningitis. The most commonly encountered organisms are gram-negative bacilli, followed by group B *Streptococcus, Listeria monocytogenes,* and others. The neonatal meningitides are believed to be acquired as a result of the delivery process, chorioamnionitis, immaturity, or iatrogenic problems (e.g., catheters, inhalation therapy equipment). The lack of a developed immune system at birth makes neonates susceptible to organisms that are normally not very virulent. These children frequently have severe parenchymal brain

---

BOX 5-2   Complications of Leptomeningeal Infection

Arterial infarction
Atrophy (late)
Basilar adhesions
Encephalitis
Epidural empyema
Focal abscess formation
Hydrocephalus
    Communicating
    Obstructive
Subdural empyema
Subdural hygroma
Vasculitis
Venous thrombosis/venous infarction
Ventriculitis

---

damage as a result of the infection that ultimately produces a multicystic-appearing brain often with hydrocephalus. The imaging findings are those of multifocal encephalomalacia leading to multiple distended intraventricular and paraventricular cysts. This pattern is also seen in *Herpes simplex* virus (HSV) type 2 viral perinatal infections (with no temporal lobe predilection like HSV-1). In children (1 month to 15 years old), *Haemophilus influenzae (H. flu)* is a common pathogen associated with upper respiratory infections and can produce a virulent meningitis with vascular infarction. Other bacteria in this group are *Neisseria meningitidis* and *Streptococcus pneumoniae.* In adults, *S. pneumoniae* and *N. meningitidis* are the most common bacterial organisms producing meningitis. *H. flu* characteristically has a high rate of subdural effusions.

Inevitably, CSF must be sampled. Clearing a patient for lumbar puncture (LP) requires spotting findings that could lead to acute herniation from the pressure gradient created by the lumbar CSF tap. Be sure that you do not see cerebellar tonsils engaged in the foramen magnum, cerebellar herniation syndromes, obliterated or trapped fourth ventricles

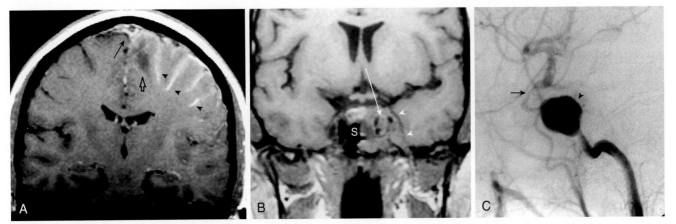

FIGURE 5-5 Complications of sinusitis **A,** Septic thrombosis of the sagittal sinus *(arrow)* occurred in this case from meningitis *(arrowheads).* A stroke in the left frontal cortex *(open arrow)* left the patient in a monoplegic vortex. **B,** Meningitis, sphenoid sinusitis (s) and cavernous sinus thrombosis *(arrowheads)* also led to cavernous internal carotid artery stenosis. The cavernous sinus funky looking flow voids portended a mycotic aneurysm *(long arrow)* on steroids. **C,** Angiogram reveals aneurysmal mycosis *(arrowhead)* and distal ICA stenosis *(arrow).*

(see Chapter 7), cerebellar masses or strokes, transtentorial downward herniation, completely effaced basal cisterns/sulci or significant subfalcine herniation. Obliteration of the superior cerebellar and quadrigeminal plate cisterns with sparing of the ambient cisterns is concerning. Better to err on suggesting use of clinical acumen in diagnosing meningitis than clearing a patient for LP and having them herniate and die from the "approved tap." Words to the wise…from the law firm of Dewey, Cheatham, and Howe.

Leptomeningitis versus Pachymeningitis

Leptomeningitis and pachymeningitis (Fig. 5-6) are entities that may be legitimately separated by their enhancement pattern on MR. Leptomeningeal enhancement follows the pia into the gyri/sulci and/or involves the meninges around the basal cisterns (because the dura-arachnoid is widely separated from the pia-arachnoid here). Pachymeningeal enhancement is thick and linear/nodular following the inner surface of the calvarium, falx, and tentorium and without extension into the sulci or involvement of the basal cisterns. One entity that simulates infectious pachymeningitis is idiopathic hypertrophic cranial pachymeningitis. This rare disorder, characterized by severe headache, cranial nerve palsies, and ataxia, peaks during the sixth decade. The clinical course is chronic with some initial improvement with steroids. Box 5-3 provides a list of conditions that can produce pachymeningeal versus leptomeningeal enhancement.

Pyogenic Brain Abscess

Cerebral abscess is most often the result of hematogenous dissemination from a primary infectious site. The various causes of cerebral abscess are listed in Box 5-4. The most frequent locations are the frontal and parietal lobes in the distribution of the middle cerebral artery. Intracranial abscess affects predominantly preadolescent (DMY) and middle-age groups (RN). In part, this is related to the incidence of congenital heart disease, intravenous (IV) drug abuse, acquired immunodeficiency syndrome (AIDS), and tympanomastoid and paranasal sinus infections. In all series there is a preponderance of male patients over female.

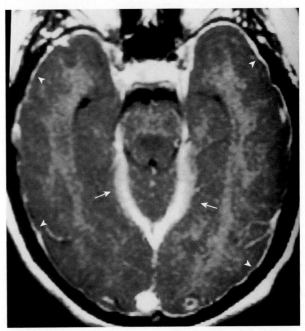

FIGURE 5-6 Pachymeningitis. Note that the very thick enhancement of the dura mater *(arrowheads)* and tent *(arrows).* This pattern is pachymeningitic, not leptomeningitic; on that you can bet the rent!

Abscesses may be unilocular or multilocular, solitary or multiple. A variety of bacterial organisms are commonly cultured from brain abscesses depending upon skin/IV sources (staphylococcus, streptococcus, enterobacteria), heart (streptococcus, staphylococcus) or lung (streptococcus, bacilli, fusobacterium) sources. In addition, numerous other pathogens can infect the brain when the immune system is compromised such as in transplant patients (candida, aspergillus, nocardia, gram-negative bacilli).

Abscess formation has been divided into four stages: (1) early cerebritis (1 to 3 days); (2) late cerebritis (4 to 9 days); (3) early capsule formation (10 to 13 days); and (4) late capsule formation (14 days and later). The cerebritis phase of abscess formation consists of an inflammatory infiltrate of polymorphonuclear cells, lymphocytes, and

---

BOX 5-3 Conditions That Produce Pachymeningeal and Leptomeningeal Enhancement

PACHYMENINGEAL

Cerebrospinal fluid leak
Idiopathic hypertrophic pachymeningitis
Sarcoidosis
Metastases including those involving the skull
Shunting
Postsurgical
Spontaneous intracranial hypotension

LEPTOMENINGEAL

Acute stroke
Leptomeningeal infection
Sarcoidosis
Subarachnoid hemorrhage
Metastatic carcinomatosis

BOX 5-4 Causes of Cerebral Abscess

DIRECT EXTENSION

Meningitis
Otomastoiditis
Paranasal sinusitis
Via dermal sinus tracts

HEMATOGENOUS DISSEMINATION

Arteriovenous shunts (Osler Weber Rendu)
Cardiac (endocarditis, infected thrombi)
Drug abuse
Pulmonary infection
Sepsis

TRAUMA

Penetrating injury
Postsurgical
Immunosuppression
No predisposing factors in 25% of cases

---

plasma cells. By the third day, a necrotic center is formed. This deliquescent (we liked the word, so check your medical dictionary) region is surrounded by inflammatory cells, new blood vessels, and hyperplastic fibroblasts. In the late cerebritis phase, extracellular edema and hyperplastic astrocytes are seen. Thus the cerebritis phase of abscess formation starts as a suppurative focus that breaks down and begins to become encapsulated by collagen at 10 to 13 days. This process continues with increasing capsule thickness.

The deposition of collagen is particularly important because it directly limits the spread of the infection. Factors that affect collagen deposition include host resistance, duration of infection, characteristics of the organism, and drug therapy. Steroids may decrease the formation of a fibrous capsule and the effectiveness of antibiotic therapy in the cerebritis phase and may reduce antibiotic penetration into the brain abscess.

Brain abscesses that are spread hematogenously usually occur at the junction of the gray and white matter. Collagen deposition is asymmetric, with the side towards the white matter and ventricle having a thinner wall, resulting in a propensity for intraventricular rupture or daughter abscess formation, which is sometimes useful in distinguishing abscess from tumor (neoplastic walls are uniformly thick). Death from cerebral abscess is due to its mass effect with herniation, abrupt hydrocephalus, and/or the development of a ventricular empyema. In the late capsule phase, there is continued encapsulation and decreasing diameter of the necrotic center.

The imaging characteristics of cerebral infection depends on the pathologic phase during which the inflammation is being examined. In the cerebritis phase, CT demonstrates low-density abnormalities with mass effect. Patchy enhancement may be present. On MR, one may see low intensity on T1WI and high signal intensity on T2WI/FLAIR (Fig. 5-7), typically centered at the corticomedullary junction as well as patchy enhancement. In the late cerebritis phase, ring enhancement may be present. The presence of ring enhancement should not unequivocally imply capsular formation. The surgeon contemplating drainage should appreciate that a firm, discrete abscess may not be present despite ring enhancement. (In patients with multiple abscesses, in eloquent locations, and poor surgical risk, conservative therapy with antibiotics alone has been advocated in conjunction with close monitoring of the clinical and imaging findings.)

On CT, the encapsulated intracerebral abscess shows a low-density center and low-density surrounding the lesion (edema). Ring enhancement is virtually always present on contrast-enhanced computed tomography (CECT) in pyogenic brain abscess. Thickness, irregularity, and nodularity of the enhancing ring should raise the suspicion that one is dealing with a tumor (most of the time) or an unusual infection (e.g., fungus). An uncommon observation on noncontrast CT is the presence of a complete slightly hyperdense ring. This noncontrast ring is most often identified in metastases, but does occur with abscesses and astrocytomas as well.

By the same token, a thin rim of low signal on T2WI and high signal on T1WI (Fig. 5-8) may characterize the wall of an abscess and would help distinguish an abscess from a necrotic tumor. This may be related to free radical formation (secondary to oxidative effect of the respiratory burst of the bacteria), hemorrhage, or other factors. At this point the DWI scan may be bright; however, some cases of cerebritis/encephalitis may also show cytotoxic edema. The low ADC is probably related to high protein, high viscosity, and cellularity (pus) within the abscess cavity.

After 2 to 3 weeks, a mature abscess appears on T1WI as a round, well-demarcated low-intensity region with mass effect and peripheral low intensity (edema) beyond the margin of the lesion. On T2WI/FLAIR, high intensity is noted in the cavity and in the parenchyma surrounding the lesion (Figs. 5-9, 5-10). Most pyogenic lesions enhance with a thin rim surrounding the necrotic center. Tiny abscesses may appear to have nodular enhancement. Multiple ring-enhancing lesions are more consistent with hematogenous dissemination of an infectious focus. Multiple rings in a single location can be seen with daughter abscesses but have also been noted with glioma (and other lesions). Box 5-5 is a partial differential diagnosis of the ring-enhancing lesion.

The vast majority of pyogenic abscesses evoke considerable edema. Remember that the vasogenic edema surrounding the pyogenic abscess will be bright on ADC maps indicating no restricted diffusion, unlike the abscess itself which is dark on ADC with restriction of diffusion.

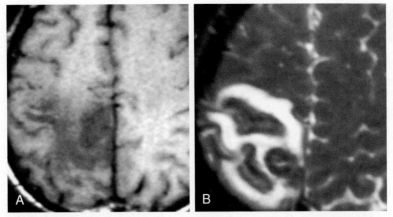

FIGURE 5-7 Early cerebritis. **A,** Low signal on T1-weighted scan, found in cerebritis in a drug abusing man. **B,** With corresponding high signal on T2, said the intern to the attending, "Now what should we do?"

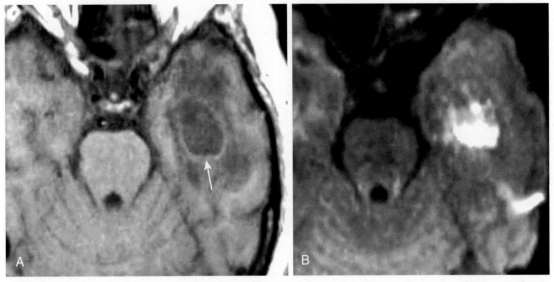

FIGURE 5-8 A, T1-weighted imaging (T1WI) hyperintense rim. The hyperintense rim on the T1WI scan *(arrow),* combined with bright diffusion-weighted imaging **(B),** is vintage. Seen together you should display your astute prowess, and declare "an intracerebral abscess."

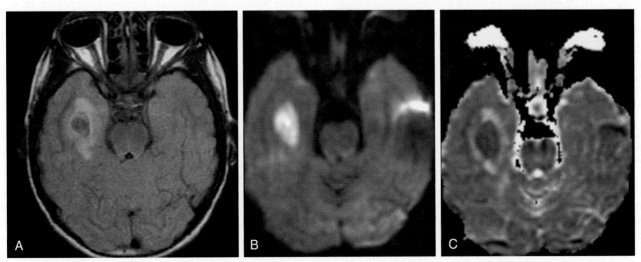

FIGURE 5-9 Mature abscess. **A,** Fluid-attenuated inversion recovery; **B,** diffusion-weighted imaging; and **C,** apparent diffusion coefficient (ADC) reveal a mass with edema, restricted water in axis Z. "Low ADC means pus in the center," says to you your neuroradiology mentor.

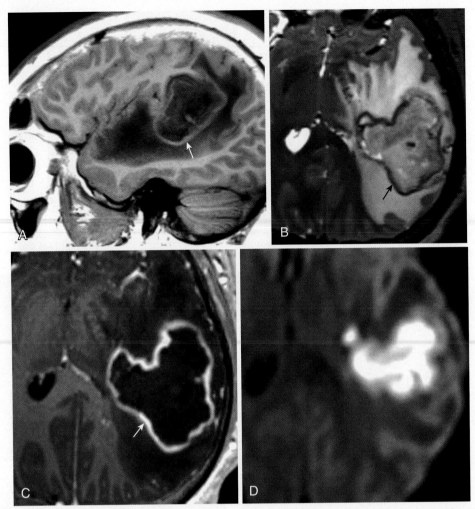

FIGURE 5-10 Abscess. **A,** Unenhanced T1-weighted imaging (T1WI) sagittal image demonstrating frontal lesion with high-intensity ring *(arrow)*. **B,** A dark border and surrounding edema on T2WI typify the thing *(arrow)*. **C,** Enhancing rim is smoother on outside than inside, which is considered a sign of abscess *(arrow)*. **D,** Bright signal on diffusion-weighted image confirms our suspicion in this patient with streptococcus.

BOX 5-5 Differential Diagnosis of Ring Enhancement

Metastasis
High-grade astrocytoma
Demyelinating plaque
Abscess
Radiation necrosis
Tuberculoma
Infarction
Lymphoma
Subacute parenchymal hematoma
Thrombosed aneurysm

A ring-enhancing lesion that does not evoke much edema and does not have restricted diffusion should steer you away from a diagnosis of abscess. A differential diagnosis, including granuloma, primary or metastatic tumor, and demyelinating disease, is more appropriately proffered in such ambiguous cases. DWI is usually positive in pyogenic abscesses. Unfortunately, many nonbacterial pathogens have not read the literature.

Occasionally, the abscess may spread into the ventricles because of lower collagen content in the medial wall, producing periventricular enhancement and/or layering debris within the ventricles. Hemorrhage is rare in acute bacterial abscesses. However, toxoplasmosis, after treatment, may show hemorrhagic byproducts. Fungal septic emboli may elicit heme.

Ventriculitis

Ventriculitis (ependymitis) can be seen as part of the spectrum of infection including meningitis as a postoperative complication (particularly related to ventricular shunting), or as an isolated finding. Ventriculitis is more commonly seen in neonates with meningitis and is a feature of cytomegalovirus (CMV) infection as well as some puerperal anaerobic infections. Organisms are introduced into the ventricle as a result of bacteremia, from abscess, trauma, or instrumentation of the ventricle or CSF spaces. The ventricles can be enlarged, and on T2WI/FLAIR high intensity can be observed surrounding the ventricles or even within the CSF. The key to the diagnosis is DWI bright layering material or subependymal/choroid plexus enhancement (Fig. 5-11). Periventricular calcification can be seen on CT after neonatal ventriculitis.

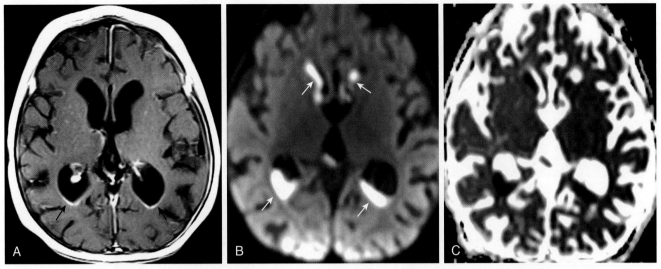

FIGURE 5-11 Ependymitis. **A,** Enhancement of the ependyma *(arrows)* leads to an infectious dilemma. **B,** Diffusion-weighted image and **(C)** apparent diffusion coefficient map show proteinaceous materials layering secretly in the ventricles *(arrows).*

Enhancement of the ependyma is also observed in lymphoma and other malignant lesions with subependymal spread.

Choroid Plexitis

Capillaries of the choroid plexus, because of their fenestrated epithelium, serve as a conduit through which infections may gain access to the brain. A second barrier between blood and CSF, the choroidal epithelium, possesses tight junctions and prevents passive exchange of proteins and other solutes. Usually, choroid plexitis is seen in association with encephalitis, meningitis, or ventriculitis. It is rarely seen as an isolated infection. Pathogens with a propensity for producing choroid plexitis include Nocardia and Cryptococcus.

Xanthogranulomatous change to the choroid plexus is bright on DWI and sometimes FLAIR but does not enhance much. The differential diagnosis of choroid plexus disease is provided in Box 5-6.

Septic Embolus

The most frequent manifestation of infective endocarditis is stroke, with *Staphylococcus aureus* by far the most common organism. However, sepsis from any cause including pulmonary arteriovenous malformations, pulmonary infection, intravenous drug abuse, infected catheters with cardiac septal defects, and occult infection may produce septic emboli to the brain. Septic emboli are associated with persistent mass effect, edema, and enhancement beyond a 6-week period. This should alert the radiologist to consider septic infarction with development of abscess formation in association with cerebral infarction. The vast majority of septic emboli we see in our population in East Baltimore are related to intravenous drug abuse, and most lead to septic emboli in the lungs, not the brain. When they pass through the lungs or are derived from septic endocarditis/valvular vegetations on the arterial side, they may result in

---

BOX 5-6 Differential Diagnosis of Choroid Plexus Lesions

ANGIOMATOUS
Klippel-Trenaunay-Weber syndrome
Primary angiomas and arteriovenous malformations of choroid plexus
Sturge-Weber syndrome

INFECTION
Cryptococcus
Nocardia
Other

INFLAMMATION
Rheumatoid nodule
Sarcoid granuloma
Xanthogranuloma

NEOPLASTIC
Astrocytoma
Choroid plexus papilloma
Ependymoma
Lymphoma
Meningioma
Metastasis
Neurocytoma
Oligodendroglioma
Primitive neuroectodermal tumor
Subependymal giant cell astrocytoma
Subependymoma

TRAUMA
Hematoma

---

brain abscess, mycotic aneurysm (these occur in distal vessels, usually the middle cerebral artery, and are less likely to hemorrhage; Fig. 5-12) or obliterative vasculitis. Mycotic aneurysm often present with intraparenchymal hemorrhage (see Fig. 5-12, *B*) leaking into the subarachnoid space in the periphery of the brain, not the circle of Willis.

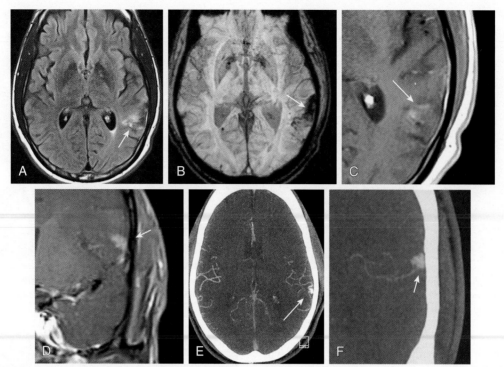

FIGURE 5-12 *Streptococcus viridans* bacterial endocarditis with septic aneurysm. **A,** Fluid-attenuated inversion recovery shows a wedge shaped abnormality in the temporal lobe *(arrow).* **B,** Bloody stroke is confirmed on susceptibility-weighted imaging probe *(arrow).* **C,** Postgadolinium there also is meningeal irritation *(arrow),* which is confirmed on the coronal T1-weighted imaging magnification (**D,** *arrows*). **E,** Axial and **(F)** magnified axial computed tomography angiography images show middle cerebral artery distal mycotic aneurysm *(arrows),* which leads to a classic *Requisites* wisdom: "If you do intravenous drugs, you risk brain-damaging bugs."

SPECIFIC INFECTIOUS PATHOGENS

Viral

See list of viruses in Table 5-2.

Herpes Simplex

Herpes simplex virus is the most common cause of fatal endemic encephalitis in the U.S. The survivors of infections with this virus have severe memory and personality problems. Early diagnosis and therapy with antiviral agents can favorably affect the outcome. Both the oral strain (type 1) and the genital strain (type 2) may produce encephalitis in human beings. Type 2 is responsible for infection in the neonatal period, presumably acquired either transplacentally or during birth from mothers with genital herpes. This strain when contracted in utero may cause a variety of teratogenic problems including intracranial calcifications, microcephaly, microphthalmia, and retinal dysplasia. It carries a high morbidity and mortality rate, with CNS infection either primary or part of a disseminated infection. Sequelae from neonatal herpes also include multicystic encephalomalacia, seizures, motor deficits, mental and motor retardation, and porencephaly. The features of intracranial neonatal herpes are different from those in adults and are summarized in Box 5-7. The early findings in neonatal herpes are subtle regions of low density on CT in various regions in the brain parenchyma including gray matter and cerebellum. These regions enlarge rapidly, with meningeal and gyriform enhancement. Later gyri may demonstrate strikingly high density on noncontrast CT possibly from laminar necrosis.

Calcification can appear between 17 and 21 days after disease onset and can be variable in location. Thalamic hemorrhage has been observed. The normal low intensity on T1WI of neonatal white matter and high intensity on T2WI/FLAIR limit the sensitivity of MR in this disease. Loss of gray-white contrast is an early abnormality that can be incorrectly interpreted as poor quality images. DWI may show gyriform high signal. The MR correlate of the high density in the cortex on CT is hypointensity on T2WI/FLAIR. The cause of these characteristic cortical changes is not understood, but increased cortical blood volume (with deoxyhemoglobin), calcification, laminar necrosis, and other associated paramagnetic ions may be possible causes.

The type 1 virus is responsible for the fulminant necrotizing encephalitis seen in children and adults. It has an incidence of one to three cases per million. The clinical picture (just like "maintenance of certification examination encephalopathy") is one of acute confusion and disorientation followed rapidly by stupor and coma. Seizures, viral prodrome, fever (in more than 95% of cases), and headache are common presentations. Focal neurologic deficits such as cranial nerve palsies are found in less than 30% of cases. Those patients with left temporal disease become symptomatic earlier because of their language impairment and thus may have more subtle imaging findings at the time of presentation.

The pathologic findings are stereotypic. The virus asymmetrically attacks the temporal lobes, insula, orbitofrontal region, and cingulate gyrus. Approximately one third of CNS herpes infections are due to primary infection

TABLE 5-2 Characteristic Targets of Viral Encephalitides

Location	Favored Virus	Useful Hint
Medial temporal lobe, insula, cingulum	HSV I	May bleed, bilateral
Holohemispheric, cortical	HSV 2	Neonatal, cystic encephalomalacic
MCA infarcts	VZV	Vasculitis
Brain stem and cerebellum	EBV	Mono symptoms, nodes, nasopharyngeal cancer
Brain stem, deep gray matter	EEE	Whinny
Deep gray matter, brain stem	Japanese	Giant panda sign
White matter, atrophy	HIV	Opportunistic infections
Periventricular, ependymitis	CMV	Congenital migrational anomalies
Subcortical white matter	JC virus	Immune compromised
Small vessel infarcts in white matter	Nipah	Batty
Myelitis, central brain and deep gray matter	West Nile virus	
Substantia nigra	St Louis encephalitis	
Basal ganglia, brain stem	Rabies	Tick bite
Cortical edema, basal ganglia disease	Measles	Rash

CMV, Cytomegalovirus; *EBV*, Epstein-Barr virus; *EEE*, eastern equine encephalitis; *HIV*, human immunodeficiency virus; *HSV*, herpes simplex virus; *MCA*, middle cerebral artery; *VZV*, varicella zoster virus.

BOX 5-7 Features of Neonatal Herpes Simplex Infection

Atrophy
Encephalomalacia with cysts (late)
Hydrocephalus
Increased density in cortical gray matter
Intracranial calcification from punctate to gyriform
Microcephaly
Microophthalmia
Multiple cysts

(usually in persons younger than 18 years of age), whereas two thirds are the result of reactivation confirmed by the presence of preexisting antibodies. A possible explanation for the focality and latency of herpes simplex type 1 may depend on its known residence in the trigeminal ganglia. This latent virus under certain circumstances becomes reactivated and spreads along the trigeminal nerve fibers, which innervate the meninges of the anterior and middle cranial fossae. A diffuse meningoencephalitis with a predominant lymphocytic infiltration is seen. Histologically, there is marked necrosis and hemorrhage, with loss of all neural and glial elements. The result in untreated cases, if the patient survives, is an atrophic cystic parenchyma. Laboratory diagnosis is dependent on polymerase chain reaction (PCR) on CSF for herpes.

In type 1 herpes encephalitis, MR findings within the first 5 days of the disease show high intensity on T2WI/FLAIR and either positive or negative DWI in the temporal and inferior frontal lobes and progressive mass effect including the cingulate region, whereas CT findings at this time may be subtle to nonexistent (Fig. 5-13). Negative diffusion images may indicate some possibility for reversibility. The earliest CT abnormalities are low-density areas in the temporal lobe and insular cortex. Hemorrhage may be identified especially with susceptibility-weighted imaging (SWI) sequences. The areas of abnormality in the temporal lobe and insula abruptly end at the lateral putamen, which is characteristically spared. MR is frequently able to detect asymmetric bilateral temporal lobe involvement even as only unilateral disease is seen on CT. This picture is virtually pathognomonic of herpes simplex infection.

Gyriform enhancement may be visualized; however, enhancement varies with severity and stage of disease. Leptomeningeal enhancement has also been observed. Mass effect may persist for a considerable time (weeks). Residual abnormalities include areas of low density and parenchymal loss at the site of involvement.

Herpes simplex encephalitis is a potentially treatable encephalitis where outcome depends on early diagnosis. Untreated herpes encephalitis has a 70% mortality rate, with only 2.5% of survivors returning to a normal functional life. MR is clearly the imaging modality of choice, and it behooves the radiologist to be alert to this condition at all times. This is one diagnosis you do not want to blow. While there are similarities to middle cerebral artery (MCA) strokes, the involvement of areas normally subsumed by anterior choroidal artery supply, the more selective cortical involvement, the bilaterality, the cingulate involvement (anterior cerebral artery territory) more subtle decrease in ADC, and the clinical symptomatology should point instead to herpes.

Herpes Zoster

Herpes varicella zoster virus (VZV) is a deoxyribonucleic acid (DNA) virus that affects elderly immunocompetent individuals but has a propensity for infecting immunosuppressed patients, particularly those with lymphoma and AIDS. It is caused by reactivation of latent varicella zoster virus that has remained dormant in cranial nerve and dorsal root ganglia since primary varicella infection (chickenpox). The virus travels along the sensory nerve from the dorsal root ganglion to the skin producing a rash (shingles) with associated severe radicular pain and allodynia (sensitivity to touch). It may be detected in the blood or CSF by methods based upon PCR.

VZV is most common in the thorax followed by the face (Ramsay Hunt syndrome-Herpes zoster oticus). One clinical presentation of intracranial involvement occurs in patients who initially have involvement of the ophthalmic division of the trigeminal nerve, herpes zoster ophthalmicus, and then develop contralateral hemiplegia several weeks later. Rarely, VZV can produce infarction of the optic nerve. Pathology reveals an occlusive granulomatous vasculitis involving the intracranial arteries and meningoencephalitis.

Cerebral angiography can demonstrate the severe vasculitis caused by herpes revealing multiple areas of segmental constriction, usually involving proximal segments

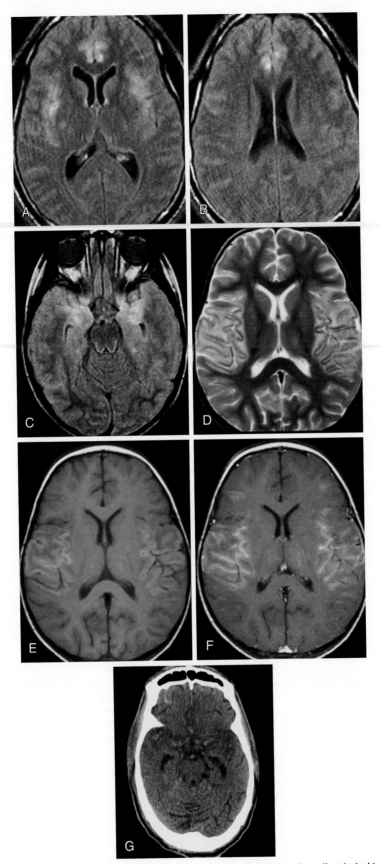

FIGURE 5-13 Herpes simplex I gallery. Case 1: **A** and **B,** Despite the patient's epileptic jerkies, this scan shows features of herpes. There is high signal that both temporal lobes share, in bilateral insula and cingulum, on fluid-attenuated inversion recovery (FLAIR). Case 2: **C,** Here we see amygdala involved on FLAIR, our first picture. By the time the T2-weighted imaging scan is done **(D),** both temporal lobe high signal is a fixture. The brightness pregadolinium suggests laminar necrosis **(E),** and everything enhances on postgadolinium—bad prognosis **(F).** Case 3: **G,** Even later one sees the low density on computed tomography, seen in the medial temporal and perisylvian regions bilaterally.

of the anterior and middle cerebral arteries without involvement of the extracranial vessels (Fig. 5-14).

The spinal form of shingles usually has no imaging findings. Rarely, Zoster may produce myelitis secondary to neurotrophic spread of the virus. This occurs 1 to 2 weeks after the appearance of the rash. Patients present with paraparesis and a sensory level. On MR there is high intensity on T2WI in the dorsal root entry zone and within a significant section of the spinal cord emanating from the root. Patients may improve (especially those that are immunocompetent) but are usually left with a residual neurologic deficit. Aggressive early treatment with acyclovir has been strongly advocated. Ipsilateral, posterior column atrophy has been described in shingles.

VZV is linked to cerebral granulomatous angiitis. Granulomatous angiitis, more recently termed primary angiitis of the CNS (PACNS; see Chapter 3), has also been reported in immunosuppressed patients, particularly those with lymphoproliferative disorders. This may be because of reactivation of varicella zoster infection from the trigeminal ganglion leading to the basal cisterns and from there the circle of Willis vessels. There is diffuse infiltration with lymphocytes, giant cells, and mononuclear cells, of small cerebral arteries and veins (<200 micrometers in diameter). Clinical manifestations include disorientation and impaired intellectual function, with up to 25% of cases progressing to death. MR shows multiple bilateral supratentorial lesions producing infarction, particularly in the deep white matter. Other presentations can be intracranial hemorrhage and a tumorlike appearance. Enhancement is commonly observed in these lesions. These areas correspond to segmental abnormalities in the blood-brain barrier secondary to vasculitis. Spherical white matter lesions have been reported to increase the specificity for VZV (see Fig. 5-14).

Postvaricella Encephalitis

In children postvaricella encephalitis can demonstrate bilateral symmetric regions of high signal intensity on T2WI, without enhancement, involving the caudate nuclei, basal ganglia, internal and external capsules, and claustrum. Associated clinical symptoms include headache, nausea, vomiting, fever, nuchal rigidity, cerebellar ataxia, and parkinsonism. This may occur 1 to 3 weeks after chickenpox with a differential diagnosis of acute disseminated encephalomyelitis (ADEM). Another MR pattern demonstrated multiple high-intensity lesions on T2WI in the gray and white matter, some of which enhance.

Epstein-Barr Virus and Infectious Mononucleosis

Epstein-Barr virus (EBV) can cause a spectrum of neurologic disorders including seizures, neuropathies both cranial and peripheral, myelitis, aseptic meningitis, and encephalitis (incidence between 0.37% and 7.3%). The brain stem and cerebellum are preferred sites of involvement in children and are characterized by high intensity on T2WI/FLAIR. The differential diagnoses in such cases involve brain stem tumor, multiple sclerosis, *Listeria* infection, and acute disseminated encephalomyelitis.

Eastern Equine Encephalitis

This mosquito-borne arboviral infection presents with a short typical viral prodrome followed by altered mental status, seizures, and focal findings. High-intensity T2WI lesions in the basal ganglia, thalamus, and brain stem have been observed early in the course of the disease (Fig. 5-15). Other less common areas of involvement are the periventricular and cortical regions. Meningeal enhancement can occasionally be identified. These findings are not specific for eastern equine encephalitis and have been described in Japanese encephalitis, measles, mumps, echovirus 25 encephalitides, and even Creutzfeldt-Jakob disease. Influenza A may cause thalamic, pontine, cortical, and subcortical abnormalities.

Japanese Encephalitis

This disease occurs throughout Asia usually in the summer and early fall. It is a fulminant disease with rapid

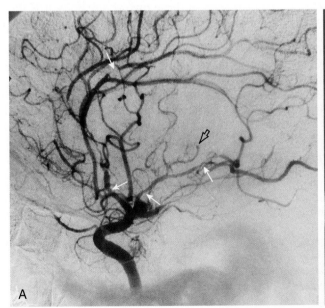
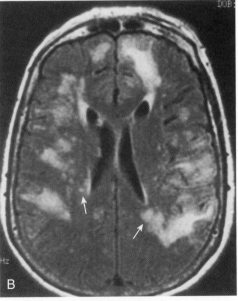

FIGURE 5-14 A, Varicella zoster virus arteritis. Internal carotid angiogram with multiple areas of vasculitic narrowing *(arrows)* and vascular occlusion *(small open arrow).* **B,** Multiple segmental areas of abnormal signal intensity at the gray-white junction, with spherical white matter lesions *(arrows)* should not cause confusion.

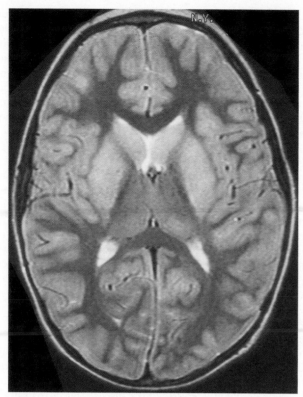

FIGURE 5-15 Eastern equine encephalitis. T2-weighted image demonstrating increased intensity in the deep gray, leading to a patient with motor sway and stray. (Courtesy R. Zimmerman, MD.)

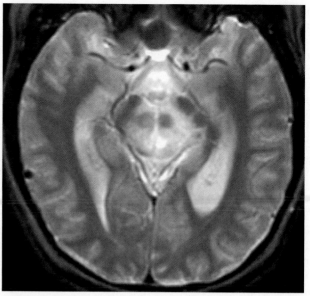

FIGURE 5-16 Giant panda sign in Japanese encephalitis. Abnormal signal in the brain stem is seen on the fluid-attenuated inversion recovery. So can you make out the giant panda bear? (Case provided by Mai-Lan Ho, UCSF.)

progression to coma. It causes 30,000 to 50,000 cases a year with up to 10,000 deaths per year. It is caused by a virus with the vector being a mosquito. Clinical signs include high fever, headache, and impaired consciousness. Neurologic signs include extrapyramidal signs (tremor, dystonia, and rigidity). At present there is no specific antiviral therapy, with associated high fatality rate, but it is important to distinguish this viral disease from herpes simplex, which is treatable. It can affect the brain stem, hippocampus, thalamus, basal ganglia, and white matter. On T2WI/FLAIR high-intensity, bilateral lesions in the thalami and putamina are reported as well as tegmental disease, which spares the red nuclei and corticospinal tract resulting in a "Giant Panda sign" in the brain stem (Fig. 5-16). Some lesions especially in the thalamus may be hemorrhagic. Enhancement is not usually observed.

Acquired Immunodeficiency Syndrome and Human Immunodeficiency Virus

Approximately 40% of patients with AIDS have neurologic symptoms during life, whereas the CNS is involved in more than 75% of autopsies. In general, MR in patients with AIDS yields little information in the asymptomatic cases. Indeed, in some cases, despite neurologic disease, the imaging may be rather unremarkable. With increasing symptoms, atrophic and white matter changes are more obvious. Divide them into encephalitis, meningitis, pure white matter changes, mass lesions (both infectious and neoplastic), and atrophy. Also, appreciate that AIDS may affect the spinal cord as well.

Acquired Immunodeficiency Syndrome Encephalopathy and Atrophy. Brain infection, clinically manifested by a subcortical dementia with cognitive, motor, and behavioral deficits, results from human immunodeficiency virus (HIV), which is the putative agent responsible for the progressive dementia complex or AIDS dementia complex, also known as AIDS encephalopathy, seen in patients during the end stage of their disease. The major imaging findings on MR are periventricular high signal abnormalities on T2WI/FLAIR in periventricular white matter and deep gray matter, and atrophy (Fig. 5-17). The periventricular predilection is in contradistinction to the subcortical white matter involvement favored by progressive multifocal leukoencephalopathy (PML; see later). No enhancement is seen in these involved regions. It is thought that the high-intensity abnormalities are the result of HIV infection directly or HIV-induced vasculopathy.

No significant correlation exists between the imaging findings and state of immunosuppression, the severity of histopathologic white matter changes, atrophy, and the severity of AIDS dementia. Protease inhibitor therapy as part of a combination highly active antiretroviral (HAART) drug therapy can result in regression or stabilization of periventricular and subcortical white matter signal intensity abnormalities seen in HIV encephalopathy (see Fig. 5-17, *D* and *E*).

HIV encephalitis can coexist with CMV, toxoplasmosis, cryptococcus infection, progressive multifocal leukoencephalopathy, and primary CNS lymphoma. Detection of mass lesions argues against the diagnosis of HIV alone. CMV in HIV-seropositive patients may produce a progressive encephalopathy characterized by dementia, which yields no specific findings on CT or MR. Atrophy, periventricular or subependymal enhancement, and periventricular low density on CT or high intensity on T2WI/FLAIR may be associated with CMV infection but can be seen in the plethora of other diseases affecting the brain in AIDS.

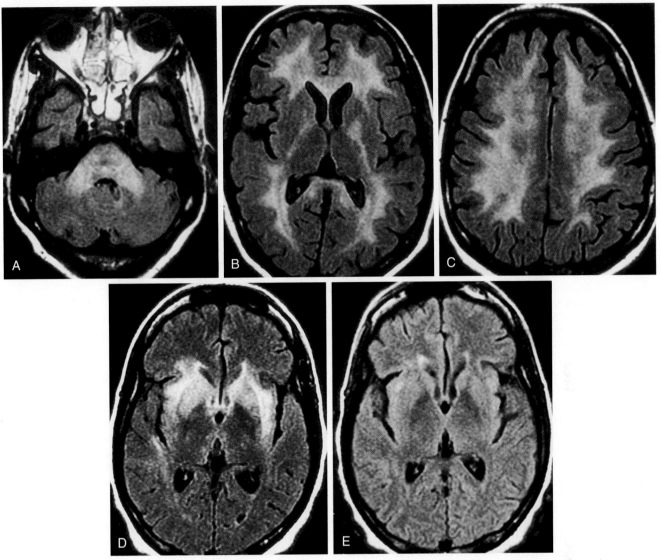

FIGURE 5-17 Pastiche of acquired immunodeficiency syndrome (AIDS). Case 1: **A,** Brain stem abnormal signal intensity is well seen on the fluid-attenuated inversion recovery. **B,** Diffuse bilateral symmetric white matter involvement is present everywhere. **C,** Centrum semiovale disease generally spares subcortical U white matter. List progressive multifocal leukoencephalopathy in your differential platter. Case 2: **D,** Basal ganglia involvement was present in this woman with dementia from AIDS. **E,** After highly active antiretroviral therapy, note that the gray matter disease fades.

Meningitis in Patients with Human Immunodeficiency Virus. Meningitis is sometimes manifested by enhancement of the meninges, which may be associated with vasculitic changes and ischemia. Common pathogens include cryptococcus, toxoplasmosis, tuberculosis, and CMV.

Meningovascular syphilis occurs rarely in patients with AIDS and should be in the differential diagnosis of vasculitis and stroke syndromes in this patient population. There is widespread thickening of the meninges and perivascular spaces with a lymphocytic infiltration. Angiographic findings of syphilis are those of segmental constriction and occlusion of the supraclinoid carotid artery, the proximal anterior and middle cerebral segments, and involvement of the basilar artery (this large and middle-sized vascular involvement is termed Heubner arteritis). Smaller vessels can also be involved (Nissl arteritis). A spectrum of parenchymal abnormalities has been described, including enhancing nodules (cerebral gumma) with variable intensity patterns and an appearance of mesiotemporal high intensity on T2WI that could be confused with herpes encephalitis.

White Matter Lesions. Progressive multifocal leukoencephalopathy (PML) is described in Chapter 6. It is very difficult to distinguish this condition from primary HIV changes, from lymphoma, and from other infections in patients with AIDS. The key finding in PML is high signal intensity on T2WI/FLAIR in the peripheral white matter (although PML can involve the cortex), usually, but not always without enhancement and mass effect; obviously a nonspecific picture. PML characteristically involves the subcortical U-fiber as opposed to HIV or CMV, which tend to involve the white matter more centrally. PML has been described in patients with primary immunologic deficiencies (hyperimmunoglobulinemia M, hypogammaglobulinemia) or secondary immunologic deficiencies (HIV infection or immunosuppressive therapies).

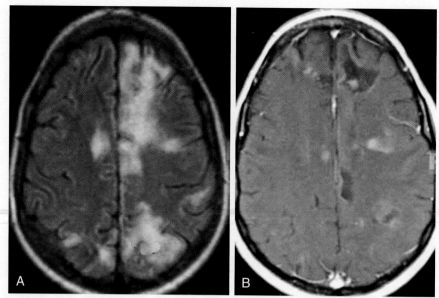

FIGURE 5-18 Immune reconstitution inflammatory syndrome. **A,** This patient with progressive multifocal leuko-encephalopathy was down-HAARTed (highly active antiretroviral therapy). His condition worsened and his white matter departed. On gadolinium **(B)** it even showed enhancement, which suggested demyelination advancement. But when his HAART was stopped, no worries, the signal intensity and enhancement dropped.

Immune reconstitution inflammatory syndrome (IRIS) can make a nonenhancing ho-hum PML case into a fulminant inflammatory demyelinating disorder (Fig. 5-18).

Cerebral Infarction. The frequency of clinical cerebral infarction in AIDS is between 0.5% and 18%, whereas the autopsy frequency is between 19% and 34%. The most common location for infarction is the basal ganglia. The causes for infarction include altered vasoreactivity from HIV in combination with drug use (particularly cocaine), HIV vasculitis, infection (varicella-zoster, CMV, tuberculosis, cryptococcosis, syphilis, and toxoplasmosis), marantic endocarditis, disseminated intravascular coagulopathy, or hypoxia. The vascular disease associated with HIV can produce bright signal on T1WI in the basal ganglia.

Lymphoma. Primary CNS lymphoma of the brain has an incidence in AIDS patients of about 6%. It is also seen in other types of immunosuppressed patients, particularly allograft recipients (heart, kidney). This entity and its imaging characteristics has been described in Chapter 2. The tumor may be isodense, hyperdense, or hypodense (in AIDS) on unenhanced CT. It is unusual for this lesion to demonstrate central necrosis except in AIDS. Most lesions enhance in a solid or ring fashion, although nonenhancing lymphoma has been reported. DWI often shows restricted diffusion.

Lymphoma versus Toxoplasmosis. It is important to differentiate lymphoma from toxoplasmosis because the former benefits from radiation therapy with more than a fourfold increase in survival rates over the untreated. In addition, it is believed that delaying therapy to embark on a course of antitoxoplasmosis therapy diminishes the benefit of radiation therapy. Certain features might enable one to distinguish CNS lymphoma and toxoplasmosis, and although it is wise not to bet the ranch, trends do emerge. CT is more specific than MR with respect to the diagnosis of lymphoma. High-density masses on noncontrast CT (not as frequent in AIDS lymphoma) and periventricular lesions with subependymal spread are findings suggestive

of lymphoma. Encasement of the ventricle does not occur with toxoplasmosis but can be seen in patients with lymphoma. On MR, lymphoma has a tendency to be isointense to hypointense relative to white matter on T2WI/FLAIR, whereas toxoplasmosis is more likely to be hyperintense on T2WI/FLAIR. Toxoplasmosis can have a low-intensity ring surrounded by high intensity on T2WI/FLAIR with ring enhancement (Fig. 5-19). Both lesions may present with solid or multiple lesions, have lesions of various sizes (although toxoplasmosis abscesses tend to be more numerous and smaller than lymphoma), and either ring-enhance or display solid enhancement. Location of the lesions does not separate entities, although toxoplasmosis has a propensity for the basal ganglia and does not spread in a periventricular pattern or typically involve the ependyma, whereas lymphoma has a penchant for the periventricular region and subependymal spread. Toxoplasmosis has a predilection for hemorrhage, particularly in the healing phase, which is not the case with CNS lymphoma (except after steroids or radiation therapy). These diseases can coexist. Thallium-201 SPECT has been reported to show abnormal uptake in lymphoma but not in toxoplasmosis. Improved specificity may be gained by performing a quantitative analysis of the scans to exclude lesions with activity not greater than that of scalp. MR perfusion measurement is another technique that has been suggested to differentiate lymphoma (increased perfusion) from toxoplasmosis (decreased perfusion). Despite what we have just said, clinicians usually treat for toxoplasmosis first and image both before and after (very cost effective).

Spinal Cord Lesions. AIDS-associated myelopathy (vacuolar myelopathy) consists of spinal cord white matter vacuolation and lipid-laden macrophages (its autopsy incidence varies from 3% to 55%). Clinical findings include the insidious onset of urinary incontinence, progressive spastic-ataxic paraparesis, and sensory loss. Viral myelitis (herpes simplex type 2, CMV, and varicella zoster) has also been implicated in patients with AIDS. Toxoplasmosis can

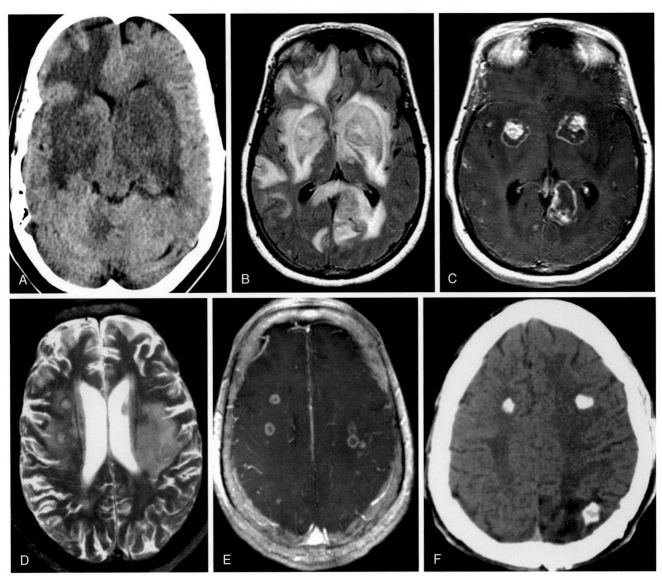

FIGURE 5-19 A, Acute toxoplasmosis on computed tomography (CT) with basal ganglia low density seen on fluid-attenuated inversion recovery **(B)** as hyperintensity. Lower signal areas amidst the brightness show enhancement on postgadolinium T1WI **(C),** hence lightness. **D,** In a different patient, lots of edema seen associated with these lesions on T2 demonstrates postgadolinium ring enhancement **(E)** in a few. **F,** Burnt out healed toxo looks on CT like calcified rocks-o.

rarely produce a myelitis. The thoracic spinal cord is most frequently involved in AIDS-associated myelopathy generally manifesting late in the disease. Findings on MR of atrophy (most frequent) with or without high signal on T2WI/FLAIR, which usually does not enhance in AIDS cases, should evoke this differential.

Pediatric Acquired Immunodeficiency Syndrome. At this time, only 1% to 2% of AIDS occurs in the pediatric population, yet in 80% of these cases CNS involvement including acquired microcephaly, diffuse cerebral atrophy, calcifications, and HIV encephalitis develops. Scattered areas of gliosis correspond in part to focal white matter abnormalities on T2WI/FLAIR. Calcifications are uncommon before the age of 1 year or when the neurologic examination yields negative results and are a prominent feature of HIV infection in children but not in adults. The calcifications are located in the basal ganglia, periventricular, frontal white matter, and cerebellum (Fig. 5-20).

Vascular ectasia, stenoses, and aneurysmal dilatation of the circle of Willis occurs with neonatal AIDS. The incidence of symptomatic vascular disease in pediatric AIDS is a little over 1%; however, at autopsy, 25% of cases were observed to have vascular lesions caused by hypoperfusion, thromboembolic disease, and infectious vasculitis. Strokes on imaging are seen in 17% of pediatric patients associated with an 8% rate of vascular stenosis. Infections and lymphoma are very uncommon in the pediatric patients, whereas cerebral atrophy, basal ganglia calcification, white matter high intensity on T2WI/FLAIR, and hemorrhage are the predominant findings in this population.

Sinusitis and mastoiditis are common coexisting infections in HIV-positive children. Adenopathy and adenoidal hypertrophy, as one would expect, are seen on the scans too.

Cytomegalovirus. Cytomegalovirus (CMV), a member of the herpes virus family, causes diseases in both the normal and immunosuppressed patient. It is the most frequent cause of fetal and neonatal viral infection. Transplacental transmission occurs as a result of either recurrent or primary maternal infection. It commonly exists in a latent form in the adult population. Reactivation of latent maternal infection is associated with a 3.4% risk of congenital fetal infection, whereas primary CMV infection poses a 30% to 50% risk of intrauterine infection. The mechanism of injury is ischemia because of insufficient fetal circulation probably secondary to placentitis and secondary chronic perfusion insufficiency. Injury before 18 weeks results in agyria, and between 18 and 24 weeks, polymicrogyria (see Chapter 8).

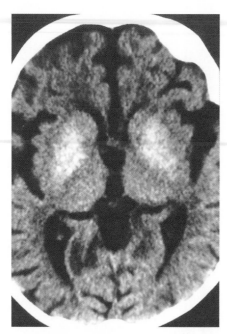

FIGURE 5-20 Pediatric acquired immunodeficiency syndrome (AIDS). Calcification in the basal ganglia and atrophy on computed tomography, in a child with AIDS contracted neonatally.

CMV has a propensity for the ependymal and subependymal regions. CNS abnormalities associated with CMV include microcephaly, ocular defects, and deafness. Other findings in infants with CMV infection include atrophy, cerebellar hypoplasia, focal white matter lesions, hippocampal abnormalities, ventriculomegaly, hydranencephaly, porencephaly, paraventricular cysts, and gyral anomalies including complete lissencephaly, pachygyria, microgyria, and localized cortical dysplasia (Fig. 5-21). Subependymal calcifications of CMV are more obvious on CT. Periventricular low density on CT or high intensity on T2WI/FLAIR, although nonspecific, has also been reported in CMV (Fig. 5-22).

Bilateral periventricular calcifications have been described in CMV and in infants with toxoplasmosis and with bacterial meningitis complicated by ventriculitis. Nevertheless, CMV calcifications are usually limited to the subependymal region, whereas neonatal toxoplasmosis has calcifications not only in the periventricular region but also throughout the brain, especially the basal ganglia. Herpes simplex has rarely been associated with calcifications and rubella has not, although microscopic calcification in relationship to blood vessels can be found pathologically. In the differential diagnosis of periventricular calcifications, tuberous sclerosis should also be considered, although calcifications of the subependymal nodules occur in childhood.

Zika Virus

Although it's been around for decades, the Zika flavivirus (transmitted to humans by mosquito bite) has gained recent worldwide attention. Most people infected by the virus experience flu-like symptoms with or without conjunctivitis which resolves. Unfortunately, there is increasing evidence that pregnant women when infected by Zika virus can transmit it transplacentally to the fetus. A dramatic rising incidence of infants born with microcephaly

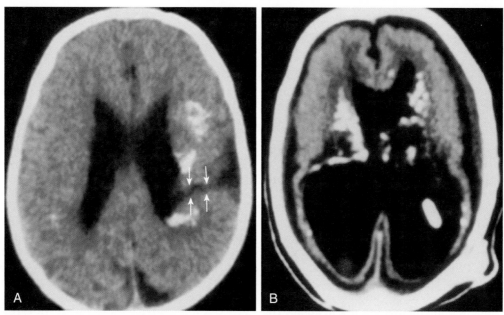

FIGURE 5-21 Cytomegalovirus (CMV) infection. **A,** Computed tomography shows periventricular calcification and schizencephalic cleft *(arrows)*. **B,** CMV calcifications, pachygyria, and hydrocephalus left this patient deaf.

in Brazil in 2015 has been linked to the Zika virus infection in utero. Imaging findings, although nonspecific to Zika, include parenchymal calcifications, ventriculomegaly, white matter hypoattenuation, disruption in cortical gyral development, and cerebellar hypoplasia. The developmental and cognitive impacts of these changes on these infants remains to be seen.

Mycobacterial/Bacterial

Tuberculosis

The incidence of tuberculosis (TB) has markedly increased in recent years, particularly in conjunction with AIDS and the emergence of drug-resistant strains of the bacillus. Approximately 5% to 10% of cases of tuberculosis have CNS involvement; however, between 4% and 19% of patients with AIDS have coexisting CNS tuberculosis. Intracranial tuberculosis has two related pathologic processes: tuberculous meningitis and intracranial tuberculoma. The two conditions are separate clinical entities, with only 10% of patients with tuberculoma having tuberculous meningitis. Tuberculous meningitis is more commonly seen in children associated with primary infection (55% of cases). Hydrocephalus after tuberculosis meningitis occurs in 81% of kids, 32% of adults and strokes occur in 38% and 20% respectively. Intracranial involvement with TB follows development of an initial focus of disease, which is usually pulmonary but may occur in the abdomen or genitourinary tract. Hematogenous dissemination of the bacilli seeds the leptomeninges and brain parenchyma. The tuberculoma is a small (1 to 3 mm in diameter) nodule, and can coalesce to form a large lesion. The intracranial tuberculoma produces symptoms from mass effect and associated edema and is the most common manifestation in adults, seen in 73% of cases (although 50% of kids also have tuberculomas). Symptoms are related to the location of the lesion. Signs and symptoms include seizures, raised intracranial pressure, and papilledema; however, patients are usually noted to have a disparity between minimal symptoms and rather significant

lesion burden. Worldwide, tuberculoma is the most common brain stem lesion, and in underdeveloped countries, tuberculoma may account for a significant incidence of intracranial mass lesions (15% to 50%). Tuberculomas may be solitary or multiple. They can occur in the supratentorial and infratentorial regions.

Tuberculomas may be dormant for years. The infection may completely resolve, or the tuberculoma may rupture into the subarachnoid space, discharging its necrotic debris and causing tuberculous meningitis and more properly a meningoencephalitis affecting the brain parenchyma and blood vessels. The basal cisterns are most affected by the exudative meningitis, centered in the interpeduncular cistern with spread into the prepontine, sylvian, and superior cerebellar cisterns. Obstruction to the normal CSF flow results in hydrocephalus. Inflammatory changes in the blood vessels caused by this process lead to an arteritis and infarction most frequently seen in the basal ganglia and internal capsule. Cranial nerve palsy may also occur as a result of the basilar meningoneuritis.

Clinical features of tuberculous meningitis in adults include confusion, fevers, headache, lethargy, and meningismus. LP reveals hypoglycorrhachia, increased protein, pleocytosis (predominantly lymphocytes), and negative smears for organisms. This progresses in a subacute fashion with stupor, coma, decerebrate rigidity, cranial nerve palsy, and possible stroke. It is interesting to note that 19% of patients with tuberculous meningitis have no evidence of extrameningeal active disease at the time of diagnosis. The tuberculin skin test is often negative early in the disease. In children, the most common symptoms include nausea, vomiting, and behavioral changes with a medical history that is usually unremarkable.

Imaging features depend on the stage of the infection. In tuberculous meningitis, the basal and sylvian cisterns are poorly visualized without contrast because of the dense exudate. On FLAIR, the basal cisterns can have increased intensity compared with normal, because of the

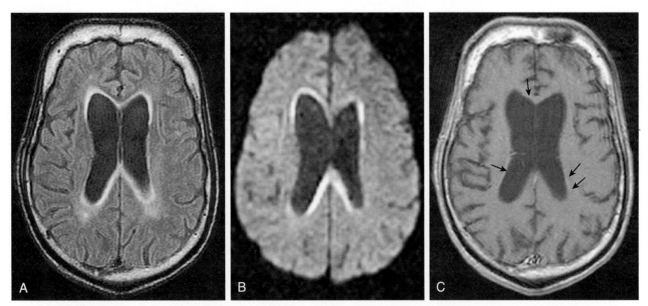

FIGURE 5-22 Cytomegalovirus infection in a patient with acquired immunodeficiency syndrome. **A,** Axial fluid-attenuated inversion recovery scan shows dilated ventricles and a peripheral rim of brightness. **B,** This periependymal hyperintensity is seen on diffusion-weighted imaging as lightness. **C,** The superficial enhancement *(arrows)* suggests ependymitis.

thick proteinaceous exudate. The cisterns enhance uniformly and intensely, and this enhancement can extend into the hemispheric fissures and over the cortical surfaces. In severe cases this is evident even on CT (Fig. 5-23). Rarely, calcification can be appreciated (particularly on CT) in the basal meninges. Periventricular low density has been observed on CT. Sequelae of tuberculous meningitis include hydrocephalus and infarction secondary to panarteritis of the vessels in the basal cisterns.

The intracranial tuberculoma appears as a nodule that ranges from low to high density on CT. These nodules may be solitary but are commonly multiple and are associated with mass effect and edema. Calcification is uncommon (less than 20%). As it enlarges, the tuberculoma may adhere to dura, causing hyperostosis and thus masquerades as a meningioma. The typical tuberculoma would appear as a nodule with a small central area (caseous necrosis) of high signal on T2WI/FLAIR or low density on CT. It is interesting to note that on MR, high intensity may be observed in the wall of tuberculomas on T1WI and low intensity on T2WI/FLAIR. This low-intensity rim has also been seen in toxoplasmosis and aspergillosis (see low intensity on T2W image in Fig. 5-19 toxoplasmosis case). Surrounding the nodule is high intensity on T2WI/FLAIR or low density on CT. This represents edema with mass effect. Enhancement is in rings (with irregular walls with variable thickness leading to a "crenated" look) or nodules (which may have punctate nonenhancing centers) (Fig. 5-24). Noncaseating tuberculomas are often high intensity on T2WI with nodular enhancement, whereas caseating lesions are isointense to hypointense on T2WI and ring enhance.

Follow-up imaging after appropriate antibiotic therapy shows that the tuberculous nodules decrease in size and

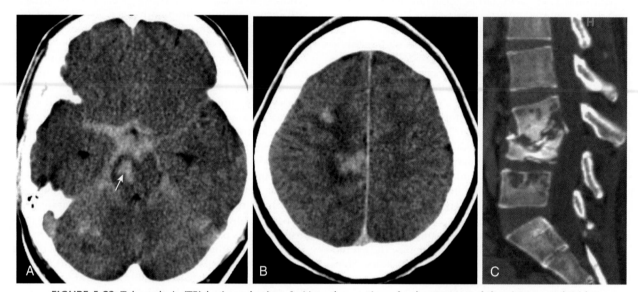

FIGURE 5-23 Tuberculosis (TB) brain and spine. **A,** Note the coating of enhancement of the cisterns and midbrain. A focus in right brain stem makes the eyes untrain *(arrow).* **B,** Above this level are additional enhancing foci, which raise the specter of red snapper tuberculi. **C,** Note the contiguous L3-5 spread, and combine that with what's up in the head. With the lumbar spine evidence of disease of Pott, you'd better consider TB as more likely than not. ("TB or not TB causing congestion?" Yes, dear friends "phthisis" the question.)

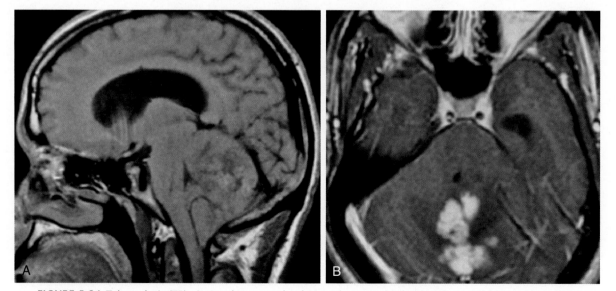

FIGURE 5-24 Tuberculosis (TB). **A,** Another example of TB in the posterior fossa, with tonsils below where they oughta. **B,** The lesion enhances and compresses the fourth, so look for big vents as you survey up north.

may show small areas of punctate calcification on CT at the tuberculoma site.

Atypical mycobacterial infections occur in HIV-positive patients; however, there are no specific imaging findings that distinguish these pathogens.

Listeria Monocytogenes

Listeriosis is a food-borne illness and the organism can be present in uncooked meats and vegetables, unpasteurized milk and cheese, and smoked seafood. Involvement of the brain elicits meningitis, meningoencephalitis, and rarely brain abscess. *Listeria* has a particular predilection for the brain stem, producing abscess, or encephalitis (rhombencephalitis). Rhombencephalitis (inflammation of the brain stem and cerebellum) has a broad differential diagnosis (Box 5-8). The principal problem lies in the difficulties encountered in laboratory diagnosis. CSF specimens often contain

---

BOX 5-8 Differential Diagnosis of Rhombencephalitis

Acute disseminated encephalomyelitis
Behçet disease
Brain stem tumor
Coccidioidomycosis
Legionnaire disease
Listeria monocytogenes
Lyme disease
Mycoplasma infection
Rickettsia
Tuberculosis
Viral diseases
    Adenovirus
    Arbovirus
    Influenza A
Whipple disease

---

only a few *Listeria* organisms, which are easily mistaken for diphtheroids (gram-positive rods) and dismissed as a contaminant or confused with pneumococcus in the CSF.

The diagnosis of *Listeria* infection should be considered in patients with impaired cellular immunity (including chemotherapy, other related predisposing conditions such as diabetes, alcoholism, renal transplantation, and AIDS) with meningitis or brain abscess when the initial cultures are negative (Fig. 5-25).

Rickettsia

Rocky Mountain Spotted Fever. This is a seasonal disease seen between late spring and summer in the South Atlantic region of the United States with clinical manifestations of headache, fever, seizures, hearing impairment, neuropathy, altered mental status, lethargy, and coma. A rash is only noted in 50% of cases. Patients older than 15 years of age tend not to have the cutaneous manifestations. Mortality of 20% is reported in untreated cases. The CNS manifestations include meningoencephalitis and vasculitis. Imaging abnormalities are uncommon, and include basal ganglia infarction, diffuse cerebral edema, diffuse meningeal enhancement, and dilatation of the perivascular spaces in the basal ganglia. Enhancement has been seen in the distal spinal cord, conus medullaris, and cauda equina. Normal imaging studies are associated with better prognosis.

Spirochetal

Lyme Disease. Lyme disease is an inflammatory disease that can involve multiple systems in the body. It is caused by a spirochete *(Borrelia burgdorferi)* and transmitted most commonly by the deer tick *(Ixodes dammini)*. Three clinical stages have been described: stage 1, constitutional symptoms and an expanding skin lesion (erythema chronicum

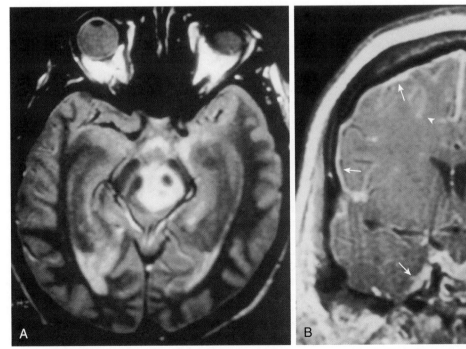

FIGURE 5-25 A, *Listeria monocytogenes* rhombencephalitis. Fluid-attenuated inversion recovery with high intensity in brain stem is present in this patient with altered cellular immunity. **B,** Leptomeningeal *(arrowheads)* and pachymeningeal *(arrows)* enhancement may occur with impunity.

migrans); stage 2, cardiac and neurologic problems (meningitis, radiculoneuropathy) occurring weeks to months after initial infection; and stage 3, arthritis and chronic neurologic problems that are noted from months to years after infection. These stages may overlap or occur alone. Approximately 10% to 15% of patients with Lyme disease have CNS involvement. Facial nerve palsy can result either from meningitis or as a mononeuritis multiplex. In endemic areas (northeastern United States), it may account for two thirds of childhood facial nerve palsy. It may be associated with optic neuritis. Diagnosis is based on clinical findings and serology, although the yield of culture is very low.

On MR, findings include (1) most commonly a normal scan; (2) high signal abnormalities in the white matter on T2WI/FLAIR, which can vary in size from punctate to large mass lesions; and (3) contrast-enhancing parenchymal lesions, meninges, labyrinth, and cranial nerves (Fig. 5-26). Other abnormalities reported include hydrocephalus, high intensity in the pons, the thalamus and basal ganglia. There may be a predilection for subcortical high-intensity abnormalities on MR in the frontal and parietal lobes. Lyme disease should be considered in the differential diagnosis of multiple sclerosis, acute disseminated encephalomyelitis, and vasculitis.

Lyme encephalopathy has been characterized as a cognitive disturbance of mild to moderate memory and learning problems sometimes accompanied by somnolence months to years after the onset of infection.

Fungal

Fungal infections can produce meningitis, granuloma formation, and rarely encephalitis. We shall discuss some specific entities below.

Cryptococcosis

Cryptococcus is a yeast with a polysaccharide capsule that distinctively stains with India ink. Sensitive immunologic diagnosis can also be made by detecting cryptococcal antigen or anticryptococcal antibody in the CSF or blood. Pathologic findings include meningitis, meningoencephalitis, or granuloma formation. Hematogenous dissemination from an occult pulmonary focus is the usual vector into the CNS. *Cryptococcus* ranks third behind HIV and toxoplasmosis as a cause of CNS infection in AIDS. CNS cryptococcosis develops in up to 11% of patients with AIDS.

MR may be normal (why do you think it is called "crypto-"?), or can demonstrate a spectrum of abnormalities. Recognition of dilated Virchow-Robin spaces in young immunosuppressed patients should raise a red flag concerning the possibility of intracranial cryptococcus (Fig. 5-27). These dilated spaces filled with gelatinous cysts ("pseudotumor"), in and adjacent to the basal ganglia and the corticomedullary junction, may or may not enhance. These are seen as numerous bilateral small foci of high signal intensity on T2WI/FLAIR along the perivascular spaces but can also be seen, rarely, in coccidioidomycosis and candidiasis and glycogen storage diseases. Gumming up of the arachnoid villi by crypto leads to its most common manifestation—hydrocephalus. Diffuse confluent basal cisternalleptomeningeal enhancement is not usually present in cryptococcal meningitis (though this is not a hard and fast rule), differentiating it from tuberculosis or bacterial meningitis.

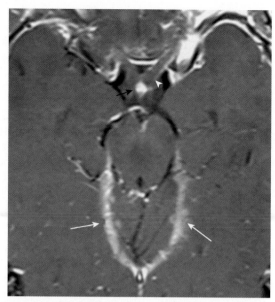

FIGURE 5-26 Lyme disease. Enhancement of the optic nerves *(arrowhead)*, stalk *(black arrow)*, and meninges *(white arrows)* should suggest the possibility of infection by *Borrelia burgdorferi*.

A rare cryptococcal pattern is that of multiple miliary enhancing parenchymal and leptomeningeal nodules with involvement of the choroid plexus in the trigone as well as the spinal cord and spinal nerve roots. Cryptococcomas occur in 4% to 11% of patients with cryptococcal meningitis presenting as a solid mass or as disseminated lesions predominately in the midbrain and basal ganglia. Contrast is needed to detect such leptomeningeal nodules.

Other less common fungal infections affecting HIV-positive patients include *Candida, Aspergillus, Histoplasma,* and *Mucor*.

Coccidioidomycosis

Coccidioidomycosis (Valley Fever) is an endemic fungus infection in the southwestern United States and northern Mexico. The spore is inhaled, and a primary pulmonary focus develops. Hematogenous dissemination to the CNS occurs within a few weeks or months, but dissemination years later has been reported. It can involve the calvarium or skeleton. The intracranial infection may be manifested pathologically by a thick basilar meningitis with meningeal and parenchymal granulomas. The intraaxial granulomas have been reported to have a propensity for the cerebellum. Vasculitis producing occlusion has also been noted. The disseminated cerebral form of coccidioidomycosis occurs predominantly in white men, and diagnosis is confirmed by CSF serology or culture. It is one of the most common causes of eosinophilic meningitis.

On MR, dilated Virchow-Robin spaces similar to *Cryptococcus* (see previous discussion) can be identified. On PDWI/FLAIR, there is increased signal in the cisterns. Enhancement is noted in the basal meninges, cisterns, and sulci. When ependymitis is present, enhancement is noted along the ventricular margin. The CT picture is of basal arachnoiditis with obliteration and distortion of the cisterns. These may show increased density before contrast. Associated with this active arachnoiditis is communicating hydrocephalus. Focal areas representing infarction secondary to vasculitis and enhancing nodules can occasionally be observed.

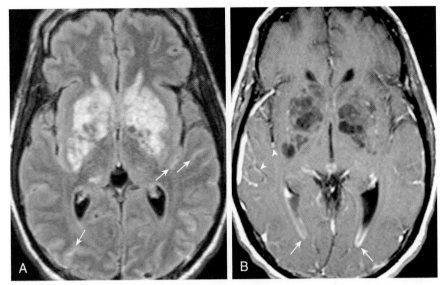

FIGURE 5-27 **A,** Cryptococcal meningitis. Axial fluid-attenuated inversion recovery image demonstrates enormous hyperintense Virchow-Robin spaces. **B,** Axial contrast-enhanced T1 scan shows the same finding in multiple basal ganglia places. If enlarged and mass-like, think gelatinous pseudotumors; have them search for Crypto in bodily humours. Bonus findings: *Arrows* in **A** indicate nonsuppression of cerebrospinal fluid signal (oh, my!) implying meningeal disease in the sulci. Abnormal enhancement along sulci and ventricular margins (*arrowheads* and *arrows* in **B,** respectively) indicate meningitis and, you guessed it, ventriculitis!

Mucormycosis

The prognosis in mucormycosis is directly related to early recognition of this disease. It affects patients with abnormalities in host defenses, including altered cellular immunity. Particularly prone to this pathogen are diabetic patients with ketoacidosis or debilitated patients with burns, uremia, or malnutrition. It is also seen in HIV patients with a history of drug abuse presenting with basal ganglionic lesions. It has been reported that persons taking the iron chelating agent deferoxamine, such as patients undergoing dialysis, are at increased risk for mucormycosis. Deferoxamine apparently abolishes the normal fungistatic effect of serum on *Mucor*. This fungus is usually inhaled and rapidly destroys the nasal mucosa, forming black crusts (classic eschar). It may then spread into the paranasal sinuses (with or without bone destruction), orbit, and the base of the skull, or may extend through the cribriform plate, resulting in involvement of the anterior cranial fossa. It has a high frequency (50% in some series) of intracranial extension. Clinical symptoms include facial pain, bloody nasal discharge, dark swollen turbinates, chemosis, exophthalmos, as well as cranial nerve palsy progressing rapidly to stroke, encephalitis, and death.

Mucor infection characteristically presents as a rim of soft-tissue thickening within the nasal cavity and along the walls of the paranasal sinus. CT features of sinonasal *Mucor* infection include sinus opacification, air fluid concentrations, increased density or calcification and obliteration of the nasopharyngeal tissue planes (Fig. 5-28). On MR, low intensity may be present in the sinuses on T1WI and T2WI/FLAIR. In some cases, bony destruction is present. Orbital extension from the ethmoid sinuses can produce proptosis and chemosis, and thrombosis of the superior ophthalmic vein, with extension through the orbital apex and subsequent thrombosis of the cavernous sinus. *Mucor* can extend into the infratemporal fossa and pterygopalatine fossa from the maxillary sinus.

Mucor has a striking tendency to proliferate along and through vascular structures producing arteritis with aneurysm, pseudoaneurysm, abscess formation, vascular occlusion, and infarction. It most frequently affects the cavernous portion of the internal carotid but has been seen in the basilar artery. When this virulent fungus extends intracranially (after extension into the orbit, cribriform plate, or the deep facial structures), low-density CT abnormalities are noted particularly in the anterior cranial fossa but may be present in any part of the brain. These regions show mass effect and enhancement (see Fig. 5-28). *Mucor* may also cause large-vessel cerebral infarction with the associated CT and MR findings. Intracranial abscess, consisting of a low-density mass without significant vasogenic edema, evidence of leptomeningitis, or well-defined ring enhancement on CT, has been reported.

Nocardiosis

Nocardia, an aerobic fungus resembling *Actinomyces,* is associated with a state of compromised immunity, especially in the setting of steroid therapy. However, it may infect patients with normal immunity as well. *Nocardia* complicates a spectrum of diseases that includes pulmonary alveolar proteinosis, sarcoidosis, ulcerative colitis, and intestinal lipodystrophy. Hematogenous dissemination occurs from a pulmonary focus into the CNS. This usually results in brain abscess formation; meningitis is rare. The onset of symptoms is often insidious. Nocardial lesions show an enhancing capsule commonly containing multiple loculations (Fig. 5-29). The diagnosis should be considered in the appropriate clinical setting because *Nocardia* is relatively sensitive to the sulfonamides.

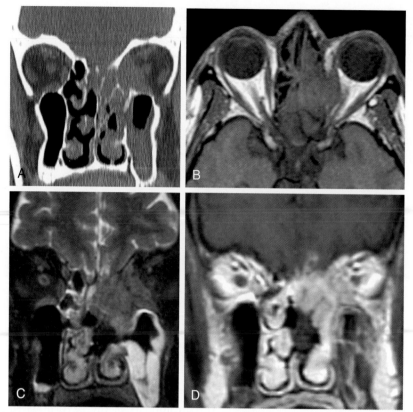

FIGURE 5-28 Mucormycosis. **A,** Coronal reconstructed computed tomography shows soft tissue in nasal cavity and left orbit. The cribriform plate bone looks like something absorbed it. **B,** Axial T1-weighted imaging (T1WI) confirms orbital spread, better search for dread in the head. **C,** Coronal T2WI shows typical dark infection, now look in the superior direction. **D,** Yes there is spread of enhancement through the skull basis, so typical of Mucor—from sinus to orbit to brain it races. Tissue in the orbit, destruction of nose, meningitis—it's Mucor! Case closed.

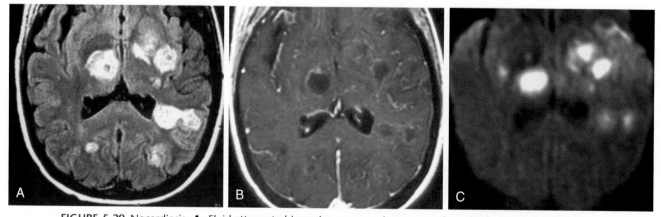

FIGURE 5-29 Nocardiosis. **A,** Fluid-attenuated inversion recovery image reveals multiple lesions in this immunosuppressed patient. **B,** Following contrast, observe the rim enhancement and third ventricular effacement. **C,** Diffusion-weighted image shows apparent diffusion coefficient reduction suggesting abscesses with brain destruction.

Aspergillosis

In contradistinction to *Nocardia*, where a well-formed capsule is usually apparent, intracranial aspergillosis may or may not demonstrate ring enhancement. This ubiquitous fungus primarily infects the immunocompromised host. It may gain entry into the CNS by inoculation, hematogenous dissemination (most often from a pulmonary focus), or direct extension from the paranasal sinuses.

Pathologically, aspergillosis involves the brain in an aggressive form, producing meningitis and meningoencephalitis with subsequent hemorrhagic infarction. Less malignant presentations include solitary cerebral abscess or isolated granulomas. In aggressive aspergillosis, one histologically visualizes invasion of blood vessels with secondary thrombosis and infarction.

MR shows high-intensity lesions on T2WI/FLAIR and at times on T1WI (Fig. 5-30). If the site of origin is the paranasal sinus, one may see decreased intensity on T2WI/FLAIR and/or bright signal on T1WI secondary to calcification or manganese accumulation. There may be evidence of infarction with

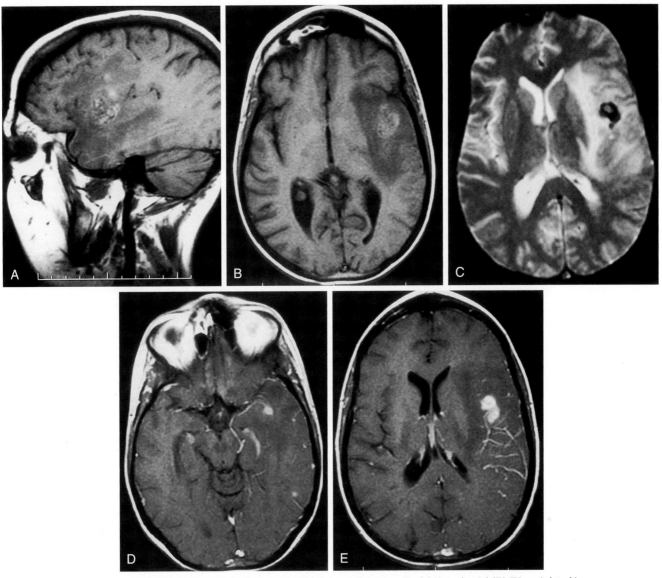

FIGURE 5-30 Aspergillosis. Hemorrhage is seen as high intensity on sagittal **(A)** and axial **(B)** T1-weighted imaging slices, and low intensity on the gradient echo devices **(C)**. **D** and **E** show enhancing nodules at temporal lobe gray-white junctions, likely knocking out speech and some hearing functions.

or without enhancement. The MR findings are nonspecific but may cause curiosity when the lesions do not enhance, as occurs sometimes. This may be from the invasion of blood vessels restricting delivery of the contrast to the site. Basal ganglia lesions are common with fungi and they, like TB, can have a crenated appearance. CT abnormalities are subtle and include areas of low density with minimal mass effect, poor contrast enhancement, and usually no ring formation. In fact, the presence of true ring enhancement militates against the most aggressive meningoencephalitic variety of aspergillosis. The relatively benign CT picture contrasts sharply with the consumptive nature of this infection. The lack of correlation between the radiographic and pathologic findings is related to the rapidity of the destructive process (inability to form an effective capsule) and to suppression of enhancement by concomitant steroid therapy.

Candidiasis

Candida is the most common cause of autopsy-proved non-AIDS cerebral mycosis. It has a propensity for neutropenic patients who are receiving steroids. *Candida* reaches the CNS by hematogenous dissemination through the respiratory or gastrointestinal tracts. These various pathologic presentations consequently generate different images that include hydrocephalus (leptomeningeal disease), enhancing nodules with edema (granuloma), calcified granuloma, infarction, and (micro) abscess formation. The infectious presentation depends on the state of the host's natural defenses. The inability to mount an effective localizing cell-mediated immune response when challenged favors an aggressive infection rather than granuloma formation.

And you thought *Candida* was just another "yeast infection."

Parasitic

Toxoplasmosis

Toxoplasma gondii is a ubiquitous protozoan parasite (20% to 70% of the American population is seropositive) infecting the CNS in approximately 10% of patients with

AIDS and also affecting adults with compromised cellular immunity (particularly defects in the lymphocyte-monocyte system). Acute toxoplasmosis (see previous) in immunocompromised patients often occurs from reactivation of remotely acquired latent infection. Lesions have a propensity for the basal ganglia, corticomedullary junction, white matter, and periventricular region. As opposed to congenital toxoplasmosis, calcification is not common, although it has been reported after therapy. Occasionally, these lesions may be hemorrhagic. On MR, multiple lesions of high intensity on T2WI/FLAIR associated with vasogenic edema and ring or nodular enhancement are seen on T1WI (see Fig. 5-19). Prompt response to appropriate antibiotic therapy can distinguish toxoplasmosis from lymphoma.

Congenital toxoplasmosis occurs during maternal infection. Its manifestations include bilateral chorioretinitis associated with hydrocephalus (secondary to ependymitis producing aqueductal stenosis) and intracranial calcifications particularly in the basal ganglia and cortex.

Cysticercosis

This parasite is endemic in parts of Mexico, Central and South America, Asia, Africa, and Eastern Europe. Human beings are the only known definitive hosts (in which the parasite undergoes sexual reproduction) for the adult tapeworm *(Taenia solium)* and the only known intermediate hosts (in which the larval or asexual stage is present) for the larval form *(Cysticercus cellulosae),* which prospers in the CNS. The parasite is acquired by ingestion of insufficiently cooked pork containing the encysted larvae (one of the benefits of keeping kosher or being vegetarian). The larvae develop into the adult tapeworm in the human intestinal tract. The oncospheres (active embryo) released from ova of the adult tapeworm by gastric digestion burrow through the intestinal tract to reach the bloodstream, which carries them to the CNS and to other regions, where they form cysticerci. Humans may also directly ingest the eggs (as an intermediate host) in contaminated food, through self-contamination by the anus-hand-mouth route (yuck!), or by regurgitation of ova. Infestation of the CNS produces many neurologic problems, the most common being seizures and headaches seen in 30% to 92% of patients. After several weeks, a cystic covering and a scolex develops and lodges in the CNS. The interval between the probable date of infection and the first distinctive symptom varies from less than 1 year to 30 years, with the average being approximately 4.8 years.

The cysticerci vary in size from pinpoint to 6 cm in diameter. They are located in the brain parenchyma, subarachnoid space, ventricles, or rarely intraspinal locations. Symptoms are related to the site of the parasites. It is not until the larvae die that an acute inflammatory reaction is incited and patients become symptomatic. The parenchymal cysts have a propensity for the cortical and deep gray matter, whereas the subarachnoid cysts can produce basal meningitis, hydrocephalus and mass lesions, particularly in the suprasellar cistern, cerebellopontine angle cistern, sylvian cistern, and cerebral arteritis (in over 50% of cases with subarachnoid lesions) with subsequent infarction usually in the middle cerebral or posterior cerebral distribution. Intraventricular cysts may be free or attached to the wall and become symptomatic when ventricular drainage is obstructed.

The acute encephalitic phase of the infection is more common in children and is characterized by either multiple diffuse nodular lesions (85%) or localized lesions. In most cases of encephalitis, cysts are located throughout the parenchyma, although they can be localized to one region. The encephalitis phase of the disease lasts from 2 to 6 months, with edema persisting after the enhancement disappears. Diffuse brain edema and ring or nodular enhancement patterns are present. This phase can be fatal.

CT shows calcification in the brain parenchyma. These are characteristic, with a slightly off-center spherical calcification of 1 to 2 mm in diameter, representing the scolex, surrounded by a partially or totally calcified sphere (7 to 12 mm in diameter). The calcification only occurs in the dead larvae.

A spectrum of MR and CT findings is associated with active cysticercosis (Fig. 5-31). Four stages of cyst formation have been described with imaging that roughly parallel these changes. In the vesicular stage, the larvae are alive and the cyst contains clear fluid. Edema is minimal, and the cyst is surrounded by a thin capsule. On MR, the fluid appears isointense to CSF on all pulse sequences, and the eccentric scolex (which appears as a mural nodule) can be identified (see Fig. 5-31, *A* and *B*). CT in this stage shows a circumscribed cyst with a density similar to CSF and a denser scolex.

In the second (colloidal vesicular) and third (granular nodular) stages, the fluid in the cysts becomes turbid and the larvae die, with associated thickening of the capsule. In some cases, there is a strong inflammatory reaction and encephalitis can occur as the cysts move into the colloidal vesicular stage. In the third stage, the cyst shrinks between one third and one fourth its original size and begins to calcify. In patients without significant reaction to the parasite, no enhancement may occur and edema can be minimal. The cyst in these stages can be isointense to hyperintense on T1WI. On T2WI/FLAIR the fluid is isointense to high intensity and the cyst wall is difficult to identify or may be hypointense. The lesions uniformly enhance as dense nodules or small ring areas. Enhancement is thought to be the result of larval death, reaction to antigen, and release of metabolic products with associated blood-brain barrier abnormality.

In the fourth stage (nodular calcified stage), focal calcifications are seen. CT is more specific than MR, with the brain parenchyma containing calcified dead larvae and low-density cysts (see Fig. 5-29, *D*). These low-density cysts may or may not enhance. On MR, the lesion is seen as hypointense nodule with susceptibility artifact on T2WI so it can be confused with cavernous malformations. The time it takes for the parasite to evolve through all of the different stages is from 2 to 10 years with an average of approximately 5 years (slightly longer than the normal political parasite). The dead cyst often elicits a lot of edema despite its small size.

Ventricular cysticercosis is best noted on MR, where lesions are of higher intensity than CSF on T1WI and the cyst wall may be identified. They have the same density as CSF on CT. The eccentric scolex can be seen as a nodule isointense to brain. The fourth ventricle is the most common location for these lesions. Intraventricular cysticercosis (7% to 20% of intracranial cysticercosis) produces obstructive hydrocephalus and also causes a granular ependymitis by provoking an inflammatory reaction secondary to toxic

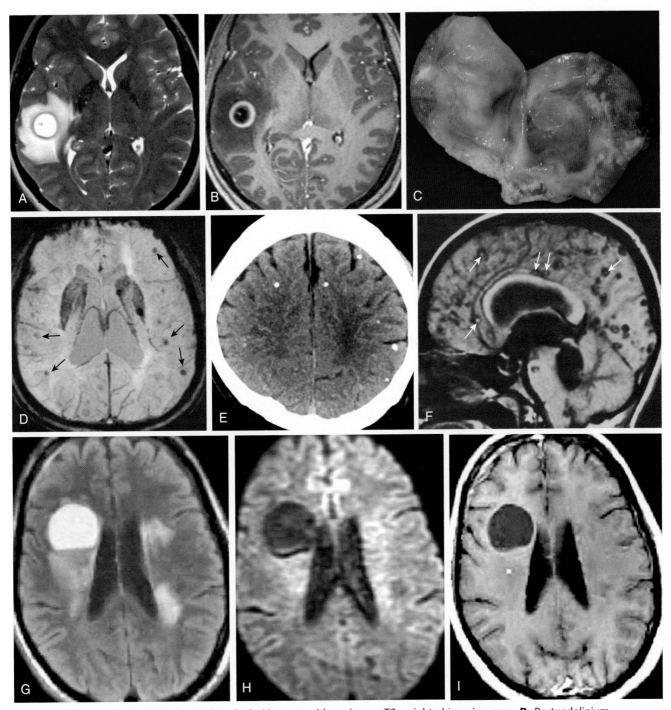

FIGURE 5-31 Cysticercosis. Case 1: **A,** Live cyst with scolex on T2-weighted imaging scan. **B,** Postgadolinium ring in pork-eating man. **C,** Live cysticercus in the pathology pan. Case 2: **D,** Focal parenchymal signal loss foci *(arrows)* on scans for susceptibility. **E,** Multiple dead calcifications seen on computed tomography. Focal parenchymal signal loss foci on scans for susceptibility. **F,** Innumerable lesions *(arrows)* including one in the fourth, suggests cysticercosis for what that is worth. On fluid-attenuated inversion recovery **(G),** diffusion-weighted imaging **(H),** and postcontrast T1 **(I),** the nonrestricting cyst says, "Check me out, hon!"

substances released from the permeable cyst wall of the dead larvae. The fourth ventricle is the most common ventricle affected. The cyst may be adherent to the wall of the ventricle producing a subependymal rim of high intensity on T2WI/FLAIR (see Fig. 5-29, *F*). The scolex may be visualized as a small dot within what looks like a dilated ventricle.

Cysticercosis has been reported rarely to cause obstructive hydrocephalus by presenting as a mass

(cysticercotic cyst) of the septum pellucidum. Larvae can have a racemose (grapelike cluster) form that occurs in the subarachnoid spaces (particularly the basal cisterns, sylvian fissures) and ventricles. The diagnosis of racemose cysticercosis should be considered in situations where there is a subarachnoid cyst associated with leptomeningeal enhancement in a patient from an endemic region.

Spinal cysts are seen in 3% of patients, mostly in the thoracic region, and can cause arachnoiditis and spinal cord compression. Inflammation of the meninges is often more severe in spinal cysticercosis. The entity can involve the subarachnoid space, the epidural space, or be intramedullary with similar findings to the spectrum of lesions within the brain. Differential diagnosis here involves arachnoid cyst and epidermoid.

The diagnosis of neurocysticercosis is usually made with the aid of serologic testing for specific antibody in serum, CSF, or both by enzyme-linked immunosorbent assay (ELISA). Treatment includes antiparasitic, antiseizure, and steroid medication. Surgical resection is risky, and reserved only for those patients with hydrocephalus, intraventricular cysts, or intraspinal cysts.

> Cysticercosis
> Produces neurosis,
> Convulsions,
> Revulsions,
> And missed diagnosis!

Echinococcosis

Echinococcosis (Hydatid disease) is another great disease brought to you by the dog tapeworm. Clinical signs and symptoms include seizures, raised intracranial pressure, and focal neurologic deficits. It is endemic in the Middle East, South America, Eurasia, and Australia caused by ingestion of dog feces that include ova of dog tapeworm *(Echinococcus granulosus)*. The ova hatch in the gastrointestinal tract and embryos are spread throughout the body. The embryo matures into a cystic larva (hydatid cyst), which are commonly large and unilocular although other cystic configurations may be seen. Cerebral involvement is seen in 2% to 5% of cases. The most common location is the parietal lobe. The cystic component has an intensity similar to CSF. Severe inflammatory reaction in the brain occurs if the cyst ruptures (surgeons need to be careful here). Scolices and aggregates of scolices (hydatid sand) can be seen on MR within the cyst. Calvarial echinococcal cyst can extend into the intracranial cavity.

Prion Disease

These entities are described in the neurodegenerative disorder chapter. Suffice it to say that Creutzfeldt-Jacob disease and Bovine spongiform encephalopathy, the most common of these disorders are being diagnosed serologically or through CSF samples more readily these days. However, the imaging features of deep gray matter and superficial cortical gray matter areas of restricted diffusion with or without FLAIR signal intensity abnormalities are becoming sufficiently well-described that we neuroradiologists may be the ones suggesting surveillance for these CSF and serologic markers. Score one for the observant and well-informed NeuroRad guys and gals.

NONINFECTIOUS INFLAMMATORY DISEASES

Sarcoidosis

Sarcoidosis, a systemic granulomatous disease of unknown etiology, primarily occurs in the third and fourth decades of life, affecting the nervous system clinically in approximately 5% of cases of systemic sarcoid

and up to 16% in autopsy series. A very small percentage of patients have only the CNS disease without systemic manifestations. CNS sarcoidosis is an exclusionary diagnosis because even a positive biopsy represents only a nonspecific reaction to a variety of diseases.

Granulomatous intracranial disease has two principal patterns. The more common presentation is a chronic basilar leptomeningitis with involvement of the hypothalamus, pituitary stalk, optic nerve, and chiasm. The convexities may also be involved. Patients may have unilateral or bilateral cranial nerve (particularly nerve VII) palsy, or endocrine or electrolyte disturbances. In these patients, communicating hydrocephalus develops and signs of meningeal irritation may be present. The granulomatous process frequently spreads from the leptomeninges to the Virchow-Robin spaces, invading and thrombosing affected blood vessels (arteries and/or veins) and producing a granulomatous angiitis similar to primary angiitis of the CNS. The CSF abnormalities are elevated lymphocytes and protein with hypoglycorrhachia. These findings are nonspecific and not diagnostic for this disease.

The second pattern is parenchymal nodules. These granulomatous masses are usually associated with extensive arachnoiditis and microscopic granulomas throughout the brain parenchyma. These masses may be calcified and avascular. The sarcoid nodules produce signs and symptoms of an intracranial mass. High-intensity white matter can be observed in sarcoid, and this can be indistinguishable from multiple sclerosis.

The imaging picture depends on whether there is leptomeningitis, granulomas in the brain, or a mass lesion. On MR, high signal on T2WI/FLAIR in the parenchyma and at the gray-white junction has been recognized. Intraparenchymal sarcoid masses may or may not enhance. Diffuse, focal, and gyriform enhancement of the leptomeninges and basilar enhancement in the hypothalamic region can be observed. An interesting observation is that the enhancement can follow the Virchow-Robin spaces and appear linear (Fig. 5-32). Hydrocephalus, either obstructive or communicating, is common.

Nodules may be calcified, have slightly increased density, or be isodense, and can occur throughout the brain parenchyma with a marked predilection for the skull base, pituitary, pons, hypothalamus, and periventricular region. The nodules are not usually associated with edema. Sarcoid nodules do not cavitate as frequently as tuberculous nodules.

Sarcoid can masquerade as a brain tumor, pituitary lesion, meningioma (with similar imaging features), vasculitis, multiple sclerosis, and even angiographically as a subdural mass. Extraaxial sarcoid is particularly a common mimic of meningioma, with a dural-based mass, occasional hyperostosis, and meningeal enhancement. Up to 25% of patients can have ophthalmic manifestations including anterior uveitis (most common manifestation), posterior uveitis, lacrimal gland infiltration, optic nerve/sheath involvement, retrobulbar masses, exophthalmos, extraocular muscle thickening, and infiltration of the visual pathways. It can mimic Tolosa-Hunt syndrome.

Sarcoid may affect the spinal cord with leptomeningeal coating of the cord or appear as an intramedullary mass. Rarely sarcoid can inflame the cauda equina, causing a polyradiculopathy and demonstrating nodularity and thickening of the nerve roots (Box 5-9). Leptomeningeal sarcoid must be distinguished clinically from

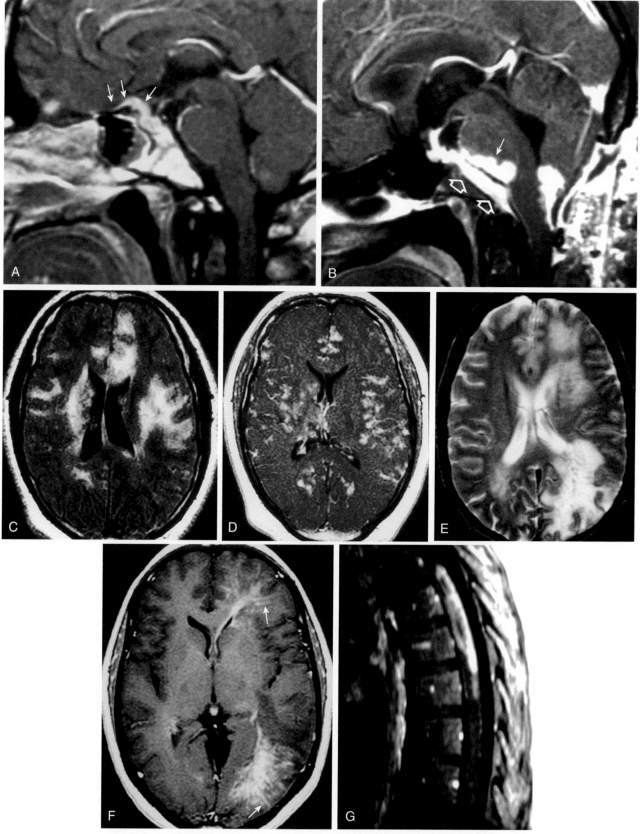

FIGURE 5-32 A mélange of sarcoidosis. **A,** Case 1: Sagittal-enhanced T1-weighted imaging scan of the sella region shows a mass that looks like a meningioma lesion *(arrows)*. Despite the dural tail and blistered bone, it was "sarcoid" the pathologist intoned. **B,** Case 2: With leptomeningeal *(arrow)*, dural *(open arrows)*, and brain stem parenchymal growth, again "This is sarcoidosis" the pathologist quoth. **C,** Case 3: Abnormal intensity to the white matter and cerebrospinal fluid, including avid enhancement of all **(D)** left the neuroradiologist bereft. Fortunately, the pathologist had no such morosis, stating "Yet another case of sarcoidosis!" **E,** Case 4: In this case asymmetric white matter changes were found. **F,** The bizarre perivascular enhancement *(arrow)* did confound. Dr. Path opined, "This is not lymphoma, I see a non-caseating granuloma." Case 5: In this case we saw superficial cord enhancement in the spine, and this time brain and spine sarcoidosis we did correctly opine **(G).**

BOX 5-9   Differential Diagnosis of Radicular Enhancement

Arachnoiditis
Charcot-Marie-Tooth
CSF metastases
Dejerine Sottas
Disc herniation with root inflammation
Guillain-Barré
Infection
    CMV
    Herpes Zoster
    Lyme
    TB
Neurofibroma
Sarcoid and other granulomatous diseases
Schwannoma

CMV, Cytomegalovirus; *CSF,* cerebrospinal fluid; *TB,* tuberculosis.

carcinomatous, lymphomatous, and infectious meningitis. Dramatic response has been noted in some cases with steroid therapy, which can produce a complete disappearance of the enhancing lesions. Sarcoid has replaced syphilis as the great mimic. When you have no idea, think sarcoid.

Granulomatosis with Polyangiitis

The term *granulomatosis with polyangiitis* (GPA) has supplanted Wegener granulomatosis much to the chagrin of Friedrich Wegener, the German pathologist who described the condition. Unfortunately, he joined the Nazi party in 1932 and rumors in the late twentieth century suggested he may have engaged in human experimentation in that capacity. Hence the name change.

GPA, also called antineutrophil cytoplasmic antibody (ANCA)–associated granulomatous vasculitis, is a disease characterized by necrotizing granulomas that can involve multiple organs of the body (upper and lower respiratory tracts, kidneys, orbits, heart, skin, and joints). In 2% to 8% of cases, the meninges and brain can be affected. Manifestations include meningeal thickening and enhancement either diffuse or focal in morphology in the brain and spinal cord, infarction, nonspecific white matter abnormalities on T2WI/FLAIR, intraparenchymal granuloma, and atrophy. It produces a vasculitis that results in peripheral neuropathy (frequent), myopathy, intracerebral hemorrhage, subarachnoid hemorrhage, arterial and venous thrombosis. It can involve the pituitary gland and stalk either from direct extension of extracranial disease or as a remote granuloma.

Behçet Disease

Behçet disease is a multisystem immune-related vasculitis, with CNS involvement in 5% to 10% of cases, characterized by exacerbations and remissions. The diagnosis requires recurrent oral ulcerations and two of the following to establish a definite diagnosis: recurrent genital ulcerations, skin or eye lesions, or a positive pathergy test. The pathology has a clear venous vasculitic component in the CNS particularly the brain stem.

Three patterns of neurologic manifestations are observed: a brain stem syndrome, a meningoencephalitic syndrome, and an organic confusion syndrome. It occurs along the Silk Road from Japan to the Mediterranean, and the Middle East but can be seen in North America, particularly in persons from these ethnic backgrounds. On MR, multiple high T2 signal intensity lesions can be seen with variable enhancement. The most common finding is a mesodiencephalic junction lesion with edema extending along tracts in the brain stem and diencephalon (Fig. 5-33). Another pattern is involvement of the pontobulbar region. These lesions can simulate brain stem tumors. Other regions affected include basal ganglia, spinal cord, cerebral hemispheres, and rarely the optic nerve. Enhancement may occur as a ring or nodule, and multiple lesions may be present. Hemorrhage can occasionally be detected in these lesions. The radiologist should appreciate that Behçet disease is associated with venous thrombosis in more than one third of cases, although it rarely occurs in the dural sinuses. Brain stem atrophy has been reported in chronic cases. The vast majority of patients are less than 50 years of age. It is thus difficult to separate these lesions from those of multiple sclerosis, other vasculitides such as primary arteritis of the CNS, and inflammatory diseases such as sarcoid.

Whipple Disease

Whipple disease, a chronic granulomatous disorder with a propensity for the gastrointestinal tract, can primarily involve the CNS (20% of cases). Clinical findings include cognitive changes leading to progressive dementia, ophthalmoplegia, seizures, myoclonus, gait disturbance, and hypothalamic dysfunction. It can be fulminant, progressing to coma and death. The brain lesions are located in the gray matter including the basal part of the telencephalon, the hypothalamus, and the thalamus. However, it has also been reported to involve other regions of the brain including the chiasm, posterior fossa, and the spinal cord.

LAST BUT NOT LEAST

Rasmussen Encephalitis

This is seen in children with a mean age of 6 to 8 years. Focal motor seizures are followed over time by progressive loss of ipsilateral motor function associated with cognitive decline. The etiology has been attributed to a viral-induced autoimmune mechanism. Early in the disease, imaging can be normal although cerebral swelling has been reported. High-intensity lesions on T2WI in the basal ganglia and periventricular white matter have also been observed. Because the therapy of this disease is hemispherectomy, make certain you comment on basal ganglia involvement. Removal of the basal ganglia gives a worse motor prognosis although not removing affected tissue can result in persistent seizures (bad news either way). Late in the disease, diffuse atrophy either progressive or nonprogressive has been noted. The frontal or frontotemporal regions are most commonly involved with a predominant unilateral distribution (Fig. 5-34).

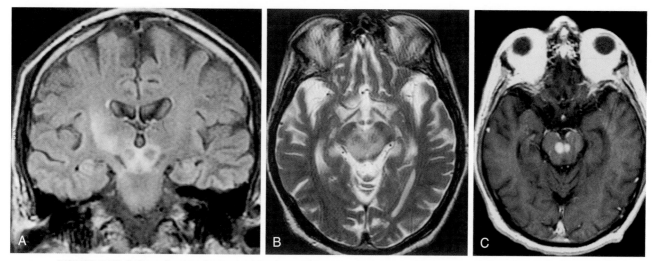

FIGURE 5-33 Behçet disease. **A,** Coronal fluid-attenuated inversion recovery portrays the brain stem involvement at the meso-diencephalic junction. **B,** Diffuse spread in the midbrain-pons region was seen on axial T2-weighted imaging leading to extraocular muscle dysfunction. **C,** Enhancement was noted in red nuclei and leptomeninges. Aphthous ulcers, uveitis, and genital lesions sealed the deal. It was Behçet, lest you forget.

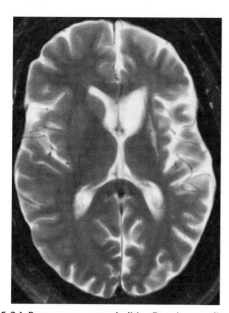

FIGURE 5-34 Rasmussen encephalitis. Despite medication, this patient's seizures failed to cease. T2-weighted imaging showed left-sided volume decrease. Note the dilated left frontal horn and reduced deep gray. Rasmussen's will do this; "Hemispherectomy!" the neurosurgeon did say.

This is another disease in which SPECT scanning may be more sensitive than MR revealing large regions of diminished cerebral perfusion. Crossed cerebellar diaschisis (secondary to disruption of the corticopontocerebellar system) has been reported. This appears as an area of diminished perfusion of the cerebellar hemisphere contralateral to the affected cerebral hemisphere.

FORTUNE COOKIE

Man who eat contaminated meat
Better examine his sheet
If disease exist in his bed
Better also examine his head

Chapter 6

White Matter Diseases

White matter diseases are heterogeneous conditions linked together because they involve the same substrate. Magnetic resonance imaging (MRI) is quite sensitive and, when combined with age and other pertinent clinical information, provides a reasonable amount of specificity for white matter lesions. We are going to start by dividing white matter diseases into demyelinating (Box 6-1) and dysmyelinating diseases (Box 6-2).

The oligodendrocyte is the cell responsible for wrapping the axon concentrically to form the myelin sheath, and although we speak of white matter diseases as those that affect myelin, in actual fact, there is now a great deal of evidence that myelin is not the only brain tissue damaged in "demyelinating diseases" and axons and neurons (and even gray matter) are also commonly affected. Our appreciation of the nature of these disorders has improved dramatically with more precise histopathology and new magnetic resonance methodology.

We can further divide these diseases based on their presumed etiology (see Box 6-1). We shall consider each of them shortly, but first we need to give all of our perspicacious readers a dysesthetic diversion—dysmyelinating conditions.

Dysmyelinating disorders involve intrinsic abnormalities of myelin formation or myelin maintenance because of a genetic defect, an enzymatic disturbance, or both. These diseases are rare, usually seen in the pediatric or adolescent population, and often associated with a bizarre appearance on magnetic resonance (MR). Some diseases such as adrenoleukodystrophy have characteristics of both demyelinating and dysmyelinating processes (although in Box 6-2 it is operationally listed in the dysmyelinating category). The term leukodystrophy is used interchangeably with dysmyelinating diseases and represents primary involvement of myelin.

PRIMARY DEMYELINATING DISEASE

Multiple Sclerosis

Multiple sclerosis (MS), first described by Charcot in 1868, is the most common demyelinating disease encountered in clinical practice (as well as in imaging). Affecting over 2 million worldwide, it is the leading cause of nontraumatic neurologic disability in young and middle-aged adults.

MS has a peak age range of 30 years with a female predominance; however, it can occur in children and adolescents (3% to 12%) and in those over 50 (10% to 20%). The authors feel that age is a state of mind and the authors' immaturity level as evidenced by the attempts at so-called humor suggests a prolonged "at-risk" stage.

Symptoms range from isolated cranial nerve palsy, optic neuritis, and vague sensory complaints including

---

BOX 6-1   Demyelinating Diseases Categorized by Presumed Etiology

AUTOIMMUNE (IDIOPATHIC)

Multiple sclerosis (MS)
Monophasic demyelination (clinically isolated syndrome [CIS])
Acute disseminated encephalomyelitis (ADEM)
Acute hemorrhagic leukoencephalitis
Optic neuritis (may also be a manifestation of MS)
Acute transverse myelitis (may also be a manifestation of MS)
Neuromyelitis optica (NMO)

VIRAL

Progressive multifocal leukoencephalopathy
Subacute sclerosing panencephalitis
Human immunodeficiency virus (HIV)–associated encephalitis

VASCULAR (HYPOXIC/ISCHEMIC)

Migraines
Cerebral autosomal dominant arteriopathy with subcortical infarcts and leukoencephalopathy (CADASIL)
Leukoaraiosis
Postanoxic encephalopathy
Posterior reversible leukoencephalopathy (PRES)
Subcortical vascular dementia (aka subcortical arteriosclerotic encephalopathy/Binswanger disease/subcortical leukoencephalopathy)

METABOLIC/NUTRITIONAL

Osmotic demyelination
Marchiafava-Bignami
Combined systems disease (B$_{12}$ deficiency)

TOXIC

Radiation
Toxins
Drugs
Disseminated necrotizing leukoencephalopathy

TRAUMA

Diffuse axonal injury

BOX 6-2   Most Common Dysmyelinating Disorders (alphabetical)

Adrenoleukodystrophy
Alexander disease
Canavan disease
Krabbe disease
Metachromatic leukodystrophy
Pelizaeus-Merzbacher disease
Sudanophilic leukodystrophy

paresthesias and numbness, to paresis and paraplegia of limbs, and myelopathy. Subtle and obvious changes in intellectual capacity are also identified in patients with MS.

The diagnosis of MS has recently been revisited at the International Panel of the Diagnosis of MS in Dublin, 2010 (Table 6-1). The diagnosis can now be made based on clinical findings alone (2 events, 2 objective clinical findings) or in combination with clinical findings and MR findings. For MR, things have been simplified in that, if two or more of four possible locations of plaques are identified (those being the periventricular region, subcortical region, infratentorial region, and spinal cord), the MR scan meets the criteria of "dissemination in space" (DIS). If nonenhancing plaques and enhancing plaques coexist on the same scan or a second scan shows new enhancing or nonenhancing plaques, the criteria for dissemination in time (DIT) have been met. The full slate of possibilities for diagnosing MS is described in Table 6-1.

As noted in Table 6-2, two new entities have been described; one is clinically isolated syndrome (CIS), where only a single episode which may be unifocal clinically is identified. Eighty percent of patients with CIS at 15 years will meet diagnostic criteria for MS if they have positive MR findings but only 20% at 15 years get diagnosed with MS if they have a normal MR when staged as CIS. Seventy-seven percent of patients presenting with an isolated brain stem syndrome have been reported to have asymptomatic supratentorial white matter abnormalities. Progression to MS occurs in about 57% of patients with isolated brain stem syndrome and in 42% of patients with spinal cord syndrome.

The second entity is the radiologically isolated syndrome (RIS). In RIS, the patients have fulfilled the three of the four 2010 McDonald criteria for MS: (1) at least one gadolinium-enhancing lesion or nine T2 hyperintense lesions if there is no gadolinium enhancing lesion; (2) at least one infratentorial lesion; (3) at least one juxtacortical lesion; (4) at least three periventricular lesions. However, clinically they have not had an MS episode. In this situation, one third progress to a diagnosis of MS in 3 years and the rate increases significantly if there are enhancing plaques.

TABLE 6-1 2010 McDonald Criteria for Multiple Sclerosis Diagnosis

Clinical Findings	Magnetic Resonance Findings Required
• 2 or more attacks (relapses) • 2 or more objective clinical lesions	None
• 2 or more attacks • 1 objective clinical lesion	Dissemination in space (DIS) demonstrated by the presence of 1 or more T2 lesions in at least 2 of 4 of the following areas of the central nervous system: periventricular, juxtacortical, infratentorial, or spinal cord
• 1 attack • 2 or more objective clinical lesions	Dissemination in time (DIT), demonstrated by simultaneous presence of asymptomatic gadolinium-enhancing and nonenhancing lesions at any time, or a new T2 and/or gadolinium-enhancing lesion(s) on follow-up magnetic resonance imaging, irrespective of its timing with reference to a baseline scan

TABLE 6-2 Clinically Isolated Syndrome versus Radiologically Isolated Syndrome

Entity	Criteria	Progression
Clinically isolated syndrome (CIS)	• 1 attack • 1 objective clinical lesion Type 1 CIS: clinically monofocal, at least one asymptomatic magnetic resonance imaging (MRI) lesion Type 2 CIS: clinically multifocal, at least one asymptomatic MRI lesion Type 3 CIS: clinically monofocal, MRI may appear normal; no asymptomatic MRI lesions Type 4 CIS: clinically multifocal, MRI may appear normal; no asymptomatic MRI lesions Type 5 CIS: no clinical presentation to suggest demyelinating disease, but MRI is suggestive (RIS)	80% progress to multiple sclerosis (MS) if magnetic resonance criteria met in 15 years, 20% if not If optic neuritis as first symptom, 50% progress to MS
Radiologically isolated syndrome (RIS)	1. Ovoid, well-circumscribed, and homogeneous foci with or without involvement of corpus callosum 2. T2 hyperintensities measuring >3 mm and fulfilling the McDonald criteria (at least 3 out of 4) for dissemination in space 3. Central nervous system (CNS) white matter anomalies inconsistent with vascular pattern No historical accounts of remitting clinical symptoms consistent with neurologic dysfunction MRI anomalies do not account for clinically apparent impairments in social, occupational, or generalized areas of function MRI anomalies not because of direct physiologic effects of substances (recreational drug abuse, toxic exposure) or medical condition Exclusion of individuals with MRI phenotypes suggestive of leukoaraiosis or extensive white matter pathology lacking involvement of corpus callosum The CNS MRI anomalies are not better accounted for by other disease processes	33% progress to MS in 3 years, 80% if enhancing foci coexist

Laboratory tests for MS including visual, auditory, and somatosensory evoked responses can confirm the presence of lesions, but they are nonspecific and provide no clue to the cause of the abnormal finding. Approximately 70% of patients with MS have elevated cerebrospinal fluid (CSF) levels of IgG, and approximately 90% have elevated oligoclonal bands.

MS is a disease characterized by a variety of clinical courses. Terminology about clinical classification can be confusing and even contradictory. Relapsing remitting (RR) MS is the most common course of the disease initially occurring in up to 85% of cases. At the beginning, exacerbations are followed by remissions. However, over years additional exacerbations result in incomplete recovery. Within 10 years, 50% (and within 25 years, 90%) of these cases enter a progressive phase, termed secondary progressive (also termed relapsing progressive) MS. During this phase, deficits are progressive without much remission in the disease. Less commonly, the disease is progressive from the start, termed "primary progressive MS." These patients (5% to 10% of the MS population) may present at a later age with progressive neurologic findings including paraparesis, hemiparesis, brain stem syndromes, or visual loss, and typically have a more severe disability. They may have occasional plateaus and temporary improvements, but do not have distinct relapses. Progressive-relapsing MS, a rare clinical course, is defined as progressive disease with clear acute relapses, with or without full recovery, and with the periods between relapses characterized by continuing progression. Benign MS describes those cases, where after initial clinical symptomatology, there is no clinical progression over, approximately, a 10- to 15-year course. Conversely, a rapid progressive disease leading to significant disability or death in a short time after the onset has been termed malignant MS.

The Kurtzke Expanded Disability Status Scale (EDSS) is often used to assess the deficits of patients with MS. The EDSS relies on an assessment of gait and eight functional systems: pyramidal motor function, cerebellar, brain stem, sensory, bowel and bladder, visual, cerebral or mental, and other. EDSS scores below 4.0 have normal gait and scores are determined by functional system deficits, if any. People with EDSS scores of 4.0 and above have some degree of gait impairment. Scores between 4.0 and 9.5 are determined by both gait abilities and the eight functional system scores.

Magnetic Resonance Findings

Brain Lesions. Fluid-attenuated inversion recovery (FLAIR) and T2-weighted imaging (T2WI) are the most useful MR sequences for identifying MS plaques. On both of these sequences MS plaques are seen as individual flame-shaped or confluent high-intensity areas in the white matter. MS lesions have a predilection for certain regions of the brain including the periventricular region, corpus callosum (best visualized on sagittal FLAIR images; Fig. 6-1), subcortical region (best seen on FLAIR), optic nerves and visual pathways, posterior fossa, the interface between the corpus callosum and septum pellucidum, and the cervical region of the spinal cord (Fig. 6-2). However, MS lesions can and do occur in any location in the brain. This includes the cortex (6%), where white matter fibers track up to the superficial cortical cells, and the deep gray matter (5%), best seen on FLAIR. Caution to those of you who rely so heavily on FLAIR imaging and skip T2WI fast spin echo images—FLAIR imaging does not detect lesions in the posterior fossa, brain stem, and spinal cord as well as T2WI scans. In addition, very hypointense lesions on T1WI (see "black holes" below) may look similar to CSF on FLAIR, that is, not bright (like some of your professors) and thereby be overlooked (see Fig. 6-2).

High-intensity lesions at the callosal-septal interface (sagittal MR either proton density [PD] or FLAIR) have been suggested to have 93% sensitivity and 98% specificity in differentiating MS lesions from vascular disease (see Fig. 6-1). The shape of these MS plaques may be variable. However, ovoid lesions are believed to be more specific for MS. Their morphology has been attributed to inflammatory changes around the long axis of a medullary vein (Dawson fingers; Fig. 6-3).

On T1WI, plaques are isointense or hypointense regions, whereas on T2WI, the lesions are high intensity (see Fig. 6-2, *A*). Uncommon hypointense lesions on T1WI that approximate CSF intensity have been termed "black holes" and have been reported to be associated with areas of greatest myelin loss. The volume of black

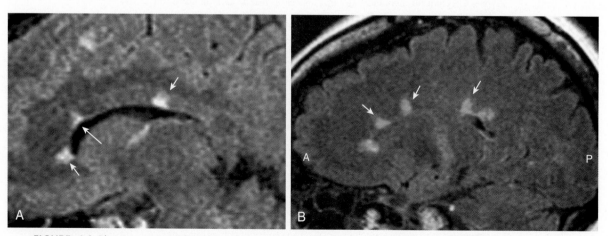

FIGURE 6-1 Plaques proximate the pellucidum parasagittally. Sagittal fluid-attenuated inversion recovery (FLAIR) images (**A** and **B**) with high-intensity lesions emanating from the callosal-septum pellucidum interface.

holes correlates most closely with disability as determined by the EDSS.

Postcontrast MR images in patients with plaques may show no enhancing plaques, or a wide variety of patterns of enhancement indicating active demyelination. These include solid nodular, solid linear, complete ring, open ring/arcs, and punctate enhancement. All different types may coexist (Fig. 6-4). Meningeal irritation near demyelination can cause meningeal enhancement as well. Cranial nerves (besides the optic nerve) can also enhance in MS as can

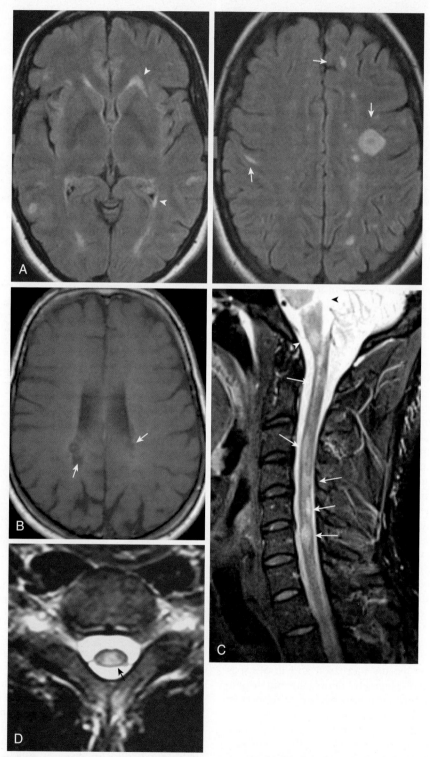

FIGURE 6-2 Mundane multiple manifestations of multiple sclerosis (MS). **A,** Axial fluid-attenuated inversion recovery (FLAIR) scan shows multiple hyperintense periventricular *(arrowheads)* and subcortical white matter lesions *(arrows).* **B,** The T1-weighted image shows low intensity lesions in periventricular sites *(arrows)* indicative of the so-called "black holes" that correlate well with disability scores. The rim of high signal, thought to be due to paramagnetic deposition, is also sometimes seen in MS. **C,** Infratentorial brain stem *(arrowheads)* and cord demyelinating plaques *(arrows)* are evident. **D,** Lesions favor the posterior cord *(arrow).*

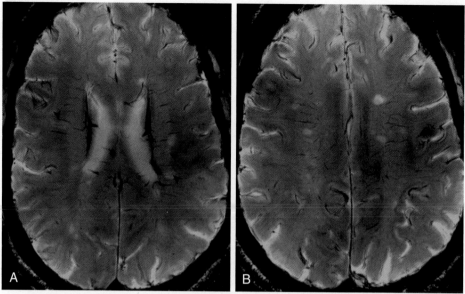

FIGURE 6-3 A, Perivenular preference of plaques portrayed perfectly. Perivenular multiple sclerosis on 7 Tesla gradient echo magnetic resonance images. Note both the ovoid appearance of the lesions and the veins *(arrowheads)* actually course through the middle of the ovoid lesion. **B,** At the ventricular junction, these perpendicular lesions are what is referred to as Dawson fingers. (Compliments of Robert I. Grossman.)

superficial brain stem enhancement, but it does mean that other diagnoses should be considered (Fig. 6-5). The normal window of enhancement is from 2 to 8 weeks; however, plaques can enhance for 6 months or more. Enhancement cannot be viewed as "an all or none" phenomenon, rather it is dependent on the time from injection to imaging, the dosage of contrast agent, the magnitude of the blood-brain barrier (BBB) abnormality, and the size of the space where it accumulates. Delayed imaging (usually 15 to 60 minutes following injection) increases the detection of enhancing MS lesions (and decrease patient throughput and revenue). Triple doses of gadolinium (0.3 mmol/kg) or a single dose (0.1 mmol/kg) with magnetization transfer (MT) to suppress normal brain can increase the number of detectable MS lesions (and fill the coffers of drug companies, while bankrupting radiology departments).

Lesions may display mass effect that can mimic a tumor (tumefactive MS) and have been associated with seizures (Fig. 6-6). There are several hints that aid in suggesting this diagnosis including the history, which is usually acute or subacute onset of neurologic deficit(s) in a young adult, and other white matter abnormalities unassociated with the mass lesion, but characteristic of MS, such as in the periventricular zone, spinal cord, or callosal-septal interface. Tumefactive MS lesions often have a leading edge of enhancement and an incomplete "horseshoe-shaped" ring. Perfusion in tumors is usually increased and in MS it is typically not. Veins are displaced by neoplasms but course through MS lesions. There have also been rare reports of hemorrhage into demyelinating lesions.

Other findings include atrophy of the brain and spinal cord. The greater the loss of myelin and axons, the more likely there is atrophy and that the lesion becomes hypointense on T1WI. High intensity on unenhanced T1WI can be observed infrequently, most often in the periphery of the plaque (see Fig. 6-2, *B*). The cause of this phenomenon is unknown, but hypotheses include a small amount of paramagnetic accumulation from hemorrhage, myelin catabolites including fat,

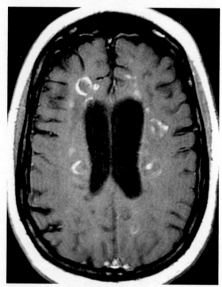

FIGURE 6-4 Images of enhancement in idiopathic oligodendroglial inflammation (multiple sclerosis). This one image shows all patterns of multiple sclerosis enhancement from open ring to ring to linear to nodular. It's all there is, one unlucky patient!

free radical production from the inflammatory response, or focally increased regions of protein.

Increased iron deposition (in the thalamus and basal ganglia) producing low intensity on T1 and T2* or susceptibility-weighted sequences has been reported in patients with long-standing MS. This latter finding is nonspecific, having been described in a variety of different conditions including Parkinson disease, Wilson disease, multisystem atrophy, and other degenerative conditions.

With MS, one must recognize that there is disease in the normal appearing white matter (NAWM) which is beyond the resolution of standard imaging and contrast techniques. There are many new MR methods (diffusion

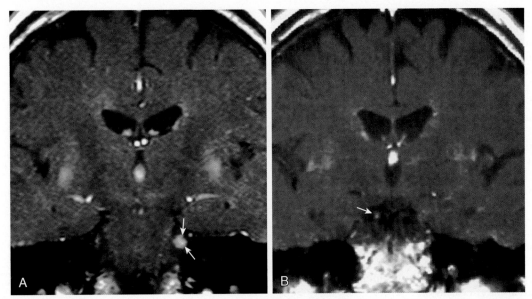

FIGURE 6-5 Increased amplification (and enlargement) of innervators. **A,** Coronal postcontrast T1-weighted imaging revealing enhancement of cranial nerve V *(white arrows).* **B,** Coronal enhanced image with enhancement of right cranial nerve III *(arrow).* Focus on the cranial nerves here, not on the pulsation artifact creating pseudolesions in the parenchyma! (Courtesy Robert Quencer, MD, Miami, Florida.)

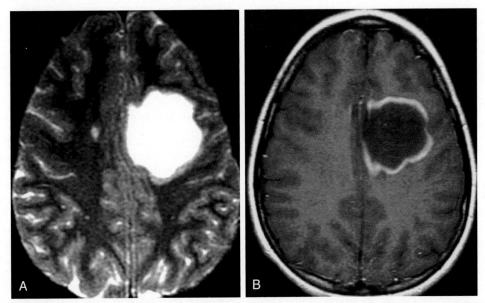

FIGURE 6-6 Demyelination deluding doctors into diagnosing deadly disease. **A,** Axial T2-weighted imaging (T2WI) with large mass compressing the left medial frontal gray matter presumed to be a glioma. **B,** T1WI with open ring of enhancement medially classic for tumefactive multiple sclerosis and unusual for glioblastoma multiforme, lymphoma, or metastasis.

tensor imaging [DTI], magnetization transfer imaging [MTI], magnetic resonance spectroscopy [MRS], susceptibility-weighted imaging [SWI]) that clearly demonstrate that the NAWM in MS is not normal, that is, there are lesions that we cannot detect by conventional MR. This is important, as the extent of disease in MS patients is generally greater than the visible T2 lesion load.

New Methods

Magnetic Resonance Spectroscopy. *N*-acetyl aspartate (NAA) is decreased in MS. This indicates that MS is more than a white matter disease. Axonal-neuronal loss occurs both early and often. It is suggested that the neuronal loss is associated with irreversible neurologic impairment. Choline (Cho) is increased and is associated with membrane (myelin) breakdown, inflammation, and remyelination. Creatinine (Cr) may also be decreased. Lipid resonances have been observed in acute lesions using short echo-times (TE ≤30 ms). Myo-inositol and lactate have also been reported to be present in some lesions.

Magnetization Transfer

Magnetization transfer (MT) results from the transfer of magnetization from protons attached to rigid macromolecules (such as myelin) to free water protons. The effect is observed by noting a decrease in intensity on MR images performed with an off resonance pulse. Injury resulting in demyelination causes a decrease in MT. Usually MT effects are noted as one minus the ratio of the image intensity with a saturation pulse on divided by the intensity with the saturation pulse off (1- MTs/MTo). This is termed the magnetization transfer ratio (MTR). Thus, low MTR equates with myelin loss.

There is decreased MT in MS, including in plaques and in NAWM. Correlation with clinical disability has been found using MT. Histogram analysis using MT enables interrogation of the entire brain.

Diffusion

Diffusion is higher in subacute and chronic MS plaques compared with NAWM, because with white matter breakdown, the tissue becomes more like water, which has maximal diffusion. Vasogenic edema shows increased diffusion and increased apparent diffusion coefficient (ADC), which is the most common signature for MS. Mean diffusivity is also increased in NAWM. However, one may see reduced ADC in early acute plaques, perhaps because of (1) cytotoxic edema, (2) myelin fragments and reduced fiber tract organization restricting water diffusion, or (3) activated microglia and oligodendrocyte apoptosis before infiltration of blood borne inflammatory cells.

Some neurologists have advocated using diffusion-weighted imaging (DWI) scans to show "different-aged lesions" to fulfill DIT criteria, even in the face of not having enhancing and nonenhancing plaques concurrently.

Spinal Cord Disease

MS can affect the spinal cord alone (5% to 24%) or, more commonly, both the brain and the spinal cord. Approximately 60% of spinal cord lesions occur in the cervical region (see Fig. 6-2, *D*). Lesions can be single or multiple. Spinal cord MS tends toward the cord periphery or center, that is, does not involve the entire cord cross-section and is said, in 90% of cases, to extend less than two vertebral body segments in length. Spinal cord swelling associated with lesions occurs in 6% to 14% of cases, whereas atrophy ranges from 2% to 40%. Most lesions in the spinal cord do not demonstrate enhancement; it takes a pretty talented (and rare) neurologist to predict a cord lesion in MS and have that correlated with an enhancing plaque at that cord level. Rarely see it, but most neurologists have also never won the lottery. Same odds. To wit, no correlation is usually found between spinal cord lesion load and EDSS, but clinical disability has been correlated with spinal cord atrophy.

Clinical Differential Diagnosis of Spinal Cord Lesions

When confronted with a clinical condition that is suggestive of MS and you see high-intensity or enhancing spinal cord lesions, what is the differential diagnosis? It should include vascular lesions, particularly dural arteriovenous malformation producing venous hypertension and subsequent venous infarction, as well as other vascular malformations and arterial lesions. In addition, collagen vascular diseases such as lupus can produce myelitis, and other inflammatory diseases such as sarcoid and acute disseminated encephalomyelitis also involve the spinal cord. Other considerations also include intrinsic spinal cord neoplasms and infections, both viral (including human immunodeficiency virus [HIV])

> **BOX 6-3   Differential Diagnosis of Spinal Cord Lesions**
>
> Idiopathic acquired transverse myelitis
> Autoimmune demyelination
> NMO
> ADEM
> MS
> Spinal cord tumor (primary or metastatic)
> Syringohydromyelia
> Acute infarction
> Vascular lesions including dural arteriovenous malformation, infarction
> Infectious processes (toxoplasmosis, vacuolar myelopathy in AIDS, herpes zoster)
> Lupus
> Trauma (hematomyelia)
> Diffuse leptomeningeal coating of the spinal cord from sarcoid, lymphoma, or other tumors

ADEM, Acute disseminated encephalomyelitis; *MS,* multiple sclerosis; *NMO,* neuromyelitis optica.

---

and bacterial, which can all masquerade as spinal MS. An appropriate history, cerebrospinal fluid analysis, and careful examination of the MR are important in differentiating these lesions (also see Differential Diagnosis of Multiple Sclerosis Lesions on Magnetic Resonance). Subacute combined degeneration of the spinal cord caused by vitamin B12 deficiency involves the spinal cord posterior columns symmetrically and is associated with a peripheral neuropathy. Box 6-3 lists the differential diagnosis of an enlarged T2 hyperintense spinal cord lesion.

Multiple Sclerosis Syndromes

Many syndromes are associated with MS. Balo disease (concentric sclerosis) represents a histologic MS lesion with alternating concentric regions of demyelination and normal brain (Fig. 6-7). Rarely, a similar pattern may be observed on T2WI scans. Diffuse sclerosis (Schilder disease) is an acute, rapidly progressive form of MS with bilateral relatively symmetric demyelination. It is seen in childhood and rarely after the age of 40 years. It is characterized by large areas of demyelination that are well circumscribed, often involving the centrum semiovale and occipital lobes. The Marburg variant of MS is defined as repeated relapses with rapidly accumulating disability producing immobility, lack of protective pharyngeal reflexes and bladder involvement.

Optic Neuritis

There are numerous causes of optic neuropathy. The most frequent, usually seen in the elderly, is acute ischemic optic neuropathy, which occurs in patients with atherosclerotic risk factors, particularly diabetes. This is more of a vascular insult than an inflammatory one. When one refers to optic "neuritis," we are usually thinking about infectious (viral usually), inflammatory (e.g., sarcoidosis), or demyelinating disorders (Box 6-4). It has been said that a person with an optic neuritis at first presentation has a 50% chance of carrying the diagnosis of MS in 5 years; that rate increases if there are MS-like plaques on their brain MRI scan but is less if the brain MRI scan is negative. All told, 80% of MS patients have an episode of optic neuritis at some point in their lives. Thus, the entity

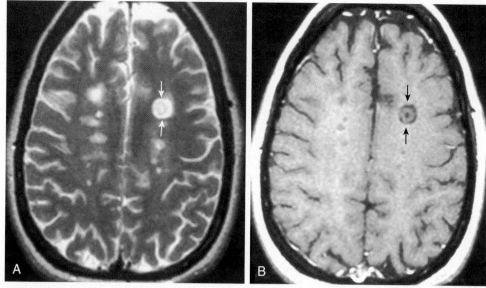

FIGURE 6-7 Bilobed ball of biconvex Balo. **A,** Left frontal high-intensity lesion. There are alternating rings of different intensities typical of Balo concentric sclerosis *(arrows).* **B,** Note the rings of intensity on this postgadolinium T1-weighted imaging *(arrows).*

BOX 6-4 Causes of Optic Neuritis

INFECTIOUS

Viral: HSV, CMV, HIV, varicella
Bacterial: Lyme, syphilis, TB, spread from sinusitis
Protozoan: toxoplasmosis

INFLAMMATORY

Sarcoidosis
Lupus
Wegener
"Pseudotumor" (idiopathic orbital inflammation)

DEMYELINATION

Idiopathic isolated optic neuritis
Clinically isolated syndrome
MS
ADEM
NMO

ADEM, Acute disseminated encephalomyelitis; *CMV,* cytomegalovirus; *HIV,* human immunodeficiency virus; *HSV,* herpes simplex virus; *MS,* multiple sclerosis; *NMO,* neuromyelitis optica; *TB,* tuberculosis.

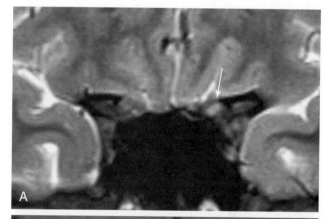

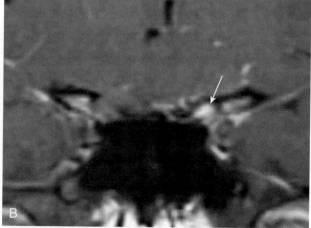

FIGURE 6-8 Optic neuritis. **A,** Coronal T2-weighted imaging with high intensity in the left optic nerve *(arrow).* **B,** Enhanced image of intracranial optic nerves with left optic nerve enhancement *(arrow).*

is considered separate from MS, but may overlap MS at the same time (50/50). For best visualization of optic neuritis, fat-suppressed scans through the orbit using FLAIR or T2 weighting is best (Fig. 6-8). This nulls the orbital fat and/or the CSF around the nerve. That said, there have been studies suggesting that optic nerve enhancement may actually precede signal intensity abnormalities on T2WI scans in some cases of optic neuritis.

Differential Diagnosis of Multiple Sclerosis Lesions on Magnetic Resonance

There is a baseline incidence of high-intensity abnormalities in the brains of healthy individuals. This number varies depending upon the exact report and the cohort's age. An increase in number of these nonspecific white matter lesions with increase in age has been described in asymptomatic people. Some people have adopted the adage that you are "allowed" one white matter "ditzel" for every decade of life, given no symptoms. This allows one author to ignore 5.6 tiny punctate foci in his subcortical white matter (then again, he is a migraineur—see later).

The following lesions are truly not primary demyelinating processes but may simulate MS in their MR appearance. These conditions manifest lesions with or without enhancement and occur in a similar patient population to that of MS.

Lyme Disease

An important infection that may produce symptoms similar to MS is Lyme disease. It is discussed in the infection chapter; however, Lyme disease can have high-intensity lesions on T2WI scans that enhance. The cranial nerves and/or meninges may also demonstrate enhancement. It can present with acute central nervous system (CNS) manifestations including transverse myelitis.

Vasculitides

Vasculitides (see Chapter 3) including primary angiitis of the central nervous system, polyarteritis nodosa, Behçet disease, syphilis, Wegener granulomatosis, Sjögren syndrome, sarcoidosis, and lupus should be in the differential diagnosis of MS both clinically and radiographically. Lesions of multiple ages which may or may not enhance occur in the vasculitides. Usually there are more deep and cortical gray matter lesions than MS.

Migraine Lesions

Migraine sufferers often have small, less than 3 mm, white matter lesions in the subcortical locations predominantly in frontal and parietal lobes. The paucity of periventricular and posterior fossa and spinal cord lesions should help distinguish the pattern of juxtacortical migrainous lesions from MS patterns (Fig. 6-9, *B*). The lesions are not flame shaped and do not have a Dawson fingers look. The cause of these high signal abnormalities remains a mystery, although platelet aggregation is increased during attacks and some investigators hypothesize that these lesions could be the result of microemboli. They could also be the consequence of primary neuronal damage related to migraine pathophysiology.

Virchow-Robin Spaces

Virchow-Robin spaces are invaginations of the subarachnoid space into the brain associated with leptomeningeal vessels (Fig. 6-10). Dilated perivascular spaces occur in rather characteristic locations, typically in the basal ganglia, around the atria, near the anterior commissure, in the corona radiata, centrum semiovale, periinsular region, and in the middle of the brain stem, medial and posterior to the reticular portion of the substantia nigra. Usually, they follow the intensity of CSF, being hypointense to brain on T1WI and FLAIR, and hyperintense on T2WI scans. FLAIR scans are best at discriminating perivascular spaces from white matter lesions because the perivascular spaces remain isointense to CSF, whereas MS and other lesions are hyperintense on these pulse sequences (unless there are cystic black holes in MS).

Virchow-Robin spaces (see Fig. 6-9) tend to enlarge with age and hypertension as the vessels within the space become more ectatic. This has been termed état criblé, which is defined as dilatation of perivascular spaces, usually with thinning and pallor of the perivascular myelin associated with shrinkage, atrophy, and isomorphic gliosis around the vessel.

Trauma

Diffuse axonal injuries (see Chapter 4, Diffuse Axonal Injury section) induce high signal abnormalities on T2WI scans at the gray-white junctions, brain stem, corpus callosum, and internal capsules, but an appropriate history of trauma should be present. Patients with subclinical unreported trauma or in head-banging athletes with or without minor concussions (e.g., soccer players who head the ball frequently, football players, boxers, lacrosse players who get their "bell rung") show abnormal diffusivity in their brains and may have tiny white matter foci at gray-white junctions or in the corpus callosum that could look like MS. Perform SWI; if punctate blood products coexist it may be from head trauma, not MS.

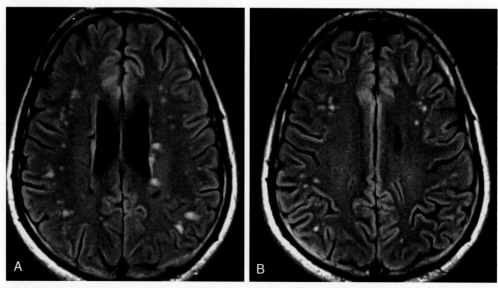

FIGURE 6-9 Mundane multiple miscellaneous markings might mitigate for migraines. **A,** Nonspecific white matter lesions on fluid-attenuated inversion recovery (FLAIR) scans could be attributable to accelerated atherosclerosis leukoaraiosis in this 50-year-old. Some would include posttraumatic or postinflammatory residua in the differential diagnosis. **B,** This pattern of juxtacortical punctate lesions is typical of migraine sufferers.

Neuromyelitis Optica

Neuromyelitis optica (NMO; formerly called Devic disease) represents either an acute variant of MS or a separate demyelinating disease (Fig. 6-11). The disease consists of both transverse myelitis and bilateral optic neuritis. Symptoms may occur simultaneously or be separated by days or weeks. The clinical manifestations can be severe. The importance of making this diagnosis is that the treatment is with monoclonal antibody agents (i.e., Rituximab) rather than the typical MS cocktail of steroids and/or interferon (the latter of which

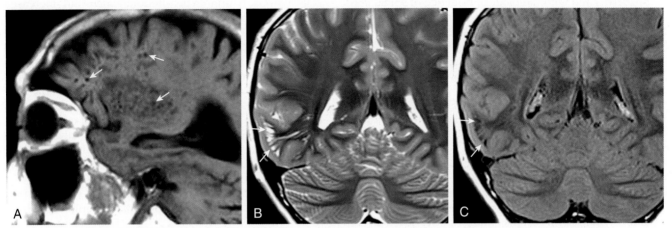

FIGURE 6-10 Very variable vasculophilic venous Virchow-Robin (VR) spaces. **A,** Sagittal T1-weighted imaging (T1WI) reveals low-intensity VR spaces in the basal ganglia *(arrows)*. **B,** In a different patient, coronal T2WI shows the high signal intensity VR spaces *(arrows)* in the subcortical region of the right temporal lobe, a characteristic appearance in a less common location. **C,** Fluid-attenuated inversion recovery (FLAIR) image showing VR spaces have similar intensity to cerebrospinal fluid *(arrows)*.

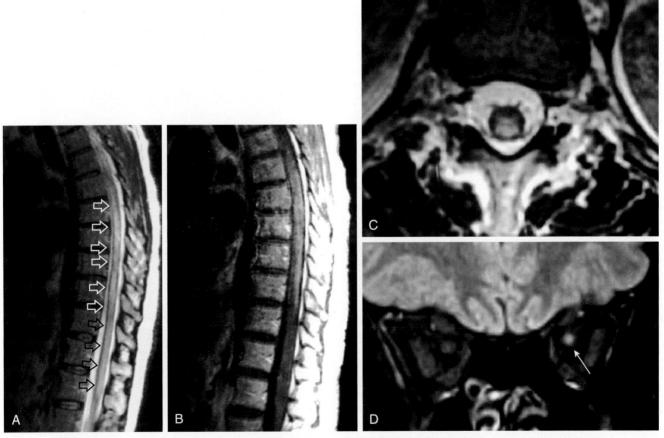

FIGURE 6-11 Devic demyelinating disease de-eponymed to neuromyelitis optica (NMO). **A,** High intensity on T2 sagittal image involves more than 7 vertebral body segments *(arrows)*. **B,** Enhancement throughout the upper area of abnormality is seen on postcontrast T1-weighted image. **C,** Axial T2-weighted imaging with central hyperintensity. **D,** Fat-suppressed coronal fluid-attenuated inversion recovery (FLAIR) shows bright left optic nerve *(arrow)* representing the optic neuritis element of NMO.

may exacerbate NMO). NMO-IgG/aquaporin-4 antibodies are the culprits that attack the astrocytic water channels in the periependymal/perivascular regions of the brain.

NMO major criteria are listed as:
- Optic neuritis in one or both optic nerves
- Transverse myelitis, clinically complete or incomplete, but associated with radiological evidence of spinal cord lesion extending over three or more spinal segments on T2WI MRI images and hypointensity on T1WI scans when obtained during acute episode of myelitis
- No evidence for sarcoidosis, vasculitis, clinically manifest systemic lupus erythematosus or Sjögren syndrome, or other explanation
The minor criteria state:
- Most recent brain MRI scan of the head must be normal or may show abnormalities not fulfilling McDonald diagnostic criteria, including:
- Nonspecific brain T2 signal abnormalities not satisfying McDonald criteria
- Lesions in the dorsal medulla, either in contiguity or not in contiguity with a spinal cord lesion, hypothalamic and/or brain stem lesions, "linear" periventricular/corpus callosum signal abnormality, but not ovoid, and not extending into the parenchyma of the cerebral hemispheres in Dawson finger configuration
- Positive test in serum or CSF for NMO-IgG/aquaporin-4 antibodies

Final Thoughts

It is critical for the radiologist to understand that the diagnosis of MS is based on the clinical signs and symptoms of the patient. The principal role of imaging is still to (1) confirm the clinical suspicion of MS, (2) suggest plausible alternative diagnoses for the patient's neurologic complaints, and (3) assess for evidence of DIS and DIT to fulfill MS criteria. MR alone cannot make the diagnosis of MS, so neurologists can still bring home the baguettes; although managed care and bundled payments have made them 3-day-old loaves.

SECONDARY DEMYELINATING DISEASES

Allergic/Autoimmune

Acute Disseminated Encephalomyelitis

Acute disseminated encephalomyelitis (ADEM) was originally described as a monophasic demyelinating disease but frequent recurrences and gray matter involvement have led to a change in the gestalt that is ADEM. It has also been shown that patients who develop ADEM are at higher risk for developing MS, perhaps because of a jazzed up immune system. The usual history is of a child with a recent viral infection, vaccination, respiratory infection, or exanthematous disease of childhood. Before using this as a poor excuse not to vaccinate your kids, keep in mind that ADEM has been identified most frequently with antecedent measles, varicella, mumps, and rubella infection, but is not limited to these viruses. Today, Epstein-Barr virus, cytomegalovirus, or mycoplasma pneumoniae respiratory infections are the most common precipitants, but others include myxoviruses, herpes group, and HIV. It may also be idiopathic. A latency period of 7 to 14 days or even longer has been described.

The suspected etiology is based on an allergic or autoimmune (cell-mediated immune response against myelin basic protein) cross-reaction with a viral protein. Symptoms are similar to a single episode of acute MS. Clinical syndromes of acute transverse myelitis, cranial nerve palsy, acute cerebellar ataxia (acute cerebellitis), or optic neuritis are well described. The diagnosis is usually made by history and CSF, which may demonstrate increase in white cells with a lymphocytic predominance and increased myelin basic protein. ADEM may have a mortality rate of up to 30%, and steroid therapy is commonly rendered.

On imaging, the lesions, which may be multiple and large, are high intensity on T2WI scans and may enhance in a nodular or ring pattern (Fig. 6-12). No new lesions should appear on MR after approximately 6 months from the start of the disease. There may, however, be incomplete resolution of lesions. ADEM can enlarge the spinal cord or brain stem (and appear as a mass lesion), but it usually is seen in the cerebrum. Gray matter lesions can also be identified.

The distribution of lesions in ADEM varies depending upon the age of onset, as noted in Table 6-3.

Although rare, at the fulminant end of the spectrum of ADEM is acute hemorrhagic leukoencephalitis (Hurst

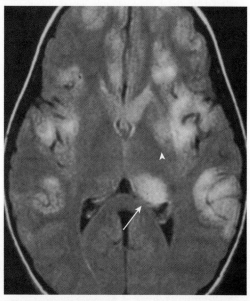

FIGURE 6-12 Any imbecile ought to opine on acute disseminated encephalomyelitis (ADEM). Fluid-attenuated inversion recovery (FLAIR) image shows high signal in the thalamus *(arrow)*, putamen *(arrowhead)*, and juxtacortical bilateral white matter in this patient with ADEM.

TABLE 6-3 Frequency of Lesions in Acute Disseminated Encephalomyelitis: Adults versus Children

Location	Percentage Adults	Percentage Children
Subcortical white matter	60%	93%
Cerebral cortex	10%	80%
Periventricular white matter	43%	77%
Brain stem	37%	57%
Deep gray matter	63%	50%
Cerebellum	50%	14%
Spinal cord	50%	10%

disease) associated with diffuse multifocal perivascular demyelination and hemorrhage confined to the cerebral white matter with strict sparing of the subcortical U-fibers.

Viral

Progressive Multifocal Leukoencephalopathy

Progressive multifocal leukoencephalopathy (PML) is a demyelinating disease with a known viral etiology. It is caused by the JC (the initials of John Cunningham, the VA patient with PML who donated his brain and from whom the new human polyomavirus was isolated postmortem) virus infecting the oligodendrocyte and is associated with the immunosuppressed state. Box 6-5 lists the conditions that have been associated with PML. Although originally described as having a propensity for the parietooccipital region, PML can occur anywhere in the brain (including the posterior fossa), particularly in patients with AIDS, and may be solitary or multifocal with eventual widespread confluence. PML may present

---

BOX 6-5 Conditions Associated with Progressive Multifocal Leukoencephalopathy

Acquired immunodeficiency syndrome (AIDS)
Autoimmune disease
Cancer
Immunosuppressive therapy
Immune reconstitution inflammatory syndrome (IRIS)
Lymphoproliferative disorders
Myeloproliferative disorders
Nontropical sprue
Sarcoid
Sepsis
Transplantation
Tuberculosis
Whipple disease

---

on MR as a focal region of low intensity on T1WI with high intensity on T2WI, most often without enhancement (so, what else is new?) (Fig. 6-13). Histologically, there are multifocal demyelinating plaques involving the subcortical U-fibers with sparing of the cortical ribbon and the deep gray matter. In a patient with HIV and subcortical white matter abnormality, favor PML over other HIV related diseases. This disease peaked at the height of the acquired immunodeficiency syndrome (AIDS) epidemic, with 2% to 7% of AIDS patients acquiring PML. Although less well recognized, PML may present as a mass lesion. It may demonstrate enhancement in about 10% of cases.

Highly active antiretroviral therapy (HAART) may improve the prognosis of PML. However, a new disease process, immune reconstitution inflammatory syndrome (IRIS), has developed in association with HAART. The use of combined antiretroviral therapy markedly improves immune function and prognosis in HIV-infected patients; however, PML may transiently worsen with antiretroviral therapy, despite a recovery of the immune system. This manifestation is believed to be a result of IRIS. It requires the patient to discontinue HAART for a period of time to get over the fulminant IRIS episode. IRIS is seen on MRI as enlarging white matter PML lesions, often with enhancement and mass effect, and frequently also involving deep gray matter (Fig. 6-14).

Human Immunodeficiency Virus

The patient with AIDS has a propensity for multiple CNS infections as well as changes related to intrinsic HIV infection. The most common imaging manifestation of HIV encephalopathy is parenchymal volume loss; however, patchy and multifocal T2 and FLAIR hyperintensity (sparing the subcortical U-fibers) can also be seen in both cerebral hemispheres. The diagnosis of HIV encephalopathy can be confounded by the propensity of this population to develop neoplasms and PML. Patients with the AIDS

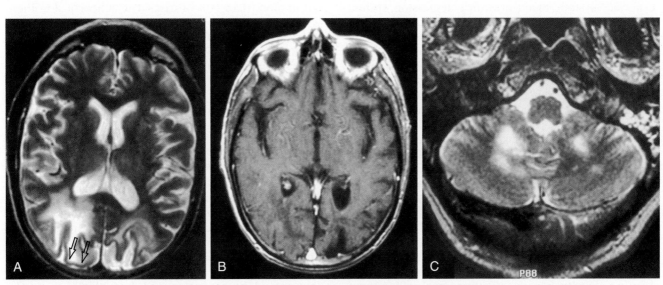

FIGURE 6-13 Prototypical portrayal of progressive multifocal leukoencephalopathy (PML) predominantly posterior. **A,** Note the absence of mass effect and the involvement of the subcortical U-fibers *(open arrows)* on axial T2-weighted imaging (T2WI) with high intensity in the parieto-occipital white matter. **B,** T1 enhanced axial image with no enhancement in the regions of signal abnormality. **C,** Another patient T2WI with posterior fossa involvement with PML.

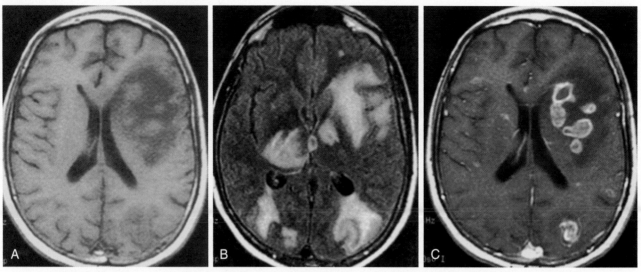

FIGURE 6-14 Unusual advanced example of immune reconstitution inflammatory syndrome (IRIS), accompanied with extensive enhancement, edema, and aggressiveness. **A,** Axial T1-weighted image (T1WI) with left sided hypointense mass compressing the lateral ventricle. **B,** Axial fluid-attenuated inversion recovery (FLAIR) image with multiple high-intensity lesions with marked edema and irregularity. **C,** Axial postcontrast T1WI with multiple ring enhancing lesions—very unusual for progressive multifocal leukoencephalopathy but in keeping with IRIS.

dementia complex develop atrophy and regions of demyelination in the white matter. This most commonly affects the supratentorial compartment particularly the deep gray and white matter.

Vascular

Age-Associated Leukoaraiosis

You should appreciate that the elderly accumulate white matter abnormalities that result, at some level, from ischemia secondary to injury to the long penetrating arteries of the brain. They have been correlated with several clinical factors including silent stroke, hypertension (especially systolic blood pressure), depression, spirometry, and income (the poor shall inherit the earth, but they might have to hire the rich to run it because of the preponderance of unidentified bright objects [UBOs] in their brain). The greater the amount of white matter abnormality, the more likely the individual is to have impaired cognitive function and gait. However, mild periventricular white matter abnormality probably has little clinical significance. Foci of white matter disease may be termed "age associated leukoaraiosis" rather than "small vessel ischemic *disease.*" This is particularly true in the age of patients having access to their medical records and imaging reports. Better they should read "age associated leukoaraiosis" than fear they have had "ischemia" (aka strokes)…or NOT.

Binswanger Disease

Binswanger disease (subcortical arteriosclerotic encephalopathy [SAE]), described in 1894, is a demyelinating disease equally affecting men and women generally more than 55 years old. It is associated with hypertension (approximately 98% of patients) and lacunar infarction. These features distinguish it from MS. Patients may have acute stroke followed by declining mental status or slower insidious mental status changes with decreased levels of mentation, dementia, psychiatric disturbances, seizures, urinary incontinence, and gait disturbance. MR reveals broad regions of high-intensity abnormalities in the white matter of the frontal-parietal-occipital regions into the centrum semiovale. Tissue damage in Binswanger disease is more severe than in the high-intensity regions in the nondemented patient, but not as severe as in infarction. Histopathology in Binswanger disease usually displays demyelination with relative axonal sparing, in association with arteriosclerosis, narrowing of white matter arteries and arterioles. Lacunar infarction is present in more than 90% of cases. The subcortical U-fibers, which have a dual blood supply from the involved medullary arteries and uninvolved cortical arteries implicating an ischemic etiology, are spared. Binswanger disease differs from multiinfarct dementia because of its distinctive white matter involvement and the absence of consistent focal stroke syndromes. This again is a disease that needs the appropriate history to aid in focusing on the correct diagnosis.

Cerebral Autosomal Dominant Arteriopathy with Subcortical Infarcts and Leukoencephalopathy

Cerebral autosomal dominant arteriopathy with subcortical infarcts and leukoencephalopathy (CADASIL) is an inherited arterial disease caused by mutations of *Notch 3* gene on chromosome 19. Its onset is in the fourth decade of life, with a mean age of death at 59 years. The disease is characterized by migraines, recurrent transient ischemic attacks, strokes, dementia, depression, pseudobulbar palsy, and hemiplegia or quadriplegia. It affects the frontal lobes, temporal lobes, and insula. Hyperintensity on T2WI scans is observed in the white matter, particularly periventricular and deep white matter, basal ganglia, and brain stem. CADASIL

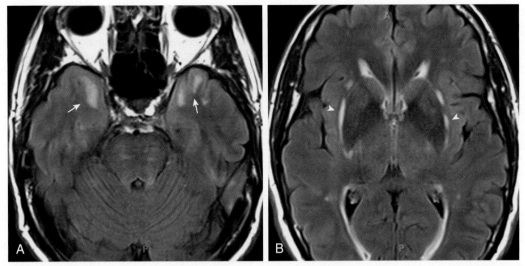

FIGURE 6-15 Cerebral autosomal dominant arteriopathy with subcortical infarcts and leukoencephalopathy (CADASIL) characterized by capsule and confluent cotemporal cephalopathy. **A,** Axial fluid-attenuated inversion recovery (FLAIR) showing diffuse high intensity in the white matter of the anterior temporal lobes *(arrows)*. **B,** Higher image with involvement of the striatocapsular region *(arrowheads)*.

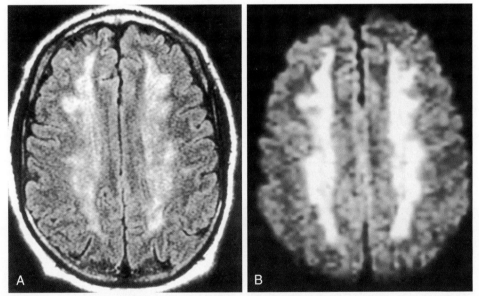

FIGURE 6-16 Anoxic encephalopathy affecting all association areas equally and inducing apparent diffusion coefficient (ADC) attenuation. **A,** Fluid-attenuated inversion recovery (FLAIR) image with diffuse high intensity throughout the centrum semiovale. **B,** Corresponding diffusion-weighted imaging scan with high intensity (ADC was low) representing cytotoxic edema.

typically involves the subcortical U-fibers where lacunar infarcts can be seen and is relatively selective for the anterior inferior temporal lobes and inferior frontal lobes as well as subinsular regions (Fig. 6-15). External capsule involvement is also a characteristic feature. The frequency of MR lesions increases dramatically with the age of the patient. Both symptomatic and asymptomatic (but with *Notch 3* mutation) patients have MR lesions. Diffusion tensor measurements have revealed increased diffusivity and concomitant loss of diffusion anisotropy in CADASIL patients that can be correlated with clinical impairment. This diffusional abnormality is hypothesized to be the result of neuronal loss and demyelination.

Postanoxic Encephalopathy

Postanoxic encephalopathy occurs after an anoxic episode severe enough to produce coma. The patient recovers in 24 to 48 hours and then precipitously declines within a 2-week period, progressing from confusion to coma and death. This is most likely an allergic demyelination caused by exposure to a myelin antigen during the hypoxic period. Pathologic changes are most prominent in the white matter with demyelination and necrosis.

MR demonstrates high signal on T2WI scans throughout the white matter, particularly involving the corpus callosum, subcortical U-fibers, and internal/external capsules. Low intensity on T2WI scans has also been observed in the thalamus and putamen. Diffusion images are positive (Fig. 6-16).

Carbon monoxide exposure can produce a similar MR and clinical picture. In carbon monoxide poisoning, however, there is a propensity for symmetric globus pallidus and/or hippocampal Ammon's horns lesions. The differential diagnosis of bilateral basal ganglionic lesions is provided in Box 6-6.

BOX 6-6   Bilateral Basal Ganglionic Lesions

Acquired immunodeficiency syndrome (AIDS)
Aminoacidopathies
Calcium phosphate dysmetabolism
Canavan disease
Cockayne syndrome
Fahr disease
Pantothenate kinase-associated neurodegeneration (PKAN)
Hepatic encephalopathy
Huntington disease
Hypoglycemia
Hypothyroidism
Infarction
Ischemia
    Artery of Percheron
    Vein of Galen thrombosis
Leigh disease
Multisystem atrophy
Neoplasms (lymphoma, multicentric glioma)
Neurofibromatosis
TORCH infections
Toxins
    Lead intoxication
    Methanol
    Mercury
    Organophosphates
    Toluene
    Methanol poisoning (putamen)
Wilson disease

TORCH, Toxoplasmosis, other infections, rubella, cytomegalovirus, herpes simplex.

Posterior Reversible Encephalopathy Syndrome

Posterior reversible encephalopathy syndrome (PRES; Box 6-7) is controversial not only in name, but also in etiology. Patients have a variety of symptoms including consciousness impairment, headache, seizures, confusion, drowsiness, nausea, vomiting, and visual disturbances. Most of the patients have a recent history of elevated blood pressure. Conditions associated with this syndrome include malignant hypertension, toxemia of pregnancy, renal disease, immunosuppressive drugs (cyclosporine, tacrolimus, ARA-A and ARA-C), lupus, antiphospholipid antibody syndrome, hepatorenal syndrome, organ transplantation, dialysis, cocaine use, and porphyria.

There are a diverse group of etiologies that produce high-intensity abnormality on T2WI scans primarily in the cortex and subcortical white matter of the occipital and parietal regions (Fig. 6-17). The abnormality is usually reversible, and can extend into other regions including the temporal and frontal lobes, pons, and cerebellum. Diffusion imaging can distinguish reversible changes from early ischemia. Enhancement may or

---

BOX 6-7   Posterior Reversible Encephalopathy Syndrome

Hypertension
Cyclosporine
ARA-A/ARA-C
FK 506 (tacrolimus)
DMSO
Eclampsia/Preeclampsia
SLE, cryoglobulinemia, hemolytic uremic syndrome
Cisplatinum

DMSO, Dimethyl sulfoxide; *SLE,* systemic lupus erythematosus.

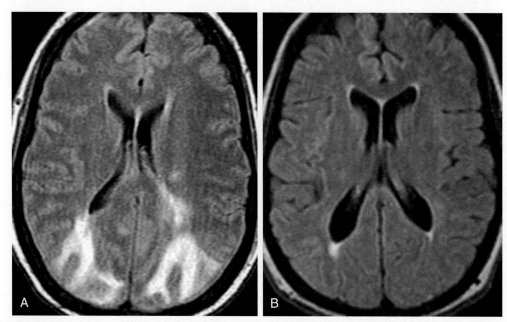

FIGURE 6-17 Posterior reversible encephalopathy syndrome pattern predominantly posteriorly and peripherally from preeclampsia purging postpartum. **A,** Before childbirth the patient had hypertension, headaches, and drowsiness associated with high signal bilaterally in the white matter. **B,** Two weeks after delivery, the abnormality had largely resolved and patient was normotensive and asymptomatic.

may not be present and the pattern of enhancement is variable.

The most likely underlying problem is the inability of the posterior circulation (sparsely innervated by sympathetic nerves) to autoregulate in response to acute changes in blood pressure. This leads to hyperperfusion and blood-brain barrier disruption with escape of fluid from the intravascular compartment into the interstitium (subcortical edema) but without infarction of the brain (classical explanation). The common substrate is elevated blood pressure from baseline.

In preeclampsia-eclampsia there is data to suggest that the brain edema is associated with abnormalities in red blood cell morphology and elevated lactate dehydrogenase and not with hypertension level. These findings signify microangiopathic hemolysis resulting from endothelial damage. The cause of the endothelial damage is thought to be from circulating endothelial toxins or antibodies against the endothelium. Indeed, in many of the conditions listed in Box 6-5, there is evidence of endothelial dysfunction or damage.

An entity that is related to PRES is reversible cerebral vasoconstriction syndrome (RCVS). This may have similar features to PRES on MR, but occurs as a sudden illness with recurrent "thunderclap headaches." It is manifested by nonaneurysmal subarachnoid hemorrhage over the convexities with vascular narrowings and dilatations in peripheral and/or central cerebral vessels (Fig. 6-18). It may not have the same underlying predisposing conditions such as pregnancies, illnesses and drugs as PRES, but the primary finding remains posterior white matter abnormal signal, generally without infarction. Enhancement

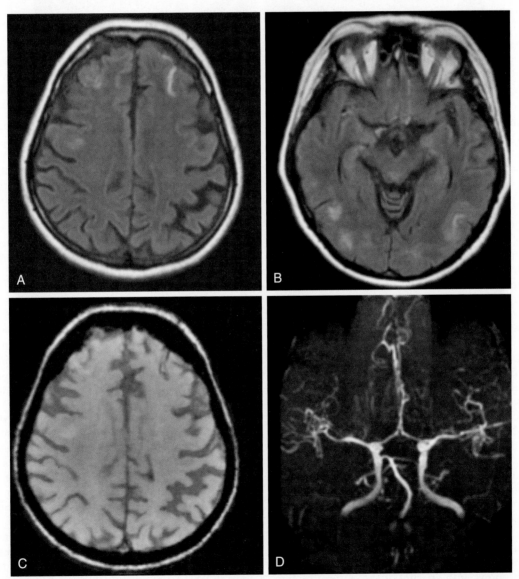

FIGURE 6-18 Reversible cerebral vasoconstriction syndrome resembling posterior reversible encephalopathy syndrome (reversible, rheologic, reviviscent). Thirty-eight-year-old female patient with reversible vasoconstriction syndrome (postpartum) and focal left frontal subarachnoid hemorrhage (fluid-attenuated inversion recovery **[A and B]**; susceptibility-weighted imaging **[C]**), cortical and subcortical edema **(A and B).** Time-of-flight magnetic resonance angiography **(D)** reveals the typical narrowing and irregularities of the intracranial vessels. (Compliments of Jennifer Linn, MD.)

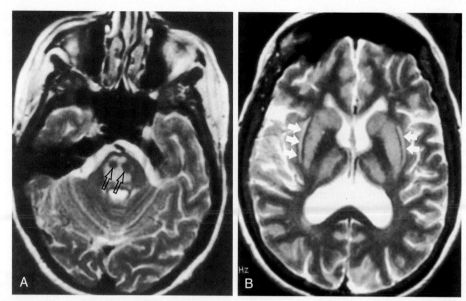

FIGURE 6-19 Osmotic insult inducing abnormal appearance in encephalon and other essential nuclei. **A,** T2-weighted imaging (T2WI) with high intensity in the pons with sparing of the periphery and the descending corticospinal tracts *(open arrows)* from osmotic demyelination aka central pontine myelinolysis. **B,** T2WI with extrapontine involvement of the caudate nuclei, putamen, thalami, and claustrum *(arrows)*.

is more common than in PRES (10%) but is still largely uncommon. Diffuse, uniform wall thickening in RCVS with negligible-to-mild vessel wall enhancement which resolves over weeks to months is typical.

Toxic/Metabolic Derangements

Osmotic Demyelination or Central Pontine Myelinolysis

Central pontine myelinolysis is a demyelinating disorder recognized in alcoholic, debilitated, or malnourished persons most typically in the setting of rapid correction of hyponatremia. It has also been associated with chronic renal failure, liver disease, diabetes mellitus, dysequilibrium syndrome (complication to rapid dialysis), and the syndrome of inappropriate antidiuretic hormone secretion. In the pediatric population, it has been associated with orthotopic liver transplantation, acute myelogenous leukemia, Hodgkin disease, Wilson disease, and craniopharyngioma. The usual scenario is that the patient is admitted with a low serum sodium level, which is enthusiastically corrected by the overzealous intern in mid-July looking to be rewarded by the attending staff for prompt attention to the numbers. Unfortunately, the patient deteriorates subacutely (usually within a few days) with extrapyramidal motor symptoms, pseudobulbar palsy, quadriparesis, and coma. The intern decides to go into radiology. This condition can be fatal, although with increasing awareness of this diagnosis, patients may survive, often with significant neurologic impairment.

On imaging, there is classically high T2 signal intensity within the pons with sparing of the outermost tegmentum and a peripheral rim of ventral pontine tissue. Many times, two central symmetric isointense structures encircled by the abnormal high intensity may be observed. These are the descending corticospinal tracts, which may be spared. What is most interesting is that the disease may also or exclusively involve extrapontine

structures including the thalamus; putamen; caudate nuclei; internal, external, and extreme capsules; claustrum; amygdala; and cerebellum (Fig. 6-19). The deep gray matter structures can be involved bilaterally out to the insular cortex. The cortex and subcortical regions may also be involved. The lesions are not associated with mass effect or enhancement. Demyelination is noted without an inflammatory response, with sparing of the blood vessels, most nerve cells, and axons. The propensity of extrapontine involvement suggests that a more appropriate term, osmotic demyelination, be used to describe this process.

Alcoholism

Alcoholic patients and others with nutritional deficiencies may sustain demyelination of the corpus callosum. This may be considered a variant of extrapontine myelinolysis and has the eponym of Marchiafava-Bignami disease (MBD). (The Italian wine industry took offense to this eponym and so Fig. 6-20 is from France.) It should be considered in cases of sudden encephalopathy in alcoholic patients. The acute form presents with seizures, neurologic dysfunction (muteness, diffuse muscular hypertonia with dysphagia), and coma and is usually fatal. A subacute form displays the sudden onset of dementia progressing to the chronic vegetative state. The chronic form is distinguished by progressive dementia and a disconnection syndrome.

On imaging, it is most commonly characterized by demyelination and spotty focal, mostly posterior necrosis of the corpus callosum, described as MBD, type B. However, in Type A the pyramidal tract and entire corpus callosum is involved. It may also be extensive and involve other brain regions. Hemorrhage and hemosiderin deposition have also been reported. MBD is depicted as a low-signal abnormality on T1WI (particularly in the sagittal plane; edema and cystic changes) and high signal on T2WI. Diffusion images are positive in the acute form. The acute

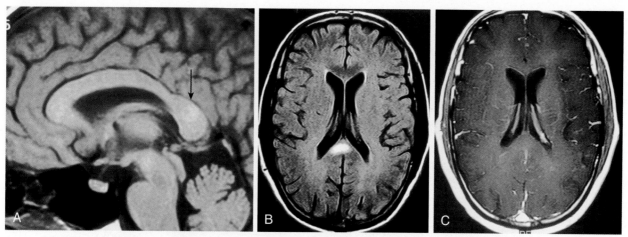

FIGURE 6-20 Marchiafava-Bignami. **A,** T1 sagittal image with swollen splenium of corpus callosum *(arrow).* **B,** Axial fluid-attenuated inversion recovery (FLAIR) image with high intensity in the splenium. **C,** Postgadolinium T1-weighted image showing no enhancement in the splenium.

BOX 6-8   Drugs, Toxins, and Conditions Associated with White Matter Abnormalities

DRUGS

Drug abuse
    Methamphetamine
    Cocaine
    Heroin (inhalational)
Isoniazid
Metronidazole (Flagyl)
Chemotherapeutic agents
    Actinomycin D
    Cisplatinum
    Cytosine arabinoside
    Adenine arabinoside
    Cyclosporine
    Methotrexate

OTHER

Hypertensive encephalopathy
Eclampsia
Radiation

---

variety may reveal swelling and enhancement, whereas the chronic variety demonstrates atrophy.

Wernicke Encephalopathy

This results from thiamine deficiency and could also be associated with alcoholism, starvation, postbariatric surgery, and anorexia/bulimia. The hallmark is severe memory impairment with anterograde amnesia. There is atrophy of the mammillary bodies and there may be high intensity in the mamillothalamic tracts. In the acute setting this may show restricted diffusion and enhancement of the periaqueductal gray matter, mammillary bodies, and medial thalamic nuclei. The white matter disease is not as dramatic as the central deep gray matter findings. If the entity is not recognized, the deficits may become permanent.

Toxic Lesions

Toxins can produce white matter abnormalities (Box 6-8). Methanol poisoning causes optic nerve atrophy with

necrosis of the putamen and subcortical white matter. The caudate and hypothalamus are less commonly involved. Methanol can produce hemorrhage as well as peripheral white matter lesions. Ethylene glycol toxicity affects the thalamus and pons. Toluene effects include atrophy of the cerebrum, corpus callosum, and cerebellar vermis. It is associated with high signal on T2WI in the white matter, poor gray-white differentiation, and low signal in the basal ganglia and thalamus. Carbon monoxide poisoning produces low or high intensity on the T1WI and high intensity on T2WI in the globus pallidus. It also affects the white matter and hippocampus. These may demonstrate restricted diffusion in the acute exposure.

Wallerian Degeneration

Wallerian degeneration can result in high signal abnormalities and atrophy in the white matter and is defined as antegrade destruction of axons and their myelin sheaths secondary to injury of the proximal axon or cell body. It has many causes including infarction, hemorrhage, white matter disease, trauma, MS, and neoplasia. On MR, it is relatively easy to demonstrate wallerian degeneration by noting high intensity on proton density–weighted imaging (PDWI)/T2WI that follows a particular white matter pathway, most commonly the corticospinal tract after a middle cerebral artery (MCA) stroke or surgical resection affecting the motor strip (Fig. 6-21).

Hypertrophic Olivary Degeneration

A curious entity, hypertrophic olivary degeneration, may lead to enlargement of the medullary olive in association with ipsilateral red nucleus and central tegmental tract lesions or contralateral dentate nucleus pathology (Fig. 6-22). The enlarged olive may last years…great for long-term martini drinkers.

Radiation Changes

MR is particularly sensitive to radiation changes in the brain. Commonly found in the irradiated fields are high signal abnormalities on T2WI and atrophy, both of which conform to the radiation portal.

Radiation alone can produce demyelination. It can also generate, usually when associated with chemotherapy

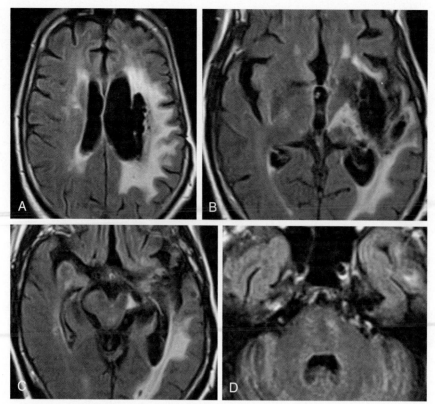

FIGURE 6-21 Wallerian degeneration after left hemispheric infarction. **A-D,** Serial scans from superior to inferior shows left middle cerebral artery stroke with high signal in corona radiata **(A)**, capsule **(B)**, cerebral peduncle **(C)**, and central pons **(D).**

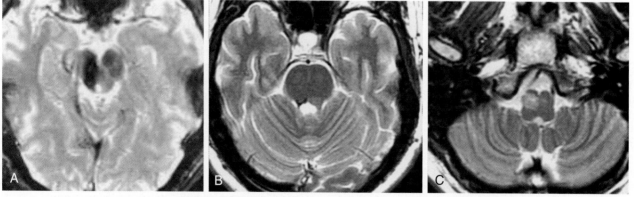

FIGURE 6-22 Heme heralds hypertrophy in hemi-medulla with hypertrophic olivary degeneration. **A,** Hemorrhage in the red nucleus affects the Guillain Mollaret triangle. **B,** Some hemosiderin is seen in the central tegmental tract. **C,** This leads to hypertrophic olivary degeneration in the ipsilateral medulla.

(particularly methotrexate), a mineralizing microangiopathy. This is best characterized on computed tomography (CT), where calcification is identified as high density in the basal ganglionic region, the anterior cerebral–middle cerebral and middle cerebral–posterior cerebral watershed zones, and the cerebellum. On MR, this calcification usually displays low intensity on T2WI, particularly on T2*WI. Other radiation-induced findings include hemosiderin deposition, telangiectasias/cavernomas, and BBB disruption.

Disseminated Necrotizing Leukoencephalopathy

Disseminated necrotizing leukoencephalopathy (DNL) is a demyelinating disease first described in children undergoing cranial and/or spinal radiation in combination with intrathecal methotrexate for leukemia. Subsequently, it has been reported after combination radiation and chemotherapy, most commonly in adult leukemia. In these patients, a progressive neurologic disorder develops. It is characterized by decreasing mental status and neurologic changes including seizures, usually progressing to coma and death. Associated with this neurologic syndrome is the appearance of marked low density in the white matter on CT. Pathologic findings of DNL (Fig. 6-23) include axonal swelling, multifocal demyelination, coagulation necrosis, and gliosis. These changes have a predilection for the periventricular region and centrum

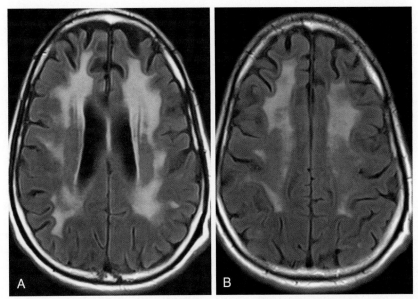

FIGURE 6-23 Disseminated necrotizing leukoencephalopathy denotes diffuse dendritic damage done. Note confluent white matter disease involving the periventricular zone **(A)** as well as the deep **(B)** white matter. The disease begins centrally and progresses peripherally especially after intrathecal methotrexate therapy.

semiovale with sparing of the U-fibers. On MR, diffuse regions of high intensity on T2WI may be visualized throughout the white matter. There have been reports of enhancement in these lesions.

Chemotherapeutic white matter injury has been reported more frequently in children. It can cause white matter changes that are indistinguishable from radiation. Transient white matter high signal abnormalities that are not precursors of DNL may also develop in children undergoing chemotherapy for acute lymphocytic leukemia, but may resolve with cessation of chemotherapy. This is well described with methotrexate therapy. A special form of this injury occurs when chemotherapy is instilled directly into the brain by an indwelling ventricular catheter, resulting in the production of focal necrosis. In such instances, high intensity on T2WI MR or low density with focal enhancement can be seen adjacent to the catheter. Thus, chemotherapy and/or radiation therapy can result in white matter changes characterized by high intensity on T2WI. Focal mass lesions produced by radiation and/or chemotherapy may enhance.

DYSMYELINATING DISEASES

Dysmyelinating diseases are much rarer than the previously described conditions. How rare are they? As rare as the truth from your used car dealer. Their appearance, particularly on MR, is usually much more bizarre. One often notes diffuse high intensity on T2WI throughout the white matter.

Metachromatic Leukodystrophy

Metachromatic leukodystrophy (MLD) is the most common of these uncommon dysmyelinating diseases. It is an autosomal recessive chromosome 22 *ARSA* gene disease that results from a deficiency of the enzyme arylsulfatase A, which hydrolyzes sulfatides to cerebrosides, resulting in accumulation of ceramide sulfatide within macrophages and Schwann cells. Metachromatic-staining lipid granules (sulfatides) are found within neurons, and diffuse myelin loss is observed in central and peripheral nerves. There is symmetrical demyelination with characteristic sparing of the subcortical U-fibers.

Several varieties of MLD are characterized by age at onset (late infantile, juvenile, and adult), which probably relates to the degree of enzyme deficiency. The diagnosis is confirmed by documenting decreased arylsulfatase activity in peripheral leukocytes or urine. Clinical findings include peripheral neuropathy, psychosis, hallucinations, delusions, impaired gait, and spasticity. There is symmetric diffuse high signal on T2WI scans throughout the white matter and the cerebellum (Fig. 6-24). Enhancement is the exception. In the adult form, multifocal lesions in the white matter with a propensity for the frontal lobes, and atrophy with ventricular dilatation have been noted.

Adrenoleukodystrophy

Adrenoleukodystrophy is a x-linked (*ALD* gene chromosome Xq28) or autosomal recessive (neonatal) peroxisomal disorder associated with cerebral degeneration and adrenal cortical insufficiency (which may be clinically inapparent). It is caused by a deficiency of acyl-CoA synthetase, preventing peroxisomal breakdown of very long-chain fatty acids with accumulation of these fatty acids in the white matter, adrenal cortex, as well as in plasma and red blood cells. This disease possesses some characteristics of a demyelinating disease with prominent perivascular inflammation and extensive demyelination, although it is classified as a dysmyelinating disease. Impaired hearing and vision, abnormal skin pigmentation, hypotonia, difficulty in swallowing, behavioral difficulties, and seizures are the most common clinical manifestations of the disease.

There have been several phenotypes of this disease and variations in the imaging characteristics of

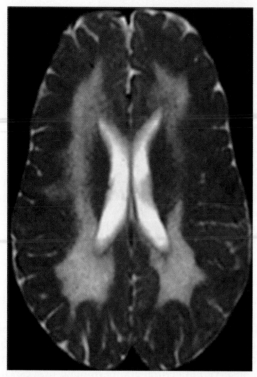

FIGURE 6-24 Markedly marred myelin mainly in the middle in metachromatic leukodystrophy (MLD). T2-weighted imaging revealing widespread symmetrical high intensity throughout the white matter found centrally with sparing of the U-fibers. Note the tiny black dots perforating the high signal white matter, which is termed the "tigroid stripes" of MLD.

adrenoleukodystrophy. One type starts in the parietooccipital region and progresses anteriorly to involve the temporal and frontal lobes together with the corpus callosum. The disease can also progress from anterior to posterior. The advancing edge of the lesion represents the region of active demyelination and enhances, whereas the nonenhanced regions are gliotic (Fig. 6-25). Patients have been described with enhancement of major white matter tracts including the corticospinal, spinothalamic, visual (including the lateral geniculate body), auditory, and dentatorubral pathways. Other findings include calcifications in the trigone or around the frontal horns, mass effect in the advancing region of demyelination, and isolated frontal lobe involvement. On MR, there may be relative sparing of the subcortical U-fibers. Spinal cord disease can be seen with degeneration of the entire length of the corticospinal tracts and cord atrophy. Stem cell transplantation is an effective treatment.

Alexander Disease

Alexander disease is a progressive leukodystrophy characterized by fibrinoid degeneration of astrocytes and diffuse Rosenthal fibers in the subependymal, subpial, and perivascular regions. It has been divided into three clinical subgroups. The infantile group has seizures, spasticity, psychomotor retardation, and macrocephaly associated with extensive demyelination. The juvenile group (7 to 14 years old) demonstrates progressive bulbar symptoms with spasticity, and the adult group may have similar clinical appearance to MS or may be asymptomatic. The latter two groups have preservation of neurons and less myelin loss.

On CT, hyperdensity has been described in the caudate nucleus and diffuse hypodensity in the white matter and in the internal and external capsules. Enhancement has been observed early in the course of the disease. T2WI signal

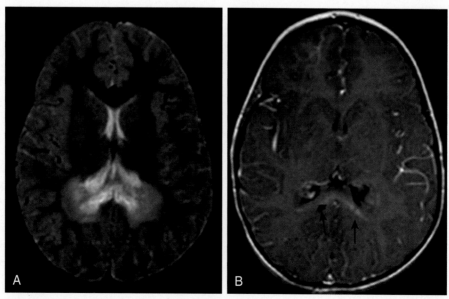

FIGURE 6-25 Adrenoleukodystrophy affecting occipital areas advancing anteriorly along an enhancing edge. **A,** T2-weighted image showing high intensity in the splenium of the corpus callosum. **B,** Enhancement is seen usually at the advancing edge of the white matter disease *(arrows).*

intensity changes have a frontal predominance. Brain stem atrophy and decreased intensity in the basal ganglia have also been reported.

Canavan Disease

Canavan disease (spongiform degeneration) is an autosomal recessive leukodystrophy resulting from a deficiency in the enzyme *N*-acetylaspartoacylase from the *ASPA* gene on chromosome 17. The age at disease onset is from 2 to 4 months. Hallmarks including a large brain (macrocephaly), hypotonia, and failure to thrive are followed by seizures, optic atrophy, and spasticity. Death usually occurs by the age of 5 years.

Proton spectroscopy has shown high levels of NAA, which is synthesized in the mitochondria, and may be the carrier for acetyl groups across the mitochondrial membrane. *N*-acetylaspartoacylase cleaves NAA into acetate and aspartate in the cytosol, and deficiencies in this enzyme can interfere with the supply of acetate for fatty acid synthesis and myelination. This is one of the few diseases that spectroscopy has made a contribution to and thus is included for those minutiae-hungry persons.

On routine MR imaging, the deep gray matter and adjacent subcortical U-fiber white matter are most often involved. Symmetric involvement of white matter (high signal on T2WI) and ventriculomegaly have been described (Fig. 6-26).

Alexander and Canavan diseases should also be considered in the differential diagnosis of macrocephaly. Megalencephalic leukoencephalopathy with subcortical cysts, an entity associated with the *MLC1* gene of chromosome 22, presents in the first year of life and, unlike Alexander and Canavan disease, preserves cognition and demonstrates cysts in the anterior temporal and frontoparietal lobes.

Krabbe Disease

Krabbe disease (globoid cell leukodystrophy) is an obscure entity (1 in 100,000 to 200,000 live births). Krabbe disease is caused by an autosomal recessive *GALC* gene chromosome 14 mutation leading to a deficiency of a lysosomal enzyme (beta-galactocerebrosidase) normally present in white and gray matter. Clinically, the leukodystrophy has been classified into an early infantile, late infantile (early childhood), late childhood, and even possibly an adolescent or adult onset group with galactosylceramide deficiency. The infantile form is the most frequent with age at onset within the first 6 months of life. These patients have seizures, hypotonia followed by spasticity, and progressive psychomotor retardation. It is characterized by globoid cell infiltration and demyelination.

Increased density on CT (some of which represents calcification) in the basal ganglia, thalami, corona radiata, and cerebellar cortex, and hyperintensity on T2WI throughout the cerebral (parietal lobes in particular) and cerebellar white matter have been noted. Late in the disease, atrophy and high density in the corona radiata have been identified. Enlargement of the optic nerves in association with patchy high intensity in the white matter on T2WI has been reported. The differential diagnosis of bilateral symmetric thalamic lesions is provided in Box 6-9. Symmetric linear signal abnormalities on T1WI and T2WI have been reported from the periventricular to subcortical regions in the centrum semiovale. This represents demyelination

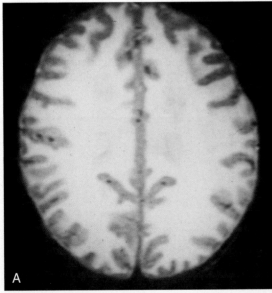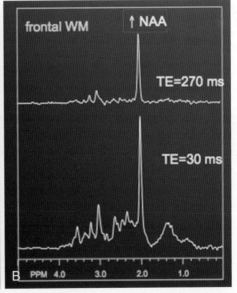

FIGURE 6-26 Megalencephaly and myelinopathy and magnetic resonance spectroscopy mark this malady. **A,** In this patient with Canavan disease virtually all the white matter is abnormal on this T2-weighted image. This patient, and the authors after they made the correct diagnosis, had a big head (megalencephaly). Well, it beats microcephaly, our faithful readers.... **B,** Note the elevated *N*-acetyl aspartate (NAA) peak at low and high TE values in Canavan. Compare with how low choline and creatine are (peaks at 3.2 and 3.0 ppm, respectively). *PPM,* Parts per million; *WM,* white matter.

with axon preservation associated with globoid cell infiltration and gliosis.

Sudanophilic Leukodystrophy/Pelizaeus-Merzbacher Disease

Sudanophilic leukodystrophy is a degenerative myelin disorder distinguished by accumulation of sudanophilic material in the brain. It has a variable age at onset, usually within the first months of life (but can be seen from neonatal to late infancy), with slow progression. The disease is divided into three groups, classic (slowly progressive with death in young adulthood), connatal (more severe, with death in the first decade of life), and transitional (less severe than connatal, with average age at death of 8 years). These differ in their onset and clinical severity. There appears to be failure of myelin maturation. This may be related to a deficiency of proteolipid apoprotein and reduced amount of other myelin proteins (necessary for oligodendrocyte differentiation). Pelizaeus-Merzbacher disease may present as an x-linked form of sudanophilic leukodystrophy because of mutation of the proteolipid protein *(PLP)* gene (Fig. 6-27). Clinical findings include bizarre eye

movements/rotatory nystagmus, head shaking, psychomotor retardation, hypotonia, and cerebellar ataxia.

On CT, atrophy, low density in the white matter, and pinpoint periventricular calcifications have been reported. MR may be normal early on, but later they reveal cerebral, cerebellar, brain stem, and upper cervical cord atrophy. On MR, diffuse symmetric increased signal in the white matter (more or less complete lack of myelination/hypomyelination) associated with low intensity on T2WI in the lentiform nucleus, substantia nigra, dentate nuclei, and thalamus has been described and is probably related to increased iron deposition. A "tigroid" pattern has been noted that corresponds to histopathologic findings in which small regions of normal neurons and myelin sheaths are scattered in the diffusely abnormal white matter. The corpus callosum is atrophic and undulating.

Vanishing White Matter Disease/Childhood Ataxia with Central Hypomyelination

Vanishing white matter (VWM) disease/childhood ataxia with central hypomyelination (CACH) is a recently described disease seen in children and teenagers and is believed to be a disorder related to the guanine nucleotide exchange factor for the eukaryotic initiation factor 2B. The clinical findings consist of prominent ataxia, spasticity, optic atrophy, and relatively preserved mental capabilities. The disease follows a chronic progressive course with decline being associated with episodes of minor infections and head trauma. The cortex is relatively normal. However, beneath the cortex the white matter is largely destroyed with the exception of some sparing of the U-fibers. There is cystic degeneration from the frontal to the occipital region with the temporal lobe being least involved. In noncystic regions of the brain, there is diffuse and severe myelin loss. MR demonstrates regions of white matter that have a signal intensity similar to CSF on all pulse sequences, and the brain has a swollen appearance with cystic degeneration around the periventricular region. Cerebellar atrophy is present and may be severe, particularly in the vermis. Symmetric high intensity in the pontine tegmentum, cerebellar white matter and central tegmental tract on T2WI has also been observed.

BOX 6-9   Bilateral Symmetric Thalamic Lesions

Acute febrile encephalopathy
Artery of Percheron infarction (single trunk that supplies paramedian thalamic arteries)
Cytoplasmically inherited striatal degeneration
Deep venous occlusion, infarction
Glioma
Hypoxic ischemic encephalopathy
Krabbe disease
Kernicterus
Leigh disease
Measles encephalitis
Molybdenum deficiency
Near-drowning
Reye syndrome
Sandhoff disease (GM$_2$ gangliosidosis)
Wernicke-Korsakoff syndrome
Wilson disease

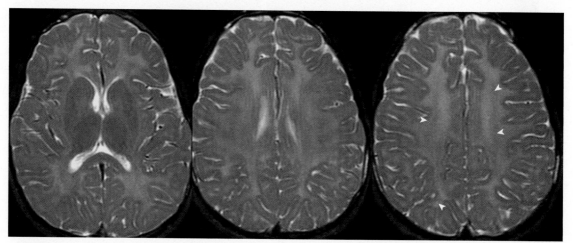

FIGURE 6-27 Paucity of proteolipids produces poor myelin proteins. There is absent white matter myelination and pinpoint dots of low intensity *(arrowheads)* in the white matter yielding another example of a disease characterized by tigroid appearance. Normally white matter would be dark on T2-weighted imaging.

MRS findings included decreased NAA, normal or slightly elevated Cho, and the presence of lactate.

Vacuolating Megalencephalic Leukoencephalopathy with Subcortical Cysts

This is also referred to as van der Knaap syndrome and is due to the *MLC* gene isolated to chromosome 22. Clinical manifestations are seizures, ataxia, macrocephaly, spasticity, and mental retardation presenting in the first year of life. MRI shows cysts at the temporal tips, which can get extreme in size, and frontoparietal regions with white matter disease, sparing U-fibers and gray matter (Fig. 6-28). Rarely there is absence of macrocephaly.

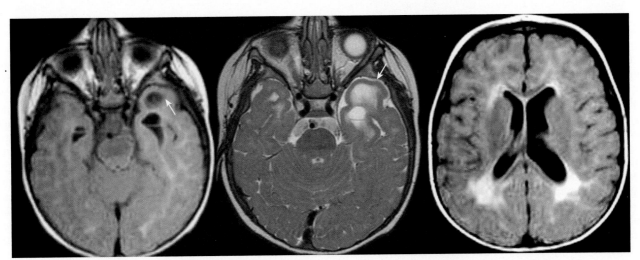

FIGURE 6-28 Subcortical cysts coincident with craniomegaly. Note the temporal lobe cysts *(arrow)*, the white matter dysmyelination, and the megalencephaly that characterize vacuolating megalencephalic leukodystrophy with subcortical cysts of van der Knaap syndrome.

Chapter 7

Neurodegenerative Diseases and Hydrocephalus

The debate over whether a patient has atrophy or hydrocephalus has consumed an enormous number of pages in the neuroradiologic literature and whole sessions of national meetings. Why the fuss? Well, the implications as far as prognosis and treatment are vastly different between the two, so accurate distinction is essential!

Atrophy reflects the loss of brain tissue, be it cortical, subcortical, or deep. With the loss of cell bodies in the cortex (gray matter), axonal wallerian degeneration occurs with white matter atrophy or demyelination. Selective atrophy of the white matter may also occur with perivascular small-vessel insults. Remember that certain drugs (steroids) or metabolic states (dehydration, alcoholism) may cause an appearance of increased cerebrospinal fluid (CSF) spaces (suggesting atrophy) but are potentially reversible (Box 7-1).

BOX 7-1   Causes of Reversible Diminished Brain Volume

Alcoholism (limited reversibility)
Chemotherapy
Dehydration/shock
Marijuana
Radiation therapy
Starvation/anorexia
Steroid use

There is a spectrum of normal brain parenchymal volume for any age, and therefore, until you have a good sense of what the normal brain looks like at all ages, be hesitant to label a brain "atrophic." It is preferable to use the terms age-related changes or parenchymal volume appropriate for age when the findings are within your range of normal. On the other hand, one might say "parenchymal volume loss greater than expected for age" in those cases where you want to raise the issue to a level of concern. Remember also that men have more prominent sulci at most ages than women.

Hydrocephalus reflects expansion of the ventricular system from increased intraventricular pressure, which is in most cases caused by abnormal CSF hydrostatic mechanics. Hydrocephalus may be due to three presumed causes: (1) overproduction of CSF; (2) obstruction at the ventricular outlet level; or (3) obstruction at the arachnoid villi level, leading to poor resorption of CSF back into the intravascular space. Although atrophy and hydrocephalus often share the finding of dilatation of the ventricular system, the prognostic and therapeutic implications of the two are markedly different. Whereas generally there is no treatment for atrophy, hydrocephalus can often be treated with well-placed ventricular or subarachnoid space shunts and/or removal of the obstructing or overproducing lesion.

Computed tomographic (CT) or magnetic resonance imaging (MRI) findings that suggest hydrocephalus over atrophy are summarized in Table 7-1. The presence of

---

TABLE 7-1   Differentiation of Hydrocephalus and Atrophy

Characteristic	Hydrocephalus	Atrophy
Temporal horns	Enlarged	Normal except in Alzheimer disease
Third ventricle	Convex Distended anterior recesses	Concave Normal anterior recesses
Fourth ventricle	Normal or enlarged	Normal except with cerebellar atrophy
Ventricular angle of frontal horns on axial scan	More acute	More obtuse
Mamillo-pontine distance	<1 cm	>1 cm
Corpus callosum	Thin, distended, rounded elevation Increased distance between corpus callosum and fornix	Normal or atrophied Normal fornix-corpus callosum distance
Transependymal migration of cerebrospinal fluid	Present acutely	Absent
Sulci	Flattened	Enlarged out of proportion to age
Aqueductal flow void	Accentuated in normal-pressure hydrocephalus	Normal
Choroidal-hippocampal fissures	Normal to mildly enlarged	Markedly enlarged in Alzheimer disease
Sella	Erosion of floor and ballooning of sella	Normal

dilatation of the chiasmatic, infundibular and suprapineal recesses of the third ventricle, rounding of the frontal horns, convexity to the third ventricle, expansion of the temporal horns, effacement of sulci, enlargement of ventricles out of proportion to sulcal dilatation, periventricular smooth high signal representing transependymal CSF flow (best seen on fluid-attenuated inversion recovery [FLAIR]), marked accentuation of the aqueductal signal void, narrowing of mamillo-pontine distances, and associated papilledema are indicative of hydrocephalus (Fig. 7-1).

The corpus callosum may be compressed against the rigid falx in long-standing hydrocephalus. Clefts of abnormal signal in the body of the corpus callosum, scalloping of its dorsal surface, and tethering of pericallosal vessels can be seen in cases of hydrocephalus due to aqueductal

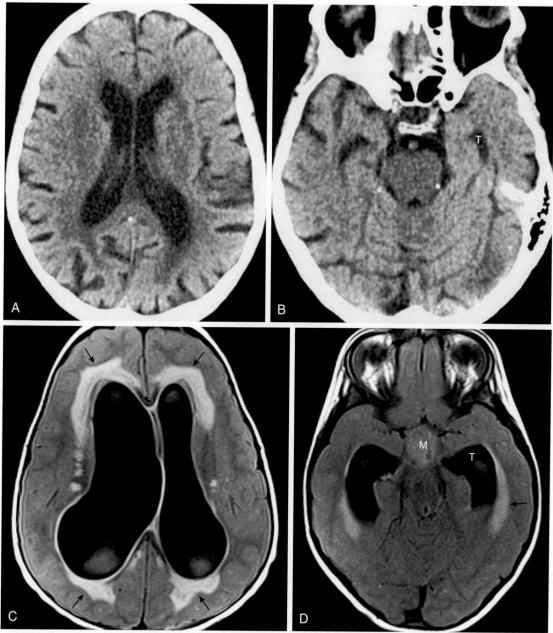

FIGURE 7-1 Atrophy versus hydrocephalus? **A,** Axial computed tomography image from an 85-year-old patient shows moderate enlargement of the lateral ventricles, with commensurate moderate dilatation of the cerebral sulci, indicating atrophy. Central white matter ischemic injury and right caudate lacune coexist. **B,** In this same patient, the temporal horns (T) are relatively nondilated compared with the remainder of the lateral ventricles. **C,** Axial fluid-attenuated inversion recovery (FLAIR) image in a different patient shows marked enlargement of the lateral ventricles but with effacement of the cerebral sulci, indicating hydrocephalus. Note the rounded margins of the frontal horns. There is increased signal *(arrows)* around the margins of the lateral ventricles, indicating transependymal flow of cerebrospinal fluid (CSF). **D,** Axial FLAIR image in same patient shows marked enlargement of the temporal horns (T) further supporting diagnosis of hydrocephalus. Transependymal flow of CSF is again demonstrated. The cause of the hydrocephalus is a large mass (M) within the third ventricle with suprasellar extension.

stenosis possibly owing to the impact of the towering corpus callosum against the falx. The damage may be due to arterial or venous vascular compromise.

HYDROCEPHALUS

Overproduction of Cerebrospinal Fluid

It has been theorized that patients with choroid plexus papillomas and choroid plexus carcinomas have hydrocephalus based on the overproduction of CSF. Increasingly, this hypothesis has come into question because it is believed that some cases of hydrocephalus may in fact be due to obstruction of the arachnoid villi or other CSF channels secondary to adhesions from tumoral hemorrhage, high protein levels, or intraventricular debris. This is particularly true with fourth ventricular choroid plexus papillomas, which generally tend to obstruct the sites of egress of the CSF in the foramina of Luschka and Magendie. In the cases of lateral ventricle choroid plexus papillomas (particularly in the pediatric population), the overproduction of CSF may be the cause of hydrocephalus.

Noncommunicating and Communicating Hydrocephalus

Obstructive hydrocephalus may be separated into non-communicating and communicating forms. Noncommunicating forms are due to abnormalities at the ventricular outflow levels (Box 7-2). Communicating hydrocephalus is due to abnormalities at the level of the arachnoid villi or blockage at the incisura of the foramen magnum. Noncommunicating types often need brain surgery to remove offending agents; communicating types respond best to shunts.

Causes of Obstructive Hydrocephalus

Colloid Cyst

The classic cause of obstruction at the foramina of Monro is the colloid cyst (Fig. 7-2). This is typically located in the anterior region of the third ventricle. On unenhanced CT, the lesion is high in density. Magnetic resonance (MR) often shows a lesion that is high intensity on T1-weighted imaging (T1WI) and T2WI. The signal of colloid cysts is variable, depending on the protein concentration, presence of hemorrhage, and other paramagnetic ion effects.

Congenital Aqueductal Stenosis

The cerebral aqueduct is one of the narrowest channels through which the CSF in the ventricles must flow. Congenital aqueductal stenosis is just one of the obstructers of the aqueduct. This is most commonly an X-linked recessive disorder seen in early childhood, although it can present at any age. Children typically have enlarging head circumferences and dilatation of the lateral and third ventricles, but with a normal-appearing fourth ventricle. Aqueductal webs, septa, or diaphragms may also obstruct the exit of CSF from the third ventricle. Brain stem, tectal, and pineal region lesions can also result in aqueductal stenosis caused by extrinsic mass effect on the aqueduct.

BOX 7-2	Causes of Noncommunicating Hydrocephalus

OBSTRUCTION OF LATERAL VENTRICLES
Colloid cyst

Tumors
Choroid plexus papilloma (children)
Ependymoma
Meningioma
Neurocytoma
Subependymal giant cell astrocytoma

OBSTRUCTION OF THIRD VENTRICLE
Aqueductal web, fenestration, diaphragm
Congenital aqueductal stenosis (autosomal recessive)
Clot

Tumors
Craniopharyngioma
Ependymoma
Hypothalamic glioma
Pineal neoplasms
Vein of Galen aneurysm

OBSTRUCTION OF FOURTH VENTRICLE
Tumors
Astrocytoma
Choroid plexus papilloma (adults)
Ependymoma
Medulloblastoma (PNET)

OBSTRUCTION OF ANY SITE
Arachnoid cyst
Complications of hemorrhage, infection, synechiae
Cysticercosis
Hematoma
Meningioma
Metastasis
Primary brain neoplasm

PNET, Primitive neuroectodermal tumor.

Sagittal MR is very helpful for distinguishing extrinsic mass compression from an intrinsic aqueductal abnormality (Fig. 7-3). Aqueductal stenosis may also be diagnosed on CSF flow (phase contrast) MR imaging. Phase contrast MR with a velocity encoding set to 10 to 15 cm/sec may be the best way to assess aqueductal patency. Application of two gradients of equal magnitude but opposite direction can produce signal from moving protons moving in either direction, whereas stationary protons do not produce signal. Biphasic flow indicated by sequentially bright and dark signal (indicating to and fro flow) should be seen on the CSF flow scan (Fig. 7-4).

Clots or Synechiae

Other intrinsic causes of ventricular obstruction include clots or synechiae resulting from trauma or chronic infection. Patients with a large amount of subarachnoid hemorrhage may demonstrate obstruction caused by clot formation anywhere within the ventricular system. Synechiae may be due to fibrous adhesions after ventriculitis or meningitis. An infectious cause of ventricular

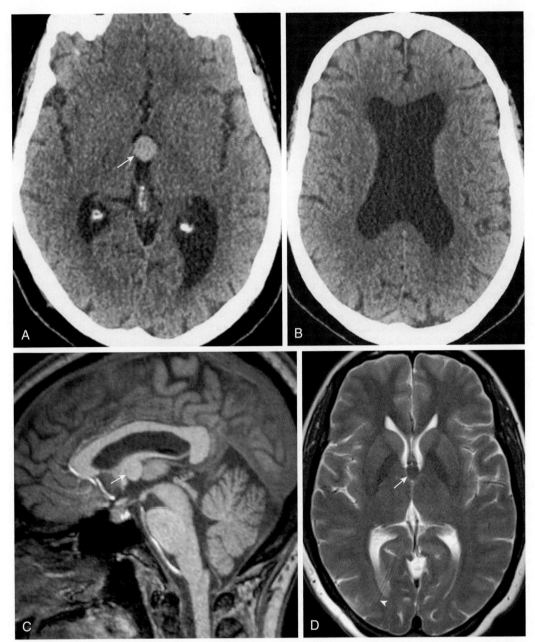

FIGURE 7-2 Colloid cyst. **A,** Axial computed tomography image shows classic appearance of a colloid cyst as a rounded hyperattenuating lesion within the anterior aspect of the third ventricle *(arrow).* **B,** More superiorly, there is enlargement of the lateral ventricles because of obstructive hydrocephalus by the colloid cyst. **C,** Sagittal T1-weighted image (T1WI) in a different patient shows a hyperintense rounded mass in the anterior aspect of the third ventricle *(arrow).* **D,** The same lesion is hypointense on T2WI *(arrow).* Note that the patient has been shunted *(arrowhead)* to decompress the lateral ventricles which now appear normal in size.

obstruction is cysticercosis. Occasionally, only parts of the ventricular system may be affected, resulting in compartmentalized dilatation of the obstructed components of the ventricular system.

Masses or Tumors

Masses of the pineal gland are the most common causes of extrinsic obstruction of the aqueduct. These generally compress the aqueduct from posteriorly and cause dilatation of the lateral and third ventricular system. Tectal gliomas also obstruct the aqueduct early in their course (Fig. 7-5). Occult cerebrovascular malformations may occur there as well. The base of the aqueduct may be obstructed

by tumors of the posterior fossa such as medulloblastomas (primitive neuroectodermal tumors [PNETs]) or ependymomas.

Any of the pediatric and adult posterior fossa tumors may obstruct the fourth ventricle and/or lower portion of the aqueduct. Ependymomas are one of the classic intraventricular tumors to infiltrate the foramina of Magendie and Luschka and may cause hydrocephalus from outflow obstruction. Cerebellar astrocytomas, medulloblastomas, or hemangioblastomas may compress the fourth ventricle extrinsically.

Other tumors that may obstruct parts of the ventricular system include meningiomas, neurocytomas, astrocytomas,

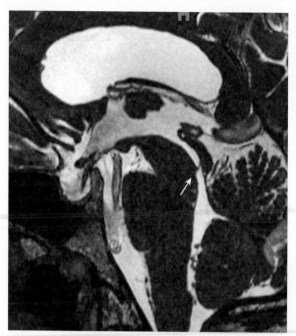

FIGURE 7-3 Aqueductal web. This high-resolution sagittal constructive interference steady-state T2-weighted image shows a fine web *(arrow)* crossing the inferior aqueduct. The ventricles are not tremendously enlarged, because of lateral ventriculostomy catheter (not shown).

choroid plexus papillomas, oligodendrogliomas, subependymal giant cell tumors, arachnoid/ependymal cysts, craniopharyngiomas, and (epi)dermoids.

Hematomas and Infarcts

A well-placed hematoma can compress the ventricles and lead to occlusion at the foramen of Monro, aqueduct, or fourth ventricle. These can be seen in the settings of trauma, acute subdural and/or epidural posterior fossa hematomas, and with hypertensive bleeds. Posterior fossa strokes are notorious for bringing about the downfall of the patient by eliciting acute hydrocephalus as the fourth ventricle is compressed and obliterated by mass effect. At the same time, excessive mass effect in the posterior fossa can lead to downward herniation of tonsils and subsequent obstruction of CSF flow at the foramen magnum.

Trapped Ventricles

The so called "trapped ventricle" may occur when the egress of CSF is obstructed, either from intrinsic or extrinsic masses. For example, in the trauma setting, a large subdural hematoma (SDH) can compress the ipsilateral lateral ventricle, but because of midline shift and outflow obstruction at the foramen of Monro, the contralateral lateral ventricle can abnormally dilate. Not uncommonly, periventricular hypoattenuation on CT or hyperintensity on T2/FLAIR around the margins of the trapped ventricle can be seen, indicating transependymal flow of CSF.

Trapping of the third ventricle is uncommon. Selective enlargement of the third ventricle must be distinguished from the presence of an intraventricular ependymal/arachnoid cyst and third ventricle squamopapillary craniopharyngiomas.

Isolation of the fourth ventricle may occur when the aqueduct of Sylvius, foramen of Magendie, and Luschka are occluded. The fourth ventricle becomes "trapped" and will expand as CSF production by the choroid plexus continues unabated (Fig. 7-6). This expansion may compress the cerebellum and brain stem and lead to posterior fossa symptoms. Many of these cases are due to fibrous adhesions with or without earlier hemorrhage.

Causes of Communicating Hydrocephalus

Infection, Hemorrhage, Tumors

The arachnoid villi are sensitive, delicate structures that may get gummed up by insults of several causes, resulting in communicating hydrocephalus (Box 7-3). Think of them as the little fenestrations in your bathtub drain; the whole tub will overflow if these tiny conduits are obstructed. The most common causes of obstruction include infectious meningitis, ventriculitis, ependymitis, subarachnoid hemorrhage, and carcinomatous meningitis. As the CSF becomes more viscous with a higher protein concentration, the arachnoid villi lose their ability to reabsorb the fluid. This causes hydrocephalus with dilatation of the ventricular system.

Do not let a normal appearance to the fourth ventricle dissuade you from considering communicating hydrocephalus. The fourth ventricle is the last ventricle to dilate, possibly because of its relatively confined location in the posterior fossa, surrounded as it is by the thick calvarium and sturdy petrous bones. Thus it is not uncommon to see dilated lateral and third ventricles but a normal-sized fourth ventricle and have communicating hydrocephalus. Still the most sensitive indicator will be the enlargement of the temporal horns and/or anterior recesses of the third ventricle—without that you probably do not have hydrocephalus. The hunt for a source of the ventricular dilatation should not stop at the aqueductal level with this pattern.

As with any cause of hydrocephalus, there may be periventricular high signal intensity on MR, very nicely demonstrated with FLAIR scanning. This is due to transependymal CSF migration into the adjacent white matter leading to interstitial edema (dark on diffusion-weighted imaging [DWI]). This is most commonly seen at the angles of the lateral ventricles and, because of its smooth and diffuse nature, can usually be distinguished from the focal periventricular white matter abnormalities associated with atherosclerotic small vessel ischemic disease. Be aware that there may normally be mild high intensity at the angles of the ventricle (ependymitis granulosa) in middle-aged patients.

Adult Hydrocephalus

Normal-pressure hydrocephalus (NPH) or adult hydrocephalus has a classic triad of clinical findings; the recent onset of gait apraxia, dementia, and urinary incontinence (Box 7-4). Half of patients have no known prior insult (idiopathic NPH), while the other half carry a remote history of prior infection or hemorrhage (nonidiopathic NPH). Imaging shows enlarged ventricles from communicating hydrocephalus with particular enlargement of the temporal horns. There may be evidence of transependymal CSF leakage on MR or CT. MR often shows accentuation of the cerebral aqueduct flow void (Fig. 7-7), but this can be seen as a normal finding as well. These patients may respond to

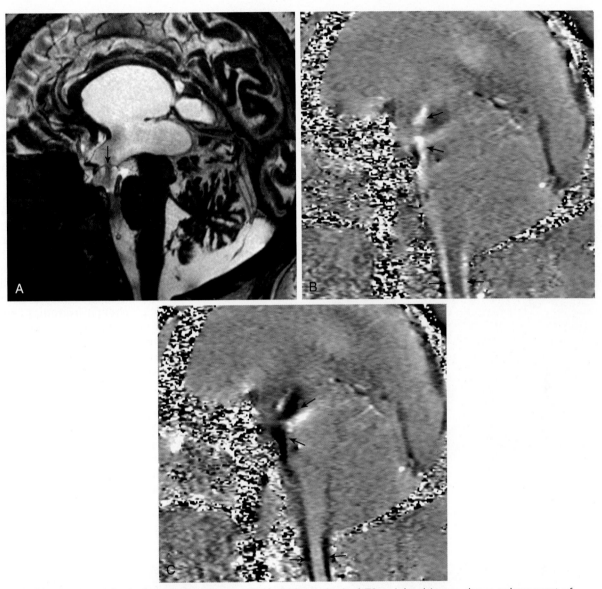

FIGURE 7-4 Biphasic flow of cerebrospinal fluid (CSF). **A,** Sagittal T2-weighted image shows enlargement of the lateral and third ventricles and effacement of the cerebral aqueduct in this patient with aqueductal stenosis. The patient underwent third ventriculostomy procedure, whereby a defect is intentionally created at the floor of the third ventricle to allow for CSF flow between the third ventricle and suprasellar cistern *(arrow)*. Note CSF pulsation artifact through the defect, indicating CSF flow. Phase contrast magnetic resonance imaging CSF flow study shows bright **(B)** and dark **(C)** signal indicating robust biphasic CSF flow at the level of the ventriculostomy defect as well as ventrally and dorsally at the foramen magnum *(arrows)*.

shunting or endoscopic third ventriculostomy procedures with amelioration of their clinical symptoms, making this a treatable cause of dementia (although the gait disturbance is more readily responsive to treatment). Because there is a chance at the possibility of return of function with a shunt, it is important to at least consider the diagnosis of NPH when ventricles appear larger than expected (Fig. 7-8). The most accurate predictors of a positive response to shunting are (1) absence of central atrophy or ischemia, (2) gait apraxia as the dominant clinical symptom, (3) upward bowing of the corpus callosum with flattened gyri and ballooned third ventricular recesses, (4) prominent CSF flow void, and (5) a known history of intracranial infection or bleeding (nonidiopathic NPH).

In patients with suspected NPH, an indium 111-DTPA (diethylenepentaacetic acid) study is sometimes ordered.

The agent is instilled in the CSF through a lumbar puncture. Normally, the tracer is resorbed over the convexities without ventricular reflux within 2 to 24 hours. In cases of communicating hydrocephalus and NPH, reflux of the tracer into the ventricles is seen with lack of tracer accumulation over the convexities 24 to 48 hours after instillation (Fig. 7-9). Patients who demonstrate this scintigraphic appearance allegedly have a better response to shunting than patients with normal or equivocal indium findings.

The rate of clinical improvement after shunting of patients with NPH is still only 50%. Prominence of the CSF flow void in patients with this condition has led some investigators to use phase contrast MR techniques to measure the flow through the cerebral aqueduct. A stroke volume of greater than 42 mL has been shown to be predictive of better response to shunting. The specific parameters inherent in this measurement are related to scanner field strength and pulse sequences, so they are not necessarily transferrable to your own scanner, but the point made is that greater flow through the aqueduct means a better chance for shunt improvement.

The best predictor of therapeutic response to shunting seems to be the patient's response to trials of large volume CSF drainages and/or a multiday trial of lumbar CSF drainage tube placement.

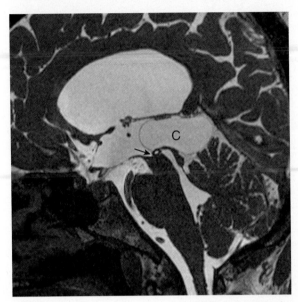

FIGURE 7-5 Arachnoid cyst. A large arachnoid cyst (C) in the superior vermian cistern focally occludes the superior aspect of the cerebral aqueduct *(arrow)* and causes hydrocephalus of the lateral and third ventricles, seen as bowing of the corpus callosum and dilatation of the third ventricle's anterior recesses.

External Hydrocephalus

Another benign cause of hydrocephalus from arachnoid villi malfunction is "external hydrocephalus," also referred to as "benign macrocephaly of infancy," "benign enlargement of the subarachnoid spaces in infants (BESS)," "benign extraaxial collections of infancy," "extraventricular obstructive hydrocephalus," and "benign subdural effusions of infancy." This may be due to immaturity of the arachnoid villi with a decreased capacity to absorb CSF. BESS is typically seen in children less than 2 years old who have a rapidly enlarging head circumference. Transient developmental delay may be present at the time of presentation; however, the clinical and imaging findings usually resolve by the time the child is 3 to 4 years old and the head circumference returns to normal. Prematurity, a history of intraventricular hemorrhage, and some genetic syndromes predispose to "external hydrocephalus."

CT and MR show dilatation of the subarachnoid spaces over the cerebral convexities and normal or slightly enlarged ventricular system (Fig. 7-10). The differential diagnosis includes chronic subdural hygromas and atrophy caused by previous injury. Sulcal dilatation and vessels coursing through the subarachnoid spaces indicate enlarged subarachnoid spaces, whereas displacement of vessels towards the surface of the brain imply the presence of a subdural collection. Atrophy is not usually associated with an enlarging head circumference. Beware, patients with BESS have an increased rate of SDHs presumably because the enlarged subarachnoid space causes stretching of the bridging veins in the subdural space. The conundrum is if nonaccidental trauma (NAT; now referred to as inflicted injury) is suspected in these patients because SDHs of different ages and volume loss may be present in NAT victims. Different aged SDHs would suggest NAT—along with additional findings of retinal hemorrhages and fractures (see Chapter 4).

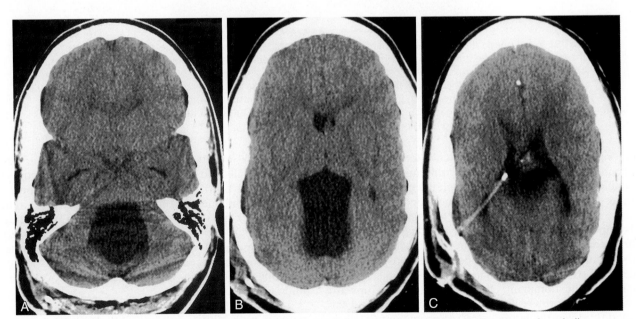

FIGURE 7-6 Trapped fourth ventricle. **A,** The fourth ventricle is markedly enlarged with effacement of cerebellar sulci. Note, however, how small the temporal horns and frontal horns are. **B,** A more superior cut shows the ballooning upward of the fourth. Did you catch the edge of the skull film finding? **C,** Yes, there was a shunt present that decompressed the ventricles above, but the aqueductal obstruction coupled with Magendie and Luschka outflow occlusion trapped the fourth ventricle.

The Almighty Failed Shunt

Shunt failure accounts for a large number of unenhanced CT scans in pediatric neuroradiology. The typical scenario is a child with a ventriculoperitoneal shunt in place who presents with nausea, vomiting, irritability and/or fever. This occurs in 30% of individuals in their first year with a shunt and in 50% of subjects within the first 6 years after shunt placement. Shunt infection occurs at a rate of about 10% in the first year.

Having prior imaging on hand is very valuable in the diagnosis of shunt failure. Evaluation should include a careful comparison of the entire ventricular system between the current exam and the prior, with search for interval change in size. Remember that there may be compartmentalization of the ventricular system in more complicated cases of brain tumors, ventricular hemorrhages, and infections, such that portions of the ventricular system may show interval change in size rather than the entire ventricular system. Next, contemplate the principal mechanisms responsible for shunt failure: (1) obstruction of the catheter tip in the ventricular system; (2) malfunction of the valve; (3) kinks in the tubing; (4) obstruction at the distal end of the catheter (i.e., intraperitoneal, intracardiac); and (5) component disconnection. The valves come in a variety of pressure settings for various resistances.

BOX 7-3   Causes of Communicating Hydrocephalus

OBSTRUCTION OF ARACHNOID VILLI

Hemorrhage
    Subarachnoid hemorrhage from aneurysm
    Traumatic intraventricular or subarachnoid hemorrhage
Infectious meningitis
Noninfectious inflammatory meningitis (sarcoidosis)
Carcinomatous meningitis
Chemical meningitis (fat, arachnoiditis, intrathecal medications)
Increased venous pressure from arteriovenous shunt, vein of Galen malformation
Venous thrombosis

OBSTRUCTION AT THE SKULL BASE

Chiari malformations
Achondroplasia
Dandy-Walker cysts
Arachnoid cysts at foramen magnum

OTHER CONDITIONS

External hydrocephalus
Normal-pressure (adult) hydrocephalus

BOX 7-4   Confirmatory Tests for Normal-Pressure Hydrocephalus

History: classic clinical triad with gait apraxia, dementia, urinary incontinence
Cerebrospinal fluid (CSF) withdrawal trial
Indium-labeled CSF study with ventricular reflux, no flow over convexities at 24 to 48 hours
Marked aqueductal flow void
Trial of ventricular shunting (acid test)
Trial of lumbar drainage (72 hours)

Shuntograms in which 2 to 3 mL of nonionic contrast are injected into the shunt reservoir may be revealing. Normally, the contrast clears from the shunt tube within 3 to 10 minutes. In adults, it may take 10 to 15 minutes to clear. The first step in the evaluation is withdrawing CSF from the shunt valve. If CSF cannot be withdrawn, the ventricular catheter is obstructed or the valve is faulty. If contrast refluxes from valve to ventricle, the valve is faulty as this is supposed to be

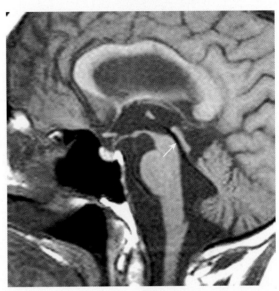

FIGURE 7-7 Normal-pressure hydrocephalus with accentuation of aqueductal flow void. Sagittal T1-weighted image shows generalized enlargement of the ventricular system and prominent cerebrospinal fluid flow void within the cerebral aqueduct *(arrow)*.

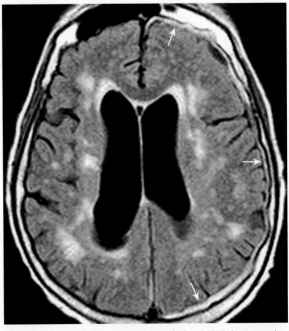

FIGURE 7-8 Normal-pressure hydrocephalus (NPH) (same patient as in Figure 7-7). Ventricular enlargement out of proportion to sulcal dilatation without an obstructing lesion is the sine qua non of NPH. The periventricular high signal on this fluid-attenuated inversion recovery (FLAIR) scan is of limited differentiating value because small-vessel ischemic changes abound in this age group. Note the incidental subdural hematoma overlying the left cerebral hemisphere *(arrows)*.

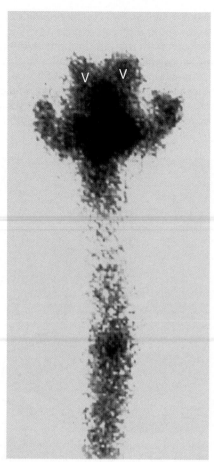

FIGURE 7-9 Indium-DTPA study in a patient with normal-pressure hydrocephalus. Coronal indium-labeled study shows lack of ascension of the radiotracer over the convexities with reflux of the tracer into the lateral ventricles (V). This study was taken 48 hours after intrathecal tracer insertion.

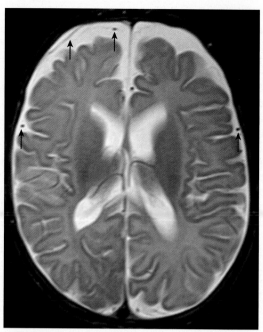

FIGURE 7-10 External hydrocephalus. Note the prominence of the extraaxial spaces overlying the frontal lobes in this infant with external hydrocephalus. The sulci are not effaced, as would be seen in the setting of bilateral subdural collections. Cortical vessels *(arrows)* cross the extraaxial spaces, not displaced toward the surface of the brain (as would be seen with subdural fluid collections). The ventricular system is also prominent, not uncommonly seen in this condition.

a one-way valve to prevent "dreck" from the flowing backwards to the ventricles. If the contrast does not flow freely out but after pumping the valve it seems to work, there is probably incomplete obstruction of the shunt system and/ or a malfunctioning valve-pressure system. If there is no spillage intraperitoneally even after pumping the valve, or if what spills gets loculated, clearly, there is a problem with the end of the shunt system. Ventricular catheter obstruction, valve malfunction, and distal obstruction are the most commonly seen causes of shunt failure.

Third ventriculostomies, where a small hole is made at the floor of the third ventricle that allows communication between the floor of the third ventricle and the suprasellar cistern, has proven to be effective in relieving hydrocephalus (see Fig. 7-4). This is most useful for those obstructions distal to the third ventricle as it bypasses the obstructed region. These are often placed through the use of a fiberoptic endoscope and/or three-dimensional reconstructions with image-guided navigation. Expect the reduction in ventricular size to appear within a couple of weeks of the procedure—not as rapidly as with lateral ventricular shunts. Flow through the third ventriculostomy may be visualized with phase contrast flow studies at a velocity encoded at 5 mL/min.

At those institutions who have ready access to MR, radiologists are encouraging the use of single shot fast spin

echo MRI (timing in at 1 to 2 seconds per slice) for evaluation of shunted kids to avoid the repeated irradiation of children's heads with CT. No more frying of the brain with CT for suspected shunt failure—that's a good thing!

Slit Ventricle Syndrome

Sometimes shunted patients are symptomatic but have tiny ventricles. These are usually patients who have had long-term shunt problems. The thought here is that the ventricles and/or brain lose compliance and cannot expand despite the fact that they are under high pressure. This is probably caused by the decreased compliance in the lateral ventricular walls owing to fibrosis in chronically shunted patients. Alternatively, they may be small owing to marked reduction in the pressure because of chronic overdrainage or leakage. In younger patients, the calvarium will thicken because of the overdrainage of the CSF. Then, if the cranial sutures close early, the brain has nowhere to go as it grows, being confined by the calvarium. This in turn leads to small ventricles and no capacity to accommodate normal variations in intracranial pressure. In this way, you can have small ventricles yet high pressure. The flow out of the shunt will be limited.

Pseudotumor Cerebri

Pseudotumor cerebri, or idiopathic benign intracranial hypertension or idiopathic intracranial hypertension, is included here because it is also a disease of abnormal CSF mechanics. The abnormality may be due to decreased absorption of CSF at the arachnoid villi, increased water content in the brain, or increased resistance to drainage because of venous obstruction. Patients with this disorder are typically obese (95% of patients), black (62%), and young or middle-aged women. They have frequent headaches, cranial nerve VI palsies, papilledema, or visual field deficits on examination. The disease may occur in association with pregnancy, endocrine abnormalities, medications, or intracranial venoocclusive disease. One should exclude a dural venous malformation, venous stenosis, or venous thrombosis as potential causes for the elevated pressure by means of computed tomography venography (CTV) or magnetic resonance venography (MRV). On physical examination, the patients may have other signs of increased intracranial pressure. The lumbar puncture demonstrates extreme elevations of CSF pressure (we've seen cases as high as 60 cm water—like a geyser—normal is 10 to 20 cm water).

On imaging, the ventricles are either normal in size or slightly small. The cerebral subarachnoid space volume is larger in patients with pseudotumor cerebri compared with age-matched control subjects. Most MR studies in patients with pseudotumor cerebri are normal, but calculated measurements of white matter intensity on T2WI may show subtle increases over normal control subjects. The venous sinuses and veins may be small and may enlarge after spinal fluid drainage. In some cases venous sinus stenosis or venous compromise may yield the underlying cause of the condition. If an orbital study is performed, reverse cupping of the optic disk corresponding to the papilledema may be noted, and this finding correlates well with the degree of vision loss. The optic nerve sheath complex is also enlarged and more tortuous in pseudotumor cerebri. These patients have a higher rate of expanded, empty sellas (greater than the 30% rate of partially empty sellas seen in the normal population). These patients also develop other areas of spontaneous CSF-filled outpouchings of the dura (meningoceles) in 11% of cases and secondary CSF leaks, both frequently around the Meckel cave and the petrous apex.

Treatment consists of repetitive lumbar punctures to drain fluid but often the disease remits spontaneously. Occasionally, CSF shunting or lumbar drain placement is required for those patients with intractable headaches and visual impairment. Diuretics and carbonic anhydrase inhibitors may reduce CSF pressures as well. The prognosis is generally good when the disorder is treated expediently.

ATROPHY

Neurodegenerative Disorders

The disorders that demonstrate supratentorial atrophy are often associated with dementia (Table 7-2).

Alzheimer Disease

Of the disorders in Table 7-2, Alzheimer disease (dementia Alzheimer type, or DAT) is one of the most notorious and common, accounting for 60% to 90% of the dementing disorders with progressive memory loss often with personality changes and impaired cognition. DAT affects 2 to 4 million Americans and 8% of the population older than 65 and 30% of those over 85 years old. Women are more commonly affected by a 2:1 margin. Olfaction is one of the first senses to show some effects of the disease, but this is

rarely tested. Depression often coexists. Late in the course the patient becomes severely impaired, myoclonic, vegetative, and weak. Current treatments are limited in effectiveness and focus on slowing progression of disease.

Senile plaques, seen as amorphous material in the cerebral cortex, and neurofibrillary tangles in the nerve cells in the form of tangled loops of cytoplasmic fibers, are the diagnostic histopathologic features of DAT. Disease progression from the entorhinal cortex to the hippocampus to the neo-cortex is the rule. Curiously, these same pathologic findings are seen in adult patients with Down syndrome, Parkinson disease, and "punch-drunk" fighters ("dementia pugilistica").

The main finding on CT and MR scanning of DAT is diffuse cortical atrophy, often more prominent in the mesial temporal lobes (Fig. 7-11). Temporal horn dilatation is seen in more than 65% of patients with DAT. Increases in ventricular size, sulcal size, sylvian fissure size, and total CSF volume are noted in patients with DAT compared with age-matched control subjects. The subiculum and entorhinal cortex of the parahippocampal region appears to be most severely affected in DAT. On longitudinal studies, the rate of atrophic change in patients with DAT is much faster than in normal persons.

One study found that the measure with the best sensitivity in discriminating DAT patients from control subjects was the width of the temporal horn. If one combines the width of the temporal horn, width of the choroid fissure, height of the hippocampus, and the interuncal distance into a compound factor one can discriminate patients with mild DAT from control subjects with 86% sensitivity. There may, in fact, be a continuum of progressive hippocampal atrophy between normal elderly people, those with mild cognitive impairment (MCI), and those with DAT. In a similar fashion there is a stepwise increase in the number of micro-hemorrhages seen on susceptibility-weighted scans between normal, MCI and DAT subjects. People have suggested that the amyloid deposition, the same amyloid in amyloid angiopathy, may be the source of these dark susceptibility-weighted imaging (SWI) foci.

There has been considerable investigation concerning the presence of deep white matter and periventricular white matter areas of high signal intensity on the T2WI scans in patients with DAT. There is a nonstatistical trend toward more small foci of white matter abnormality in patients with clinically diagnosed probable DAT.

On MR spectroscopy, reduced levels of *N*-acetyl aspartate (NAA) and increased levels of myoinositol characterize DAT. The NAA levels are significantly reduced in frontal, temporal, and occipital cortex of DAT patients, presumably because of neuronal loss. Being able to distinguish between normal aging and DAT with magnetic resonance spectroscopy (MRS) runs at the mid 80% range; distinguishing between DAT and other dementias drops the accuracy to the mid 70% range.

Some investigators have used dynamic susceptibility contrast-enhanced MR perfusion imaging in diagnosing DAT and have found that relative values of temporoparietal regional cerebral blood volume (as a percentage of cerebellar rCBV) were reduced by a factor of 20% bilaterally in the patients with DAT compared with normals. Using left and right temporoparietal rCBV as index measures, specificity was 96% and sensitivity was 95% in moderate DAT and 88% in mild DAT.

Positron emission tomographic (PET) scanning has demonstrated decreased oxygen utilization and decreased regional cerebral blood flow in frontal, parietal, and temporal lobes in patients with DAT (Fig. 7-12). The findings are most striking in the posterior temporoparietal lobes. On single-photon emission computed tomography (SPECT) brain studies, patients with DAT have reduced cerebral blood flow as measured by parietal-to-cerebellar and parietal-to-mean cortical activity. The severity of symptoms may correspond to the reduction in uptake of technetium hexamethylpropyleneamine oxime (HMPAO).

Pittsburgh Compound B Imaging

The usefulness of amyloid imaging using ^{11}C-labeled Pittsburgh compound B (^{11}C-6-OH-BTA-1, also known as ^{11}C-PIB or PIB), a thioflavin-T derivative, was first demonstrated in mice models of Alzheimer disease in 2003, with encouraging results, including rapid entry into the brain, quick labeling of amyloid deposits, rapid clearance of nonspecific binding, and prolongation of specific binding. Shortly after, the first human study using PIB showed marked retention of the radiotracer in DAT patients compared with controls, namely in areas of association cortex known to contain large amounts of amyloid deposits. On the other hand, MCI patients repeatedly showed PIB uptake values that were

TABLE 7-2   Neurodegenerative Causes of Atrophy

Entity	Distinguishing Imaging Findings	MRS Features	Distinguishing Clinical Findings
Alzheimer disease	Temporal lobe predominance Increased hippocampal-choroidal fissure size Hippocampal and amygdala atrophy, global atrophy	Decreased NAA Increased myoinositol, phosphomonoester (PME)	Severe memory loss Speech and olfaction affected early No early myoclonus or gait disturbance Course in years
Frontotemporal dementias	Behavioral variant: anterior temporal and frontal atrophy, hemispheric asymmetry Primary progressive aphasia variant perisylvian and insular atrophy, especially of superior temporal gyrus	NKY	In general, speech Behavioral variant: prominent changes in personality and interpersonal relationships Primary progressive aphasia variant word-finding difficulties, difficulties with writing and comprehension
Pick disease	Severe frontal and mild temporal lobe predominance Caudate atrophy, sparing of parietal cortex	Increased PME, phosphodiester (PDE)	Cognition, personality severely affected Abulia and apathy
Creutzfeldt-Jakob disease	Abnormal intensity in basal ganglia and thalami	Normal early, late reduction in NAA and increased myoinositol (in hamsters)	Transmission of prion Rigidity Myoclonus Course in months
Parkinson disease	Substantia nigra decreased in size	Normal spectrum or decreased NAA and elevated lactate in demented patients	Rigidity, bradykinesia, tremor
Multisystem atrophy	Variants include olivopontocerebellar degeneration with olive, pons, cerebellar; and putaminal atrophy, striatonigral degeneration with smaller midbrain	NKY	Parkinsonism, autonomic dysfunction, cerebellar gait disorders
Progressive supranuclear palsy	Midbrain, collicular atrophy, increased putaminal iron	Decreased NAA/Cr	Ophthalmoplegia, pseudobulbar palsy, rigidity
Cortical-basal ganglionic degeneration	Atrophy of the paracentral structures, superior parietal lobule knife blade atrophy, dilated central sulcus asymmetry characteristic	NKY	Rigidity of limbs Alien limb phenomenon Personality disorders Myoclonus
Lewy body dementia	Brain stem, substantia nigra cortical atrophy	NKY	Cognitive difficulties, hallucinations
AIDS	Atrophy High intensity in basal ganglia Superimposed infection PML and lymphoma	Decreased NAA/Cho Decreased NAA/Cr Increased Cho/Cr Increased myoinositol/Cr All phosphorus levels lower	Young age Risk factors Positive HIV
Multi-infarct dementia	White matter and deep gray lacunae Strokes of different ages Central pontine infarcts	Decreased NAA	Stuttering course with discrete events Stroke risk factors Early gait disturbance
Amyotrophic lateral sclerosis	Hypointensity in motor cortex (T2WI) Hyperintensity in corticospinal tract (T2WI) Anterior horn cell atrophy	Decreased NAA/Cr Decreased NAA/Cho Decrease PCr/Pi	Weakness Atrophy Spasticity Preserved cognition

AIDS, Acquired immunodeficiency syndrome; *Cho*, choline; *Cr*, creatine; *HIV*, human immunodeficiency virus; *MRS*, magnetic resonance spectroscopy; *NAA*, N-acetyl aspartate; *NKY*, not known yet; *PCr*, phosphocreatine; *Pi*, inorganic phosphate; *PML*, progressive multifocal leukoencephalopathy; *T2WI*, T2-weighted image.

either control-like or DAT-like. The striking finding, though, was that of "cognitively normal" controls with higher than normal PIB uptake, sometimes not significantly different from uptake in DAT patients. This raised the possibility that PIB may be sensitive for detection of preclinical Alzheimer disease state (that is, before onset of symptomatology). Recently, a 2-year longitudinal follow-up of 26 MCI patients studied with PIB demonstrated that 5 of 13 PIB-positive MCI patients converted to clinical DAT, whereas none of the 10 PIB-negative patients did. Other amyloid binding agents from other pharmaceutical companies are also being used to evaluate patients currently. These include

a fluorinated derivative of PIB, [18F]flutemetamol, [18F] AV-45 florbetapir, and florbetaben [41].

Frontotemporal Dementia

The term frontotemporal dementia (FTD) is used to classify patients with focal cortical atrophy affecting the frontal and/or temporal lobes. Clinically, patients with FTD have peculiar behaviors (hyperorality, hypersexuality, lack of personal awareness, apathy, and perseverations) with personality shifts, inappropriate social conduct, and psychiatric overtones. This entity is thought to represent a constellation of diseases that lead to widespread cortical atrophy. Frontotemporal dementia occurs in younger age groups compared with Alzheimer disease, usually presenting in the age range of 40 to 75 years and affecting men and women equally.

On imaging, differentiation with DAT is difficult; however, in general, the frontal and anterior temporal lobes are more atrophic in frontal variant FTD than in DAT. Technetium HMPAO SPECT scans will show hypoperfusion (and hypometabolism) in the ventromedial frontal lobes with frontal variant FTD.

FTD was originally called Pick disease, but this term is now reserved for a more specific subtype of FTD. Pick disease causes cerebral atrophy and manifests clinically with memory loss, confusion, cognitive and speech dysfunction, apathy, and abulia. As in DAT, there is an anterior temporal lobe predominance to the atrophy; however, inferior frontal lobe changes are also present, sometimes focally (Fig. 7-13). The posterior cortex and parietal lobe in general are spared. Pathologically, there is severe atrophy of the anterior frontal and temporal lobes with swollen nerve cells and spherical intracytoplasmic inclusions (Pick bodies). Some studies have shown concomitant caudate atrophy. No treatment is available, and so the patient's cognition spirals downward within months to years. Fortunately, Pick disease is uncommon.

In the spectrum of frontotemporal dementias is primary progressive aphasia (PPA), a rare dementing disorder with language impairment as the dominant deficit (Fig. 7-14).

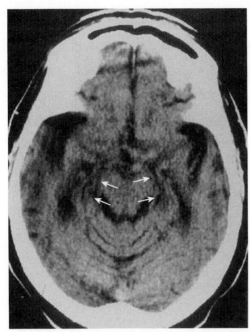

FIGURE 7-11 Dementia, Alzheimer type. Axial unenhanced computed tomography scan shows dilation of the choroidal-hippocampal fissure complex *(arrows)* with dilation of the adjacent temporal horns caused by temporal lobe atrophy.

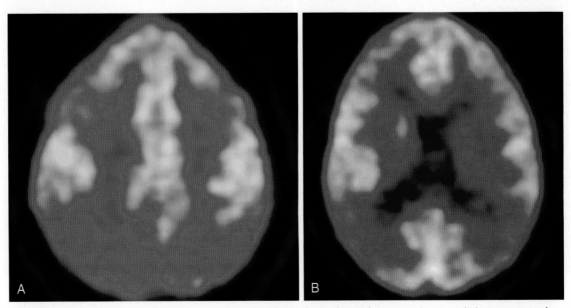

FIGURE 7-12 Dementia, Alzheimer type. Images from fluorodeoxyglucose–positron emission tomography (FDG-PET) show decreased activity in the parietal **(A)** and temporal **(B)** regions.

There are three subtypes that have been described based on specific speech and language features. These include nonfluent or agrammatic PPA, semantic PPA, and logopenic PPA.

In nonfluent progressive aphasia, one sees cognitive diminution over a period of years with word finding difficulties leading over time to mutism. Behavioral changes are less prominent. Atrophy is seen in a perisylvian location rather than the hippocampal region. "Knife blade shaped gyri" (knife blade atrophy) are described in the anterior aspect of the superior temporal gyrus with a widened sylvian fissure and insular atrophy.

Patients with semantic dementia (previously called temporal variant FTD) have difficulties with naming, word comprehension, object recognition, semantic relatedness of objects, and interrelations between words and meanings. Syntax, executive functioning, and phonology are spared. Short-term memory is usually intact, but long-term memory is affected. The technetium SPECT scans in semantic dementia show hypoperfusion of one or both temporal lobes. Temporal lobe atrophy is more dramatic with semantic FTD than with frontal variant FTD. The temporal pole and inferolateral gyri (including the parahippocampal gyri) are more affected than the hippocampi, marking a distinction from DAT. The left temporal lobe is more often affected than the right with hemispheric asymmetry not unusual.

In the logopenic variant of PPA, word retrieval and sentence repetition deficits are major features. Speech is

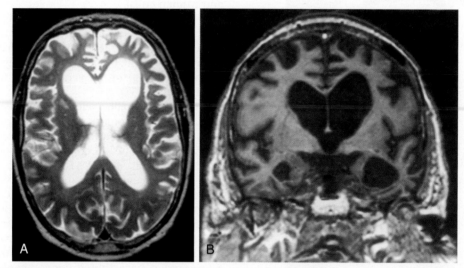

FIGURE 7-13 Frontotemporal dementia, frontal variant. **A,** Axial T2-weighted image (T2WI) is remarkable for the dilated subarachnoid space over the frontal lobes, signifying atrophy. The frontal horns of the lateral ventricle enlarge to fill the void. The atrophy is striking in its frontal predominance. **B,** The coronal T1WI again shows the striking frontal and anterior temporal atrophy. The patient was in his late 50s.

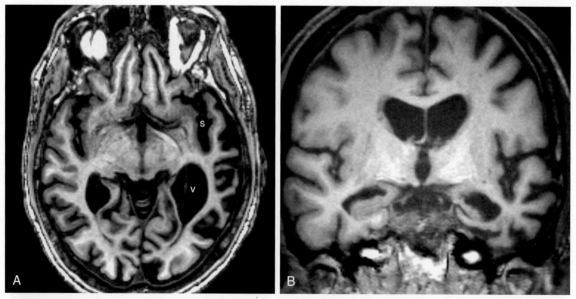

FIGURE 7-14 Primary progressive aphasia (PPA). **A,** Axial T1-weighted image (T1WI) shows generalized volume loss with superimposed marked atrophy of the left temporal lobe in this patient with PPA. Note the expanded sylvian fissure (s) and ex vacuo enlargement of the atrium of the left lateral ventricle (v). **B,** Coronal T1W1 in this same patient again demonstrates the striking asymmetry of left sided temporal volume loss.

slowed with frequent pauses. On imaging, volume loss in the left temporoparietal junction including posterior temporal, supramarginal and angular gyri is seen.

More recently, motor disorders associated with frontotemporal dementia have been described. These include corticobasal syndrome, progressive supranuclear palsy, FTD with parkinsonism, and FTD with amyotrophic lateral sclerosis. Each of these variants of FTD will be characterized by their primary features later in this chapter.

Creutzfeldt-Jakob Disease

Creutzfeldt-Jakob disease (CJD) is a rare dementing disorder that often affects a younger population than DAT. It is caused by prions (small proteinaceous infectious agents devoid of deoxyribonucleic acid [DNA] and ribonucleic acid [RNA]) and is related to diseases that were described in New Guinea. When seen in brain-eating cannibals the disease is called kuru, but it is called scrapie when found in New Guinea sheep. Scrapie in European sheep has been thought to be responsible for the spread to cattle in the form of bovine spongiform encephalopathy. There is a familial form of CJD (fCJD) that accounts for about 5% of cases.

Clinically sporadic CJD (sCJD) is characterized by cognitive decline, behavioral changes, vision disturbance, anxiety, jerky movements, and memory loss. Electroencephalogram (EEG) findings that suggest CJD are spike and wave complexes that occur about once every second. Elevated levels of the 14-3-3 protein (a regulatory molecule that binds to signaling proteins kinases, phosphatase, and transmembrane receptors), neuron-specific enolase (NSE), and total tau protein (T-tau) in the CSF support the diagnosis.

CJD is a human disease that occurs as a rapidly progressive dementing disorder. CT studies may be normal (80%) or may show rapidly progressive atrophic changes (20%) in the brain. On MR, diffusion-weighted scans are the most sensitive pulse sequences and along with FLAIR images, shows abnormal high signal most commonly in the cortex and/or deep gray matter (Fig. 7-15). With time, cerebral atrophy and symmetric high signal intensity foci in all of the basal ganglia, thalami, occipital cortex (the Heidenhain variant), and white matter may develop.

Mad Cow Disease

Mad cow disease (bovine spongiform encephalopathy) was first recognized in 1986 and is characterized by the cows being apprehensive, hyperesthetic, and uncoordinated with progressive mental status deterioration. The brains of these cattle revealed spongiform encephalopathy. It was caused by feeding cattle with infected offal (animal tissue discarded by slaughterhouses), which contained the prions from sheep with scrapie. The transmission to humans led to variant CJD (vCJD). As opposed to sCJD where one finds cortical and lentiform nucleus hyperintensities, with bovine spongiform encephalopathy the MR abnormalities are usually limited to bilateral thalamic pulvinar hyperintensity on DWI and FLAIR (hockey stick sign). Additional involvement of the putamen and caudate may be seen as well, and cortical involvement may coexist. EEG and CSF analysis for 14-3-3 protein is useful to assist in suggesting the diagnosis. Because vCJD also involves the lymph nodes, spleen, tonsil and appendix, a tonsil biopsy may be recommended to cinch the diagnosis. Moooo!

Parkinson Disease

Parkinson disease (PD) is characterized by bradykinesia, shuffling gait, rigidity, and involuntary tremors. Dementia, when or if it occurs at all, is a late phenomenon. If dementia occurs within the first year of onset of the movement disorder, the diagnosis of Lewy Body Disease (see later) is favored. PD is the disease that has reduced the most

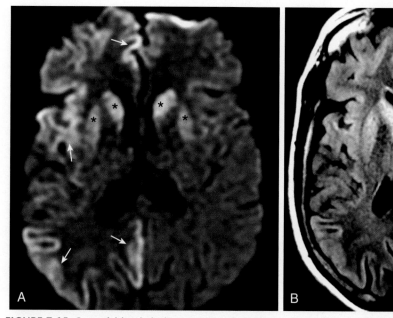

FIGURE 7-15 Creutzfeldt-Jakob disease (CJD). **A,** Diffusion-weighted and **(B)** fluid-attenuated inversion recovery (FLAIR) magnetic resonance (MR) study reveals peripheral cortical high intensity *(arrows)* reflecting the most common manifestation of CJD on MR. The location may be variable, but all lobes may be involved. In this case, the caudate heads and putamina are also involved bilaterally as might be seen in variant CJD *(asterisks)*.

famous boxer of all time, Muhammad Ali, from "floating like a butterfly" to shuffling like a man many years his senior. The lesion in PD has been localized to the dopaminergic cells of the pars compacta of the substantia nigra. Treatment consists of dopamine stimulation therapy (levodopa), bromocriptine, anticholinergics (benztropine), piperidyl compounds (trihexyphenidyl), and/or tricyclic antidepressants. Research into surgical implantation of fetal substantia nigra or stem cells has shown some promise but remains experimental at this time.

Although no statistically significant differences in signal intensity or size of the substantia nigra pars compacta have been identified on conventional field strength MR, a trend toward decreased width of the pars compacta has been noted (Fig. 7-16). At 7T MR, signal loss (probably from iron accumulation) can be seen on SWI within the pars compacta, with degree of signal loss approaching that of the pars reticulata, which normally contains twice the iron content of the pars compacta. On spectroscopy, one sees a diminution in NAA (what else?) and a significant increase in the ratio of lactate to NAA, especially in those PD patients with dementia. Fluorodeoxyglucose (FDG) PET is typically normal, but occasionally there can be diminished uptake in the parietooccipital cortex.

More and more neurosurgeons are employing image guidance for stereotactic pallidotomy for patients with PD and/or dystonia. The surgeons combine our ability to provide anatomic three-dimensional guidance with microelectrode electrophysiologic recording to identify the ventral internal globus pallidus. Then they burn a 100 to 200 mm³ hole in the brain and when the smoke clears, voila!, less rigidity and less bradykinesia. It is not uncommon to see hemorrhage and edema after this procedure, even extending into the optic tract or internal capsule (structures in proximity to the posterolateral internal globus pallidus), but this is usually asymptomatic when done right. The globus pallidus internus (GPi) can be targeted for stimulator implantation and is then corroborated intraoperatively by microelectrode identification of movement-related kinesthetic cell firing. The subthalamic nucleus may also be targeted. Chronic stimulation of the subthalamic nucleus can reduce parkinsonian movement disorders.

Still others are using the gamma knife radiosurgical technique to perform thalamotomies and pallidotomies for movement disorders. One should expect to see abnormal signal in the target (globus pallidus interna or ventralis intermedius thalamic nucleus). A ring-enhancing focus with vasogenic edema is typically seen at the 3-month mark after the radiosurgery, but therapeutic benefit usually begins at 1 month.

Atypical Parkinson Disease

Atypical PD includes the entities of multisystem atrophy, progressive supranuclear palsy, corticobasal ganglionic degeneration, and dementia with Lewy bodies. In each of these entities there is a movement disorder.

Dementia with Lewy Bodies

Dementia with Lewy bodies (DLB) is an entity probably related to Parkinson disease in which Lewy bodies are found not only in deep gray matter structures, but diffusely in the brain, including the cortex. DLB presents in older subjects than PD and accounts for 25% of dementing disorders. While dementia may be primarily associated with Parkinson disease, clinicians distinguish these entities based on the fact that Parkinson disease dementia shows parkinsonism preceding the dementia by 1 year or more. Dementia preceding or accompanying parkinsonism is DLB (dementia with Lewy bodies). PD dementia is termed "subcortical" (psychomotor slowing, difficulty concentrating, impaired retrieval), but DLB is a cortical dementia (aphasia, anomia, apraxia, visuospatial problems, memory deficits). Histopathologically, they may look the same. Clinically patients with DLB show

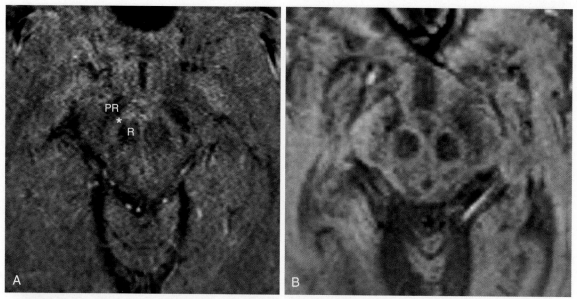

FIGURE 7-16 Parkinson disease. **A,** Normal appearance of the midbrain on susceptibility-weighted imaging shows hypointense signal within pars reticulata (PR) of the substantia nigra as well as the red nucleus (R). The pars compacta is normal in width *(asterisk)*. **B,** In this patient with Parkinson disease, note that the width of the pars compacta is decreased compared with normal, and this appears more pronounced on the right compared with the left (yes, it is subtle).

fluctuating cognitive impairment, visual hallucinations, and parkinsonism. Cognitive deficits are usually in memory, attention, executive function, and visuospatial and visuoconstructional abilities.

On MRI, a greater degree of cortical atrophy in DLB distinguishes it from PD. Brain stem and substantia nigra atrophy may be evident. SPECT scans show decreased striatal uptake and on FDG PET, generalized decreased uptake is seen most pronounced in the occipital lobes.

Multiple System Atrophy

Shy-Drager syndrome, olivopontocerebellar atrophy, and striatonigral degeneration can be lumped into the disorder known as multisystem atrophy (MSA). MSA is associated with autonomic system dysfunction that presents in middle age and simulates PD (tremors, cogwheel rigidity, bradykinesia, and ataxia) in some aspects. MSA-striatonigral (MSA-P) type displays more parkinsonian features (rigidity, bradykinesia, postural instability) and autonomic dysfunction (impotence, incontinence, orthostatic hypotension), whereas MSA-olivopontocerebellar atrophy type (MSA-C) is characterized by dominance of cerebellar dysfunction. Patients may have autonomic abnormalities of temperature regulation, sweat gland function, and maintenance of the blood pressure (orthostatic hypotension). Dementia is usually not a prominent component of MSA.

Brain stem, putaminal, and cerebellar atrophy may be seen on MRI, and the posterolateral putamina may show iron-related decreased signal intensity on T2WI. The width of the middle cerebellar peduncle and the midbrain are reduced in MSA and measurements of these structures can lead to presumptive diagnosis. T1 and T2 shortening may also be found in the cortex and globus pallidus. Neuromelanin may also account for some signal changes in the putamina.

The "hot cross buns" sign (Fig. 7-17) has been described in this disorder in which there is a cruciform linear area of high signal on T2WI in the pons with tiny round darker areas within the checkerboard of the cross. This is relatively specific for multisystem atrophy, though the cause of this finding has yet to be determined.

Progressive Supranuclear Palsy

Progressive supranuclear palsy (PSP) resembles PD in its manifestations (rigidity, bradykinesia) but also expresses a severe supranuclear ophthalmoplegia (impaired downward gaze), gait disorder, dysarthria, postural instability and pseudobulbar palsy a few years after the onset of parkinsonian symptoms. Patients present with hyperextension of the neck and contracted facial muscles giving a "surprised look" to the face. PSP is another dementing disorder associated with personality changes of uncontrolled emotions, social withdrawal, and depression.

On CT and MR, dilatation of the third ventricle, atrophy of the midbrain, and enlargement of the interpeduncular cistern is seen. MR may show decreased width of the pars compacta of the substantia nigra, atrophy of the superior colliculi, and high intensity to the periaqueductal gray matter. Measurement of the superior cerebellar peduncle is a sensitive parameter for PSP—the width is reduced to 2 mm in greater than 80% of patients with the disease. Volume loss in the midbrain contributes to the "hummingbird sign" on sagittal imaging of the brain stem (Fig. 7-18). Increased iron may be found in the putamen so that it appears more hypointense than the globus pallidus on T2WI (the opposite of normal patients).

Corticobasal Ganglionic Degeneration

Corticobasal ganglionic degeneration (CBGD) presents with postural instability, dystonia, akinesia, apraxia, myoclonus, bradykinesia, and limb rigidity. This disease demonstrates neuronal loss in the substantia nigra, frontoparietal cortex, and striatum.

MR will demonstrate symmetric or asymmetric thinning of precentral and postcentral gyri with central sulcus dilatation. The superior parietal lobule and superior frontal gyrus seem to be at particular risk for volume loss

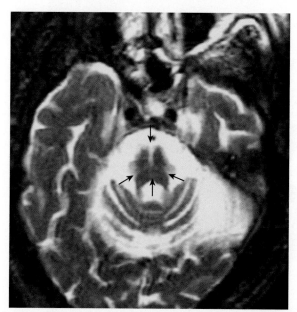

FIGURE 7-17 Hot cross buns sign. On this axial T2-weighted image, note the four quadrants of dark signal *(between arrows)* that make up the hot cross buns sign in the pontine region in this patient with multisystem atrophy. Note the marked atrophy of the pons.

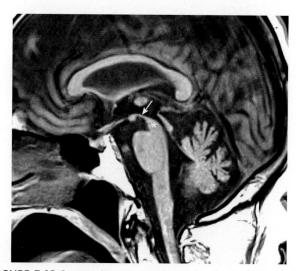

FIGURE 7-18 Progressive supranuclear palsy. Sagittal T1-weighted image shows atrophic midbrain, with characteristic configuration of a hummingbird head in profile. The *arrow* indicates the beak and the *asterisk* indicates the head of the hummingbird.

(knife-blade atrophy again) as well, whereas the temporal and occipital lobes are less involved. Parasagittal involvement is prominent. These features clearly distinguish this dementing disorder from DAT or FTD. Atrophy of the basal ganglia and midbrain may be subtle. Subcortical gliosis seen as high intensity on T2WI may be a clue to this diagnosis. One feature typical of CBGD is the asymmetry in the parasagittal and paracentral atrophy between the two hemispheres. This also is not a common feature of DAT, PD, MSA, DLB, PSP, or ABCDEFG (ok, that last one we made up). FTD does show asymmetry, but the anterior frontal and temporal lobe involvement is more distinctive. High signal on T1WI of the subthalamic nuclei may also suggest CBGD.

Acquired Immunodeficiency Syndrome Dementia Complex

The acquired immunodeficiency syndrome (AIDS) dementia complex has become an increasingly common cause of

dementia. Once a superimposed cytomegalovirus (CMV) infection occurs, the patient may develop a more fulminant decline in mentation.

Although the overall manifestations of AIDS are best described in the chapter on inflammation, note that a young patient with cerebral atrophy and mild to marked white matter hyperintensity without mass effect may well have AIDS dementia complex (Fig. 7-19). MR spectroscopic imaging often reveals diffuse decreases in NAA and elevated choline in AIDS dementia complex. These spectroscopic findings may precede mental status deterioration and are present in children and adults with AIDS. Myoinositol/creatine ratios may actually rise (similar findings to DAT). The magnetization transfer ratios of human immunodeficiency virus (HIV)–related white matter disease are higher than those of progressive multifocal leukoencephalopathy. As the viral load in HIV-positive individuals increases, the water diffusivity increases as well and the white matter anisotropy decreases.

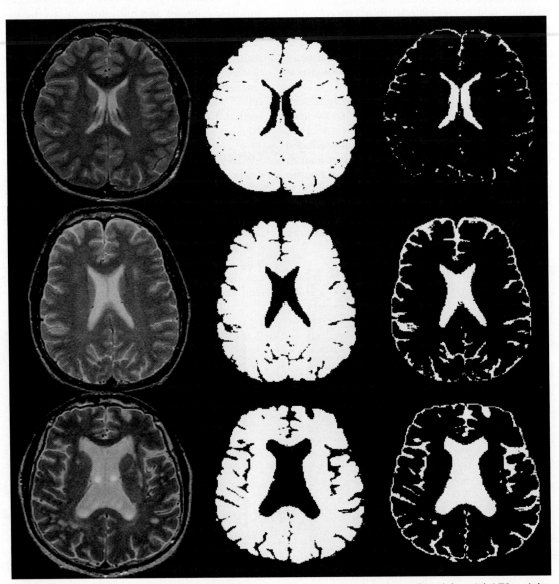

FIGURE 7-19 Spectrum of human immunodeficiency virus (HIV) brain with top row *(from left to right)* T2-weighted image, brain parenchymal segmentation, and cerebrospinal fluid (CSF) segmentation. The *top row* is from a normal patient. Middle row is in asymptomatic HIV patient, and lower row is in acquired immunodeficiency syndrome (AIDS) dementia complex. Observe how the CSF spaces increase and the parenchyma volume decreases.

Protease inhibitors may cause reversal in the cognitive decline associated with HIV encephalopathy and may also result in the regression of periventricular white matter and basal ganglia signal intensity abnormalities. However, the MR imaging response to the highly active antiretroviral therapy (HAART) is often delayed compared with the clinical response. In fact, the MR findings may progress for the first 6 months before regressing or stabilizing with time. Basal ganglia and brain stem manifestations seem to respond to the greatest degree.

Vascular Dementias

Multi-infarct dementia (MID) is an entity seen in patients with long-standing hypertension and/or other risk factors for atherosclerotic disease. The patients are demented and manifest unusual personalities, pseudobulbar effect and/or palsies, incontinence, and ataxia. MID is characterized clinically by a progressive, episodic, stepwise downward course. There may be intervals of clinical stabilization or even limited recovery. People have also called this disease Binswanger disease or subcortical arteriosclerotic encephalopathy, but some neurologists distinguish the stepwise decline and combined gray and white matter disease (MID) from Binswanger, which they claim is slowly progressive and exclusively involves white matter. Even if there really is a clinical difference between Binswanger disease and MID, pathophysiologically, the mechanism is the same—arteriolosclerosis. Lacunar infarcts and central pontine ischemic foci may coexist. Histologically, one sees myelin and axonal loss, astrocytosis, and areas of infarction in deep white matter. MID is characterized on imaging studies by multiple areas of white matter infarction accompanied by severe deep gray matter lacunar disease caused by atherosclerosis of deep penetrating arteries (Fig. 7-20). Overall, cerebral blood flow is diminished.

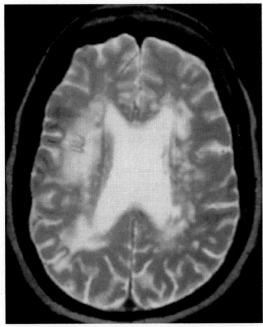

FIGURE 7-20 Multiinfarct dementia. Multiple white matter ischemic foci and a right middle cerebral artery cortical infarct can be found on T2-weighted image in this demented patient with hypertension and diabetes.

Cerebral autosomal dominant arteriopathy with subcortical infarcts and leukoencephalopathy (CADASIL) is another source of vascular dementia. This disorder, linked to the *notch* gene on chromosome 19, presents with severe lacunar disease and subcortical white matter ischemic changes in the frontal and anterior temporal lobes and reduced perfusion to affected areas. Clinically, the patients present with presenile dementia and migraine headaches in the third to fourth decade of life. CADASIL shows a striking pattern of subcortical lacunar infarcts and confluent high signal on FLAIR in the frontal and anterior temporal lobes; this is rarely seen in any other entities and suggests this unusual diagnosis (see Figs. 3-41 and 6-15).

Amyotrophic Lateral Sclerosis

Amyotrophic lateral sclerosis (ALS; also known as Lou Gehrig disease) is a degenerative disease of upper motor neurons, generally manifested in the spinal cord, but also affecting the full extent of the corticospinal tract. The disease causes relentless loss of motor strength in facial, limb, and diaphragmatic musculature with atrophy and hyperreflexia. Death usually occurs from pulmonary infections as airway competency and respiratory muscle integrity are lost. This disease spares the patient from dementia, which may be more tragic because the patient is cognizant of the deadly downward course, which occurs over 3 to 6 years. There are sporadic (95%) and familial forms (5%).

The most common finding seen on brain MR is high signal intensity of the corticospinal tract at the level of the internal capsules. Occasionally, one sees extension of the atrophy or wallerian degeneration along the full length of the corticospinal or bulbospinal tract into the brain stem, cerebral peduncles, internal capsule, corona radiata, and Betz cells of the cortex (Fig. 7-21). This is best seen on coronal FLAIR scans. The more severe the MR findings, the more rapid the progression and severe the disease. Sometimes (around 40% to 60% of cases) the signal intensity of the precentral gyrus is decreased on T2WI, thought to be because of iron deposition (Fig. 7-22). Atrophy of the anterior horn cell region of the spinal cord may be evident. One spectroscopic study of the precentral gyrus region revealed a strong correlation between reductions in NAA and glutamate levels and increases in choline and myoinositol levels with increasing ALS disease severity.

ALS fits under the overall rubric of motor neuron diseases, is the most common form, and affects both upper and lower motor neurons. The other varieties of motor neuron diseases include (1) primary lateral sclerosis (PLS) affecting the upper motor neurons, (2) progressive muscular atrophy (PMA) affecting lower motor neurons of the spinal cord, and (3) progressive bulbar palsy (PBP) affecting lower motor neurons of the brain stem. Selective atrophy of the affected sites may be seen (atrophy in the precentral gyrus region and corticospinal tract hyperintensity in PLS, anterior cord atrophy in PMA, and brain stem atrophy on PBP).

Dyke-Davidoff-Masson Syndrome

In Dyke-Davidoff- Masson syndrome, hemiatrophy of one hemisphere is present. The calvarium may be thickened on that side, the petrous ridge and sphenoid wing

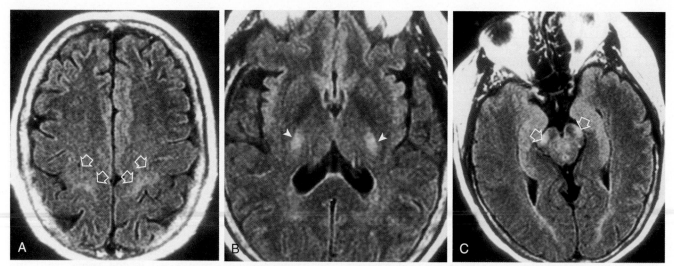

FIGURE 7-21 Amyotrophic lateral sclerosis (ALS). Follow the high signal on these fluid-attenuated inversion recovery (FLAIR) scans from the white matter of the motor strip anterior to the central sulcus bilaterally *(arrows)* **(A),** to the posterior limb of the internal capsule *(arrowheads)* **(B),** to the cerebral peduncles *(arrows)* **(C)** in this patient with ALS.

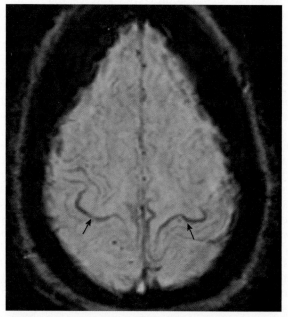

FIGURE 7-22 Amyotrophic lateral sclerosis (ALS). Axial susceptibility-weighted imaging scan shows signal loss along the cortical margins of the precentral gyrus bilaterally *(arrows)* in this patient with ALS.

elevated, and the frontal sinus may be grossly enlarged on the side of the atrophy (Fig. 7-23). This denotes a cerebral injury that occurred early in life or in utero. Usually, a middle cerebral vascular-ischemic cause is invoked. Dyke-Davidoff-Masson syndrome can also be seen with Sturge-Weber syndrome.

Porencephaly

Porencephaly refers to an area of focal encephalomalacia that communicates with the ventricular system, causing what appears to be a focally dilated ventricle (Fig. 7-24). When the dilated ventricle abuts the inner cortex of the skull without significant brain tissue superficially, the CSF

pulsations within the ventricle may cause remodeling of the bone. Sometimes synechiae may actually be within the ventricle, creating a ball-valve effect that leads to progressive enlargement of the ventricle and expansion of the skull. Alternatively, when there is intervening brain tissue and CSF pulses are not transmitted to underlying structures, the skull may thicken in a Dyke-Davidoff-Masson fashion. The causes of porencephaly are manifold and include trauma, infection, and perinatal ischemic injury.

Miscellaneous

Atrophy may also be caused by previous infections, long-standing multiple sclerosis, and extensive traumatic injury to the brain.

Deep Gray Nuclei Disorders

The diseases that affect the deep gray matter nuclei (Table 7-3) are fascinating because of the movement disorders they produce. If you ever wonder what role the caudate nucleus or globus pallidus has in coordination, observe a patient with a degenerative deep gray matter disease. Smooth motor control is dependent on these structures; it is a wonder that larger blood vessels were not devoted to their supply.

Huntington Chorea

Huntington chorea is a dementing disorder manifested by selective atrophy of the caudate nuclei. Patients have involuntary choreoathetoid movements and severe memory impairment. The disorder has an autosomal dominant inheritance pattern and is expressed in young adulthood. The Huntington gene on the short arm of chromosome 4 has been identified and there are genetic tests available for this disorder. This gene may affect the processing of the Huntington protein, which may be the mediator of this basal ganglionic disease. Histochemical analysis may show increased deposition of iron in the caudate and putamen.

On imaging studies, the frontal horns of the lateral ventricles are dilated and rounded as a result of caudate

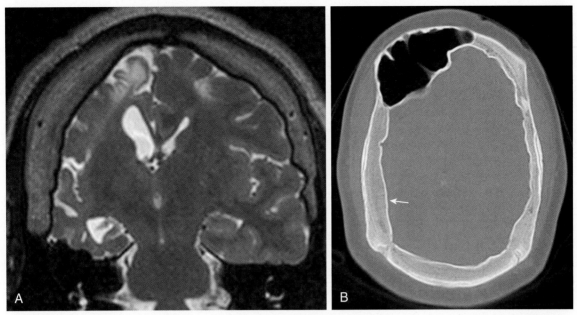

FIGURE 7-23 Dyke-Davidoff-Masson syndrome. **A,** Coronal T2-weighted image shows encephalomalacia of the right cerebral hemisphere and associated enlargement of the right lateral ventricle due to in utero insult. **B,** Axial computed tomographic image in bone windows in same patient shows associated thickening of the right skull *(arrow)* relative to the left in addition to enlargement of the right frontal sinus. Often one sees petrous ridge and sphenoid wing elevation with sinus enlargement ipsilateral to the hemiatrophy.

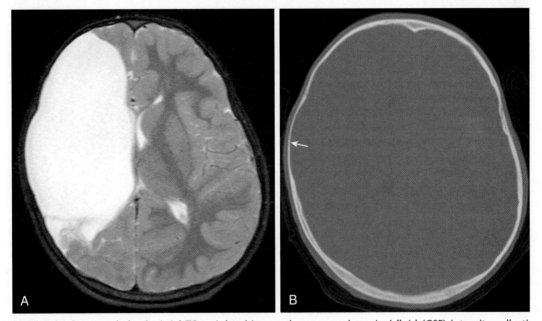

FIGURE 7-24 Porencephaly. **A,** Axial T2-weighted image shows a cerebrospinal fluid (CSF) intensity collection communicating with the right lateral ventricle with no significant overlying cortical tissue. **B,** On the axial computed tomographic image you should note that the adjacent skull *(arrow)* is remodeled because of the long-term transmission of CSF pulsations. (Courtesy K. Chang, MD.)

atrophy. Increased signal intensity in the putamen and globus pallidus has been described in the juvenile form of Huntington disease, and frontal atrophy is usually present (Fig. 7-25). Hypometabolism in the caudate on PET and hypoperfusion on SPECT have been reported.

Pantothenate Kinase-Associated Neurodegeneration

Pantothenate kinase-associated neurodegeneration (PKAN; formerly Hallervorden-Spatz syndrome) is a familial autosomal recessive disorder associated with abnormal

involuntary movements, spasticity, and progressive dementia. Characteristic accumulation of iron-containing compounds is present in the globus pallidus, red nuclei, and substantia nigra pathologically. This is seen on MR as decreased intensity on T2WI. Reports of high signal intensity in the globus pallidus and white matter have also been reported on T2WI, presumably caused by gliosis and/or demyelination with axonal swelling. In addition, cortical and sometimes caudate atrophy is present. In the presence of a young adult with bradykinesia, muscle rigidity, and

TABLE 7-3 **Deep Gray Nuclei Disorders**

Disorder	Distinguishing Imaging Features	MRS Features	Distinguishing Clinical Features
Huntington disease	Caudate atrophy Abnormal signal in lentiform nuclei	Increased lactate	Familial transmission, autosomal dominant Choreiform movements
Wilson disease	Abnormal intensity in basal ganglia Cerebellar atrophy	Decreased NAA/Cho Increased lactate/NAA	Kayser-Fleischer copper rings in globes Coexistent liver disease Ataxia, dysarthria
Pantothenate kinase-associated neurodegeneration	Basal ganglia, red nuclei low-intensity Globus pallidus atrophy	Increased glutamate-glutamine/Cr Decreased PME Increased PCr/Pi	Dementia in young adults Rigidity, bradykinesia, toe-walking, hyperreflexia Familial transmission, autosomal recessive
Leigh syndrome	Favors putamen, other basal ganglia structures, brain stem tegmentum	Increased lactate	Childhood onset Failure to thrive Lactic acidosis
Fahr	Dense symmetric calcifications and atrophy in basal ganglia, thalami, dentate nuclei, subcortical white matter		Autosomal dominant and autosomal recessive inheritance patterns Presents in mid-late adulthood Psychosis, dementia, cognitive impairment, gait abnormality, movement disorders

Cho, Choline; *Cr,* creatine; *MRS,* magnetic resonance spectroscopy; *NAA, N*-acetyl aspartate; *PCr,* phosphocreatine; *Pi,* inorganic phosphate; *PME,* phosphomonoester.

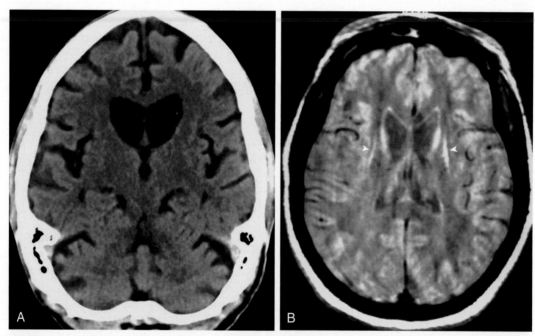

FIGURE 7-25 Huntington disease. **A,** Axial unenhanced computed tomography demonstrates caudate atrophy with ballooning of the frontal horns of the lateral ventricles. **B,** Proton density–weighted image in a different patient shows high signal intensity in the caudate nuclei bilaterally and in the putamina *(arrowheads)*. Again seen is frontal horn dilatation caused by the atrophy of the caudate nuclei.

choreoathetoid movements of the body with ataxia, the presence of dramatic iron deposition within the basal ganglia with a central spot of high signal in the globus pallidus seen on T2WI MR should suggest the diagnosis of PKAN (Fig. 7-26). This pattern is that of the "eye of the tiger," although this has a limited differential diagnosis (Box 7-5).

Wilson Disease

Wilson disease (hepatolenticular degeneration) is an autosomal recessive disorder caused by abnormal ceruloplasmin metabolism with deposition of copper in the liver and brain. Patients are seen in early adulthood with dysarthria, dystonia, and tremors. Rigidity and ataxia may follow. Copper deposition in the cornea accounts for the classic

Kayser-Fleischer rings seen on slit-lamp ophthalmologic examination.

On CT, the patients have atrophy involving the caudate nuclei and brain stem, with nonenhancing foci of hypodensity in the basal ganglia and thalami. MR findings in Wilson disease include T2 hypointensity in the lentiform nucleus and thalami, possibly caused by a paramagnetic form of copper and/or associated iron deposition. Still others have noted bilateral and symmetric deep gray areas of abnormally increased signal on T2WI, in the outer rim of the putamen (Fig. 7-27), ventral nuclear mass of the thalami, and the globus pallidus. In cases of copper toxicosis, one may also see bright hypothalami and bright anterior pituitary gland on T1WI. Signal abnormality may

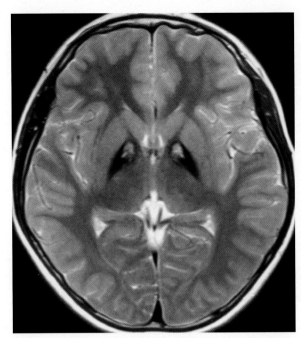

FIGURE 7-26 Pantothenate kinase-associated neurodegeneration (PKAN). Axial T2-weighted image at the basal ganglia level shows excess iron deposition within the globus pallidus bilaterally (dark signal), with central high T2 signal, the "eye of the tiger" sign.

BOX 7-5   Differential Diagnosis of "Eye of the Tiger" Sign

Carbon monoxide poisoning
Leigh disease
Neurofibromatosis
Pantothenate kinase-associated neurodegeneration (PKAN) syndrome
Parkinson disease
Progressive supranuclear palsy
Senescent basal ganglia calcification
Shy-Drager syndrome
Toxins
Wilson disease

---

also be seen in the brain stem, particularly the tegmentum of the midbrain, the red nuclei, and pons. Atrophy is common, especially in the caudate nuclei. Scattered high signal intensity foci in the white matter, pathologically correlating to demyelination, may be present as well. These high intensity foci tend to occur in the corticospinal tract, dentatorubrothalamic tract, and pontocerebellar tract.

Bringing structure and function together, some investigators have shown that abnormalities of the basal ganglia and pontocerebellar tracts depicted on MR images correlate with pseudoparkinsonian signs, whereas abnormal dentatothalamic tracts correlate with cerebellar signs.

Fahr Disease

Fahr disease (familial cerebral ferrocalcinosis) is an inherited disorder with both autosomal dominant and recessive inheritance patterns. The condition is characterized by calcium deposition in arterial, capillary, and venous walls, without abnormality of calcium metabolism. Patients present in mid to late adulthood with progressive symptoms including psychosis, movement disorders, cognitive decline and dementia. Unfortunately, there is no treatment available.

On CT, dense symmetric calcification and atrophy in the basal ganglia, thalami, dentate nuclei, and subcortical white matter is seen. These areas can show high T1 and low T2 signal on MRI. Additional areas of T1 and T2 hyperintensity in the white matter of the centrum semiovale separate from calcified regions may also be seen.

Metabolic and Toxic Disorders

Hepatic Failure

High signal intensity in the basal ganglia on T1WI has been described in patients with hepatic encephalopathy, portosystemic shunting without hepatic encephalopathy, and disorders of calcium-phosphate regulation, and in patients receiving parenteral nutrition therapy (Figs. 7-28, 7-29). With hepatic failure, one may see increased signal intensity in the globus pallidus and putamina, the midbrain, and the anterior pituitary gland. T2 values as well as T1 values are shortened in cases of hepatic encephalopathy, particularly in the globus pallidus. Atrophy is usually seen elsewhere as well, especially affecting the cerebellum. Deposition of manganese, which bypasses the detoxification in the liver, has been postulated as the source of the high signal intensity on T1WI in the globus pallidus. The high signal finding is reversible after correction of the underlying hepatic disorder and/or a return to oral alimentation. Do not mistake high T1 signal in the posterior limb internal capsule reflecting normal myelinated brain for evidence of liver failure in the newborn patient!

With short echo time stimulated echo acquisition mode (STEAM) sequences on MRS, patients with hepatic encephalopathy have decreased myoinositol (mI) to creatine (Cr) and choline to creatine ratios and elevated glutamine-glutamate to creatine ratios compared with normal. These ratios will normalize or even overshoot the other way with correction of the hepatic dysfunction. The mI/Cr ratio seems to be the most sensitive (80% to 85%) indicator of hepatic encephalopathy. The MRS (and clinical) response return toward normal before reversal of the bright basal ganglia sign, which is delayed 3 to 6 months. Other causes of bilateral increased signal intensity on T1WI of the basal ganglia are found in Box 7-6.

Other Metabolic Derangements

Elevated accumulation of ammonia in the brain (hyperammonemia) can be seen in patients with acute liver failure and in patients with inborn errors of metabolism. In patients with severe acute elevations of ammonia, increased T2 signal and diffusion restriction is seen in the basal ganglia as well as insular and cingulate cortices. On MRS, the toxic metabolite glutamate-glutamine can be detected. Resolution of imaging findings can be seen within months of correction of the metabolic derangement.

Hyperglycemia can have a widely variably clinical presentation, ranging from increased thirst to seizures, stupor and coma. In patients with hyperglycemia, increased attenuation in the caudate and globus pallidus can be seen on CT, and on MRI, these structures show increased signal on T1W1. The signal abnormality is usually bilateral but can be unilateral as well (Figure 7-30).

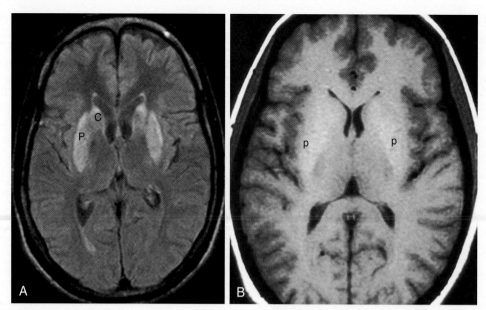

FIGURE 7-27 Wilson disease. **A,** Fluid-attenuated inversion recovery (FLAIR) images are particularly valuable in showing the bright signal in the caudate heads (C) and putamina (P) in this patient with Wilson disease. **B,** The T1-weighted (T1WI) scan also shows high signal intensity in the putamen (P). Although the differential diagnosis is quite broad, the presence of Kayser-Fleischer rings in this individual clinched the diagnosis of Wilson disease.

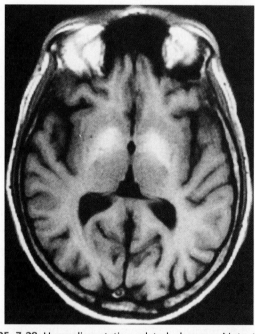

FIGURE 7-28 Hyperalimentation-related changes. Note the bilateral increased signal intensity in the globus pallidus on this T1-weighted image. The apparent cause of the hyperintensity is thought to be related to manganese metabolism, which is abnormal in patients undergoing hyperalimentation.

Patients with profound hypoglycemia can present with seizures and coma. In severe cases, increased T2 signal and diffusion restriction can be seen in the cortex, hippocampi and basal ganglia. In less severe cases, such signal changes can be seen in the corpus callosum, internal capsule, and corona radiata and can be reversible.

Demyelination seen in the setting of central pontine myelinolysis is discussed in Chapter 6. It is included here as a reminder that extrapontine myelinolysis can also be seen in the setting of rapid correction of hyponatremia and manifests on MRI as increased T2 signal in the globus pallidus, putamen, caudate, internal and external capsules and thalamus. Increased T2 signal in the cerebellum can also be seen. Diffusion restriction in involved structures is variable.

Drugs and Toxins

Several toxic substances have been implicated in basal ganglionic and diffuse white matter lesions. These are enumerated in Table 7-4. Anoxic injury and carbon monoxide poisoning affect the globus pallidi (Fig. 7-31). Recent reports have noted that abuse of "ecstasy," a popular recreational party drug, also known as 3,4-methylenedioxymethamphetamine (MDMA), can cause bilateral globus pallidus necrosis, possibly because of prolonged vasospasm. In a similar vein, cocaine can induce ischemia, vasoconstriction, vasculitis, hypertensive white matter changes, intraparenchymal hematomas, subarachnoid hemorrhage, and even moyamoya vasculopathy. "Crack" cocaine has an even higher rate of drug-induced ischemic strokes, secondary to vasoconstriction, platelet dysmetabolism, and episodic hypertension. Heroin inhalation ("chasing the dragon") can result in toxic spongiform encephalopathy with high T2 signal in the posterior cerebral and cerebellar white matter, cerebellar peduncles, splenium of the corpus callosum, and posterior limbs of the internal capsules.

Basal Ganglionic Calcification

Although basal ganglionic calcification may be a normal senescent process, there are endocrinologic/metabolic (hyperparathyroidism [Fig. 7-32], hypoparathyroidism, pseudohypoparathyroidism, pseudopseudohypoparathyroidism, and hypothyroidism), mitochondrial (Kearns-Sayre; mitochondrial encephalomyopathy, lactic acidosis, and stroke-like episodes [MELAS]; and myoclonic epilepsy with ragged red fibers [MERRF]), idiopathic (Fahr disease), congenital (Cockayne syndrome, Down syndrome, trisomy 13), phakomatoses (neurofibromatosis and tuberous sclerosis),

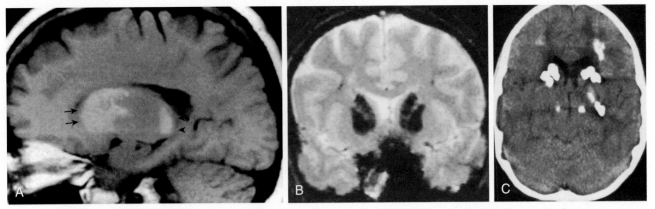

FIGURE 7-29 Basal ganglia changes in patients with pseudohypoparathyroidism. **A,** Sagittal T1-weighted image (T1WI) shows bright signal intensity in the head of the caudate *(arrows),* with similar changes in the posterior portion of the thalamus *(arrowheads).* **B,** On a gradient echo T2WI the striking hypointensity in the basal ganglia is caused by the paramagnetic accumulation. **C,** Axial unenhanced computed tomography in a different patient shows dense calcification within the basal ganglia, thalami, and periventricular white matter.

BOX 7-6   Causes of Hyperintense Basal Ganglia on T1-Weighted Imaging

Senescent basal ganglia calcification
Calcium-phosphorus dysmetabolism
   Hyperparathyroidism
   Pseudohyperparathyroidism
   Pseudopseudohyperparathyroidism
   Hypoparathyroidism
Hepatic encephalopathy
Hyperglycemia
Acquired immunodeficiency syndrome (AIDS)
Hyperalimentation therapy (manganese)
Hemorrhage (hypertension)

Hypoxic-ischemic encephalopathy
Kernicterus
Neurofibromatosis type 1
Wilson disease
Carbon monoxide poisoning
Pantothenate kinase-associated neurodegeneration (PKAN) syndrome
Langerhans cell histiocytosis
Fucosidosis
Japanese encephalitis

Data from Lai PH, Chen C, Liang HL, Pan HB. Hyperintense basal ganglia on T1-weighted imaging. *AJR Am J Roentgenol.* 1999;172:1109-1115.

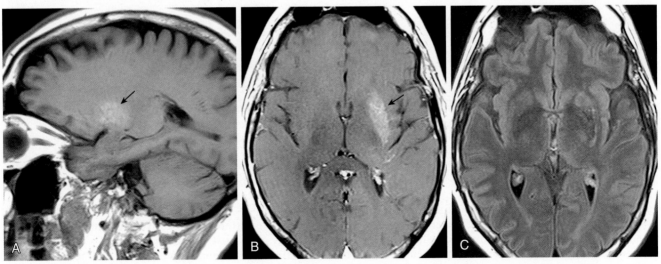

FIGURE 7-30 Hyperglycemia in a patient presenting with hyperglycemia (fasting blood glucose >300) and right upper extremity choreiform movements. **A,** Sagittal T1-weighted image (T1WI) shows high signal in the basal ganglion *(arrow).* **B,** Axial postcontrast T1WI shows the high signal to be a unilateral phenomenon *(arrow)* without associated enhancement. Note that there is no parenchymal signal change evident of fluid-attenuated inversion recovery image **(C).**

TABLE 7-4   Manifestations of Toxic Insults to the Basal Ganglia

Toxin	Basal Ganglia Findings	Other Imaging Findings
Alcohol	Limited until portosystemic shunts, then hyperintense lentiform nuclei on T1WI	Cerebellar (superior vermis) and cortical atrophy, signal changes in periaqueductal gray matter, hypothalamus in Wernicke encephalopathy
Carbon monoxide	Bilateral globus pallidus injury	Diffuse anoxic ischemia to brain, white matter injury sparing subcortical fibers, injury to the hippocampus and cerebellum
MDMA	Globus pallidus necrosis	
Methanol	Putaminal necrosis	Edema of brain, petechial hemorrhages
Mercury	None	Cerebellar, calcarine cortex atrophy
Manganese (see total parenteral nutrition, portosystemic shunts)	Hyperintense globus pallidus in T1WI	Diffuse injury when severe
Lead	Late basal ganglia calcification	Swelling, then atrophy of cerebrum (favoring limbic system) and cerebellum
Cyanide	Globus pallidus infarctions	Diffuse swelling of cerebrum early after ingestion, demyelination
Reye syndrome (aspirin)	Abnormal basal ganglionic signal	Diffuse swelling, increased gray-white differentiation
Hydrogen sulfide	Basal ganglia infarction	Diffuse swelling
Toluene	Basal ganglia increased intensity	White matter disease in infratentorial and supratentorial compartments, diffuse white matter change showing obvious brain atrophy, including hippocampal atrophy and thinning of the corpus callosum
Typewriter correction fluid (trichloroethane)	Lesions in basal ganglia and cortex similar to those observed in patients with methanol and carbon monoxide poisoning	

MDMA, 3,4-methylenedioxy-methamphetamine.

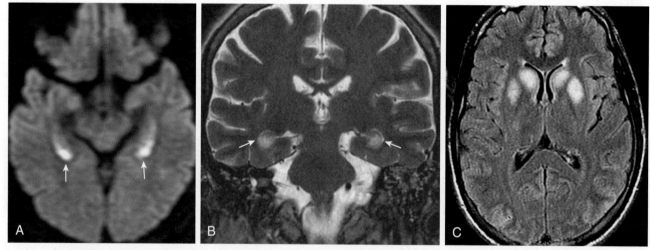

FIGURE 7-31 Anoxic injury. **A,** Axial diffusion-weighted image shows symmetric hyperintensity (with restriction on apparent diffusion coefficient map, not shown), along the hippocampal formations bilaterally *(arrows)*, an area very susceptible to anoxia. **B,** Coronal T2-weighted image shows high signal and swelling in the same location bilaterally *(arrows)*. **C,** On this fluid-attenuated inversion recovery (FLAIR) image in a different patient, one sees symmetric signal abnormality in basal ganglia, also highly susceptible to anoxic injury. (Compliments of Brian Fortman.)

infectious (toxoplasmosis, rubella, cytomegalovirus, herpes simplex [TORCH] infections; cysticercosis; and AIDS, among others), traumatic (dystrophic calcifications), ischemic (anoxia, carbon monoxide poisoning), and iatrogenic (radiation therapy, methotrexate therapy, foreign body reaction to retained shunt catheters) causes of this finding.

Cerebellar Atrophy

Atrophy of the cerebellum has many causes (Box 7-7). We review the most common below.

Alcohol Abuse

Chronic ingestion of excessive amounts of alcohol is probably the most common cause of cerebellar atrophy. The vermis appears to be more commonly involved than other parts of the brain with alcohol abuse, but the whole cerebellum suffers (Fig. 7-33). When one adds poor nutrition and thiamine deficiency to the alcohol abuse, the midbrain, mammillary bodies, and basal ganglia are damaged. This condition is known as Wernicke encephalopathy. By that point, the patient has tremors, delusions, confabulation, ophthalmoparesis,

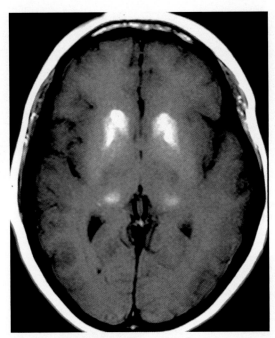

FIGURE 7-32 Hyperparathyroidism. The caudate heads, putamina, and thalami are bright on this T1-weighted image. One should consider causes of calcium-phosphorus dysmetabolism, especially given the absence of globus pallidus involvement. The corollary on computed tomography would be calcification.

ataxia, and confusion. When severe amnesia accompanies Wernicke encephalopathy, Korsakoff syndrome is invoked, although the two diseases are caused by the same factors. Although chronic alcoholism is the most common cause of Wernicke encephalopathy, other possible etiologies include bariatric surgery, gastric bypass, hyperemesis gravidarum, prolonged infectious-febrile conditions, carcinoma, anorexia nervosa, and prolonged voluntary starvation.

In addition to generalized cortical and cerebellar vermian atrophy seen on CT and MR, high signal intensity areas in the periaqueductal gray matter of the midbrain (40%), the paraventricular thalamic regions (46%), the mamillothalamic tract, and in the tissue surrounding the third ventricle on T2WI can be seen (Fig. 7-34). Reversible thalamic/pulvinar lesions in the dorsal medial nuclei have also been reported. These areas may or may not enhance (in some cases the enhancement may be dramatic, almost sarcoid-like), may have high signal on DWI scans, and may be associated with mammillary body atrophy. In fact, mammillary body enhancement or abnormal signal on DWI and/or FLAIR may be the sole manifestation of Wernicke encephalopathy.

Long-Term Drug Use

Infratentorial atrophy is also a common finding after long-term use of certain drugs. Phenytoin (Dilantin) and phenobarbital are the classic drugs that produce cerebellar atrophy. Patients may have reversible nystagmus, ataxia, peripheral neuropathies, and slurred speech. The cerebellar degeneration becomes irreversible after long-term administration. It is manifested by dilatation of the fourth ventricle and cerebellar sulci. Phenytoin may also cause thickening of the skull when used early in life for a long period.

TOXIC

Alcohol abuse
Long-term use of phenytoin (Dilantin), phenobarbital, high-dose cytarabine
Mercury poisoning
Poor nutrition
Steroid use
Thallium poisoning

VASCULAR

Strokes
Vertebrobasilar insufficiency

HEREDITARY

Ataxia-telangiectasia
Dandy-Walker complex
Down syndrome
Friedreich ataxia
Hereditary ataxia
Joubert syndrome
Olivopontocerebellar degeneration
Shy-Drager syndrome
Vermian hypoplasia

INFECTIOUS

Creutzfeldt-Jakob disease
Cytomegalovirus/rubella

TREATMENT RELATED

Postoperative
Radiation therapy

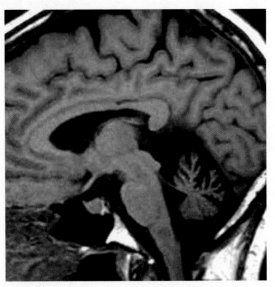

FIGURE 7-33 Cerebellar atrophy. Sagittal T1-weighted image shows enlargement of the cerebellar sulci and decrease in the size of the cerebellum.

Olivopontocerebellar Degeneration

Olivopontocerebellar degeneration is a disease that occurs in young adults and is associated with cerebellar and brain stem atrophy. The disease may occur sporadically or may be transmitted in an autosomal dominant inheritance. Patients generally have truncal and limb ataxia. Problems with speech, tremors, nystagmus,

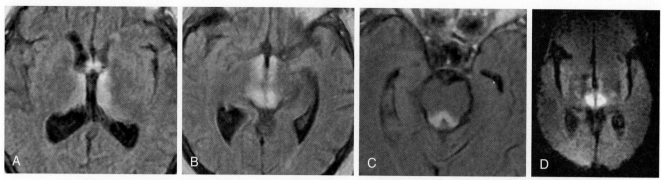

FIGURE 7-34 Wernicke encephalopathy. **A-C,** Three fluid-attenuated inversion recovery (FLAIR) images show high signal in periaqueductal gray matter, medial thalami, and hypothalamus. **D,** Acute cytotoxic edema is demonstrated on the diffusion-weighted image in the periaqueductal gray matter region. (Compliments of Mauricio Castillo.)

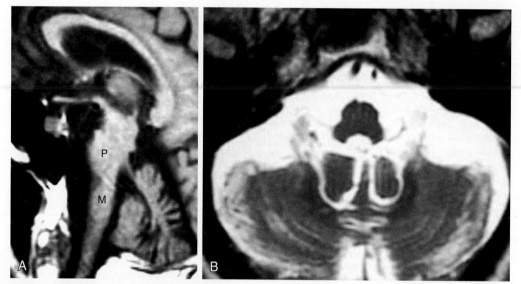

FIGURE 7-35 Olivopontocerebellar degeneration. **A,** Sagittal T1-weighted image (T1WI) shows marked cerebellar atrophy with flattening of the belly of the pons and the enlarged inferior cerebellar cistern. **B,** Axial T2WI reveals atrophy of the pons as it swims in the prepontine cistern.

rigidity, brain stem dysfunction, and mild mental impairment may ensue. On imaging, the pons and inferior olives are strikingly small, as are the cerebellar peduncles (Fig. 7-35). Cerebellar atrophy, particularly of the vermis, is marked. Remember that this is sometimes considered in the spectrum of Parkinson-plus syndromes and multisystem atrophy.

Friedreich Ataxia

Friedreich ataxia is an autosomal disorder associated with cerebellar atrophy that presents in the second decade of life. It has autosomal dominant and recessive forms. Cardiac anomalies may coexist, particularly affecting the conduction pathways. Patients are usually seen in late childhood with lower extremity ataxia, kyphoscoliosis, and tremors of the upper extremity. Areflexia in the lower extremities, scoliosis, deafness, optic atrophy, and dysarthria may occur. In addition to cerebellar volume loss, atrophy of the cervicomedullary junction with a decreased anteroposterior diameter of the upper part of the cervical cord has been reported. Signal intensity in the cerebellum and cord is normal.

Ataxia Telangiectasia

Ataxia telangiectasia (Louis-Bar syndrome) is an autosomal recessive disorder characterized as one of the phakomatoses. Hallmarks of the disease include cerebellar vermian and hemispheric atrophy and telangiectatic lesions on the face, mucosa, and conjunctiva. It occurs in 1 in 20,000 to 100,000 live births and has been mapped to chromosome 11. An associated abnormality of the immune system (predominantly affecting immunoglobulin A) causes recurrent sinus and lung infections. Leukemia or non-Hodgkin lymphoma may develop, and the patients usually die of the disease in childhood. If the patients survive to adulthood, they may be besieged by breast, gastric, CNS, skin, and liver malignancies. Increased sensitivity to radiation effects and progeria are manifestations of the disorder as well. The typical imaging findings are vermian/holocerebellar atrophy and increased T2 white matter signal intensity. Rarely, intracranial occult cerebrovascular malformations may be present. These are manifest as dark foci of hemosiderin on T2* scans of the brain and may be due to perivascular hemorrhages.

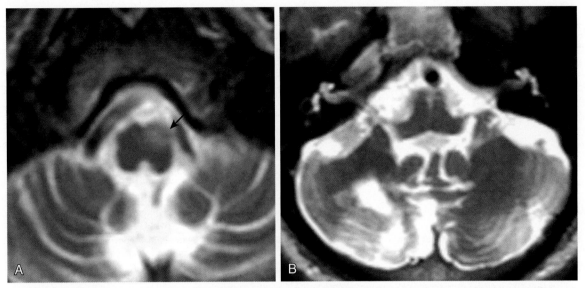

FIGURE 7-36 Hypertrophic olivary degeneration. **A,** On this axial T2-weighted image, there is focal increased signal and slight enlargement within the left ventral medulla *(arrow)* at the level of the inferior olivary nucleus. **B,** The hypertrophic olivary degeneration has occurred due to insult in the right dentate nucleus.

Ischemia

Chronic vertebrobasilar atherosclerotic disease may result in parenchymal volume loss and gliosis seen on imaging from multiple prior infarctions. Usually, supratentorial manifestations of atherosclerotic ischemic disease (lacunar infarctions, white matter gliosis) coexist and are the tip-off to the disease.

Hypertrophic Olivary Degeneration

Hypertrophic olivary degeneration (HOD) occurs as a result of ischemic, traumatic, neoplastic, or vascular insult to the components of the "triangle of Guillain and Mollaret," that is, the red nucleus, inferior olivary nucleus and contralateral dentate nucleus. These structures are connected by the superior cerebellar peduncle (dentate to red nucleus), the central tegmental tract (red nucleus to inferior olivary nucleus), and the inferior cerebellar peduncle (inferior olivary nucleus to dentate nucleus—the olivodentate tract). Lesions in the first two tracts can lead to neuronal degeneration of the inferior olivary nucleus. Initially this leads to high T2 signal (first 2 months) and then hypertrophy with increased T2 signal (6 months to 3 to 4 years) of the inferior olivary nucleus of the medulla (Fig. 7-36). In the later stages, olivary hypertrophy resolves and increased T2 signal persists. Contralateral cerebellar atrophy can occur although this is more commonly seen with involvement of the olivodentate fibers. Patients typically present with a palatal tremor termed "palatal myoclonus."

Paraneoplastic Syndromes

Cerebellar atrophy is one of the manifestations of paraneoplastic syndromes. These may be associated with neuroblastoma, Hodgkin disease, and ovarian, gastrointestinal, lung, and breast cancers. An autoimmune mechanism is implicated because antibodies (anti-Yo antibodies) to cerebellar antigens (predominantly directed against Purkinje cells) may be identified. The cerebellar degeneration precedes discovery of the primary tumor in up to 60% of cases.

METABOLIC DISORDERS

The number of metabolic disorders that affect the brain is probably in the hundreds. We touch on the major metabolic disorders and general imaging manifestations here.

Mucopolysaccharidoses

Clinically, the mucopolysaccharidoses (MPS) are characterized by mental retardation, peculiar facies, and musculoskeletal deformities. The mucopolysaccharidoses are associated with brain atrophy and abnormalities of the white matter on imaging (Table 7-5). The diagnoses of these disorders are usually based on biochemical and/or chromosomal evaluation. The mucopolysaccharidoses are usually manifested on MR as diffuse cribriform or cystic-appearing areas of abnormal signal intensity in the white matter (low on T1WI, high on T2WI) and hypodensity in the white matter on CT (Fig. 7-37). There often is dilatation of the perivascular spaces that are seen in the striatocapsular regions and occasionally in the corona radiate white matter. At the foramen magnum, thickening of the dura (or mucopolysaccharide deposition) may cause medullary compression in some of the mucopolysaccharidoses. Some of these disorders are associated with enlarged heads and thickened skulls. MPS III, Sanfilippo disease, is unique among the mucopolysaccharidoses in that the non-CNS abnormalities are relatively minor. Arachnoid cysts are included among the intracranial findings with MPS disorders. It is presumed that build-up of glycosaminoglycans in the leptomeninges account for the arachnoid membrane obstruction and ball valve effect that can lead to such cysts, as well as optic nerve sheath dilation.

Lipidoses and Other Storage Diseases

The gangliosidoses are diseases that manifest white matter abnormalities and cortical atrophy (Table 7-6). Lacunar and white matter small vessel infarctions may occur

TABLE 7-5   Imaging Findings in Mucopolysaccharidoses

Eponym	Deficient Enzyme	Inheritance	CNS Findings
Hurler (MPS I)	α-l-Iduronidase	AR	Thickened meninges, atrophy, kyphosis, atlantoaxial subluxation, ligamentous thickening causing cord compression, cribriform or cystic areas within white matter or basal ganglia, delayed myelination, hydrocephalus, vertebra plana
Scheie (MPS IS)*	α-l-Iduronidase	AR	Pigmentary retinopathy
Hunter (MPS II)	Iduronate sulfatase	XR	Macrocrania; enlarged Virchow-Robin spaces; cribriform or cystic areas within white matter, corpus callosum, or basal ganglia; delayed myelination; thickening of dura matter, especially in the spine; communicating hydrocephalus
Sanfilippo (MPS III)	Sulfamidase	AR	Atrophy, cribriform or cystic areas within white matter or basal ganglia, delayed myelination, retinal degeneration
Morquio (MPS IV)	Galactosamine-6-sulfate sulfatase	AR	Ligamentous thickening leading to cord compression, odontoid hypoplasia, atlantoaxial subluxation, vertebra plana, white matter high intensity
Maroteaux-Lamy (MPS VI)	*N*-Acetyl galactosamine-sulfatase B	AR	Meningeal thickening, hydrocephalus, perivascular gliosis, cord compression from ligamentous thickening, atlantoaxial subluxation
Sly (MPS VII)	Glucuronidase	AR	None or hydrocephalus, white matter high intensity

AR, Autosomal recessive; *CNS,* central nervous system; *MPS,* mucopolysaccharidosis; *XR,* X-linked recessive.
*Initially classified as MPS V.

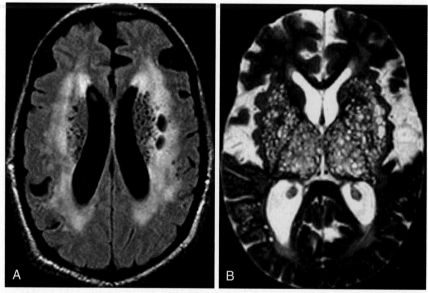

FIGURE 7-37 Hunter mucopolysaccharidosis. **A,** Axial fluid-attenuated inversion recovery (FLAIR) scan shows dilated perivascular spaces (dark) and abnormal corona radiata white matter (bright). **B,** On T2-weighted image, the perivascular spaces of the deep gray matter are highlighted in a cribriform pattern.

in Fabry and Gaucher disease. Of the lipidoses, cerebellar ataxia and cerebellar atrophy are more common in Tay-Sachs disease and Niemann-Pick disease. Bilateral thalamic abnormalities with hyperdensity on CT is characteristic of Sandhoff disease, but may also be seen with Sandhoff disease, Tay-Sachs disease, Fabry disease, and toxic exposures (Fig. 7-38).

Mitochondrial Defects

Leigh Disease

Leigh disease (subacute necrotizing encephalomyelopathy) may be the prototype of the mitochondrial enzymatic disorders (Table 7-7). The disease is thought to be due to a deficiency in the enzymes associated with pyruvate breakdown; its accumulation leads to lactic acid build-up. It is manifested clinically by motor system abnormalities,

ataxia, nystagmus, ophthalmoplegia, spasticity, psychomotor retardation, cranial palsies, and metabolic acidosis. Leigh disease is a neurodegenerative disorder seen in children under 2 years old and expressed on MR by abnormally high signal intensity of the putamina bilaterally on T2WI (Fig. 7-39). Other areas may also show abnormal intensity including the caudate nuclei, globus pallidi, thalami, and brain stem. Lower brain stem involvement correlates with loss of respiratory control, a potentially fatal complication of the disease. Diffuse supratentorial white matter T2 hyperintensity may accompany the deep gray matter findings. These areas do not enhance. MR may also detect midbrain and spinal cord atrophy in this disorder. The telltale laboratory finding is metabolic acidosis with increased lactate levels; the lactate may be detectable with proton or phosphorous MR spectroscopic examination.

TABLE 7-6   Imaging Findings in Lipidoses, Gangliosidoses, and Other Storage Diseases

Disorder	Deficiency	Inheritance	CNS Findings
Ceroid lipofuscinosis (Batten disease)	Several proteins and enzymes identified	AR	Cerebellar or less often cortical atrophy, periventricular gliosis, optic atrophy, thickened dura, thick skull
Fabry disease	α-Galactosidase A	XR	Multiple infarcts, vascular stenoses, and thromboses; basal ganglia lacunae
Farber disease (lipogranulomatosis)	Acid ceramidase	AR	Atrophy, hydrocephalus
Fucosidosis	α-l-Fucosidase	AR	Demyelination, gliosis, atrophy, high-intensity T1 in basal ganglia, thickened skull, craniostenosis, poorly developed sinus, short odontoid, platyspondyly, anterior beaking, scoliosis, vacuum disk, square vertebrae
Gaucher disease	Acid β-glucosidase	AR	Minimal to no atrophy, vertebral body collapse, dementia, infarcts
GM₁ gangliosidosis	β-Galactosidase	AR	White matter disease, late atrophy, macroglossia, organomegaly, dementia (1-3 yr), seizures, gibbus deformity, beaked vertebrae, bright putamen caudate atrophy on T2WI
Krabbe disease (globoid cell leukodystrophy)	Galactocerebroside β-galactosidase	AR	Increased CT attenuation in cerebellum, thalami, caudate nuclei; decreased signal on T2WI of cerebellum; optic atrophy; small atrophic brain, demyelination, and intracranial optic nerve enlargement
Mannosidosis	α-Mannosidase	AR	Low density in parieto-occipital white matter, atrophy, thickened skull, lenticular opacities, brachycephaly, craniostenosis
Niemann-Pick disease	Sphingomyelinase	AR	Normal vs areas of demyelination, gliosis, slight atrophy, small corpus callosum
Pompe disease	Acid maltase	AR	Gliosis, macroglossia
Sandhoff disease	Hexosaminidase A and B (gangliosidase)	AR	Atrophy, thalamic hyperdensity on CT, possibly caused by calcification; diffuse white matter disease
Tay-Sachs disease (GM₂ gangliosidosis)	Hexosaminidase A (gangliosidase)	AR	Megalencephaly early; atrophy late, especially of optic nerves and cerebellum; demyelination; high-density thalami, deep gray abnormal signal, large caudate

AR, Autosomal recessive; *CNS,* central nervous system; *CT,* computed tomography; *GM₁,* monosialotetrahexosylganglioside; *T2WI,* T2-weighted imaging; *XR,* X-linked recessive.

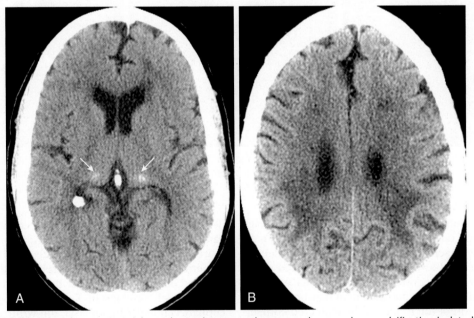

FIGURE 7-38 Fabry disease. **A,** Axial unenhanced computed tomography scan shows calcification isolated to the ulvinar region of the thalmi *(arrows),* which is felt to be a characteristic imaging feature of this condition. **B,** More superiorly, chronic ischemic changes are present in the periventricular white matter. (Courtesy K. Chang, MD.)

Mitochondrial Encephalomyopathy, Lactic Acidosis, and Strokelike Episodes Syndrome

MELAS syndrome represents an unusual form of mitochondrial dysmetabolism. Imaging studies demonstrate multiple areas of cortical high signal intensity abnormality on T2WI, which often crosses a traditional cerebral artery distribution (Fig. 7-40). Although the lesions may be bright on DWI scans, they usually do NOT show reduced apparent diffusion coefficient (ADC) values and are therefore not acute infarcts. The patients may have cortical blindness. Strangely

TABLE 7-7 Imaging Findings in the Mitochondrial Disorders

Disorder	Inheritance	Deficient Enzyme/Mutated Gene	CNS Findings
MELAS syndrome	Maternal	*A3243G* mutation	Pseudostrokes, demyelination in cord
MERRF syndrome	Maternal	*A8344G, T8356C*	White matter demyelination, especially superior cerebellar peduncles, posterior columns, gliosis
Kearns-Sayre syndrome	Spontaneous	MtDNA	Diffuse white matter disease, calcified basal ganglia, more cerebellar than cortical atrophy, retinopathy, ophthalmoplegia, microcephaly
Alpers disease	AR	Cytochrome c-oxidase	Microcephaly, posterior cortical encephalomalacia, atrophy
Leigh disease	AR	Pyruvate dehydrogenase	Periaqueductal gray and putaminal abnormal signal intensity, swollen caudate, low density in putamina on CT, demyelination, lactate peak on MRS
Menkes kinky hair disease	XR	Copper metabolism ATPase, *Menkes/OHS* gene	Subdural effusions, irregular vascular atrophy, gliosis, increased wormian bones, infarcts, tortuous dilated vessels
Zellweger syndrome (cerebrohepatorenal syndrome)	AR	Peroxisome assembly, *PXR1* gene	Abnormal neuronal migration (polymicrogyria, heterotopias, pachygyria), white matter disease, optic atrophy, decreased *N*-acetyl aspartate level, macrocephaly, open sutures, ventricular dilatation
Refsum disease	AR	Phytanic acid-2-hydroxylase	Demyelination of spinal cord tracts, atrophy, abnormal signal intensity, dentate nucleus

AR, Autosomal recessive; *CNS*, central nervous system; *CT*, computed tomography; *MELAS*, mitochondrial encephalomyopathy, lactic acidosis, and strokelike episodes; *MERRF*, myoclonic epilepsy with ragged red fibers; *MRS*, magnetic resonance spectroscopy; *XR*, X-linked recessive.

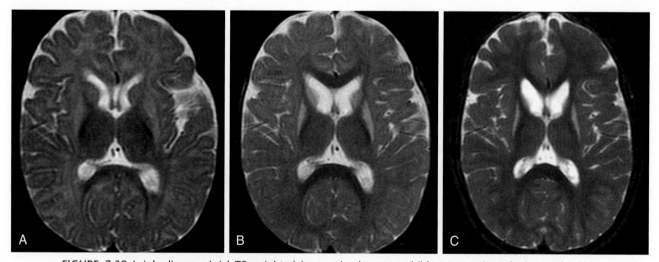

FIGURE 7-39 Leigh disease. Axial T2-weighted images in the same child at approximately 3 months **(A)**, 8 months **(B)**, and 2 years of age **(C)** show bilateral symmetric putaminal and caudate high signal intensity, with progressive volume loss of these structures over time.

enough, these lesions may disappear with time, leaving only minimal sulcal dilatation in their place. The abnormality is associated with serologic and spectroscopic evidence of elevated lactic acid levels. The lesions associated with MELAS have also been reported in the putamina, caudate nuclei, and thalami. Contrast enhancement may be present. Although resolution of the lesions is the expected course, one may see new lesions appearing as well. Ultimately, atrophy favoring the posterior cerebrum is expected.

Other mitochondrial encephalomyopathies include Kearns-Sayre syndrome (which also affects the basal ganglia) and MERRF syndrome. MERRF has a higher rate of brain stem focal lesions than MELAS or Kearnes-Sayre.

Kearns-Sayre syndrome affects the orbits with retinitis pigmentosa, ophthalmoplegia, extraocular muscle weakness (and possible atrophy), and affects the heart, resulting in cardiac conduction deficits and a cardiomyopathy. Basal ganglia calcification occurs in Kearns-Sayre syndrome more frequently than in MELAS or MERRF syndrome. Hyperintense basal ganglia, cerebral and cerebellar atrophy, and diffuse white matter hyperintensity are the most common MR findings in Kearns-Sayre syndrome. Microcephaly, cerebellar hypoplasia, and white matter demyelination may occur as well.

Zellweger syndrome is an autosomal recessive disorder with the defective gene on chromosome 8 characterized by hypomyelination, microgyric and pachygyric cortical malformations that are most severe in the perisylvian and perirolandic regions as well as caudothalamic groove cysts. Soft-tissue calcifications may be present. Dolichocephaly and sutural widening and macrocephaly are seen. A marked decrease in NAA is found at MRS.

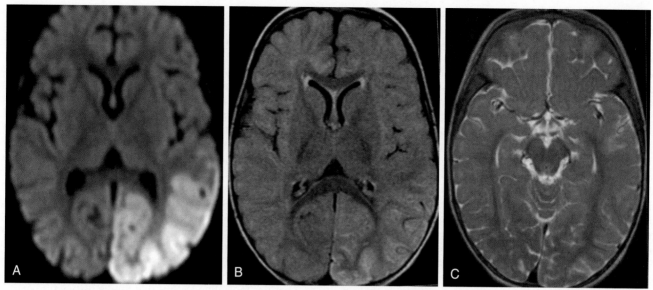

FIGURE 7-40 Mitochondrial encephalomyopathy, lactic acidosis, and strokelike episodes (MELAS) syndrome. **A,** Axial diffusion-weighted image shows cortically based diffusion hyperintensity in the left occipital and posterior temporal lobe. **B,** Axial fluid-attenuated inversion recovery (FLAIR) and **(C)** axial T2-weighted imaging show high signal intensity in the swollen cortex in the region of diffusion abnormality. Note that the abnormality is not confined to a single vascular territory but rather spans the right middle cerebral artery and posterior cerebral artery distributions without affecting the basal ganglia.

TABLE 7-8   Imaging Findings in Amino Acid Disorders

Disorder	Inheritance	Deficient Enzyme	Brain Findings
Hyperglycinemia	AR	Glycine metabolism	Microcephaly, atrophy, dysmyelination, corpus callosum anomalies
Maple syrup urine disease	AR	Branched-chain amino acid decarboxylase	Abnormal neuronal migration, acute edema after birth followed by atrophy at few months, demyelination of posterior white matter, dorsal midbrain edema, deep gray matter swelling, deep cerebellum edema; MRS shows branched-chain amino acids
Methylmalonic academia	AR	CoA carboxylase or mutase	Early swelling then atrophy, decreased density in white matter, basal ganglia (especially globus pallidus) injury, delayed myelination
Phenylketonuria	AR	Phenylalanine hydroxylase	White matter demyelination posteriorly (often optic radiations), neuronal migrational abnormalities, atrophy, cerebellar and brain stem high signal foci, basal ganglia calcification, atrophy
Propionic acidemia	AR	CoA carboxylase	Atrophy, decreased density in white matter
Glutaric acidemia type II	AR	Acyl CoA dehydrogenase	Globus pallidus and white matter hyperintensities on T2WI, macrocephaly, atrophy (frontal/temporal), hydrodense basal ganglia, bat-wing dilatation of sylvian fissures, temporal arachnoid cysts, subdural hematomas

AR, Autosomal recessive; *MRS,* magnetic resonance spectroscopy; *T2WI, T2*-weighted imaging.

Aminoacidopathies and Other Enzyme Deficiencies

As a group, the amino acid metabolic disorders (Table 7-8) are usually seen in children and are manifested neurologically by developmental delay, mental retardation, seizures, and vomiting (when not diagnosed early by screening tests). The normal maturation of the brain is often delayed, and myelination may be affected. Brain atrophy is seen in later stages. Usually dietary manipulations and/or replacement therapies are effective in gaining some return of function and control of the disease.

Krabbe disease, also referred to as globoid cell leukodystrophy (GLD), is due to galactocerebroside–beta-galactocerebrosidase deficiency. The gene for this disease has been mapped to chromosome 14. Cerebrosides from catabolized myelin cannot be degraded to galactose and ceramide in this condition and leads to harmful build-up of galactosylsphingosine cerebroside and psychosine in the brain parenchyma. When the disease presents before age 2 (the infantile form), the pyramidal tracts, cerebellar white matter, deep gray matter, posterior corpus callosum, and parietooccipital white matter are usually involved (as they are the first areas to myelinate). The U-fibers are generally spared. Hyperdensity of the thalami may be seen on CT. MRS findings include abnormally elevated choline and myoinositol (thought to be secondary to myelin breakdown products or phospholipid membrane metabolism). There is a marked decrease in relative anisotropy of the affected white matter tracts identified on diffusion-weighted scans. Atrophy ultimately develops.

In the late-onset group, the same locations may be involved except for the cerebellar white matter and deep gray matter.

The corticospinal tract is involved consistently (as it is in the infantile form). Splenial disease is also common in this form. The upper motor tract, corresponding to the lower extremity region, is affected to a greater extent than the regions that subserve the face and arms. Cerebral atrophy is present in early but not late disease. Optic atrophy is present in both varieties. Treatment trials of stem cell transplantation have been encouraging. Successful treatment has corresponded with a trend toward increased relative anisotropy.

Glutaric acidemia may be suggested when one finds macrocephaly with open opercula and abnormally high signal in the basal ganglia, especially the globus pallidus. Patients may present with encephalitis. Other imaging findings include atrophy, subarachnoid space dilatation anterior to the temporal lobes and basal ganglia volume loss. Bleeding abnormalities can occur in some patients with this condition, and intracranial hemorrhages (often SDHs) of variable age can be appreciated on imaging. It can be challenging to distinguish the subdural hemorrhages in this condition from those occurring in the setting of nonaccidental trauma (inflicted injury).

Homocystinuria is caused by cystathionine beta synthase deficiency, mapped to chromosome 21. The resultant high plasma levels of homocysteine result in multiple thrombotic events, including strokes. Premature atherosclerosis occurs. The thromboses may be either arterial or venous. The presence of lens subluxations might suggest a diagnosis of Marfan syndrome but those with homocystinuria are "up and in" and those with Marfan disease are "down and out." Other neuroradiologic findings may include optic atrophy, osteoporosis, cataracts, scoliosis, and biconcave vertebral bodies.

In maple syrup urine disease (secondary to abnormal decarboxylation of branched chain amino acids), head ultrasound may show symmetric increase of echogenicity of periventricular white matter, basal ganglia (mainly pallidi), and thalami. CT and MR show diffuse edema in similar locations as well as in the cerebellum and capsular regions. Delayed myelination is present. MRS may show a peak at 0.9 ppm representing the branched chain amino acid peaks.

Propionic acidemia (PPA) and methylmalonic acidemia (MMA) are inherited disorders of the tricarboxylic cycle caused by enzymatic defects, which can lead to metabolic acidosis. These diseases are detected within the first month of life as babies have feeding difficulties, muscular hypotonia, choreoathetosis, microcephaly, and seizures. The disease occurs more frequently in patients from Saudi Arabia where it is transmitted as an autosomal recessive disorder (though can be seen with biotin and/or cobalamin deficiency also). Atrophy and abnormal white matter signal with delayed myelination are seen at conventional MRI. Basal ganglionic lesions (typically in the globus pallidus in MMA and in globus pallidus, caudates, and putamina in PPA) after 1 year of age may appear. The etiology is unclear, and sometimes they resolve. Lactate peaks may be seen on proton MRS. These two entities look alike except that low-density white matter on CT is present in PPA but not in MMA. Delays in myelination are more evident in MMA.

Lysosomal Disorders

Juvenile neuronal ceroid lipofuscinosis is a lysosomal neurodegenerative disorder caused by the accumulation of lipopigment in neurons. The neonatal form of this disease is associated with microcephaly and visual deficits. Volume loss of the CNS, most prominently in the cerebellum, hypointense thalami and hyperintense periventricular white matter on T2WI, and proton MRS spectra revealing reduced NAA and increased myoinositol and glutamate/glutamine characterize this disorder. Periventricular high-signal rims are seen after 1 year of life. Hypointensity of the thalami and basal ganglia and peritrigonal CSF-like hyperintensities on T2WI scans are also seen.

Chapter 8

Congenital Anomalies of the Central Nervous System

TIMING OF FORMATION OF CONGENITAL LESIONS

If you think your ex utero life is complicated, wait till you see what goes on in utero! Understanding development of the central nervous system (CNS) aids in understanding congenital brain and spine anomalies and coexistence of multiple anomalies. Although there is some overlap, to put things simply, the CNS goes through six major developmental events that include (1) primary neurulation (3 to 4 weeks of gestation); (2) promesencephalic development (2 to 3 months of gestation); (3) neuronal proliferation (3 to 4 months of gestation); (4) neuronal migration (3 to 5 months of gestation); (5) organization (5 months of gestation to postnatal years); and (6) myelination (birth to postnatal years). Above mentioned termination periods of each event means that the malformation of that particular event may have its onset before the event is over.

Primary Neurulation

Neurulation is a series of inductive events that takes place in the dorsal aspect of the embryo and results in development of the brain and spinal cord. Primary neurulation involves events that are related to formation of the brain and spinal cord exclusive of its most caudal region. The first fusion of neural folds occur in the lower medulla. Closure generally proceeds rostrally and caudally, although it is not a simple zipper-like process. Disorders of primary neurulation are cranioschisis, anencephaly, myeloschisis, encephalocele, myelomeningocele, and Chiari II malformations. Various malformations of neural tube closure are accompanied by axial skeleton, meningovascular and dermal covering abnormalities.

Secondary neurulation occurs later than primary neurulation with sequential processes of canalization and retrogressive differentiation. Disorders of secondary neurulation result in malformations of the lower sacral and coccygeal segments of the neural tube affecting the conus medullaris, filum terminale, and cauda equina. Lipomas of the filum, short thickened tethered fila and tethered cords, and caudal regression syndrome with sacral agenesis are also listed under this category.

Promesencephalic Development

The primary inductive relationship is between the notochord-prechordal mesoderm and forebrain and takes place ventrally in the rostral end of the embryo, thus some authors use the term *ventral induction*. Formation of the face and forebrain occur during this event, so malformations of the brain are usually accompanied by facial anomalies (such as cyclopia and probosci). The spectrum of malformations can be as severe as aprosencephaly, to perhaps clinically occult callosal malformations. Three major sequential events (and related malformations) are promesencephalic development (atelencephaly), premesencephalic cleavage (holoprosencephaly), and midline promesencephalic development (agenesis of corpus callosum, agenesis of septum pellucidum, septooptic dysplasia, hypothalamic dysplasia).

Cerebellar development also occurs at the 5- to 15-week period of fetal development. Just as there is cerebral hemispheric development from neural tissue derived from the germinal matrix around the lateral ventricles, so too is cerebellar development dependent on the germinal matrix about the fourth ventricle. Because hemispheric development occurs before vermian development, and the superior vermis forms before the inferior vermis, it is rare to see hemispheric anomalies without vermian maldevelopments and superior vermian lesions without associated inferior vermian anomalies. Inferior vermian hypoplasia may be seen in isolation. Those entities associated with vermian dysgenesis include the Dandy-Walker malformation, Joubert syndrome, and rhombencephalosynapsis.

Disorders of hemispheric development include hemispheric hypoplasia, but this is a rare bird more commonly seen with other supratentorial and vermian anomalies.

Neuronal Proliferation

All neurons and glia are derived from the ventricular and subventricular zones of the germinal matrix. Disorders of neuronal proliferation can result in small or large brain (microcephaly or macrocephaly). Keep in mind the neurocutaneous syndromes when assessing macrocephaly or hemimegalencephaly...stay tuned, more on this later in the chapter.

Neuronal Migration

The neurons migrate from the ventricular and subventricular zones of the subependyma to their final residence for life. Initially neurons migrate by translocation of the cell body followed by two basic varieties of cell migration: radial and tangential. Radial migration leads to projection neurons of the cerebral cortex and deep nuclei in the cerebrum, and Purkinje cells in the cerebellum. Tangential migration leads to interneurons of the cerebral cortex and internal granule layer of the cerebellar cortex. The layering of the neurons are inside out: early arriving neurons are deep whereas late arriving neurons are in the superficial

Posterior to anterior
Central to peripheral
Caudal to cephalad
T1-weighted imaging to T2-weighted imaging
Immature to mature

aspect of the cortex. Disorders of neuronal migration include schizencephaly, lissencephaly (pachygyria), polymicrogyria, heterotopia, and focal cerebrocortical dysgenesis. Tip: Commisural anomalies (e.g., corpus callosum agenesis/dysgenesis) and septum pellucidum anomalies can accompany these disorders, so keep a look out!

Neuronal Organization

A series of complex processes take place: subplate neuron differentiation; alignment, orientation and layering of cortical neurons; establishment of synaptic contacts; cell death; proliferation and differentiation of glia. Primary disorders are mental retardation, autism, and syndromes such as Fragile X, Rett, Down, and Angelman. Potential disturbances such as those seen in premature infants (germinal matric hemorrhage spectrum, periventricular leukomalacia) can occur.

Myelination

The structure of myelin is rich in lipid and protein. In the CNS, myelin is primarily found in white matter, although it is also present in gray matter, but in smaller quantities. Myelination of the brain begins during the fifth fetal month, progresses rapidly for the first 2 years of life, and then slows markedly after 2 years. In contrast, fibers to and from the association areas of the brain continue to myelinate into the third and fourth decades of life. In general, myelination has a predictable maturation, with progressing from caudal to cephalad, posterior to anterior, and central to peripheral directions (Box 8-1). As an example, myelination progresses from the brain stem to the cerebellum and basal ganglia, and then to the cerebral hemispheres. In a particular location, the dorsal region tends to myelinate before the ventral regions. The process of myelination also relates to functional requirements such that the somatosensory system in neonates myelinates earlier than motor and association pathways. A rapid growth of the myelinated white matter volume is observed between birth and 9 months of age.

Anatomic magnetic resonance imaging (MRI) sequences (T1- and T2-weighted) are very helpful in assessment of myelination. There is reduction in T1 and T2 relaxation times with continued white matter maturation that corresponds with reduction in tissue water as well as the interaction of water with myelin lipids. On T1-weighted imaging (T1WI), most of the white matter in the newborn brain is hypointense compared with gray matter, and the appearance is similar to that of T2-weighted imaging (T2WI) in adults. Some of the exceptions include the dorsal brain stem pathways (medial lemnisci, medial longitudinal fasciculi) and the posterior limb of the internal capsules, which are myelinated at birth and show hyperintensity on T1. In general, the white matter myelinates earlier on T1WI than on T2WI for reasons that have to do with water

TABLE 8-1 Timing of Myelination

Anatomic Structure	T1WI	T2WI
PLIC (posterior portion)	36 GW	40 GW
Median longitudinal fasciculus	25 GW	29 GW
Superior cerebellar peduncles	28 GW	27 GW
Middle cerebellar peduncles	Birth	Birth-2 mo
PLIC (anterior portion)	Birth-1 mo	4-7 mo
Anterior limb internal capsule	2-3 mo	5-11 mo
Cerebellar white matter	Birth-4 mo	3-5 mo
Splenium of corpus callosum	3-4 mo	4-6 mo
Genu of corpus callosum	4-6 mo	5-8 mo
Occipital white matter (central)	3-5 mo	9-14 mo
Frontal white matter (central)	3-6 mo	11-16 mo
Occipital white matter (peripheral)	4-7 mo	11-15 mo
Frontal white matter (peripheral)	7-11 mo	14-18 mo
Centrum semiovale	2-4 mo	7-11 mo

GW, Gestational weeks; *PLIC,* posterior limb of internal capsule; *T1WI,* T1-weighted imaging; *T2WI,* T2-weighted imaging.

Disorders of the forebrain development
    Holoprosencephaly
    Commissural anomalies
    Septooptic dysplasia
    Pituitary anomalies
Cortical malformations
    Proliferation disorders
        Primary microcephaly with simplified gyral pattern
        Hemimegalencephaly
    Migrational Disorders
        Lissencephaly
            Pachygyria
            Subcortical band heterotopia
        Nodular heterotopia
        Cobblestone brain
        Focal cortical dysplasia
        Hamartomas
    Organizational disorders
        Polymicrogyria
        Schizencephaly

content versus lipid content. Diffusion tensor imaging (DTI) can visualize and quantify neural tracts relating to their myelination: fractional anisotropy (FA) increases with age (myelination), whereas apparent diffusion coefficient (ADC), axial diffusivity (AD), and radial diffusivity (RD) decreases, with a turning point around 6 years of age. Note that developmental stages differ by topography and tract of interest, and changes on T1WI and T2WI differ. Myelination can be assessed using these milestones as outlined in Table 8-1. After 2 years of age, delayed myelination can be detected only when quite severe.

SUPRATENTORIAL CONGENITAL LESIONS

It is useful to separate congenital disorders of the brain into those involving the supratentorial structures (Box 8-2)

Anomalies limited to the cerebellum
   Dandy-Walker malformation
   Blake pouch cyst
   Mega cisterna magna
   Rhombencephalosynapsis
   Other
     Isolated vermian hypoplasia
     Arachnoid cyst in the posterior fossa
Anomalies involving the cerebellum and brain stem
   Joubert syndrome
   Pontocerebellar hypoplasia
   Cerebellar disruptions
Chiari malformations I-III

---

and those involving the infratentorial structures (Box 8-3), because radiologists think in terms of anatomy rather than embryology. Naturally, disorders can affect both spaces, but as an initial classification scheme, this may be helpful. In addition to supratentorial and infratentorial compartments, separating the lesions into those that are cystic and those that are solid can provide some help in arriving at a differential diagnosis. Clinical information is always useful in distinguishing among various congenital abnormalities, because several disorders also have associated cutaneous, ocular, or metabolic abnormalities.

Disorders of Forebrain Development

Holoprosencephaly

Holoprosencephaly refers to a constellation of disorders characterized by failure of cleavage or differentiation of the prosencephalon, resulting in a fused appearance of the prosencephalonic structures (Fig. 8-1). These abnormalities are associated with rather severe mental retardation, microcephaly, hypotelorism, and abnormal facies. The finding of a solitary median maxillary central incisor is also indicative of holoprosencephaly. The olfactory bulbs and tracts are usually absent with lack of development of olfactory sulci and flat gyri recti. The range of this disorder is subclassified as alobar, semilobar, and lobar holoprosencephaly (with decreasing severity; Table 8-2). However, you should be aware that the term holoprosencephaly represents a continuum of forebrain malformations, and that no clear distinction exists between categories.

Lobar holoprosencephaly presents with near complete cleavage/separation of the frontal lobes (see Fig. 8-1). Formation of the frontal horns of the lateral ventricles can be dysplastic but present. Sylvian fissures are near normal but formation of an interhemispheric fissure and cerebral falx is incomplete. The deep cerebral nuclei are fully formed. Tip: A fully formed third ventricle, at least partially present frontal horns, and presence of the posterior body and splenium of corpus callosum classifies the anomaly under lobar holoprosencephaly.

The most severe form is alobar holoprosencephaly (Fig. 8-2), also the most common one, yet rarely encountered in clinical practice because most infants are stillborn. No interhemispheric fissure, falx, or significant separation of the hemispheric structures is identified. A crescentic/horseshoe-shaped monoventricle continuous with a large

dorsal cyst usually occupies most of the cranium. The basal ganglia and thalami are fused, and the septum pellucidum and corpus callosum are absent. The anterior cerebral arteries in these cases are nearly always azygous.

Between the two extremes of lobar and alobar holoprosencephaly is semilobar holoprosencephaly, in which there is partial development of the falx and the interhemispheric fissure (with partial separation of the lateral ventricles). The basal ganglia and thalami are at least partially fused.

Common differential diagnoses include hydranencephaly and severe congenital hydrocephalus. If no cortical mantle is discernible around the dilated cerebrospinal fluid (CSF) space centrally, the diagnosis is hydranencephaly, especially if all that can be seen is a nubbin of occipital or posterior temporal cortex remaining with a falx present (Fig. 8-3). If you see a well-formed falx, cortical mantle, and separated ventricles with a septum pellucidum, suggest severe hydrocephalus. If the thalami are fused, septum pellucidum or falx is absent, a cortical mantle is seen, and the ventricles have lost their usual shape, diagnose alobar holoprosencephaly. Holoprosencephalies can be associated with a number of clinical syndromes (Box 8-4).

Commissural Anomalies

The corpus callosum is the largest of the three midline commissures; others are the anterior commissure and the hippocampal commissure. Agenesis of the corpus callosum is one of the most commonly observed features in the malformations of the brain and is a part of many syndromes. The classic, more descriptive callosal segments include the lamina rostralis, genu, body, isthmus, and splenium. However, from a functional and developmental anatomical point of view, the corpus callosum is divided into two segments by the isthmus: a prominent anterior frontal segment that carries the fibers of the frontal cortex and white matter, and a smaller posterior splenial segment that carries the fibers of the primary visual as well as the posterior parietal and medial occipitotemporal cortices and white matter. This helps understanding of two important concepts: (1) myelination proceeds posterior to anterior reflecting the fact that myelination of the primary cortical areas connected through the isthmus and splenium precedes myelination of the body, genu, and rostrum, and (2) in holoprosencephalies, the posterior callosum is present, because the temporal and occipital lobes are separated in those cases.

The septum pellucidum forms in close association with the anterior corpus callosum. The cavum is not truly apparent before 20 gestational weeks. This is important to understand partial agenesis/dysgenesis of the genu of the corpus callosum in cases with septooptic dysplasia. The interhemispheric glial bridge provides a support for the first pioneer axons to cross at about gestational weeks 12 to 13. The corpus callosum develops within a very short time during week 13 and at week 14 it is virtually complete, although still short. The shape of the corpus callosum is essentially final by week 20; however, its sagittal cross-sectional area is only 5% of what it will be in a mature brain (Fig. 8-4). The corpus callosum enlarges together with the connectivity and the tangential growth of the cortex.

It has long been assumed that the corpus callosum develops from front to back, and the partial commissural agenesis, most commonly posterior splenial agenesis, would be lesser

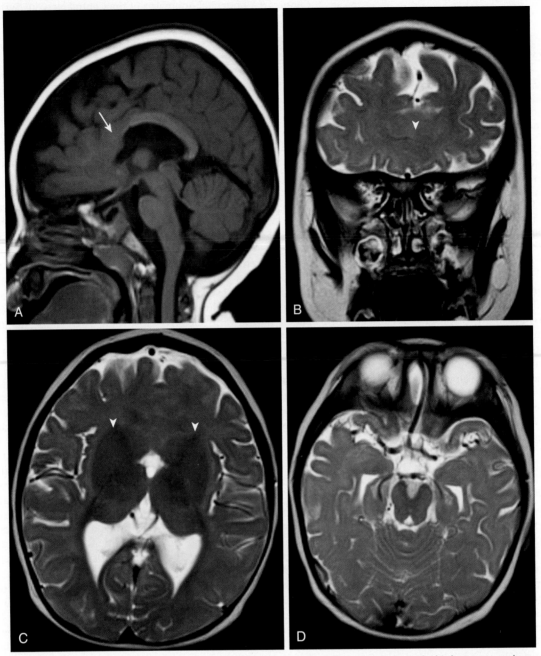

FIGURE 8-1 Lobar holoprosencephaly. **A,** Sagittal T1-weighted imaging (T1WI) shows lack of rostrum and genu of corpus callosum *(arrow)* whereas the posterior body and splenium are present. **B,** Coronal T2WI scan shows lack of most anterior portion of the falx cerebri and lack of cleavage in the frontal lobes *(arrowhead)*. **C,** Axial T2WI scan shows rudimentary frontal horns *(arrowheads)*. Note normal position of the sylvian fissures. **D,** The anterior cerebral artery is azygous as shown in this axial T2WI scan.

TABLE 8-2 **Holoprosencephaly Variants**

Feature	Lobar	Semilobar	Alobar
Facial deformities (cyclopia)	None	None to minimal	Yes
Falx cerebri	Anterior tip missing or dysplastic	Partially formed	Absent
Thalami	Separated	Partially fused	Fused
Interhemispheric fissure	Formed	Present posteriorly	Absent
Dorsal cyst	No	Yes, if thalamus is fused	Yes
Frontal horns	Yes, but unseparated	No	No
Septum pellucidum	Absent	Absent	Absent
Vascular	Normal	Normal except rudimentary deep veins	Azygous anterior cerebral artery, absent venous sinuses and deep veins
Splenium of corpus callosum	Present	Present without genu or body	Absent
Third ventricle	Normal	Small	Absent
Occipital horns	Normal	Partially formed	Absent

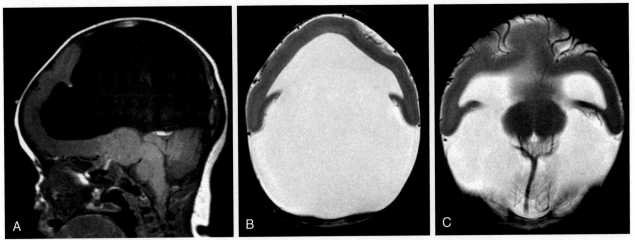

FIGURE 8-2 Alobar holoprocencephaly. **A,** Sagittal T1-weighted imaging (T1WI) demonstrates thin layer of cortex surrounding a large monoventricle which communicates with a large dorsal cyst. **B,** Axial T2WI scan shows the crescentic shaped monoventricle communicating with the dorsal cyst. Septum pellucidum and falx cerebri are absent. **C,** The basal ganglia and thalami are a small mass of fused gray matter which also fuses with the cerebral peduncles.

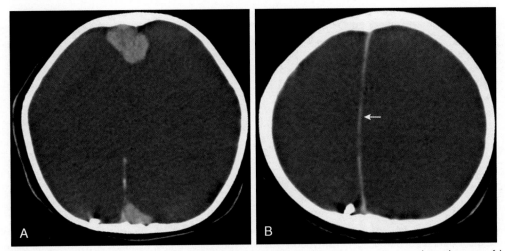

FIGURE 8-3 Hydranencephaly. **A,** Axial computed tomography image shows near complete absence of brain parenchyma except for small portions of anteroinferior frontal lobes and medial occipital lobes. **B,** The falx cerebri is present *(arrow)*.

BOX 8-4 Associations with Holoprosencephaly

Caudal agenesis
DiGeorge syndrome
Fetal alcohol syndrome
Kallmann syndrome
Maternal diabetes
Trisomy 13, 15, 18

---

forms of agenesis. It is now understood that by determining the missing portions, one cannot reliably predict whether the partial agenesis is a sequela of destructive event versus a developmental anomaly. One needs to evaluate all interhemispheric commissures, the isthmus, the connection of anterior segment to isthmus, and the connection of the posterior segment to isthmus to conclude whether or not the corpus callosum is developmentally anomalous.

Sagittal midline T1WI and/or T2WI are ideal in this evaluation. In the most typical partial agenesis, the entire corpus callosum is present, but short in anteroposterior (AP) diameter. Alternatively along the spectrum of callosal dysgenesis, the anterior and posterior segments may be present without the connecting isthmus (no connection of the fornix to the splenium), the splenium may be hypoplastic, or the corpus callosum can be completely absent (Fig. 8-5).

In fact, in classic corpus callosum agenesis, the commissural fibers are not completely genetically absent but rather heterotopic. The noncrossing fibers make an angle and form a parasagittal bundle in the medial aspect of the lateral ventricles, forming the Probst bundles. The lateral ventricles are parallel, frontal horns are pointed and crescentic in shape and occipital horns are enlarged (colpocephaly), with a high-riding third ventricle (if no corpus callosum, there is nothing to hold the third ventricle in place). The cingulate gyrus is everted, and the cingulate sulcus does not form resulting in radiating/disorganized appearance of the sulci in the interhemispheric region (Box 8-5; Fig. 8-6).

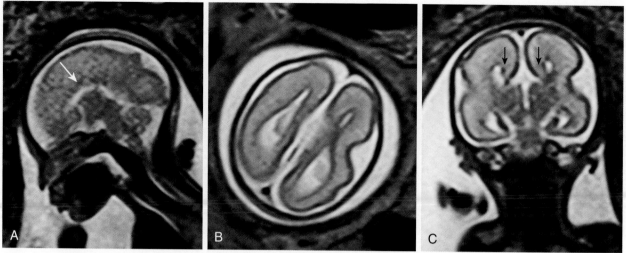

FIGURE 8-4 Agenesis of corpus callosum in an early second trimester fetus. **A,** Sagittal half-Fourier-acquisition single-shot turbo spin-echo (HASTE) demonstrates lack of corpus callosum in the midline *(arrow).* **B,** Axial HASTE shows the pointed frontal horns and dilated occipital horns of the lateral ventricles (Colpocephaly). The lateral ventricles are parallel. **C,** Crescentic shape of the frontal horns secondary to medially located bundles of Probst *(black arrows).* The cingulate gyri are everted.

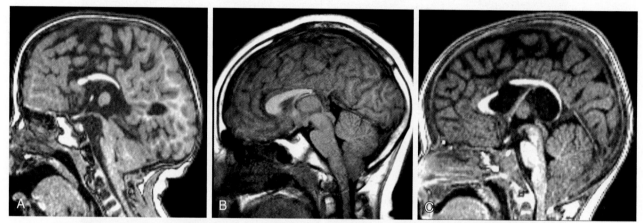

FIGURE 8-5 Examples of partial agenesis/hypogenesis of corpus callosum. **A,** Reduced anteroposterior diameter of the corpus callosum with lack of rostrum, genu and splenium of corpus callosum. Note the small posterior fossa, effaced fourth ventricle, flattened/elongated brain stem, inferior descent of the tonsils and tectal beaking in this patient with Chiari II malformation. **B,** The splenium of corpus callosum is absent. Note the inferior descent of the cerebellar tonsils below the level of foramen magnum with peg-like configuration in this patient with Chiari I malformation. **C,** The anterior and posterior segments of the corpus callosum are disconnected with the lack of isthmus. Note the Chiari II malformation with milder degree of findings described in **A.**

BOX 8-5	Findings in Agenesis of Corpus Callosum

Pointed, crescent-shaped frontal horns
Colpocephaly
High-riding enlarged third ventricle
Incomplete development of hippocampal formation
Interhemispheric cyst or lipoma
Medial impingement of Probst bundle on ventricles
No cingulate sulcus
Radially oriented fissures (eversion of cingulate gyrus) into the high-riding third ventricle
Septum pellucidum absent or widely separated

Other midline abnormalities may be associated with agenesis of the corpus callosum, the most common being interhemispheric cysts or a midline lipoma (both of which are related to meningeal dysplasia). Interhemispheric cysts may be communicating (Fig. 8-7) or noncommunicating with the ventricles; these cysts are important to identify especially in those cases with ventriculomegaly. As expected, the cyst would follow the same density/signal intensity as the CSF, unlike the lipoma, which has fat density/signal intensity (Fig. 8-8).

Callosal development is closely associated with development of the cortex so associations of dysgenesis with heterotopias, agyria, pachygyria, holoprosencephaly, septooptic dysplasia, cephaloceles, Chiari I and II malformations, and Dandy-Walker syndrome are not uncommon. Trisomy 13, 15, and 18, fetal alcohol syndrome, Meckel syndrome (occipital encephalocele, microcephaly, polycystic kidneys, polydactyly) are also associated with agenesis of the corpus callosum (and also with holoprosencephaly). Aicardi syndrome is characterized

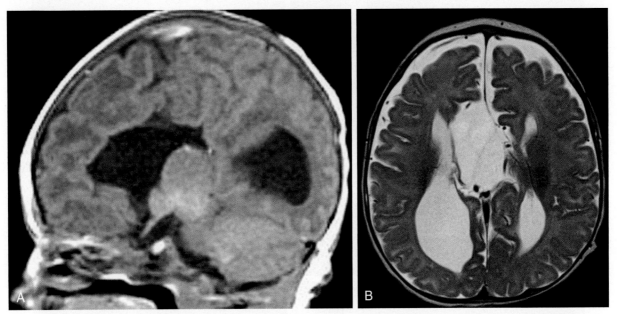

FIGURE 8-6 Agenesis of corpus callosum with interhemispheric cyst. **A,** Sagittal T1-weighted imaging (T1WI) shows agenesis of corpus callosum. Note interhemispheric sulci radiating into the third ventricle with everted cingulate sulcus. **B,** Axial T2WI scan shows interhemispheric cyst. Lateral ventricles are parallel with asymmetric dilation of the right lateral ventricle.

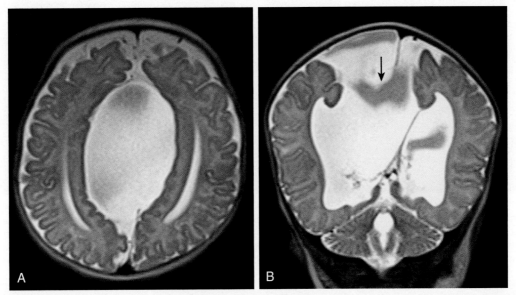

FIGURE 8-7 Large midline interhemispheric cyst with agenesis of corpus callosum. **A,** Axial half-Fourier-acquisition single-shot turbo spin-echo (HASTE) image demonstrates a large midline interhemispheric cyst, which communicates with the right lateral ventricle in the coronal HASTE image. **B,** Cerebrospinal fluid flow artifact confirms communication in addition to the displaced septum *(arrow).*

by the triad of agenesis of the corpus callosum, infantile (or even neonatal) spasms usually without typical hypsarrhythmia (abnormal interictal pattern seen on EEG in patients with infantile spasms), and severe neurological and mental impairment. The imaging features are partial/total callosal agenesis, a marked asymmetry between the hemispheres, an interhemispheric cyst, polymicrogyria, periventricular or subcortical nodular heterotopias, and choroid plexus cysts or papillomas, posterior fossa cyst, choroidal ocular lacunae and ocular colobomata (Fig. 8-9).

Interhemispheric lipomas occur secondary to meningeal dysplasia believed to result from an abnormal differentiation of the meninx primitiva. These are usually associated with dysplastic vessels and may calcify. Many (up to 50%) are associated with agenesis of the corpus callosum. The most common sites of lipomas are the interhemispheric fissure (50%), the quadrigeminal plate cistern, the pineal region, hypothalamic region, and the cerebellopontine angle cistern. These lesions typically do not grow and are symptomatic only because of other associated congenital anomalies and/or mass effect on

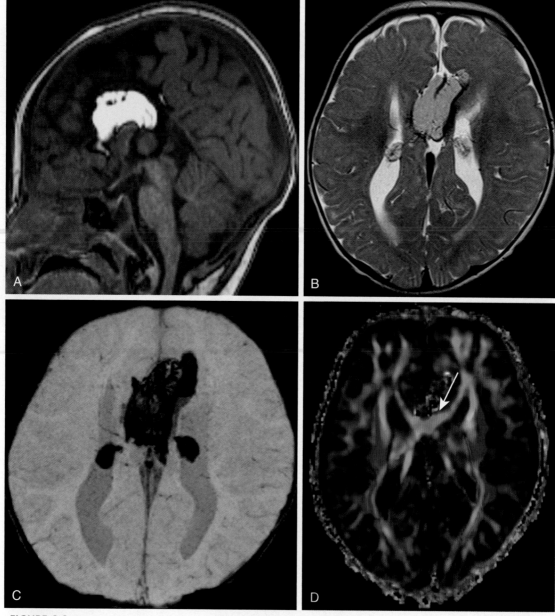

FIGURE 8-8 Agenesis of corpus callosum with midline lipoma. **A,** Sagittal T1-weighted imaging (T1WI) shows T1 bright lobular mass following the same signal as the subcutaneous fat representing a lipoma. **B,** Axial T2WI scan shows that this lipoma is vascular and extends into the lateral ventricles. **C,** Axial minimum intensity projection of susceptibility-weighted imaging shows extensive susceptibility covering the lipomas indicating mineralization/calcification in this lipoma. **D,** Fractional anisotropy map demonstrate at least partial presence of genu of the corpus callosum *(arrow)*.

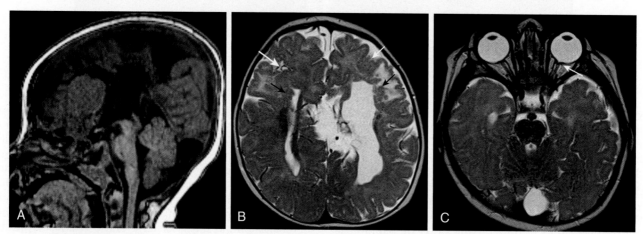

FIGURE 8-9 Aicardi Syndrome. **A,** Agenesis of corpus callosum demonstrated on the sagittal T1-weighted imaging (T1WI) scan. Note the relatively small vermis and small cyst in the posterior fossa confirmed on image **C**. **B,** Axial T2WI scan shows asymmetric left lateral ventriculomegaly, bifrontal polymicrogyria *(white arrows)*, bilateral subependymal heterotopias *(black arrows)*, and subcortical abnormal myelination. Note the abnormal T2 signal of the subcortical white matter in the region of polymicrogyrias. **C,** Note the coloboma *(arrow)* in the left globe on this T2WI scan.

neighboring structures. Treatment tends to be conservative because vessels often course through the lipoma, making surgical removal that much harder. Lipomas have attenuation values on computed tomography (CT) that are in the negative range, usually −30 to −100 Hounsfield units (HU), and are isodense to subcutaneous fat. High signal intensity on T1WI and bright on fast spin echo T2WI, the fat of the lipoma can be corroborated by the presence of a chemical shift artifact along the frequency encoding direction. This causes one edge of the lesion to be highlighted as very bright on T2WI and the other edge along the frequency-encoded axis to be darkened. Alternatively, a fat-suppression pulse sequence can verify the fat by demonstrating signal diminution after it is applied.

Septooptic Dysplasia

Septooptic dysplasia (de Morsier syndrome) consists of hypoplasia of the optic nerves and hypoplasia/absence of the septum pellucidum along with hypothalamic-pituitary axis abnormality. Seizures may coexist. Effects on the visual pathway may range from blindness to normal vision, nystagmus to normal eye movements. Absence of the septum pellucidum causes a squared-off appearance to the frontal horns of the lateral ventricles (Fig. 8-10). MRI appearances are categorized in two groups, one with a high incidence of malformations of cortical development (especially schizencephaly and heterotopias), and partial absence of the septum pellucidum; the second with overlapping features of mild lobar holoprosencephaly with complete agenesis of the septum pellucidum. Hypoplasia of the anterior falx can be seen in the second group. When the septum pellucidum is partially absent, usually the anterior portion is present. This is best seen on coronal magnetic resonance (MR). Agenesis of the corpus callosum and white matter hypoplasia may be associated with this abnormality. In general, patients with septooptic dysplasia demonstrate small hypoplastic optic nerves and a small optic chiasm resulting from the dysplastic optic

pathways. In some cases that dysplasia may be limited to the optic disc and the nerves/chiasm may not be small. Pituitary abnormalities such as ectopic posterior pituitary gland can be seen.

Pituitary Anomalies

Rathke cleft cysts are embryologic remnants of Rathke pouch, the neuroectoderm that ascends from the oral cavity to the sellar region to form the pituitary's anterior lobe and pars intermedia. These cysts are lined with a single layer of cuboidal or columnar epithelial cells and may arise within the sella, the suprasellar region, or most commonly both. They are not very common lesions. The cysts can compress normal posterior or anterior pituitary tissue to cause symptoms of hypopituitarism, diabetes insipidus, headache, and visual field deficits but patients are usually asymptomatic. The cysts are well-defined masses that are located between the adenohypophysis and neurohypophysis. Signal characteristics may vary on MRI: high or low signal intensity on T1WI (depending of amount of proteinaceous contents), high signal intensity on T2WI, and lack of postcontrast enhancement are common (Fig. 8-11). These can appear hypodense on CT. Intracystic, yellow waxy solid nodules have recently been reported in Rathke cysts containing cholesterol and/or mucinous proteins probably accounting for the bright signal on T1WI in some cysts. The differential diagnosis is a craniopharyngioma or hemorrhagic pituitary gland; first check out the location—if the lesion is between the adeno- and neurohypophysis and lacks calcification, solid component, and enhancement, then favor Rathke cleft cyst.

Cortical Malformations
Proliferation Disorders

Primary Microcephaly with Simplified Gyral Pattern. These children are microcephalic because of the reduced proliferation of neurons and glia in the germinal

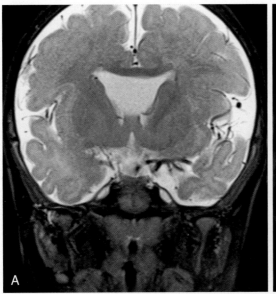
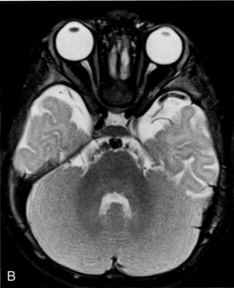

FIGURE 8-10 Septooptic dysplasia. **A,** Coronal T2-weighted imaging (T2WI) shows absence of septum pellucidum, and squared off appearance of the frontal horns. **B,** Bilateral hypoplastic optic nerves are demonstrated on this axial T2WI scan.

zones or increased apoptosis. The head circumference is 3 or more standard deviations below the norm. Neuroimaging shows the primary and secondary sulcations; however, the tertiary sulcations are not formed. The primary sulci are shallow grooves on the surface of the brain that become progressively more deeply infolded and that develop side branches, designated secondary sulci. Gyration proceeds with the formation of other side branches of the secondary sulci, referred to as tertiary sulci. The volume of the white matter in the cerebral hemispheres is reduced. This is a heterogenous group of disorders classified under six subgroups of increasing severity of neuroimaging findings. Groups 5 and 6 have the simplest looking gyral pattern and are named microlissencephaly (Fig. 8-12).

Hemimegalencephaly. Hemimegalencephaly is hamartomatous overgrowth of all or part of a cerebral hemisphere with defects in neuronal proliferation, migration, and organization, and thus this seems to be an abnormal stem cell proliferation disorder. This is a heterogenous disorder which can occur in isolation, or more commonly associated with cutaneous abnormalities and hemihypertrophy. Hemimegalencephaly may occur in a variety of syndromes including epidermal nevus syndrome, Proteus syndrome, neurofibromatosis type 1 (NF-1), Soto syndrome, tuberous sclerosis, and Klippel-Trenaunay-Weber syndrome, to name a few (Box 8-6). These children are typically macrocephalic at birth. Patients have seizures, hemiplegia, developmental delay, and abnormal skull configurations.

On MRI and CT, part or all of the affected cerebral hemisphere appears moderately to markedly enlarged. Typically, the gyri are broadened, thickened, and dysplastic; however, various degrees of poly/pachygyria and dysplasia of the cortex can be seen. White matter volume is increased and

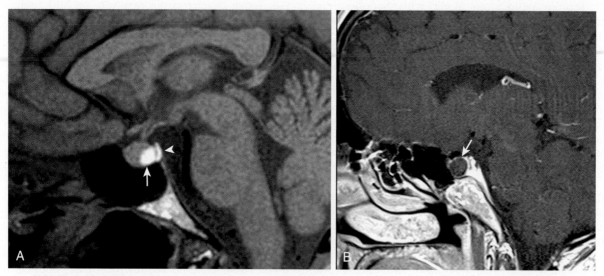

FIGURE 8-11 Rathke cleft cyst. **A,** Precontrast sagittal T1-weighted imaging (T1WI) demonstrates hyperintense lesion in the pituitary gland *(arrow)*. Note the normal precontrast bright T1 signal of the dorsum sellae *(arrowhead)*. **B,** Postcontrast T1WI in a different patient nicely demonstrates nonenhancing cyst in the sella representing the Rathke cleft cyst *(arrow)*.

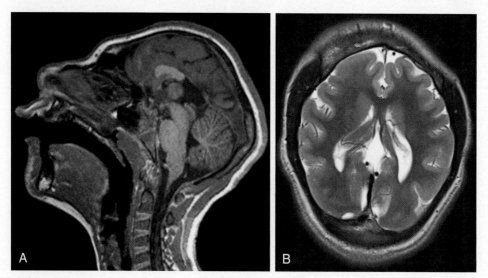

FIGURE 8-12 Microcephaly with simplified gyral pattern. **A,** Sagittal T1-weighted imaging (T1WI) shows the markedly small head size in relation to the face. Note the hypogenesis of the corpus callosum mostly affecting the posterior segment. **B,** Note the too few gyri and shallow sulci in this axial T2WI scan.

Neurofibromatosis type 1
Dyke-Davidoff-Masson syndrome/Sturge-Weber association
McCune Albright syndrome
Soto syndrome
Tuberous sclerosis
Klippel-Trenaunay-Weber syndrome
Proteus syndrome
Epidermal nevus syndrome

usually reveals heterogeneous signal either because of heterotopic gray matter or delayed myelination. Characteristically, the ipsilateral ventricle is enlarged with dysmorphic appearance of the frontal horn (Fig. 8-13). This unique feature of ventricular dilatation on the side of the enlarged hemisphere separates congenital hemimegalencephaly from other infiltrative lesions. Interestingly, the radiological appearance may change over time. Patients with hemimegalencephaly of one hemisphere in infancy have been reported to develop atrophy of the affected hemisphere at 1 year of age. Reduced tracer uptake has been noted

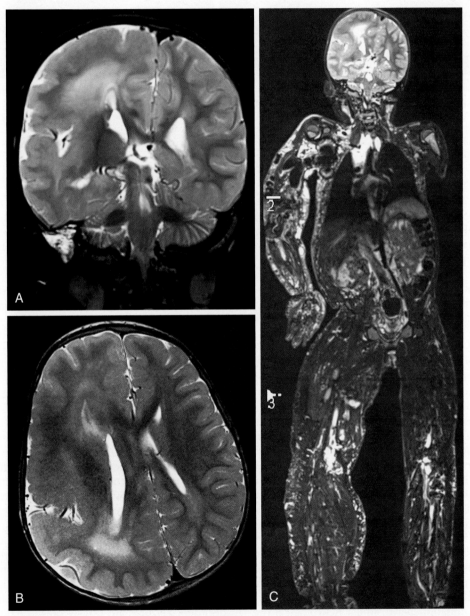

FIGURE 8-13 Unilateral hemimegalencephaly in a patient with congenital lipomatous overgrowth, vascular malformations, and epidermal nevi (CLOVE) syndrome. **A** and **B,** Coronal T2 and axial T2-weighted imaging demonstrate marked enlargement of the right cerebral hemisphere, diffuse pachygyria of the right cerebral cortex, mild asymmetric enlargement of the right lateral ventricle and straightening of the right frontal horn. In addition, the T2 signal of the white matter is diffusely abnormal and gray-white matter distinction is fuzzy. **C,** Stitched coronal short tau inversion recovery (STIR) sequence of the whole body demonstrates overgrowth in the right upper and lower extremities and torso and to a lesser degree in the left lower extremity with extensive venous and lymphatic malformations in this patient with CLOVE syndrome.

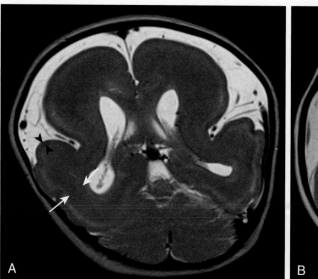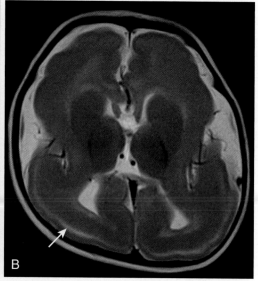

FIGURE 8-14 Lissencephaly. Note the vertically oriented sylvian fissures giving the brain a figure eight or hourglass appearance. **A,** Coronal T2-weighted imaging shows the thin cortical outer layer *(black arrowheads)* separated from arrested neurons by a normal white matter in the cell sparse zone. The thick layer of arrested neurons are outlined between the *two white arrows.* The thin layer of cortex is outlined by the *black arrowheads.* **B,** Very few shallow sulci in the frontal lobes. The T2 bright cell-sparse zone *(arrow)* is quite prominent in the agyric posterior parietooccipital lobes.

on iodoamphetamine single-photon emission computed tomography (IMP-SPECT) , which has been attributed to episodes of status epilepticus. Although rare, associated enlargement and dysplasia of the cerebellum and brain stem can be seen, a condition named as total hemimegalencephaly.

Anatomical or functional hemispherectomy may be indicated in cases with intractable seizures, if the contralateral hemisphere is normal. Therefore, careful evaluation of the contralateral hemisphere is critical in these patients.

Migrational Disorders

Lissencephaly. The term lissencephaly means "smooth brain" with an observed paucity of gyral and sulcal development of the surface of the brain. Agyria is a global absence of gyri along with thickened cortex and is the same entity as "complete lissencephaly," whereas "pachygyria" refers to partial involvement in a hemisphere. Multiple genetic mutations have been identified in lissencephaly, and one of the most well-known is a defect of *LIS1* gene at locus 17p13.3. Miller-Dieker syndrome also has a mutation in chromosome 17, and presents with a characteristic facies and lissencephaly on imaging. Mutation in the *DCX* gene (also known as *XLIS*) in chromosome 22 is known as X-linked lissencephaly and accounts for about 75% of lissencephaly mutations. Appendicular and oropharyngeal spasticity develop with maturation of the CNS. The X-linked gene *doublecortin (XLIS)* predisposes to lissencephaly in boys and band heterotopia in girls (protected by two X chromosomes).

LIS1 and *DCX* mutations have similar neurological presentations, namely hypotonia at birth. Classic lissencephaly has characteristic neuroimaging features: smooth brain surface, diminished white matter and shallow/vertically oriented Sylvian fissures (thus the figure of eight or hourglass appearance on axial images). The thin outer cortex is separated from the thick deeper cortical layers by a zone of white matter, called as "cell-sparce zone." The trigones

and occipital horns of the ventricles are dilated likely because of underdevelopment of the calcarine sulcus. The brain stem is small, likely secondary to lack of normally formed corticospinal and corticobulbar tracts (Fig. 8-14).

Pachygyria. Agyria/pachygyria typically results from abnormal neuronal migration. Compare this with polymicrogyria, which is a malformation of cortical development and involves interruptions in late neuronal migration and cortical organization. Keep in mind that the sulci in polymicrogyria are abnormal and do not correspond to any normal described in textbooks, whereas the sulci in pachygyria are normal in their location and can be identified neuroanatomically, although shallow and less in number. Patients with congenital cytomegalovirus (CMV) infection have high rates of pachygyria (Fig. 8-15). As opposed to hemimegalencephaly, white matter volume is decreased. The gray matter appears thicker because of poor sulcation. Abnormal myelination may coexist.

Heterotopia. Heterotopia is disorganized brain tissue, and it is usually gray matter that is located in the wrong place. Heterotopias form when migration of the neuroblasts from the periventricular region to the pia is thwarted, possibly because of damage to the radial glial fibers, which orient migrating neurons. The classification of heterotopias is usually divided into two varieties; nodular and band types.

Band Heterotopia. The "band heterotopia" (double cortex) may present at any age, but usually in childhood with variable degrees of developmental delay and mixed seizure disorders. Seizures are usually on the milder side. The *DCX* mutation is the most common genetic abnormality with a strong female preponderance (>90%). As the name implies, band heterotopia is a homogenous band of gray matter located between the cortex and white matter, separated from the ventricles and cortex by normal appearing white matter. The band heterotopia may be complete or partial. Partial frontal lobe bands are more common in females whereas posterior

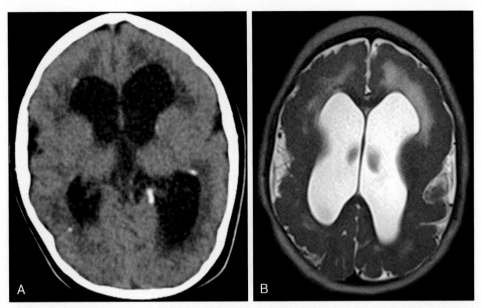

FIGURE 8-15 Bilateral pachygyria in a patient with cytomegalovirus infection. **A,** Axial computed tomographic (CT) image shows diffuse thickening of the cortex bilaterally with low density of the white matter. Note the periventricular calcifications. **B,** Axial T2-weighted imaging of the brain demonstrate diffuse thickening of the cortex and shallow sylvian fissures. The volume of the white matter is reduced. The increased T2 signal of the white matter corresponds to the decreased density in the CT representing delayed myelination.

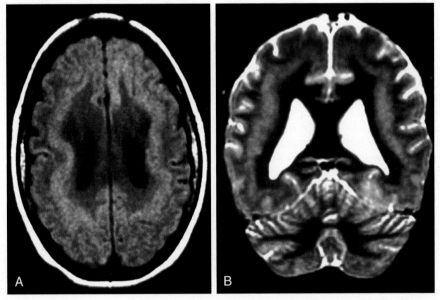

FIGURE 8-16 Band heterotopia. **A** and **B,** Coronal and axial T1-weighted imaging demonstrate a thin layer of gray matter in the subcortical white matter of bifrontal lobes representing a partial band heterotopia.

involvement is more common in males. The overlying cortex may be normal or malformed. Look for gray, white, gray, white, ventricle as the pattern for band heterotopias—it is the second gray that is the rub (Fig. 8-16).

Nodular Heterotopia. Under nodular heterotopias, you will find subependymal and subcortical variants. Patients with subependymal heterotopias can be divided in two groups: the larger group of patients have symmetric and fewer subependymal heterotopias usually confined to the trigones and temporal/occipital horns (Fig. 8-17). These are rarely familial but can be seen in cases with Chiari II

malformations, callosal anomalies and cephaloceles. The smaller group of patients may have familial either X-linked or autosomal recessive patterns of inheritance. Mutations in several genes (such as *Filamin*-1 gene on the long arm of the X chromosome) may cause subependymal heterotopias. Subependymal heterotopias appear as smooth, ovoid masses that are isointense to gray matter on all imaging sequences (see Fig. 8-17). The long axis of the heterotopias is parallel to the ventricular wall and the heterotopias do not show evidence of enhancement on postcontrast images nor do they have perilesional edema. They can grow exophytically,

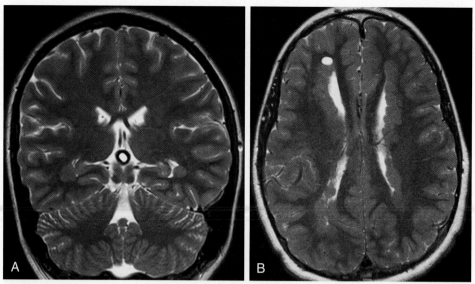

FIGURE 8-17 Examples of nodular heterotopias. **A,** Coronal T2-weighted imaging (T2WI) shows nodular heterotopia outlining the temporal horns of bilateral ventricles. **B,** Axial T2WI scan shows bilateral extensive nodular masses outlining the lateral ventricles, following the same signal of the cortex.

extending into the ventricle, and if big enough may have mass effect on the ventricle (see Fig. 8-17). Hyperintensity on T1WI may be due to dystrophic microcalcifications and these may show hyperdensity at CT.

Patients with subcortical heterotopias often have abnormal sulcation patterns superficial to the heterotopia. The hemisphere ipsilateral to the site of the subcortical heterotopia may be smaller with thinning of the overlying cortex. Subcortical heterotopias may be nodular or curvilinear in shape. The nodular variety of subcortical heterotopias are usually identified in a periventricular or subcortical location, whereas the diffuse (or laminar) heterotopias are seen more commonly in or close to the cortex. Imaging features are similar to that of subependymal heterotopias. Encephaloceles, holoprosencephaly, schizencephaly, Chiari malformations, and agenesis of the corpus callosum may coexist with gray matter heterotopias. If you see one, actively search for other anomalies!

Cobblestone Malformations. Congenital muscular dystrophies (CMD) are included in this group of malformations, characterized by hypotonia at birth, generalized muscle weakness, joint contractures, and other CNS anomalies. Fukuyama type was the first described in this group, an autosomal recessive condition prevalent in Japan. Walker-Warburg syndrome and muscle-eye-brain disease have brain and ocular anomalies.

There is considerable overlap in pathologic and imaging findings. Walker-Warburg syndrome patients have cobblestone lissencephaly, congenital hydrocephalus, severe congenital eye malformations (microphthalmus, retinal dysplasia, persistent hypoplastic primary vitreus, optic nerve hypoplasia), and may have occipital encephaloceles. The cortex is thick with very few sulci. Cobblestone lissencephaly refers to the quite distinctively irregular gray-white matter junction possible reflecting the disorganized neurons interrupting the white matter. More severe cases have pontine hypoplasia with a distinctive kink at the mesencephalic-pontine junction because of cerebellar hypoplasia/dysplasia. Fukuyama

Congenital Muscular dystrophy is characterized by frontal polymicrogyria and temporooccipital cobblestone cortex. In addition, the cerebellum is dysplastic and with subcortical cysts. Muscle-eye-brain disease shows similar features to that of above two described entities; however, the severity is somewhat intermediate.

Focal Cortical Dysplasia

Type 1 Focal Cortical Dysplasia (Without Balloon Cells). Cortical dysplasia may be a source of seizures and motor deficits. The findings may be very subtle, manifested as mild focal volume loss, thickening of cortex, abnormal sulcation, and blurring of gray-white matter junction (Fig. 8-18). The CSF cleft overlying a cortical dimple is a specific sign of cortical dysgenesis. The signal intensity of the dysplastic cortex and underlying white matter may change with age; therefore, multiple studies may be required to find the focal cortical dysplasia (FCD) in young children. Cortical dysplasia can be associated with other conditions such as neuroglial tumor (dysembryoplastic neuroepithelial tumors) and hippocampal sclerosis.

Focal Cortical Dysplasia with Balloon Cells. FCD with balloon cells (a histopathologic term) consists of a cortex with abnormal lamination, abnormal cells (dysplastic neurons and balloon cells), and focal signal abnormality that extends from the gray-white matter junction to the ventricular surface (which led to the term transmantle cortical dysplasia). Cortical thickening, hyperintense subcortical white matter on T2WI (likely because of hypomyelination and astrogliosis), and radial bands from the ventricle to the cortex are seen (Fig. 8-19). These bands may be of white or gray matter signal intensity. The gray-white junction is blurred.

A frontal lobe location favors a balloon cell dysplasia, whereas a temporal lobe (especially medial temporal lobe) location is more suggestive of a neoplasm. Rarely, one may see anomalous venous drainage from areas of cortical dysplasia. Both radiologic and histologic findings may overlap with cortical tubers seen in tuberous sclerosis, therefore genetics and pediatric neurology consult may be advisable in these cases.

Hamartomas. Hamartomas represent an abnormal proliferation of disorganized but mature cells, usually a combination of neurons, glia and blood vessels in an abnormal location. Whereas the heterotopias are due to anomalous neuronal migration, hamartomas are a nonneoplastic proliferation of brain tissue. There is a propensity for hamartomatous formation in the hypothalamus typically located between the mammillary bodies and the tuber cinereum of the hypothalamus (Fig. 8-20). Boys are more commonly affected than girls. These patients typically present with precocious puberty (before 2 years of age in boys and slightly later in girls) and gelastic (laughing spells) seizures. However, occasionally, visual disturbances may be present, because the hypothalamic hamartoma involves the optic pathways.

Because the tissue that makes up a hamartoma is essentially normal brain substance, the hamartoma is isodense with the gray matter on CT. On MR, the hamartoma is isointense to gray matter on T1WI and variable on T2WI. The lesion is identified as a bulbous protrusion of the hypothalamic region in the midline. The hamartomas have a normal blood-brain barrier and are not expected to show enhancement on either CT or MR. Occasionally, mass effect may be associated with the hamartoma as evidenced by displacement of the inferior portion of the third ventricle.

After the hypothalamic region, the next most common location for hamartomas is the cerebral cortex-subcortical region. Occasionally, hamartomas may be seen in a periventricular location. Fetus in fetu refers to duplication of brain structures, usually seen as an extraaxial frontal region mass. The signal intensity approaches that of normal brain. Case studies of ectopic brain in the nasopharynx or pterygopalatine fossa have also been reported.

Organizational Disorders

Polymicrogyria. Polymicrogyria is a malformation of cortical development which results from interruptions in later neuronal migration and neuronal organization. Histologically, there is derangement of the six-layered lamination of the cortex with an associated derangement of sulcation.

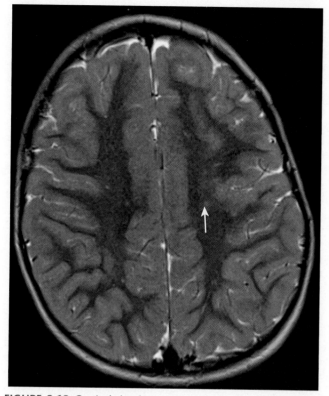

FIGURE 8-18 Cortical dysplasia, type I. Note the localized blurring of the gray-white matter junction in the posterior left frontal lobe in the medial aspect representing focal cortical dysplasia. In addition few tiny foci of gray matter signal is seen in the white matter in the vicinity of this focal cortical dysplasia suggesting additional subcortical heterotopias *(arrow)*.

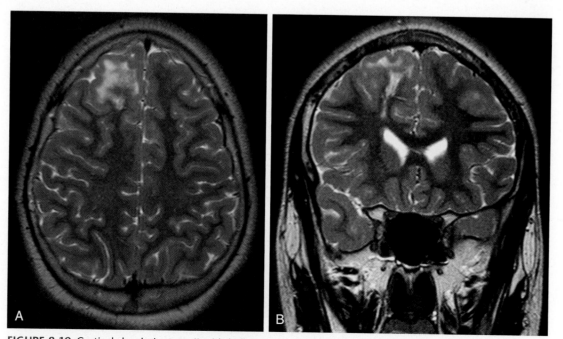

FIGURE 8-19 Cortical dysplasia, type II with balloon cells. **A,** Axial T2-weighted imaging (T2WI) shows a focal area of increased T2 signal in the subcortical white matter of the right frontal lobe radiating to the superolateral margin of the right frontal horn in this coronal T2WI **(B).**

Therefore, in polymicrogyria, no normal sulci are seen. Patients may present with developmental delay, focal neurologic signs/symptoms or epilepsy at any age. CMV infection, in utero ischemia, or chromosomal mutations can be associated with polymicrogyria. It can be focal, multifocal or diffuse; it can be unilateral or bilateral. Posterior sylvian fissure and frontal lobe are common locations for this sneaky abnormality.

On imaging, one can see excessive numbers of small, disorganized cortical convolutions. The cortex appears thickened. The white matter thickness is normal (remember that it is increased in hemimegalencephaly and smaller in agyria). To distinguish pachygyria from polymicrogyria, check the cortex; the cortex is less thick with

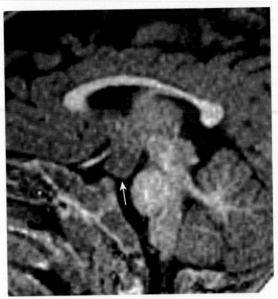

FIGURE 8-20 Hypothalamic hamartoma. Sagittal T1-weighted imaging shows pedunculated brain tissue extending caudally from tuber cinereum of the hypothalamus.

polymicrogyria than with true pachygyria (>5 mm thick). On thin section images polymicrogyria is "bumpier" than pachygyria. Cortex may appear buckled or with an inward folding. Another distinguishing feature between pachygyria and polymicrogyria is the possible presence of abnormal deep white matter in the latter (Fig. 8-21). Keep in mind that the degree of myelination affects the imaging appearance. In unmyelinated areas, the polymicrogyria looks thin, whereas in myelinated areas the thickness may reach 5 mm or more with a smoother looking outer cortex. The polymicrogyria tends to involve the frontal and parietotemporal lobe, whereas pachygyria tends to involve the temporooccipital lobes. An association with developmental venous anomalies (anomalous venous drainage) is noted with polymicrogyria as with other dysplastic cortices.

Congenital bilateral perisylvian (opercular) syndrome is recognized as an entity in which there is polymicrogyria involving the opercular cortex associated with abnormal sylvian fissure sulcation (Fig. 8-22). The inheritance pattern is heterogenous. Patients with congenital bilateral perisylvian syndrome disorder have seizures, congenital pseudobulbar paresis, and developmental delay. The abnormal sylvian fissure may have cortical thickening on either side of it. Schizencephaly may also be present.

Bilateral symmetrical frontoparietal polymicrogyria syndrome and bilateral medial parietooccipital polymicrogyria syndrome have also been described.

Schizencephaly. Schizencephaly is gray matter lined cleft that extend from the pial covering of the cortex to the ependymal lining of the lateral ventricle. They are difficult to classify, because the malformation likely occurs during neuronal cell proliferation, migration, and organization. Give the toss-up, we left them in the organization section. Both genetic and acquired causes are considered in the etiology.

The abnormality can be unilateral or bilateral, with closed lips or open lips. Bilateral involvement (~%40) is

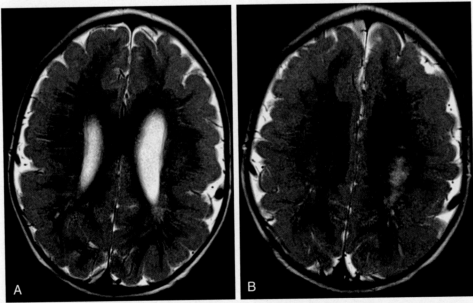

FIGURE 8-21 Pachygyria. **A,** Axial T2-weighted imaging demonstrates bilateral diffuse thickening of the frontoparietal cortices. Although cortex is thickened, the normal anatomy of the sulci can be identified. **B,** Note the abnormally increased T2 signal in the left posterior centrum semiovale, which may indicate abnormal myelination.

associated with seizures, worse developmental delay, and developmental dysphasia. Motor dysfunction is more common with frontal lobe schizencephaly, open lipped varieties, and wider gaps in the open lips.

The cleft has a dysplastic gray matter lining with abnormal lamination that looks like polymicrogyria and is usually seen in the supratentorial space (near the precentral and postcentral gyrus) coursing to the lateral ventricles. Interestingly, the common locations follow that of polymicrogyria. Schizencephaly occurs most commonly in the frontal (44%), frontoparietal (30%), and occipital (19%) lobes. The gray-white matter junction is irregular along the clefts. The lips of the cleft may be opposed ("closed lipped") or gaping ("open lipped") (Fig. 8-23). The closed lipped variety may be missed if the clefts are tightly apposed, but a dimple at the ventricle-cleft interface should suggest the diagnosis. The gyral pattern of the cortex adjacent to the schizencephaly is usually abnormal.

Optic nerve hypoplasia is seen in one third of the patients. Optic nerve hypoplasia with high absence of septum pellucidum can result in some patients getting categorized under septooptic dysplasia.

Schizencephaly is often associated with FCD, gray matter heterotopias, agenesis of the septum pellucidum (~80%), and pachygyria. If bilateral clefts are present, then the septum pellucidum is absent. The lining of the cleft with dysplastic gray matter, best seen on MR, is the differential point in this lesion and distinguishes it from ischemic/encephalomalacic abnormalities, which are usually lined by white matter (Fig. 8-24). Prenatal open lip clefts may become closed lip postnatally in 50% of cases. The inner surface of the cleft is pia-lined and communicates with the ependyma of the ventricle, which differentiates the lesion from an enlarged sylvian fissure because of volume loss such as those seen in premature infants.

Anencephaly

Anencephaly is failure of anterior neural tube closure. Anencephaly (Fig. 8-25) is now a prenatal diagnosis because most women are screened relatively early in pregnancy for elevated levels of serum a-fetoprotein, a marker of neural tube defects. The diagnosis by obstetric ultrasound (US) is made when the cranial vault is seen to be small, with only the fetus's face and posterior fossa well seen. Only a nubbin of tissue is seen at the skull base on ultrasound, and amniotic fluid α-fetoprotein levels are elevated. These

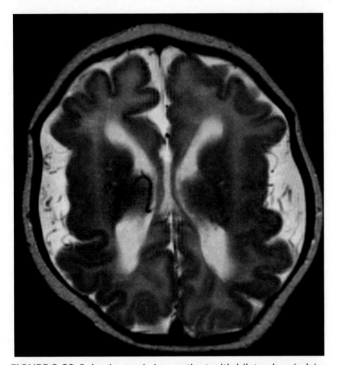

FIGURE 8-22 Polymicrogyria in a patient with bilateral perisylvian syndrome. Axial T2-weighted imaging shows thick and bumpy insular cortex where gray-white matter differentiation is blurry. The sylvian fissures are open because of abnormal opercularization.

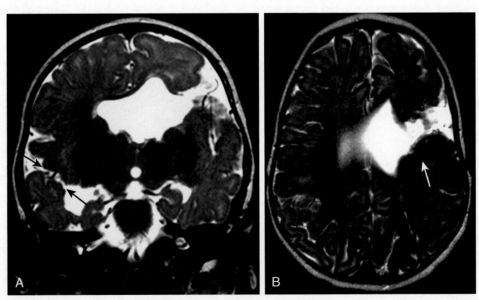

FIGURE 8-23 Schizencephaly. **A,** Coronal T2-weighted imaging (T2WI) shows a cleft extending from the cortical surface of the left frontal lobe to the frontal horn of the left lateral ventricle lined with dysplastic gray matter representing an open lip schizencephaly. Note the additional closed lip schizencephaly in the right temporal lobe *(black arrows)*. Septum pellucidum is absent. **B,** Dysplastic gray matter lining of the left frontal open lip schizencephaly is better depicted in the axial T2WI. Note thickened cortex lining the cleft *(arrow)*.

babies die soon after delivery. An association with spinal dysraphism exists.

INFRATENTORIAL ABNORMALITIES

Inherited (genetic) or acquired (disruptive) causes are identified in congenital abnormalities of the posterior

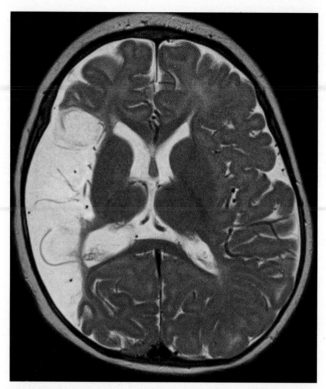

FIGURE 8-24 Intrauterine middle cerebral artery (MCA) infarct in a 7-month-old demonstrates cystic encephalomalacia in the right MCA territory. Note the lack of dysplastic gray matter lining along the borders of the encephalomalacia. In addition the remaining frontal and occipital lobes demonstrates reduced parenchymal volume, typical of intrauterine territorial infarcts.

fossa. It is important to differentiate the disruptive causes, since the chance of recurrence in other offspring is very limited. Disruptive abnormalities most commonly result from prenatal infections, hemorrhage and ischemia. If one identifies a unihemispheric abnormality of the cerebellum, disruptive causes should be considered.

For practical purposes, posterior fossa malformation are classified based on the neuroimaging pattern: (1) predominantly cerebellar, (2) cerebellar and brain stem, (3) predominantly brain stem, and (4) predominantly midbrain. We will focus on the first two in this classification because they cover the most common malformations of the posterior fossa. Evaluation of vermis is of great importance, so make good use of your mid-sagittal images! Always pay attention to the size and form of the fourth ventricle, and everything else around it.

Anomalies Limited to the Cerebellum

The cerebellum consists of the vermis and two cerebellar hemispheres. The cerebellum may be hypoplastic (small volume), dysplastic (abnormal foliation and architecture of the cerebellar white matter), or a combination of both, involving the entire cerebellum, limited to the vermis only, or limited to the cerebellar hemispheres only.

Dandy-Walker Malformation

This is the most common malformation of the posterior fossa and is typically sporadic, with a very low risk of recurrence. These days, most cases are prenatally diagnosed. Macrocephaly and symptoms of increased intracranial pressure (ICP) manifest before 1 year of age in most patients.

The key neuroimaging features are (1) hypoplasia of the cerebellar vermis (rarely agenesis), typically involving the inferior portion with the remaining vermis elevated and upwardly rotated, and (2) dilation of the cystic-appearing fourth ventricle which may fill the entire posterior fossa (Fig. 8-26). The fourth ventricle is not filled with an arachnoid cyst, it is just enlarged because of vermian hypoplasia. The cerebellar hemispheres display normal morphology

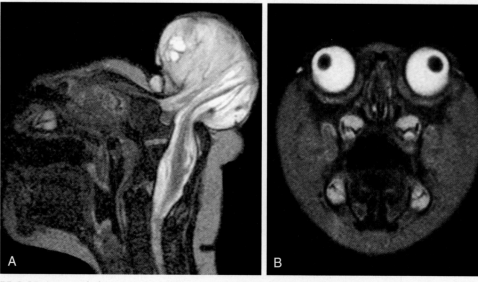

FIGURE 8-25 Anencephaly. Sagittal T2 **(A)** and coronal T2 **(B)** weighted imaging shows lack of cranial vault. Tangle of disorganized neuronal elements and glia are noted along with rudimentary brain stem.

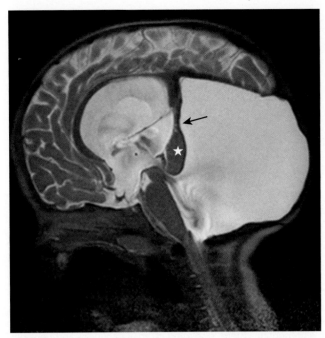

FIGURE 8-26 Dandy-Walker malformation. Sagittal half-Fourier-acquisition single-shot turbo spin-echo (HASTE) image demonstrates a hypoplastic vermis *(star)* with upward rotation *(arrow)* and cystic dilation of the fourth ventricle enlarging the posterior fossa. Note the elevation of the tentorium, ventral displacement of the brain stem and supratentorial hydrocephalus. The cerebral aqueduct is patent with cerebrospinal fluid flow artifact through it.

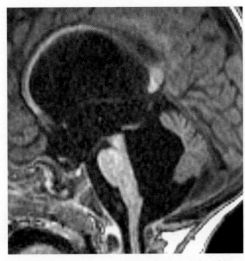

FIGURE 8-27 Blake pouch cyst. Sagittal T1-weighted image in the midline shows enlargement of the fourth ventricle, which communicates with a cystic infravermian compartment representing the Blake pouch cyst. The vermis is normal. Note the tetraventricular hydrocephalus.

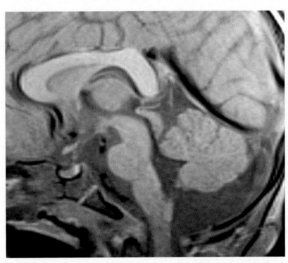

FIGURE 8-28 Mega cisterna magna. Sagittal T1-weighted imaging in the midline shows mild enlargement of the posterior fossa with scalloping of the occipital bone. Normal vermis, no hydrocephalus.

although they are displaced anterolaterally. The posterior fossa is usually enlarged, and the torcular Herophili and transverse sinuses are elevated. Additional abnormalities such as agenesis/dysgenesis of the corpus callosum, occipital encephalocele, polymicrogyria and subependymal/subcortical heterotopias can be seen in up to 50% of patients. Hydrocephalus is seen in about 90% of patients. The more malformations you find in the entire brain, the poorer the prognosis. Coexisting cardiovascular, urogenital or skeletal abnormalities also influence the prognosis negatively.

Neuroradiologists have to make the utmost effort not to fall into the confusing terminologies such as "Dandy-Walker variant," "Dandy-Walker Spectrum," or "Dandy-Walker Complex" when they find one or two things wrong with the posterior fossa. Rather, recognize the true "Dandy-Walker malformation" and remain with descriptive terminology for the rest of the cases.

Blake Pouch Cyst

The Blake pouch is a normal embryologic structure that normally should fenestrate to form the foramen of Magendie, and when it does not, it can result in hydrocephalus. Lack of fenestration in the Blake pouch results in absence of communication between the fourth ventricle and subarachnoid space. Patients demonstrate tetraventricular hydrocephalus. The size of the posterior fossa is normal. The vermis has normal size and form (Fig. 8-27).

The typical neuroimaging findings are a retrocerebellar or infra-retrocerebellar cyst. The choroid plexus is displaced inferior to the cerebellar vermis. Supratentorial abnormalities other than hydrocephalus are usually absent. A Blake pouch cyst occurs sporadically without known risk of recurrence.

Mega Cisterna Magna

As the name implies, mega cisterna magna refers to an enlarged cisterna magna (measuring >10 mm in midsagittal slice). The vermis is of normal size and shape, and so is the fourth ventricle. The posterior fossa may demonstrate variably enlarged size (Fig. 8-28). The enlargement is confined to the subarachnoid space only. Normal communication of CSF between the fourth ventricle in cisterna magna and cervical subarachnoid spaces are seen on CSF flow studies. This is usually an incidental finding, of no clinical consequence. There is no known risk of recurrence in future pregnancies.

Rhombencephalosynapsis

This malformation is characterized by the absence of a vermis with continuity across midline of the cerebellar white matter and folia, dentate nuclei, and superior cerebellar peduncles which results in a key-hole shaped

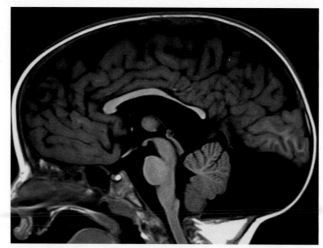

FIGURE 8-29 Retrocerebellar arachnoid cyst. Sagittal T1-weighted imaging shows a posterior fossa cyst which is isointense to cerebrospinal fluid. Note mass effect on the normally formed vermis, normal fourth ventricle, and scalloping of the occipital bone.

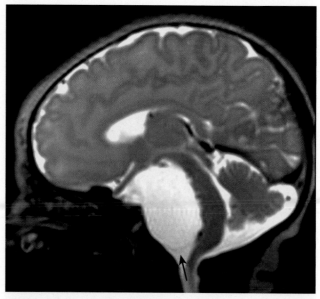

FIGURE 8-30 Prepontine arachnoid cyst. Sagittal half-Fourier-acquisition single-shot turbo spin-echo (HASTE) shows an extraaxial cyst in the prepontine cistern, displacing cerebral peduncles and pons posteriorly. The inferior cyst wall is nicely delineated in this image *(arrow)*.

fourth ventricle. Coronal T2WI best depicts the continuation of horizontal folial pattern. It is sporadic, therefore risk of recurrence in future pregnancies is low. Most cases are nonsyndromic, except for Gomez-Lopez-Hernandez syndrome and in patients with VACTERL (vertebral anomalies, anal atresia, cardiac defects, tracheoesophageal fistula, renal anomalies, and limb anomalies). Ataxia, abnormal eye movement, and delayed motor development are the most common clinical presentation.

Other

Isolated Vermian Hypoplasia. This is a distinct entity not to be confused with Dandy-Walker malformation. Vermian hypoplasia refers to partial absence in the inferior vermis. More than 75% of these patients have a favorable outcome. Prenatal diagnosis is most reliable after 18 gestational weeks, because incomplete development of the inferior vermis may be physiologic before then.

Arachnoid Cyst in the Posterior Fossa. About 10% of the arachnoid cysts occur in the posterior fossa. The duplication of the arachnoid membrane produce these CSF filled cysts. They can be retrocerebellar (Fig. 8-29), supravermian, prepotine (Fig. 8-30), or anterior/lateral to the cerebellar hemispheres. No communication is seen with the fourth ventricle. Arachnoid cysts occur sporadically without known risk of recurrence. These cysts can be asymptomatic, or present with macrocephaly and increased intracranial pressure, if the CSF flow is obstructed. Neuroimaging findings are similar to their supratentorial counterparts, revealing a well-defined, smooth-contoured extraaxial cyst that follow the CSF density/signal on all images. Internal proteinaceous content may result in lack of signal suppression on fluid-attenuated inversion recovery (FLAIR). No restricted diffusion is seen. Remodeling/thinning of the adjacent bone can be seen, occurring as a result of repetitive CSF pulsations. Be cognizant of the association of the arachnoid cyst with acoustic schwannomas in the cerebellopontine angle cistern.

To summarize the most common cystic lesions in the posterior fossa: the vermis is hypoplastic in Dandy-Walker malformation and inferior vermian hypoplasia; the fourth ventricle is enlarged in Dandy-Walker malformation, inferior vermian hypoplasia, and Blake pouch cysts; posterior fossa is enlarged in Dandy-Walker malformation, and variably in mega cisterna magna; hydrocephalus is seen in Blake pouch cysts, in most patients with Dandy-Walker, and sometimes in posterior fossa arachnoid cyst; and finally bony scalloping is seen with posterior fossa arachnoid cysts, and may be seen in cases with mega cisterna magna.

Anomalies Involving the Cerebellum and Brain Stem

Joubert Syndrome

The clinical presentation of Joubert syndrome is characterized by hypotonia, ataxia, oculomotor apraxia, neonatal breathing dysregulation, and variable degree of intellectual disability. Autosomal recessive inheritance is seen in all cases with the exception of *OFD1* mutation where inheritance is X-linked. The classic neuroimaging finding is the "molar tooth sign," which consists of elongated and thickened superior cerebellar peduncles, a deep interpeduncular fossa, and vermian hypoplasia (Fig. 8-31). Diffusion tensor imaging can show the absence of decussation of superior cerebellar peduncles, which may imply an underlying defect in axonal guidance. Other additional abnormalities can be seen such as dysmorphic tectum and midbrain, and thickening and elongation of the midbrain, as well as a small pons. Supratentorial involvement can be seen in 30% of cases showing callosal agenesis/dysgenesis, cephaloceles, hippocampal malrotation, neuronal migrational disorders and ventriculomegaly. Renal (nephronophthisis), liver (congenital hepatic fibrosis), ocular (coloboma), and skeletal (polydactyly) abnormalities can also be seen. Renal and liver involvement leads to a poorer prognosis. There has been no correlation between neuroimaging findings and genotype, therefore neuroimaging alone doesn't help further genotypic classification.

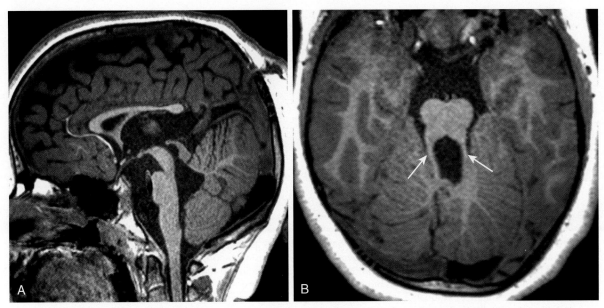

FIGURE 8-31 Joubert syndrome. **A,** Sagittal T1-weighted imaging (T1WI) shows hypoplastic vermis, upward and posterior displacement of the fastigium, mild enlargement of the fourth ventricle, and the narrow pontomesencephalic isthmus. Note the atretic parietal encephalocele. **B,** Axial T1WI scan shows the molar tooth sign, with elongated parallel superior cerebellar peduncles *(arrows).*

The molar tooth sign can also be seen in oro-facial-digital syndrome type VI. Presence of a hypothalamic hamartoma allows differentiation of oro-facial-digital syndrome type VI from Joubert syndrome.

Pontocerebellar Dysplasia

This is a group of autosomal recessive neurodegenerative disorders with prenatal onset. The disease is characterized by hypoplasia of the cerebellum and pons. In cases with prenatal onset, there is progressive atrophy of the already hypoplastic cerebellum. Ten subtypes have been identified. On coronal images the appearance of the cerebellum resembles a "dragon fly" with small volume/flattened cerebellar hemispheres and the relatively preserved vermis representing the body. This morphologic appearance is not specific for pontocerebellar hypoplasia and can be seen in the setting of insults (such as those seen in extreme prematurity), and neurometabolic diseases.

A few words on predominantly brain stem malformations: Pontine tegmental cap dysplasia is characterized by flattened ventral pons, partial absence of the middle cerebellar peduncles, vermian hypoplasia, a molar-tooth like pontomesencephalic junction, and absent inferior olivary prominence. Horizontal gaze palsy with progressive scoliosis is a rare autosomal recessive disease, characterized by butterfly shaped medulla and prominent inferior olivary nuclei. The pons is hypoplastic with a dorsal midline cleft.

Cerebellar Disruptions

Cerebellar maturation and development is complex, starting in the midst of the first trimester and ending about 2 years of age. There is rapid growth (30-fold increase in the surface area of the cerebellar cortex) of the cerebellum between 28 gestational weeks and term. This rapid growth is dependent on high levels of energy supply resulting in increased vulnerability especially between 24 and 32 gestational weeks (Fig. 8-32). The cerebellum

is vulnerable to metabolic, toxic, and infectious insults as well as hemorrhage and ischemia; however, in the immediate prenatal, perinatal, and postnatal period it is resilient to hypoxic ischemic injury. Cerebellar injury occurs in 20% of preterm infants born at less than 32 gestational weeks. Hemorrhage and ischemia may result in parenchymal volume loss.

Chiari Malformations

Chiari Malformations were initially described by Chiari in 1891 as three major malformation of the hindbrain.

Chiari I Malformation

Chiari I malformation is caudal cerebellar tonsillar ectopia, measuring 5 mm or more below the level of foramen magnum (on the sagittal midslice of the brain, draw a horizontal line between the tip of the basion and opisthion and measure the craniocaudal length of the cerebellar ectopia perpendicular to that line). Chiari I malformation may result from multiple different processes. In some cases, there is underdevelopment of the posterior fossa from paraaxial mesoderm that forms the occipital somites resulting in a cranial base dysplasia. The posterior fossa may be relatively small with short/flattened clivus (Fig. 8-33). There may be basilar invagination, or a missing odontoid tip. Other craniovertebral anomalies may accompany these cases. However this is seen in only some of the cases with Chiari I malformations. Other cases do show evidence of tonsillar ectopia without the rest of these skull base abnormalities. There is no hydrocephalus, and the fourth ventricle is of normal size and configuration. The symptoms are similar in either form ranging from asymptomatic to suboccipital headaches, retroorbital pressure or pain, clumsiness, dizziness, vertigo, tinnitus, paresthesias, muscle weakness, and lower cranial nerve symptoms (i.e., dysphagia, dysarthria, sleep apnea, tremors, especially in children younger than 3

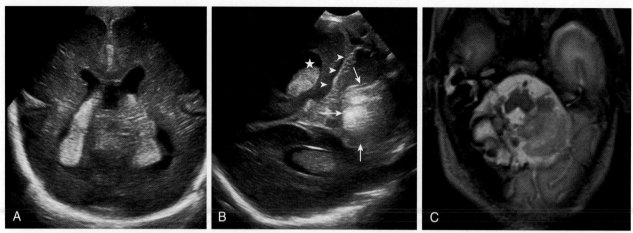

FIGURE 8-32 Cerebellar hemorrhage in a 26 gestational week premature infant. **A,** Coronal image through the anterior fontanelle demonstrates simple gyral pattern which is age appropriate. Prominent size of choroid is normal for age. **B,** Transtemporal view demonstrates hemorrhage in the fourth ventricle and cerebellar parenchyma (*star:* occipital horn; *arrowheads* outline the tentorium; *white arrows:* blood filled fourth ventricle). **C,** Term equivalent age magnetic resonance imaging of the brain showing marked volume loss and hemosiderin staining in the right cerebellar hemisphere representing disruptive cerebellar injury secondary to hemorrhage.

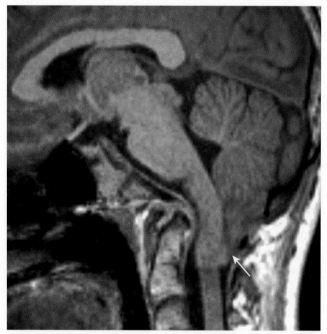

FIGURE 8-33 Chiari I malformation. Sagittal T1-weighted imaging shows inferior descent of the cerebellar tonsils below the level of the foramen magnum *(arrow).* The cerebellar tonsils have a peg-like shape. Note the short clivus, posterior tilt of the odontoid process of C2, and kinking at the craniocervical junction.

years of age). The entity usually presents in the second or third decade of life and women outnumber men by a 3:1 ratio. Spine imaging should be performed in these cases because the incidence of syringohydromyelia is 20% to 65% (Fig. 8-34). Posterior indentation of the dens is associated with higher incidence of syringohydromyelia.

Although, anatomical MRI evaluation and clinical presentation is sufficient in most cases, CSF flow studies are useful where the measurements are borderline or a mismatch is seen between symptoms and measurements.

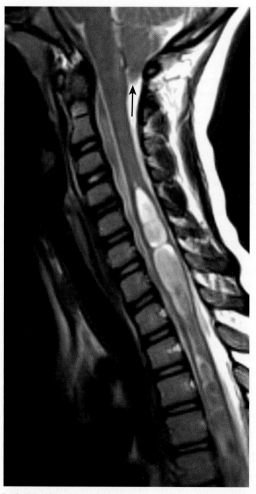

FIGURE 8-34 Chiari I malformation with syrinx. Sagittal T2-weighted imaging shows segmented syrinx in the distal cervical and upper thoracic spinal cord. Note the cerebrospinal fluid flow artifact in the syrinx. The cerebellar tonsils are low-lying *(arrow).*

BOX 8-7 Findings in Chiari II Malformation

INFRATENTORIAL FINDINGS

Myelomeningocele
Small posterior fossa
Tonsils and medulla below foramen magnum
Beaking of tectum
Caudal displacement of medulla
Cerebellum wrapped around brain stem
Petrous bone scalloping
Fourth ventricle compressed, elongated, trapped, and low
Syringohydromyelia
Cervicomedullary kinking
Low torcular
Dysplastic tentorium
Scalloping of clivus posteriorly
Absent hypoplastic arch of C1

SUPRATENTORIAL FINDINGS

Hydrocephalus
Callosal hypogenesis
Lückenschädel skull
Falx hypoplasia
Interdigitation of gyri along widened interhemispheric fissure
Fused, enlarged massa intermedia
Colpocephaly
Malformations of cortical development
Caudate hypertrophy, bat-wing lateral ventricles
Biconcave third ventricle

Patients with symptoms almost always have abnormal CSF flow studies. Foramen magnum obstruction may lead to increased systolic spinal cord motion, impaired spinal cord recoil, and impaired diastolic CSF motion anteriorly at the C2-C3 level and in the posterior fossa. Impaired systolic and unimpaired diastolic flow may also be seen just below the foramen magnum. Some neurosurgeons perform sub-occipital decompression procedures on patients with a variety of clinical symptoms including headaches, vertigo, weakness, and fibromyalgia who have imaging findings of abnormal CSF motion with tonsillar ectopia. Hyperdynamic movement of the tonsils with posterior movement of the medulla in patients with Chiari I malformations has been reported.

Chiari II Malformation

Chiari II malformation is a complex malformation that involves abnormal development of the hindbrain, spine and skull base. Chiari II (the original Arnold-Chiari malformation) anomalies occur in 0.02% of births and affect girls twice as often as boys. The majority of cases are diagnosed prenatally with ultrasonography and amniocentesis (elevated alpha fetoproteins). The cerebellar tonsils, vermis, fourth ventricle, and brain stem are herniated through the foramen magnum, and the egress from the fourth ventricle is obstructed (Box 8-7). A kink may be present at the cervicomedullary junction. Hydrocephalus may occur prenatally and usually increases after closure of the myelomeningocele (which often takes place within the first 48 hours of life). The frontal horns of the lateral ventricles are squared off, the fourth ventricle is compressed, and the aqueduct is stretched inferiorly. The tectum of the midbrain is beaked. The massa intermedia and the caudate heads are abnormally enlarged. The superior cerebellum towers superiorly through a widened tentorial incisura because the whole posterior fossa is too small. The rest of the cerebellum may literally wrap around the brain stem. Associated abnormalities of the corpus callosum is seen in 70% to 90% of the cases predominantly affecting the splenium (Fig. 8-35). Other supratentorial malformations such as heterotopias and abnormal gyral patterns are common. The gyral pattern is usually abnormal in the medial occipital lobes, having the appearance of multiple small gyri, not to be confused with polymicrogyria, since the cortical thickness is not increased. This appearance is called *stenogyria*. Falx cerebri is usually fenestrated.

Virtually all patients have myelomeningoceles. The hindbrain findings of Chiari II are best explained by the theory of McLone and Knepper. In this theory, the hindbrain abnormality results from a small posterior fossa with low tentorial attachment in the setting of a rostral-caudal pressure gradient (secondary to the myelomeningocele). A lipoma of the filum may coexist. The anchoring of the distal portion of the craniospinal axis may account for the downward herniation of intracranial contents in this disorder. Rarely, the tonsils may necrose secondary to compression of critical vessels at the foramen magnum. Plain radiography or CT of the skull shows irregularity of the surfaces of the inner and outer table of the skull called as lacunar skull or *luckenschadel* appearance (Fig. 8-36).

Chiari III Malformation

Chiari III malformations are associated with herniation of posterior fossa contents through a posterior spina bifida defect at the C1-C2 level. Nearly all Chiari III encephaloceles contain brain tissue, usually the cerebellum, although even the brain stem can herniate through the defect. Heterotopias, agenesis of the corpus callosum, anomalies of venous drainage, and syringohydromyelia are commonly associated in this extremely rare malformation.

Meningoencephaloceles in the occipital region are most commonly seen with Chiari III malformations (Fig. 8-37). In the United States, parietal (10%) and frontal (9%) meningoencephaloceles are the next most common locations after occipital region. Vietnamese and Southeast Asian patients have a propensity for nasofrontal (Fig. 8-38) or sphenoethmoidal meningoencephaloceles.

Symptoms of Chiari II and III malformations are developmental delay, ataxia, vertical nystagmus, headache, cranial nerve VI through XII findings, and occasionally, central canal syndromes caused by syringohydromyelia. At autopsy, disordered neuronal tissue is present in the brain stem and cerebellum. Thus, it is not clear whether the symptoms are entirely because of compression at the foramen magnum or because of dysplastic neurons. Decompression of the foramen magnum often does not improve symptoms in most individuals with Chiari II, suggesting that disorganized medullary tissue may be etiologic in some cases.

CONGENITAL HYDROCEPHALUS AND ARACHNOID CYSTS

Amongst the congenital malformations of the brain, Chiari II malformation and Dandy-Walker malformation are the most common causes of hydrocephalus. Chiari II malformation accounts for 40% of congenital hydrocephalus in children. From 70% to 80% of children with Dandy-Walker malformation have hydrocephalus. These malformations are discussed in detail in their respective sections above.

Aqueductal Stenosis

Aqueductal stenosis causes lateral and third ventricular enlargement without fourth ventricular dilatation. Aqueductal stenosis can be developmental or acquired, seen in around 20% of cases with hydrocephalus. During the normal maturation of CNS, the aqueductal size gradually decreases until birth, reaching a mean cross-sectional diameter of 0.5 mm² at birth. The aqueductal stenosis is focal usually at the level of the superior colliculi or intercollicular sulcus.

Aqueductal stenosis can be benign or associated with a tumor such as tectal/tegmental gliomas. T2-weighted and FLAIR imaging is quite helpful in diagnosing these low-grade tumors, as they appear as focal round/oval masses centered at the level of the aqueduct. They typically do not enhance. Acquired causes of aqueductal stenosis are numerous and include clots from subarachnoid hemorrhage and fibrosis after bleeds or infections such as those seen in the preterm infants with germinal matrix spectrum hemorrhages. In those cases, isolated (trapped) fourth ventricle can also be seen. Aqueductal gliosis is a postinflammatory process secondary to perinatal hemorrhage and infection, and is becoming more prevalent as these newborns have increasing survival rates. Differentiation of

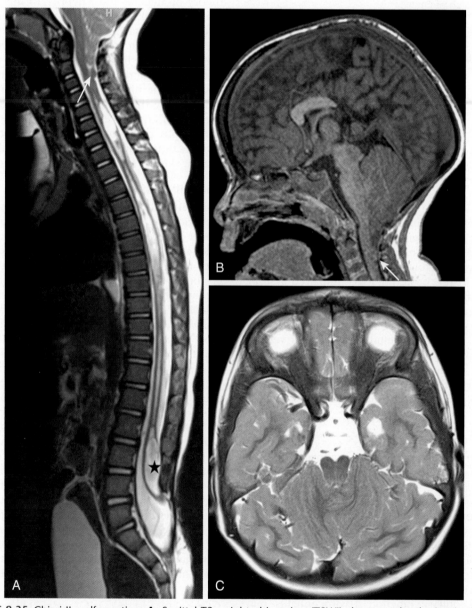

FIGURE 8-35 Chiari II malformation. **A,** Sagittal T2-weighted imaging (T2WI) shows cord tethering in this patient status postmyelomeningocele repair. Note the segmented syringohydromyelia and dilated ventriculus terminalis *(black star)*. **B,** Sagittal T1WI scan shows small posterior fossa, effacement and inferior displacement of the fourth ventricle, beaking of the tectal plate, enlargement of the massa intermedia, and inferior descent of the cerebellar tonsils reaching the level of C3 *(arrow, as also seen on* **A***)*. Note the shortened anteroposterior diameter of the corpus callosum in this patient with agenesis of the posterior segment of the corpus callosum. **C,** Axial T2WI scan shows towering of the cerebellum.

aqueductal stenosis and gliosis is not possible by imaging only, as they appear identical on all modalities.

The congenital causes may be due to webs, septa, or membranes. The aqueductal web is a thin membrane in the distal aqueduct (Fig. 8-39). X-linked aqueductal stenosis is a hereditary type stenosis with variable symptoms such as mental retardation, aqueductal stenosis, spasticity of lower extremities, and clasped adducted thumbs. Pathologic studies showed malformations of cortical development in addition to hydrocephalus. Few MRI reports showed fusion of the thalami, small brain stem, and diffuse hypoplasia of cerebellar white matter.

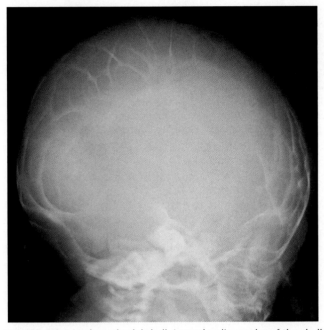

FIGURE 8-36 Luckenschadel skull. Lateral radiography of the skull in a 1-day-old infant with Chiari II malformation with lacunar skull appearance, secondary to dysplasia of membranous skull vault. Note areas of apparent thinning in the calvarium.

Arachnoid Cyst

If you had to have a congenital lesion in the brain, this would be the lesion of choice. The arachnoid cyst is the most common congenital cystic abnormality in the brain. It consists of a CSF collection within layers of the arachnoid. The cyst may distort the normal brain parenchyma (Fig. 8-40).

The arachnoid cyst is typically a serendipitous finding and is usually asymptomatic. The most common supratentorial locations for an arachnoid cyst are (in decreasing order of frequency) (1) the middle cranial fossa, (2) perisellar cisterns, and (3) the subarachnoid space over the convexities. Infratentorially, arachnoid cysts commonly occur in the (1) retrocerebellar cisterns (see Fig. 8-29), (2) cerebellopontine angle cistern, and (3) quadrigeminal plate cistern. Intraventricular cysts are rare but favor lateral and third ventricles (Fig. 8-41).

CT scans demonstrate a CSF density mass that typically effaces the adjacent sulci and may remodel bone. The mass measures from 0 to 20 HU (fluid density of CSF) and shows no enhancement. In those difficult cases where an arachnoid cyst and a dilated subarachnoid space must be distinguished, MRI, especially high-resolution T2WI (such as constructive interference in steady state [CISS]/fast imaging employing steady-state acquisition cycled phases [FIESTA] sequence) can be very helpful in delineating the cyst walls.

On MRI, the most common appearance is that of an extraaxial mass that has signal intensity identical to CSF on all pulse sequences. Occasionally, the signal intensity may be greater than that of CSF on proton density–weighted imaging (PDWI) scans because of the stasis of fluid within the cyst as opposed to the pulsatile CSF of the ventricular system and subarachnoid space. FLAIR and diffusion-weighted imaging (DWI) scans usually show a dark mass similar in intensity to CSF (again pulsation effects may cause some higher intensity). Rarely, the fluid within the arachnoid cyst may be of higher protein content than that of the CSF, accounting for the difference in the signal intensity.

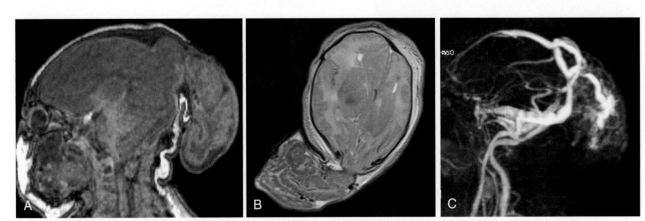

FIGURE 8-37 Chiari III malformation. **A,** Sagittal T1-weighted imaging (T1WI) shows a large calvarial defect with protrusion of occipital and posterior parietal brain through the defect. Note the retraction of the brain stem and posterior fossa toward the encephalocele. **B,** Axial T2WI scan shows that the calvarial defect is to the left of midline, and the encephalocele contains dysplastic brain tissue. Maldevelopment of most of the visualized cortices and subependymal heterotopias noted. **C,** Three-dimensional reconstruction of postcontrast magnetic resonance venography shows aberrant deep draining veins and ectopic venous sinuses, critical information for the neurosurgeons. Hypogenesis of the corpus callosum noted.

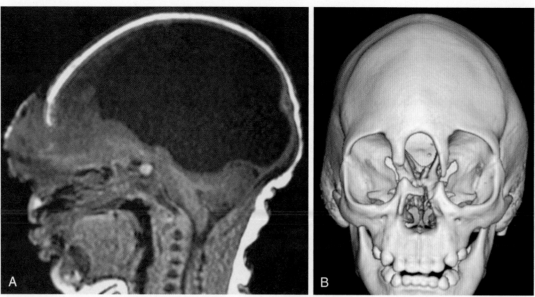

FIGURE 8-38 Frontonasal encephalocele. **A,** One-day-old infant with a large defect in the frontonasal region with extension of the dysplastic brain tissue. Note the ventriculomegaly, and dysplastic brain stem. **B,** Three-dimensional surface rendered image of head computed tomography at 15 months of age demonstrates the large bone defect in the frontonasal region.

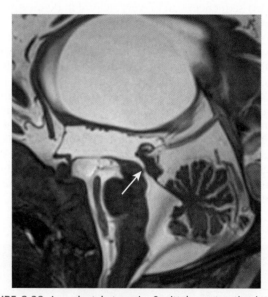

FIGURE 8-39 Aqueductal stenosis. Sagittal constructive interference in steady state (CISS) demonstrates a thin web *(arrow)* in the distal cerebral aqueduct. Supratentorial hydrocephalus is noted. The anterior and inferior recesses of the third ventricle are dilated.

The differential diagnosis of an arachnoid cyst is limited and generally revolves around three other diagnoses: a subdural hygroma, dilatation of normal subarachnoid space secondary to underlying atrophy or encephalomalacia, and epidermoid. Although subdural hygromas have been thought to be due to chronic CSF leaks through traumatized leptomeninges, in most cases the trauma results in sufficient blood deposited within the "hygroma" so that the signal intensity on T1WI and FLAIR is different from that of CSF. In addition, subdural hygromas are typically crescentic in shape, whereas

arachnoid cysts tend to have convex borders. Both efface sulci and show mass effect. In contradistinction, dilatation of the subarachnoid space secondary to underlying encephalomalacia does not demonstrate mass effect and the adjacent sulci are enlarged. Another distinguishing feature is the fact that the cerebral veins in the subarachnoid space are seen to course through the CSF in the case of underlying encephalomalacia (see Fig.8-24), as opposed to the subdural hygroma and arachnoid cyst, where the veins are displaced towards the surface of the brain. An epidermoid may simulate an arachnoid cyst on CT and T2WI; however, FLAIR and DWI should show higher signal intensity than CSF because of elevated protein content. In fact, T1WI will often also be brighter than CSF.

A feature often seen in association with arachnoid cysts that may suggest the diagnosis is bony scalloping. The bone may be thinned or remodeled, probably because of transmitted pulsations and/or slow growth. This would not be seen with hygromatous collections or atrophy. However, the finding may be seen occasionally in epidermoids or with porencephaly, where the ventricular pulsations may be transmitted through the porencephalic cavity to the inner table of the skull. The underlying brain parenchyma may appear hypoplastic, yet with normal function. Absence of soft-tissue intensity or density, calcification, or fat distinguishes arachnoid cysts from those of the dermoid-epidermoid line.

CONGENITAL INFECTIONS

With respect to in utero infections, the age of the fetus at the time of the insult plays a critical role in prognosis. The infections in the first and second trimester result in congenital malformations, whereas the third trimester infections result in destructive lesions. Another unique feature

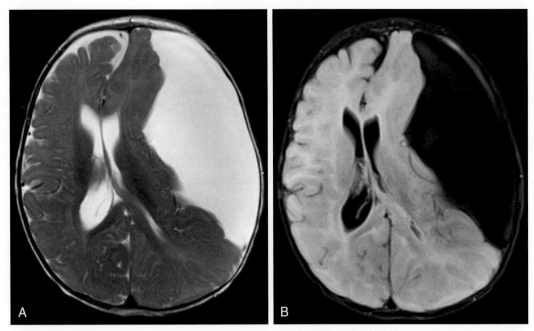

FIGURE 8-40 Arachnoid cyst. Large extraaxial cystic lesion following the same signal as cerebrospinal fluid on axial T2-weighted **(A)** and fluid-attenuated inversion recovery (FLAIR) **(B)** images. Note the midline shift, bowing of the falx, effacement of the left lateral ventricle and hypoplastic appearance of the left cerebral hemisphere.

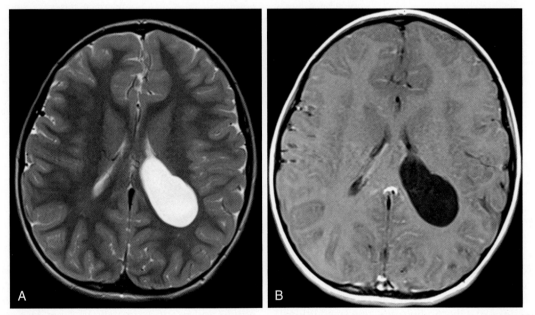

FIGURE 8-41 Intraventricular arachnoid cyst. Axial T2-weighted imaging (T2WI) **(A)** and postcontrast T1WI **(B)** scans demonstrate a cyst in the left lateral ventricle.

of prenatal infection is the altered biological response of the fetal brain to injury; the immature brain doesn't respond to injury by astroglial reaction. The transmission of the infection is either ascending from the cervix (bacterial infections), or transplacental (TORCH: toxoplasmosis, other infections, rubella, cytomegalovirus, herpes simplex). In many instances, these infections lead to in utero death. Other infants are seen in the perinatal period with failure to thrive, hydrocephalus, and/or seizures (Table 8-3). The transmission may be at the time of passage through the vaginal canal during delivery.

The neuroimaging findings are variable depending on the timing and severity of injury. Microcephaly, intracranial calcifications, pachygyria/agyria, neuronal migrational anomalies, white matter abnormalities/delayed myelination, and cysts can be seen. It is important to remember that intracranial calcifications are not specific for congenital infections, and that ischemic/metabolic diseases may result in dystrophic calcifications. Linear branching hyperechogenicities can be seen in the basal ganglia with transfontanellar head US. This entity is called mineralizing vasculopathy, and can be seen in congenital infections,

TABLE 8-3   In Utero Infections

Characteristic	Rubella	HSV (Type 2 > 1)	Toxoplasmosis	CMV	HIV
Frequency	0.0001% of neonates	0.02% of neonates	0.05% of neonates	Most common; 1% of neonates	Growing exponentially
Clinical manifestations	Hearing loss, mental retardation, autism, speech defects	Skin lesions	Usually fetal death; developmental delay, seizures	Hearing loss, psychomotor retardation, visual defects, seizures, optic atrophy	Asymptomatic at birth, later presentation, developmental delay, late spastic paraparesis, ataxia
Ocular changes	Cataracts, glaucoma, pigmentary retinopathy	Chorioretinitis, microphthalmos	Chorioretinitis	Chorioretinitis	
Neuronal migrational anomaly	Rare	None	None	Frequent (polymicrogyria, heterotopia, hydranencephaly, lissencephaly, pachygyria), cerebellar hypoplasia	
Head size	Microcephaly	Microcephaly unless hydrocephalus	Microcephaly	Microcephaly	
Parenchymal changes	Necrotic foci, delayed myelination	Hydranencephaly; patchy areas of low density in cortex, white matter; vast encephalomalacia; cortical laminar necrosis, no predilection for temporal lobe	Hydrocephalus from aqueductal stenosis, intracranial calcifications	Hemorrhage, especially at germinal matrix, loss of white matter, delayed myelination, cortex, subependymal cysts around occipital horns, cerebellar hypoplasia, atrophy	Glial, microglial nodules in basal ganglia, brain stem, white matter; demyelination, atrophy, corticospinal tract degeneration
Vessels	Vasculopathy	Can infect endothelial cells	Infarctions	Vasculopathy, vasculitis with calcifications	
Calcifications	Basal ganglia, cortex	71%, periventricular, basal ganglia, parenchyma	Frequent (40%), periventricular, can have cortical calcifications	Perivascular in basal ganglia, cerebellum	Basal ganglia
Non–central nervous system	Patent ductus arteriosus, pulmonic stenosis, rash, hepatosplenomegaly	Hepatosplenomegaly rash	Hepatosplenomegaly	Neck adenopathy, oral candidiasis	

CMV, Cytomegalovirus; *HIV*, human immunodeficiency virus; *HSV*, herpes simplex virus.

trisomies, prenatal drug exposure, congenital heart disease and variety of anoxic/toxic injuries, as well as a normal variant (far more commonly). Therefore mineralizing vasculopathy is a nonspecific finding, and unless additional abnormalities are seen in the brain, congenital infection should not be strongly considered.

The most common viral infection is CMV, seen in approximately 1% of all births in the USA (see Fig. 8-15). Hepatosplenomegaly and petechiae can be the first signs of disease. Lissencephaly, delayed myelination, marked ventriculomegaly and significant periventricular calcifications are presumed to be a result of early first trimester CMV infections, whereas those infected in the middle of the second trimester typically have polymicrogyria, less pronounced ventriculomegaly and less pronounced cerebellar hypoplasia. Microcephaly caused by atrophy can also be seen with CMV infections (27%). White matter damage (with increased water content) can be seen at any gestational age. Periventricular cysts, usually around the occipital poles, may also be present.

Toxoplasmosis is caused by a protozoan *toxosplasma gondii* and is approximately 10 times less common than CMV. The principal CNS findings are chorioretinitis, hydrocephalus and seizures. The inflammatory infiltration of the meninges is seen with granulomatous lesions in the brain thus leading to obstructive hydrocephalus. Unlike CMV, malformations of cortical development are uncommon. Toxoplasmosis calcifies most frequently of the congenital infections (71% of the time in one series). With treatment of congenital toxoplasmosis, 75% of cases show diminution or resolution of the intracranial calcifications by 1 year of age. If treatment does not occur, is delayed, or is inadequate, the intracranial calcifications may increase. The status of the calcifications often mirrors neurologic function. On MR, periventricular and subcortical white matter injury is seen as high signal intensity on T2WI.

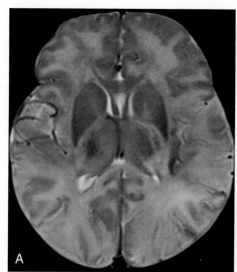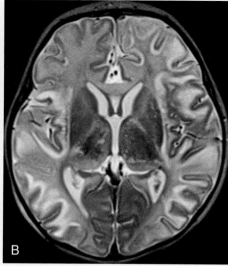

FIGURE 8-42 Neonatal herpes infection: herpes simplex virus type 2 (HSV-2) encephalitis. Axial T2-weighted imaging at age 24 days of life **(A)** and 33 days of life **(B)**. **A,** Diffuse marked increase in T2 signal of the entire supratentorial white matter and cortex with relative sparing of the bifrontal cortices. The thalami are swollen with markedly increased T2 signal. **B,** Within 7 days, rapid subacute-chronic changes in the brain parenchyma with volume loss and laminar necrosis, in the above described regions.

Herpes simplex infection may be acquired as the child passes through the birth canal. The imaging may have a similar appearance to CMV, but microcephaly and microphthalmia are more prevalent. Neuroimaging shows diffusely increased echogenicity or high T2 signal with head US or MRI respectively (Fig. 8-42). Diffusion-weighted imaging is the sequence that demonstrates injury the earliest, with areas of restricted diffusion. Contrast enhancement is usually in a meningeal pattern. Progression to chronic encephalomalacic changes (usually cystic) is very rapid within few weeks. Hemorrhagic foci may be present in the basal ganglia and cortical laminar necrosis may be seen as high signal on T1WI.

Rubella infection is extremely rare in Western countries because of maternal screening during pregnancy. It may lead to cataracts, chorioretinitis, glaucoma, and cardiac myopathies. Deafness caused by sensorineural injury is very common. Microcephaly and seizures may lead to medical attention. Calcifications in the periventricular white matter and basal ganglia are usually seen as a sequela to the ischemia from vasculopathy.

Congenital HIV infection is a significant healthcare problem. Transmission of disease to the fetus is about 30% in untreated mothers. Maternal treatment and cesarean section can reduce transmission to fetus to less than 2%.

Neuroimaging of congenital HIV shows meningoencephalitis, atrophy and calcific vasculopathy. Diffuse calcification is seen throughout the brain, not limited to the periventricular region or the basal ganglia. Microcephaly may develop. Congenital HIV infection is also associated with arteritis, fusiform aneurysms, and arterial sclerosis with vascular occlusion. One can see diffuse dilatation of circle of Willis vessels in these children.

Patients with congenital acquired immunodeficiency syndrome (AIDS) rarely present with neurological symptoms in the neonatal period. Ninety percent of HIV-infected infants get AIDS within the second year of life and thus, they may develop AIDS encephalitis, infections, and lymphoma at early age. The AIDS encephalitis is characterized by atrophy, diffuse white matter hyperintensity on T2WI, and basal ganglia vascular calcification. Progressive multifocal leukoencephalopathy (PML), toxoplasmosis, and tuberculosis are rare in children with AIDS.

Congenital syphilis leads to seizures and cranial nerve palsies in infancy. Radiological manifestations include optic atrophy, tabes dorsalis, meningitis, and vasculitis with enhancing meninges and perivascular spaces. Vasculitis may lead to infarctions.

PHAKOMATOSES

The phakomatoses refer to a group of congenital malformations mainly affecting the structures of ectodermal origin, the nervous system, skin (thus the neurocutaneous disorders), retina, globe and its contents. The visceral organs can be involved as well but to a lesser degree. *Phakos* is Greek for mother-spot, probably reflecting the cutaneous manifestation. Although many diseases have been described under the heading of phakomatosis, the most common and classic ones will be discussed in this chapter including neurofibromatosis, tuberous sclerosis, von Hippel-Lindau disease, and Sturge-Weber syndrome. The quintessential lesion is the neurogenic tumor, the tuber, the hemangioblastoma, and the angioma respectively. Hereditary hemorrhagic telangiectasia, ataxia-telangiectasia, neurocutaneous melanosis, basal cell nevus syndrome, Wyburn-Mason syndrome, and Parry-Romberg syndrome are also classified as phakomatoses. Read on, tender student.

Neurofibromatosis

Neurofibromatosis Type 1

Neurofibromatosis type 1 (von Recklinghausen disease, or NF-1; Table 8-4) is a disease of childhood that occurs with autosomal dominant inheritance. Overall, the incidence of neurofibromatosis is approximately 1 in 3000 to 5000 people in the general population. NF-1 appears to be transmitted on the long arm of chromosome 17 (17q11).

The NF-1 gene produces a protein called neurofibromin and appears to be a tumor suppressor gene that is inactivated in patients with this disease. The phenotypic appearance is quite variable, both clinically and radiologically. NF-1 is diagnosed if there are two or more of the following findings: (1) six or more café-au-lait spots, (2) two or more

TABLE 8-4 Neurofibromatosis Type 1 versus Neurofibromatosis Type 2

Feature	NF-1	NF-2
Chromosome involved	17	22
Optic gliomas	Yes	No
Acoustic schwannomas	No	Yes
Meningiomas	No	Yes
UBOs in deep gray matter, cerebellum	Yes	No
Incidence	1/4,000	1/50,000
Skin findings	Many	Few
Spinal gliomas	Astrocytoma	Ependymoma
Skeletal dysplasias	Yes	No
Lisch nodules (iris hamartomas)	Yes	No, but sublenticular cataracts
Dural ectasia	Yes	No
Age at presentation (yr)	<10	10-30
Vascular stenoses	Yes	No
Plexiform neurofibromas	Yes	No
Malignant change	Yes	No
Sphenoid wing absence	Yes	No
Hydrocephalus	Yes, obstructed/ stenotic aqueduct	No
CNS hamartomas	Yes	No
Paraspinal neurofibromas	Yes	Yes
Meningocele	Yes, lateral thoracic	No

CNS, Central nervous system; *NF-1,* neurofibromatosis type 1; *NF-2,* neurofibromatosis type 2; *UBOs,* unidentified bright objects.

Lisch nodules (hamartomas) of the iris, (3) two or more neurofibromas or one or more plexiform neurofibromas, (4) axillary/inguinal freckling, (5) one or more bone dysplasias or pseudarthrosis of a long bone, (6) optic pathway glioma, or (7) a first-degree relative with the diagnosis of NF-1. The following features are common characteristics of the disease: gliomas of the optic pathway, kyphoscoliosis, sphenoid wing dysplasia, vascular dysplasias (moyamoya syndrome), nerve sheath tumors, macrocephaly, cognitive impairment (wide range of learning disabilities). Additional findings that one might see in patients with neurofibromatosis include spinal dural ectasia, lateral thoracic meningoceles, aqueductal stenosis, and syringomyelia.

The most common brain abnormality in NF-1 is the optic pathway glioma, typically of the low-grade pilocytic variety. These optic gliomas present in childhood yet may have little effect on vision until they are large. Fifteen percent of patients with NF-1 have optic pathway gliomas and only half of the affected patients are symptomatic. Optic gliomas can be confined to a single optic nerve, involve both optic nerves, the optic chiasm, or rarely the optic radiations (Fig. 8-43). Involvement of the optic chiasm and optic radiations indicates a poorer prognosis. Optic chiasm and hypothalamic involvement may be associated with precocious puberty. On imaging, one may see enlargement of the optic nerves or chiasm. Enhancement is variable. These tumors are generally low-grade pilocytic astrocytomas, are slow growing and are usually watched; treatment is withheld until the patients become progressively symptomatic. Cerebellar, brain stem, and cerebral astrocytomas are also seen with NF-1. Patients with NF-1 also have increased incidence of astrocytomas of the spinal cord.

In addition, as noted on MR, the patients have multiple high signal intensity foci on T2WI or FLAIR scans that appear in the cerebellar peduncles or deep gray matter of the cerebellum, the brain stem (especially the pons), the basal ganglia (especially the globus pallidus), and the white matter of the supratentorial space (Fig. 8-44). There is considerable debate concerning what these areas of high signal intensity on T2WI represent. At this time most people believe they are due to myelin vacuolization, or myelin dysplasia. Typically these lesions do not enhance,

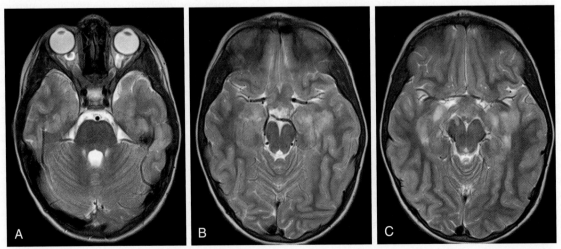

FIGURE 8-43 Neurofibromatosis type 1. Optic pathway gliomas. Axial T2-weighted imaging demonstrates thickening and tortuosity of bilateral intraorbital optic nerves **(A)**, thickening of prechiasmatic optic nerves and optic chiasm **(B)**, and bilateral optic tracts **(C)**.

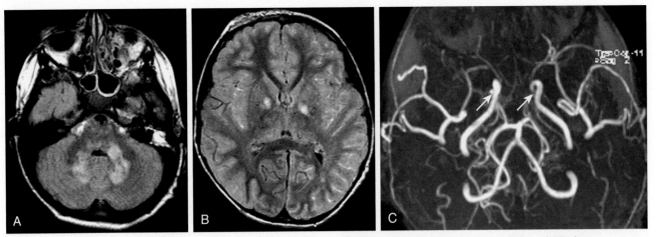

FIGURE 8-44 Neurofibromatosis type 1. Axial fluid-attenuated inversion recovery (FLAIR) images **(A),** and **(B)** demonstrates FLAIR signal hyperintensities in bilateral dentate nuclei and globus pallidi, the most common places for unidentified bright objects. **C,** Three-dimensional reconstruction from magnetic resonance angiography demonstrates occlusion of distal internal carotid arteries (paraclinoid ICAs are indicated with *arrows*), bilateral middle cerebral arteries, and anterior cerebral arteries in this patient with moyamoya. The collaterals are not well displayed in this image.

and demonstrate normal or near normal proton magnetic resonance spectroscopy (H1 MRS) findings to those of normal brain, with lack of mass effect/edema. The basal ganglia foci may also be bright on T1WI.

Follow-up studies of these high-intensity foci in NF-1 patients suggest that, although present in the basal ganglia and brain stem in the younger decades with relatively high frequency, they often decrease in size with age. Lesions in the cerebellar white matter and dentate nuclei are found in younger patients and are rare after the third decade. Rarely, they may enlarge in adulthood. When these high-intensity foci enhance or grow larger over time, the possibility of neoplasm must be raised. Short-term follow-up scans are indicated.

The presence of a plexiform neurofibroma strongly suggests NF-1. A plexiform neurofibroma consists of sheets of collagen and Schwann cells that spread in an aggressive manner, insinuating themselves in a cylindrical fashion around a nerve (Fig. 8-45). These lesions tend to involve the scalp, neck, mediastinum, retroperitoneum, cranial nerve V, and orbit. The lesions are soft and elastic, and probably account for the "elephantiasis" of neurofibromatosis.

Sarcomatous degeneration of neurofibromas occurs in about 5% of patients with peripheral nerve sheath neural tumors. The more neurofibromas one has, the higher the likelihood of malignant degeneration, usually occurring in mid-adulthood. Plexiform neurofibromas have a higher rate of malignant change, much more so than schwannomas.

Neurofibromatosis Type 2

Neurofibromatosis type 2 (NF-2) appears to be transmitted on chromosome 22q12 and is approximately one tenth as common as NF-1. This is a different entity than NF-1. The patients have fewer skin lesions compared with NF-1, and the pathognomonic imaging sign of NF-2 is bilateral vestibular schwannomas (Fig. 8-46). Diagnosis is made if the patient has one of the following features: (1) bilateral vestibular schwannomas; (2) a first-degree relative with NF-2 and either a single vestibular schwannoma or any two of the following: schwannomas, neurofibromas, meningiomas, or gliomas. Cranial nerve V is

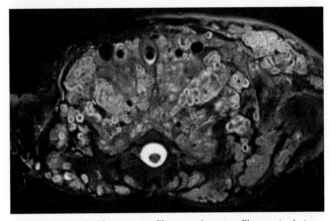

FIGURE 8-45 Plexiform neurofibromas in neurofibromatosis type 1. Axial T2-weighted imaging: large, infiltrative neurofibromas infiltrating the entire neck, to a point where normal soft tissues cannot be identified. Note the target sign in the neurofibromas.

the second most common site of schwannomas in NF-2. Sensory roots are affected more commonly than motor roots. The patients also have increased incidence of meningiomas and rarely have other glial tumors (ependymomas). The nickname MISME (multiple inherited schwannomas, meningiomas, and ependymomas) attests to the fact that neurofibromas are not a feature of NF-2. A posterior sublenticular capsular cataract in a young patient is also typical of this disorder.

Unlike adults, hearing loss is an uncommon presentation in childhood. Instead, seizures and facial nerve palsy are more common.

On MRI, enhancing masses located in and around the cerebellopontine angle or extending into the internal auditory canal are seen. The lesions generally are slightly hyperintense on T2WI. Occasionally, dural ectasia of the cranial nerve VIII root sleeve can enlarge the internal auditory canal, producing the same type of bony flaring seen with vestibular schwannomas. Arachnoid cysts may accompany the vestibular schwannomas. In addition, meningiomas may occur at this location, although the parasagittal regions predominate.

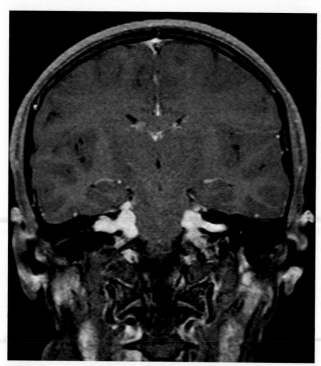

FIGURE 8-46 Neurofibromatosis type 2. Coronal postcontrast T1-weighted image shows bilateral symmetric enhancing vestibular schwannomas expanding the internal auditory canals and extending into the cerebellopontine angle cisterns.

The characteristic spinal manifestations of NF-2, multiple paraspinal nerve sheath tumors, intraspinal meningiomas, and intramedullary spinal cord tumors, can be seen in 63% to 90% of cases. In one report 53% of patients had intramedullary lesions, 55% intradural extramedullary tumors (88% were nerve sheath tumors and 12% meningiomas), and 45% both intramedullary and extramedullary masses (Fig. 8-47). Of those with intramedullary masses, over half had multiple ones. Multiple nerve sheath tumors (both schwannomas [75%] and neurofibromas [25%]) are seen in the cauda equina and may be intradural and/or extradural. Of the intramedullary tumors, ependymomas predominate, but astrocytomas and intramedullary schwannomas may occur. Syringohydromyelia can be seen, secondary to the altered CSF dynamics in the setting of either intramedullary spinal cord tumors or extramedullary masses.

See Table 8-4 summarizing the differences between NF-1 and NF-2. Although NF-1 and NF-2 are two distinct entities, in practice, cases with overlapping features can be seen, which at some point lead to the term NF-3, along with other forms of NF (believed to be variations of NF-1).

Tuberous Sclerosis

Tuberous sclerosis (TS; Bourneville disease) (Box 8-8) is an autosomal dominant disorder that involves multiple organ systems. The *TSC1* gene localized on the long arm of chromosome 9 (9q34) and the *TSC2* gene localized to chromosome 16 (16p13) have been identified in patients with TS. It arises in 1 in 6000 to 15,000 live births. The characteristic findings with tuberous sclerosis are adenoma sebaceum (60% to 90% of cases), mental retardation (50%), and seizures (60% to 80%). All three findings occur in only one third of cases. Patients also may have retinal hamartomas (50%), shagreen patches (20% to 40%), ungual fibromas (20% to 30%), rhabdomyomas of the heart (25% to 50%), angiomyolipomas of the kidney (50% to 90%), cystic skeletal lesions, as well as the intracranial manifestations. Tuberous sclerosis is like a shotgun blast: it can hit all parts of the body.

The intracranial manifestations include supratentorial periventricular subependymal hamartomas/nodules (the most common lesion seen in virtually all cases), cortical and subcortical peripheral tubers (the most characteristic lesion seen in 94% of patients on MR), white matter hamartomatous lesions (Fig. 8-48), and subependymal giant cell astrocytomas (6% to 16%) (Fig. 8-49). Cerebellar lesions such as cortical tubers or subependymal hamartomas have been reported in 10% of patients. Patients may have cortical heterotopias and ventriculomegaly as well. Eighty-eight percent of periventricular subependymal nodules are calcified (increasing with age), whereas only 50% of the parenchymal hamartomas are calcified. The frequency of cortical tubers and white matter lesions is highest in the frontal lobes followed by the parietal, occipital, temporal, and cerebellar regions. Tubers may expand gyri or show central umbilication and are bright on long TR sequences.

Patients with cerebellar tubers are older than those with cerebral tubers, have more extensive disease, and may have focal cerebellar volume loss associated with these tubers. White matter lesions may appear on MR as curvilinear or straight thin bands radiating from the ventricles (88%), wedge-shaped lesions with apices near the ventricle (31%), or tumefactive foci of abnormal intensity (14%). Subependymal nodules (31%), cortical tubers (3%), and white matter lesions (12%) may show enhancement on MR. FLAIR imaging is particularly good at spotting subcortical small tubers, even more so than T2WI scans. It should be noted, however, that the number, size, and location of tubers seems to be unrelated to the neurologic symptoms in adults. A greater number of tubers occur in children with infantile spasms, seizures before 1 year of age, and mental disability. The pathogenesis of the various lesions of tuberous sclerosis is thought to be due to abnormal radial-glial migration of dysgenetic giant cells that are capable of astrocytic or neuronal differentiation. Multipotential cells with multipotential problems.

In neonates, head US can identify the subependymal hamartomas. In infants, the subependymal hamartomas are hyperintense on T1WI and hypointense on T2WI (because of unmyelinated white matter), the opposite of what is seen in adults. They are NOT gray matter, therefore not isointense to gray matter and therefore should not be confused with subependymal nodular heterotopias. White matter anomalies are more visible in infants. However, cortical tubers are more difficult to identify in infants.

Tuberous sclerosis has an association with subependymal giant cell astrocytomas (SGCA aka SEGA lesions) that generally occur around the foramina of Monro (see Fig. 8-49). As opposed to subependymal tubers, these lesions enhance commonly and uniformly, are large, grow with time, cause obstructive hydrocephalus, and have a lower rate of calcification. It is believed that these tumors arise from subependymal nodules. Progressive enlargement and/or greater than 12 mm diameter in an enhancing subependymal hamartoma strongly suggests SGCA. They occur in approximately 5% to 10% of patients who have tuberous sclerosis. Their malignant potential is small.

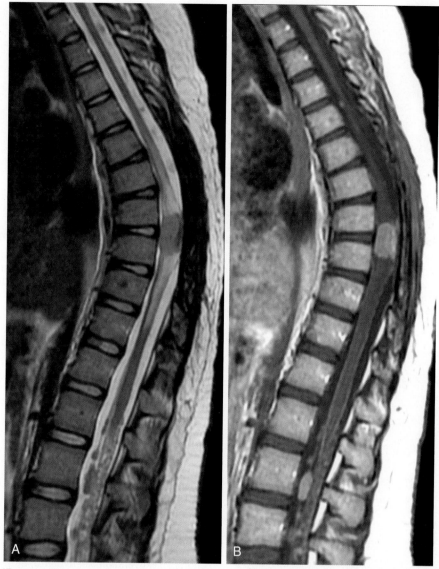

FIGURE 8-47 Neurofibromatosis type 2, ependymomas. **A,** Sagittal T2-weighted imaging (T2WI) demonstrates an expansile intramedullary mass in the mid thoracic spinal cord representing an ependymoma. Note additional multifocal nodular masses along the cauda equina representing schwannomas. **B,** Postcontrast T1WI scan shows enhancement in all the lesions.

BOX 8-8 Tuberous Sclerosis Findings

CLINICAL

Adenoma sebaceum
Ash-leaf spot
Café-au-lait spots
Mental retardation
Retinal hamartomas
Retinal phakoma
Seizures
Shagreen patches
Subungual fibromas

CENTRAL NERVOUS SYSTEM IMAGING FINDINGS

Atrophy
Calcified optic nerve head drusen
Cortical tubers
Giant cell astrocytomas
Intracranial calcifications
Periependymal nodules
Radial glial fiber hyperintensity/myelination disorder

NON–CENTRAL NERVOUS SYSTEM IMAGING FINDINGS

Angiomyolipomas of kidneys
Aortic aneurysm
Hepatic adenomas
Pulmonary lymphangiomyomatosis
Rhabdomyomas of heart
Renal cell carcinoma
Renal cysts
Skeletal cysts, sclerotic densities, periosteal thickening
Upper lobe interstitial fibrosis, blebs, pneumothorax, chylothorax
Vascular stenosis

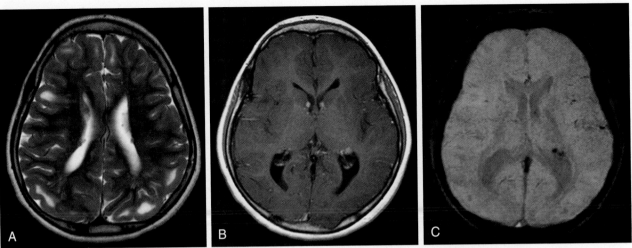

FIGURE 8-48 Tuberous sclerosis (TS). **A,** Axial T2-weighted imaging (T2WI) shows multifocal areas of increased T2 signal in the subcortical white matter representing cortical tubers, the most characteristic lesion in TS. Note the T2 dark subependymal nodules. **B,** Postcontrast T1WI scan shows enhancing subependymal nodules in the region of the foramen of Monro and also in the left occipital ependyma. **C,** Note the signal drop in susceptibility-weighted imaging associated with the left occipital subependymal nodule representing calcification.

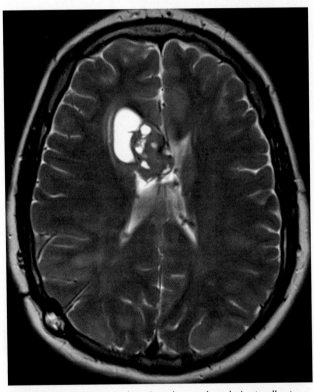

FIGURE 8-49 Tuberous sclerosis, subependymal giant cell astrocytomas. Axial T2-weighted imaging shows a subependymal giant cell astrocytoma (>1 cm) with solid and cystic components. Note smaller T2 dark subependymal nodules. Patient is shunted.

Susceptibility-weighted imaging should be part of the MRI protocol in patients with TS because calcification can be seen in the most common and characteristic lesions.

Sturge-Weber Syndrome

Sturge-Weber syndrome (encephalotrigeminal angiomatosis; Box 8-9) is associated with facial port wine stain with

---

BOX 8-9   Findings in Sturge-Weber Syndrome

CLINICAL

Accelerated myelination
Choroidal angioma
Glaucoma-buphthalmos
Hemiparesis, hemiplegia
Mental retardation
Scleral telangiectasia
Seizures
Trigeminal angioma (capillary telangiectasia); port wine nevus in cranial nerve V-1 distribution
Visceral angioma

RADIOGRAPHIC

Anomalous venous drainage to deep veins
Choroid plexus angioma or hypertrophy ipsilateral to angiomatosis
Dyke-Davidoff-Masson syndrome
Elevated petrous ridge, sphenoid wing
Enlarged frontal sinuses
Hemihypertrophied skull
Hemiatrophy
Intracranial calcification (tram-tracks)
Pial angioma

---

hemiplegia and seizure disorder. Infantile spasms develop in 90% by 1 year of age. Other manifestations include congenital glaucoma and buphthalmos, choroidal or scleral hemangiomas, and mental retardation. This disease usually occurs sporadically with equal male and female incidence. The cutaneous vascular malformation in the face is usually in the V-1 distribution, which is in fact a capillary malformation. As the patient ages, thickening of the capillary malformation is not uncommon. Ipsilateral to the vascular angioma within the brain, one often identifies enlargement and enhancement of the choroid plexus, suggesting an angioma.

Leptomeningial capillary-venous angiomatosis is also seen in Sturge-Weber syndrome. On susceptibility-weighted imaging, Sturge-Weber syndrome is usually detected by

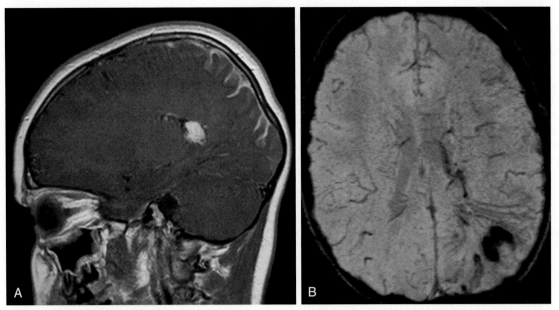

FIGURE 8-50 Sturge Weber. **A,** Sagittal postcontrast T1-weighted imaging shows thickening and enhancement of the parietooccipital leptomeninges secondary to leptomeningeal angiomatosis. **B,** Axial susceptibility-weighted imaging shows signal drop in the subcortical white matter representing calcification. In addition note the dilated deep intramedullary veins. Overall, left cerebral hemisphere is smaller compared with the right, secondary to atrophy.

cortical calcification (tram-line) in the ipsilateral occipital, parietal, or temporal lobe underlying the leptomeningeal vascular malformation with associated hemiatrophy (Fig. 8-50). This can be bilateral in 20% of the cases. The abnormality is truly a pial vascular abnormality, which demonstrates underlying patchy gliosis and demyelination. The calcification is within the cortex and not the meninges, arises in the fourth layer of the cortex, and is thought to result from a localized steal phenomenon from the affected cortex due to the leptomeningeal angiomatosis. Calcification, however, is not seen until about the second year of life. Susceptibility-weighted imaging demonstrates the cortical calcification to best advantage over conventional MRI sequences. Atrophy is present on the ipsilateral side, as is enlargement of the choroid of the lateral ventricles.

MR detects a greater extent of disease than CT as in any of the phakomatoses, particularly on enhanced T1WI and susceptibility-weighted imaging. In this instance the pia enhances dramatically, giving a true demonstration of the degree of the vascular abnormality. Cortical enhancement caused by ischemic injury may also be present. Abnormally low signal intensity within the white matter on T2WI is probably related to abnormal myelination from ischemia. Bilaterality of the vascular lesion is often seen on MR.

On angiography an increase in the number and size of the medullary veins with decreased cortical veins (anomalous venous drainage) and a capillary stain are usually seen in the parietal and occipital lobes. There is slow cerebral blood flow, and the ipsilateral cerebral arteries are generally small. Keep your eyes on your patients' eyes! In nearly 50% of patients with Sturge-Weber syndrome you will find abnormal ocular enhancement, be it because of choroidal hemangiomas or inflammation from glaucoma. Visualization of ocular hemangiomata is increased with bilateral intracranial disease, extensive facial nevi, and ocular glaucoma.

The Wyburn-Mason syndrome may be a forme fruste of Sturge-Weber syndrome. Patients have a facial vascular nevus in nerve V distribution, retinal angiomas, and a midbrain arteriovenous malformation. To make things a bit more confusing some have postulated that Sturge-Weber syndrome with involvement of the viscera and extremities is called Klippel-Trenaunay-Weber syndrome (hemihypertrophy, cutaneous angiomas, slow-flow vascular malformations, varices, and/or anomalous venous drainage). But we beg to differ, and believe that these are different entities.

von Hippel-Lindau Disease

CNS hemangioblastomas combined with retinal hemangioblastomas (67%), cysts of the kidney, pancreas, and liver, renal cell carcinomas, islet cell tumors and adenomas form von Hippel-Lindau disease (VHL). Diagnosis is based on a hemangioblastoma of the CNS or retina and the presence of one VHL associated-tumor or a previous family history. Pheochromocytomas of the adrenal gland may also be present, linking the multiple endocrine neoplasia syndromes with VHL. Endolymphatic sac tumors have also been described with this entity. There also appears to be a link to neurofibromatosis. The disorder appears to be transmitted through autosomal dominant inheritance in 20% of cases on chromosome 3p25, the *VHL* gene. Twenty percent of patients with hemangioblastomas of the cerebellum have VHL. Cerebellar hemangioblastomas ultimately develop in 83% of patients with VHL.

Multiple CNS hemangioblastomas are the sine qua non of this syndrome, and they may arise in the cerebellum (most commonly), the medulla, the spinal cord, or less commonly, supratentorially. The lesions may be cystic, solid, or combined. The classic description is a highly vascular, enhancing mural nodule associated with a predominantly cystic mass in the lateral cerebellum.

Spinal hemangioblastomas represent about 3% of spinal cord tumors. One third are associated with von Hippel-Lindau disease. Of all spinal hemangioblastomas, 80% are single, 20% multiple (almost all associated with VHL), 60% intramedullary, 11% intramedullary and extramedullary, 21% intradural but purely extramedullary, and 8% extradural. Their location is usually characterized by the surgeons as "sub-pial." They are more commonly found in the thoracic cord than the cervical cord. Most hemangioblastomas seen with VHL are 10 mm or less in size and may be intramedullary or along the dorsal nerve roots. If you are expecting to see vascular flow voids to make the diagnosis of spinal hemangioblastoma, don't hold your breath. It is extremely rare to see flow voids on MR in hemangioblastomas under 15 mm in size; you will have to wait for the tumors to grow. On the other hand, a syrinx may be present in 40% to 60% of cases (often out of proportion to the small size of the tumor).

Hereditary Hemorrhagic Telangiectasia

Hereditary hemorrhagic telangiectasia (also known as HHT or Osler-Weber-Rendu syndrome) is an entity consisting of mucocutaneous telangiectasias and visceral arteriovenous malformations. One of the many presenting symptoms may be epistaxis secondary to sinonasal mucocutaneous telangiectasias. However, one should not forget that 5% of patients with HHT have a cerebral arteriovenous malformation (AVM) and 2% of AVMs are associated with HHT. When a patient with HHT has one cerebral AVM, there is a 50% chance of a second AVM in the brain elsewhere. The AVMs may have a nidus (seen more frequently in adults) or may actually represent a fistula (more common in kids). There can be additional AVMs in the lungs, liver, spine, gastrointestinal tracts and pancreas.

Neurocutaneous Melanosis

Another relatively obscure syndrome put in the classification of the phakomatoses is neurocutaneous melanosis. Neurocutaneous melanosis is usually a sporadic disease discovered in children because of hydrocephalus from gummed up arachnoid villi. Multiple cranial neuropathies may develop. Melanoblasts from neural crest cells are present in the globes, skin, inner ear, sinonasal cavity, and leptomeninges and are the source of this disorder. It is characterized by cutaneous nevi and melanotic thickening of the meninges. Diffuse enhancement of the meninges of the brain and spine (20%) is seen; the melanin may (Fig. 8-51) or may not be detected on the preenhanced T1WI. Hydrocephalus, cranial neuropathies, and syringohydromyelia may develop.

Although malignant degeneration of the skin lesions is very uncommon, malignant transformation of CNS melanosis occurs in up to 50% of cases. When this occurs, parenchymal or intramedullary infiltration is the hallmark.

PHACES Syndrome

Presence of hemangiomas in the head and neck with intracranial pathologies qualifies PHACES syndrome (posterior fossa malformations, hemangiomas, arterial anomalies, cardiac anomalies and aortic coarctation, eye anomalies, and sternal clefting and/or supraumbilical raphe) under the category of phakomatosis. The vascular anomaly of the soft

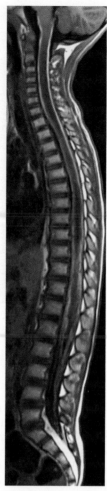

FIGURE 8-51 Neurocutaneous melanosis. Noncontrast sagittal T1-weighted imaging shows T1 shortening and thickening of the leptomeninges covering the spinal cord and cauda equina secondary to melanin deposition.

tissues in the head and neck is large (>5 cm) and called segmental infantile hemangioma (previously called as capillary hemangioma or strawberry hemangioma). The cerebellar anomalies are the most common structural brain abnormality. Dandy-Walker malformation, cerebellar hypoplasia, and cortical dysplasias can be seen. The most common anomalies identified on neuroimaging are vascular in origin such as persistence of the trigeminal artery, and aplasia/hypoplasia of internal or external carotid arteries, vertebral arteries or posterior cerebral arteries (Fig. 8-52).

CONGENITAL SPINAL ANOMALIES

Development of the spinal canal and its contents follow four distinct and somewhat overlapping processes: (1) gastrulation and development of the notochord, (2) primary neurulation, (3) segmentation with formation of the somites, and (4) secondary neurulation (caudal cell mass). We will discuss the most common congenital spinal anomalies based on these processes.

Disorders of Primary Neurulation

Separation of the neural tube from the overlying ectoderm during closure of the neural tube is a process called

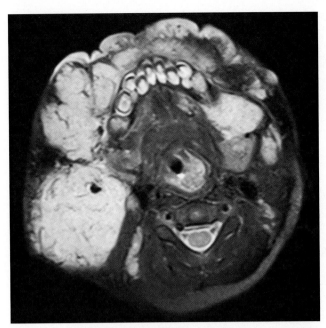

FIGURE 8-52 PHACES (posterior fossa malformations, hemangiomas, arterial anomalies, cardiac anomalies and aortic coarctation, eye anomalies, and sternal clefting and/or supraumbilical raphe). Large multifocal T2 bright solid masses infiltrating the lower chin and right parotid gland represent infantile hemangiomas. Note smaller ones on the left side.

disjunction. After disjunction the ectoderm closes in the midline, dorsal to the closed neural tube. The perineural mesenchyma migrates into the space between the closed neural tube and ectoderm inducing the formation of meninges, bony spinal column, and paraspinous musculature.

Open Spinal Dysraphism

Open spinal dysraphism refers to the exposed abnormal neural tissue with leakage of the CSF. The skin, muscle, and bone are deficient with variable degree of severity. In closed spinal dysraphism, the malformed neural tube is covered by mesoderm and ectodermal elements such as subcutaneous fat and skin (see Box 8-8). Spinal bifida occulta refers to presence of defective/incomplete closure of the posterior osseous elements only. The skin is covered, the spinal cord is unremarkable. Typically no other CNS malformations are identified.

Complete/segmental nondisjunction of the cutaneous ectoderm from neural ectoderm results in formation of myeloceles and myelomeningoceles. Many of these disorders are detected prenatally by serologic or amniocentesis tests (elevated alpha-fetoprotein level) and US. The clinical and neurological symptoms are believed to arise from two major issues: the neural placode being less functional because of deranged neuroarchitecture, and long-lasting intrauterine exposure of the neural tissue to the amniotic fluid.

The myelocele and the myelomeningocele are the most common forms of open dysraphism, with myelomeningoceles more frequently encountered than myeloceles. All children with open spinal dysraphism are believed to have a Chiari II malformation and variable degree of hydromyelia is observed (Fig. 8-53). The most common location is the lumbar level; however, it can occur in the cervical (Fig. 8-54) and thoracic levels as well. Although quite rare,

anterior dysraphism can occur, most commonly in the sacral region. The Currarino triad refers to (1) anorectal malformation, (2) bony sacral defect, and (3) presacral mass (e.g., anterior sacral meningocele).

A myelocele refers to herniation of the neural placode (a flat plate of neural tissue) through the bony defect such that it lies flush with the surface of the skin of the back. Very little CSF is evident, which is continuous with the subarachnoid space, and only a layer of arachnoid is present at the ventral surface of the myelocele. Both the ventral and dorsal nerve roots arise from the neural placode. A myelomeningocele is identical to a myelocele except for expansion of the ventral subarachnoid space resulting in posterior displacement of the neural placode and stretching of the nerve roots.

Myelomeningoceles are typically repaired soon after birth, often without any preoperational imaging, because it is quite clear with visual inspection only.

Closed Spinal Dysraphism

Skin covered or closed spinal dysraphisms include lipomyelocele and lipomyelomeningoceles (Box 8-10). The prognosis is far better compared with open spinal dysraphism because the neural tissue is covered by the subcutaneous fat and the skin. They result from premature disjunction of the neural ectoderm from skin ectoderm. Then the mesenchymal tissue remains in touch with the neural tube, leading to excess production of fat. The size of the lipomas is variable and the extension can be exclusively intradural or intradural with subcutaneous continuation.

Similar to open spinal dysraphism, the classification depends on the level of involved elements; if the neural tissue is seen at the same level as the spinal canal and skin surface, then the malformation is called lipomyelocele, if the neural tissue is pushed beyond the level of the spinal canal and skin, then the malformation is called lipomyelomeningocele (Fig. 8-55). There may be associations with other malformations such as cloacal malformations. No CSF leakage, no Chiari II!

Both the open and closed dysraphisms are usually associated with a low-lying, tethered conus medullaris below the L2-L3 disk. This brings up the discussion about the normal location of the conus. In several series involving more than 1000 cases, the normal termination of the conus is above the L2-3 disk level; and less than 2% of normal humans have conus below the L2-3 disk level. Therefore, conus at or below the level of L2-L3 disk space should prompt the search for tethering mass, bony spur or thick filum.

The goal of therapy is to (1) enclose the defect into the intraspinal compartment so that infection is prevented, and (2) free up the distal end of the neural tube so that tethering does not occur as the patient grows. If tethering does occur, then downward herniation of intracranial contents through the foramen magnum is a possibility. One of the complications of the surgery is production of fibrous tissue, which may re-tether the dysraphic tissue. On postoperative examination, you may not see a difference in the location of the conus medullaris from the preoperative location. Nonetheless, symptomatic improvement usually occurs even with no change in the site of the bottom of the cord. Presumably, CSF pulsations return to normal and the cord and hindbrain can move again.

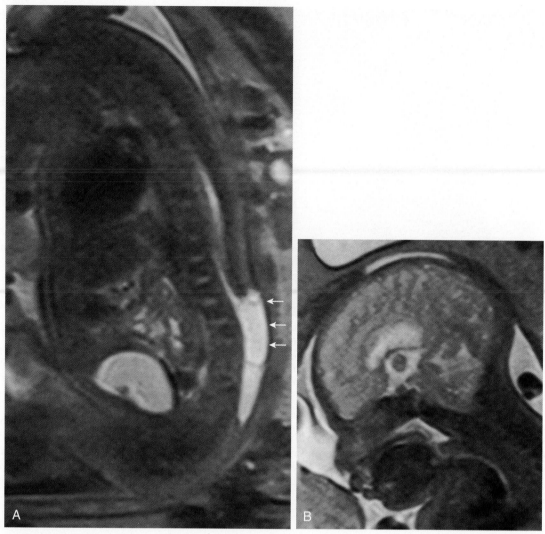

FIGURE 8-53 Myelomeningocele: third trimester fetus. **A,** Sagittal half-Fourier-acquisition single-shot turbo spin-echo (HASTE) through the thoracolumbar spine shows a lumbosacral spina bifida with dorsal myelomeningocele *(arrows)*. **B,** Sagittal HASTE through the brain confirms Chiari II malformation.

Dorsal Dermal Sinus

These represent an intermediate malformation between open and closed spinal dysraphism but closer to the closed type. An epithelial-lined tract remains between the skin surface and the spinal canal. The sinus can attach at the dura without transgressing through it, resulting in dural tenting dorsally at the attachment site. If the dorsal dermal sinus tract courses through the dura, it may terminate in a dermoid/epidermoid in 50% of cases (Fig. 8-56), or they may terminate on the spinal cord, conus medullaris, filum terminale or a nerve root. Focal/incomplete disjunction of the skin ectoderm and neuroectoderm is believed to be the cause. These are most frequently seen in the lumbosacral and occipital regions (because these are the last portions of the neural tube to close).

They are skinny and small; therefore, high-resolution thin and consecutive slices (such as CISS) should be obtained on MRI, in addition to the conventional sequences. Only minor bony changes such as hypoplastic spinous process, single bifid spinous process, or laminar defect are noted if any.

Disorders of Secondary Neurulation

Disorders of secondary neurulation affect the lowest part of the spinal cord: the conus medullaris and filum terminale (approximately 10% of the entire CNS). Secondary neurulation starts after completion of primary neurulation about day 48 of gestation.

Fibrolipoma of the Filum Terminale

Lipomas of the filum terminale are another form of spinal dysraphism that may cause tethering of the cord. These lesions are identified as having fat density on CT or signal intensity on MR (Fig. 8-57). Small amounts of fat (less than 2 mm in transverse diameter) in the filum are usually asymptomatic in kids and may be termed "fibrolipomas." Thoracic lipomas outnumber lumbosacral ones. Alternatively, the filum terminale may be markedly thickened and tethered without fatty infiltration (the adult tethered spinal cord syndrome). The conus medullaris may be dragged downward by the thick filum. The normal filum is less than 1 mm in cross-sectional diameter.

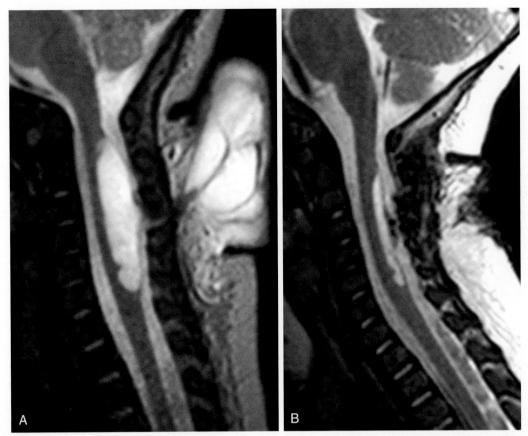

FIGURE 8-54 Cervical myelocele. Sagittal T2-weighted imaging of the cervical spine demonstrates the presurgical **(A)** and postsurgical **(B)** cervical myelocele. Note the syringohydromyelia protruding through the bony defect **(A)**.

BOX 8-10   Classification of Spinal Defects

OPEN SPINAL DYSRAPHISM
Myelomeningocele (99%)
Myelocele (1%)

CLOSED SPINAL DYSRAPHISM
With a Subcutaneous Mass
Lipomyelocele
Lipomyelomeningocele
Meningocele
Myelocystocele

Without a Subcutaneous Mass
Filar lipoma
Neuroenteric cyst
Dermal sinus
Diastematomyelia

Previously the upper limit of normal filar thickness was considered to be 2 mm; however, with the advent of higher-resolution thinner slice imaging, now 1 mm in considered as the upper limit of normal. Associated findings with a taut filum terminale (adult tethered cord syndrome) are kyphoscoliosis, midline bony defects in the lumbosacral part of the spine, and fatty infiltration of the filum. Patients may develop conus ischemia, the dreaded sphincteric dysfunction, foot deformities, and abnormal gait.

Caudal Regression Syndrome

In caudal regression syndrome you may see a trunculated/blunt ending spinal cord in the caudal spine (usually in the lower thoracic level) associated with absence of matching vertebral osseous elements. Caudal regression is most commonly seen in children of diabetic mothers. The condition can be seen in association with urogenital anomalies such as pulmonary hypoplasia (secondary to renal insufficiency) and imperforate anus. Caudal regression can also be seen as part of OEIS (omphalocele, exstrophy, imperforate anus, spinal defects) syndrome.

Variable amounts of vertebral body fusion and spinal stenoses may be seen, but the Aunt Minnie is the club- or wedge-shaped distal cord terminating above L1 (Fig. 8-58). Typically, the sacrum and coccyx are absent, and depending on the severity, higher-level vertebral bodies may be missing. The malformation can be either high and abrupt, or low and tethered. Additional findings are hypoplastic musculature and dislocated hips.

Sacrococcygeal Teratoma

Sacrococcygeal teratomas are the most common tumor in neonates and can be visualized on prenatal ultrasound. They are thought to arise from the rests of totipotential cells in the caudal cell mass. Approximately two thirds are mature teratomas. Girls are affected four times more than boys. Classification is based on the location; type I, almost completely outside the pelvis, with a minimal presacral extension, and type IV, almost completely in the pelvis, with type II and III in between types I and IV (Box 8-11).

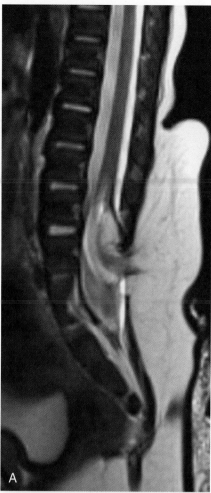

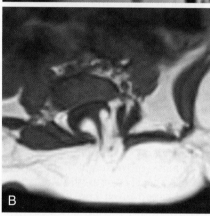

FIGURE 8-55 Lipomyelomeningocele. **A,** Sagittal T2-weighted imaging shows a tethered cord with a T2 and T1 bright (as seen in **B**) intradural lipoma extends from the subcutaneous fat through the bony spina bifida into the dorsal subarachnoid space **(B).**

The deeper the sacrococcygeal teratoma in the pelvis, the worse the prognosis. Surgical resection should involve the coccyx. Usually it is seen on imaging as a mixed solid and cystic mass; however, purely cystic forms are seen as well.

Anomalies of Notocord Development

Diastematomyelia

Diastematomyelia (split cord malformation) refers to a complete or incomplete (anterior only or posterior only)

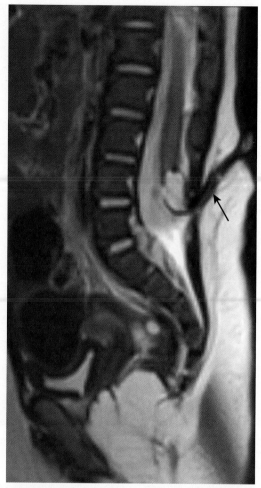

FIGURE 8-56 Dermal sinus tract. Tethered cord terminating at L4-L5 disk space level with an intradural lipoma. See the downward oblique orientation of the dermal sinus tract *(arrow)* extending from the skin surface to the lipoma.

longitudinal split in the cord. The hemicords may be symmetric or asymmetric; however, each involves a central canal, as well as a ventral and dorsal horn. The split may also involve the dura so that there may be two dural sacs; however, more commonly, two hemicords within one enlarged sac is seen. Usually this abnormality occurs in the lower thoracic-upper lumbar region and is associated with bony abnormalities 85% of the time including spina bifida, widened interpediculate distances, hemivertebrae, and scoliosis. Hairy skin patches occur in 75% of cases. The separation of the spinal cord into two hemicords may be due to a bony spur, cartilaginous separation, or fibrous bands (Fig. 8-59). When this occurs, generally the cord reunites below the cleft. There often is some asymmetry in the size of the hemicords. A bony spur causing the diastematomyelia is more commonly associated with two separate dural sacs than a fibrous split. Associated tethering of the conus medullaris, myelomeningoceles (31% to 46%), and hydrosyringomyelia can be seen. Almost all cases have some form of formation/segmentation anomaly of the spinal column.

All these neural tube defects are best evaluated with MR in conjunction with CT. Although CT is useful in detecting the bony canal abnormalities and the bony spur between the diastematomyelia, MR is superior in

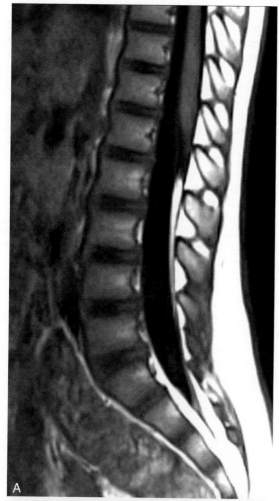

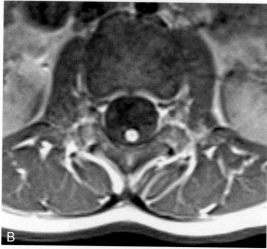

FIGURE 8-57 Filar lipoma. Sagittal **(A)** and axial **(B)** T1-weighted imaging shows thickening of the filum terminale with bright T1 signal representing fatty infiltration. The thick fatty filum lies in the dorsal aspect of the thecal sac, sometimes only seen in axial images. Note normal location of the conus medullaris.

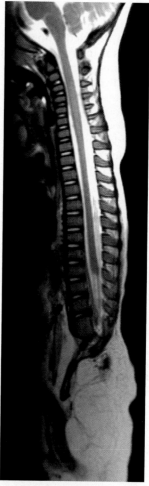

FIGURE 8-58 Caudal regression syndrome. Sagittal T1-weighted imaging shows a shortened spine. The lowest vertebra is T12. Note the blunted conus terminus at the level of T9.

BOX 8-11   Grading of Sacrococcygeal Teratoma by Extent

I: Protrudes predominantly externally with or without presacral component
II: Protrudes externally and internally with intrapelvic mass
III: Protrudes internally with pelvic and abdominal mass, minimal external component
IV: Protrudes only internally in presacral space

---

locating the distal portion of the conus medullaris and identifying the thickening of the filum terminale, the fatty component to the dysraphic state, and the presence of hemicords. Obviously, the presence or absence of a hydromyelia within the spinal cord is better identified with MR than with CT. A full examination of the patient who has spinal dysraphism should include imaging evaluation of the skull base to assess for a spinal dysraphism and the spine to detect hydromyelia, to assess the position of the conus medullaris, to identify the structures protruding through the bony defect, and to evaluate for block-vertebrae or hemivertebrae.

Neuroenteric Cysts

These are usually unilocular, single, smooth cysts most commonly seen in the cervical and thoracic regions. In most cases, the signal intensity follows that of the CSF; however, milky, xanthochromic content may alter the signal intensities on MR. Typically, enteric cysts are located in the ventral or ventrolateral aspect of the cord in an intradural, extramedullary location; however, 10% to 15% of the cases may have an intramedullary component. Enteric cysts may be extraspinal, within the mediastinum or abdomen. Dorsal enteric diverticulum refers to an anomaly where the intraabdominal contents extend through a diastematomyelia and/or malformed vertebral bodies.

Neuroenteric cysts arise from failure of cleavage between the endodermic bronchial or gastrointestinal tract and the spinal system. The persistent connection (via the canal of Kovalevsky) may be manifest as an intradural extramedullary cyst in the spinal canal (most commonly), but in some cases (<20%) there can be a coexistent thoracic bronchoenteric cyst. Over 50% have vertebral segmentation anomalies including but not limited to hemivertebrae, block vertebral bodies, Klippel-Feil syndrome, butterfly vertebrae and fused vertebrae. Bony remodeling from the longstanding cyst may be present.

Formation and Segmentation Anomalies of the Spinal Column

Development of the vertebral column involves three major stages: membranous development, chondrification and ossification. From a practical standpoint, these anomalies refer to either abnormally formed vertebra (such as butterfly or hemivertebrae), or abnormally unsegmented vertebrae (such as block vertebra; Fig. 8-60). These can be solitary findings or can be seen in other syndromes such as in VACTERL and OEIS.

Although scoliosis is idiopathic in the majority of cases (>90%), the reason for imaging is to exclude or identify formation/segmentation anomalies of the spine. Kidney anomalies are more common in these patients, so take a peek at the kidneys as you review the spine!

The Klippel-Feil anomaly refers to incomplete segmentation of multiple cervical spine bodies. It may be associated with the Chiari malformations and syringohydromyelia. The C2-C3 and C5-C6 levels are the most common sites of segmentation anomalies.

Other Bony Disorders. Many bony disorders are associated with spinal anomalies. Hypoplasia and/or incomplete development of the C1 arch and/or odontoid process is a common occurrence and is usually asymptomatic. However, occipitalization of the C1 vertebral body may be associated with atlantoaxial instability.

Dural Ectasia

Dural ectasias can be seen in Marfan syndrome, Loey-Dietz syndrome and NF-1. If you measure the diameter of the dural sac and correct for vertebral body diameters, you will find that the dural sac ratios in these patients are increased compared with controls, particularly at L3 and S1.

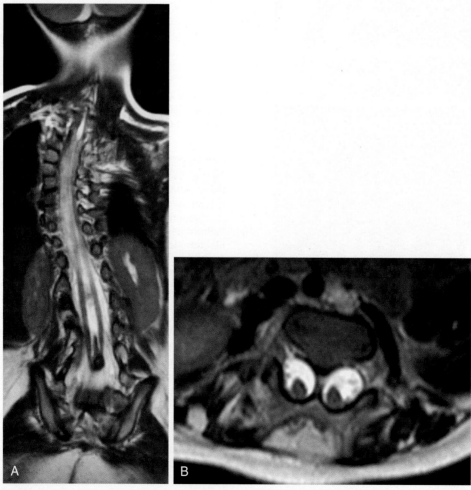

FIGURE 8-59 Diastematomyelia. **A,** Coronal T2-weighted imaging (T2WI) shows scoliosis. Note the two hemicords in the distal thoracolumbar region. The T2 dark bony spur is seen at the lumbosacral junction. **B,** Axial T2WI scan shows the bony spur extending between the two hemicords.

Associated findings of dural ectasia include widening of the canal, thinning of adjacent bone, enlargement of neural foramina, increased interpediculate distance, scalloping of the posterior vertebral body, and meningocele formation.

Dural ectasia, after excluding neurofibromatosis as a cause, is considered a principal criterion for Marfan syndrome. Genetic mutations associated with Marfan syndrome have been mapped to chromosome 15's long arm (*fibrillin-1* gene). Other manifestations include aortic dilatation, dissection, or coarctation, lens dislocation (up and out), arachnodactyly, pectus excavatum, tall stature, osteopenia, dolichocephaly, pes planus, scoliosis, ligamentous laxity, glaucoma, mitral valve regurgitation, pulmonary cysts, and blue sclera.

Just remember that scalloping of the posterior vertebral body is not unique to dural ectasia, because it is also seen in patients with achondroplasia where the spinal canal is narrowed from reduced interpediculate distance.

Miscellaneous
Syringohydromyelia

A cavity in the cord may be due to central canal dilatation (hydromyelia) or a cavity eccentric to the central canal (syrinx). On detailed radiological and pathological exam, most cases are found to involve both, thus the term

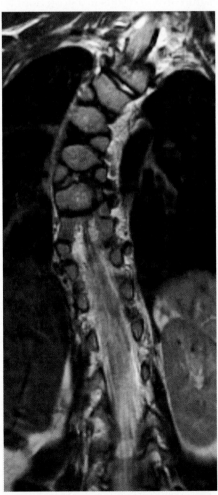

FIGURE 8-60 Formation and segmentation anomalies. Coronal T2-weighted imaging shows multilevel hemivertebrae with lack of proper segmentation (no visible disc space).

syringohydromyelia is the most commonly used terminology among neuroradiologists.

Most hydromyelic cavities are associated with congenital spinal and hindbrain anomalies such as Chiari malformations and myelomeningoceles. Most syrinxs are also congenital but may arise as a result of spinal cord trauma, ischemia, adhesions, or neoplasms.

Ventriculus Terminalis

During the development of the spinal cord, the central canal is widest at the conus; which is referred to as ventriculus terminalis. This may persist into infancy and even into adulthood. Typically, this is an incidental finding without associated clinical deficits. This appears as an ovoid, nonenhancing, smooth dilation of the central canal with the signal intensity of cerebrospinal fluid on all pulse sequences.

Lateral Thoracic Meningoceles

These are most commonly seen with NF-2, but can be seen in Marfan syndrome. Lehman syndrome includes wormian bones, hypoplastic atlas, and malar hypoplasia with lateral meningoceles. Ehlers-Danlos can also be associated with lateral meningoceles.

SKULL ANOMALIES

Craniostenosis

At birth, the sagittal, coronal, lambdoid, and metopic sutures are open. Occasionally, one will see a suture that crosses from one lambdoid suture to the other, the persistent mendosal suture. *Craniostenosis,* or *craniosynostosis,* refers to abnormal early fusion of one or more of the sutures of the skull (Table 8-5). In 75% of cases, only one suture or part of a suture is fused; in 25% of cases, more than one suture is affected. This leads to abnormal head shapes and a palpable ridge at the site of fusion. Males are affected much more often than females. These disorders, if severe and early in development, can cause abnormal growth of the brain. Microcephaly may occur.

Syndromes associated with early sutural closure include Crouzon syndrome, Apert syndrome, hypophosphatasia, and Carpenter syndrome. Endocrinologic abnormalities, including rickets, hyperthyroidism, and hypophosphatasia can cause craniosynostosis. Premature closure of the sagittal suture, the most common variety (1 in 4200 births), produces a head that cannot grow side to side, so the head looks long and thin, or scaphocephalic, also

TABLE 8-5 Craniostenosis

Type	Suture Involved	Head Shape
Dolichocephaly	Sagittal	Long and thin
Brachycephaly	Coronal	Round and foreshortened
Turricephaly	Lambdoid	High-riding top
Plagiocephaly	Any unilateral suture	Asymmetric
Trigonocephaly	Metopic	Anteriorly pointed head
"Harlequin eye"	Unilateral coronal	One eye points upward

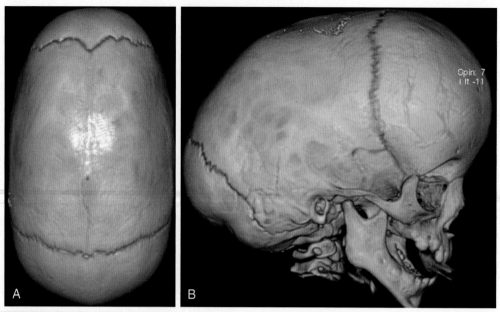

FIGURE 8-61 Scaphocephaly on three-dimensional surface rendered computed tomographic images. **A,** Supero-anterior view shows narrow head with reduced biparietal diameter. **B,** Lateral view shows elongated head with relatively increased anteroposterior diameter, secondary to early closure of the sagittal suture.

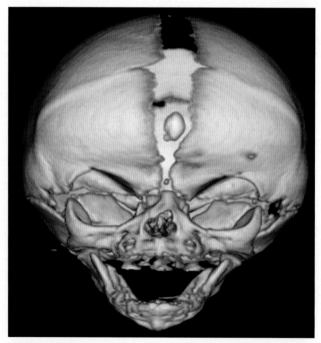

FIGURE 8-62 Brachycephaly. Three-dimensional surface rendered image, anterior view shows early closure of bilateral coronal sutures in this young infant. Note the bony ridge along the early closed coronal sutures. The anterior fontanelle and sagittal suture are still open. Note the elliptical orbits, known as "harlequin eye" deformity.

termed *dolichocephalic* (Fig. 8-61). If the coronal suture fuses early, the head is short and fat, or brachycephalic (Fig. 8-62). Sometimes only one portion of a coronal or lambdoid suture fuses too early, leading to an asymmetric skull with plagiocephaly. The forehead and orbital rim (eyebrow) have a flattened appearance on that side. This gives a harlequin "winking" eye. Lambdoid suture closure can lead to turricephaly, with a high-riding vertex. Unilateral

lambdoid suture fusion results in a posterior flattened plagiocephaly. Metopic sutural closure causes trigonocephaly with a ridge that runs down the forehead like a triceratops. Communicating hydrocephalus and tonsillar herniation are seen in some patients with complex craniosynostosis.

Surgery is attempted for cosmetic reasons or when elevations of ICP are dangerous secondary to the growth of the brain against the noncompliant calvarium. The elevated ICP may lead to reduction in brain perfusion. If brain growth is stunted because of craniostenosis, operative intervention is also indicated. Chronic venous hypertension because of jugular foramen stenosis has been proposed as an etiology for the increased ICP.

Osteopetrosis

Osteopetrosis may be inherited as an autosomal dominant or recessive condition, with the latter being the more virulent form. Calvarial involvement is frequent in both types, with clinical manifestations mostly relating to involvement of skull base foramina. Therefore, cranial nerve palsies, optic atrophy, and stenoses of the carotid and jugular vessels may be present. One may see diffuse bone thickening and increased marrow density on CT or decreased intensity on MRI. MR may depict optic nerve sheath dilation, tonsillar herniation, ventriculomegaly.

Achondroplasia

Achondroplasia is the most common cause of dwarfism and is characterized by a small foramen magnum but a large head. The macrocephaly may be secondary to hydrocephalus. The hydrocephalus may be due to venous outflow obstruction at the jugular foramen level, leading to elevated venous pressure and reduced flow in the superior sagittal sinus. The lack of resorption in the arachnoid villi because of these pressure effects produces larger CSF volume in the ventricles, increased ICP, maintenance of widened calvarial sutures, and an enlarged head.

Other features of achondroplasia include a short clivus, platybasia, and a J-shaped sella. Congenital spinal stenosis because of short pedicles is another feature of this disease.

Wormian Bones

Wormian bones (secondary ossification centers within sutural lines) are seen in osteogenesis imperfecta, cleidocranial dysplasia, cretinism, pyknodysostosis, Down syndrome, hypothyroidism, progeria, and hypophosphatasia. Add normal variant to this list and you've got it all covered!

Basilar Invagination and Platybasia

Basilar invagination refers to upward protrusion of the odontoid process into the infratentorial space. Two lines must be learned: (1) the McGregor line, extending from the posterior margin of the hard palate to the undersurface of the occiput, and (2) the Chamberlain line, from the hard palate to the opisthion (the midportion of the posterior margin of the foramen magnum). If the dens extends more than 5 mm (or half its height) above these lines, basilar invagination is present. Paget disease, rickets, fibrous dysplasia, osteogenesis imperfecta, hyperparathyroidism, osteomalacia, achondroplasia, cleidocranial dysplasia, and Morquio syndrome are among the causes of this finding. In adults, think rheumatoid arthritis. *Basilar impression* is the term sometimes used when the finding is found secondary to bone-softening diseases.

Basilar invagination should not be confused with platybasia. *Platybasia* literally means "flattening of the base of the skull" and is said to be present when the basal angle formed by intersecting lines from the nasion to the tuberculum sellae and from the tuberculum along the clivus to the anterior aspect of the foramen magnum (basion) is greater than 143 degrees. Platybasia is seen with Klippel-Feil anomalies, cleidocranial dysplasia, and achondroplasia.

MISCELLANEOUS

Although not congenital malformations, there are a few entities that should be included in this chapter. We will briefly touch base on germinal matrix-intraventricular hemorrhage in the preterm infant and hypoxic ischemic encephalopathy in the term infant. Additionally, the topic of temporal lobe seizures and mesial temporal sclerosis will be discussed here.

Germinal Matrix-Intraventricular Hemorrhage

Germinal matrix-intraventricular hemorrhage (GM-IVH) is the most common neonatal intracranial hemorrhage and characteristically seen in preterm infants. High incidence of prematurity and increasing incidence of survival in the preterm infants especially in very low birth weight preterms (<1500 g) highlight the importance of this entity. The capillary bed in the germinal matrix is rich with arterial blood receiving supply from anterior and middle cerebral arteries and internal carotid artery (through the anterior choroidal artery). The venous system drains blood from cerebral white matter (intramedullary veins), choroid

plexus (choroidal veins), and thalamus (thalamastriate veins) all coming to a confluence at the caudathalamic notch to form the terminal vein. Finally, the end drainage is into the vein of Galen. The site of origin for germinal matrix hemorrhage is the subependymal germinal matrix, which is the source of cerebral neuronal precursors; this provides glial precursors that become cerebral oligodendroglia and astrocytes. The pathogenesis of GM-IVH is multifactorial including fluctuating cerebral blood flow, rapid volume expansion, decreased hemoglobin, decreased blood glucose, increased cerebral venous pressure, respiratory disturbances, tenuous vascular supply, vulnerability to hypoxic injury and deficient extravascular support.

Head US is the best image neuroimaging modality for evaluation of GM-IVH given lack of radiation and bed-side serial imaging capability over time. Germinal matrix-IVH severity can be graded (Table 8-6): Grade 1 GM-IVH is hemorrhage limited to the germinal matrix with little to no extension to the ventricular system; Grade 2 GM-IVH is intraventricular hemorrhage without enlargement of the ventricular system, Grade 3 GM-IVH is intraventricular hemorrhage with enlargement of the ventricular system. Periventricular hemorrhagic infarction (PVHI) is not an extension of intraventricular hemorrhage. Although it is causally related to GM-IVH, the pathogenesis is primarily obstruction of flow to the terminal vein typically ipsilateral to the GM, indicating that the primary injury is venous ischemia secondary to venous obstruction, which later develops hemorrhagic conversion. Typically PVHI is either unilateral (67%) or asymmetrical (33%). Approximately 80% of the cases are associated with large IVH.

It is important to distinguish PVHI from periventricular leukomalacia (PVL; see next). PVL is usually bilateral and symmetrical, nonhemorrhagic and ischemic injury to the white matter.

Periventricular Leukomalacia

PVL is an injury to cerebral white matter, classically with two forms: focal and diffuse. The focal form is defined as distinguishable cystic focal lesions in the white matter consisting of localized necrosis deep in the periventricular white matter. In diffuse PVL, more commonly seen than focal PVL, focal necroses are microscopic in size and not readily seen by neuroimaging. This form of PVL, which accounts for the vast majority of cases, is termed noncystic PVL. Diffuse/noncystic form of PVL is characterized by marked astrogliosis and microgliosis. The decrease in premyelinating oligodendrocytes is counteracted by an increase in oligodendroglial progenitors, which lack the capacity for full differentiation to mature myelin-producing

TABLE 8-6 Grading of Germinal Matrix Hemorrhage

Grade	Location of Hemorrhage, Findings
1	Limited to GM
2	GM and IVH
3	GM and IVH with hydrocephalus
PVHI	Hemorrhagic conversion of the periventricular white matter venous ischemia/infarction

GM, Germinal matrix; *IVH,* intraventricular hemorrhage; *PVHI,* periventricular hemorrhagic infarction.

cells, resulting in hypomyelination with ventriculomegaly as later sequelae result.

Neuronal/axonal disease is an important counterpart of PVL. The regions of involvement include the cerebral white matter (axons and subplate neurons), thalamus, basal ganglia, cerebral cortex, brain stem, and cerebellum. On imaging, decreased volume of neuronal structures such as the thalamus, basal ganglia, cerebral cortex, and cerebellum is seen, as early as term-equivalent age, as well as later in childhood, adolescence, and adulthood.

Hypoxic-Ischemic Encephalopathy

Hypoxic-ischemic encephalopathy (HIE) is the leading cause of disability and death in the newborn despite neuroprotective therapeutic treatment options such as cooling. HIE is characterized by a deficit in oxygen supply either because of hypoxemia (diminished oxygen in blood supply) or ischemia (diminished amount of blood perfusing the brain) in the neonate. Ischemia is the more important part in the mechanism of HIE. Hypoxemia usually results in impaired cerebral autoregulation, leading to ischemia. Severity and duration of ischemia/hypoxia as well as the gestational age of the infant are principal factors of the final neuropathology.

The etiology may be antepartum, intrapartum, or postpartum insults. Most common intrapartum events leading to HIE are disturbances to the placenta or cord such as acute placental abruption or cord prolapse, uterine rupture, prolonged labor with transverse arrest, and difficult forceps extractions. Postnatal factors such as severe persistent fetal circulation, severe apneic spells, and congenital heart disease make up only 10% of the cases and affect preterm infants more commonly. Apgar scores are diminished at birth with profound metabolic acidosis, hypotonia, lethargy, seizures, coma and other system involvement (kidneys, liver, etc.).

Volpe describes four major patterns of injury based on animal model studies where the type of insult has been correlated with the neuroimaging findings. Areas of involvement have been described as cerebral (cortex and subcortical white matter usually parasagittal/watershed distribution), deep nuclear (thalamus and basal ganglia especially putamen), brain stem (inferior colliculus and tegmentum), cerebral white matter (periventricular and central white matter).

In severe prolonged injury, cerebral, deep nuclear and brain stem involvement is seen. In moderate prolonged or intermittent injury cerebral and/or deep nuclear involvement is seen. In severe brief injury, brain stem and deep nuclear involvement is seen. Mild/moderate, prolonged injury involves the cerebral white matter. These findings are usually bilateral and symmetrical.

Head US shows generalized brain edema in the white matter and/or basal ganglia reduced resistive indices in the circle of Willis in the first few days of life. MRI shows areas of restricted diffusion in the acute phase (first 6 days of life), in the above-described regions with variable degree of increased T2 signal. The findings of diffusion restriction may be present up to 8 to 10 days in those neonates who had cooling treatment, since there is delayed pseudonormalization of ADC attributed to cooling. Recognition of injury on head CT can be very challenging in the watery neonatal brain, not to mention the disadvantage of the radiation. Progression of injury to chronic stages happens much faster in the neonate compared with adults.

Age of the neonate is quite critical. Hint: Do not regard lack of dark T2 signal in the posterior limb of the internal capsule in a 34 gestational week born preterm as injury, because it has not myelinated yet!

Temporal Lobe Seizures and Mesial Temporal Sclerosis

Temporal lobe seizures are usually partial (that is, there is no loss of consciousness), as opposed to generalized seizures (where there is loss of consciousness). If consciousness is altered the term used is *complex partial seizure;* if no alteration, neurologists call it *simple partial seizure.* The causes of temporal lobe seizures are manifold (Box 8-12). Sources of seizure foci include (in the young patient) neoplasms, vascular malformations, gliotic abnormalities, and malformations of cortical development (add strokes to this list for older adults). Occipital lobe epileptogenic foci are most often because of developmental abnormalities (e.g., FCD, heterotopias, hamartomas, migrational anomalies) or tumors (usually gliomas).

Thirty percent of temporal lobe seizures are idiopathic, with no cause, 40% have an underlying visible cause (symptomatic), and 30% are cryptogenic, where the cause may be unknown but a lesion probably exists. Although 60% of all seizure disorders are well-managed with medications, some causes, including mesial temporal sclerosis (MTS), often require surgical intervention. Mesial temporal sclerosis accounts for as many as 50% of subjects undergoing temporal lobe surgery.

When the localization of a seizure focus is problematic, the neurosurgeons may place stainless steel or platinum alloy subdural grid contact arrays on the brain surface. This is a last resort when ictal scalp recordings have failed and when functional cortical mapping is required before surgery. Placement of depth electrodes into brain parenchyma is the final stage of invasive localization. The placement of these grids and electrodes is not without risk; be wary of the complications of extraaxial hematomas or effusions, venous thromboses, subfalcine or transtentorial herniations, tension pneumocephalus, intraparenchymal hemorrhages, and cerebral infections.

MTS, also known as hippocampal atrophy, is like the Holy Grail of neuroimaging because this entity is a common source for poorly controlled seizures in adolescents and young adults. Although most cases present in or after adolescence, the roots of MTS are controversial, with possible association with febrile seizures in infancy. The mechanism for congenital, nonfebrile development of MTS is mysterious, but may be due to a perinatal ischemic event

---

BOX 8-12 Causes of Temporal Lobe Epilepsy

Mesial temporal sclerosis (hippocampal sclerosis)
Ganglioglioma
Astrocytomas of all types
Oligodendroglioma
Cavernous hemangioma
Cortical dysplasia
Heterotopia
Vascular malformations
Ischemia
Trauma

because of compression of arteries during delivery, intra-uterine hypoxia, hypoxia secondary to status epilepticus, neurotoxic effects of excessive glutamate production, or hypoglycemia. Nonetheless, MTS is considered a progressive disease and the imaging findings of selective atrophy may progress with age. Histopathologically, one sees neuronal loss and gliosis in affected patients.

To make the imaging diagnosis of MTS, go back to the anatomy of the hippocampus. Going from the superomedial to inferolateral surface of the hippocampus, one passes from the fimbria to the alveus to the dentate fascia and then Ammon's horn. After passing around the hippocampal fissure, one proceeds into the subiculum (the superomedial border) to the parahippocampal gyrus medially and the entorhinal and piriform cortex inferiorly. The collateral sulcus separates the parahippocampus above from the occipitotemporal gyrus below. Ammon's horn, or cornu ammonis, has four zones of granular cells. CA1 is also called the vulnerable sector because it is the most sensitive area of the brain (along with the globus pallidus) to anoxia and is the main site of disease for MTS. CA2 (dorsal resistant zone) and CA3 (resistant Spielmeyer sector) are thought to be more resistant to anoxic damage. CA4 (end folium) is partially affected by anoxia. Sclerosis of CA1, and to a lesser extent CA4, is the etiology of MTS and has been linked to cerebral ischemia and febrile seizures. The selective vulnerability of CA1 may be due to overactivity of glutamate receptors and increased concentration of intracellular calcium ions in these granular cells.

On MRI, hippocampal volume loss is the salient feature of MTS. Atrophy of the mesial temporal lobe may affect the amygdala (12%), hippocampal head (51%), hippocampal body (88%), and hippocampal tail (61%). Hyperintense signal on T2WI scans may involve the amygdala (4%), hippocampal head (39%), hippocampal body (81%), and hippocampal tail (49%) or the entire hippocampus (44%). Coronal oblique FLAIR or T2WI through the temporal lobes can show signal intensity changes and inversion recovery T1WI scans (phase-sensitive inversion recovery) can show hippocampal atrophy; these sequences will tell you which temporal lobe is abnormal in the vast majority of cases (Fig. 8-63). The imaging findings are more often

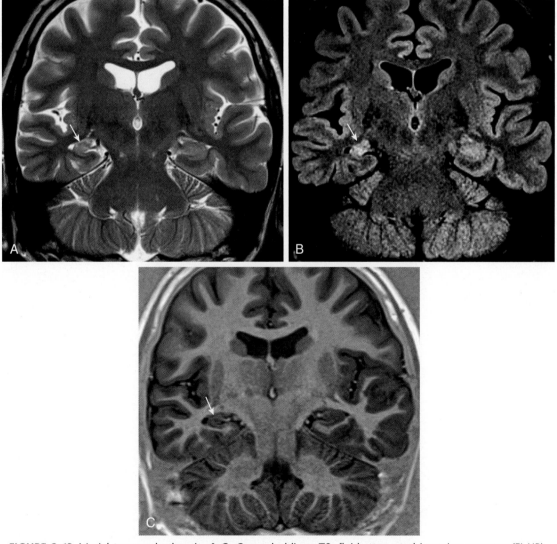

FIGURE 8-63 Mesial temporal sclerosis. **A-C,** Coronal oblique T2, fluid-attenuated inversion recovery (FLAIR), and TI inversion recovery images show elevated T2 signal **(A, B)** and diminished volume **(C)** of the right hippocampus compared with the left. To keep things in perspective, this case is *very* obvious, so you can imagine the challenge of making the diagnosis in more subtle cases.

subtle than glaringly obvious. Keep in mind that there is a normal asymmetry with respect to hippocampal volumes: the left hippocampus is typically 5% to 8% smaller than the right. Also be forewarned that bilateral involvement occurs in approximately 20% of cases, which can make a tricky diagnosis even trickier. Another finding in the spectrum of MTS is the loss of the normal cortical interdigitations of the hippocampal head. The sensitivity of this finding is approximately 90%; it may be present even when atrophy and signal intensity changes are absent in the medial temporal lobe. Additionally, the fornix and mammillary body ipsilateral to the side of MTS may be atrophic secondary to the decreased input to crossing fibers and limbic contributions (but these findings may also be present in cases of temporal lobe resections, strokes, and tumors).

In those patients with ipsilateral hippocampal atrophy, surgical removal of MTS is 90% effective in eliminating seizures. In adults with temporal lobe epilepsy who have resection of their temporal lobe, 65% of the time, the pathology shows MTS. Twenty percent of cases show normal MR structural scans, which just goes to show that the diagnosis of this condition should not be based on imaging alone (it takes a village!).

One can obtain greater and greater layers of sophistication in the analysis of the temporal lobes using quantitative volumetry, T2 relaxometry, magnetic resonance spectroscopy (MRS) (decreased N-acetyl aspartate [NAA], also reported as NAA/choline ratio <0.8 and NAA/creatine ratio <1.0), positron emission tomography, and single-photon emission computed tomography. If a patient is scanned during or within 24 hours of acute temporal lobe seizures, one finds lactic acid build up or lipid peaks on the proton MRS studies of the affected temporal lobe. Or you can call for the electroencephalography (EEG) results and be done with it. For those cases with equivocal EEG findings or bilateral disease, careful scrutiny of the images will stand you in good stead. There is a whole cottage industry of volumetric programs that will determine which hippocampus is smaller; they can be extremely accurate, but the trained eye is pretty good too.

One potential obstacle associated with the work-up for seizures should be recognized. If you image the patient immediately after a seizure or even during status epilepticus that is unapparent to the clinicians, you may see meningeal enhancement and high signal intensity in the seizing temporal lobe. This may imply a more diffuse process than is actually there (encephalitis or gliomatosis) and in fact, this "abnormality" may resolve completely on MR in a few days. This is probably because of the increased blood flow to the seizing temporal lobe; that is, a perfusion effect. Even DWI scans may be transiently positive, so keep an open mind when reviewing scans on patients with temporal lobe seizures!

Chapter 9
Orbit

Time to play with the eye-ball—at first base, anatomy. The orbit is a cone-shaped structure made up of seven bones (Fig. 9-1). The roof is formed by the orbital plate of the frontal bone anteriorly and the lesser wing of the sphenoid bone posteriorly. The lateral wall of the orbit is composed of the zygomatic bone anteriorly and the greater wing of the sphenoid posteriorly. The orbital floor consists primarily of the orbital plate of the maxillary bone; however, the zygoma forms part of the antero-lateral floor, whereas the palatine bone is at the posterior aspect of the floor. The bones of the medial wall are the lacrimal (anterior), lamina papyracea (ethmoid), and sphenoid (posterior). The orbit is bordered superiorly by the anterior cranial fossa, medially by the ethmoid sinus, inferiorly by the maxillary sinus, posteriorly by the middle cranial fossa, and laterally by the temporal fossa. Fortunately, the supreme almighty neuroradiologist has decided not to change the anatomy for this fourth edition. She also told us to place orbital trauma in Chapter 4 (Head Trauma).

ANATOMY

Globe

The globe usually approaches a sphere in shape with a diameter of approximately 2.5 cm, containing three enveloping layers: (1) sclera, (2) uvea, and (3) retina (Fig. 9-2). The most peripheral outer layer is the sclera, composed of collagen-elastic tissue. Covering the sclera anteriorly is the conjunctiva, a clear mucous membrane. The sclera is continuous with the cornea and beneath the sclera is the vascular pigmented layer termed the uveal tract, composed of the choroid, ciliary body, and the iris. The inner layer of the globe is the retina, which is continuous with the optic nerve. It can be further separated into an inner sensory layer containing photoreceptors, ganglion cells, and neuroglial elements, and an outer layer of retinal pigment epithelium, which is adjacent to the basal lamina of the choroid (Bruch membrane).

The iris separates the anterior chamber from the posterior chamber. Clear aqueous humor circulates between the two chambers. The ciliary body lies between the iris and choroid, contains muscles attached to the lens by the suspensory ligament that control the curvature of the lens, and secretes the aqueous humor. Posterior to the lens is the posterior segment filled with a jelly-like substance, the vitreous body (humor). What would the globe (or this book) be without its humor? Hemorrhage may occur within these different compartments—anterior hyphema within the anterior chamber and the rarely seen eight ball hyphema of the posterior chamber (rarely seen because the posterior chamber is so tiny).

There are four potential areas where bad humor accrues: politics, sex, religion, and this book …oops, back to anatomy…. The space between the base of the vitreous and the sensory retina is the posterior hyaloid space (Fig. 9-3). The space between the layers of the retina (sensory retina and retinal pigment epithelium) is the subretinal space, and between the choroid and the sclera is the suprachoroidal space. When the posterior hyaloid membrane separates from the sensory retina it is termed a posterior vitreous detachment. It is curvilinear in shape and anterior to the retina and separate from the optic disc. Total retinal detachments are V shaped with the apex pointing to the optic disc and the arms at the ora serrata (the anterior terminations of the retina which are located at approximately 10 o'clock and 2 o'clock on the globe). In fact, retinal separation is probably a better term as the two layers of the retina, neurosensory and retinal pigment epithelium, separate. Choroidal detachments are generally limited by the vortex veins or posterior ciliary arteries and usually do not reach the optic disc. In addition, choroidal detachments extend anteriorly beyond the limitations of the ora serrata, since the choroidal epithelium goes up to the ciliary body, occasionally detaching it. Sub-Tenon space is located between the sclera and the fibrous membrane (Tenon capsule) adjacent to the orbital fat extending from the ciliary body to the optic nerve. Hemorrhages in this space, most often from trauma, conform to the curvilinear shape of the eyeball.

Foramina

Three major foramina are in the orbit: (1) the optic canal, (2) the superior orbital fissure, and (3) the inferior orbital fissure (see Fig. 9-1). Table 9-1 lists the structures that traverse these foramina. The optic canal is formed by the lesser wing of the sphenoid bone, closely approximating the anterior clinoid process. The shape of the canal is horizontally oval at its intracranial entrance, round at its midportion, and vertically oval at its orbital end. It often is bordered by ethmoid or sphenoid sinus aerated cells which puts the nerve at risk if there is a dehiscence or if an errant functional endoscopic sinus surgery (FESS) probe goes awry. The superior orbital fissure is formed from the greater and lesser wings of the sphenoid and is separated from the optic canal by a thin strip of bone, the optic strut. The inferior orbital fissure lies between the orbital plate of the maxilla and palatine bones, and the greater wing of the sphenoid. The inferior orbital fissure communicates with the pterygopalatine fossa. A few other foramina are in the orbit, including the anterior and posterior ethmoidal foramina just medial to the optic canal. These transmit the anterior and posterior ethmoidal

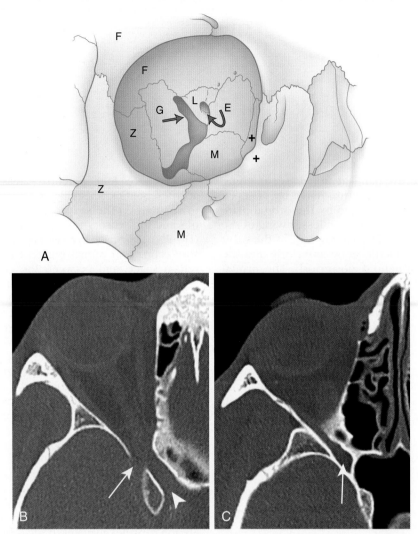

FIGURE 9-1 A, The "eyes" have it: bones of the orbit. A diagram of the bony orbit as seen from the front. The bones of the medial wall are the lacrimal (+), ethmoid (lamina papyracea [E]), and sphenoid (lesser wing [L]). The orbital roof is formed by the orbital plate of the frontal bone (F) anteriorly and the lesser wing (L) of the sphenoid bone posteriorly. The lateral wall of the orbit is composed of the zygomatic bone (Z) anteriorly and the greater wing of the sphenoid (G) posteriorly. The orbital floor consists primarily of the orbital plate of the maxillary bone (M); however, the zygoma (Z) forms part of the anterolateral floor while the palatine bone (not seen) is at the most posterior aspect of the floor. The superior orbital fissure is identified *(straight arrow)* as well as the optic canal *(curved arrow).* **B,** Axial computed tomographic (CT) image shows the superior orbital fissure *(arrow)* and optic canal *(arrowhead)* from the imaging perspective. **C,** Axial CT shows the inferior orbital fissure *(arrow).*

arteries, which can bleed like stink if the FESS surgeon is not careful. There are infraorbital nerves and supraorbital nerves which traverse their respective canals and transmit sensory nerve branches of the trigeminal nerve as well. These are sources of perineural spread from cutaneous cancers back to the CN V ganglia. The nasolacrimal duct on the inferomedial surface of the orbit communicates with the inferior meatus and can serve as a pathway for nasal tumors to extend directly into the orbit. The soft tissues of the orbit are principally composed of the lacrimal sac and gland, six extraocular muscles, optic nerve, orbital fat, and many vascular structures.

Extraocular Muscles

The extraocular muscles include the medial, superior, inferior, and lateral rectus, which originate from the annulus of Zinn at the optic foramen and insert on the globe. The superior oblique and inferior oblique muscles have separate origins (superomedial to the optic foramen and orbital plate of the maxilla, respectively). The levator palpebrae superioris muscle arises above the superior rectus and inserts into the upper lid. The four rectus muscles are classically thought to be connected by an intramuscular fibrous membrane creating the so-called *intraconal space* (Fig. 9-4). This anatomic concept may not truly exist beyond the globe. However, for radiologic purposes this boundary serves as a useful landmark in categorizing and diagnosing orbital lesions. Thus, intraconal lesions are associated with the optic nerve, its vessels, and orbital fat; conal lesions involve the muscles; and extraconal disease includes the bony orbit, peripheral fat, and extraorbital structures like the paranasal sinuses, skull, or brain.

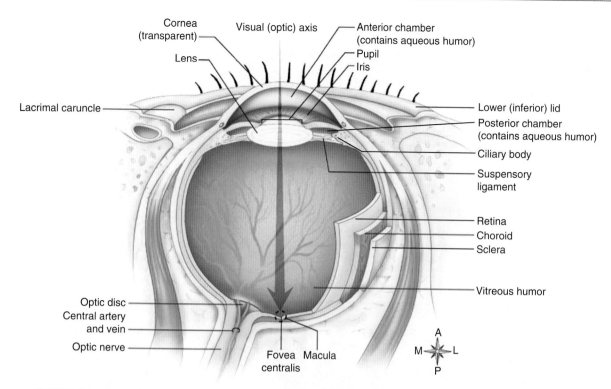

FIGURE 9-2 Eye-socenter section through left eyeball, viewed from above. (From Thibodeau GA, Patton KT. *Anatomy and Physiology.* 7th ed. St. Louis: Mosby; 2009:515. Used with permission.)

The orbit is fortuitously lined with adipose tissue that is well organized and divided by fibrovascular septa. This fat produces the contrast, enabling orbital structures to stand out on magnetic resonance (MR) or computed tomography (CT), and of course decreases the conspicuity of enhancing lesions on MR unless fat-saturation MR techniques are used. The orbital septum is a reflection of orbital periosteum (periorbita) inserting on the tarsal plate of the eyelid. The periorbita is an excellent barrier to neoplastic or inflammatory disease emanating from the sinuses. The lacrimal gland is an almond-shaped structure in the anterosuperolateral portion of the orbit lying in the lacrimal fossa at the level of the zygomatic process of the frontal bone. The fat functions as a shock absorber for squash balls and other orbit-seeking foreign bodies.

Optic Nerve

The optic nerve is a white matter tract that is sheathed by the leptomeninges and dura mater. The subarachnoid space of the optic nerve sheath is continuous with the intracranial subarachnoid space (see Fig. 9-4) providing a pathway for spread of infection, inflammation, tumors, or hemorrhage. This explains the finding of papilledema in instances of increased intracranial pressure.

The course of the optic nerve may be divided into four segments: intraocular, intraorbital, intracanalicular, and intracranial. The intraorbital nerve has a sinuous course in both horizontal and vertical planes. After the optic nerves leave the canals, they ascend at an angle of approximately 45 degrees as prechiasmatic nerves, meet to form the chiasm beneath the floor of the third ventricle, and proceed posterolaterally to enter the brain as the postchiasmatic optic tracts. The chiasm is approximately 10 mm above the diaphragma sellae.

The caveat that should be kept in mind when the optic nerve is evaluated, particularly in the setting of trauma (e.g., optic nerve avulsion). Hypodensity, hyperintensity, or thinning of segments of the optic nerve must be demonstrated on reformatted or true oblique views parallel to the segment in question before the finding is considered pathologic.

Vascular Structures

The important vascular structures in the orbit are the ophthalmic artery and its branches, and the superior and inferior ophthalmic veins. The ophthalmic artery usually arises from the internal carotid artery just after emerging from the cavernous sinus. On entering the orbit via the optic canal, the ophthalmic artery is initially inferolateral to the nerve. The orbital arteries radiate from the apex and diverge through the orbital fat to pierce the orbital septa. The main trunk of the ophthalmic artery divides into two relatively independent systems: the retinal vascular system (central retinal artery), which supplies the optic nerve and inner aspect of the sensory retina, and the ciliary vascular system.

The ophthalmic artery constitutes a major anastomotic pathway between the internal and external carotid artery. These anastomoses are listed in Box 9-1 and are particularly important in extracranial occlusive vascular disease. The choroidal plexus of the eye, supplied by the short posterior ciliary arteries, is seen on the lateral conventional catheter arteriogram as a thin crescent (choroidal

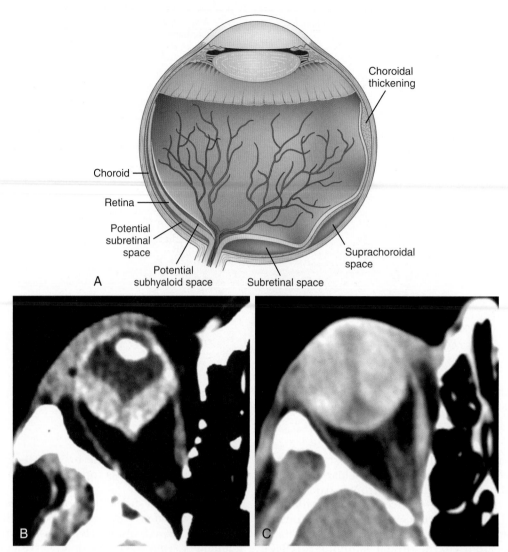

FIGURE 9-3 Eye-solated collections of fluid beneath the membranes. **A,** Schematic representation of the various potential spaces for fluid/hemorrhage accumulation in setting of ocular detachments. **B,** Axial computed tomographic (CT) image shows typical V-shaped appearance of retinal detachment with the apex at the optic nerve head and anterior extent to the level of the ora serrata (at 10 and 2 o'clock). **C,** Axial CT image shows typical configuration of choroidal detachment. The margin of the detached choroid extends to the expected location of the vortex veins.

TABLE 9-1   Contents of Major Orbital Foramina

Orbital Foramen	Contents
Optic canal	Optic nerve, sympathetic fibers, ophthalmic artery
Superior orbital fissure	Cranial nerves III, IV, first division of CN V, and CN VI; superior ophthalmic vein; sympathetic fibers; orbital branch of middle meningeal artery
Inferior orbital fissure	Second division of CN V, infraorbital artery and vein, inferior ophthalmic vein

crescent) with an anterior concavity. In patients older than 30 years, delay in appearance of the choroidal blush (0.6 seconds) should suggest hemodynamically significant disease of the internal carotid artery.

Anatomic variations of the ophthalmic artery are important when embolizations and carotid surgery are planned. The ophthalmic artery can originate from the middle meningeal artery (meningolacrimal artery), or branches of the meningeal artery may connect with the ophthalmic artery. Rarely the ophthalmic artery can arise more inferiorly from the cavernous internal carotid artery (ICA) and can enter the orbit through the superior orbital fissure. Occlusion of the external carotid during carotid endarterectomy or embolization of the middle meningeal artery could result in visual loss when these vascular anomalies are present.

There is an extensive venous network throughout the orbit arranged in a circular fashion following the fibrous septa of the orbital connective tissue. The superior ophthalmic vein is formed by the angular, nasofrontal, and supraorbital veins, which converge at the superolateral aspect of the nose. The superior ophthalmic vein (1.5 to 3 mm in diameter on MR) runs under the orbital roof in its anterior and posterior segments. Its middle segment pierces the muscle cone to course under the superior rectus in a connective tissue septum (superior ophthalmic vein hammock) and over the ophthalmic artery. Posteriorly, it enters the superior orbital fissure and then

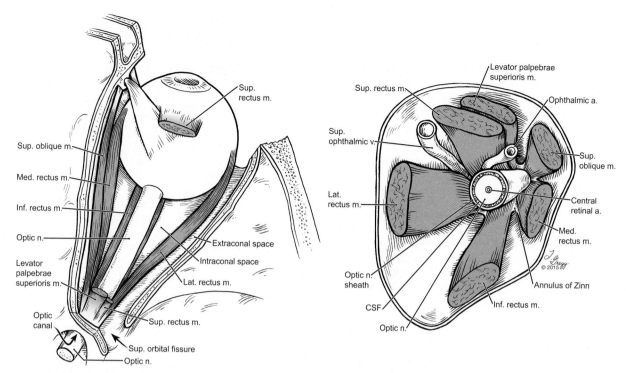

FIGURE 9-4 Global spaces of the orbit. This diagram shows the classic separation of the sites of pathology in the orbit into ocular disease (the globe), intraconal, conal, and extraconal spaces including the lacrimal sac and gland. *CSF*, Cerebrospinal fluid. (© 2015 Lydia Gregg.)

---

drains into the cavernous sinus. Its position may cause it to be obstructed by an enlarged inflamed superior rectus muscle (from Graves or pseudotumor). The inferior ophthalmic vein starts as a plexus of small veins beneath the globe and courses posteriorly above the inferior rectus muscle to join the superior ophthalmic vein or to enter the cavernous sinus directly. Inferiorly, a second branch passes through the inferior orbital fissure to drain into the pterygoid venous plexus. A medial ophthalmic vein, observed in about 40% of cases, runs in the nasal extraconal space and joins the superior ophthalmic vein near the superior orbital fissure. These veins are all valveless with numerous anastomoses, which can serve as a conduit for facial infections (zit-popper syndrome) to proceed intraorbitally, into the cavernous sinus, and intracranially.

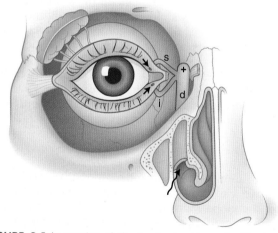

FIGURE 9-5 In-tear-ior of the lacrimal system. **A,** The lacrimal puncta in the medial aspect of the lid margin *(small arrows)* drains via the superior (s) and inferior (i) canaliculi into the lacrimal sac (+) via the sinus of Maier. The lacrimal sac lies in the medial inferior wall of the orbit (lacrimal fossa) and drains via the valve of Krause into the nasolacrimal duct (d) through the valve of Hasner into the nasal cavity beneath the inferior turbinate *(squiggly arrow)*.

These veins can become engorged and cause edema/mass effect when the patient has a dural malformation or carotid cavernous fistula.

Lacrimal System

Tears produced by the lacrimal system and its glands track medially and flow through the lacrimal puncta in each lid margin. Nasolacrimal drainage follows superior and inferior canaliculi, which are situated along the medial part of the lids and drain via the sinus of Maier into the lacrimal sac (Fig. 9-5). The lacrimal sac is a membranous tissue situated in the medial inferior wall of the orbit (lacrimal fossa) and drains via the valve of Krause into the nasolacrimal duct and through the valve of Hasner into the nasal cavity beneath the inferior turbinate (more trivia). Obstruction of this drainage system can result in (1) dacryocystoceles in children, (2) epiphora (excessive tearing), or (3) dacryocystitis (infection of the nasolacrimal sac) or dacryoadenitis (lacrimal gland inflammation).

Visual Pathway

Just a brief word concerning the visual pathway (see Chapter 1 for more detail). Lesions of the prechiasmatic optic nerve generally produce monocular visual loss, those involving the optic chiasm cause bitemporal or heteronymous field defects, and those behind the chiasm result in visual loss restricted to one side of the visual field (homonymous hemianopsia).

IMAGING CONSIDERATIONS: INDICATIONS FOR MAGNETIC RESONANCE, COMPUTED TOMOGRAPHY, AND ORBITAL ULTRASOUND

CT is the initial imaging modality of choice for most traumatic situations involving the orbit. Fractures are easier to detect with CT, and bone lesions are usually better visualized with CT. Furthermore, orbital metallic foreign bodies are a contraindication to MR. One of the most important attributes of CT is its ability to detect calcification. This is particularly crucial in the diagnosis of retinoblastoma, where calcification appears to be a hallmark of the tumor (but many other lesions may calcify). MR on the other hand is recommended for looking at orbital masses that affect the optic nerve, may have subarachnoid seeding or intracranial extension, or are intraconal or intramuscular. Inflammation may be assessed by both modalities (although CT has advantages with paranasal sinus sources), but demyelinating or ischemic optic nerve lesions are best evaluated with MR.

A word about ultrasonography…surrender. At this juncture, leave it to the ophthalmologist to assess the integrity of the globe following trauma, to look for retained foreign bodies, to demonstrate choroidal rupture, retinal detachment, flow dynamics of vessels in the orbit, some ocular tumors, and collections under the ocular membranes. We do a superior job with CT and MR, but let's "overlook" ocular ultrasonography and throw the eye docs a financial bone.

A Few Technical Points

CT is an excellent method for detection and localization of intraocular and intraorbital foreign bodies (Fig. 9-6). The minimum detectable intraocular particle size for steel is 0.06 mm³, and 1.82 mm³ for auto window glass. Multidetector computed tomography (MDCT) with reformatted images will save radiation dosage to the eye, reduce radiation-induced cataracts (ooops! So much for that financial bone we offered last paragraph), and still be more sensitive than conventional thick section multiplanar scanning for detection of metallic foreign bodies.

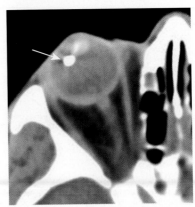

FIGURE 9-6 Iron in eye. This metallic foreign body *(arrow)* pierced the globe after flying off of a steel wheel in the boys' "shop class." That'll teach him to wear his safety glasses. The right lens is probably injured as well.

Interestingly, small fragments of wood (a source of orbital infection) may be variably detectable on CT. Wood with lead paint appears high density. Freshly cut wood has a high water content and may be difficult to distinguish from orbital soft tissues on CT or MR. Hardwoods such as oak also have higher CT attenuation values than do softwoods. When wood dries out the water content is replaced by gas content, and the CT values can appear more like air or fat. Extremely dry wood and balsa wood look like air on CT; on T1-weighted imaging (T1WI), dry wood is hypointense. Wide bone window settings can show the reticulated matrix of wood, but on soft-tissue windows wood may be mistaken for air. "Woodeyes" are bad. "Woodies" are great.

MR provides excellent soft-tissue detail. The use of fast spin echo or constructive interference in steady state (CISS) techniques with increased matrix size (512 × 512) can provide beautiful T2-weighted imaging (T2WI) of the optic nerve and demonstrate intrinsic lesions such as optic neuritis particularly when performed coronally, perpendicular to the structure of interest. These images enable differentiation of the nerve and sheath. The major problem with MR in contrast to CT is related to the longer scan duration and concomitant image degradation from motion.

For diagnostic purposes, it is convenient to divide orbital disease into ocular and retrobulbar components. The latter region is subdivided into intraconal, conal, and extraconal lesions. We will consider each location and develop a differential diagnosis based on anatomic region.

> There was an eye doctor named Job
> Who devoted his life to the globe
> He treated glaucoma,
> Retinoblastoma,
> And any ocular microbe

OCULAR LESIONS

There is a remarkable amount of pathology that can affect this small but critical structure, including degenerative, infectious, and neoplastic processes. Imaging can be very useful, particularly when lid swelling limits a complete ophthalmologic examination. Even if the diagnosis is known, imaging can play an important role in determining the extent of ocular disease not appreciable by ophthalmologic examination alone.

Ocular Calcifications

CT is exquisitely sensitive to ocular calcification and density differences such as those exhibited by the ocular lens. The normal lens is the densest part of the globe. It gets denser in senescent cataracts but acute traumatic cataracts lead to lens edema and a less dense lens (say that 10 times fast—LOL). Contemporary cataract surgery usually includes placement of an intraocular lens. These can also be identified on CT or T2WI as a wafer thin implant or absent native lens. In orbital trauma one must be concerned about the position, integrity and density of the lens. The lens may also be dislocated in a variety of medical conditions. Box 9-2 provides a list of these conditions as well as the common position of these dislocations.

Box 9-3 lists the causes of ocular calcification (Fig. 9-7). It should be appreciated that the lens appears dense on CT because of its high protein content. On T1WI and T2WI, most pure calcifications (without paramagnetic ions) are hypointense and do not change in intensity as one increases the echo time. Susceptibility differences produce hypointensity so that calcium, which has altered diamagnetic susceptibility compared with tissue, appears hypointense. This is sometimes useful in detecting calcium on an MR in a patient with retinoblastoma. This hypointensity is not specific for calcium and other entities including blood products (deoxyhemoglobin, intracellular methemoglobin, hemosiderin), air and paramagnetic substances are hypointense on gradient echo images.

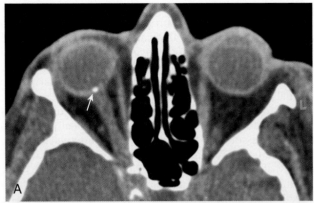

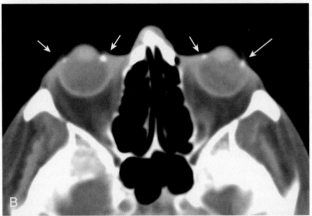

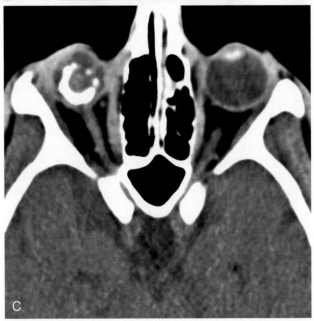

FIGURE 9-7 Icalc? A new Apple product or short for ocular/orbital calcifications. **A,** Punctate calcification on computed tomography (CT) at the junction of the head of the optic nerve and right globe *(arrow)*. This is a drusen (body). **B,** CT with senescent calcifications in insertions of medial and lateral rectus tendon sheaths *(arrows)*. **C,** In this case, phthisis bulbi occurred from trauma to the eye resulting in shrunken globe with calcification.

BOX 9-2   Lens Dislocations

Marfan: up and out
Homocystinuria: down and in
Ehlers-Danlos
GEMSS syndrome (glaucoma, lens ectopia, microspherophakia, stiffness, and shortness)
Trauma

BOX 9-3   Causes of Ocular Calcification

DEGENERATIVE

Cataracts
Optic nerve drusen
Senile calcification of insertion of muscle tendons in the globe

CHRONIC/POST-INSULT

Phthisis bulbi
Retinal detachment
Retrolental fibroplasia

NEOPLASTIC

Astrocytic hamartoma
Neurofibromatosis
Tuberous sclerosis
von Hippel-Lindau syndrome
Choroidal osteoma
Retinoblastoma

INFECTIOUS

Cytomegalovirus
Herpes simplex
Rubella
Syphilis
Toxoplasmosis
Tuberculosis

Senescent calcifications of the globe, sometimes called limbus calcifications, are usually found at the insertion sites on the globe of the medial and lateral rectus muscles. They have also been termed scleral plaques. Trochlear calcifications are seen at the tendon site of the superior oblique muscle. Diseases that cause hypercalcemia may also lead to earlier and more prominent benign globe calcifications. Another benign calcification is the optic drusen, which appears as a punctate calcification located at the optic nerve insertion at the globe, involving the optic disc, and sometimes leading to an appearance ophthalmoscopically as papilledema. There is evidence to suggest that in the normal course of drusen there is progressive enlargement over years. This condition has an incidence of up to 2% of the population. Drusen may be familial (typically bilateral) or may be associated with other conditions such as retinitis pigmentosa. They may be unilateral or bilateral and are located at the junction of the optic nerve and globe.

Phthisis bulbi is the end stage of globe injury, be it from infection, inflammation, trauma, or autoimmune disease. The globe is nonfunctional, small, and it eventually calcifies.

Pediatric Ocular Diseases

There are a number of conditions that can affect the pediatric eye which deserve special attention in this section, probably the most important of which is retinoblastoma. Recognition of other pathologies is also important, if only to exclude the dreaded retinoblastoma from the differential diagnosis.

Retinoblastoma

Retinoblastoma is the most common intraocular tumor of childhood. The vast majority of lesions occur as sporadic mutations (90%) whereas approximately 10% are familial. Familial retinoblastoma is an autosomal dominant trait, with variable penetrance and a propensity for bilaterality and multifocality (same globe; 30% of cases). The retinoblastoma gene has been localized on chromosome 13 at the q14 locus. Sporadic retinoblastomas are usually unilateral, occurring later than the inherited form. Children with monocular disease are more likely to have a single tumor and negative family history. There is a high incidence of nonocular tumors in the hereditary form (the "oncogene") with this predisposition for cancer leading to a 59% 35-year mortality rate (remember they get the first tumor at 2 to 3 years of age). These include midline, primitive neuroectodermal, pineal tumors, osteogenic sarcoma, soft-tissue sarcoma, malignant melanoma, basal cell carcinoma, and rhabdomyosarcoma.

Patients with retinoblastoma usually (98%) present before the age of 3 years with leukokoria, strabismus, decreased vision (particularly in bilateral lesions), retinal detachment, glaucoma, ocular pain, or signs of ocular inflammation. Leukokoria (white pupil) is due to the inability of light to be reflected off the retina, and it is caused by any opaque tissue that interferes with the passage of light through the globe. Thus, leukokoria itself is a nonspecific sign with many causes including retinoblastoma (a little less than half of all comers with leukokoria, but mostly kids), congenital cataract, retinopathy of prematurity, chronic retinal detachment associated with retrolental fibroplasia, choroidal hemangioma, retinal astrocytoma, persistent hyperplastic primary vitreous, Coats disease, and *Toxocara canis*

infection. Only the most radiologically relevant lesions are considered in the subsequent sections.

Accurate radiologic diagnosis of retinoblastoma is crucial for timely treatment and survival. Although ophthalmoscopy is usually able to visualize the ocular abnormalities, imaging is of critical importance in revealing retrobulbar disease including other tumors, metastasis, and invasion of the optic nerve posterior to the lamina cribrosa where the meninges insert on the optic nerve (associated with a poor prognosis) (Fig. 9-8). Retinoblastoma may spread directly into the subarachnoid space via the optic nerve (occasionally producing spinal subarachnoid implants), hematogenously, or via lymphatics and has a propensity to hemorrhage. Metastases from retinoblastoma occur within the first 2 years after treatment.

Pineoblastoma, although usually associated with bilateral retinoblastoma, has been noted with unilateral retinoblastoma and may be detected simultaneously with, after, or even before the ocular lesion. The pineal gland in lower animals contains photoreceptors and has been termed the "third eye," so that pineal tumors (usually pineoblastoma) associated with bilateral retinoblastoma have been named "trilateral" retinoblastoma (see Fig. 9-8, *C*). For those erudite readers the pineal gland is also considered the "third testicle" because of the origin of germ cell tumors. Ectopic retinoblastoma has also been reported in the parasellar or suprasellar regions.

The CT findings are those of calcifications in the posterior portion of the globe with extension into the vitreous (see Fig. 9-8, *A*). The calcification may be homogeneous or irregular and occurs in 95% of patients. In children younger than 3 years of age, ocular calcification is highly suggestive of retinoblastoma; conversely, its absence in this same group makes the diagnosis of retinoblastoma highly unlikely. These lesions minimally enhance on CT, obscured by the ocular calcification; however, contrast is useful to appreciate the intracranial extension (better assessed with MR).

MR findings in retinoblastoma include moderately high intensity on T1WI and hypointensity on T2WI, with the noncalcified portion of the lesion less hypointense than the calcification. We speculate that the high intensity on T1WI may be caused by calcium or by some coexistent paramagnetic substance, tumor protein, or hemorrhagic component. Associated retinal detachment may be high intensity on T1WI and of variable intensity on T2WI, depending on the protein content and whether there is associated hemorrhage in the subretinal fluid.

In patients with suspected retinoblastoma, CT and MR are complementary examinations. CT detects calcification better, whereas MR is more sensitive to tumoral extension into the nerve and intracranial space, and to secondary lesions. Enhanced MR images better show the tumor and subarachnoid/meningeal/perioptic spread (the latter best on coronal postcontrast fat-suppressed images).

A number of other conditions can occasionally simulate retinoblastoma and should be distinguished from it, often presenting with leukokoria. Most of these lesions can be differentiated by history, clinical presentation, and examination.

Persistent Hyperplastic Primary Vitreous

Persistent hyperplastic primary vitreous (PHPV) is caused by persistence of various portions of the primary vitreous (embryonic hyaloid vascular system) with hyperplasia of the associated embryonic connective

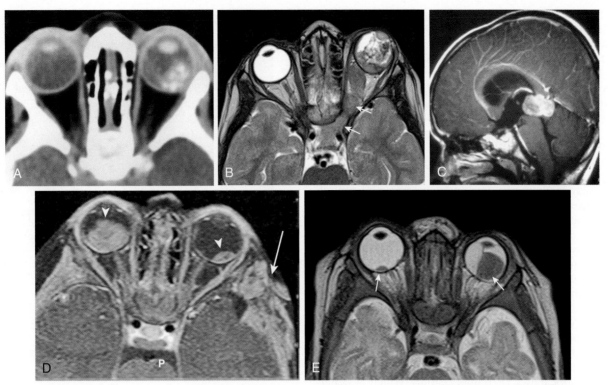

FIGURE 9-8 Eye? B-eye? Tr-eye? Quad-eye? **A,** On computed tomography, retinoblastoma calcifications are noted extending from the posterior retina into the vitreous on the left. **B,** T2-weighted imaging (T2WI) shows retinoblastoma that has spread along the optic nerve *(arrows)* through the optic canal into the intracranial space. **C,** Note the trilateral retinoblastoma/pineoblastoma in a patient with bilateral retinoblastoma. **D,** This patient with bilateral retinoblastomas *(arrowheads)* also has a soft-tissue sarcoma *(arrow)* in the left temporalis muscle seen on post-gad fat-sat T1WI. **E,** Bilateral retinoblastomas *(arrows),* larger on the left, well seen on T2WI.

tissue, and may present as leukokoria in the neonate. The globe usually demonstrates microophthalmia, although this can be minimal, and the lens can be small with flattening of the anterior chamber. Initially, a funnel-shaped mass of fibrovascular tissue (including the hyaloid artery) is present in the retrolental space and runs in an S-shaped course (termed Cloquet canal) between the back of the lens and the head of the optic nerve. PHPV is vascular, prone to repeated hemorrhages, and may vary in size. CT demonstrates generalized increased density in the globe and enhancement of the intravitreal tissue after intravenous contrast. Unlike retinoblastoma, PHPV does not calcify (this is key!). The CT/MR diagnosis is facilitated by observing the tubular fetal tissue between the lens and head of the optic nerve of the persistent canal (Fig. 9-9). A blood-vitreous layer may be seen as a result of posterior hyaloid detachment. On MR, vitreal intensity is variable.

Coats Disease

Coats disease is a vascular anomaly of the retina that may be clinically difficult to distinguish from retinoblastoma. This lesion is usually unilateral. These telangiectatic vessels leak serum and lipid into the retina (lipoproteinaceous effusion) and subretinal space. Patients are usually seen when retinal detachment occurs in the first 2 years of life, although the age at discovery may be as late as 6 to 8 years with approximately two thirds of cases in boys. The CT appearance

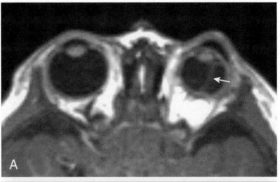

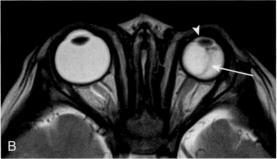

FIGURE 9-9 Small ball? Eye thinks so. **A,** In this patient with persistent hyperplastic primary vitreous (PHPV), note the presence of Cloquet canal in the left globe *(arrow).* **B,** The T2-weighted imaging scan best shows the persistent canal *(arrow)* and the associated dysmorphic lens *(arrowhead)* in this somewhat small globe.

is of high density in the vitreous related to the exudate. Calcification is rare, again distinguishing it on CT from retinoblastoma. On MR, Coats disease is hyperintense on T1WI and T2WI, whereas retinoblastoma tends to be high intensity on T1WI but low intensity on T2WI. Coats disease enhances along the leaves of the detached retina and at the sites where the retina reinserts whereas retinoblastoma enhances in a masslike fashion.

Toxocara Canis Infection

Toxocara canis infection begins when a child ingests soil contaminated with dog feces containing the ova of the nematode *Toxocara canis*. (Your children should ask the maître d' for their dirt sans worms.) The ova hatch in the gastrointestinal tract and then migrate throughout the body. The ocular *T. canis* infection results when larvae of *T. canis* within the eye die, producing an inflammatory reaction (just like cysticercosis in the brain) with vitreous opacification and retinal detachment. Ocular lesions occur months to years after initial infection. This chronic endophthalmitis causes unilateral leukokoria. On CT, diffuse high density is identified within the globe without calcification. Thick enhancement of the sclera has been observed. On MR, these lesions are reported to be high intensity on T1WI and T2WI. This is somewhat different from the appearance of retinoblastoma, which is generally of lower intensity on T2WI. The intensity patterns are related to the protein content and/or associated hemorrhage in the globes.

Retinopathy of Prematurity

Formerly called retrolental fibroplasia, retinopathy of prematurity (ROP) is the result of abnormal vascular development of the retina. Risk factors are low birth weight (<1 kg) and low gestational age, indicators of prematurity, but episodes of hyperoxia, sepsis, blood transfusions, and hypercarbia can contribute to its occurrence. Prolonged oxygen therapy in premature infants is a major risk factor. ROP can present as bilateral leukokoria because of traction retinal detachments. CT findings are of increased density in the posterior portion of the globe, which may represent calcification (Fig. 9-10). Occasionally, a thin line representing the detached retina can be identified beneath the high density.

Hamartomas

Astrocytic (glial) hamartomas are observed in cases of tuberous sclerosis, neurofibromatosis, and rarely von Hippel-Lindau disease (see Chapters 2 and 8). On CT, a focal region of calcification is noted in the retina. This can potentially be confused with retinoblastoma. These hamartomas grow slowly and do not metastasize. A hallmark of von Hippel-Lindau disease is retinal angiomatosis (which is not identified angiographically). Hemangioblastomas of the retina and retrobulbar optic nerve and orbit have also been reported in von Hippel-Lindau disease. The primary differential diagnosis of such optic nerve lesions is angioblastic meningioma and optic nerve glioma (seen in neurofibromatosis type 1 [NF-1]). Half the patients with optic nerve hemangioblastoma have other associated lesions of von Hippel-Lindau disease. Choroidal angiomas are reported in Sturge-Weber syndrome (tomato ketchup retina) (Fig. 9-11).

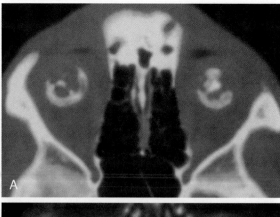
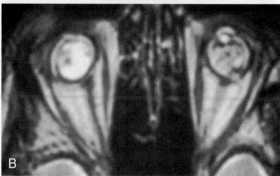

FIGURE 9-10 Two small I's from too much O-2. Retinopathy of prematurity may present with small dense eyes, sometimes with calcifications. **A,** Bilateral shrunken calcified globes are seen in this patient. They also are abnormally in signal intensity on T2-weighted imaging **(B)** from the calcification and chronic retinal detachments.

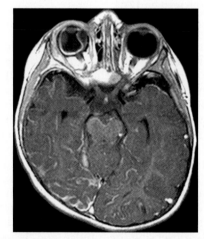

FIGURE 9-11 Choroidal angioma of the right globe in patient with Sturge-Weber syndrome. Note abnormal enhancement of the choroid along with choroidal detachment. Leptomeningeal angiomatosis is present over the right hemisphere and left occipital lobe, with volume loss of the right cerebral hemisphere compared to the left, typical of this syndrome.

Adult Ocular Diseases

The majority of primary and metastatic ocular neoplasms involve the choroid or retina. The radiologist can play an important role in the diagnosis of ocular neoplasms and in recognizing some of the nonneoplastic mimics of these conditions. Imaging is essential in detection of episcleral spread of tumor or for showing lesions spreading posteriorly along the optic nerve

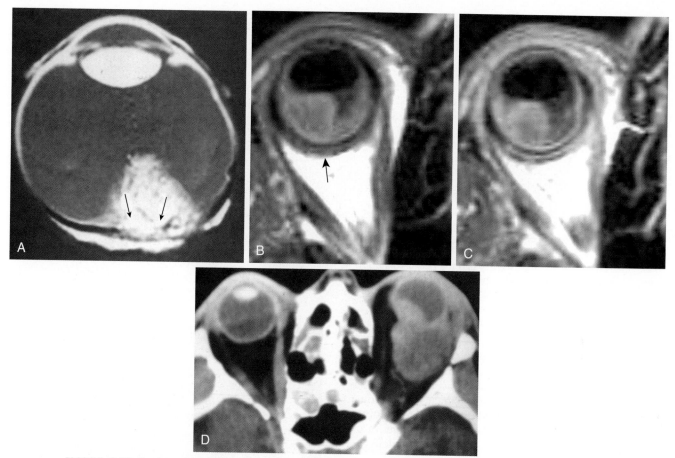

FIGURE 9-12 Ocular melanoma—not a pigment of your eye-magination. **A,** T1-weighted imaging (T1WI) scan of enucleated globe with melanotic melanoma. Note the high intensity of the tumor (paramagnetic effect of melanin) and the episcleral extension *(arrows).* In a different patient, precontrast **(B)** and postcontrast **(C)** T1-weighted images demonstrate how the precontrast high signal from melanin can make the enhancement difficult to appreciate. **D,** Computed tomography of melanoma (in vivo) with posterior episcleral/orbital extension.

sheath complex. Furthermore, there are circumstances in which the ophthalmologist may have a limited view of the globe and retina because of cataract formation and/or corneal, vitreous opacities. Don't let the globe become a radiologic blind spot. Hyuk.

Melanoma

Melanomas of the uveal tract are the most common intraocular malignancy in adults. They are rare in African Americans but when they occur tend to be larger, more pigmented, and more necrotic. Conditions predisposing to uveal melanoma include congenital melanosis, nevus of Ota, blue nevi, ocular melanocytosis, oculodermal melanocytosis, and uveal nevi. Unlike most other tumors, melanotic melanomas are hyperintense on T1WI and hypointense on T2WI. Free radicals known to exist in melanin are responsible for the associated T1 and T2 shortening by the proton-electron dipole-dipole proton relaxation enhancement mechanism (Fig. 9-12). The degree of T1 and T2 shortening appears to be related to the melanin content. Amelanotic melanomas have MR characteristics similar to those of other tumors (hypointense on T1WI and hyperintense on proton density–weighted imaging [PDWI]/T2WI). Other lesions with similar intensity patterns that could be confused with melanotic melanoma include fat and subacute hemorrhage in the intracellular methemoglobin state. Use

of fat saturation images can easily separate fat from melanotic tumor. Associated with the melanoma may be retinal detachments, with subretinal proteinaceous effusions, that appear as high intensity on T2WI.

On CT, uveal melanomas are high density on noncontrast images and enhance. MR is superior to CT in visualizing retinal detachment, for noting associated vitreous changes, and for differentiating uveal melanoma from choroidal hemangioma (based on intensity characteristics) and choroidal detachment (based on enhancement). Extraocular invasion is important to detect as it is associated with a poorer prognosis and has different therapeutic implications. Ocular ultrasound and CT may be particularly useful here. Choroidal melanoma has a propensity for metastasis to the liver and lung. Melanoma is one of the best reasons not to show off your exposed body at the beach.

Choroidal Hemangioma

This is a vascular hamartoma detected in middle-aged and elderly individuals. There are two varieties: a circumscribed or solitary form, unassociated with other abnormalities, and a diffuse angiomatous lesion associated with Sturge-Weber or facial nevus flammeus. The solitary lesion can be mistaken for choroidal melanoma. On MR, these lesions are isointense to slightly high intensity and hyperintense on T2WI. They enhance avidly.

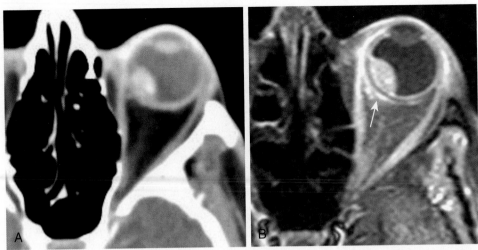

FIGURE 9-13 "Eye see it!" said the neuroradiologist. **A,** Axial computed tomographic (CT) image shows unsuspected choroidal metastasis in a patient with known metastatic colon cancer. **B,** Axial postcontrast T1-weighted image shows the postscleral extention of this metastasis *(arrow)* to better advantage compared to CT. Fat suppression is key.

Medulloepithelioma

This tumor, formally termed dictyoma (we will pass on this joke), arises from the primitive unpigmented epithelium lining of the ciliary body. It occurs in pediatric as well as adult patients. Its MR appearance is similar to melanoma. In children it is in the differential diagnosis of retinoblastoma whereas in adults think melanoma. These tumors are benign or just locally invasive but they too may calcify!

Metastases

Metastatic breast, lung, gastrointestinal and skin cancers, lymphoma, and leukemia are the most common nonmelanoma ocular malignancies seen in adults. Many metastatic neoplasms to the globe affect the uveal tract and therefore show eccentric thickening of the uveoscleral rim and retrobulbar extension, which can be detected on CT and MR, and may not be clinically apparent (Fig. 9-13). Choroidal lymphoma and leukemia can appear on MR to be similar to uveal amelanotic melanoma, may cause retinal detachment and may be bilateral (rare for melanoma).

Choroidal Osteoma

Choroidal osteoma is a benign tumor seen in young women. These lesions are usually unifocal, demonstrate ossification, and are juxtapapillary in location. CT is the imaging modality of choice in this situation. Punctate calcification in the posterior pole of the globe, usually on the temporal side of the optic nerve, may be easily identified. An astrocytic hamartoma could look identical.

Ocular Infections

Both CT and MR are nonspecific in identifying specific pathogens in the setting of ocular infections. However, imaging can be useful in establishing a diagnosis when the clinical picture is unclear and can help in defining the extent of the infection. Imaging is useful in ocular pseudomonas infections to rule out posterior scleritis, which demonstrates posterior scleral thickening and enhancement. In patients positive for human immunodeficiency virus (HIV), cytomegalovirus (CMV) infection–related arcuate retinal enhancement can be seen.

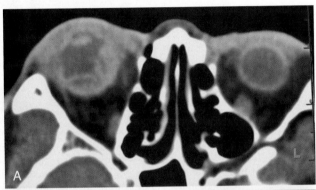

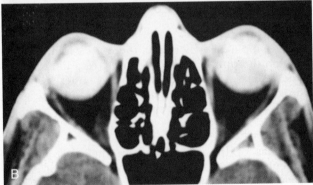

FIGURE 9-14 C the CMV (cytomegalovirus)? **A,** The patient with acquired immunodeficiency syndrome (AIDS) has a retinal detachment associated with the CMV infection of the chorioretinal membranes. **B,** Hemorrhagic CMV. Computed tomography reveals bilateral vitreous hemorrhages in a patient with AIDS.

CMV is the most common opportunistic infection of the retina and choroid. Up to one fourth of patients with acquired immunodeficiency syndrome (AIDS) have CMV. The chorioretinitis may be hemorrhagic and can produce high density on CT or various MR appearances depending upon the stage of blood (Fig. 9-14). Other infectious pathogens affecting the retina in AIDS include toxoplasmosis, herpes simplex, and herpes zoster.

Endophthalmitis is an acute suppurative infection of the aqueous or vitreous humor, acquired either by

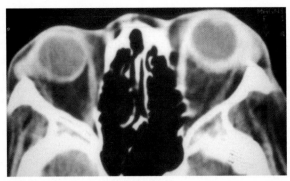

FIGURE 9-15 Visible inflection affecting vision. Computed tomography with left posterior enhancement of episcleral region and shaggy borders extending into retrobulbar fat signifying posterior scleritis (and orbital cellulitis).

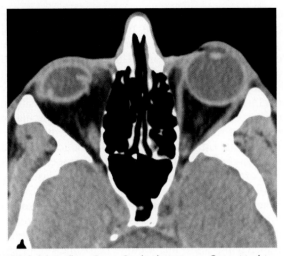

FIGURE 9-16 Deflate-Gate: Ocular hypotony. Computed tomography following perforation of the right globe. The globe has the appearance of a flat tire when compared to the left, which is characteristic of this condition. Note the focal disruption of the globe contour on the medial globe margin, also supporting the diagnosis of globe rupture.

direct inoculation (e.g., trauma, foreign body, or iatrogenic) or hematogenously from other infected source (e.g., cardiac). Aggressive treatment is warranted to prevent the possibility of permanent blindness. On CT, heterogeneous or diffuse increased attenuation within the affected globe can be seen, along with thickening and enhancement of the sclera. Proteinacous content within the globe can show elevated signal on T1- and T2-weighted images.

Scleritis can be infectious (bacterial, fungal, or viral) or autoimmune (collagen vascular diseases) in origin. It is separated into acute and chronic categories. Findings include thickening and enhancement of the posterior sclera and uveal layers (Fig. 9-15). The thickening may be nodular or diffuse. The nodular variety can be mistaken for melanoma; the diffuse type can mimic lymphoma.

Intraocular Hemorrhage Following Subarachnoid Hemorrhage

Retinal hemorrhage occurs in 18% to 41% of adults with subarachnoid hemorrhage and may be as high as 70% in children. If vitreous hemorrhage also occurs the condition is termed *Terson syndrome*. Retinal hemorrhages in nonaccidental trauma may be related. Fundal hemorrhage is believed to be caused by a sudden rise in intracranial pressure transmitted into the distal optic nerve sheath. This pressure may compress the central retinal vein and cause the intraocular hemorrhage. Intraocular hemorrhage has also been seen in HIV patients with CMV retinitis.

Abnormalities of Globe Morphology

Ocular Hypotony and Choroidal Detachments. Ocular hypotony is defined as low tension in the globe secondary to surgery, trauma, or glaucoma therapy, and may be observed on CT as uveoscleral infolding also referred to as the "flat tire" or "umbrella" sign (Fig. 9-16). This is most often seen after perforation and can be reversed by instillation of saline, silicone oil, or sodium hyaluronate (Healon), which expand the globe. Just two more words: double perforation. This term is used by ophthalmologists and means a perforation through the anterior and posterior portions of the globe. The posterior perforation is one reason to perform an imaging study; it usually cannot be repaired (poor outcome). The

most common cause of the double perforation is the BB sized shot. Ocular hypotonia may be the cause of serous choroidal detachments.

Table 9-2 compares the various ocular detachments radiologist commonly encounter. Hemorrhagic choroidal detachment may be observed after penetrating injury or intraocular surgery and has a lenticular morphology. This is bad news because it means there has been rupture of a choroidal vessel, which has an associated poor prognosis. Serous choroidal detachment (benign prognosis) is crescentic or ring shaped, and the curvilinear choroid can be identified. Serous choroidal effusions beneath the detached choroid appear as convex regions of low density outlined by the detached choroid. Hemorrhagic choroidal effusions are high density and do not change with position. Inflammatory choroidal effusions secondary to uveitis and posterior scleritis are high density but may change in location with changes in head position. Uveoscleral thickening and enhancement can be seen in inflammatory choroidal detachment.

MR is beneficial in distinguishing hemorrhagic choroidal detachment from serous choroidal detachments. On MR, hemorrhagic choroidal detachment can display a variety of intensity patterns related to age of the hemorrhage. Serous choroidal detachment has been noted to be high intensity on T1WI and T2WI.

Globe Tenting. Globe tenting occurs when the posterior globe configuration changes so that it appears as a conical or tented structure. It is caused by an intraorbital mass lesion, either acute or subacute, producing proptosis with tethering of the globe by the stretched optic nerve. This differs from the appearance of a staphyloma, described below, by the posterior globe shape that subsumes an angle of less than 130 degrees. Ocular tenting has been reported in acute trauma, inflammatory processes, carotid cavernous fistula, subperiosteal abscess, hemorrhage into lymphangioma, and varix. Tenting with acute proptosis is an indication for emergent surgical decompression.

Coloboma. A coloboma of the eye is a congenital defect in any ocular structure. It is classified as typical or atypical depending on location and derivation. Atypical colobomas occur in the iris. The typical coloboma is a cone-shaped or notch deformity that occurs in the inferior medial portion of the globe. It is caused by incomplete closure of the choroidal fissure. This can result in an abnormal elongated or malformed globe or ocular cyst at the site of the fetal optic fissure closure. It may be associated with orbital cysts, midline craniocerebrofacial clefting including sphenoidal encephalocele, agenesis of the corpus callosum, and olfactory hypoplasia, as well as cardiac abnormalities, retardation, genital hypoplasia, and ear anomalies. It is a part of the CHARGE syndrome (coloboma, heart defects, atresia of the choana, retarded growth and development, and ear anomalies). Other syndromes having coloboma include Lenz microophthalmia syndrome, Meckel syndrome, trisomy 13, Goldenhar syndrome, Rubinstein-Taybi syndrome, Aicardi syndrome, and Waardenburg anophthalmia syndrome.

On imaging, there is a cone-shaped or notch-shaped deformity, usually of the posterior globe, that may involve the optic nerve, with eversion of a portion of the posterior globe (Fig. 9-17). Other structures including retina, choroid, iris, and lens can be affected. The sclera is normal. Colobomas usually arise sporadically and are unilateral, although rarely an autosomal dominant form can exist, with about 60% of those affected having bilateral coloboma.

Staphyloma. Posterior staphyloma and coloboma may be detected by CT or MR and are superficially similar (Fig. 9-18). Staphylomas are acquired defects in the sclera or cornea. They are lined with uveal tissue. Posterior staphyloma is associated with increasing globe size in patients with axial myopia and is usually on the temporal side of the optic disc. With posterior staphyloma one may demonstrate outward bulging (which is more diffuse than the cone-shaped deformity in coloboma and triangulation in globe tenting) of the posterior portion of the globe with uveoscleral thinning and lack of enhancement. Posterior staphyloma can be a cause of proptosis.

The Big or Small Globe. Occasionally, one is confronted with an image of an enlarged globe (Box 9-4). The most common cause of this is axial myopia. The most common cause of axial myopia is reading too many books in your training (leading to axial myopia), so throw out the other neuroradiology texts besides the requisites and protect yourself. Other conditions causing enlarged globes include congenital glaucoma (buphthalmos), collagen disorders (e.g., Marfan syndrome), posterior staphyloma (see Fig. 9-18), coloboma (see Fig. 9-17), juvenile glaucoma in patients with Sturge-Weber syndrome, Proteus syndrome, and neurofibromatosis.

Small globes (microophthalmia) have been reported as an isolated event or may be associated with conditions such as craniofacial anomalies, congenital rubella, persistent hyperplastic primary vitreous, retinopathy of prematurity, and phthisis bulbi. Anophthalmia represents complete absence of ocular tissue.

RETROBULBAR LESIONS

Retrobulbar lesions may be conveniently divided into intraconal, conal, and extraconal locations. The differential diagnosis of lesions by virtue of their anatomic origin is provided in Boxes 9-5 through 9-8. Remember that intraconal lesions are associated with the optic nerve, its vessels, and orbital fat; conal lesions involve the muscles; and extraconal lesions are those centered in the bony orbit or extending into the orbit from the paranasal sinuses, skull, or brain. MR, by virtue of its superior anatomic and multiplanar imaging, is the imaging modality of choice for localizing orbital lesions.

Intraconal Lesions

For lesions of the optic nerve, see Box 9-6.

Enlarged Cerebrospinal Fluid Space

An enlarged perioptic subarachnoid space is easily detected by MR and may occur congenitally, as a result of optic atrophy, or in situations where raised intracranial pressure exists (such as idiopathic [benign] intracranial hypertension, also known as pseudotumor cerebri). Symptoms include headaches, transient visual obscuration, and intracranial pulsation noises. In cases of raised intracranial pressure, the globe at its junction with the optic nerve can be indented (reverse cupping) by the transmitted raised intracranial pressure, providing imaging evidence of

TABLE 9-2 Types of Ocular Detachments

Detachment	Separated Layers	Shape	Extent	Association
Retinal	Sensory retina from retinal pigment epithelium—subretinal space	V shaped with apex at optic disc (total)	To ora serrata	Retinoblastoma, Coats disease, *Toxocara* endophthalmitis, diabetes, melanoma, choroidal hemangiomas, following trauma, senile macular degeneration, or persistent hyperplastic primary vitreous (PHPV)
Choroidal-serous or hemorrhagic	Between choroid and sclera—suprachoroidal space	Linear, crescentic, or ring-shaped-serous convex-hemorrhagic	Leaves do not extend to the optic disc because the posterior choroid is anchored by short posterior ciliary arteries and nerves	Ocular hypotony, trauma, surgery, inflammatory choroidal lesions, melanoma
Posterior hyaloid space	Between posterior hyaloid membrane and sensory retina	Thin, semilunar, gravitational layering	Variable	Macular degeneration, PHPV, posterior vitreous detachment (PVD)

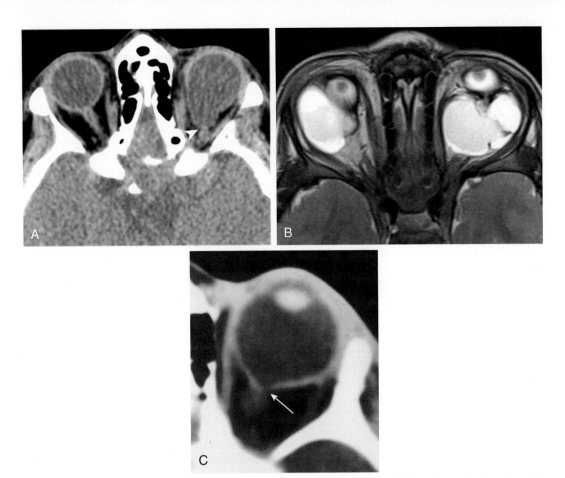

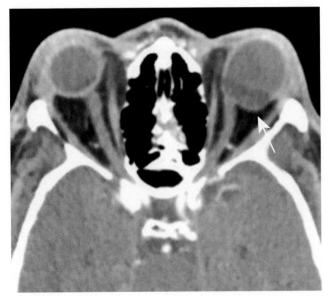

FIGURE 9-17 This case caught my eye. **A,** Coloboma. Left-sided cone-shaped deformity at the junction of the optic nerve head and globe *(arrowhead).* **B,** Bilateral orbital cysts on axial T2-weighted imaging in a patient with colobomas. **C,** Classic in-cupping of the cyst *(arrow)* to the optic nerve insertion of this coloboma.

FIGURE 9-18 Keep an eye on this one. Computed tomography of posterior staphyloma. Notice the uveoscleral thinning in the posterior lateral margin of the left globe *(arrow).*

BOX 9-4 Causes of Macroophthalmia and Microophthalmia

MACROOPHTHALMIA

Aniridia
Axial myopia
Buphthalmos
Congenital glaucoma
Ehlers-Danlos
Homocystinuria
Intraocular masses
Marfan
Neurofibromatosis 1
Proteus
Staphyloma
Sturge-Weber

MICROOPHTHALMIA

PHPV
Trisomy 13 (Patau syndrome)
Surgery
Fetal rubella, varicella, herpes simplex, infection
Holoprosencephaly
Phthisis bulbi
Retinopathy of prematurity
Radiation
CHARGE syndrome
Infant of diabetic mother
Fetal alcohol syndrome
Pseudohypoparathyroidism

CHARGE, Coloboma, heart defects, atresia of the choana, retarded growth and development, and ear anomalies; *PHPV,* persistent hyperplastic primary vitreous.

BOX 9-5   Intraconal Lesions

Lymphangioma
Lymphoma
Hematoma
Metastasis
Optic nerve lesions: Glioma, meningioma, hemangioblastoma,
   schwannoma, neuritis, sarcoid
Orbital pseudotumor
Rhabdomyosarcoma
Schwannoma
Vascular: Cavernous hemangioma, varix, venous vascular
   malformation

BOX 9-6   Causes of Optic Nerve Sheath Complex Enlargement

TUMORS

Leukemia, lymphoma
Hemangioblastoma
Meningioma
Metastasis
Optic nerve glioma
Schwannoma

NONNEOPLASTIC CAUSES

Central retinal vein occlusion
Cysticercosis
Graves disease
Idiopathic intracranial hypertension (pseudotumor cerebe27ri)
Increased intracranial pressure
Normal variant with enlarged subarachnoid space
Optic neuritis
Orbital pseudotumor
Sarcoidosis
Toxoplasmosis
Traumatic hematoma of optic nerve
Tuberculosis

BOX 9-7   Conal Lesions

Acromegaly
Carotid-cavernous fistula, thrombosis, or dural malformation
Cellulitis
Hematoma or swelling from trauma
Lymphoma, leukemia
Metastasis
Myositis (from adjacent sinusitis)
Orbital pseudotumor
Rhabdomyosarcoma
Sarcoid
Thyroid ophthalmopathy
Wegener

BOX 9-8   Extraconal Lesions

Bone lesions including fibrous dysplasia, Langerhans histiocytones,
   adjacent meningioma
Capillary hemangioma
Cellulitis, abscess
Dermoid, epidermoid/teratoma
Fractures, hematoma
Granulomatous disease
Lacrimal gland tumor and other lacrimal gland pathologies
Lacrimal sac abnormalities
Lymphangioma
Lymphoma/leukemia
Meningioma
Metastases
Mucocele
Orbital cyst
Pseudotumor
Rhabdomyosarcoma
Schwannoma/plexiform neurofibroma
Sinus carcinoma

---

papilledema (Fig. 9-19). Marked enhancement of the optic nerve heads has also been observed and thought to be secondary to breakdown of the blood retinal barrier following sudden rise in intracranial cerebrospinal fluid (CSF) pressure. This may be seen in papillitis and acute intracranial pressure elevations. There may even be restricted diffusion at the optic nerve head. Rarely one may see dural ectasia of the optic nerve sheath associated with NF-1.

Optic Nerve Atrophy

The end result of a variety of insults to the optic nerve is optic atrophy. The nerve loses both function and its pink color from loss of vascular supply (optic nerve pallor). On MR, the nerve looks atrophic (Fig. 9-20). Causes of optic nerve atrophy include congenital optic nerve hypoplasia, macrophthalmos, compressive lesions including pituitary tumors and craniopharyngioma, herpes zoster, multiple sclerosis/optic neuritis, trauma, glaucoma, ischemic optic neuropathy, toxin exposure and nutritional deficiencies.

When you see optic nerve hypoplasia in children think De Morsier (septooptic dysplasia) syndrome, which has associated poor vision, hypotelorism, absence of the septum pellucidum, and hypopituitarism.

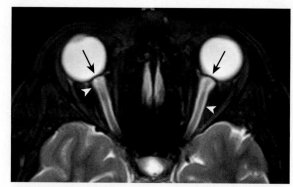

FIGURE 9-19 Eye-popping pressure. T2-weighted imaging shows reverse cupping of the optic nerve insertion to the globe *(arrows)* as well as prominence to the optic nerve sheath complex *(arrowheads)* indicative of elevated intracranial pressure and clinically appreciated papilledema.

Optic Neuritis

Optic neuritis is an inflammatory lesion of the optic nerve clinically associated with pain, decreased visual acuity, abnormal color vision, and afferent pupillary defect. Optic neuritis occurs in up to 80% of patients with multiple sclerosis (MS) and may be the first clinical manifestation of MS. In patients with their first attack of optic neuritis, up to 65% have asymptomatic white matter abnormalities in the brain on MR and 50% in 5 years are diagnosed with multiple sclerosis. Other less common diseases associated with optic neuritis include ischemia, vasculitis, sarcoid, systemic lupus erythematosus, syphilis, Lyme disease, viral infection, toxoplasmosis, tuberculosis, chemotherapy (cis-platinum), and radiation therapy (Fig. 9-21). MR is the best imaging modality for optic neuritis. Visualization is quite technique dependent. Fast spin echo coronal T2WI (best performed with fat and/or CSF suppression) will reveal high intensity of the involved nerve. The nerve is almost always normal in size. The use of enhancement is particularly useful, especially with fat saturation for the orbital portion of the nerve. Perivenous inflammation is responsible for the enhancement, which can be seen in approximately 50% of patients with acute optic neuritis. Enhancement can also be observed in the intracranial portion of the optic nerve, sometimes even in the absence of T2WI signal abnormality.

Neuromyelitis optica (aka NMO or Devic disease) is characterized by demyelination in the optic nerves and/or chiasm as well as long segment lesions in the spinal cord. Most of the patients have specific immunoglobulin G (IgG) antibodies to the aquaporin 4 water channel on astrocytic foot processes. The disease may be severe, fulminant, and monophasic. The treatment is monoclonal antibody treatment (rituximab) to deplete the B cells.

Optic Nerve and Visual Pathway Glioma

Optic nerve glioma (ONG) is classified as a juvenile pilocytic astrocytoma and represents about two thirds of all primary optic nerve tumors, 1.5% to 3% of all orbital tumors, and up to 1.5% of all intracranial tumors. Optic gliomas have a mean age at presentation of 8.5 years (5 years old in NF-1 and 12 years old in the absence of neurofibromatosis), with more than 80% occurring before 20 years of age. Optic nerve lesions tend to occur more often in females, whereas chiasmal tumors are present equally in males and females. These lesions have a growth phase during childhood when they become symptomatic. Most of the tumors then stabilize with indolent progression in about 40% of children. Patients may be asymptomatic and those with symptoms tend to have strabismus, visual loss, and afferent pupillary defect followed by proptosis. Decreased visual acuity occurs late in the course of optic nerve glioma (distinguishing it from an optic sheath meningioma described below). In 10% to 38% of cases (25% average)

there is an association with NF-1 (von Recklinghausen disease).

Approximately one fourth of visual pathway gliomas are confined to the intraorbital optic nerve. The rest occur in the intracranial portions of the optic nerve, the chiasm, or along the remainder of the visual pathway. Prepubertal children have a higher incidence of glioma restricted to the optic nerve. Forty percent of chiasmatic gliomas extend into the hypothalamus, which significantly increases the morbidity and mortality rate. As opposed to children, in whom the lesions are slow growing and benign, the rare optic nerve glioma in the adult is commonly malignant (ONG GB OMG!!), often extending intracranially, and is associated with a very poor prognosis (Fig. 9-22).

MR is the imaging modality of choice for this lesion. It beautifully defines the tortuous enlarged nerve that usually appears isointense on T1WI and of isointensity to high intensity on T2WI. MR is especially advantageous for lesions at the orbital apex and in the optic canal. Prominent CSF spaces around a normal nerve are differentiated from tumor in that they follow the intensity of CSF, whereas tumor may be higher intensity on fluid-attenuated inversion recovery (FLAIR) and lower intensity on T2WI. These lesions demonstrate variable enhancement. MR not only depicts the orbital lesion but is also very useful in discerning the associated cerebral abnormalities. These include enlargement and abnormal signal in the intracranial portion of the optic nerves, chiasm, and optic tracts.

CT reveals an enlarged optic nerve and canal (if the lesion extends through the optic canal). The nerve enlargement may be fusiform or nodular. Enhancement is variable, and calcification is rare unless the patient has been previously treated with radiation. The mass usually cannot be radiographically separated from the optic nerve. Optic nerve gliomas may have cystic components associated with them. These result from ischemia, radiation therapy, or mucin deposition. The nerve-tumor complex may display kinks or buckling, perhaps secondary to the mucin content of the tumor (which would make it more pliant) and perineural axial growth.

Optic Nerve Sheath Meningioma

Meningiomas may arise primarily from the meninges of the orbital optic nerve or extend secondarily from the cranial meninges. They originate from meningothelial cells within the meninges. Orbital optic nerve sheath meningiomas present with the insidious onset of visual

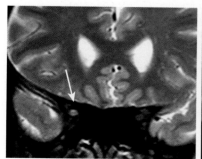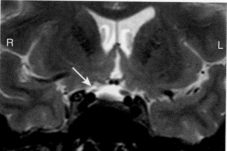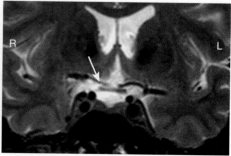

FIGURE 9-20 Eye-solated atrophy on the right. Serial coronal T2-weighted imaging shows high signal and small volume of the right optic nerve *(arrows)* compared with the left as far back as the chiasm.

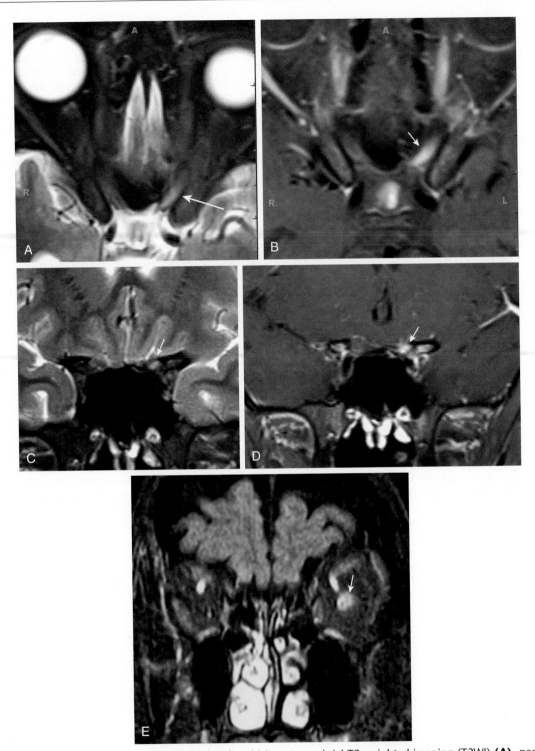

FIGURE 9-21 What a mess of MS (multiple sclerosis) in my eye. Axial T2-weighted imaging (T2WI) **(A)**, postcontrast T1 **(B)**, coronal T2WI **(C)**, and postcontrast T1WI **(D)** show abnormal signal in the optic nerve *(arrows)*, which enhances at the left optic canal. **E,** If you do fluid attenuation coupled with fat suppression as in this case, you can see the bright left optic neuritis really well.

loss, optic atrophy, mild proptosis, and opticociliary shunts (dilated veins from the optic nerve head, observed in 32% of cases). These lesions have a predominance for middle-aged women but may also be seen in children with NF-1 or NF-2 in adults (possibly bilaterally). Patterns of optic nerve enlargement may include diffuse tubular appearance, fusiform mass, or globular perioptic enlargement. In many cases, the nerve may be visualized as a separate entity from the mass of tumor. This helps differentiate the nerve sheath meningioma from the optic nerve glioma, where in most instances the nerve cannot be separated from the tumor.

Bilateral optic nerve sheath meningiomas are rare lesions that most often arise via spread from the region of the tuberculum sellae or planum sphenoidale and grow forward along both optic nerves. The risk for bilateral blindness

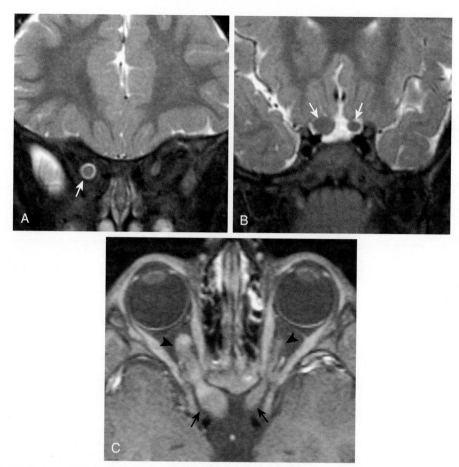

FIGURE 9-22 Bad eyesight b-eye-laterally. OMG it's an ONG (optic nerve glioma)! **A,** Coronal T2-weighted image (T2WI) shows marked enlargement of the right optic nerve *(arrow)* compared to the left. Often the T2WI signal of the nerve is normal and, owing to the low grade, the tumors may not enhance. **B,** More posteriorly at the level of the suprasellar cistern, coronal T2WI shows both optic nerves are markedly enlarged *(arrows)*, right more than left. **C,** Postcontrast T1WI shows enhancement along the involved optic nerve segments within the orbit *(arrowheads)* and in the suprasellar cistern *(arrows)*.

makes this a "can't miss" diagnosis. Beware how "en plaque" these may be. Associated findings in this situation include pneumosinus dilatans of the sphenoid sinus and blistering of the planum sphenoidale. As opposed to optic nerve gliomas, which do not calcify, optic nerve sheath meningiomas may demonstrate calcification (20% to 50% of cases).

CT findings include a high-density mass that demonstrates enhancement (Fig. 9-23). Bony changes may be present in the region of the optic nerve canal with bone erosion and occasionally hyperostosis. The optic canal is commonly normal, although it may be large or small. On axial CT the appearance after contrast has been termed the "sandwich sign" (also called the "tram track" sign) with the nerve (the "bologna") sandwiched between the tumor (the "bread"). Calcified meningiomas can have a similar appearance on unenhanced images.

MR is exceptionally useful for orbital apex and intracanalicular lesions, whereas detection on CT might be difficult in this region because of the high density of adjacent bone. MR reveals isointensity/or slight hypointensity on T1WI and slightly high intensity on T2WI (see Fig. 9-23). Optic nerve sheath meningiomas enhance, and such enhancement is particularly evident with fat saturation techniques. The enhancement on axial MR has been analogized to tram tracks in the axial plane or a doughnut in the coronal plane. The differential diagnoses of axial tram-tracking include pseudotumor, sarcoid, metastatic disease, lymphoma, and leukemia. MR is exceptionally useful for orbital apex and intracanalicular lesions, whereas detection on CT might be difficult in this region because of the high density of adjacent bone.

Slow-Flow Vascular Malformations

Venous Malformations. Congenital hemangiomas are tumors that are visible at birth, whereas infantile hemangiomas are seen after a few weeks of life. The congenital hemangiomas may demonstrate a rapidly involuting congenital hemangioma (RICH) or noninvoluting congenital hemangioma (NICH) progression. Infantile hemangiomas are tumors that have a growth phase and an involutional phase. These tumors are the most common childhood tumors of the orbits. They favor girls and premature infants and present at a mean age just under 1 year old. They may appear in the skin, the subcutaneous tissues, and the extraconal >> intraconal orbit (Fig. 9-24). As they involute between ages 2 and 8 they may undergo fibrofatty change. The lesions often show an artery feeding it and are opacified in the arterial phase of a dynamic MR enhanced scan.

Venous vascular malformations (VVM), previously called cavernous hemangiomas, are a nonneoplastic entity, and slowly enlarge over time. VVMs are the most common primary orbital vascular mass, consisting of dilated endothelial lined vascular channels encompassed by a fibrous pseudocapsule. They occur in females more commonly than males and present in the second to fifth decade of life with the slow onset of proptosis (which commonly goes unnoticed) and visual disturbance. Growth can be exceedingly slow and may stop after a certain time period; however, rapid

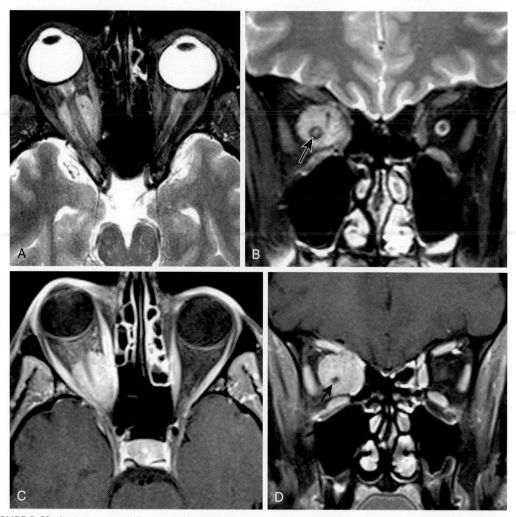

FIGURE 9-23 Optic nerve with frosting in the "cone": optic nerve meningiomas. **A,** Axial and **(B)** coronal T2-weighted images show hyperintense mass circumferentially investing the right optic nerve. There is proptosis of the right globe as a result. Note the elevated T2 signal within the nerve due to compressive effect by the mass. **C,** Axial and **(D)** coronal postcontrast T1-weighted images show homogeneous contrast enhancement of the mass. Note the nonenhancing optic nerve encased by the tumor (*arrow* in **D**). Meningiomas can be much sneakier than this case, arising from the planum sphenoidale or clinoid process before insinuating along the optic nerve sheath (see Fig. 10-32).

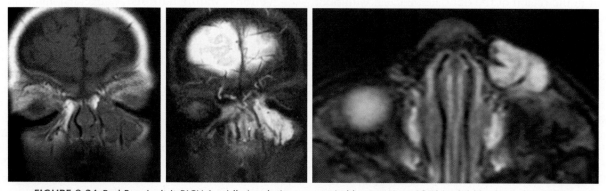

FIGURE 9-24 Red Eye: Isn't it RICH (rapidly involuting congenital hemangioma)? This child had a red raised lesion along the left lower eyelid that extended into the orbit that was present at birth. Fortunately, over months it involuted, hence a RICH.

growth has been observed during pregnancy. CT shows a smoothly marginated high-density round or oval intraconal mass that densely enhances with or without phleboliths (that do not occur in infantile hemangioma). Rarely may they be extraconal. Commonly, there is evidence of orbital bone expansion. MR depicts a smoothly marginated, intraconal mass that has no flow voids and demonstrates marked enhancement. VVMs are isointense to muscle and hypointense to fat on T1WI and high intensity on T2WI (Fig. 9-25). On early postcontrast images they enhance in a patchy

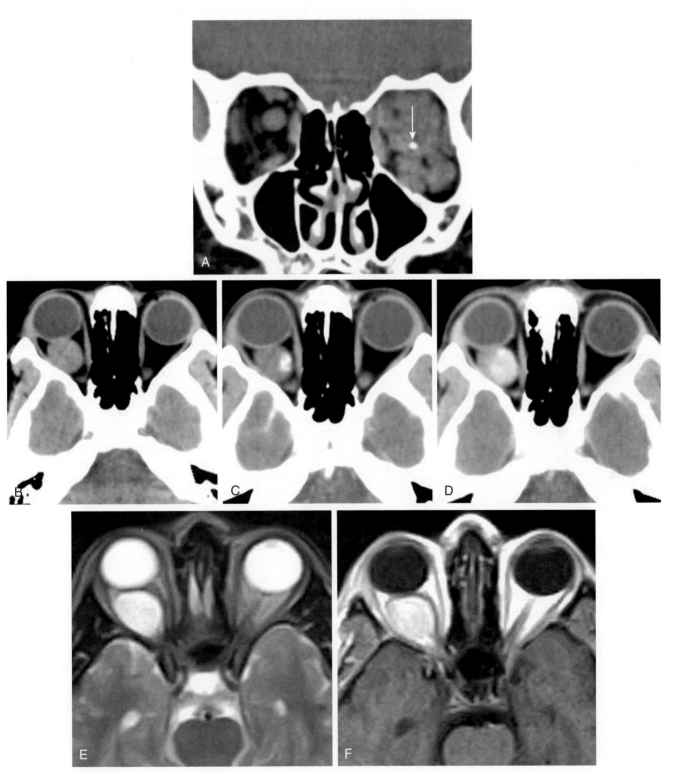

FIGURE 9-25 Phlabulous phlebolith in a VVM (venous vascular malformation). **A,** Note the calcification *(arrow)* among the numerous intraconal vascular channels that define this as a venous vascular malformation. More classic VVM in a different patient on noncontrast **(B)**, followed by early **(C)** and delayed **(D)** contrast-enhanced computed tomographic images seen as a well-defined intraconal enhancing mass that accrues contrast over time. The same lesion is bright on T2 **(E)** and shows avid enhancement on postcontrast T1 **(F)** images.

fashion but totally fill in within 30 minutes. They opacify in the later venous phase of the dynamic MRA.

Histopathologic differentiation between lymphatic and venous channels is extremely difficult. Thus an orbital varix may simulate an orbital hemangioma at pathology. There may be combined venolymphatic malformations of the orbit, which have both venous and lymphatic components. There has been a reported association of these lesions with intracranial vascular malformations.

Orbital Varix. The orbital varix is a venous malformation in which there is dilatation of an otherwise normal venous system (primary) or abnormal venous channels (secondary) affecting the superior and/or inferior ophthalmic veins, resulting in intermittent proptosis associated with retrobulbar pain. The spectrum can range from a single dilated venous structure to multiple varicosities. Varices are associated with a history of proptosis in conjunction with Valsalva maneuvers, such as straining (this can be a bathroom metamucil diagnosis) or coughing, that increase venous pressure. The increased venous pressure is transmitted through the valveless jugular veins to the dilated intraorbital venous network. These lesions are considered congenital venous vascular malformations (primary) but can also be related to orbital trauma or associated with orbital or intracranial arteriovenous malformations or carotid-cavernous fistula. Plain CT reveals a high-density intraconal mass. Enhanced CT of the patient in the prone position with Valsalva maneuver often results in significant

change in the size of orbital "mass" (Fig. 9-26). Flow in the varix is detected as low intensity on spin echo sequences in a vascular-shaped structure. With cessation of blood flow and thrombosis, the appearance is variable, depending on the stage of the clot. MR is the best noninvasive imaging modality for detecting flow and the cessation of flow.

Lymphangioma/Lymphatic Vascular Malformation. Lymphangiomas consist of dysplastic vascular channels (lymphatic or venous), loose connective tissue, and smooth muscle bundles. They are most likely vascular hamartomas that arise from an anlage of vascular mesenchyme. Lymphangioma of the orbit is generally an extraconal mass that may be located anteriorly or posteriorly; however, lymphangiomas may occur in any orbital space. These congenital lesions are not well encapsulated. The mean age at presentation in one series was 23 years but lymphangiomas can be seen in the newborn to adult patient. These tumors have a tendency to hemorrhage with the acute onset of symptoms. CT exhibits orbital expansion, irregular margins crossing anatomic boundaries and increased density on unenhanced CT scan but variable enhancement following contrast administration. Peripheral enhancement can be demonstrated in cystic regions of the lesion. In addition to the morphologic characteristics, MR can show evidence of hemorrhage into multiple or single cysts. These appear as high signal intensities on T1WI, and fluid levels are identified (Fig. 9-27).

High-Flow Vascular Malformations: Carotid Cavernous Fistula and Dural Vascular Malformation

Carotid-cavernous fistula is a communication between the intracavernous portion of the carotid artery and the venous cavernous sinus. These most often result from significant trauma, either nonpenetrating or penetrating. Another cause is the spontaneous rupture of a cavernous carotid aneurysm, usually in an elderly patient. Spontaneous carotid cavernous fistula has been reported in osteogenesis imperfecta (even more imperfecta than you knew), Ehlers-Danlos, and pseudoxanthoma elasticum.

As such they are high-flow lesions. Mostly there are single points of communication between the carotid artery

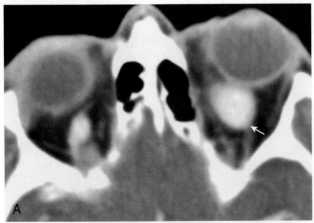

FIGURE 9-26 Popeye with a proptotic eye. **A,** Note both the mild left proptosis as well as the enlarged vascular structure in the left eye *(arrow)*. **B,** On the coronal view one can see that this vessel represents the enlarged superior ophthalmic vein in a venous varix *(arrow)*. The normal sized right vein can be seen in comparison.

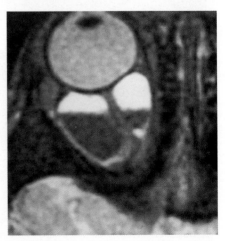

FIGURE 9-27 Tearful lesion in the eye: lymphangioma. Axial proton density–weighted imaging in a young man who had acute proptosis demonstrates hemorrhagic fluid levels as is typically seen with lymphangioma.

and cavernous sinus (particularly the traumatic ones), sometimes multiple connections, sometimes ICA and/or external carotid artery (ECA) connections, and sometimes an intervening nidus of a true arteriovenous malformation. In all, the clinical presentation is of pulsating exophthalmos, orbital bruit, and/or motility disturbance with dilated conjunctival vessels, and glaucoma. Carotid-cavernous fistula causes enlargement of the extraocular muscles, proptosis, and dilatation of the superior ophthalmic vein secondary to congestion and edema from orbital venous hypertension. Computed tomography/computed tomography angiography/computed tomography venography nicely show these findings with enlargement of the superior ophthalmic vein (Fig. 9-28). Irregularity or absence of superior ophthalmic vein enhancement suggests partial or complete thrombosis. The superior ophthalmic veins can also be enlarged in patients with diffuse cerebral swelling. In such cases the veins are enlarged bilaterally as opposed to carotid cavernous fistula where they are usually unilaterally enlarged. Periorbital swelling and blurring of the globe's margin related to conjunctival edema or pulsations can be observed on CT. The cavernous sinus may be distended and bowed convex to the middle cranial fossa. MR shows similar findings and can indicate, with flow techniques, either flow or its absence in the superior and inferior ophthalmic veins. Cavernous carotid artery aneurysms as a potential source of the fistula can be visualized with MR. Ultrasound is also useful and demonstrates reversal of blood flow.

Dural malformations involving the cavernous sinus are different from direct carotid-cavernous fistula in many ways. They are not usually associated with a traumatic event and are low flow processes. Venous thrombosis has been linked to the development of these malformations. Symptoms are generally not as fulminant as the direct carotid-cavernous fistula, although if venous pressure is increased, periocular pain, third, fourth, and sixth nerve palsies, and visual loss (secondary to acute glaucoma) can rapidly occur. If the venous drainage is posterior, that is, out the inferior petrosal sinus, you may have a "white-eyed" cavernous shunt (that is, without the typical proptosis and chemosis of a cavernous carotid fistula). The sixth nerve palsy may result from venous compression in Dorello canal. Supply is from branches of the meningohypophyseal artery, ascending pharyngeal artery, middle meningeal artery, accessory meningeal artery, artery of the foramen rotundum, or other meningeal tributaries, and supply can be bilateral in patients with unilateral symptoms. If there is partial or complete thrombosis of the venous drainage of the orbit including the superior ophthalmic vein, patients' symptoms could be temporarily paradoxically worsened although this is ultimately therapeutic.

Both lesions are amenable to interventional embolization. The preferred treatment of direct carotid-cavernous fistula is the use of detachable balloons, coils, stents, or liquid embolic agents flow-directed through the fistula into the cavernous sinus, tamponading the hole in the carotid. The approach may be intraarterial or retrograde venous. Another approach, reserved for cases in which the balloon cannot be placed safely through the fistula, involves trapping of the fistula above and below its origin with balloons and/or surgical intervention. Dural fistulas respond to particulate or liquid embolic material. The trick here is to promote venous thrombosis so that the fistula occludes. At times these fistulas spontaneously thrombose, although other therapies include angiography and carotid massage (spa therapy for vessels). Partial embolization can also result in complete thrombosis at times. However, dural malformations can recur and may be difficult to cure. Emergency intervention is required for acute visual

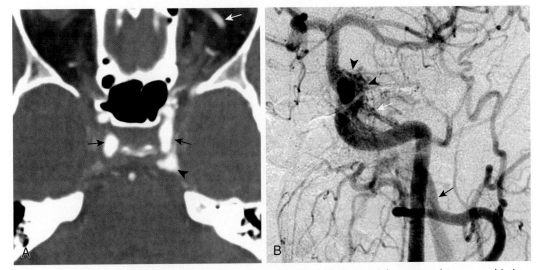

FIGURE 9-28 Take the red eye home: cavernous carotid fistula (CCF). **A,** Axial computed tomographic image from computed tomography angiography acquired in arterial phase shows normal arterial enhancement in the cavernous carotids bilaterally *(black arrows)* and abnormal enhancement in the left cavernous sinus *(arrowhead)* of similar (arterial) attenuation indicating presence of a CCF. While you might normally see some cavernous sinus enhancement on arterial phase images, it should never be the same density as the arteries. The superior ophthalmic vein is dilated and enhancing on the left *(white arrow)* compared to the right. **B,** Conventional catheter angiogram in the same patient following left common carotid injection again shows the abnormal arterialized enhancement in the cavernous sinus *(arrowheads)* and there is early venous draining in the petrosal sinus *(black arrow)*. This malformation was supplied both by internal carotid and external carotid *(white arrow)* branch feeders.

deterioration, thrombosis of the superior ophthalmic vein with increased collateral orbital blood flow, or the presence of cortical venous drainage (because of risk of intracranial hemorrhage).

Metastases and Lymphoma

Metastases and lymphoma can involve any part of the orbit, presenting as intraconal, conal, or extraconal lesions. Metastatic scirrhous breast carcinoma and neuroblastoma in children have a propensity for the extraconal space and enophthalmos and are discussed in the extraconal lesion section.

Neurogenic Tumors

Neurogenic tumors can arise from orbital branches of cranial nerves III, IV, V, VI, sympathetic and parasympathetic nerves, and the ciliary ganglion. They include schwannomas, neurofibromas, and amputation neuromas. Schwannomas appear similar to those in other locations with cystic and solid components and a propensity to enhance (Fig. 9-29). Neurofibromas may be localized, diffuse or plexiform, all three associated with neurofibromatosis with plexiform being pathognomonic.

Conal Lesions

Box 9-7 enumerates lesions that primarily involve the muscles (conal lesions).

Thyroid Ophthalmopathy

Thyroid eye disease (Graves orbitopathy, thyroid eye disease, thyroid orbitopathy) leading to proptosis can be present in euthyroid or hyperthyroid persons. Women have a 4:1 predominance over men. It may occur before, during, or after treatment and has a subacute onset, extending for months. CT exhibits prominent extraocular muscle enlargement and exophthalmos (greater than two thirds of the globes project anterior to a line connecting lateral orbital wall anteriormost borders [interzygomatic line]. There are normal muscle insertions on the globe, fusiform enlargement of the muscle belly, and protrusion of the orbital fat. In patients with clinical thyroid ophthalmopathy, 85% have bilateral involvement, 5% have unilateral involvement, and 10% have normal muscles. The most frequently affected muscles are the inferior and medial recti, although the most common single pattern is enlargement of all muscles. The lateral rectus muscle is rarely if

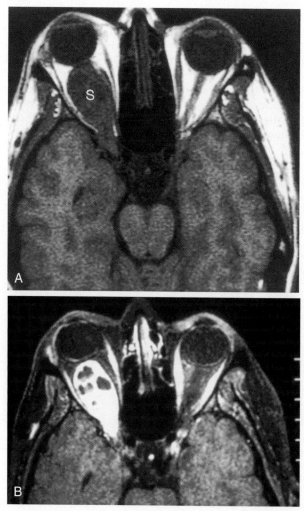

FIGURE 9-29 You've got some nerve! Schwannoma of the orbit (nerve unknown). **A,** Unenhanced T1-weighted image (T1WI) shows intraconal mass in the right orbit (S) causing proptosis. **B,** Postcontrast T1WI shows inhomogeneous contrast enhancement of the lesion due to presence of cystic components which don't enhance.

ever involved alone and is the last to be affected when all muscles are involved. Try this mnemonic for remembering the frequency of muscle involvement in Graves—"I'm slo(w)"—inferior, medial, superior, lateral, oblique muscle involvement.

Coronal scanning is the method of choice for assessing muscle thickness (Fig. 9-30). Orbital fat may be increased alone in 8% of cases with forward displacement of the orbital septum, whereas approximately 46% of cases involve both muscle and fat. Fatty infiltration of extraocular muscles may also be seen. The lacrimal glands and eyelids may be swollen. Increased orbital fat, producing exophthalmos, also has been reported to be caused by exogenous steroids and Cushing disease. The eyebrow fat pad can be involved producing bulky eyebrows. The superior ophthalmic vein can be distended. Progressive optic neuropathy is a serious complication of Graves orbitopathy and is seen in 5% of patients. There is an association of optic neuropathy and intracranial prolapse of orbital fat through the superior orbital fissure. Other features include increased volume of orbital contents and lacrimal gland displacement or apical crowding with dilatation of the superior ophthalmic vein and optic nerve sheath (just eyeball it). At present, MR may not add much more useful information than CT.

Many of the signs and symptoms of Graves orbitopathy including lid lag, diplopia, limited extraocular muscle movements, proptosis, and optic nerve compression result from the periorbital fibrosis that develops when the inflammation resolves. Orbital fat volume may actually be decreased relative to the total orbital volume in some patients with optic neuropathy. The T2 relaxation times of the extraocular muscles in Graves have been reported to be prolonged most likely secondary to increased water content as a manifestation of inflammation. Potentially, this can be used to distinguish patients with inflammatory changes (long T2) from those of fibrosis (shorter T2). This can have therapeutic implications, that is, treatment of inflammation versus no treatment for fibrosis.

Multiplanar MR or CT is important in assessing the degree of optic nerve compression by the enlarged muscles at the orbital apex. It is easy to appreciate the extent of optic nerve compression, particularly on the coronal image. In extreme cases where vision

is threatened, orbital decompression is performed with partial removal of the floor or medial wall of the orbit. This is currently being performed by an endoscopic approach, in-fracturing the orbital floor and/or lamina papyracea.

The differential diagnosis of enlarged muscles is similar to that of conal lesions (obviously).

Orbital Inflammatory Pseudotumor

There have been almost as many names for this entity (nonspecific orbital inflammation, inflammatory pseudotumor, orbital pseudotumor, orbital inflammatory syndrome, idiopathic orbital inflammation) as locations of the disease in the orbit (sclera, conjunctiva, muscle, nerve, sheath, gland, sac, etc.). Orbital pseudotumors may mimic a variety of pathologic states so that the appropriate history (rarely found on radiology request slips) is essential for making the correct diagnosis. The clinical features include restriction of ocular motility, chemosis, lid swelling, and pain. It is a cause of unilateral exophthalmos. These findings usually have a rapid onset and respond to steroids, although there is also a chronic form with progressive fibrosis and a mild or poor response to steroids. In such cases, chemotherapy or radiation therapy is used.

Acute orbital pseudotumor is an inflammatory condition that may be the result of an autoimmune condition involving the lacrimal gland (dacryoadenitis), lacrimal sac, the extraocular muscles, the connective tissue surrounding the dura of the optic nerve, the orbital fat, the epibulbar connective tissue, and the sclera. The disease has been recently categorized as IgG4-mediated inflammation, since perivascular lymphocytic infiltration rich in IgG4-positive plasma cells and elevated serum IgG4 are found. Systemic diseases associated with orbital pseudotumor include Wegener granulomatosis, polyarteritis nodosa, sarcoidosis, as well as autoimmune conditions such as lupus erythematosus, dermatomyositis, and rheumatoid arthritis. Related and associated fibrotic processes include retroperitoneal fibrosis, sclerosing cholangitis, Riedel thyroiditis, and mediastinal fibrosis. The term *multifocal fibrosclerosis* is used as a collective description of these disorders.

Pseudotumor may present as a lacrimal mass, a diffuse, ill-defined retrobulbar enhancing mass obscuring fascial

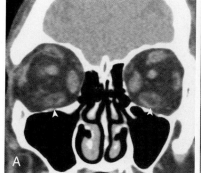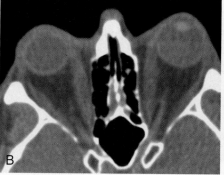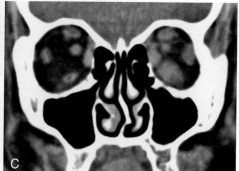

FIGURE 9-30 A TED talk about TED (thyroid eye disease)? **A,** Thyroid eye disease manifests proptosis and enlargement of extraocular muscles. Sometimes there is fatty infiltration of the muscles as seen in these enlarged inferior recti muscles *(arrowheads)*. In this case the medial recti are also large and the disease is bilateral. **B,** Classic medial rectus involvement. **C,** Yes, even for unilateral proptosis, TED is the most common diagnosis. TED gets around.

planes, or simply as thickened muscles and/or their tendons and sheaths. Over 70% of cases display proptosis. There may be a subtle increase in the density of orbital fat (dirty fat) or optic nerve thickening. In contrast to Graves disease the contour of the enlarged muscles may not be smooth and the tendinous insertions may be affected. Unfortunately, this is not true for all orbital pseudotumors! Indeed, some pseudotumors may have smooth muscles with uninvolved tendons. At times pseudotumor may be bilateral and may show intracranial extension. On T1 and T2WI pseudotumor has a tendency to be low intensity, whereas metastatic lesions have a longer T2 (high intensity) (Fig. 9-31).

Tolosa-Hunt

Tolosa-Hunt syndrome is an idiopathic inflammatory condition similar to orbital pseudotumor that involves the cavernous sinus and orbital apex. It presents with painful ophthalmoplegia. With respect to the orbit, inflammatory tissue can be identified in the orbital apex in the majority of cases. Pathologically lymphocytes and plasma cells infiltrate the involved region and there is thickening of the dura mater. On CT, the low density of orbital fat is replaced by soft tissue. This is a subtle but important finding in lesions of the orbital apex. On MR, soft tissue is isointense to muscle and can be recognized at the orbital apex and/or the cavernous sinus. This tissue enhances and has signal intensity changes identical to lymphoma, sarcoid, and meningioma, which are in the differential.

Sarcoidosis

Sarcoidosis commonly involves the orbit. Up to 25% of all patients with sarcoid have ophthalmic involvement. Uveitis is the most common manifestation of orbital sarcoidosis but other sites of involvement include the lacrimal gland, optic nerve/sheath, chiasm, muscles, and retrobulbar tissue producing proptosis (Fig. 9-32). Isolated orbital disease without pulmonary findings is rare; when limited to the orbit it usually affects the lacrimal glands.

Granulomatosis with Polyangiitis

Granulomatosis with polyangiitis (formerly Wegener granulomatosis) is characterized by granulomatous inflammation, tissue necrosis, and vasculitis that involve arteries, veins, and capillaries. It may involve the orbit secondarily from the paranasal sinuses, or may present as primary orbital disease. Up to 54% of all patients with the systemic condition have neurologic involvement and half have disease in the orbit. The most common ocular manifestations are keratitis and scleritis, whereas the orbital involvement produces pain, proptosis, erythematous eyelid edema, and limitation of extraocular movements caused by conal and intraconal spread. The ocular and orbital processes may coexist. Orbital inflammation may cause painful swelling, proptosis, nasal-lacrimal obstruction, or dacryocystitis. Antineutrophil cytoplasmic antibodies are highly sensitive indicators of the disease.

On MR, this condition has been reported to be hypointense relative to orbital fat on T2WI and enhance homogeneously. The classic lesion is a homogeneously enhancing

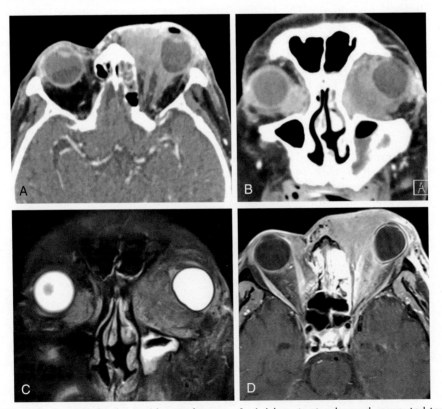

FIGURE 9-31 Looky here, looky there with pseudotumor. **A,** Axial contrast-enhanced computed tomographic image shows bulky enhancing mass centered in the preseptal tissues/postseptal extraconal medial orbit on the left, distorting the globe and displacing it laterally. **B,** Yikes! On coronal view, there's another mass in the inferomedial orbit on the right. **C,** These orbital masses are intermediate signal on T2 and show avid enhancement on postcontrast T1-weighted images **(D).**

mass with associated sinus disease and bone destruction, but if everything were classic, radiology would be pretty dull and you wouldn't rate the big bucks. The anterior segment is usually involved more than the posterior segment. The differential diagnosis includes polyarteritis nodosa and lymphomatoid granulomatosis, Beçhet disease, primary central nervous system vasculitis, lymphoproliferative disorders, sarcoidosis, Churg-Strauss syndrome, and infectious/inflammatory or neoplastic meningeal infiltration.

Orbital Lymphoma

Lymphoma may occur as a primary orbital tumor or may be associated with systemic lymphoma. It is generally seen in older persons presenting with slowly progressive painless periorbital swelling and low-grade proptosis. CT reveals a diffuse infiltrative mass that destroys the normal orbital architecture so that anatomic structures cannot be defined. On CT, one may see increased density on the precontrast scan. The lesions can extensively infiltrate muscles and/or the lacrimal gland. Lymphoma molds to the contour of the orbit and its structures; bone

destruction is rare. Extension beyond the orbit is unusual. The superior rectus is the extraocular muscle most often involved. Distinguishing lymphoma from orbital pseudotumor on MR is problematic because signal intensities and location are similar but the latter is painful painful painful (Fig. 9-33).

Extraconal Lesions

The extraconal lesions (see Box 9-8) arise from structures outside the muscle cone such as the lacrimal gland, peripheral fat, sinuses, or adjacent bony orbit. Bilateral orbital masses suggest the diagnosis of systemic inflammatory, leukemic, lymphoproliferative, histiocytic, or metastatic diseases.

Orbital Infection

The periorbita (periosteum), which lines the orbit, is reflected anteriorly on the tarsal preseptal (anterior to septum) space and globe proper. The orbital septum is a fibrous tissue attached to the outer bony orbital periphery continuous with the orbital periosteum, that acts as

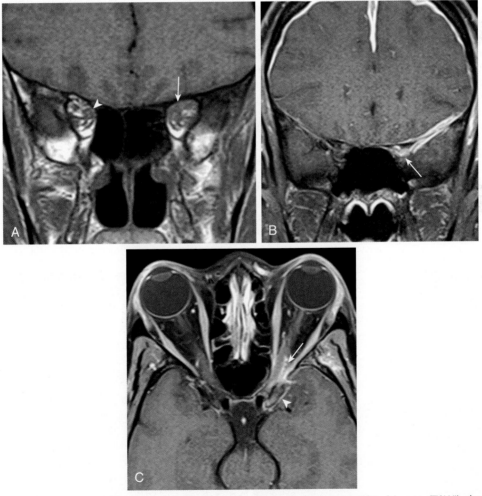

FIGURE 9-32 Orbital sarcoidosis: a diagnosis to cry about. **A,** Coronal T1-weighted image (T1WI) shows T1 hypointense soft-tissue infiltration at the left orbital apex *(arrow)*. Compare to the normal expected fatty signal surrounding the optic nerve sheath at the apex on the right *(arrowhead)*. **B,** Postcontrast T1WI shows abnormal enhancement of this tissue as it invests the optic nerve sheath on the left *(arrow)*. Note extensive dural-based enhancement intracranially. **C,** Axial postcontrast T1WI again shows the abnormal enhancement along the optic nerve sheath complex on the left *(arrow)* and adjacent dural enhancement around the left optic strut *(arrowhead)*.

a barrier to the spread of infection, and can be seen on high-resolution MR. The superior aspect can be observed descending from the superior orbital rim and fusing with the levator aponeurosis before reaching the superior aspect of the tarsus. The inferior aspect of the septum ascends from the inferior orbital rim towards the lower lid tarsus. It divides the orbit into the superficial anterior preseptal space and the deep postseptal space. The clinical manifestations of preseptal cellulitis are swelling and erythema of the skin and subcutaneous tissues of the eyelids. There is no evidence of proptosis, disturbances of ocular motility, or chemosis. CT/MR demonstrates no abnormalities posterior to the orbital septum; preseptal periorbital cellulitis is treated medically as an outpatient.

Orbital (postseptal) cellulitis is located within the bony orbit and deep to the orbital septum. It presents with painful ophthalmoplegia, proptosis, chemosis, and decreasing visual acuity. There is edema and inflammation without discrete abscess formation within the orbit. The causes of orbital inflammatory disease include sinus infection (particularly in the pediatric population), bacteremia, skin infection (secondary to trauma, insect bite, and impetigo), or foreign body. CT has an advantage over MR in the diagnosis of orbital cellulitis because of its ability to visualize the air-filled sinuses and demonstrate foreign bodies. CT can distinguish between preseptal cellulitis, orbital cellulitis, and subperiosteal infection. While CT in *preseptal cellulitis* shows swelling and obliteration of the preseptal soft tissues without extension deep to the orbital septum (Fig. 9-34), in *orbital (postseptal) cellulitis*, the orbital tissue planes are poorly defined and there may be either an intraconal or an extraconal soft-tissue mass. On fat-saturated T1WI MR, one can detect enhancement in the preseptal and postseptal tissue, particularly the extraocular muscles or retrobulbar fat.

Nodular coalescence or focal organized collection with ring enhancement suggests discrete abscess. A soft-tissue mass extending from the bony wall of the orbit with displacement of muscle and preservation of a thin strip of extraconal fat implies subperiosteal infection, a known complication of rhinosinusitis. There is marked swelling, chemosis, proptosis, and limitation of motility particularly to the side of the subperiosteal abscess. The common location of these lesions is in the medial orbit subjacent to the ethmoid air cells. Most often, this requires surgical intervention.

Orbital infection can produce venous thrombosis of the orbital veins with extension into the cavernous sinus. It is important to image the brain in cases of orbital cellulitis and abscess and of subperiosteal infection. Occasionally foci of infection are seen as a frontal epidural abscess, subdural empyema, or an intraparenchymal abscess. Intracranial complications are associated with a 50% to 80% mortality rate.

Opportunistic infections, a hallmark of AIDS, often involve the anterior and posterior segments of the eye. Rarely do they involve the orbit. Spread from contiguous sinusitis, usually the ethmoid and maxillary sinuses, is the most common source of orbital cellulitis in immune compromised hosts. Think orbital aspergillosis in HIV patients with proptosis, pain, visual loss, and ophthalmoplegia. These fungal infections may be very aggressive, invade the orbit, and lead to cavernous sinus inflammation and thrombosis.

Herpes zoster ophthalmicus, a grouped vesicular eruption, occurs along the first division of cranial nerve V. It is observed in both immunosuppressed and nonimmunosuppressed (usually elderly) individuals. When seen in patients younger than 45 years old, think about HIV. This can be a virulent infection where the virus grows along the optic nerve and vessels producing infarction of the optic nerve and large vessel vasculitis in the brain.

Lacrimal Sac Lesions

The anatomy of the lacrimal sac region was briefly discussed at the beginning of this chapter. The lacrimal sac is a preseptal structure lying anterior to the medial orbital septum. Obstruction of the sac or the nasolacrimal duct may lead to dilatation and inflammation (dacryocystitis).

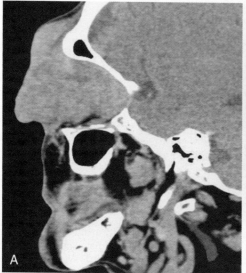
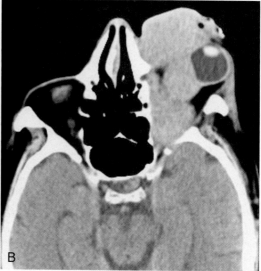

FIGURE 9-33 "Rimphoma" orbiting the orbit. **A,** Massive tumor without gross destruction may suggest lymphoma over other cancers. Note in **(B)** that the globe is misshapen without invasion.

Trauma can interrupt the normal lacrimal drainage, producing epiphora and dacryocystitis. Lacrimal sac dilatation is observed on CT or MR, and preseptal swelling, and/or cellulitis can be noted (Fig. 9-35). Enhancement is seen in the walls of the dilated sac, producing a ring configuration on axial images. Tumors of the lacrimal sac are uncommon, with the most common malignancies being squamous cell carcinoma and transitional cell carcinoma. Bone erosion with a mass lesion occurs

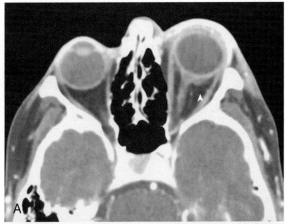

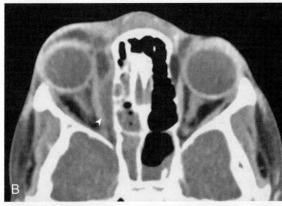

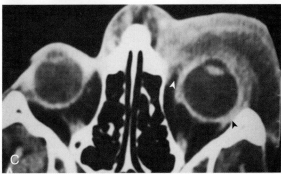

FIGURE 9-34 Manifestations of orbital infection. **A,** Postseptal orbital cellulitis. Computed tomography shows soft tissue behind the globe *(arrowhead)* as well as diffuse thickening over the left globe. Disease behind the orbital septum makes this orbital cellulitis. **B,** Subperiosteal abscess. Appreciate the lateral extension of the ethmoid sinusitis into the subperiosteal region fluid collection *(arrowhead)* with lateral displacement of the medial rectus muscle and orbital fat. There is also significant preseptal swelling. **C,** Preseptal periorbital cellulitis. Despite the huge amount of inflammation depicted it is all anterior to the medial *(white arrowhead)* and lateral *(black arrowhead)* attachments of the orbital septum. Hence preseptal cellulitis.

with these cancers. Lymphoma and minor salivary gland lesions also occur here. The most common benign tumor is the papilloma, which has a distinct incidence of malignant degeneration.

Lacrimal Gland Lesions

The lacrimal gland is histologically similar to the salivary gland and can be involved by a comparable spectrum of pathologic entities. Lymphoid and inflammatory lesions comprise about 50% of lacrimal masses whereas epithelial tumors represent the other 50%. Inflammatory conditions (dacryoadenitis) such as Mikulicz (nonspecific swelling of the lacrimal and salivary glands in association with conditions such as sarcoid, tuberculosis, and leukemia) and Sjögren syndrome (lymphocytic infiltration and enlargement of the lacrimal and salivary glands associated with connective tissue diseases), lymphoid hyperplasia, and acute dacryoadenitis appear as glandular enlargement with a homogeneous density on CT. Generalized contrast enhancement can be seen. Sarcoid can commonly involve the associated lacrimal gland and is associated with enlargement and enhancement (see Fig. 9-32). Dacryoadenitis may also be caused by viral infections.

Common presentations of patients with lacrimal gland tumors (Box 9-9) include a palpable lacrimal fossa mass or proptosis. Epithelial tumors include the benign pleomorphic adenoma (most common benign tumor). Pleomorphic adenomas may have cystic spaces without lytic destruction of adjacent bone. With long-standing lesions, remodeling, lysis, or excavation of bone may be present. Minimal or moderate enhancement can be observed.

Malignant adenoid cystic carcinoma (most common epithelial malignant tumor) presents with pain, diplopia, and visual loss during a 3- to 6-month period. Bone involvement is common. The overall 5-year survival is 21% and death is usually secondary to intracranial spread. This tumor has a propensity for perineural spread proceeding into the cavernous sinus.

Other tumors involving the lacrimal gland include lymphoma, mucoepidermoid carcinoma, adenocarcinoma, malignant mixed cell tumors, squamous cell carcinoma, undifferentiated (anaplastic carcinoma, sebaceous carcinoma, and metastasis). Mucoepidermoid carcinoma displays high density on plain CT that markedly enhances. Along with sebaceous carcinoma, it may reveal high intensity on T1WI. The other lesions have variable CT/MR characteristics.

A mass lesion in the lacrimal fossa that does not produce bony erosion is most likely lymphoid (lymphoma or benign pseudolymphoma) or inflammatory whereas epithelial neoplasms generally involve bone. Lymphoma demonstrates diffuse homogeneous involvement of the lacrimal gland that molds to the bony orbit or globe. MR appearances of these lesions are nonspecific.

Mucoceles

Mucoceles are expansile sinonasal lesions than can present with extraconal mass effect, diplopia, and proptosis. The common locations are the frontal and ethmoidal sinuses. Patients have a history of sinusitis, allergy, or trauma. CT manifests an isodense smooth mass (with an

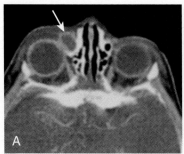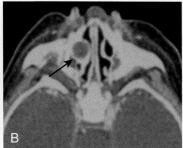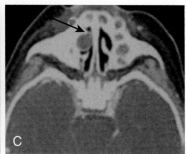

FIGURE 9-35 Ee eye ee eye ow! Dacrocystitis. **A-C,** Peripherally enhancing fluid collection *(arrows)* arising from the nasolacrimal duct canal in this little one indicates infected dacrocystocele. Note associated preseptal soft-tissue swelling.

enhancing rim) centered in the sinus. Bowing and thinning of the bony margins of the orbit can be seen. MR shows the anatomic limits of the lesion (Fig. 9-37). Mucoceles have variable intensities depending on the protein concentration and viscosity but commonly are observed to be high intensity on T1WI and T2WI. They demonstrate peripheral enhancement as opposed to neoplasms, which have solid enhancement. These are discussed further in Chapter 12.

Dermoids

Dermoid cysts (Fig. 9-37) are the most common benign congenital lesion of the orbit (infantile hemangiomas are the most common congenital tumors), accounting for 1% to 2% of all orbital masses. They usually present in the first decade of life but can be subclinical until adulthood, where they may present by rupturing, inducing granulomatous inflammation, and scar formation. They arise from epithelial rests, most often in the superolateral portion of the orbit at the frontozygomatic suture near the lacrimal fossa, but are not true lacrimal gland tumors. They can also arise medially, inferiorly, or posteriorly. These lesions are extraconal and displace the globe medially and inferiorly. CT exhibits either no enhancement or a thin enhancing margin with a low-density fatty center (like a tootsie pop). Bony scalloping is present, and partial marginal calcification can sometimes be identified. On MR, the diagnosis is clinched by high signal on T1WI in this region, which suppresses with fat saturation techniques. Fat may be seen floating in cystic fluid on T1WI. The differential diagnosis of orbital cysts is provided in Box 9-10.

Orbital Rhabdomyosarcoma

Orbital rhabdomyosarcoma is the most common primary malignant orbital tumor of childhood as well as the most common site of head and neck rhabdomyosarcomas. Mean age is 7 to 8 years with 90% occurring before the age of 16 years. Children present with rapidly progressive painless exophthalmos although about 10% may have headache or periorbital discomfort. On examination, a mass may be palpable and ecchymosis, conjunctival chemosis, and ophthalmoplegia may be present. Primary orbital tumor arises from primitive orbital mesenchymal elements whereas secondary involvement occurs from the extraocular muscles, nasopharynx, or paranasal sinuses. It is isodense or slightly high density and uniformly enhances. It is usually seen behind the globe (50%), although other locations—above the globe (25%), below the globe (12%)—are

---

BOX 9-9 Lacrimal Gland Enlargement

INFLAMMATION
Collagen vascular lesions
Dacryoadenitis
Mikulicz/Sjögren
Other granulomatous disorders
Pseudotumor
Sarcoidosis
Wegener granulomatosis

NEOPLASMS
Germ cell tumors: dermoid, epidermoid
Lymphoma, leukemia
Metastasis
Minor salivary gland (epithelial) tumors
Sarcoma

---

common. Bone destruction has been reported with intra-orbital extension of paranasal rhabdomyosarcoma. The tumor appears as a homogeneous well-defined enhancing mass without definite density/intensity characteristics. MR reveals the extent of tumor spread, and this has important prognostic implications.

Fibrous Histiocytoma

Although less than 1% of all primary orbital tumors, fibrous histiocytoma is the most common primary orbital mesenchymal tumor in adults. The clinical presentation includes proptosis, diplopia, and a palpable mass. The tumor is seen in patients in their forties and fifties. They are mostly benign or locally aggressive; however, malignant forms also occur. Histologically, the lesions are combinations of histiocytes and fibroblasts. These lesions may be discrete, smooth masses that can be intraconal or extraconal and usually enhance. They may also display an infiltrating pattern; however, the behavior of the lesion is not necessarily related to its appearance. Other rare mesenchymal orbital tumors include fibroma, fibromatosis, fibrosarcoma, solitary fibrous tumor, leiomyoma, leiomyosarcoma, lipoma, liposarcoma, and mesenchymal chondrosarcoma.

Metastases

Metastases account for approximately 10% of orbital neoplasms with an average survival after detection of approximately 9 months. Patients complain of diplopia, ptosis,

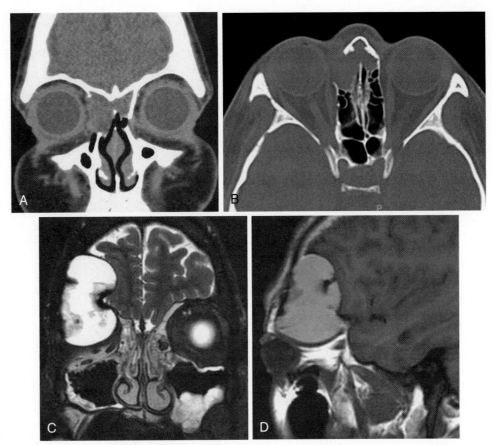

FIGURE 9-36 Sinusitis does not start with an (I) eye! Computed tomography of mucocele. **A,** Frontoethmoidal mucocele with thinning of the bony wall of the sinus. **B,** Note the dehiscent wall of the ethmoid sinus. **C,** Different patient with right frontal mucocele demonstrating bright signal on T2-weighted imaging (T2WI) and high signal on T1WI **(D)** with encroachment on the globe.

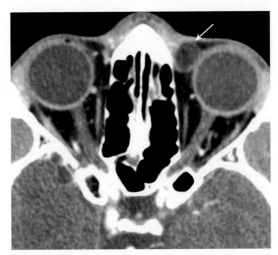

FIGURE 9-37 Computed tomographic image of orbital dermoid: a sight for sore eyes. There is a smoothly marginated nonenhancing fluid-attenuating mass *(arrow)* abutting the medial left globe. If we saw fat in the lesion, the diagnosis would be a no-brainer. However, identification of fat is not always appreciable on imaging, especially when it's microscopic.

proptosis, eyelid swelling, pain, and visual loss. Breast and lung cancers account for over 50% of orbital metastases. Breast is thought to initially involve orbital fat whereas prostate goes to bone. Most metastatic lesions with time

---

BOX 9-10   Differential Diagnosis of Orbital Cysts

Dacryocystocele
Dermoid cyst
Easter-cele
Encephalocele
Epidermal inclusion cysts
Hematic cyst
Lymphangioma
Meningocele
Mucocele
Teratoma

---

produce diffuse lesions; however, thyroid, carcinoid, and renal cell may remain as discrete nodules.

The bones of the orbit harbor most metastases and the greater wing of the sphenoid is the most common site, affecting the lateral extraconal space. CT depicts the orbital and cranial soft-tissue components of the lesion. Metastatic lesions are isodense or high density on unenhanced scan and may enhance. Although CT is generally nonspecific for histologic findings, a picture of an infiltrative retrobulbar mass and enophthalmos is characteristic of scirrhous breast carcinoma. Metastatic disease may account for 7% of cases of extraocular muscle enlargement found on CT. Isolated enlargement of the lateral

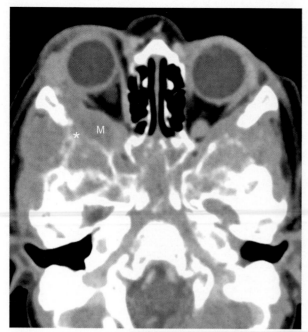

FIGURE 9-38 Naughty neuroblastoma metastases. Sutural based metastases centered at the sphenozygomatic suture bilaterally (indicated with *asterisk* on right) are associated with bulky extraosseous soft-tissue tumor extending into the extraconal orbit bilaterally (indicated with *M* on right). There is resulting proptosis on the right. Note extraosseous tumor extension into the masticator space bilaterally and mottled appearance of the central skull base due to additional bony metastases.

rectus should be thought to be secondary to metastasis, orbital pseudotumor or infection, as it does not occur in Graves.

Metastatic neuroblastoma is second to rhabdomyosarcoma as the most frequent malignant orbital tumor in childhood with 8% presenting initially with orbital lesions. Metastasis to the bone displaces or elevates the periosteum, producing a smooth extraconal mass (Fig. 9-38). The neuroblastoma metastasis may be found at bony sutures/growth plates. These lesions can be high density on CT or display MR characteristics of blood secondary to intratumoral hemorrhage. Neuroblastoma can be distinguished from rhabdomyosarcoma by its high-density values and its lack of preseptal extension (which is much more common in rhabdomyosarcoma). In children, also think of Ewing sarcoma (especially with sudden proptosis and orbital hemorrhage), which commonly involves the sphenoid wing and has extensive soft-tissue multicompartmental components involving the middle cranial fossa, the posterior lateral portion of the orbit, and the soft tissues of the temporal region. Oh, and eosinophilic granuloma, more punched out….

Sphenoid Wing Meningioma

Sphenoid wing meningioma (Fig. 9-39, *A* and *B*) presents as a mass associated with hyperostosis mass displacing the muscles and causing proptosis. It can have a sizable component both in the orbit and intracranially. Proptosis from extraosseous tissue, a blistered appearance to the bone, or purely intraosseous mass can manifest as meningiomas. A number of fibro-osseous lesions also give a somewhat similar

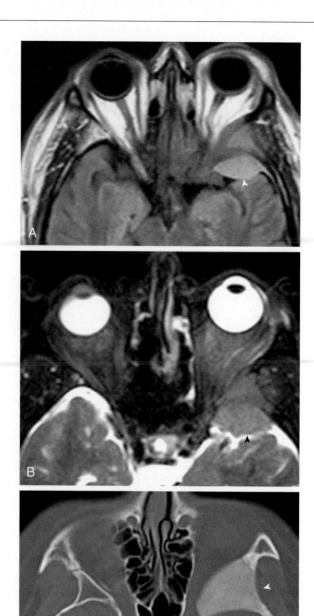

FIGURE 9-39 Winging it. Sphenoid wing meningioma (*arrowheads*) on **(A)** fluid-attenuated inversion recovery (FLAIR) and **(B)** T2-weighted imaging shows an intracranial component as well as an osseous component. Both meningiomas and fibrous dysplasia (**C**, *arrowhead*) can lead to proptosis when the sphenoid wing is affected.

radiologic appearance including ossifying fibroma (which has discrete bony margins and is monostotic), osteoma, sclerotic metastases such as prostate, and fibrous dysplasia (which has poorly defined margins and may be polyostotic).

Primary Bone Lesions

In fibrous dysplasia, normal bone is replaced by immature bone and osteoid in a cellular fibrous matrix. Pain, swelling, and disfigurement occur. Malignant transformation is rare and is associated with previous radiation. In about 20% of cases, craniofacial bones are involved. Most orbital lesions are monostotic and affect the sphenoid wing (see Fig. 9-39, *C*) but the disease entity may involve multiple skull bones and cross suture lines. This disease can occur in adults as well as adolescents and children.

The typical appearance is that of "ground glass" on CT. Differential diagnosis includes Paget disease and meningioma, which produces homogeneous thickening of bone, and often displays soft-tissue involvement rarely seen in fibrous dysplasia.

Primary bone lesions of the orbital walls include eosinophilic granuloma and other fibro-osseous lesions of the orbit including osteoma, ossifying fibroma, osteoblastoma, osteosarcoma, osteoclastoma, brown tumor of hyperparathyroidism, aneurysmal bone cyst, and giant cell reparative granuloma. Use the clinical information including patient's age and presentation to help guide your differential. Pinpointing a specific diagnosis is not always possible, but as with bone lesions elsewhere, a discussion about the extent of the lesion, zone of transition, presence of extraosseous soft-tissue component, and presence of remodeling versus destruction can help narrow the differential and guide clinical management. Excuse the superficial treatment of bone tumors, but what do you expect from neuroradiologists?

Chapter 10
Sella and Central Skull Base

ANATOMY

True story: in order to understand imaging of the skull base you are going to have to appreciate the normal anatomy first. We define the skull base as the region from the upper surface of the ethmoid bone and orbital plate of the frontal bone to the occipital bone. Central to the skull base is the sphenoid bone—the main attraction, so to speak. The bone itself has the appearance of a bat with its wings extended (Fig. 10-1). The feet of the bat are the medial and lateral pterygoid processes, the head being the body of the sphenoid bone, and wings being the greater and lesser wings of the sphenoid. The body of the sphenoid bone is just behind the cribriform plate of the ethmoid bone.

The roof of the sphenoid sinus, the planum sphenoidale, is anterior to the sella turcica and connects the two lesser wings of the sphenoid. The posterior aspect of the planum sphenoidale is termed the *limbus* of the planum sphenoidale. Just posterior to the limbus is the chiasmatic groove; then a bony prominence, the tuberculum sellae; and then the sella turcica (Fig. 10-2). The pituitary gland sits in the sella turcica, which is bounded anteriorly by the chiasmatic groove (the optic chiasm is not located here; however, the lateral portions of the sulcus lead to the optic canals), the tuberculum sellae, and the anterior clinoid processes (part of the lesser wing of the sphenoid), onto which the tentorium cerebelli attaches. The posterior boundary of the sella is the dorsum sellae, from which arises the posterior clinoid processes, onto which the tentorium and petroclinoid ligaments (from the petrous apex) also insert.

Inferior to the dorsum sellae is the clivus, which extends inferiorly to the foramen magnum. Anteriorly, the clivus merges with the sphenoid sinus. Its lateral margins are the petrooccipital fissures. Beneath the sella is the sphenoid sinus, which is usually separated asymmetrically by a vertical bony septum. The high-riding ethmoid sinus variant, Onodi cell, can occasionally be seen above the sphenoid sinus and be intimately associated with the anterior clinoid and optic canal.

The sphenoid sinus displays a wide range of normal variations including asymmetric expansion of its lateral recess into the pterygoid plate or the greater wing of the sphenoid bone. The sinus wall adjacent to the groove for the carotid artery can be quite thin normally.

The lateral surface of the sphenoid body joins with the greater wings of the sphenoid and the medial pterygoid plates. The superior margin of the junction of the sphenoid body with the greater wings of the sphenoid is the carotid sulcus, over which the carotid artery runs. The inner surface of the greater wings of the sphenoid forms part of the floor of the middle cranial fossa and the posterior part of the lateral wall of the orbit.

The pterygopalatine fossa (PPF) is an important conduit for the spread of tumor and infection in and around the skull base. This region can be easily recognized on axial computed tomography (CT; Fig. 10-3). The PPF is bounded anteriorly by the maxillary bone, anteromedially by the perpendicular plate of the palatine bone, and posteriorly by the base of the pterygoid process. The numerous communications of the PPF with other skull

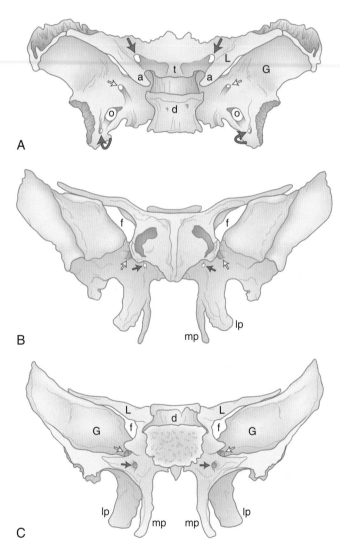

FIGURE 10-1 Diagram of the sphenoid bone. **A,** Superior view. **B,** Anterior view. **C,** Posterior view. Anterior clinoid (a), tuberculum sellae (t), optic canal *(large arrows),* foramen spinosum *(curved arrows),* foramen ovale (o), foramen rotundum *(small open arrows),* dorsum sellae (d), lesser wing of sphenoid (L), greater wing of sphenoid (G), vidian canal *(small closed arrows),* medial pterygoid plate (mp), lateral pterygoid plate (lp), superior orbital fissure (f).

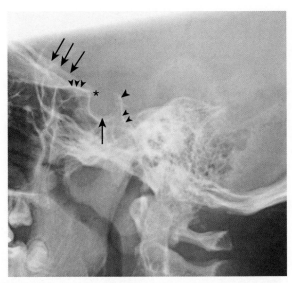

FIGURE 10-2 Lateral radiograph of the sella. You can appreciate the floor of the anterior cranial fossa *(triple black arrows)*, the planum sphenoidale *(triple black arrowheads)*, the anterior clinoid process *(asterisk)*, the sella turcica *(single black arrow)*, the dorsum sellae *(large single arrowhead)*, and the clivus *(double black arrowheads)*.

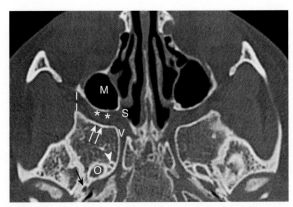

FIGURE 10-3 Anatomy of pterygopalatine fossa (PPF). Axial computed tomography scan shows the pterygopalatine fossa *(asterisks)* bounded anteriorly by the posterior wall of the maxillary sinus (M), and bounded posteriorly by the base of the pterygoid process *(white arrows)*. The PPF communicates medially with the sphenopalatine foramen (S), and laterally with the masticator space through the pterygomaxillary fissure *(dashed line)*. Also indicated on this image are vidian canal (V), foramen ovale (O), foramen spinosum *(black arrow)*, and foramen of Vesalius *(arrowhead)*.

base foramina allow for spread of disease into the orbit, nasal cavity, infratemporal fossa, and hard palate, as well as intracranially. Specifically, the PPF communicates with the orbit via the inferior orbital fissure. The PPF communicates medially with the sphenopalatine foramen (entering the posterosuperior nasal fossa); laterally with the pterygomaxillary fissure (leading to the masticator space); superoposteriorly with the foramen rotundum (and therefore the Meckel cave and the cavernous sinus); inferoposteriorly with the vidian canal (which communicates with the region of the foramen lacerum); and inferiorly with the greater and lesser palatine canals and foramina (to the hard palate).

TABLE 10-1   Major (and Some Minor) Foramina at the Base of the Skull and Their Contents

Foramen	Contents
Superior orbital fissure	Cranial nerves III, IV, first division of V, and VI; orbital branch of middle meningeal artery; sympathetic nerve; recurrent meningeal artery, superior ophthalmic vein
Optic canal	Optic nerve, ophthalmic artery
Inferior orbital fissure	Infraorbital artery, vein, and nerve (branch of second division of cranial nerve V)
Foramen rotundum	Second division of cranial nerve V, artery of foramen rotundum, emissary veins
Foramen ovale	Third division of cranial nerve V, lesser petrosal nerve, accessory meningeal artery, emissary veins
Foramen spinosum	Middle meningeal artery and vein, recurrent branch of third division of cranial nerve V, lesser superficial petrosal nerve
Foramen lacerum	Meningeal branch of ascending pharyngeal artery, nerve of pterygoid canal
Foramen of Vesalius	Emissary vein from cavernous sinus to pterygoid plexus
Vidian canal	Vidian artery and nerve
Jugular foramen	Pars nervosa: cranial nerve IX, inferior petrosal sinus Pars vascularis: Cranial nerves X and XI; jugular bulb
Hypoglossal canal	Cranial nerve XII, hypoglossal persistent artery (in rare instance when it is present)
Pterygopalatine fossa	Pterygopalatine ganglia (V_2); pterygopalatine plexus
Foramen magnum	Medulla oblongata; vertebral artery, anterior spinal artery, posterior spinal artery.

Table 10-1 lists important foramina at the base of the skull and their contents. You should know these inside and out, upside down and right-side up, in your sleep and when you wake...you get the picture. Let us start from below and work our way up.

The hypoglossal canal courses obliquely within the occipital bone (Fig. 10-4). Through it runs the hypoglossal nerve and, occasionally, the persistent hypoglossal artery (a primitive connection between the proximal cervical internal carotid artery at approximately C1-C2 level and the proximal basilar artery). The meningeal branch of the ascending pharyngeal artery as well as a small emissary vein (anterior condyloid) arising from the inferior petrosal sinus may inconstantly also run through this foramen. The jugular tubercles separate the hypoglossal canal from the jugular foramen with the two regions being about 8 mm apart on the inner surface of the skull. Intracanalicular enhancement on magnetic resonance (MR) is always present representing multiple emissary venous radicles, and linear filling defects in the enhancing regions are the hypoglossal nerve rootlets. In addition, dural enhancement can be seen along the margins of the entrance of the canal and

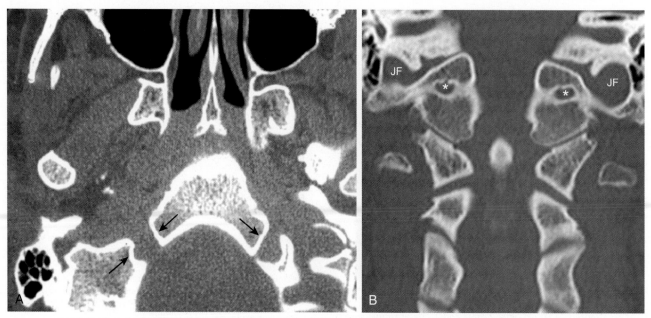

FIGURE 10-4 A, The hypoglossal foramen is outlined by *arrows*. **B,** Coronal reconstruction from cervical spine computed tomography shows the hypoglossal canal indicated by *asterisks*. It is separated from the jugular foramen (JF) along its superolateral margin by the jugular tubercles (often referred to as the eagle's head on coronal images).

<table>
<tr><td>

BOX 10-1   Lesions Involving the Hypoglossal Nerve and Canal

Schwannoma
Meningioma
Metastasis
Chordoma
Large glomus jugulare neoplasm
Perineural spread of tumor
Myeloma

</td></tr>
</table>

anteriorly into the carotid space. Box 10-1 lists the lesions involving the hypoglossal canal.

The jugular foramen is demarcated by the petrous portion of the temporal bone anterolaterally and by the occipital bone posteromedially (Fig. 10-5). It is divided into two parts, the pars nervosa (anteromedial) and the pars vascularis (posterolateral), by a bony or fibrous septum (jugular spur). Cranial nerve IX runs lateral to the inferior petrosal sinus within the pars nervosa portion of the jugular foramen. The inferior petrosal sinus runs posterolaterally along the petrooccipital fissure to the pars nervosa and then into the jugular vein (within the pars vascularis). The pars vascularis is the larger of the two compartments and contains cranial nerves X and XI in a common sheath medial to the jugular bulb, which is also in the pars vascularis (yes, it is ironic that two nerves run through the pars vascularis versus the pars nervosa). The jugular bulb is the confluence between the sigmoid sinus and the jugular vein. It is usually larger on the right side. The petrous portion of the carotid artery is anterolateral to the pars nervosa.

The internal auditory canal is just superior to the jugular foramen. It contains cranial nerves VII and VIII. These nerves are discussed in detail in Chapter 11 (Temporal

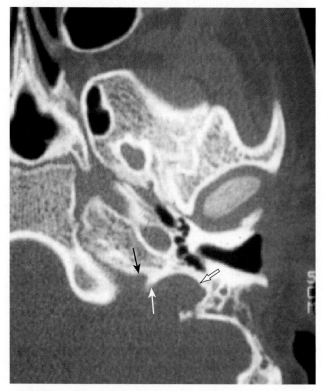

FIGURE 10-5 Jugular foramen. Note the pars nervosa anteromedially *(black arrow),* and the pars vascularis posterolaterally *(open arrow).* Between them is the jugular spur *(white arrow).*

Bone). Other skull base foramina in the temporal bone including vestibular and cochlear aqueducts are also described in this same chapter.

The inferior petrosal sinus can be visualized on contrast-enhanced CT or MR (Fig. 10-6). The basilar venous plexus

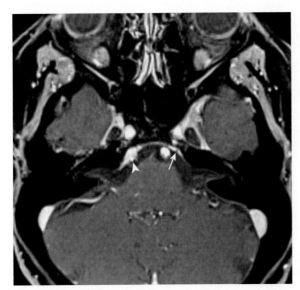

FIGURE 10-6 Inferior petrosal sinus. On contrast-enhanced magnetic resonance imaging, the inferior petrosal sinus is behind the clivus and enhances. The sixth cranial nerve can be seen as a "filling defect" *(arrow)* within the enhancing left inferior petrosal sinus as it courses the Dorello canal. There is an enhancing dural-based mass at the level of the Dorello canal on the right *(arrowhead)* felt to represent meningioma, accounting for this patient's clinical presentation of right abducens palsy.

connects the superior portions of the inferior petrosal sinuses. Dorello canal is located just below the petrous apex and is a conduit for cranial nerve VI to reach the cavernous sinus. The canal is located within the inferior petrosal sinus and can be observed on contrast-enhanced axial MR as an unenhanced line crossing the enhancing sinus obliquely. There may be asymmetry and differences in size in this structure.

The abducens nerve (cranial nerve VI) exits the pontomedullary sulcus, courses through the subarachnoid space, enters the Dorello canal, and passes into the cavernous sinus running just lateral to the intracavernous internal carotid artery. Exiting the cavernous sinus, it then enters the orbit through the superior orbital fissure and terminates on the lateral rectus muscle.

The dorsal meningeal artery (from the meningohypophyseal trunk), or a branch of it, may also run through the Dorello canal. It is located between two dural layers and demarcates an interdural venous confluence. The cranial nerve VI courses in this venous confluence and is separated from blood by a dural and/or arachnoidal sheath. The posterior portion of the cavernous sinus, the lateral basilar sinus along the clivus, and the superior petrosal sinus drain this region, which then forms the inferior petrosal sinus draining into the jugular bulb.

There are conditions that produce abducens palsy precisely because of fixation of the nerve in the Dorello canal. These include nerve injury caused by brain stem shifts from trauma or mass lesions, and Gradenigo syndrome (cranial nerve VI palsy associated with inflammatory lesions of the petrous apex and facial pain caused by involvement of cranial nerve V as it crosses the petrous apex). Increased venous pressure in the PVC from carotid-cavernous fistula and dural malformations may compress and injure the nerve.

The foramen lacerum is not a true foramen and the carotid artery does not run through it. Rather, the carotid artery runs over the fibrocartilage (making up the endocranial floor of the foramen lacerum) on its way to the cavernous sinus.

The greater superficial petrosal nerve (GSPN) is a branch of the facial nerve that innervates the lacrimal glands and mucous membranes of the nasal cavity and palate. It is a mixed nerve containing sensory and parasympathetic fibers. The parasympathetic fibers exit the brain stem as the nervus intermedius of cranial nerve VII. The GSPN courses anteromedially from the geniculate ganglion and exits the facial hiatus in the petrous bone. It passes under the gasserian ganglion in the Meckel cave and goes forward to the region of the foramen lacerum. Here it merges with the deep petrosal nerve, arising from the sympathetic carotid plexus, and forms the vidian nerve. This nerve runs anteriorly in the vidian canal with the parasympathetic fibers synapsing in the pterygopalatine ganglia and the sensory fibers passing through the ganglion to the nasal cavity and palate. The vidian canal connects the pterygopalatine fossa anteriorly to the foramen lacerum posteriorly and transmits the vidian artery (see Fig. 10-3). The vidian artery, a branch of the maxillary artery, joins the carotid artery in its petrous segment.

The foramen of Vesalius (just saying it out loud makes you feel fancy, doesn't it?) is an inconstant emissary foramen that can be seen anterior and medial to foramen ovale. Besides the emissary vein, the ascending intracranial branch of the accessory meningeal artery can enter the middle cranial fossa through the foramen of Vesalius or the foramen ovale (see Fig. 10-3).

On either side of the sella is the cavernous sinus, a trabeculated venous plexus containing cranial nerves III, IV, VI, and the first and second divisions of V (Figs. 10-7, 10-8). These are located in the lateral portion of the sinus. Cranial nerves III, IV, and the first and second divisions of V are in the lateral wall of the cavernous sinus and maintain that order from superior to inferior in the coronal plane. Cranial nerve VI is medial in the cavernous sinus but lateral to the cavernous carotid artery.

Now pay attention because this is important! Cranial nerve V exits the ventral pons as separate motor and sensory roots at the "root entry zone," an area often compressed from above by the superior cerebellar arteries, and less commonly by other basilar branches, which may cause symptoms of trigeminal neuralgia. The roots run forward together through the prepontine cistern and exit through the porus trigeminus of the petrous apex. These roots enter the trigeminal cistern (the space containing cerebrospinal fluid [CSF]), which is in the Meckel cave, a dural invagination at the posterior aspect of the cavernous sinus. The dural layers of the Meckel cave demonstrate thin peripheral enhancement. In addition, a discrete semilunar enhancing structure within the inferolateral aspect of the Meckel cave representing the gasserian (aka trigeminal) ganglion has been observed to enhance suggesting the lack of a blood-nerve barrier. The gasserian ganglion is a meshwork of sensory neural fibers permeated by CSF from the trigeminal cistern. On CT or magnetic resonance imaging (MRI) the CSF in the trigeminal cistern is clearly visualized, and with high resolution MR the nerve fibers can be seen. The three

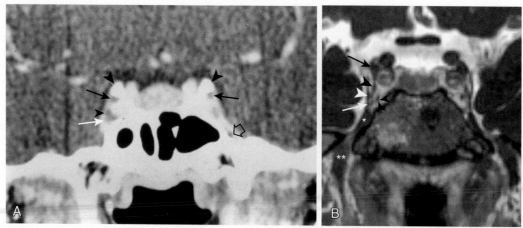

FIGURE 10-7 A, Cranial nerves in cavernous sinus. Enhanced computed tomography in coronal plane shows cranial nerves in cavernous sinus. The cranial nerves appear as filling defects within the enhancing cavernous sinuses. Cranial nerve III *(black arrows)* is directly under the anterior clinoid process *(large arrowheads)*. Also identified on the patient's right side are cranial nerves IV *(arrowhead)* and the first division of cranial nerve V *(white arrow)*. On the left side, the second division of cranial nerve (CN) V is marked *(open arrow)*. **B,** Coronal contrast-enhanced constructive interference in steady state image shows the cranial nerves are dark filling defects within the enhancing hyperintense cavernous sinuses. Cranial nerves III *(black arrow)*, IV *(black arrowhead)*, V1 *(white arrowhead)*, V2 *(white arrow)*, and VI *(double black arrowheads)* are indicated. Also seen are more distal portions of V2 headed toward foramen rotundum *(single asterisk)* and V3 *(double asterisk)* extending inferiorly from foramen ovale.

sensory divisions of the trigeminal nerve leave the gasserian ganglion, with the first and the second divisions running in the lateral wall of the cavernous sinus to exit the superior orbital fissure (along with cranial nerves III, IV, and VI and the superior ophthalmic vein) and foramen rotundum, respectively. The motor root passes under the gasserian ganglion and, after it exits foramen ovale, combines with its sensory root counterpart to form the mandibular nerve.

The superior and inferior ophthalmic veins drain into the cavernous sinus via the superior and inferior orbital fissures, respectively; however, there are many variations of this venous drainage pattern. The cavernous sinus is formed by two layers of dura mater. The periosteal layer forms the floor and most of the medial wall, and the meningeal layer (dura propria) forms its roof, lateral wall, and the upper part of its medial wall. The lateral wall may have two layers of dura: a deep layer, which ensheathes cranial nerves III and IV and first and second divisions of cranial nerve V, and a superficial dural layer. In addition, like most other venous structures in the body, the cavernous sinus has many variations and much controversy about its exact internal venous anatomy. It has been reported that the true cavernous sinus (a large venous channel surrounding the internal carotid artery) exists in only 1% of patients. In the other instances the cavernous sinus is formed by numerous small veins including (1) the veins of the lateral wall, (2) the veins of the inferolateral group, (3) the medial vein, and (4) the vein of the carotid sulcus. In point of fact, enhancement within the cavernous sinus can be asymmetric because of these anatomic variations, and this asymmetry alone should not prompt a diagnosis of cavernous sinus thrombosis without other supportive imaging findings. Fat in the cavernous sinus is a frequent occurrence.

The cavernous sinus enhances dramatically within 30 seconds after contrast injection for CT or MR. The cavernous sinus drains into the superior and inferior

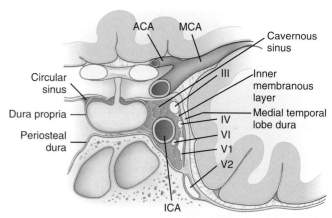

FIGURE 10-8 Cavernous sinus diagram in coronal view. The cavernous sinus is in blue. Cranial nerves (III, IV, V₁, V₂, VI). *ACA,* Anterior cerebral artery; *ICA,* internal carotid artery; *MCA,* middle cerebral artery.

petrosal sinuses. Many venous connections exist between the cavernous sinuses around the sella. The basilar venous plexus, the largest intercavernous connection, lies within the dura behind the clivus connecting the two cavernous sinuses and the superior and inferior petrosal sinuses. The coronary sinus is located along the roof of the sphenoid sinus and joins the two cavernous sinuses. There are also venous communications between the cavernous sinus and the pterygoid plexus of veins via emissary veins in the foramen ovale and foramen rotundum, and through the inconstant foramen of Vesalius (oh so fancy!). These basilar foramina can be a path (and can demonstrate enlargement) for nasopharyngeal tumors coursing into the cavernous sinus.

The cavernous sinus can be subdivided into an intracavernous and interdural compartments. Cavernous sinus tumors that arise interdurally (within the lateral wall) such as schwannomas of the cranial nerves, epidermoid tumors, melanomas, and cavernous angiomas have smooth contours, oval shape, and displace the intracavernous portion of the internal carotid without encasement or narrowing. Intracavernous lesions include pituitary macroadenomas, meningiomas, hemangiopericytomas, and ganglioneuroblastomas. These lesions tend to encase and the meningiomas may narrow the internal carotid artery. The cavernous sinus may be compressed but not obliterated by interdural lesions whereas it may be obliterated by intracavernous tumors.

The pituitary gland is surrounded by a dural bag with the medial wall of the cavernous sinus being the lateral extent of the dural bag. The gland is divided into anterior and posterior lobes. The anterior lobe of the pituitary (adenohypophysis) is divided into the pars tuberalis, pars intermedia, and the pars distalis. The pars tuberalis consists of thin anterior pituitary tissue along the median eminence and anterior infundibulum. Rarely, suprasellar adenomas and other suprasellar pituitary tumors may originate from this tissue, and it may function after hypophysectomy. The pars intermedia lies between the pars distalis and the posterior lobe of the pituitary. It is noted to contain small cysts (pars intermedia cysts, colloid cysts) and may be the origin of Rathke cleft cysts.

The adenohypophysis secretes prolactin (from lactotrophs), growth hormone (from somatotrophs), thyroid-stimulating hormone (from thyrotrophs), follicle-stimulating hormone and luteinizing hormone (from gonadotrophs), and corticotropin (ACTH) precursor and melanocyte-stimulating hormone (from corticotrophs).

The neurohypophysis is composed of the neural (posterior) lobe, the infundibular stem, and the median eminence. Besides storing antidiuretic hormone and oxytocin, the neural lobe also contains nonsecreting cells termed pituicytes. Their exact role is uncertain. The posterior lobe of the pituitary has a direct blood supply from the inferior hypophyseal artery, a branch of the meningohypophyseal trunk arising from the cavernous carotid. The superior hypophyseal arteries, arising from the supraclinoid internal carotid arteries and posterior communicating arteries (usually not visualized on angiography) supply a plexus around the base of the hypophyseal stalk and median eminence and then supply the anterior lobe of the pituitary indirectly through the pituitary portal system. The implications of this quaint blood supply are that, on dynamic imaging, the posterior pituitary and infundibulum enhance immediately because of their direct blood supply, whereas the anterior pituitary is slightly delayed because of the portal system. The indirect blood supply to the anterior lobe of the pituitary makes it susceptible to ischemia, which can be seen in cases of autoinfarction of pituitary tumors and in postpartum pituitary necrosis (Sheehan syndrome). The venous drainage of the pituitary is into the cavernous sinuses.

The diaphragma sellae is the sheet of dura forming a roof over the sella turcica overlying the pituitary gland. The diaphragm has a central hiatus of variable size through which the infundibulum passes. The portion of the hypophysis located just below the diaphragm is concave superiorly like the region just around the stem of an apple and creates the hypophyseal cistern. This cistern is an expansion of the chiasmatic cistern and is separated from the interpeduncular and prepontine cisterns by the

membrane of Liliequist. This membrane is just below the floor of the third ventricle and is entered to approach a basilar tip aneurysm.

The infundibulum arises from the tuber cinereum (a prominence of the inferior portion of the hypothalamus) and courses in an anterior inferior direction towards the pituitary gland. It is an important landmark in pituitary anatomy, marking the anterior portion of the posterior pituitary gland.

The suprasellar cistern is superior to the diaphragma sellae. This cistern contains the circle of Willis with anterior cerebral arteries, anterior and posterior communicating arteries, and the tip of the basilar artery. Anteriorly, the cistern is bounded by the inferior frontal lobes and the interhemispheric fissure, laterally by the medial portions of the temporal lobes, and posteriorly by the prepontine and interpeduncular cisterns. Lying central in the suprasellar cistern is the optic chiasm, which is anterior to the infundibular stalk. The normal chiasm is about 3 to 4 mm posterosuperior to the tuberculum sellae. In some circumstances the chiasm can overlie either the tuberculum sellae (prefixed optic chiasm, seen in 9% of cases) or the dorsum sellae (postfixed optic chiasm, seen in 11% of cases). Such anatomic anomalies are important with respect to visual symptoms and surgical approach to suprasellar lesions.

The hypothalamus forms the ventral and rostral part of the wall of the third ventricle. The chiasmatic and infundibular recesses of the third ventricle project inferiorly into these respective structures (chiasm and infundibulum). Posterior to the infundibular stalk are the anteroinferior third ventricle and mammillary bodies. The tuber cinereum is the lamina of gray substance from the floor of the third ventricle (hypothalamus) between the mammillary bodies and the optic chiasm.

IMAGING OF THE NORMAL PITUITARY GLAND AND THE PARASELLAR REGION

MR has several advantages over CT in imaging the sellar region; however, CT does have some use. MR can demonstrate the relationship of pituitary lesions to the optic chiasm and cavernous sinuses. It has the capability of distinguishing solid, cystic, and hemorrhagic components of lesions. Preservation or absence of the expected flow voids within major cerebral vasculature in the sellar region can be observed on MRI. Calcification, although usually imaged as low intensity on T1-weighted imaging (T1WI) and T2-weighted imaging (T2WI), is better seen on CT. Bony septa in the sphenoid sinus are also better visualized on CT. This may be important if a transsphenoidal surgical approach is being considered.

In general, sagittal and coronal T1WI before and following gadolinium administration with thin sections (<3 mm) is all that is necessary to image the pituitary gland. Dynamic postcontrast scanning with serial imaging as the gadolinium infuses the pituitary gland has been shown to identify an additional 20% of microadenomas over static imaging. T2WI can occasionally add additional information for the differential diagnosis by providing the intensity characteristics of a particular lesion. This is not necessarily essential in typical cases of "rule out microadenoma."

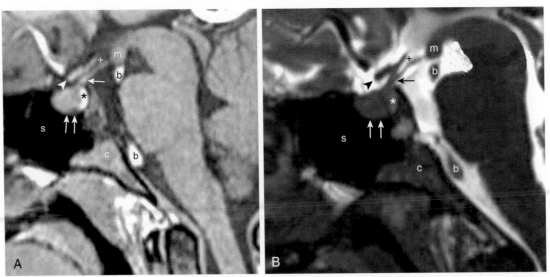

FIGURE 10-9 Normal sellar anatomy. Sagittal T1-weighted **(A)** and sagittal T2 constructive interference in the steady state (CISS) **(B)** images show the posterior pituitary bright spot *(asterisk)*, infundibular recess (+), infundibulum *(arrow)*, chiasm *(arrowhead)*, mammillary body (m), basilar artery (b), clivus (c), floor of sella *(double arrows)*, and sphenoid sinus (s).

In women, the maximal height of the pituitary has been reported as 9 mm whereas in men it is 8 mm. In children younger than 12 years of age, the gland should be 6 mm or less, with its upper surface flat or slightly concave. The gland changes shape and size during puberty and pregnancy and lactation up to 12 mm because of physiologic hypertrophy. After the first postpartum week, the gland rapidly returns to normal. In teenaged girls, it may measure up to 10 mm in height, and convex upper margins may be identified. This can be noted in teenaged boys but appears to be less striking. Convexity has been observed in children with precocious puberty. Similar to some other "organs," the gland gradually decreases in size after the age of 50 years.

Intensity is important in MR diagnosis. The anterior lobe of the pituitary gland is isointense to brain on T1WI and T2WI (Fig. 10-9). However, in children younger than 2 months of age the pituitary is rounder, larger, and of high intensity on T1WI. This is most likely related to its high level of metabolic and hormonal function during early infancy, although it has been suggested that the high-intensity results from an increase in the bound fraction of water molecules caused by hormone secretion. Hyperintensity on T1WI during pregnancy has also been noted. Reversible hyperintensity has been reported in patients receiving parenteral nutrition (as seen with the basal ganglia secondary to manganese deposition). Iron can accumulate in the anterior lobe of the pituitary gland in patients with hemochromatosis and produce low intensity on T2WI and gradient echo T2* weighted images (Fig. 10-10).

The posterior pituitary gland is high intensity on the T1WI and of lower intensity on the T2WI (see Fig. 10-9). It is much more conspicuous in younger people and becomes less conspicuous with increase in age. The precise cause of the high signal in the posterior of the pituitary is probably related to the carrier protein (neurophysin) stored in the neurosecretory granules of the posterior pituitary, intracellular lipid in glial cell pituicytes, water interactions with paramagnetic substances, or low molecular weight

molecules such as vasopressin or oxytocin. Posterior to the posterior pituitary is a rim of hypointensity, representing cortical bone of the dorsum. Posterior to this hypointense margin is the hyperintensity of fatty marrow in the clivus. High signal intensity has also been observed in the infundibular stalk on fluid-attenuated inversion recovery (FLAIR) images presumably related to the fluid rich component (prolonged T2) in the pituitary stalk. The high intensity of the posterior pituitary gland has been noted to be absent in patients with diabetes insipidus.

After intravenous contrast injection, enhancement is promptly noted on T1WI in the infundibulum and the cavernous sinuses with more delayed enhancement at the anterior pituitary. Remember that the posterior pituitary is already high intensity, so that any enhancement would be difficult to ascertain. The initial enhancement gradually fades in 20 to 30 minutes or more. The pituitary and cavernous sinuses generally enhance to a similar extent.

INTRASELLAR LESIONS

See list of intrasellar lesions in Box 10-2.

Congenital Lesions of the Pituitary

Classically, it is taught that the anterior lobe develops from a Rathke pouch, a diverticulum of the primitive buccal (oral) cavity. The posterior lobe originates from neuroectoderm and migrates inferiorly from the hypothalamus. A Rathke pouch starts growing toward the brain during the fourth week of gestation. By the eighth week, the connection with the oral cavity disappears and the pouch is in close contact with the infundibulum and posterior lobe of the pituitary. A Rathke cyst is most often located as an intrasellar cyst between anterior and posterior lobes and may grow to the suprasellar location. The remnants of embryological tract can persist in the form of the craniopharyngeal canal (which can be visualized on axial CT with bone windows as a

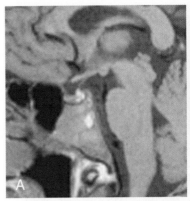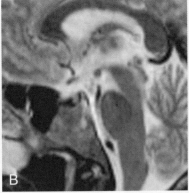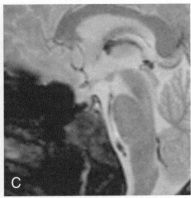

FIGURE 10-10 Hemochromatosis affecting the pituitary. **A,** Sagittal T1-weighted image (T1WI) of the pituitary. Note the bright posterior pituitary and the isointense anterior aspect. **B,** T2WI revealing marked anterior pituitary hypointensity representing iron deposition. **C,** Gradient-echo image emphasizing T2* shows increased susceptibility. Bone marrow abnormality also reflects iron deposition. (Courtesy Y. Miki, MD.)

small foramen in the sphenoid in up to 9% of children younger than 3 months and up to 0.5% of adults) or as ectopic pituitary tissue in the nasopharynx or sphenoid sinus. The craniopharyngeal canal extends from the floor of the sella through the sphenoidal septum into the vomer. Ectopic craniopharyngioma can therefore arise anywhere along this pathway, but the classic location is in the suprasellar cistern.

Congenital abnormalities of the pituitary include aplasia, hypoplasia, ectopia or duplication (Fig. 10-11). These have been observed to occur alone or with a variety of different developmental syndromes, including septooptic dysplasia; holoprosencephaly; anencephaly, sphenoidal encephalocele; Kallmann syndrome; Pallister-Hall syndrome; CHARGE syndrome (coloboma, heart defects, atresia of the choana, retarded growth and development, and ear anomalies); and 17q, 18p, or 20p chromosomal deletions. For this reason, when congenital abnormalities of the pituitary are observed, it is important to evaluate the brain parenchyma and face for additional congenital malformations and/or anomalies.

Intracranial ectopic pituitary adenoma occurs most frequently in the suprasellar cistern most often contiguous with the pituitary stalk. These lesions result from cells of the pars tuberalis located above the diaphragma sellae or from aberrant pituitary cells. Rarely, a connection with the stalk is not demonstrated and all that is observed is a T1 hyperintense mass in the suprasellar cistern.

Pituitary dwarfism, produced by diminished levels of growth hormone, presents as delayed skeletal maturation, slow growth, short stature, and delayed dentition. More males than females have growth hormone deficiency, and isolated growth hormone deficiency can progress to multiple pituitary hormonal deficiencies. In most cases with only isolated growth hormone deficiency, the infundibulum may be thin or truncated (most common), normal or absent, and the adenohypophysis is either normal or small. Isolated growth hormone deficiency may in some cases just be associated with abnormalities intrinsic to the pituitary cells producing growth hormone or perhaps to partial transections of the infundibulum. Ectopic posterior pituitary glands with isolated growth hormone deficiency are uncommonly seen.

Patients with multiple pituitary hormonal deficiencies, however, do tend to have a small or absent anterior

---

BOX 10-2 Intrasellar Lesions

Congenital
    Rathke cleft cyst
    Pituitary aplasia
    Hypoplasia
    Ectopia
    Duplication
Empty sella
Pituitary adenoma
Pituitary hyperplasia secondary to end-organ failure (primary hypothyroidism)
Lymphocytic adenohypophysitis
Meningioma
Pituitary apoplexy (infarct, hemorrhage)
Rathke cleft cyst
Craniopharyngioma
Aneurysm
Arachnoid cyst
Chordoma
Choristoma
Sarcoid and other granulomatous processes
Metastasis
Infection
Pituitary stone

---

pituitary and/or stalk, with the neurohypophysis being either ectopic or absent. This occurs more commonly after a breech delivery, neonatal hypoglycemia, and PROP1 AG deletions. The absent infundibulum and ectopic location of the normal posterior pituitary bright spot are identified on T1WI near the median eminence (ectopic posterior pituitary bright spot) (see Fig. 10-11). Thus the presence of a thin stalk is very indicative of isolated growth hormone deficiency, whereas its absence strongly suggests multiple pituitary hormonal abnormalities. To evaluate the pituitary stalk, contrast enhancement is essential.

Disruption of the infundibulum, which is distal to the ectopic neurohypophysis, interferes with the hypothalamohypophyseal portal system and anterior pituitary function. In the setting of ectopic posterior pituitary, the neurohypophysis functions normally. This is because communication still exists between the neurohypophysis and the hypothalamus, so that diabetes insipidus is not present.

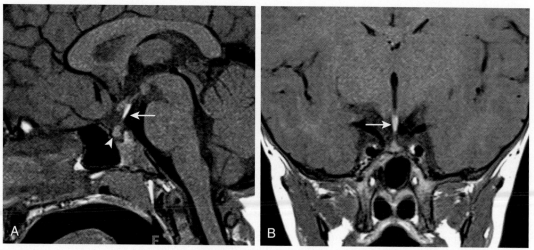

FIGURE 10-11 Ectopic posterior pituitary. **A,** Noncontrast sagittal and **(B)** coronal T1-weighted images show high signal intensity in an ectopic posterior pituitary *(arrow)* along the infundibulum. The normal pituitary bright spot within the sella is absent. Note the rounded configuration of the pituitary gland *(arrowhead)* and shallow configuration of the sella.

Transections lead to bright signal above the cut, because of accumulations of neurosecretory granules.

Pituitary Adenoma

Autopsy series indicate that the pituitary gland can be a reservoir for the "incidentaloma," including asymptomatic microadenomas (14% to 27% of cases), pars intermedia (Rathke) cysts (13% to 22%), and occult metastatic lesions (about 5% of patients with malignancy). This means that clinical input is critical in assessing small lesions of the pituitary because many "normal" patients may have small, insignificant nonsecretory abnormalities visualized on CT or MR. On the other hand, serum prolactin levels of more than 200 ng/mL are highly specific for prolactin-secreting adenomas. Intermediate levels of prolactin (30-70 ng/mL) may be on the basis of glandular compression from nonprolactinoma lesions.

Pituitary microadenomas (<10 mm) are generally hypointense compared with the normal gland on T1WI and display a variable intensity on T2WI. On CT, the microadenoma is of low density compared with the normal gland with or without enhancement. In about 75% of cases, microadenomas present because of symptoms from the hormones they secrete. Nonhormonally active lesions become symptomatic because of their size (macroadenomas), producing headache, visual disturbances (classically bitemporal visual defects), cranial nerve palsy, and CSF rhinorrhea. The diagnosis of pituitary microadenoma can be made without contrast (Fig. 10-12), but microadenomas are more conspicuous with administration of intravenous contrast. In most cases with contrast-enhanced MRI, on T1WI the microadenoma appears hypointense relative to the normally enhancing pituitary gland (Fig. 10-13), especially using dynamic MR after contrast (rapid imaging of the same region repeated for a short time: six to eight thin sections every 30 seconds, for 3 to 5 minutes, typically in the coronal plane). The dynamic images obtained within the first minute appear to provide the greatest contrast between enhancing normal gland and pituitary adenoma that does not initially enhance. If a delayed scan (20 minutes after the injection of contrast) is performed, the tumor

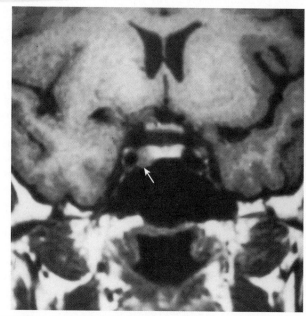

FIGURE 10-12 Microadenoma. Unenhanced coronal magnetic resonance shows lower intensity lesion *(arrow)* compared with normal pituitary.

may appear hyperintense relative to the normal glandular tissue.

Macroadenomas (>1 cm) are obvious abnormalities on unenhanced T1WI. Coronal MR beautifully shows the relationship of the macroadenoma to the optic chiasm, third ventricle, and cavernous sinuses (Fig. 10-14). Macroadenomas have roughly the same signal characteristics as microadenomas; however, they have a propensity for hemorrhage and infarction because of their marginal blood supply. Thus, these tumors can possess a variable intensity pattern. Treatment with bromocriptine increases the likelihood of intratumoral hemorrhages which may be asymptomatic or may be associated with the syndrome of pituitary apoplexy (Fig. 10-15). Cystic regions in macroadenomas produce low intensity on T1WI and high intensity on T2WI. Pituitary

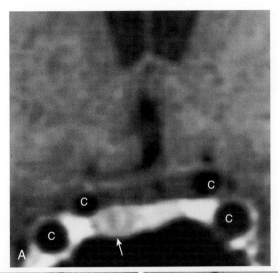

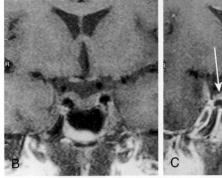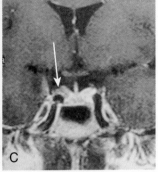

FIGURE 10-13 Postcontrast imaging of microadenoma. **A,** The normal pituitary gland enhances more than the microadenoma *(arrow)*. Note that the carotid artery flow voids (C). **B,** In a different patient, the microadenoma is not perceptible on unenhanced coronal T1-weighted image of the pituitary. **C,** After intravenous contrast, the basophilic adenoma *(arrow)* is now appreciable and the lesion even extends into but not entirely through the cavernous sinus.

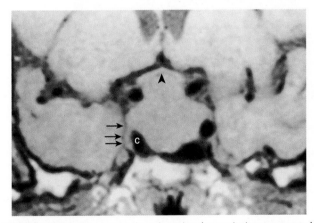

FIGURE 10-14 Cavernous sinus invasion by a pituitary macroadenoma. There is extra soft tissue lateral to the cavernous carotid artery flow void (c). Lateral dural margin of the cavernous sinus *(arrows)* is bowed. Note mass effect upon the floor of the optic chiasm, which is displaced superiorly *(arrowhead).*

adenomas associated with hemorrhage may be confused with craniopharyngiomas or Rathke cysts on MR.

Detection of cavernous sinus invasion may be made with either unenhanced or dynamic enhanced sequences.

If the pituitary tumor remains medial to a line drawn through the mid diameter of the cavernous carotid artery, cavernous sinus invasion is *not* present. If the tumor extends lateral to a line drawn along the lateral edge of the cavernous carotid artery wall, cavernous sinus invasion *is* present (see Fig. 10-14). When the tumor is between those lines, all bets are off, but markedly elevated hormonal levels usually mean the cavernous sinus is violated. If the tumor infiltrates the venous compartment inferomedial to the basal turn of the cavernous carotid artery or greater than 50% of the intracavernous artery is encased, it means that cavernous sinus invasion has occurred. Noninvasion can be assured if (1) there is intervening normal pituitary tissue between the adenoma and cavernous sinus, or (2) less than 25% of the intracavernous carotid artery is encased.

Pituitary macroadenomas can become very large, to the point that it may be difficult to discern whether the tumor is in fact of pituitary or extrapituitary origin. Such adenomas can result in significant mass effect on the chiasm and adjacent brain parenchyma, resulting in obstructive hydrocephalus and herniation syndromes (Fig. 10-16).

Adenomas have rarely been reported at extrasellar sites not in continuity with the sella, including the sphenoid sinus, nasal cavity, petrous bone, and third ventricle.

Pituitary Apoplexy

Pituitary apoplexy is a syndrome that appears suddenly with combinations of ophthalmoplegia, headache, visual loss, and/or vomiting. This syndrome occurs as a consequence of pituitary hemorrhage and/or ischemia most commonly in the setting of a preexisting pituitary tumor. Sheehan syndrome presents as hypopituitarism as a consequence of pituitary ischemia in the peripartum period. Pituitary hemorrhages follow the pattern of intraparenchymal hemorrhages, with acute hemorrhage revealing hypointensity on T2WI and subacute hemorrhages exhibiting high intensity on T1WI (Fig. 10-17). As opposed to most simple intracranial hemorrhages, in which hemosiderin is deposited in the walls of the cavity, pituitary hemorrhage is not associated with hemosiderin deposition.

Metastatic Disease

Metastatic lesions can occur in the pituitary with a frequency reported from 1.8% to 12% of all pituitary lesions; in clinical practice, however, these are rarely identified. The most common primary tumor is breast followed by gastrointestinal carcinoma (Fig. 10-18). Lymphoma can also enlarge the stalk. Usually, one cannot distinguish metastatic lesions from adenomatous disease. Obviously, clinical history of known primary tumor and lesion multiplicity helps suggest the diagnosis of metastatic disease.

Abscess

Abscess can form in the pituitary just as in other parts of the brain. This can occur after surgery but also in situations that predispose to infection, including sinusitis. These are uncommon lesions, and the patient is seen with symptoms such as fever and headache. The abscess produces

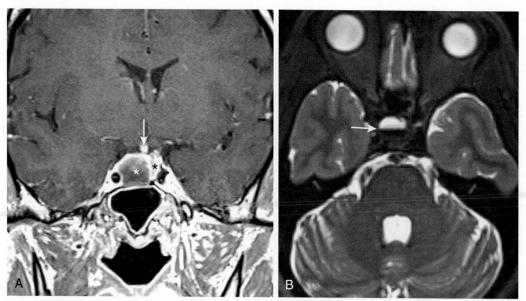

FIGURE 10-15 Macroadenoma with hemorrhage. **A,** Coronal postcontrast T1-weighted image shows a large macroadenoma *(white asterisk)* in the right aspect of the sella. The tumor is hypo-enhancing relative to the enhancing pituitary tissue in the left aspect of the sella *(black asterisk)*. The infundibulum *(arrow)* is displaced towards the left at its inferior extent. **B,** Fluid level within the adenoma on T2-weighted image *(arrow)* indicates intratumoral hemorrhage. In this case, the patient was asymptomatic.

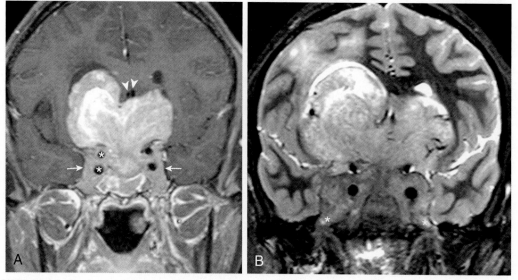

FIGURE 10-16 Giant macroadenoma. **A,** Postcontrast coronal T1-weighted imaging (T1WI) shows enhancing tumor centered within the sella invading the cavernous sinuses bilaterally *(white arrows),* with preservation of the internal carotid artery flow voids *(asterisks)*. There is significant mass effect upon the inferior frontal lobes. The A2 segments of the anterior cerebral arteries are displaced superiorly by the mass *(arrowheads)*. **B,** Coronal T2WI shows edema within the right frontal lobe because of mass effect from this giant macroadenoma. There is left ward midline shift. Note that the tumor has invaded the Meckel cave bilaterally (the normally T2 hyperintense signal within the cave is absent because of tumor invasion, indicated by *asterisks*) and is headed into foramen ovale bilaterally.

compression of surrounding structures. Infection can extend to involve the skull base, leptomeninges, cavernous sinuses, orbits, brain parenchyma, and circle of Willis, so keep your eyes peeled in these situations.

Granulomatous Disease

The pituitary gland can infrequently (except at the Boards) be affected by granulomatous diseases such as giant cell granuloma, Langerhans cell histiocytosis (LCH), and

sarcoidosis. In the case of giant cell granuloma, a pituitary mass is present with associated hypopituitarism and rarely diabetes insipidus. These lesions cannot be differentiated from other pituitary lesions; however, there is an association between giant cell granulomas and granulomas in the adrenal glands and liver. Sarcoidosis can produce intrasellar or suprasellar mass lesions that can masquerade as pituitary adenomas or as meningiomas. Tuberculosis can affect the sella in a similar fashion. Careful review for additional evidence of intracranial tuberculous involvement including

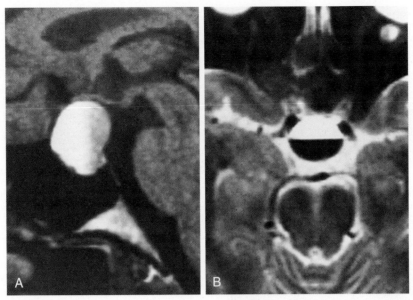

FIGURE 10-17 Pituitary apoplexy. Sagittal T1-weighted image (T1WI) **(A)** of large hemorrhage into a pituitary tumor. There is a fluid level, which is best appreciated on the axial T2WI **(B),** with the hypointense dependent level containing deoxyhemoglobin. In this case, the patient was acutely symptomatic.

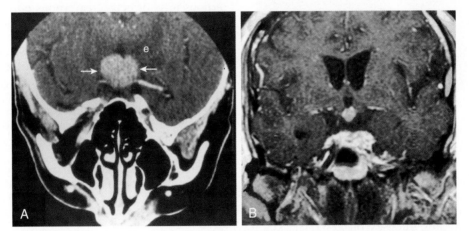

FIGURE 10-18 Sellar and suprasellar metastasis. **A,** Enhanced coronal computed tomography demonstrates suprasellar mass *(arrows)* and associated edema (e). **B,** A coronal postcontrast T1-weighted image in a patient with metastatic hepatocellular carcinoma shows enlargement and enhancement of the pituitary stalk in this patient who presented with diabetes insipidus.

parenchymal and leptomeningeal disease can help distinguish tuberculosis from other pituitary lesions. LCH is the most common cause of infundibular thickening in children; loss of the posterior pituitary bright spot can be seen in these cases.

Lymphocytic Hypophysitis

This is an uncommon inflammatory disease of the pituitary gland that can also involve the infundibulum. It is seen in young women during late pregnancy or in the postpartum period. However, it can also occur in nonpregnant women and men of all ages. The condition has also been termed lymphocytic infundibuloneurohypophysitis (if the neurohypophysis and infundibulum are involved). There is enlargement and enhancement of the pituitary gland and

thickening of the stalk (Fig. 10-19). Endocrinologic abnormalities can include all anterior pituitary hormonal functions, and when the infundibulum and neurohypophysis are affected, diabetes insipidus can ensue. The inflammation has been reported to occasionally extend into the cavernous sinus. Dynamic enhanced MR studies indicate that the blood supply to the posterior pituitary is compromised by the inflammatory process. The enlargement may regress spontaneously or with steroids.

Posterior Pituitary Tumors

Although unusual, tumors of the posterior pituitary gland occur and can often be diagnosed on MR (Fig. 10-20). Pituicytomas and granular cell tumors (choristoma/myoblastoma of the posterior pituitary) may produce visual or

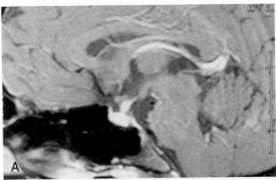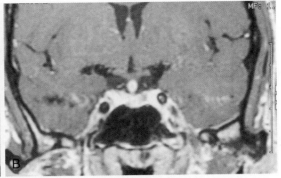

FIGURE 10-19 Lymphocytic adenohypophysitis. Sagittal **(A)** and coronal **(B)** postcontrast T1-weighted images are remarkable for a prominent pituitary gland and vigorous enhancement with enlargement of the stalk. It should not be convex outward.

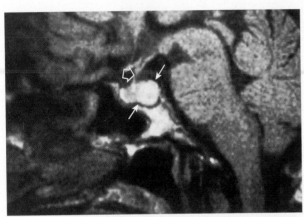

FIGURE 10-20 Pituicytoma. Sagittal T1-weighted image demonstrates this posterior pituitary tumor *(arrows)*. Observe that the mass is behind the infundibulum *(open arrow)* and thus is located in the posterior pituitary and is high intensity.

endocrinologic disturbances. They are of variable intensity on T1WI, proton density–weighted images (PDWI) and T2WI, and they enhance. However, some reports indicate a characteristic low intensity on T2WI. The key to the diagnosis is the sagittal MR, which localizes the lesion to the posterior pituitary. These tumors have also been reported in the suprasellar region and third ventricle.

Intrasellar Meningioma

An intrasellar meningioma can simulate the appearance of a pituitary tumor; however, careful attention can show that these masses are not arising from the pituitary gland proper. Intrasellar meningiomas can arise from the diaphragma sellae. The diagnosis can be suggested if the diaphragma sellae is visualized. In lesions originating in the suprasellar cistern the diaphragm should be depressed, whereas with intrasellar lesions the diaphragm is elevated. Careful observation also reveals slightly different intensity on MR between the meningioma and the inferior pituitary tissue. A dural tail and homogeneous enhancement may be visualized with meningioma, distinguishing it from pituitary adenoma. Meningiomas can also arise from the planum sphenoidale and infiltrate the sella (and optic canals) from anterosuperiorly (Fig. 10-21). When these tumors

involve the cavernous sinuses, narrowing of the internal carotid artery (ICA) flow void may be seen. Just as with macroadenomas, these tumors can compress the optic chiasm with suprasellar extension.

Aneurysm

Internal carotid artery aneurysms arising from the cavernous ICAs or ectatic cavernous portions of the carotid arteries may produce sellar enlargement and mass effect (Fig. 10-22). Pulsation artifact related to cavernous ICA aneurysms can raise suspicion for vascular etiology of the apparent sellar/parasellar mass and should prompt further evaluation by computed tomographic angiography (CTA) or magnetic resonance angiography (MRA). About 50% of cases of persistent trigeminal artery (which arises from the carotid and courses posteriorly, penetrating the sella, before joining the basilar artery) have an intrasellar course, potentiating iatrogenic injury. Surgeons need to be informed by us and cognizant of these anatomic variations or else, after the transsphenoidal hypophysectomy, the *radiologist* may be on the receiving end of an epistatic event.

Empty Sella Syndrome

CSF is easily noted in the empty or partially empty sella because of a patulous diaphragma sella and extension of the suprasellar arachnoid space inferiorly (Fig. 10-23). The sellar floor is often expanded and downwardly displaced. This may be seen with aging and in patients with pseudotumor cerebri (idiopathic intracranial hypertension). Other findings associated with pseudotumor include an enlarged tortuous optic nerve sheath, bulging of the optic nerve head (papilledema), and meningoceles in other locations. Cases have been reported in which the appearance of the empty sella was observed to be reversible following treatment of the intracranial hypertension.

Occasionally, the suprasellar visual system is noted to herniate into the sella. In fact, this results from traction from previous adhesions and from arachnoiditis after surgery for macroadenomas. These patients seldom have symptoms. On coronal and sagittal images the herniated chiasm and floor of the third ventricle are noted in the sella.

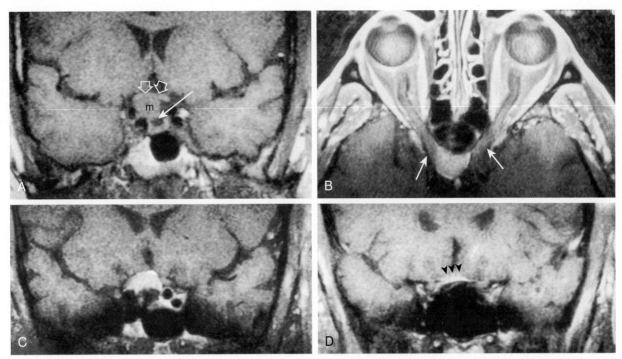

FIGURE 10-21 Intrasellar meningioma. **A,** Unenhanced T1-weighted image shows a mass (m) off the diaphragma sella *(arrow).* The diaphragm is slightly depressed. Chiasmatic compression is present *(open arrows).* **B,** The mass enhances. Its relationship to the two optic nerves *(arrows)* is worrisome on this axial scan. **C,** The bulk of the lesion enhances, but so does the dural tail *(arrowheads)* extending anteriorly along the planum in **D.**

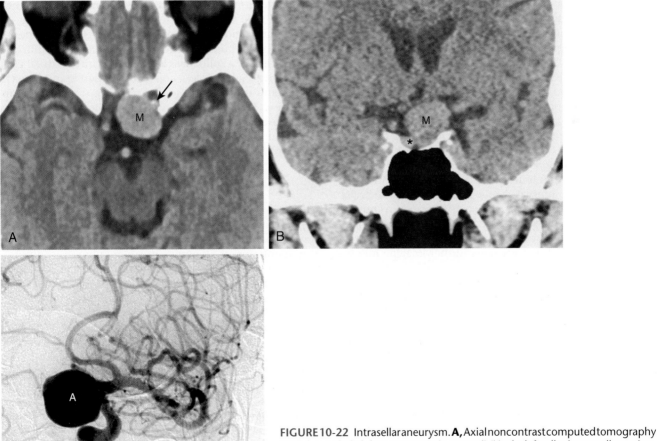

FIGURE 10-22 Intrasellar aneurysm. **A,** Axial noncontrast computed tomography (CT) shows slightly hyperattenuating mass (M) in the left sellar/suprasellar region, intimately associated with the left cavernous internal carotid artery (ICA; *arrow*). **B,** Coronal reconstruction from the same CT confirms left sellar/suprasellar location of the mass (M) and shows its relationship to the pituitary gland *(asterisk).* **C,** Conventional catheter angiography confirms presence of a large aneurysm (A) arising from the cavernous ICA.

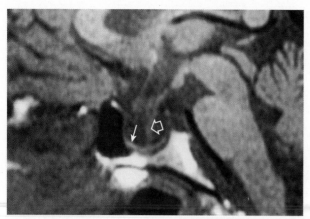

FIGURE 10-23 Partially empty sella. Sella is expanded. Note the residual pituitary tissue *(solid arrow)* and infundibulum *(open arrow).*

Postoperative Pituitary

The postoperative MR examination may reveal what appears to be persistent mass effect even though there is clinical evidence of chiasmatic decompression. Immediately following surgery, a peripherally enhancing cystic cavity can be seen at the site of previously demonstrated adenoma; this cystic cavity will decrease in size over the next few months (Fig. 10-24). Additional postoperative changes include fat or other packing material in the sphenoid sinus and persistent enhancement at the operative tract. Recognize that tumors with cavernous sinus invasion are often not completely resected; what one should expect is decompression of the midportion of the tumor where access via the transsphenoidal approach is possible to relieve indentation on the optic chiasm, pituitary stalk, third ventricle, and hypothalamus. Remaining normal glandular tissue is inconstantly identified. It may be difficult to distinguish residual normal pituitary tissue from residual tumor and/or granulation tissue when reviewing the postoperative images. Comparing directly with the preoperative scan can certainly be helpful in this regard; however, identifying these different tissue types is less important than assessing for interval change in tissue bulk in the sella over time and compressive effect on adjacent structures. Do not irritate your endoscopic neurosurgeon by calling residual tumor in every adenoma case. "Tissue is present in the operative cavity that warrants further imaging follow-up" (ka-ching, ka-ching).

Occasionally, the suprasellar visual system is noted to herniate into the sella. In fact, this results from traction from previous adhesions and from arachnoiditis after surgery for macroadenomas. These patients seldom have symptoms. On coronal and sagittal images, the herniated chiasm and floor of the third ventricle are noted in the sella.

SUPRASELLAR LESIONS

See list of suprasellar lesions in Box 10-3.

Infundibular Lesions

The infundibulum should generally not be larger than the basilar artery at the level of the clivus. Granulomatous diseases (LCH, sarcoid, tuberculosis) can enlarge the infundibulum (Box 10-4). In many cases, these lesions involve the hypothalamus as well. Granulomas of the infundibular-hypothalamic region can be isointense to brain and enhance (Fig. 10-25). Diabetes insipidus is a common finding in lesions affecting the stalk and hypothalamus. Other lesions that can involve the infundibulum include metastasis, lymphoma, germinoma, craniopharyngioma, Rathke cleft cyst, and secondarily, prolactinoma. In the end, the pituitary adenoma is the most common lesion to affect the infundibulum with displacement, foreshortening, thickening, or effacement.

Rathke Cleft Cyst

Rathke cleft cysts arise from a Rathke pouch and may be found in the anterior sellar region (25%), suprasellar region (5%) or both (70%). They are benign lesions lined with cuboidal or columnar epithelium and may contain mucus. Rathke cleft cysts may cause visual disturbances, pituitary insufficiency, or diabetes insipidus. These lesions have variable intensities on both T1WI and T2WI (Fig. 10-26). Their MR intensity is related to the contents of the cyst (e.g., protein, paramagnetic). An intracystic nodule having high signal intensity on T1WI and low signal intensity on T2WI is seen in three fourths of Rathke cysts. Fluid-fluid levels also denote a Rathke cyst. The principal differential diagnosis of Rathke cleft cyst is craniopharyngioma. CT may be useful here because of its sensitivity to calcification as compared with MR. In general, Rathke cleft cysts do not calcify whereas a high percentage of craniopharyngiomas do. Rathke cleft cysts have a smooth contour with homogeneous signal intensity within the lesion without a solid mass associated with it. Rathke cleft cysts enhance much less consistently than craniopharyngioma, and if they do, it's usually with just a rimlike configuration. This rim enhancement has been attributed in some cases to displaced pituitary tissue. They do not have the solid or soft tissue nodular enhancement seen with craniopharyngiomas.

Arachnoid Cyst

Approximately 15% of all arachnoid cysts arise in the suprasellar region. It is hypothesized that they arise developmentally as a result of a lack of perforation of the membrane of Liliequist. If the membrane is imperforate, normal CSF flow anterior to the pons can produce a wind sock, which can subsequently close off and become a true cyst or may remain as an arachnoid pouch. Such a cyst produces mass effect on adjacent structures, including hypothalamus, chiasm, and brain stem. These cysts can grow and produce hydrocephalus. Age at presentation is variable, from childhood to the second or third decade of life. The density and intensity of these cysts are those of CSF (Fig. 10-27). They are not associated with enhancement or calcification.

If the arachnoid cyst invaginates into the third ventricle, it can be mistaken for an ependymal cyst (neuroepithelial cyst) of the third ventricle or an enlarged third ventricle on the basis of aqueductal stenosis (which you should see on the MR). The key here is to determine whether you can separate the third ventricle from the lesion (if you can, it is most likely an arachnoid cyst). Cisternography or high

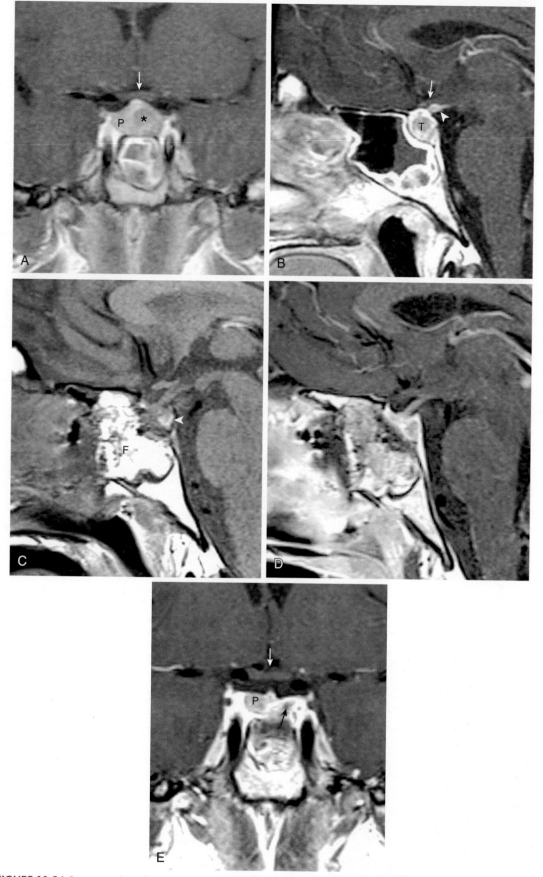

FIGURE 10-24 Postoperative sella. **A,** Coronal postcontrast T1-weighted imaging (T1WI) shows a relatively hypoenhancing adenoma *(asterisk)* in the left aspect of the sella with mass effect upon the inferior margin of the optic chiasm *(arrow)*. Normal pituitary tissue is seen in the right aspect of the sella (P). **B,** Sagittal postcontrast T1WI shows the same mass (T), resulting in superior displacement of the infundibulum *(arrowhead)* and chiasm *(arrow)*. **C,** Following transsphenoidal resection of the adenoma, sagittal T1WI shows fat packing material in the sphenoid sinus (F) and a small amount of hemorrhage at the site of tumor *(arrowhead)*. Note the more anatomically appropriate positioning of the chiasm and infundibulum on this image as well as on **D,** sagittal postcontrast T1WI. **E,** Postcontrast coronal T1WI in the same patient shows surgical cavity at site of resected adenoma *(black arrow)*, residual normal pituitary tissue in the right aspect of the sella (P), and resolution of mass effect upon the chiasm *(white arrow)*.

BOX 10-3   Suprasellar Lesions

Pituitary adenoma
Meningioma
Aneurysm
Arachnoid cyst
Infundibular lesions (see Box 10-4)
Chiasmatic or hypothalamic glioma
Rathke cleft cyst
Craniopharyngioma
Dermoid
Enlarged third ventricle extending into cistern
Epidermoid
Germinoma
Hamartoma of tuber cinereum
Lipoma
Lymphoma
Cavernous hemangioma
Sarcoid and other granulomatous processes
Teratoma
Cysticercosis or echinococcal cyst

BOX 10-4   Infundibular Lesions

GRANULOMATOUS DISEASES

Erdheim-Chester disease
Langerhans cell histiocytosis
Sarcoid
Tuberculosis

TUMORS

Pituitary adenoma
Rathke cyst
Craniopharyngioma
Germinoma
Lymphoma
Metastasis
Astrocytoma

OTHER

Stalk transection
Lymphocytic hypophysitis

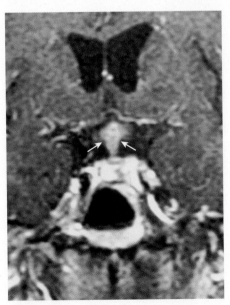

FIGURE 10-25 Sarcoid involving the infundibulum. Coronal enhanced T1-weighted image shows an enlarged thickened infundibulum *(arrows)* in a patient with intracranial sarcoid.

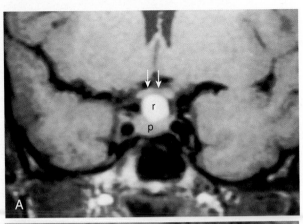

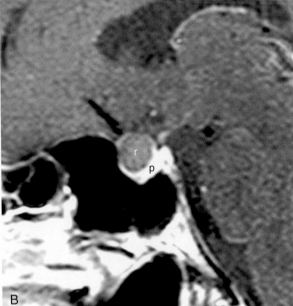

FIGURE 10-26 Rathke cleft cyst. **A,** Coronal T1-weighted image (T1WI) of Rathke cleft cyst (r), which is in the suprasellar cistern. It is high intensity on T1WI. The pituitary (p) and optic chiasm *(arrows)* are identified. **B,** Sagittal postcontrast T1WI in a different patient with Rathke cleft cyst shows cystic lesion (r) without enhancing component within the suprasellar cistern, sitting just above normal pituitary tissue (p).

resolution thin section constructive interference in the steady state (CISS) T2WI can be helpful in separating these lesions. Prompt filling with contrast following intrathecal contrast injection indicates a dilated third ventricle as opposed to an arachnoid cyst, which will show delayed contrast filling. The sagittal MR is also particularly useful in trying to sort this out, but at times only the pathologist will know for sure.

Other Developmental Lesions

The epidermoid lesion can have the same density and intensity on CT and T1W and T2W MR as CSF. In the vast majority of cases it is of slightly higher density and intensity than CSF. It may be intradural or extradural, and can arise in the third ventricle and in the parasellar region around the gasserian ganglion, where it can erode the petrous apex. It can have a variegated appearance, which is best appreciated on CT after intrathecal

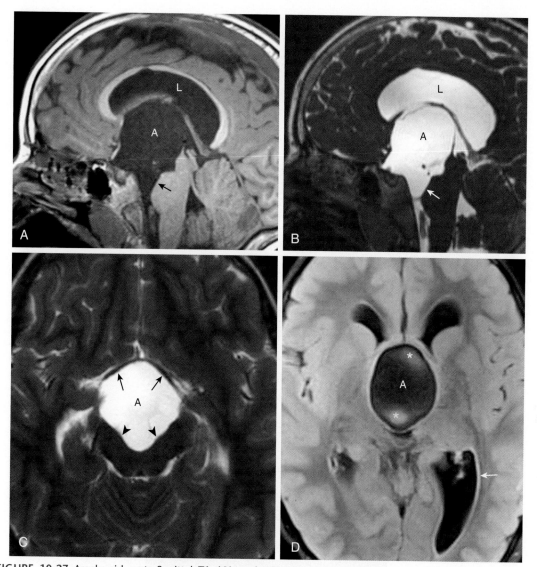

FIGURE 10-27 Arachnoid cyst. Sagittal T1 **(A)** and T2 constructive interference in the steady state (CISS) **(B)** magnetic resonance images through the midline show an extraaxial cerebrospinal fluid (CSF) signal lesion (A) centered within the suprasellar cistern. Note the flattening on the ventral pons because of mass effect *(arrow)* and enlargement of the lateral ventricle (L) because of obstructive hydrocephalus. **C,** Axial T2-weighted imaging shows mass effect and lateral displacement of the bilateral A1 segments of the anterior cerebral arteries *(arrows)* and splaying of the bilateral cerebral peduncles *(arrowheads)* by the arachnoid cyst (A). **D,** Axial fluid-attenuated inversion recovery (FLAIR) image shows presence of CSF pulsation artifact *(asterisks)* within the arachnoid cyst (A). Note enlarged lateral ventricles and transependymal flow of CSF *(arrow)* because of obstructive hydrocephalus.

contrast (with contrast accumulating in its interstices). Epidermoids insinuate throughout the suprasellar region and can grow behind the clivus, thereby pushing the brain stem posteriorly. Occasionally, epidermoids may have rim calcification. These lesions rarely if ever enhance. Fortunately, the diagnosis of these lesions has been made rather simple with the advent of FLAIR and diffusion-weighted imaging, where they are brighter than CSF (Fig. 10-28). Restricted diffusion is from the highly proteinaceous contents. Differential may include suprasellar craniopharyngioma with largely cystic components.

Dermoids, lipomas, and teratomas can also arise in the suprasellar cistern. Teratomas and dermoids can grow to be large, compressing the third ventricle and adjacent structures. On MR, high intensity on T1WI from fat can be observed in dermoids, lipomas, and teratomas. Lipomas

in this region are usually well circumscribed (Fig. 10-29). Teratomas may contain dense calcification or ossification, which has been observed to be in the central portion of the lesion. Dermoids are midline lesions containing fat, squamous epithelium (as do epidermoids), hair follicles, sweat glands, and sebaceous glands (Fig. 10-30). Their capsules are thick with peripheral calcification.

Craniopharyngioma

Craniopharyngiomas may be seen in children and adults. They comprise between 1.2% and 3% of all intracranial tumors. In children, they account for a greater percentage of tumor cases, but more than 50% of craniopharyngiomas occur in adults. Two histopathologic subtypes exist and include adamantinomatous and squamous/papillary (Table 10-2). These subtypes have different clinical profiles,

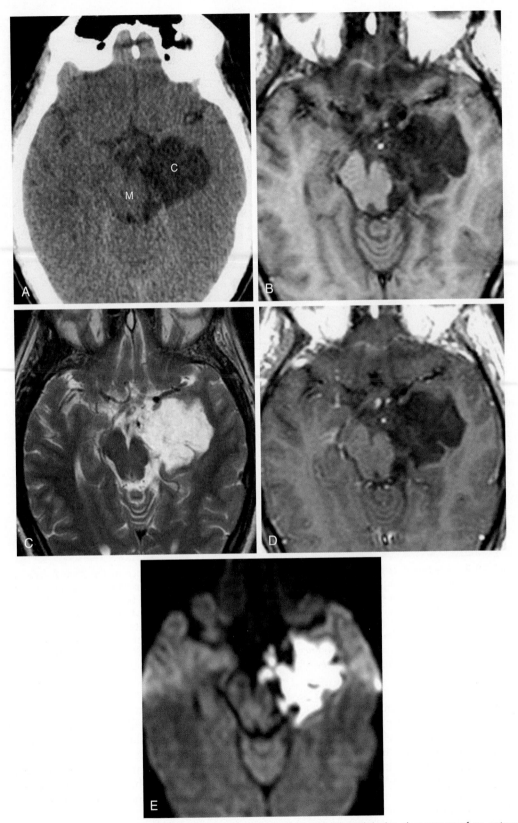

FIGURE 10-28 Epidermoid. **A,** Unenhanced computed tomography shows lobulated contours of an extraaxial lesion isodense to cerebrospinal fluid (CSF) (C) centered in the left middle cranial fossa, with extension into the left parasellar region. Note mass effect upon the left aspect of the midbrain (M). No calcification is seen. **B,** Axial T1-weighted imaging (T1WI) and **(C)** axial T2WI show the lesion to have signal intensity similar to CSF. **D,** Postcontrast T1WI shows that the lesion does not enhance. So far this lesion could represent an arachnoid cyst. However, the diffusion-weighted scan **(E)** nails the diagnosis of an epidermoid.

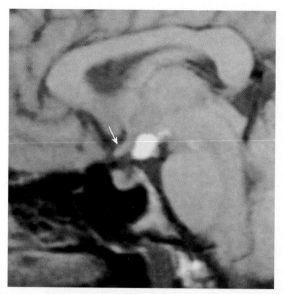

FIGURE 10-29 Lipoma in the suprasellar cistern. A high-intensity mass is seen on this sagittal T1-weighted image, representing a lipoma in the suprasellar cistern. Optic chiasm *(arrow)* is anterior to the lipoma. Note incidental Chiari I malformation.

TABLE 10-2   Histopathologic Types of Craniopharyngioma

	Adamantinomatous	Squamous-Papillary
Location	Suprasellar	Intrasellar/suprasellar or suprasellar
Age	Children, occasionally adults	Adults
Tissue structure	Predominantly cystic	Predominantly solid
T1 without contrast	Hyperintense cysts typical	If ever, hypointense cysts
Shape	Lobulated	Spherical
Encase vessels	Yes	No
Tumor recurrence	+++	+
Calcifications	+++	+

+, Possible; +++, very likely.

imaging characteristics, and prognoses but in practice they can be difficult to distinguish on imaging.

Craniopharyngiomas are usually centered in the suprasellar (20%), suprasellar and sellar (70%), sellar (10%), or infrasellar regions (<1%), but they can be extensive, including the anterior fossa, middle fossa, posterior fossa, retroclival region, and the third ventricles. Other rare origins include the lateral ventricle, sphenoid bone, nasopharynx, cerebellopontine angle, and pineal gland. A good rule is that if the tumor looks bizarre and has a component at the base of the skull, think craniopharyngioma. They arise from metaplasia of squamous epithelial remnants (Rathke pouch) of the adenohypophysis and anterior infundibulum, or from ectopic embryonic cell rests of enamel organs.

Three imaging hallmarks of craniopharyngioma have been identified (although an individual lesion may have none or all these characteristics): (1) calcification, (2) cyst formation, and (3) enhancement (solid, nodular) (Fig. 10-31). T1 bright signal intensity, young age, and calcification on CT favor the adamantinomatous type over the papillary type of craniopharyngiomas. Calcification may be nodular or rimlike, occurring in approximately 80% of cases; for this reason, CT can be a useful adjunct to MRI in making the diagnosis as calcifications are better appreciated by CT. Cystic regions are observed in about 85% of cases; the cyst can show low density on CT, and variable intensity on MR. The intensity on T1WI ranges from hypointense to hyperintense and is usually high intensity on T2WI. The high intensity on T1WI appears to be due to methemoglobin and/or high protein (>9000 mg/dL). Some ultra-high-protein craniopharyngiomas are low intensity on T1WI. This has been attributed to the increased viscosity associated with such elevated protein levels. The solid portion of the tumor enhances.

The papillary intraventricular variety of craniopharyngioma is unusual, probably originating from the pars tuberalis that extends to the tuber cinereum in the floor of the third ventricle. These lesions do not extend beneath the floor of the third ventricle (i.e., they are not in the suprasellar space). Other features distinguishing these lesions include their incidence in adults and their male preponderance.

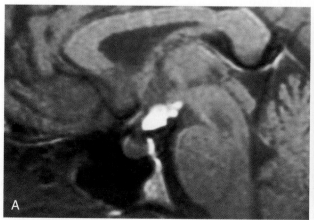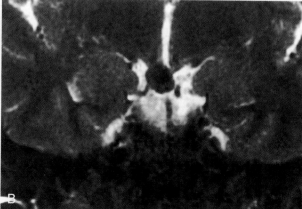

FIGURE 10-30 Dermoid in suprasellar region. **A,** The fat of the dermoid is T1 hyperintense. **B,** Note decreased intensity of fat on this fat-suppressed T2-weighted image.

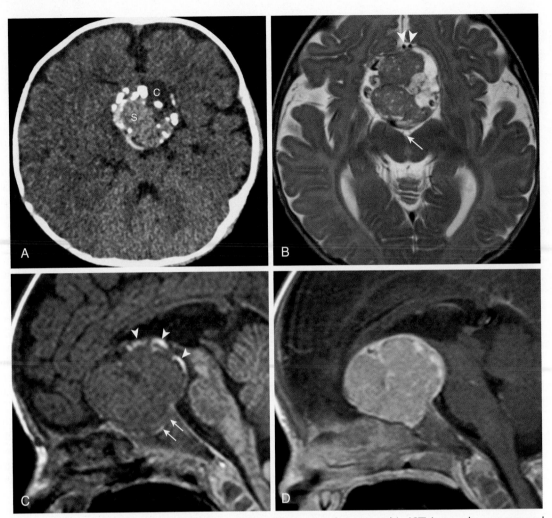

FIGURE 10-31 Craniopharyngioma. **A,** Axial unenhanced computed tomographic (CT) image shows a suprasellar mass with hyperattenuating solid (S) and hypoattenuating cystic (C) components, with chunky peripheral and central calcifications. **B,** Axial T2-weighted imaging (T2WI) shows the same lesion containing low signal (solid) and high signal (cystic) components. Note widening of the interpeduncular cistern *(arrow)* and anterior displacement of the A2 segments of the anterior cerebral artery *(arrowheads)* because of mass effect. **C,** Sagittal T1WI shows remodeling of the sella *(arrows)*, indicating slow-growing lesion. Curvilinear hyperintense signal along the posterior margin of the tumor *(arrowheads)* probably reflects calcified components seen on CT. Again seen is posterior displacement of the midbrain because of mass effect. **D,** Heterogeneous contrast enhancement of the tumor is demonstrated on postcontrast sagittal T1WI.

Hormonal and visual disturbances are rare, again because of their intraventricular location. There is a lower incidence of calcification or cyst formation (as opposed to the run-of-the-mill adamantinomatous craniopharyngioma), with uniform enhancement. This solid intraventricular lesion has no specific hallmarks. The differential diagnosis of such third ventricular lesions besides the rare papillary craniopharyngioma is important and includes cavernous malformation of the third ventricle, choroid plexus papilloma, ependymoma, pilocytic astrocytoma, and meningioma.

Fusiform dilatation (pseudoaneurysm formation) of the supraclinoid carotid artery has been reported postoperatively in craniopharyngioma; however, to date no cases of subsequent subarachnoid hemorrhage have been reported. Treatment and retreatment with cyst puncture/drainage and local instillation of antineoplastic drugs for recurrences are common. This is a sticky tumor that is hard to get rid of without injury to critical normal structures like the optic apparatus, pituitary stalk, and hypothalamus.

Meningioma

Meningiomas in this region arise from the tuberculum sellae, anterior clinoid processes, diaphragma sellae, planum sphenoidale, and upper clivus. Progressive visual loss is the most frequent complaint. Meningiomas are extraaxial in location, buckling the gray-white interface. CT can reveal hyperostosis, blistering of the tuberculum and planum sphenoidale, and erosion of the dorsum sellae.

On T1WI, meningiomas are isointense to slightly hypointense to brain (Fig. 10-32). On T2WI, they are isointense to slightly hyperintense to brain. These lesions are generally not cystic. They enhance dramatically. On angiography, they have a characteristic stain from ophthalmic and carotid meningeal branches. Remember meningiomas commonly narrow the carotid artery. Careful imaging is critical because these tumors may be thin and at times difficult to separate from normally enhancing pituitary tissue or dura.

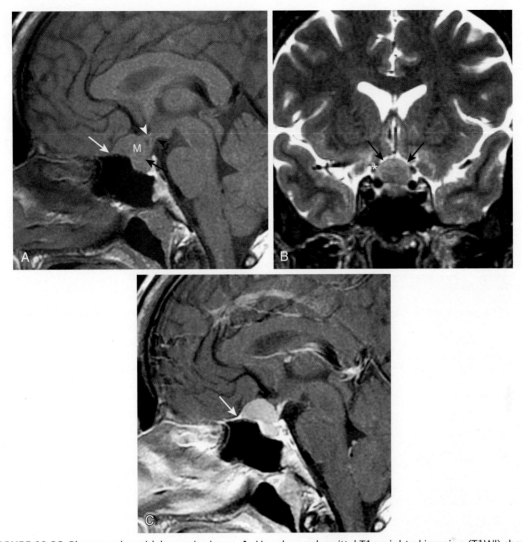

FIGURE 10-32 Planum sphenoidale meningioma. **A,** Unenhanced sagittal T1-weighted imaging (T1WI) shows a mass (M) in the suprasellar region. However, there is a subtle tissue plane present between the mass and pituitary gland *(black arrow),* indicating that this tumor is not arising from the pituitary gland. There is dural based attachment on the planum sphenoidale *(white arrow),* indicating dural origin of this meningioma. The infundibulum is displaced posteriorly *(black arrowhead),* and the optic chiasm is elevated superiorly *(white arrowhead)* by the mass. **B,** Coronal T2WI shows the tumor to be very slightly hyperintense to brain *(arrows).* Note lateral displacement of the A1 segments of the anterior cerebral arteries *(asterisk).* **C,** Following contrast administration, the dural attachment is again seen *(arrow),* and the tumor is shown to enhance slightly less than the normal pituitary gland.

Chiasmatic and Hypothalamic Astrocytoma

Chiasmatic/hypothalamic astrocytomas (gliomas) present as mass lesions in the suprasellar cistern. They are typically isointense to brain on T1WI and high intensity on T2WI (Fig. 10-33). This high intensity may be noted throughout the visual pathway and is of uncertain significance. Enhancement is variable, and calcification in the nonirradiated tumor is rare. At times, a cystic component (fluid attenuation/intensity) is present with the tumor and they are commonly the grade 1 pilocytic varieties. Bilateral optic nerve astrocytomas are associated with neurofibromatosis. Hypothalamic astrocytomas and gangliogliomas may be difficult to distinguish from chiasmatic lesions; a normal chiasm, with an inhomogeneous mass in the floor of the third ventricle and suprasellar cistern, suggests a hypothalamic as opposed to a chiasmatic astrocytoma.

These lesions, because of their location, are rarely resected. Fortunately they are slow-growing low-grade astrocytomas so delay, delay, delay until one is forced to radiate and/or operate. Radiating the optic chiasm…not a great option.

Hamartoma of the Tuber Cinereum

Hamartomas of the tuber cinereum are known to cause central precocious puberty and gelastic seizures (spasmodic laughter). These lesions are congenital nonneoplastic heterotopias. The tuberoinfundibular tract probably carries releasing hormones that modulate gonadotropins. The mechanism for precocious puberty is neurosecretion by the hamartoma of luteinizing hormone–releasing hormone. Seizures may result from connections between the lesion and the limbic system

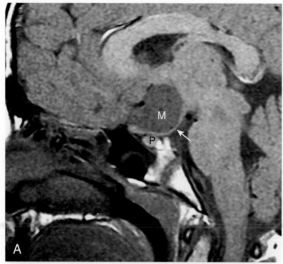

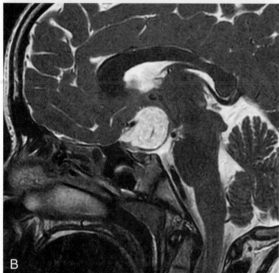

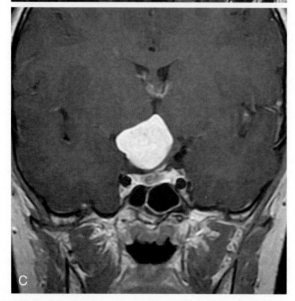

FIGURE 10-33 Pilocytic astrocytoma of chiasm-hypothalamic origin. **A,** Sagittal T1-weighted imaging (T1WI) shows a large low signal intensity mass (M) centered in the region of the floor of the third ventricle with extension inferiorly into the suprasellar region. Note inferior displacement Sella of the optic pathway *(arrow)*. The pituitary (P) is not involved. **B,** Sagittal T2WI demonstrates the lesion to be hyperintense. **C,** Enhancement of this degree is typical of pilocytic astrocytomas but does not imply high grade.

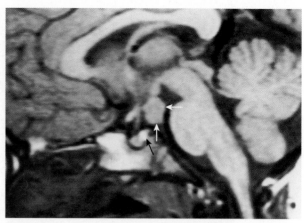

FIGURE 10-34 Hamartoma of the tuber cinereum. Pedunculated mass extending from the tuber cinereum *(white arrows)* is isointense to brain. Note the high signal in the posterior pituitary gland *(black arrow).*

or possibly, if present, associated brain abnormalities including callosal dysgenesis, optic malformation, heterotopias, and microgyria. These findings might indicate an insult in the first month of gestation. These tumors can be completely resected when they are pedunculated with amelioration of symptoms.

The anatomic location of these hamartomas (just anterior to the mammillary bodies) together with a signal intensity similar to gray matter on T1WI and most often higher intensity than gray matter on PDWI and particularly T2WI, strongly supports this diagnosis (Fig. 10-34). Morphologically, hamartomas may be pedunculated or broad based, ranging in size from 0.4 to 4 cm in diameter. The typical hamartoma of the tuber cinereum does not calcify, enhance, contain fat, or have cysts. Histologically, they are composed of neurons supported by normal microglia. These lesions tend to be stable over time.

Can we distinguish this lesion from craniopharyngioma and hypothalamic glioma? It would be unusual for the former to be isointense to brain on T1WI. CT is also more sensitive for calcification as seen in craniopharyngioma. The hypothalamic glioma and the craniopharyngioma are both heterogeneous lesions compared with the homogeneous appearance of the hamartoma. The absence of enhancement greatly favors hamartoma. Morphology, location, and clinical history usually make this a straightforward Aunt Minnie diagnosis.

Aneurysm

The diagnosis of aneurysm on MR/MRA and CTA has already been discussed in the vascular chapter (Chapter 3). Partially thrombosed giant aneurysms contain a flow void and signal intensities from the various stages of clot on MRI (Fig. 10-35). Pulsation artifact extending through and beyond the margins of the sella should raise suspicion for aneurysm on MRI. Angiography is required to characterize the exact site of origin and identify the neck. Petrous apex cholesterol granulomas may mimic giant clotted petrous carotid aneurysms. *Hopefully you can find the normal/ displaced ICA flow void nearby.* In the suprasellar cistern the mixed intensity and calcification of a craniopharyngioma may lead to confusion with a partially thrombosed

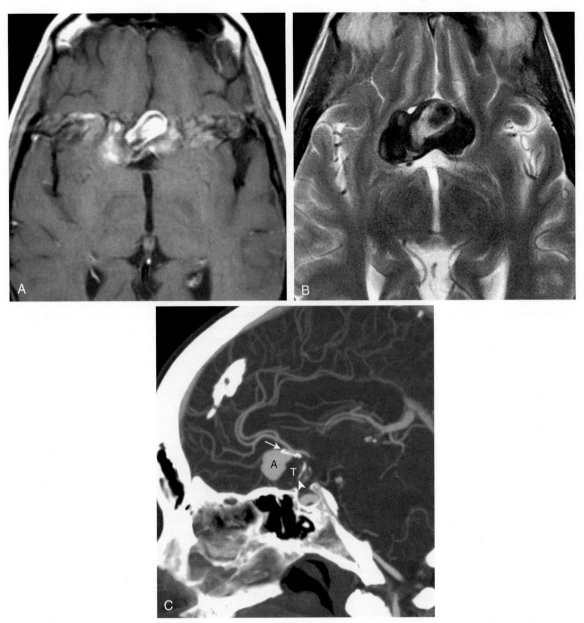

FIGURE 10-35 Giant suprasellar aneurysm. **A,** Axial T1-weighted imaging (T1WI) shows pulsation artifact extending along the image in the phase-encoding direction at the level of a suprasellar mass, indicating that this mass is aneurysmal in nature. **B,** Axial T2WI shows the suprasellar mass with mixed signal indicating that the aneurysm is partially thrombosed. **C,** Sagittal maximum intensity projection reconstruction from computed tomography angiography confirms the presence of suprasellar aneurysm (A) arising from the A1-A2 junction. Note the nonenhanced thrombosed component (T) displacing vessels posteriorly *(arrowhead)*. Calcifications are present along the base of the aneurysm *(arrow)*.

calcified aneurysm/pseudoaneurysm. We repeat... *Hopefully you can find the normal/displaced ICA flow void nearby.*

Germinoma

These lesions are seen in children and young adults. Presentation is variable, but common findings include diabetes insipidus, hypopituitarism, and optic chiasm compression. They arise from primitive germ cells in the suprasellar region. They appear as mass lesions, which may be locally invasive. On CT, germinomas are high density on unenhanced scan and uniformly enhance (Fig. 10-36).

The density is similar to lymphoma and probably the result of the increased tumor protein. These lesions do not calcify. Coexisting pineal masses may be identified. They can metastasize by subarachnoid seeding. On MR, there is less specificity between this lesion and other soft-tissue tumors in this region. They melt under the influence of radiation therapy (or chemotherapy) in a matter of days to weeks.

PARASELLAR LESIONS

Parasellar lesions are those centered within the cavernous sinus, Meckel cave, overlying dura, and adjacent

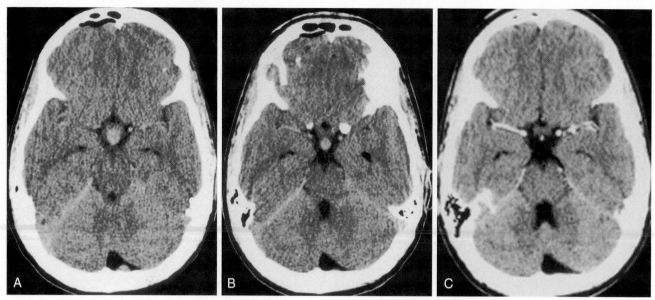

FIGURE 10-36 Germinoma. **A,** Noncontrast computed tomography (CT) demonstrates high density in a suprasellar mass representing a germinoma. **B,** Noncontrast CT inferior to **A** shows high density in a markedly enlarged infundibulum containing germinoma. **C,** Postcontrast CT following radiation therapy. Observe the normal infundibulum.

BOX 10-5   Cavernous Sinus Lesions

Pituitary adenoma
Meningioma
Metastasis (including perineural spread of tumor)
Schwannoma (cranial nerves III, IV, V, and VI)
Vascular lesions (ectatic carotids, carotid-cavernous fistula, cavernous carotid aneurysm, cavernous hemangioma, and cavernous sinus thrombosis)
Lymphoma
Infection
Chondrosarcoma
Chordoma
Idiopathic inflammatory disease (Tolosa-Hunt syndrome)
Plasmacytoma

---

bones (Box 10-5). We will try not to reiterate the imaging characteristics of lesions that have already been commented on earlier in this chapter and in other chapters in the book. Rather, we will focus on what other characteristics in this particular location aid in making the diagnosis.

Aneurysm

Aneurysms may erode and undermine the anterior clinoid processes, and are associated with either obvious flow voids or layers of thrombus. You can make the diagnosis on T1WI, and this can be supplemented by CTA, MRA, and conventional angiography. MRI/MRA has in most cases taken the guesswork out of this lesion. Cavernous sinus aneurysms produce mass effect on the intracavernous cranial nerves. When they rupture, they create carotid-cavernous fistula as opposed to intradural aneurysms arising more distally from the carotid and its branches, which produce subarachnoid hemorrhage.

Carotid-Cavernous Fistula or Dural Malformation

Carotid-cavernous fistula or dural malformation can enlarge the cavernous sinus. The fistula may be a result of trauma or extradural aneurysm rupture. Associated with a prominent cavernous sinus is usually an enlarged superior ophthalmic vein or other orbital veins. On CTA obtained in the arterial phase, early or increased venous contrast enhancement within the affected cavernous sinus relative to the contralateral side can be seen. Occasionally, dilated intercavernous sinus collateral veins can be identified. This can be seen on an angiogram when one internal carotid artery is injected and the contralateral cavernous sinus and its tributaries are opacified (see Fig. 3-61).

Thrombosis of the Cavernous Sinus

This one is a tricky diagnosis to make, even among the best of us. Thrombosis of the cavernous sinus may occur as part of a septic process, spread from invasive fungal sinusitis, retrograde extension from superior ophthalmic vein thrombosis, may be associated with spontaneous dural malformations or may result from an interventional or surgical procedure (Fig. 10-37). The last situation most commonly occurs in conjunction with carotid-cavernous fistula or dural malformation. On MR without enhancement, high intensity may be seen in the subacutely occluded cavernous sinus. However, many times the sinus may be only partially occluded. Evaluating postcontrast images alone on MR is not very useful because nonthrombosed regions of the sinus enhance, and subacute clot is also high intensity. Absence of gadolinium enhancement of the cavernous sinus without bright signal suggests acute thrombosis. On CT, an irregularly enhancing sinus can be detected with lack of contrast opacification within the thrombosed portion of the sinus. An enlarged superior ophthalmic vein, periorbital swelling, or thickening of the

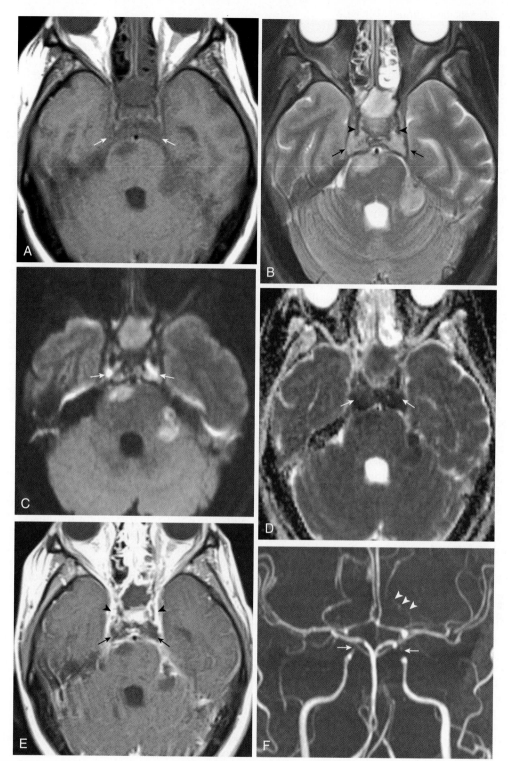

FIGURE 10-37 Cavernous sinus thrombosis in the setting of meningitis. **A,** Axial T1-weighted imaging (T1WI) shows absence of small amount of T1 hyperintense fat signal that can often be seen in the cavernous sinuses *(arrows)*. **B,** Axial T2WI shows lateral bulging of the cavernous sinus lateral margins on both sides *(arrows)* as well as abnormal signal within the posterior aspects of the cavernous sinuses bilaterally. There is also abnormal signal within the right ventral pons and left cerebellopontine angle (CPA) cistern. Note marked narrowing and irregular contours of the cavernous carotid flow voids anteriorly *(arrowheads)* and complete absence of flow voids posteriorly within the cavernous sinuses. Diffusion restriction is seen within the posterior aspects of the cavernous sinuses bilaterally on diffusion-weighted imaging (**C,** *arrows*), and apparent diffusion coefficient map (**D,** *arrows*), indicating purulent material. Diffusion restriction is also seen within the right ventral pons and left CPA cistern. **E,** Axial postcontrast T1WI shows expansile filling defects within the cavernous sinuses bilaterally *(arrows)* with preserved normal enhancement within the cavernous sinuses anteriorly *(arrowheads)*. Note abnormal leptomeningeal enhancement along the pons in this patient with meningitis. Abnormal signal within the pons represents extension of inflammation into the parenchyma, and abnormal signal within the left CPA signal represents loculated meningeal deposit of purulent material. **F,** Maximum intensity projection reconstruction from three-dimensional time-of-flight magnetic resonance angiography shows abrupt occlusion of the cavernous internal carotid arteries bilaterally *(arrows)* as well as severe irregularity of the left posterior cerebral artery *(arrowheads)*, indicating arterial vasculitis secondary to meningitis.

extraocular muscles should send your eyes searching for clot in the cavernous sinus. Be careful not to confuse filling defects on CT or high intensity on T1WI in the cavernous sinus with fat seen in normal persons.

Cavernous Sinus Meningioma

In general, cavernous sinus meningiomas follow the lateral margin of the cavernous sinus and may extend posteriorly along the tentorial margin with a dove's tail appearance on contrast enhanced imaging studies (Fig. 10-38). The lateral margin of the sinus is smoothly bulged. Meningiomas can encase or distort the cavernous portion of the internal carotid artery. Their intrinsic characteristics on CT, MR, or angiography are similar to those of meningiomas in other locations. Meningiomas in rare instances can extend through the base of the skull. After the orbit, extracranial meningiomas are seen in the paranasal sinuses, the nasopharyngeal parapharyngeal space, the jugular foramen and the foramen magnum.

Trigeminal Nerve Lesions

Trigeminal schwannomas are rare tumors (<0.4% of brain tumors) arising from the intracranial portion of the trigeminal nerve. They can be based predominantly in the middle cranial fossa, in the region of the gasserian ganglion/Meckel cave (most common) or the posterior fossa (next most common), or may be dumbbell shaped (least common), involving both regions. The peak incidence of these tumors is in the fourth decade of life, with an equal prevalence between men and women. The most frequent symptom is trigeminal nerve dysfunction, including pain, numbness, and paresthesia.

On MR, they are smooth masses, isointense on T1WI and high intensity on T2WI, with avid enhancement (Fig. 10-39). Regions of "cystic" change may be observed in the enhancing mass. They may grow through skull base foramina, producing smooth enlargement, and can be traced from the pons, prepontine cistern, and gasserian ganglia into the cavernous sinus. Sagittal MR is useful in following the course of the tumor along the nerve. On CT, particularly with bone windows, erosion can be appreciated at the petrous apex. The differential diagnosis of petrous apex erosion includes epidermoid tumor, giant petrous apex cholesterol cyst (granuloma), cerebellopontine angle schwannomas, meningioma, mucocele, chordoma, metastasis, osteochondroma, and chondrosarcoma. Most of these lesions are readily distinguished from trigeminal schwannoma. Other findings associated with the slow growth of the tumor are smooth erosion of the floor of the middle fossa and enlargement of the foramen ovale, foramen rotundum, and superior orbital fissure.

Plexiform neurofibromas may diffusely infiltrate the fifth cranial nerve and extend into the deep spaces of the face via the various affected nerves exiting their foramina. Malignancies (malignant peripheral nerve sheath tumors [MPNST]) may occur as with plexiform neurofibromas in patients with neurofibromatosis type 1 most commonly. Other rare lesions of the proximal portion of cranial nerve V include lipoma, epidermoid, metastasis, and inflammatory disease.

Imaging studies are performed in the work-up of trigeminal neuralgia. In such cases the surgeons are looking

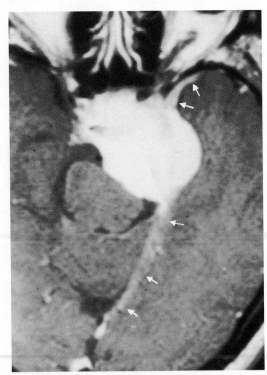

FIGURE 10-38 Cavernous sinus meningioma. A large homogeneously enhancing cavernous sinus mass is identified. The lesion bulges the sinus laterally and presents with enhancement along its anterior and posterior (dove's tail) dural margins *(arrows)*.

for causes of tic douloureux/trigeminal neuralgia, including (1) mass lesions such as a schwannoma; (2) vascular lesions (large looping arteries or veins, arteriovenous malformations, cryptic malformations); and (3) other causes (such as multiple sclerosis). Arterial and venous vascular compression upon the trigeminal nerve can be appreciated on high-resolution T2-weighted MRI (Fig. 10-40). MRA head performed concurrently can aid in diagnosis as well. The most common vessels to compress the trigeminal nerve and require microvascular decompression are (1) the superior cerebellar artery, (2) a vein (perimesencephalic), (3) anterior inferior cerebellar artery, and (4) vertebral artery.

Perineural Spread of Tumor

Metastases occur via perineural, subarachnoid, and hematogenous spread. Such dissemination indicates a poor prognosis. Head and neck tumors may demonstrate perineural spread through the foramina at the skull base and into the brain. Adenoid cystic carcinoma, basal cell carcinoma, squamous cell carcinoma, lymphoma, mucoepidermoid carcinoma, melanoma, and schwannoma all have a propensity for this kind of infiltration (Fig. 10-41). Skin, minor/major salivary glands, and the sinonasal cavity may be the site of origin. Enlargement or asymmetry of any basal neural foramina in the appropriate clinical setting should alert you to this possibility of perineural tumor extension (Fig. 10-42). Infections such as actinomycosis, Lyme disease, and herpes zoster can also demonstrate perineural involvement.

The third division of cranial nerve V, because of its extensive neural network about the head and neck, is

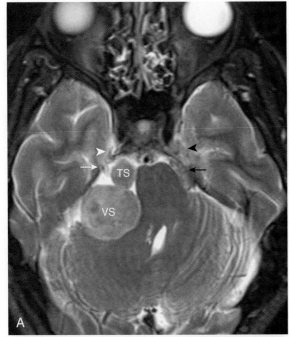

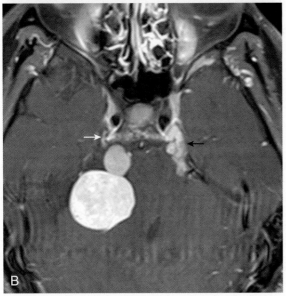

FIGURE 10-39 Bilateral trigeminal schwannomas. **A,** In this patient with neurofibromatosis type 2, axial T2-weighted imaging (T2WI) shows multiple bilateral extraaxial masses hyperintense to brain. On the left, an elongated lobulated mass extends along the prepontine course of the trigeminal nerve *(black arrow)* and into the Meckel cave with effacement of the normal fluid signal present within the cave *(black arrowhead),* compatible with trigeminal schwannoma. On the right, a rounded trigeminal schwannoma (TS) arising from the proximal segment of the nerve displaces the more distal prepontine segment of the trigeminal nerve laterally *(white arrow).* An additional lesion along the trigeminal nerve is suspected within right Meckel cave given partial absence of expected fluid signal *(white arrowhead).* Posterior to this is a large vestibular schwannoma (VS), compressing and displacing the brain stem to the left. **B,** These schwannomas enhance homogeneously on postcontrast T1W1. Note the enhancement of schwannomas within left Meckel cave *(black arrow)* and right Meckel cave *(white arrow).*

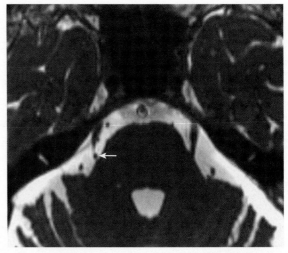

FIGURE 10-40 Neurovascular compression. In this patient presenting with right-sided trigeminal neuralgia, axial high-resolution T2 constructive interference in steady state (CISS) image shows distal superior cerebellar artery *(arrow)* contacting and laterally displacing the prepontine segment of the right trigeminal nerve at the root entry zone level.

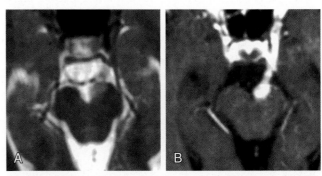

FIGURE 10-41 Lymphoma of cranial nerve III. Infiltration of the third cranial nerve is best viewed on the postcontrast scan **(B),** but the enlargement is even evident on the T2-weighted scan **(A).**

particularly prone to perineural spread of lesions (second only to the V_2 maxillary nerve). Such spread may be antegrade or retrograde from peripheral involvement. The findings that can be noted with perineural invasion along the third division of cranial nerve V include (1) thickening of the nerve; (2) concentric enlargement of the foramen ovale; (3) replacement of the normal CSF density and intensity in the trigeminal cistern (region of Meckel cave) by a soft-tissue mass; (4) atrophy of the ipsilateral masticator muscles; (5) high signal on T2WI in the masticator muscles; and (6) avid enhancement of the denervated muscles.

Tolosa-Hunt Syndrome

The Tolosa-Hunt syndrome is an idiopathic inflammatory disease of the cavernous sinus. It has been characterized clinically as (1) steady, gnawing, retroorbital pain; (2) defects in cranial nerves III, IV, and VI, and the first division of cranial nerve V, with less common involvement of the optic nerve or sympathetic fibers around the cavernous carotid artery; (3) symptoms lasting days to weeks; (4) occasional spontaneous remission; (5) recurrent attacks; and (6) prompt response to steroid therapy. Reports have

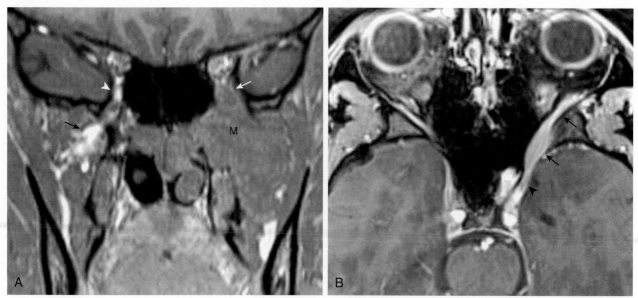

FIGURE 10-42 In diagnosing perineural tumor spread, fat is our friend. **A,** Coronal T1-weighted image (T1WI) shows a tumor mass centered in the left pterygopalatine fossa/masticator space (M) with obliteration of the normal fat in this space as shown on the normal right side *(black arrow).* There is extension of tumor into the inferior orbital fissure (IOF), which is widened, and normal fat in this space is obliterated by the tumor *(white arrow).* Compare to the normal IOF on the right *(arrowhead).* **B,** Contrast-enhanced fat-suppressed T1WI shows abnormal enlargement and enhancement within the IOF *(arrows)* with extension along V2 toward the cavernous sinus *(arrowhead).* Yup, it's a juvenile nasopharyngeal angiofibroma. Yikes!

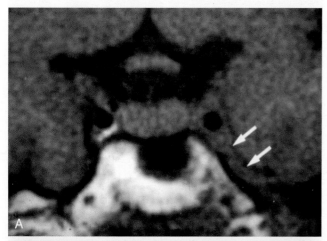

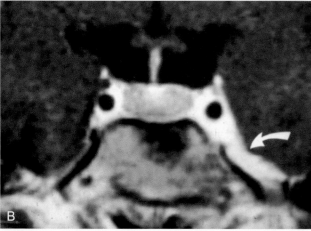

FIGURE 10-43 Tolosa-Hunt syndrome. Unenhanced **(A)** and enhanced **(B)** scans show the infiltrative pseudotumor *(arrows)* entering the cavernous sinus. In many cases, an orbital apex component coexists.

demonstrated narrowing of the cavernous carotid on angiography and irregularity or thrombosis of the superior ophthalmic vein and thrombosis of the cavernous sinus on orbital venography. On MR, a spectrum of findings has been observed including (1) normal scan; (2) abnormal signal (isointense with muscle on T1WI and isointense with fat on T2WI) and/or mass lesions in the cavernous sinus (enlargement of the cavernous sinus); (3) extension of the lesion into the orbital apex; (4) thrombosis of the cavernous sinus and/or superior ophthalmic vein; and (5) enhancement that can extend into the orbital apex (identified with fat saturation techniques) and along the floor of the middle cranial fossa (Fig. 10-43). The pathologic substrate of this syndrome is debated, with some investigators believing that it represents a granulomatous inflammation of the cavernous sinus and others finding no granulomas but rather nonspecific inflammatory changes. With a prompt response to steroid therapy, it is difficult in most cases, to justify biopsy. This is analogous to the idiopathic inflammatory condition of orbital pseudotumor and has now been lumped into immunoglobulin G (IgG)4-related disease (IgG4RD) inflammatory processes. These IgG4RD account for autoimmune pancreatitis, retroperitoneal fibrosis, orbital pseudotumor, and sclerosing sialadenitis (Mikulicz disease).

The differential diagnosis of Tolosa-Hunt syndrome includes sarcoid, meningioma, lymphoma, metastatic, and perineural spread of tumor into the cavernous sinus, and infections such as actinomycosis.

Chondrosarcoma

Chondrosarcoma is a rare neoplasm that is said to arise from embryonal rests, endochondral bone, or cartilage and is located at the skull base, in the meninges, or in the brain. To involve the parasellar region, this tumor can arise from

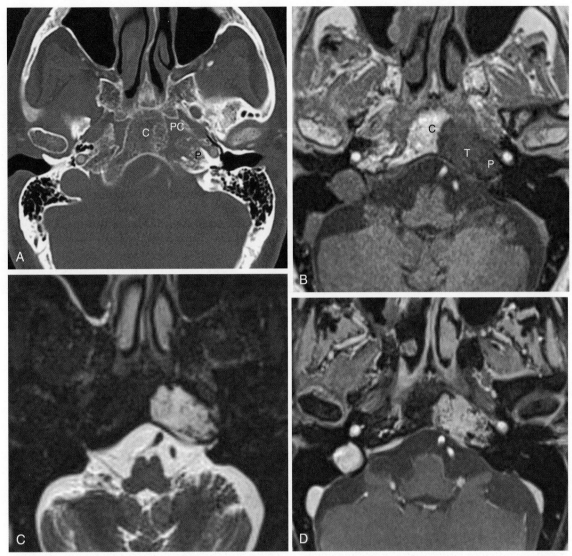

FIGURE 10-44 Chondrosarcoma. **A,** Axial computed tomographic image shows lytic destructive lesion centered at the left petroclival synchondrosis (PC) with lytic change involving the left lateral aspect of the clivus (C) and medial aspect of the left petrous apex (P). **B,** Axial T1-weighted imaging (T1WI) shows the tumor (T) again centered in the petroclival synchondrosis with abnormal marrow signal within the left aspect of the adjacent clivus (C) and petrous apex (P). **C,** Axial T2WI shows characteristic high T2 signal of this lesion. Hypointense foci reflect calcified matrix. **D,** The tumor enhances on postcontrast T1W1.

BOX 10-6   Parasellar Lesions Containing Calcifications

Atherosclerosis-dolichoectasia
Dural calcification/petroclinoid ligament
Aneurysm
Craniopharyngioma
Meningioma
Chondrosarcoma
Chordoma
Granuloma
Teratoma

BOX 10-7   Infrasellar and Base of Skull Lesions

Mucocele
Metastasis
Pituitary adenoma
Meningioma
Paget disease
Fibrous dysplasia
Cholesterol granuloma
Craniopharyngioma
Chondrosarcoma
Chordoma
Plasmacytoma/multiple
   myeloma

Developmental lesions
   (cephalocele)
Eosinophilic granuloma
Epidermoid
Infection
Juvenile angiofibroma
Malignant otitis externa
Nasopharyngeal carcinoma
Sphenoid sinus carcinoma

---

the petroclival synchondrosis or petrosphenoid fissure and extend anteriorly towards the cavernous sinus. Chondrosarcoma usually occurs in patients in their second to fourth decade of life. Usually, patients have a long history of headache and cranial nerve problems (particularly cranial nerve VI), which are sometimes intermittent.

CT reveals calcification in more than 60% of cases in a stippled, finely speckled, amorphous, or ringlike configuration. Pure lytic bone destruction may also occur. Contrast-enhanced CT demonstrates enhancing neoplastic tissue.

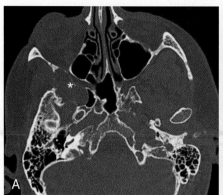
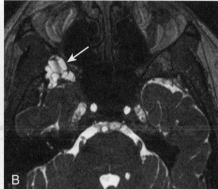
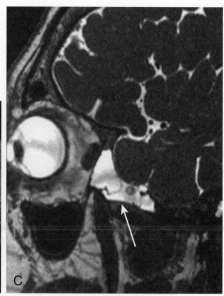

FIGURE 10-45 Meningocele. **A,** Lytic lesion in the sphenoid wing at first, on computed tomography, seems to be a primary bone lesion *(asterisk)*. The axial **(B)** and sagittal **(C)** T2-weighted image show the cerebrospinal fluid signal *(arrow)* and communication with the subarachnoid space indicative of a skull base meningocele.

CT is probably more specific for this tumor because of its sensitivity to calcium. MR shows low to intermediate intensity on T1WI and high intensity on T2WI with heterogeneity. The lesions will enhance, but the degree (and the high intensity on T2WI) is tempered by the amount of calcified matrix, which will not show enhancement (Fig. 10-44). The key here is appreciating whether there is calcification. Chondrosarcoma should be in the differential diagnosis of calcified enhancing parasellar masses (Box 10-6). They present 15 to 20 years earlier in life than adult patients with chordomas but can look similar. The site of origin is more commonly the petrooccipital synchondrosis or petrosphenoid fissure rather than the midline clivus or sella (which is more characteristic of chordoma).

INFRASELLAR AND BASE OF SKULL LESIONS

See list of infrasellar and base of skull lesions in Box 10-7.

Developmental Lesions

Basal meningoencephaloceles comprise approximately 10% of all cephaloceles. They include sphenopharyngeal (through the sphenoid body), sphenoorbital (through the superior orbital fissure), sphenoethmoidal (through the sphenoid and ethmoid bones), transethmoidal (through the cribriform plate), and sphenomaxillary (through the maxillary sinus) meningocephaloceles. These anomalies produce round, smooth erosion in the bone of the particular anatomic region. Their intensity on MR depends on the contents of the cephalocele, which includes meninges, brain which may be dysplastic, and/or CSF. High-resolution MRI can elegantly show the contents of meningoencephalocele, while thin section CT images can identify the skull base defect through which the tissue herniates (Fig. 10-45). At the skull base, cephaloceles can present as a pharyngeal mass producing airway obstruction and CSF rhinorrhea, and can be a source for recurrent meningitis.

Juvenile Angiofibroma

Juvenile angiofibroma is seen almost exclusively in young male patients, accounts for approximately 0.5% of head and neck neoplasms, and is the most common benign tumor of the nasopharynx. These tumors may originate in the posterolateral wall of the nasal cavity/sphenopalatine foramen. Characteristically, they grow through the pterygopalatine foramina and may extend into the infratemporal fossa. Other regions of spread include the sphenoid sinus, cavernous sinus, and paranasal sinuses. Unusual locations for these lesions include the parapharyngeal space and the pterygoid muscle region. The tumor can be seen with epistaxis, nasal obstruction, or facial deformity. These masses are extremely vascular, accounting for the common presentation of epistaxis. Preoperative embolization followed by surgery is the present approach to these lesions.

Historically, plain film diagnoses relied on anterior displacement of the posterior maxillary sinus wall, but these days we have better tools at our disposal. MR reveals the extent of the lesion and careful evaluation of the skull base and pterygopalatine fossa is essential in determining the true extent of the tumor (Fig. 10-46). CT can play a complementary role in showing expansion of soft-tissue effacement of the normal fat within the involved skull base foramina. There are no specific intensity characteristics to the tumor itself, although flow voids may be seen in highly vascular lesions. The tumor enhances avidly. Angiography reveals a dense tumor stain, and depending on its location, recruits blood vessels from around the base of the skull, including the inferolateral trunk and branches of the internal maxillary artery, including meningeal arteries.

Chordoma

Chordomas occur at sites of notochordal remnants and constitute less than 1% of all intracranial tumors.

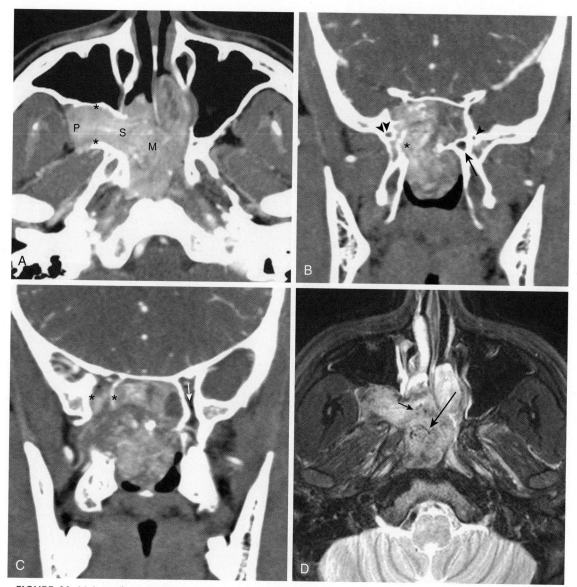

FIGURE 10-46 Juvenile angiofibroma. **A,** The axial contrast-enhanced computed tomography image demonstrate a heterogeneously enhancing mass (M) involving the nasopharynx, posterior nasal cavity bilaterally, and right pterygopalatine fossa (PPF). The PPF (indicated as *space between asterisks*) is markedly widened. Tumor arrived in the PPF through the sphenopalatine foramen (S), also widened, and is spreading to the infratemporal fossa through the pterygomaxillary fissure (region indicated by P). **B,** Coronal reconstruction from the same computed tomographic exam shows widening of the right foramen rotundum *(double arrowheads)* indicating perineural spread, compared with the normal left side *(single arrowhead)*. Vidian canal is completely disrupted on the right *(asterisk)* because of tumor extension compared with the normal canal on the right *(arrow)*. **C,** On this coronally reconstructed image, the right inferior orbital fissure is expanded (indicated *between asterisks*) because of tumor infiltration into the orbit; expected fat within the fissure is obliterated by tumor. The normal left inferior orbital fissure containing fat is indicated on the left *(arrow)*. **D,** The mass shows characteristic serpentine flow voids on T2-weighted imaging, which speaks to the hypervascular nature of this tumor *(arrows)*.

Approximately 35% to 40% of chordomas are cranial (peak age between 20 and 40 years), 50% are in the sacrum (peak age 40 to 60 years, more common in men), and 15% are in the spine. Intracranially, the most common site of origin is the clivus. Other sites of origin include the basioccipital and parasellar regions, and rarely the paranasal sinuses. Chordomas are considered a low-grade malignancy but can be aggressive and extensive, spreading into the nasopharynx or the prepontine regions. Patients have headache, visual disturbances, and usually cranial nerve VI palsy. These signs and symptoms have a variable duration and can remit and recur spontaneously for months to years. Although most chordomas are histologically benign, they are locally aggressive, with a poor overall prognosis. Complete surgical removal is rarely possible; thus, partial resection is performed followed by radiation therapy. Radiation with proton beam or linear accelerator therapy has been advocated because chordomas are resistant to standard radiation treatment protocols.

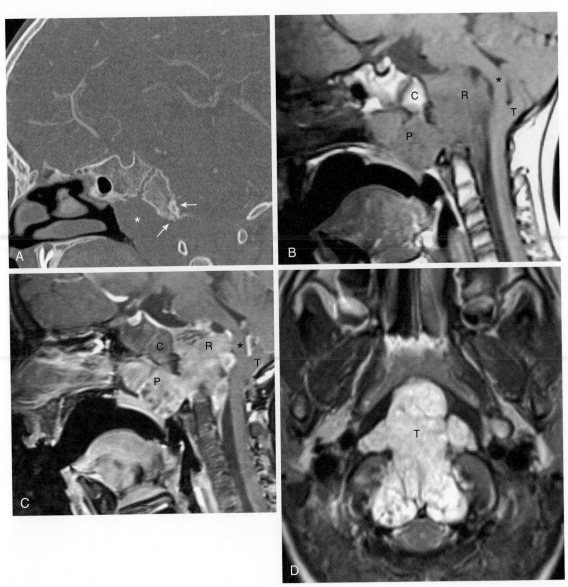

FIGURE 10-47 Clivus chordoma. **A,** Sagittal reconstruction from contrast-enhanced computed tomography shows destructive process centered at the base of the clivus, involving both anterior and posterior margins of the clivus *(arrows).* Large prevertebral soft-tissue component is appreciated *(asterisk),* resulting in near complete occlusion of the nasopharyngeal and oropharyngeal airway. Sagittal precontrast **(B)** and postcontrast **(C)** T1-weighted imaging (T1WI) shows the truncated appearance of the clivus (C) due to destructive tumor mass. Prevertebral (P) and retroclival (R) soft-tissue components are shown here. The retroclival component compresses and posteriorly displaces the medulla *(asterisk)* and there is crowding of the cerebellar tonsils (T) because of mass effect. **D,** Axial T2WI shows characteristic high signal within this lesion (T). Note that the tumor is centered at midline, as opposed to off-midline, which would be more characteristic of chondrosarcoma.

CT and MR are complementary in imaging these lesions (Fig. 10-47). CT demonstrates a soft-tissue mass, bone destruction, and calcification. Chordomas can be very large, and multiplanar imaging is essential for delineation of tumor extent. On T1WI, chordomas are isointense to brain parenchyma, although some regions of low intensity (calcification) and high intensity can be identified. On T2WI, high intensity is the rule. Chordomas enhance. The key to the diagnosis is location, bone destruction, and calcification. Clival lesions can be tricky diagnostically. Assess the CT carefully for any attachment of the lesion to the clivus. Sometimes this can be subtle; however, when identified, it favors chordoma. Midline location can help favor chordoma over chondrosarcoma, which tends to be paramidline.

Metastatic Lesions

The most frequent metastatic lesions to the base of the skull are from prostate, lung, and breast primaries. Prostate metastases are osteoblastic and can simulate meningioma-like bone reaction, whereas lung and breast are more commonly lytic.

Granulomatosis with polyangiitis (formerly Wegener granulomatosis) and inflammatory pseudotumor (IgG4RD) can be hypointense on T2WI and can mimic malignant neoplasms of the skull base, in particular, lymphoma.

Other Infrasellar Lesions

Sellar macroadenomas typically extend superiorly into the suprasellar region, but occasionally they may be seen extending inferiorly into the infrasellar region, with involvement of the sphenoid sinus, clivus, and even nasopharynx. The imaging appearance is similar to their supratentorial counterparts. Recognition that the lesions arise from the sella can help aid in the diagnosis.

Nasopharyngeal carcinoma, sphenoid metastases, and chordomas arising in the sphenoid sinus can extend to involve the sella in a secondary fashion. Sphenoid sinus mucoceles are expansile slow-growing lesions that have a variable intensity pattern but have a tendency for high intensity on T1WI because of proteinaceous contents; these lesions do not enhance centrally but a thin rim of contrast enhancement can be seen along lesion margins. A thin margin of bone is usually present, best appreciated on CT.

Fibrous dysplasia is a common developmental bone lesion to affect the central skull base. On CT, ground glass density is seen in affected regions, which may be expanded. Lytic changes may be seen within the lesion. When extensive, the associated bone enlargement can narrow skull base foramina and can compress critical soft-tissue structures, such as within the orbit. On MRI, these lesions show a mixed signal intensity on T1WI and T2WI with variable contrast enhancement. These lesions can appear aggressive on MRI, and CT can help specify the diagnosis.

Craniopharyngioma can present as an infrasellar mass. Similar to the suprasellar appearance of this tumor, craniopharyngiomas may calcify in 30% to 50% of cases, show cyst formation, and enhance. Intensity on MR as described earlier can be variable, and that's all we have to say about that (Forrest Gump, anyone?).

Chapter 11

Temporal Bone

The detailed bony anatomy of the vestibulocochlear structures of the temporal bone makes computed tomography (CT) the primary method of evaluating the erosive and inflammatory lesions of the temporal bone. Magnetic resonance imaging (MRI) also has an important role in the evaluation of temporal bone pathology, particularly when evaluating vascular and neoplastic processes affecting this region. A good understanding of the anatomy of the temporal bone as well as clinical symptoms is essential in reviewing temporal bone imaging studies, as many pathologies in this region can be overlooked without this necessary information. For this reason, normal anatomy for each region in the temporal bone in addition to the lesions that can be seen in these regions is reviewed in this chapter.

EXTERNAL AUDITORY CANAL

Normal Anatomy

The external auditory canal (EAC) is derived from the first branchial groove and is an ectodermal structure. First and second branchial arches contribute to the cartilaginous EAC and auricle. Where the ectodermal EAC joins the endoderm of the first branchial pouch is the medial boundary of the EAC, the tympanic membrane. The EAC is made of fibrocartilage laterally and bone medially. The posterior border of the glenoid fossa housing the temporomandibular joint serves as part of the anterior wall of the EAC. Can you hear me now?

Congenital Anomalies

Atresia and Hypoplasia

Congenital anomalies of the EAC are rather common, more so than middle ear abnormalities. The degree of congenital deformities of the EAC runs the gamut from total atresia to hypoplasia or stenosis of the EAC, to microtia (a small auricle of the ear) (Fig. 11-1, Box 11-1). Thalidomide embryopathy accounted for a large blip in EAC dysplasias in the early 1960s—rubella infections still account for some these days.

The degree of microtia tends to correlate with the extent of stenosis of the EAC. Middle ear malformations also vary with the severity of the auricular anomalies. Pneumatization of the middle ear often follows the degree of microtia; 2/3 of patients with major microtia (absence of normal auricular structures) and anotia have reduced pneumatization of the middle ear and mastoid. This is important to note because the surgeon needs to know how much drilling is necessary and what size space he/she has to work with to reconstruct the middle or inner ear structures. In addition, it is important to know whether the inner ear structures are normal. Thirteen percent of patients with microtia have dysplastic inner ear structures—usually a hypoplastic lateral semicircular canal.

In minor microtia (pocket ear, absence of upper helix, absence of tragus, miniear, clefts, and so forth), 50% of patients have a dysplastic malleus-incus complex, whereas with major microtia, 67% are dysplastic and 30% have absent ossicles altogether. The incudostapedial joint is commonly abnormal (>65%) in both minor and major microtia. Abnormalities of the stapes coexist in up to 70% of cases.

Bilateral EAC atresia is seen in 23% to 29% of cases. There is often concomitant abnormality in the temporomandibular joint, with EAC anomalies manifested by flattening or absence of the glenoid fossa. Dysplasia of the mandibular condyle (remember both the EAC and mandibular condyle are formed from first branchial apparatus structures and are associated with the fifth cranial nerve) and defects of the zygomatic arch may be seen. At the extreme, atresia of the EAC may be associated with hypoplasia of the malleus, fusion of the malleus and incus, or other anomalies of middle ear structures (in greater than 50% of cases).

With a stenotic EAC, the slope of the EAC may be more vertically angulated. In addition, keratinous plugs and cholesteatomas may form. Occasionally, there is an associated anomaly of the inner ear (11% to 30%); this is rare because inner ear structures are derived from the neuroectoderm as opposed to the branchial system (but the insult that takes out the branchial apparatus can also do damage to neuroectoderm). In patients with EAC atresia, the horizontal segment of the facial nerve may have an aberrant course in close proximity to the stapes footplate and may be anteriorly located in its descending portion. The facial nerve may also be dehiscent in its tympanic course. Evaluation of the facial nerve position, carotid and jugular anomalies, middle ear and mastoid air cell pneumatization, ossicular deformities, meningoencephaloceles, sigmoid sinus location, and other features is important for preoperative planning to prevent accidental surgical injury.

Surgery to correct the EAC is fruitless if the inner ear is nonfunctional. Surgery for external ear anomalies is difficult because it often requires grafts of bone and cartilage as well as drilling new canals. Thus, correction just for cosmetic purposes is usually delayed until after adolescence, when growth has slowed down. However, if the ear anomaly leads to learning disabilities (e.g., bilateral hearing loss from EAC atresia) it may be treated before schooling (Table 11-1).

First Branchial Cleft Cyst

Another congenital lesion that can be seen at the EAC is the first branchial cleft cyst. The branchial arches and clefts are the precursors of the soft tissues and bones of the neck, and when they fail to involute appropriately embryologically, clefts, fistulae, and sinuses can form in relatively typical locations in the neck soft tissues. First branchial cleft cysts occur around the ear and/or in the neighboring parotid gland (Fig. 11-2). A fistula from a first branchial cleft anomaly may drain to the EAC at

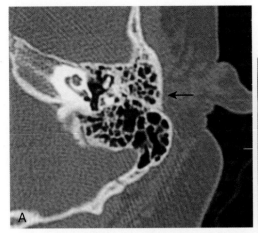
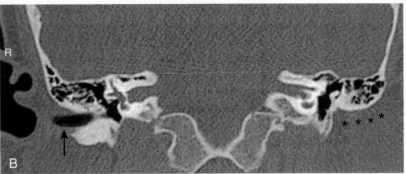

FIGURE 11-1 External auditory canal (EAC) atresia. **A,** Axial computed tomography demonstrates absence of EAC on the left side *(arrow).* The ossicles are deformed. **B,** The finding is confirmed on the coronal image, where one can see the normal right EAC *(arrow),* the absent EAC *(asterisks),* and the lateralized ossicles on the left.

BOX 11-1 Associations with Microtia

Idiopathic
External auditory canal atresia
Down syndrome
CHARGE syndrome
Crouzon syndrome
Treacher Collins syndrome

CHARGE, Coloboma, heart defects, choanal atresia, retardation, genitourinary abnormalities, ear abnormalities.

TABLE 11-1 Checklist for Evaluating for External Auditory Canal Atresia or Stenosis

Item	Rationale
Inner ear structures	No sense fixing the outer, middle ear if no sensorineural function
Stapes	Implies inner ear anomaly, requires implant
Oval and round window	Access to perilymph, endolymph
Middle ear space	Need place to put, repair ossicles, transmit sound
Facial nerve	Do not want to injure nerve
Ossicular anatomy	How many need to be replaced? Are there functioning joints?
Carotid artery, jugular vein	Makes for a bloody mess if they are anomalous and in the way

the bone-cartilage interface (Arnot type 2 first branchial anomalies). These are much less common than the second branchial cleft cyst. Although these are congenital lesions, they typically present in middle age, with recurrent drainage from the EAC. On imaging, an organized fluid density/signal mass can be seen in and around the parotid gland. Peripheral rim enhancement can be seen if the cyst becomes infected.

Calcifications of the External Ear Organs

Calcifications of the external ear and/or pinna occur in a variety of congenital and acquired lesions. Box 11-2 lists these entities.

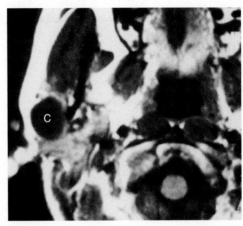

FIGURE 11-2 First branchial cleft cyst. Axial T1-weighted image shows a cystic mass (C) in the right parotid gland near the external auditory canal. The first branchial cleft anomaly can communicate with the external ear.

BOX 11-2 Causes of Calcified External Ear Organs

Idiopathic
Hyperparathyroidism
Diabetes mellitus
Pseudogout
Gout
Radiation
Frostbite
Relapsing polychondritis
Ochronosis
Acromegaly
Hypoparathyroidism
Addison disease

Inflammatory Lesions

Malignant Otitis Externa

The most severe inflammatory condition affecting the EAC is malignant otitis externa, a *Pseudomonas* infection of the EAC seen in elderly diabetic patients (93% of cases). Immunocompromised patients are also at risk, presenting with purulent discharge from the ear. The infection usually begins at the junction of the cartilaginous and bony portion of the EAC along the fissures of Santorini, which lead to the parapharyngeal space. The infection can spread to the infratemporal fossa, the nasopharynx, the parapharyngeal space, the adjacent bone, the temporomandibular joint, the middle and inner ear structures, and intracranially in the extradural space. Palsies of cranial nerves VI, VII, and IX through XII can indicate extension of the disease at the skull base and neural foramina. Venous sinus thrombosis is a complication. The process may mimic an aggressive neoplasm in many respects and often is difficult to control with antibiotics.

CT will identify soft tissue in the EAC, bony erosion of the EAC walls and skull base. MRI shows edema within the EAC mucosa, the parapharyngeal fat may be obliterated as the infection extends anteriorly and medially, and the tissue planes around the carotid sheath may be infiltrated (Fig. 11-3). Marrow involvement is evidenced by edema (high T2 signal and low T1 signal) and enhancement on MRI but CT shows later development of bony erosion. Affected cranial nerves can show abnormal enhancement. Technetium (Tc) 99m bone scans and gallium 67 citrate scans show uptake of radiotracers and may be useful to assess activity of disease.

Keratosis Obturans

This condition, caused by plugs of keratin in the EAC, is very painful and is usually seen bilaterally in middle-aged adults with histories of bronchiectasis and/or sinusitis. Temporal bone CT shows soft tissue opacifying and expanding the EAC without erosive change. The tissue does not enhance. These plugs can be surgically excised.

Swimmer's Ear and Surfer's Ear

Swimmer's ear (acute external otitis) is usually attributable to *Pseudomonas* infection. Rarely, mastoiditis and/or osteomyelitis may complicate swimmer's ear. The diagnosis of acute external otitis is made clinically, and imaging is typically not required unless complications are suspected.

Patients who swim in cold water are prone to exostoses of the EAC (surfer's ear) (Fig. 11-4). Patients present with conductive hearing loss and CT can show smoothly marginated broad-based circumferential sclerosis of the bony EAC resulting in narrowing of the canal. These are benign and typically require no treatment, unless hearing loss is severe. A more pedunculated bony overgrowth at the junction of the cartilaginous and bony EAC can be diagnosed as osteoma, which also presents with conductive hearing loss, is benign and usually requires no treatment. Some authors believe exostoses and osteomas are the same entity.

Neoplasms

See list in Box 11-3.

Benign Neoplasms

Benign masses of the external ear include hemangiomas, nevi, ceruminomas (adenomas of the ceruminous glands), polyps, and salivary gland tumors. All of these present as soft-tissue masses that may expand the EAC without destruction. All may enhance. Venous vascular malformations (VVMs), while not true neoplasms, can also occur in the external ear and could be mistaken for hemangioma, a true neoplasm. VVMs may have phleboliths and have more gradual enhancement than hemangiomas on dynamic imaging.

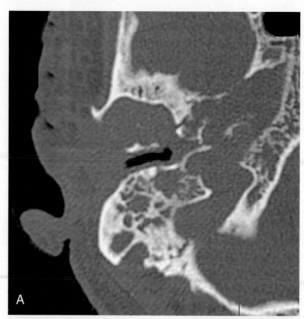

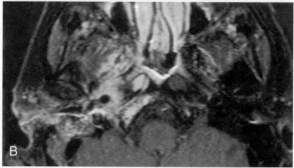

FIGURE 11-3 Malignant otitis externa. **A,** Axial computed tomographic image shows thickening of the external auditory canal (EAC) mucosa, mastoid opacification, and destructive changes at the mastoid and petrous portions of the temporal bone. **B,** The fat-suppressed T1-weighted imaging scan shows enhancing tissue in the EAC, mastoid, petrous tip, right side of clivus, right longus colli muscle, and parapharyngeal space.

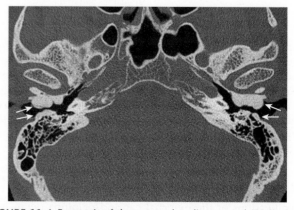

FIGURE 11-4 Exostosis of the external auditory canal (EAC). Axial computed tomographic image reveals dense EAC exostoses bilaterally *(arrows)*. Note that the hyperostosis is also seen along the posterior margin of the EAC.

Malignant Neoplasms

Squamous Cell Carcinomas. Squamous cell carcinomas of the EAC are the most common neoplastic processes in this location. This is essentially a skin cancer with the associated risk from sun exposure. The lesion may invade the middle ear and temporal bone. Cartilaginous invasion of the external ear or middle ear extension portends poor prognosis and must be treated aggressively. The 5-year prognosis of EAC squamous cell carcinomas without middle ear disease is 59% but is 23% if the cancer extends into the middle ear. Deep lesions in the bony canal have the worst prognosis (Fig. 11-5). Pain occurs early because of periosteal spread or extension to the temporomandibular joint, where trismus may also arise. Facial nerve involvement also occurs early in the course. Metastatic adenopathy within the parotid and periparotid nodes that drain the EAC can be seen in conjunction with the primary tumor.

CT can identify bony erosion and destruction along the osseous margins of the EAC and adjacent temporomandibular joint and mastoid air cells. Intracranial extension is better evaluated by contrast-enhanced MRI. While distinguishing the tumor from obstructed opacified air cells can be difficult, on MRI, the tumor tends to be darker on T2 images, whereas obstructed air cells are T2 bright, making MRI a more reliable tool in assessing the margins of the tumor. The tumor enhances solidly. Secretions enhance on the periphery.

Other Primary Malignant Tumors. Other skin tumors such as basal cell carcinomas or melanomas may also affect the external ear in the same manner as squamous cell carcinomas. They can demonstrate perineural spread via cranial nerves V and VII. Lymphatic drainage may be to intraparotid, retropharyngeal, occipital, and skull base lymph nodes. Kaposi sarcoma in individuals who are human immunodeficiency virus (HIV) positive can affect the ear. In children, rhabdomyosarcomas and lymphomas may present as external ear masses.

Direct extension into the EAC from a primary tumor outside the EAC can also be seen. Carcinomas of the parotid gland commonly invade the temporal bone and EAC, and perineural growth may occur along cranial nerve VII.

Metastases. Metastases rarely affect the EAC portion of the temporal bone.

| BOX 11-3 | External Auditory Canal Masses |

Cerumen impaction
Foreign body
Exostosis
Hemangioma
Polyp
Basal cell carcinoma
Squamous cell carcinoma
Melanoma
Papilloma
Malignant otitis externa
Keratosis obturans
Epidermoid
Metastasis
Rhabdomyosarcoma
Minor salivary gland neoplasm
Ceruminoma
Chondroid neoplasm
Langerhans cell histiocytosis
Nevi

THE MIDDLE EAR AND FACIAL NERVE

Normal Anatomy

OK, take a deep breath...this will not be as bad as you think. The middle ear or tympanic cavity is often divided into a superior attic or epitympanic recess, the mesotympanum at the level of the tympanic membrane, and

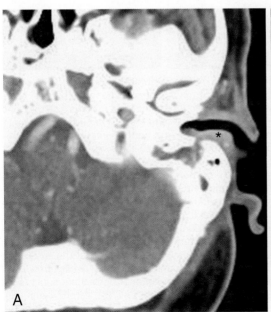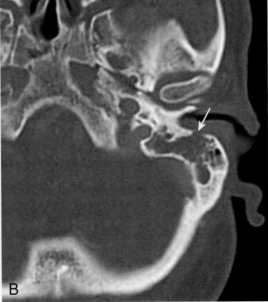

FIGURE 11-5 External auditory canal (EAC) squamous cell cancer. **A,** Axial contrast-enhanced computed tomography (CT) image in soft-tissue windows shows mucosal based thickening along the posterior wall of the EAC *(asterisk).* **B,** Close review of CT in bone windows at the same level shows associated focal bony destruction *(arrow),* making this lesion suspicious for aggressive process. Biopsy revealed squamous cell carcinoma.

the hypotympanum lying inferior and medial to the tympanic membrane (Fig. 11-6) kind of like the top floor (attic), main level (mesotympanum), and basement (hypotympanum) of your dream house. Some of the contents of the middle ear cavity and eustachian tube are derived from the first branchial pouch. The first branchial arch forms the bodies of the malleus and incus and the short process of the incus. The second branchial arch forms the superstructure (capitulum and crura) of the stapes and long process of the incus as well as the manubrium of the malleus. The first branchial pouch invaginates into the eustachian tube, mesotympanum, and mastoid air cells.

The tympanic membrane is the lateral border of the middle ear. It has a thin anterosuperior portion known as the pars flaccida and a tougher posteroinferior pars tensa. The tympanic membrane slants down and inward so that the posterosuperior wall is shorter than its anteroinferior wall. The umbo is the inward puckering of the tympanic membrane at the attachment of the handle of the malleus. The malleus has a head, which articulates with the body of the incus; a neck, an anterior process that attaches by ligaments to the wall of the mesotympanum and to the tensor tympani muscle; a lateral process; and a manubrium, which connects to the tympanic membrane. The tensor tympani muscle attaches to the upper manubrium and neck of the malleus. The named portions of the incus are its body, short process, and long process. The short process attaches by ligaments to the posterior tympanic cavity wall, whereas the long process parallels the manubrium posteromedially before bending medially and articulating with the

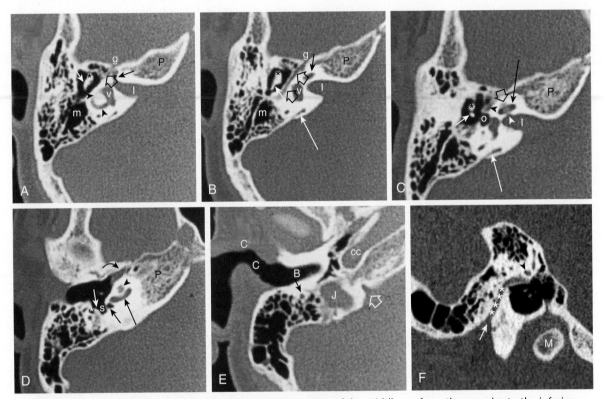

FIGURE 11-6 Computed tomography of the normal anatomy of the middle ear from the superior to the inferior region. **A,** Axial view of labyrinthine portion of the facial nerve *(black arrow),* geniculate ganglion of facial nerve (g), proximal portion of the horizontal segment of the facial nerve *(open arrow),* head of malleus *(asterisk),* and short process of incus *(white arrow).* A joint is barely seen between the incus and malleus (incudomalleal joint), vestibule (v) with the lateral semicircular canal *(black arrowheads),* mastoid (m), nonpneumatized petrous apex (P), and internal auditory canal (IAC; I). **B,** Axial view of middle turn of the cochlea *(black arrow),* geniculate ganglion of facial nerve (g), horizontal segment of the facial nerve *(open arrows),* head of malleus *(asterisk),* short process of incus *(small white arrow),* vestibule (v), mastoid (m), nonpneumatized petrous apex (P), IAC (I), and vestibular aqueduct *(large white arrow).* **C,** Axial view of middle turn of the cochlea *(large black arrow),* apical turn of the cochlea *(open arrowhead),* neck of malleus *(asterisk),* long process of incus *(small white arrow),* oval window (o), tensor tympani muscle *(small black arrowhead),* cochlear aperture *(white arrowhead),* nonpneumatized petrous apex (P), IAC (I), and vestibular aqueduct *(large white arrow).* **D,** Axial view of the hypotympanum. Basal turn of the cochlea *(large black arrow),* apical turn of the cochlea *(black arrowhead),* nonpneumatized petrous apex (P), round window niche *(small black arrow),* superior aspect of the eustachian tube and tensor tympani tendon *(curved black arrow),* sinus tympani (s), pyramidal eminence *(white arrow),* and facial nerve recess *(asterisk).* **E,** Axial view of the region inferior to the hypotympanum. Cochlear aqueduct *(open arrow),* cartilaginous portion of external auditory canal (EAC; C), bony portion of EAC (B), carotid canal (cc), jugular bulb (J), and descending portion of facial nerve *(black arrow).* **F,** Sagittal oblique reconstruction of the right temporal bone shows the descending (mastoid) segment of the facial nerve *(asterisks)* and the stylomastoid foramen *(arrow).* The tympanic segment of the facial nerve travels just underneath the lateral semicircular canal *(arrowhead).* Mandibular condyle is noted (M).

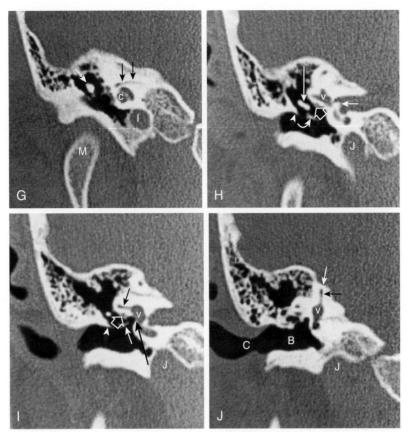

FIGURE 11-6, cont'd **G-J,** Coronal images from anterior to posterior. **G,** Internal carotid artery (I), cochlea (c), facial nerve coursing over cochlea *(black arrows),* head of malleus *(white arrow),* and mandibular condyle (M). **H,** Crista falciformis of IAC *(small white arrow),* jugular bulb (J), vestibule (v), head and neck of malleus *(large white arrow),* tympanic portion of the facial nerve *(open arrow),* scutum *(arrowhead),* and incus *(curved arrow).* **I,** Jugular foramen (J), vestibule (v), incudostapedial joint *(white arrow),* tympanic portion of the facial nerve *(open arrow),* scutum *(arrowhead),* oval window *(long black arrow),* and lateral semicircular canal *(short black arrow).* **J,** Jugular foramen (J), arcuate eminence *(white arrow),* superior semicircular canal *(black arrow),* vestibule (v), cartilaginous portion of EAC (C), and bony portion of EAC (B).

stapes via its lenticular process. The stapes has a head (capitulum), which articulates with the lenticular process of the incus, an anterior crus, a posterior crus, and a footplate.

The footplate of the stapes covers the oval (vestibular) window. The stapedius muscle arises from the pyramidal eminence and attaches to the head of the stapes. This muscle dampens sound by preventing excessive stapedial vibration. This explains the hyperacusis with seventh nerve palsies, as this muscle is innervated by a branch of the facial nerve (Figs. 11-7, 11-8).

The scutum is a sharp, bony excrescence seen best on coronal images forming the superomedial margin of the EAC (inferolateral attic wall) from which the tympanic membrane descends. It protrudes from the roof of the epitympanic cavity, the tegmen tympani. The air space between the scutum and the middle ear ossicles is called Prussak space and is the first area filled by a pars flaccida cholesteatoma.

The tensor tympani muscle courses parallel to the eustachian tube lateral to the neck of the malleus. The petrosquamous suture connects the lateral tegmen to squamous temporal bone and transmits veins to the intracranial space; this can be a source of spread of infection.

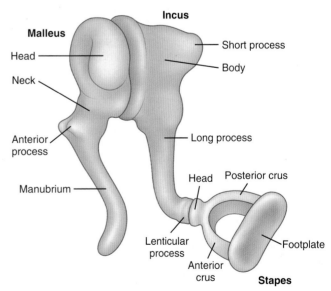

FIGURE 11-7 Schematic shows ossicles including their named parts and articulations.

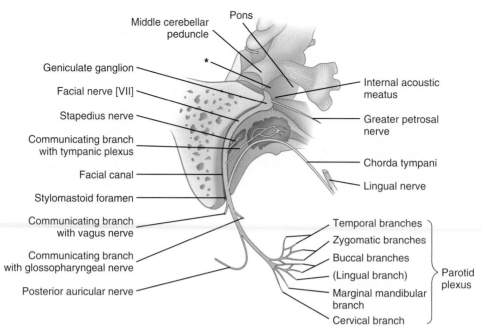

FIGURE 11-8 Facial nerve branches. Schematic shows branches of facial nerve including the greater (superficial) petrosal nerve (for lacrimation and salivation), the stapedius nerve (to dampen sound), the tympanic plexus, the chorda tympani (for taste), and infratemporal branches for facial muscles. (From Putz R, Pabst R. *Sobotta Atlas of Human Anatomy, Vol. 1: Head, Neck, Upper Limb.* Philadelphia: Lippincott Williams & Wilkins; 2001. Fig. 652, p 370.)

BOX 11-4	Segments of Cranial Nerve VII

Cisternal (cerebellopontine angle cistern)
Intracanalicular
Labyrinthine (fallopian canal)
Geniculate ganglion (first genu)
Horizontal (tympanic)
Intramastoid (second genu, descending portion)
Intraparotid

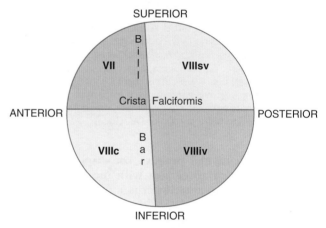

FIGURE 11-9 Internal auditory canal (IAC) subdivisions. Diagram demonstrating Bill bar and crista falciformis separating the IAC into four quadrants with cranial nerve VII and cochlear (VIIIc), superior vestibular (VIIIsv), and inferior vestibular (VIIIiv) divisions of cranial nerve VIII.

In the inferoposterior portion of the middle ear cavity, four important structures are visible on axial scans. They are, medially to laterally, the round window niche, the sinus tympani, the pyramidal eminence, and the facial nerve recess (see Fig. 11-6). The sinus tympani and facial nerve recess are indentations in the bone; the pyramidal eminence is a bony hillock separating the two. The stapedius muscle belly and tendon lie in the pyramidal eminence. These recesses are important in that disease at these sites can be difficult to see on direct inspection during middle ear surgery; recognition of disease at these levels on imaging before surgery is key to minimizing residual tissue.

The facial nerve courses through the middle ear after exiting from the internal auditory canal (IAC) (Box 11-4). The IAC is separated into superior and inferior sections by the transverse crista falciformis and into anterior and posterior quadrants by Bill bar (Fig. 11-9). Cranial nerve VII is found in the anterosuperior portion of the IAC ("Seven-Up"). The facial nerve has a labyrinthine segment coursing anterosuperiorly and laterally in the fallopian canal from the IAC to the geniculate ganglion. The geniculate ganglion is superior to the cochlea (see Figs. 11-7, 11-8). Here it gives off the greater superficial

petrosal nerve, which contributes to lacrimation and salivation. From the geniculate ganglion, the facial nerve forms its first genu and runs posteroinferolaterally on the undersurface of the lateral semicircular canal and above the oval window niche in its horizontal (or tympanic) segment. The facial nerve then makes its second turn (the second genu) to course inferiorly in the mastoid bone in its descending (or mastoid) segment before exiting at the stylomastoid foramen. Along its course it gives innervation to the stapedius muscle and, just above the stylomastoid foramen, to the chorda tympani for taste to the anterior two thirds of the tongue. The chorda tympani doubles back on itself running superiorly and reenters the mesotympanum before exiting

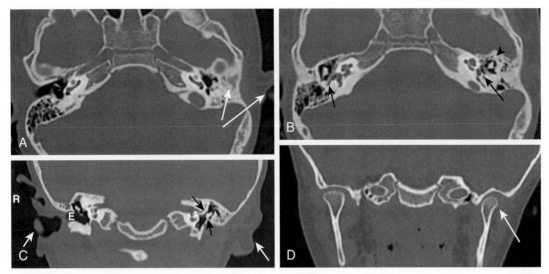

FIGURE 11-10 Left microtia and external auditory canal (EAC) atresia. **A,** Axial computed tomographic (CT) scan through the temporal bone shows an absent left EAC and malformed auricle *(arrows).* **B,** The ossicles on the left are fused and there is no incudomalleolar joint *(arrowhead).* The position of the facial nerve second genu on the left is anterior to the right *(arrows),* which puts it in potential danger during the ossicular reconstruction procedure. Note decreased air space of left middle ear. **C,** Coronal reconstruction of axial CT images shows the maldeveloped left pinna with microtia (compare *white arrows*), the absence of an EAC on the left (see normal EAC indicated by E on the right), with a poorly aerated left middle ear and fused ossicles *(black arrows).* **D,** When the EAC and ossicles are congenitally malformed, the temporomandibular joint *(arrow)* on the same side is also typically maldeveloped as seen on the left here—shallow and maloriented.

anteriorly via the petrotympanic fissure to join the lingual nerve.

The middle ear cavity connects via the eustachian tube to the nasopharynx at the torus tubarius. This explains the frequent coexistence of serous otitis media and/or mastoiditis with nasopharyngeal carcinoma or adenoidal hypertrophy. The eustachian tube is a conduit for spread of lesions in both directions (e.g., malignant otitis externa from ear to nasopharynx and carcinoma from nasopharynx to middle ear cavity). It is also a conduit of cash from parents to pediatricians as the adenoids obstruct the tube leading to recurrent otitis media and numerous office visits. The presence of unilateral otomastoid effusions on imaging should prompt a search for nasopharyngeal mass as a cause of the effusions.

Congenital Anomalies
Hypoplasia and Fusion
The middle ear is a site of congenital dysplasias that may be associated with EAC stenosis or atresia (Fig. 11-10). Ossicular fusion, hypoplasia, or maldevelopment can occur and may coexist with anomalies of the facial nerve as it runs through the middle ear cavity. Isolated middle ear congenital abnormalities are not as common as those of the EAC or inner ear. When they occur in the absence of external ear anomalies, the distal incus (especially the long process) and the stapes are most commonly affected in concert, followed by the stapes alone and incus alone. These components of the ossicular chain are derived from the second branchial arch. Absence, hypoplasia, and fixation to the attic may be seen. A malleus bar is not the best place to grab a drink but rather is a bony fixation of the neck to the posterior tympanic cavity.

Epidermoids
Epidermoids are ectodermal/epithelial rests that may arise in a variety of locations within the temporal bone. To prevent confusion with acquired cholesteatomas, use the term epidermoids rather than the older terms congenital cholesteatomas, epidermoidomas, or primary cholesteatomas. The temporal bone is one of the most common skull base sites of congenital epidermoids. The classic locations for these lesions include the petrous apex, Körner septum (the petrosquamous suture), mastoid air cells, eustachian tube opening, geniculate ganglion region, and middle ear including epitympanum junction, incudostapedial joint, sinus tympani, petrous apex, and facial nerve recess. Intracranial locations for epidermoids are in the cerebellopontine angle cistern and the other posterior fossa cisterns.

Patients present with hearing loss, vertigo, or facial nerve palsy. On direct visual inspection, these lesions are pearly white just like your smile will be when you're done with this chapter.

On imaging, these lesions are nonenhancing, hypointense on T1-weighted imaging (T1WI), and hyperintense on T2-weighted imaging (T2WI). As in the brain, they are as bright as cerebrospinal fluid (CSF) on T2WI but are more intense than CSF on T1WI. They are bright on fluid-attenuated inversion recovery (FLAIR) scanning and bright on diffusion-weighted imaging (DWI) scans. On CT, they appear as noninvasive, low density, expansile, well-circumscribed lesions in the temporal bone with scalloped margins (Fig. 11-11). Although in the middle ear they may be whitish lesions that simulate acquired cholesteatomas, these lesions are not associated with perforation of the tympanic membrane, and patients have no history of antecedent ear infections or previous surgeries. The scutum is usually intact. Epidermoids may be solid or cystic. Treatment is surgical excision.

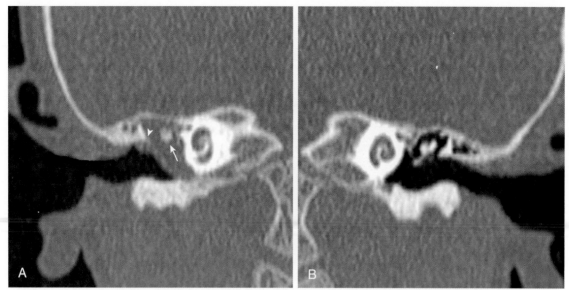

FIGURE 11-11 Epidermoid in middle ear cavity. **A,** Coronal computed tomographic image in 5-year-old child presenting with hearing loss on the right shows expansile tissue in the middle ear, resulting in medialization of the ossicles, demineralization of the malleolar head *(arrow)*, with intact scutum *(arrowhead)*. The tympanic membrane was intact on direct inspection, but there was no question about the presence of the characteristic pearly white mass behind it. **B,** Compare these findings with the normal left side.

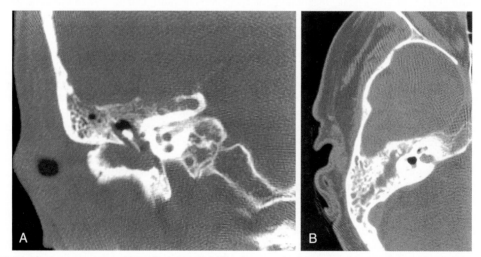

FIGURE 11-12 Acute otomastoiditis. **A,** An air-fluid level is seen in the epitympanic space with complete opacification of the mesotympanum on this coronally acquired computed tomographic image. The mastoid air cells are also opacified. **B,** Air in the vestibule and in the cochlea (pneumolabyrinth), seen in this patient with otomastoiditis, implies an open connection between the middle ear and the inner ear.

Inflammatory Lesions

Otitis Media

Inflammatory disease of the tympanic cavity is common in children. Opacification of the epitympanic recess with thickening of the tympanic membrane is often seen in patients with otitis media. The most frequent causes of otitis media are *Streptococcus, Moraxella catarrhalis, Haemophilus influenzae,* and *Pneumococcus.* Obstruction or impairment of the function of the eustachian tube from nasopharyngeal lymphoid or mucosal hypertrophy caused by upper respiratory tract infections is often responsible for this condition in children. Otitis media generally responds well to antibiotics.

Imaging is typically not required in the diagnosis of otitis media, unless complications are suspected or the

diagnosis is not clear. The middle ear is filled with fluid density and intensity on CT and MR, respectively, in uncomplicated otitis media (Fig. 11-12). Rarely, ossicular erosions, usually a marker for acquired cholesteatoma, can occur in association with acute otitis media. When they occur, they usually affect the long process of the incus and can lead to conductive hearing loss. The erosions appear on CT as tiny, lytic, punched-out areas in the ossicle. Although pneumolabyrinth is more commonly seen in cases of barotrauma or in the postoperative setting, occasionally air can be seen in the inner ear structures with infections, presumably from a labyrinthine fistula.

Another complication of chronic otitis media is ossicular fixation. This may cause a conductive hearing loss and may be fibrous (soft tissue around the ossicles) or

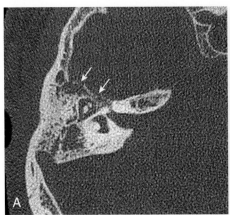
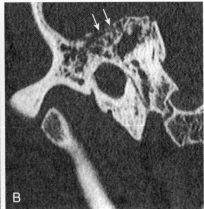
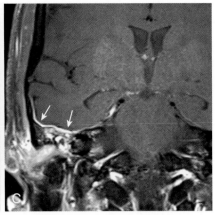

FIGURE 11-13 Coalescent mastoiditis. **A,** Axial and **(B)** coronal computed tomographic images show destruction of mastoid air cell septations and areas of dehiscence along the anterior and superior walls of the mastoid bone *(arrows)*. **C,** Coronal postcontrast T1-weighted image shows abnormal dural-based enhancement along the floor of the middle cranial fossa *(arrows)*, likely reactive, as a consequence of breach of the mastoid roof.

tympanosclerotic (calcification around ossicles or ossicular ligaments).

The finding of fluid opacification in the middle ear is very nonspecific. Middle ear effusions may be present both with infectious and noninfectious processes. The process may be due to a pressure phenomenon and can be seen in patients after air travel in poorly pressurized compartments. Radiation therapy may also be a cause. Alternatively, look for the long-standing nasogastric tube as a predisposing factor that can also lead to mastoid/middle ear air cell opacification.

Mastoiditis

Infection may travel from the middle ear via the aditus ad antrum (the narrow channel connecting the middle ear cavity to the mastoid antrum) to the mastoid air cells. Mastoiditis may occur secondary to otitis media. Again, imaging is not required unless complications are suspected. Coalescence of mastoiditis portends a poor prognosis because it represents bony infection with destruction rather than mucositis. β-Hemolytic streptococci and pneumococci are usually the pathogens involved. CT reveals opacification of air cells; bone destruction constitutes a criterion for coalescent mastoiditis. The fluid is bright on T2WI. Occasionally, air-fluid levels within the small mastoid air cells can be seen. Middle ear and/or petrous apex opacification may coexist (Fig. 11-13).

Complications of acute otomastoiditis include sigmoid sinus thrombosis, thrombophlebitis, epidural abscesses, meningitis, subperiosteal abscesses, fistulas, and osteomyelitis. Cerebellar or temporal lobe encephalitis is uncommon. A Bezold abscess is an inflammatory collection that occurs within the soft tissues inferior to the mastoid tip as the infection spreads from the bone to the adjacent soft tissue (Fig. 11-14). It can spread down the plane of the sternocleidomastoid muscle to the lower neck.

Acquired Cholesteatomas

Long-standing otomastoiditis may result from chronic eustachian tube dysfunction. This may lead to recurrent otitis media and acquired cholesteatomas; these two entities are easier to distinguish in textbooks than in real life, where imaging characteristics may overlap (Table 11-2). Acquired cholesteatomas are erosive collections of

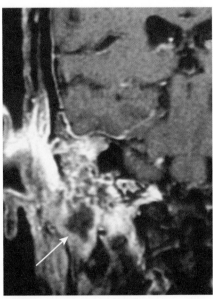

FIGURE 11-14 Bezold abscess. Coronal postcontrast T1-weighted image shows ring-enhancing abscess *(arrow)* just below the opacified mastoid tip.

keratinous debris from an ingrowth of stratified squamous epithelium through a perforated tympanic membrane.

Usually the diagnosis of cholesteatoma is made by the clinician based on clinical presentation and "pearly white" appearance of the lesion on otoscopic evaluation. CT imaging is useful to better understand the full extent of the lesion for surgical planning. The cholesteatoma most often arises from a perforation in the pars flaccida of the tympanic membrane. Once the pars flaccida has been violated, the inflammatory process proceeds into Prussak space, which is located lateral to the ossicles and medial to the scutum in the epitympanic space. The key features in identifying a lesion as a cholesteatoma are the presence of mass effect, bony erosion, and/or expansion (Fig. 11-15). One often sees a soft-tissue mass causing erosion of the scutum and medial displacement of the malleus and incus with pars flaccida cholesteatomas. The head of the malleus and body of the incus are the areas most susceptible to erosion by a pars flaccida cholesteatoma;

TABLE 11-2 **Cholesteatoma versus Otitis Media**

Feature	Cholesteatoma	Otitis Media
Middle ear opacified	Yes	Yes
Scutum	Eroded	Normal
Ossicular erosion	Yes	Infrequent
Ossicular displacement	Yes	No
Expansion of aditus ad antrum	Sometimes	No
Lateral semicircular canal fistula	Sometimes	No
Gadolinium enhancement	Rare	Rare
T2-weighted image signal intensity	Intermediate	Bright
Tympanic membrane retracted	Yes	No
Diffusion-weighted imaging signal	Bright	Low unless really purulent
Tegmen tympani erosion	Sometimes	No
Facial nerve canal dehiscence	Sometimes	No

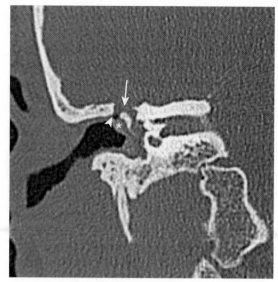

FIGURE 11-16 Tegmen tympani erosion. This example of a cholesteatoma shows erosion of the roof of the epitympanic space *(arrow)* and the scutum *(arrowhead)*. There is opacification of Prussak space *(asterisk)* and entire middle ear cavity including the hypotympanum.

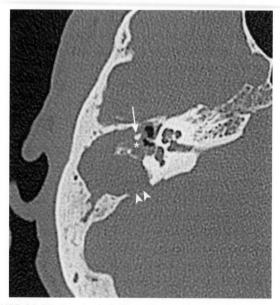

FIGURE 11-15 Acquired cholesteatoma. This rampant cholesteatoma has eroded the ossicles as well as the posterior wall of the mastoid air cells *(arrowheads)*. Remnants of malleus and incus are present *(arrow)*, and the middle ear cavity *(asterisk)* is opacified.

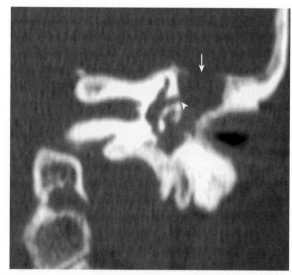

FIGURE 11-17 Lateral semicircular canal dehiscence from cholesteatoma. Coronal reconstruction of temporal bone computed tomographic image shows cholesteatoma focally eroding the lateral margin of the lateral semicircular canal *(arrowhead)* as well as the tegmen typmani *(arrow)*.

lysis of all the ossicles is uncommon. From Prussak space the lesion often spreads through the aditus ad antrum, expanding its waist as the inflammatory process proceeds into the mastoid air cells.

Pars tensa cholesteatomas are much less common than pars flaccida cholesteatomas. They arise from perforations through the posterosuperior-most portion of the pars tensa, which is the inferior portion of the tympanic membrane. From this location the sinus tympani, pyramidal eminence, and facial recess may be expanded and/or eroded. Pars tensa cholesteatomas present with a mass in the hypotympanum of the middle ear, erosion of the long process of the incus or stapes, epitympanic spread, and ossicular displacement. The scutum is usually intact.

Complications of cholesteatomas include a perilymphatic (perilymph/labyrinthine) fistula from the middle ear into the semicircular canals (4% to 25% of cases), with the lateral semicircular canal most commonly affected. This may be identified as a dehiscence in the bony labyrinth with a soft-tissue mass expanding the region of the oval window or lateral margin of the lateral semicircular canal. Alternatively, cholesteatomas may erode the tegmen tympani (the roof of the epitympanic space) and subsequently invade the intracranial compartment (Fig. 11-16). Another area of potential erosion is the lateral or inferior wall of the tympanic portion of the facial nerve (Fig. 11-17). If there is dehiscence or skeletization of the facial nerve canal or sinus tympani, the surgeon must know

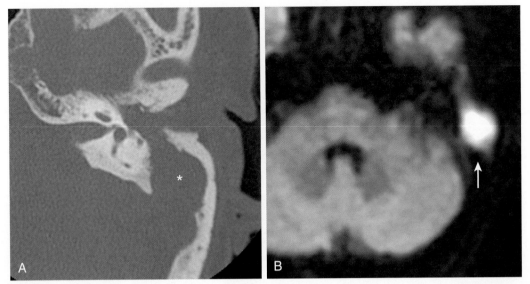

FIGURE 11-18 Recurrent cholesteatoma. **A,** Axial computed tomographic image in a patient with recurrent ear drainage many years following cholesteatoma resection shows expansile opacification within the mastoid bone *(asterisk)* with complete absence of normal mastoid bony architecture (automastoidectomy). Opacification is also present in the middle and external ears. **B,** The expansile soft tissue shows diffusion restriction *(arrow)*, indicating recurrent cholesteatoma.

this preoperatively so that removal of the cholesteatoma is done in a careful manner so as not to injure the underlying structures.

MRI is not typically required in the primary diagnosis of cholesteatoma, but can be helpful in distinguishing recurrent cholesteatoma from granulation tissue in patients who have undergone surgical resection in the past and have continued symptoms. On MR, cholesteatomas are hypointense on T1WI, intermediate on T2WI, and do not enhance, as opposed to granulation tissue (postoperative), which does enhance. Recurrent cholesteatomas will show diffusion restriction; granulation tissue does not (Fig. 11-18).

Other Inflammatory Conditions of the Middle Ear

Granulomatosis with polyangiitis (formerly Wegener granulomatosis) can attack the eustachian tube and from there invade the nasopharynx or skull base.

Langerhans cell histiocytosis affects the temporal bones of children as eosinophilic granulomas in isolation or as part of the wider spectrum of disease (histiocytosis X, Langerhans granulomatosis). Eosinophilic granuloma has a propensity for involving the mastoid portion of the temporal bone. Hearing difficulties without pain may be the initial complaint. CT shows lytic lesions in the involved regions. The classic appearance in the skull of eosinophilic granuloma is a well-defined lytic lesion with beveled edges involving the outer table more than the inner table. On MRI, the lesion is dark on T1WI and bright on T2WI, and enhances. When seen, be sure to check for other lesions elsewhere in the skull, skull base, and orbits as multiplicity can be seen (Fig. 11-19).

Another cause of lysis of the temporal bone is osteoradionecrosis. This has been described most commonly after irradiation for nasopharyngeal carcinoma.

Surgery for Otomastoiditis

Many operations are performed for chronic inflammatory conditions of the middle ear and mastoid air cells.

The simple mastoidectomy spares the EAC and ossicular chain but removes the offending mastoid air cells. Another term for the simple mastoidectomy is the canal wall up mastoidectomy and it preserves the posterior wall of the EAC; canal wall down mastoidectomies take this border down (Fig. 11-20). The modified radical mastoidectomy preserves the ossicles but removes the mastoid air cells and EAC. The radical mastoidectomy removes the mastoid air cells and most of the ossicular chain but preserves the stapes. The incus is the most commonly diseased ossicle; if the stapes can be preserved, a partial prosthesis can be used (partial ossicular replacement prosthesis [PORP] versus total ossicular replacement prosthesis [TORP], which includes stapes down to the footplate). The postoperative mastoid cavity is usually filled with bone chips, fascia, or fat. In all these surgeries, the facial nerve is preserved at all cost.

Ossiculoplasty or tympanoplasty is a procedure to restore the conductive capability of the ossicular chain usually after damage from cholesteatoma, chronic otitis media, or congenital malformation. There are at least five types of tympanoplasties. In type 1 the procedure spares all the ossicles and the graft rests on the malleus, in type 2 the graft rests on the incus, in type 3 the graft rests on the head of the stapes, in type 4 the graft connects to the footplate of the stapes, and in type 5 the stapes is removed. Type 3 PORP is the most common form of tympanoplasty. A stapedotomy refers to a procedure in which a tiny wire connects the long process of the incus to the stapes footplate with only the superstructure resected. A stapedectomy removes the footplate of the stapes and an implant is placed from the incus through the oval window into the labyrinth. A variant of the stapedectomy-stapedotomy procedure is one in which the posterior crus of the stapes footplate is preserved.

Autografts (from the host) for ossicular replacement are difficult to sculpt and are currently rarely used. Homografts

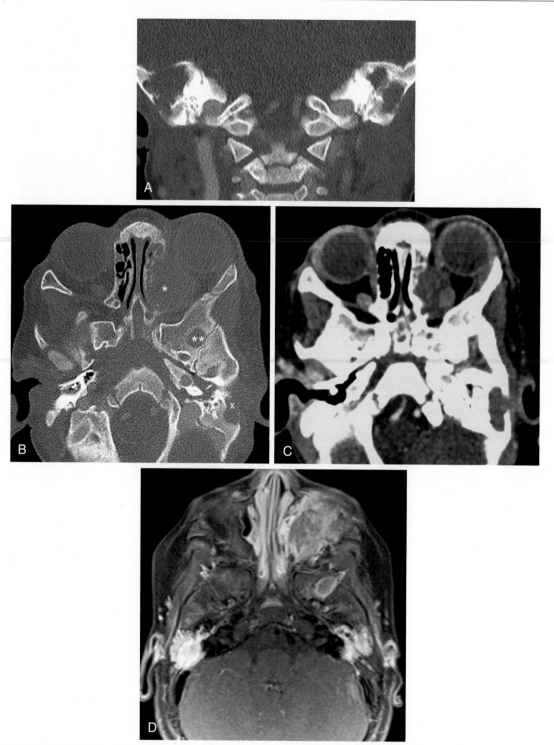

FIGURE 11-19 Eosinophilic granuloma. **A,** Coronal computed tomographic (CT) image in a pediatric patient shows lytic lesions within the mastoid portions of the temporal bones bilaterally. There is dehiscence of the mastoid roof on the right, and thinning/dehiscence along the lateral margins of the mastoid bones bilaterally. **B,** Axial CT image shows the lytic lesion in the left mastoid bone (X) extending into the middle ear; destructive, expansile mass involving the left ethmoid sinus and orbit *(asterisk);* and additional lytic lesion in the left sphenoid wing *(double asterisks).* (In case you are wondering, the high-density structure on the right at the level of the tympanic membrane represents a previously placed tympanostomy tube.) Enhancing soft-tissue components of these same lesions can be seen on **(C)** axial postcontrast CT image and **(D)** axial postcontrast fat-suppressed T1-weighted image.

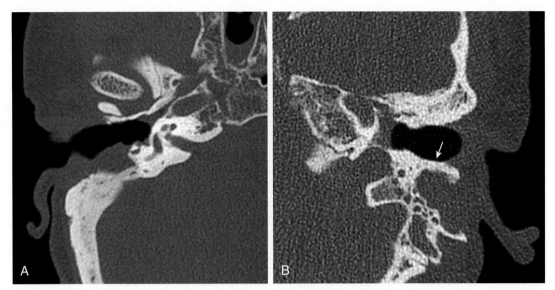

FIGURE 11-20 Up or down? **A,** Axial computed tomographic (CT) image shows canal wall down mastoidectomy defect with surgical absence of the posterior wall of the external auditory canal (EAC). There is accumulation of debris/granulation tissue in the mastoidectomy "bowl." **B,** Axial CT image shows typical appearance of canal wall up mastoidectomy defect with intact posterior wall of the EAC *(arrow).*

from cadavers (for the incus) are not commonly used at present, because of an unspoken fear of transmitting infectious agents. Synthetic ossicular prostheses (Proplast and Plastipore prostheses) have gained favor. Some ossicular prostheses that an erudite neuroradiologist should be familiar with include the stapes prosthesis, the incus interposition graft (for incudostapedial joint disease), the Applebaum prosthesis (a synthetic prosthesis from long process of incus to capitulum of stapes), the Black oval top synthetic prosthesis (from tympanic membrane to capitulum of stapes or oval window), the Richards synthetic prosthesis (from tympanic membrane to capitulum of stapes or oval window), and the Goldenberg prosthesis (from tympanic membrane to capitulum of stapes or oval window, or stapes to malleus or footplate to malleus) (Fig. 11-21). The latter three ossicular replacement prostheses may be total (TORP) or partial (PORP) depending on whether the stapes superstructure is preserved. They therefore extend from the tympanic membrane to the stapes capitulum (PORP) or footplate (TORP). The PORP is the most commonly used synthetic ossicular prosthesis and the hydroxyapatite head of the prosthesis attaches to the tympanic membrane and the Plastipore shaft to the capitulum of the stapes.

Failure of ossicular replacement prostheses may be due to recurrent otitis media, recurrent cholesteatoma, reparative granulomas (foreign body reactions), ossicular subluxation or dislocation, adhesions, fracture of the prosthesis, granulation tissue, recurrence of otospongiotic bone, excessive postoperative bony reaction, or extrusion of the prosthesis. Subluxation of the prosthesis accounts for 50% to 60% of the cases in which there is postoperative hearing loss after ossicular replacement surgery. When PORPs or TORPs are subluxed, it usually occurs at its distal site either with the stapes capitulum or oval window, respectively. In the case of the stapes prosthesis, look for the subluxed prosthesis inferior and posterior to the oval window (Fig. 11-22). The wire between the prosthesis and the incus may also migrate inferiorly.

Failed stapedectomy cases may also be caused by persistent perilymphatic fistulas. Evidence for this complication includes air in the labyrinth or persistent fluid in the middle ear. Extrusions of prostheses may occur through holes in the tympanic membrane from surgery or inflammatory perforations. The prosthesis may be found in these cases in the EAC or on the pillow in the morning. Extrusions rarely occur into the vestibule. The stapes prosthesis should not extend beyond 0.25 mm into the oval window. Too far is not as bad as having an air gap between the stapes prosthesis and the oval window, resulting in conductive hearing loss.

The postoperative examination of a patient who has undergone temporal bone surgery is fraught with difficulties. How can one tell whether the soft tissue seen in the operative cavity is due to recurrent or residual cholesteatoma, scar tissue, acute inflammation, or non-cholesteatomatous granulation tissue on CT? It is nearly impossible unless the recurrent mass is focal and erosive, indicating recurrent/residual cholesteatoma. If the postoperative soft tissue enhances, it is more likely to be granulation tissue than cholesteatoma. Fortunately DWI MRI shows recurrent cholesteatoma as bright with low apparent diffusion coefficient, whereas granulation and acute inflammation is usually dark on DWI. Remember to provide landmarks for the surgeon to assess the mass endoscopically or by biopsy. Identifying areas of dehiscence in the base of the skull or near vascular structures (sigmoid sinus, carotid canal, and jugular bulb) may be the most important role of imaging if reoperation is contemplated.

Benign Neoplasms
Glomus Tympanicum

Glomus tympanicum tumors are fascinating entities that have interesting anatomic ramifications and pathologic manifestations. Ringing in the ears (tinnitus) is a hallmark of these lesions (Box 11-5). The glomus tympanicum is the most common neoplasm of the inferior part of the middle ear, and presents most commonly in middle-aged women.

This tumor arises from the glomus bodies (neural crest tissue) along the tympanic branch of cranial nerve IX known as the Jacobson nerve. This nerve runs from the inferior ganglion of cranial nerve IX, through the inferior tympanic canaliculus (between jugular foramen and carotid canal), to the middle ear. The Jacobson nerve forms the tympanic plexus that may cover the cochlear promontory before rejoining as the lesser superficial petrosal nerve. Branches of the tympanic plexus may be found near the round window, eustachian tube egress, tensor tympani tendon, and along the inferior tympanic canaliculus.

The glomus tympanicum, because it produces symptoms relatively early, is usually seen as a small soft-tissue mass bulging behind the tympanic membrane (Box 11-6), which enhances markedly. Its location is variable within the medial aspect of the middle ear. Although classically it presents at the cochlear promontory (Fig. 11-23), it may also reside anterior to the promontory, beneath the cochleariform process and the semicanal of the tensor tympani, inferior to the promontory, in the recess beneath the basal turn of the cochlea.

The tumors are classified on the basis of their spread from the cochlear promontory, to the middle ear, the mastoid, and the EAC or carotid canal. As opposed to cholesteatomas and other middle ear masses, the glomus tympanicum does not erode the ossicles but engulfs them, best appreciated on high-resolution thin-section CT with multiplanar reconstructions. Because a glomus tympanicum is so small, it is usually identified on MR as an enhancing soft-tissue mass of intermediate signal intensity without large vessels. Although glomus tympanicum tumors gain blood supply from the inferior tympanic artery branch of the ascending pharyngeal artery, they rarely require preoperative embolization because of their small size.

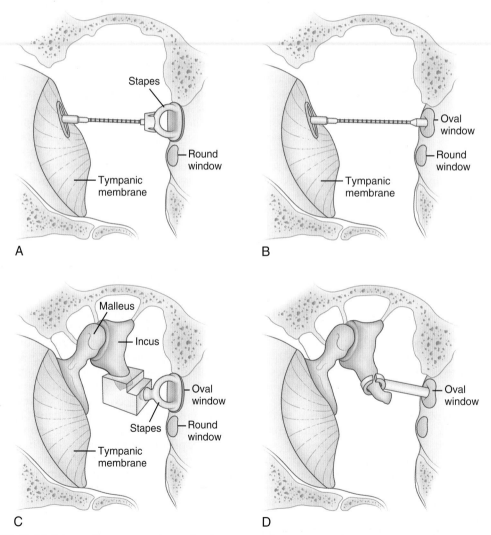

FIGURE 11-21 Schematic representations of various types of ossicular prostheses (by no means all-inclusive!). **A,** Partial ossicular replacement prosthesis (PORP) extends from the tympanic membrane to the stapes capitulum. **B,** Total ossicular replacement prosthesis (TORP) extends from the tympanic membrane to the oval window. **C,** The L-shaped configuration of the Applebaum prosthesis is depicted, with one end of the prosthesis extending over the partially resected long process of the incus and the other extending over the capitulum of the stapes. **D,** In the case of stapedectomy, the stapes is resected and an implant extends from the incus to the oval window.

Glomus jugulare tumors, described later, can grow of sufficient size to present in the middle ear where glomus tympanicums reside, hence the term glomus jugulotympanicum for those at the jugular foramen *and* middle ear cavity. Stay tuned....

Hemangiomas

A hemangioma behind the tympanic membrane is another cause of a vascular tympanic mass. It enhances markedly and has nonspecific density and intensity characteristics unless flow voids are seen. Facial nerve hemangiomas (see later) are another source of a vascular retrotympanic mass. In some cases these are likely VVMs.

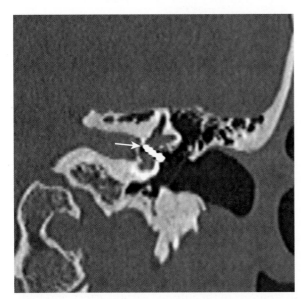

FIGURE 11-22 Subluxed stapes prosthesis. Coronal computed tomographic image shows a displaced stapes prosthesis extending through the oval window deep into the vestibule *(arrow)*. Ouch! The prosthesis should normally not extend 0.25 mm deep to the oval window.

Variations in Vascular Anatomy

These lesions are not tumors but mimic glomus tumors and may present similarly on otoscopic examinations as a retrotympanic vascular mass; they are therefore included in this section. The aberrant internal carotid artery passes through the middle ear cavity and runs anteromedially to the horizontal portion of the cavernous carotid canal (Fig. 11-24). This is easily recognized as a flow void on MR but may be visualized as an enhancing mass of uncertain cause on CT.

BOX 11-5   Causes of Pulsatile Tinnitus

Idiopathic
Idiopathic intracranial hypertension (pseudotumor cerebri)
Ménière disease
High-grade stenosis of internal carotid artery*
High-riding jugular bulb
Paraganglioma (glomus jugulare, glomus tympanicum, glomus jugulotympanicum)*
Hemangioma/venous vascular malformation*
Cholesterol granuloma
Meningioma
Aberrant carotid artery/persistent stapedial artery*
Dural vascular malformations, fistulas*
Carotid aneurysms
Otosclerosis
Paget disease

---

*May be "objective tinnitus."

BOX 11-6   Vascular Intratympanic Masses

High-riding/dehiscent jugular bulb or diverticulum
Paraganglioma (glomus jugulare, glomus tympanicum, glomus jugulotympanicum)
Hemangioma/venous vascular malformation
Cholesterol granuloma/chronic hypervascular inflammatory tissue
Meningioma
Aberrant carotid artery/persistent stapedial artery
Petrous carotid aneurysms

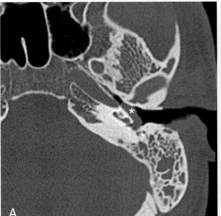
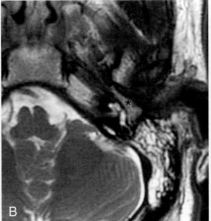
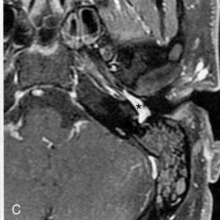

FIGURE 11-23 Glomus tympanicum. **A,** Axial computed tomographic image shows small soft-tissue mass in the middle ear *(asterisk)*. While nondescript in its appearance, in the setting of pulsatile tinnitus, glomus tympanicum is the major diagnostic consideration. The location, overlying the cochlear promontory *(asterisk)* is perfect for glomus tympanicum. **B,** Axial T2-weighted image (T2WI) shows a slightly hyperintense to brain mass in the middle ear *(asterisk)*. Note opacification of mastoid air cells attributable to obstruction of the eustachian tube by the tumor. **C,** Postcontrast fat-suppressed T1WI shows the mass to be avidly enhancing *(asterisk)*.

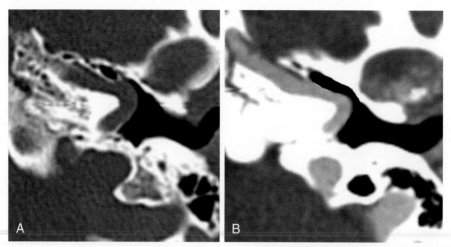

FIGURE 11-24 Aberrant left internal carotid artery. **A,** Axial computed tomographic (CT) image in bone windows shows extension of the petrous internal carotid artery (ICA) into the middle ear. **B,** If you're not convinced that was the ICA, here's a contrast-enhanced CT showing the expected enhancement in the vessel to alleviate all doubt. Tell your surgeon to resist the temptation to biopsy!

The persistent stapedial artery is a rare anomaly (<0.5% of patients). This artery appears transiently in embryologic development as a branch of the hyoid artery (derived itself from the second aortic arch) connecting the external and internal carotid artery. Ultimately, it regresses to form pieces of the caroticotympanic artery. When it persists it may present as a pulsatile middle ear mass. Because it subsumes the role of the middle meningeal artery, its imaging findings include absence of the foramen spinosum, a soft-tissue mass along the horizontal portion of the tympanic facial nerve, and an additional branch leading from the petrous carotid artery. It enters the middle ear to create the obturator foramen of the stapes (between the crura) and leaves the middle ear near the geniculate ganglion (where it can simulate a schwannoma, albeit a bloody one). This anomaly may coexist with an aberrant internal carotid artery (60%). Associations with trisomy 13, 15, 21, thalidomide exposure, anencephaly, and neurofibromatosis have been reported.

Jugular bulb variations are very common and can be associated with pulsatile tinnitus. Do not be alarmed by asymmetry in jugular foraminal size; that is the rule, not the exception. Until you see erosion or soft-tissue destruction, do not call it a glomus tumor. Remember also that the jugular foramen has two parts: a pars vascularis, containing the internal jugular vein and cranial nerves X and XI, and a pars nervosa with cranial nerve IX and continuation of the inferior petrosal sinus. The jugular bulb wall may be dehiscent or nondehiscent; and the jugular vein may rise high into the middle ear cavity (Fig. 11-25). The nondehiscent jugular bulb has preservation of the bony plate of the top of the jugular foramen, whereas the dehiscent jugular bulb shows no bony margin. Most people use either the inferior rim of the tympanic annulus or the IAC as the uppermost limit for a normal jugular bulb. High jugular bulbs may extend into the middle ear or into the petrous bone near the endolymphatic sac. Diverticuli of the jugular bulb may also enter the middle ear. They are also seen as retrotympanic vascular red masses.

Dural Vascular Malformations

Another entity that may cause tinnitus is a dural vascular malformation. These commonly affect the transverse sinus and in some cases may be due to previously thrombosed veins or sinuses, resulting in abnormal collateral flow of arteries and veins around the thrombosis. If no soft-tissue mass in the middle ear or skull base is seen on cross-sectional imaging in patients with objective tinnitus, angiography may be indicated to detect this lesion (Fig. 11-26). At times arteriography is required to distinguish between other lesions that cause pulsatile tinnitus including aneurysms and arteriovenous malformations of the temporal bone, high-flow turbulence associated with stenoses of the internal carotid artery, and high jugular bulbs. Please remember though, the most common neurologic cause of subjective tinnitus is pseudotumor cerebri.

Physicians may separate patients into those with objective pulsatile tinnitus (heard by both the physician and patient) and those with subjective tinnitus (heard only by the patient). Objective pulsatile tinnitus is usually evaluated with angiography of the ipsilateral carotid and vertebral arteries initially, with MR or CT to follow. Subjective tinnitus, by contrast, is best evaluated noninvasively initially with CT or MR. Objective pulsatile tinnitus is usually caused by true vascular malformations, stenoses, or aneurysms. Subjective tinnitus is usually seen in patients with glomus tumors, aberrant internal carotid arteries, and jugular vein abnormalities.

Facial Schwannomas

Facial schwannomas may arise within the IAC, the labyrinthine portion of the facial nerve, the tympanic portion of the facial nerve, the geniculate ganglion, the mastoid portion of the facial nerve, or within the parotid gland. If you identify expansion of the facial nerve canal, you may make the diagnosis of facial schwannoma; however, these lesions often cannot be separated from lesions that occur in neighboring structures such as vestibular schwannomas within the IAC or glomus tumors within the tympanic cavity (Box 11-7). All three of these

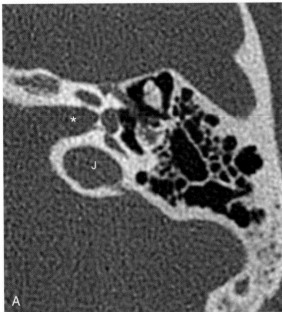

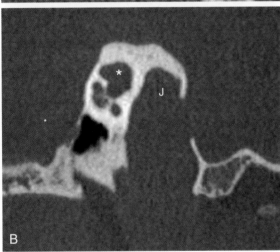

FIGURE 11-25 High-riding jugular bulb. **A,** Axial computed tomographic (CT) image shows the jugular bulb (J) at the level of the internal auditory canal *(asterisk)* in this patient presenting with tinnitus. **B,** Sagittal CT reconstruction in the same patient shows that the roof of the jugular bulb (J) approaches that of the internal auditory canal *(asterisk).*

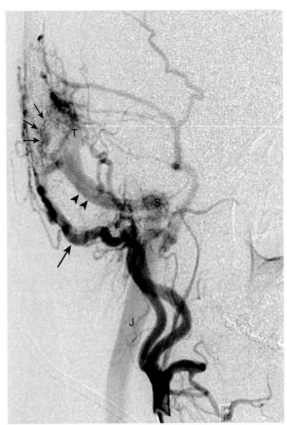

FIGURE 11-26 Dural arteriovenous fistula on conventional catheter angiogram. In this patient presenting with tinnitus, selective catheterization of the external carotid artery demonstrates an enlarged occipital artery *(large arrow)* and multiple fistulous connections (one of which is indicated by *small arrows*) to the venous drainage system. Note the early enhancement within the torcular (T), transverse sinus *(arrowheads),* sigmoid sinus (S), and jugular vein (J).

BOX 11-7	**Intratympanic Facial Nerve Lesions**

Normal enhancement misread as a lesion
Bell palsy
Cholesteatoma
Fracture
Infection (viral, Lyme disease, spread from otitis media or malignant otitis externa)
Schwannoma
Hemangioma/venous vascular malformation
Sarcoidosis
Perineural malignant spread
Meningioma

---

lesions demonstrate enhancement (Fig. 11-27). If you are fortunate enough to see the flow voids of blood vessels within the glomus tumor, then you may be able to make the differential diagnosis.

Facial Hemangiomas

Facial nerve hemangiomas can occur in the IAC, around the geniculate ganglion, and at the posterior genu of the facial nerve canal. Bone erosion with a soft-tissue mass is seen by CT and involved marrow space may have a "honeycomb" bony trabecular internal architecture—pretty, eh? (Fig. 11-28). Again, although called hemangiomas, these are likely VVMs. These tumors often present with slowly evolving facial paresis and twitching. Hemangiomas of the facial nerve are said to be as common as schwannomas of the temporal bone facial nerve—in other words, pretty rare. The borders of a hemangioma are indistinct compared

with the well-demarcated schwannoma. The distinction is important because a schwannoma is rarely resected without sacrificing the nerve, but a hemangioma can sometimes be separated from the nerve.

Facial Nerve Enhancement

The facial nerve normally may show some enhancement at the geniculate ganglion and in its horizontal and descending portions. The reason for the normal enhancement is

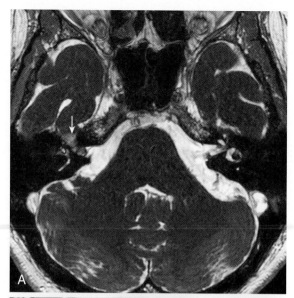

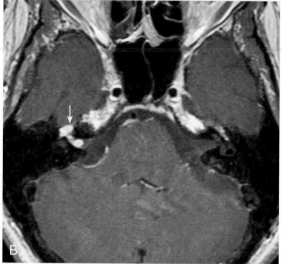

FIGURE 11-27 Facial nerve schwannoma. **A,** Axial T2 constructive interference in steady state image shows expansile mass within the lateral aspect of the internal auditory canal extending along the facial nerve canal on the right *(arrow)*. Compare with normal appearance on the left. **B,** Postcontrast T1-weighted image shows enhancement of this mass involving the canalicular and labyrinthine segments as well as genu of the right facial nerve *(arrow)*.

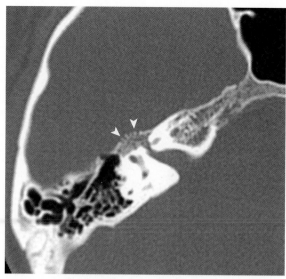

FIGURE 11-28 Hemangioma of the facial nerve. Axial computed tomography demonstrates an expansile mass *(arrowheads)* with permeative change in the bone surrounding the facial nerve canal. Internal architecture should suggest the diagnosis of hemangioma.

the prolific circumneural arteriovenous plexus around the nerve. The plexus is not present in the IAC or intralabyrinthine and extracranial portions of the nerve. Therefore, enhancement of the facial nerve in the cerebellopontine angle cistern, IAC, in the labyrinthine portion, and in the parotid gland, by contrast, is always abnormal. The differential diagnosis of nerve VII enhancement includes schwannomas, Lyme disease, lymphoma, hemangioma, sarcoidosis, Bell palsy, viral neuritis, Ramsay-Hunt syndrome (a herpetic infection), Guillain-Barré syndrome, and perineural tumor spread. Of the two most common lesions, Bell palsy is thought to account for 80% of facial nerve paralysis, with schwannomas comprising only 5%.

In Bell palsy, the region of pathology most commonly is located at the narrowest canal that the facial nerve must go through in its course—at the distal intrameatal

segment of the IAC. Smooth, linear, abnormally intense contrast enhancement of the distal intrameatal segment, indicating peripheral inflammatory nerve palsy is a common finding indicative of a viral (presumed Bell) palsy of the facial nerve. This may also be found after facial nerve trauma related to breakdown of the blood/peripheral nerve barrier associated with nerve degeneration and regeneration after traumatic stretching of the greater superficial petrosal nerve. Scar formation along the nerve may also produce thickening and intense enhancement of the affected nerve segments. Labyrinthine segment enhancement proximal to the geniculate ganglion is another finding commonly seen in Bell palsy. Please note that most people believe that imaging of patients with acute facial nerve palsy should be limited to "atypical Bell palsy," that is if the symptoms persist longer than 4 weeks, the paralysis is progressive, other cranial nerves are affected, pain is a prominent feature, or hemifacial spasm is present.

Another important source of facial nerve dysfunction is malignant perineural spread. Parotid malignancies, especially adenoid cystic carcinoma and lymphoma, have a propensity for tracking up the facial nerve and presenting with Bell palsy. Squamous cell carcinomas of the parotid or face also present this way. Perineural tumor spread along the facial nerve is seen as enlargement and enhancement of the nerve itself, secondary enlargement of the stylomastoid foramen, or loss of normal fatty tissue surrounding the nerve within the foramen (Fig. 11-29).

Malignant Neoplasms

Malignancies of the middle ear are uncommon. Usually these masses are derived from EAC structures and secondarily invade the middle ear. This is particularly true of dermal squamous cell carcinomas and rhabdomyosarcomas. Primary rhabdomyosarcomas of the middle ear may arise from muscular cells along the eustachian tube opening. Rarely, ectopic salivary gland tissue may be present

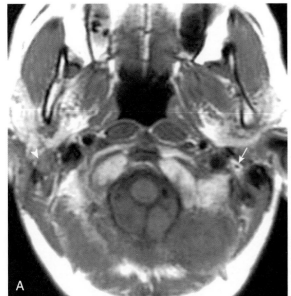

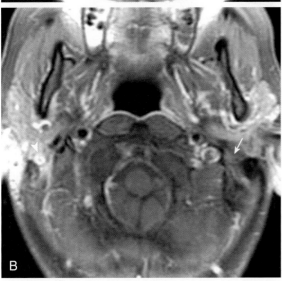

FIGURE 11-29 Perineural tumor spread in a patient with adenoid cystic carcinoma of the right parotid gland. **A,** Axial T1-weighted image (T1WI) shows normal appearance of the facial nerve as it exits the stylomastoid foramen on the left *(arrow)*. Note hypointense nerve surrounded by normal hyperintense fat signal within the foramen. On the right, the fat within the foramen is effaced by infiltrative soft tissue *(arrowhead)*, indicating perineural spread of tumor along the right facial nerve. **B,** Axial postcontrast T1WI shows abnormal enhancement in the right stylomastoid foramen *(arrowhead)*, confirming perineural tumor spread. Note normal postcontrast appearance of the stylomastoid foramen on the left *(arrow)*.

within the middle ear cavity and neoplasms such as adenocarcinoma or adenoid cystic carcinoma may arise in such ectopic salivary tissue. The most common primary tumors to metastasize to the temporal bone are lung and breast cancer.

INNER EAR AND PETROUS APEX

Normal Anatomy

The cochlea and semicircular canals comprise the principal components of the inner ear (Fig. 11-30). The osseous labyrinth (otic capsule) includes the vestibule, semicircular canals, and cochlea. The membranous labyrinth includes the perilymph (within the scala vestibuli and tympani of the cochlea) and the endolymph (within the cochlear duct, semicircular canals, and vestibular aqueduct). The cochlea has apical, middle, and basal turns. The apical turn amplifies low tones and the basal turn amplifies high tones.

The vestibule is the common chamber to which the semicircular canals join. There are lateral, superior, and posterior semicircular canals. The superior and posterior semicircular canals share a nonampullated end known as the crus communis. The bony ridge over the superior semicircular canal is called the arcuate eminence. The stapes articulates to the vestibule via the oval window. The vestibule contains the posteriorly located utricle and the round, anterior saccule, the sense organs responsible for balance. The endolymphatic sac within the bony vestibular aqueduct courses posterolaterally from the vestibule and dilates into a blind-ending sac (see Fig. 11-6, *B* and *C*).

The cochlea is filled with perilymph but also has an endolymph channel. The cochlea has a base and a cupula (or apex), and is divided by a bony central canal known as the modiolus. At the scala vestibuli, perilymph from the vestibule communicates with that of the cochlea. At the end of the scala tympani along the basal turn, the perilymphatic space connects to the round window. The scala tympani and vestibuli join at the helicotrema of the cupula. The cochlear aqueduct extends from the scala tympani posteromedially to drain perilymph into the subarachnoid space of the posterior fossa. It is seen on sections below the IAC (see Fig. 11-6, *E*).

Congenital Anomalies

Although many causes of sensorineural deafness are derived from acquired disorders of the inner ear, several different types of congenital anomalies cause inner ear dysplasias and hearing loss. Underlying causes for these anomalies include thalidomide exposure, congenital rubella or cytomegalovirus infection, and eponymical genetic disorders.

Cochlear Abnormalities

The otic capsule structures develop very early, between the third and tenth weeks of gestation. Congenital inner ear abnormalities vary in degree of severity, ranging from mild dysplasias to complete aplasia depending on when during embryogenesis the failure of normal development occurred. For example, complete labyrinthine aplasia occurs as a result of failed formation of the otic placode in the third gestational week, whereas milder deformities such as incomplete partition type II (IP-II) occur at the seventh gestational week.

At present, most proscribe to the classification scheme of Sennaroglu and Saatchi, in which malformations of the cochleovestibular apparatus are categorized by decreasing severity (Table 11-3). Specifically, labyrinthine aplasia (aka Michel anomaly) consists of complete absence of the cochlea and vestibule. Cochlear aplasia is characterized by absence of the cochlea, but with the vestibule present (although usually not normal) (Fig. 11-31). The common cavity deformity is manifested by a cystic cavity without differentiation into the cochlea and vestibule. Cystic cochleovestibular anomaly (also known as incomplete

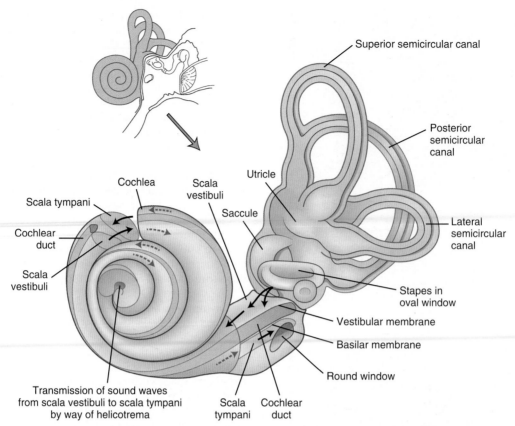

FIGURE 11-30 Schematic shows the vestibulocochlear apparatus and its numerous components.

TABLE 11-3   Congenital Vestibulocochlear Anomalies

Malformation	Features
Labyrinthine aplasia (Michel deformity)	Absent cochlea and vestibule
Cochlear aplasia	Absent cochlea
Common cavity deformity	Cystic cavity without differentiation into cochlea or vestibule
Incomplete partition type I	Cochlea lacks entire modiolus, with cystic cochlea and vestibule
Cochlear hypoplasia	Separate but small cochlea and vestibule
Incomplete partition type II (Mondini malformation)	1.5 turns of cochlea with cystic apical and middle turns, dilated vestibule and vestibular aqueduct

As proposed by Sennaroglu L, Saatchi I. A new classification for cochleovestibular malformations. *Laryngoscope* 2002;112:2230-2241.

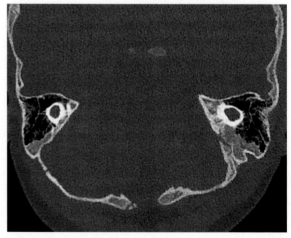

FIGURE 11-31 Cochlear aplasia. Axial computed tomography shows complete absence of the cochlea with abnormal cystic-appearing vestibule bilaterally.

partition type I, or IP-I) consists of a cystic cochlea and cystic vestibule (Fig. 11-32). Cochlear hypoplasia is manifested by small but separate cochlea and vestibule. With IP-II (aka Mondini dysplasia), the cochlea has only 1.5 turns with cystic middle and apical turns and the vestibule and vestibular aqueduct can be enlarged (see large vestibular aqueduct syndrome, discussed later).

Cochlear nerve deficiency is a cause of sensorineural hearing loss and consists of diminished caliber of absence of the cochlear nerve. This condition can be congenital or acquired (attributable to atrophy). Making this diagnosis is important because cochlear implants are contraindicated.

On MRI, sagittal reconstructions from high-resolution T2 images (i.e., constructive interference in steady state) can show the small or absent nerve in the anterior inferior aspect of the IAC. On CT, the cochlear nerve canal is small in caliber as it enters the basal turn of the cochlea (cochlear aperture stenosis).

If the cochlear aqueduct is congenitally enlarged, hearing may be impaired. Enlargement is difficult to define with the cochlear aqueduct as it narrows from medially

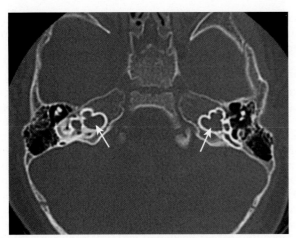

FIGURE 11-32 Incomplete partition type I. Axial computed tomography shows cystic-appearing cochlea with absent modiolus bilaterally *(arrows)* as well as cystic-appearing vestibule bilaterally.

to laterally and may be nearly invisible in its petrous and lateral segments. A medial orifice size of greater than 1.5 mm should be considered abnormal and a midportion diameter of greater than 1.2 mm is suspicious. An enlarged cochlear aqueduct, because it represents a communication between the scala tympani and the subarachnoid space, has been implicated in children with recurrent meningitis and ear infections. This anomaly has also been associated with the poststapedectomy "gusher," so named because CSF and perilymphatic fluid intermingle in spurts.

Vestibular Aqueduct Abnormalities

Enlargement and flaring of the vestibular aqueduct greater than 1.5 to 2 mm (large vestibular aqueduct syndrome) is the most common cause of congenital sensorineural hearing loss (Fig. 11-33) and can be seen as part of the IP-II spectrum. The midpoint of the normal duct should have a transverse diameter equal to or less than 1.5 mm in size or about the same size as the adjacent posterior semicircular canal. Enlarged vestibular aqueducts cause high-frequency hearing loss and may be seen in isolation or in association with abnormal cochlea spiralization (76%), cystic vestibules (31%), or abnormal semicircular canals (23%). One may see the dilated endolymphatic sac on MR as brighter than CSF on T1WI as a result of high-protein, hyperosmolar fluid. CHARGE syndrome (coloboma, heart defects, choanal atresia, retardation, genitourinary abnormalities, ear abnormalities), Pendred syndrome, and congenital cytomegalovirus infections may predispose to enlarged vestibular aqueducts. A narrowed vestibular aqueduct less than 0.5 mm in diameter may be seen with Ménière disease.

Semicircular Canal Abnormalities

Maldevelopment of the semicircular canals is another form of otic dysplasia. The lateral semicircular canal is most often affected in the form of hypoplasia because the lateral semicircular canal is the last to form embryologically (superior first, posterior second, lateral last). If the posterior (next most common) or superior (least

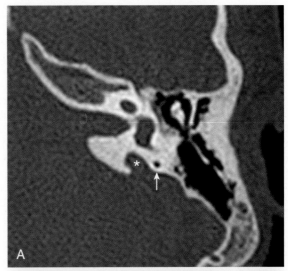

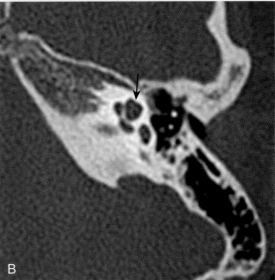

FIGURE 11-33 Large vestibular aqueduct syndrome with incomplete partition type II. **A,** Axial computed tomographic image shows enlargement of the vestibular aqueduct (the bony housing of the endolymphatic sac, *asterisk*). The normal vestibular aqueduct should be similar in diameter to the adjacent posterior limb internal capsule *(arrow)*. **B,** In this same patient more superiorly, the cochlea shows abnormal morphology, with cystic appearance of the apical and middle turns *(arrow)*.

common) semicircular canal fails to develop, the lateral semicircular canal must also be affected. Compensatory enlargement of the vestibule will occur. Hearing loss occurs rather than imbalance. If one sees an isolated semicircular canal deformity without cochlear anomalies it implies that the defect occurred after 8 to 9 weeks of gestation; by that time the cochlea has completely developed. Aplasia of the semicircular canals can be seen in CHARGE syndrome.

Superior semicircular canal dehiscence is characterized clinically by sound and/or pressure induced vertigo. Oscillopsia (the perception that stationary objects are moving) and vertigo evoked by loud noises is Tullio phenomenon, a frequent clinical finding (but is also seen in otosyphilis, Ménière disease, perilymphatic fistulas, and Lyme disease).

Although the defect has been described in congenital and syndromic conditions, there is an increased prevalence of the defect among older age groups suggesting that this is more commonly an acquired condition. Thin section imaging afforded by multidetector helical CT scanning allows for the imaging finding, that of a dehiscence of the bone overlying the superior semicircular canal, which is pretty subtle unless submillimeter sections are performed. A bilateral finding of dehiscence or thinning, at the least, is not unusual. Furthermore, the imaging finding may be seen in patients without the typical clinical presentation and is of doubtful clinical significance in these cases. Surgical exploration of the middle cranial fossa has confirmed the CT findings in selected patients; covering up the gap may lead to symptom relief.

Congenital Internal Auditory Canal Abnormalities

Incomplete partition type III (IP-III) is an X-linked deformity manifested by bulbous dilatation of the IAC laterally along with enlargement of labyrinthine segments of the facial and superior vestibular nerve canals. The bone between the IAC and basal turn of the cochlea can be deficient or absent.

Atresia or stenosis of the IAC is often associated with absent cranial nerves VIII and less commonly VII. Finding a nerve is important in candidates who are considered for cochlear implants. High-resolution MRI would be the study of choice. Often the facial nerve will leave the IAC early or aberrantly when cranial nerve VIII is aplastic or hypoplastic.

Congenital Syndromes

Inner ear dysplasias are rather common in patients with Down syndrome with hypoplastic inner ear structures, vestibular malformations, and deficiencies of the lateral semicircular canal reported. Additionally, fusion of the lateral semicircular canal and vestibule, large vestibular aqueduct syndrome, cochlear nerve deficiency, and IAC stenosis or duplication can be seen.

In patients with achondroplasia, one can see (1) poorly developed mastoid air cells; (2) upward tilting ("towering") of the petrous ridges and IACs; (3) rotation of the cochlea and ossicles; (4) changes of chronic otomastoiditis; and (5) narrowing of the skull base and foramen magnum.

Additional congenital syndromes associated with inner ear anomalies include (but are not limited to) neurofibromatosis, osteogenesis imperfecta, Apert syndrome, and Treacher Collins syndrome.

Labyrinthine Disease
Perilymphatic Fistula

A perilymphatic fistula is an abnormal connection between the subarachnoid space and the perilymphatic space of the inner ear. The usual sites of the fistula in children are at the oval and round windows often with associated stapes superstructure malformations. Spread of middle ear infections to the meninges or of meningitis to the inner or middle ear can occur through perilymphatic fistulas. Labyrinthitis ossificans can result as a sequela of such spread. Congenital sources include enlarged vestibular aqueducts,

Mondini malformations, and Michel anomalies. Acquired causes of perilymphatic fistulas include cholesteatomas, chronic otitis media, and trauma.

Labyrinthine Ossification

After chronic middle or inner ear infections, temporal bone trauma, cholesteatoma, bacterial meningitis, mumps, or labyrinthectomy, labyrinthitis ossificans may develop. Meningitis can cause labyrinthitis ossificans through the spread of the infection from the subarachnoid space to the scala tympani by means of the cochlear aqueduct. Fibroblasts in the labyrinth are induced by the inflammatory state to produce fibrosis, and they may differentiate into osteoblasts to form ossific deposits in the cochlea. This is another cause of a "dead" (deaf) ear (often with vertigo) that is best evaluated with CT. Bony replacement of the labyrinthine portion of the inner ear with dense sclerosis is identified on CT (Fig. 11-34, Box 11-8). Imaging findings include cochlear stenosis (approximately 40%), cochlear fibroossific change (perhaps better seen with high-resolution T2WI MR as the fibrous obliteration may not be evident on CT), and cochlear ossification (>30%). Obliterative changes in the semicircular canals and vestibule are not uncommon in association with cochlear labyrinthitis ossificans. Osseous obliteration at the round window niche may lead to inadequate cochlear implant insertion; in general, the further into the cochlear turns that a multichannel electrode can be inserted, the better the quality of hearing.

Otospongiosis

Otospongiosis is another cause of sensorineural hearing loss that is usually bilateral (80%) and seen most frequently in young to middle-aged women. Otospongiosis suggests the pathophysiology in which endochondral bone is replaced by spongy bone. In the early phases, one identifies a lytic lucent erosion of the labyrinthine margins of the oval window, the round window niche, and/or the cochlea. In later stages (otosclerosis), the bone again becomes hyperattenuating and the diagnosis is difficult to make. Fenestral otospongiosis most frequently affects the anterior margin of the oval window (fissula ante fenestram). In the cochlear (retrofenestral) form of otospongiosis, the middle and basal turns of the cochlea are most frequently involved, showing areas of demineralization (Fig. 11-35). A "double ring" (lucent) sign caused by resorption of bone immediately around the membranous cochlea may be seen as a result of the normal basal turn lucency paralleled by otospongiosis. In the late phases of this disease increased bony density caused by recalcification is visualized. Osteogenesis imperfecta has an identical appearance on temporal bone CT. The differential diagnosis also includes otosyphilis and rarely fibrous dysplasia and Paget disease.

Fenestral otospongiosis is more common than cochlear otospongiosis. This is a disease of young adulthood; 70% of cases occur in patients 18 to 30 years old. It typically involves the oval window (80% to 90%) border with the anterior crus of the stapes and the round window niche (30% to 50%). Bilaterality is seen in up to 85% of patients with fenestral otospongiosis. The stapes is essentially glued in position to the oval window, preventing

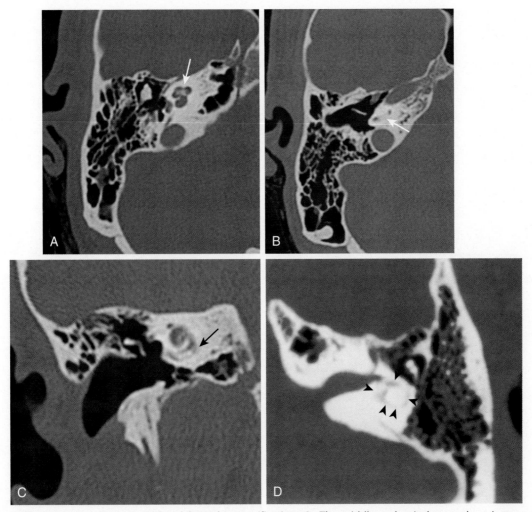

FIGURE 11-34 Montage of cochlear labyrinthine ossification. **A,** The middle and apical turns show increased bony obliteration *(arrow)*. That is too much for a normal modiolus. **B,** The basal turn *(arrow)* was also involved. **C,** The coronal CT is definitive. The *arrow* is on basal turn ossification. **D,** Labyrinthitis ossificans. The vestibule and semicircular canal *(arrowheads)* show the same obliterated appearance resulting from labyrinthitis ossificans.

BOX 11-8   Causes of Labyrinthine Ossification

Meningitis complication
Chronic otitis media complication
Chronic labyrinthitis (bacterial)
Labyrinthine fistula (infection/cholesteatoma/trauma)
Trauma/hemorrhage
Labyrinthectomy
Otospongiosis
Paget disease
Sickle cell disease

---

transmission of sound and resulting in conductive hearing loss. The oval window niche is narrowed with fenestral otospongiosis with plaques of bone anteriorly. The density of the fenestral plaques is variable. Although it is usually seen as hyperdense to the normal oval window membrane, it is only rarely (15%) as dense as the otic capsule. The surgery of choice for fenestral otospongiosis is a small fenestral stapedotomy or total stapedectomy. With a total stapedectomy, a prosthesis must be inserted

into the oval window. Metal, Teflon, and wire devices are commonly used.

A cochlear implant may be required in patients with cochlear otospongiosis or other causes of sensorineural hearing loss. This operation is a surgical procedure that consists of inserting multichannel electrodes through the round window into the cochlea with the distal end along the basal membrane of the cochlea where the auditory nerve transmits the sound. Preoperative evaluation with CT is sometimes ordered to ensure that the facial nerve is in its normal anatomic position (to prevent injury at surgery) and that cochlear patency is present. If the patient has bilateral hearing loss and one cochlear implant is being inserted, the surgeon will place it in the cochlea that is (more) patent. A list of what the surgeon needs to know before implantation is given in Box 11-9.

Of children who receive cochlear implants, nearly 50% have deafness secondary to meningitis. Congenital lesions and viral infections account for most of the rest of the cases. The cochlea of patients with a history of meningitis may be obstructed (labyrinthitis ossificans) and this

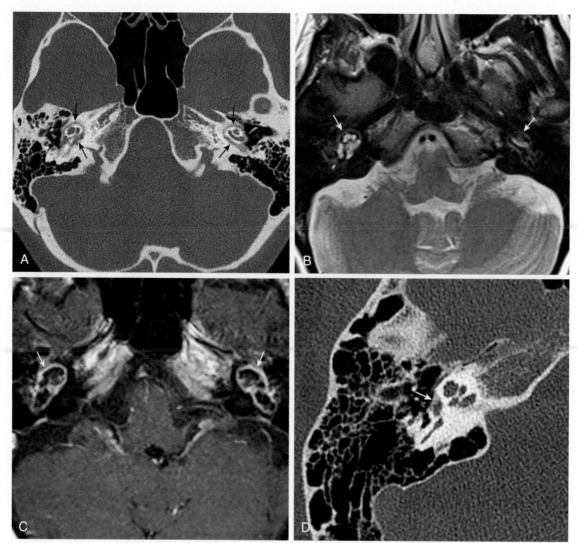

FIGURE 11-35 Otospongiosis. **A,** Axial computed tomographic image shows abnormal lucency in the bone surrounding the cochlea bilaterally in this patient with cochlear otospongiosis *(arrows)*. **B,** Axial T2-weighted image (T2WI) shows abnormal hyperintense signal and **(C)** axial postcontrast T1WI shows abnormal enhancement within the demineralized bone surrounding the cochlea and vestibular structures *(arrows)*. **D,** Demineralization at the fissula ante fenestram *(arrow)* is seen with fenestral otospongiosis with identical imaging findings seen in patients with osteogenesis imperfecta (OI). This patient has OI.

BOX 11-9   Precochlear Implant Computed Tomography Evaluation

Bilateral acoustic schwannomas (major differential diagnosis)
Bilateral obliterative labyrinthine ossification (little hope for multichannel implant)
Cochlear patency (Does bone obliterate labyrinth? Is round window open?)
Concurrent middle ear infections (need for preoperative antibiotics)
Congenital cochlear anomalies and coexistent vestibular disease (risk of "gusher" ear)
Determine location of facial nerve, carotid artery, sigmoid sinus
Fractures, unsuspected trauma
Hypoplasia of internal auditory canal (Is there a nerve?)
Size of middle ear cavity (for access)

---

leads to a higher rate of implant placement failure. Look for cochlear stenosis (basal turn most commonly affected), cochlear ossification, and round window ossification as predictors for suboptimal placement.

Ménière Disease

Ménière disease (endolymphatic hydrops) is a condition characterized by episodic vertigo, hearing loss, tinnitus, and ear pressure and is felt to occur secondary to abnormal endolymphatic pressure. Various reports suggesting that the absence of visualization of the endolymphatic sac on high-resolution three-dimensional T2WI scans is indicative of Ménière disease have surfaced. Some even believe that the stage of the disease can be assessed with this technique, the sac becoming visible once again when Ménière disease is quiescent. Eventually, we will have the resolution to see the dilated scala channels in the cochlea with

Chronic osteitis from infection
Lack of pneumatization at childhood (hypoplasia)
Blastic metastases
Paget disease
Fibrous dysplasia
Adjacent meningioma
Osteopetrosis
Engelmann disease
Otosyphilis

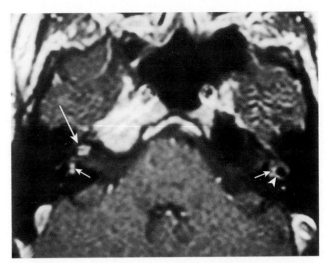

FIGURE 11-37 Labyrinthitis. T1-weighted image shows enhancement of the right cochlea *(large arrow)* and left and right vestibule *(small arrows)* in this patient with viral labyrinthitis. Left lateral semicircular canal also enhances abnormally *(arrowhead).*

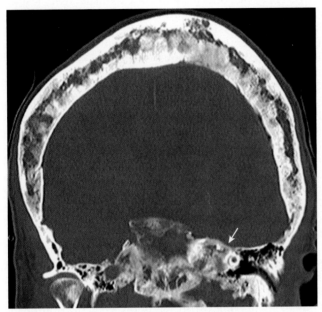

FIGURE 11-36 Paget disease of the temporal bone. Axial computed tomography reveals diffuse increased bone density with thickening throughout the base of the skull. Note the predominance in the petrous apex *(arrow)* with relative sparing laterally, especially around the right labyrinth.

bulging membranes. Our own experience is that you may see endolymphatic sac enhancement on MRI in patients with Ménière disease.

Paget Disease

Another of the bone-producing lesions in the inner ear is Paget disease (Box 11-10). Paget disease may cause either sensorineural or conductive hearing loss. The increased vascularity of involved temporal bone can also account for the clinical presentation of pulsatile tinnitus. In its early phases one identifies a diffuse lytic process involving the bony labyrinth; however, in the late phases increased density is seen. The lytic phase appears to begin medially in the petrous apex and to progress laterally (Fig. 11-36). Otospongiosis is in the differential diagnosis.

Fibrous Dysplasia

Fibrous dysplasia may affect the temporal bone, causing increased density in a ground-glass manner. The mastoid portion is affected most commonly, and the involvement may lead to conductive hearing loss.

Postoperative after schwannoma resection
Labyrinthitis (viral, bacterial, luetic, Lyme disease, sarcoidosis, sickle cell disease)
Posttrauma with hemorrhage into labyrinth
Autoimmune labyrinthitis (antibodies to cochlear antigens)
Labyrinthine schwannoma
Cogan syndrome (interstitial keratitis, vestibuloauditory abnormality, vasculitis)

Labyrinthitis

Enhancement of the labyrinth on MR in patients with sudden hearing loss and vertigo has been described as suggestive of labyrinthine infection (Fig. 11-37). Cochlear enhancement or vestibular apparatus enhancement may occur and often correlates with electronystagmogram findings and clinical symptoms. Labyrinthitis may be attributable to viral, bacterial, luetic, or idiopathic causes. Asymptomatic patients do not show labyrinthine enhancement. Autoimmune labyrinthitis occurs when antibodies to cochlear antigens form; there is often enhancement of the cochlea bilaterally on MR. The differential diagnosis includes labyrinthine schwannoma. Other causes of labyrinthine enhancement are listed in Box 11-11.

Petrous Apex Lesions

Petrous Apicitis

Petrous apicitis is a nondestructive inflammatory condition of the aerated petrous apex (Box 11-12). Pneumatization of the petrous apex is present in 30% to 35% of people, thus petrositis can develop in these (unfortunate) persons, often after what was thought to be successful mastoidectomy surgery for inflammatory disease. Associated with this condition is Gradenigo syndrome, which causes pain in the distribution of cranial nerve V, a VI nerve palsy, and otorrhea. The lesion appears as opacification of the petrous air cells, typically of low signal intensity on T1WI and high intensity on T2WI. If chronic infection persists, the signal intensity of the apicitis may change with the higher protein content and viscosity, causing high signal on T1WI and/or lower signal intensity on T2WI. The dura near the gasserian ganglion (Meckel cave) may enhance on MR. Cranial nerve VI is affected as it passes through the Dorello canal, a bony passageway leading from the tip of the temporal bone to the cavernous sinus containing nerve VI and the inferior petrosal sinus. Gradenigo syndrome may also occur in the absence of pneumatized petrous air cells.

Cholesterol Granulomas

Cholesterol granulomas are lesions that typically arise in the petrous apex of the temporal bone. The inciting

event in the genesis of cholesterol granulomas (also known as chocolate cysts, cholesterol cysts, epidermoids, and blue-domed cysts) seems to be a small blood vessel rupture with recurrent hemorrhage in the petrous apex. This may be caused by negative pressures occurring in the petrous air cells, resulting from chronic obstruction. This elicits a foreign body reaction by the mucosa of the air cells, causing giant cell and fibroblastic proliferation and cholesterol crystal deposition with subsequent recurrent subclinical hemorrhages. The lesion expands as the host response perpetuates itself. Eventually, the patient develops cranial nerve findings (usually V or VIII). These lesions present as expansile lytic lesions within a pneumatized petrous apex filled with soft-tissue debris on CT. The cholesterol granuloma is lined by fibrous connective tissue as opposed to acquired cholesteatomas, which are encapsulated by stratified squamous epithelium.

Cholesterol granulomas and cholesteatomas are commonly confused solely because they share four syllables, but they look nothing alike on MR. A cholesterol granuloma has high signal intensity on all pulse sequences because of the hemorrhagic products and/or the cholesterol debris (Fig. 11-38), whereas acquired cholesteatomas are not bright on T1WI. Sometimes cholesterol granuloma blood products are dark on T2WI, but they are classically bright on T1WI. The differential diagnosis includes a mucocele of the petrous apex, petrous apicitis, or a hemorrhagic bony metastasis.

If you are accustomed to using fast spin echo T2WI scans on MR beware of this pitfall: both petrous apex fat and a cholesterol granuloma will look the same on fast spin echo scans, that is, bright on both T1WI and T2WI. Only by identifying expansion of the bone or by applying fat suppression to the sequence will you be able to get out of this quandary.

BOX 11-12   Petrous Apex Masses

Petrous apicitis
Petrous apex fat thought to be cholesterol granuloma
Cholesterol granulomas
Mucocele of petrous apex
Chondroid lesions
Chordoma
Cranial nerve V schwannomas
Epidermoid
Paraganglioma
Meningioma
Metastases
Myeloma
Endolymphatic sac tumor
Petrous carotid aneurysms
Eosinophilic granuloma (Langerhans cell histiocytosis)

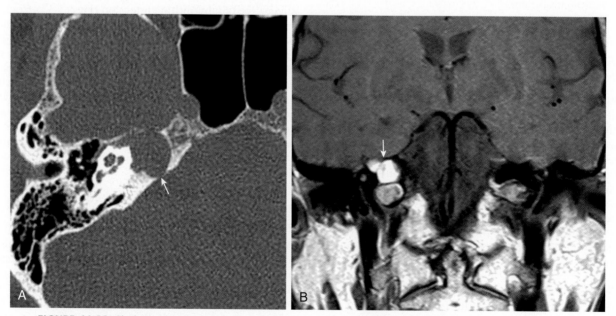

FIGURE 11-38 Cholesterol granuloma. **A,** Axial computed tomographic image shows expansile lucent lesion within the right petrous apex *(arrow)*. **B,** The diagnosis becomes clear on this unenhanced coronal T1-weighted image, which shows the lesion to be bright due to blood products within *(arrow)*.

Other Causes of Hearing Loss

The work-up of acute hearing loss usually yields an abundance of cases of viral or immune-mediated disease, Ménière disease, vascular disorders, syphilis, neoplasms (vestibular schwannomas), multiple sclerosis, and/or perilymphatic fistulas. Sickle cell disease is associated with intralabyrinthine hemorrhages that may present with sudden hearing loss.

Dural malformations, neoplasms, or other vascular lesions that may cause chronic recurrent hemorrhage may lead to superficial (hemo)siderosis of the central nervous system. This is an unusual cause of hearing loss in which chronic bleeding leads to hemosiderin deposition on the brain stem and nerves running through the basal cisterns. Cranial nerve VIII is particularly sensitive to the effects of hemosiderin deposition. The characteristic MR appearance is a thin, dark rim around the surface of the brain stem and cerebellum on T2WI, although susceptibility-weighted imaging is more sensitive for these changes.

Benign Neoplasms of the Inner Ear

The benign masses associated with the inner ear—the glomus tympanicum, facial schwannoma, and IAC schwannomas—have all been described earlier. Intralabyrinthine schwannomas are rare tumors but enhance markedly on MR. They are usually situated close to the round window niche. They are an unusual cause of hearing loss and, as opposed to the more typical vestibular schwannomas, appear to arise from the cochlear (not the vestibular) branch of cranial nerve VIII. Usually the patients are thought to have Ménière disease because of the associated vertigo. The tumors are associated with neurofibromatosis II.

Cranial nerve V schwannomas may also occur along the petrous apex. These lesions begin in an extraosseous location but may erode the medial petrous bone near the trigeminal impression. They enhance solidly or heterogeneously.

Malignant Neoplasms of the Inner Ear

There are relatively few primary malignant lesions of the inner ear. Squamous cell carcinoma is probably the most common malignancy to affect the inner ear by direct extension. It can arrive there via (1) the EAC, (2) the middle ear, (3) the back of the nasopharynx along the eustachian tube, (4) a cholesteatoma, or (5) the parotid gland. Hematogenous metastases may occur in the inner ear, but direct invasion by carcinoma is more common. Rarely, neurofibrosarcomas, rhabdomyosarcomas, lymphomas, or malignant hemangiopericytomas may occur in this location. Perineural spread of malignancies along the facial nerve may lead to destructive processes affecting the inner ear.

Endolymphatic Sac Tumors

These tumors of the endolymphatic sac were previously called adenomatoid papillary tumors and were initially thought to be thyroid metastases because of the papillary histology. More recently, their site of origin has been reevaluated and it currently appears that they need not be of endolymphatic sac origin. Some arise from the top of the jugular bulb, the mucosa of the aerated cells around the jugular bulb, or the mastoid air cells. The tumors are characterized by aggressive bony destruction and calcified matrix on CT and bright signal on T1WI, possibly from hemorrhage (Fig. 11-39). Tumors larger than 2 cm may have flow voids owing to branches of the external carotid artery that supply this hypervascular tumor. The orientation of the tumor, parallel to the posterior margin of the petrous temporal bone, simulates the vestibular aqueduct. There is a strong association with von Hippel–Lindau disease. Approximately 11% to 16% of patients with von Hippel–Lindau disease have an endolymphatic sac tumor, and of these one third are bilateral. Most endolymphatic sac tumors are associated with von Hippel–Lindau disease. Scan the cerebellum and spine for additional hemangioblastomas when this diagnosis is suspected.

JUGULAR FORAMEN LESIONS

Lesions that occur in this location include glomus tumor, neurofibroma or schwannoma, meningioma, superior spread of nasopharyngeal carcinoma, and metastatic disease (Box 11-13, Table 11-4). Occasionally, a glomus jugulare extends from the skull base superiorly into the middle ear cavity and may simulate a localized glomus tympanicum (thereby called a glomus jugulotympanicum). The glomus tumor, neurofibroma, schwannoma, and metastatic lesions often erode/remodel bone in the jugular foramen. CT is again best at demonstrating bony erosion or foraminal enlargement. The glomus tumor can occlude the jugular bulb and characteristically grows into the jugular vein (unusual for the other masses in the region). All these lesions enhance, and when no flow voids are in the lesion, differentiation may be difficult. Meningiomas that extend into the jugular foramen usually demonstrate a dural base and an enhancing tail.

Glomus jugulare most commonly arise from the adventitia of the jugular vein in the jugular foramen, although glomus bodies also accompany the auricular branches of the vagus nerve (Arnold nerve) or the tympanic branch of the glossopharyngeal nerve (Jacobson nerve). A hereditary form of paragangliomatosis is associated with multiple glomus tumors including jugulare, vagal, carotid body, and tympanicum tumors. Overall, multiple paragangliomas occur in 15% of patients with a glomus tumor. Paragangliomas rarely will metastasize.

On imaging, the glomus jugulare has a typical MR salt-and-pepper appearance on T2WI and enhanced T1WI (Fig. 11-40). This is seen as flow voids within the tumor, surrounded by tumor substance. Time-of-flight magnetic resonance angiography can show prominent flow-related signal within the mass. On CT, this mass erodes the jugular foramen of the temporal bone with permeative lytic change in the affected marrow spaces. The mass may grow inferiorly into the jugular vein or may grow from the jugular bulb region into the sigmoid and transverse sinuses. Alternatively, the mass may cause thrombosis of

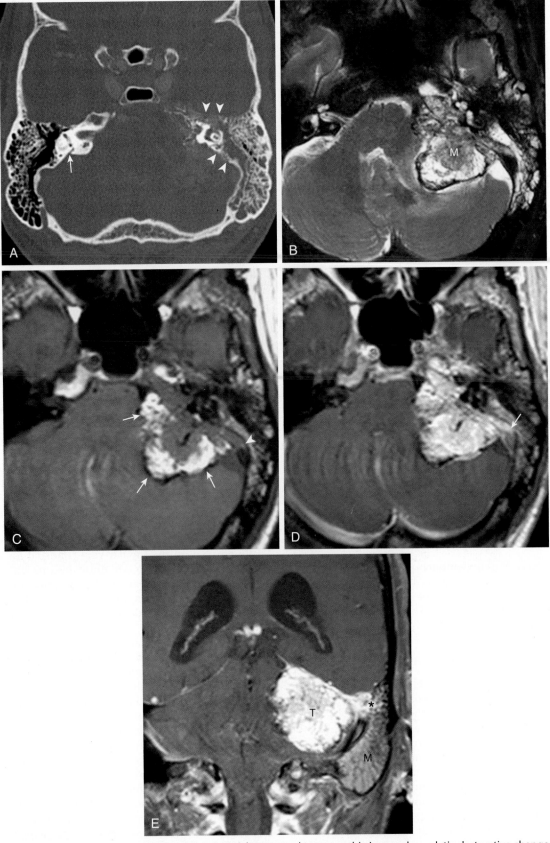

FIGURE 11-39 Endolymphatic sac tumor. **A,** Axial computed tomographic image shows lytic destructive change involving the anterior and posterior margins of the mastoid portion of the left temporal bone *(arrowheads)* as well as petrous apex. The vestibular aqueduct is not appreciable due to bone destruction at this level. Note normal appearance of the vestibular aqueduct on the right *(arrow)* for comparison. **B,** Axial T2-weighted image (T2WI) shows the associated bulky mixed signal extraosseous soft-tissue mass (M) effacing the cerebellopontine angle cistern and compressing the brachium pontis and cerebellar hemisphere on the left. Intraosseous tumor involvement of the petrous apex and mastoid bone is also seen. **C,** Axial T1WI reveals that the mass is partially bright along its peripheral margins *(arrows)*, and that the mass invades the adjacent sigmoid sinus *(arrowhead)*. **D,** Post-contrast axial and **(E)** coronal T1WI show enhancing tumor with focal invasion of the sigmoid sinus *(arrow in **D**)*. In **(E)**, the tumor (T) shows marked enhancement and the mastoid portion involved with tumor *(asterisk)* can be distinguished from the proteinaceous opacification within the adjacent mastoid air cells (M).

the adjacent venous sinuses. Mass within the vessel may be distinguished from bland thrombosis by the presence of enhancement in the former.

On conventional angiography, the glomus tumor is evidenced by a hypervascular mass, often supplied by ascending pharyngeal branches, with a persistent stain. Because these tumors secrete norepinephrine, α-adrenergic blocking drugs might be required (for the patient) during arteriograms. On the other hand, β-blockers might be helpful (for the angiographer) to keep it together in these cases.

Beware of the jugular foramen! Sometimes slow flow will have intermediate signal intensity on T1WI, bright signal on T2WI, and "enhancement" of the turbulent flow. Look for phase ghosting artifacts but do not be shy about asking for an MR venogram to assure this is the artifact of a slow turbulent jugular vein rather than a mass.

BOX 11-13   Jugular Foramen Masses

Enlarged jugular bulb/diverticulum
Paraganglioma
Nasopharyngeal carcinoma spread
Schwannoma
Meningioma
Metastasis
Chondroid lesion

TRAUMA

Multidetector-row CT scanning has really helped to increase detectability of temporal bone fractures, even very subtle ones. Fractures of the temporal bone can range from simple appearing nondisplaced fractures to more complex injuries. Occult fractures should be suspected even if a clear fracture plane is not seen when secondary signs of injury are present, including opacification of mastoid air cells, adjacent pneumocephalus, and pneumolabyrinth. Keep in mind that temporal bone fractures can be seen with diastatic (widened) sutures, which also need to be reported.

At present, fractures are best described as "otic capsule sparing" and "otic capsule violating" injuries with regard to involvement of the cochlea and labyrinth in the setting of temporal bone injury. Integrity of these structures should be commented upon when reviewing a case of temporal bone fracture, for prognostication. Injury to the facial nerve can occur usually from local effects at the geniculate ganglion rather than transection. Sensorineural hearing loss can also be seen as a complication, and transection of the cochlear nerve can occur at the IAC apex (Fig. 11-41).

Involvement of the EAC and glenoid fossa of the temporomandibular joint is very common. When the middle ear is involved, ossicular dislocation can be seen (Fig. 11-42). The incudostapedial joint, being the weakest of the middle ear articulations, is most commonly affected. This is detected by seeing a fracture of the long process of the

---

TABLE 11-4   Differential Diagnosis of Jugular Foramen Masses

Entity	T2WI Intensity	Enhancement	Calcification/ Bone Erosion	MR Technique Helpful	Angiographic Appearance	CT Density
Glomus jugulare	Salt and pepper	Marked: downward dip on dynamic enhancement	Erosion	Dynamic enhancement, MR venogram	Hypervascular with arteriovenous shunting, stain	Hyperdense
Schwannoma	Hyperintense; may have cystic degeneration	Moderate: upward slope on dynamic scanning	No; bone remodeled	Traditional	Hypovascular	Isodense
Metastasis	Hyperintense	Moderate	Erodes bone and infiltrates	Traditional	Most often hypovascular— exceptions include hypervascular metastases such as renal, thyroid	Isodense
Chondroid lesions	Hyperintense with mottling (secondary to calcification)	Moderate	Yes; calcified matrix and bone erosion	Gradient echo for calcification	Hypovascular	Areas of high density from matrix
Enlarged jugular bulb	Flow effects	Varies with technique, turbulence	No	MR venogram	Venous phase	Vascular
Nasopharyngeal carcinoma	Hyperintense to intermediate	Moderate	Erodes bone and infiltrates, look for perineural spread	Traditional	Most often hypovascular	Isodense
Meningioma	Isointense to slightly hyperintense	Marked	Osteolysis versus hyperostosis	Traditional	Hypervascular; persistent stain on all tests	Slightly hyperdense

CT, Computed tomography; *MR,* magnetic resonance; *T2WI,* T2-weighted image.

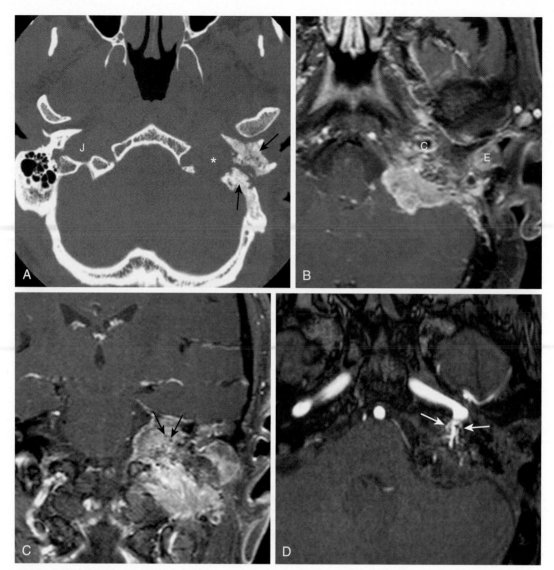

FIGURE 11-40 Glomus jugulare. **A,** Axial computed tomographic image shows marked expansion of the left jugular foramen *(asterisk),* compared with the normal foramen on the right (J). Note permeative marrow change in the adjacent mastoid bone on the left *(arrows),* which is very characteristic of glomus jugulare. **B,** Axial post-contrast T1-weighted image (T1WI) shows enhancing tumor arising from the expanded jugular foramen with extension into the posterior fossa, encasement of the carotid artery (C), and extension into the external auditory canal (E). **C,** Coronal postcontrast T1WI again shows enhancing tumor centered within the jugular foramen with extension below the skull base along the course of the jugular vein. Flow voids are present *(arrows),* contributing to characteristic salt and pepper appearance of this tumor. **D,** Three-dimensional time-of-flight magnetic resonance angiography shows high-flow shunting to the tumor *(arrows).*

incus with separation of more than 1 mm from the stapes head posterolaterally. Incudal dislocations are the source of nearly 80% of the cases of posttraumatic conductive hearing defects attributable to the ossicles.

Disruption of other vital structures including adjacent venous sinuses and carotid canal must be reported, followed by recommendation for angiographic evaluation preferably by CT to assess for acute arterial and venous injury including dissection, pseudoaneurysm, vasospasm, cavernous-carotid fistula, traumatic occlusion, and thrombosis (Fig. 11-43).

Be sure to evaluate the intracranial contents for associated extraaxial hemorrhages and parenchymal contusions in the temporal lobe. Other complications of temporal bone fractures include otorrhea, meningitis, traumatic meningoencephaloceles, perilymphatic fistula, and CSF leakage. Disruption of the jugular foramen can result in cranial nerve IX, X, and XI palsies and injury to the petrous apex can cause stretch injury to the abducens nerve as it enters the Dorello canal.

One more thing: beware of normal sutures, fissures, and vascular channels in the temporal bone that can mimic the appearance of a fracture! To avoid this error, firstly, know your anatomy, but when in doubt, check the contralateral side for symmetry (if it is symmetrical, it is probably not a fracture) and soft tissues for scalp swelling (if there is swelling, it probably is a fracture) to help you out when you are in a pickle.

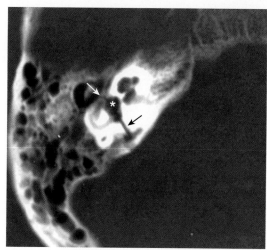

FIGURE 11-41 Temporal bone fracture violating otic capsule. Linear fracture *(black arrow)* through the temporal bone extends across the vestibule *(asterisk)* on the right side perpendicular to the plane of the petrous ridge. One can see that the plane of the fracture would cross the horizontal portion of the facial nerve *(white arrow)*. Fluid in the mastoid air cells is probably hemorrhage from the fracture.

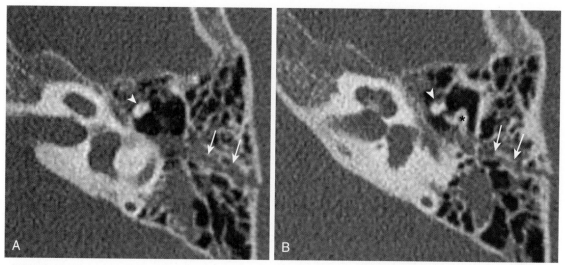

FIGURE 11-42 Ossicular dislocation. Axial computed tomographic images **(A)** and **(B)** show nondisplaced fracture through the temporal bone *(arrows)*. There is associated opacification within the mastoid air cells. Note that the head of the malleus *(arrowhead)* seems to be "falling off" of the short process of the incus *(asterisk)*. This is incudomalleolar dislocation, most often found in the setting of trauma.

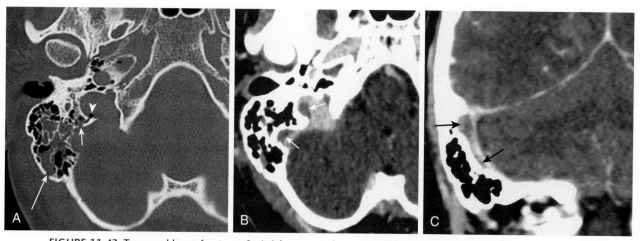

FIGURE 11-43 Temporal bone fracture. **A,** Axial computed tomographic (CT) image shows minimally displaced fracture through the mastoid portion of the temporal bone *(large arrow)*, which extends anteromedially to the jugular foramen. There is focal disruption of the bone adjacent to the expected course of the sigmoid sinus *(small arrow)* as well as focus of gas within the jugular foramen *(arrowhead)*. These imaging findings should raise suspicion for acute venous injury and prompt further imaging work-up. **B,** Axial and **(C)** coronal images from subsequently performed CT venogram show nonocclusive filling defects within the sigmoid sinus and jugular vein, indicating acute venous thrombosis *(arrows)*.

Introduction to Head and Neck Chapters

Imaging of the head and neck can be divided into three main regions: sinonasal, mucosal, and extramucosal spaces. Within each of these are subdivisions based on anatomic landmarks, which allow a more organized approach to head and neck imaging. The anatomy of this area is fearsome to most radiologists in training (or out of training). Therefore, the next three chapters emphasize the appropriate parlance of the head and neck specialist.

Although it is important to know the imaging characteristics that may suggest a limited diagnosis in head and neck lesions, the more essential role of the radiologist is to provide answers to the specific questions of the otorhinolaryngology clinician that will help in the staging of the lesion and/or treatment of the patient. For this reason, rather than emphasizing the differential diagnosis for a lesion with a given set of density or intensity characteristics, these chapters emphasize the relevant clinical issues associated with a given disease. Because many lesions occur in various locations along the aerodigestive system, we have tended to avoid repeating the imaging characteristics at each site. With this approach in mind, this section of the book begins with a border zone between the brain and the head and neck: the sinonasal cavity.

Have no fear, young Jedi knight. The more you study this anatomy, the easier it gets. Remember, only the cribriform plate separates "Nosage" from "Knowledge."

Chapter 12
Sinonasal Disease

To appreciate the pathogenesis of sinusitis, you must understand the normal anatomic pathways of mucociliary clearance in the paranasal sinuses (Fig. 12-1). The cilia within the maxillary sinus propel the mucous stream in a starlike pattern from the floor of the maxillary sinus toward the ostium situated superomedially. In approximately 30% of patients, a second accessory ostium to the maxillary sinus is present inferior to the major opening. From the maxillary sinus ostium, mucus from the maxillary antrum (the maxillary antrum and maxillary sinus are synonymous) gets swept superiorly through the infundibulum, which is located lateral to the uncinate process and medial to the inferomedial border of the orbit. The uncinate process, a sickle-shaped bony extension of the lateral nasal wall extending anterosuperiorly to posteroinferiorly, is rarely (<2.5% of patients) pneumatized itself. Occasionally, the uncinate process attaches to the lamina papyracea (the medial wall of the orbit). If it does so, the infundibulum does not have a superior opening, thus creating a blind pouch, the recessus terminalis. The hiatus semilunaris is a slitlike air-filled space anterior and inferior to the largest ethmoid air cell, the ethmoidal bulla, and right above the uncinate process. Mucus is passed through the hiatus semilunaris posteromedially via the middle meatus, a channel between the middle turbinate and the uncinate process, into the back of the nasal cavity to the nasopharynx where it is subsequently swallowed (blech!) or expelled (yuck!).

The ostiomeatal complex (OMC) refers to the maxillary sinus ostium, the infundibulum, the uncinate process, the hiatus semilunaris, the ethmoid bulla, and the middle meatus; this is the common drainage pathway of the frontal, maxillary, and anterior ethmoid air cells.

The frontal sinuses drain inferomedially via the frontal recess (previously termed the frontoethmoidal recess or frontonasal duct). The frontal recess connotes the common drainage of the frontal sinus and the anterior ethmoid air cells. The frontal recess is the space between the inferomedial frontal sinus and the anterior part of the middle meatus. The frontal sinus and the anterior ethmoid air cells usually drain directly into the middle meatus via the frontal recess, or less commonly into the superior ethmoidal infundibulum, before passing to the middle meatus.

Several ethmoidal air cells have specific names. The most anterior ethmoid air cells located anterior, lateral, and below the frontal recess are termed *agger nasi* cells. They are present in more than 90% of patients (Fig. 12-2). The *ethmoidal bulla* is the term used for the ethmoid air cell directly above and posterior to the infundibulum and hiatus semilunaris. A very large ethmoidal bulla can obstruct the infundibulum and hiatus semilunaris, and lead to interference with the drainage of the maxillary and anterior ethmoid sinuses. When anterior ethmoid air cells are located inferolateral to the bulla, along the inferior margin

of the orbit protruding into the maxillary sinus, they are termed Haller cells, also known as maxilloethmoidal cells or infraorbital cells. They are seen in 10% to 45% of patients. When greatly enlarged, Haller cells may narrow the infundibulum or maxillary sinus ostium (Fig. 12-3). Between the ethmoidal bulla and the basal lamella (the lateral attachment of the middle turbinate to the lamina papyracea of the orbit) is the sinus lateralis. The sinus lateralis, comprising the suprabullar and retrobullar recesses, may open into the frontoethmoidal recess or into a space posterior to the bulla, the hiatus semilunaris posterioris.

The posterior ethmoid air cells are located behind the basal lamella of the middle turbinate and drain via the superior meatus, the supreme meatus, or other tiny ostia just under the superior turbinate, which ultimately drain along with the sphenoid sinuses into the sphenoethmoidal recess of the nasal cavity (Fig. 12-4), from which the secretions pass to the nasopharynx. In some patients the most posterior ethmoid air cell may pneumatize into the sphenoid bone, superior to the sphenoid sinus. This is termed an *Onodi cell*. Its importance is that, if perforated surgically, it may lead intracranially or into the optic canal.

The roof of the ethmoid sinus is termed the fovea ethmoidalis, which merges medially with the cribriform plate.

The nasal cavity typically has three sets of turbinates: the superior, middle, and inferior turbinates. Occasionally, a fourth superiormost turbinate can be seen, the supreme turbinate. An aerated middle turbinate, which usually communicates with the anterior ethmoid air cells, is termed a concha bullosa and is seen in approximately 34% to 53% of patients. Most people believe that, unless huge, the presence of a concha bullosa does not predispose to chronic sinusitis. Significant pneumatization of the inferior or superior turbinates is much less common (<10% of patients). Reversal of the characteristic medially directed curve of the middle turbinates is described as "paradoxical."

The nasal septum is the midline structure between the right and left turbinates. The nasal septum is composed of three parts: a cartilaginous anteroinferior portion; a bony posteroinferior portion known as the *vomer;* and a superoposterior bony portion, the perpendicular plate of the ethmoid bone. The nasal septum is aerated only rarely. Nasal septal deviation, however, is common, and bony spurs often develop at the apex of the deviation. Spurs may cause the sensation of nasal obstruction.

The nasolacrimal duct courses downward from the lacrimal sac bordering the medial canthus, where it is in close association with agger nasi air cells. Inflammation of agger nasi cells may be associated with epiphora because of this close relationship. The duct subsequently runs in the anterior and inferior portions of the lateral nasal wall. Its ending opens below the inferior turbinate at the inferior meatus.

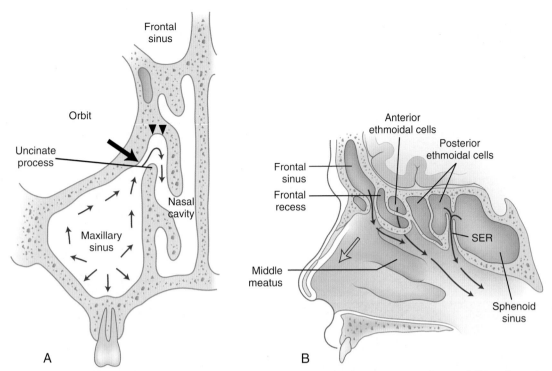

FIGURE 12-1 **A,** On this coronal view schematic, mucociliary clearance passes from the antral floor along the walls of the maxillary sinus toward the main maxillary sinus ostium. It then passes lateral to the uncinate process in the infundibulum *(fat black arrow)* into the hiatus semilunaris *(arrowheads)* and then to the middle meatus. **B,** Note the anterior to posterior flow of mucus from the frontal, ethmoid, and sphenoid sinuses on sagittal view schematic. The frontal recess, middle meatus, and sphenoethmoidal recess (SER) are the respective egresses from the sinuses. (Figure modified from Zinreich SJ, Kennedy DW, Rosenbaum AE, et al. Paranasal sinuses: CT imaging requirements for endoscopic surgery. *Radiology.* 1987;163:769-775.)

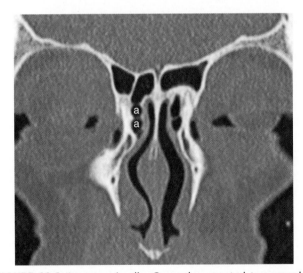

FIGURE 12-2 Agger nasi cells. Coronal computed tomographic image shows bilateral agger nasi air cells (labeled *a* on right).

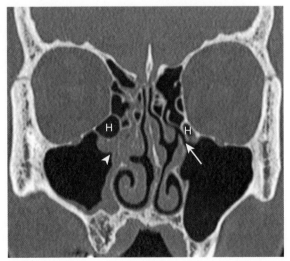

FIGURE 12-3 Haller cell. Coronal computed tomographic image demonstrates Haller cells (H) bilaterally. On the left, the air cell is small enough that it does not narrow the left maxillary sinus ostium *(arrow)*. However, on the right side, the ostium is occluded *(arrowhead)* because of the presence of the larger Haller air cell.

PARANASAL SINUS DEVELOPMENT

Remember the development of the sinuses by the mnemonic: Embattled Military Fought Saddam (ethmoid, maxillary, frontal, sphenoid).

The ethmoid sinuses are present at birth. Rapid expansion of the ethmoid air cells occurs during ages 0 to 4 years and again with the adolescent growth spurt from 8 to 12 years. The ethmoid sinus is usually the source of infection in childhood sinusitis. This may also lead to orbital cellulitis or subperiosteal orbital abscesses in children.

The maxillary antrum is also present, although small at birth, and growth continues to age 14 years.

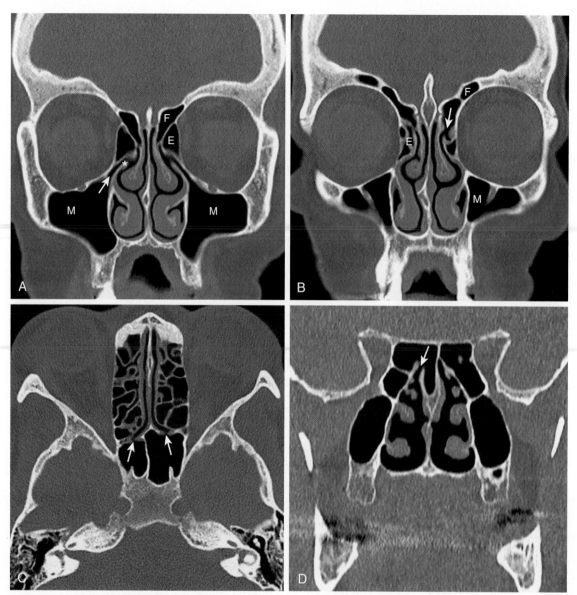

FIGURE 12-4 Major drainage pathways of the paranasal sinuses. **A,** Coronal computed tomography (CT) reconstruction shows the maxillary sinus ostium *(arrow)* and uncinate process *(asterisk)* facilitating drainage from the maxillary sinus (M) into the nasal cavity. Frontal (F) and ethmoid (E) sinuses are also indicated. **B,** More posteriorly in the same patient, the frontal recess is indicated *(arrow)*, facilitating drainage from the frontal sinus (F) and ethmoid air cells (E) toward the middle meatus. Maxillary sinus is indicated (M). On axial **(C)** and coronal **(D)** CT, the sphenoethmoidal recess *(arrows)* leads from the sphenoid and posterior ethmoid sinus into the nasal cavity.

The frontal sinus is not present at birth but pneumatization evolves from age 1 to 12 years. Growth over the orbital roof usually occurs from ages 4 to 8 years. Frontal sinusitis before age 4 years is therefore rare.

The sphenoid sinus begins its pneumatization at approximately age 2 years and the growth is slower and more delayed than the other sinuses. The ultimate size of the sphenoid sinus is quite variable.

Note that the OMC is developed at birth and functional endoscopic sinus surgery (FESS) is an option for chronic childhood sufferers of sinusitis. Similarly, the mucosa in the infant is somewhat redundant and up to 60% of asymptomatic infants can have complete or near complete opacification of their sinuses. Therefore, mucosal thickening should not be assumed to be due to sinusitis.

IMAGING MODALITIES

Sinonasal imaging has progressed methodically as each new generation of imaging modality has encroached on the domain of the former generation. Although plain films once served as the most commonly ordered study to evaluate the sinonasal cavity, computed tomography (CT) has now supplanted plain films, because the endoscopic sinus surgeon has required greater anatomic precision. During the 1970s and 1980s, FESS replaced the more traditional Caldwell-Luc and maxillary antrostomy procedures for treating chronic sinusitis.

FESS is done via an intranasal endoscope rather than with an external approach, so surgeons must know where they are at all times, to prevent complications such as orbital or intracranial entry, particularly when they are operating posteriorly in the sinonasal cavity. CT serves as the road map for this procedure. The goal of FESS is to maintain the normal mucosa of the sinonasal cavity and to preserve the natural pathway of mucociliary clearance. FESS does not attempt to strip the mucosa clean as was sometimes performed in a Caldwell-Luc procedure, so mucociliary motility is preserved. Also, rather than creating an alternate egress of mucus from the maxillary sinus, as in an inferior maxillary antrostomy (the Caldwell-Luc procedure), FESS enlarges the natural ostia and passageways of the paranasal sinuses. Whereas in the past maxillary and frontal sinusitis were thought to be the primary processes in patients with chronic sinusitis, it is now believed that these sinuses are secondarily obstructed because of disease in the OMC. The classic theory of FESS is that disease at the ostium and inferior infundibulum obstructs the maxillary sinus, whereas disease in the middle meatus and posterior infundibulum obstructs the frontal and anterior ethmoid air cells. Therefore, surgery is directed toward removing potential obstacles to mucociliary clearance at the OMC. Therefore amputation of the uncinate process, enlargement of the infundibulum and maxillary sinus ostia, and creation of a common unobstructed channel for the anterior ethmoid air cells are common practices in FESS. Usually FESS also includes complete or partial ethmoidectomies.

There are four potential dehiscent areas where the FESS surgeon can get into anatomic (and medicolegal) trouble: (1) cribriform plate with inadvertent entry intracranially leading to cerebrospinal fluid (CSF) leak/brain damage, (2) lamina papyracea with puncture into the orbit leading to expanding intraorbital hematoma, (3) sphenoethmoidal junction with the optic canal with potential optic nerve injury, and (4) sphenoid sinus lateral wall with possible carotid artery injury/pseudoaneurysm. Make sure you check each case for intact walls in these four areas.

For the FESS surgeon, coronal CT is ideal because it simulates the appearance of the sinonasal cavity from the perspective of the endoscope. At present, coronal reconstructions of axial CT data from multidetector spiral CT scans are just as good as direct coronal images and may be able to eliminate some dental amalgam artifact that otherwise is present. To eliminate the effects of reversible sinus congestion, patients undergoing CT for evaluation of chronic sinus disease are best scanned 4 to 6 weeks after medical therapy and not during an acute infection. Some radiology departments also administer nasal spray decongestants or antihistamines to reduce reversible mucosal edema before the patients are placed in the scanner. Even with this preparation, surgeons claim that in approximately 10% of cases with "normal CT scans" they find endoscopic evidence of significant sinusitis. This goes to show that imaging diagnosis of acute sinus inflammation is not always accurate, and the diagnosis of sinusitis remains a clinical one.

Magnetic resonance (MR) examination of the sinonasal cavity can be performed in a standard head coil or, for more precise anatomic resolution, with a surface coil placed over the anterior part of the face. Both T1-weighted imaging (T1WI) and T2-weighted imaging (T2WI) are required because of the variability of signal intensity of sinonasal secretions caused by protein concentration. Fat-suppressed contrast-enhanced T1WI are employed for the evaluation of complicated sinusitis or for suspected neoplastic disease. Differentiating tumors from infections of the sinonasal cavity may be best achieved with enhanced MR: infected mucosa enhances in a peripheral manner, whereas tumors usually enhance solidly and centrally (Fig. 12-5).

CONGENITAL DISORDERS

The piriform aperture of the nasal cavity can be narrowed congenitally. A width of less than 11 mm is indicative of congenital nasal piriform aperture stenosis. Abnormal dentition and a midline bony inferior palatal ridge are associated imaging findings. The most common anomalies that result in infantile airway compromise include posterior choanal stenoses and atresias, dacryocystoceles, and stenosis of the piriform nasal aperture.

Choanal Atresia

Choanal atresia is usually diagnosed in infancy because neonates are obligate nose breathers as they suck on a bottle or breast (Box 12-1). The child presents with respiratory distress. Although the diagnosis can be suggested by the inability to pass a nasogastric tube through the nose, imaging is necessary to determine whether the obstruction is membranous (15% of cases) or bony (85%) and whether other congenital central nervous system (CNS) or non-CNS anomalies are associated (50%). In addition to the narrowed posterior choana, look for thickening of the vomer (Fig. 12-6). The posterior choanal opening should be greater than 0.5 cm in width in neonates and 1 cm in adolescents. The vomer should measure less than 0.3 cm in children younger than 8 years old. Rather than atresia, some patients have mere stenosis of the passageway. Often, unilateral choanal atresia may escape detection into adulthood. Patients may be unaware of the hyposmia often associated with this disorder. A dacryocystocele or piriform aperture stenosis may mimic choanal atresia clinically.

Congenital Sinonasal Masses

There are several congenital lesions of the sinonasal cavity, including congenital encephaloceles, dermoid/epidermoid cysts, sinus tracts, and nasal gliomas (Table 12-1). These lesions occur as an abnormality in the process of invagination of the neural plate. In embryogenesis, the dura of the brain contacts the dermis at the nasion region as the neural plate retracts. Normally, the dermal connection regresses; when it does not, one of the lesions cited earlier may develop. A CSF connection to the intracranial contents is maintained with meningoencephaloceles, whereas the connection is fibrous only with a nasal glioma (Figs. 12-7, 12-8). Nasal gliomas are *not* neoplasms but congenital anomalies. They are extranasal more commonly than intranasal. (What an oxymoron—nasal gliomas are usually extranasal and are not gliomas!) Most patients with dermal sinus tracts have a pit in the middle of the nose. Dermoid cysts occur more commonly than tracts; however, tracts may cause more severe symptoms because of their intracranial connection in 25% of cases. Thus meningitis, osteomyelitis, and intracranial abscesses

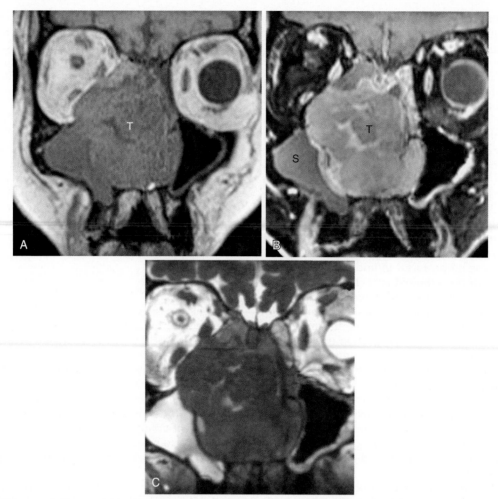

FIGURE 12-5 Differentiation of tumor versus inflammation on magnetic resonance imaging. **A,** Coronal T1-weighted imaging (T1WI) shows a bulky enhancing mass (T) within the nasal cavity. The mass is invading the right inferomedial orbit. It is unclear whether the maxillary sinus next door is full of tumor or secretions blocked within the sinus due to obstruction at the ostiomeatal complex. **B,** Coronal enhanced fat-suppressed T1WI demonstrates solid enhancement within this neoplasm (T). Note that much of the material within the right maxillary sinus does not enhance, indicating backed-up secretions (S). Thus with contrast, the border of the tumor is well delineated. **C,** In this case, there is also marked signal difference between the tumor and the secretions in the maxillary sinus on coronal T2WI, but the distinction on T2 is not always quite so clear, as a result of proteinaceous material that can develop within sinus contents.

BOX 12-1   Entities Associated with Choanal Atresia

Achondroplasia
Fetal alcohol syndrome
CHARGE syndrome (coloboma, heart defects, choanal atresia,
    retardation, genitourinary abnormalities, ear abnormalities)
Crouzon syndrome
Apert syndrome
Holoprosencephaly
Treacher Collins syndrome
Amniotic band syndrome
Thalidomide embryopathy

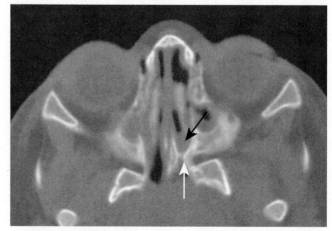

FIGURE 12-6 Membranous and bony choanal atresia. On the left side the nasal passageway narrows, the vomer is thicker on that side, and both a bony (posterior; *white arrow*) and soft tissue (anterior; *black arrow*) plug is seen. No luck passing the tube on the left side.

TABLE 12-1   Congenital Nasal Lesions

Lesion	Imaging Findings	Clinical Examination	Treatment
Nasal glioma	Soft-tissue mass with characteristics of the brain No connection or fibrous connection to the brain or CSF	Intranasal or extranasal mass	Excision
Dermoid/epidermoid	Variably cystic/soft-tissue mass ±Sinus tract Inflammatory changes	Nasal dimple Sinus tract	Exploration and excision
Encephalocele	Connection to CNS with associated defect in skull bone Brain or meninges included Other CNS anomalies	Pulsatile mass Dural covering	Patch dura Reduce brain tissue

CNS, Central nervous system; *CSF,* cerebrospinal fluid.

may occur in the setting of dermoid tracts. Epidermoids do not have hair/fat/other dermal elements and are more cystic.

Hypoplastic Maxillary Antrum

Congenital hypoplasia of the maxillary sinus occurs in 9% of patients. Bony changes that suggest the diagnosis of a hypoplastic antrum are listed in Box 12-2. Causes of *over*-expansion of paranasal sinuses are listed in Box 12-3. In the differential diagnosis of sinus hypoplasia is the "silent sinus syndrome" or maxillary sinus atelectasis. In this entity, ostial obstruction from chronic sinusitis leads to chronic negative pressure, which leads to hypoventilation, which, over time, reduces the volume of the sinus, hence "sinus atelectasis." Patients present with enophthalmos (not sinus symptoms strangely enough) as the orbital floor becomes depressed, the maxillary walls retract centripetally, and expanded retromaxillary fat fills the space left by the atelectatic sinus. CT shows the retracted maxillary sinus walls in association with a small volume, opacified sinus and hypoglobus with enophthalmos (Fig. 12-9).

INFLAMMATORY LESIONS

Sinusitis

Sinusitis ranks as one of the most common afflictions in the United States. It is estimated that more than 31 million people in the United States are affected by sinus inflammatory disease each year and that 16 million visits to primary care physicians annually are for sinusitis and its complications. Adults average two to three colds per year, and 0.5% of viral upper respiratory infections are complicated by sinusitis. Overall, Americans spend more than $150 million per year for over-the-counter cold and sinusitis medicines, $100 million of which is for antihistamine medications.

Most cases of acute sinusitis are related to an antecedent viral upper respiratory tract infection. With mucosal congestion as a result of the viral infection, apposition of mucosal surfaces results in obstruction of the normal flow of mucus, which results in retention of secretions, creating a favorable environment for bacterial superinfection. The ethmoid sinuses are most commonly involved in sinusitis, possibly because of their position in the "line of fire" as inspired particles collide with and irritate the fragile ethmoid sinus lining.

The bacterial pathogens responsible for acute sinusitis include *Streptococcus pneumoniae, Haemophilus influenzae,* β-hemolytic streptococcus, and *Moraxella catarrhalis.* In the chronic phase *Staphylococcus, Streptococcus,* corynebacteria, *Bacteroides,* fusobacteria, and other anaerobes may be responsible. The fungi that may infect the sinuses include *Aspergillus, Mucor, Bipolaris, Drechslera, Curvularia,* and *Candida.*

The diagnosis of acute sinusitis is a clinical one, and imaging is typically not required. Imaging can be performed to evaluate for recurrent sinus infections or atypical sinus infections, often with the intention of surgical intervention.

Anatomic Considerations

Several issues must be addressed when a patient's CT is evaluated before FESS. Is the uncinate process apposed to the medial orbital wall (an atelectic infundibulum)? If so, its vigorous removal may result in orbital penetration. Are there areas of dehiscence in the lamina papyracea, or do the orbital contents protrude into the ethmoid sinus (both of which may lead to unintentional orbital entry from the ethmoid sinus)? Defects in the lamina papyracea have been reported in 5% to 10% of autopsy specimens. Because orbital hematomas are the most common orbital complication of FESS, it is important to identify any gaps in the lamina papyracea. CT is the best means for identifying the thin medial bony wall of the orbit.

Are there areas of dehiscence in the cribriform plate and sphenoid sinus walls (Fig. 12-10)? Remember that there is an attachment of the middle turbinate to the cribriform plate and that the carotid canal courses along the sphenoid sinus walls. If the surgeon tugs too hard, the cribriform plate will fall and down will come CSF and/or the brain through the new foramen in the skull base. The potential for intraorbital, intracranial, carotid, or optic nerve perforation at the time of surgery depends on these anatomic variants, found in 4% to 15% of patients. Three percent of people have optic nerves that are in contact with the posterior ethmoid wall—most course along (90%) or through (6%) the sphenoid sinus. There have been limited reports of optic nerve transection during sphenoethmoidectomy from an intranasal approach, and dehiscence of the sphenoid wall may be a predisposing factor. An intersinus septum in the sphenoid sinus that attaches to the carotid canal is important to recognize preoperatively and is typically best identified in the axial view (Fig. 12-11). Overvigorous removal of such an intersinus septum during surgery may result in carotid laceration, a real bloody mess.

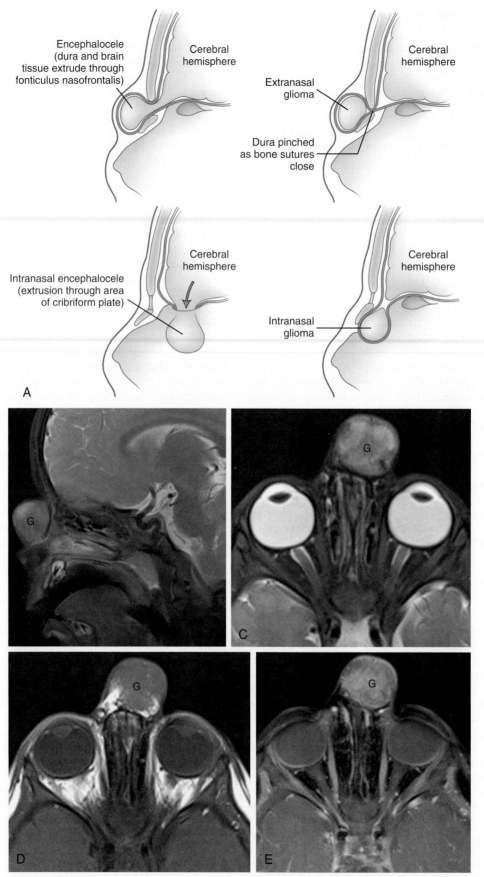

FIGURE 12-7 Congenital nasal lesions. **A,** Schematic representation of encephaloceles *(left)* and nasal gliomas *(right).* Sagittal T2 **(B),** axial T2 **(C),** and axial precontrast **(D)** and postcontrast **(E)** T1-weighted images show an extranasal soft-tissue mass (G) looking kind of brainy with mild enhancement; there is no attachment to the intracranial structures. Again, not nasal and not glioma = nasal glioma (go figure!). (**A** from Gorenstein A, Kern EB, Facer GW, Laws ER Jr. Nasal gliomas. *Arch Otolaryngol.* 1980;106:536. Copyright © Mayo Clinic.)

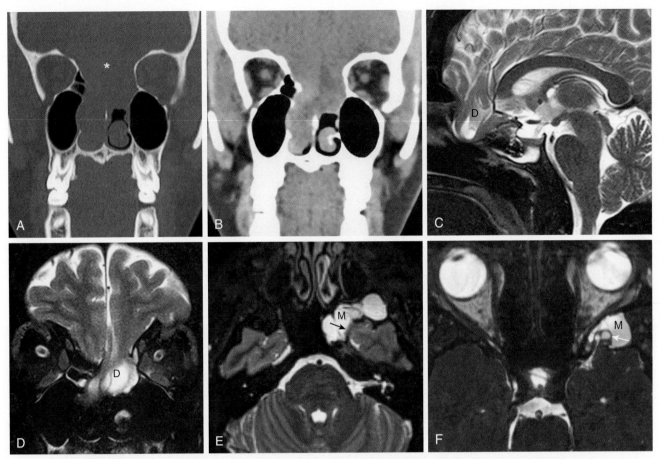

FIGURE 12-8 Meningoencephaloceles galore. Contrast the previous case with this one. An intranasal meningoencephalocele is seen on coronal computed tomography in bone **(A)** and soft-tissue **(B)** windows. There is a large deficiency at the cribriform plate *(asterisk),* allowing for herniation of brain tissue into the nasal cavity. Sagittal imaging **(C)** and coronal **(D)** T2-weighted imaging (T2WI) confirm the herniation of brain tissue (D) into the nasal cavity. Note the T2 hyperintensity within the herniated tissue, indicating dysplastic brain. **E,** And one more…on this axial T2WI, you might be fooled into calling this a lateralized sphenoid sinus filled with snot but it's not (heh heh). Note the focal deficiency of the middle cranial fossa transmitting a small amount of brain tissue *(arrow),* making this a meningoencephalocele (M). **F,** For you nonbelievers out there, slightly more superiorly in this same patient, axial T2 constructive interference in steady state imaging shows another defect along the middle cranial fossa, transmitting clearly dysplastic brain tissue *(arrow)* into the aerated but opacified sphenoid wing, again cinching the diagnosis of meningoencephalocele (M).

BOX 12-2   Findings in a Hypoplastic Maxillary Sinus

Downward sloping antral roof
Enlarged ipsilateral nasal cavity
Lower ipsilateral orbital floor
Thickened lateral antral wall

BOX 12-3   Overpneumatized Sinuses

Normal variant
Marfan syndrome
Turner syndrome
Acromegaly
Pneumosinus dilatans from meningioma
Dyke-Davidoff-Masson syndrome (usually unilateral with ipsilateral
   atrophic brain)
Homocystinuria

Appearance on Imaging

In addition to commenting on the normal anatomic variations in the CT report, the radiologist should identify areas of mucosal thickening and sinus passageway opacification. It has come to be accepted that the location of sinusitis is more important in producing a patient's symptoms than the extent of the sinusitis. Therefore, a subtle area of opacification in the infundibulum of the OMC may cause more pain and discomfort than nearly complete opacification of the maxillary sinus with a mucous retention cyst and/or polyp.

OMC opacification correlates well with the development of sinusitis (Fig. 12-12). The positive predictive value of infundibular opacification for the presence of maxillary sinus inflammatory disease is approximately 80%. When the middle meatus is opacified, the maxillary and ethmoid sinuses show inflammatory change in 84% and 82% of patients, respectively. The specificity of middle meatus opacification for maxillary or ethmoid sinus disease is

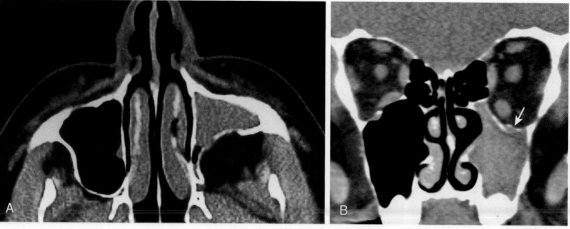

FIGURE 12-9 Silent sinus syndrome. **A,** Note the retracted posterior wall of the left maxillary antrum. Fat fills the vacuum. The sinus is nearly completely opacified. **B,** On this coronal computed tomographic image in a different patient, the left maxillary sinus is completely opacified and smaller than the right. Note the slightly thickened walls of the left maxillary sinus from chronic inflammation. The orbital floor of the left is depressed *(arrow)* compared to the normal right side. The sinus has become "atelectatic."

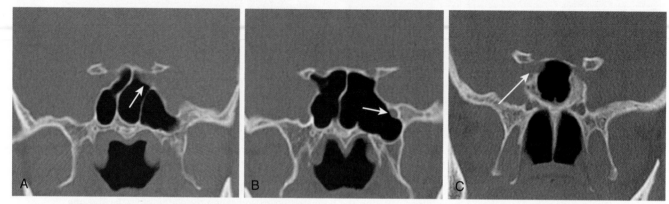

FIGURE 12-10 Potential sites of disaster: dehiscent areas in the sinuses. **A,** Along the left optic canal in the sphenoid sinus *(arrow)*. **B,** Along the left maxillary nerve *(arrow)*. **C,** Along the right optic nerve *(arrow)*.

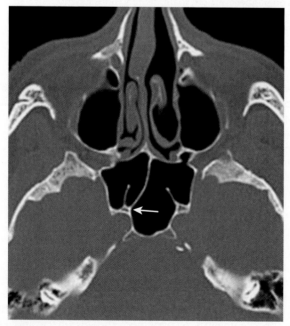

FIGURE 12-11 Sphenoid sinus septa. On this axial computed tomographic image, the right septae in the sphenoid sinus attach to the medial wall of the right internal carotid artery *(arrow)*. Overvigorous removal during sphenoid sinus surgery can cause a laceration in the carotid wall (ouch!).

more than 90%. These findings support the contention that obstruction of the narrow drainage pathways leads to subsequent sinus inflammation.

Some head and neck radiologists categorize recurrent inflammatory sinonasal disease into five patterns: (1) infundibular, (2) ostiomeatal unit, (3) sphenoethmoidal recess, (4) sinonasal polyposis, and (5) sporadic or unclassifiable disease. The infundibular pattern is seen in 26% of patients and refers to isolated obstruction of the inferior infundibulum, just above the maxillary sinus ostium. Limited maxillary sinus disease often coexists with this pattern, whereas the ostiomeatal unit pattern, seen in 25% of cases, often has concomitant frontal and ethmoidal disease. The ostiomeatal unit pattern is designated when middle meatus opacification is present. Sphenoethmoidal recess obstruction occurs in 6% of cases and leads to sphenoid or posterior ethmoid sinus inflammation. When the sinonasal polyposis pattern is present, enlargement of the ostia, thinning of adjacent bone, and opacified sinuses are usually seen in conjunction with nasal polypoid disease.

The presence of air-fluid levels and/or frothy secretions is more typically associated with acute sinusitis than with chronic inflammatory disease, however this finding is by no means specific for acute sinusitis. Fluid levels can often be seen within the paranasal sinuses

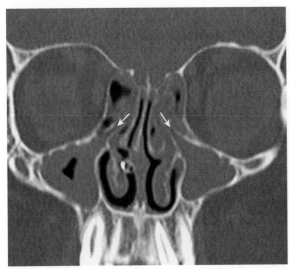

FIGURE 12-12 The ostiomeatal complex in sickness. Note that both maxillary sinus ostia and infundibula *(arrows)* are opacified in this individual. Associated ethmoid and bilateral maxillary sinus inflammation is present. There is nasal tubing in the right nasal cavity.

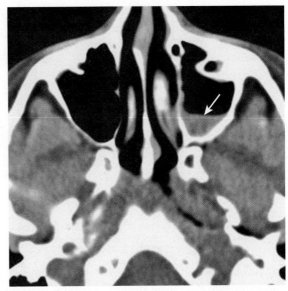

FIGURE 12-13 Acute sinusitis. An air-fluid level can be seen in a number of clinical scenarios, but if the clinician is concerned about acute sinus inflammation, the air-fluid level would be a salient imaging sign *(arrow)*.

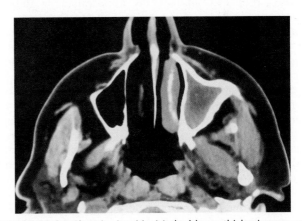

FIGURE 12-14 Chronic sinusitis. Marked bony thickening around the opacified left maxillary sinus signifies osteitis from chronic sinus disease.

of intubated patients and are not necessarily infected. Drowning victims or trauma patients can also present with fluid levels in the sinus, but these sinuses are not necessarily infected. However, in cases of clinically suspected acute sinusitis, air-fluid levels and/or complete opacification of a sinus is present in 63% of cases (Fig. 12-13). Of course, acute sinusitis may be superimposed on chronic changes. The findings suggestive of chronic sinusitis include mucosal thickening, bony remodeling, polyposis, mucous retention cysts, and bone thickening secondary to osteitis from adjacent chronic mucosal inflammation (Fig. 12-14).

Hyperdense secretions on CT may be due to four main causes: (1) inspissated secretions, (2) fungal sinusitis, (3) hemorrhage in the sinus, and (4) calcification. The hyperdense sinus may be the only clue to fungal sinusitis and is an important feature to note. However, chronically inflamed sinuses infected with bacteria occasionally show hyperdense contents on CT, particularly in patients who have very long-standing disease, polyposis, or cystic fibrosis. A hyperdense sinus often corresponds to a hypointense sinus on T2WI (Fig. 12-15).

If you measure the Hounsfield units (HU) you can gain some specificity to the "hyperdense sinus." Those with HUs greater than 2000 have a 93.3% chance of maxillary sinus aspergillosis. The mean CT density of the sinus concretions without aspergillosis is 778 HU. Make sure this hyperdensity is not reflective of hemorrhage secondary to a traumatized sinus.

The following features of the "calcified" sinus mass have been reported. A single discrete hyperdensity is most likely to be an inflammatory mass (aspergilloma, rhinolith), but multiple discrete calcifications could be seen in tumors (enchondromas, inverted papillomas, meningiomas) or inflammatory lesions. If the intrasinus calcification is central, fine, and punctate-like, it is most likely attributable to fungi. If the calcification is peripheral, curvilinear, and eggshell-like, it is probably nonfungal. If the process is diffusely hyperdense with a well-defined margin, think of a fibroosseous lesion but if it is poorly marginated, consider

a high-grade sarcoma (chondrosarcoma, osteosarcoma). Although calcification is not unusual in inverted papillomas it is more likely to be residual bone, not calcifications. By contrast, esthesioneuroblastomas have intrinsic calcifications.

A rhinolith (stone in the nose) is often attributable to a foreign body that has become lodged in the nose and has slowly calcified. This may occur in the setting of traumatic or recreational nasal septum injuries.

While you are evaluating coronal CT scans for sinusitis, do not forget to look at four other related areas: the teeth, sella, nasopharynx, and temporomandibular joints. Maxillary periodontal disease can incite sinus inflammation (odontogenic sinusitis), and therefore thinning and/or dehiscence of the maxillary alveolus adjacent to an inflamed carious tooth necessitates reporting. The sinus disease will not resolve until the offending tooth is removed.

Malocclusions or temporomandibular joint degenerative change may indicate a maxillofacial pain syndrome

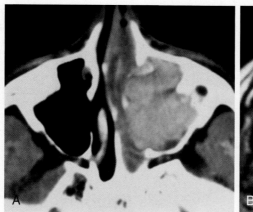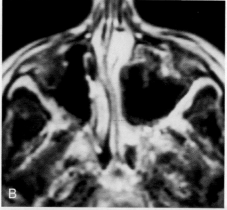

FIGURE 12-15 Fungal sinusitis. **A,** Unenhanced computed tomographic scan shows a hyperdense opacified left maxillary antrum. **B,** On T2-weighted imaging, the secretions were black, suggesting inspissation, paramagnetic accumulation, or fungal sinusitis. Cultures grew *Drechslera* fungi.

that may simulate pain from sinusitis. Look for narrowing, osteophytes, or sclerosis at the joint.

Beware of the incidental sellar lesions and nasopharyngeal carcinomas that might be appreciable on routine sinus CT. When these tumors extend into paranasal sinuses, they can simulate the appearance of ho-hum sinus opacification—don't be fooled! On that note, do not forget to evaluate the imaged portions of the brain for intracranial pathology—all kinds of sneaky buggers might be lurking there and will remain undetected unless you look for them.

Screening sinus CT images are often performed before bone marrow transplantation (BMT) in patients with hematologic malignancies. Of those showing severe sinusitis on initial screening CT scans, two thirds experience clinical sinusitis post-BMT. The presence of sinusitis on pre-BMT studies is also associated with a trend toward overall decreased survival.

Trauma often affects the walls of the paranasal sinuses. Remember that the floor of the orbit serves as the roof of the maxillary sinus, so orbital floor fractures cause blood levels in the sinuses. Similarly, the medial wall blowout fracture affects the ethmoid sinus and orbital roof fractures may affect supraorbital ethmoid cells or the frontal sinus. Direct blows to the forehead may cause inward displacement fractures of the frontal sinuses (Fig. 12-16). Skull base fractures may cross the sphenoid sinus. These air-filled spaces are not necessarily the best buttresses against trauma, being very thin and with nothing but air between them and the directed force.

The sensitivity of MR to mucosal thickening accounts for the visualization of the normal nasal cycle. There is cyclical passive congestion and decongestion of each side of the nasal turbinates, nasal septum, and ethmoid air cell mucosa that rotates from side to side over the course of 1 to 8 hours in humans. Thus, 1 to 2 mm of ethmoid sinus mucosal thickening may not be due to inflammatory disease but may reflect the normal intermittent congestion of the nasal cycle. Be aware of the nasal cycle so you do not overread increased mucosal signal from the turbinates and ethmoid sinus as disease on MR. Even if you use 5 mm of thickness of the mucosa or greater as a determinant of pathology on MR, incidental sinus inflammation is present in more than 32% of asymptomatic patients coming to MR scanning.

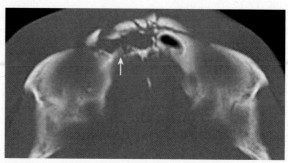

FIGURE 12-16 Trauma to the frontal sinus. When a fracture goes through both walls of the frontal sinus, the potential for a cerebrospinal fluid leak, meningitis, pseudomeningoceles, and intracranial hemorrhagic complications increases. Note extensively comminuted fractures of the frontal sinus with focal defect at the posterior wall of frontal sinus *(arrow)*.

Sinonasal secretions are often, but not always, bright on T2WI and dark on T1WI. The change in signal intensity of sinonasal secretions is based on protein concentration and mobile water protons. The changes are probably a result of the increased cross-linking of glycoproteins in hyperproteinaceous secretions, leading to fewer available free water protons and more bound protons to glycoproteins. As protein concentration increases, the signal intensity on T1WI of sinus secretions changes from hypointense to hyperintense, to hypointense again. On T2WI, hypoproteinaceous watery secretions are initially bright, but as the protein concentration and viscosity increase, the signal intensity on T2WI decreases. Therefore, you can get four confusing intensity patterns for sinus secretions: (1) hypointense on T1WI and hyperintense on T2WI in the most liquid form, (2) hyperintense on T1WI and hyperintense on T2WI in the mild-to-moderate proteinaceous form, (3) hyperintense on T1WI and hypointense on T2WI in a highly proteinaceous form, and (4) hypointense on T1WI and hypointense on T2WI when the secretions are in a nearly solid form (Fig. 12-17).

Because air and very hyperproteinaceous secretions appear as a signal void, this is a potential pitfall. You may think the sinus is aerated but it is not. Other potential problems in which hypointensity may be encountered on both T1WI and T2WI include osteomas, odontogenic lesions,

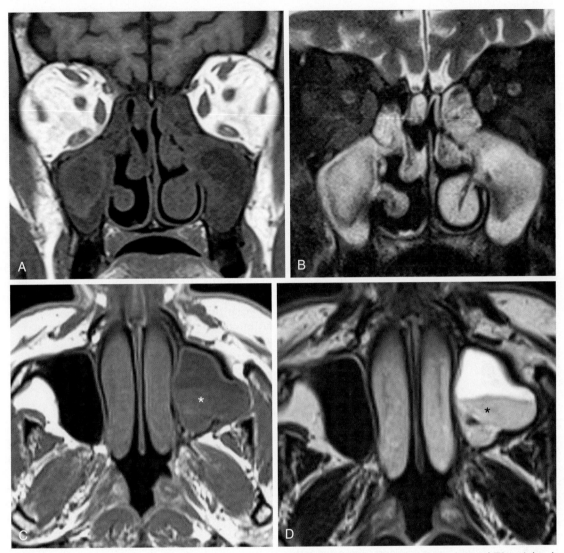

FIGURE 12-17 Proteinaceous sinonasal secretions on magnetic resonance imaging. **A,** On coronal T1-weighted imaging (T1WI) and **(B)** coronal T2WI, the maxillary sinus contents show mixed signal intensity indicating variable concentrations of proteinaceous secretions. In a different patient, **(C)** axial T1- and **(D)** axial T2-weighted images show simpler fluid layering non-dependently within the left maxillary sinus, and more proteinaceous contents layering dependently *(asterisks)*. In general, signal from lowest protein content to highest protein content progresses from dark T1-bright T2 to bright T1-bright T2 to bright T1-dark T2, and finally dark T1-dark T2 signal intensity.

osteochondromas, fibrosis, and fungal sinusitis mycetomas. The low signal of fungal sinusitis is thought to result from the paramagnetic effects of either iron or manganese metabolized by the fungi (Box 12-4).

Because obstructed sinus secretions may have any combination of signal intensities, it may become difficult to distinguish inflammation from neoplasm solely based on intensity patterns. For this reason, the presence of peripheral rim enhancement is a reassuring finding that suggests inflammation rather than neoplasm (which enhances centrally).

Post-Functional Endoscopic Sinus Surgery Scanning

There is one classification for types of FESS that is worth remembering; see Table 12-2. Although postoperative studies will show open OMCs in patients with chronic sinusitis and polyposis, the overall extent of sinus opacification before and after FESS is more often than not minimally changed (Fig. 12-18).

BOX 12-4   Sinonasal Lesions with Signal Voids on Magnetic Resonance

Teeth
Inspissated secretions
Fungus balls
Osteomas
Sinoliths
Hemorrhage
Odontogenic lesions
Chondroid lesions
Amyloidomas
Fibrosis

CT is the study of choice in the acute setting of orbital complications after FESS, which usually occur within 48 hours after surgery. CT is reliable in diagnosing an orbital hematoma, assessing for optic nerve compression or inadvertent resection, and planning therapy

TABLE 12-2 **Types of Functional Endoscopic Sinus Surgery**

Type	Uncinate Resected?	Ethmoidectomy	Maxillary Sinus Surgery
I	Yes	Only agger nasi cells	No
II	Yes	Bulla, anterior ethmoidectomy, frontal sinus opening	No
III	Yes	Bulla ethmoidectomy, frontal sinus opening	Antrostomy
IV	Yes	Anterior and posterior ethmoidectomy	Antrostomy
V	Yes	Anterior and posterior ethmoidectomy plus sphenoidectomy	Antrostomy

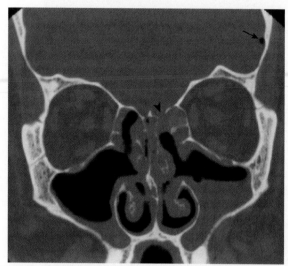

FIGURE 12-18 Postoperative study after uncinectomies and ethmoidectomies. Despite the surgery to open the ostia and ethmoid sinuses, this patient continued to have acute maxillary sinusitis attacks (note the air-fluid levels on the coronal prone study) and chronic mucosal changes in the ethmoid sinuses. Extra credit: Did you catch the dehiscent area in the left fovea ethmoidalis *(arrowhead)?* This may be a manifestation of the disease or the surgery. The dot of intracranial air *(arrow)* is the second edge of the film diagnosis.

to relieve increased intraorbital pressure. Often, however, if vision is deteriorating rapidly, this complication is treated on the basis of clinical findings. Acute hematomas are usually hyperdense on CT; occasionally, diffuse orbital fat edema may be the most salient finding. Usually, orbital hematomas are due to transection of an ethmoidal artery at FESS. The artery then retracts into the orbit and continues to bleed. The intraorbital pressure rises with the expanding hematoma, leading to compromise of the flow of the retinal artery and ischemia of the optic nerve. Decompressive canthotomies are emergently performed (within hours), or irreversible nerve damage will occur.

Other complications include trauma to the medial wall of the orbit. While removing the middle turbinate, the lateral attachment to the lamina papyracea (the basal lamella) may be yanked off and injure the medial wall of the orbit. Contusion to the medial rectus muscle, fat herniation into the sinus, and orbital hematomas may follow. Orbital emphysema may be seen.

CSF leaks resulting from trauma to the cribriform plate and damage from overvigorous removal of the superior attachment of the middle turbinate at the fovea ethmoidalis are other potential FESS complications. This may lead to postoperative meningitis, epidural abscess, or pneumocephalus. Fortunately, FESS surgeons are getting good at patching their own dural holes with mucosal free grafts or septal flaps.

Osteoplastic frontal sinus grafts are often placed during frontal sinus obliterative procedures when FESS has been unsuccessful in opening the frontal recess. Postoperatively, patients may develop pain that may be secondary to neuromas, mucoceles, or recurrent sinusitis if the sinus is not plugged completely. Most surgeons use fat to obliterate the sinus. MR imaging with fat suppression can evaluate the frontal sinus after obliteration with adipose tissue to look for mucoceles and to differentiate viable adipose tissue from fat necrosis in the form of oil cysts. The amount of fat intensity in the sinus obliteration decreases over time as fibrosis occurs.

Less Common Inflammatory Conditions of the Sinuses

Granulomatous Diseases

Sarcoidosis may affect the sinonasal cavity and is one of the sources of nasal septal perforation (along with cocaine abuse, syphilis, leprosy, and granulomatosis with polyangiitis [formerly Wegener granulomatosis]) (Fig. 12-19). The entities of lethal midline granuloma, Stewart syndrome, midline granuloma syndrome, polymorphic reticulosis, lymphomatoid granulomatosis, and pseudolymphoma have recently been reclassified into cases of granulomatosis with polyangiitis and/or non-Hodgkin T-cell lymphoma. The generic term of "midline destructive lesions of the sinonasal tract" has also been applied to these entities (Box 12-5). Both granulomatous disease processes and non-Hodgkin T-cell lymphoma may present with nasal septal perforations, bony/skull base erosions, and soft-tissue masses in the sinonasal cavity.

Granulomatous disease processes show intracranial involvement in up to 5% of cases, usually manifesting with cranial neuropathies. Meningeal inflammation, vasculitis, and hypophysitis may also coexist.

In cases of non-Hodgkin T-cell lymphoma, bony erosion may be seen, particularly involving the medial wall of the maxillary antrum or the hard palate. Relative preservation of the sinus morphology in the setting of a sinus soft-tissue mass can favor lymphoma over carcinoma, which results in frank destruction of the sinus walls.

Foreign body granuloma, polyarteritis nodosa, lupus, and hypersensitivity angiitis may also present with destructive sinonasal masses. Pseudotumor, an idiopathic inflammatory disease characterized by fibroblasts, histiocytes, and inflammatory cells, can simulate an aggressive sinus process. Pseudotumor has been more recently classified under IgG4-related disease, and can be associated with rhinosinusitis with or without bony destruction; patients have high levels of serum IgG4 levels and respond well to glucocorticoid treatment.

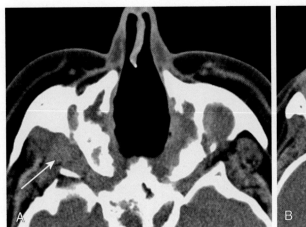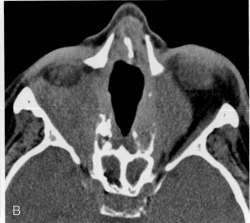

FIGURE 12-19 Septal perforations. **A,** This patient had granulomatosis with polyangiitis (formerly Wegener granulomatosis). The nasal septum is gone and the patient had infiltration of the orbits and the soft tissue around the right maxillary antra *(arrow)*. **B,** Similar findings with septal and lateral nasal wall destruction as well as orbital soft tissue extension are seen in this case of sarcoidosis.

BOX 12-5   Causes of Nasal Septal Perforation

Cocaine
Iatrogenic
Foreign body reaction
Sarcoidosis
Granulomatosis with polyangiitis (formerly Wegener granulomatosis)
Lymphoma/prelymphoma
Tuberculosis
Syphilis
Rhinoscleroma (Klebsiella)
Leprosy

Complications of Sinusitis

Although MR has no role in the evaluation of uncomplicated sinusitis, it is very useful in evaluating intracranial and intraorbital complications, including meningitis, thrombophlebitis, extraaxial empyemas, intracranial abscesses, and perineural or perivascular spread of infection (Fig. 12-20). Nonsuppression of CSF signal in the meningeal spaces on fluid-attenuated inversion recovery, diffusion restriction, and abnormal meningeal enhancement following contrast administration can indicate meningeal infection. Coronal scans are particularly good for identifying meningitis adjacent to frontal or ethmoidal sinusitis.

Osteomyelitis of the sinus marrow space can be complicated by spread of infection into the overlying scalp. This is most commonly seen centered in the frontal sinus, the so-called "Pott puffy tumor." This is not a tumor at all, but actually infection within the scalp overlying the inflamed sinus. Contrast-enhanced images with CT or magnetic resonance imaging (MRI) can show an organized fluid collection within the scalp. When this finding is made, careful attention to the intracranial compartment is imperative, as intracranial extension of inflammation can also occur. This includes infection of the extraaxial spaces, meninges, and brain parenchyma. MRI is more sensitive in the evaluation of intracranial involvement than CT.

Orbital complications of sinusitis may include preseptal and postseptal cellulitis, subperiosteal abscesses (the most common intraorbital complication in pediatric patients), and phlegmon. Particularly in pediatric patients, spread of infection through the inflamed sinus into the orbit can occur without dehiscence of the bone, and occurs along vascular channels (Fig. 12-21). Carefully evaluate the orbital fat for abnormal soft-tissue infiltration adjacent to the inflamed sinus. Contrast-enhanced images can show presence of organized abscess (peripherally enhancing fluid collection) or phlegmon (solidly enhancing soft tissue). Assessment for mass effect on orbital structures is important, including stretching of the optic nerve and proptosis.

Thrombophlebitis

Thrombophlebitis is an important complication to recognize as a consequence of sinusitis. Although thrombosed veins and venous sinuses may occur as a result of adjacent sinusitis, it is more commonly seen with mastoiditis. Sigmoid sinus venous thrombosis and venous infarction may complicate mastoiditis and petrous apicitis. In the paranasal sinuses, the most common venous channel involved is the cavernous sinus, and thrombosis is usually associated with sphenoid sinus inflammation. CT venography and/or contrast-enhanced MR venography are equally effective in making this diagnosis but beware of hyperintense thrombi, which may simulate flow on two-dimensional time-of-flight venograms. Enlargement of the superior ophthalmic vein within the orbit can help indicate presence of venous obstruction at the level of the cavernous sinus. Keep in mind that bilateral enlargement of these veins can be seen physiologically with Valsalva maneuvering, so do not overcall this finding if the clinical context does not fit.

Fungal Sinusitis

MR is of particular value in a patient who has an aggressive fulminant fungal sinus infection such as mucormycosis or aspergillosis. These fungal infections in their fulminant or invasive form have a propensity for invasion into the orbit, the cavernous sinus, and the neurovascular structures. Intracranially, fungi may cause vascular insults such as vascular thrombosis or cerebral infarcts. Mucormycosis

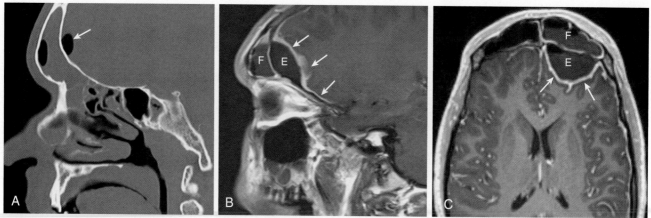

FIGURE 12-20 Epidural abscess associated with sinusitis. **A,** Sagittal computed tomography shows opacification within the frontal sinus with air-fluid level indicating acute sinusitis. Thickening of the frontal sinus walls indicates element of chronic inflammation. This could be run of the mill acute on chronic frontal sinusitis except for the air bubble intracranially *(arrow)* indicating intracranial extension of infection. **B,** Sagittal and **(C)** axial postcontrast T1-weighted images show the frontal sinus opacification (F), as well as an organized peripherally enhancing fluid collection within the subjacent epidural space (E). Note thickened, enhancing dura *(arrows)*.

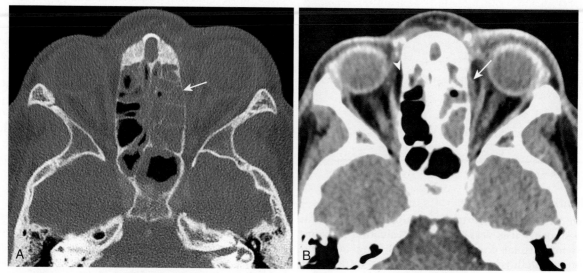

FIGURE 12-21 Orbital inflammation from sinusitis. **A,** Axial bone targeted computed tomographic image shows near complete opacification of the left ethmoid air cells with preservation of the integrity of the lamina papyracea *(arrow)* in this 12-year-old patient. **B,** Axial postcontrast image in soft-tissue windows at the same level shows developing subperiosteal phlegmon in the medial left orbit *(arrow)*. Compare this with the normal orbital fat on the other side *(arrowhead)*.

in particular appears to spread intracranially along the vessels. Prompt detection of this complication can lead to life- or vision-saving therapy.

Periantral soft-tissue infiltration either anterior or posterior to the maxillary sinus should suggest the possibility of invasive fungal sinusitis in the appropriate clinical setting. Early detection may save the patient from a significant face-ectomy, so scout the walls of the antrum for adjacent fatty infiltration in the fat outside the sinus suggesting the "horse has left the barn."

There are different levels of aggressiveness to mycotic infections. Sinonasal mycotic infection is similar in this regard to pulmonary fungal disease. Extramucosal fungal infection may be manifested as polypoid lesions (i.e., fungal saprophytic growth on retained secretions in patients with atopy) or as fungus balls. These are benign conditions usually caused by *Aspergillus*. They are often dense on CT

as well but are much more benign-looking and localized than allergic fungal sinusitis (AFS). Fungus balls are usually rounded masses, perhaps with a lamellated appearance. They have high signal on T1WI and low on T2WI (Fig. 12-22).

AFS, a nonvirulent form of the disease, is characterized on CT by increased attenuation within multiple paranasal sinuses. Bilateral involvement is frequently present. Complete opacification with expansion, erosion, or remodeling and thinning of the sinuses and/or expansion of the ostia are signature features of AFS (Fig. 12-23). The signal intensity on T2WI is usually low. Patients often carry history of asthma.

Steroid therapy and local excision are sufficient treatment for extramucosal fungal infections such as AFS or fungus balls.

Infiltrating fungal sinusitis occurs in an immunocompetent host but is not as aggressive as the fulminant infection seen in the immunocompromised person. The fulminant disease is the lethal form of infection and may be caused by *Mucor* or *Aspergillus*. Invasive fungal sinusitis is often hyperdense on CT and hypointense on T1WI and T2WI on MR. Wide, local excision and intravenously administered antifungal drugs are required for extirpation of the invasive mycotic infections. Orbital exenteration, hyperbaric oxygen treatment, and radical surgical therapy often are necessary for fulminant cases, and even then the prognosis is grim.

Cystic Fibrosis

Cystic fibrosis is a common autosomal recessive disease characterized by viscous secretions and hence chronic sinusitis (>90%) and polyposis (46%). Mucociliary clearance is inhibited by the thick grungy secretions. Sinonasal development in this childhood disease is often retarded by chronic infections. The mucosa is often polypoid in its appearance and mucocele formation is not infrequent. The triad of frontal sinus hypoplasia, medial bulging of the lateral nasal wall (indicative of maxillary sinus polyposis), and ethmoid opacification in a child is highly suggestive of cystic fibrosis. Add in some high density of the secretions and polyps on the turbinates and you are home free. Hypoplasia of the sphenoid sinus is present also to such a degree that if you do see pneumatization of the basisphenoid, you should question the diagnosis of cystic fibrosis.

Mucocele

Another complication of sinusitis is a mucocele. Although CT best demonstrates the bony distortion and thinning of sinus walls with mucocele formation and the remodeling of the osseous structures suggesting chronicity, MR can best detect its interface with intracranial or intraorbital structures, which can be compressed and distorted when mucoceles become large. When infected, the term used is mucopyocele. The density of the secretion on CT is usually highly indicative of chronicity and inspissation. The signal intensity of mucoceles may vary considerably based on protein content. Mucoceles are most common in the frontal (65%) and ethmoid (25%) sinuses, with maxillary (10%) and sphenoid sinus mucoceles the least common. A peripheral rim enhancement pattern is seen on MR, but can also be seen with infected mucoceles (mucopyocele) (Fig. 12-24). This pattern is useful for distinguishing mucoceles and obstructed secretions from neoplasms such as inverted papillomas or malignancies, which may also remodel bone but show solid enhancement. Tumors, fibro-osseous bone lesions, trauma, postoperative scarring, and hematomas can cause mucoceles.

Postoperative Encephaloceles and Cerebrospinal Fluid Leaks

Sinus encephaloceles are better distinguished from sinonasal inflammation by MR than by CT. Encephaloceles may occur from congenital defects or may arise in the postoperative setting when the sinus has been violated (Fig. 12-25). In the postoperative encephalocele that can occur after anterior skull base surgery, rhinorrhea is the presentation that elicits the imaging study, but this may occur as late as 18 months after surgery. Rhinorrhea may be due to a dural tear or a postethmoidectomy encephalocele. Dural tears occur most commonly on

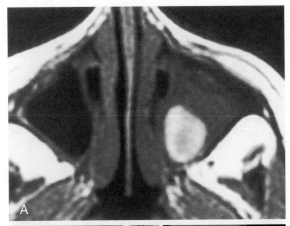

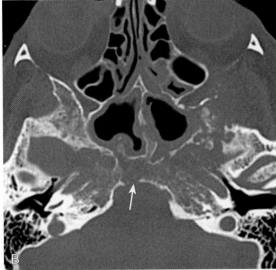

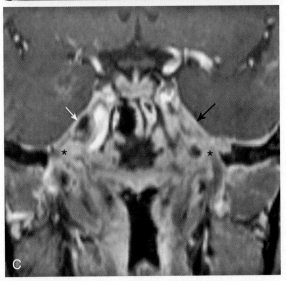

FIGURE 12-22 Two forms of *Aspergillus*. **A,** The hyperintense focus in the opacified left maxillary antrum may have been an inspissated mucus retention cyst, but they are usually of water intensity. Au contraire, this proved to be an aspergilloma. **B,** This is not run of the mill sphenoid sinus mucosal thickening. On closer inspection, there are destructive changes involving sphenoid sinus walls and septum. Destruction and permeative lytic change of the central skull base, sphenoid wings, clivus *(arrow)*, and petrous apices are also present in this patient with invasive fungal sinusitis. **C,** Coronal postcontrast T1-weighted imaging in the same patient as **(B)** shows abnormal enhancement at foramen ovale bilaterally *(asterisks)* resulting from soft-tissue invasion by aspergillosis. Note extension of infection into the left Meckel cave *(black arrow)* compared with the not-yet-violated right Meckel cave *(white arrow)*.

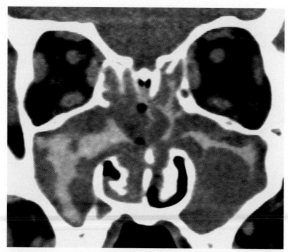

FIGURE 12-23 Allergic fungal sinusitis. On this unenhanced coronal computed tomographic image, note the expanded paranasal sinuses that are nearly completely opacified with mixed but predominantly hyperattenuating material, a typical appearance for allergic fungal sinusitis.

a patient's right side because most surgeons are right handed, making the right sinonasal cavity more difficult to surgerize.

Persistent postoperative leaks caused by dural tears are probably best assessed with a combination of studies. Nuclear medicine studies in which indium–diethylene triamine pentaacetic acid is injected intrathecally are highly accurate for confirming leaks when active flow is present, even at a slow rate. Pledgets are placed in the nose on both sides and assessed for "counts" (which will declare themselves if a leak is present) after removal of the pledgets. However, the more precise anatomic localization of site of leak afforded with CT with or without intrathecal contrast instillation is more useful to the clinician and is gradually replacing nuclear medicine studies. These scans are best performed with the patient in the prone position. If active CSF leakage is rapid, the CT may show the intrathecally injected contrast dye in the adjacent sinus or at the site of intracranial bony defect (Fig. 12-26). Perform these studies with both soft tissue and bone windows for optimal detection and localization of the leak.

If there is the possibility of the brain herniating through the dehiscent area, MR is valuable. It may also distinguish postoperative scar tissue, recurrent polyps, or inflammation from meninges, CSF, and the brain. Some people have advocated using a heavily T2-weighted MR study, "MR cisternography," to detect the site of CSF leaks. In many ways, direct endoscopic visualization of intrathecally instilled fluorescein actively leaking from the cribriform plate is the most reliable test in the hands of a skilled endoscopist. (Often the ear, nose, and throat surgeons instill an agent for nuclear medicine cisternography at the same time.) A CT scan before or after this test will often solve the CSF leak riddle: where's the leak?

Polyposis

One of the manifestations of allergic sinusitis is sinonasal polyposis. Polyps may also occur in the absence of allergies and are caused by nonneoplastic hyperplasia of inflamed mucous membranes.

From an imaging standpoint, this entity is somewhat problematic because the lesion, although benign, may demonstrate aggressive bony distortion. On CT, the findings suggestive of polyposis include enlargement of sinus ostia, rounded masses within the nasal cavity, expanded sinuses or portions of the nasal cavity, thinning of bony trabeculae, and, less commonly, erosive bone changes at the anterior aspect of the skull base (Fig. 12-27). Most cases of sinonasal polyposis show bilateral changes and nasal involvement (>80%).

Because aggressive skull base erosion might suggest a malignancy, it would be useful if specific findings on MR or CT distinguished polyposis from cancer. Unfortunately, this is not always possible. Although the signal intensity of most sinonasal polyps on T2WI is bright, this overlaps with some of the neoplasms seen in the sinonasal cavity, including sarcomas, adenoid cystic carcinomas (Fig. 12-28), and other, less common, minor salivary gland neoplasms. Polyps usually enhance peripherally because they represent hypertrophied mucosa, but, occasionally, when the mucosa folds repetitively on itself, they may enhance solidly like neoplasms.

Polyps occur in 1.3% of the general population and 16.2% of patients with chronic sinusitis. They are commonly associated with aspirin intolerance, nickel exposure (4%), cystic fibrosis (10% to 20%), asthma (30%), allergic rhinitis (25%), and Kartagener syndrome (Box 12-6). The latter refers to the autosomal recessive disorder otherwise known as immotile cilia syndrome in which patients have recurrent sinusitis, situs inversus, bronchiectasis, and infertility.

Steroids may retard the growth or decrease the size of sinonasal polyps. Depending on the severity of the patient's symptoms, intranasal or oral steroid treatment may be prescribed for sinonasal polyposis. Unfortunately, the disease often keeps coming back. The patient is then faced with a choice of pulsed intermittent steroid therapy or surgery. For cystic fibrosis children, this often means recurrent surgery.

Antrochoanal (Killian) polyps arise in the maxillary sinus but may protrude into the nasal cavity or the nasopharyngeal airway (Fig. 12-29). The polyp usually extends through the accessory ostium of the sinus and projects posteriorly through the posterior choana to the nasopharynx. Sphenochoanal polyps can usually be found between the nasal septum and the middle turbinate and track through the sphenoethmoidal recess. These polyps are smooth hypodense masses that often remodel bone and enlarge the maxillary sinus accessory ostium on CT, features indicative of slow sustained growth. On MR, they are very bright on T2WI and have a variable amount of usually peripheral enhancement. Choanal polyps are seen in children and young adults with a strong male preponderance.

Retention Cysts

Mucous retention cysts develop as a result of obstruction of small seromucinous glands. In most cases, it is impossible to distinguish a polyp from a retention cyst. The distinction is of little clinical relevance. These cysts are often not addressed unless obstructive in size. Retention cysts are smooth-domed homogeneous lesions usually found in the maxillary sinus and often in the dependent portion. They are bright on T2WI.

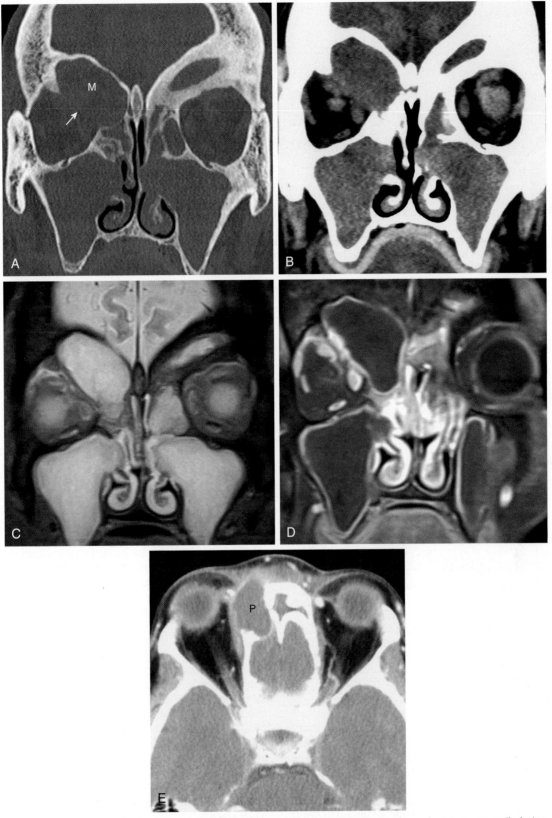

FIGURE 12-24 Mucocele. **A,** Coronal computed tomography (CT) in bone windows shows an expansile lesion within the right frontal sinus (M) resulting in dehiscence of the superomedial orbital wall *(arrow)*. There is also complete opacification and osteitis of the included portions of the left frontal, bilateral ethmoid, and bilateral maxillary sinuses. **B,** Coronal CT in soft-tissue windows shows the lesion to be near fluid density. Note the displacement of the superior and medial rectus muscles by this lesion. Although very often mucoceles contain proteinaceous contents, in this case **(C)** coronal T2-weighted imaging (T2WI) and **(D)** postcontrast T1WI show the lesion to be of fluid signal with only mucosal enhancement present (similar to the other paranasal sinuses). **E,** On this axial contrast-enhanced CT of a different patient, a similar lesion shows peripheral contrast enhancement and overlying soft-tissue swelling, consistent with pyomucocele (P).

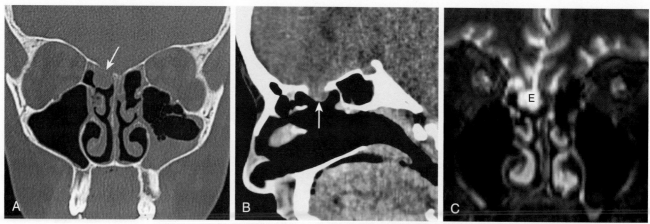

FIGURE 12-25 Postoperative encephalocele. **A,** Coronal computed tomography (CT) shows a focal bony defect at the right cribriform plate *(arrow)* following functional endoscopic sinus surgery. **B,** Sagittal CT in soft-tissue windows shows that the tissue herniating through the defect is isodense to the brain *(arrow)*. **C,** Coronal T2 demonstrates herniation of the brain and meninges (E) through the defect at the cribriform plate.

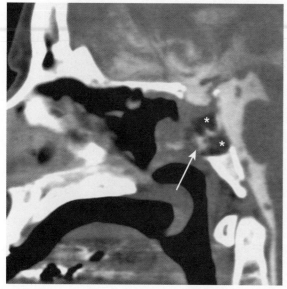

FIGURE 12-26 Cerebrospinal fluid leak. Sagittal computed tomographic (CT) image from CT cisternogram in a patient complaining of continuous clear nasal drainage following transsphenoidal resection for pituitary tumor shows clear seepage of contrast *(arrow)* through the fat-packing material *(asterisks)* in the surgical bed into the posterior nasal cavity (sniffle, sniffle).

BENIGN NEOPLASMS

Of the benign neoplasms to affect the sinonasal cavity, osteomas, papillomas, schwannomas, and juvenile angiofibromas are most commonly seen. The stereotypical benign neoplasm expands and remodels the bone as a result of its slow, nonaggressive growth. This is contrasted with malignancies that destroy bone and invade in an ill-defined, poorly marginated manner. Four types of skull base erosion have been described in relation to sinonasal masses. These are (1) resorption of the central skull base, (2) enlargement of the foramina of the central skull base, (3) thinning of the central skull base, and (4) displacement of the bone. The first two are more common with malignant lesions and the last two with benign tumors. Some malignancies,

however, may show bony remodeling, and some benign tumors aggressively destroy bone.

Osteomas

Osteomas are usually identified in the frontal sinus and may infrequently be a source for recurrent headache and/or recurrent sinusitis (Fig. 12-30). Occasionally, the osteoma (or osteochondroma) results in mucocele formation and/or pneumocephalus as the posterior wall of the frontal sinus is breached (Fig. 12-31). The classic history associated with a frontal sinus osteoma narrowing the sinus opening is a patient who has severe sinus pain associated with takeoffs from airplane flights. Osteomas are benign masses that often are completely invisible on MR because of the presence of dense compact sclerotic bone making up the mass that is indistinguishable from adjacent air within the sinus. By contrast, they are easily identified on CT as markedly hyperdense bony masses protruding in the sinus. Remember that a patient with colonic polyps and osteomas may have Gardner syndrome.

Papillomas

Papillomas come in many different varieties, the most common of which (around 75%) is the inverted papilloma. This benign neoplasm is remarkable for the coincidence of squamous cell carcinoma in approximately 15% of cases. Inverted papillomas may show a rather aggressive bone destruction pattern and have been known to cross the cribriform plate into the anterior cranial fossa. The lesion typically arises from the lateral nasal wall, nasal septum, or medial maxillary sinus, and it accounts for approximately 4% of all tumors of the nasal cavity.

Staging classification has been proposed to include: Stage I, limited to the nasal cavity alone; Stage II, limited to the ethmoid sinuses and medial and superior portions of the maxillary sinuses; Stage III, extension to the lateral or inferior aspects of the maxillary sinuses or extension into the frontal or sphenoid sinuses; and Stage IV, spread outside the nose and sinuses.

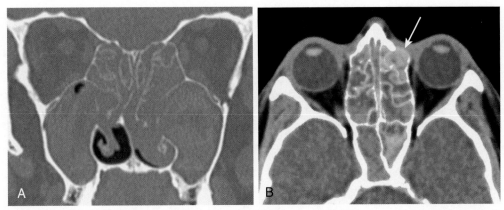

FIGURE 12-27 Sinonasal polyposis. **A,** The coronal computed tomographic image filmed in bone windows belies the extensive high-density secretions seen best in soft-tissue settings **(B).** Note the expansion of the sinus walls with a focal area of dehiscence *(arrow)* where the polyp has skeletonized the bone. This type of hyperdensity may characterize fungal infection or inspissated secretions, but the expanded bone suggests polyps (with or without fungus).

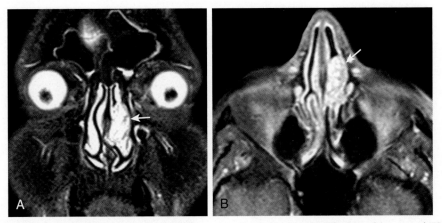

FIGURE 12-28 A, A hyperintense mass in the left nasal cavity *(arrow)* is identified on this coronal T2-weighted scan. You're probably thinking polyp, right? **B,** The enhancement is solid on postcontrast T1-weighted image. Therefore, it is not a polyp. Pleomorphic adenoma? In the nasal cavity? Yup!

BOX 12-6 Associations with Sinonasal Polyps

Aspirin intolerance
Cystic fibrosis
Immotile cilia syndrome
Kartagener syndrome
Peutz-Jeghers syndrome

CT may demonstrate a nonspecific-enhancing mass along the lateral nasal wall. The specific diagnosis is not evident except when, as in approximately 20% of cases, it contains stippled calcium. On MR, the lesion is typically isodense to muscle on T1WI and is isointense to hypointense on T2WI. Most of the other benign polypoid masses are bright and homogeneous in intensity on T2WI. The lesion enhances in a solid manner and in approximately 50% of cases is heterogeneous in both its signal intensity and enhancement. A convoluted "cerebriform" appearance to the lesion on T2WI and postenhancement is typical of inverted papilloma (Fig. 12-32). Necrosis may be an indicator of coexistent squamous cell carcinoma.

The inverted papilloma can erode the skull base in a manner similar to aggressive cancers; because its signal intensity characteristics overlap those of malignancies, there is no way to preoperatively predict the diagnosis. The lesion is particularly problematic to surgeons who are required to treat it as if it were a malignant neoplasm by an aggressive surgical approach. Unfortunately, the recurrence rates are approximately 22% to 40% despite aggressive operations. Recurrences may be distinguished from postoperative thickening on dynamic-enhanced MR; recurrent inverted papillomas have earlier and greater enhancement than granulation tissue.

Enchondromas

Enchondromas are rare neoplasms of the sinonasal cavity. On CT, they often have "popcorn" calcification, different from the "stippled" calcification of inverted papillomas. The nasal septum is one of the favored sites of chondroid lesions.

Meningiomas

One percent of meningiomas occur outside the CNS, presumably from embryologic arachnoid rests. The most common sites for these "extradural" meningiomas are the sinonasal cavity. The upper nasal cavity, the ethmoid sinuses,

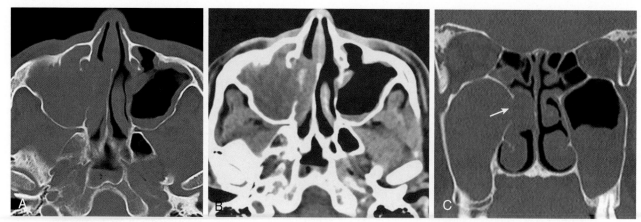

FIGURE 12-29 Antrochoanal polyp. Axial computed tomography (CT) in bone **(A)** and soft-tissue **(B)** windows shows a fluid density expansile lesion opacifying the right maxillary sinus and extending through the ostium into the right nasal cavity. **C,** Coronal CT from the same patient shows widening of the right ostium *(arrow)* by this antrochoanal polyp.

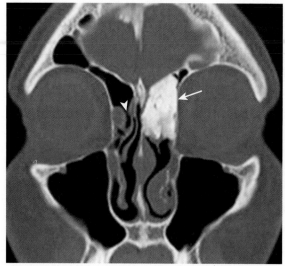

FIGURE 12-30 Osteoma. Coronal unenhanced computed tomography shows sclerotic lesion with narrow zone of transition sitting at and occluding the left frontal recess *(arrow)*. Note the patent frontal recess on the right *(arrowhead)* for comparison. This is a characteristic look and location for sinus osteoma and, in this case, is large enough to cause recurrent frontal sinusitis as a result of its obstructive nature.

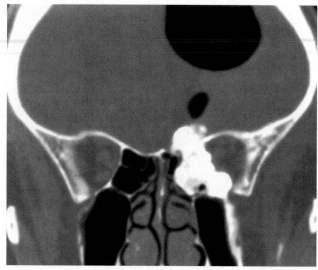

FIGURE 12-31 Osteochondroma and pneumocephalus. Classically, osteomas are said to be the tumors that can cause pneumocephalus; however, in this case, the osteochondroma originated in the ethmoid sinus, perforated the anterior cranial fossa, and led to a communication with the brain—hence pneumocephalus.

and the frontal sinuses are the most common sinonasal sites. As with all meningiomas they are usually slightly hyperdense masses on CT, intermediate in intensity on MR pulse sequences, and will avidly enhance. A dural attachment may be present or absent, best seen on postcontrast T1WI.

Fibrous Dysplasia

Fibrous dysplasia is one of the most common bone lesions to affect the paranasal sinuses. Monostotic and polyostotic forms are both common and can affect the walls of the sinuses and skull base or the ethmoid lattice and nasal cavity. Fibrous dysplasia has a classic ground-glass appearance on CT (Fig. 12-33). On MRI, the picture can be a little more perplexing, with variable signal intensity and enhancement (although most commonly

dark on all pulse sequences). Rarely large vascular flow channels can be seen. Bony expansion by these lesions can lead to obstruction of the sinus drainage pathways and cause sinusitis. Orbital and intracranial involvement by fibrous dysplasia can cause a whole host of other problems, too.

MALIGNANT NEOPLASMS

When evaluating a sinonasal malignancy, identifying the histopathologic nature of the tumor is not always possible. Rather than trying to name that tumor, it is more useful to our surgical colleagues to define the extent of the malignancy, including identifying primary tumor focus, sites of invasion (such as the skull base, orbit, brain, and vessels), mass effect on critical structures, evidence of perineural tumor spread, and lymphadenopathy.

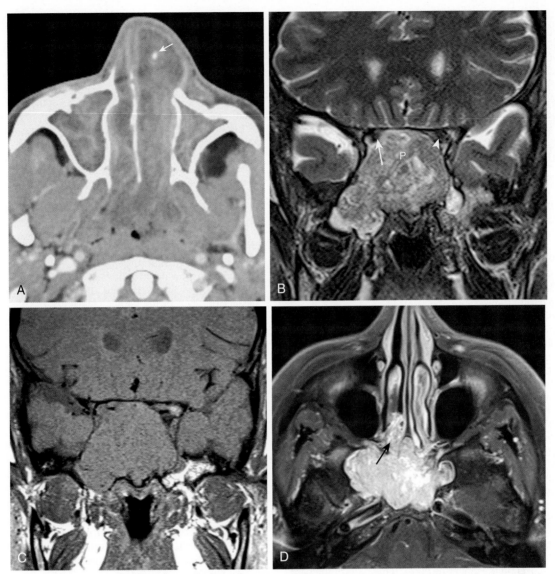

FIGURE 12-32 Inverted papilloma. **A,** On this contrast-enhanced axial computed tomographic image, an expansile mildly enhancing mass occupying the nasal cavity on the left bulges anteriorly through the nasal vault and posteriorly to the nasopharynx. There is tiny focus of calcification *(arrow)*, which could lead you to suspect this to be an inverted papilloma. Inspissated hyperintense secretions are present in the bilateral maxillary sinuses. **B,** Coronal T2-weighted imaging (T2WI) in another patient shows a mass (P) occupying and expanding the sphenoid sinus (admittedly a somewhat unusual location for an inverted papilloma). There is a "cerebriform" pattern to the signal intensity of the lesion, which has been described with inverted papillomas. The right optic nerve is slightly compressed because of the expansile nature of this lesion *(arrow)*, compared with the normal optic nerve appearance on the left *(arrowhead)*. **C,** On coronal T1WI the lesion is isointense to muscle, but with contrast administration the lesion enhances avidly. **D,** Note insinuation of the papilloma through the sphenoethmoidal recess into the posterior nasal cavity *(arrow)*. Based on imaging alone, we cannot readily distinguish this lesion from a sinus cancer...the diagnosis of inverted papilloma is rendered by the pathologist in this case.

CT and MR play complementary roles in the evaluation of sinonasal malignancies because of CT's superiority in defining bony margins and MR's superior soft-tissue resolution, multiplanar capability, and ability to define intracranial, meningeal, or intraorbital spread. One of the advantages of MR is its ability to distinguish sinus neoplasm from postobstructive secretions. Skull base invasion is another area where MR has gained ascendancy. MR provides more information than CT with regard to demonstrating dural invasion, cavernous sinus infiltration, tumor and muscle differentiation in the infratemporal fossa and masticator space, optic nerve identification amid adjacent tumor, and

tumor separation from the internal carotid artery and cavernous sinus.

When you encounter a sinonasal mass that is eroding intracranially, you should consider carcinoma, olfactory neuroblastoma, sarcoma, lymphoma, metastasis, sinonasal polyp, and inverted papilloma in your differential diagnosis. Necrosis, hemorrhage, or calcification in carcinomas, olfactory neuroblastomas, or sarcomas may cause signal heterogeneity. Malignancies tend to show a broad, flat base of skull base erosion; benign conditions have a rounded, polypoid intracranial excrescence. The T system of staging sinus cancers is summarized in Boxes 12-7 and 12-8.

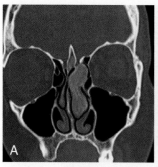
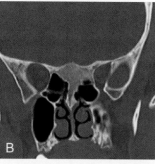

FIGURE 12-33 Polyostotic fibrous dysplasia. **A,** The left middle turbinate and ethmoid strut is ground-glass in appearance and expansile. **B,** There is also involvement of the floor of the anterior cranial fossa and nasal septum.

BOX 12-7   T System of Staging Maxillary Sinus Cancer

TX: Primary (main) tumor cannot be assessed.
T0: No evidence of primary tumor.
Tis: Cancer cells are limited to the innermost layer of the mucosa (epithelium). These cancers are known as carcinoma in situ.
T1: Tumor is only in the tissue lining the sinus (the mucosa) and does not invade bone.
T2: Tumor has begun to grow into some of the bones of the sinus, other than into the bone of the back part of the sinus.
T3: Tumor has begun to grow into the bone at the back of the sinus (called the posterior wall) or the tumor has grown into the ethmoid sinus, the tissues under the skin, or the side or bottom of the eye socket.
T4a: Tumor is growing into other structures such as the skin of the cheek, the front part of the eye socket, the bone at the top of the nose (cribriform plate), the sphenoid sinus, the frontal sinus, or certain parts of the face (the pterygoid plates or the infratemporal fossa). This is also known as moderately advanced local disease.
T4b: Tumor has grown into the throat behind the nasal cavity (called the nasopharynx), the back of the eye socket, the brain, the tissue covering the brain (the dura), some parts of the base of the skull (middle cranial fossa or clivus), or certain nerves. This is also known as very advanced local disease.

From Edge SB, Byrd DR, Compton CC, et al, eds. *AJCC Cancer Staging Manual.* 7th ed. New York: Springer; 2010.

Squamous Cell Carcinoma

Squamous cell carcinomas account for 80% of the malignancies affecting the paranasal sinuses and 80% occur in the maxillary antrum. Seventy-five percent of patients affected are older than 50 years and there is a male preponderance. Occupational exposure to nickel and chrome pigment and the use of Bantu snuff and cigarettes have been implicated as risk factors.

Squamous cell carcinoma (relatively hypointense on T2WI) should be distinguished from most inflammation (hyperintense on T2WI). Inflammation and neoplasm can be distinguished in 95% of cases on the basis of T2WI and enhancement (Fig. 12-34). Even when the sinus secretions become increasingly inspissated and the signal intensity on T2WI decreases, the neoplasm can be distinguished from the obstructed secretions by its typical heterogeneity as opposed to the typically smooth homogeneous appearance

---

BOX 12-8   T System of Staging Ethmoid Sinus Cancer

TX: Primary (main) tumor cannot be assessed.
T0: No evidence of primary tumor.
Tis: Cancer cells are only in the innermost layer of the mucosa (epithelium). These cancers are known as carcinoma in situ.
T1: Tumor is only in the nasal cavity or one of the ethmoid sinuses, although it may have grown into the bones of the sinus.
T2: Tumor has grown into other nasal or paranasal cavities, and may or may not have grown into nearby bones.
T3: Tumor has grown into the side or bottom of the eye socket, the roof of the mouth (palate), the cribriform plate (the bone that separates the nose from the brain), and/or the maxillary sinus.
T4a: Tumor has grown into other structures such as the front part of the eye socket, the skin of the nose or cheek, the sphenoid sinus, the frontal sinus, or certain bones in the face (pterygoid plates). This is also known as moderately advanced local disease. Cancers that are T4a are usually resectable (meaning they can be removed with surgery).
T4b: Tumor is growing into the back of the eye socket, the brain, the dura (the tissue covering the brain), some parts of the skull (the clivus or the middle cranial fossa), certain nerves, or the nasopharynx (throat behind the nasal cavity). This is also known as very advanced local disease. These tumors are not resectable (they cannot be removed with surgery).

From Edge SB, Byrd DR, Compton CC, et al, eds. *AJCC Cancer Staging Manual.* 7th ed. New York: Springer; 2010.

---

of sinus secretions. This is also true in the case of mucoceles, which may occur after or in association with sinus neoplasms. However, do not count on this trend all the time—low intensity on T2WI is an inconstant finding in sinonasal malignancies. The signal intensities of nonsquamous cell tumors (especially minor salivary gland tumors, sarcomas, and lymphoma) may show overlap with inflammation.

Squamous cell carcinomas of the sinonasal cavity enhance in a solid manner as opposed to a peripheral rim of enhancement in sinus secretions and/or mucoceles. Unfortunately, lymphomas, undifferentiated carcinomas, inverted papillomas, and some sarcomas may have identical signal intensity and enhancement characteristics as squamous cell carcinoma (although the cerebriform convoluted pattern predominates in inverted papillomas). Therefore, among sinonasal tumors, specific histologic diagnoses are elusive. The hallmark of imaging malignancies of the sinonasal cavity is bony destruction, seen in approximately 80% of scans of sinonasal squamous cell carcinomas at initial presentation. Maxillary sinus carcinomas are confined to the maxillary antrum in only 25% of cases at presentation.

With MRI, contrast-enhanced imaging is particularly useful for demonstrating epidural or meningeal invasion of neoplasms. Often, enhanced coronal scans with fat-suppression techniques are necessary to identify enhancement amid the abundant skull base fatty marrow. It should be noted that meningeal enhancement need not necessarily imply neoplastic invasion; just as in cases of meningioma, the dura may enhance because of reactive fibrovascular changes alone. Discontinuous dural enhancement without intervening hypointense epidural margination favors neoplastic invasion. Nodular enhancement and enhancing tissue greater than 5 mm thick imply neoplasm over reactive changes.

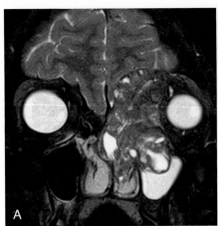
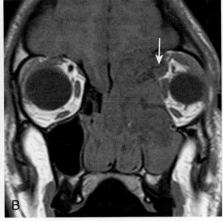
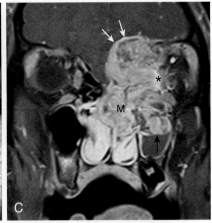

FIGURE 12-34 Poorly differentiated squamous cell carcinoma. **A,** Coronal T2-weighted imaging (T2WI) shows an aggressive, expansile mass centered within the left nasal cavity and maxillary ostium with extension laterally into the orbit and superiorly into the cranial vault. **B,** On coronal T1WI the mass is isointense to muscle. Infiltration of the orbital fat on the left can be seen, with invasion of the superior oblique muscle *(arrow)*, effacement of medial orbital fat, and displacement of other orbital structures laterally. Note the violation of the cribriform plate on the left. **C,** The mass (M) enhances heterogeneously with contrast on this coronal T1WI. Note the thick and slightly nodular dural-based enhancement at the anterior skull base *(white arrows)*, indicating dural invasion. Orbital invasion is again demonstrated *(asterisk)*. The margin of enhancing tumor within the maxillary sinus *(black arrow)* can be distinguished from backed-up fluid within the sinus, which does not enhance.

Sinonasal Undifferentiated Carcinoma

Sinonasal undifferentiated carcinomas (aka SNUC) are very aggressive malignancies associated with a very poor prognosis. These most commonly occur in the ethmoid sinus and show early bone destruction. Involvement of the adjacent structures of the nose, skin, orbit, and calvarium is common even at presentation. Their imaging appearance is similar to that of a squamous cell carcinoma gone wild, often with necrosis. The tumor enhances heterogeneously. Histopathologically, a high mitotic rate, tumor necrosis, and prominent vascularity are seen. Dural metastases occur at a high rate with this neoplasm.

Minor Salivary Gland Cancers

Minor salivary gland tumors are the next most common malignancy to affect the sinonasal cavity after squamous cell carcinoma. The minor salivary gland tumors represent a wide variety of histologic types including adenoid cystic carcinoma, mucoepidermoid carcinoma, and adenocarcinoma. Of these tumors, adenoid cystic carcinoma is the most common variety. Its signal intensity may be high or low on T2WI, possibly related to the degree of tubular or cribriform histologic pattern as well as cystic spaces, necrosis, and tumor cell density. Tissue specificity is not readily achievable with MR or CT except perhaps in some melanomas.

Contrast is of particular use with adenoid cystic carcinomas, which have a propensity (60%) for perineural spread (Fig. 12-35). With sinonasal cavity malignancies, always attempt to trace the branches of cranial nerve V via the pterygopalatine fossa, foramen rotundum, foramen ovale, and orbital fissures to identify perineural neoplastic spread. Retrace your steps. Check the hard palate for spread down the greater and lesser palatine foramina.

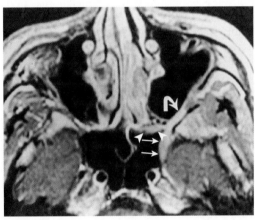

FIGURE 12-35 A tricky case. In a patient with adenoid cystic carcinoma, follow-up exam including axial-enhanced T1-weighted image demonstrates enhancing tumor traveling along the foramen rotundum on the left side *(arrows)* and extending through the pterygopalatine fossa *(arrowheads)* and pterygomaxillary fissure *(curved arrow)*. This is perineural spread of adenoid cystic carcinoma.

Melanoma

Melanoma is a tumor that is usually identified in the nasal cavity (two to three times more common than in the paranasal sinuses) and is sometimes associated with melanosis, in which there is diffuse deposition of melanin along the mucosal surface of the sinonasal cavity (Fig. 12-36). Therefore, multiplicity of lesions may suggest melanoma as a diagnosis. The nasal septum is the most common site of malignant melanoma, followed by the turbinates. When melanoma contains melanin, there is paramagnetism, which causes T1 and T2 shortening, accounting for high signal intensity on T1WI and low signal intensity on T2WI. However, an amelanotic melanoma may have low intensity on T1WI and bright signal intensity on T2WI. The presence

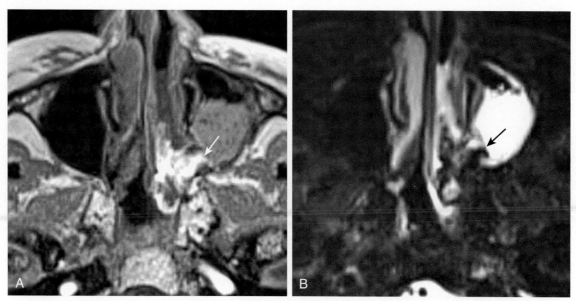

FIGURE 12-36 Sinonasal melanoma. Bright on T1-weighted imaging (T1WI) **(A)** and dark on T2WI **(B),** this nasal cavity melanoma *(arrow)* was very well localized. Sinonasal melanomas span the gamut from tiny discolored mucosal lesions identified incidentally for epistaxis to much larger and more aggressive invasive masses. Remember that they need not be bright on T1WI; that is merely a reflection of quantity of melanin.

of hemorrhage associated with the melanoma, a common occurrence because of the coincidence of epistaxis, may further obfuscate the signal intensity pattern. Melanoma is another tumor that has a propensity for perineural spread. It also readily metastasizes via hematogenous routes.

Adenocarcinoma

Adenocarcinomas of the paranasal sinuses have a predilection for the ethmoid sinuses and appear more commonly in woodworkers. This tumor also tends to have low signal intensity on T2WI but may have high signal intensity in a small percentage of cases.

Olfactory Neuroblastoma

A calcified malignancy high in the nasal cavity or ethmoid vault might represent olfactory neuroblastoma (esthesioneuroblastoma) (Fig. 12-37). This tumor arises from olfactory epithelium in the nasal vault from cells derived from the neural crest. Olfactory neuroblastomas have a bimodal peak seen both in males age 11 to 20 years and in middle-aged adults (sixth decade of life). Patients present with a history of nasal obstruction, epistaxis, or decrease in olfactory function.

As with squamous cell carcinoma, olfactory neuroblastomas typically have low signal intensity on T2WI. Tumoral cysts at the peripheral margins of the intracranial mass have been described and are virtually pathognomonic for this malignancy. Esthesioneuroblastomas have a particular propensity for crossing the cribriform plate to enter the intracranial space (35% to 40%). Stage A tumors are confined to the nasal cavity. Stage B tumors show nasal cavity and paranasal sinus involvement. Stage C tumors show skull base, orbital, and intracranial extension and/or distant metastasis.

When intracranial extension is identified, a craniofacial approach with a neurosurgical-otorhinolaryngologic

team is required. In this type of surgery the frontal lobe is retracted to gain optimal exposure to the cribriform plate so that the tumor can be removed en bloc. A fascia lata or galeal pericranial graft is placed between the brain and resected dura, and is sutured closed. This is followed by skin grafting under the dural surface. Craniofacial resections have decreased the recurrence rates of not only olfactory neuroblastoma but also other upper nasal vault–cribriform plate tumors such as adenocarcinomas, squamous cell carcinomas, and sarcomas.

Olfactory neuroblastomas can have lymphatic and hematogenous metastases. Recurrence rates are greater than 50%.

Sarcomas

Sarcomas of the sinonasal cavities are very rare, with chondrosarcoma the most common. Again, the histologic diagnosis is probably better suggested by CT based on characteristic whorls of calcification. It arises most commonly along the nasal septum (in the cartilaginous portion). Their aggressiveness is variable.

Rhabdomyosarcomas are not uncommon in the sinonasal cavity (particularly in children), although they can be seen in the orbit, pharynx, and temporal bone more commonly. They are usually of the embryonal cell type and often have a benign appearance to the manner in which they erode bone; these lesions may expand and remodel the bone rather than destroy it. Most are homogeneous in CT density, T1 and T2 signal intensity, and contrast enhancement. Intratumoral hemorrhage is not a rarity.

Lymphoma

Non-Hodgkin lymphoma occurs in the paranasal sinuses and may also have variable signal intensity. It is characterized by homogeneous signal intensity without necrosis

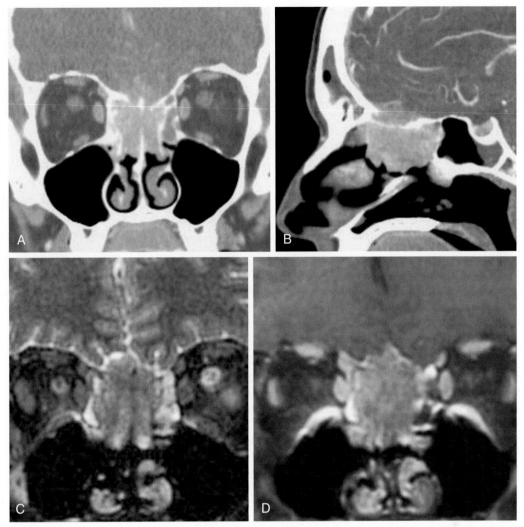

FIGURE 12-37 Olfactory (esthesio)neuroblastoma has a propensity for violating the cribriform plate. **A** and **B,** This is demonstrated in this case with absence of bone and enhancing tissue entering the anterior cranial fossa from the superior nasal cavity on contrast-enhanced coronal and sagittal images, respectively. **C,** On T2-weighted image (T2WI), the tumor has intermediate signal intensity (as with most small-cell tumors). **D,** The postcontrast T1WI shows nodular enhancing tissue extending through the cribriform plate.

and is associated with cervical lymphadenopathy. Nasal lymphoma often presents with nasal obstruction (80%), nasal discharge (64%), and epistaxis (60%). Septal perforations occur. Most (75%) are of T-cell lineage as opposed to nasopharyngeal carcinoma, which is more commonly of B-cell clonality (69%). Advanced age and bulky disease are associated with reduced survival. T-cell lymphomas occur mostly in the nasal cavity and ethmoid sinus. Of the B-cell lymphomas of the sinonasal cavity (25%) most arise in the maxillary sinus. Five-year survival rates for T-cell nasal lymphomas are less than 40%. Five-year survival rates of B-cell lymphomas are better than T-cell lymphomas.

Nasal natural killer cell lymphomas in posttransplant patients have recently been reported. This appears in the overall spectrum of posttransplant lymphoproliferative diseases but is one of the more aggressive varieties. Nasal T-cell/natural killer cell lymphoma presents with obliteration of the nasal passages and maxillary sinuses, erosion of the maxillary alveolus or hard palate, and/or invasion of the orbits and nasopharynx in more than 50% of cases. Necrosis and erosion of the nasal septum is typical of this entity.

Neuroendocrine Tumors

Neuroendocrine tumors of the sinonasal cavity can be divided into those referred to as typical (well differentiated), atypical (moderately differentiated), and small cell neuroendocrine (poorly differentiated) carcinomas. Small cell varieties are most common. Paranasal sinus neuroendocrine carcinomas expand and destroy sinus walls. They show intermediate T2WI signal intensity and enhance.

Metastases

Metastatic disease to the paranasal sinuses is extremely rare. Of the primary tumors that metastasize to the sinuses, renal cell carcinoma is the most common. This is a very vascular tumor that also has a propensity for hemorrhage. Depending on the stage of hemorrhage, the renal cell carcinoma metastasis may have variable density and signal intensity.

A quick aside about renal cell carcinoma: it is the most common tumor to metastasize to the nasal cavity, larynx, and skin. After lung and breast carcinoma, renal cell carcinoma is the most common tumor to metastasize to the head and neck in general. Approximately 15% of patients with hypernephroma have metastases to the head and neck, and the most common site is the thyroid gland.

Myeloma may affect the walls of the paranasal sinuses as can plasmacytomas. These are lytic lesions that can have soft-tissue masses associated with them (Fig. 12-38).

Keep in mind that malignant disease within the paranasal sinuses might not have originated in the sinus but in fact may have spread directly from other spaces in the head and neck. There is a high rate of sinonasal involvement by nasopharyngeal carcinoma. In a Hong Kong series of 150 cases of nasopharyngeal carcinoma, extrapharyngeal spread occurred at rates of 63% for the skull base, 56% for the parapharynx, 53% for the nasal cavity, 17% for oropharynx, 27% for the sphenoid sinus, 14% for the ethmoid sinus, 5% for the orbit, and 5% for the maxillary antrum. Spread to the maxillary sinus and the orbit portended the worst prognosis.

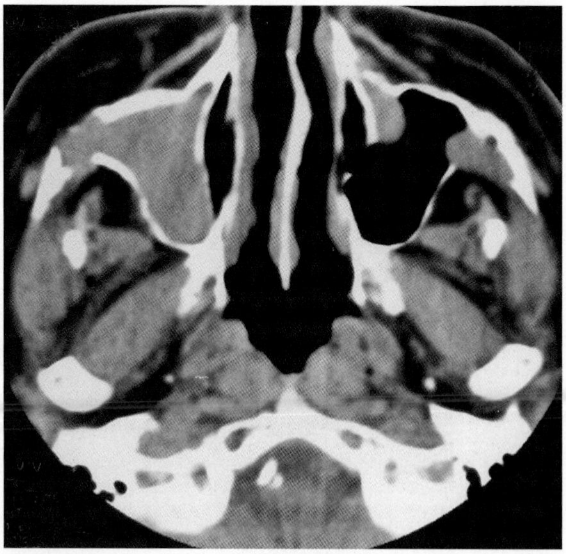

FIGURE 12-38 Multiple myeloma. See the numerous punched-out lytic lesions of the sinus walls in this patient with myeloma.

LYMPH NODE DRAINAGE OF NEOPLASMS

The lymph node drainage of sinonasal neoplasms is not terribly well understood. Generally, however, the drainage of sphenoid and posterior ethmoid sinus malignancies is to retropharyngeal lymph nodes and from there to the high jugular chain. Maxillary sinus cancers drain to the submandibular lymph node chains. At presentation, only 9% to 14% of sinonasal cancers have spread to the lymph nodes; this figure has justified routine radiation therapy after surgery for lesions in this area. Limited supraomohyoid neck dissections (see Chapter 13) may also be performed.

OPERABILITY OF SINONASAL TUMORS

What makes a sinonasal tumor inoperable, and what should you be particularly cognizant of when reviewing these cases? See Boxes 12-7 and 12-8 for stage 4b disease! There are five general criteria for nonresectability: (1) distant metastases, (2) optic chiasm invasion, (3) extensive cerebral involvement, (4) bilateral carotid infiltration, and (5) very poor general health. In some university settings cavernous sinus and/or chiasm invasion no longer constitute contraindications for surgery.

Chapter 13

Mucosal and Nodal Disease of the Head and Neck

The approach to this chapter on mucosal disease is divided along disease categories combining all regions except for the discussion of malignant masses. Because the basis for discussing cancers with clinicians is rooted in the TNM staging of the American Joint Commission on Cancer, which is itself subdivided by anatomic location, the chapter will analyze squamous cell carcinoma based on sites of origin.

An old adage avers that adding alliteration to an authoritative authorship augments educational attainment; all of us authors agree. Can you find 10 examples in this chapter?

ANATOMY

The anatomy of the mucosal layer of the head and neck will be described along general regions of interest.

Nasopharynx

The nasopharynx is broadly defined as that area of the mucosal surface that encompasses the walls of the aerodigestive tract above the soft and hard palate and extends to the skull base. Below the nasopharynx lies the oral cavity anteriorly and the oropharynx posteriorly. The nasopharynx is lined by stratified squamous and ciliated columnar epithelium and includes the mucosa overlying the eustachian tube orifice, the cartilaginous portion of the eustachian tube (torus tubarius), and the posterolateral pharyngeal recesses known as the fossa of Rosenmüller (Fig. 13-1). The mucosa of the nasopharynx is separated from the deeper retropharyngeal space by the pharyngobasilar fascia. The pharyngobasilar fascia forms a rather stiff barrier to the spread of mucosal diseases but it has bilateral openings, the sinus of Morgagni, to emit the eustachian tubes. The buccopharyngeal fascia is deep to the pharyngobasilar fascia and serves as another of the fascial barriers from nasopharynx to retropharyngeal and parapharyngeal spaces.

On either side of the eustachian tube orifice lie, anterolaterally, the tensor veli palatini (innervated by cranial nerve V_3) and posteromedially the levator veli palatini muscle (innervated by cranial nerve X), deep to the mucosa (see Fig. 13-1). These muscles elevate and tense the soft palate (into which they insert), preventing nasal regurgitation during swallowing. Between these muscles is a slip of fat (typically obliterated in early nasopharyngeal carcinomas) and posterolateral to these muscles lies the fat-filled parapharyngeal space, another common place to which cancer spreads. Fixate on fat—an effective friend for finding pharyngeal foulness.

The nasopharynx also houses the adenoidal lymphoid tissue. The amount of adenoidal tissue present depends on the age of the patient, because it atrophies by the fourth decade of life. In a young adult the normal adenoidal tissue and/or lymphoid hyperplasia may simulate a lymphoma or an exophytic squamous cell carcinoma. In patients who are positive for human immunodeficiency virus (HIV), on average, the width of the adenoidal tissue is twice that of age matched controls (Fig. 13-2). The normal variation in adenoidal thickness requires vigilance for deep invasion, infiltration of the parapharyngeal fat, concomitant middle ear and/or mastoid opacification, and/or skull base erosion or obscuration of the planes between the tensor and levator veli palatini muscles to definitively suggest tumor. Lymphoid tissue here and in the palatine and lingual tonsils is usually slightly hyperintense on T1-weighted imaging (T1WI) and hyperintense on T2-weighted imaging (T2WI) and enhances. These imaging characteristics would be unusual for a squamous cell carcinoma but could occur in a lymphoma. The adenoids, the palatine (also known as faucial) tonsils, and the lingual tonsils make up Waldeyer ring of lymphoid tissue (Fig. 13-3). All of these regions may show lymphoid hyperplasia in cases of infection or because of exposure to chronic irritants (cigarette smoke, alcohol, chewing tobacco, and mothers-in-law).

Minor salivary gland tissue is present throughout the aerodigestive system and is relatively abundant in the nasopharynx, oropharynx, and oral cavity. The hard and soft palate has the highest concentration of minor salivary glands (and consequently the highest rates of minor salivary gland neoplasms). Squamous epithelium lines the vast majority of the aerodigestive system mucosa of the head and neck.

Oropharynx

The oropharynx includes the posterior third of the tongue (also known as the tongue base); the vallecula; the palatine tonsils and tonsillar fossa; the posterior and superior pharyngeal walls from the level of the soft palate down to the pharyngoepiglottic folds; the uvula; and the soft palate (Fig. 13-4). The circumvallate papillae of the tongue separate the oral tongue (a part of the oral cavity) anteriorly from the oropharynx posteriorly. The hard palate is part of the oral cavity, but the soft palate is part of the oropharynx. Besides the palatine tonsils, the oropharynx also contains the lingual tonsillar tissue seen at the base of the tongue.

Oral Cavity

The oral cavity includes the lips, the anterior two thirds of the tongue, the buccal mucosa, the gingiva, the hard palate, the retromolar trigone, and the floor of the mouth. For the radiologist the phrase "the floor of the mouth" should be equated with the mylohyoid musculature (which constitutes the inferior sling of the mouth) and the sublingual

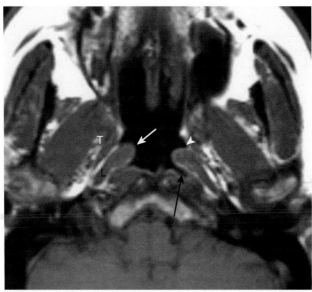

FIGURE 13-1 Normal nasopharyngeal anatomy. The torus tubarius *(white arrow)*, eustachian tube orifice *(white arrowhead)*, and fossa of Rosenmüller *(black arrow)* are labelled. The tensor veli palatini (T), and levator veli palatini (L) are seen.

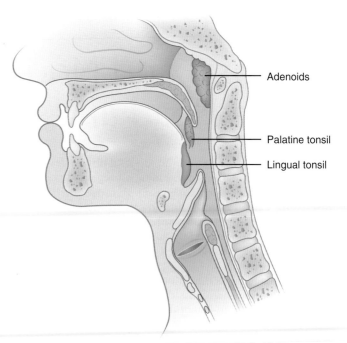

FIGURE 13-3 Waldeyer ring. Although the adenoids and lingual tonsils are basically midline structures, the palatine tonsils are found bilaterally framed by the pharyngeal faucial arches.

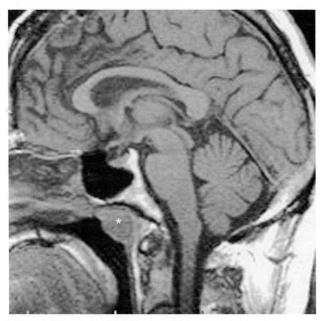

FIGURE 13-2 Adenoidal hypertrophy. The width of the nasopharyngeal adenoid tissue *(asterisk)* is enlarged in human immunodeficiency virus (HIV) patients and is one of the findings to check for on sagittal T1-weighted imaging scans of the brain. Diminished marrow signal intensity, posterior triangle nodes, and parotid cysts may be other signs of early HIV.

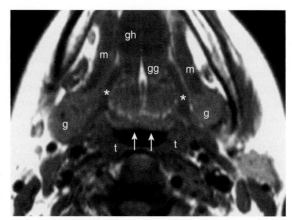

FIGURE 13-4 Normal anatomy of the oropharynx. On this axial T1-weighted imaging scan, one can identify the base of the tongue with lingual tonsil tissue *(arrows)* and the palatine tonsils (t). Also identifiable on this scan are the submandibular glands (g), the sublingual space extending from the submandibular glands anteriorly, and the midline fatty lingual septum with posterior aspect of the genioglossus muscles (gg) on either side. Muscles on either side of the sublingual space are the mylohyoid muscles (m) laterally and the hyoglossus *(asterisks)* medially. Geniohyoid muscle (gh) makes up the bulk of the tissue anteriorly in the tongue usually below genioglossus, partially included here.

space, between the mylohyoid muscle and the hyoglossus muscle (Fig. 13-5). The lingual nerve from the trigeminal nerve and the hypoglossal nerve run together from the floor of the mouth into the tongue base and sublingual space and are important for the surgeon to identify to maintain tongue function. Radiologists must identify whether tumor is in the sublingual space, to alert the surgeon regarding the potential for invasion or sacrifice of these nerves. The chorda tympani from the facial nerve supplies taste to the anterior two thirds of the tongue and its branches join that of the lingual

nerve. The glossopharyngeal nerve supplies taste to the posterior third of the tongue.

Hypopharynx

The hypopharynx is the junction between pharynx and larynx and includes three major subsites: the pyriform sinus, the postcricoid region (pharyngoesophageal junction), and the posterior pharyngeal wall above the inferior border of the cricoid cartilage (Fig. 13-6). The top of

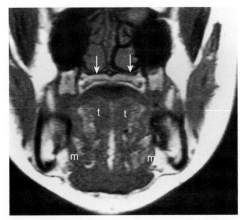

FIGURE 13-5 Normal anatomy of the oral cavity. The hard palate *(arrows)*, the anterior two thirds of the tongue (t), and the gingival surfaces of the mandible constitute portions of the oral cavity. The oral cavity also includes the floor of the mouth, seen as the mylo-hyoid (m) muscular sling inferolaterally.

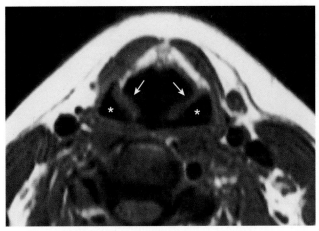

FIGURE 13-6 Anatomy of the hypopharynx. Axial T1-weighted imaging demonstrates the air-filled pyriform sinuses *(asterisks)*, which are delineated anteromedially by the aryepiglottic folds *(arrows)*. Other subsites of the pharynx in this location include the posterior and lateral pharyngeal walls.

the hypopharynx is at the epiglottic level, and its inferior border is the pharyngoesophageal junction. The mucosa over the posterior surface of the cricoarytenoid joints is part of the hypopharynx. The anteromedial border of the pyriform sinus is the aryepiglottic fold, a structure of the supraglottic larynx.

The pyriform sinus is best evaluated with the patient undergoing Valsalva maneuver because this distends the airway down to its inferiormost portion, the pyriform apex. This maneuver may be employed during barium studies or ultrafast computed tomography (CT) scans. The apposition of mucosal surfaces in the pyriform sinus often makes lesion localization difficult. You may not be able to distinguish extension to the adjacent aryepiglottic fold, the lateral pyriform sinus mucosa, or the posteromedial mucosa without such maneuvers to maximize distention.

Larynx

The larynx is broadly separated into the supraglottis, the glottis, and the subglottis (Fig. 13-7). Each of these areas

is dealt with individually by the head and neck oncologic surgeon, although lesions often cross these boundaries of the larynx (transglottic cancers). The supraglottis includes the subsites of the false vocal cords, the arytenoids, the infrahyoid and suprahyoid epiglottis, and the aryepiglottic folds. The glottis includes the true vocal cords, the anterior and posterior commissures, and the vocal ligament extending from the arytenoid cartilage. The laryngeal ventricle is said to separate the supraglottis and glottis, but is itself a part of the supraglottis. The subglottis begins 1 centimeter below the laryngeal ventricle and extends to the first tracheal ring. The dominant structure of the subglottis is the cricoid cartilaginous ring.

The larynx is anchored on a framework composed of the hyoid bone, the epiglottis, the thyroid cartilage, the cricoid cartilage, and the arytenoids, each of which has an integral function. Of these, the complete ring of the cricoid cartilage is the indispensable strut for preservation of airway patency. The major role of the epiglottis is to protect the airway during swallowing. From the inferior portion of the arytenoid cartilage the vocal ligament stretches to the thyroid cartilage anteriorly and supports the vocal cord. The lower cricoarytenoid joint is the marker for the level of the true vocal cord (Fig. 13-8). The true vocal cords meet in the midline at the anterior commissure. This junction should be no more than 1 to 2 mm of thickness. The posterior commissure refers to the mucosa between the two vocal processes on the anterior surface of the arytenoid cartilage.

On the lateral side of the laryngeal mucosal surface is the paraglottic space, which contains fat, lymphatics, and small muscles. At the false cord level the paraglottic space is dominated by fat, whereas at the true cord level it contains the thyroarytenoid muscle, which makes up the bulk of the true vocal cord, parallels the vocal ligament, and marks the glottic level. Thus, you can tell if you are at a supraglottic level by seeing fat deep to the mucosa—at the glottic level, it is muscle that is seen submucosally. The cricoarytenoid muscle moves the arytenoids to narrow or open the glottic airway for speech.

The vagus nerve innervates the larynx through two branches: the recurrent laryngeal nerve and the superior laryngeal nerve. The only muscle supplied by the latter is the cricothyroid muscle, and its paralysis causes only minor changes in the voice. The vagus descends from the medulla through the jugular foramen into the carotid sheath. The vagus follows the carotid sheath inferiorly with the recurrent laryngeal nerve looping under the aortic arch on the left and the subclavian artery on the right, before ascending in the tracheoesophageal groove. The branches of the recurrent laryngeal nerve perforate the cricothyroid membrane to supply the functioning intrinsic musculature of the larynx.

CONGENITAL LESIONS

Although the location of the following congenital lesions varies as they ascend or descend in the neck, they will be presented here more or less in a top-down order.

Tornwaldt Cyst

The most common congenital lesion of the nasopharynx (and BTW the head and neck) is the Tornwaldt cyst

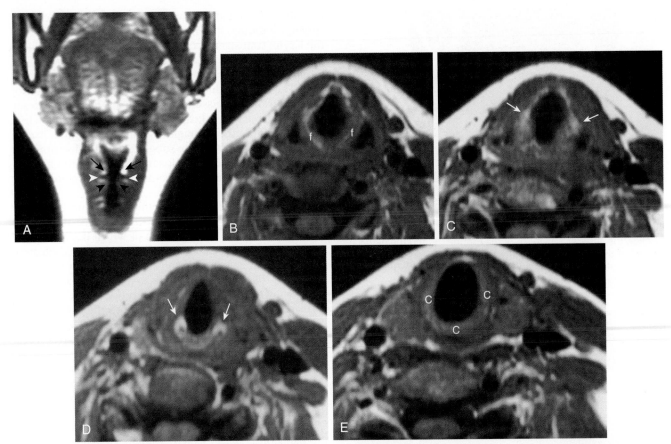

FIGURE 13-7 Normal laryngeal anatomy. **A,** Coronal T1-weighted imaging (T1WI) shows the laryngeal ventricle *(white arrowheads),* which separates the false cord *(black arrows)* above from the true cord below *(black arrowheads).* Note that at the false cord level the paraglottic tissue is high intensity from fat, whereas below the ventricle the soft tissue is muscular from the thyroarytenoid muscle, which makes up the bulk of the true cord. One centimeter below the ventricle is the margin of the subglottic region. **B,** Axial T1WI demonstrates the supraglottic structures including the aryepiglottic fold (f) extending posteroinferiorly toward the arytenoid region. **C,** At the false cord level one can again identify the fat *(arrows)* in the paraglottic space. **D,** At the true cord level the paraglottic tissue is made up of the thyroarytenoid musculature. The landmark for the true cord level is the cricoarytenoid joint *(arrows).* **E,** Subglottic region is marked by the full cricoid (C) ring.

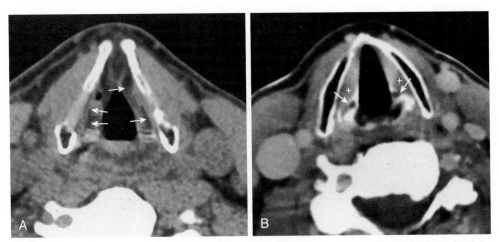

FIGURE 13-8 False cord nad true cord levels. **A,** The false cord is characterized by low-density fat *(arrows)* in the paraglottic space. **B,** The true cord is located at the lower cricoarytenoid joint *(arrows).* The thyroarytenoid muscle (+) makes up the "paraglottic" tissue and this muscle attaches to the vocal process of the arytenoid.

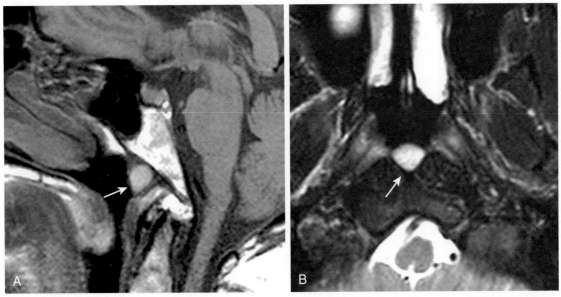

FIGURE 13-9 Tornwaldt cyst. **A,** Sagittal T1-weighted imaging (T1WI) demonstrates a hyperintense mass *(arrow)* high in the nasopharynx. **B,** On the T2WI the Tornwaldt cyst *(arrow)* is in the midline and bright on this sequence as well.

(Fig. 13-9). This results from apposition of the mucosal surfaces of the nasopharynx in the midline as the notochord ascends through the clivus to create the neural plate. A Tornwaldt cyst usually is hypodense on CT. The intensity on magnetic resonance imaging varies with the protein content but is usually bright on T1WI and T2WI. The cyst is usually well defined and characteristically occurs in the midline, although it may be seen off midline in a small percentage of cases. The cysts become infected on rare occasions and may be a source of persistent halitosis.

Tornwaldt cysts should be distinguished from mucus retention cysts, which are also often seen within the nasopharyngeal mucosa, and are T1 dark, T2 bright with peripheral rim enhancement.

Bony Congenital Lesions

Of the congenital cysts in the maxilla, the nasopalatine cyst (incisive canal cyst) is the most common. This cyst usually arises in the (para) midline incisive canal and slowly expands the maxilla and hard palate. Usually the cyst is painless and the teeth are unaffected. Therefore, it is not considered an odontogenic cyst.

Cherubism can occur in patients with familial fibrous dysplasia of the mandible, neurofibromatosis, Noonan syndrome (male Turner syndrome), and Ramon syndrome (cherubism, gingival fibromatosis, epilepsy, and mental retardation). The ground glass appearance and "bubbliness" akin to fibrous dysplasia are typical of this bilateral bony lesion (Fig. 13-10).

Branchial Apparatus Fistulas

Branchial cleft cysts are not considered to be "mucosal lesions"; however, when they drain to the mucosal surface they may present as such. Second branchial cleft fistulas may drain to the palatine tonsil from their typical location deep to the sternocleidomastoid muscle but superficial to the carotid sheath (Bailey type II second branchial cleft

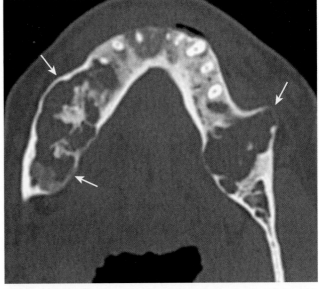

FIGURE 13-10 Cherubism. Computed tomography shows bilateral bubbly lesions of the mandible *(arrows)* in a patient with cherubism. The expansion without destruction is typical of this entity. The maxilla is also commonly affected.

cyst [BCC]). Those arising in the parapharyngeal space and potentially draining to the mucosa are classified as Bailey type IV second BCC.

The pyriform sinus is a drainage site for third branchial cleft cysts, and the pyriform sinus apex may be a site of sinus tracts leading from fourth branchial cleft cysts. The third branchial cleft sinus tract passes between the common carotid artery and vagus nerve to the anterior border of the inferior sternocleidomastoid muscle. The fourth branchial cleft sinus tract passes around the great vessels and the aortic arch on the left side. Pyriform sinus fistulas from the fourth branchial apparatus may pass to the thyroid gland causing acute suppurative

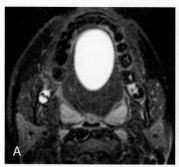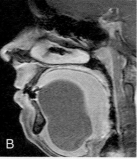

FIGURE 13-11 (Epi) Dermoid cyst of tongue. **A,** Axial T2-weighted imaging (T2WI) is dominated by the very bright midline mass in the oral tongue. It is too anterior to be a thyroglossal duct cyst and not lateral or inferior enough to be a ranula. **B,** The sagittal postcontrast T1WI confirms the cystic nature of the mass by virtue of the absence of enhancement.

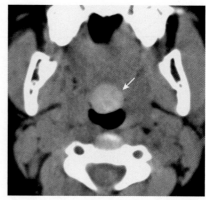

FIGURE 13-12 Lingual thyroid. This axial coronal computed tomographic image shows a mass *(arrow)* in the base of the tongue that is hyperdense on a noncontrast scan. Think thyroid that thoroughly thickens the throat thereabouts.

thyroiditis or may open to the skin surface. The vast majority of these fistulas are left sided. Recurrence rates are about 40%.

Developmental Cysts

Epidermoids and less commonly dermoids and lymphangiomas (usually multilocular) may occur in the aerodigestive system favoring the oral tongue and floor of the mouth. Epidermoid cysts usually occur in the sublingual space and have fluid density and signal intensity (Fig. 13-11). The dermoids contain fat or fluid or other dermal appendages with density and intensity that is characteristically heterogeneous. However, more often than not, these microscopic elements are not appreciable on imaging, and both epidermoid and dermoid need to be included in the differential diagnosis. If you perform diffusion-weighted imaging (DWI) on your head and neck study you may see that epidermoid lesions may have reduced apparent diffusion coefficient (ADC). The "sac of marbles" sign refers to geometric designs within the cyst.

Dermoid/epidermoid cysts should be distinguished from thyroglossal duct cysts (TGDCs), which may also occur in the tongue base or floor of mouth. The classic distinguishing feature is that TGDCs are embedded within the strap musculature. TGDCs are much more common entities than lingual thyroid glands but are usually infrahyoidal. This entity is covered in the thyroid gland section of Chapter 14.

Lingual Thyroid Gland

Ectopic thyroid tissue may be seen embedded in the posterior tongue because of arrested descent from the foramen cecum (the origin of the thyroglossal duct), which is located at the junction of the two sides of the circumvallate papillae along the anterior edge of the base of the tongue. This thyroid tissue is typically in the midline, is hyperdense on noncontrast computed tomography (NCCT) because of its natural iodine content, and enhances following contrast administration. After identifying the lingual thyroid tissue, the radiologist must next image the neck for any additional thyroid gland residua (Fig. 13-12). In 80% of cases the lingual thyroid tissue is the only functioning thyroid tissue in the body. These ectopic tissues are subject to the

same pathologies seen in the thyroid gland proper, including tumors. Incomplete descent of the thyroid gland elsewhere in the lower part of the neck is an uncommon congenital anomaly.

Venous Vascular Malformations

Venous vascular malformations (VVMs), with or without evidence of phleboliths, can occur anywhere in the neck, favoring the oral cavity and oropharynx for mucosal sites. They are most commonly seen in the skin and subcutaneous tissue, but populate the entire aerodigestive tract and nonmucosal spaces (Fig. 13-13). Previously, these were called hemangiomas, but this term is now relegated to those neoplastic lesions that tend to spontaneously regress in childhood or in response to steroids. Forty percent of VVMs are in the head and neck, followed by the extremities and the trunk. VVMs are often dense on NCCT and enhance avidly. These mucosal lesions grow with age and are evident because of their coloration on endoscopy.

Airway Lesions

Webs can occur in the larynx, usually at the true cord level. This is due to incomplete recanalization of the airway tube. As one would expect, the children present with airway obstruction and dysphonia. This entity is associated with subglottic stenosis in one third of cases.

When subglottic narrowing is seen in an infant, the differential diagnosis is usually between a capillary hemangioma and idiopathic subglottic stenosis. Stridor suggests subglottic stenosis: scoping shows stricture sans submucosal swelling. On the other hand, the subglottic capillary hemangioma is a benign neoplasm that generally occurs in 1- to 2-year-olds. Fifty percent of patients with subglottic hemangiomas have cutaneous hemangiomas as well. This is a benign tumor that is easy to diagnose endoscopically (Fig. 13-14) because of its characteristic location and red coloration. It often responds to steroid therapy, benign neglect (because it regresses with age) and/or laser therapy. In general, excisional treatment is not required except in those cases where extensive laryngeal narrowing is produced by the mass. On plain films the subglottic hemangioma causes an asymmetric narrowing of the airway of the subglottis. CT and MR reveal a smooth,

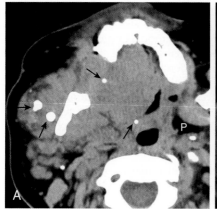
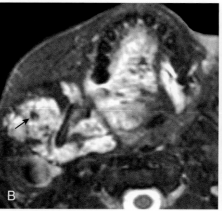
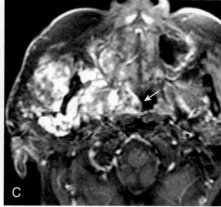

FIGURE 13-13 Venous vascular malformation: VVM (Va Va vooM!) **A,** Axial computed tomographic (CT) image without contrast gives us the sense that there is bulky soft-tissue mass investing the masticator space (masseter and pterygoid musculature) and oral cavity. Note effacement of the parapharyngeal fat on the right (normal space on the left indicated by *P*). There are chunky phleboliths *(arrows)* scattered throughout the lesion. **B,** Axial T2-weighted image (T2WI) makes the extent of the lesion more clear, given its intrinsic high T2 signal. Note the dark signal from the phlebolith *(arrow)*, much more difficult to discern compared to CT. **C,** The lesion enhances on postcontrast T1WI. There is mass effect from this lesion on the nasophayngeal airway *(arrow)*.

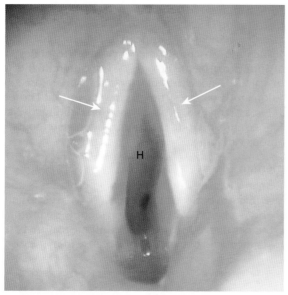

FIGURE 13-14 Subglottic hemangioma. This endoscopic view shows the subglottic hemangioma (H) below the true cords *(white arrows)*. The airway below the true cords is compromised. (Courtesy David Tunkel, MD.)

enhancing mass that is asymmetric in the subglottis and that compromises the airway. In adults hemangiomas are usually supraglottic lesions.

Tracheolaryngomalacia is another neonatal entity that can cause airway narrowing and that is usually outgrown with time. The children have inspiratory stridor because the floppy laryngeal cartilages collapse under the effects of negative inspiratory pressure.

Narrowings of the oropharynx can cause or be a result of obstructive sleep apnea (OSA) and Pickwickian syndrome. Pickwickian syndrome refers to morbid obesity, hypoventilation, and polycythemia. The narrowing of the airway may be due to obesity, redundancy of mucosal and muscular tissue, hypertrophy of lymphoid tissue, or on a congenital basis (hence discussed herein). In general, cross-sectional diameters of the oropharynx are reduced in patients with OSA at the soft palate, base of tongue, or uvula levels. The normal cross-sectional airway measurement should be about 100 mm^2. Patients with OSA often have values at or below a width of 50 mm^2 and may correspond with oxygen desaturation episodes. Treatment options for OSA are varied and include behavioral modification such as weight reduction, sleep hygiene, intraoral appliances that advance the mandible, positional training, and continuous positive airway pressure (CPAP) applied via a nasal mask. Uvulopalatopharyngoplasty or other surgical interventions are reserved for those who fail conservative therapy and who experience significant deoxygenation/hypoxia in their sleep.

LESIONS OF THE AERODIGESTIVE SYSTEM AND LYMPH NODES

Inflammatory Pharyngeal Disease

Typically, pharyngeal mucosal based thickening is seen in patients presenting with fever and sore throat. Because this is an easily made clinical diagnosis in the proper setting, no imaging is typically required. Nonetheless, occasionally one may find an associated peritonsillar abscess in a patient who does not respond to antibiotics for tonsillitis. The typical microorganisms that cause a tonsillar abscess include *Streptococcus pneumoniae, Streptococcus viridans,* and occasionally gram-negative anaerobes. Often one is dealing with multiple bugs at once.

If a fistula in the tonsil is identified, consider a branchial cleft anomaly as a possible cause. Typically, the second branchial cleft fistulas drain to the tonsil and may or may not be associated with a cystic lesion in the soft tissues of the neck near the angle of the mandible. Actinomycosis is an unusual cause of peritonsillar fistulas.

Imaging is also performed to look for complications including organized abscess formation or retropharyngeal extension of inflammation (Fig. 13-15). Pharyngitis and retropharyngitis can lead to Grisel syndrome—torticollis with rotatory subluxation of C1 on C2 secondary to an adjacent inflammatory mass. More often, one sees retropharyngeal lymphadenitis because of pharyngitis or tonsillitis.

Calcifications in the adenoids and tonsils (tonsilloliths) are not uncommon after infections. They simply symbolize sequelae of successful solution of such sickness. These are of no import and do not implicate granulomatous infections.

Abscesses have low-density centers and a peripheral rim of enhancement on CT. Adjacent inflammation of the fat may be due to neighboring cellulitis, myositis, and fasciitis in the neck. On MR, inflammatory lesions are very bright on T2WI because of the marked amount of edema and swelling associated with lesions such as abscesses. As in the brain, the DWI of a neck abscess may show restricted diffusion. A peritonsillar location is the most common site of abscess in children, followed by the retropharyngeal region. More often, what one is calling a retropharyngeal abscess is really necrotizing retropharyngeal lymphadenitis caused by pharyngitis or tonsillitis and need not be drained surgically. If the collection crosses the midline, call it an abscess. If paramedian along the expected location of retropharyngeal nodes, call it necrotizing adenitis. A phlegmon can be considered an immature abscess that can be arrested with early appropriate therapy.

Submucosal cysts that occur in the nasopharynx (Box 13-1) as sequela of previous infections are most often seen around the fossa of Rosenmüller. These cysts usually have low signal intensity on T1WI and high signal intensity on T2WI. Lymphoid hyperplasia can occur as a response to inflammation in adjacent regions. When this is seen, check for pharyngitis or tonsillitis in children, mononucleosis in hormonally active teenagers, HIV infection in adults, and cigarette butts in the ashtrays of your waiting room. Calcifications in the lymphoid tissue of Waldeyer ring, including the adenoids, are not uncommon. They simply symbolize sequelae of successful sanatory of such sickness.

Adenitis

Lymphadenitis refers to inflammation of lymph node(s). This can be reactive, as a response to adjacent infectious process such as pharyngitis. Primary inflammation/infection of the lymph node(s) can also occur. Enlargement, abnormal enhancement, intranodal cystic/necrotic change, calcifications, and perinodal stranding should alert you to the presence of lymphadenopathy, although the cause is not always clear without the clinical context and biopsy. Children seem to have a predilection for dramatic adenitis. Posterior triangle reactive adenopathy often accompanies the ubiquitous middle ear infections of young childhood.

Tuberculous Adenitis

The classic cause of an inflammatory cervical adenitis is tuberculous adenitis (scrofula), usually seen in Southeast Asians. The patients have painless posterior neck masses with or without systemic symptoms. The source of the infection is usually contaminated milk associated with *Mycobacterium bovis,* causing a subclinical pharyngitis. In the United States *Mycobacterium tuberculosis* is the most common cause. Atypical mycobacteria (*Mycobacterium scrofulaceum* especially) may also cause tuberculous adenitis. Concomitant pulmonary tuberculosis is relatively uncommon. The disease manifests as bilateral low-density necrotic lymph nodes, often in the level V posterior triangle distribution (Fig. 13-16). The nodes often have ringlike thick enhancement and appear multiloculated. Adjacent fat planes are obscured or edematous. The nodes often calcify after treatment (Box 13-2). The differential diagnosis of calcified nodes should include tuberculosis, other granulomatous diseases (fungi, sarcoidosis, and Thorotrast granulomas), amyloid, treated lymphoma, anthracosilicosis, and metastatic thyroid carcinoma, adenocarcinoma, and squamous carcinoma.

Castleman Disease

Castleman disease (angiofollicular hyperplasia) is a nodal disease that can be seen in the chest (70% of cases) and the head and neck (10%). Usually the patient is asymptomatic and younger than 30 years old. The unique feature of these nodes is their avid enhancement because of hypervascular

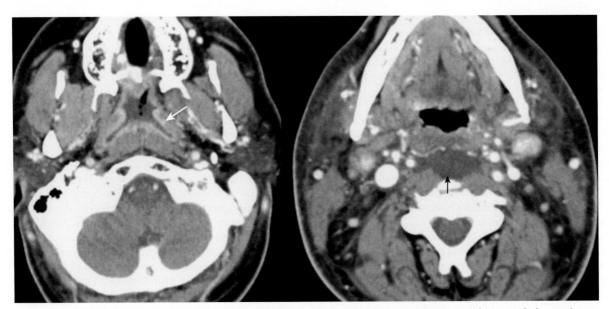

FIGURE 13-15 Pharyngitis with retropharyngeal edema. The patient has a rip-roaring nasopharyngeal pharyngitis evidenced by the enhancing mucosa *(white arrow)*. There was resultant retropharyngeal edema *(black arrow)* without an abscess. An abscess would have a better defined and enhancing peripheral border.

stroma (Fig. 13-17). The nodes are nonnecrotic. There may be a stellate area of nonenhancement in the center of the enhancing nodes.

Mononucleosis

Mononucleosis, caused by Epstein-Barr virus (EBV) infection, is another inflammatory source of multiple enlarged nonnecrotic lymph nodes. The differential diagnosis includes but is not limited to acquired immunodeficiency syndrome (AIDS) and sarcoidosis. In AIDS and mononucleosis the lymphoid tissue of Waldeyer ring (adenoids, palatine tonsils, and lingual tonsils) may be hypertrophied and bilateral nodes are the rule (Fig. 13-18). If intraparotid lymphoepithelial cysts are present, HIV infection is more likely. If the parotid glands are enlarged and infiltrated diffusely, sarcoidosis may be suggested but a chest radiograph with bihilar adenopathy may be the best clue. Otherwise the differential diagnosis may rely on serology or the appropriate (kissing) history.

Cat Scratch Fever

Another "zebra" that can cause bilateral lymph nodes, including intraparotid nodes, is cat scratch fever. There

---

BOX 13-1   Inflammatory Lesions of the Nasopharynx

Adenoidal hypertrophy (e.g., HIV)
Pharyngitis
Infected Tornwaldt cyst
Mucous retention cyst/submucosal cysts
Infectious mononucleosis
Retropharyngeal abscess/adenitis

HIV, Human immunodeficiency virus.

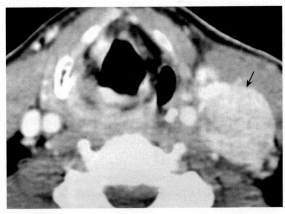

FIGURE 13-17 Castleman disease. The nodal mass *(arrow)* on this enhanced computed tomographic scan enhanced very avidly. Such an enhancing node is suggestive of angiofollicular hyperplasia (Castleman disease), thyroid carcinoma, Kaposi sarcoma, some lymphomas, and Kimura disease.

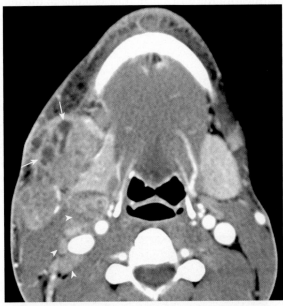

FIGURE 13-16 Tuberculous adenitis. The necrotic nodes *(arrows)* in the right neck with soft-tissue stranding and irregular margination begs the diagnosis of tuberculosis adenitis especially in a young person. Other right neck nodes show enlargement and heterogeneous enhancement *(arrowheads)*, and are also involved. Enhanced computed tomography shows rim enhancing nodes in the left neck infiltrated with "red snappers."

BOX 13-2   Causes of Calcified Lymph Nodes

Tuberculous adenitis
Granulomatous (nontuberculosis) infections
Rarely inactive (burnt-out) inflammatory nodes
Thyroid metastases
Mucinous adenocarcinoma metastases
Amyloid
Radiated lymph nodes
Thorotrast

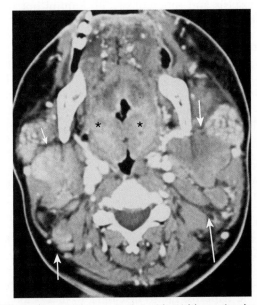

FIGURE 13-18 Large lymph nodes with Waldeyer ring lymphoid hyperplasia. Note the enlarged lymph nodes *(arrows)* in the neck and the enlargement of Waldeyer ring tissue in the palatine tonsils *(asterisks)*. Differential diagnosis includes acquired immunodeficiency syndrome, amyloidosis, mononucleosis, lymphoma, sarcoidosis, or pharyngeal carcinoma with malignant lymphadenopathy among other infectious/inflammatory processes. This is where the clinical context needs to come in, and biopsy if that is inconclusive.

often is edema surrounding the nodes. The etiologic agent in this infection has recently been characterized as a gram-negative intracellular bacterium *(Bartonella henselae)*. If you quiz the clinician about a cat scratch history and actually come up with the diagnosis, you will be a star in his or her eyes forever. It is a self-limited disease that may manifest regional lymphadenopathy, fever, and malaise, but can progress to encephalopathy and neuropathy after a cat scratch or flea bite. Parinaud oculoglandular syndrome, characterized by unilateral conjunctivitis with polypoid granuloma of the palpebral conjunctiva, and preauricular, parotid, and periparotid lymphadenopathy can be caused by Bartonella infections. The diagnosis is confirmed by positive serologic tests or positive polymerase chain reaction assays to the bacterium.

Sinus Histiocytosis

Sinus histiocytosis may cause massive nodes. The eponym is Rosai Dorfman disease and the disease presents with painless, bilateral, cervical lymphadenopathy accompanied by fever, leukocytosis, and elevated serum inflammatory markers. The nodes usually resolve as the disease is self-limited.

Kikuchi Fujimoto Disease

Kikuchi disease, histiocytic necrotizing lymphadenitis, predominantly affects young adults of Asian ethnicity. The etiology is unclear, but is probably viral, and the patients present with adenopathy, fever, and leukopenia. It is associated with large necrotic and nonnecrotic, enhancing and nonenhancing adenopathy. The CT appearance of Kikuchi disease can simulate lymphoma and other nodal diseases that have necrosis, such as metastasis and tuberculosis. Be Kareful not to Miss Kikuchi!

Kimura Disease

Kimura disease is another import from Asia. It is a chronic inflammatory process with associated diffuse hypervascular adenopathy in the cervical chains favoring the submental and submandibular regions, eosinophilia, and a predilection for Asian men aged 10 to 30. The salivary glands may be swollen and tender. The nodes are round, solid, hypoechoic, and homogeneous. They have hilar hypervascularity on power Doppler scans (hey, since this is an Asian import, the sonographic features are what is best known about the disease).

Posttransplant Lymphoproliferative Disorder

Posttransplant lymphoproliferative disorder (PTLD) is one of those cross-over diseases between a reactive node, a benign neoplasm, and a malignant neoplasm. The aggressiveness of the entity ranges widely from benign lymphoid hypertrophy to myeloma, to monoclonal lymphomas, to polyclonal lymphomas, all from the EBV leading to B-cell proliferation. Basically, you can find enlarged nodes in the abdomen, chest, and neck (in descending order of frequency) in patients after whole organ transplants (heart, liver, kidney, or bone marrow). The inciting event appears to be an infection by Epstein Barr virus, leading to a proliferation of B cells. If the T cells are deficient, as in transplant patients or HIV infections,

this B-cell mass production gets revved up. Although this may result in a polyclonal lymphoproliferation, if unabated, a monoclonal dominant spike may appear. In some of these patients, lymphoma develops; in others the adenopathy resolves with manipulation of medications to produce a reduction in the degree of immunosuppression. Chemotherapy may be required in some. PTLD can affect the adenoids and tonsils and is usually seen after solid organ transplantations (lung is most common). The onset of this process may range from months to years after the transplant.

Cysts

Mucous retention cysts (MRC) that occur in the aerodigestive system as a sequela of previous infections or obstruction of minor salivary glands can be seen anywhere in the head and neck mucosa. In the nasopharynx, they are often seen laterally around the fossa of Rosenmüller. These cysts usually have low signal intensity on T1WI and high signal intensity on T2WI. Midline MRCs and Tornwaldt cysts with similar signal intensities are indistinguishable. Occasionally, one will also see cysts deep to the mucosa that are along the pathway of the eustachian tube heading to the middle ear cavity. Rarely, these lesions may grow large enough to cause airway compromise. Vallecular cysts can also be seen (Fig. 13-19).

Laryngotracheal Papillomatosis

Laryngotracheal papillomatosis refers to a viral infection that is seen most commonly in children and is probably due to transmission of the human papilloma (DNA) virus types 6 and 11 at the time of passage through the birth canal. The true cords are the most common laryngeal site. This disease is currently treated with laser surgery and steroids, but the papillomas commonly recur. When the

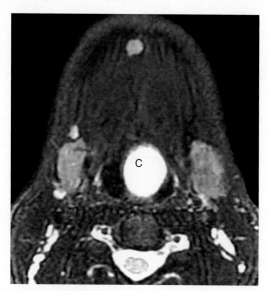

FIGURE 13-19 Vallecular cyst. While mucous retention cysts may abound in all sites of the aerodigestive system, a cyst in the vallecula (C) may cause a globus sensation and may lead to clinical presentation. Note that it is very bright on T2-weighted imaging.

branches of the trachea and bronchi are affected, postobstructive pneumonias and respiratory compromise may occur.

Macroglossia

Inflammation/edema of the tongue may occur with acute allergic reactions, and is one of the complications of iodinated contrast injections of which all radiologists should be aware. Treatment is dependent on severity and airway compromise but have the epi-pen, Benadryl, and steroids ready. Diffuse macroglossia may be present in patients with hypothyroidism, amyloidosis, Down syndrome, glycogen storage diseases, mucopolysaccharidoses, and neurofibromatosis among other conditions (see Box 13-3). Macroglossia makes one a masticatory misfit with a maddening maelstrom of mucous mash and malocclusive malfeasance.

Odontogenic Lesions

Numerous odontogenic abnormalities may be found within the oral cavity, the most common of which is the lytic lesion known as the radicular (periapical) cyst (Table 13-1). This is a lucent lesion of either the mandible or the maxilla, and is associated with an infected tooth. Look for associated bony dehiscence and extension of inflammation into the adjacent soft tissues. Associated subperiosteal or soft-tissue abscesses can be better appreciated with intravenous contrast. The second most common odontogenic cystic lesion is the dentigerous cyst. This cyst is associated with an unerupted tooth and is usually seen in the mandible, particularly around the molar region. Both the radicular cyst and the dentigerous cyst are usually unilocular, as opposed to the keratocystic odontogenic tumor (previously odontogenic keratocyst) and the ameloblastoma, which are benign but aggressive multilocular cystic lesions most commonly affecting the mandible (Fig. 13-20). Keratocystic odontogenic tumors (KCOT) are associated with the basal cell nevus (Gorlin) syndrome, in which patients have proliferative falcine calcification, multiple basal cell carcinomas of the skin, scoliosis, ribbon-shaped ribs, central nervous system tumors, and keratocysts of the mandible.

Sclerotic dental lesions also span the spectrum of inflammatory, benign neoplastic, and malignant lesions (Table 13-2). Dental disease is daunting because of dumbfounding and redundant distributions and depictions of different dense and destructive diagnoses (Fig. 13-21).

Inflammatory Laryngeal Disease

Almost all of us have experienced the inconvenience (and serendipitous benefits) of laryngitis. This condition is usually associated with a viral upper respiratory tract infection and is benign and self-limited. Chronic laryngitis may actually be due to laryngeal nodules, a nonneoplastic reaction to chronic voice abuse.

Supraglottitis

We use the term supraglottitis because most cases of "epiglottitis" will affect the aryepiglottic folds and even

---

BOX 13-3   Conditions Associated with Macroglossia
Acromegaly
Amyloidosis
Beckwith-Weidemann syndrome
Down syndrome
Glycogen storage diseases
Hemangioma/venous vascular malformation
Hypothyroidism
Idiopathic/familial macroglossia
Infants of diabetic mothers (transiently)
Lingual thyroid tissue
Mucopolysaccharidoses

TABLE 13-1   Benign Lytic Dental Lesions

Lesion	Imaging Appearance	Typical Clinical Findings
Ameloblastoma	Multiloculated (60% of cases) lytic lesion often associated with dentigerous cysts; hyperostotic margins; cortex eroded or penetrated	85% in mandibular molar area with expanded jaw; painless, male predominance
Brown tumor	Lytic lesion with erosion of lamina dura; ill-defined borders	Hyperparathyroidism
Central odontogenic fibroma	Multilocular lesion with sclerotic borders	Expanded mandible
Cherubism	Bilateral, symmetric multilocular lucencies (soap bubble) in posterior mandible; expanded cortex without perforation; simulates fibrous dysplasia	Painless; bilateral enlargement of lower part of face; angelic appearance; autosomal dominant inheritance; regression after adolescence
Dentigerous (follicular) cyst	Lytic, lucent, expansile lesion adjacent to unerupted tooth; spares cortex; sclerotic margins	Unerupted asymptomatic third molar or canine tooth
Giant cell granuloma	Well-defined multilocular lucency with sclerotic margins involving mandible	Asymptomatic in children and young adults
Globulomaxillary cyst	Lucent lesion between lateral incisor and canine in maxilla	Asymptomatic
Incisive canal cyst (nasopalatine duct cyst)	Lucent lesion in midline hard palate with hyperostotic borders at canal	Usually asymptomatic
Keratocyst (primordial cyst)	Unilocular or multilocular expansile lucent lesion; erodes cortex but does not perforate it	Recurrent posterior mandibular lesion with thin walls; associated with basal cell nevus syndrome
Radicular (periapical) cyst	Lytic; lucent at apex of erupted tooth; loss of lamina dura; hyperostotic borders	Carious, tender nonvital tooth

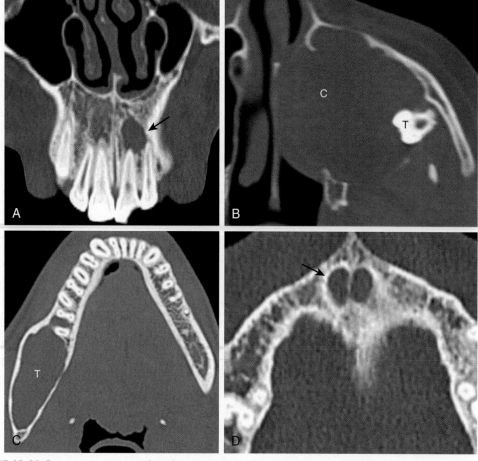

FIGURE 13-20 Four most common dental cysts. **A,** Radicular cyst *(arrow)* is at the root of a carious tooth and is oriented along the vertical plane of the maxilla or mandible. **B,** Dentigerous cyst (C) is associated with the crown of an unerupted tooth (T). **C,** An odontogenic keratocyst (now called keratocystic odontogenic tumor) may be oriented horizontally along the long plane of the mandible (T). Note that the lesions splays the dentition, rather than arising from the teeth, and causes expansion of the medullary compartment with cortical remodeling. **D,** Nasopalatine cyst *(arrow)* is a congenital cyst of the incisive canal and will therefore be midline in the maxilla.

TABLE 13-2   Sclerotic Dental Lesions

Lesion	Imaging Appearance	Typical Clinical Findings
Adenomatoid odontogenic tumor	Calcified well-defined lesions with thick capsule; associated with impacted tooth; involves crown of tooth	Teenagers with impacted maxillary front teeth; painless; female predominance
Cementoblastoma	Circular radiodensity attached to a mandibular tooth with pencil-thin border; surrounded by lucency; radial spicules	Expanded mandible with vital tooth; occurs first to third decades of life
Chondrosarcoma	"Moth-eaten" appearance with chondroid whorls; may be lucent or dense	Maxillary swelling; painful in adults
Ewing sarcoma	Onion-skinning; destructive lesion; poorly defined	5-25 years of age with painful mandibular mass, fever, rapid growth, loose teeth
Fibrous dysplasia	Ground-glass appearance, homogeneous in later stages	Focal painless mass; slow growth; posterior maxilla
Garré sclerosing osteomyelitis	Predominantly sclerotic bony lesion; hot on scintigrams; often with periosteal reaction and apical lucency	Bony-hard cortical swelling of mandible; carious molars; nonvital teeth
Lymphoma	"Moth-eaten," sclerotic bone	Often systemic symptoms
Metastases	Dense or lytic permeative lesions	Lung, breast, prostate, colon, kidney, thyroid primary tumors; loose teeth; often painful
Odontoma	Compound: Miniature teeth (enamel) within maxilla with peripheral lucent zone Complex: Irregular opaque mass in mandibular molar region	Young patient with mass between canines or in mandible; painless; young adults or children
Osteoma	Dense benign excrescence; well-defined	Associated with Gardner syndrome with colonic polyps, supernumerary teeth, cysts; seen as a torus on palate; painless, slow-growing
Osteosarcoma	Sclerotic or lytic; poorly defined with opaque spicules with sunray appearance; resorbs roots	Maxillary or mandibular mass; rapid growth; painful; loose teeth
Paget disease	Dense thickened bone with risk of osteosarcoma; cotton-wool appearance in maxilla; loss of lamina dura	Elderly patient with dentures no longer fitting commonly in maxilla
Pindborg tumor	Multiple small calcifications within lytic lesion associated with impacted teeth	Mass in posterior mandible

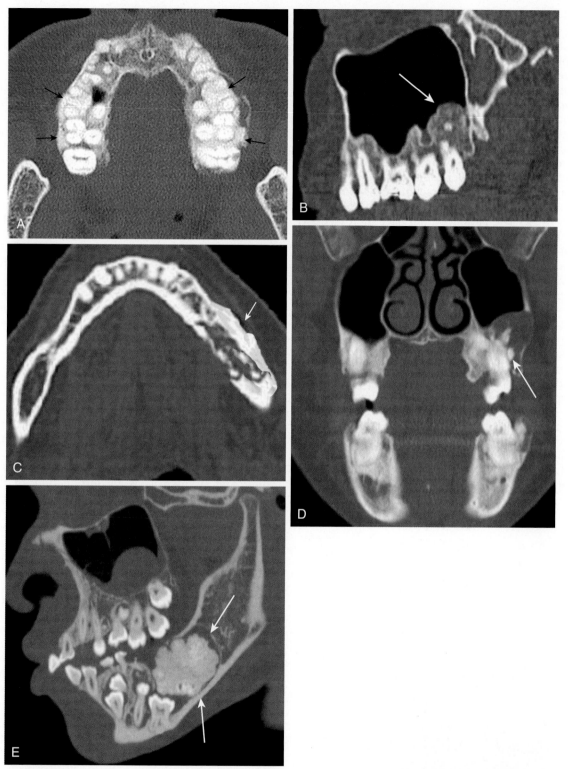

FIGURE 13-21 Common sclerotic dental lesions. **A,** Axial computed tomographic image shows periapical cemental dysplasia, which occurs near the roots of teeth with dense sclerotic appearance without bony destruction *(arrows).* **B,** Fibrous dysplasia is more common in the maxilla than the mandible. Note expansile nature of this lesion with ground glass matrix. **C,** Sclerosing osteomyelitis with thickened outer bone *(arrow)* as well as lytic lingual surface mandible is from chronic infection. Odontomas may be simple **(D)** or compound **(E)** depending upon the amount of tooth elements associated with the mass.

the superior aspects of the arytenoids. In some cases the soft palate and prevertebral swelling may be the predominant factor in creating upper respiratory symptoms. Epiglottitis is a life-threatening illness that generally is seen in 2- to 4-year-old children. This is a bacterial infection of the epiglottis that is usually caused by *Haemophilus influenzae*. The epiglottis is markedly thickened (thumb-shaped), and dilatation of the airway above the epiglottis is present. Manipulation of the epiglottis in the acute setting may cause diffuse laryngeal edema, producing acute respiratory compromise. Fifty percent of infants with this disorder ultimately require intubation. Epiglottitis is generally a clinical diagnosis with imaging limited to a confirmatory upright lateral plain film (Fig. 13-22). The patients often have increased stridor when they are placed supine, and therefore cross-sectional imaging is not recommended—unless you have a gutsy radiologist. Furthermore, you want to keep the patient in a location close to where an emergency tracheostomy can be performed, which is usually not in an outpatient imaging center.

Streptococcus may predominate as an organism in adult supraglottitis. The infection is milder in adults and is less likely to cause acute respiratory arrest or obstruction. Adult supraglottitis is a more indolent infection than pediatric epiglottitis because adults can tolerate more supraglottic and prevertebral swelling than children. In adults, the ratio of the width of the epiglottis to the anteroposterior width of C-4 should not exceed 0.33 (sensitivity, 96%; specificity, 100%), the prevertebral soft tissue to C-4 should not exceed 0.5 (sensitivity, 37%; specificity, 100%) and the hypopharyngeal airway to the width of C-4 should be less than 1.5 (sensitivity, 44%;

specificity, 87%) or epiglottitis should be suspected. Look for aryepiglottic folds enlargement and arytenoid swelling.

Croup

Epiglottitis must be distinguished from croup. Croup is a viral infection that occurs in children younger than those affected by epiglottitis. Parainfluenza 1 or 2 and influenza A are the most common pathogens. Croup is characterized by edema of the glottis and subglottis such that the normal contours of the laryngeal ventricle and true cords are obliterated, and a smooth steeple-shaped laryngeal airway is produced. Croup is a much less morbid disease than epiglottitis and should be distinguishable on the basis of plain films (Table 13-3). Wooof, wooof—sounds like the classic barking cough of croup.

Granulomatous inflammatory conditions, including tuberculosis (TB), sarcoidosis, giant cell reparative granulomas, granulomatosis with polyangiitis (formerly Wegener granulomatosis). Langerhans histiocytosis, and candidiasis may affect the larynx. Of these, TB has a propensity for invading the cartilage. Relapsing polychondritis can affect the cricoarytenoid joints.

Pharyngoceles and Laryngoceles

Pharyngoceles may occur in patients with chronic increased intrapharyngeal pressure such as in horn blowers (Dizzy Gillespie), shofar blowers (Hymie Gillespie), glass blowers (John Gillespie), and shaft blowers (Monica Lewinspie). These lesions are usually air filled but occasionally may become obstructed and fill with fluid. The laryngocele is an outpouching of the laryngeal ventricle caused by obstruction of the ventricular saccule. The saccule is a superolateral extension of the ventricle into the paraglottic space. A laryngocele may be filled with either air or fluid (so-called saccular cyst) and is seen frequently in people playing wind instruments who have chronic increased intraglottic pressure. The laryngocele is characterized as being internal, mixed, or external in its location. The internal laryngocele remains confined by the thyrohyoid membrane, whereas the external laryngocele protrudes through the thyrohyoid membrane. By definition the lesion arises within the laryngeal ventricle so that an isolated external laryngocele is almost unheard of; most lesions are in fact mixed laryngoceles, which have both a component

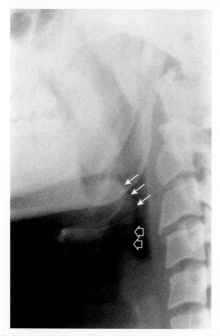

FIGURE 13-22 Epiglottitis. The epiglottis *(arrows)* and aryepiglottic folds *(open arrows)* are markedly thickened in this patient with epiglottitis. The case is unusual in that it occurred in an adult (note the degenerative change in the cervical spine). Make this diagnosis on an upright lateral plain film, one of the last indications for plain radiography in neuroradiology.

TABLE 13-3 Differentiation of Croup from Epiglottitis

Feature	Croup	Epiglottitis
Organism	Parainfluenza virus	Haemophilus influenza (children); streptococcus (adults)
Radiographic findings	Steeple-shaped subglottis	Thickened epiglottis, arytenoids, and aryepiglottic folds
Region of larynx	Glottic/subglottic	Supraglottic
Fever	No	Yes
Age at onset (years)	<2	>2
Dysphagia	No	Yes

internal and external to the membrane. Occasionally the laryngocele becomes infected and it is then termed a *pyolaryngocele*.

Occasionally, a laryngocele may arise because of neoplastic processes (Fig. 13-23). To exclude a carcinoma at the saccule, endoscopy to evaluate the ventricle is recommended in patients with laryngoceles. Imaging studies should allow careful scrutiny of lesions in this region, which may be blind to endoscopy because of the overhanging shelf of the false cord.

Ranula

An inflammatory lesion of the oral cavity is the ranula (mucus escape cyst). The name is attributed to the cool shape of these lesions, likened to a frog (ranula means frog in Latin). This is a (pseudo) cystic lesion that is due to obstruction of the sublingual gland duct or minor salivary gland of the floor of the mouth, which causes a backup of secretions and subsequently spillage of secretions in the adjacent tissues. The obstruction may be due to previous infection, trauma, or calculi. Depending on whether the

ranula protrudes through the mylohyoid musculature, the lesion may be termed simple (above the mylohyoid) or plunging (through the muscle). Do not poo-poo this minutia as it makes a difference to the surgeon—the simple ranula is approached via an intraoral resection, whereas the plunging ranula requires a submandibular or transcervical or combined intraoral and cervical resection. The ranula's wall usually enhances. The ranula may have a pointed edge along its anterior extent in the sublingual space at the site of obstruction (Fig. 13-24). Rupture of a ranula can cause an encapsulated mucus-containing infection in the deep tissue of the neck.

Trauma

Crush injuries of the adult larynx usually involve fractures of the thyroid cartilage and/or cricoarytenoid joint dislocation. The classic scenario is an impact of the larynx against the steering wheel of a car. A mucosal tear may occur as an isolated finding or in combination with the cartilaginous fracture. Pneumomediastinum or subcutaneous air or air leakage into the paraglottic spaces may be present. The

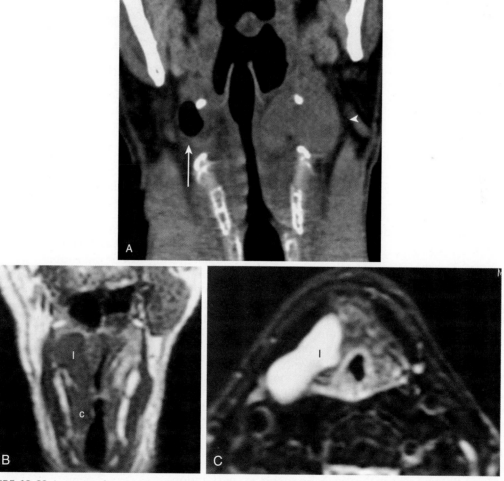

FIGURE 13-23 Laryngoceles. **A,** Bilateral laryngoceles, fluid-filled on the left *(arrowhead),* which is mixed, and air-filled on the right *(arrow),* which is external only. **B,** This patient had a transglottic squamous cell carcinoma (c) that obstructed the saccule of the laryngeal ventricle on the right side, shown on coronal T1-weighted image. Internal laryngocele (l) developed in the paraglottic soft tissue. **C,** T2-weighted imaging (T2WI) demonstrates the fluid-filled laryngocele (l), also known as a saccular cyst, in the paraglottic tissues. This should not be confused with squamous cell carcinoma, which would not be as bright on the T2WI.

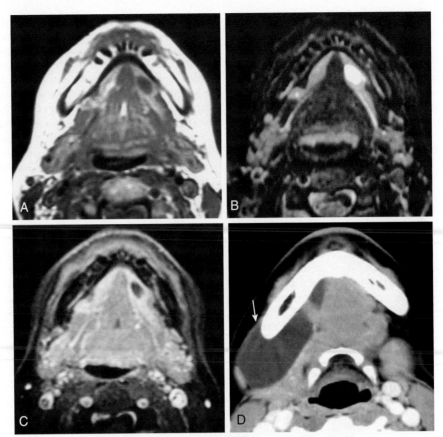

FIGURE 13-24 Simple ranula. **A,** Note the cute little lesion in the sublingual space on the left on this T1-weighted image. **B,** It has fluid intensity on T2-weighted imaging. **C,** Only peripheral enhancement is seen. This is a quintessential simple ranula, probably from a sublingual duct obstruction. **D,** This right-sided ranula *(arrow)* has perforated laterally and inferiorly and therefore constitutes a plunging/diving ranula.

primary danger to the patient is loss of airway patency, particularly if swelling occurs or the cricoid cartilage is affected. Late sequelae include infection and chondritis as an exposed cartilaginous surface is a neat culture medium for bacteria. Crushed cartilages from car crashes can create crepitus, chondritis, and chondromalacia culminating in chili con carne for cricoids.

Vocal Cord Paralysis

In the investigation of vocal cord paralysis, you should follow the course of the vagus and recurrent laryngeal nerve on the scans and include the skull base and the superior mediastinum. Remember the anatomy from a couple pages back? Or do you not have the vagus node-tion of what we are talking about? The recurrent laryngeal nerve loops around the subclavian artery on the right side and the aortic arch on the left. Separate causes of vagal neuropathy into those lesions above the hyoid bone and those below. Metastases to the jugular foramen, glomus tumors, schwannomas, nasopharyngeal carcinomas, and chordomas account for the majority of skull base causes of vocal cord paralysis (Fig. 13-25). Because slips of the vagus also supply the pharynx at the suprahyoid level, some swallowing and gag reflex dysfunction should be associated with lesions of the high vagus. If you find pharyngeal muscle atrophy or a deviated uvula on your imaging studies, insist the etiology is an upper vagus nerve lesion—remember that the gag is

sensory IX and motor X and can be politically motivated. Cranial nerves IX, XI, and XII travel with the vagus in its uppermost segment; therefore lesions affecting the vagus superiorly, usually (95%), affect one or all of these cranial nerves as well.

Below the hyoid bone, an isolated vocal cord paralysis may be seen without pharyngeal effects. The common lesions causing lower vagus abnormalities are related to lesions in and around the carotid sheath including squamous cell carcinoma, thyroid masses, glomus vagale tumors, schwannomas, posttraumatic dissections and pseudoaneurysms, intraoperative injury, and lymphadenopathy. In the mediastinum lymphoma, bronchogenic carcinoma, lymphadenopathy, patent ductus arteriosus, mitral stenosis (because of the associated pulmonary artery dilatation), aneurysms, and arterial dissections may cause recurrent laryngeal nerve paralysis. Situated in the tracheoesophageal groove as it ascends to the larynx after exiting the mediastinum, the nerve is susceptible to lesions in the thyroid gland (cancer, goiter, and trauma), the parathyroid tissue (adenoma and carcinoma), and esophagus (Zenker diverticulum, perforation).

Several imaging findings suggest the presence of vocal cord paralysis: (1) atrophy of the cord (thyroarytenoid muscle); (2) dilatation of the laryngeal ventricle; (3) medial orientation of the vocal cord; (4) dilatation of the ipsilateral pyriform sinus and vallecula; (5) medial orientation and angulation of

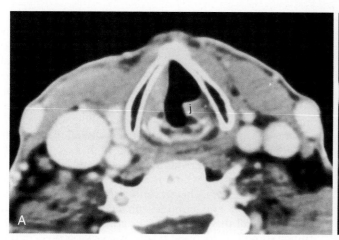
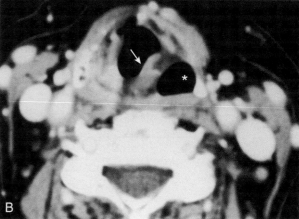

FIGURE 13-25 Left-sided vocal cord paralysis. **A,** Note the cord atrophy suggesting chronic vocal cord paralysis on the left side. In addition, there is medial orientation of the cricoarytenoid joint (j) and the remaining vocal cord tissue. The source of this patient's vocal cord paralysis was not evident because the resident forgot to scan inferiorly to the aortic arch, where the nerve circles before doubling back in the tracheoesophageal groove. That same resident has become a very successful imaging center entrepreneur. Case 2. **B,** With time, the left piriform sinus *(asterisk)* dilates as the left aryepiglottic fold *(arrow)* slides medially.

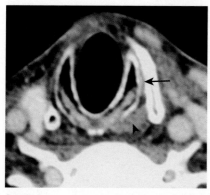

FIGURE 13-26 Right vocal cord paralysis. Axial computed tomographic scan in a patient with a right-sided glomus vagale show atrophy of the right posterior cricoarytenoid (PCA) muscle and right cricothyroid muscle relative to the normal left side *(arrowhead indicates normal left PCA muscle; arrow, normal left cricothyroid muscle).* (From Romo LV, Curtin HD. Atrophy of the posterior cricoarytenoid muscle as an indicator of recurrent laryngeal nerve palsy. *AJNR* 1999;20:467-471.)

the aryepiglottic fold; and (6) anteromedial deviation of the arytenoid cartilage (see Fig. 13-25). Recently, a new finding has been brought to our attention, that of posterior cricoarytenoid (PCA) muscle atrophy as this is one of the intrinsic muscles of the larynx innervated by the recurrent laryngeal nerve. Atrophy of the PCA muscle is found in over half of patients with vocal cord paralysis but less than 5% of radiologists can identify this muscle (Fig. 13-26). Curious.

Zenker Diverticulae

These occur near the cricopharyngeus muscle and protrude posteriorly. They are best demonstrated by gastrointestinal radiologists and barium. They increase in frequency and size with age and develop at Killian's dehiscence, an area of weakness between the cricopharyngeus and the inferior pharyngeal constrictor muscles. The patient may complain of dysphagia, bad breath, aspiration pneumonia, or regurgitation of undigested food. A Killian-Jamieson

---

> **BOX 13-4   Staging of Juvenile Nasopharyngeal Angiofibroma**
>
> IA—Tumor limited to posterior nares or nasopharyngeal vault
> IB—Extension into one or more paranasal sinuses
> IIA—Minimal lateral extension through sphenopalatine foramen into medial pterygomaxillary fossa
> IIB—Full occupation of pterygomaxillary fossa, displacing posterior wall of antrum forward; superior extension eroding orbital bones
> IIC—Extension through pterygomaxillary fossa into cheek and temporal fossa
> III—Intracranial extension

pharyngoesophageal diverticulum protrudes laterally and is typically smaller than a Zenker diverticulum.

BENIGN NEOPLASMS OF THE AERODIGESTIVE TRACT

Juvenile Nasopharyngeal Angiofibroma

The classic benign nasopharyngeal tumor is the juvenile nasopharyngeal angiofibroma (JNA; also see Chapter 10). This is a tumor that is characterized by its high vascularity and its propensity for bleeding. The typical patient is a male adolescent who has recurrent epistaxis. This lesion is seen nearly exclusively in males. The lesion has been said to arise from the nasopharynx, the sphenopalatine foramen (a medial egress from the pterygopalatine fossa), the pterygopalatine fossa itself, the vidian canal, and/or the nasal cavity. The tumor, although benign, often has aggressive growth with spread via the pterygopalatine fossa into the infratemporal fossa and the intracranial compartment. A grading system for the juvenile angiofibroma has been derived that forces one to learn some of the skull base foramina (Box 13-4). Remember that the sphenopalatine foramen leads to the nasal cavity, the vidian canal to the foramen lacerum, carotid canal, and skull base, the inferior orbital fissure to the orbit, and the pterygomaxillary fissure to the infratemporal fossa. Extension into adjacent structures is exceedingly common (Table 13-4).

On CT, the tumor is isodense to muscle before contrast administration but demonstrates marked enhancement. By the time it is diagnosed the tumor usually spans the nasal cavity, the nasopharynx, and the pterygopalatine fossa with bowing of the posterior wall of the maxillary antrum anteriorly and widening of the pterygomaxillary fissure. JNA used to be a classic plain film (base view) diagnosis in which the posterior wall of the maxillary sinus is seen to be displaced anteriorly. It may still be seen on residents' boards as such, so be wary if you pop up a plain film case on the computerized test for neuroradiology boards.

On MR, the salient feature of the angiofibroma is the abundance of flow voids from its high vascularity. The tumor has a characteristic "salt (tissue) and pepper (flow voids)" appearance, which is also described with glomus tumors (Fig. 13-27). The salt and pepper may be "served" both on T2WI and contrast-enhanced T1WI. When the tumor is small, this finding may be absent. The extent of the tumor should be mapped relatively easily on postcontrast

TABLE 13-4   Spread of Juvenile Angiofibromas

Site of Spread	%
Pterygopalatine fossa	89
Infratemporal fossa	85
Sphenoid sinus	61
Maxillary sinus	43
Intracranial	20

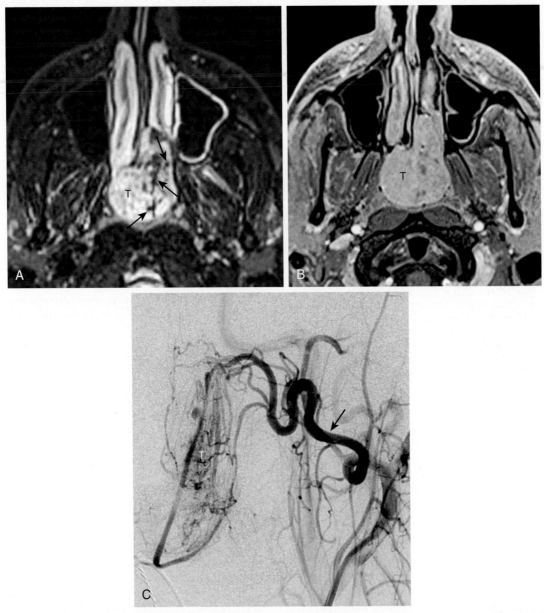

FIGURE 13-27 Juvenile angiofibroma. **A,** Axial T2-weighted image (T2WI) in a young male presenting with epistaxis shows a large mass in the nasopharynx extending into the posterior aspect of the left nasal cavity (T). The mass shows numerous serpentine flow voids *(arrows)*. **B,** On this postcontrast T1WI the mass (T) is shown to enhance avidly. **C,** Conventional catheter angiogram shows marked enlargement of the internal maxillary artery *(arrow)*, which provides the major arterial supply to the tumor (T). Preoperative embolization can reduce the morbidity of a surgical excision.

fat-suppressed MR. One should pay particular attention to the skull base foramina and the numerous exits from the pterygopalatine fossa (Fig. 13-28).

On angiography, these tumors are highly vascular, and are usually supplied by branches of the ascending pharyngeal artery and the internal maxillary artery. Often bilateral

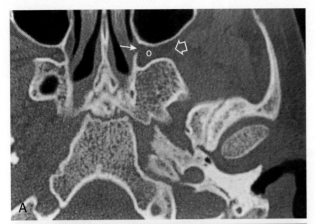

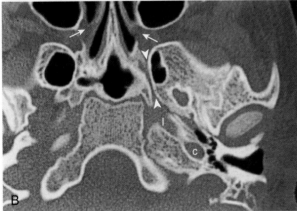

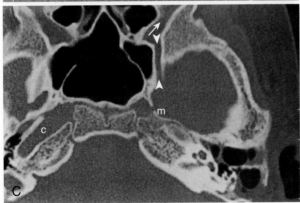

FIGURE 13-28 Anatomy of the pterygopalatine fossa. **A,** The openings from the pterygopalatine fossa (o) into the nasal cavity via the sphenopalatine foramen *(arrow)* and into the infratemporal fossa via the pterygomaxillary fissure *(open arrow)* are labeled. **B,** *Arrowheads* denote the vidian canal, another egress from the pterygopalatine fossa. Foramen lacerum (l) will become the floor of the internal carotid artery in its horizontal course. *Arrows* indicate the sphenopalatine foramina. **C,** *Arrowheads* in more superior section outline the foramen rotundum while the *arrow* shows the inferior orbital fissure. These are the upper terminations of the pterygopalatine fossa. V$_2$ will pass through rotundum to get to the Meckel cave (m). Horizontal petrous carotid artery (c)—lacerum would be just below.

supply as a result of rich pharyngeal anastomoses is present. Recruitment from internal carotid artery tributaries including petrous and cavernous carotid branches may occur, with extension of the tumor through the base of the skull. Embolization of the tumor may be useful preoperatively to reduce blood loss.

Recurrence rates are as high as 60% because the tumor tends to infiltrate the pterygoid bone, base of skull, and sphenoid bone. It can reactivate from there. Tumor recurrences can be treated with primary radiation therapy.

Neurogenic Tumors

Schwannomas of the hypoglossal or lingual nerve present as well as defined masses in the submucosal space that are intermediate in signal intensity on T2WI and show enhancement. There may be consequential atrophy of the muscles they innervate with fatty replacement. Another uncommon location for schwannomas in children or adults is in the supraglottis. The aryepiglottic fold and arytenoid region are most commonly involved, presumably because of superior laryngeal nerve involvement by the tumor. These neoplasms appear as exophytic or submucosal masses that enhance.

Benign Odontogenic Tumors

Ameloblastoma

The ameloblastoma is a benign neoplasm of the oral cavity that is hard to remove completely and has a high rate of recurrence (Fig. 13-29). These characteristics may also be seen with keratocysts, myxomas, and aneurysmal bone cysts. Ameloblastoma is the second most common odontogenic tumor and arises in the mandible in 81% of cases. The molar region is affected in 70% of cases. The lesion is painless unless superinfected and has scalloped margins, multiloculation, and expanded cortical surfaces. Solid and cystic components are seen on MR with frequent mural nodules. Enhancement is marked in the periphery, not in the cysts. High intensity on T1WI sometimes occurs and may be due to hemorrhage or cholesterol crystal accumulation.

Osseous Tumors

Benign lesions of the oral cavity are common and include bony lesions such as osteomas and tori palatini, which are benign bony outgrowths from the hard palate that present as ossified masses. Tori mandibuli occur along the inner cortex of the mandible.

Papilloma

The laryngeal papilloma is a true neoplasm (as opposed to a polyp) and, although benign, has a small association with squamous cell carcinoma. It appears as a small focal lesion on the vocal cord.

Pleomorphic Adenoma

Minor salivary gland tumors occur within the aerodigestive system especially along the soft palate. In this location benign tumors approximate malignant minor salivary gland tumors in nearly equal abundance. The most common benign tumor of the soft palate minor salivary glands

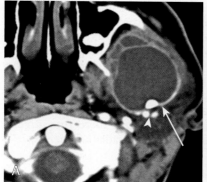
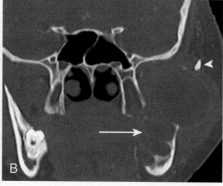
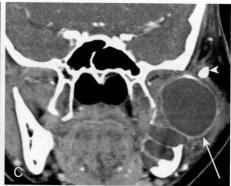

FIGURE 13-29 Ameloblastoma. **A-C,** This mass arose in the mandible and has multiple cystic loculations *(arrows)*; residual tooth remnant *(arrowheads)* erodes bone.

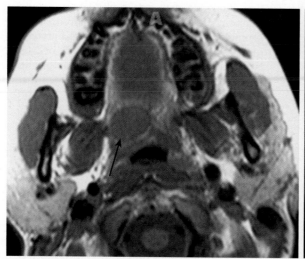
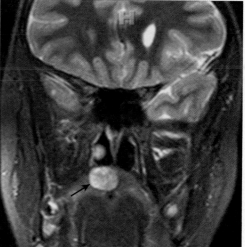

FIGURE 13-30 Pleomorphic adenoma. Axial T1-weighted imaging (T1WI) and T2WI show a well-defined mass of the soft palate *(arrows)*. The relative brightness of the lesion on T2 is typical of pleomorphic adenoma. It is certainly not specific, however, and biopsy/surgical resection is necessary for the diagnosis.

is the pleomorphic adenoma (Fig. 13-30). Pleomorphic adenomas are usually very bright on T2WI. They enhance.

Other Rare Birds

Other rare benign tumors of the aerodigestive mucosa include granular cell tumors, leiomyomas, and rhabdomyomas. Most extracardiac rhabdomyomas arise from pharyngeal constrictor muscles, the floor of the mouth, and the tongue base—not so much the larynx, although these can happen too. Paragangliomas and lipomas infrequently frequent the larynx.

MALIGNANT NEOPLASMS: SQUAMOUS CELL CARCINOMA

TNM Staging

The TNM staging system is used to classify head and neck cancers. T refers to primary Tumor extent, N to regional lymph Nodes, and M for distant Metastases. On the basis of the TNM stage of a tumor, treatment regimens are planned, prognoses are predicted, and treatment efficacy is assessed. The TNM staging of squamous cell carcinoma is used throughout the aerodigestive system.

Nasopharyngeal Carcinoma

The latest American Joint Commission on Cancer staging guide considers T1 nasopharyngeal cancer to be confined to the nasopharynx. Tumor spread to the soft tissues of the oropharynx or nasal fossa is graded T2. For T2b or not 2b, the parapharyngeal fat is the question. If there is no parapharyngeal spread, call it T2a; if the fat is infiltrated, it would be T2b. T3 designates invasion into bony structures and/or paranasal sinuses, whereas intracranial, infratemporal fossa, masticator space, hypopharyngeal, orbital, or cranial nerve involvement merits a T4 lesion. The World Health Organization classifies nasopharyngeal carcinomas into three types; type 1 keratinizing squamous cell carcinoma, type 2 nonkeratinizing carcinoma (with or without lymphoid stroma), and type 3 basaloid squamous cell carcinoma (Table 13-5). Of these types, type 1 and 2 are the most common varieties of carcinoma and all types of nasopharyngeal carcinomas account for 80% of nasopharyngeal malignancies, followed by lymphoma, minor salivary gland tumors, and rhabdomyosarcoma.

A bimodal distribution appears in the incidence of nasopharyngeal carcinoma. A younger age group (15 to 25 years old) is seen with undifferentiated nasopharyngeal

TABLE 13-5 Nasopharyngeal Carcinoma Subtypes (OK, Lymphoma Too)

Feature	Major Risk Factors	Radiation Sensitivity	Frequency	AIDS Association	Site of Origin	Survival	Growth
Keratinizing squamous cell carcinoma	Smoking, alcohol, radiation induced, Caucasian		2nd	Yes		Fair	Deep
Nonkeratinizing squamous cell carcinoma (differentiated and undifferentiated)	Highest EBV risk, Southeast Asian, nitrosamines in salted food, burning incense sticks	Highest	1st	Yes	Fossa of Rosenmüller	Very good	Deep
Basaloid squamous cell carcinoma	EBV, HPV		3rd	No		Poor	Deep
Lymphoma NHL B cell and T/NK cell	HIV		Rare	Yes	Adenoidal tissue	Very good	Into airway, exophytic

EBV, Epstein-Barr virus; *HIV,* human immunodeficiency virus; *HPV,* human papillomavirus; *NHL,* non-Hodgkin lymphoma.

carcinoma in Southeast Asia, particularly in the Chinese population where it accounts for 18% of all cancers. Chinese Americans have a sevenfold increased risk of nasopharyngeal carcinoma over non-Chinese counterparts. The Westerners who get nasopharyngeal carcinoma are usually older adults, from 40 to 60 years old, who have more squamous and nonkeratinizing carcinomas. Exposure to the EBV has been suggested to lead to an increased incidence of carcinoma of the nasopharynx, particularly in the nonkeratinizing/undifferentiated varieties. Late membrane protein (LMP) 1, LMP2, and Epstein-Barr virus nuclear antigen (EBNA) 1 proteins are coded by the EBV genome incorporated into the nasopharyngeal carcinoma cells and are now the target of systemic immunotherapies directed at this tumor. In addition, some investigators say that AIDS may predispose to nasopharyngeal carcinoma. In these instances the patient is younger than those in the second peak of nasopharyngeal carcinoma, the late-middle-aged adult. The latter patients often carry risk factors of cigarette smoking and alcohol abuse, although the association with nasopharyngeal carcinoma is less strong than elsewhere in the aerodigestive system.

Nasopharyngeal carcinoma is notorious for demonstrating minimal mucosal disease while having a large submucosal component, which is best evaluated by cross-sectional imaging techniques. In fact, often the lesion may be completely inapparent to the endoscopist because of the frequent apposition of mucosal surfaces around the fossa of Rosenmüller, which hides the primary tumor (Fig. 13-31). In this case fat is your friend; its loss between tensor and levator veli palatini muscles or in the parapharyngeal tissue suggests malignancy. Whenever new serous otitis media occurs in an adult, think nasopharyngeal carcinoma and be sure that you scrutinize this area.

Nasopharyngeal carcinoma is also notorious for its spread beyond the mucosal surface of the head and neck, into the parapharyngeal space (65% to 84% of cases), the retropharyngeal space (40%), the masticator space (15%), the carotid space (23%), the perivertebral space (15%), and the bony skull base. The sphenoid sinus (27%), nasal cavity (22%), and ethmoid sinus (18%) are also frequently invaded. Situated as it is near the base of the skull, nasopharyngeal carcinoma also is notorious for direct and/or perineural spread through the various skull base foramina to the intracranial space (31% to 48%).

On CT, nasopharyngeal cancers demonstrate density similar to that of muscle. Because of the low differentiation between muscle and the tumor, subtle nasopharyngeal carcinomas may be difficult to diagnose. Infiltration of the parapharyngeal space is the most reliable (but a T2b) sign of nasopharyngeal carcinoma. Looking just at the mucosal surface is fraught with difficulty because of the frequent asymmetry in the aeration of the fossa of Rosenmüller. For every early nasopharyngeal carcinoma diagnosed because of tiny mucosal asymmetries, there are probably three cases in which only mucosal apposition or secretions are seen at endoscopy. This is a classic "overcall" region. For this reason one must emphasize analysis of the parapharyngeal and intermuscular fat for obliteration. Loathful lesions lead to lipid loss.

Invasion intracranially or along the cranial nerves is difficult to diagnose early with CT. One must identify enlargement of the nerves, enlargement of skull base foramina, or infiltration of the bone. Scans with contrast and bone windows may help identify the skull base disease on CT. By the time the skull base foramina have enlarged, however, the tumors are probably far advanced. In addition, the evaluation of the base of the skull and/or meninges is problematic with CT because of difficulties associated with beam-hardening artifact. For this reason and because of its superior soft-tissue resolution, MR is the recommended study for evaluation of nasopharyngeal carcinoma. The involvement of cranial nerve V from the pterygopalatine fossa up to the cavernous sinus is much better seen with unenhanced T1WI scans and/or fat suppressed enhanced T1WI scans. If intracranial extension is seen, do not forget to evaluate the orbits, because perineural tumor extension from the cavernous sinus can occur here too. Skull base involvement through the sinus of Morgagni along the eustachian tube or to the foramen lacerum is also well demonstrated on unenhanced T1WI.

On T1WI, the carcinoma in the nasopharynx may be isointense to the muscles. On T2WI, carcinoma of the nasopharynx is often of low to intermediate signal intensity but is always brighter than the dark signal of muscle (see Fig. 13-31, *B*). However, some tumors may be very bright on T2WI, making differentiation with mucosal edema very difficult. To make matters worse, the normal nasopharyngeal mucosa, the adenoidal tissue, and some carcinomas and lymphoma enhance. In the performance of postcontrast MR of the nasopharynx, fat saturation should be used to better evaluate the invasion of skull base foramina and fat.

Treatment of nasopharyngeal carcinoma is primarily nonsurgical. Early stage disease (Box 13-5) is treated with radiation therapy alone with 3-year survival rates in the 90% range. More recently, intensity modulated radiotherapy (IMRT) has been used to great effect in treating nasopharyngeal carcinoma. This technique employs variable intensity beams through static multileaf collimators that shape the radiation field. Tumor tissue can, in this way, get doses up to 70 to 80 gray while vital intracranial and spinal structures are spared from the harmful rays. Sequential tomography with multiple arcs and multiple fields of radiation beams may also be employed for T4 disease. These techniques, with chemotherapy, have rendered nasopharyngeal carcinoma an eminently curable disease at a local level; unfortunately for the patient, the tumor has a relatively high propensity for hematogenous spread. Bulky disease, small exophytic recurrences after radiotherapy, and persistent nodal disease may occasionally be managed surgically. These tumors have traditionally not been part of the bailiwick of the surgeon. However, a few centers are starting to operate primarily on patients with localized nasopharyngeal cancers.

Recent studies have suggested that tumor volume may be a better predictor of radiation/chemotherapy response than T staging, at least for high-grade nasopharyngeal carcinomas. The cumulative survival for T3 and T4 patients with less than 30 mL of tumor (>80%) is better than that for patients with 30 to 60 mL of tumor (45%), and for those with more than 60 mL of tumor (<10%) at 5-year follow-up. Presumably, hypoxia and radioresistance occur in large tumor volume cancers, making cure less likely.

With concomitant chemotherapy added (cis-platinum and fluorouracil), stage II disease survival rates get bumped up by about 5% to 10% over radiotherapy alone into the 85% to 90% range. Stage III and IV nasopharyngeal carcinomas are also treated with chemotherapy and concomitant radiotherapy. Undifferentiated cancers seem to respond the best. However, do not expect the images of the nasopharynx to normalize. Au contraire, you may expect to see residual asymmetry and mucosal apposition forever more (quoth the Baltimore Raven).

First echelon nodal spread from nasopharyngeal carcinoma is to retropharyngeal and high jugular (level II) lymph nodes. Nodal spread is assumed with nasopharyngeal

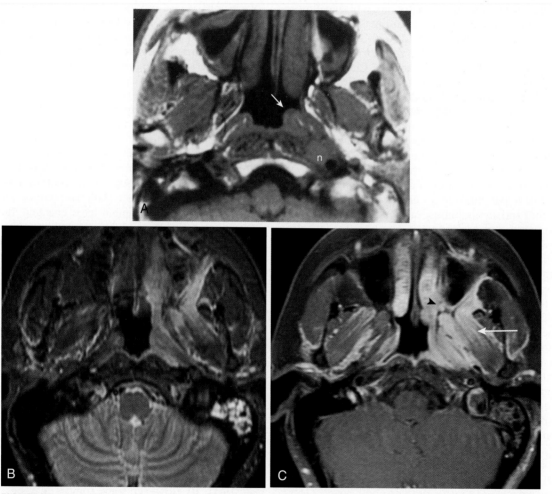

FIGURE 13-31 Nasopharyngeal carcinoma with a retropharyngeal lymph node. **A,** T1-weighted imaging (T1WI) demonstrates enlargement of the torus tubarius *(white arrow)* on the left side resulting from a nasopharyngeal carcinoma in this 20-year-old Southeast Asian man. Retropharyngeal lymph node (n) is identified medial to the carotid artery on the left side. **B,** This T2WI scan shows infiltration of the left fossa of Rosenmüller with spread along the anterior mucosal surface as well as the deep musculature. Note fluid backed up in mastoid air cells due to eustachean tube outlet obstruction. **C,** On postcontrast scans, the enhancement of the medial pterygoid muscle *(arrow),* the tissue along the outer wall of the maxillary antrum, and into the pterygopalatine fossa *(black arrowhead)* is evidence of T4 disease.

carcinoma because it occurs at a rate of about 80%. Disease in the nodes and the size of the nodes show less correlation with nasopharyngeal carcinoma than other head and neck carcinomas. Even a tiny NPCA spreads to the nodes early; that is probably why the T staging of NPCA is not based on size criteria the way oral cavity and other pharyngeal cancers are.

Oropharyngeal Squamous Cell Carcinoma

As opposed to the nasopharynx, which is classified by spread, staging of the oropharynx for squamous cell carcinoma is based largely (sic) on size criteria (Box 13-6).

BOX 13-5 Staging of Nasopharyngeal Carcinoma

T1—Tumor confined to the nasopharynx, or tumor extension to oropharynx and/or nasal cavity without parapharyngeal extension
T2—Tumor with parapharyngeal extension
T3—Tumor involves bony structures and/or paranasal sinuses
T4—Tumor with intracranial extension and/or involvement or cranial nerves, infratemporal fossa, hypopharynx or orbit, or masticator space

From Edge SB: *AJCC Cancer Staging Manual.* 7th ed. New York: Springer; 2009. Used with permission.

BOX 13-6 T-Staging of the Oropharynx

T1—Tumors 2 cm or less in greatest dimension
T2—Tumor >2 cm but not >4 cm in greatest dimension
T3—Tumor >4 cm in greatest dimension or extension to lingual surface of epiglottis.
T4a—Tumor invades the larynx, deep/extrinsic muscle of tongue, medial pterygoid, hard palate, or mandible
T4b—Tumor invades lateral pterygoid muscle, pterygoid plates, lateral nasopharynx, or skull base or encases carotid artery

From Edge SB: *AJCC Cancer Staging Manual.* 7th ed. New York: Springer; 2009. Used with permission.

With the T4a lesion, tumor invades larynx, deep/extrinsic muscle of tongue, medial pterygoid, hard palate, or mandible, whereas T4b classification is generally considered nonresectable disease and includes invasion of the lateral pterygoid muscle, pterygoid plates, lateral nasopharynx, skull base or carotid artery.

The most common locations for a primary squamous cell carcinoma of the oropharynx are the anterior portion of the tonsil (Fig. 13-32) and the tongue base. Tonsillar carcinomas have a propensity for invasion into the tongue musculature beyond the circumvallate papillae, thereby spreading across the anatomic boundary between the oropharynx and oral cavity. Tonsillar carcinomas also may spread laterally and superiorly to invade the retromolar trigone.

The staging criteria for T1 to T3 lesions necessitate measuring all lesions of the oropharynx to assist the surgeon in the proper therapy for a given oropharyngeal squamous cell carcinoma. Rulers rule. In addition, it is critical to assess for invasion into the mandible or maxilla or the soft tissues of the neck. Invasion of the deep muscles and soft tissues of the neck may be equally well visualized with MR and CT; however, the increased soft-tissue resolution on MR with T1WI and T2WI and enhanced fat-suppressed imaging makes MR the preferable examination, if the patient can tolerate it. On CT, there is less conspicuity between tumors and muscle because both have similar densities.

Other issues that must be addressed with respect to oropharyngeal carcinomas are listed below.

Preepiglottic Fat Invasion

From the base of the tongue, a tumor may extend into the vallecula, the air space anterior to the epiglottis but posterior to the tongue base. Once the vallecula is infiltrated, the critical space to be evaluated by the radiologist is the preepiglottic fat. When the preepiglottic fat space is infiltrated with tumor, it is surgically impossible to remove the fat while maintaining the integrity of the epiglottis'

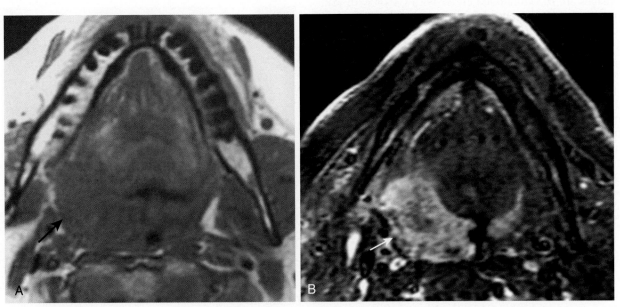

FIGURE 13-32 Tonsil carcinoma. **A,** T1-weighted imaging (T1WI) scan of tonsillar squamous cell carcinoma shows a hypointense mass centered in the right palatine tonsil *(arrow)* with partial effacement of the right parapharyngeal fat. These tumors often invade anterolaterally to enter the retromolar trigone, medially to enter the soft palate, and anteriorly to invade the tongue base. **B,** Tonsil and tongue cancers are very well-depicted on fat-suppressed T2WI *(arrow)*. They are as subtle as a Donald Trump campaign speech.

petiole (inferior stem). A tumor of the base of the tongue that does not involve the preepiglottic fat can be resected without requiring a portion of the supraglottic larynx to be included in the operative specimen. On the other hand, if the preepiglottic fat is invaded, the patient will often require an epiglottectomy, partial supraglottic laryngectomy or, worse, a total laryngectomy. The combination of base of tongue surgery with laryngeal surgery often leads to a poor quality of life in which both swallowing and speaking are compromised.

The preepiglottic fat is best assessed with sagittal T1WI scans and axial T1WI scans in which soft-tissue signal intensity within the fat suggests cancerous involvement. Occasionally, adjacent inflammation or peritumoral edema or partial volume effects might simulate preepiglottic fat invasion. Massive preepiglottic fat disease also often makes radiation therapy undesirable because of the bulk of the tumor, thereby limiting curability. Preepiglottic perforation predisposes to palpable nodal pathology, predictably portending poor prognosis.

Mandibular Invasion

Although periosteal invasion is nearly impossible to identify by any current imaging technique, cortical invasion is equally well seen on CT and MR with T1WI showing loss of the hypointense cortex, fat-suppressed T2WI showing high signal infiltration, and postcontrast fat suppressed T1WI showing enhancement of the cortical defect. Infiltration of the bone marrow of the mandible or the maxilla is better appreciated with MR, appearing as low signal on T1WI infiltrating the normally high-intensity fatty marrow in an adult. Scans will identify tumor adjacent to the mandible, cortical erosions, infiltration of marrow fat, and/or tumor on both sides of the mandible. Particularly when a lesion arises on the alveolar surface of the mandible, single plane imaging may be insufficient to determine mandibular invasion. For this reason, when the issue of mandibular involvement is raised, axial and coronal images are recommended either with CT or with MR. Depending on the extent of involvement, the oral cavity/oropharyngeal cancer adjacent to the mandible is treated differently. If tumor abuts the mandible but is not fixed to the periosteum or mandible, the periosteum is resected as the margin. For tumor fixed to or superficially invading the periosteum and/or cortex, inner cortex resection (marginal corticectomy) can be performed for margin control. Marginal resection can be performed for superficial alveolar (oral cavity) cancers. Once the cortex has been violated or marrow has been infiltrated, a more extended mandibular resection is required for cure as primary radiation incurs the risk of osteoradionecrosis at doses high enough to sterilize the bone disease. In most cases, microvascular osteomusculocutaneous free flaps are used to replace the bone and to achieve a cosmetic result in which facial deformity is not evident.

Prevertebral Muscle Invasion

If a cancer is fixed to the prevertebral musculature (longus capitus-longus colli complex) the patient is deemed unresectable (Fig. 13-33). Although the imaging findings of high signal intensity on T2WI scans in the muscles, contrast enhancement of the muscle, or nodular infiltration of the muscles would suggest neoplastic infiltration, in fact these findings have not been very reliable. The surgical evaluation at the time of panendoscopy or open exploration remains the gold standard, despite the fact that in rare instances a plane can be found between tumor and the prevertebral musculature. Because the prevertebral musculature is so close to the spinal canal and spinal cord, there are some issues with regard to curative radiotherapy in individuals who have infiltration in this location. At the very least the radiologist should suggest the possibility of violation of the prevertebral musculature when the aforementioned findings are present and/or there is obliteration of the retropharyngeal fat stripe by cancer. At that point the surgeon must take over and evaluate for fixation at panendoscopy or at surgery (trying to create a plane in the retropharyngeal space between tumor and longus muscles). Sending surgeons searching for spurious squamous cells in this space squanders seconds and subsidizes civil suits.

Pterygopalatine Fossa Invasion

Extension to the pterygopalatine fossa or to other avenues of the fifth cranial nerve raises the possibility of perineural spread of the cancer to the skull base. "Losing the tumor at the skull base" because of spread along the cranial nerves happens infrequently with squamous cell carcinoma when the radiologist cautions the surgeon about this possibility; on the other hand, this is typical for adenoid cystic carcinoma where remote perineural spread is almost the norm. Once again, MR appears to have the advantage in evaluating the nerves over CT, showing abnormal enlargement and enhancement of the nerve. Foraminal enlargement and bony infiltration is a reliable finding on CT, but is pretty uncommonly seen. If one sees infiltration of the fat of the pterygopalatine fossa, atrophy of the muscles innervated by the trigeminal nerve, or abnormal enhancement in the Meckel cave on CT, perineural invasion is implied.

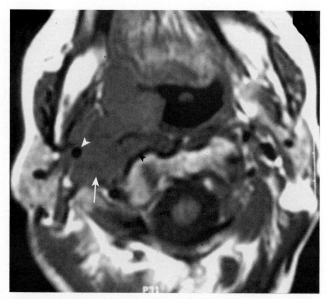

FIGURE 13-33 Spread of oropharyngeal cancer. This tumor has spread posterolaterally to the prevertebral musculature *(black arrowhead)*, to the carotid artery where the mass *(arrow)* encases the medial 180 degrees of the carotid *(white arrowhead)*. It is also in the parapharyngeal tissues.

Perineural spread of tumor may be antegrade or retrograde and may show "skip lesions" radiographically.

Pterygomandibular Raphe Invasion

The pterygomandibular raphe stretches from the medial pterygoid muscle's insertion on the medial pterygoid plate to the mylohyoid ridge of the mandible. It serves as the origin or insertion of the buccinator muscle and the pharyngeal constrictor muscles. The pterygomandibular raphe is one of the boundaries between the oral cavity and the oropharynx as it effectively divides the anterior tonsillar pillar and the retromolar trigone. The retromolar trigone, a portion of the oral cavity, is the area behind the maxillary teeth but in front of the coronoid process and ascending ramus of the mandible. Tumor can spread along this plane superiorly to the temporalis muscle, medially into the pterygomandibular space where the lingual and inferior alveolar nerves run, or inferiorly into the floor of the mouth. If tumors spread anteriorly from the medial pterygoid plate they enter the pterygopalatine fossa.

Bilateral or Deep Invasion of the Tongue Base

Tongue base lesions also have characteristic problems inherent to their location. Because the neurovascular bundle (including the hypoglossal nerve, the lingual nerve, and the lingual artery) enters from the base of tongue, infiltration of these structures makes surgical excision, while preserving function, more difficult. These same neurovascular structures then course along the styloglossus-hyoglossus complex and the floor of the mouth to supply the anterior aspect of the tongue. If the base of the tongue is infiltrated bilaterally, there is a relatively limited chance that a patient will be able to have functioning tongue available after surgical removal of the tumor. The surgical guideline to abide by is that you need 25% of the base of the tongue and one hypoglossal nerve and one lingual artery to have a functioning tongue.

The tongue is one of the most critical organs of the aerodigestive system, being instrumental in both swallowing and phonation. Without a tongue or a viable reconstruction, the patient is left with a permanent feeding tube and, in many cases, a tracheostomy as the airway becomes unprotected. Part of the emphasis in head and neck surgery is the creation of sophisticated flaps that can be constructed to allow swallowing (like a seal) without bolus formation. For this reason it is important for the radiologist to identify spread of tumor across the fatty lingual septum that defines the midline of the tongue. If unilateral disease only is present, tongue function will often be sufficient for near normal lifestyle. Patients who undergo hemiglossectomies can form a bolus and can speak reasonably intelligibly if not a little garbled. Nowhere else in the head and neck is bilaterality of disease more important with respect to patient quality of life. Once the base of the tongue has been resected, neurovascular grafting cannot be achieved to preserve function in the anterior aspect of the tongue. If the midline of the base of the tongue has been violated to any significant degree by cancer, the possibility of having a complete resection with adequate margins while maintaining a functioning tongue is remote. If a small mobile portion of the base of the tongue is preserved, functional recovery with flap reconstruction is much improved. In some institutions, total glossectomies are never performed, leaving radiotherapy and chemotherapy with or without brachytherapy implants, the only options for cancer cure. With large pectoralis, rectus abdominis free flaps, or bulky microvascular flaps, one can usually get the patient swallowing again after total glossectomy so that lifelong feeding by gastrostomy tubes is not required. The risk of aspiration into the larynx is also reduced when bulky flaps are used.

Some programs follow the adage that a total glossectomy means a laryngectomy. Although this is certainly true in patients who have already had radiation, have supraglottic spread of oropharyngeal tumors, have huge disease, or are respiratorily compromised (defined as unable to walk two flights of stairs chased by a health maintenance organization (HMO) regulator without getting winded), in the advanced cancer centers this need not be the case in all comers. The pectoralis or bulky free flap allows the patient to swallow with a head tilt and the supraglottis protects the airway. The patient is taught to cough after swallowing to prevent the *petit pois* passing by the petiole and to protect the laryngeal passageway.

The tongue undergoes predictable changes after it has been denervated (usually as a result of primary tumor resections, neck nodal surgery, anterior horn cell disease [ALS], and occasionally from primary neoplasms of the hypoglossal nerve). In the first few months one will see high signal on T2WI in the ipsilateral hemitongue and a relatively normal T1WI. After about 5 months, fatty infiltration of the tongue, manifesting as high signal on T1WI (and low density on CT), and volume loss ipsilaterally may be seen. The same pattern is seen with denervated muscles of mastication.

Lymphadenopathy

Tonsillar carcinomas are also the most common source of occult primary tumors that present as cervical adenopathy alone (the dreaded "carcinoma of unknown primary"). These microscopic cancers may be deep within the crypts of the lymphoid tissue and may be invisible to both endoscopy and imaging. Other sites for carcinomas of unknown primary that should be assiduously studied include the nasopharynx, piriform sinus, base of tongue, and chest.

The appearance of oropharyngeal squamous cell carcinoma is similar to that elsewhere in the aerodigestive system on both CT and MR. It should be noted, however, that the base of tongue drains to bilateral lymphatic systems, and therefore one must critically assess both sides of the neck for associated lymphadenopathy. Primary drainage is to upper jugular and submandibular lymph node chains. Cystic nodes are possible with a tonsillar carcinoma primary. This is particularly true of human papillomavirus (HPV)–positive cancers. Fortunately, even with nodal disease HPV-positive cancers have a decent prognosis.

Prognosis

The prognosis for oral cavity and oropharyngeal carcinoma for all-comers is about 50% for 5-year survival. Obviously, the smaller the tumor, the better the prognosis. Surgery, radiation therapy, and chemotherapy all play roles in treatment with the latter generally reserved for more advanced, widespread, or unresectable disease. Markers for a worse prognosis include expression of mutated tumor suppressor gene *p53*, enhancement of oncogene *cyclin D1*, and high levels of vascular endothelial growth factor.

The latter works to increase radioresistance and increase hematogenous metastases and leads to lower survival and disease-free rates.

Oral Cavity Squamous Cell Carcinoma

The risk factors for oral cavity squamous cell carcinoma include smoking, alcohol abuse, HPV exposure, and chewing tobacco. Chewing betel nuts has also been associated with an increased risk of squamous cell carcinoma. Surprisingly, the lower lip is the second most common site of squamous cell carcinoma in the head and neck, after the skin. Radiologists are not aware of this fact, because lip lesions usually are not imaged. Lower lip cancers are many times more common than upper lip tumors because of the sun exposure, gravitational effects and the reluctance to apply SPF #30 to one's lips.

The presentation of patients with oral cavity cancer may be delayed because people often assume that the lesions in the mouth are due to trauma from biting or chewing rather than from a neoplastic proliferation. Denial also plays a large role. Therefore it is not uncommon for a patient to have the minor complaint of "ill- fitting dentures" or a nonhealing ulcer of the oral cavity, only to find that the lesion is an aggressive, deeply infiltrative squamous cell carcinoma.

The T staging of oral cavity cancer is very similar to that of the oropharynx and is divided by size criteria from T1 to T3 (Box 13-7). A stage T4 tumor shows infiltration to cortical bone, deep muscles, maxillary sinus, skin (T4a) or masticator space, pterygoids, skull base, and carotid artery (T4b).

The critical questions that the radiologist must address with oral cavity cancer are as follows: (1) Is there mandibular or maxillary invasion (Fig. 13-34)? (2) Is there extension into the retromolar trigone and pterygomandibular raphe? (3) Is there extension into the oropharynx? (4) Is there spread into laryngeal soft tissues, necessitating a laryngectomy? (5) Is there extension across the midline of the tongue requiring total glossectomy versus a radiation-chemotherapy cocktail? (6) Is there perineural spread of the tumor? Life without a tongue is for the birds.

The maxilla is more commonly involved with oral cavity and specifically retromolar trigone cancers than with oropharyngeal cancers. Rarely, a soft palate cancer may affect the maxilla and tonsillar cancers may spread to the retromolar trigone and from there infiltrate the maxilla. Partial maxillectomies are relatively well tolerated by patients as long as appropriately tailored obturators are constructed, which separate the nasal cavity from the oral cavity and oropharynx. Otherwise, regurgitation of food products into the nasal cavity and/or phonation difficulties such as velopharyngeal insufficiency may arise from this common cavity. After the maxilla, the tumor may grow into the maxillary sinus or the pterygopalatine fossa.

Although bilaterality is important with all tongue lesions, it is the base of the tongue that is the critical component to the swallowing mechanism. Therefore you can be reasonably functional after oral tongue resections as long as you have most of your posterior tongue still wiggling around after the surgeon's knife filets you. Phonatory issues also are different between oral tongue and

---

BOX 13-7 **T Staging of the Lip and Oral Cavity**

TX—Primary tumor cannot be assessed

T0—No evidence of primary tumor

Tis—Carcinoma in situ

T1—Tumor 2 cm or less in greatest dimension

T2—Tumor >2 cm but not >4 cm in greatest dimension

T3—Tumor >4 cm in greatest dimension

T4 (lip)—Tumor invades through cortical bone, inferior alveolar nerve, floor of mouth, or skin of face, i.e., chin or nose

    T4a (oral cavity)—Tumor invades adjacent structures (e.g., through cortical bone, into deep [extrinsic] muscle of tongue [genioglossus, hyoglossus, palatoglossus, and styloglossus], maxillary sinus, skin of face)

    T4b—Tumor invades masticator space, pterygoid plates, or skull base and/or encases internal carotid artery

Note: Superficial erosion alone of bone/tooth socket by gingival primary is not sufficient to classify a tumor as T4.

From Edge SB: *AJCC Cancer Staging Manual.* 7th ed. New York: Springer; 2009. Used with permission.

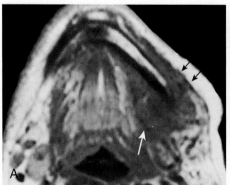 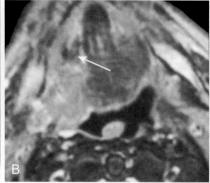 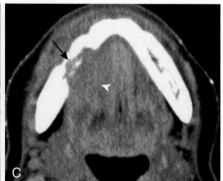

FIGURE 13-34 Oral cavity cancer. **A,** Axial T1-weighted imaging (T1WI) depicts a soft-tissue mass along the floor of the mouth *(arrow)* that has a component *(small black arrows)* superficial to the mandible. Note that the normal T1 hyperintense marrow signal is preserved although the lesion sweeps around its posterolateral edge and there is poor definition of the mandibular cortex. Corticectomy could cure confined cortical carcinoma. **B,** Cancers are well seen on T2WI against the backdrop of the intrinsic musculature of the tongue. In this case the hyoglossus muscle, an extrinsic muscle, is infiltrated *(arrow),* making the tumor T4a by the American Joint Committee on Cancer staging. **C,** This cancer *(arrowhead)* leads to erosion of the lingual surface of the mandible *(arrow),* also T4a.

oropharyngeal tongue resections. The tip and anterior portion of the tongue are more important with creating certain consonant sounds such as Ts, Ds, Gs, Js, and Zs. With total glossectomies you take out not only the tongue, but a lot more consonants and vowel sounds.

Nodal disease is less frequent with superficial oral cavity primary cancers than oropharyngeal ones. The exact numbers from different series vary widely, but roughly 30% of patients with oral cavity cancers have nodes at presentation whereas the percentage for oropharyngeal cancers runs approximately 65%. Nodal spread impacts significantly on patient outcome (reducing 5-year survival by 50%), emphasizing the importance of identifying pathologic nodes in all patients with cancer. Drainage of the anterior two thirds of the tongue goes to the submandibular lymph nodes and from there to the high internal jugular chain. Because reactive lymph nodes in the submandibular region may grow to more than 1 cm in size, submandibular nodes are not suggested to be neoplastic until they are more than 1.5 cm in diameter.

With oral cavity cancers, the issues of depth of skin invasion, pterygomandibular raphe invasion, maxilla invasion, and pterygopalatine fossa invasion (the latter secondary to retromolar trigone cancer) remain important. If the disease is limited or superficial, transoral resection with reconstruction by skin grafting, local flaps, or healing by secondary intention can be used. More extensive skin grafting may be required with oral cavity cancers that invade superficially than the oropharyngeal cancers, which tend to occur in the deeper tissues of the head and neck.

The two pulse sequences that are of the greatest use with MR of the tongue are (1) T2WI, because tumors are brighter than the very dark intrinsic musculature of the tongue, and (2) the fat-suppressed enhanced scans, which often demonstrate marked enhancement in tongue cancers. Mandibular marrow invasion can be diagnosed by hypointense soft-tissue infiltration across the dark cortical margin of the mandible into the normally bright signal intensity fat on T1WI. On CT, this diagnosis requires demonstration of direct bony erosion.

Floor-of-mouth cancers may cause obstruction of the submandibular duct. This causes an enlarged, edematous, painful, submandibular gland that may simulate inflammation caused by calculous disease and lead to delayed diagnosis. For the evaluation of a painful salivary gland, unenhanced CT is recommended to exclude calculi.

One must also be cognizant of the role of the nasopalatine nerves, greater and lesser palatine canals, inferior alveolar canal, and pterygopalatine fossa as avenues for the possible spread of cancers along nerves. Ultimately, the foramen rotundum and foramen ovale should be assessed with imaging to insure that intracranial extension of tumor along the cranial nerves has not occurred.

The combined metachronous (lesions that will develop) and synchronous (two lesions at the same time) rate with squamous cell carcinoma of the oral cavity is 40%, and therefore these patients are followed up closely for the rest of their lives with panendoscopy for the possibility of the second tumor. In addition, because of the uniform presentation of the carcinogens to the oral cavity, there is the possibility of "field cancerization," in which multiple sites of carcinoma in situ, severe dysplasia, and invasive carcinoma may be present within the oral cavity at presentation.

Hypopharyngeal Squamous Cell Carcinoma

The staging of hypopharyngeal cancer depends on the number of subsites that are invaded, the size of the lesion, as well as the presence or absence of fixation of the hemilarynx (Box 13-8). Once again, whip out that measuring stick because you must make distinctions between tumors: less than 2 cm, 2 to 4 cm, and over 4 cm in size. Size matters!

If the tumor invades adjacent structures such as the thyroid or cricoid cartilage, or extends out into the soft tissues of the neck, the lesion is considered a T4a cancer. Once you get used to the T4b classifications you will recognize that for most sites it means invasion of the prevertebral fascia, encasement of the carotid artery, or involvement of mediastinal structures.

The anatomy of the hypopharynx gets somewhat confusing because the anteromedial margin of the pyriform sinus is the lateral aspect of the aryepiglottic fold, which is considered a portion of the supraglottic larynx. The anterolateral wall of the pyriform sinus is the posterior wall of the paraglottic space more inferiorly. Because tumors in the pyriform sinus often obliterate the space between the lateral/pharyngeal mucosa of the aryepiglottic fold and the mucosa of the pyriform sinus, it is often difficult to determine whether a tumor is supraglottic (arising from the aryepiglottic fold), or hypopharyngeal (arising along the lateral aspect of the pyriform sinus). Endoscopists have a much better appreciation of this distinction because they are able to slide the endoscope medial to the tumor if it is a lateral pyriform sinus cancer (hypopharyngeal) or lateral to the tumor if the lesion is arising from the aryepiglottic fold (supraglottic). However, endoscopy is limited in the evaluation of very large tumors that obscure the pyriform sinus apex; this may be an area where either cross sectional imaging with reconstructions or barium studies may be of particular use.

Most hypopharyngeal cancers (60%) arise in the pyriform sinus with the remainder evenly split between postcricoid and posterior pharyngeal locations. The pyriform sinus cancers do not behave well—they metastasize early, invade aggressively into the soft tissue of the neck, and

---

BOX 13-8 T Staging of the Hypopharynx

T1—Tumor limited to one subsite of hypopharynx and ≤2 cm in greatest dimension

T2—Tumor invades more than one subsite of hypopharynx or an adjacent site, or measures >2 cm but not >4 cm in greatest diameter without fixation of hemilarynx

T3—Tumor >4 cm in greatest dimension or with fixation of hemilarynx

T4a—Tumor invades thyroid/cricoid cartilage, hyoid bone, thyroid gland, esophagus, or central compartment soft tissue*

T4b—Tumor invades prevertebral fascia, encases carotid artery, or involves mediastinal structures

Subsites: Postcricoid region, pyriform sinus, aryepiglottic fold (pharyngeal wall), posterior pharyngeal wall (to cricoarytenoid joint).
From Edge SB: *AJCC Cancer Staging Manual.* 7th ed. New York: Springer; 2009. Used with permission.
*Note: Central compartment soft tissue includes prelaryngeal strap muscles and subcutaneous fat.

present late (because this area is clinically silent; Fig. 13-35). There seems to be a particular affinity for pyriform sinus cancers to spread through the thyrohyoid membrane or cricothyroid membrane into the neck where they may encircle the carotid arteries. This is not good, as most deem cancerous carotid encasement a criterion for cancelling curative cutouts.

As in laryngeal carcinoma, one of the major issues regarding hypopharyngeal tumors is the invasion of cartilage. The superior aspect of the thyroid cartilage is particularly vulnerable to hypopharyngeal cancer. At this time most radiologists believe that MR has higher sensitivity but lower specificity than CT in the evaluation of early laryngeal cartilage invasion. MR may be able to detect the more subtle cartilaginous invasion before through-and-through disease is present. For this indication T1 enhanced fat-suppressed MR appears to be a particularly valuable pulse sequence. In the later stages infiltration of the strap muscles superficial to the cartilage is seen equally well on MR and CT.

Watch for prevertebral muscle invasion with posterior pharyngeal wall hypopharyngeal carcinomas. Findings include bright signal intensity on T2WI and enhancement of the longus musculature. Watch for Plummer-Vinson syndrome references (glossitis, anemia, and cervical esophagus or hypopharyngeal webs) in patients with postcricoid carcinomas.

Pyriform sinus cancers also have a high rate of metastasis to the adjacent lymphatics, and lymphadenopathy in some cases is reported to occur in 75% of patients at presentation. Jugular lymph nodes in levels II, III, and IV are the primary drainage sites. Necrosis in these nodes seems more common with a hypopharyngeal primary tumor.

The critical questions the radiologist must answer are: (1) Is there cartilaginous invasion? (2) Is there extension into deep tissues of the neck? (3) Is lymphadenopathy present? (4) Is there evidence of infiltration into the posterior paraspinal musculature? (5) Is there extension into the larynx or paraglottic space?

Most patients require a resection of supraglottic structures and pharyngectomy for pyriform sinus cancers. Occasionally, when the paraglottic space is infiltrated, a total laryngectomy and pharyngectomy is necessary.

Squamous Cell Carcinoma of the Larynx

An analysis of malignancies of the larynx requires discussion by anatomic subsites because of the different surgical issues relevant to each site; therefore, we will explore supraglottic, glottic, and subglottic cancers separately.

Supraglottic Squamous Cell Carcinoma

Staging of supraglottic carcinoma is based on subsites of the supraglottis, cord mobility, and deep invasion (Box 13-9).

Because laryngeal conservation therapy is a hot area in head and neck surgery, the indications for supraglottic laryngectomy as opposed to total laryngectomy must be reviewed. In a supraglottic laryngectomy, the epiglottis, false vocal cords, aryepiglottic folds, and preepiglottic fat are totally removed. If tumor extends to or below the laryngeal ventricle, a horizontal supraglottic laryngectomy, in which the surgical cut is through the plane of the ventricle

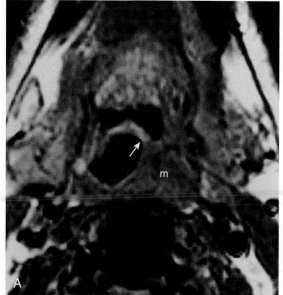

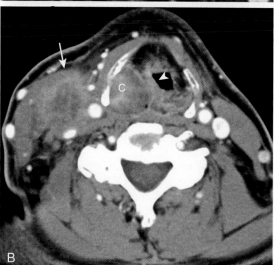

FIGURE 13-35 High hypopharyngeal lesion. **A,** This mass (m) is located at the top of the pyriform sinus and invades the lateral pharyngeal wall below the level of the epiglottis *(arrow).* Because the lesion is apposed to the lateral aspect of the epiglottis, it is difficult to determine whether this is a supraglottic lesion of the aryepiglottic fold or a lesion in the high pyriform sinus. Fortunately, the endoscopist can determine the lesion's origin because if the scope passes medially, it suggests a pyriform sinus origin; if the scope passes laterally, it suggests aryepiglottic fold origin. **B,** This hypopharyngeal cancer (C) has infiltrated the paraglottic soft tissues and pushes the aryepiglottic fold *(arrowhead)* medially. Note the lymph node *(arrow)* with extracapsular spread of tumor. Bad prognostic sign!

removing all laryngeal structures above but preserving thyroid cartilage cannot be performed (Fig. 13-36). Therefore, the presence of tumor at the upper margin of the true vocal cord is a critical branch in the surgical decision-making tree. In addition, if there is cartilaginous, postcricoid, or anterior commissure invasion, a supraglottic laryngectomy is not possible. The incidence of thyroid cartilage invasion is much higher than that of hyoid bone, arytenoid cartilage, or cricoid cartilage invasion, but thyroid cartilage extension usually occurs when the tumor is transglottic.

T1—Tumor limited to one subsite of supraglottis with normal vocal cord mobility

T2—Tumor invades mucosa of more than one adjacent subsite of supraglottis or glottis, or region outside the supraglottis (e.g., mucosa of base of tongue, vallecula, medial wall of pyriform sinus) without fixation of the larynx

T3—Tumor limited to larynx with vocal cord fixation and/or invades any of the following: postcricoid area, preepiglottic tissues, paraglottic space, and/or minor thyroid cartilage erosion (e.g., inner cortex)

T4a—Tumor invades through the thyroid cartilage and/or invades tissues beyond the larynx (e.g., trachea, soft tissues of neck including deep extrinsic muscle of the tongue, strap muscles, thyroid, or esophagus)

T4b—Tumor invades prevertebral space, encases carotid artery, or invades mediastinal structures

Subsites: False cords, arytenoids, suprahyoid epiglottis, infrahyoid epiglottis, aryepiglottic folds.
From Edge SB: *AJCC Cancer Staging Manual.* 7th ed. New York: Springer; 2009. Used with permission.

Other contraindications to supraglottic laryngectomy (be it supracricoid or horizontal) include involvement of both arytenoid cartilages, arytenoid fixation (which implies cricoid cartilage invasion to the "blades"), or extensive bilateral involvement of the base of tongue and/or preepiglottic fat (Fig. 13-37).

Understand that a supraglottic cancer may have extensive submucosal glottic and subglottic invasion but look totally normal by endoscopy. The route of spread is submucosally into the paraglottic space, where the normal fat is readily permeable for the invasion of tumor. Epiglottic carcinomas like to spread to the preepiglottic fat and from there can grow inferiorly to affect the petiole and anterior commissure of the glottis. Paraglottic spread is the mode du jour for aryepiglottic fold and false cord tumors. Invasion to the postcricoid region is usually limited to those affecting the posterior commissure or interarytenoid region. Because of these modes of deep spread, cancers can grow extensively invisible to endoscopy and masquerade as a relatively limited neoplasm. Coronal MR may be very helpful in deciding therapy by defining the extent of

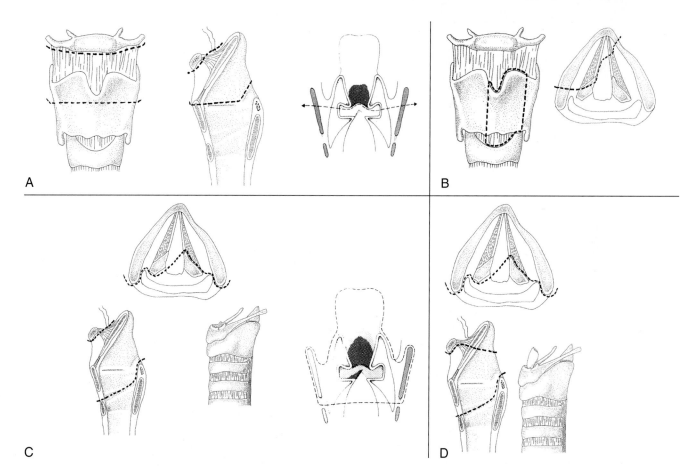

FIGURE 13-36 A, The horizontal supraglottic laryngectomy removes the epiglottis and the upper margin of the thyroid cartilage. The plane of section crosses just above the laryngeal ventricle. The true vocal cords are preserved and the voice is good. **B,** The vertical hemilaryngectomy (VHL) removes a significant portion of the thyroid cartilage unilaterally and may extend anteriorly across the midline. If the VHL does not cross the midline, it may be called a frontolateral VHL; if it captures portions of the contralateral vocal cord, with or without the arytenoids, it may be termed an anterolateral VHL. The epiglottis is spared. **C,** The supracricoid laryngectomy with cricohyoidopexy removes all of the thyroid cartilage and epiglottis. One of the arytenoids is usually removed ipsilateral to the tumor. **D,** If a portion of the epiglottis is spared, the reconstruction is called a cricohyoidoepiglottopexy. The hyoid bone is spared with all voice conservation procedures. (From Maroldi R, Battaglis G, Nicolai P, et al. The appearance of the larynx after conservative and radical surgery for carcinomas. *Eur Radiol* 1997;7:418-431.)

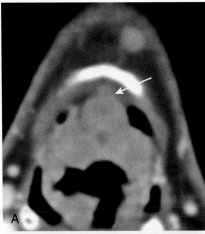

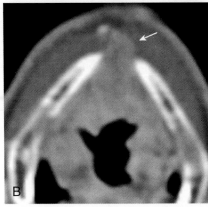

FIGURE 13-37 Supraglottic carcinoma considerations. **A,** Note the preepiglottic fat infiltration *(arrow)* from this epiglottis carcinoma. **B,** Cartilaginous invasion through the anterior commissure and into adjacent soft tissues from this epiglottis tumor also renders it unacceptable for horizontal supraglottic laryngectomy.

the tumor, because the tumor is well outlined by paraglottic fat at the false cord and ventricle levels. In contrast, the soft tissue lateral in the paraglottic space at the true cord level is the thyroarytenoid muscle and is readily separable from the fat above. The true cord is below the laryngeal ventricle.

Of all the laryngeal cancers, supraglottic and transglottic squamous cell carcinomas have the highest frequency of nodal metastases at presentation. This is because of the abundant lymphatics associated with the supraglottis as opposed to the relatively sparse lymphatics of the glottis.

As noted previously, the distinction between supraglottic and hypopharyngeal lesions gets blurred when the tumor extends from the aryepiglottic fold to the pyriform sinus and/or the posterior pharyngeal wall. Often it is difficult to determine the site of origin of a cancer that has spread in this fashion, but in either case an extended resection (partial laryngectomy), taking the pyriform sinus, may be required (Table 13-6).

The debate on whether CT or MR is the better method for evaluating the larynx for cartilaginous invasion has not been settled. The irregular ossification-calcification of the thyroid cartilage is particularly troublesome, and you will note that the T4 staging for supraglottic and glottic laryngeal carcinoma refers to thyroid cartilage invasion. Thus, the elastic cartilage of the epiglottis and the hyaline cartilage of the arytenoids are not considered in TNM staging,

TABLE 13-6 Supraglottic Surgical Procedures

Lesion	Minimal Supraglottic Procedure
Small suprahyoid epiglottic lesion	Laser epiglottectomy
Infrahyoid epiglottic lesion (into preepiglottic fat)	Supraglottic laryngectomy
Unilateral arytenoid involvement	Extended supraglottic laryngectomy, taking arytenoid, or supracricoid laryngectomy
Tongue base involvement (limited)	Extended supraglottic laryngectomy, taking unilateral tongue base
Aryepiglottic fold involvement	Extended supraglottic laryngectomy, taking unilateral piriform sinus
True cord involvement or close to true cords	Supracricoid laryngectomy with cricohyoidopexy
Piriform sinus involvement, unilateral	Supracricoid hemilaryngopharyngectomy
Extension deep into subglottis, T4 lesions, or extensive piriform sinus involvement	Total laryngectomy

although they often are invaded with cancer. Cancer is said to preferentially invade ossified cartilage, infiltrating the bone marrow with ease compared with nonossified cartilage.

On CT, the absolute findings diagnostic of thyroid cartilage invasion are through-and-through erosion with extension of the mass into the strap muscles. Sclerosis of the cartilage may be a harbinger of cartilaginous invasion; in a recent study only 50% of patients with this finding had invasion, with the remainder showing only perichondrial involvement (Fig. 13-38). A cautionary note however: areas of arytenoid sclerosis occur in 16% of normal subjects especially in the body of the arytenoid and more commonly in women. Other CT findings include lysis or destruction of cartilage or obliteration of the marrow space of ossified cartilage. Unfortunately CT tends to be relatively insensitive to cartilage invasion.

On MR, involvement of the strap muscles is a sure sign of thyroid cartilage invasion, but enhancement of tissue in the fat suppressed walls of the cartilage and brightening of intracartilaginous tissue on fat suppressed fast spin echo T2WI suggest early invasion (Fig. 13-39). Using these criteria, one finds that MR is very sensitive (85%) to cartilage invasion. Unfortunately, this is at a cost of decreased specificity as we now know that there is a ton of peritumoral inflammation around that can simulate neoplasm and can brighten signal on T2WI and cause enhancement. The patient and clinician end up in the radiologic nightmare; order a CT and you may underestimate disease, get caught with residual tumor, and have a worse prognosis, but save a voice box. Choose MR and you may remove a "clean" larynx, render the patient "speechless," but have a better chance of tumor free margins.

Glottic Squamous Cell Carcinoma

The staging of squamous cell carcinoma of the glottis (Fig. 13-40) is also based on cord mobility and fixation as

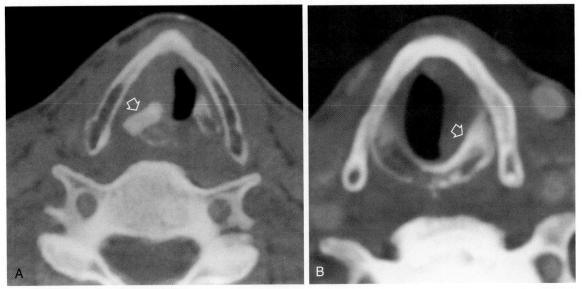

FIGURE 13-38 Cartilage sclerosis. **A,** This glottic carcinoma has led to arytenoid sclerosis *(arrow)* on the right side. Although that is pretty obvious, the density change in the anterior half of the thyroid cartilage may be entirely normal. **B,** Do you buy the left cricoid sclerosis *(arrow)* in this patient with left vocal cord carcinoma?

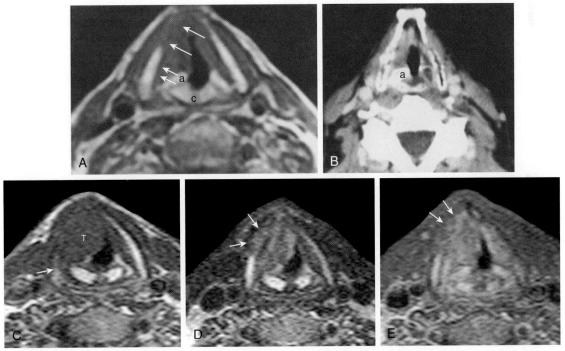

FIGURE 13-39 Cartilaginous invasion by laryngeal carcinoma. **A,** Axial T1-weighted imaging (T1WI) reveals invasion of the anterior aspect of the right thyroid cartilage *(large arrows)* by a glottic cancer. Note that the high signal intensity of the cartilaginous marrow and the low signal intensity of the edge of ossified cartilage *(small arrows)* are preserved posteriorly. True cord level is well demonstrated at the crico(c)-arytenoid (a) joint. **B,** Axial computed tomographic image in a different patient shows sclerosis of the arytenoid cartilage, which is worrisome for invasion. Invasion of the thyroid cartilage is really hard to know here. **C,** T1WI in yet another patient shows a laryngeal tumor (T) invading the right thyroid cartilage with replacement of the right thyroid cartilage with low signal *(arrow)*. Compare this signal to the normal left side. **D,** Axial T2W1 and **(E)** postcontrast fat-suppressed T1WI show tumor extension through the thyroid cartilage into the strap musculature *(arrows)*.

well as extension into adjacent soft tissue and/or cartilage (Box 13-10).

Because patients with squamous cell carcinoma of the glottis are seen early because of voice changes, the prognosis associated with this lesion is better than that elsewhere in the larynx. The cancer is usually detected at an earlier stage. If you are hoarse for a while, you usually seek out medical attention sooner rather than later.

Just as the principle issue in treatment of supraglottic laryngeal carcinoma is its superior-inferior extension, the critical features in analyzing primary glottic carcinoma are its transverse and anterior-posterior dimensions. One of the potential surgical treatments for primary glottic carcinoma is the vertical hemilaryngectomy (unilateral removal of the cord and supraglottic structures, sparing the contralateral side). This surgery leaves the patient with a working voice and is the preferred modality of treatment in the appropriate patient population. However, consideration of this therapy is contingent on the tumor not extending contralaterally beyond the anterior third of the opposite vocal cord (Table 13-7). Posterior extension into the arytenoid cartilages or the cricoid cartilage is also a contraindication to the vertical hemilaryngectomy. If the tumor extends superiorly or inferiorly into the supraglottic or subglottic region, the possibility of a vertical hemilaryngectomy is lowered.

Glottic carcinomas may spread from the anterior commissure, via the neurovascular perforations through the thyrohyoid membrane to the soft tissues of the neck, to the paraglottic space, or to the subglottic compartment. This is difficult to visualize at office endoscopy and sometimes the presence of subglottic disease is initially borne out by imaging studies showing thickening in this location.

Supracricoid laryngectomies remove the supraglottic structures and the entire thyroid cartilage. This surgery requires at least one freely mobile arytenoid with no interarytenoid tumor and a clean cricoid cartilage for providing this option to the patient. Pharyngeal involvement and subglottic spread also are contraindications in most settings. Effectively, what happens is that the arytenoid apposes to the tongue base to allow speech and sphincteric function.

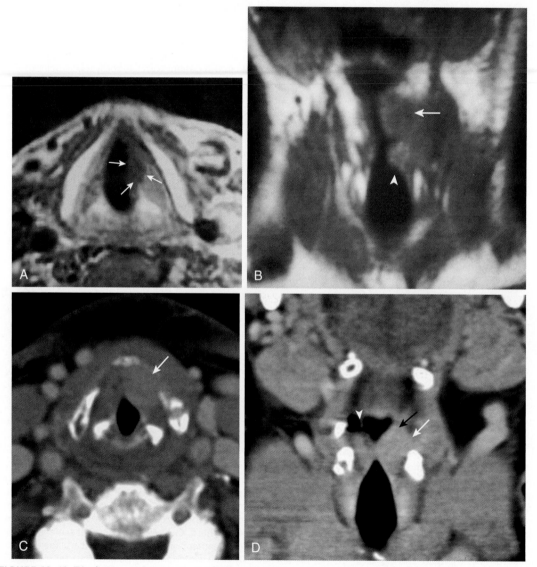

FIGURE 13-40 T1 glottic carcinoma. **A,** Axial T1-weighted imaging delineates a small focal mass *(arrows)* in the left true vocal cord. Mass did not impair cord mobility and was localized to the left side, thereby classifying it as a T1a lesion. **B,** This case endoscopically appeared to be a small true vocal cord lesion *(arrowhead).* However, deep to the endoscopist's view, there was extensive paraglottic spread upward *(arrow)* rendering a T3 designation. **C,** Axial computed tomographic image shows a left glottic cancer that has invaded the paraglottic soft tissues *(arrow).* It abuts on the nonossified thyroid cartilage. **D,** Coronal reconstruction shows that, in the paraglottic tissues the left vocal cord cancer *(white arrow)* has ascended above the laryngeal ventricle level *(arrowhead)* and therefore this part of the cancer *(black arrow)* is considered "transglottic."

This requires a "pexy" procedure in which the cricoid is suspended from the hyoid bone.

As with all laryngeal carcinomas, the presence of cartilaginous invasion or extension into the adjacent soft tissues is very important. As stated previously, the risk of lymphatic invasion is much lower with glottic carcinoma. Only when the tumor extends to the supraglottis or subglottis is expectant treatment of the lymph nodes required.

T1 and T2 glottic carcinomas may be treated with radiation therapy alone in some instances. The cure rates for T1 glottic carcinoma with either surgery or radiation are approximately 90%. However, the quality of the voice is generally better after focal localized radiation therapy. The radiation therapy usually does not include the lymph node chains because of the sparse lymphatics involved.

Subglottic Squamous Cell Carcinoma

Subglottic squamous cell carcinomas constitute less than 5% of all laryngeal carcinomas. These lesions are very difficult to identify at indirect or fiberoptic endoscopy because visualization of the undersurface of the vocal cord or the proximal trachea may be obscured by the shadow of apposed true cords. The best views are obtained under supported direct laryngoscopy in the operating room.

The staging of subglottic laryngeal carcinoma is similar to that in the glottis and supraglottis and deals primarily with cord mobility, cartilaginous destruction, or extension to the soft tissues of the neck (Box 13-11). The prognosis for subglottic lesions is generally poor because by the time the diagnosis is made, the lesion is generally fairly extensive. It may infiltrate the trachea, esophagus, and even the thyroid gland. Remember to look at the paratracheal, esophageal, anterior visceral (Delphian level VI), and upper mediastinal nodal chains when you find a subglottic cancer. The mucosa of the subglottis is pencil thin. Any nodularity or thickening must be suspected of being cancerous on any imaging modality (Fig. 13-41).

The major issue when examining a patient for subglottic carcinoma is whether there is paraglottic, glottic and/or supraglottic extension at the time of diagnosis. Because surgical resection of subglottic carcinomas nearly always requires cricoid cartilage resection, total laryngectomy is usually performed. Take it from the noted head and neck

---

BOX 13-10 T Staging of Glottic Cancer

T1—Tumor limited to the vocal cord(s) (may involve anterior or posterior commissure) with normal mobility
    T1a—Tumor limited to one vocal cord
    T1b—Tumor involves both vocal cords
T2—Tumor extends to supraglottis and/or subglottis, and/or with impaired vocal cord mobility
T3—Tumor limited to the larynx with vocal cord fixation and/or invades paraglottic space, and or minor thyroid cartilage erosion (e.g., inner cortex)
T4a—Tumor invades through the thyroid cartilage and/or invades tissue beyond the larynx (e.g., trachea, soft tissues of neck including deep extrinsic muscle of the tongue, strap muscles, thyroid, or esophagus)
T4b—Tumor invades prevertebral space, encases carotid artery, or invades mediastinal structures

Subsites: True cords including anterior commissure, posterior commissure.
From Edge SB: *AJCC Cancer Staging Manual.* 7th ed. New York: Springer; 2009. Used with permission.

BOX 13-11 T Staging of Subglottic Cancer

T1—Tumor limited to the subglottis
T2—Tumor extends vocal cord(s) with normal or impaired mobility
T3—Tumor limited to larynx with vocal cord fixation
T4a—Tumor invades cricoid or thyroid cartilage and/or invades tissues beyond the larynx (e.g., trachea, soft tissues of neck including deep extrinsic muscles of the tongue, strap muscles, thyroid, or esophagus)
T4b—Tumor invades prevertebral space, encases carotid artery, or invades mediastinal structures

From Edge SB: *AJCC Cancer Staging Manual.* 7th ed. New York: Springer; 2009. Used with permission.

TABLE 13-7 Glottic Surgical Procedures

Lesion	Minimal Glottic Procedure
Very small midcord lesion	Excisional biopsy
Moderate-sized midcord lesion	Laser cordectomy
Large midcord lesion	Open cordectomy with false cord imbrication
T1a, whole unilateral cord	Vertical hemilaryngectomy
T1b, bilateral cord	Supracricoid laryngectomy with cricohyoidoepiglottopexy
T3, unilateral subglottic extension, piriform sinus involvement	Near-total laryngectomy
T4	Total laryngectomy

T2, T3, and selected T4 (e.g., involving but not through thyroid cartilage) glottic cancer can be resected with supracricoid laryngectomy with cricohyoidoepiglottopexy or cricohyoidopexy.

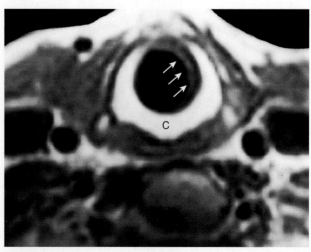

FIGURE 13-41 Subglottic carcinoma. Axial T1-weighted imaging demonstrates thickening of the mucosa *(arrows)* along the left subglottic region. At this level one should not see any thickness to the mucosa of the subglottic larynx. Note the high signal of the cricoid cartilage (c).

surgeon, Johnnie Cochran: "If over 1.0 cm of subglottic extension is in doubt, the larynx must come out."

Below the inferior surface of the cricoid cartilage the tumors are referred to as tracheal lesions rather than subglottic laryngeal lesions. Tracheal carcinomas are much less common than laryngeal carcinomas and enter the realm of pulmonary radiology. We will stop here.

To detect recurrences of squamous cell carcinoma, nuclear medicine positron emission tomography (PET) studies and a prayer may be useful. Pray that the clinicians actually get the baseline CT or MR study you recommend at 8 weeks after all therapies (Fig. 13-42). Because that is not likely, rely on fluorodeoxyglucose (FDG) PET scans for differentiating recurrent squamous cell carcinoma from posttreatment changes if the CT findings are not clear cut (Fig. 13-43). As expected, growing tumors are hot. The limitations of these studies are predominantly attributable to lower resolution, normal variations in activity, variable uptake shortly after therapy, and the nuclear medicine doctors not having read the anatomy portion of this chapter well enough to challenge "true" neuroradiologists. Combined PET-CT scanners are ideal for detecting primary tumor or nodal recurrences. PET-CT studies may also predict response to therapy. Be careful of scans within 6 weeks of treatment; false-positive studies abound.

Chondroradionecrosis

One of the most catastrophic of iatrogenic mishaps to affect the larynx is chondroradionecrosis. This may occur in some unusual individuals after standard doses of radiotherapy for laryngeal neoplasms, but is, thankfully, rare to see. The hallmarks of chondroradionecrosis are soft-tissue swelling of the larynx, sloughing of the arytenoid cartilage with subluxation, fragmentation, sclerosis, and collapse of the thyroid cartilage, and/or the presence of gas bubbles around the cartilage (Fig. 13-44). Super infection often occurs. Invariably, this entity requires a total laryngectomy, which often was the surgical extreme that the therapists were hoping to avoid by going with radiation therapy for the cancer.

Carcinosarcoma

A neoplasm that has characteristics of both squamous cell carcinoma and sarcoma is known as the spindle cell sarcoma or carcinosarcoma. This lesion is rarely seen elsewhere in the aerodigestive tract besides the larynx. Its prognosis is similar to that of squamous cell carcinoma in the larynx, and therefore the histologic distinction does not generally alter therapy or prognostic consideration. Radiation therapy in carcinosarcoma produces similar control rates to irradiated patients with similar volume disease with the more typical squamous cell carcinoma.

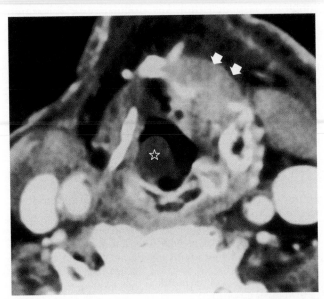

FIGURE 13-42 Recurrent carcinoma after radiation. Note the soft-tissue mass *(arrows)* along the left thyroid cartilage in this patient with recurrent squamous cell carcinoma. Incidentally, an unusual mucous retention cyst *(star)* protrudes into the lumen of the larynx.

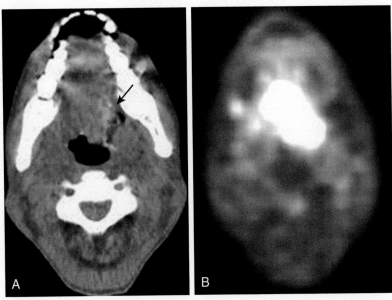

FIGURE 13-43 Recurrent cancer on positron emission tomography (PET). **A,** After chemoradiation there was still a necrotic mass in the oral tongue and tongue base *(arrow)* favoring the left side. **B,** The PET study showed diffuse infiltration of the tongue base evidenced by avid fluorodeoxyglucose uptake.

Other Mucosal Based Malignancies

Lymphoma

Squamous cell carcinoma does not have a monopoly on bad news in the nasopharynx. Because of the abundant lymphoid tissue associated with the adenoids of the nasopharynx, lymphoma also occurs here. It is usually in the form of non-Hodgkin lymphoma and affects young adults. Its appearance is identical to that of carcinoma except that necrosis is uncommon in lymphoma. Otherwise density and intensity characteristics overlap. Remember that B-cell lymphomas are part of the spectrum of PTLD to affect the adenoidal region and can cause a necrotic mass. Nodes are present in 80% of patients with "extranodal" lymphoma. Lymphoma in the nasopharynx has the opportunity to spread to the skull base and cranial nerve foramina.

Because of the presence of faucial and lingual tonsillar tissue along the tongue base, the incidence of lymphomatous involvement of the oropharynx is significant. Once again these lesions may appear as exophytic masses and can be challenging to distinguish from tonsillar lymphoid hypertrophy. Both lymphoma and lymphoid hyperplasia may be relatively high in signal intensity on T2WI and may show enhancement, thereby making their differentiation very difficult (Fig. 13-45). The presence of associated lymphadenopathy may not even be helpful. Biopsy may be necessary. This quandary is an issue in AIDS patients with suspected lymphoma.

Hodgkin disease rarely affects Waldeyer ring and afflicts adults of younger age than non-Hodgkin lymphoma.

Minor Salivary Gland Malignancies

The extensive minor salivary gland tissue in the nasopharynx accounts for neoplasms such as adenoid cystic carcinoma (the most common minor salivary gland malignancy), mucoepidermoid carcinoma, acinic cell carcinoma, adenocarcinoma, pleomorphic low-grade adenocarcinoma, and undifferentiated carcinoma. Adenoid cystic carcinomas may be particularly difficult to treat because of their tendency to spread perineurally (seen in 60% of individuals). The skull base foramina are easy targets for nasopharyngeal adenoid cystic carcinomas, and from there the tumor marches up the nerves to the intracranial compartment. Perineural progression portends a poor prognosis and potentially produces partial paresis plus paresthesia.

Nearly half of minor salivary gland tumors of the oropharynx are pleomorphic adenomas with the other half being malignant. Minor salivary gland malignancies of the oropharynx are usually due to adenoid cystic carcinoma. The most common site of occurrence is the soft palate.

It is nearly impossible to distinguish among the major histologic types of noncarcinomatous tumors that involve the nasopharynx. Age is probably the best discriminator, with lymphomas occurring more commonly in the young adult. Lymphoma frequently has a homogeneous bland appearance even when the tumor is large. Its signal intensity may be variable but usually is indistinguishable from that of nonnecrotic carcinoma. Lymphoma often grows in a more exophytic pattern, with less deep infiltration than carcinoma. Minor salivary gland tumors have a variable pattern of signal intensity on MR. Occasionally, one may identify a very bright minor salivary gland tumor on T2WI, which usually represents either adenoid cystic carcinoma or low-grade mucoepidermoid carcinoma. Perineural spread will suggest the former.

Plasmacytomas

Plasmacytomas can occur in the nasopharynx, palatine tonsils, and base of tongue. They are usually oval in shape and have similar intensity to muscle on T1WI and are only slightly hyperintense on T2WI. They enhance notably, often with a heterogeneous center.

Sarcomas

In children the predominant tumor of the nasopharynx is rhabdomyosarcoma, an eminently treatable tumor with chemotherapy (Fig. 13-46). Intracranial extension is common with rhabdomyosarcomas, and rhinorrhea may be a presenting symptom. Of the tumors of the nasopharynx that may be hemorrhagic, rhabdomyosarcoma is the most common. Rhabdomyosarcoma rapidly races retropharyngeally requiring a referral for radiotherapy. The survival rate with combined chemotherapy and radiation is more than 50% at 5 years. Rhabdomyosarcomas may also occur within the tongue. Rhabdos run rampant in rug rats requiring radiation for remissions. Also in the sarcomatous category, hemangiopericytomas and synovial sarcomas may occur in the oropharynx.

The malignancies other than squamous cell carcinoma that affect the vocal cords include the chondroid tumors. The cricoid cartilage is associated with chondromas and chondrosarcomas. These lesions are identified on CT by their "popcorn-like" whorls of calcification caused by the chondroid

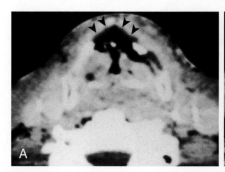
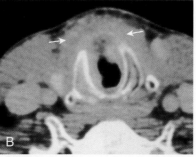

FIGURE 13-44 Chondronecrosis of the larynx. **A,** Loss of the anterior portions of the thyroid cartilage bilaterally is coupled with extralaryngeal air in the preglottic and paraglottic soft tissues *(arrowheads)*. Fragmentation is typical. Note the absence of fat planes characteristic of the irradiated neck. **B,** At the cricoid level the necrotic tissue *(arrows)* is seen projecting anteriorly into the strap muscles. (**A** compliments of Hugh Curtin, MD, Boston, MA.)

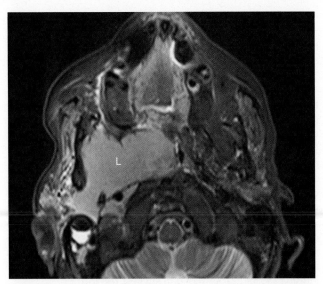

FIGURE 13-45 Oropharyngeal lymphoma. Note how homogeneous this oropharyngeal lymphoma (L) is affecting the palatine tonsil and spilling into the parapharyngeal, masticator, carotid, and retropharyngeal spaces.

matrix of the tumor (Fig. 13-47). Distinguishing a benign from a malignant chondroid tumor is sometimes difficult; in the larynx the chondrosarcomas are usually low-grade and slow growers. They are often treated with piecemeal resection. Chondrosarcomas account for less than 1% of laryngeal malignancies, are seen in the 40 to 60 age range and predilect males by 5 to 10 to 1. Over 70% arise in the cricoid cartilage.

Another rare malignancy seen in the pharynx is the synovial sarcomas. A well-defined yet heterogeneous mass with septations, hemorrhage, cysts, calcification, or multilocularity should raise suspicion of a synovial sarcoma. These lesions *do not* arise from synovial structures. This is a misnomer. Saying "synovial sarcoma" should not suggest a synovial source, simply cell shape.

MALIGNANT ADENOPATHY

Although the lymph nodes are not strictly a part of the mucosal system of the head and neck, it seems more relevant to discuss the nodes in the chapter dealing with squamous cell carcinoma. The most common cause of nodal disease in an adult is a squamous cell carcinoma metastasis from a mucosal primary tumor.

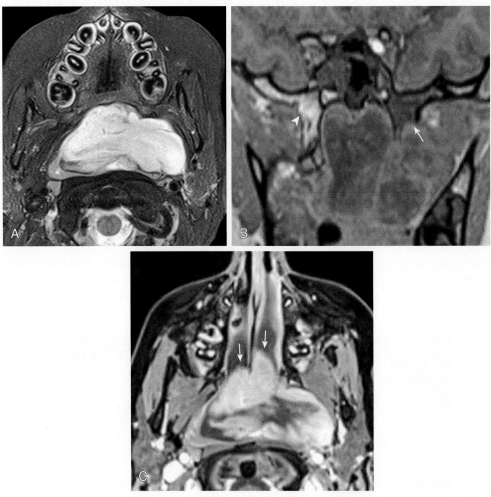

FIGURE 13-46 Rhabdomyosarcoma of the nasopharynx. **A,** Axial T2-weighted image (T2WI) shows a bulky high signal pharyngeal mass obliterating the airway. **B,** Coronal T1WI shows loss of the normal fatty signal and loss of cortical definition indicating invasion of the pterygoid plates *(arrow)*. Compare to normal right side *(arrowhead)*. **C,** The tumor enhances heterogeneously on postcontrast T1WI. Extension into the posterior nasal cavity bilaterally is seen *(arrows)*.

The nomenclature of the lymph nodes of the neck has undergone a recent change, which is supported by various societies of head and neck surgeons. This nomenclature identifies the lymph nodes by a grading system from I to VII (Fig. 13-48). Lifelong labor in learning levels of lymph nodes leads to learned lymphatic luminaries. Level I lymph nodes include the submental (Ia) and submandibular (Ib) lymph node chains. Level II lymph nodes include the internal jugular lymph node chain above the hyoid bone that are anterior or immediately adjacent to the jugular vein (IIa) or posterior to the jugular vein deep to the sternocleidomastoid muscle (IIb). The IIa nodes include the jugulodigastric node, a node that sits along the posterior belly of the digastric muscle at the upper pharyngeal level. Level III lymph nodes involve the jugular chain between the hyoid and cricoid cartilage, and level IV lymph nodes involve the jugular chain below the cricoid cartilage. Level V is designated as all the lymph nodes of the posterior triangle of the neck (deep and posterior to the sternocleidomastoid muscle and above the clavicle). If above the cricoid they are designated Va, if below Vb. Level VI lymph nodes are those which previously were identified in the anterior jugular and visceral chain in front of the thyroid gland. Finally, level VII nodes are in the superior mediastinum region.

The N staging of lymph nodes in the TNM classification is noted in Box 13-12 (aerodigestive sites) and Box 13-13 (for the nasopharynx). M staging is M0 for no distant metastases and M1 for distant metastases.

The critical numbers to remember are less than or equal to 3 cm for N1 nodes, 3 to 6 cm for N2 lymph nodes, and greater than 6 cm for the N3 classification. Nodal infiltration from carcinoma of the nasopharynx is present in 80% of

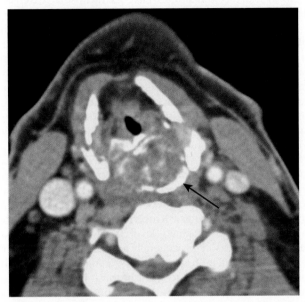

FIGURE 13-47 Chondrosarcoma of the larynx. Note the cricoid origin *(arrow)*, popcorn-like appearance with whorly matrix that characterize this tumor.

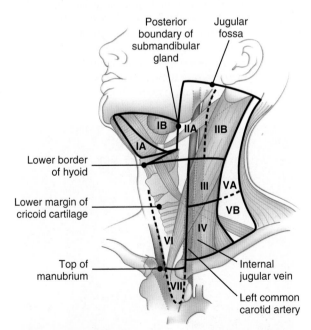

FIGURE 13-48 Nodal Classification. The revised nodal classification scheme includes an A and B designation for nodes in groups I, II and V. IA refers to submental and IB submandibular nodes. In II the separation depends on the relationship to the carotid sheath/jugular vein with IIA anterior or along and IIB behind, but both above the hyoid bone as opposed to level III nodes between hyoid and cricoid cartilage. With the posterior triangle classification V, the upper nodes above the cricoid cartilage are designated VA and the lower ones are VB.

BOX 13-12 N (Nodal) Staging Criteria for Oral Cavity, Oropharynx, Hypopharynx, Paranasal Sinuses, and Larynx

NX—Regional lymph nodes cannot be assessed
N0—No regional lymph node metastasis
N1—Metastasis in single ipsilateral lymph node, ≤3 cm in greatest dimension
N2—Metastasis in single ipsilateral lymph node, >3 cm but ≤6 cm in greatest dimension, or multiple ipsilateral lymph nodes, none >6 cm in greatest dimension, or bilateral or contralateral lymph nodes, non >6 cm in greatest dimension
   N2a—Metastasis in single ipsilateral lymph node >3 cm but ≤6 cm in greatest dimension
   N2b—Metastasis in multiple ipsilateral lymph nodes, none >6 cm in greatest dimension
   N2c—Metastasis in bilateral or contralateral lymph nodes, none >6 cm in greatest dimension
N3—Metastasis in lymph node >6 cm in greatest dimension

From Edge SB: *AJCC Cancer Staging Manual.* 7th ed. New York: Springer; 2009. Used with permission.

BOX 13-13 Regional Lymph Nodes (N): Nasopharynx

NX—Regional lymph nodes cannot be assessed
N0—No regional lymph node metastasis
N1—Unilateral metastasis in cervical lymph node(s), <6 cm in greatest dimension, above the supraclavicular fossa, and/or unilateral or bilateral retropharyngeal lymph nodes <6 cm in dimension
N2—Bilateral metastasis in lymph node(s), <6 cm in greatest dimension, above the supraclavicular fossa*
N3—Metastasis in a lymph node(s) and/or to supraclavicular fossa
   N3a—>6 cm in dimension
   N3b—Extension to the supraclavicular fossa

From Edge SB: *AJCC Cancer Staging Manual.* 7th ed. New York: Springer; 2009. Used with permission.
*Note: Midline nodes are considered ipsilateral nodes.

patients and is bilateral in 50%. The first echelon of drainage is to lateral retropharyngeal nodes. From there, the high jugular (level II, jugulodigastric) and posterior triangle (level V) chains are next to be infiltrated. The nodal staging for the nasopharynx is "special"; N0 no nodes, N1 unilateral supraclavicular node 6 cm or less, N2 bilateral supraclavicular nodes 6 cm or less, and N3 nodes greater than 6 cm or into the supraclavicular region. Issues other than pure size criteria include the presence (or absence) of central nodal necrosis, which suggests tumor no matter what the size.

BOX 13-14 Features of Nodes that Lead to 50% Drop in Prognosis

1. Any adenopathy
2. Bilateral adenopathy
3. Extracapsular spread of tumor
4. Fixation to vital structures (e.g. carotid artery, prevertebral musculature)

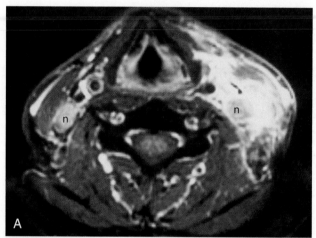

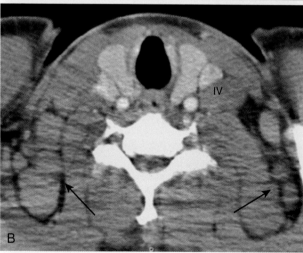

FIGURE 13-49 Extranodal spread of tumor. **A,** Fat-suppressed, enhanced T1-weighted imaging readily depicts the spread of neoplasm from the lymph node (n) into the soft tissues of the neck. The presence of extracapsular spread portends a worse prognosis and problems obtaining "clean margins" at surgery. These nodes are level III nodes because they are below the hyoid bone. **B,** This patient has nodes below the cricoid cartilage classified as level IV and level VB *(arrows).* Because they are bilateral, score this as N2c disease.

When the lymph nodes are being evaluated, it is important to analyze whether extracapsular extension (spread outside the confines of the lymph node to the adjacent soft tissue) is present because it reduces the 5-year prognosis by 50% and often leads to local treatment failure (Box 13-14). The incidence of extracapsular spread of tumor correlates with the nodal size. Histopathologically, 23% of nodes less than 1 cm, 53% of nodes 2 to 3 cm, and 74% of nodes greater than 3 cm in transverse diameter have extracapsular extension (Fig. 13-49). Measure those nodes!

The clinical evaluation of the lymph nodes relies on the gift of palpation. The numbers argue that even in the best of hands, over 20% of patients with N0 necks clinically have occult nodal metastases. Our mission is to improve on that number. Most patients with T2 cancers undergo a neck dissection even if they are N0. Of the positive specimens from neck dissections of N0 patients (around 30% of cases), 25% reveal metastases less than 3 mm in size. Can we honestly expect to find these lesions?

Some lymph nodes that are located deep in the head and neck are not clinically palpable. Retropharyngeal lymph nodes, which commonly accompany nasopharyngeal carcinoma, are impossible to detect clinically (Fig. 13-50). Sometimes, deep parapharyngeal lymph nodes and/or lymph nodes in the jugular digastric chain (level 2) may not be readily palpable. For this reason it is generally accepted that 15% to 30% of all malignant lymph nodes escape clinical detection. The role of the radiologist is to identify these lymph nodes and the very small lymph nodes that demonstrate central nodal necrosis that may be dismissed by the clinician (and unsophisticated radiologist) as insignificant.

Most head and neck radiologists agree that CT is the easiest method for identifying the lymph nodes in the neck, provided that a large bolus of contrast has been delivered to the blood vessels to distinguish them from lymph nodes. In addition, there is some evidence that central nodal necrosis may be more readily identified by CT than by MR.

There has been considerable debate about the appropriate size criteria to identify lymph nodes as enlarged and therefore

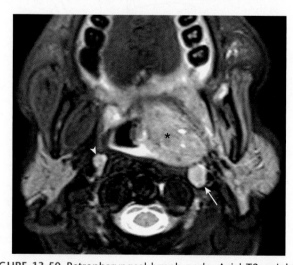

FIGURE 13-50 Retropharyngeal lymph node. Axial T2-weighted image shows a bulky pharyngeal mass (yup, you guessed it, squamous cell carcinoma), clearly visible on endoscopy. What the clinician can't see is the asymmetrically enlarged left retropharyngeal lymph node *(arrow)* indicating intranodal metastasis. Compare to more normal size on the right *(arrowhead).*

infiltrated with neoplasm. Again and again size does matter in the head and neck. Masters of the mucosa must measure masses. But, for a change, smaller is better. The debate centers on millimeters, some favoring 10 mm, some 12 mm, and some 15 mm as the maximum normal transverse diameter of a reactive node. Most would agree that at greater than 15 mm (longest measurement in the axial plane) the node should be considered suspicious until proved otherwise. Retropharyngeal nodes should not exceed 8 mm in diameter (see Fig. 13-50). More leeway is given to the highest jugular node (jugulodigastric node) and submandibular nodes, but read on, faithful student, and you will see that despite these guidelines, radiology "ain't" cutting the mustard in predicting neoplastic infiltration of nodes.

A radiology diagnostic oncology group (RDOG) study looked at how well we in imaging do at predicting neoplastic invasion in nodes. The answer is that we are abysmal. In this large multiinstitutional study of thousands of nodes they found that using the 10 mm cut off, CT was only 39% specific and 88% sensitive for detecting disease and MR did not fare much better (48% and 81% respectively). With use of a 10-mm size or central necrosis to indicate a positive node, CT had a negative predictive value (NPV) of 84% and a positive predictive value (PPV) of 50%, and MR imaging had an NPV of 79% and a PPV of 52% (Table 13-8).

RDOG found that one had to use a 5-mm cutoff to achieve a 10% error rate of a negative study (the negative predictive value), but that left a false-positive rate of 56%. Why are we even doing these studies then? Good question! In many cases, whether we say there are nodes or not, the surgeons go ahead and treat the necks. They want to pick up the 20% to 30% of cases with microscopic disease. They know that some primary tumors, such as oropharyngeal, hypopharyngeal, and supraglottic cancers have such a high rate of nodal involvement (in the 60% range) that they are simply going to treat regardless of what they feel, you see, and Aunt Minnie wants. Furthermore, they tend to take out the nodes because it affords them better access to the primary tumor with that fatty nodal stuff removed. It is in the way! Get it out! They realize that the incidence of malignant cervical lymphadenopathy with squamous cell carcinoma varies with the primary tumor from site to site (Table 13-9).

The sonographers, led by our European and Asian counterparts are strong advocates for the use of ultrasound (US) in the evaluation of lymph nodes. Power Doppler US will show round nodes with no hilar flow and peripheral parenchymal nodal flow in most metastatic nodes. Hilar vessels with branching suggest lymphadenitis whereas peripheral vessels suggest metastases. If you combine a short axis diameter of 5 to 7 mm with absence of hilar blood flow, the sensitivity can rise to near 90% and specificity to around 95%. If that isn't good enough you can use ultrasound to provide guidance for fine needle aspiration of nodes for cytologic evaluation. This technique is very accurate (with a specificity approaching 99%) but requires the patience and expertise that many Americans do not possess.

The shape of the lymph node may also be a secondary indication of whether it is involved by tumor. More rounded lymph nodes have a greater chance of being neoplastic than those which have a kidney bean shape. Most authors agree that a lymph node that has central necrosis no matter what its size should be considered malignant until proved otherwise, unless the patient has active tuberculosis or has been treated with radiation in the past for lymphoma. A lymph node with fat centrally is benign and may signify previous inflammation and/or radiation. Heterogeneous contrast enhancement, calcifications, asymmetric clustering, and perinodal stranding should help improve your odds in detecting malignant adenopathy.

So what is the big deal about nodes? Remember Dave's 50% rule of lymph nodes. Having a lymph node with your primary tumor cuts your 5-year prognosis by 50%, having bilateral nodes cuts it another 50%, having extracapsular spread of tumor cuts it another 50%, having fixation to vital structures (carotids, vertebral arteries, paraspinal muscles, transverse processes, brachial plexi) cuts it another 50%, and 50% of our school age children have lymph nodes palpable 50% of the time.

Malignant Adenopathy: Other Considerations

Another important issue to consider when you are dealing with internal jugular chain lymph nodes is the presence or absence of carotid invasion by lymphadenopathy. Presently, no ideal radiographic study can predict whether the carotid artery is invaded by tumor when there is contiguous lymphadenopathy. It has generally been accepted that if less than 270 degrees of the circumference of the carotid artery is encircled by squamous cell carcinoma, the likelihood of carotid wall invasion is small. This suggests that the surgeon may be able to "scrape" the tumor off the carotid artery. However, when the tumor encircles the carotid artery greater than 270 degrees, the carotid wall is

TABLE 13-8 Lymph Node Size and Pathology

Size	CT Sensitivity	CT Specificity	MR Sensitivity	MR Specificity
5 mm	98%	13%	92%	20%
8 mm	95%	22%	87%	31%
10 mm	88%	39%	81%	48%
12 mm	74%	67%	66%	72%
15 mm	56%	84%	51%	86%

Data from Curtin HD, Ishwaran H, Mancuso AA, Dalley RW, Caudry DJ, McNeil BJ. Comparison of CT and MR imaging in staging of neck metastases. *Radiology.* 1998;207:123-130.
CT, Computed tomography; *MR,* magnetic resonance.

TABLE 13-9 Primary Sites of Squamous Cell Carcinoma versus Presence of Lymphadenopathy

Primary Site	Nodes at Presentation (% of cases)	Bilateral Nodes (% of cases)	Nodes after Therapy (% of cases)	Nodal Drainage Levels
Nasopharynx	72-85	25-33	15-20	I-V
Oropharynx	66-76	20-25	15-20	II, III, 5
Hypopharynx	59-65	9-13	15-20	II-V
Supraglottic larynx	55-80	21	10-20	II-IV
Oral cavity	3-36	8-12	8-28	I-III
Sinonasal	10-17	<5	16	I-III
Glottis	3-7	<5	<5	II-IV

nearly always invaded and the carotid artery may need to be sacrificed. The surgeon may spend a great deal of time palpating the lymph nodes to assess clinically whether they are fixed to the carotid artery; however, imaging, particularly MR, may be more useful than the clinical examination. At least we radiologists would like to believe that. Carotid angiography may not be as useful because the lumen of the artery may be entirely normal whereas its adventitia and media may be infiltrated with cancer. On the other hand, the arteriogram is invaluable in determining whether the patient can tolerate carotid sacrifice at the time of surgery via reversible balloon occlusion testing. By the time the carotid is encased, the long-term prognosis is abysmal and even sterling surgeons with sharpened scalpels shan't stunt the Stygian survival.

Extension of lymph nodes into the posterior musculature of the neck, paraspinal soft tissue and/or the base of the skull also makes surgical resection much more difficult if not impossible. The presence of these findings suggests that the tumor may be extensive beyond visible dimensions and may require heroic surgery for a small chance of cure.

In general, clinicians use relatively rigid clinical staging criteria to determine whether the lymph nodes are to be removed. In many instances, to remove the possibility of microscopic disease, the surgeon operates on a clinically N0 neck. The alternative treatment option is to treat these patients with postoperative radiation therapy to eliminate microscopic lymph node disease. Often, however, the clinicians hold radiation therapy in abeyance because of the 15% likelihood of a metachronous primary squamous cell carcinoma in a patient with head and neck cancer. Radiation may be held in reserve to treat the secondary primary tumor. This may be a rationale for a surgical approach to disease in a younger patient. In some cases radiation is used before surgery to reduce the overall bulk of the disease so that the surgical removal is easier. Nonetheless, no study has shown that cure rates with surgery after radiation therapy are significantly better than those with radiation therapy after surgery.

Occasionally, a patient has malignant lymphadenopathy without a clinically apparent primary tumor. The most common head and neck sites for an occult primary neoplasm invisible to endoscopy are the nasopharynx, the tonsil, the pyriform sinus apex, and the tongue base. Therefore, in those patients who have unknown primary tumors, these

areas should demand greater attention by the radiologist. The work-up should include either CT or MR as well as a barium swallow and upper gastrointestinal examination. In patients who have cervical adenopathy with an unknown primary tumor by clinical and endoscopic examination, imaging will identify the primary tumor in 25% of cases. In many instances the primary tumor is never discovered. Occasionally, the patient who has lymphadenopathy in the lower part of the neck may have an esophageal, tracheal, bronchial, or pulmonary primary tumor.

The results of using FDG PET to identify the primary sites of cancers when malignant adenopathy is present without a clinical or radiographic source for it have been mixed. PET potentially predicts primaries proactively in pretty high percentages; 25% to 50% of otherwise occult primary cancers are discovered with PET. Performing bilateral tonsillectomies with the neck node dissection probably reveals an equal number of primary tumors.

Nodal Metastases from Thyroid Carcinoma

Lymphadenopathy from papillary thyroid carcinoma is unique in many senses (Fig. 13-51). In 50% of cases the nodal disease is the presenting sign of the cancer and in 25% of cases the thyroid cancer may be occult to palpation and imaging. Thyroid papillary carcinoma lymph nodes may be (1) cystic, (2) calcified, (3) highly vascular, (4) highly enhancing, (5) colloid containing, and therefore (6) bright on T1WI, (7) tiny, (8) retropharyngeal, (9) hemorrhagic, and (10) worthy of their own separate paragraph. Metastatic lymphadenopathy from papillary thyroid carcinoma can be present even in the setting of microscopic and radiographically undetectable primary disease in the thyroid gland. The nodes may be treatable with iodine therapy. Remember also that papillary carcinoma, though known for its regional lymphatic spread, may also produce a miliary pattern in the lungs and so make sure you flip to the lung windows once you hit the apices on your CT scans. For once, do not just think like a neuroradiologist.

Lymphoma

The most common noninflammatory cause for a unilateral neck mass in a patient between 20 and 40 years old is

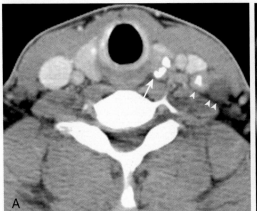
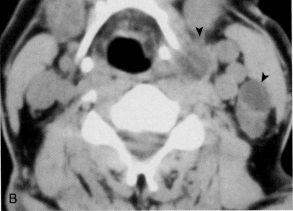

FIGURE 13-51 Metastatic papillary thyroid cancer to lymph nodes. **A,** Axial contrast-enhanced computed tomographic image shows a calcified nodule in the left thyroid gland *(arrow)* with calcified *(single arrowhead)* and cystic *(double arrowheads)* metastatic lymph nodes. **B,** A different patient with cystic nodal metastases *(arrowheads)* from papillary thyroid cancer.

lymphoma. Non-Hodgkin lymphoma (NHL; 75% of lymphomas) is more common in the head and neck than Hodgkin lymphoma (25%), and usually Waldeyer ring is also involved with NHL. Fifty percent of patients with Waldeyer ring NHL have nodes at presentation. When one has isolated nodal disease in the neck or mediastinal involvement, the odds shift more in favor of Hodgkin disease (HD). A history of mononucleosis is present in many patients with HD and may play a role in the etiology. Classically HD proceeds in an organized fashion through the lymphatic chains, from one echelon to the next. NHL is more likely to jump around and have pulmonary, osseous, and noncontiguous nodal involvement. The head and neck is the second most common site of NHL after the gastrointestinal tract (although franchises of the NHL may be found throughout the United States and Canada).

Either gallium scanning or thallium scanning may be used to identify lymph nodes with Hodgkin disease.

POSTTREATMENT IMAGING

Do not just shrug your shoulders and murmur something about your inability to distinguish normal postoperative or postradiotherapeutic changes from recurrent disease when these cases come across your screen. You can contribute some information about the possibility of recurrence, particularly if you have a baseline posttreatment study when edema has resolved (after approximately 8 weeks). Ninety percent of head and neck cancer recurrences take place within the first 12 months and 96% within 2 years. For this reason patients are not considered cured of their tumor until 5 years have passed and no evidence of tumor is identified. Recurrences are usually split evenly between those at the primary site, those in the nodes, and those at both the primary site and the nodes. Therefore, the survey for residual or recurrent disease should be performed frequently within the first 2 years, usually at 6-month intervals, and must cover the primary site and the cervical lymph nodes. Growth of tissue after the 8-week posttreatment scan should make one worry about recurrence. Focal exophytic (into the extramucosal deep soft tissue) or endophytic (into the aerodigestive airway) soft-tissue bulges should be histologically examined for recurrence. Morphology is the key here. MR intensity, CT density, and enhancement features are useless, particularly after radiotherapy. Growth with time is the only sure-fire sign to hang your hat on for recurrence! Unfortunately, we rarely have a baseline scan to work with after therapy despite entreaties (and donuts) sent to our clinical colleagues.

Before you are able to identify recurrences, you must be familiar with the terminology of operations on the neck. Neck dissections are performed to remove suspected adenopathy. The nodes may have been identified by palpation or imaging preoperatively, or sometimes the nodes are removed in an N0 neck because of a high rate of microscopic disease with that particular primary site (e.g., supraglottic larynx, base of tongue, and tonsil). If the sternocleidomastoid muscle, the spinal accessory nerve (XI), the submandibular gland, and the jugular vein are removed with level I to V nodal chains, the procedure is called a radical neck dissection. The side of the neck is flattened, and minimal tissue is left besides skin, arteries, and the anterior viscera. A modified neck dissection usually spares either the spinal accessory nerve (allowing a functional

trapezius muscle), the sternocleidomastoid muscle, and/or the internal jugular vein. A supraomohyoid neck dissection removes level I to III nodes above the cricoid cartilage and usually spares the sternocleidomastoid muscle, internal jugular vein, and spinal accessory nerve. This dissection is commonly performed for N0 primary tumors high in the aerodigestive system, with a low to intermediate rate of occult metastases. An anterior neck dissection may be performed for thyroid cancer to remove anterior visceral chain nodes (level VI). Bilateral neck dissections may also be performed. It is important to remember that once the normal pathway of lymphatic drainage from a primary site has been removed surgically, nodal metastases may subsequently occur in unusual locations for that particularly primary tumor. After both nodal chains have been removed, watch out for dermal lymphatic drainage of tumor cells because the nodes are no longer available to sequester the neoplasm. This is identifiable as subcutaneous stranding and thickening on CT or MR. Unfortunately this may be indistinguishable from the edema associated with acute radiation treatment.

A word about head and neck cancer radiation: it is vitally important for the patient to receive the radiotherapy for the cancer with as little interruption as possible, so do not overcall lumps that may delay therapy. A recent study has shown that with treatments less than 55 days apart, 56 to 65 days apart, and more than 66 days apart, the 5-year survival rates were 56%, 46%, and 15%, respectively in patients undergoing sequential chemotherapy and radiation therapy. Treatment interruptions during definitive radiation therapy are a no-no. Try to soft call anything in this time range.

Most radiologists have a hard time understanding flaps and grafts beyond shaking down your local equipment vendors. For this reason we reiterate some of the terminology of these surgical procedures. Surgical reconstruction of defects in the head and neck after operation requires the interposition of flaps and grafts. Flaps are usually separated into several categories: site (local, regional, or distant), tissue (cutaneous, fasciocutaneous, musculocutaneous, or osteomusculocutaneous), and blood supply (random, axial, pedicled, or free). Modern techniques of inserting osteointegrated implants into bone grafts (often distant osteomusculocutaneous free flaps of the fibula) afford the patient an opportunity to have a dental surface capable of chewing. After radiation therapy and/or in individuals who have carious teeth, marrow changes may occur that might simulate tumoral infiltration but may actually represent radiation fibrosis, osteoradionecrosis (Fig. 13-52), osteomyelitis, or periodontal disease.

Surgeons have a whole host of skin (cutaneous), muscle (myo-), bone (osteo-), and fat (lipo-) plugs that they are capable of inserting these days. These are usually defined with the prefixes provided, so that an osteomyocutaneous flap is one that contains bone, muscle, and skin.

The site from which the flap is taken precedes the description of what is taken. A "free" flap is not one for which there has been no charge (far from it!). It really is one that is "far from it," at a distant site and wholly separated from the original surgical defect. So a fibular free flap may be used to reconstruct the mandible, or a radial forearm free (cutaneous) flap frequently fixes florid flaws in the floor of the mouth. Free flaps require anastomoses of

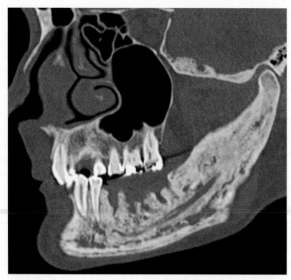

FIGURE 13-52 Osteoradionecrosis. Sagittal reconstructed computed tomographic image shows mixed lucent and sclerotic change in the mandible from bone destruction. Pathologic fractures can coexist. The absence of a soft-tissue mass in conjunction with history of oral cancer and radiation treatment help make the diagnosis of osteoradionecrosis. (Courtesy Nafi Aygun, MD.)

blood vessels (and sometimes nerves) from the donor site to the surgical bed with microvascularization techniques.

A local or regional flap requires rotation or "pedicling" a piece of adjacent tissue into a surgical gap without disrupting the original blood supply of the flap. Thus a local muscular (myocutaneous) temporalis flap may be rotated down to fill the gap of an infratemporal fossa resection, or a pedicled pectoralis major flap may reconstruct a large base of tongue or floor of mouth defect. Often on imaging the most striking feature of these flaps is the dominance of the fat in the graft. Granted, some of the appearance varies with patient habitus, nationality, whether Oprah is staying on her diet, and the type of graft used (in Americans the rectus abdominis graft usually has more fat than a radial forearm flap), but the muscle of the graft often atrophies to a greater extent than the fat. A graft that has been irradiated often has a denser appearance on CT and MR than one that has not been irradiated and may appear as a sheet of fibrous-like tissue without good tissue planes.

FUTURE CANCER RISK

A general adage among head and neck clinicians is that tumors recur at surgical margins along the periphery of the tumor, whereas they recur centrally within the tumor bed when radiation therapy is the primary modality of treatment. Recurrent squamous cell carcinoma may be very difficult to identify when small and this has been another area where FDG PET has shown value. Nonetheless, for the rest of their lives patients with head and neck cancers generally are surveyed for the possibility of a secondary malignancy. If a surgeon has a few of these patients as long-term survivors, the office schedule can be filled for years. As stated previously, the incidence of synchronous or metachronous oral cavity cancers in a patient who has had a previous cancer is approximately 40%. In patients who have head and neck cancers outside the oral cavity, the metachronous rate is still 15%.

Another fascinating fact: 19% of patients with head and neck carcinoma who undergo thoracic CT have one or more primary or metastatic neoplasms discovered. Synchronous primary lung cancers are twice as common as metastatic lesions to the lungs. Of the second lesions, less than one third are discovered on chest radiographs. The primary sites of the head and neck cancers that have second lesions are fairly evenly distributed between larynx, pharynx, and oral cavity and most have stage IV lymphadenopathy. Most series of patients with head and neck cancers report a second lung primary incidence of 0.8% to 6% and lung metastases incidence of 5% to 7%. Bottom line: Recommend a chest CT in patients with head and neck squamous cell carcinomas to pick up additional lesions so that the head and neck surgery will not be for naught.

Do not be too morose after reading this chapter with its emphasis on cancer and head and neck surgery. By chewing tobacco, smoking and excessive drinking, humans abuse their mucosa and the mucosa strikes back in a vicious way: squamous cell carcinoma. Although beauty may be only skin (mucosa) deep, cancer may go clear down to the bone. So be good to your mucosa. Kick the habit!

For a quick pick-me-up, read the next chapter (extramucosal diseases, Chapter 14). There are a lot more benign tumors and interesting infections there.

Our authors' outstanding alliterations are over, allowing uninterrupted ease in education.

Chapter 14

Extramucosal Diseases of the Head and Neck

As opposed to the mucosal diseases of the head and neck, in which the differential diagnosis usually revolves around squamous cell carcinoma 90% of the time, the extramucosal space allows the radiologist to exercise finely honed skills in forming a differential diagnosis. Rumination, differentiation, pontification and "clinical correlation" may follow—and not just in this chapter.

The reader should understand that more comprehensive multivolumed dry texts dealing with head and neck imaging are readily available for reference (at a price guaranteed to shock the trainee back to undergrad years). This chapter will be bargain basement, kick the tires, head and neck radiology, giving you 95% of what you need to interpret cases well—which is probably better than 95% of the radiologists practicing now and at a 95% discount compared with those multiauthored texts gathering dust on your reference shelf (that we have used to fill in holes in our chapter below).

In the suprahyoid and infrahyoid neck, layers of the deep cervical fascia serve to encapsulate regions of the anatomy that lend themselves to specific analysis. Although these layers of fascia are rarely visualized, they do represent a subtle barrier that restricts the free movement of pathology from one area of the neck to the other. While it is true that some entities, both malignant (e.g., squamous cell carcinoma antigen [SCCA]) and benign (e.g., venous vascular malformations) easily cross the deep cervical fascia boundaries and other lesions are multispatial in their distribution (e.g. lymphoma), this framework is quite useful as a learning tool for studying the head and neck.

SALIVARY GLANDS

Anatomy

We shall begin the discussion of the extramucosal lesions of the head and neck with the salivary glands, to get the juices flowing (pun intended). Not much is known about the salivary glands because they are so secretive. Haha, just kidding!

In the context of deep cervical fascia spatial anatomy, the major salivary glands are encapsulated in the parotid space and the submandibular space.

The three major salivary glands are the paired parotid, the submandibular, and the sublingual glands. The parotid gland is located superficially under the tissue below and about the ear and extends over the ascending ramus of the mandible. The parotid "space," enclosed in deep cervical fascia, includes the gland, branches of the external carotid artery and retromandibular vein, cranial nerve VII, and a few branches of the auriculotemporal rami of cranial

nerve V_3. The parotid gland's consistency changes with age. As you get older, your gland (and other body parts we have found) gets fattier. The size and the consistency of the gland also depend on body habitus; in someone who gives the gland a good workout and grows corpulent, the gland tends to be bigger and fattier.

A portion of the parotid gland extends deep to the plane of the facial nerve (identified on computed tomography [CT] by the stylomandibular tunnel from the styloid process to the ascending ramus of the mandible/retromandibular vein) and is termed the deep lobe of the parotid. The superficial lobe extends from just under the skin and often has an accessory tongue of tissue that passes over the masseter muscle. For the parotid gland, the "lobes" are an arbitrary distinction dividing the gland into anatomic sections based on the facial nerve and useful for teaching first year residents who have just passed the "lung" rotation (with its clearly anatomically defined lobes). The importance of differentiating the deep and superficial portions of the parotid stems from the different surgical approaches to tumors (Fig. 14-1). If a lesion is in the superficial portion of the parotid gland, it is usually approached from an external periauricular incision, the facial nerve is dissected deep to the mass to ensure its safety, and the lesion is plucked from the gland superficial to the nerve. If a mass in the deep portion of the parotid is well defined and noninfiltrative, it is also approached from the same incision and the facial nerve is dissected and then lifted up to shell out the deep lobe mass. However, if the lesion is infiltrating through the deep portion and may encase the nerve, the approach may be combined with a parapharyngeal space approach via a neck incision below the ear. Unfortunately, with infiltrating deep lobe lesions the facial nerve may be surgically sacrificed (if it has not already been sacrificed by the tumor itself). Good news... grafts can get function back. Floppy facies need not come standard.

The parapharyngeal fat is a good marker for telling what space a lesion is in (Table 14-1) in the deep neck by its displacement. Deep parotid space lesions displace the parapharyngeal fat anteromedially and maintain a fat plane between the lesion and the mucosal surface of the pharynx.

The parotid duct is termed the Stensen duct, and it passes over the masseter muscle before curving medially to insert in the cheek at the second maxillary molar. Because the parotid gland is late in its encapsulation, lymphoid tissue lies within it. For that reason, the parotid gland is the only salivary gland that has the potential for lymphadenitis, metastases to intraglandular lymph nodes, lymphoma, and autoimmune lymphocytic disorders.

The submandibular space (Box 14-1) encompasses the tissue below the mucosa of the floor of the mouth yet above the fascia connecting the mandible to the hyoid bone. As such, it contains the sublingual compartment with the mylohyoid muscle, sublingual gland, portions of the submandibular gland, the associated ducts, and the corresponding neurovascular structures. The sublingual space communicates with the submandibular space along the posterior margin of the mylohyoid muscle.

The submandibular gland is located in the floor of the mouth, deep to the angle of the mandible (Fig. 14-2). The submandibular gland secretes seromucinous saliva as opposed to the parotid gland, which secretes serous saliva. In addition, the pH of the saliva produced by the submandibular gland is more alkaline and the fluid is more viscous. The duct of the submandibular gland is called the Wharton duct, and it drains on either side of the frenulum of the floor of the mouth. The duct has a tighter orifice but is wider than the Stensen duct and is more easily traumatized in the mouth. The duct of the submandibular gland courses anteriorly and superiorly before reaching its orifice.

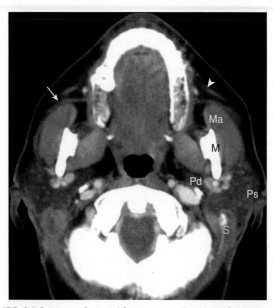

FIGURE 14-1 Normal parotid anatomy. Computed tomography (CT): The plane between the styloid process (S) and the mandible (M) defines the superficial portion of the parotid gland (Ps) from its deep portion (Pd) on CT. The parotid duct *(white arrow)* on the right is distinguished from the zygomaticus muscle *(white arrowhead)* on the left. Note parotid tissue superficial to the masseter muscle (Ma).

The sublingual gland is located in the sublingual space medial to the mylohyoid muscle in the floor of the mouth. The size of the sublingual gland ranges from being inapparent on imaging studies to a readily identifiable almond sized structure. The sublingual gland has many draining ducts (known as ducts of Rivinus) opening below the tongue. If a dominant sublingual duct connects to Wharton duct, it is called the duct of Bartholin (don't get the Bartholin duct in the mouth mixed up with the Bartholin duct of the vagina....nuf said!). The saliva produced by the sublingual gland is also seromucinous. It is the smallest of the major salivary glands and has the fewest lesions associated with it, but it does get in the path of floor of the mouth squamous cell carcinoma.

Minor salivary glands are found scattered throughout the aerodigestive system but abound in the oral cavity (especially the hard and soft palate). Minor salivary glands can also be found in the oropharynx, nasopharynx, sinonasal cavity, the parapharyngeal space, the larynx, the trachea, the lungs, and even into the middle ear and eustachian tube. These glands secrete mucinous saliva. Minor salivary glands do not have readily identifiable ducts; however, they are the source of the many mucous retention cysts that are seen as benign lesions in the aerodigestive system. Neoplasms of minor salivary glands have been addressed in the mucosal chapter (Chapter 13); suffice it to say pleomorphic adenomas and adenoid cystic carcinomas are the most common benign and malignant histologies respectively.

Now that the hors d'oeuvres have been served and your appetite has been whetted, you are ready to digest diseases of the salivary lesions.

Congenital Disorders

Branchial Cleft Cyst

First branchial cleft cysts (BCCs) classically occur in the parotid gland (Box 14-2) or around the external auditory canal (EAC). Several classifications of first BCCs have been developed including Arnot type I (intraparotid cyst) and Arnot type II (cyst in the anterior neck that may drain with a tract through the deep portion of the parotid gland to the EAC; Figs. 14-3 and 14-4). Work type I first BCCs are anterior and inferior to the pinna and may communicate with skin anterior and medial to the EAC in the preauricular zone. Work type II BCCs are more common, arise at the angle of the mandible (where we typically see second BCCs), and extend to

TABLE 14-1 How to Place a Lesion

Lesion Location (Space)	Displacement of Parapharyngeal Fat	Displacement of Styloid Musculature	Displacement of Longus Musculature
Parotid deep portion	Anteromedial	Posterior	Posterior
Masticator	Posteromedial	Posterior	Posterior
Carotid	Anterior	Anterior	Posterior
Mucosal	Posterolateral	Posterior	Posterior
Retropharyngeal	Anterolateral	Anterior	Posterior
Perivertebral	Rarely displaced	Anterior	Anterior
Parapharyngeal	N/A	Posterior	Posterior

the junction of the bony and cartilaginous EAC. They may rarely drain into the EAC. Both types of first BCCs may be intimately associated with the facial nerve which gives the ear, nose, and throat (ENT) surgeons fits. Because second BCCs are so much more common, they probably outnumber first BCCs even in the periparotid location (Table 14-2).

Although these lesions usually have fluid density and intensity characteristics, if infected or traumatized, their appearance may be a bit more complex (see Fig. 14-3). Adjacent inflammation may be present, and the whole complex may simulate an infiltrative process. Fistulization to the bone-cartilage junction of the external ear may occur with first branchial cleft anomalies.

Venous Vascular Malformations

Venous vascular malformations (formerly inaccurately called hemangiomas because they are *not* tumors) may occur in the submandibular space as well as in other spaces of the head and neck. On CT, they may show phlebolithic calcification in their later stage (as opposed to monolithic ossification of the Stone Age), but invariably one identifies a lobulated, heterogeneously enhancing mass. Areas of enhancement that are as extensive/intensive as that of the neighboring jugular vein are not unusual. If you perform a fine-needle aspiration (FNA) of this lesion you will be faced with the bane of the aspirator's existence—recurrent samples of "nothing but blood." You will be doing the study in vein (Ha! Pun-ishment again!). Do not bother. Interestingly enough, there is a 20% coincidence rate of developmental venous anomalies (venous angiomas) of the brain in those subjects with head and neck venous malformations.

BOX 14-1   Submandibular and Sublingual Space Compartments and Contents

SUBLINGUAL SPACE

Mylohyoid muscle
Portions of submandibular gland
Sublingual gland
Wharton duct
Lingual artery
Lingual, hypoglossal, glossopharyngeal nerves

SUBMANDIBULAR SPACE

Submandibular gland
Facial artery branches
Level IA lymph nodes

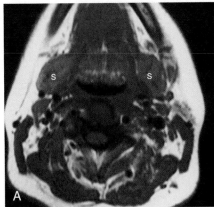

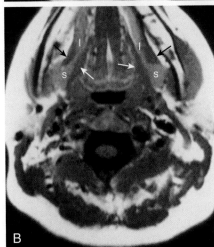

FIGURE 14-2 Normal submandibular-sublingual anatomy. **A,** Submandibular glands (s) can be seen on this T1-weighted image in the neck. Note that there normally is some heterogeneity to the gland because of its hilum and ductal system. **B,** Superior portion of the submandibular gland (s) can be seen on this section, which also nicely demonstrates the sublingual gland tissue (l). Note that the sublingual space is bounded by the mylohyoid musculature *(black arrows)* laterally and the styloglossus-hyoglossus complex *(white arrows)* medially.

BOX 14-2   Cystic Masses in the Parotid Gland

Abscesses
Branchial cleft cysts (BCCs)
Lymphoepithelial lesions
Pleomorphic adenomas and Warthin tumors
Pseudocysts
Sialectasis
Sialoceles
Simple cysts

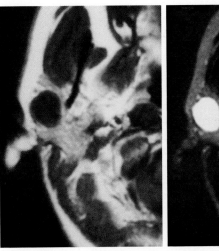

FIGURE 14-3 First branchial cleft cyst (BCC). This intraparotid cyst with a sharply defined wall is dark on T1-weighted imaging (T1WI) (to the left) and very bright on T2WI (image on the right). There is no way of knowing whether this is an inflammatory cyst, a posttraumatic sialocele, or a BCC based on imaging alone.

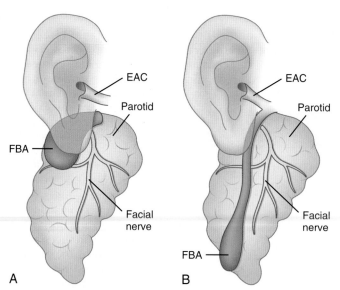

FIGURE 14-4 First branchial cleft cyst's Arnot types. **A,** Type I first branchial cleft anomaly (FBA). The cyst is located in the parotid gland. There is no communication with the external auditory canal (EAC). **B,** Type II first FBA. The proximal portion of the anomaly communicates with the EAC. The cyst tract typically extends inferiorly within the deep lobe of the parotid gland. The main portion of the cyst is usually located inferior to the parotid gland. Consequently, these masses may present as submandibular masses. (From Mukherji SK, Fatterpekar G, Castillo M, et al. Imaging of congenital anomalies of the branchial apparatus. *Neuroimaging Clin North Am.* 2000;10:76-77.)

TABLE 14-2 **Branchial Cleft Anomalies**

Features	BCC type 1	BCC type 2	BCC type 3	BCC type 4
% of all BCCs	7	90	2	1
Location	Inferoposteromedial to pinna Angle of mandible to EAC Periparotid	Anterior to mid SCM, deep to ICA Angle of mandible At CCA bifurcation Parapharyngeal	Anterior to lower SCM, superficial	Low, anterior to SCM Left side in 90% Follows recurrent laryngeal nerve
Sinus drainage	Ear, skin	Tonsillar fossa, skin	Pyriform sinus	Pyriform apex
Etiology	Persistent cleft	Failure of arch to proliferate, persistent cervical sinus	Persistent cervical sinus	Persistent cervical sinus

BCC, Branchial cleft cyst; *CCA,* common carotid artery; *EAC,* external auditory canal; *ICA,* internal carotid artery; *SCM,* sternocleidomastoid muscle.

Inflammatory Lesions

Calculous Disease

By our estimation, calculous disease (in the nonmathematical sense) is the most common benign condition to affect the salivary glands by a factor of the $(\pi^3 \ !3)^4$. Because the submandibular gland secretes a more mucinous, viscous, alkaline saliva and the wider duct must drain in an uphill direction with greater possibility of stasis, calculi occur most commonly in the Wharton duct (Fig. 14-5). In fact, submandibular gland calculi outnumber those in the parotid gland by a "calculation" of four to one. Sublingual gland calculi and minor salivary gland calculi are extremely uncommon. Although most of the calculi associated with the salivary glands are radiopaque, a small percentage of calculi (20%) may not be so radiodense as to be visible on plain films. Viva La CT!

Patients who have calculous disease usually have painful glands, exacerbated by chewing foods that precipitate salivation. If the clinician suspects calculous disease, the usual work-up includes plain films to evaluate for large radiopaque calculi. If no calculi are identified by plain films, or if the patient has a fever associated with a painful salivary gland and an abscess is suspected, CT should be done because it is more sensitive for the detection of calcification and inflammatory masses (Fig. 14-6). This may be one of the rare indications for unenhanced (to see a small calculus and distinguish it from a vessel) and enhanced (to look for signs of inflammation/abscess) CT examinations of the neck. Remember that magnetic resonance (MR) imaging is less sensitive to calcified or noncalcified tiny calculi; larger calculi can be seen as low-intensity areas on T2-weighted scans but flow voids of vessels may simulate calculi.

Calcifications of the parotid glands are also associated with human immunodeficiency virus (HIV) infection, chronic kidney disease, alcoholism, elevated alkaline phosphatase, and autoimmune disease, including sarcoidosis and systemic lupus erythematosus.

In the instance of a calculus that is not radiopaque on CT, one might be forced to perform conventional sialography (cannulation of the salivary duct with injection of contrast). MR sialography can employ either a single shot fast spin echo heavily T2-weighted sequence that highlights the ducts alone or a high-resolution three-dimensional fast spin echo or constructive interference in steady state (CISS) T2-weighted sequence. Sialography is useful in demonstrating strictures of the ducts after

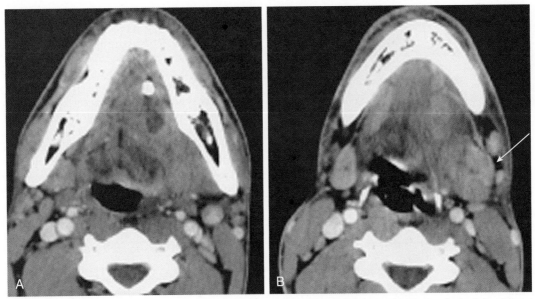

FIGURE 14-5 Submandibular sialadenitis. **A,** A large stone with a dilated Wharton duct is seen in the left sublingual space. **B,** Note the swollen submandibular gland *(arrow)* with the effaced sublingual space fat planes.

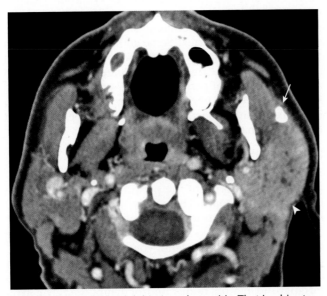

FIGURE 14-6 Parotid sialolithiasis and parotitis. That is a big stone *(arrow)* in a small Stenson duct. Note how much larger and relatively hyperenhancing the inflamed left parotid gland *(arrowhead)* is when compared with the right.

passage of a calculus and intraluminal filling defects from nonopaque stones. The sialogram suggests the diagnosis of autoimmune/chronic sialadenitis if there are pruned, truncated main ducts with punctate/globular collections peripherally in the glandular parenchyma.

Treatment of sialolithiasis (a catchy seven-syllable word for salivary gland stones) generally consists of supplying sour solutions to stimulate salivation to spew said stone (say that 10 times fast without spitting!). Transoral resection and sialodochoplasty can be performed for isolated distal duct (close to ampulla) sialoliths. For proximal or glandular stones, the surgeon may decide to treat the patient with resection of the gland. This is often the preferred treatment with recurrent bouts of sialolithiasis

with disruptive sialadenitis. A cervical (submandibular) approach may be taken with sialoliths that extend beyond the mylohyoid (in the proximal duct).

Sialosis

Sialosis is a painless enlargement of the parotid glands that has been associated with numerous causes, including (1) diabetes, (2) alcoholism, (3) hypothyroidism, (4) medications including phenothiazines and some diuretics, (5) obesity, (6) starvation, (7) radiation, and (8) the dreaded "idiopathic" causes. This usually is a bilateral and symmetric process that may resolve when the underlying cause has been removed. It should be noted, however, that the normal range of parotid gland size and consistency is varied and often it is difficult to state definitively that the glands are larger than normal. On imaging studies the glands with sialosis generally have a CT density and signal intensity on T2-weighted imaging (T2WI) slightly greater than that of normal fatty parotid glands. The glands in sialosis usually are not as bright on T2WI as glands that are infected. You may see distortion of the facial contour by the enlarged glandular tissue.

Sialadenitis

Sialadenitis refers to glandular inflammation, whereas sialectasis refers to dilatation of ductal spaces. Sialadenitis is often associated with sialectasis. The most common cause of these conditions is calculous disease. The inflamed gland is enlarged and hyperenhancing on CT and MR imaging compared with the contralateral normal side and periglandular stranding can be seen. Microabscesses within the parotid tissue may be seen in a person who has sialectasis and/or sialadenitis. They are identified on CT as areas of low density with peripheral rim enhancement and on MR as areas of very bright signal intensity on T2WI of the salivary glands. Microabscesses are often multiple and may be a source of painful parotid glands with fever. Abscesses may also develop around the mandible, sublingual gland,

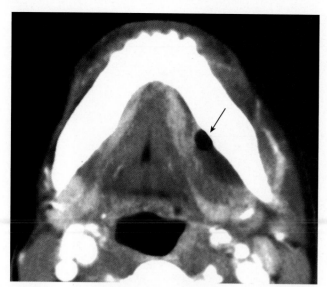

FIGURE 14-7 Abscess. An abscess in the sublingual space is usually a result of carious teeth or treatment for such. A pocket of air and fluid is marked by the *arrow*. This is what happens when dentists do not wash their instruments of torture properly.

or submandibular gland in association with dental infections (Fig. 14-7).

A wide variety of inflammatory conditions may affect the salivary glands. Although mumps may be the most common infection to affect the salivary glands (specifically the parotid glands), imaging is unnecessary; the diagnosis is a clinical one. Other viral etiologies include HIV, coxsackie, and influenza viruses. Bacterial infections are uncommon and are usually due to *Streptococcus, Haemophilus,* and *Staphylococcus* species.

Other etiologies of acute parotitis include granulomatous (tuberculosis, candida, cat scratch fever) and idiopathic causes (e.g., postpartum parotitis). Poor dental hygiene may contribute to the development of infections affecting the submandibular, sublingual, and parotid glands. The minor salivary glands rarely show inflammatory change other than mucous retention cysts from local obstruction. Vallecular retention cysts, believed to arise from salivary tissue at the tongue base, can get huge.

Sialodochitis

Sialodochitis refers to inflammation of the main salivary ductal system. A number of autoimmune conditions may cause sialadenitis and sialodochitis. Mikulicz disease, or in the most recent classification, Sjögren type 1 disease (poor Dr. Mikulicz is getting squeezed out), is an autoimmune disorder that causes chronic sialadenitis and sialodochitis and leads to fibrous salivary gland tissue (primarily of the minor glands) with resultant dry mouth. This disorder usually affects middle-aged women. They have a higher rate of stones also so do the scan "dry" (Ha!). When the disease is associated with a collagen-vascular disease (most commonly rheumatoid arthritis more so than systemic lupus erythematosus) and there is involvement of the lacrimal glands, the disorder is classified as Sjögren syndrome (Sjögren type 2). Sjögren syndrome is an autoimmune disorder that causes dry eyes, dry mouth, and arthritis. Patients with

Sjögren syndrome have a 10-fold increased risk of lymphoma, which may have its first manifestations in the parotid glands. The lymphoma is usually of the non-Hodgkin variety and may also affect any other area of the head and neck (Fig. 14-8). These patients are often scanned to survey for the possibility of lymphoma.

Sjögren syndrome is characterized on conventional or MR sialography by punctate, globular, cavitary, or destructive appearance of the ducts of the parotid glands. Tiny pools of contrast may be seen in the gland. With increasing severity of disease there is greater and greater replacement of glandular tissue with fat. Thus, some have suggested that the severity of fat deposition correlates well with the impairment of salivary flow in Sjögren patients. Sjögren syndrome is also associated with the presence of lymphoepithelial cysts and nodules akin to those seen in patients with HIV-associated parotid lesions (see following discussion).

The cross-sectional imaging appearance of parotid glands in patients who have Sjögren disease may range from normal to a dried up, scarred down, atrophic gland, to one with lots of large and/or tiny cysts and nodules within it, to one with a dominant mass within it. This looks very much like acquired immunodeficiency syndrome (AIDS)–related parotid disease (see later discussion).

The glands with autoimmune sialadenitis generally are denser on CT than normal. Some authors have described a "salt and pepper" appearance (there is that same old food analogy) to the gland on both T1-weighted imaging (T1WI) and T2WI in 46% of patients with Sjögren syndrome, presumably reflecting fibrosis and lymphocytic infiltration intermixed with sialectasis. Biopsy should be performed on any dominant mass to rule out lymphoma.

Lymphoepithelial Lesions

Since the ascent of AIDS in the young population, lymphoepithelial lesions of the parotid gland have become much more common. These may include purely cystic lesions or solid lesions of lymphoid aggregates (Fig. 14-9). Therefore, in a younger patient with multiple lesions in the parotid gland, you should consider lymphoepithelial lesions as opposed to multiple Warthin tumors. The differential diagnosis also includes multiple intraparotid lymph nodes and/or lymphoma. The lymphoepithelial lesions of the parotid have been associated with HIV seropositivity, and the presence of these abnormalities may predate the infection that classifies the patient as having AIDS. Associated findings with the HIV-related parotid disease include diffuse generalized lymphadenopathy in the neck and prominence of adenoidal and tonsillar tissue. Bone marrow signal intensity on T1WI may be lower than normal. When the lymphoepithelial lesion is cystic, it has low density on CT and signal intensity characteristics of cerebrospinal fluid on T1WI and T2WI. However, the lymphoepithelial solid nodules may have a more variable density and signal intensity on cross-sectional imaging. They are not distinguishable from Sjögren-related benign lymphoepithelial lesions (BLELs).

Sialocele

A sialocele refers to a collection of saliva that communicates with the parent duct in a manner similar to that

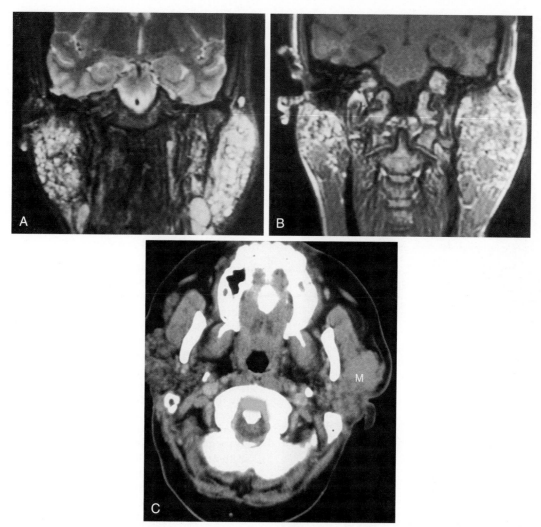

FIGURE 14-8 Sjögren syndrome. **A,** The coronal T2-weighted imaging (T2WI) reveals many tiny benign lymphoepithelial lesions (BLELs) as well as nodes inferiorly and bilaterally. **B,** Dominant masses are seen on the T1WI, but which do you biopsy? Fine-needle aspiration revealed lymphoid aggregates. **C,** Lymphoma of the parotid gland and Sjögren syndrome in another patient with rheumatoid arthritis. The axial computed tomography scan shows a mass (M) in the left parotid gland diffusely infiltrating its superficial and deep portion. The right gland is not normal, with mixed pattern of glandular density and fatty replacement. Note widening of the anterior atlantodental interval due to ligamentous laxity from the patient's rheumatoid arthritis.

of a pharyngocele or a laryngocele filled with fluid (Fig. 14-10). The most common cause of sialoceles is penetrating trauma, although blunt trauma may also cause disruption of the duct and leakage of salivary contents into the parenchyma and outside the gland. This most commonly occurs in the parotid gland, either from a punch to the side of the face or from a stab wound. The entity is distinguished from a pseudocyst because it communicates with the parent duct and is not lined by fibrous tissue.

Ranula

This is a postinflammatory pseudocystic lesion that results from obstruction and focal rupture of either the sublingual or submandibular duct and that produces a cystic mass either confined by the mylohyoid muscle (simple ranula, epithelial lined) or extending to the submandibular region (a plunging ranula, not epithelial lined) by passing through the mylohyoid dehiscence known as the "boutonniere" (see Fig. 13-24). The name ranula is derived from the word "rana," which means frog in Latin, since the shape of the pseudocyst

has been likened to that of a frog. A ranula has also been termed a "mucous escape cyst," a mucous retention cyst, and a mucocele of the sublingual gland or neighboring minor salivary glandular tissue. The simple ranula is usually addressed transorally, but may be treated with resection or, in some cases, marsupialization. The lingual and hypoglossal nerves must be carefully identified during the operation. A plunging ranula may be excised through a transcervical submandibular incision with a neck dissection. This allows complete resection of the cyst and will help spare the lingual and hypoglossal nerve. Alternatively, the surgeon may excise the sublingual gland transorally and pack the cyst or place a drain in it. By treating the gland, some believe the plunging cyst will resolve on its own.

Retention Cysts

Retention cysts are very common benign "masses" that result from inflammation and obstruction of minor salivary gland ducts and therefore may be seen throughout the aerodigestive system's mucosal surface.

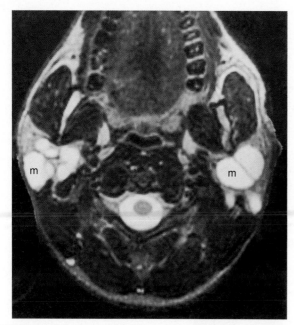

FIGURE 14-9 Lymphoepithelial lesions associated with human immunodeficiency virus (HIV). Parotid glands have multiple high intensity masses (m) on this fat-suppressed fast spin echo T2-weighted imaging, typical of lymphoepithelial cysts in this HIV-positive man.

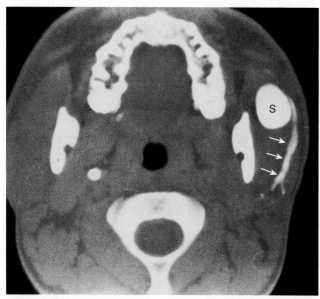

FIGURE 14-10 Sialocele. Axial computed tomography after left parotid sialography demonstrates opacification of a sialocele (s). One can see the normal parotid duct *(arrows)* coursing to and communicating with the sialocele on the left side. This patient had been punched in the left side of the face.

Miscellaneous Inflammatory Disorders

Sarcoidosis may also affect the parotid gland, usually manifesting as bilaterally enlarged glands with multifocal nodules. Gallium uptake on nuclear medicine scans may be striking. Multiple tiny calcifications in the parotid glands may be seen in sarcoidosis, Sjögren, tuberculosis, and lupus.

A mucus plug in the duct may also cause a painful swollen gland (Kussmaul disease). An inflammatory pseudomass associated with calcifications in the gland is termed

---

the "Kuttner tumor," a focal firmness of the submandibular gland caused by chronic sialadenitis from sialolithiasis.

Benign Neoplasms

There is an adage that salivary glands are unlike National Football League middle linebackers—the larger they are, the less malignant they are. Thus, the rate of malignancy increases from 20% to 25% in the parotid gland to 40% to 50% in the submandibular gland and 50% to 81% in the sublingual glands and minor salivary glands. The larger the size of the gland, the higher the rate of benign versus malignant tumors, at least in adults. In contradistinction to adults, the larger the gland of origin of the tumor in children, the more likely the tumor will be malignant. In children, 90% to 95% of salivary tumors occur in the parotid and 5% occur in the submandibular and sublingual glands. Sixty-five percent of salivary gland neoplasms in children are benign.

Although CT and MR dominate the evaluation of the major salivary glands for neoplasms, do not count sonography out completely, particularly if you live on the fish-and-chips/paella/bratwurst/borscht side of the Atlantic. It has been reported that 95% of space-occupying lesions of the major salivary glands can be completely delineated by sonography. All salivary gland neoplasms are hypoechoic to normal glandular tissue. Ultrasound (US) correctly assesses whether a lesion is benign or malignant in 90% on the basis of definition of the margins of the tumor, but 28% of malignant lesions may be misinterpreted as being benign. Sonography differentiates extraglandular from intraglandular lesions with an accuracy of 98% (all mistakes were periparotid lymph nodes). The authors hope that this last paragraph will help the overseas sales of our book despite France's arguments with our foreign policy (now back to costly CT and MR!).

Pleomorphic Adenoma

Nearly 80% of benign parotid neoplasms are pleomorphic adenomas (Box 14-3). Pleomorphic adenomas, also known as benign mixed tumors (BMT), occur most commonly in middle-aged women. Most pleomorphic adenomas are well-defined lesions that commonly appear solid and round. Pleomorphic adenomas are well identified on both CT and MR against the fatty background of the normal adult's parotid gland (Fig. 14-11). On CT, the lesions have density similar to that of muscle and demonstrate mild to moderate enhancement. On dynamic MR and CT scanning, a time to peak of less than 120 seconds and washout ratio of less than 30% suggests malignant tumors over benign with 91% accuracy. This appears to correlate with microvessel count and cellularity. An overlap of peaks between malignancies and Warthins occurred but not with washout ratios.

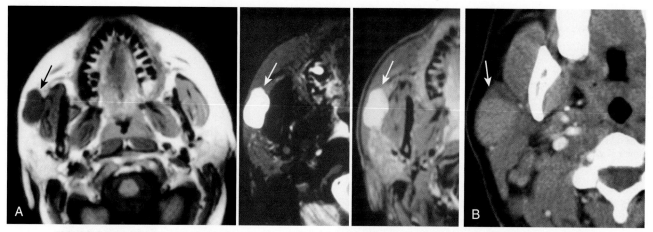

FIGURE 14-11 Pleomorphic adenoma. **A,** T1-weighted imaging (T1WI), T2WI, and enhanced T1WI scans from left to right show a typical pleomorphic adenoma *(arrows),* well-defined, very bright on T2WI, and enhancing homogeneously. **B,** The well-defined nature is also seen on a computed tomography scan. On delayed scanning the pleomorphic adenoma accumulates more and more contrast even as the vascular opacification fades (not shown).

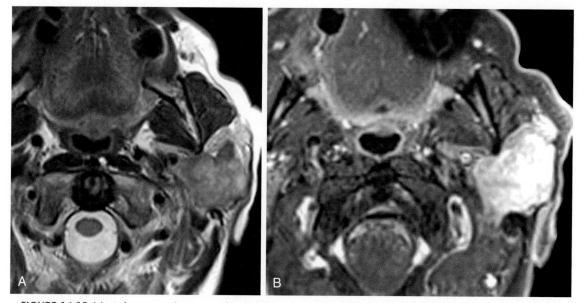

FIGURE 14-12 Meet the ex: carcinoma ex pleomorphic adenoma, that is. **A,** Axial T2-weighted image (T2WI) shows a heterogeneous signal mass replacing most of the left parotid gland. **B,** The mass enhances on postcontrast fat-suppressed T1WI. On resection, carcinoma was seen arising from a pleomorphic adenoma histopathologically.

With a delay, one may see an increase in the degree and homogeneity of enhancement in parotid pleomorphic adenomas. On MR, the lesions are best identified on T1WI amid the bright signal of the parotid fat; however, they are usually seen on T2WI as very bright lesions (add that to your 80% rule; 80% of bright lesions in the parotid are pleomorphic adenomas). Additional MR findings include a complete capsule (often low intensity on T2WI) and a lobulated contour. Pleomorphic adenomas inconstantly have cystic degeneration or calcification within them. Because the incidence of calcification is so much lower in other types of parotid tumors, the presence of calcification nonetheless suggests pleomorphic adenoma.

Pleomorphic adenomas are the most common tumor of the parotid gland in both its superficial and deep portions. In fact, pleomorphic adenomas are the most common benign tumor of the submandibular, sublingual, and minor salivary glands. If you guess pleomorphic adenoma, you will have a better batting average than Ted Williams.

They are multicentric in 0.5%, but as they account for 80% of benign tumors and benign tumors are 80% of all tumors, this means that approximately 3% of multiple parotid neoplasms are pleomorphic adenomas.

If left alone to grow, a pleomorphic adenoma will degenerate into or coexist with a carcinoma ex pleomorphic adenoma in a significant (10% to 25%) percentage of cases within 25 years. Yelled the ENT surgeon, "Cut it out!" It is not known whether a carcinoma is present within the pleomorphic adenoma from its outset (malignant mixed tumor) or whether this is a manifestation of malignant transformation (carcinoma ex pleomorphic adenoma), but most pathologists favor the latter theory (Fig 14-12). The luminal epithelial cells undergo malignant change more commonly than the myoepithelial cells. In any case, it is because of this carcinomatous association that pleomorphic adenomas are removed in their entirety with a cuff of normal salivary tissue in most instances.

Rarely, they may metastasize (metastasizing pleomorphic adenoma). The average age of carcinoma ex is 20

BOX 14-4   Multiple Masses in the Parotid Gland

Lymph nodes (normal or abnormal)
Lymphoepithelial lesions
Lymphoma
Pleomorphic adenomas
Metastases
Warthin tumors
Sarcoidosis
Sjogren syndrome
Acinic cell carcinomas

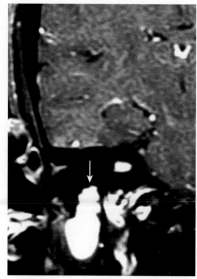

FIGURE 14-14 Facial nerve schwannoma. This schwannoma ascends up the intramastoid portion of the facial nerve on this post-gadolinium T1-weighted coronal scan.

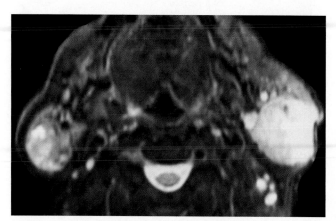

FIGURE 14-13 Warthin tumor. This T2-weighted image shows bilateral masses with heterogeneous mixed signal.

years older than pleomorphic adenoma, lending support to the theory of malignant degeneration of a benign tumor.

Monomorphic adenomas (myoepitheliomas, oncocytic adenomas, canalicular adenomas) look very much like pleomorphic adenomas but are more commonly seen in submandibular glands. They are well defined and enhance.

Warthin Tumor

Warthin tumors are also known as cystadenoma lymphomatosum. Guess what? They can be cystic or lymphoma-like in their appearance. These tumors are nearly exclusive to the parotid gland and are the most common multiple and bilateral tumors in the parotid (Box 14-4). As opposed to pleomorphic adenomas, which are generally seen in middle-aged women, Warthin tumors are most commonly seen in elderly men. Warthin tumors may have a tumoral cyst and favor the parotid's tail (think dirty old men chasing tail). These lesions are entirely benign and show no evidence of malignant transformation. Therefore, if an FNA identifies a lesion as a Warthin tumor, surgeons may conservatively watch the tumor rather than remove it, although that may break their heart. On MR, the lesions are well seen hypointense masses on T1WI contrasting with the native high signal intensity of the parotid gland. However, the signal intensity on T2WI is often heterogeneous and variable and may overlap that of the bright signal of pleomorphic adenomas (Fig. 14-13) or the darker intensity of malignancies of the parotid gland (see following section). Warthin tumors, like oncocytomas, have increased uptake on technetium 99m pertechnetate nuclear medicine scans. Therefore, if an FNA is equivocal or nondiagnostic, you could recommend a nuclear medicine technetium scan to make the diagnosis of Warthin tumor.

Oncocytoma

The oncocytoma is a relatively rare benign tumor almost exclusively seen in the parotid gland (of the salivary glands). These lesions have been called the vanishing parotid mass because, while they are hypointense on T1WI, they are isointense to the native parotid gland on fat-saturated T2 and postcontrast T1 imaging. The tumors may also take up technetium on nuclear medicine scans.

Hemangioma

Hemangiomas are the most common salivary gland tumors seen in children, girls more than boys. Hemangiomas enhance brightly and are very hyperintense on a T2WI. Cutaneous hemangiomas may coexist. Congenital capillary hemangiomas represent 90% of parotid gland tumors in neonates. They will undergo spontaneous resolution by adolescence in most cases. Low-flow vascular malformations of lymphatic and venous origin are the next most common benign lesions to affect pediatric parotid gland. The third most common benign tumor of the pediatric parotid gland is the pleomorphic adenoma.

Other Benign Tumors

Lipomas, schwannomas (Fig. 14-14), and neurofibromas may also be seen in association with the salivary glands, most commonly the parotid. Lipomas have fat density and present in children. The neurogenic tumors generally follow cranial nerves V (submandibular, sublingual, and parotid glands) or VII (parotid glands). Be careful to trace these tumors along the expected course of the nerves, even to their foramina. What lies at the bottom of the ocean and twitches? A nervous wreck.

Malignant Neoplasms

Patients with parotid malignancies usually have a palpable, discrete, painless mass (98% of cases). Other presentations include facial nerve dysfunction (24%) and cervical adenopathy (6%). Facial nerve paralysis associated with a parotid mass usually means a malignancy is present. Of

Classification of Salivary Gland Malignancies

Primary Tumor (T)
TX—Tumor that cannot be assessed
T0—No evidence of primary tumor
T1—Tumor <2 cm in diameter without extraparenchymal extension
T2—Tumor >2 cm but <4 cm in diameter without extraparenchymal extension
T3—Tumor >4 cm and/or tumor having extraparenchymal extension*
T4a—Tumor invades the skin, mandible, ear canal, and/or facial nerve
T4b—Tumor invades skull base and/or pterygoid plates and/or encases carotid artery
Nodal Involvement (N)
NX—Regional lymph nodes cannot be assessed
N0—No regional lymph node metastasis
N1—Metastasis in single ipsilateral lymph node, 3 cm in greatest dimension
N2—Metastasis in single ipsilateral lymph node, >3 cm but <6 cm in greatest dimension, or multiple ipsilateral lymph nodes, none >6 cm in greatest dimension, or bilateral or contralateral lymph nodes, none >6 cm in greatest dimension
   N2a—Metastasis in single ipsilateral lymph node >3 cm but <6 cm in greatest dimension
   N2b—Metastasis in multiple ipsilateral lymph nodes, none >6 cm in greatest dimension
   N2c—Metastasis in bilateral or contralateral lymph nodes, none >6 cm in greatest dimension
N3—Metastasis in lymph node >6 cm in greatest dimension
Distant Metastasis (M)
MX—Presence of distant metastasis cannot be assessed
M0—No distant metastasis
M1—Distant metastasis

*Note: Extraparenchymal extension is clinical or macroscopic evidence of invasion of soft tissues. Microscopic evidence alone does not constitute extraparenchymal extension for classification purposes.

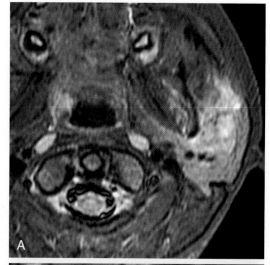
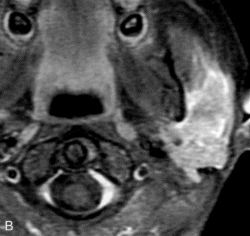

FIGURE 14-15 Mucoepidermoid carcinoma. **A,** Axial T2-weighted imaging (T2WI) scan shows a poorly defined heterogeneous signal lesion in the parotid gland focally invading the adjacent masseter and pterygoid musculature. **B,** Axial fat-suppressed contrast-enhanced T1WI shows heterogeneous enhancement in the mass as well as the invasive component in the masseter muscle. Invasive mass in the parotid in a kid? ...Best guess is mucoepidermoid carcinoma, which this turned out to be.

the malignancies to cause a facial nerve paralysis, adenoid cystic carcinoma and undifferentiated carcinoma predominate. The staging of malignant salivary gland lesions is outlined in Box 14-5.

The differentiation of deep or superficial parotid malignancies is critical from the standpoint of the extent of dissection needed to separate the nerve from the tumor or to gain access to the tumor, the attendant risk to the facial nerve, and in the case of tumors extending into the parapharyngeal space, the need for a cervical approach with or without mandibulotomy. Demonstration or suspicion of direct invasion of the nerve at the stylomastoid foramen (or above) prods the surgeon to plan for transmastoid (temporal bone surgery) identification of the facial nerve to control disease and prevent tumor spillage. The superficial parotidectomy thereby becomes skull base surgery with its attendant risks (to the other cranial nerves, venous sinuses, carotid artery, and temporomandibular joint function) and morbidity. If the skull base is invaded, the cartilaginous auditory canal may have to be addressed and possibly resected. A radical mastoidectomy is contemplated and even the ascending ramus of the mandible may be removed.

The main role of the radiologist in the evaluation of malignant parotid masses, because there are no specific imaging findings to suggest a specific histologic diagnosis, is to describe the location of the lesion (deep versus superficial), assess for lymphadenopathy or extraglandular spread, and detect perineural tumor spread that would render a diagnosis of malignant neoplasm more likely and influence the surgical/therapeutic approach. But, like political pundits, we enjoy speculating on the ultimate (histologic/political) winner (diagnosis) and the route to the endpoint. Don't beat around the Bush, Rand-er a diagnosis, and tell the pathologist to "Trump that!"

Mucoepidermoid Carcinoma

Mucoepidermoid carcinomas account for 30% of all salivary gland malignancies, and is the most common malignant lesion of the parotid gland (60% of them occur in the parotid gland; Fig. 14-15). Mucoepidermoid carcinomas, like squamous cell carcinoma, can be graded from low to high, and the prognosis varies with the grade. Mucoepidermoid carcinoma is the most common pediatric salivary gland malignancy. Thirty-five percent of salivary gland neoplasms in children are malignant and of these, 60% are mucoepidermoid carcinomas. Primary salivary gland lymphoma does occur in the parotid gland as can leukemic infiltration in children.

In the parotid gland, a lesion's morphology may be misleading as far as predicting benignity versus malignancy. Some pleomorphic adenomas have tentacles with an irregular margin. By the same token, some mucoepidermoid carcinomas are well defined by a pseudocapsule and do not appear to be invasive. Therefore, you cannot rely on definition of margins to distinguish cancer from benign tumors. Just like shape cannot differentiate one Jenner from the true Jenner.

On CT, most tumors of the parotid gland have a density equal to muscle—no help there. Low-grade mucoepidermoid carcinomas may have high T2W signal intensity, but high-grade, poorly differentiated mucoepidermoid carcinomas and parotid malignancies of other histologies are usually low in intensity on T2WI. A useful adage is that if the T2W signal intensity is low, the lesion should be biopsied for probable malignancy (careful, Warthin tumors may be dark on T2WI and simulate a high-grade mucoepidermoid carcinoma). If the lesion is bright on T2WI, you can assume it is a pleomorphic adenoma and proceed accordingly without presurgical sampling. MR also does well in demonstrating the perineural, vascular, and dural invasion that may be present with parotid malignancies. Malignancies enhance, have low apparent diffusion coefficient (ADC) values, and have *lower* blood flow and blood volume (compared with benign parotid tumors in contradistinction to brain tumors).

Adenoid Cystic Carcinoma

Adenoid cystic carcinoma (ACC) is the second most common primary malignancy of the parotid gland and the most common tumor of the submandibular, sublingual, and minor salivary glands. Similar to the mucoepidermoid carcinoma, variable intensity occurs with the T2WI (Fig. 14-16). Noncystic ACC masses in the parotid gland have muscular CT density, low intensity on T1WI, and mild to moderate enhancement. ACCs are notorious for their propensity for perineural spread (50% to 60%) and persistence despite "complete surgical removal." They are like dust mites; you cannot get rid of them, they keep multiplying, and you learn to live with them. An ACC of the parotid gland may spread via the ramifications of cranial nerve VII retrograde into the temporal bone or may spread via the auriculotemporal branches of cranial nerve V to the Meckel cave region through the foramen ovale (see Fig. 14-16, *B*). In the other salivary glands, adenoid cystic carcinoma generally demonstrates perineural extension along the branches of the second and third divisions of cranial nerve V. This cancer is a relentless, slow-growing tumor whose prognosis is generally measured in terms of decades rather than 5-year survival rate because of its prolonged course.

Squamous Cell Carcinoma

Squamous cell carcinoma may be seen within the parotid gland. Sometimes it is difficult to tell whether the squamous cell carcinoma is present secondarily because of invasion of lymph nodes from a primary site outside the parotid or is intrinsic to the parotid gland (Fig. 14-17). How it gets there is mysterious because squames are not intrinsic to glandular tissue; they are presumably caused by metaplasia of the ductal columnar epithelium into

squamous cells. This same difficulty lies with lymphoma of the parotid—is it a primary parotid tumor or secondary spread? When multifocal in the parotid, it is generally accepted that the squamous cell carcinoma is probably within lymph nodes in the parotid gland. A search for a primary tumor is undertaken in the overlying skin and ear as primary sites that drain to the parotid gland. Squamous cell carcinoma does not generally occur in submandibular, sublingual, or minor salivary glands as a primary site, although it certainly spreads to them from adjacent mucosal surfaces. Squamous cell carcinomas are virtually always hypointense on T2WI unless necrosis coexists.

Adenocarcinoma

Adenocarcinomas may also arise within the parotid gland, sublingual gland, submandibular gland, or minor salivary glands. This lesion generally has a worse prognosis than that

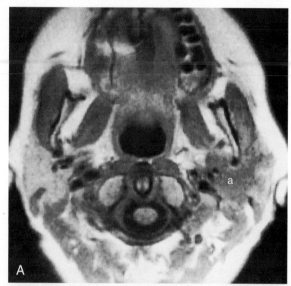

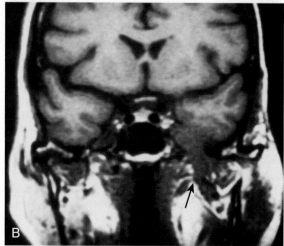

FIGURE 14-16 Adenoid cystic carcinoma of the parotid gland. **A,** T1-weighted imaging of the left parotid shows infiltrative mass (a) extending into both superficial and deep portions of the gland. Note how well the mass is identified by the replacement of the normal high intensity parotid tissue. Better watch cranial nerve VII on this lesion. Mind your facials! **B,** Woops! There it goes up the foramen ovale (*arrow*) coursing along the mandibular nerve. How did it get there? The auriculotemporal branch of V₃, which is in close proximity to VII branches. Fooled you!

of mucoepidermoid carcinoma and adenoid cystic carcinoma (Box 14-6). This tumor is derived from the glandular tissue itself as opposed to ductal tissue. Signal intensity is variable, depending on mucinous, cystic, or solid contributions, but usually dark on T2WI and enhance. Remember that some adenocarcinomas (carcinoma ex) occur from malignant degeneration of pleomorphic adenomas.

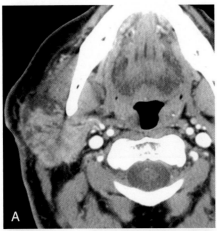

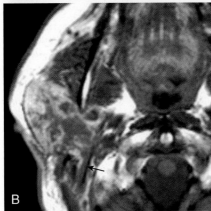

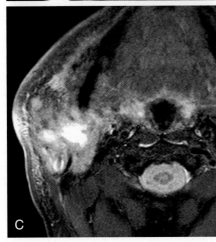

FIGURE 14-17 Squamous cell carcinoma of the parotid gland. **A,** Postcontrast axial computed tomographic image in a patient with facial swelling and history of right periauricular squamous cell carcinoma shows an infiltrative mass nearly completely replacing the right parotid gland. **B,** Axial T1-weighted image (T1WI) and **(C)** T2WI show the same infiltrative lesion. Note that some T1 dark tumor is sneaking back toward the stylomastoid foramen *(arrow),* very worrisome for facial nerve extension.

The polymorphous low-grade adenocarcinoma of the salivary gland (PLAC) is a low-grade neoplasm that predominantly occurs in the minor salivary glands of the oral cavity (mucosa of the soft and hard palates, in the buccal mucosa, and in the upper lip) but can be seen in the parotid gland. It has a much more benign prognosis and course than the traditional butt-kicking adenocarcinoma.

Acinic Cell Carcinoma

Acinic cell carcinoma (Fig. 14-18) is the most common primary multifocal malignancy to affect the parotid gland. Its prognosis is intermediate between that of mucoepidermoid carcinoma and adenocarcinoma of the parotid gland. This tumor is seen almost exclusively in the parotid gland, and the incidence of bilateral acinic cell carcinomas is approximately 1% to 3%. No specific imaging features other than its multifocality suggest this diagnosis. Consider it the spitting image of adenocarcinoma (pun-time!)

Undifferentiated Carcinoma

Undifferentiated carcinoma is also seen as a salivary gland tumor. This lesion has a very poor prognosis but fortunately is rarely seen.

Metastatic Adenopathy

The incidence of nodal metastases in untreated parotid cancers is low except in T3 or T4 lesions. Although mucoepidermoid carcinoma and squamous cell carcinoma metastasize to nodes in 37% to 44% of cases, the other histologic types are only associated with lymphadenopathy in 5% to 21%, with adenoid cystic carcinoma being the lowest. ACC "plucks" your nerves instead of nodes.

MASTICATOR SPACE

Anatomy

The masticator space is defined by layers of the deep cervical fascia and encompasses the muscles of mastication (the medial and lateral pterygoid, masseter, and temporalis muscles) as well as the condyle and ascending ramus of the mandible, branches of the external carotid artery and third division of cranial nerve V, and venous branches from the jugular system (Fig. 14-19). Of the muscles of mastication, the small lateral pterygoid has a primary function of opening the mouth, whereas the bulky medial pterygoid, masseter, and temporalis muscles serve to keep the mouth closed. Given that the Lord has given us three huge

---

BOX 14-6 **Five-Year Prognosis of Parotid Malignancies (Percentage of Parotid Malignancies)**

GOOD

Acinic cell carcinoma (12%)
Adenoid cystic carcinoma (6%-14%)
Low-grade mucoepidermoid carcinoma (25%)

POOR

High-grade mucoepidermoid carcinoma (20%)
Adenocarcinoma (8%)
Squamous cell carcinoma (1%-7%)
Carcinoma ex pleomorphic adenoma (9%)
Undifferentiated carcinoma (9%)

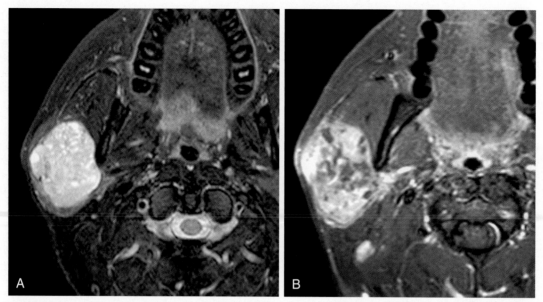

FIGURE 14-18 Acinic cell carcinoma. **A,** Axial T2-weighted image (T2WI) shows a well circumscribed mass in the right parotid gland. So you're thinking pleo, right? **B,** The mass enhances heterogeneously on T1WI. This turned out to be acinic cell carcinoma. Thank goodness for FNAs!

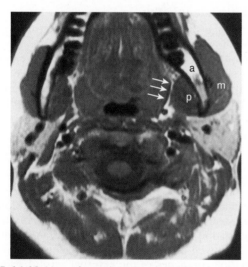

FIGURE 14-19 Normal masticator space. Masseter muscle (m), pterygoid muscle (p), and angle of the mandible (a) are well visualized on this T1-weighted imaging scan. Note that a masticator space lesion would displace parapharyngeal fat *(arrows)* medially and predominantly posteriorly.

muscles to keep our yappers shut and only one small one to open our mouths, you can see why we feel politicians and lawyers are defying the rightful laws of nature and should have their lateral pterygoids removed.

How do you identify a lesion as being within the masticator space? Chew on this: When a pterygoid masticator space lesion is present, the parapharyngeal fat is displaced posteromedially and may be infiltrated along its anterolateral aspect (see Table 14-1).

Inflammatory Lesions

Most of the inflammatory lesions of the masticator space relate to infections of odontogenic origin (Fig. 14-20). Therefore, abscesses around the teeth, osteomyelitis of the mandible, and cellulitis associated with carious teeth are the prime offenders in this category (Fig. 14-21). Look for lucency

surrounding the suspected tooth root. You could also see dehiscence of the adjacent maxillary or mandibular cortex, and there just might be an accompanying subperiosteal abscess. These might be small and difficult to really appreciate without intravenous contrast, but you should still look. Why? If not addressed, the infection can become quite extensive, involving other soft tissues of the face and neck, causing airway compromise (Ludwig angina), thrombophlebitis, and vision loss (from orbital extension). It is important to let the clinician know of the offending tooth, because no matter how the abscess gets drained, the problem won't be solved until that tooth is dealt with. Dearest students, please floss regularly.

Occasionally, cellulitis and myositis of the masticator space develop as a result of penetrating injuries, superficial facial infections, sinus tracts, or adjacent parotitis. Most inflammatory lesions of the masticator space can be readily identified on CT or MR. On CT, look for thickening of the adjacent muscle, infiltration of the nearby fat, subcutaneous tissue, or skin with a stranding pattern to it, or a tract leading to the skin or mucosal surface. If an abscess has developed, it will appear in a fashion similar to abscesses elsewhere, with a low-density center and a peripheral enhancing rim, often bordering on the bad tooth. On MR, inflammatory lesions, because of their high water content from edema, are very bright on T2WI. There are some exceptions to this rule, namely actinomycosis (because of sulfur granule deposition) and fungal infections (because of paramagnetic iron and manganese accumulation).

Bruxism and Atrophy

Bruxism may develop in patients who constantly gnash their teeth. This is a fancy word for (usually) bilateral enlargement of the muscles of mastication. It occurs most commonly idiopathically but may develop as a result of malocclusion, excessive chewing, or clenching the teeth (as occurs when radiologists are called to witness stand). Rarely, this may be a unilateral phenomenon. Conditions that can result in an enlarged appearance of the muscles of mastication are listed in Box 14-7.

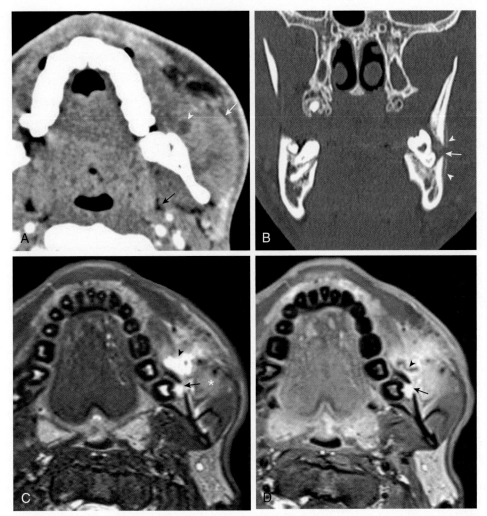

FIGURE 14-20 A plea to brush regularly. **A,** Axial contrast-enhanced computed tomography (CT) shows marked fullness of the left masseter muscle *(white arrow)* with low density, possibly necrotic component *(arrowhead)*. There is also fullness of the pterygoid musculature resulting in displacement of the parapharyngeal fat *(black arrow)*. Could this be a tumor? **B,** Coronal CT shows periapical lucency surrounding the left mandibular molar tooth, with dehiscence of the buccal cortex *(arrow)* and periosteal reaction along the mandibular ramus *(arrowheads)*. Hmmm, this looks like a case of dental infection gone really bad, with extraosseous extension into the masticator soft tissues, abscess in the masseter, and osteomyelitis of the mandible. **C,** Axial T2-weighted and **(D)** postcontrast T1-weighted images show the sinus tract from mandible to overlying soft tissues *(arrows)*, as well as the abscess *(arrowhead)* and edema *(asterisk* in C) in the masseter muscle.

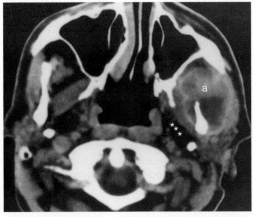

FIGURE 14-21 Abscess in masticator space. Axial computed tomography shows an abscess (a) centered on the left mandible. Note that the parapharyngeal fat *(asterisks)* is displaced postero-medially by this masticator space infection.

BOX 14-7   Infiltrative or Expansile Lesions of the Muscles of Mastication

Bruxism
Neurofibromatosis
Beckwith Weidemann syndrome
Glycogen storage disease
Hypothyroidism
Lipodystrophy
Proteus syndrome

Alternatively, one may see atrophy of the muscles of mastication when one has a cranial nerve V abnormality (Fig. 14-22). This may be caused by lesions in the peripheral branches of the third division of cranial nerve V. Perioperative injury may cause denervation atrophy as can trauma. With denervation atrophy one may see volume loss, fatty infiltration, enhancement after gadolinium administration

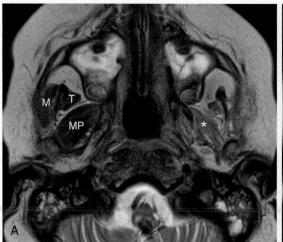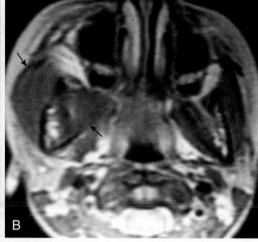

FIGURE 14-22 A, Axial T2-weighted image in a young child with metastatic medulloblastoma involving the trigeminal nerve intracranially. There is marked atrophy of the left muscles of mastication *(asterisk)* due to V3 disease. Compare to the normal bulk of the medial pterygoid (MP), and the insertions of the temporalis (T) and masseter (M) muscles on the right. **B,** The muscles *(arrows)* on the right in a different patient are larger than the left. This could be from malocclusion or a glycogen storage disease infiltration.

and/or high signal intensity on T2W scans in the muscles of mastication (in reverse order chronologically). Neurogenic tumors may also induce these imaging characteristics.

Fibrous Dysplasia

Fibrous dysplasia causes enlargement of the mandible and maxilla with expansion of the outer cortex of the bone, generally without erosion through the bone. McCune-Albright syndrome refers to precocious puberty, café-au-lait spots, and polyostotic fibrous dysplasia, (which may affect the mandible unilaterally) in girls. When a patient has bilateral bubbly bone lesions with a ground-glass appearance in the maxilla or mandible, consider either fibrous dysplasia or cherubism, which is a related disorder (see Fig. 13-21, *B*). Both of these disorders tend to expand during adolescence, are possibly related to hormonal influences, and either regress or stabilize in young adulthood.

The skull and facial bones are involved in 10% to 25% of cases of monostotic fibrous dysplasia and in 50% of polyostotic fibrous dysplasia. The mandible, maxilla, and calvarium are frequently involved.

Temporomandibular Joint Syndrome

The temporomandibular joint (TMJ) falls under the rubric of the masticator space. The TMJ may be the source of chronic facial or head pain. TMJ (or chronic maxillofacial pain) syndrome is seen in women nine times more frequently than men, is often precipitated by a traumatic event, and is very poorly understood, although, because it often regresses around menopause, is felt to be hormonally mediated. Presently, most imaging is performed with MR where the disk, condyle, glenoid fossa, and surrounding soft tissues may be evaluated. The disk has an anterior triangular band and a larger posterior band, which are joined in the middle by an intermediate zone (Fig. 14-23). The posterior band is attached to the posterior joint by the retrodiskal tissue, or bilaminar zone. The disk should be centered over the condylar head in open- and closed-mouth positions on sagittal proton density–weighted (PDW) MR with the posterior

margin of the posterior band between the 11- and 12-o'clock point of the condyle in the closed-mouth position. The joint capsule itself has an anterosuperior compartment and an inferior compartment, which usually do not communicate, allowing effusions in multiple compartments. The lateral pterygoid muscle opens the jaw and inserts on the anterior portion of the disk. The medial pterygoid, temporalis, and masseter muscles close the jaw (except in lawyers).

Anterior meniscal dislocations are the most common type in patients with TMJ complaints. In this setting, the disk's posterior band is dislocated anteriorly from directly over the condyle. This posterior edge of the posterior band displacement is more than 10 degrees anterior to the 12-o'clock position, more like the 9- to 10-o'clock position. It may be far in front of the condyle in the closed-mouth position on sagittal PDW imaging. The dislocation may reduce on opening (often with a clicking sound—the "opening click") and may redislocate on closing (a closing click). Thus, scanning is performed in closed and open mouthed views. Alternatively, the disk may remain anteriorly dislocated even on opening ("anterior displacement without recapture" is the phrase used in TMJ MR imaging). The location of the disk in front of the condyle may subsequently restrict the joint's motion (a closed-lock situation).

A medial or lateral component in addition to the anterior dislocation (rotational dislocation) is not uncommon, said to occur in 25% to 30% of cases. Isolated medial and lateral (termed "sideways") dislocations are relatively rare; however, with a transverse component the possibility of nonsurgical reduction of the dislocation is decreased (from 46% to 9%). For this reason a coronal MR in the closed mouth position should be performed and the disk must be assessed for medial or lateral offset from the condyle even as the disk is viewed anterior to that condyle in rotational dislocation. Posterior dislocations are extremely rare.

A stuck disk is one that does not move in open- or closed-mouth positions. It may or may not be anteriorly displaced, but it is usually fibrosed in and immobile. This may be associated with pain and disability.

Approximately 35% of subjects *without* TMJ symptoms have anterior dislocations, but 78% of patients

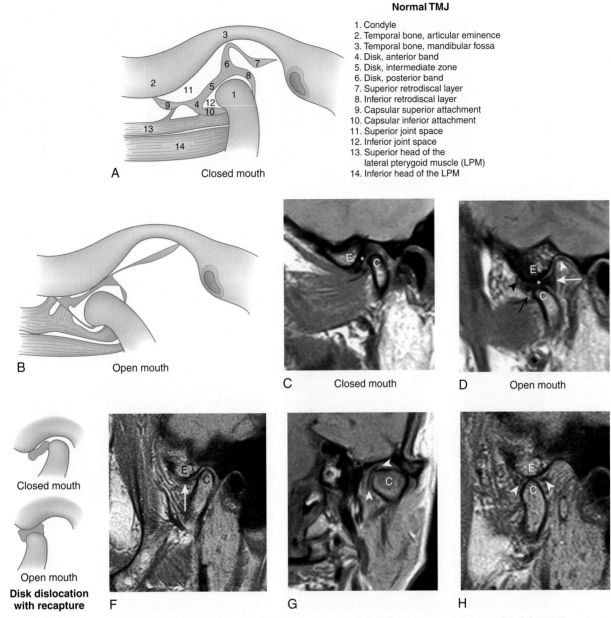

Normal TMJ

1. Condyle
2. Temporal bone, articular eminence
3. Temporal bone, mandibular fossa
4. Disk, anterior band
5. Disk, intermediate zone
6. Disk, posterior band
7. Superior retrodiscal layer
8. Inferior retrodiscal layer
9. Capsular superior attachment
10. Capsular inferior attachment
11. Superior joint space
12. Inferior joint space
13. Superior head of the lateral pterygoid muscle (LPM)
14. Inferior head of the LPM

A — Closed mouth

B — Open mouth

C — Closed mouth

D — Open mouth

E — Disk dislocation with recapture (Closed mouth / Open mouth)

F G H

FIGURE 14-23 Normal temporomandibular joint (TMJ) anatomy and disk derangements. Schematic of the TMJ in sagittal oblique plane demonstrates normal anatomic relationships in the **(A)** closed- and **(B)** open-mouth positions. Note how the mandibular condyle translates forward anteriorly on mouth opening to allow for disk recapture. **C,** Sagittal oblique proton density (PD) image in closed-mouth position shows the mandibular condyle (C), articular eminence (E), and normal disk position *(asterisk)* relative to these structures. **D,** Sagittal oblique PD image in the open-mouth position shows the normal position of the articular disk *(asterisk)* juxtaposed between the articular eminence (E) and condyle (C), indicating appropriate recapture of the disk. The posterior band *(white arrow),* superior retrodiskal later *(white arrowhead),* as well as superior *(black arrowhead)* and inferior capsular *(black arrow)* attachments, are shown. **E,** Schematic representation of anterior disk dislocation on closed mouth view with appropriate recapture despite the dislocation on mouth opening. **F,** PD image in closed-mouth position shows a thinned irregular anteriorly displaced disk *(arrow)* relative to the condyle (C). Note relationship of the disk to the articular eminence (E). **G,** A coronal T1-weighted image shows that the disk *(between arrowheads)* is also medially dislocated relative to the condylar head (C). **H,** On mouth opening, despite its degenerative appearance, this disk *(between arrowheads)* does recapture appropriately between the condyle (C) and eminence (E).

Continued

with TMJ symptoms have disk displacements. The degree of disk displacement is greater in symptomatic than asymptomatic individuals and recapture of the displaced disk is less frequent in patients than asymptomatic volunteers (where it is the norm). Bilateral involvement is twice as common in patients as in volunteers. Clearly, the baseline incidence of this disease is very high.

What causes the pain? The size of TMJ effusions correlates well with subjective pain scales. To demonstrate an effusion, one performs a T2WI in the sagittal plane for bright fluid (see Fig. 14-23, *K*). Other authors have focused on the retrodiskal tissues and believe that inflammation (again bright on the sagittal T2WI) there may irritate nerve fibers, accounting for the pain. The TMJ is to facial pain what the lumbar spine is to back pain. Shrug and call what you see.

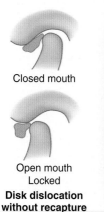

Closed mouth

Open mouth
Locked

**Disk dislocation
without recapture**

I

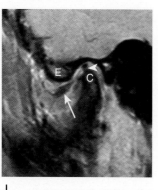

J

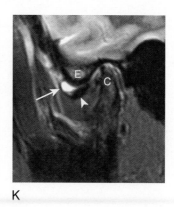

K

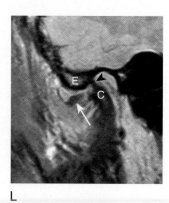

L

FIGURE 14-23, cont'd I, Schematic representation of anterior disk dislocation on closed-mouth view without appropriate recapture on mouth opening. Note how the condyle shows very limited anterior translation, giving the "locked" appearance. **J,** PD image shows anteriorly displaced irregular disk on closed-mouth position *(arrow)*. The eminence (E) and condyle (C) position are noted. Note the osteophyte at the joint space *(arrowhead)* that's got to be very annoying. **K,** Sagittal oblique T2 image shows the same anteriorly displaced disk *(arrowhead)*, condyle (C), and eminence (E) on closed-mouth view. There is joint fluid present *(arrow)*, which can correlate with patient pain. **L,** On mouth opening, the pesky osteophyte *(arrowhead)* limits anterior translation of the condyle (C), and the displaced disk *(arrow)* does not recapture. Eminence indicated by E. **(A** and **B** redrawn with permission from Tomas X, Pomes J, Berenguer J, et al. MR imaging of temporomandibular joint dysfunction: a pictorial review. *Radiographics.* 2006;26:765-778.)

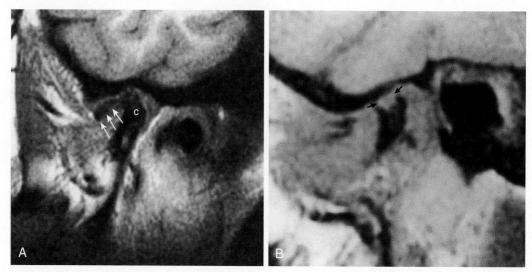

A

B

FIGURE 14-24 Avascular necrosis of the temporomandibular joint (TMJ). **A,** Sagittal T1-weighted imaging (T1WI) shows decreased intensity in the mandibular condyle (c) in this patient with chronic TMJ syndrome. Note that the disk *(arrows)* is anteriorly dislocated in this closed mouth view. This patient had avascular necrosis of the condyle, confirmed by wedge resection. **B,** Erosive synovitis of the TMJ. Note the erosion *(arrows)* of the condylar head on the sagittal T1WI in this patient with rheumatoid arthritis.

MR is not able to detect disk perforations other than showing secondary findings of joint effusion and meniscal displacements. Arthrography or arthroscopy is the only means of demonstrating perforations. Perforations usually occur at the junction of the bilaminar zone with the posterior band. They occur more frequently with a transverse component to meniscal dislocations.

Initial treatment for TMJ syndrome is splinting, analgesics, behavior modification, and muscle relaxants. If the patient fails to respond to this noninvasive therapy and has meniscal abnormalities, plication of the disk into a more normal location or meniscal removal with or without prosthesis may be attempted. Adhesions may be lysed if present under arthroscopy. Complications related to the surgery including foreign body reactions, granular cell reactions,

or avascular necrosis of the condylar head (which may also occur even without surgery). The latter may be identified as low signal intensity on all pulse sequences in the condylar head. Sclerosis is seen on CT (Fig. 14-24).

In the long term, degenerative changes in the joint may develop and include narrowing, osteophytic bird-beaking of the condyle, eburnation, ankylosis, joint mice/loose bodies, and fragmentation.

Rheumatoid arthritis may also affect the TMJ. Usually erosions of the condyle are seen with a soft-tissue component (see Fig. 14-24, *B*). Effusions and meniscal perforations also complicate the rheumatoid joint. Other arthritides include septic arthritis, gout, and pseudogout of the TMJ.

The TMJ may be affected by such lesions as pigmented villonodular synovitis (characterized by hemosiderin or other blood products seen as dark signal on T2* MR), synovial chondromatosis (Fig. 14-25), chondroid masses (Fig. 14-26) and tumoral calcinosis.

Benign Neoplasms

Hemangiomas and Vascular Malformations

In 1982, a new classification of hemangiomas and vascular malformations was proposed by Mulliken and Glowacki and has since been adopted. In this new design, infantile hemangiomas have been relegated to those benign neoplasms of endothelial cells that may or may not be present at birth. Congenital hemangiomas are visible at birth, whereas infantile hemangiomas are seen after a few weeks of life. The former

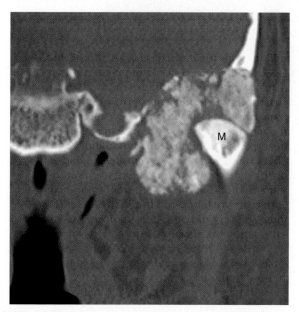

FIGURE 14-25 Calcified lesion of temporomandibular joint. This temporomandibular joint lesion shows a calcified matrix and destruction of the temporal bone making up the glenoid fossa. The differential diagnosis would include synovial chondromatosis, a chondrosarcoma, a chondroblastoma, calcium pyrophosphate dihydrate deposition disease, tumoral calcinosis, and Brown tumor, in decreasing order of likelihood. Final diagnosis: synovial chondromatosis.

may demonstrate a rapidly involuting congenital hemangioma (RICH) or noninvoluting (NICH) progression. Both may have a rapid proliferative phase shortly after birth, and 50% completely resolve by 5 years of age and 72% by 7 years of age. Such hemangiomas occur more frequently in Caucasians and girls. They may be in the skin, subcutaneous tissues, or deep spaces of the head and neck. If the lesions fail to involute on their own, they may be prodded to do so with steroids, alpha interferon, or chemotherapy. The masses are low signal on T1WI, bright on T2WI, and enhance. As they involute, fibrofatty infiltration occurs and they enhance less. The infantile hemangioma is the most common pediatric, benign neoplasm of the masticator space. This generally begins in the superficial tissues but may extend deeply into the muscles and adjacent fat of the masticator space. Kasabach-Merritt syndrome is characterized by infantile hemangiomas associated with consumptive coagulopathy, high-output congestive heart failure, and respiratory distress with or without splenomegaly.

Vascular malformations are the other lesions that before 1982 were called "hemangiomas" and have been reclassified. They may be primarily venous, lymphatic, or combination lesions when they are low-flow ones. They are comprised of dysplastic vessels, are present at birth, grow during fast growth phase of the child, and persist. The majority of these lesions are the previously referred to venous vascular malformations and therefore are seen as slow-flow vascular channels on MR. They too are dark on T1WI, bright on T2WI, and enhance (see Fig. 14-26). These lesions may show calcifications/phleboliths on CT. Maffucci syndrome may include venous malformations and enchondromatosis. Klippel-Trenaunay-Weber syndrome has mixed capillary-venous-lymphatic malformations with hemihypertrophy. Percutaneous sclerosis may be employed as treatment for venous vascular malformations, even using ethanol under general anesthesia to control pain.

The more lymphatic the lesion the more cystic it may appear but the cystic components usually have brighter signal than CSF on T1W and fluid-attenuated inversion recovery (FLAIR) scans. Fluid-fluid levels are more common in lymphatic malformations than venous vascular malformations.

Purely lymphatic malformations, that is, the cystic hygroma family of lesions, are increased in patients with Turner syndrome, Noonan syndrome, Down syndrome,

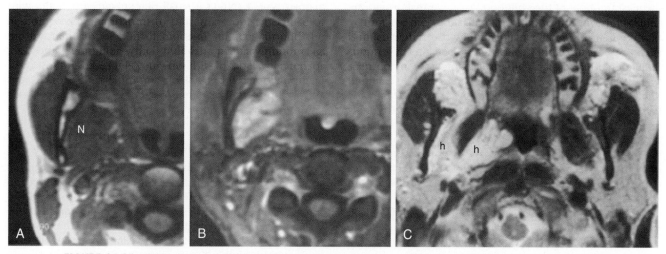

FIGURE 14-26 Venous vascular malformation of the masticator space. Note how this mass (N) that is isointense to muscle in **A** enhances so dramatically in **B. C,** Different case, and same diagnosis: T2-weighted imaging shows infiltrating venous vascular malformations (h).

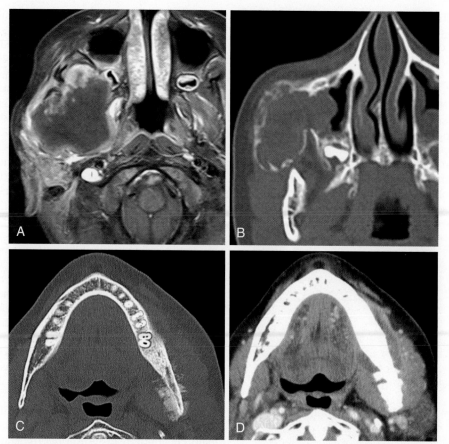

FIGURE 14-27 Sarcomas of the mandible. **A,** Ewing sarcoma arising from the right mandible extends into the masticator space on this postcontrast axial T1-weighted image. **B,** Axial computed tomography (CT) in a different patient with Ewing sarcoma shows expansile destructive mass arising from the right zygoma. **C,** Axial CT in bone window in another patient shows marked scleosis of the left mandible with striking periosteal reaction. This was an osteosarcoma that developed in a patient who underwent radiation therapy for oral squamous cell carcinoma (double bummer). **D,** CT in soft-tissue window in this same patient shows soft-tissue tumor involvement of the masseter and pterygoid muscles (compare to size on the normal right side) and fluid component along the buccal and lingual surfaces of the mandible.

and trisomy 13. They frequent the posterior triangle and axilla most commonly but can occur anywhere, detected before age 2 in 90% of cases.

High flow vascular malformations contain an arterial component and may contain a nidus or a fistula. They are the least frequent of the childhood vascular malformations but may be seen with Osler-Weber-Rendu syndrome. These are basically arteriovenous malformations of the head and neck and are uncommon.

Ameloblastoma

A relatively common benign bony tumor of the mandible is the ameloblastoma. It is labeled benign because it does not metastasize, but it is highly aggressive locally and may have malignant growth patterns. Ameloblastomas present as solitary masses of the mandible or maxilla. They are usually multiloculated and have septations within the lesion. There may be an extraosseous component to this lesion as it expands beyond bony confines. The lesion is seen as a lytic process within the mandible on CT with fine septations running within it (see Fig. 13-29). On MR, it is high intensity on T2WI and may have enhancing septa within it as well as solid components, distinguishing it from other benign odontogenic cysts.

Neurogenic Tumors

Other benign neoplasms of the masticator space include schwannomas and neurofibromas derived from the branches

of cranial nerve V that permeate the masticator space. Isolated schwannomas are the most common neurogenic tumor of the masticator space. Plexiform neurofibromas may be extensive and may be a cause of a diffuse masticator space mass and are one of the criteria defining neurofibromatosis type 1 (NF-1). A small percentage of patients with neurofibromatosis may have malignant neurogenic tumors, which may affect the masticator space. Usually, the inferior alveolar canal of the mandible is eroded and expanded by malignant neurofibromas of V_3, and the tumor may ascend into and through the skull base (foramen ovale) toward the Meckel cave to the gasserian ganglion.

Malignant Neoplasms

Osteosarcoma and Ewing Sarcoma

Most of the primary malignancies of the masticator space relate to tumors of the mandible such as osteosarcoma and Ewing sarcoma (Fig. 14-27). Osteosarcomas of the craniofacial region account for less than 10% of all osteosarcomas. When they occur, they favor the mandible (50%) and the maxilla (25%). Most are lytic in appearance; the mandibular ones may be blastic. Some occur in sites of Paget disease or radiation portals. Familial retinoblastomatosis may predispose to osteosarcomas (and other head and neck sarcomas) by virtue of the chromosome 13 oncogene. These osteosarcomas often occur in the portals of the retinoblastoma.

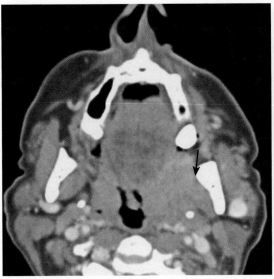

FIGURE 14-28 Squamous cell carcinoma extending into the masticator space. This left tonsil/lateral base of tongue squamous cell carcinoma grew into the pterygoid musculature *(arrows),* thereby invading the masticator space.

Both of these lesions may demonstrate periosteal reaction and either a dense hyperostotic destructive mass or a lytic process in the mandible. Osteosarcomas may have bony matrix. Ewing sarcomas occur in a younger first and second decade age group than osteosarcomas, which affect patients in their third and fourth decades of life. Rarely, ameloblastomas may demonstrate malignant potential as well.

For all mandibular or maxillary lesions, it is important to assess for cortical (erosions, defects) versus marrow (replacement of fat density/intensity) involvement. Once the bone is involved, one must be cognizant of the mental and inferior alveolar foramina to assess for perineural invasion. Always look at the pterygopalatine fossa (V2) and foramen ovale (V3) fat planes to assess for malignant spread along the nerves.

Squamous Cell Carcinoma

Squamous cell carcinoma of the mucosa is the most common malignancy to secondarily invade the masticator space. Typically, the squamous cell cancers arise in the region of the oral cavity (especially the retromolar trigone), oropharynx, or nasopharynx, where extension to the pterygoid musculature is not uncommon. Alternatively, oral cavity tumors may spread into the mandible (Fig. 14-28). Buccal cancers may invade the masseter muscle. Only the temporalis muscle is relatively spared from squamous cell carcinoma spread (except from skin cancer penetration).

Metastasis

Metastases to the bones of the masticator space are not uncommon and typically arise from breast, kidney, or lung cancer. Thyroid cancer may also be a source of metastatic involvement of the mandible. The most common metastasis to the mandible is an adenocarcinoma from the breast, and mandibular involvement occurs five times more commonly than maxillary involvement. Metastases to the mandible are more common than primary mandibular bony tumors. In children, neuroblastomas may go to the bones or soft tissue of the masticator space (Fig. 14-29).

Soft-Tissue Sarcoma

In addition to malignant lesions of the bone and spread from adjacent mucosal surfaces, some soft-tissue sarcomas

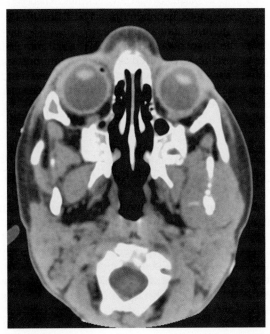

FIGURE 14-29 Metastasis from neuroblastoma to mandible. There is a soft-tissue mass centered about the left mandible on this scan. Note that this is a child. Diagnosis: neuroblastoma.

affect the masticator space. Rhabdomyosarcomas are the most common of these tumors and are generally seen in children. Rhabdomyosarcomas have a propensity for intratumoral hemorrhage and are generally bright on T2WI (Fig. 14-30). Fibrosarcomas and osteosarcomas have also been reported in the head and neck, often in association with retinoblastoma. This association may be related to the oncogene found with bilateral retinoblastomas; fibrosarcomas may be dark on T2WI. Malignant neurofibrosarcomas and synovial sarcomas may also affect the masticator space.

Lymphoma

Lymphoma in the facial nodes may present as a superficial masticator space lesion, but lymphoma may also permeate the

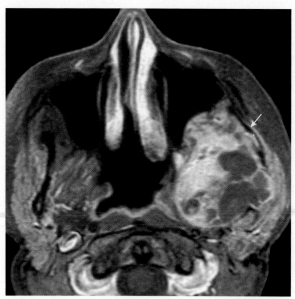

FIGURE 14-30 Rhabdomyosarcoma of masticator space: Postcontrast T1-weighted imaging scan shows diffuse infiltration of the left masticator space by rhabdomyosarcoma. Note the remodeling of the left mandible *(arrow)* and posterolateral wall of the maxillary sinus.

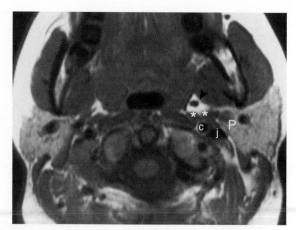

FIGURE 14-31 Normal parapharyngeal space. Axial T1-weighted imaging shows the high signal intensity of the prestyloid parapharyngeal space fat *(arrowhead)*, anteromedial to the deep lobe of the parotid gland (P). Separating the prestyloid and poststyloid parapharyngeal space is the styloid musculature *(asterisks)*. Directly behind the styloid musculature, one can identify the carotid artery (c) and jugular vein (j) within the carotid space. On computed tomography, the styloid process may be the best anatomic landmark to separate the two spaces.

mandible and/or infiltrate the muscles of mastication primarily. This is most commonly seen as non-Hodgkin lymphoma.

You got your juices flowing with salivary lesions. You have chewed on a few masticator processes. Now you will be served the foie gras, a fatty infiltrated region, the parapharyngeal space.

PARAPHARYNGEAL SPACE

Anatomy

The parapharyngeal space is classically separated into a prestyloid compartment and a poststyloid compartment (Fig. 14-31). The poststyloid parapharyngeal space contains the carotid sheath and has been called the carotid space by many authors. When cognoscenti refer to the parapharyngeal space, however, they are generally referring to the prestyloid portion. The fascia of the stylopharyngeus, styloglossus, and tensor veli palatini muscle separates prestyloid and poststyloid spaces, but one can also use the styloid process for this boundary as seen on CT. Only fat, lymphatics, and very small branches of the internal maxillary artery, ascending pharyngeal artery, and mandibular (V$_3$) nerve lie within the parapharyngeal space. Occasionally, you may find ectopic minor salivary gland tissue; however, the space, like our midsections, is dominated by its fat.

The parapharyngeal fat is a readily mobile and therefore is readily displaced and infiltrated by adjacent disease. By observing the direction of displacement of this fat, you can identify a lesion as arising from one of the deep spaces of the head and neck (see Table 14-1). Bet you never thought fat could be so useful. Just do not let it metastasize to your head.

Congenital Disorders

Other than a rare second branchial cleft cyst (BCC) (see Table 14-2) and lymphangioma, congenital lesions of the parapharyngeal space are infrequent (Box 14-8). The tract of a second branchial cleft may extend to the oropharyngeal tonsillar

---

BOX 14-8 Differential Diagnosis of Prestyloid Parapharyngeal Space Lesions

Direct spread of adjacent tumors
Lymph nodes (normal and abnormal)
Branchial cleft cysts
Neurogenic tumors
Paragangliomas
Parapharyngeal space abscesses
Parotid deep lobe tumors
Pleomorphic adenomas

---

fossa and thus a second BCC may rarely arise here. Cystic schwannomas would be in the differential diagnosis, but concurrent enhancing solid tissue in schwannomas, not branchial cysts, should steer you to the correct Mike Trout home run diagnosis. Just do not bet the farm on it like Pete Rose, or you will never make it to the Neuroradiology Hall of Fame.

Inflammatory Lesions

Intrinsic inflammatory disease of the parapharyngeal space is also rare. Infections may spread secondarily from (1) mucosal infections such as tonsillitis or pharyngitis, (2) masticator space lesions such as odontogenic abscesses, and (3) parotid infections. Adenitis related to any of these primary infections may coexist.

Benign Neoplasms

Primary lesions arising within the prestyloid parapharyngeal space are extremely uncommon and are usually found incidentally. Because only fat, ectopic minor salivary gland tissue, and lymph nodes are present in this region, the most common intrinsic lesions of the prestyloid parapharyngeal space are enlarged lymph nodes, either inflammatory or neoplastic. Of primary benign tumors, minor salivary gland pleomorphic adenomas are most common, and sometimes it is hard to distinguish those exophytic from the parotid deep

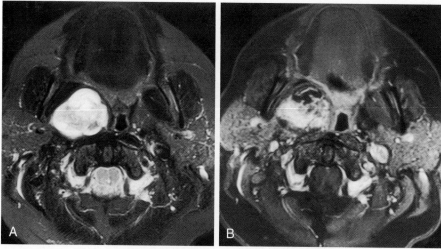

FIGURE 14-32 Parapharyngeal space pleomorphic adenoma. If you transplanted this mass to the parotid gland, this would be a "no-brainer" pleomorphic adenoma. After all, it is bright on T2-weighted imaging **(A)** and enhances **(B).** The parapharyngeal space location should not dissuade you. Pleomorphic adenomas arise from minor salivary gland tissue there. (Case courtesy Stuart Bobman, MD, Naples, FL.)

lobe versus those primarily arising from the parapharyngeal space. If the tumor is surrounded by parapharyngeal space fat, easy peasy. But if no fat is seen, hardy party (Fig. 14-32). As elsewhere, pleomorphic adenomas are very bright on T2WI and have well-defined, lobulated margins.

After salivary gland tumors (40% to 50%), neurogenic tumors (17% to 25%), glomus tumors (10% to 14%), and lipomas may primarily arise here. As noted, parotid masses in the deep lobe are the most common benign neoplasm to invade the parapharyngeal space secondarily. Remember, fat is your friend (we're all about that bass!). If the parapharyngeal fat is displaced anteromedially, suggest the diagnosis of a deep lobe parotid mass. If the parapharyngeal fat is seen between the lesion and the deep lobe of the parotid, then the lesion is either mucosal in origin or arose within the ectopic tissue of the prestyloid parapharyngeal space.

Malignant Neoplasms

Secondary invasion of the parapharyngeal space occurs often, usually spreading from a mucosal space carcinoma. Of the various sites of the mucosal space, the nasopharynx and tonsils represent the most frequent sources of secondary invasion of the parapharyngeal space (Fig. 14-33). Nasopharyngeal carcinoma has been described to invade the parapharyngeal space in 65% of cases at the time of diagnosis. Because this tumor has a propensity for submucosal growth as opposed to exophytic growth, the infiltration of the parapharyngeal fat may be the only indicator of a nasopharyngeal primary cancer. From the tonsil, lateral growth leads to the parapharyngeal fat. A tumor deep within the crypts of the tonsil may escape the endoscopist's attention and the radiologist may be the only one who can identify the primary tumor, on the basis of its infiltration of the parapharyngeal fat.

Synovial sarcomas can arise in the parapharyngeal space. These masses are not derived from the synovium and are unrelated to joints. Good name, eh? Trust the pathologists to confuse the poor radiologists, but we return the favor with our 25-gauge FNAs with four lonely spindle cells (sigh—"no-diagnosis"). Synovial sarcomas are distinguished by their propensity for fluid levels, intratumoral

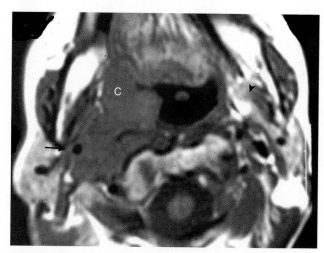

FIGURE 14-33 Malignant spread of cancer into parapharyngeal space. Axial T1-weighted imaging demonstrates a tonsillar carcinoma (c) growing laterally into the parapharyngeal space which is obliterated (*arrowhead* shows right parapharyngeal space fat). The tumor abuts on the right internal carotid artery (*arrow*).

hemorrhage, calcification, and cysts. They may be well defined and have excellent prognoses . . . for sarcomas.

Only malignant lymphadenopathy and minor salivary gland cancers, usually adenoid cystic carcinomas, qualify as intrinsic malignancies of the parapharyngeal space. Minor salivary gland malignancies are far outnumbered by mucosal SCCA tumors that invade the area secondarily.

If you are still awake after spitting, chewing, and fattening up, you will find out next what the foie gras may do to your carotid space.

CAROTID SPACE

Anatomy

The carotid space (poststyloid parapharyngeal space) deserves to be treated as a separate entity because of the numerous unique tumors associated with the carotid sheath and because this space is partially encapsulated by

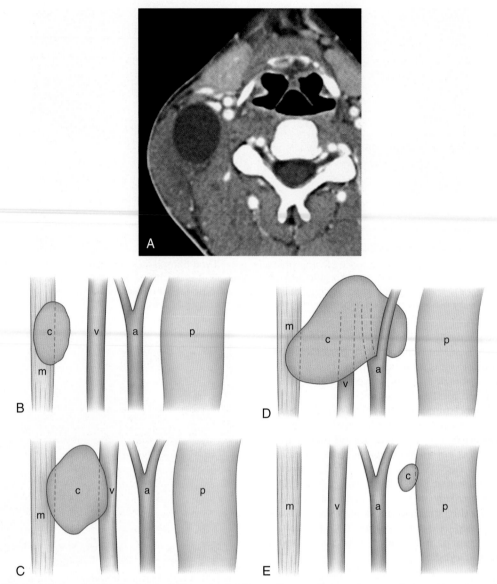

FIGURE 14-34 Branchial cleft cyst. **A,** Bailey type 2. What will you call a cystic mass deep to the sternocleido-mastoid muscle, which approaches the carotid sheath and has a thin rim? A second branchial cleft cyst, if you are smart! **B,** Type I second branchial cleft cyst anomaly. The cyst lies superficial to the anterior border of the sterno-cleidomastoid muscle beneath the cervical fascia. *a,* Common carotid artery; *c,* cyst; *m,* sternocleidomastoid; *p,* pharynx; *v,* internal jugular vein. **C,** Type II second branchial cleft cyst anomaly. The cyst abuts the carotid sheath and is varyingly adherent to the internal jugular vein. **D,** Type III second branchial cleft cyst anomaly. The cyst passes characteristically medially between the internal and external carotid arteries and extends toward the lateral wall of the pharynx. **E,** Type IV second branchial cleft cyst anomaly. This is a columnar-lined cyst located deep to the carotid vessels abutting the pharynx. (**B-E** from Mukherji SK, Fatterpekar G, Castillo M, et al. Imaging of congenital anomalies of the branchial apparatus. *Neuroimaging Clin North Am.* 2000;10:86-87.)

deep cervical fascia. The fascia around the carotid sheath is complex in that it may be incomplete in parts or absent in some people, and is uniformly intact only below the bifurcation. Its "incompetency" allows the carotid artery to veer into the retropharyngeal compartment. The normal contents of the carotid space include the carotid artery, internal jugular vein, vagus nerve (X), sympathetic nervous plexus, branches of the ansa cervicalis/hypoglossi (C1-3 roots), and cranial nerves IX, XI, and XII. Lymph nodes abound around and within the carotid sheath. The carotid space courses down the entire length of head neck and begins at the skull base. Superiorly, the carotid space is separated from the prestyloid parapharyngeal space by the styloid musculature, (styloglossus and stylopharyngeus)

just anterior to the carotid sheath but posterior to the parapharyngeal fat (see Fig. 14-31).

Congenital Disorders
Branchial Cleft Cyst and Lymphangioma

Second BCCs may be intimately associated with the carotid bifurcation. These cysts probably arise because of incomplete closure of the cervical sinus of His (not Hers, misogynist). If they connect to the external skin or the pharyngeal mucosa, they are called branchial cleft sinus tracts. If they connect from pharynx to skin, they are considered fistulae (see Table 14-2). They may occur anywhere along the path from the palatine tonsil to the supraclavicular region but are most commonly found near the angle of the mandible lateral to the carotid sheath as Bailey type II second BCCs (Fig. 14-34). Bailey type I is located anterior to the surface of the sternocleidomastoid muscle, deep to the platysma. Bailey type III may extend between the internal and external carotid artery and may send a tail to the mucosal space. Bailey type IV is actually in the pharyngeal mucosa or submucosa and is deep to the carotid arteries.

In a similar fashion, lymphangiomas may infiltrate the contents of the carotid space. These lesions are seen in children and/or neonates and may have variable signal intensity on MR because of their protein content and/or propensity for hemorrhage. Lymphangiomas develop as a result of sequestration of lymphatic sacs. They may be combined with vascular malformations in a lymphangiohemangioma. Septations and/or fluid-fluid levels are present in most lymphangiomas, and the signal intensity on T1WI may simulate muscle (75%) or fat (25%). Heterogeneity within the lesion is present in more than 80% of cases. Most (75%) occur in the posterior triangle of the neck, with the axilla the next most common site.

Inflammatory Lesions

Cervical Adenitis

Inflammation that involves the carotid space usually comes from adjacent adenitis or extension from mucosal space infections. The source of pericarotid cervical adenitis may be mucosal or odontogenic; tuberculous adenitis; or suppurative adenopathy associated with any pharyngitis can also occur here. Diffuse adenopathy in the neck should make you think of AIDS-related illnesses, mononucleosis, sarcoidosis, sinus histiocytosis, and lymphoma (see Chapter 13).

Vascular Inflammatory Lesions

Primary carotid space inflammatory processes include thrombophlebitis of the internal jugular vein (Fig. 14-35). Lemierre syndrome is an infection of the oropharynx with internal jugular vein thrombophlebitis and possible septic pulmonary emboli. This may complicate pharyngitis. Iatrogenic causes secondary to venous line placement or surgery may also account for some cases of thrombophlebitis. Often, one will see a halo of edema around the thrombosed vein and with fat-suppressed MR, enhancement of the vessel wall and the perivenous soft tissue. Typical scenario: a patient with leukemia with an internal jugular line left in just a tad too long.

Pseudoaneurysms of the carotid artery may present as a neck mass, usually after traumatic dissection. Somber is the surgeon who rushes to perform a biopsy of this "lesion" before consulting the radiologist. The cause of a pseudoaneurysm may be parapharyngeal space abscesses, syphilis, fibromuscular dysplasia, Ehlers-Danlos syndrome, trauma, surgery, neoplastic "blow-out" (Fig. 14-36), or idiopathic. Accelerated atherosclerosis may appear as thickening around the carotid bifurcation, and occasionally you may see intramural hemorrhage into an ulcerated carotid plaque. This may be the result of all that foie gras in this chapter.

Regarding carotid dissection and pseudoaneurysms derived from the knife and gun club, before acquiring the CT angiogram or conventional angiogram, a few clues may help predict whether a carotid sheath vessel has been damaged/dissected. Prevertebral soft-tissue swelling, bullet fragments, and metal foreign bodies adjacent to major vessels are useful

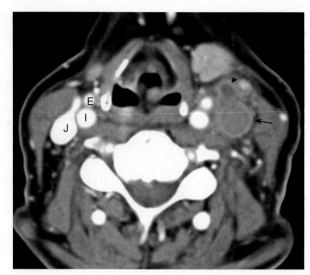

FIGURE 14-35 Thrombophlebitis. This axial contrast-enhanced computed tomographic image shows an occluded left internal jugular vein *(arrow)* and external jugular vein *(arrowhead)*. There is enhancement of the vessel margins and stranding of the soft tissues surrounding these vessels. The normally enhancing jugular vein (J), internal (I), and external (E) carotid arteries are indicated on the patient's right side.

but nonspecific radiographic signs. Flaps in the vessel and widening of the lumen may be detected too.

Carotidynia is a syndrome associated with tenderness, swelling, or increased pulsations over the carotid artery with pain in the ipsilateral neck. This is a self-limited syndrome usually lasting less than 2 weeks. Although very few imaging studies have explored this entity, a recent report has noted enhancing soft tissue around the distal common carotid artery and bifurcation region that resolves with nonsteroidal antiinflammatory medication. Differential diagnosis includes giant cell arteritis, dissection, fibromuscular dysplasia, Takayasu arteritis, and wall hematoma.

Benign Neoplasms

Schwannoma

The majority of carotid space masses are benign. Of these, one classic lesion is the schwannoma which can arise from cranial nerves IX, X, XI, XII, the sympathetic plexus, or cervical spine nerve roots. Situated posterior to the carotid artery, vagus nerve lesions tend to displace the carotid artery and parapharyngeal fat in an anterior direction. Schwannomas of the vagus nerve are usually well-defined, rounded structures hypodense to muscle on CT and enhance moderately. The lesions are circumscribed on T1WI because of the high signal intensity fat around the carotid sheath and around the parapharyngeal space. Whereas on an enhanced CT the border between the schwannoma and carotid artery or jugular vein may be indistinguishable, on MR it is possible to identify the flow voids of the carotid artery and jugular vein as opposed to the enhancing solid tumor (Fig. 14-37). On T2WI, the signal intensity of schwannomas is variable, depending on its content of Antoni A and Antoni B tissue. Occasionally, schwannomas may be cystic and demonstrate characteristic density and intensity features for cyst fluid. Schwannomas also may hemorrhage within themselves. Clues that

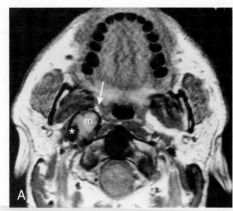
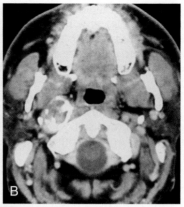

FIGURE 14-36 Pseudoaneurysm of the right internal carotid artery. **A,** Axial proton density image shows a right carotid space mass (m) that has variable signal intensity. Note the peripheral rim of signal void and the postero-lateral area of dark signal *(asterisk).* Is that because of flow void, calcification, or hematoma? Only the surgeon will know for sure if you are unfortunate enough to recommend biopsy. Note also that the parapharyngeal fat *(arrow)* is displaced anteromedially by this carotid space lesion. **B,** Axial computed tomography shows that the majority of the mass is calcified in its rim. There is partial enhancement centrally, indicating that this pseudoaneurysm is only partially thrombosed.

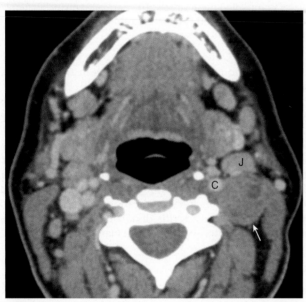

FIGURE 14-37 Schwannoma in carotid space. Axial computed to-mography shows a low-density schwannoma *(arrow).* The carotid artery (C) is displaced anteromedially and the jugular vein (J) ante-riorly by the mass. Note that this tumor is not enhancing as much as one would typically expect of a schwannoma. That can happen in the neck. Why? We haven't the vagus.

the lesion may arise from a specific nerve other than the vagus include tongue atrophy (XII), sternocleidomastoid or trapezius atrophy (XI), neural foraminal enlargement (cervical root origin) or Horner syndrome (sympathetic plexus origin). Uvular deviation suggests the vagus, but IX may affect smaller pharyngeal muscles.

Glomus Tumor

Like schwannomas that may derive from multiple origins, the paragangliomas of the carotid sheath may originate as glomus jugulare, glomus vagale, or carotid body tumors. Carotid body tumors (paragangliomas) usually arise at the carotid bulb and are the most common paraganglioma in the neck. Therefore, they tend to splay the internal and external carotid arteries away from each other. These

tumors are highly vascular and demonstrate dramatic enhancement (Fig. 14-38). If sequential CT or MR images during contrast infusion are performed, you can readily distinguish the schwannoma from the carotid body tumor (Table 14-3). The glomus tumor dynamic contrast curve shows rapid uptake of contrast, an early dip (MR), a high peak contrast, and rapid washout, whereas the schwan-noma has a slower wash-in and a lower peak. The other lesions with the glomus-type dynamic curve include hem-angiomas, aneurysms, arteriovenous malformations, and angiofibromas. Meningiomas have a rapid uptake phase and a high peak but have persistent contrast accumulation with a prolonged washout.

On unenhanced MR, you may identify numerous flow voids within a carotid body tumor; however, quite fre-quently these tumors do not demonstrate the characteristic salt-and-pepper appearance (again!) of vessels that is seen with glomus tumors elsewhere (Fig. 14-39). Nonetheless, a lesion arising at the carotid bifurcation that demonstrates marked enhancement should be considered to be a glomus tumor until proved otherwise. Conversely, tumors that do not avidly enhance are *not* glomus tumors. Angiography demonstrates the high vascularity of the glomus tumor (despite the absence of flow voids on MR) as opposed to the schwannoma, as well as its characteristic persistent staining and early arteriovenous shunting. Remember to check for the metanephrine secretion (seen in 3% to 5% of paragangliomas) before performing the angiogram, or else you may find the blood pressure monitor (and neurointer-ventionalist) going through the roof.

Glomus jugulare tumors also may grow through the skull base to involve the carotid space. In most cases, when they do so, they infiltrate the lumen of the jugular vein. Erosion of the jugular spine is the sine qua non at the skull base.

Glomus vagale (glomus intravagale tumor or vagal para-glioma) tumors may also present in the poststyloid parapharyngeal space and may displace the carotid artery anteriorly as opposed to splaying it at the bifurcation (Fig. 14-40). The tumor is derived from the nodose gan-glion, one of the vagal ganglia in the upper neck, which lies within the carotid sheath. The most common level of

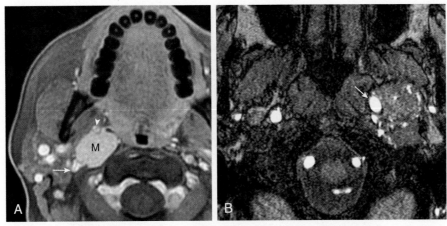

FIGURE 14-38 Carotid body tumor. **A,** Axial postcontrast T1-weighted image shows an avidly enhancing mass (M) at a high right carotid bifurcation, splaying the internal carotid artery (ICA) *(arrowhead)* and external carotid artery *(arrow)*. **B,** Three-dimensional time-of-flight magnetic resonance angiography shows a displaced left ICA *(arrow)* due to a mass in the carotid space. The presence of the numerous vessels in the mass showing flow-related signal speaks to the hypervascular nature of this tumor.

TABLE 14-3 Differentiation of Carotid Space Glomus Tumors from Schwannomas

Feature	Schwannoma	Glomus Tumor
Carotid displacement	Pushes anteriorly	Splays internal and external carotid arteries apart and displaces them anteriorly
Contrast uptake	Slow uptake	Rapid uptake dynamically, early dip
Flow voids on MR	No	Yes
Vascularity on angiography	Variable	Hypervascular
Density on unenhanced CT	Usually hypodense	Usually isodense
Morphology	May have cysts	Speckled

CT, Computer tomography; *MR,* magnetic resonance.

involvement in the neck is near the angle of the mandible and above the hyoid bone. Cranial nerves IX to XII may be affected or the patient may present with a Horner syndrome. Glomus vagale tumors are relatively uncommon, being overshadowed by the carotid body tumors. These two tumors can sometimes be differentiated on the basis of the displacement of the external carotid artery (ECA) as glomus vagales displace the ECAs anteriorly with the internal carotid artery (ICA; and the jugular vein posteriorly), whereas carotid body tumors will push the ECA posterolaterally away from the ICA. The distinction between a vagus schwannoma and the glomus vagale tumor has to be made on the basis of flow voids and/or characteristic vascular flow curves on dynamic imaging, as opposed to the displacement of the carotid blood vessels.

Treatment of glomus vagale tumors almost always results in sacrifice of the vagus nerve and subsequent vocal cord paralysis. Look for the medialized ipsilateral true vocal cord (thyroplasty) after such surgery. Implants, silastic or otherwise, may be seen in the vocal cord.

A small percentage of patients have a familial incidence of glomus tumors where the multiplicity rate may be as high as 30%. These patients also may demonstrate

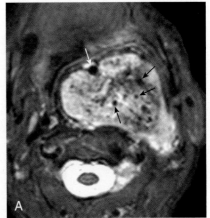

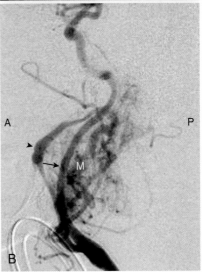

FIGURE 14-39 Glomus tumor. **A,** Axial T2-weighted image shows an enormous mass in the deep left neck—it's so big it's really hard to tell where it's originating from. The displaced internal carotid artery flow void *(white arrow)*, numerous serpentine flow voids in the lesion proper *(black arrows)*, and "salt-n-pepper" appearance helps us a lot in the diagnosis of glomus tumor. Yikes, where's the airway? **B,** Conventional catheter angiogram of the left common carotid in lateral projection (A, anterior; P, posterior) shows the avidly enhancing mass (M) displacing the internal *(arrow)* and external *(arrowhead)* arteries anteriorly, typical of glomus vagale.

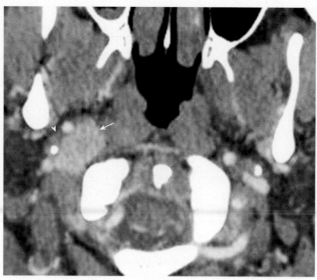

FIGURE 14-40 Glomus vagale. Perfect location for glomus vagale tumor *(arrow)*: at C1 level, in carotid sheath, behind carotid artery and styloid musculature *(arrowhead)*, enhancing nearly as much as vessels.

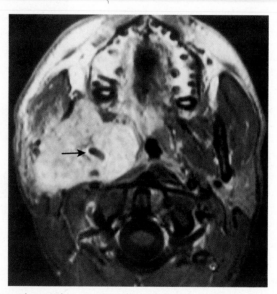

FIGURE 14-41 Carotid space adenopathy encasing the carotid artery. When the carotid artery is encircled by tumor *(arrow)*, you can bet that the surgeons will call the mass unresectable. If the tumor envelops less than 270 degrees of the vessel's circumference it may be saved.

bilateral involvement, which is said to occur in 5% to 14% of patients with carotid body tumors. Syndromic paragangliomas are also associated with succinate dehydrogenase germline mutations. Some of these are also associated with pheochromocytomas. There are rare reports that head and neck paragangliomas are linked to multiple endocrine neoplasia type 2 and von Hippel–Lindau disease.

Indium 111 octreotide scintigraphy enables distinction of glomus tumors from schwannomas and other masses of the carotid space as uptake occurs in the former but not the latter. False-positive cases can be seen in other neuroendocrine-like lesions such as medullary thyroid carcinomas, thyroid adenomas, Merkel cell tumors, and carcinoid lesions. However, for the detection of multicentric paragangliomas (greater than 1.5 cm—it *is* a nuclear medicine study after all) in patients with familial paragangliomatosis

this is a useful nukes exam (score one for the guys glowing in the dark outside of North Korea).

Malignant Disease

Lymphadenopathy

Lymphadenopathy from squamous cell carcinomas ranks as the third most common carotid space lesion after schwannomas and glomus tumors (Fig. 14-41). The lymph nodes are part of the internal jugular chain (levels II, III, and IV) and may be associated with either inflammatory or neoplastic processes involving the mucosal space. Lymph nodes may be present anterior, posterior, medial, and lateral to the carotid sheath and therefore may displace the blood vessels in any possible direction. Lymphomas may also involve the carotid sheath nodes.

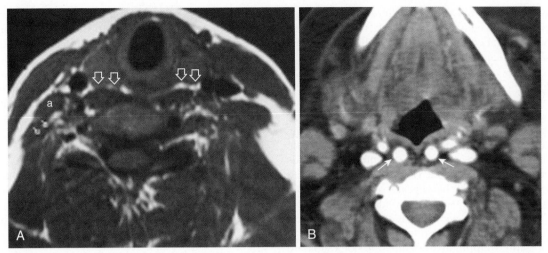

FIGURE 14-42 Retropharyngeal space. **A,** The normal fat of the retropharyngeal space is indicated with closed arrows. The prevertebral musculature is located posterior to the retropharyngeal space. As a prequel to the brachial plexus section in this chapter, also noted on this image are the anterior scalene muscle (a) and the brachial plexus coursing posteriorly to this muscle (open arrows). **B,** Note the medial location of both internal carotid arteries (arrows) in the retropharyngeal space. Careful, spine surgeons, take note before you start your anterior approach spinal fusion!

Often, one is called on to determine whether the carotid artery wall is invaded by spread from mucosal primaries or malignant lymphadenopathy. This conundrum has already been discussed in Chapter 13. Although no absolute criteria have been developed to determine carotid wall invasion, involvement of greater than 270 degrees of its circumference by neoplasm strongly suggests the wall is infiltrated with tumor, particularly if you are dealing with squamous cell carcinoma (see Fig. 14-41).

Mucosal Disease Spread

The carotid artery and space may also be invaded from extension of mucosal cancers rather than lymphadenopathy. When a tumor surrounds the carotid sheath contents by over 270 degrees, carotid invasion is implied. Intraluminal tumor is another not-so-subtle sign. Carotid blow-out in encased vessels figures into the therapeutic planning of the mucosal space squamous cell carcinoma. Radiation oncologists are generally reluctant (no guts, no glory) to radiate the bed of a tumor where the carotid artery may rupture. The incidence of rupture during radiotherapy is relatively small but is many times greater if the patient has already undergone surgery in the neck. Rarely, the interventionalist may occlude the diseased carotid artery before an attempt at complete surgical resection or radiotherapy.

RETROPHARYNGEAL SPACE

The retropharyngeal space may be likened to the intermezzo between the pharynx and the perivertebral space. Eat on.

Anatomy

The retropharyngeal space is a potential space defined by the deep cervical fascia. It is located deep to the pharyngeal mucosa and anterior to the longus colli and capitis muscles. The normal contents of the retropharyngeal space are fat and lymph nodes above the hyoid bone and fat alone below the hyoid bone, with the occasional

carotid artery dashing in there (Fig. 14-42). The retropharyngeal space extends from the base of the skull to the upper thoracic spinal level and is a site for spread from pharyngeal or esophageal lesions. The "danger space" is a term used to refer to a potential space associated with the retropharyngeal space arising from splitting of layers of the deep cervical fascia's deep layer into the alar fascia, ventral to the perivertebral space. Whereas the middle layer of the deep cervical fascia fuses with the deep layer at T6, the split in the deep layer may track to the level of the diaphragm. Some people have studied the danger space and the deep layers of the cervical fascia ad nauseum, all because they have seen two or three lesions there in their storied careers. Do not succumb to this minutia, stick with the common diseases to know well: low back pain, headaches, sinusitis, Alzheimer disease, stroke, and politics.

Characteristically, the parapharyngeal fat is displaced in an anterolateral fashion by retropharyngeal space lesions. However, one must introduce a separate structure, the muscular longus colli and capitis complex, for the differentiation of a retropharyngeal mass from a prevertebral mass. A retropharyngeal mass remains anterior to the longus musculature, whereas a perivertebral space mass displaces the muscle anteriorly or is intrinsic to them.

Congenital Disorders

Retropharyngeal Carotid Artery

Of the benign congenital lesions to affect the retropharyngeal space, the medially deviated internal carotid artery is a potentially dangerous normal variant (see Fig. 14-42, *B*). When the carotid artery is located in the retropharyngeal space, it may simulate a deep submucosal mass to the endoscopist looking from within. He or she may be tempted to perform a deep biopsy to identify the source of the bulge in the pharyngeal mucosa. This may lead to a catastrophic pulsatile complication. Malpractice lawyers will appear like vultures circling carrion. Similarly, beware of the spine surgeon who, doing an anterior approach to the cervical spine, is unaware of a retropharyngeal medialized carotid or a tortuous medially oriented vertebral artery in the foramen transversarium intervening between the neurosurgeon's drill and the vertebral column. "Kaching, kaching" gloats the plaintiff's malpractice lawyer.

Inflammatory Lesions

Retropharyngeal abscesses, suppurative (necrotizing) adenitis, and cellulitis are usually sequelae of pharyngitis (adenoidal or tonsillar infections), sinusitis, or intrinsic lymphadenitis (in children). Initially, the infection spreads to a retropharyngeal lymph node (Fig. 14-43). From there a diffuse cellulitis of the retropharyngeal space and/or lymphedema may occur as the capsule of the node is violated. Most people believe that what we radiologists called unilateral retropharyngeal abscesses in yesteryear (the 20th century) actually represented suppurative adenitis and not a separate inflammatory collection. By the same token, it is important to understand that the retropharyngeal fat may become quite edematous with adjacent inflammatory masses. Therefore, one should not jump to the conclusion that low density in this space represents an abscess. Until the collection is loculated or has a ring enhancement picture, hold off on sending in the clowns to drain the collection or it may be you who is shown to be the joker. Infected fluid density that is ill defined may be a phlegmon.

Internal carotid artery thickening, spasm, and even thrombosis may accompany retropharyngitis and/or lymphadenitis in children. Neurologic findings may be absent or subtle despite the carotid changes.

Benign Neoplasms

Lipomas, fibromyxomas, and hemangiomas may occur in the retropharyngeal space. To make these calls, look for fat in the lipomas (Fig. 14-44), an oval shaped mass for fibromyxoma, or highly enhancing tissue for a hemangioma.

Malignant Neoplasms

Lymphadenopathy

Just as in the prestyloid parapharyngeal space, malignant lesions primarily arising in the retropharyngeal space are very uncommon. The most common malignant condition is lymphadenopathy associated with nasopharyngeal or oropharyngeal cancers. In the normal child, one may identify retropharyngeal lymph nodes associated with infections; however, lymph nodes greater than 0.8 to 1.0 cm in the adult are pathologic in almost all cases (see Figs. 13-31 and 14-45). Lymph node metastasis from nasopharyngeal squamous cell carcinoma is the most common malignant lesion of the retropharyngeal space. This is the first echelon of spread of nasopharyngeal cancer before the high jugular lymph node chain. Lymph node enlargement in the retropharyngeal space may also be caused by lymphoma. The parotid gland may also drain to retropharyngeal nodes.

Other sources of lymphadenopathy in the retropharyngeal space include papillary carcinoma of the thyroid gland and malignant melanoma. For this reason, an examination of the thyroid gland for the possibility of malignancy must extend to the skull base to include retropharyngeal lymph nodes.

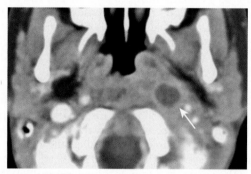

FIGURE 14-43 Retropharyngeal adenitis. Positioned as it is anterior to the longus colli muscles and lateral to the pharynx, this represents a necrotic suppurative left retropharyngeal node *(arrow)*. Note the associated pharyngitis and loss of fat planes. Is the left carotid artery smaller than the right? Vasospasm!

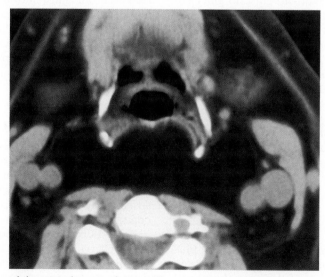

FIGURE 14-44 Lipoma of the retropharyngeal space. The computed tomography shows a mass anterior to the longus muscles, which has fat density. It is posterior to the pharyngeal musculature and resides in the retropharyngeal space.

Contiguous Spread from Mucosal Disease

Contiguous spread from nasopharyngeal carcinoma may also lead to invasion of the retropharyngeal space and displacement of the longus colli muscles posteriorly. Retropharyngeal space infiltration is present in 40% of patients with nasopharyngeal carcinoma at the time of diagnosis. If the posterior deep cervical fascia of the retropharyngeal space and the longus colli musculature are involved with oropharyngeal or hypopharyngeal cancer, the likelihood of surgical cure is markedly diminished. For nasopharyngeal cancers, it also decreases cure rates of chemoradiation protocols (remember nasopharyngeal carcinoma is nonsurgical).

Rhabdomyosarcoma

Rhabdomyosarcomas also occur in the retropharyngeal space. Rhabdomyosarcomas usually invade the retropharyngeal space from their primary site in the nasopharynx. Of the various types of rhabdomyosarcomas, the embryonal histologic type is the most common to affect the head and neck.

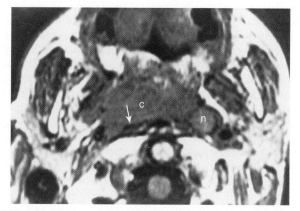

FIGURE 14-45 Nasopharyngeal carcinoma with retropharyngeal infiltration. Axial T1-weighted imaging shows infiltration of the retropharyngeal space by nasopharyngeal carcinoma (c). Again, the longus colli muscles *(arrow)* are seen to be flattened against the vertebral body. Lateral retropharyngeal lymph node (n) involved by tumor is also seen on the left side.

Leukemia/Lymphoma

Although non-Hodgkin lymphomas are a more frequent source for lymphatic infiltration of the retropharyngeal space, acute leukemia may diffusely infiltrate this region. Check for low signal on T2WI and low ADC values.

PERIVERTEBRAL SPACE

Anatomy

To clump the remaining portions of the neck into one catchall category, you will note the use of the phrase the perivertebral space. In this way, lesions in front of (prevertebral) and within, behind, and on the side of the spine (posterolateral neck) can be captured in this section. It is an arbitrary distinction, but lesions here are diverse enough that they defy organization...and this chapter is already testing your attention span. Play through.

The perivertebral space includes the longus colli-capitis muscle complex, the paraspinal musculature, the vertebral body, the posterior triangles of the neck, the neurovascular structures within the spinal canal, and the brachial plexus. Perivertebral space lesions displace the longus colli musculature anteriorly and, when large enough, displace the parapharyngeal fat anterolaterally. The deep cervical fascia of the perivertebral space encircles the paraspinal and prevertebral muscles, the vertebral bodies, the nerves, and vessels, and it divides the space into two compartments by attaching to the transverse processes.

Congenital Disorders

The classic congenital mass in the perivertebral space is the cystic hygroma located in the posterior triangle of the neck. Congenital lymphatic slow flow vascular lesions of the head and neck (in decreasing size) include cystic hygromas, cavernous lymphangiomas, and capillary lymphangiomas. Most cystic hygromas are apparent at birth (50% to 60%) and are seen most commonly in the neck (75%) and axilla (20%) (Fig. 14-46). Cystic hygromas may transilluminate on ultrasound. Occasionally, they may be diagnosed in utero in the setting of polyhydramnios. They are usually hypodense and

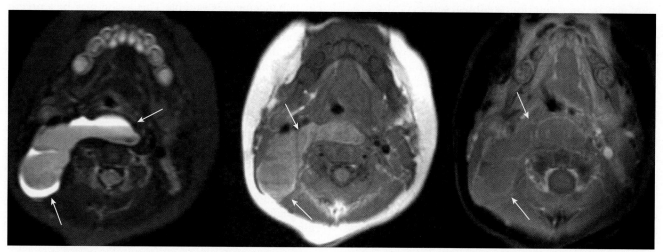

FIGURE 14-46 Lymphangioma. This lesion *(arrows)* is bright on T2-weighted imaging (T2WI) and T1WI and goes from a retropharyngeal location to the posterior neck. It does not enhance. Note the fluid-fluid level on the T2WI. Given a child, the best diagnosis is a lymphatic malformation.

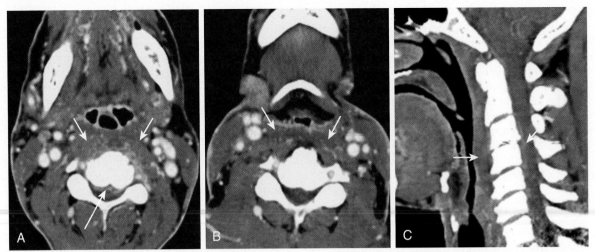

FIGURE 14-47 Prevertebral phlegmon. **A,** A prevertebral inflammatory collection is present derived from cervical spine diskitis and osteomyelitis at the C3-4 level. Retropharyngeal edema is seen in **B** *(arrows)*. **C,** prevertebral and dreaded intraspinal extension of inflammation is also seen *(arrows)*.

multiloculated on CT and have variable intensity on T1WI and T2WI because of their potential for proteinaceous-chylous-hemorrhagic content. The mass may present acutely as a result of spontaneous or posttraumatic intralesional hemorrhage with fluid-fluid levels or infection. The cystic hygroma is not invasive but is infiltrative, being compressible and distorted by arteries. The cause of these lesions is thought to be obstruction of the primitive lymphatic channels that are derived from the venous system early in gestation. Associations with Turner syndrome, Noonan syndrome, Down syndrome, and fetal alcohol syndrome are well described.

Inflammatory Lesions

The most common benign condition of the perivertebral space is spondylosis of the vertebral bodies. When a vertebral body osteophyte is large enough, it may simulate a submucosal pharyngeal or retropharyngeal mass and produce dysphagia. Large anterior osteophytes are not uncommon in elderly patients, even without a history of cervical spine trauma. In cases of diffuse idiopathic skeletal hyperostosis and/or ossification of the posterior longitudinal ligament, there may be large anterior osteophytes that coexist.

Infections of the perivertebral space center on diskitis and osteomyelitis (Fig. 14-47). These are unusual in patients with normal immune responses except in those patients in whom surgery on the cervical spine has been performed, in bacteremic patients, or in intravenous drug abusers. The radiographic findings of diskitis and osteomyelitis are discussed fully in Chapter 16.

Rotatory subluxation of the atlantoaxial joint may coexist with retropharyngitis in the entity known as Grisel syndrome.

Esophageal perforation may lead to mediastinitis and infection of the perivertebral space.

Acute calcific prevertebral tendinitis has findings of calcifications within the tendons of the longus colli muscles and therefore represents a perivertebral process. However, there may be retropharyngeal effusions associated with this entity. The patient has neck pain and a fever and swollen prevertebral muscles and improves with conservative management.

Nodular fasciitis occurs in the neck. It may present as a lateral neck mass and be misdiagnosed clinically as a lymph node. It may enhance brightly but is usually intermediate in intensity on T2WI. Most people feel this represents a low-grade soft-tissue tumor, not an inflammatory process.

Necrotizing fasciitis was in "fasciion" a few years back. Imaging findings consist of (1) diffuse thickening and infiltration of the skin and subcutaneous tissue (cellulitis); (2) diffuse enhancement and/or thickening of the superficial and deep cervical fasciae (fasciitis); (3) enhancement and thickening of the platysma, sternocleidomastoid muscle, or strap muscles (myositis); and (4) fluid collections in multiple neck compartments. This disease could result in sloughing of tissue, gas containing abscesses, and pulmonary manifestations of adult respiratory distress syndrome, mediastinitis, and pneumonia.

Aggressive fibromatosis (desmoid tumor) may also infiltrate the muscles of the perivertebral space/posterior triangle. Although there are reports that this lesion is always dark on T1W and T2W scans and is isodense to muscle on CT, other studies have shown isointensity on T1WI and hyperintensity on T2WI relative to adjacent normal muscle. Enhancement is variable. This lesion may be a precursor to or in the family of malignant fibrous histiocytoma.

Fibrosing inflammatory pseudotumor can affect the perivertebral space and spread to the skull base. As expected from its appearance in the orbit, the lesion shows characteristic MR findings of bone destruction and hypointensity on T2WI. Enhancement on MR images is weak.

Benign Neoplasms

The benign neoplasms of the perivertebral space are relatively uncommon. They include lipomas, schwannomas from the cervical nerve branches, hemangiomas of the musculature, and benign bony tumors. By sheer numbers, hemangiomas dominate the benign bone neoplasm category, but as they are most often confined to the vertebral bodies and rarely affect the head and neck surgeon, they will not be addressed here.

Lipomas can occur anywhere in the head and neck but have a predilection for the lateral subcutaneous tissues, the

Chapter 14 Extramucosal Diseases of the Head and Neck **513**

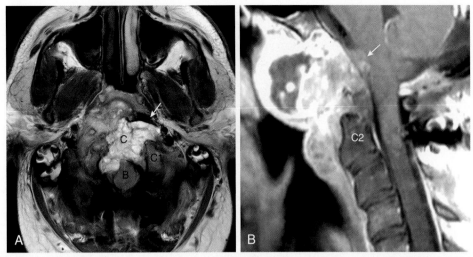

FIGURE 14-48 Chordoma of the perivertebral space. This T2-weighted imaging scan demonstrates a high-intensity chordoma (C), arising at the level of the atlas (C1), anteriorly displacing and stretching the longus coli muscle on the left *(arrow)* and infiltrating the muscle on the right with elevation of the nasopharyngeal soft tissues. Posterior extension at the craniocervical junction results in displacement and distortion of the brain stem (B).

supraclavicular fossa, and the posterior perivertebral space. The differential diagnosis will include liposarcomas, which may or may not have nonfatty soft tissue associated with them. Another bizarroma is Madelung disease, an entity with massive lipomatosis of the posterior neck. Excess fat can be seen predominantly in the posterior part of the neck, under the trapezius and sternomastoid muscles, in the supraclavicular fossa, between the paraspinal muscles, and in the anterior part of the neck (suprahyoid and infrahyoid). The patients may present with respiratory symptoms secondary to tracheal compression, neuropathies, weakness, macrocytic anemia, and venous stasis.

Chordoma

Of the histologically benign bony neoplasms, chordomas preferentially affect the cervical perivertebral space. The characteristic cell type is the physaliferous cell. Chordomas, though histologically benign, should be considered malignant tumors because they tend to invade aggressively into the skull base and metastasize in 7% to 20% of cases. Patients usually have cranial nerve symptoms (nerve VI most commonly; then nerves III, V, VII, and VIII) as the tumor invades the Dorello canal (VI), the cavernous sinus (III, IV, V, VI), or skull base/IAC (VII, VIII). Chordomas occur at the sacrum, the clivus, the C1-C2 region, in that order of frequency. The tumor is destructive and lytic and is often associated with calcifications. It displaces the longus colli musculature anteriorly and may cross the boundaries of the C1-C2 region. The signal intensity of these lesions often is bright on T2WI (Fig. 14-48). The tumor demonstrates minimal enhancement. The differential diagnosis includes chondrosarcoma, which has similar features but hopefully is located off of the midline.

Malignant Neoplasms
Metastasis

Malignancies of the perivertebral space center on the vertebral bodies. Metastases from blood-borne sources are the most common lesions in the vertebral body, often

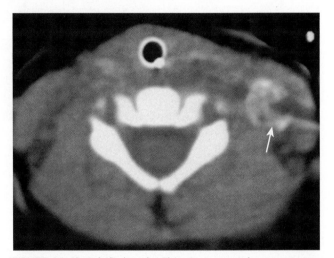

FIGURE 14-49 Calcified node. This one stumped us—a neonate with a palpable and calcified nodule in the left side of the neck *(arrow)*. Consider it to be paraspinal and you may get the correct diagnosis: metastatic neuroblastoma.

from breast, lung, or kidney primary tumors. Primary vertebral body malignancies include osteosarcomas and Ewing sarcomas in addition to plasmacytomas. Multiple myeloma may affect the cervical spine as well. Invasion by Pancoast tumors extending from the lung apex may also present as a perivertebral or supraclavicular mass and may cause a brachial plexopathy (see following discussion).

Soft-tissue malignant masses in the perivertebral musculature include lymphoma, rhabdomyosarcoma, malignant fibrous histiocytoma, and neurofibrosarcoma (MPNST). Less commonly found soft-tissue neoplasms include hemangiopericytomas and synovial sarcomas. The imaging features of these with salient points (rhabdo: heterogeneous/parameningeal, malignant fibrous histiocytoma: dark on T2WI, MPNST along nerves, hemangiopericytoma: avidly enhancing, synovial sarcoma heterogeneous) have been previously described.

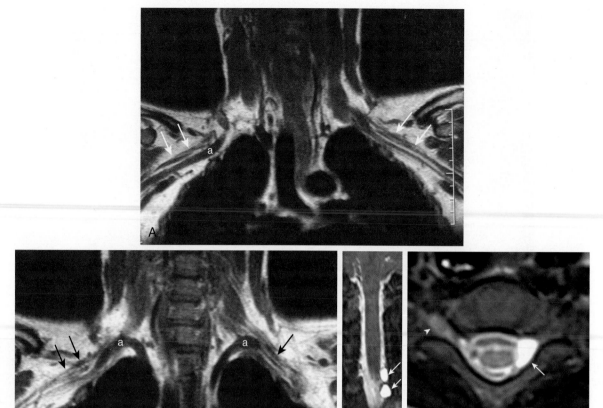

FIGURE 14-50 Brachial plexus anatomy. **A,** Note the relationship of the brachial plexus *(arrows)* to the subclavian artery (a) on these coronal scans (i.e., posterior and superior for the most part). **B,** A more posterior section shows plexus with artery anterior. **C,** Coronal and **(D)** axial T2-weighted images show left-sided pseudomeningoceles *(arrows)* due to avulsion of the left C7 and C8 nerve roots from birth trauma. The expected signal of the C7 nerve root within the neural foramen is present on the right *(arrowhead)* but absent on the left.

Lymphadenopathy

Lymphadenopathy is perhaps the most common perivertebral space lesion. The primary site of abnormality may be within or below the head and neck (Fig. 14-49). Lymphoma may also account for nodes in the posterior triangle (level V) of the neck or may present as a supraclavicular mass. Classically, Hodgkin disease manifests as a posterolateral neck mass.

BRACHIAL PLEXUS

The brachial plexus is considered a part of the perivertebral space. The brachial plexus is derived from the C5-T1 cervical nerve roots, which pass inferolaterally to the axilla for supply to the upper extremity. The mnemonic for remembering the anatomy here is "Rad Techs Drink Cold Beer." The Roots merge into Trunks (upper, middle, and lower) at the scalene muscular triangle. The Trunks divide into Divisions (anterior and posterior), which then form Cords (lateral, posterior, and middle) at the Clavicle. The cords form the Branches in an unnecessarily complicated manner. Usually His designs are simpler. Suffice it to say the brachial plexus runs between the anterior and middle scalene muscles and then with the subclavian artery to the level of the clavicles (Fig. 14-50). At that point, the plexus runs with the axillary artery, posterior to the larger axillary vein.

Congenital Lesions

Lymphatic and venous malformations are the most common congenital lesions to affect the brachial plexus. The posterior neck is the most common site for cystic hygromas, followed by the axilla. In both of these locations, the brachial plexus may be affected. Consider whether the patient has a "webbed neck" to suggest a radiographic diagnosis of Turner syndrome on the neck CT study. Lymphangiomas infiltrate and encase the brachial plexus and adjacent vessels, but usually they do not enhance, unlike venous vascular malformations. Multiloculation, fluid-fluid levels, bright fluid on T1WI are all ancillary findings highly suggestive of lymphatic malformations.

Two types of perinatal injuries may affect the brachial plexus: an Erb palsy and a Klumpke palsy. Both of these may occur as a result of shoulder dystocia at the time of delivery. In the former, the avulsion of nerve roots occurs at the C5-C6 level and the intrinsic muscles of the hand are unaffected, though the shoulder is weak (see Fig. 14-50). In the latter, the C7, C8, and even T1 roots are torn and therefore the hand muscles are affected. The Klumpke patient may have a coincident Horner syndrome secondary to involvement of the sympathetic nervous system structures and/or stellate ganglion opposite the C7-T1 level.

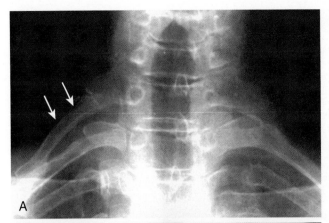

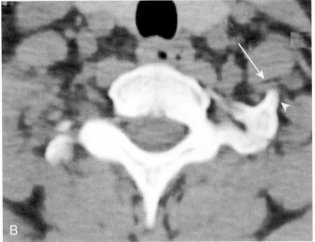

FIGURE 14-51 Cervical rib. **A,** There is a complete cervical rib on the right *(arrows)* and an incomplete one on the left. The patient had a left brachial plexopathy. At surgery, a fibrous band across the left brachial plexus from the cervical rib stump to the manubrium was present. **B,** Left cervical rib attaching to C7 transverse process with an exostosis-like process *(arrowhead)* compressing the brachial plexus *(long arrow).*

Traumatic injuries in the adults are usually from "breaking a fall by a motocross racer" injuries (i.e., Harley Davidson syndrome). Avulsions of nerve roots ensue. Look for absent nerve roots arising from the cord, dilated CSF spaces, pseudomeningoceles, displacement of the cord contralateral to the side the avulsed nerve roots took place, and, in 10% to 14% of cases, concurrent injury to the spinal cord.

Cervical ribs also can cause a brachial plexopathy (Fig. 14-51). The incomplete cervical ribs often have a band leading from them to the clavicle that traps the plexus and therefore are more often symptomatic than the complete cervical ribs. Either way a thoracic outlet syndrome may be produced. Women are affected more than men. They are unilateral in the majority of cases.

Inflammatory Lesions

Brachial plexitis (neuralgic amyotrophy) most commonly is viral in origin (Fig. 14-52). However, this may also occur as a complication of radiotherapy, especially after treatment of supraclavicular adenopathy and/or breast cancer

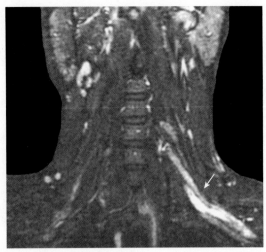

FIGURE 14-52 Brachial plexitis. Fat-suppressed T2-weighted imaging shows left sided brachial plexus *(arrow)* inflammation that was viral in etiology. This may or may not show enhancement.

axillary nodes. Radiation brachial plexitis is characterized by its symmetry, high signal on fat-suppressed T2WI, and variable enhancement.

Nodular fasciitis is a nonspecific sclerosing inflammatory condition that can lead to loss of planes around the brachial plexus. It is often low in intensity on T2WI.

Benign Neoplasms

Lipomas and neurogenic neoplasms predominate in this category. Both are late to produce symptoms and are easily characterized by density (lipoma/fat) or shape (fusiform/schwannoma). Neurofibromas of the brachial plexus outnumber schwannomas and one may have plexiform neurofibromas or MPNSTs associated with NF-1 in this locale.

Malignant Neoplasms

Primary malignancies that affect the C5-T1 nerves include malignant fibrous histiocytoma and other lesions akin to fibrosarcoma, liposarcomas, and, in children, rhabdomyosarcomas and neurofibrosarcomas (in the NF-1 population). Soft-tissue lymphomas may also infiltrate here.

Secondary invasion may be caused by contiguous involvement or lymphadenopathy. Direct invasion by Pancoast tumors or chest wall sarcomas may lead to a brachial plexopathy. Often there is an associated Horner syndrome.

The most common sources of lymphadenopathy producing a brachial plexopathy are primary tumors and lymphomas of the breast, lung, esophagus, head, and neck. Usually an infiltrative nodal mass erases the planes between the brachial plexus and the scalene muscles (Fig. 14-53). Sometimes the plexopathy only appears after radiotherapy, but the radiation oncologists will claim it was there from the outset, just overlooked. "Should have had a neurology consult…" "Shouldn't have fried that supraclavicular fossa…"

Now that you have digested the spaces of the head and neck, it is time to stimulate the head and neck hormones with a discussion of the thyroid and parathyroid glands.

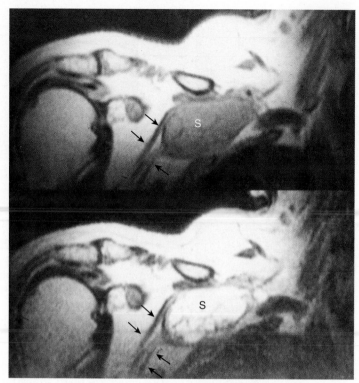

FIGURE 14-53 Brachial plexopathy pathology. A malignant schwannoma (s) on unenhanced (above) and enhanced (below) coronal T1-weighted imaging (T1WI) can be seen intertwined with the brachial plexus nerve roots (*arrows*).

THYROID GLAND

Anatomy

A plethora of lesions, both benign and malignant, affect the thyroid gland. Head and neck radiologists and pulmonary radiologists have laid claim to the thyroid gland as within their turf. If most thyroid gland abnormalities would stay in the neck (unlike goiters, retrosternal thyroids, and ectopic thyroidal tissue), head and neck radiologists would have more clout in claiming the gland as their own.

The thyroid gland has two lobes connected by a midline isthmus. The gland is located at the C5-T1 level and is encapsulated in fascia that is attached to the trachea, accounting for its movement with swallowing. A pyramidal lobe is variably present and projects upward from the isthmus. The gland is supplied by superior thyroidal arteries from the external carotid and inferior thyroidal arteries from the thyrocervical trunk off the subclavian artery.

Congenital Disorders

Agenesis and Ectopic Thyroid Tissue

Ectopic thyroid tissue is more common than total agenesis of the thyroid gland. Of the ectopic locations, 50% are in the base of the tongue (lingual thyroid—see Fig. 13-12) and 50% lie between the tongue and the normal location of the gland.

The work-up of the lingual thyroid gland requires an iodine based nuclear medicine study to search for other thyroid tissue. The ectopic tissue can also be seen on CT imaging of the neck as hyperattenuating material in an unexpected place, with absence of thyroid tissue in an expected place. In 80% of the cases, the lingual thyroid gland is the *only* source of thyroid hormone in the individual. This will influence therapy timing as thyroid hormone is important for healthy growth as a child. Imagine removing what you believe is a hemangioma of the tongue (because it enhances and you did not perform an unenhanced scan to see that it was hyperdense [from iodine] beforehand) in a child. Pathology comes back 10 days later as lingual thyroid tissue. In the meantime, the patient is in myxedema coma because you never thought to replace the thyroid hormone because you never suspected the diagnosis. Now the patient is on thyroid hormone supplementation for life from childhood. In many cases, the surgeon will leave the lingual thyroid tissue in through adolescence (unless it is obstructing the airway) and beyond if this is the only thyroid tissue in the body. It is that time you wish for struma ovarii or struma cardia (other ectopic sites). Leaving the lingual thyroid tissue in does have its drawbacks. The incidence of papillary carcinoma runs as high as 3% to 5%—about what you find at autopsy studies of cervical thyroid tissue.

Thyroglossal Duct Cyst

Thyroglossal duct cysts (TGDCs) represent the most common nonodontogenic cysts in the head and neck below Tornwaldt cysts. They account for 70% of congenital neck masses. TGDCs are due to a remnant of the duct along the pathway of embryologic descent of the thyroid gland and may occur anywhere from the foramen cecum of the tongue to the natural location of the thyroid gland. TGDCs

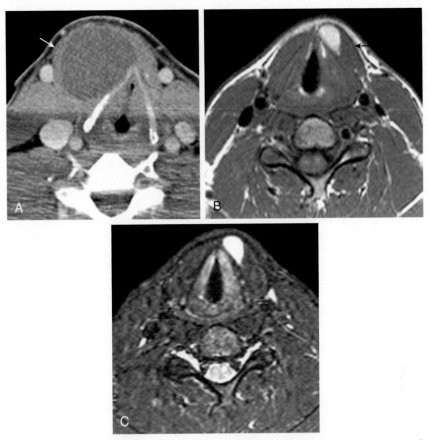

FIGURE 14-54 Thyroglossal duct cyst (TGDC). **A,** Axial computed tomography demonstrates a cystic mass at the level of the thyroid cartilage and displacing it toward the left. The cyst is embedded in the strap musculature *(arrow),* making the definitive case for TGDC. **B,** Axial T1 and **(C)** T2-weighted images in a different patient show a complex cystic lesion (resulting in bright T1 signal). Once again, this lesion is embedded in the strap musculature *(arrow* in **A**), making TGDC a slam dunk diagnosis. The T1 high signal is most likely due to thyroglobulin.

are most often infrahyoid (65%) but can occur at the hyoid level (15%) or in the suprahyoid soft tissue (20%). A cystic lesion embedded in the strap muscles (thyrohyoid, genio-thyroid, and so on) is a TGDC with little question about it. For this, you can stick your neck and TGDC out. Unlike ectopic thyroid tissue located between the tongue and lower neck, TGDCs do not typically contain functioning thyroidal tissue though microscopic rests are present in some cases. Three fourths of TGDCs are in the midline. For math aficionados, this means that 25% are off midline (Fig. 14-54). Cysts look like cysts, but sometimes they may have high protein. That leads to bright signal on T1WI. Rim enhancement is not uncommon.

On US, the appearance is more variable. The cysts may be anechoic (28%), homogeneously hypoechoic with internal debris (18%), solid appearing (28%), and hetero-geneous (28%). The majority show posterior back wall enhancement (88%), and only half show a classic typical thin wall.

Carcinoma in a TGDC is rare, arising in approximately 1% of cases. When it occurs, its histologic appearance is usually papillary, although squamous and follicular variet-ies have also been reported. As one might expect, medul-lary carcinoma does not occur in TGDCs, as these tumors arise from the parafollicular cells, not thyroid derivatives. Thyroglossal duct carcinoma can be suspected when

mural nodules, calcifications, or combinations thereof are present in the cyst.

The Sistrunk procedure, in which the tract of the thyro-glossal duct, the midline tongue base, and the midportion of the hyoid cartilage are resected, is the definitive proce-dure for treating TGDC.

If you see fat or a solid nodule associated with a cyst in the thyroid gland, consider the diagnosis of teratoma.

Autoimmune and Inflammatory Lesions

Goiter

A goiter is a diffuse enlargement of the thyroid gland often found in association with iodide deficiency. Goiters may have areas of cystic degeneration, hemorrhage, col-loid formation, or nodularity within them and therefore are difficult to analyze on the basis of nuclear medicine, US, CT, or MR characteristics (Fig. 14-55). Goiters are usu-ally evaluated by nuclear medicine studies in which a lesion is classified into three main categories: hot (hyper-accumulation of radiotracer), warm (increased uptake with suppression of background thyroidal tissue), or cold (nonfunctioning, low uptake). Warm and hot nodules, because they represent hormonally active tissue, usually have a benign cause. Cold nodules are more worrisome for malignancy. A cold nodule in a goiter is malignant in less than 14% of cases, so unless a mass in a goiter is growing, biopsy is usually not performed. Nonetheless, many thyroid nodules, discovered clinically or serendipitously while scanning for other sites, get aspirated. Once the cytologists see macrophages, colloid, and papillary cells, they call it "multinodular goiter." The surgeon may still operate if cosmesis, airway, or esophageal compromise dictate. Often a trial of hormone suppression is offered beforehand—to limited benefit.

Suppurative Thyroiditis

Suppurative thyroiditis may occur because of bacterial infection or as a complication of branchial cleft fistulae from the piriform sinus that inflame the adjacent thyroid gland. These are usually due to third or fourth branchial cleft abnormalities and may have a cutaneous opening at the clavicular level. The ratio of left to right congenital pyriform sinus fistulae is 4 to 1.

Subacute Thyroiditis

Inflammatory conditions of the thyroid gland include subacute, Hashimoto, radiation-induced, Riedel, and acute lymphocytic thyroiditis. Subacute (de Quervain) thyroiditis is due to an as yet unidentified virus. Like most thyroid disorders, de Quervain thyroiditis is more common in women. Patients are febrile, and the thyroid gland becomes swollen and tender. The disease is self-limited, and most patients are euthyroid after the acute bout, with a small percentage becoming hypothyroid.

Hashimoto Thyroiditis

As opposed to subacute thyroiditis, the other inflammatory disorders are not associated with fever or a tender thyroid gland. In Hashimoto thyroiditis, the gland is infiltrated with lymphocytes and plasma cells, and the disease is thought to be due to autoantibodies to thyroid cell proteins, especially thyroglobulins. Antimicrosomal antibodies are elevated. Although initially the patient may be euthyroid, eventually hypothyroidism sets in because of replacement of functioning tissue with lymphocytes and fibrosis. Generalized enlargement of the gland is commonly seen in Hashimoto thyroiditis. An increased rate of thyroid lymphomas has been reported in women who have this disease. Other autoimmune disorders such as Sjögren syndrome, pernicious anemia, lupus erythematosus, rheumatoid arthritis, and diabetes occur with increased frequency in patients with Hashimoto thyroiditis. Increased, but still exceptionally low.

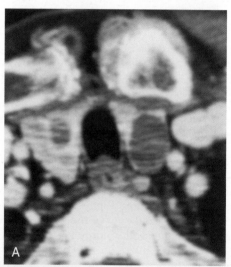

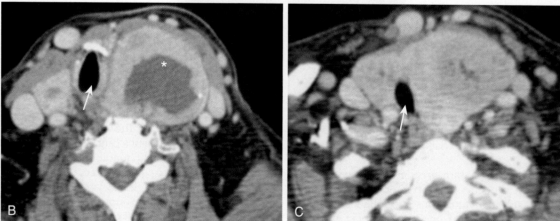

FIGURE 14-55 Thyroid lesions. **A,** Enhanced computed tomography (CT) shows multiple nodules within both thyroid lobes. The nature of these lesions cannot be determined by CT alone. These turned out to be cysts. **B,** This enlarged gland with a dominant mass *(asterisk)* is displacing the trachea *(arrow)* to the right. The diagnosis was adenoma. **C,** This enlarged gland also pushed the trachea over *(arrow)*. The left lobe mass was a papillary cancer.

Riedel Thyroiditis and Radiation-Induced Thyroiditis

Riedel thyroiditis and radiation-induced thyroiditis both lead to fibrosis of the gland in the long run and are chronic in their courses. The thyroid gland may feel firm and woody in both these disorders. Hypothyroidism occurs in up to 50% of patients. Riedel thyroiditis may be associated with retroperitoneal fibrosis, idiopathic orbital inflammation (aka pseudotumor), mediastinal fibrosis, and sclerosing cholangitis. The term multifocal fibrosclerosis has been coined for this entity. On imaging Riedel thyroiditis is homogeneously hypoechoic on US, hypodense on CT, and hypointense on all sequences with MR. Enhancement is variable.

Graves Disease

Graves disease is synonymous with diffuse toxic goiter and is a common cause of hyperthyroidism. The disease has an ophthalmopathy associated with proptosis, lid retraction, fatty infiltration, and extraocular muscle enlargement (thyroid eye disease), but this may occur whether the patient is hyperthyroid, euthyroid, or hypothyroid. The orbital manifestations of this syndrome have been discussed in Chapter 9. Graves is an autoimmune disease associated with increased levels of long-acting thyroid stimulator (LATS), antibodies to thyroid-stimulating hormone (TSH) receptors that circulate in 80% to 90% of persons with Graves disease. The gland is enlarged and is hot diffusely on nuclear medicine scans.

Diffuse nontoxic goiters and multinodular goiters usually develop as a response to increased TSH levels and are seen in euthyroid patients. The gland is enlarged, with nodularity.

Benign Neoplasms

Adenoma

The most common benign tumors of the thyroid gland are adenomas. More than 70% of solitary neoplasms in the thyroid glands are benign adenomas, either follicular or papillary. The lesion may be inhomogeneous, but the margins by and large are well circumscribed on CT or MR (see Fig. 14-55, *B*). Thyroid adenomas may be associated with calcifications; however, the same is true of thyroid carcinomas.

If you were to perform thyroid US as a screen in middle-aged women, you would find that more than one third have nodules in their thyroid glands. On autopsy specimens, 50% of thyroid glands have nodules. Fortunately, the vast majority of these are benign degenerated nodules: 4.2% of nodular glands have malignancies (see Fig. 14-55, *C*). The percentages shift toward malignancies when the patient has a history of radiation exposure especially after two or three decades.

More highly differentiated adenomas may concentrate radiotracers and appear as hot or warm nodules on thyroid scintigrams. Because they are often autonomously secreting (independent of TSH stimulation), they suppress the remaining normal thyroid tissue. If necrosis or intralesional hemorrhage develops, the nodule will become cold.

Colloid Cyst

A colloid cyst is another common lesion of the thyroid gland. This may have high density on CT and most characteristically has high intensity on T1WI. Colloid and hemorrhage may be difficult to distinguish on the basis of CT and MR. This lesion may cause a cold spot on a technetium or iodine nuclear medicine scintigram.

Malignant Neoplasms

Only 4% to 7% of histologically examined thyroid nodules are positive for cancer, but a solitary nodule in a male patient or a child has a higher rate of being neoplastic. Papillary, follicular, and mixed papillary-follicular carcinomas account for 80% of thyroid malignancies. Medullary carcinoma (10% of thyroid malignancies), undifferentiated or anaplastic carcinoma (3%), and Hürthle cell (2%) are less common cancers of the thyroid gland. All these lesions may have well-circumscribed or poorly circumscribed margins and do not have a characteristic CT or MR appearance to distinguish them from adenomas.

As in some adenomas, the lesions demonstrate a photopenic area on technetium or iodine scans. Nonetheless, only 20% of cold nodules are cancerous (the remainder are adenomas, colloid cysts, focal thyroiditis, or other benign cysts). Thyroid carcinomas may be cystic, hemorrhagic, calcified, or hyperproteinaceous (Fig. 14-56). They may spread diffusely to the lymphatics and/or hematogenously and therefore be seen with cervical lymphadenopathy or with distant metastases (Box 14-9). They may infiltrate the trachea and esophagus seen as circumferential involvement,

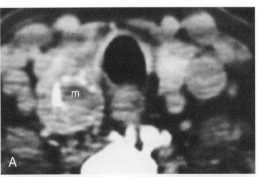

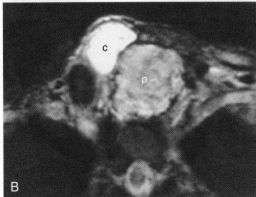

FIGURE 14-56 Thyroid carcinoma's multiple faces. **A,** Axial computed tomography demonstrates a mass (m) with peripheral calcification in the posterior portion of the right thyroid gland. Do not be fooled by the presence of calcification. It need not suggest benignity in the thyroid gland. This was follicular carcinoma. **B,** This patient had papillary carcinoma (p), which was associated with a large cyst (c). Hemorrhage, calcification, cyst formation, and lymphatic and hematogenous metastases may all be characteristic of the various types of thyroid cancer.

BOX 14-9 TNM Classification of Thyroid Malignancies

Primary Tumor (T)
TX—Primary tumor cannot be assessed
T0—No evidence of primary tumor
T1—Tumor <2 cm limited to thyroid
T2—Tumor >2 cm but <4 cm limited to thyroid
T3—Tumor >4 cm limited to thyroid or any tumor with minimal extrathyroid extension
T4a—Tumor of any size beyond capsule to subcutaneous tissues, larynx, trachea, esophagus, or recurrent laryngeal nerve
T4b—Tumor invades prevertebral fascia or encases carotid artery or mediastinal vessels
(All anaplastic carcinomas are considered T4 tumors)
    Note: All categories may be subdivided: (a) solitary tumor, (b) multifocal tumor (the largest determines the classification).
Regional Lymph Nodes (N)
NX—Regional lymph nodes cannot be assessed
N0—No regional lymph node metastasis
N1—Regional lymph node metastasis
    N1a—Metastasis to level VI nodes
    N1b—Metastasis in bilateral, midline, or contralateral cervical or mediastinal lymph node(s)
Distant Metastasis (M)
MX—Presence of distant metastasis cannot be assessed
M0—No distant metastasis
M1—Distant metastasis

endoluminal spread, enhancement of the wall or focal nodularities.

If a complete pseudocapsule (an even-thickness band of low intensity around the mass) is present on MR, this usually signifies cystic degeneration of an adenoma. If the capsule is irregular in thickness (yet continuous) or discontinuous with nodular penetrating excrescences, carcinoma is more likely. Discontinuous pseudocapsules without areas of penetration are commonly seen in adenomatous goiters. Hemorrhagic degeneration is more common in adenomas and goiters but may also be present in cancer.

Papillary Carcinoma

Papillary carcinomas are the most common malignancies of the thyroid gland, accounting for half of thyroid cancers. In a patient less than 40 years old, papillary adenocarcinomas outnumber all other cancers of the thyroid by a 5-to-1 margin. Women are more commonly affected than men, and cervical lymphadenopathy is a common presentation. Lymphatic rather than hematogenous spread is the rule. A propensity for cystic-appearing nodes has been demonstrated. The nodes may also be calcified, colloid containing, highly vascular, and tiny. The nodes may be bright on T1WI or have calcification evident on CT.

The prognosis with papillary carcinoma is the best of the thyroid malignancies and most cancers in general. Age is the most important prognostic factor in patients with papillary carcinoma. Even with distant metastases, a patient under 45 years of age can have an excellent prognosis. Among patients over the age of 45, there are several features that portend decreased survival in some cases of papillary carcinoma. Large tumors,

esophageal invasion, tracheal invasion, lymphadenopathy, and carotid encasement are unfavorable characteristics. Although those patients with incidentally discovered carcinomas and microcarcinomas have a better prognosis than clinically apparent differentiated thyroid carcinoma, bilaterality, lymph node metastasis, thyroid capsule invasion, and disease recurrence are seen more commonly in nonincidental microcarcinomas and conventional macrocarcinomas. Therefore, total thyroidectomy followed by an adequate exploration of the central neck compartment is usually recommended.

Imaging findings indicative of tracheal invasion include abnormal T2WI signal in the wall, circumferential involvement of more than 180 degrees, intraluminal masses, and displacement over 2.5 cm from the midline. Esophageal invasion by tumors is suggested if the wall has abnormal MR signal, the wall enhances, there is tumor encasement greater than 270 degrees, or if there is intraluminal tumor.

Follicular Carcinoma

Follicular carcinomas tend to spread hematogenously. The prognosis is almost as good as papillary carcinoma, with age again playing a critical factor in prognosis (younger patients have better long-term prognosis than older patients).

Medullary Carcinoma

Medullary carcinoma is a rare neoplasm of the thyroid gland, accounting for only 7% to 10% of thyroid malignancies. Ten percent are associated with the multiple endocrine neoplasia (MEN) II syndrome, and some cases are familial without other endocrine lesions (Table 14-4). MEN IIa and IIb signify syndromes with increased rates of pheochromocytoma and parathyroid hyperplasia. MEN IIb may be associated with marfanoid facies and mucocutaneous neuromas. Medullary carcinoma is derived from parafollicular cells that secrete calcitonin; elevated serum calcitonin levels may be a marker for the disease. Stippled calcification may be the clue to making this diagnosis.

Anaplastic Carcinoma

Anaplastic carcinomas are highly aggressive, large bulky lesions that occur in an older population (Fig. 14-57). On nuclear scintigraphy, the tumors do not take up iodine usually, but may be thallium or gallium avid. At the time of diagnosis, there is usually infiltration of adjacent trachea, esophagus, or extrathyroidal tissues and the prognosis is grim, very grim. Spread to the trachea is the rule with these lesions as the larger the tumor, the higher the rate of invasion. Twenty-nine percent of patients with thyroid carcinoma have tracheal invasion, but the rate is over 80% for anaplastic carcinoma. Intraluminal mass and 180 degrees or more of circumferential tracheal involvement has 100% positive predictive value. Three MR findings, soft tissue through cartilage, intraluminal mass, or circumferential involvement of over 180 degrees suggest the presence of tracheal invasion by thyroid carcinomas.

Non-Hodgkin Lymphoma

Non-Hodgkin lymphoma may also occur in the thyroid gland and is usually manifested by solitary nodules (44% to

TABLE 14-4 Multiple Endocrine Neoplasia Syndromes

Feature	MEN I	MEN II or IIa	MEN III or IIb
Eponym	Wermer syndrome	Sipple syndrome	Mucosal neuroma syndrome, Froboese syndrome
Parathyroid abnormality	Hyperparathyroidism (90% of cases) more commonly caused by hyperplasia than by adenoma	Parathyroid hyperplasia in 20%-50%	Very rare
Thyroid lesion	Goiter, adenomas, thyroiditis are rare	Medullary thyroid carcinoma (100%)	Medullary thyroid carcinoma (100%)
Pituitary lesions	Adenomas (20%-65%)	No	No
Pheochromocytoma	No	Yes (50%, bilateral)	Yes (50%)
Other manifestations	1. Pancreatic islet cell adenomas (insulinoma or gastrinoma) in 30%-80% 2. Adrenal cortex adenomas or carcinomas (30%-40%) 3. Rarely glucagonomas, VIP-omas, carcinoid tumors 4. Zollinger-Ellison syndrome	1. Pheochromocytoma 2. Scoliosis	1. Mucocutaneous neuromas (100%) 2. Marfanoid facies 3. Café-au-lait spots 4. Intestinal ganglioneuromatosis (100%)
Chromosomal linkage	Autosomal dominant, chromosome 11	Autosomal dominant, chromosome 10	Autosomal dominant, chromosome 10

MEN, Multiple endocrine neoplasia; *VIP,* vasoactive intestinal polypeptide.

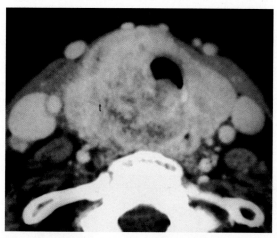

FIGURE 14-57 Dominant mass in the thyroid gland. The large size of this right thyroid mass (t), the degree of tracheal displacement, the effacement of the wall of the esophagus, and the heterogeneous density in a gland otherwise homogeneous on the left side in an elderly chap makes one write a run-on sentence and also suspicious of the correct diagnosis of anaplastic carcinoma.

86% of cases), multiple nodules (13%), or diffuse enlargement (7% to 11%). Bilateral disease has been reported in up to 53%. Tumor often invades outside the thyroid gland into the carotid sheath. The lymphoma may be primary to the thyroid gland or may result from systemic dissemination. Women are affected more than men, and this disease is seen with greatest frequency in elderly adults. Hashimoto thyroiditis may be a predisposing risk factor. Between 20% and 52% of patients are hypothyroid. On US the lesions are hypoechoic (pseudocystic), whereas on CT lymphoma is hypodense to normal thyroid tissue.

Metastasis

Renal cell carcinoma ranks third after lung and breast carcinoma in its propensity to metastasize to the thyroid gland. There are no specific imaging features for a metastasis to the thyroid gland.

TABLE 14-5 Duke's 3-Tiered System for Incidental Thyroid Nodules

CT/MRI/PET-CT Features	Recommendation
Category 1: Thyroid nodule PET avid or Thyroid nodule locally invasive or Suspicious lymph nodes	Strongly consider work-up with US for any size nodule
Category 2: Solitary thyroid nodule in patient <35 years of age	Consider work-up with US if ≥1 cm in adults. Consider work-up with ultrasound for any size in pediatric patients.
Category 3: Solitary thyroid nodule in patient ≥35 years of age	Consider work-up with ultrasound if ≥1.5 cm
Multiple nodules	Consider ultrasound with recommendations prioritized on basis of criteria (in order listed) for solitary nodule

CT, Computed tomography; *MRI,* magnetic resonance imaging; *PET-CT,* positron emission tomography–computed tomography; *US,* ultrasound.
Copyright 2016 Dr. Jenny Hoang. Image courtesy Dr. Jenny Hoang and Radiopaedia. org. Used under license.

Incidental Thyroid Nodule

What to do about the all too often incidentally noted thyroid nodule on the cervical spine CT and MRI? This is a quandary because (1) thyroid nodules are so common, (2) imaging features are nonspecific, (3) most nodules are benign, (4) the malignancies are often nonaggressive and early detection may not lead to longer survival, and (5) there is some morbidity associated with the surgical extirpation (vocal cord paralysis, hypoparathyroidism, vascular injury). We recommend the Dr. Jenny Hoang approach from Duke (Table 14-5).

PARATHYROID GLANDS

Anatomy

The parathyroid glands are usually located in a perithyroidal location (duh?). Ninety percent of patients have four glands, but there may be two to six parathyroid glands, usually situated along the upper and lower poles of the thyroid gland. Rarely, parathyroid glands are located in the upper mediastinum, within the thyroid gland, in the tracheoesophageal groove, or at the thoracic inlet. The parathyroid glands are derivatives of the third and fourth pharyngeal pouches.

Congenital Disorders

Most parathyroid cysts are due to degeneration of adenomas. Rarely, a congenital parathyroid cyst may be seen, usually within the lowermost glands.

The MEN syndromes are associated with parathyroid adenomas or hyperplasias (see Table 14-4). The gene for MEN I has been localized to the eleventh chromosome. In this syndrome, the gene must be deleted or inactivated because it normally expresses a protein, menin, which is a tumor-suppressor protein. Most patients with MEN I have hyperparathyroidism and this is the most common presenting symptom. MEN I can manifest with parathyroid hyperplasia or adenoma and one may define anterior pituitary, islet cell pancreatic, thymic, adrenal, foregut, and bronchial neoplasms as well. Remember P^{3}AT (pituitary, parathyroid, pancreas and adrenal and thyroid). Collagenomas and facial angiofibromas may also coexist.

Benign Neoplasms

When a patient is seen with hypercalcemia, a search for a parathyroid adenoma and/or parathyroid gland hyperplasia may lead to imaging. Although some surgeons operate on the parathyroid glands without preoperative imaging, feeling confident that they can identify the glands in the surgical field and can separate a normal gland from one that is hyperplastic or adenomatous, others employ imaging techniques to identify a parathyroid adenoma preoperatively. The surgical success rate for removal of a hypersecreting gland is 95% without any imaging at all at initial exploration. However, preoperative imaging is playing an increasing role these days in order to keep the surgical exploration as minimally invasive as possible.

In a patient whose hypercalcemia persists after initial surgery for resection of the parathyroid glands, imaging may be helpful because the reoperation success rate drops to as low as 62% without imaging. In these cases, aberrant or ectopic parathyroid tissue may be present. Fool me once, shame on you; fool me twice, shame on me. Get some imaging! The most common remote location for an ectopic parathyroid gland is the anterior mediastinum.

A running debate about whether CT (Fig. 14-58), US (Fig. 14-59), nuclear medicine (Fig. 14-60), or MR is the best study for identifying parathyroid adenomas continues. However, we believe that technetium sestamibi (99m technetium 2-methoxy-isobutyl-isonitrile [MIBI]) is the best functional study to perform when searching for a parathyroid adenoma. The study has accuracies between 80% and 90% for adenomas with equally high specificity when applying single-photon emission computed tomography (SPECT) technology, which has the added benefit of increased sensitivity, but an additional couple hundred buck$$ from your health maintenance organization (HMO) and health care finance administration (HCFA) agent. Unfortunately, for parathyroid hyperplasia, the sensitivity

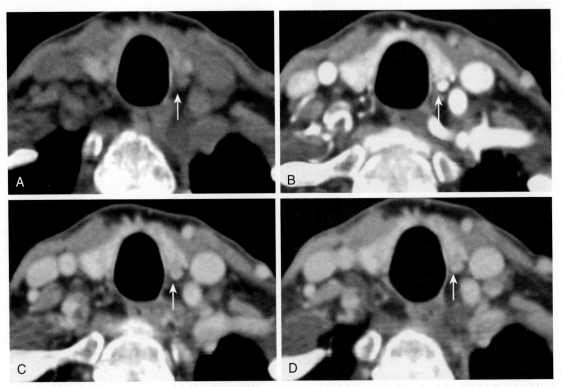

FIGURE 14-58 Parathyroid adenoma on four-dimensional computed tomography. The left sided adenoma *(arrow)* is seen as less dense than thyroid gland on unenhanced scan **(A)**, more dense than thyroid, simulating vessel enhancement on arterial phase **(B)**, and denser than lymph nodes on venous **(C)** and delayed **(D)** phases.

is only 60%, but that is still better than the competitive procedures. Ten to 25 mCi of technetium-labeled MIBI is given with images obtained in the first minutes after injection and again at 2 hours after injection of the radiotracer. If needed, Tc pertechnetate can be administered to localize the thyroid gland and even subtract the pertechnetate activity in the thyroid from the MIBI activity in the parathyroids and thyroid. Recently four-dimensional (unenhanced, arterial [25 seconds after injection], venous [60 seconds after injection], and delayed [120 seconds after injection]) CT (see Fig. 14-58) has been advocated for showing early arterial type enhancement of parathyroid hormone adenomas that can be differentiated from veins, nodes, and thyroid lesions, but the radiation dose to the neck has led to major push-back on using this for routine evaluations and should be reserved for those cases where nuclear scintigraphy fails to identify the adenoma.

In the postoperative patient with persistent hyperparathyroidism, no study is outstanding frankly. Most surgeons combine sestamibi with a cross-sectional imaging study—dealer's choice. Some surgeons use intraoperative sestamibi and a radiation detector to find the suckers. They can use intraoperative sampling of parathormone values to determine if they got the little bugger out. Ectopic adenomas in the thymus, parapharyngeal region, chest, thyroid gland, and lower neck are better detected than those in the tracheoesophageal groove, a blind spot for MR and MIBI imaging. For hyperplastic, nonadenomatous glands, MR (82%) and MIBI (90%) fare pretty well, but with hypercellular glands, the success rates of MR (77%) and MIBI (64%) drop significantly.

On MR, parathyroid adenomas are typically very bright on T2WI and can usually be readily distinguished from normal glands. In 30% of cases, the MR findings may be atypical—high intensity on T1WI or intermediate on T2WI. These atypical cases have been correlated histopathologically with cellular degeneration, intratumoral hemorrhage, fibrosis, and hemosiderin deposition. Parathyroid adenomas enhance brightly.

A problem arises when bright signal is encountered on the MR. The same appearance may be due to a lymph node, parathyroid adenoma, or even an intrathyroidal nodule. Remember, the goal of the study is to direct the surgeon to the possible location of the parathyroid adenoma.

On US, parathyroid adenomas are usually oval sonolucent masses.

Malignant Neoplasms

Rarely one may identify a parathyroid carcinoma. You might see one or two cases in your career (but not as a figure in this book because we haven't seen it, but we're still young). The distinction is based on spread to lymph nodes and rapid growth. Carcinomas account for 1% of cases of hyperparathyroidism.

CYSTIC MASSES IN THE NECK

Throughout this and the preceding chapters, you have read about numerous cystic masses that can appear in the head and neck. This section organizes these entities

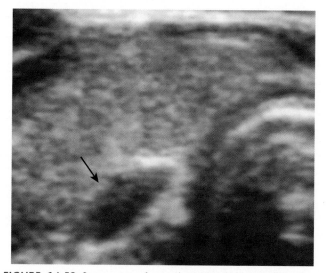

FIGURE 14-59 Sonogram of parathyroid adenoma. The *arrow* marks a parathyroid adenoma posterior to the thyroid gland on this transverse scan.

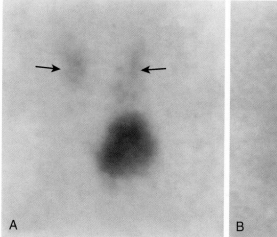
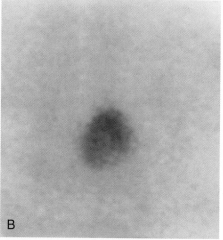

FIGURE 14-60 Parathyroid adenoma studies by technetium-sestamibi. **A,** While the initial 14-minute study shows both the normal thyroid gland (faintly seen designated by *arrows*) and the large adenoma, the 2-hour follow-up study **(B)** shows only the adenoma. Normal thyroid uptake has washed out.

for you before you hit the final summary chapter. These lesions usually present as nontender, palpable masses without associated lymphadenopathy. Rarely, they may declare themselves from superimposed infection. The specifics about each "cyst" are summarized in Table 14-6. To understand one of the most common of these cysts, the BCC, one must have an understanding of normal branchial arch derivatives (see Tables 14-2, 14-7, and 14-8).

The main differential features of these lesions are their age at onset, location, and cystic contents. The pediatric lesions include the cystic hygroma (60% evident at birth and 90% by 3 years) and the thymic cyst. In young adulthood, you will see BCCs (Fig. 14-61), thyroglossal duct cysts, dermoids, and ranulas. Older patients have thyroid and parathyroid cysts, lipomas, or cystic/necrotic lymph nodes.

The head and neck is the third most common site for dermoids after the gonads and superior mediastinum. In the head and neck region, the orbit, oral cavity, and nasal cavity account for most. A classic location of a dermoid is along the anterior floor of the mouth where they may simulate a ranula or lymphatic vascular malformation. One may see internal cobblestone architecture on MR that suggests the final diagnosis of a dermoid.

Thyroglossal duct cysts classically present in the midline and are associated with the infrahyoid strap muscles; the second BCC is most typically seen anterior to the sternocleidomastoid muscle at the carotid bifurcation level (see Fig. 14-34); the thymic cyst is low in the left neck (Fig. 14-62); the cystic hygroma is typically in the posterior triangle of the neck (see Fig. 14-46); and the thyroid, parotid, and parathyroid cysts are in their native tissue. Lower down in the neck, one may find third and fourth branchial

TABLE 14-6 Cystic Masses in the Neck

Cyst	Clinical Presentation	Imaging Appearance
Abscess	Fever, pain, systemic reaction	Thick-walled mass Infiltration, edema of subcutaneous fat Rim enhancement
BCC	Painless mass in 10- to 40-year-old patients: occasionally infected	Mass anterior to SCM, at CCA bifurcation Thin-walled unless infected Unilocular
Cystic hygroma	Painless mass in neonate	Mass posterior to SCM High intensity to T1WI (25%) because of blood, protein, fat Multiloculated Insinuates into different neck compartments Posterior triangle or submandibular
Cystic, necrotic node	Associated with papillary thyroid cancer or squamous cell carcinoma (often HPV+)	Necrotic node with thick rim Possible extracapsular extension
Dermoid	Midline cyst in tongue, oral cavity in young adult, dorsum of tongue	Fat often seen Thin wall
Epidermoid	Off-midline cyst in mouth	Cystic (no fat) Unilocular; may be in sublingual space
Laryngocele/pharyngocele	Often asymptomatic	Air- or fluid-containing sac emanating from saccule of laryngeal ventricle or pharynx Thin-walled
Lipoma	Soft, compressible mass in posterior triangle in middle-aged patient with rapid weight gain	Fat density/intensity on CT, MR
Parathyroid cyst	Painless cyst below thyroid gland, rarely hyperparathyroid as result of PTH secretion, adult	Cyst with or without nodule 95% below thyroid gland Thin-walled
Parotid cysts	Painless masses, sometimes seen in HIV-positive patients	Usually cyst intensity, density in the parotid Multiple Associated with lymph nodes, Waldeyer ring hyperplasia
Ranula	Cystic mass in floor of mouth located in sublingual space; may protrude into submandibular region, often superinfected	If not infected, thin-walled mass, homogeneous unilocular in sublingual space, possibly protruding through mylohyoid (plunging ranula) Tail pointed to obstructed sublingual or submandibular duct If infected, thick-walled
Thyroglossal duct cyst	Midline painless mass in adult	Unilocular cyst embedded in strap muscles Suprahyoid, 20%; hyoid, 14%; infrahyoid, 65% May have high protein 75% midline
Thyroid cysts	Enlarged lobe of thyroid with palpable mass and/or goiter	Colloid cysts are dense on CT, hyperintense on T1WI Follicular cysts may have high or low intensity on T1WI Often have hemorrhage
Thymic cyst	Midline cyst low in neck in child, ectopic thymus may be on left side	Anterior mediastinum or low neck cyst Multiloculated

BCC, Branchial cleft cyst; *CCA,* common carotid artery; *CT,* computed tomography; *HIV,* human immunodeficiency virus; *HPV,* human papillomavirus; *MR,* magnetic resonance; *PTH,* parathyroid hormone; *SCM,* sternocleidomastoid muscle; *T1WI,* T1-weighted imaging.

TABLE 14-7   Arch Derivatives and Anomalies

Arch	Arch Artery Derivative (Mesoderm)	Arch Nerve Derivative	Arch Cartilage Derivative	Arch Anomalies
1	Degenerates	V_3	Mandible; malleus; most of incus; tensor tympani; masticator, tensor veli palatini, anterior digastric muscles; middle ear cavity; tonsils	1. Treacher Collins syndrome; abnormal external ear, hypoplastic mandible 2. Pierre Robin syndrome: hypoplastic mandible, cleft palate, low ears, eye defects
2	Degenerates	VII	Manubrium of malleus; long process of incus; stapes; styloid process; parts of hyoid stapedius, facial, posterior digastric muscles; inferior facial canal; tonsils; parathyroid; tonsillar fossa	Persistent stapedial artery
3	Carotid arteries	IX	Part of hyoid, stylopharyngeus muscles; parathyroid; thymus; pyriform sinus	DiGeorge syndrome: absent thymus and third parathyroid gland
4	Left = aorta Right = proximal subclavian	X (superior laryngeal)	Thyroid cartilage; cuneiform cartilage, inferior pharyngeal constrictor, cricopharyngeus muscles; inferior pyriform sinus	Aberrant subclavian artery
5	This arch is MIA			
6	Pulmonary arteries	X (recurrent laryngeal)	Cricoid, arytenoids, trachea, intrinsic muscles of the larynx	

TABLE 14-8   Branchial Apparatus Derivatives: Cleft and Pouch

Branchial Apparatus	Cleft/Groove (Ectoderm) Derivative	Pouch (Endoderm) Derivative	Nerve Associated
1	External auditory canal	Eustachian tube, mastoid air cells	V_3
2	Cervical sinus of His	Palatine tonsil, tonsillar fossa	VII
3	Cervical sinus of His	Thymus, inferior parathyroid gland	IX
4	Cervical sinus of His	Ultimobranchial body, superior parathyroid	X

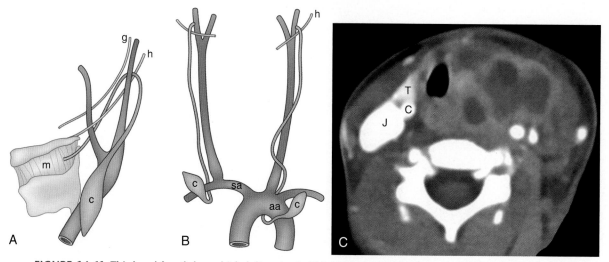

FIGURE 14-61 Third and fourth branchial cleft cysts. **A,** Third branchial cleft cyst anomaly. The epithelium-lined cyst is located posterior to the sternocleidomastoid muscle. The cyst tract typically ascends posterior to the internal carotid artery. It then courses medially to pass over the hypoglossal nerve (h) and below the glossopharyngeal nerve (g). It then pierces the posterolateral thyrohyoid membrane (m) to communicate with the pyriform sinus. **B,** Fourth branchial cleft cyst anomaly. The nonkeratinized stratified squamous epithelium-lined cysts are located anterior to the aortic arch on the left and the subclavian artery on the right, respectively. The sinus tract is seen to hook inferiorly around the adjacent vascular structures (like the recurrent laryngeal nerve) and ascends to the level of the hypoglossal nerve posterior to the common and internal carotid arteries. It then loops over the hypoglossal nerve to pass deep to the internal carotid artery. **C,** This patient had a fourth branchial cleft cyst that got recurrently infected. There is a rim-enhancing fluid collection in the left perithyroidal soft tissues, displacing the trachea to the right. The normal thyroid gland (T), carotid artery (C), and jugular vein (J) are indicated on the right. *aa,* Aortic arch; *c,* cyst; *h,* hypoglossal nerve; *sa,* subclavian artery. (**A** and **B** from Mukherji SK, Fatterpekar G, Castillo M, et al. Imaging of congenital anomalies of the branchial apparatus. *Neuroimaging Clin North Am.* 2000;10:89, 92.)

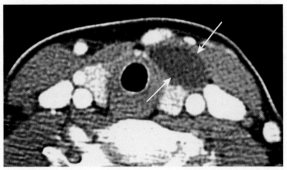

FIGURE 14-62 The differential diagnosis of this cystic mass *(arrows)* on computed tomography in the lower neck would include a third or fourth branchial cleft cyst, a cystic hygroma/lymphangioma, a thymic cyst, a thyroid cyst, necrotic lymph node, or a jugular lymphatic cyst/lymphocele. In this case, it was a thymic cyst.

apparatus lesions/cysts (see Fig. 14-61, *C*) and jugular lymphatic sacs. These may be from iatrogenic/trauma to the thoracic duct.

The presence of fat identifies lipomas and dermoids. High protein or hemorrhage may be seen in thyroid, parathyroid, lymphatic, and epidermoid cysts as well as in a cystic hygroma (macrocystic low flow lymphatic vascular malformation). Most parotid cysts and noninfected BCCs have densities and intensities that simulate simple fluid.

If a thyroglossal duct cyst has high CT density, it usually means that it has previously been infected. This reasoning also applies to other primary congenital cysts. Thick-walled lesions are usually abscesses or necrotic lymph nodes.

You have consumed the offerings we have laid out for you, and hopefully you are attempting to digest all this fabulous food for thought. We hope its transit time through your brain is slower than it would be through your intestines.

Chapter 15

Anatomy and Degenerative Diseases of the Spine

SPINAL PARLANCE

This chapter begins with a brief review of the anatomy of the spine. Imaging techniques are then discussed, followed by the normal imaging appearance of the spine on magnetic resonance (MR) imaging and computed tomography (CT). We then consider degenerative diseases of the spine. These common diseases may manifest as localized back pain, radiculopathy (pain radiating in a spinal root distribution), or myelopathy (signs of spasticity, increased tone, and increased reflexes). The radiologic differential diagnosis of spine lesions is based on localizing the lesion to a particular space.

A brief word concerning terminology. Understanding the terminology used in localizing spinal lesions is critical in framing your differential diagnosis as well as presenting yourself as a knowledgeable radiologist. So a little repetition would not hurt. The anatomic algorithm historically used by radiologists was predicated on myelographic interpretation as to whether lesions were extradural, intradural extramedullary, or intramedullary. Intramedullary lesions are indigenous to the spinal cord, tend to expand it, and narrow the subarachnoid space. They include spinal cord tumors such as astrocytoma, ependymoma, and spinal cord metastases, as well as nonneoplastic lesions such as syringohydromyelia, infections, and inflammation such as transverse myelitis, acute disseminated encephalomyelitis, or cord infarcts. Intradural extramedullary lesions are lesions outside the spinal cord but within the thecal sac. These include meningioma, neurogenic tumors, subarachnoid seedings, or vascular mass lesions such as arteriovenous malformations or varices. Those which are intradural but extramedullary (outside the cord) expand the subarachnoid space on the ipsilateral side (producing a meniscus) and shove the cord over to the contralateral side. Obviously, intradural lesions below the termination of the spinal cord are extramedullary. Lesions that compress or are intrinsic to the cord often produce myelopathy, whereas those which compress and irritate the roots cause radiculopathy.

Extradural lesions occur outside the thecal sac and may originate from the adjacent paraspinal soft tissues, disk (herniated disk), disk space (epidural infection), or the vertebrae (osteophytes, primary bone and metastatic tumors). Extradural lesions are a more common cause of compression on the cord than intradural extramedullary lesions, purely by virtue of their large numbers; that is, compressive disk herniations and bone metastases are more common than intradural lesions.

Although spinal anatomy lacks the sex appeal of other anatomic sites, it does have certain features that are intriguing and may even keep your attention (SEX SEX SEX—see, got your attention!).

The Doc ordered MR of the spine
For "Painful back of 3 weeks' time"
The HMO said,
"6 weeks in bed!"
To operate is "Medicare crime"

ANATOMY

Spinal Nerves

The bony spine is divided by region into the cervical spine containing seven vertebrae (the first two of which are rather unique and are discussed further), the thoracic spine, consisting of 12 vertebral bodies; and the lumbar spine, with five vertebral bodies. The distal spine consists of the sacrum and coccyx. The spine encases the spinal cord, which normally terminates at a variable level from approximately T12 to L2 in adults.

The spinal cord contains eight cervical, 12 thoracic, five lumbar, five sacral, and one coccygeal paired spinal nerves. These nerves are rather easily identified on CT with intrathecal contrast or high-resolution T2-weighted (T2W) MR. It is important to appreciate that C1, which is a sensory nerve, exits above the C1 and C2 interspace so that the C2 nerve exits between C1 and C2, C3 nerve exits between C2 and C3 and so on. Therefore, the C8 nerve root exits between C7 and T1. In the thoracic region, T1 exits between T1 and T2, and T12 exits between T12 and L1. In the lumbar spine the L1 root exits between L1 and L2 and so forth, so that the L5 root exits between L5 and S1. However, a funny thing happened on the way to creation. The bodies in the lumbar region became much longer. The nerve roots in this region leave the thecal sac right under the pedicle (Fig. 15-1), well above the interspace. Paracentral disk herniations in the lumbar region characteristically strike the root in the thecal sac that will exit below the interspace. This is because the disk space is inferior to the same numbered exiting root at that level. Thus, an L4-5 disk herniation most often compresses the L5 root because the L4 root is already in the foramen (above the disk). Very lateral herniated disks may compress the upper root; that is, an L4-5 lateral herniation can compress the L4 root in the foramen. Larger disks can compress many roots in the thecal sac. Furthermore, disk fragments may migrate superiorly and compress the root exiting at the appropriate interspace; that is, an L4-5 free fragment can compress the L4 root or a combination of both the L4 and L5 nerve roots.

Contrast that finding with the cervical spine where a C6-7 disk herniation in the foramen injures the C7 nerve

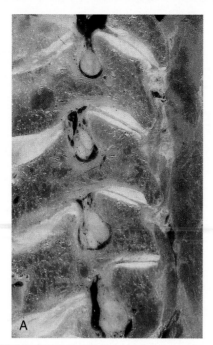

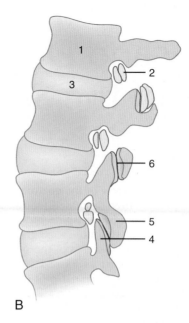

FIGURE 15-1 Nerve roots in cervical neural foramina give your arms their sensorimotor stamina. **A,** Cervical roots reside in the inferior aspect of the neural foramen. **B,** Lumbar roots (2) travel in the upper half of the foramen. 1, Vertebral body; 2, nerve root, 3, disk; 4, superior articular facet; 5, inferior articular facet; 6, facet joint. (From Firooznia H, Golimbu C, Rafii M, et al. *MR and CT of the musculoskeletal system,* St Louis: Mosby-Year Book; 1992.)

root and a paracentral one also may affect the same nerve root because of the horizontal course of the root. In the cervical region, the roots are in the lower portion of the foramen whereas in the lumbar region they are in the upper aspect of the neural foramen. In the thoracic region, disk herniations may cause myelopathic changes; however, they can also produce thoracic radiculopathy. Lesions at a given thoracic spine body level might also produce sensory symptoms one to two segments below the compression. This is because the cord ends at approximately T12 to L2 so that cord lesions result in neurologic deficits that are localized below their vertebral body anatomic location.

An anatomic variation is the conjoined nerve root, which occurs in less than 5% of patients, with L5-S1 being the most common disk space. This normal variation consists usually of two nerve roots traveling in the same dural pouch and exiting through the same or through different foramina. The problem is really the radiologist's. She or he should not mistake the conjoined root on imaging for an epidural defect with obliteration of the thecal sac below the conjoined root.

Each spinal nerve is divided into a dorsal or sensory root and a ventral or motor root. The dorsal root ganglion is a distal dilatation of the dorsal root just proximal to its joining with the ventral root to form the spinal nerve (Fig. 15-2).

Vertebrae

The generic vertebra is composed of the cylindric vertebral body, which contains cancellous bone with marrow and fat, covered by a thin layer of compact bone, and the vertebral arch or posterior elements, covered by a thick layer of compact bone (cortex), including the pedicles, laminae, superior and inferior facets, transverse processes, and spinous process (Fig. 15-3). The vertebral configuration is

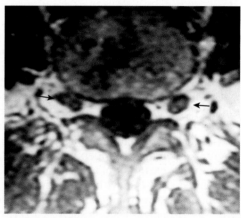

FIGURE 15-2 These ganglia of dorsal root supply sensation to the foot. Axial T1-weighted imaging demonstrates normal appearance of the dorsal root ganglia *(arrows)*. Do not mistake this for a foraminal disk herniation. Note normal appearance of surrounding hyperintense fat within the foramen.

modified in the different regions of the spine. The cervical vertebrae have their neural foramina between the transverse processes. The superior and inferior articular facets have joints between them (zygapophyseal) and form the articular pillar (Fig. 15-4). The uncovertebral (Luschka) joints (neurocentral joints, joints of Luschka) originate from the lateral posterior portion of the vertebral body, articulate with the contiguous vertebral body, and insinuate themselves between the disk and the nerve root canal. These joints have no synovium and no hyaline cartilage, but clefts can develop in the fibrocartilage with age and degeneration. The vertebral artery enters the foramen transversarium (in the cervical transverse process, naturally) at approximately C6 and travels superiorly.

The first cervical vertebra (atlas) has no body but rather just an anterior arch connected to two lateral masses and a posterior arch (Fig. 15-5). On the upper surface of the posterior arch is a groove over which the vertebral artery courses after it leaves the foramen transversarium of C1. The vertebral arteries pass through the posterior atlantooccipital membrane and course anteriorly superiorly upward through the foramen magnum. As it pierces the dura the vertebral artery may be slightly narrowed, and this caliber change can serve as a marker for the beginning of the intradural segment of the vertebral artery. The first spinal nerve exits here as well.

The second cervical vertebra, the axis, is unique, with a superior extension from its body termed the dens (odontoid process; see Fig. 15-5, *B*). The dens represents the lost vertebral body of the atlas and is usually found fractured

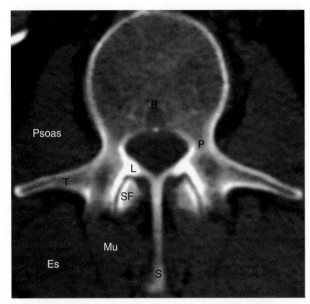

FIGURE 15-3 Lumbar vertebral anatomy shown by multiplanar computed tomography (CT). This CT of a lumbar vertebral serves as a model for a generalized vertebra. Pedicle (P), lamina (L), transverse process (T), basivertebral venous plexus (B), superior facet from vertebral below (SF) and spinous process (S) are labeled. Psoas muscle, multifidus (Mu), and erector spinae (Es) are labelled.

by residents on their first call. The articulation between the atlas and axis is composed of multiple synovial joints: one medial between the dens and the anterior arch, one on each side between the inferior articular facet of the lateral mass of the atlas and the superior facet of the axis, and multiple ones between the dens and the atlantoaxial ligaments (transverse, cruciate, and alar). Rheumatoid arthritis has a propensity for the atlantoaxial joint with pannus formation, leading to bone erosions and subluxations.

The thoracic vertebrae have an articulation on the transverse process for the rib and no foramen transversarium, whereas the lumbar vertebrae have neither a foramen transversarium nor a facet for the rib articulation. The lumbar vertebral articulations are composed of the lumbar disk and two facet joints posteriorly. The lateral recess of the lumbar spine is in the anterolateral portion of the spinal canal, with boundaries consisting of the posterior margin of the vertebral body and disk anteriorly, the medial margin of the pedicle laterally, and the superior articular facet, the medial insertion of the ligamentum flavum, the lamina, and the pars interarticularis posteriorly (Fig. 15-6).

Intervertebral Disks and Spinal Ligaments

The diskovertebral complex is composed of three components: the cartilaginous endplate, annulus fibrosis, and nucleus pulposus. The endplate consists of a flat bony disk with an elevated rim (attached ring apophysis), which produces a central depression in the endplate occupied by hyaline cartilage.

The annulus fibrosus surrounds the nucleus pulposus (the remnant of the notochord). The nucleus is eccentrically located near the posterior surface of the disk. The lamellae of the annulus are fewer in number, thinner, and more closely packed posteriorly than anteriorly. This anatomic arrangement may account for the propensity for posterior herniation. The external fibers of the annulus are connected to the bone of the vertebral bodies by Sharpey fibers, which usually cannot be distinguished by imaging. Annular fibers also merge with both anterior and posterior longitudinal ligaments. Important ligaments of the vertebral column are (1) the anterior longitudinal ligament, running along the anterior aspect of the vertebral bodies; (2) the posterior longitudinal ligament, running along the posterior aspect of the vertebral

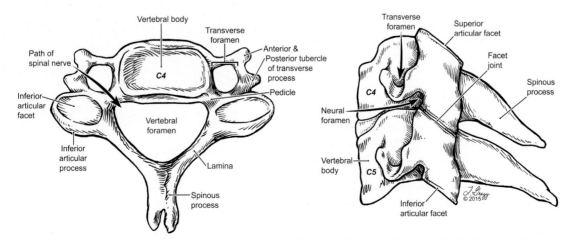

FIGURE 15-4 Cervical Vertebra illustration shows vertebral parts and relationships. (© 2015 Lydia Gregg.)

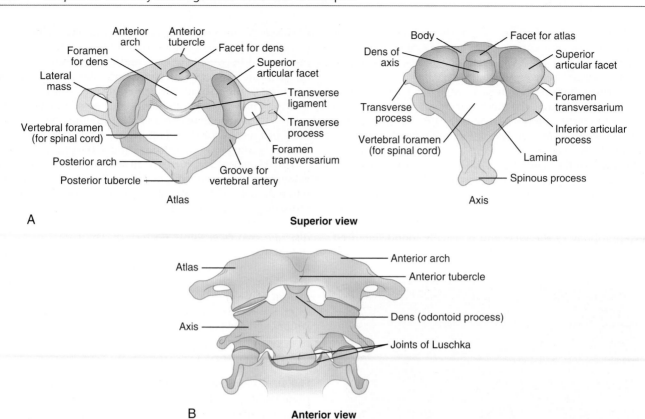

FIGURE 15-5 Schematic representation of the atlas and axis from a superior view **(A)** and anterior view **(B)**.

bodies anterior to the thecal sac; (3) the ligamentum flavum, connecting the laminae and extending from the midline laterally to the facets; and (4) the interspinous ligament, joining the superior portion of the spinous process below to the inferior portion of the spinous process above and meeting the ligamentum flavum in the midline (Fig. 15-7). The facet joints also help stabilize the spine and may be yet another contributor to spinal stenosis. There are also small ligaments in the neural foramen, which may play a role in foraminal stenosis.

Spinal Cord

The spinal cord extends from the medulla oblongata, at the level of the upper border of the atlas, to T12 to L2, where it terminates in the conus medullaris. At the apex of the conus, continuous with the pia mater (i.e., "the tender mother"), is the filum terminale, which descends initially in the thecal sac and then becomes covered with adherent dura as it leaves the thecal sac to insert in the coccyx. The cauda equina emanates from the conus medullaris and contains the nerve roots of the lumbar and sacral nerves. The spinal cord has two enlargements in its course, one in the cervical region from approximately C4 to approximately T1 and the other in the lower thoracic region from approximately T9 to T12. Do not mistake these normal expansions for a pathologic process; otherwise you will need a return ticket to your board examination. These enlargements correspond to the locations in the cord that supply the spinal nerves for the upper and lower extremities.

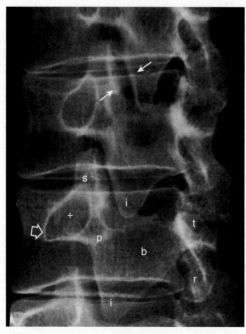

FIGURE 15-6 Anatomy on oblique scan explains the "Scottie Dog" diagram. Right facet joint is seen *(arrows)*. At the disk space below, components of the Scottie dog are identified: *face,* right pedicle (+); *neck,* right pars interarticularis (p); *ear,* right superior articular facet (s); *front leg,* right inferior articular facet (i); *body,* lamina (b); *nose,* right transverse process *(open arrow)*; *tail,* left superior articular facet (t); *rear leg,* left inferior articular facet (r). That's the straight poop on anatomy.

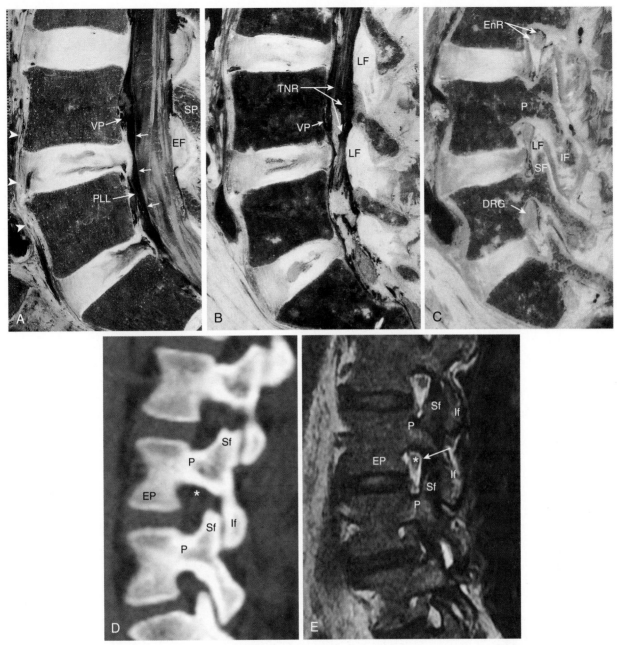

FIGURE 15-7 Cryomicrotome anatomic section **(A** to **C)** reveals lumbar spine anatomy to perfection. *Arrowheads,* Anterior longitudinal ligament; *small arrows,* dura mater. *DRG,* Dorsal root ganglion; *EF,* epidural fat; *EnR,* exiting nerve root; *IF,* inferior facet; *LF,* ligamentum flavum; *P,* pedicle; *PLL,* posterior longitudinal ligament; *SF,* superior facet; *SP,* spinous process; *TNR,* traversing nerve root; *VP,* venous plexus. **D,** Corresponding parasagittal computed tomographic scans shows dorsal root ganglion *(asterisk)* surrounded by bony margins of the pedicle (P), superior (Sf) and inferior (If) facets and the endplace of the vertebral body (EP). Similar anatomy on T2-weighted magnetic resonance **(E)** with the dorsal root ganglion in the neural foramen *(arrow)* surrounded by fat.

Blood Supply to the Spinal Cord

The blood supply to the spinal cord depends on the particular location (Fig. 15-8). In the cervical region the anterior spinal artery is formed by branches that originate from the vertebral arteries just before joining the basilar artery. The anterior spinal artery supplies the anterior two thirds/majority of the spinal cord including the anterior column of the central gray matter, the corticospinal, spinothalamic, and other tracts. In addition, paired posterior spinal arteries originate from the vertebral arteries and supply the dorsal portion of the cord; that is, the posterior columns and the posterior horn of the central gray matter. These two arterial systems do not usually have significant anastomoses between them and can lead to different myelopathic cord syndromes when compromised (Table 15-1). The anterior spinal artery is in the midline whereas the posterior spinal arteries lie off the midline (Fig. 15-9, *B*). The anterior and posterior spinal arteries rarely originate at the same level.

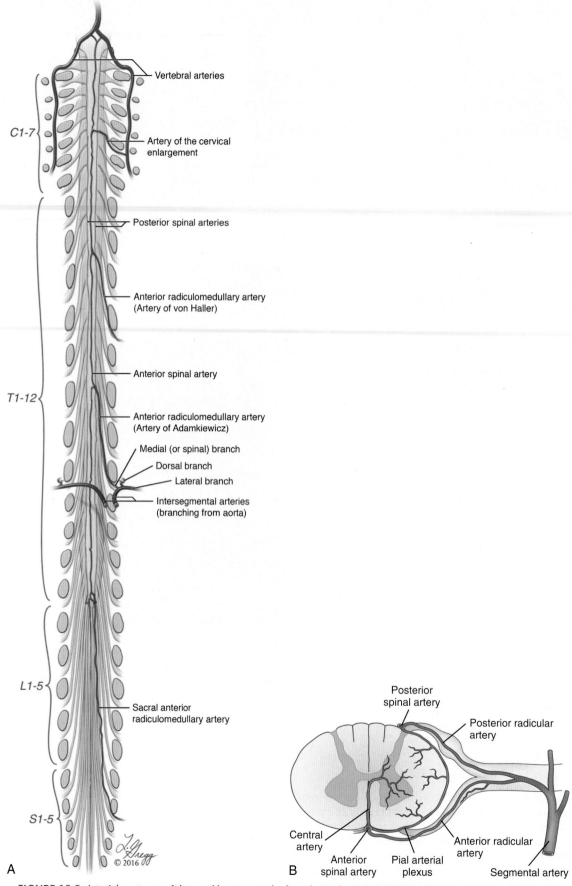

FIGURE 15-8 Arterial anatomy of the cord keeps a reader from being bored. **A,** Schematic representation. **B,** Arterial blood supply at a single segment. Both the anterior and posterior medullary arteries are illustrated. This is not the typical arrangement. It is unusual for both anterior and posterior medullary arteries to enter at the same segment in any region of the cord.

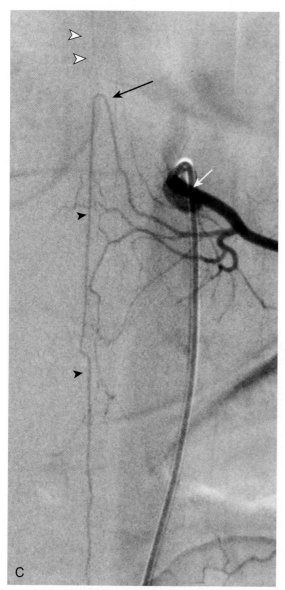

C

FIGURE 15-8, cont'd **C,** Spinal arteriogram. Injection into the T9 left intercostal artery *(white arrow)* reveals filling of artery of Adamkiewicz with its characteristic hairpin turn *(long black arrow).* Anterior spinal artery is filling faintly above *(white arrowheads)* and more emphatically below *(black arrowheads).* (**A** © 2016 Lydia Gregg; **B** from Krauss WE. Vascular anatomy of the spinal cord. *Neurosurg Clin North Am* 1999;10:10.)

The caliber of the anterior spinal artery at a particular spinal level is proportional to the metabolic demands of the spinal gray matter.

At the C3-4, C5-6, and C7-T1 levels, radicular feeders from the vertebral, ascending cervical (anterior to the transverse process), and deep cervical (posterior to the transverse process) arteries anastomose with the spinal arteries. The radicular feeders enter the thecal sac through the intervertebral foramina and divide into the anterior and posterior branches coursing with the nerve roots. Because they follow the nerve root, the spinal arteries have a sharper angle in the lumbar region than in the cervical region. However, not all spinal nerves have radicular arteries. The cervical and upper two thoracic levels comprise one vascular territory. The midthoracic region

T3-T7 is supplied by intercostal branches from the aorta, branches of the supreme intercostal arteries from the subclavians, and lumbar arteries. This region may have a tenuous blood supply. The lower thoracolumbar region to the filum terminale is supplied by the artery of Adamkiewicz. It is commonly located on the left side between T9 to L2 (85% of the time) and T5 to T8 (15% of the time). It enters the spinal canal with the nerve roots and makes a characteristic hairpin loop, giving off a small superior branch from the apex of the turn and a large descending branch, which supply the anterior spinal cord and anastomoses with the posterior spinal arteries in the region of the conus medullaris.

Uncommonly, an artery named the artery of Lazorthes arises from the common or internal iliac arteries and accompanies one of the sacral roots of the cauda equina to supply the conus medullaris. The venous blood supply is comparable to the arterial blood supply with a variable amount of anterior and posterior spinal veins running with the spinal arteries.

RADIOLOGIC WORK-UP

Imaging Fundamentals and Techniques

Although imaging techniques vary, important concepts in the spine should be appreciated. These generalizations hold true for the vast majority of situations encountered, regardless of the type of equipment available. Currently, MR is the technique of choice in most situations. However, CT, myelography, and myelography supplemented by CT (myelo-CT) all have roles in particular situations and with particular surgeons. As Sam Walton once said: There is only one boss. The customer. And he can fire everybody in the company from the chairman on down, simply by spending his money somewhere else…like on a DI book.

Magnetic Resonance Technique

Dollar for dollar, MR is the best single method for imaging the spine. Complete MR examination requires an excellent sagittal image. Take care to scan from neural foramen to neural foramen with thin sections, with the minimum interslice gap that your particular instrument permits. This sagittal sequences should include T1-weighted imaging (T1WI), T2WI, and short tau inversion recovery (STIR) scans to assess for disk/endplate/body/ligamentous signal abnormalities. Assess the foramina for bony constriction and the normal fat and nerves both in the axial and sagittal plane. The axial plane, because of the oblique orientation of the neck foramina, is better for such an evaluation in the cervical region (unless you play with oblique sagittal scans). In the lumbar region, the foramina have a vertical orientation and are filled in part with fat so they are well visualized in the sagittal plane. The nerve root is in the superior portion of the foramen under the pedicle in the lumbar zone. Lesions obliterating the fat are easily detected on T1WI. These include lateral disk herniations, facet joint hypertrophy, neurofibromas, soft tissue from inflammatory changes, postoperative scar, and spondylolisthesis.

In the cervical region, thin-section axial images are also necessary. A thin section low flip angle volumetric acquisition through the cervical spine enables sections of 1.5 mm

TABLE 15-1   Cord Syndromes and Clinical Manifestations

Syndrome	Part of Cord Affected	Manifestations	Less Common Manifestations	Typical Lesion
Anterior spinal artery syndrome	Anterior 2/3 cord bilaterally	Bilateral motor loss and pain and temperature sensation loss	Hypotension, sexual dysfunction	Cord infarct (ASA)
Incomplete ASA syndrome	Just anterior horn cells	Paraplegia	Painful diplegia	Cord infarct (ASA)
Posterior spinal artery syndrome	Posterior column injury	Loss of proprioception, vibratory sense, total anesthesia at level of injury	Mild weakness, rarely bilateral	Cord infarct (PSA)
Brown-Sequard syndrome	Hemicord injury	Ipsilateral hemiparesis and loss of proprioception, contralateral loss of pain and temperature		Demyelinating diseases
Cauda equina syndrome	Conus, cauda equina injury	Back pain, saddle anesthesia, bowel and bladder dysfunction with incontinence, sexual dysfunction	absent anal reflex, weakness in the legs, gait disturbance	Compression by bone/ disk pathology
Central cord syndrome	Central cord	Bilateral loss of pain and temperature and motor control	Bladder dysfunction, urinary retention	Trauma, cervical spondylosis, cardiac arrest, hypotension

ASA, Anterior spinal artery; *PSA,* posterior spinal artery.

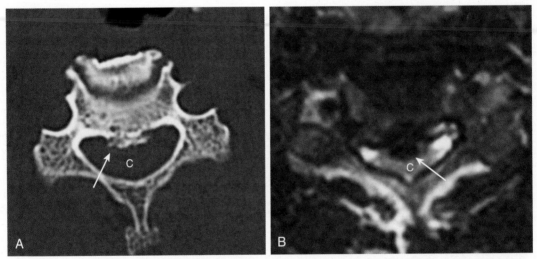

FIGURE 15-9 Cervical spine osteophyte. Computed tomography (CT) **(A)** and T2-weighted imaging **(B)** show osteophyte from posterior body cervical spine *(arrows)* compressing the spinal cord (C), which is inferred on the CT but directly visualized in cross-section on the magnetic resonance study.

or less with contrast such that the cerebrospinal fluid (CSF) is high signal, disk material is intermediate signal, and osteophytes are low signal intensity (Fig. 15-10). Look for osteophytic compression of the roots or impingement on the spinal canal. Being able to distinguish a dark osteophyte from a brighter disk is critical—otherwise you default to the lazy "disk-osteophyte complex" phrase that plagues our specialty. Disk herniations are amenable to discectomies...osteophytes are not...and require shaving... distinguish between them or else go do an MSK fellowship instead of neuroradiology.

Gradient echo MR may tend to exaggerate foraminal or bony canal stenosis of the spine, and what appears as high-grade or complete blocks on MR may not be as severe on myelography, CT myelography, or T1WI. Fast spin echo (FSE) T2WI is the routine supplemental technique used for imaging the CSF, spinal cord, and nerve roots. One advantage for FSE imaging derives from its relative insensitivity to susceptibility effects compared with gradient and

conventional spin echo techniques. Thus, spinal osteophytes are not as exaggerated on sagittal images as with the other techniques and there is better visualization of the extent of thecal sac compression, cord compression, and intrinsic cord pathology. Short tau inversion recovery images are excellent for detection of ligamentous injury and bone edema that otherwise is lost on FSE T2WI where fat and edema are both bright in the bone.

Diffusion-weighted/tensor imaging of the spinal cord has a role in the diagnosis of ischemic, inflammatory, traumatic, and demyelinating lesions of the spinal cord as well as studying the normal maturation of the cord in pediatric patients. Cord infarcts are readily detected. DTI may show that astrocytomas disrupt fibers whereas ependymomas are prone to displacing fibers.

Although MR is the best technique for demonstrating spinal cord compression, it cannot definitively judge whether a particular lesion will produce a complete block on myelography. This may be less of an issue with

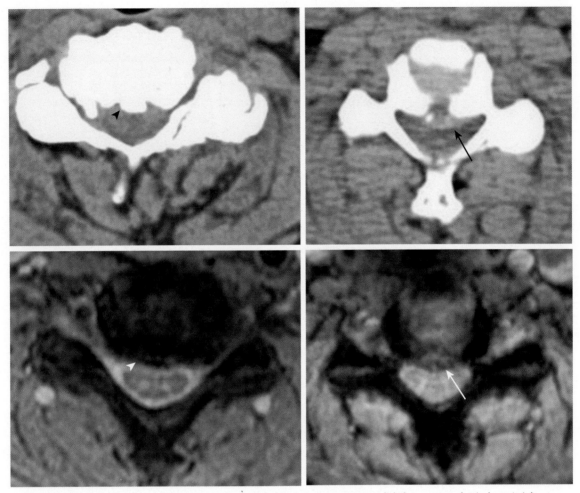

FIGURE 15-10 Contrast in gradient echo magnetic resonance separates disk from osteophytic bar. Axial computed tomographic images are above and gradient echo images below. Note the difference of brighter intensity in the disk herniation *(arrows)* versus an osteophyte of dark intensity *(arrowheads)*.

new imaging techniques, because treatment protocols are directed more toward lesions as opposed to the consequences of the lesions (the myelographic block). Indeed, the "block" has gone the way of the typewriter—only used by the old curmudgeon. Do not regret growing older. It is a privilege denied to many.

Intravenous contrast on MR imaging is necessary in the postoperative back to distinguish between scar and disk (Fig. 15-11), in infectious and inflammatory conditions of the spine to assess the full extent of disease, and in the evaluation of the spinal cord to rule out tumor. Although enhancement with fat suppression is in many cases useful in detecting metastatic disease to the vertebral bodies, it is not always necessary, especially with good quality STIR images, and many times, replacement of the fatty marrow by tumor is obvious on unenhanced images. Furthermore, because both benign and malignant marrow processes enhance, in many cases contrast enhancement of a marrow lesion will not help you make a distinction between the two.

Myelography

Although not performed as often as it once was, myelography, almost always combined with CT (myelo-CT), is still a sensitive and useful technique for disk herniation and, more importantly, osteophytic impingement

on *cervical* roots. The advantage of the myelo-CT is the exquisite bone detail superimposed upon the subarachnoid contrast. Myelo-CT unambiguously reveals extradural bony lesions compressing the subarachnoid space, roots, and spinal cord. For patients in whom MR is contraindicated, such as persons with metallic implants or cardiac pacemakers, or those who cannot tolerate MR, myelo-CT is fine...except to see intrinsic cord lesions. CT-myelography could be performed in the immediate postoperative setting if a "blocking" hematoma is suspected.

Computed Tomography

CT alone is favored over MR for evaluating metallic hardware for spine stabilization such as pedicle screws and anterior metallic plates (Fig. 15-12). So get used to those unenhanced CT postoperative "hardware scans." CT is also very useful in the trauma setting in the evaluation of acute bony injury. CT without intrathecal contrast is adequate for the lumbar region, where natural contrast exists between epidural fat, disk, and bone (Fig. 15-13). However, there is usually little contrast between the spinal cord and the subarachnoid space in the cervical and thoracic regions, so that intradural processes are suboptimally imaged without intrathecal contrast. Intramedullary lesions require MR evaluation.

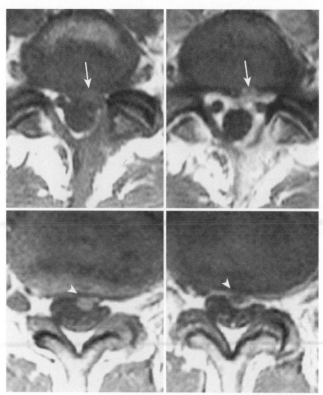

FIGURE 15-11 Scar versus residual disk requires gadolinium to mitigate misdiagnosis risk. Note that on the images above, the soft tissue in the left anterior epidural space after left hemilaminectomy shows enhancement on the top images *(arrows)*; Diagnosis: Scar! Contrast that (excuse the pun) with the images below where the soft tissue *(arrowheads)* does not enhance on the postcontrast scan; Diagnosis: Disk Herniation.

Discography

This is a most controversial topic. The procedure calls for injection of contrast material within the nucleus itself. After injection, plain radiographs are made and additional CT images can be performed. There are several observations to be made that include whether the contrast is confined by the annulus, streaks into the annulus, or leaks into the epidural space. Here you may be able to diagnose an unsuspected annular tear. Equally important is the reproduction of patient symptoms from the injection of contrast material. Thus, the injector must have enough experience to perform the technique in a reproducible manner and hopefully the injectee will have a rather standardized response. Yes, we agree it sounds more like art than science and more subjective than most studies we perform. Do these procedures add anything? For one thing, they are useful in cases of considerable pain and no definitive imaging findings. Potentially, you can find the level that is producing the symptoms. A second possible useful application is in cases with multiple disk herniations and no definitive notion of which level is the symptomatic culprit.

Low Tech: Plain Spine Films

Plain spine films are still performed and are useful particularly when looking for fractures in cases of trauma and to check alignment of the vertebral bodies, (ab)normal motion on flexion and extension for assessment of instability or fusion status, and the position of bone grafts, pedicle

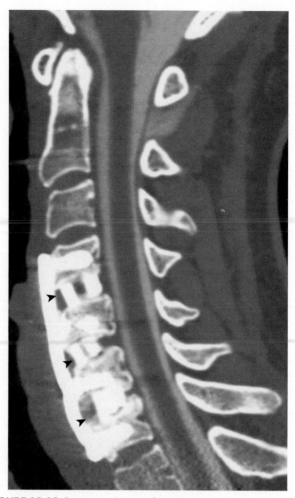

FIGURE 15-12 Postoperative myelogram–computed tomography sagittal reconstruction from thin section axial data nicely shows good alignment, fusion from C4 to C7, and use of autologous bone interbody spacers *(arrowheads)*. There is no impingement on the cord.

screws, cages, plates. Subsidence (settling/sinking of the interbody space device into the endplate) and pseudarthrosis (incomplete or absent fusion leading to a mobile and often painful segment) can be evaluated with flexion/extension plain films with or without CT.

The lateral cervical spine radiograph provides an excellent view of the odontoid process and the anterior arch of C1. The distance between the anterior aspect of the odontoid and the posterior surface of the anterior arch of C1 should not be greater than 3 mm in an adult and 5 mm in a child. Alignment of the spine and the disk spaces is easily evaluated with the lateral view. With the lateral view, you receive at no extra charge the prevertebral soft tissues and the sella turcica. Occasionally, you will note unexpected findings including an enlarged sella or prevertebral mass. A "swimmer's" view is used to study C7-T1 if it cannot be visualized on the lateral, and consists of a lateral view with the tube-side arm depressed and the film-side arm elevated. Ossification of the posterior longitudinal ligament and diffuse idiopathic skeletal hyperostosis (DISH) are identified well on the lateral radiograph. Anteroposterior (AP) films visualize the pedicles and vertebral bodies and can identify uncovertebral joint spurs. The AP view and odontoid view show the relationship of the C1 and C2 vertebrae for alignment (Fig. 15-14).

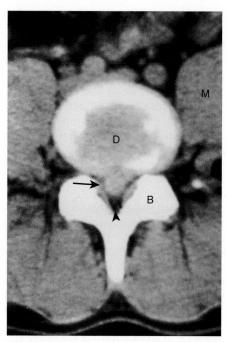

FIGURE 15-13 Herniated disk on unenhanced computed tomography reveals a disk at L2-3. Huge central herniated disk *(arrow)* is squashing the thecal sac. Herniated disk and the parent disk (D) have the same density. Note the array of densities from disk (D), bone (B), muscle (M), and posterior epidural fat *(arrowhead)*.

The right and left cervical oblique radiographs are obtained to view the neural foramina, the uncovertebral joints, and the facets (Fig. 15-15). Remember in the cervical region, the foramina are directed anterior and lateral; therefore in the right posterior oblique projection with the film behind the neck, the left foramina are being visualized. Of course, the opposite is true for the left posterior oblique film.

Thoracic AP films visualize the pedicles and vertebral bodies. Carefully determine whether there is pedicle erosion, a sign of metastatic disease, or whether the interpediculate distance is abnormal. This indicates an intraspinal lesion, or a possible congenital spinal problem. The lateral thoracic radiograph provides information on thoracic alignment, abnormal calcifications in a disk, and the state of the vertebral body and its associated disk spaces.

Information obtained from the lumbar spine AP and lateral films is similar to that of the thoracic spine; however, the oblique radiographs produce the well-known "Scottie dog," which provides an excellent view of the pars interarticularis. Fractures of the "neck" of the dog indicate spondylolysis (see Fig. 15-6). Spondylolysis is associated with spondylolisthesis (anterior slippage of a vertebral body), both of which are readily detectable on plain films, CT and MR, to be discussed later in the chapter.

NORMAL MAGNETIC RESONANCE APPEARANCE OF THE VERTEBRAL BODY MARROW

The beauty of MR is its ability to provide multiplanar images of both the bone and soft tissues of the spine. In the adult, the normal vertebral marrow generally has intermediate to high signal intensity on T1WI. On T2WI FSE images, the normal vertebral marrow is high intensity making lesion

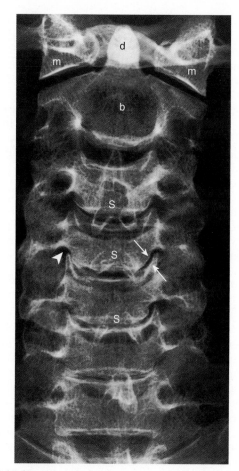

FIGURE 15-14 Anteroposterior view of cervical spine confirms the bodies are in line. Lateral masses of C-1 (m), dens (d), body of C-2 (b), bifid spinous processes (S), C3-C4 left uncovertebral joint *(arrows)*, and a neural foramen *(arrowhead)* are identified.

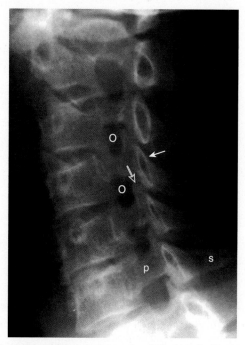

FIGURE 15-15 Oblique radiograph of the cervical spine. Right posterior oblique demonstrates the left neural foramina (O), the pedicle (p), the superior articular facet *(open arrow)*, the lamina *(arrow)*, and the spinous process (s).

detection in the vertebral body more challenging. Fat saturation in the form of STIR imaging is important when vertebral body lesions are suspected and FSE techniques are being employed. However, the saturation may not be uniform particularly when large field-of-views are employed.

In children, the marrow is lower in intensity than in adults because of the low fat content of hematopoietic marrow. In young adults, a small region of high intensity on T1WI is observed at the entry of the basivertebral veins. With aging, the hematopoietic (red) marrow is gradually converted to fatty (yellow) marrow. In older patients, this process can result in focal regions of high intensity on T1WI (focal fatty replacement) with a heterogeneous appearance. In children, the normal marrow may enhance; however, in adults normal marrow does not enhance significantly. In patients with anemia, fatty marrow is replaced by hematopoietic marrow with decreased signal intensity on T1WI.

DEGENERATIVE DISEASES (AND OTHER COMMON SPINE MALADIES)

Unlike great wines, the spinal column does not improve with age—it degenerates. Spondylosis deformans refers to the typical aging process in which the annulus fibrosis and adjacent apophyses form bone spurs/osteophytes from the endplates or adjacent joints and the nucleus pulposus degenerates. It is important to separate the process of disk degeneration from disk herniation. The pathophysiology of the degenerative process consists of dehydration of the nucleus pulposus and decreased tissue resiliency (intervertebral osteochondrosis) with decrease in the height of the disk space and endplate changes. Initially, the nucleus pulposus is soft and gelatinous; however, with aging it is replaced by fibrocartilage and the distinction between the nucleus and annulus fibrosus becomes less well defined. The cartilaginous endplate becomes fissured and more hyalinized. The annulus, which is initially attached to the anterior and posterior longitudinal ligaments, loses its lamellar configuration and develops fissures. The cracks have negative pressure so that gas, primarily nitrogen, comes out of solution and deposits in the intervertebral disk, close to the subchondral bone plate or in other locations. This is termed the vacuum phenomenon (Fig. 15-16). Early MR abnormalities representing disk degeneration include infolding of the anterior annulus and a hypointense central dot within the disk on T2WI. With time the disk degeneration is manifested on MR by decreased signal intensity on T2WI. Disk calcification commonly occurs in the elderly, and is part of the normal ageing process. It is also associated with other often systemic conditions (Box 15-1). Increased intensity on T1WI of the disk can be seen uncommonly with mild calcification associated with degeneration. As calcification increases, the intensity on T1WI decreases. Ultimately the degenerative changes permit disk material to bulge and subsequently to herniate, but disk herniation may occur in the absence of significant disk degeneration especially with acute trauma.

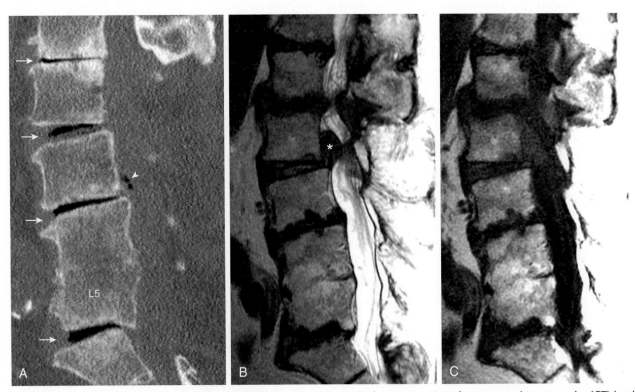

FIGURE 15-16 Vacuum cleft is disks caused by nitrogen-oxygen mix. **A,** Sagittal reconstruction of a computed tomography (CT) lumbar spine in a patient following lumbar laminectomy from L2 through L5 shows multilevel vacuum disks at L1-L2 through L5-S1 *(arrows)* with the exception of L4-L5 where bony fusion has developed across the disk level. Note disc herniation at L3-L4 into the spinal canal is evident by virtue of the associated vacuum phenomenon *(arrowhead)*. Sagittal T2 **(B)** and T1 **(C)** images from the same patient demonstrate how challenging it is to appreciate the vacuum phenomenon (dark intensity on both sequences) by magnetic resonance imaging. The giant disc herniation at L1-L2 on the other hand is obvious on T2-weighted image *(asterisk)* but not appreciable on the CT.

Disk Herniation

Degenerative processes may be accompanied by disk herniation. In such an instance, the nuclear material squeezes itself through the annular fissures. The radiologic descriptions that are associated with various imaging findings of disk herniation had been imprecise and confusing until a consortium of radiologists and orthopods and neurosurgeons agreed on a common lumbar disk nomenclature, accepted in 2001 and revised a day before we started this chapter in 2014. Of course they still did not agree whether it should be "disc" or "disk." Too fundamental.

Low back pain occurs in more than 60% of all adults and diagnosis and treatment costs the USA more than 14 billion dollars (four times more than paid out in *all* medical malpractice judgments and settlements). When to image? Health maintenance organizations (HMOs) believe you should wait 4 to 6 weeks. The radicular symptoms and motor signs tend to improve over time and waiting can save big bucks...unless you are an HMO executive in pain, in which case you get imaged immediately to know the true diagnosis (because the pain after the motor vehicle collision could be from cancer). However, if you actually have weakness and muscle loss, go see the take-no-insurance surgeon immediately (not 6 weeks later when it is permanent); do not pass "Go," do not pay $200 co-pay.

Here are the definitions that have been advocated to more precisely define degenerative lesions in the lumbar spine. An annular fissure is a separation between annular fibers, avulsion of fibers from their vertebral body insertions, or breaks through fibers that extend in a particular direction (see following discussion; Fig. 15-17). Such lesions are usually *not* associated with trauma and should *not* be called an annular tear. As to anular or annular...who gives a disk? There are three types of annular fissures: concentric, radial and transverse with the latter two being identifiable on MR. Transverse fissures are peripheral disrupting Sharpey fibers. Radial fissures start from the nucleus and course through the annulus. On MR, globular or linear high intensity can be observed on T2WI within the posterior annulus. Anterior radial annular fissures are rare but can be associated with back pain. This can be associated with

pain or segmental instability. Annular fissures can enhance and show T2 hyperintensity for years—very persistent.

A herniation is a *localized* displacement of disk material (nucleus, cartilage, fragmented apophyseal bone, or annulus) beyond the limits of the intervertebral disk space (Box 15-2). Because a "herniation" may comprise more than just nucleus pulposus, the spine nomenclature discards the abbreviation "HNP" (herniated nucleus pulposus). *"Localized"* or "focal" is most recently defined as less than 25% (90 degrees) of the periphery of the disk. If the disk material extends beyond 25% of the circumference of the disk, it is now termed "asymmetric bulge." The 2014 revision says a herniation must be less than 25% of the disk circumference—no more broad-based disk herniation in the lexicon.

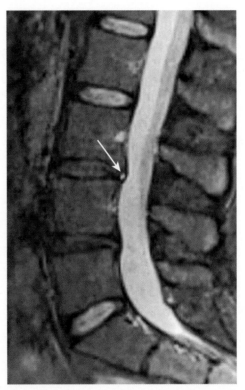

FIGURE 15-17 Annular fissure (not a tear) as seen on this sagittal short tau inversion recovery magnetic resonance imaging scan often take years to repair. High signal intensity zone (HIZ) *(arrow)* at the margin of this disk represents an annular tear. Recent studies have noted that the annular tears enhance and can remain hyperintense on T2-weighted imaging for years after their occurrence. Hence, high intensity and enhancement of annular tears do not connote acuity.

BOX 15-1   Diseases Associated with Disk Calcification

Acromegaly
Amyloidosis
Ankylosing spondylitis
Calcium pyrophosphate deposition disease (CPDD)
Chondrocalcinosis
Diabetes mellitus
Degenerative disk disease
Diffuse idiopathic skeletal hyperostosis (DISH)
Gout
Hemochromatosis
Homocystinuria
Hyperparathyroidism
Hypothyroidism
Ochronosis
Osteoarthritis
Poliomyelitis
Sequelae of disk infection

BOX 15-2   Disk Terms Used (Out-of-Vogue Terms in Parentheses)

Bulge
    Focal = <90 degrees
    Broad based = >90 degrees
Herniation (herniated nucleus pulposus [HNP]), prolapsed disk, ruptured disk
    Protrusion (prolapsed) = broad attachment at base
    Extrusion = narrow attachment at base
    Sequestrated (sequestered) = free fragment no longer communicating at base

Herniations may be described as protrusions or extrusions. Protrusions are defined as herniations that have a broader attachment at the junction with the parent disk than distally in any plane. Extrusion is defined when the junction with the parent disk is narrower than its distal portion. If there is separation of the herniated disk material from the parent disk, it is a sequestrated disk or free fragment, which is considered a type of disk extrusion (Fig. 15-18). Herniations can be likened to The Rock (Dwayne Johnson), broad neck (protrusion); Modigliani, narrow neck (extrusion); Marie Antoinette, decapitated (sequestration). If the herniated disk is displaced from the extrusion (regardless of whether it is sequestrated or not) it is termed a migrated disk. Disk migration occurs equally commonly up and down, right and left from the parent disk. If the disk herniation is covered by the outer annulus and/or the posterior longitudinal ligament (defined as dark signal on MR pulse sequences) it is considered contained. A synonym used is "subligamentous." If the disk herniation is uncovered it is termed "uncontained" (see Fig. 15-18, *D*). This distinction aids the surgeon in knowing where to look for that dag-nabbit disk fragment in the operating room. In extremely rare instances, disks have been reported to transgress the dura and lie intradurally. This description has also been used as an asset preserver in operative reports by disk jockeys after the dura has been inadvertently breached during surgery.

Further description should include the location of the herniation and the extent, if any, that a particular structure (nerve root, thecal sac, spinal cord) is compressed by the herniation. The accepted terms to be used for describing

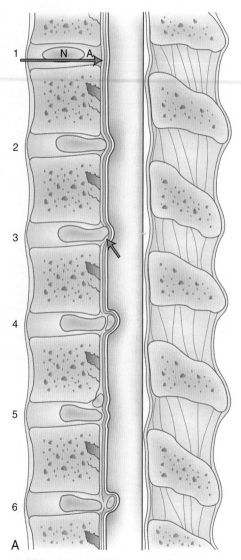

FIGURE 15-18 Disk herniations of different names led to uniform nomenclature claims. **A,** Diagram of lateral spine demonstrating spectrum of disk lesions. 1. Normal configuration of nucleus pulposus (N), intact annulus fibrosis (A), posterior longitudinal ligament *(arrow)*, and dura. 2. Bulging disk with intact annulus fibrosus. 3. Subligamentous herniated disk with rupture of the annulus fibrosus *(arrow)*. 4. Herniated disk with a free fragment. 5. Herniated disk with a free fragment that has migrated superiorly. 6. Herniated disk that has ruptured through the posterior longitudinal ligament and is against the thecal sac. In extremely rare instances the disk can continue to head east and become intradural. This is so rare that it may be seen only in the operating room when the surgeon inadvertently traverses the dura (oops!).

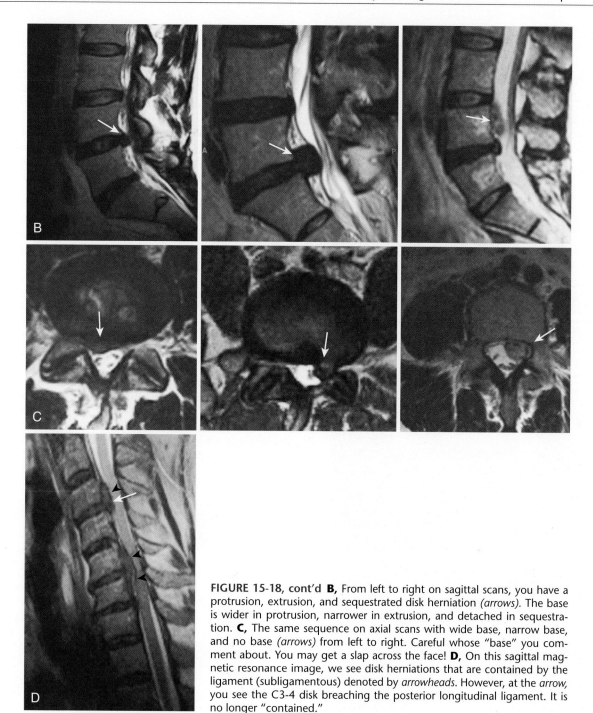

FIGURE 15-18, cont'd B, From left to right on sagittal scans, you have a protrusion, extrusion, and sequestrated disk herniation *(arrows)*. The base is wider in protrusion, narrower in extrusion, and detached in sequestration. **C,** The same sequence on axial scans with wide base, narrow base, and no base *(arrows)* from left to right. Careful whose "base" you comment about. You may get a slap across the face! **D,** On this sagittal magnetic resonance image, we see disk herniations that are contained by the ligament (subligamentous) denoted by *arrowheads*. However, at the *arrow*, you see the C3-4 disk breaching the posterior longitudinal ligament. It is no longer "contained."

disk position are (from the midline) central, right/left (para)central, subarticular (lateral recess), foraminal, and extraforaminal (far lateral; Fig. 15-19). Of course, there are anterior herniations in front of the vertebrae and intravertebral herniations affectionately known as "Schmorl nodes" by every schmegegge in radiology. The disk may also, in a craniocaudal direction, be referred to as migrating to a suprapedicular, pedicular or infrapedicular location.

The degree of canal or foraminal stenosis caused by disk disease or other perpetrators of spinal pain has been codified as mild if it narrows the spinal canal by one third or less, moderate if narrowing one third to two thirds, and marked if more than two thirds narrowed. By the same

token a disk herniation can be termed small, medium, or large, using the same criteria. Neural foraminal narrowing by degenerative changes is graded from mild to severe using the same one third, one third to two thirds, more than two thirds system. The third location we look for "stenosis" is in the lateral recess where exiting nerves, before they reach the foramen, can be affected.

Another useful way of describing herniations or osteophytes is their impact on the thecal sac or exiting nerve roots. The three terms that "tell it all" are "noncompressive," "abutting" and "compressing/displacing." Thus a herniation may be noncompressive and therefore unlikely to cause radiculopathy; it may be abutting on

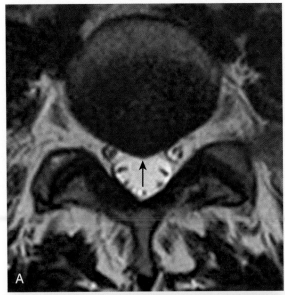

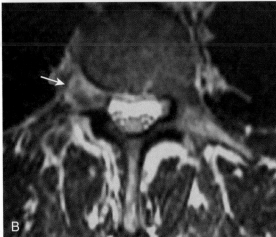

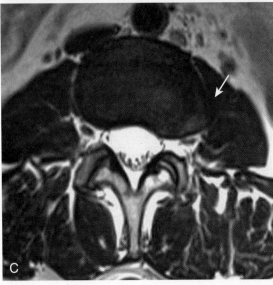

FIGURE 15-19 Disk locations accepted by committee agreed upon from central to periphery: central (**A,** *arrow*), paracentral (see Fig. 15-18, *C,* left image, *arrow*), subarticular (see Fig. 15-18, *C,* middle image, *arrow*), lateral recess (see Fig. 15-18, *C,* right image, *arrow*), foraminal (**B,** *arrow*), and far lateral, aka extraforaminal (**C,** *arrow*).

an exiting L4 nerve root implying that potentially when weight-bearing it may displace and compress the nerve root...or not; and it may, even in the supine position, compress or displace the nerve root. Thus, when it abuts the nerve root, it has no intervening tissue and is up against it, but the nerve root is in its native natural state, nondisplaced.

To describe what part of the nerve root is compressed, one can use the terms, "within the thecal sac," "as it leaves the thecal sac," "in the lateral recess," "in the neural foramen," or "exiting the foramen."

OK, let's put it all together then:

"The MR shows a moderate sized contained right central L4-5 disk extrusion that compresses the traversing right L5 nerve root as it leaves the thecal sac and leads to minimal spinal canal stenosis. There is also a noncompressive right facet joint spur that causes mild left foraminal stenosis at the same level. A small right subarticular L5-S1 uncontained disk protrusion leads to mild right lateral recess stenosis with disk material abutting on the right S1 nerve root."

Got it? Awesome, we know...and the bread and butter of private practice neuroradiology...YAWNING NOT PERMITTED.

Disk fragments may have different intensities depending on their state of hydration and the particular pulse sequence used. Degenerated disks on MR generally have lower intensity on T2WI than normal disks, and disks containing gas (vacuum disks) or calcification have very low intensity on all pulse sequences. Obviously these disks will be high density on CT (Fig. 15-20).

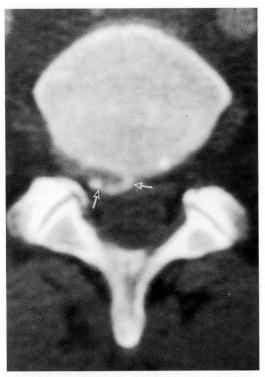

FIGURE 15-20 Computed tomography of a calcified disk *(arrows)* in the lumbar region (but more commonly seen as a thoracic lesion). Surgeon may feel a "hard disk" at operation and have trouble with standard discectomy attempt. Not soft!

Time Course

Disk herniations can occur acutely (from moving that heavy sofa across the room) or over a prolonged time course in a degenerating spine. Unfortunately, there are no specific imaging criteria for determining whether a herniation is acute or not...we have to rely on our clinical colleague's evaluation to make that determination. Disks can improve over time and spontaneous reduction of herniations, particularly those larger than 6 mm, are reported 6 to 12 months after the initial event. This is associated with clinical improvement (if you have a high pain threshold and no significant neurologic deficit to tolerate the wait). Exactly how this occurs is unknown, but investigators have hypothesized about dehydration with disk shrinkage, fragmentation, and phagocytosis as possible factors in the reduction of disk herniation. With the onset of herniation, there is an associated inflammatory response. Neovascularity occurs at the periphery of the herniated disk and the combination of inflammation and neovascularity may contribute to resorption of the disk material and better clinical outcomes. Acute epidural enhancement around a free fragment has been reported with acute disk herniation to occur in 73% of cases. The inflammatory component has also been suggested for the reason why epidural steroids have been used successfully for the nonsurgical treatment of herniated disks.

Schmorl Node

Herniation of disk material through the endplate is termed a Schmorl node (recently termed intravertebral herniation), which usually has discrete margins and intensity similar to disk material, and reveals rim enhancement. Schmorl nodes can be identified in asymptomatic patients and have been associated with infection, trauma, malignancy, osteopenia, and intervertebral osteochondrosis (Fig. 15-21). On MR, they appear as extension of disk into the vertebral body surrounded by a rim of low intensity on T2WI representing reactive sclerosis. Occasionally, a Schmorl node may be associated with bone marrow edema, which can be confused with infection or metastatic lesion. Chronic Schmorl nodes may be associated with fatty endplate changes.

If the disk herniates into the anterior ring apophysis, it is termed a limbus vertebra (seen in children and associated with back pain).

Scheuermann Disease

This degenerative disease is noted in children with the onset at puberty and has a male predominance. It consists of vertebral wedging resulting in lower thoracic kyphosis. It requires the involvement of three contiguous vertebra with wedging of more than 5 degrees. Schmorl nodes are common. The etiology is thought to be stress related through either congenitally or traumatically weakened portions of the cartilaginous endplates.

Thoracic Disk Herniation

Herniated thoracic disks have a reported incidence of 0.2% to 0.5%. They have an insidious onset with back pain, radicular paresthesias, and myelopathy. MR is the first imaging choice for demonstrating the herniated disk compressing the subarachnoid space and possibly the cord (Fig. 15-22). Occasional problems may arise in specifically diagnosing calcified herniated thoracic disks (which occur at a greater rate for some reason in the thoracic region) because of the decreased ability of MR to detect calcification. Similarly, MR cannot often distinguish between vacuum disk (hypodense on CT) and calcified disk (hyperdense on CT) both of which are dark on all pulse sequences. CT myelography is useful for confirmation of the MR findings in the situation where cord compression is suspected because of myelopathy. In patients with DISH and other enthesopathies the ligamentum flavum prominence, and anterior osteophytes can be impressive in the thoracic region.

Osteophyte Formation

The combination of loss of disk height and disk shrinkage is associated with abnormal motion, particularly in the cervical region. Loss of height (but hopefully not of stature) and tissue shrinkage are what we have to look forward to with advancing age. Abnormal stress caused by the loss of disk height produces osteophyte formation and posterior displacement of the vertebral body. Because of the abnormal stress, subluxation of the facet joints may occur. Spur formation in the cervical region takes place at the uncovertebral joints as the disks desiccate and get narrower leading to more grinding of the uncovertebral joint (UVJ) below and the vertebral body above (Fig. 15-23). Osteophyte formation at the uncovertebral joints produces foraminal

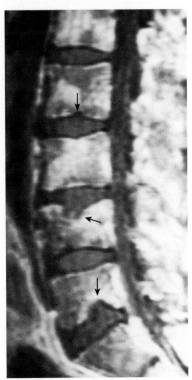

FIGURE 15-21 A Schmorl node causes the endplate to erode. Sagittal T1-weighted imaging demonstrates multiple intrabody disk herniations (Schmorl nodes) *(arrows)* in this patient who incidentally has acute myeloid leukemia (note generalized marrow heterogeneity).

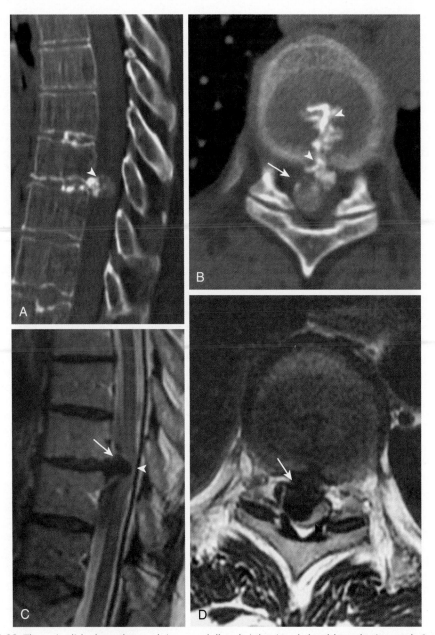

FIGURE 15-22 Thoracic disk abuts the cord; intramedullary bright signal should not be ignored. **A,** Computed tomography shows a large calcified disk *(arrowhead)* compromising the spinal canal. **B,** The axial scan depicts the calcified disk *(arrow)* as off midline to the right with a connection of calcification to the parent disk *(arrowheads)*. **C,** Corresponding magnetic resonance (MR) shows the disk herniation *(arrow)* in the thoracic region compressing the cord and causing high signal in the cord on T2-weighted imaging *(arrowhead)*. **D,** Axial T2-weighted MR shows the disk herniation *(arrow)* and the bright compressed and laterally displaced spinal cord *(arrowhead)*.

stenoses, compression of the nerve root and clinical signs of cervical radiculopathy. It should be emphasized that in the cervical spine, UVJ degenerative disease is more commonly the source of foraminal narrowing than facet joint disease, ligamentum flavum thickening, posterior body osteophytes and disk disease. They are the most common cause of cervical root compression. Fortunately, the cervical nerves generally course in the lower portion of the neural foramen whereas the UVJ spurs are at the superior part of the foramen. Otherwise we'd all be sporting limp, painful pitching arms a la Stephen Strasburg. Rarely such UVJ overgrowth can lead to stenoses of the V_2 segment of the vertebral arteries. Please note that while there are UVJs between C3-C4 and C6-C7, it is less common to have

such a joint between C7-T1 and very rare at T1-T2. It can happen, but if you are apt to call UVJ stenosis of the neural foramen at C7-T1, it is probably a misnomer. Good thing you are reading this chapter!

Large osteophytes may form at the posterior edge of adjacent vertebral bodies, with narrowing of the subarachnoid space and spinal cord compression, producing myelopathy (Fig. 15-24). A diagnostic dilemma (on MR) is the differentiation between desiccated or calcified "hard" disk and bony osteophyte. It is difficult at times to separate osteophytic compression, ossification of the posterior longitudinal ligament, and a calcified hard disk. All may produce compression of the thecal sac or roots. Desiccated/calcified disk and osteophyte appear identical and dark

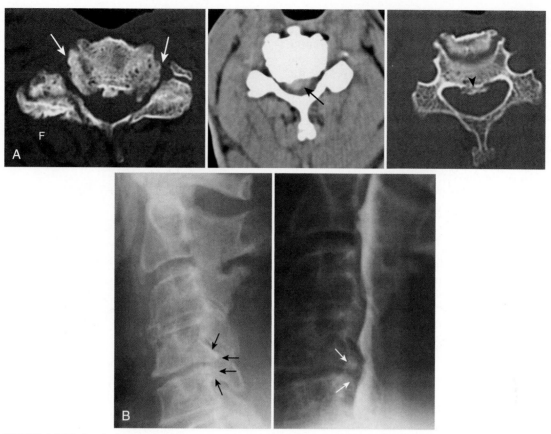

FIGURE 15-23 Cervical spine degenerative disk disease leads to nerve root agony. **A,** Note the *white arrows* denoting uncovertebral joint osteophytes narrowing the neural foramina, accompanied by facet joint disease (F) on the right. A central disk protrusion is present *(black arrow),* accompanied by ossification of the posterior longitudinal ligament *(black arrowhead).* That'll give you a neck ache! **B,** The impact of posterior osteophytes *(arrows)* on the spinal thecal sac may be better demonstrated by myelography (right image) than mere plain film (left) radiographs.

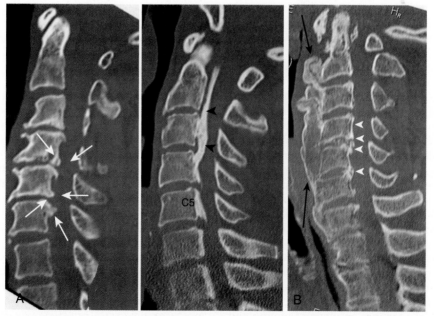

FIGURE 15-24 Ossification of the posterior longitudinal ligament (OPLL) versus osteophyte: both can cause a canal that's too tight. **A,** Contrast the osteophytes on the left hand image *(white arrows)* with the OPLL on the right image *(black arrowheads).* OPLL is more likely to be flowing and continuous and may have breaks at the disk levels such as at C4-5 and C5-6. It may be just behind the vertebral body and not at the disk as at C5. **B,** Diffuse idiopathic skeletal hyperostosis (DISH) causes enthesopathies of spinal aponeurosis. Note the appearance of osseous bridging *(arrows)* at multiple contiguous levels. This is characteristic of DISH. OPLL coexists *(arrowheads).*

on all sequences, whereas soft disks appear as a different intensity compared with bone on most gradient pulse sequences. Most of the time there is a soft-tissue component to the hard disk. On sagittal T1WI, osteophyte and disk can usually be distinguished. CT-myelo can resolve the issue if necessary.

These endplate osteophytes associated with significant degenerative disease in the spine (spondylosis deformans) and uncinate spurs result from traction stress at the osseous site of attachment of the annulus (Sharpey fibers). Indeed, the endplate is probably the most vulnerable region of the lumbar vertebral body as it bears the axial load. The nucleus pulposus in usual aging has normal turgor so that its displacement leads to traction on the Sharpey fibers of the annulus with the development of osteophytes several millimeters from the disk-overtebral junction. Osteophyte formation also occurs at the facet joints; however, this is less significant in the cervical region than in the lumbar region. In the lumbar region, osteophytic compression occurs primarily in the lateral recess and at the neural foramen. Hypertrophy of the superior articular facets in the lumbar region is most important because it lies anterior to the inferior articular facet and closer to the nerve in the lateral recess and neural foramen (Fig. 15-25).

Endplate Changes

In degenerative disease the signal intensity of the endplate vertebral marrow can be variable on conventional spin echo images with decreased intensity on T1WI and increased intensity on proton density–weighted imaging (PDWI) and T2WI (Modic type I changes), evolving at times to increased intensity on T1WI and isointense to slightly high signal on PDWI and T2WI (Modic type II changes; Fig. 15-26; Table 15-2). These have been associated with histopathologic findings representing fibrous tissue or bone marrow edema associated with acute or subacute inflammatory change (type I) and yellow marrow (type II). Hypointensity on T1WI and T2WI has also been noted and represents bony sclerosis (Modic type III). These should not be mistaken for malignant disease by unknowing residents (type 0) possessing neuroanemia.

The appearance of Modic type I changes has been associated with acute low back pain and may be a marker of recent stress or spinal instability. To wit, Modic type 1 changes are seven times more common in symptomatic low back pain patients than in asymptomatic subjects. Curiously, some have suggested that there may actually be a low-grade infection (by anaerobes and/or staphylococcus) occurring in patients with acute herniations and Modic type 1 changes. A recent study showed efficacy of Modic antibiotic (amoxicillin-clavulanate) spine therapy in patients with chronic low back pain, and new Modic type 1 changes in the vertebrae adjacent to a previously herniated disk for parameters including back pain intensity, leg pain intensity, general improvement, number of hours with pain, and physical examination range of motion and motor tests. So it's worth commenting on new Modic I type changes.

Modic type I changes are generally transient; type I changes convert into either type II changes or back to normal within 6 months following lumbar fusion, which parallels clinical improvement.

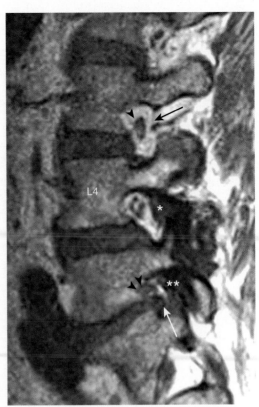

FIGURE 15-25 Foraminal stenosis can lead to lower leg ptosis. Sagittal T1-weighted image shows varying degrees of foraminal patency. L4 level is indicated. At L3-L4, the foramen is widely patent *(black arrow)*, with normal bright fat surrounding the exiting L3 nerve root *(black arrowhead)*. At L4-L5, ligamentum flavum infolding *(asterisk)* serves to narrow the posterior aspect of the foramen but the nerve root is A-OK; we can call this mild foraminal narrowing. At L5-S1, there is pronounced ligamentum flavum infolding and facet hypertrophy *(double asterisks)* along with disc herniation *(white arrow)*, which serve to nearly completely efface the fat in the foramen, resulting in moderate to severe foraminal narrowing. Note the impact upon the exiting L5 nerve root *(double black arrowheads)*, which is flattened and compressed.

Vacuum Phenomenon

Another aspect of degenerative disease (which bears repeating) is the vacuum phenomenon, with gas (nitrogen) in the disk space or facet joints. Gas from the vacuum disk that is in the spinal canal implies disk herniation (see Fig. 15-16). On CT and MR, air can be appreciated as linear low density and intensity. The presence of air in the disk space may indicate mobile spinal segment and should not be present in a fused spine.

Synovial Cysts

Synovial cysts are associated with degenerative joint disease (DJD) occurring in a characteristic location, budding from either side of the facet joint (medially or laterally) and producing a rounded posterolateral extradural mass (medial cysts), but they may occur in the posterior paraspinal tissue (lateral cysts). They occur in the lumbar region, L4-5 more commonly than L5-S1, although they have rarely been reported in other regions of the spine. Hemorrhage in these lesions, precipitating acute neurologic symptoms, has been reported, but often the hemosiderin seen in the wall of the cyst is unassociated

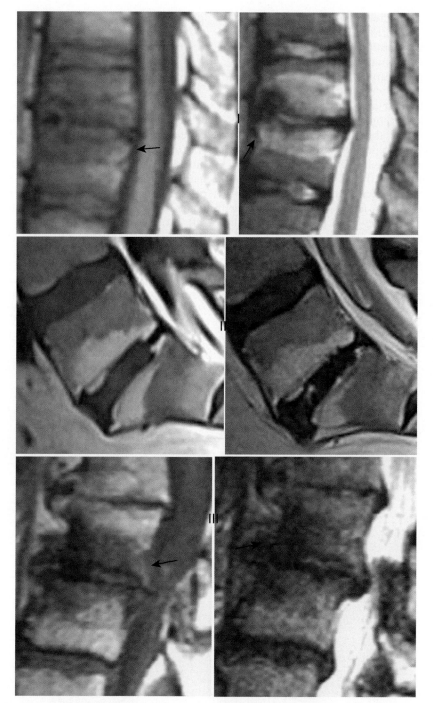

FIGURE 15-26 Modic endplate changes. Modic type I at top with dark endplate on T1-weighted imaging (T1WI) to left and bright endplate on T2WI to right. In Modic II, middle set, endplates are bright on both T1WI and T2WI. Finally, the sclerotic Modic III form has dark intensity on both T1WI and T2WI (bottom set). Endplate changes are indicated by *arrows*.

TABLE 15-2   Modic Endplate Changes

Modic Type	SI on T1WI	SI on T2WI	Explanation	Implication
I	Dark	Bright	Bone edema, possible inflammation	Could be symptomatic, infectious/inflammatory; possible segmental instability
II	Bright	Bright	Fat replacement	Usually stable
III	Dark	Dark	Sclerosis	Severe degeneration

SI, Signal intensity; *T1WI,* T1-weighted imaging; *T2WI,* T2-weighted imaging.

with acuity. This accounts for black signal on T2WI surrounding fluid signal centrally. The diagnosis is made by the characteristic location and association with degenerative disease, including disk space narrowing, eburnation, and hypertrophic changes. Peripheral rim enhancement can be seen. There is an association with spondylolisthesis and abnormal movement of the facet joint. CT can rarely detect calcification in the cyst wall and gas in the cyst, but most often the periphery is just slightly hyperdense (Fig. 15-27). These lesions can display enhancement if associated with an inflammatory process. The lesions can present with pain, usually radicular, and neurologic deficits. The major differential diagnosis is extruded posterior disk. A curious feature of synovial cysts is that sometimes they are deep to the ligamentum flavum (uncontained) suggesting somehow they squirt through a gap like Walter Payton of the Chicago Bears in his prime.

Another juxtaarticular cyst may have a connective tissue capsule and is termed a ganglion cyst. Baastrup disease represents inflammation of the interspinous ligaments usually found in the lumbar spine and isolated to one level. This may be because of excessive contact between the spinous processes resulting in their sclerosis, eburnation, and enlargement with pain maximal on extension. Many cases have associated cysts that may project in a number of directions, but they can project in the posterior epidural space in communication with the interspinous bursitis (Fig. 15-28). From there they may irritate the posteriorly

located spinal nerve roots. Treatment is to remove the offending spinous processes.

Ossification of the Posterior Longitudinal Ligament

Ossification of the posterior longitudinal ligament (OPLL) is an inflammatory degenerative condition usually associated with degenerative disease of the cervical spine. It was originally described in the Japanese (2% prevalence) but may be seen in any patient population. OPLL affects men and women typically in the fifth to seventh decades of life. It can produce compression of the spinal cord with myelopathic symptoms. The diagnosis can be tricky on MR because of its insensitivity to calcium. Thin ossifications on sagittal MR are difficult to detect, even for a spine hot shot. Detection of this lesion at times is dependent on

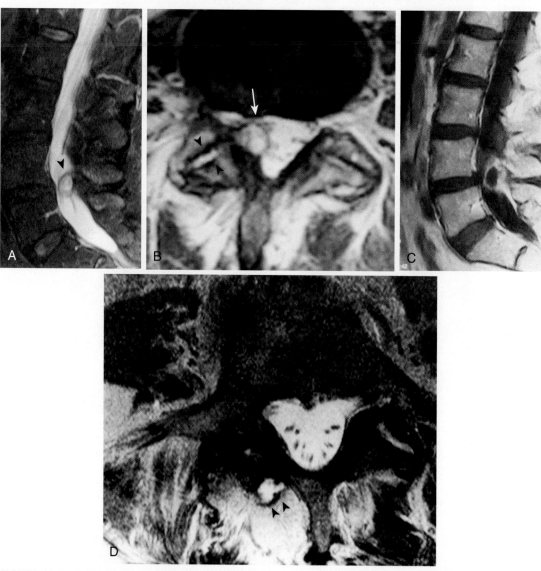

FIGURE 15-27 Synovial or juxtaarticular cyst will no longer be a postrequisites miss. **A,** Sagittal T2-weighted imaging (T2WI) shows a synovial cyst *(arrowhead)* that has a dark intensity rim and hyperintense center. **B,** The origin from the facet joint *(arrowheads)* and the effect on the right-sided nerve roots *(arrow)* is demonstrated on the T2-weighted axial scan. **C,** After contrast, the wall of the cyst enhances. Note that cyst contents in this case are dark. **D,** Synovial cysts often arise from the lateral aspect of the face joint and project into the paraspinal soft tissues as in this case *(black arrowheads).*

identification of compression of the cord. On T1WI, CSF and calcium have almost the same intensity. Axial T2WI aid in confirming extraaxial compression; however, OPLL may be superficially confused with osteophytic compression. Sometimes the ossification may truly have marrow fat and cortex in which case the bright signal marrow fat will be the giveaway on T1WI. Concurrent ligamentum flavum calcification is often present. Remember that OPLL occurs along the full course of the posterior longitudinal ligament whereas osteophytes are present only at the disk space. Calcified herniated disks usually do not occur at multiple levels. A calcified meningioma is usually round, intradural, and unlikely to extend longitudinally, as does OPLL. OPLL makes a strong case for CT and plain films to establish the diagnosis (see Fig. 15-24). OPLL occurs in 50% of patients with DISH (see following discussion). Although OPLL is a cervical spine entity, you often see thoracic spine ligamentum flavum hypertrophy and calcification with it.

Diffuse Idiopathic Skeletal Hyperostosis

DISH, or Forrestier disease, is characterized by ossification along the anterior and to a lesser extent, lateral aspect of the spine (see Fig. 15-24). In addition, hyperostosis at the sites of tendon and ligamentous attachment to bone, ligamentous ossification (e.g., OPLL), and paraarticular osteophytes in both the axial and appendicular skeleton are present. Osseous bridging of at least four contiguous vertebral bodies is one criterion for diagnosis of this condition according to Resnick but only three segments according to Utzinger. There should be preservation of the intervertebral disk space, and absence of apophyseal joints or sacroiliac inflammatory changes. The entity is most commonly observed in the lower thoracic spine but can occur anywhere.

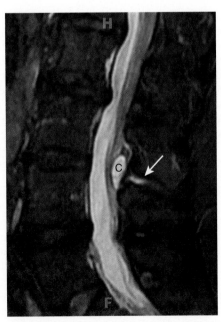

FIGURE 15-28 Baastrup disease leads to pain that on flexion recedes. Classic appearance on magnetic resonance (MR) imaging shows interspinous bursitis seen as high signal on short tau inversion recovery MR *(arrow)*, with a posterior epidural cyst (C).

The spine has a bumpy contour anteriorly, with the greatest amount of bone being deposited at the level of the intervertebral disks. Hypertrophy and ossification can be identified about the spinous process. Ossification of the ilio-lumbar and sacrotuberous ligaments with bony overgrowth of the inferior acetabular rim are findings in the pelvic region.

DISH differs from spondylosis deformans in that calcification/ossification is present in the anterior longitudinal ligament with an associated proliferative enthesopathy at the site of attachment of the anterior longitudinal ligament to the anterior vertebral body surface. The facet joints are not involved. The differential diagnosis of DISH includes ankylosing spondylitis, which is not as florid as DISH and is associated with sacroiliitis (usually the first manifestation of the disease). You can observe erosion of the superior and inferior vertebral margins, producing squaring and bridging of the vertebral bodies (bamboo spine) in ankylosing spondylitis. The facet joints are involved in ankylosing spondylitis and not DISH, and the former is also associated with HLA-B27 and osteoporosis.

Spondylolysis and Spondylolisthesis

In the lumbar region, the articular processes have an oblique orientation. The plane of the joint between the superior and inferior articular facets is from medial to lateral, from the anterior to posterior aspect of the joint. The term spondylolisthesis is defined as slippage of one vertebra onto another whereas spondylolysis is a fracture through the pars interarticularis, which may or may not be associated with vertebral slippage. Pars defects can be easily seen on plain spine films and CT with MR slightly more challenging to see the changes. Spondylolysis is the most common pathology to affect the pars with a prevalence of 3% to 10%. Single-photon emission computed tomography (SPECT) imaging is useful for detection of symptomatic acute pars defects. T1WI can detect a break in the cortical margin of the pars and STIR can see edema in the pars between the superior and inferior articular facets. Spondylolysis without spondylolisthesis is a cause of chronic low back pain particularly in children and young adults. Effectively, this can present like a stress fracture in kids from repetitive injury, although the defect is also felt by some to arise on a developmental/congenital basis.

There are many different causes of spondylolisthesis: (1) congenital spondylolisthesis, which is associated with dysplastic articular processes, abnormal orientation of articular processes, or conditions such as kyphosis, all of which may produce slippage of vertebrae (shown with widening of the spinal canal at the level of the spondylolisthesis); and (2) acquired spondylolisthesis including (a) pars interarticularis lesions produced by stress fractures with persistent defects of the pars or healing of the fracture resulting in elongation of the isthmus, (b) the effects of degenerative facet disease associated with joint instability, most often occurring at L4-5, with a higher incidence in diabetics and associated with compression of the cauda equina (shown with narrowing of the spinal canal), (c) postsurgical lesions, which may be seen in the cervical or lumbar regions, resulting from altered stress on the joints after cervical fusions or surgery for spinal stenosis, (d) acute trauma with pedicle fractures (widens the spinal canal), and (e) pathologic conditions such as osteoporosis, metastasis, infection, osteopetrosis, or arthrogryposis.

In evaluating lumbar spondylolisthesis on MR, the sagittal image is probably the most useful in identifying compression of a particular root (Fig. 15-29). Significant findings in this condition include the AP displacement of vertebrae in the midsagittal plane and the far lateral foraminal views showing altered shape of the neural foramina with the long axis converting from vertical to a horizontal configuration, loss of the foraminal fat, nerve root compression from reactive changes in the posterior longitudinal ligament, fibrocartilaginous tissue at and surrounding the site of the pars defect, and sharp angulation of the nerve root (pediculate kinking) caused by slippage. Disk herniation at the level of spondylolisthesis is unusual, but the more posterior disk is "uncovered" making it seem, to the "unrequisited" that the disk is herniated. It is really not, hence the term "pseudodisk/pseudoherniation"; it is just sticking with the appropriate more posterior endplate. Axial images therefore tend to exaggerate the spinal stenosis.

A grading system is used on the basis of the position of the posterior margin of the subluxed vertebral body compared with the posterior margin of the inferior vertebral body. When the superior body is subluxed up to one fourth of a vertebral body width on the lateral film it is termed grade I spondylolisthesis; half a vertebral body, grade II; three fourths, grade III; and a whole width is grade IV. Spondyloptosis refers to a vertebral body plopping over the lower vertebral body (ptosis: grade V).

Congenital Posterior Element Variations

Congenital anomalies of the posterior elements occur and can, at times, confuse even the professor. Facets can be catawampus ($20 word) at the same level. In this situation, the facets are sagittally (or axially) oriented. This is a particular issue when the facet on one side is medial at its posterior tip than its anterior tip and is termed facet tropism. When present, one side slips more than the other and there is vertebral rotation. The subluxation is more severe on the side of the more sagittally oriented facet. Facet tropism may be mild and asymptomatic but can produce symptoms. A facet may also be congenitally absent or hypoplastic. This results in hyperplasia of the contralateral facet with accelerated degenerative changes and stress fractures in the facet. Missing pedicles, arches, and spinous processes are a lot less exciting and more common.

Ankylosing Spondylitis

Ankylosing spondylitis (AS) is a common rheumatologic disorder with an incidence of 1.4% with young men affected more than women (at least 4:1) with HLAB27 positive in 97% of cases. The sacroiliac joints and lumbar spine are the most commonly affected locations but with time, the entire spine can become involved. Vascularization and ossification in the ligamentous attachments (enthesopathy) affecting Sharpey fibers result in classic syndesmophyte formation (bamboo spine). Erosion at the vertebral endplates in combination with ligamentous insertion ossification is the cause of squaring of the anterior vertebral concavity. The anterosuperior and anteroinferior aspects of the vertebral bodies ("shiny corner" on plain films) and

sacroiliac joints are subject to reactive changes with bony sclerosis (Romanus lesion). On MR, there is low intensity in the vertebral body on T1WI and high intensity on T2WI with enhancement. Calcification in the disk in a variety of patterns commonly occurs by a process of enchondral ossification (Fig. 15-30).

Spinal fractures (either traumatic or pseudoarthroses), progressive spinal deformity, subluxation (atlantooccipital/atlantoaxial) and rotatory instability, and spinal stenosis are common in AS. Spinal pseudoarthrosis (Andersson lesion) with associated instability is the most significant biomechanical complication of this disease. It is the result of mobile nonunion usually after occult stress fracture. There is reactive sclerosis in the vertebral bodies adjacent to a widened area of destruction across the fractured enclosed disk. Dural ectasies can also be noted in AS (see Fig. 15-30, *D*).

Spinal Stenosis

Many conditions result in compression of lumbar nerve roots and produce symptoms including radicular pain, claudication-like symptoms (which can be misdiagnosed as vascular disease), numbness, and tingling. There are many reports on normal measurements and on problems with such measurements. The authors are not obsessive-compulsive measurers. Nevertheless, some measurements are provided as a guide. The lumbar canal increases in midsagittal diameter, proceeding from superior to inferior; however, the minimal bony sagittal diameter of the lumbar central canal is approximately 11.5 to 12.42 mm (OK, maybe we are compulsive). Interpediculate measurements of less than 15 mm at L4-5 or less than 20 mm at L5-S1 and canal cross-sectional areas of less than 1.45 cm^2 are considered abnormal. The interpediculate distance increases from T12 to L5. For the radiologic etymologic entomologist, the interpedicular distance is the distance between two lice.

For people less compulsive than we the authors, one can simplify things.... If the cervical spine and the lumbar spine has a central AP width of greater than 10 mm, the patient is probably in good enough shape. The mean cervical canal diameter is generally 15 mm with SD 1.5 mm (with some variation by level), so...of course other radiologists opine that canal/body ratio is more reliable than midsagittal diameter of the cervical spinal canal in the diagnosis of cervical spinal stenosis. In the lumbar spine, mean diameters are about 13 mm with SDs 1.6 mm, but they vary by level. Oh what the heck...if more than 10 mm, let 'er go...unless you see "abutment" or "compression."

It should be pointed out that the *bony* sagittal diameter (seen on plain films or just bone windows on CT) may not be a sensitive indicator of stenosis because it does not take into account soft tissues such as ligamentum flavum, disk, fat, and facet osteophytes. The normal ligamentum flavum is 2 to 4 mm thick. It is considered to be hypertrophied (as a consequence of infolding) if it is greater than 5 mm thick in the lower spine (Fig. 15-31). Ossification of the cephalic attachment of the ligamentum flavum is occasionally noted and may be a normal anatomic variant (or, as described above, seen in concert with DISH or OPLL). It has also been described in calcium pyrophosphate dihydrate deposition disease (pseudogout), where

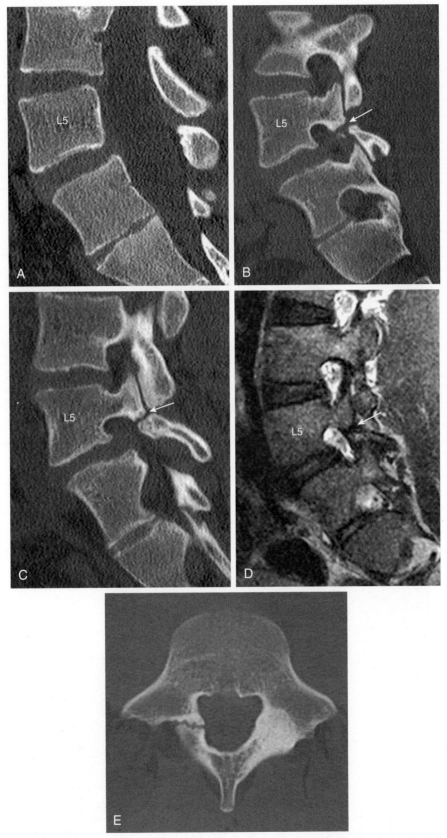

FIGURE 15-29 Spondylolysis of the spine when spondylolisthesis causes it to malalign. **A,** Midline sagittal reconstruction from lumbar spine computed tomographic (CT) image shows a grade I anterolisthesis of L5 (indicated) on S1. **B,** Right and **(C)** left parasagittal images from same exam show the bilateral L5 pars defects *(arrows).* **D,** Sagittal T2-weighted image in the same patient shows the same defect on the left *(arrow).* The defect is much harder to detect on magnetic resonance imaging (MRI) than CT, and sometimes you might think there is one on MRI only to find the pars intact on CT. So make that MRI diagnosis with caution! **E,** In another patient, a pars defect is well depicted on the axial CT. The horizontal orientation of the defect indicates that this is unlikely to be a facet joint (a first-year professor mistake).

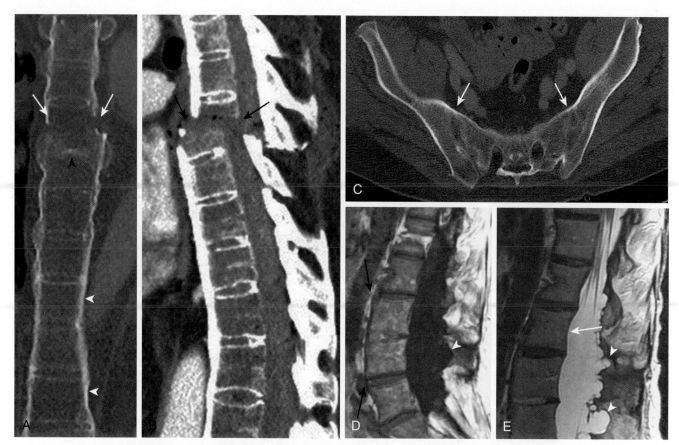

FIGURE 15-30 Ankylosing spondylitis causes squaring and spinal border brightness. **A,** Coronal reconstruction of axial computed tomography (CT) data reveals findings characteristic of ankylosing spondylitis including syndesmophytes *(white arrowheads)* and resulting bamboo spine appearance and calcification of the disk *(black arrowhead)*. Unfortunately, this patient suffered a complication of ankylosing spondylitis, a fracture *(arrows)*. **B,** The sagittal reformatted CT showed the "squaring off" of the vertebral bodies and the bridging syndesmophytes anteriorly. The fracture with displacement of the upper thoracic region is seen as well *(arrows)*. Again, some of the disks show partial posterior edge calcification. **C,** Axial CT shows the ankylosis of the sacroiliac joints *(arrows)* in this advanced case. **D** and **E,** On the magnetic imaging one can see the squaring of the vertebrae on the T1-weighted imaging (T1WI) *(black arrows)*, dural ectasias on the T2WI *(white arrowheads),* and posterior scalloping *(white arrow)*.

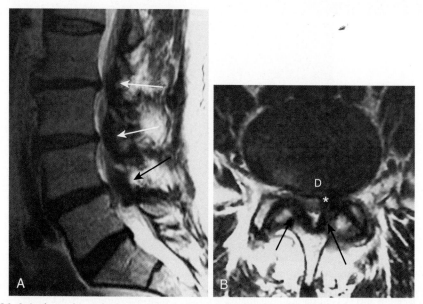

FIGURE 15-31 Spinal canal is tight from multiple "perps" in the night. **A,** Midsagittal T2-weighted scan shows multilevel spinal stenosis with limited cerebrospinal fluid in the thecal sac identified at L2-3, L3-4, and L4-5 *(arrows)*. **B,** At L3-4, the stenosis is a result of disk bulge (D) with superimposed subarticular herniation *(asterisk)* and ligamentum flavum thickening *(arrows)* associated with facet joint degenerative changes.

it has been associated with spinal stenosis. Images of canal stenosis are usually vivid, revealing degenerated disks, hypertrophied ligamentum flavum, loss of epidural fat at the pedicle and lamina level, and facet hypertrophy. Prominence of the epidural fat alone as epidural lipomatosis can compress the thecal sac, or it can combine with narrowing by ligamentum flavum, lamina, and/or facet disease. In the lower lumbar region the stenotic canal on axial scans takes on a T shape or what has been termed a trefoil shape, which results from a combination of a narrow canal and facet hypertrophy. This condition is usually the result of degenerative processes superimposed on a bony canal that is borderline normal or slightly small, and is seen in adults in their fifth to sixth decades of life. Congenitally short pedicles predispose to spinal stenosis. In cases of focal stenosis in the lumbar spine because of degenerative changes, redundancy of cauda equine nerve roots above the level of stenosis can be seen.

The width of the lateral recess in the lumbar region could be measured from the posterior aspect of the vertebral body to the most anterior aspect of the superior articular facet, but no one is that meticulous in neuroradiology. It is said that a width of 2 mm or less is considered stenotic. The nerve root buds out of the thecal sac to course in the lateral recess, then under the pedicle and out the neural foramen. Hypertrophy of the superior articular facet is the most common cause of narrowing of the lateral recess, although abnormalities of any components of the lateral recess may also compress the nerve.

Stenosis of the lumbar neural foramina can also occur. Normally the neural foramen is oval, with constriction in its inferior portion, and is bordered anteriorly by the vertebral body and disk, superiorly and inferiorly by the pedicles, and posteriorly by the pars interarticularis. Sagittal T1 MR is the best modality to view the foramina (see Fig. 15-25). The nerve and blood vessels usually lie in the superior portion of the foramen, which is normally filled with fat. The nerve may be compressed here by osteophyte or disk, or from spondylolisthesis. You should carefully evaluate the sagittal relationship between the nerve and components of the foramen to detect subtle forms of compression that may not be apparent on axial images. MR tends to overestimate the degree of spinal stenosis compared with myelography and myelo-CT. This is probably related to susceptibility effects, CSF motion, and truncation artifact.

Spinal stenosis can occur in children and adolescents with achondroplasia (Fig. 15-32), mucopolysaccharidoses, congenital lipomas, and with acquired precipitating lesions such as an acute disk herniation in combination with preexisting idiopathic spinal stenosis.

A lumbar laminectomy results in partial or complete removal of the lamina, the spinous process, the ligamentum flavum, and some of the contents of the patient's wallet. The extent of the procedure is determined by the disease and the surgical approach. On CT, loss of the lamina and absence of the ligamentum flavum indicate previous lumbar surgery. Postoperatively, on MR intermediate or high intensity on T2WI is observed in the region of the surgery. Mass effect (from the normal postoperative swelling, hemorrhage, and scar formation) can simulate preoperative disk herniation in size and signal intensity. This gradually resolves during a period of up to 6 months. Surgical disruption of the annulus appears as a line of high intensity

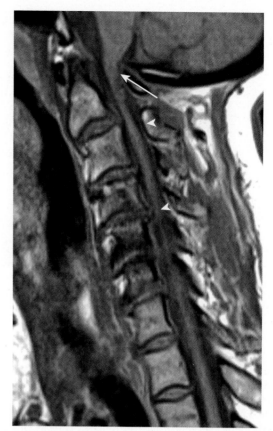

FIGURE 15-32 Achondroplastic dwarfism associated with tight foramen magnum. Sagittal T1-weighted image shows spinal stenosis at the foramen magnum *(arrow)* as well as cervical spine stenosis *(arrowheads).* In achondroplastic little people the anteroposterior and transverse diameter of the canal is often very stenotic throughout the spine. The patients can present with cervical myelopathy or with spinal claudication from lumbar stenosis (a keyhole appearing canal).

on T2WI. Scar tissue can be seen in both asymptomatic and symptomatic patients. Enhancement in the postoperative patient is dynamic over time. There is normally enhancement of the facet joint, paraspinal muscles, previously compressed nerve roots (which may enhance proximally up to the conus), the postdiscectomy disk space, and vertebral endplates. This enhancement is common and can persist for months or even longer. Sizable scarring may be identified in more than 40% of patients postoperatively, and these persons may be asymptomatic.

Postoperative Diskitis

The classic MR findings of postoperative diskitis are decreased signal intensity within the disk and adjacent vertebral body marrow on T1WI; increased signal on T2WI in the disk and adjacent marrow, often with obliteration of the intranuclear cleft; and enhancement of the disk and endplate (Fig. 15-33). A homogeneous pattern of enhancement can be identified as a horizontal band on either side of the disk space. Asymptomatic patients after discectomy may have some of these findings. They do not even have to be uniformly present in patients with confirmed infections.

The bottom line is that the MR in the postoperative patient can be erroneously suggestive of residual diskitis, particularly

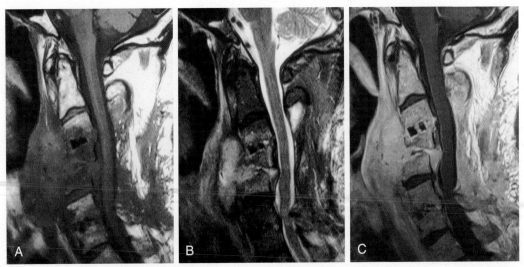

FIGURE 15-33 Postoperative diskitis can lead to surgical-legal "flight risk." **A,** Sagittal T1-weighted imaging (T1WI) scan shows loss of disk space, hypointensity of the marrow, collapse, kyphosis, epidural compression, and an impressive prevertebral mass. **B,** High signal in the disk and the adjacent marrow as well as the epidural and prevertebral mass is evident on this sagittal T2WI. **C,** After contrast, one completes the picture with enhancing disk material, endplates, epidural phlegmon, and prevertebral abscess.

within the first 6 months, even in patients who are asymptomatic. A little knowledge can be dangerous in this case; however, thorough appreciation of the normal postoperative MR findings can soothe your favorite neurosurgeon's or orthopedic surgeon's gastric mucosa (unless of course he or she operated on the wrong level, or you missed the incidental conus tumor). Without perivertebral enhancing phlegmon or frank abscess formation, contrast enhancement of the residual disc material is not very specific in the postoperative patient to rule in or to rule out diskitis.

Dural Ectasia

The phenomenon of dural ectasia is defined as enlargement of the dural sac and root sleeves anywhere along the spinal column but most often occurring in the lumbar and sacral regions. This results in scalloping of the posterior vertebral bodies, spinal canal dilatation with widening of the interpediculate distance and neural foramina, and thinning of the cortex of the pedicles and laminae. Anterior and posterior meningocele can also be seen in dural ectasia (Fig. 15-34). It is common in Marfan syndrome. It is also reported in neurofibromatosis, Ehlers-Danlos syndrome, and ankylosing spondylitis.

Myelopathy from Degenerative Disease

Cervical spondylosis compresses the spinal cord as well as its arterial and venous blood supply. Cervical myelopathy associated with high intensity on T2WI, extending over several vertebral body segments, has been reported in cases of cervical spondylosis (Fig. 15-35). This may be exacerbated if there is instability demonstrated on flexion extension films. The cause of cervical spondylomyelopathy is much debated. Pathology shows (1) edema within the cord (faint, indistinct high signal on T2WI); (2) gliosis (brighter, more well defined on T2WI); (3) demyelination; (4) cystic necrosis (bright, well defined on T2WI); and (5) necrosis, myelomalacia (seen as elevated T2 signal with

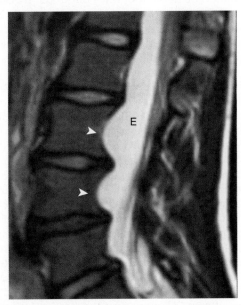

FIGURE 15-34 Dural ectasia (E): note the posterior scalloping *(arrowheads)* at L3 and L4 in this child with neurofibromatosis.

cord volume loss). The etiology is purported to be from either fixed stenosis, direct subclinical trauma, dynamic strrrrrrretch injury with motion, arterial or venous ischemia, inflammation, or posttraumatic demyelination.

Following successful stabilization and decompression, the high intensity can resolve suggesting an edematous etiology for the abnormality, but the risk of further intraoperative injury because of blood pressure changes, fluid dynamics or surgical manipulation is not to be taken lightly.

Immediate Scanning

After the lengthy surgery for spinal stenosis, you will invariably receive the late night CT scan for postoperative assessment that has become routine after hardware

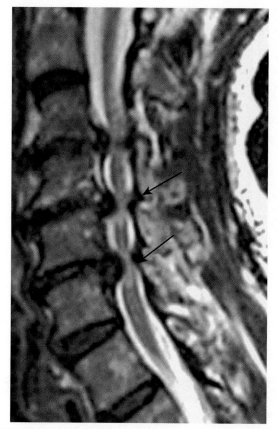

FIGURE 15-35 Spondylomyelopathy (a big word for cord injury because of degenerative disease). Note the high signal intensity in the spinal cord on this T2-weighted imaging scan with multilevel degenerative disease and spinal stenosis *(arrows).* This finding has a poor preoperative prognosis.

placement. Generally, this requires two parts to the report: one for hardware assessment and the other for immediate complications. Most surgeons are using pedicle screws in the thoracolumbar spine for posterior stabilization. If any screws violate the medial pedicle margin and encroach on the thecal sac or lateral recess they should be pointed out in a gentle way that does not instigate trouble with the sensitive (buttercup) orthopod or neurosurgeon (Fig. 15-36). If a sacral screw extends far anterior to the sacrum such that it might irritate an L5 or S1 root in the pelvis, it too should be discussed. If there are fractures along the hardware or broken screws or screws that transgress the facet joints or end up in the disk spaces, gently notify the surgeons with a courtesy call.

Intervertebral spacers of a variety of types may be placed for adding height between bodies and decompressing foramina. They should be symmetrically placed and should not overlap cortical edges on any plane.

In the cervical spine one is usually looking at lateral mass screws posteriorly at C3-C7. These screws have the potential for entering the foramen transversarium where the vertebral artery lies or the neural foramen where nerve roots abide. They should not transgress these foramina.

Anterior plates with screws may be present in the cervical region more commonly than the lumbar spine. Make sure the screws have good purchase and are not placed into the disk space. By the same token they should not transgress the spinal canal.

An assessment of spinal stability after surgery begins with flexion and extension views to look for movement and change in angulation of the spine. If there is anterolisthesis or retrolisthesis with bending or if the width of the anterior intervertebral space changes significantly with position it suggests the fusion is not solid. Values of 10 degrees for sagittal rotation and 4 mm for sagittal translation are typically used to infer instability. Other authors suggest an anterior translation greater than 6% to 8% of the vertebral body width, posterior translation greater than 9% (L1-S1), and angular displacement in flexion greater than 9 degrees (L1-L5) indicate instability. Vacuum phenomena in the disks implies movement is occurring and should not be seen at fused levels. For functional fusion, the morphologic appearance should include solid complete bridging bone across facet joints on CT scans and bone that crosses the intervertebral spaces around or through the interbody spacers (P.S. ramps are closed spacers, cages are open centrally).

Instability has been defined as loss of motion segment stiffness as shown by an abnormal response to applied loads characterized kinematically by abnormal movement in the motion segment beyond normal constraint.

Failed Back Surgery Syndrome (The Surgeon's Worst Nightmare)

A large problem confronting the clinician (and secondarily the radiologist) is recurrent or residual low back pain in the patient after lumbar disk surgery. This condition has a reported incidence of 5% to 40%, and the syndrome has been termed the "failed back" or the "failed back surgery" syndrome (FBSS). The common causes of this condition are listed in Box 15-3.

Immediately following successful low back surgery imaging reveals mass effect on the anterior spinal canal usually greater than what was there preoperatively. This diminishes after the first week and its size does not correlate with outcome. There can also be posterior compression of the dural tube secondary to placement of absorbable surgical material for hemostasis, which appears like a folded sponge surrounded by fluid. The MR appearance should not concern you unless the patient has significant referable clinical signs and symptoms. Extraspinal fluid collections reflecting hematomas or seromas can also be seen to compress the neural tube immediately postoperatively in the setting of a laminectomy procedure.

All patients have varying degrees of scar tissue at 4 months. However, scar tissue cannot necessarily be implicated as an etiology of postoperative back pain because everyone gets it and only some have chronic pain. In the failed back, it is critical to look carefully for other causes of pain including infection, residual or recurrent disk, or instability.

The diagnosis of scar versus residual/recurrent disk herniation is extremely important in this situation. Surgery is not indicated for scar (epidural fibrosis) but would be beneficial if disk can be diagnosed as the cause of the radiculopathy. Both may produce mass effect, although epidural fibrosis may cause traction of the dural tube. Nerve root displacement by a mass lesion is almost always associated with a recurrent herniated disk. Contrast is necessary in this situation. After injection, immediate scanning of scar

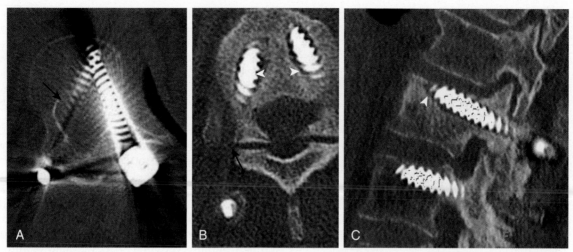

FIGURE 15-36 Pedicle screw placement (refrain from surgical debasement). **A,** The tract of the right screw *(arrow)* can be seen to violate the spinal canal. Here it can irritate nerve roots or impact the thecal sac. **B,** Not only does the right screw tract appear to be lateral to the pedicle *(black arrow)*, both screws have lucency around them *(white arrowheads)* implying loosening (in a different patient). This is usually not an immediate postoperative finding but can be seen on follow-up. **C,** If the screw violates the superior endplate *(arrowhead)* and enters the disk, it loses "purchase" and is unlikely to maintain stability. All of these misplaced screws means the patient is "screwed" as far as the stability of the spine.

demonstrates diffuse enhancement on T1WI whereas disk usually does not enhance or shows peripheral enhancement (Figs. 15-11 and 15-37). Enhancement should be performed with fat saturation, because fat and enhancement are both hyperintense on T1WI without fat suppression. Conspicuity between scar (which enhances) and fat would thus be normally lacking.

There may be immediate enhancement around a herniated disk, and fragmented disks can produce enhancement similar to scar. If scanning is delayed more than 20 minutes, disk may imbibe the contrast and therefore enhance with an appearance identical to scar, so scan immediately and with rapid sequences. Enhancement also enables detection of the nerve root (not enhancing) surrounded by enhancing scar.

Nerve roots have been noted to enhance up to 8 months postoperatively. These roots are nearly always those compressed preoperatively suggesting ongoing repair and regeneration associated with an impaired blood-nerve barrier. Relating the symptoms to root enhancement is controversial (both in the literature and in the courts). In addition, this phenomenon occurs in 5% of symptomatic patients without surgery. An alternative explanation regarding lumbar root enhancement is that it results from obstruction of small radicular veins within the endoneurium of the nerve root related to nerve root compression. It may be the radicular veins (intravascular enhancement) and *not* roots that enhance. It may occur either physiologically or pathologically and may not be specific. Sorry for all the equivocation.

Arachnoiditis

Arachnoiditis, once a very common condition after surgery is an inflammatory disorder of the spinal leptomeninges particularly affecting the nerve roots resulting in intradural adhesions. Meningitis, subarachnoid hemorrhage, and contrast myelography using previously more

---

BOX 15-3   Causes of Failed Back Syndrome

Arachnoiditis
Epidural fibrosis/granulation tissue/scarring
Immediate postoperative complications including infection, hematoma, or surgical trauma to roots
Insufficient decompression of roots by residual soft tissue or bone
Mechanical instability
New disk at level adjacent to the surgery because of changes in stress
Pseudoarthrosis
Residual or recurrent disk
Spondylolisthesis
Surgery at wrong level

---

toxic dyes can also result in arachnoiditis. It was well described with the combination of hemorrhage from myelography/surgery and retained iophendylate (Pantopaque) (about 20% of cases). Nonionic contrast agents have markedly decreased the incidence of arachnoiditis after myelography, but it still occurs. On good FSE images in the lumbar spine you should be normally able to distinguish the anterior and posterior spinal roots until they leave the thecal sac. In the lumbar region, arachnoiditis has a variable appearance ranging from (1) loss of the ability to distinguish the roots in the thecal sac and obliteration of the root sleeves (mild); (2) loss of the morphology of the thecal sac; (3) adhesion of the nerve roots to the dural tube ("empty sac"); or (4) clumping together of the nerve roots, leading to an appearance of a "pseudofilum" (Fig. 15-38). Arachnoiditis may or may not enhance. Arachnoiditis has been cited as a cause in up to 15% of cases of failed back syndrome. It can be responsible for the development of loculated arachnoid cysts particularly in the thoracic region.

Tuberculous arachnoiditis is somewhat different from other forms and involves the spinal cord, meninges, and

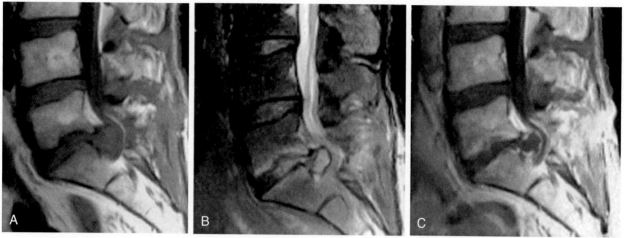

FIGURE 15-37 Postoperative failed back requiring a repeat surgical whack (hopefully not by a hack). Sagittal T1-weighted imaging (T1WI) **(A)**, sagittal T2WI **(B)**, and enhanced sagittal T1WI **(C)** show a recurrent herniated disk at L5-S1 and the spinous process defect from the laminectomy. Note that the disk does not enhance in **C**, but there is minimal peripheral granulation tissue.

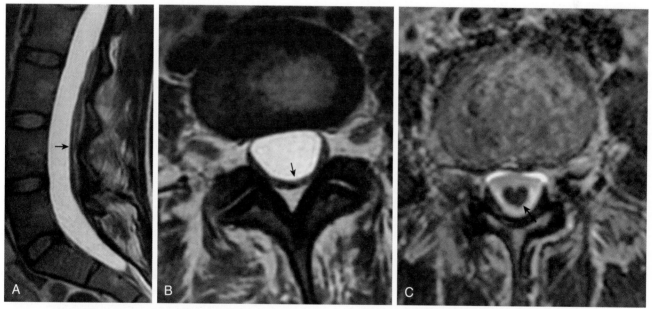

FIGURE 15-38 Pseudocord versus empty sac; both manifestations of nerve root tacked. **A** and **B**, These images show the nerve roots plastered along the posterior thecal sac such that the sac looks empty *(arrows)*. **C**, Contrast that with this section through L4-5 in a different patient where it looks like there is a spinal cord *(arrow)* present. Arnold Chiari with tethered cord? No, actually these are clumped nerve roots after surgery.

nerve roots. On MR, there is loss of distinct margins of the cord and enhancement of the leptomeninges. Enhancing nodules have also been reported. The combination of leptomeningeal thickening, nodules of the spinal cord and roots, and spinal cord involvement should suggest tuberculous arachnoiditis (although other diseases such as sarcoid and metastatic disease could produce a similar picture). The differential diagnosis of nodular or diffuse root thickening is given in Box 15-4.

Occasionally, the dura can be torn at the time of surgery (by the Greek surgeon Euripides) or can occur after trauma or even spontaneously, with leakage of CSF from the wound (postoperatively) or with the formation of an

BOX 15-4   Diffuse Root Thickening
Arachnoiditis
Carcinomatous meningitis
Chronic inflammatory demyelinating polyradiculoneuropathy
Cytomegalovirus (or other infectious agent) radiculitis
Guillain-Barré
Hypertrophic sensory motor neuropathy (Dejerine-Sottas syndrome/Charcot Marie Tooth)
Lymphoma
Neurofibromatosis I or II (i.e., neurogenic tumors)
Sarcoid
Spinal arteriovenous malformation

organized CSF collection (pseudomeningocele), which can subsequently enlarge. This collection may need to be repaired if it is progressive and/or prone to infection, and/or fistulizes to the skin. Often the surgeons seem very "ho-hum" about a pseudomeningocele as if it is an expected occurrence, but we think they are actually not keen on big ones. The differential diagnosis, particularly in the early postoperative period in the patient with pain and fever, is abscess, which might demonstrate significant peripheral enhancement. Seromas are later, smaller, and usually asymptomatic. They, like pseudomeningoceles, enhance less along the periphery or into the soft tissues than abscess.

We do not know anyone who won a Nobel Prize for making a major contribution to degenerative spine diseases. However, in practice degenerative disease will occupy more of your scanner's time than you can imagine. Here the intellectual challenges are inversely proportional to wampum potential. Unfortunately, if you have ever had back or neck pain, you would wish more Nobel Prize winners would lend their efforts to study degenerative disease of the spine than Drosophila saliva genes.

Chapter 16

Nondegenerative Diseases of the Spine

This chapter begins with infectious and noninfectious inflammatory diseases of the spine and spinal coverings. Cystic lesions of the cord followed by neoplastic diseases involving the spine and cord comprise the next two sections of this chapter. The last sections concern vascular diseases of the cord and spinal cord trauma. We have chosen this approach because it is disease oriented and in most settings the clinical findings generate the imaging algorithm. These diseases actually stimulate more intellectual discussion than "Is it a herniation or a bulge?" In fact, the diseases affecting the spine give the authors Lhermitte shivers thinking about them. So, let us decompress the spine.

INFECTIOUS AND INFLAMMATORY DISEASES

Diskitis and Osteomyelitis

Pyogenic disk space infections are usually the result of a blood-borne agent, particularly from the lung or urinary tract (or needle and syringe in the Baltimore 'hoods). The pathogen lodges in the region of the endplate in the adult or the hypervascular disk edge in the child and destroys the disk space and the adjacent vertebral bodies. The disk space infection may be acute, in which case pain is invariably present, or it may be more chronic. The symptoms start with focal back pain and progress to radicular, meningeal, and spinal cord compressive symptoms if/as the disease advances. Clinical settings include those cases in which there is a recent history of spine surgery, hematogenous dissemination from another infectious site, compromised immunity, or intravenous drug use. The most common organism is *Staphylococcus;* other common causes include *Streptococcus, Peptostreptococcus, Escherichia coli,* and *Proteus*. The sedimentation rate and C-reactive protein are elevated and there is a leukocytosis in most cases of disk space infection.

Although gallium- and indium-labeled white blood cell radionuclide imaging is highly sensitive to inflammation, magnetic resonance imaging (MRI) is the most specific method for cost-effective and time-efficient diagnosis and can, at the same time, evaluate the impact on the spinal cord. The disk space is abnormally low signal intensity on T1-weighted imaging (T1WI) and higher than normal signal intensity on T2-weighted imaging (T2WI) as a result of the associated edema (Fig. 16-1). The disk space, adjacent vertebral bodies (particularly at the endplates), and, if present, the epidural and/or paravertebral soft tissues enhance. Endplate destructive changes can be better appreciated by plain radiograph or computed tomography (CT). Unfortunately, Modic type I degenerative

marrow changes (low on T1WI and bright on T2WI) may simulate the appearance of diskitis/osteomyelitis and may enhance, making it difficult to distinguish the two entities. For this reason, high T2WI signal in the endplates of the vertebral bodies is the least reliable of the findings for diskitis/osteomyelitis (although some suggest chronic indolent infection may underlie some cases of Modic I endplates). The presence of paravertebral or intraspinal enhancing inflammatory soft tissue, if present, is very helpful in making the distinction between degenerative disease and diskitis/osteomyelitis. Look carefully for epidural or paravertebral abscesses as these may require surgical drainage.

Vertebral osteomyelitis usually occurs in the setting of disk space infection; however, osteomyelitis can occur without disk space infection from hematogenous dissemination directly to the vertebral body. The signal changes are similar to disk space infection. A three-phase technetium bone scan can show increased blood flow and persistent osseous uptake on delayed images in osteomyelitis. This examination may not be specific so that if there is a question of degenerative disease versus infection, a gallium scan or indium-111–labeled leukocyte scan may suggest an infectious process by demonstrating increased uptake and correlating it with the positive bone scan.

It is not uncommon to have spinal cord symptoms/signs without epidural compression in cases of pyogenic disk space infection or vertebral osteomyelitis. In this case, the cause may be a vasculitis of the medullary arteries and/or veins.

Granulomatous Infections

The most common granulomatous infections of the spine are tuberculosis, brucellosis, and fungal infections (blastomycosis, cryptococcosis, and coccidioidomycosis).

Tuberculosis most often affects the lower thoracic spine (Pott disease, or tuberculous spondylitis), is indolent, and is frequently associated with epidural disease, particularly paravertebral, or subligamentous abscess, or abscess of the psoas muscle. Vertebral body destruction with relative sparing of the disk space occurs late in the disease (Fig. 16-2). The posterior elements may be involved, and the infection can spread along the anterior longitudinal ligament and involve multiple levels. The disk can be of normal intensity on noncontrast magnetic resonance (MR) but usually enhances. Spinal deformity is common including gibbus deformity. Approximately 50% of patients have obvious pulmonary disease. The diagnosis of tuberculosis requires isolation of the bacillus, characteristic histopathologic changes,

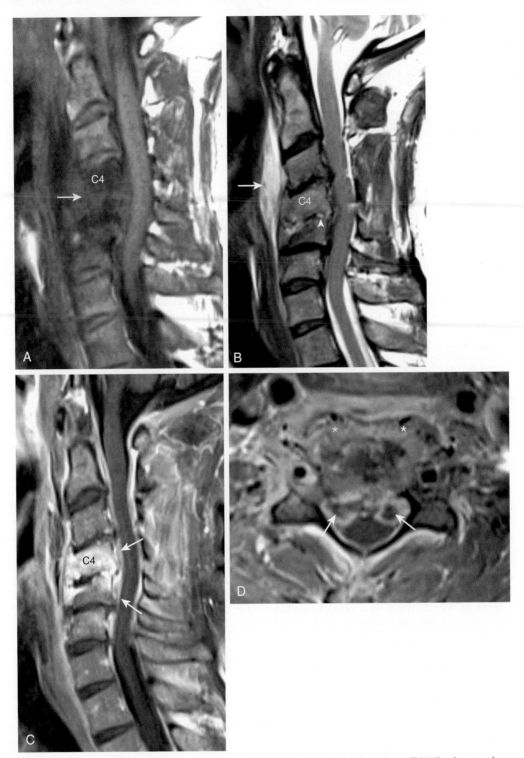

FIGURE 16-1 Cervical diskitis/osteomyelitis. **A,** Sagittal T1-weighted imaging (T1WI) shows abnormal marrow signal at C4 and C5 with loss of endplate definition *(arrow).* There is a slight kyphotic angulation centered at this level. **B,** Sagittal T2WI shows focal high signal within the C4-C5 disk space *(arrowhead),* although this alone is not sufficient to make a diagnosis of diskitis. It is the constellation of these and other findings, including prevertebral edema *(arrow),* and abnormal epidural enhancement *(arrows)* on postcontrast T1WI **(C),** that cinch the diagnosis. **D,** Note the presence of organized fluid collections within the ventral epidural space *(arrows)* resulting in canal compromise as well as thickening and enhancement of the longus colli musculature *(asterisks).*

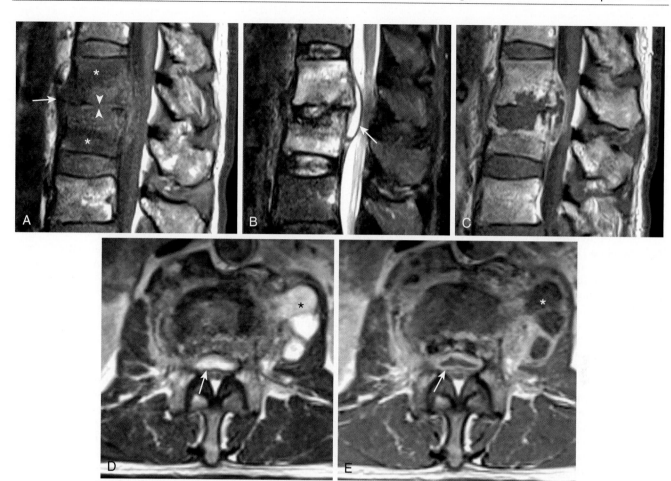

FIGURE 16-2 Tuberculous osteomyelitis. **A,** Sagittal T1-weighted imaging (T1WI) shows abnormal marrow signal involving two adjacent vertebral bodies *(asterisks)* as well as loss of endplate cortical definition *(arrowheads)*. Note prevertebral component *(arrow)* along the anterior longitudinal ligament. **B,** Sagittal short tau inversion recovery image shows associated ventral epidural fluid collection compromising the spinal canal *(arrow)*. **C,** On this sagittal postcontrast T1WI, the fluid collection shows peripheral enhancement, indicating organized abscess in communication with the intervertebral disk. **D,** Axial T2WI again shows the epidural fluid collection *(arrow)* as well as extension of inflammation and fluid collection into the left psoas musculature *(asterisk)*. **E,** Both fluid collections show peripheral rim enhancement on axial postcontrast T1WI indicating organized abscesses. Note perivertebral soft-tissue enhancement along the margins of the vertebral body.

TABLE 16-1   Salient Features of Nontuberculous Bacterial Infection, Tuberculosis, and Neoplasm

Condition	Disk Space	Paraspinal Mass	Posterior Elements	Spread
Bacterial infection	Always involved early on	May or may not be present	Very uncommon	Contiguous
Tuberculosis	Occurs in the course of the disease	Usually large	Yes	May have skip areas anteriorly
Neoplasm	Very rare	Yes	Yes	Noncontiguous

and/or response to antibiotic therapy. Table 16-1 contrasts the features of bacterial infection, tuberculosis, and neoplasm.

Intraspinal Infection

Epidural abscesses are the result of either direct extension of vertebral osteomyelitis or hematogenous dissemination from an infectious source. *Staphylococcus aureus* is the most common pathogen (45%), with the remainder being gram-negative rods, anaerobes, mycobacteria, and fungi. Patients are septic, with frequent histories of intravenous drug abuse, immunosuppression, urinary tract infections, recent surgery, and valvular infections. These patients present with tenderness over the spine, back pain, and progressive neurologic impairment. There is an association with blunt spine trauma, hematoma formation, and secondary infection.

MR is the modality of choice for imaging epidural abscesses (see Figs. 16-1, 16-2). The epidural mass has low signal intensity on T1WI and has high signal on T2WI. Uncomplicated, the cord has normal signal. There are three patterns of enhancement: (1) homogeneous enhancement, representing diffuse inflamed tissue with microabscesses (phlegmonous granulation tissue); (2) peripheral enhancement consistent with frank abscess including a necrotic center; and (3) a combination of tissue enhancement and frank abscess. Pattern (1) may be difficult to discern in the lumbar spine without the use of fat saturation techniques to distinguish between enhancement and epidural fat and venous plexus, which can be diffusely infiltrated by the inflammatory process. Precontrast T1 images can be very helpful in detecting the space occupying process in the epidural space because of effacement and/or replacement of the normal T1 bright epidural fat by abnormal tissue. Epidural abscess produces symptoms by sheer mass effect and/or septic thrombophlebitis with cord edema and infarction. Think about spinal cord infarction secondary to thrombophlebitis in the setting of epidural abscess and high signal on T2WI in the cord.

Subdural empyema has a similar presentation to epidural abscess, although it is much rarer in occurrence and may occur following lumbar puncture, spinal anesthesia, and diskography. Discrimination between the subdural and epidural compartments is usually difficult on MR. Hopefully you can see the dark dural borders around the collection in a subdural empyema as opposed to the epidural collection, which is usually bordered by epidural fat/bone/ligaments.

Meningeal inflammation can occur in conjunction with or without the intraspinal infections described above. Such pathology is not readily discernible by CT even with contrast administration. MR can show poor definition of the cord or subarachnoid space on T1WI and diffuse thick enhancement of the leptomeninges. Normal MRI does not exclude meningeal inflammation and if clinically suspected and no contraindications, lumbar puncture should be performed.

The sequela of bacterial meningitis may be arachnoiditis, which on MR or CT myelography is demonstrated by clumping of nerve roots within or around the periphery of the thecal sac. Clinically, meningeal infection may lead to symptoms of myelopathy and/or radiculopathy, which can progress to paralysis. These symptoms are related to compression of nerve roots and spinal cord by thickened dura. Infarction and cord cavitation are potential late sequelae of meningeal inflammation.

Renal Spondyloarthropathy

Patients undergoing dialysis can have changes in their spine (usually in the cervical region), which superficially may resemble infection (renal spondyloarthropathy). There is destruction of the disk space and adjacent vertebral bodies; however, the signal intensity on T2WI is usually low rather than high as expected in infection. Paravertebral inflammatory tissue is absent. Furthermore, there is no clinical evidence of infection. Amyloid deposition has been implicated as a possible cause.

Sarcoidosis

Only 6% to 8% of patients with neurosarcoidosis have spinal cord lesions with the most common location being the cervical spine. Sarcoidosis can manifest as enhancing nodules on the surface of the cord, which can appear similar to tumor nodules from subarachnoid seeding and can be associated with cord enlargement. Rarely, sarcoid nodules may be on individual nerve roots. As in the brain, intramedullary lesions result from infiltration of the Virchow-Robin spaces (Fig. 16-3). Intramedullary sarcoidosis can present as a diffuse inflammatory lesion or as a discrete mass. The lesions demonstrate intermediate to high intensity on T2WI with diffuse leptomeningeal enhancement and/or nodular enhancement. Unfortunately, patients with sarcoidosis on or off steroids have an increased rate of spinal tuberculosis or fungal infection, which obscures matters further.

Intramedullary Lesions in Acquired Immunodeficiency Syndrome

A variety of conditions affect the cord in patients with acquired immunodeficiency syndrome (AIDS); these are listed in Box 16-1 (Fig. 16-4). Vacuolar myelopathy, identified in patients with AIDS, is defined as vacuolation in the spinal white matter associated with lipid-laden macrophages typically involving the dorsal columns and lateral corticospinal tracts, although it may occur anywhere. Degeneration of the cortical spinal tract has also been reported, especially in children. Its incidence has been reported to be between 3% and 55%. The etiology is unknown but may be related to either primary or indirect effects of human immunodeficiency virus (HIV) itself, secondary infectious agents, nutritional deficiencies, or toxic effects of drugs. The vertebral marrow can also be affected by these conditions. Unfortunately, nothing is specific about the MR appearance to separate AIDS from other non-AIDS intramedullary lesions. High intensity on T2WI is seen in a spinal cord that may be large, normal, or small in size. The patients may also get infectious myelitis from viruses, toxoplasmosis, fungal agents, and bacteria.

Transverse Myelitis and Multiple Sclerosis

Transverse myelitis is a syndrome affecting the spinal cord associated with rapidly progressive neurologic dysfunction. Diseases causing this condition include acute disseminated encephalomyelitis, multiple sclerosis, connective tissue diseases (lupus, rheumatoid arthritis, and Sjögren syndrome), sarcoidosis, vascular malformations, vasculitides, or the condition may be idiopathic. The spinal cord is focally enlarged with high signal on T2WI, and may show patchy enhancement (Fig. 16-5).

With idiopathic acute transverse myelitis, the clinical course occurs over days to weeks. Pathology reveals demyelination, perivascular lymphocytic infiltrates, and necrosis. The lesion extends over multiple spinal cord segments and involves the entire cross-section of the spinal cord. There is high intensity on T2WI with variable enhancement patterns (nodular, meningeal). It has been hypothesized that this is a result of a small vessel vasculopathy (perhaps immunologically mediated), either arterial or venous, affecting gray matter as well as white matter. In some cases, the cauda equina enhances suggesting a possible relationship with Guillain-Barré syndrome.

Multiple sclerosis (MS) can affect the spinal cord and produce myelopathic signs and symptoms. It can be confined solely to the spinal cord (5% to 24% of cases), which may

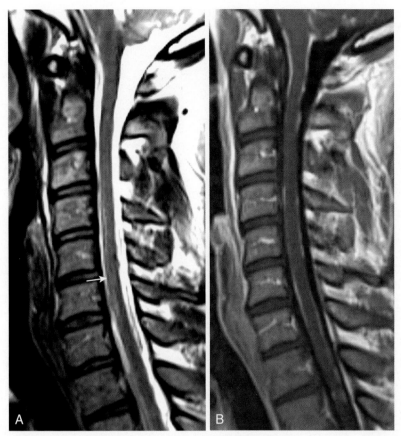

FIGURE 16-3 Sarcoidosis of cord. **A,** Sagittal T2-weighted imaging shows abnormal patchy intramedullary signal throughout the cervical spinal cord *(arrow).* **B,** On postcontrast T1WI, there is striking enhancement of the leptomeninges.

account for the "negative" brain MR findings in patients with definite clinical symptoms. MS lesions are isointense to cord or of low intensity on T1WI and high signal on T2WI. Sixty percent of the spinal cord lesions occur in the cervical region. The typical MS spinal cord lesion does not involve the entire cross-sectional area, is peripherally located, does not respect gray/white boundaries, and is less than two vertebral body segments in length (90% of the time) (Fig. 16-6). The posterior columns are favored by MS. The cord is usually normal in size with enlargement seen in only 6% to 14% of cases. The majority of lesions are patchy in configuration and rarely demonstrate enhancement, unless the patient is referred for problems specific to new spinal cord signs/symptoms. Spinal cord parenchymal loss, especially in the cervical region, can be detected over the course of the disease.

It is useful to image the brain by MRI if questions are raised concerning the cause of transverse myelitis, because asymptomatic high signal abnormalities in young adult brains help favor MS as the cause of the transverse myelitis. If you see transverse myelitis and optic neuritis, consider neuromyelitis optica (NMO, also known as Devic syndrome). There may not be intracranial demyelination and NMO shows long segment cord disease rather than short segment disease as in MS (which would more likely show intracranial white matter disease). NMO may be monophasic or polyphasic.

Lupus Myelitis

This is a recognized complication of systemic lupus erythematosus presenting as transverse myelitis with back pain,

---

BOX 16-1 Intramedullary Lesions in Acquired Immunodeficiency Syndrome

Viral (HIV, CMV, HSV, HZV) infection
Toxoplasmosis
Tuberculosis
Vacuolar myelopathy
Non-Hodgkin lymphoma
Subacute necrotizing myelopathy
Syphilis

CMV, Cytomegalovirus; *HIV,* human immunodeficiency virus; *HSV,* herpes simplex virus; *HZV,* herpes zoster virus.

---

paraparesis or quadriparesis, and sensory loss. Etiology of the lesion is vacuolar degeneration from an autoimmune process or ischemia. The spinal cord may be enlarged with high intensity on T2WI involving four to five vertebral body segments (larger than MS) of the cord (Fig. 16-7). Contrast enhancement can be detected in approximately 50% of cases.

Subacute Combined Degeneration

Subacute combined degeneration, a complication of cobalamin (vitamin B_{12}) deficiency, causes a myelopathy affecting the cervical and upper thoracic spinal cord, but it can also produce lesions in the optic tracts, brain, and peripheral nerves. Clinical findings include paresthesias of the hands and feet, loss of position and vibratory sensation, sensory ataxia, spasticity, and lower extremity weakness.

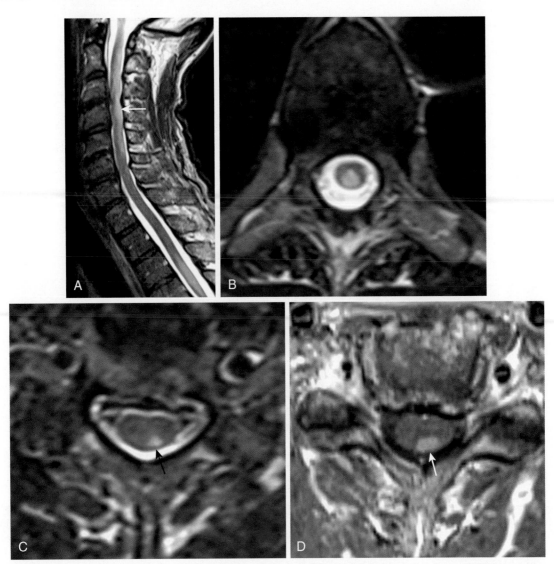

FIGURE 16-4 Acquired immunodeficiency syndrome–associated myelopathy. **A,** Long segment T2 hyperintensity predominantly affecting the dorsal cervical cord *(arrow)* is seen on this sagittal T2-weighted image (T2WI). Note additional multilevel signal abnormality at the upper thoracic cord. **B,** Axial T2WI through the thoracic cord shows clear involvement of the dorsal cord. **C,** More superiorly, this axial T2WI shows preferential involvement of the left dorsal cord *(arrow),* and this lesion enhances on **(D)** postcontrast T1WI *(arrow).*

Pernicious anemia, the inability to absorb B_{12} resulting from inactivation of intrinsic factor (secreted by gastric parietal cells), is the most frequent cause of B_{12} deficiency in the United States. Diseases that affect the terminal ileum where the B_{12}-intrinsic factor complex is absorbed (i.e., regional enteritis or tropical sprue) can produce B_{12} deficiency. More recently, this has been reported in patients having bariatric surgery (along with folate and thiamine deficiency), anorexia nervosa, veganism, and alcoholism. Pathologically, demyelination and axonal loss in the posterior and lateral spinal cord columns is seen.

On T2WI, high intensity is observed longitudinally in the posterior columns (Fig. 16-8). Following treatment this can regress and disappear if caught early enough; however, if undetected, permanent axonal loss and gliosis results. The bone marrow can also be low intensity on T1WI and T2WI associated with enhancement indicating benign hyperplasia of the hematopoietic marrow (secondary to the pernicious anemia). There are several etiologies for posterior

column disease, however, and Box 16-2 provides the differential diagnosis of lesions involving the posterior columns.

Radiation-Induced Changes

Spinal radiation converts normal bone marrow to fatty marrow (Fig. 16-9). This is manifested on MR by diffuse high signal on T1WI in the vertebral bodies and lower intensity on conventional T2WI. Box 16-3 provides a list of conditions that produce increased or decreased marrow signal on T1WI. If the cord is high signal on T2WI, it may be the result of radiation myelitis or residual tumor, if the reason for the initial therapy was a cord tumor. Cord atrophy is another manifestation of radiation therapy.

Tethered Cord

Symptoms of adult tethered cord/tight filum terminale syndrome have been associated with a precipitating traumatic

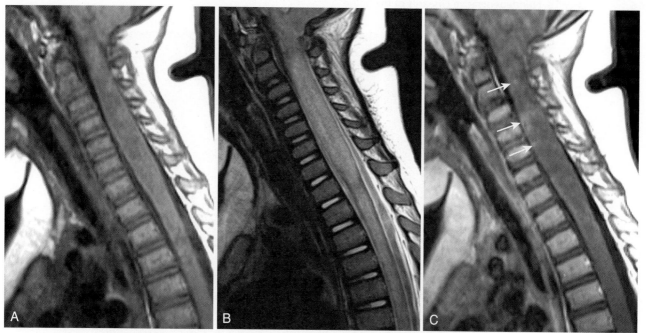

FIGURE 16-5 Transverse myelitis. **A,** Sagittal T1-weighted imaging (T1WI) shows marked enlargement of the imaged cord particularly the cervical cord. **B,** Sagittal T2WI shows generalized increased T2 signal within the cord. **C,** Following contrast administration, patchy areas of subtle enhancement can be seen on this sagittal T1WI *(arrows).*

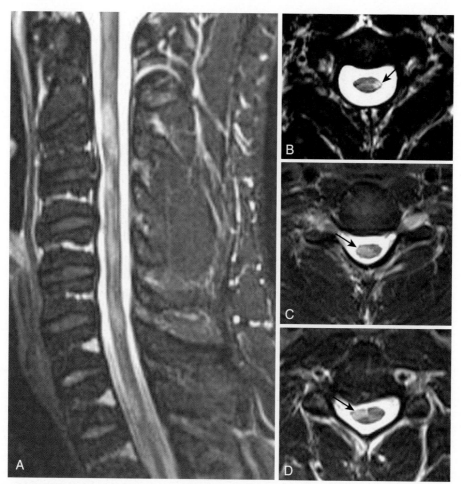

FIGURE 16-6 Multiple sclerosis. **A,** Multiple high signal intensity foci are depicted on this sagittal short tau inversion recovery image through the cervical and upper thoracic cord. **B-D,** Axial T2-weighted imaging through the cervical cord shows some of these lesions *(arrows).* Note that these typical lesions do not involve the entire cross-sectional area of the cord, are peripherally located, and do not respect great-white boundaries. Over time, cord volume loss may develop at these sites of disease.

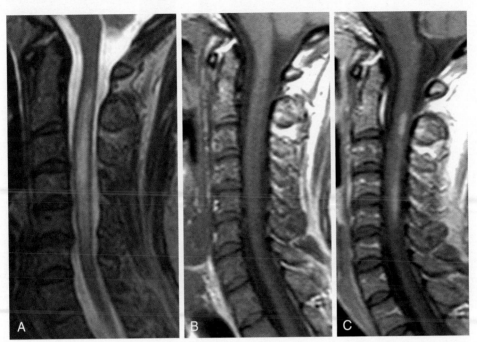

FIGURE 16-7 Lupus myelitis. **A,** Sagittal T2-weighted imaging (T2WI) shows diffusely increased T2 signal and mild enlargement of much of the cervical cord. When comparing precontrast sagittal T1WI **(B)** with the post-contrast T1WI **(C),** note that there is patchy enhancement present within this region of signal abnormality. It is difficult to know the diagnosis without the clinical history of systemic lupus erythematosus.

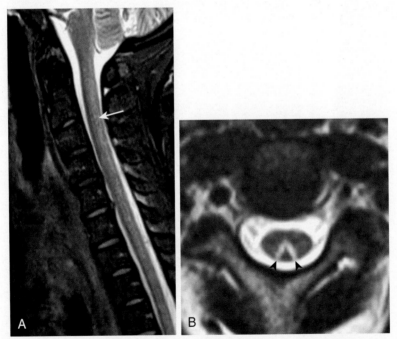

FIGURE 16-8 Subacute combined degeneration. **A,** Sagittal short tau inversion recovery image shows subtle abnormal signal predominantly in the posterior spinal cord *(arrow).* **B,** On this axial T2-weighted image in a different patient, you can see the configuration of the posterior column disease *(arrowheads).* Although vitamin B$_{12}$ and folate deficiency are the classic etiologies for this appearance (check labs and you'll look like a star!), copper deficiency should also be on the differential diagnosis.

event. This is seen in young women more often than in men and is accompanied by diffuse back and perianal pain, leg weakness, and urinary tract dysfunction. Imaging reveals the tip of the conus medullaris to be below L2, which is considered the lower limit of normal. The filum is thickened (>2 mm); however, it is sometimes difficult to distinguish between a thickened filum and a low-lying conus. The conus may be tethered by spina bifida occulta, a filum lipoma, and/or an intradural lipoma, and posteriorly displaced by fat, glial cells, and collagen. Cutaneous

BOX 16-2 Conditions Involving the Posterior Columns of the Spinal Cord

Multiple sclerosis/Devic syndrome
Cobalamin (vitamin B$_{12}$) deficiency
Copper deficiency
Folic acid deficiency
Nitrous oxide toxicity
Spinal trauma
Tumor
Tabes dorsalis (syphilis)
Toxins
Vacuolar myelopathy

BOX 16-3 Intensity Abnormalities of Vertebral Marrow on T1-Weighted Imaging

DECREASED INTENSITY

Any condition that produces edema, increased hematopoietic marrow, or fatty marrow replacement, such as infection, metastases, marrow packing disorders
Any condition that produces sclerosis of the bone (e.g., osteopetrosis)
Degenerative changes (types I and III)
AIDS
Anemia
Paget disease (variable)
Blood transfusions
Hemochromatosis (iron overload)
Chronic renal failure

INCREASED INTENSITY

Any condition that produces fatty marrow, including normal aging
Degenerative changes (type II)
Radiation
Vertebral body hemangioma
Hemorrhage
Paget disease (variable)
Fibrous dysplasia (variable)
Steroid therapy/Cushing disease

AIDS, Acquired immunodeficiency syndrome.

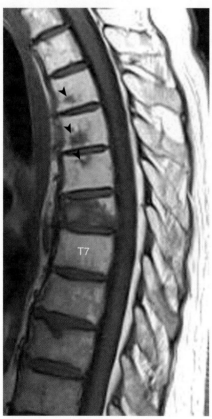

FIGURE 16-9 Radiation changes to spine. In this patient with lung cancer treated with radiation, there is relatively bright T1 marrow signal from T7 headed up. Compare to more normal heterogeneous marrow signal from T8 down. Unfortunately, this patient developed bone metastases *(arrowheads)* with more confluent metastatic disease nearly completely replacing the T6 marrow. Bummer...at least the cord was OK.

stigmata of dysraphism including hairy patch, sinus tract, or subcutaneous lipoma are common.

CYSTIC LESIONS

Arachnoid Cyst

Arachnoid cysts occur on a congenital basis or result from trauma, postoperative scarring, or inflammation. On CT myelography, you will see an extramedullary mass, which may compress the cord. These cysts can be under

pressure and produce symptoms by spinal cord compression as they get larger. Finding these lesions can be tricky because the cyst looks like CSF on MRI and CT and can "hide" within the surrounding CSF; carefully scrutinize the cord margins for any suggestion of focal mass effect or cord displacement. On MR, focal impression on the cord can be seen with intensity similar to that of CSF without enhancement (Fig. 16-10). CT myelography is helpful to make these cysts more apparent: the cyst does not opacify at the time of initial intrathecal injection and may or may not fill with contrast on delayed imaging.

As a caveat, displacement of the cord by an apparent "mass" can also be seen in cases of cord herniation syndrome, a rare condition most commonly seen in the thoracic spine, mimicking the imaging appearance of an arachnoid cyst. This can occur spontaneously, posttraumatically, or may be a postoperative complication usually becoming symptomatic on the order of years following the event. In this condition, a portion of the cord herniates through a dural defect producing symptoms. Clinical presentation includes myelopathy, radiculopathy, sensory deficits, or Brown-Séquard syndrome. Herniation secondary to previous spine surgery is very understandable particularly if the dura had been opened or torn. Here the cord herniates posteriorly through the defect. Trauma can also tear the dura and the arachnoid. Evidence of a nuclear trail sign (sclerotic changes in the vertebral endplate or disk representing the path of a previously herniated nuclear disk fragment) may suggest traumatic disk herniation resulting in ventral cord herniation. In those cases without surgery or trauma, a congenital dural defect has been implicated. MR demonstrates the cord to be small, rotated, and displaced with dilated CSF space opposite the herniation through the dura. As stated earlier, the appearance can

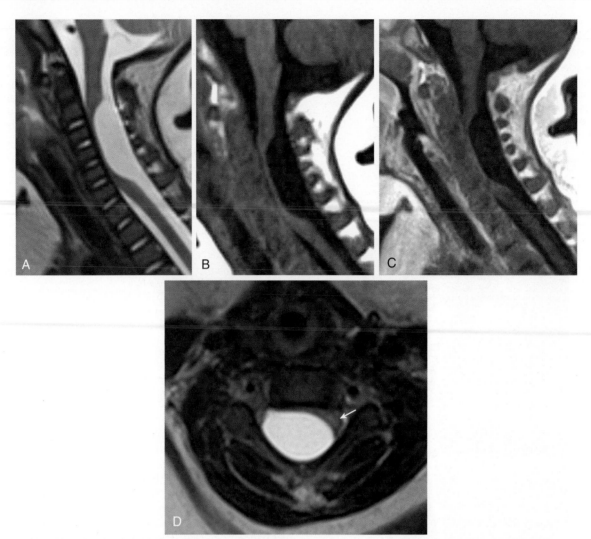

FIGURE 16-10 Arachnoid cyst. **A,** Sagittal T2-weighted imaging (T2WI) and **(B)** T1WI show dramatic ventral displacement of the cervical cord by a space-occupying lesion within the posterior spinal canal, which is isointense to cerebrospinal fluid. **C,** The lesion does not enhance following contrast administration on this axial T1WI. These findings indicate that this lesion is an arachnoid cyst. **D,** Axial T2WI shows the marked displacement and compression of the cord *(arrow)* resulting from mass effect. What a pain in the neck!

mimic that of an arachnoid cyst displacing the cord, and a CT myelogram may need to be performed to distinguish between these two entities (Fig. 16-11). High signal intensity on T2WI may be seen in the cord if strangulated. Surgery is usually beneficial.

Tarlov Cyst

Cystic dilatation of the sacral root pouches can be large and may be associated with bone remodeling (Fig. 16-12). They occur in approximately 5% of individuals and vary in size from a few millimeters to several centimeters. These are caused by a ball-valve phenomenon at the ostium of the nerve root sheath with CSF flowing into the cyst with arterial pulsation. After instillation of subarachnoid contrast these cysts may or may not fill and may or may not be symptomatic (as a cause of sciatica). Those that are symptomatic and do not readily communicate with the subarachnoid space may be

treated with fibrin glue ablation. The differential diagnosis includes meningocele, arachnoid cyst, neurofibroma, and dural ectasia.

Epidermoid Cyst

Epidermoid cysts represent less than 1% of intraspinal tumors, with a higher incidence in children. They are usually extramedullary but rarely can be intramedullary, and can be congenital or acquired. Congenital epidermoids result from displaced ectoderm inclusions occurring early in fetal life perhaps from faulty closure of the neural tube, usually in the lumbosacral region. The acquired cysts result from thecal sac instrumentation with inclusion of skin tissue in the spine. On MR, a discrete mass, which has a variable signal intensity depending on the cyst contents, is present. The lesion can calcify and rarely can be associated with peripheral enhancement. Inclusion of endoderm results in

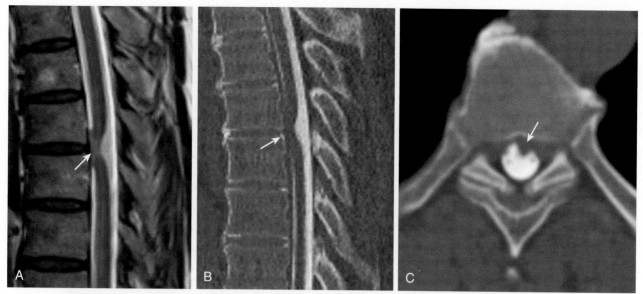

FIGURE 16-11 Thoracic cord herniation. **A,** Sagittal T2-weighted imaging shows ventral displacement of the spinal cord centered at a midthoracic disk level *(arrow)*. **B,** Sagittal and **(C)** axial images from a computed tomography myelogram again shows the cord displacement centered at a midthoracic disk level with contrast opacification of the intraspinal cerebrospinal fluid space. The cord appears adherent to the disk *(arrow)* through a dural defect that cannot be appreciated on these images. An arachnoid cyst can fill with contrast on myelogram, but opacification would progress in a more delayed manner. Time to call a surgeon!

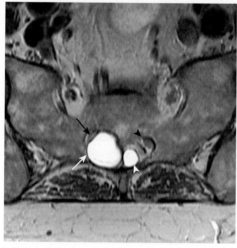

FIGURE 16-12 Tarlov cyst. On this axial T2-weighted image, there is a lobulated sacral root cyst with large foraminal component on the right *(white arrow)* and smaller canal component *(white arrowhead)*. There is remodeling of the bony sacrum due to chronic cerebrospinal fluid pulsations. Note the displaced and compressed appearance of the exiting sacral nerve root on the right *(black arrow)*, compared to the more normal appearance on the left *(black arrowhead)*.

an enterogenous cyst that is lined by mucin-secreting cells, which can produce contents with similar intensity (high signal on T1WI and T2WI) to mucoceles. They often are bright on diffusion-weighted imaging (DWI), as elsewhere in the central nervous system (CNS). These can be associated with developmental defects in the skin and vertebral bodies, and with fistulous communications to cysts in the mediastinum, thorax, or abdomen. Other cysts include dermoid cysts (which contain fatty elements) and ependymal cysts (which follow CSF intensity) (Fig. 16-13).

Syringomyelia (Syringohydromyelia) and Cysts within the Cord

In the fetus and in newborns, the central canal (in the central spinal cord gray matter) contains a small amount of CSF and is lined by ependymal cells. You should also know that the central canal extends from the obex of the fourth ventricle to the filum terminale and can be seen on T2WI even in adults from time to time (but usually <2 mm in diameter). In the region of the conus medullaris, the canal expands as a fusiform terminal ventricle, the ventriculus terminalis (fifth ventricle), which completely obliterates by approximately 40 years of age. There are probably variations in the patency of the central canal with maturation. States that alter the flow of CSF (Chiari malformations, arachnoid cysts, and adhesions) and/or that produce abnormal CSF pressure ultimately result in transmission of the fluid pressure into what is left of the central canal. How precisely this occurs is open to much speculation, but you can imagine that higher CSF pressure is transmitted along the perivascular spaces and into the interstitial spaces toward the central canal. This dilates the central canal remnant, and depending on where it is patent, is associated with the development of varieties of syringohydromyelia.

There are three types of syringohydromyelia. The first is a central canal dilatation that communicates with the fourth ventricle and is associated with hydrocephalus. These are produced by obstruction of CSF circulation distal to the outlets of the fourth ventricle. The second occurs in a region of the central canal that is dilated but does not communicate with the fourth ventricle. Here we are referring to Chiari I malformations, extramedullary intradural tumors, arachnoid cysts/arachnoiditis, cervical spinal stenosis, basilar invagination, and so forth (Fig. 16-14; Box 16-4). These abnormalities

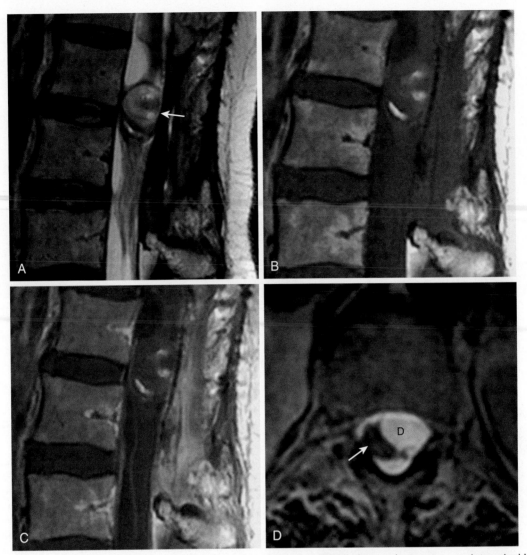

FIGURE 16-13 Lumbar dermoid. **A,** Sagittal T2-weighted imaging (T2WI) shows a heterogeneous intraspinal lesion at the L1 to L2 level *(arrow)*. Note the laminectomy defect at these levels (this lesion was partially resected in the past, so don't blame the surgeon for this one). **B,** On T1WI, high signal within the lesion indicates presence of fatty elements. **C,** On this sagittal postcontrast T1WI, no enhancement is seen. **D,** Note the mass effect and displacement of the conus *(arrow)* by the extramedullary dermoid lesion (D).

have also been termed hydromyelia. There may also be cases of asymptomatic localized dilatation of the central canal without any predisposing factors such as Chiari malformation; as long as the canal dilatation is 2 mm in diameter or less, this finding is unlikely to be of clinical significance.

The third type of syringohydromyelia differs from the first two in that it is centered eccentrically in the spinal cord parenchyma rather than the canal (Box 16-5). These are found in watershed regions of the spinal cord and are associated with direct spinal cord injury such as trauma, infection, or infarction. This is usually termed a syrinx.

There has recently been described a situation where alterations of CSF flow are recognized before syringomyelia occurs. The spinal cord becomes edematous and appears enlarged with high intensity on T2WI. This has been termed a presyrinx. This condition appears reversible if the condition producing the altered CSF pressure is treated.

Posttraumatic myelomalacia is a lesion with cord cavitation and volume loss, associated with significant spinal cord trauma including hemorrhage or infarction. Altered CSF dynamics from adhesions may predispose to development of syrinx formation in this myelomalacic cavity.

Cysts Associated with Neoplasia

Just to confuse the reader further, neoplasms of the spinal cord commonly have cysts associated with them. These cysts have been termed tumoral (or intratumoral) and nontumoral (reactive) (Fig. 16-15). Cysts rostral or caudal to the solid portion of the tumor (nontumoral) are secondary to dilatation of the central canal. They do not enhance, are not echogenic on intraoperative ultrasound, do not have septations, and usually disappear after resection of the solid lesion. Allegedly, fluid produced by the tumors dilates the central canal producing

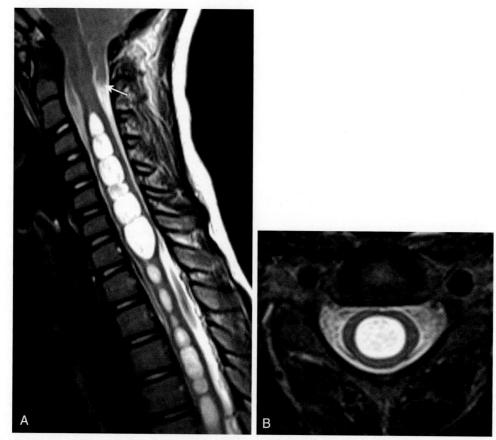

FIGURE 16-14 A, Chiari I malformation with syringohydromyelia. Sagittal T2-weighted imaging (T2WI) shows the cerebellar tonsils *(arrow)* extending well below the foramen magnum as well as a marked dilatation of the central canal throughout the imaged cord. **B,** Axial T2WI shows the cyst is centrally based, indicating dilatation of the central canal within the cord. The linked-sausage appearance is due to glial adhesions and is classic in syringohydromyelia.

the cyst. Tumoral cysts (more common in astrocytoma) are part of the tumor itself and may show peripheral enhancement.

INTRASPINAL LESIONS BY LOCATION

Intramedullary Lesions

MR has had a huge impact on the diagnosis of cord disease, including neoplastic diseases of the cord. Spinal cord tumors represent approximately 5% of CNS neoplasms. Neoplasms have three general characteristics: (1) they tend to enlarge the cord either focally or diffusely, (2) on proton density–weighted imaging (PDWI) and T2WI they produce high signal intensity, and (3) they frequently enhance. Tumors involving the cord may be primary, such as astrocytoma or ependymoma, or metastatic. It is virtually impossible to separate ependymoma from astrocytoma but we will provide tricks of the trade to improve the coin flip.

Box 16-6 provides the differential diagnosis of the enlarged spinal cord. This is an essential diagnosis to know. Memorize it! You must appreciate, however, that not all enlarged cords are neoplasms, and inflammatory and demyelinating diseases may enlarge the spinal cord as well. Especially in young patients with acute or subacute myelopathic symptoms, you should not glibly suggest that

---

BOX 16-4   Lesions Associated with Hydromyelia

Basilar invagination
Chiari malformations
Diastematomyelia
Klippel-Feil syndrome
Paget disease
Spinal dysraphism
Tethered cord syndrome

---

an enlarged spinal cord with high T2WI signal represents a tumor. Rather, think about the clinical context, assess presence of enhancement (absence unusual in tumors), get an MR of the brain, and assess whether there are additional lesions. In this patient population, corroborative findings might favor a demyelinating process, such as MS, over a spinal cord tumor.

Astrocytoma

These neoplasms make up approximately 40% of spinal tumors and usually occur in children (most common intramedullary tumor overall and in children), and adults in their third to fifth decade of life (with a mean age of 29 years). The presentation is of pain and paresthesias followed by motor signs. Males are affected more than females. The cervicothoracic cord is most commonly involved. In children, these lesions behave as grade I pilocytic astrocytomas with a good prognosis. Adults have a worse outcome related to the infiltrative nature of the lesion and more cervical predominance. Astrocytomas are graded I to IV with I being the pilocytic astrocytoma, II the low grade or fibrillary astrocytoma, III the anaplastic astrocytoma, and IV the glioblastoma. Most of the spinal cord astrocytomas are low grade (grade I and grade II nearly equal in incidence) with less than 2% being glioblastomas.

BOX 16-5 Lesions Associated with Syringomyelia

Arachnoid cyst
Arachnoiditis
Cord infarction
Myelitis
Pott disease
Spondylotic compression
Trauma
Spinal cord neoplasm
Intradural extramedullary neoplasms
Vertebral body tumors

These tumors may produce mild scoliosis, widened interpediculate distance, and vertebral scalloping but less than ependymomas. In general, cord astrocytomas are hypercellular, often large without obvious margins, and involve the full diameter of the cord, but are more eccentric than ependymomas. They are iso to low intensity on T1WI and high intensity on T2WI. The average length of involvement is seven body segments. They may have an associated cystic component (usually tumoral with small irregular or eccentric morphology), which may be appreciated on T1WI as hypointensity with rim enhancement (see Fig. 16-15). Hemorrhage is uncommon. Syrinx is more common in the pilocytic variety. Enhancement is the general rule, although it may be uneven compared with the intense enhancement of ependymoma. Malignant potential is generally less than that of brain astrocytomas. Exophytic components are sometimes present. The size of the tumor does not necessarily reflect its malignant potential. These tumors are associated with neurofibroma type 1 (NF1), are infiltrating, and not generally resectable.

Ependymoma

Ependymomas are generally more focal (involving an average of 3 to 4 vertebral body segments), although

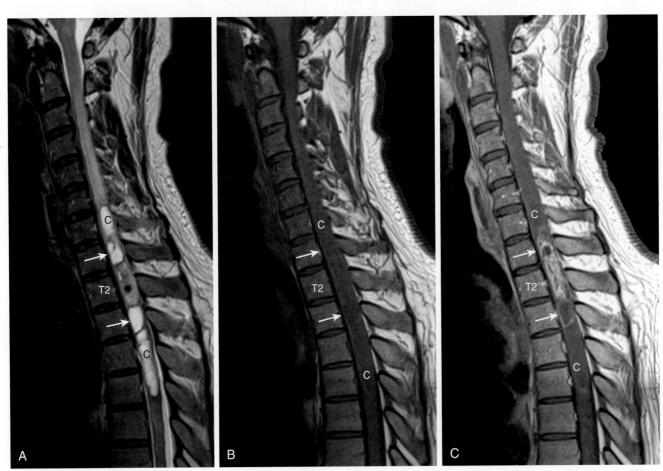

FIGURE 16-15 Neoplastic and nonneoplastic cysts in a patient with cord pilocytic astrocytoma. **A,** Sagittal T2; **B,** sagittal T1; and **C,** postcontrast sagittal T1-weighted imaging shows a mass within the cord centered at the T2 level. Low signal within the segment of tumor at the T2 level on T2-weighted imaging (T2WI) indicates small focus of hemorrhage. There are cysts above and below the level of the T2 mass. Those labeled with *arrows* show rim enhancement, indicating they are part of the tumor proper. Those labeled with "C" at the extreme ends of the mass do not show rim enhancement, and are therefore nontumoral cysts. Note edema within the cord extending to the C2 level on the T2WI.

they can be extensive (reported to involve 15 vertebral body segments), and as they arise from ependymal cells from the central canal, they tend to occupy the central portion of the cord. They account for 50% to 60% of spinal cord tumors in the third to fifth decade of life (most common intramedullary neoplasm in adults), and may be associated with neurofibromatosis type 2. Symptoms are generally mild, delaying diagnosis, with back or neck pain the most common complaint followed by sensory deficits and motor weakness, and bowel and bladder dysfunction.

They most commonly involve the cervical cord (32%) followed by the conus and cauda equina region (27%), followed by cervicothoracic (16%), thoracic (16%), thoracolumbar (5%), and cervicomedullary (3%) locations. Interestingly, although lower spinal ependymomas tend to be of lower histologic grade (World Health Organization grades I and II), recurrence rates are higher in these regions. However, tumor location does not appear to impact overall survival. They often produce a lobulated mass in the lumbosacral zone. Thus, these neoplasms can be both intramedullary and extramedullary (affecting the filum). Although unencapsulated, these glial neoplasms are well circumscribed, noninfiltrating, histologically benign with slow growth and, in certain circumstances, are totally resectable. The 5-year survival is greater than 80%. Metastatic spread to extraspinal sites is rare but includes the retroperitoneum and lymph nodes. Ependymomas outside the CNS (broad ligament and sacrococcygeal region) have a strong association with spina bifida occulta.

Ependymomas appear isointense to hypointense with regard to the cord on T1WI and may have a multinodular

---

BOX 16-6 Spinal Cord Enlargement

DEMYELINATING DISEASES

Acute disseminated encephalomyelitis
Multiple sclerosis
Transverse myelitis

INTRAMEDULLARY INFECTIONS

AIDS (see Box 16-1)

TUMOR

Metastases to cord
Primary spinal cord tumor

INFLAMMATION

Sarcoid
Systemic lupus erythematosus

SUBACUTE NECROTIZING MYELOPATHY

See Box 16-12 for causes of this condition

VASCULAR LESIONS

Acute infarction
Arteriovenous malformation
Cavernous angioma
Hemorrhage
Venous hypertension (Foix-Alajouanine syndrome)

SYRINGOHYDROMYELIA

AIDS, Acquired immunodeficiency syndrome.

---

high signal picture on T2WI that occupies the whole width of the cord. Commonly there is edema surrounding the tumor. Ependymomas have a propensity to hemorrhage, with hypointense rims ("cap sign") on T2WI, that histopathologically represent residual hemosiderin from hemorrhage (Fig. 16-16). Ependymomas may cause subarachnoid hemorrhage. Hemosiderosis has also been reported to result from ependymoma. Enhancement is common, and the tumor may have sharply defined, intensely enhancing margins. They may be associated with extensive cyst formation (in up to 84% of ependymomas—mostly nontumoral cysts but tumoral cysts may also occur), which does not usually enhance. Other associated radiographic findings include scoliosis, canal widening with vertebral body scalloping, pedicle remodeling, or laminar thinning.

Myxopapillary ependymoma (13% of all spinal ependymomas) may affect the filum with extension into the conus and the subcutaneous sacrococcygeal region. The mean age on diagnosis is 35 years and the lesion is more common in males. The presentation is of low back pain, leg pain, lower extremity weakness, and bladder dysfunction. These lesions are multilobulated, usually encapsulated, mucin-containing tumors that may hemorrhage and calcify. On MR, they are usually isointense on T1WI, high intensity on T2WI, and enhance (Fig. 16-17). However, as a result of calcification or hemorrhage they may be high or low intensity on T1WI and T2WI.

Hemangioblastoma

Hemangioblastomas are vascular lesions that may involve the cervical and thoracic spinal cord. They are the third most common primary intramedullary spinal neoplasm (1% to 7% of all spinal cord neoplasms). The tumor is composed of a dense network of capillary and sinus channels. They diffusely widen the spinal cord and may have both a cystic and solid component, with the solid component enhancing intensely. These tumors are associated with considerable edema (Fig. 16-18). Hemangioblastoma may be solid in 25% of cases.

Spinal cord hemangioblastomas may be an isolated lesion, but one third of cases are associated with von Hippel–Lindau disease and are multiple. von Hippel–Lindau disease has been isolated to a germline mutation of the *VHL* gene on chromosome 3. It has been suggested that one third of patients with von Hippel–Lindau disease have a spinal hemangioblastoma. Cord hemangioblastomas are more commonly thoracic than cervical and 43% have associated cysts.

On imaging, hemorrhagic components may be seen in the tumor nodule. There may be multiple lesions in the spine and they may be eccentric, at times appearing extramedullary, or pial based, and they can occur on the nerve roots. With this presentation, the question of dropped hemangioblastoma seed from a cerebellar lesion versus multiple primary spinal cord or nerve root lesions is always raised. Particularly in the setting of von Hippel–Lindau disease, favor multiple lesions, not seeds. Spinal angiography shows dilated arteries, a tumor stain, and draining vein. On MR, intratumoral flow voids can be identified when the lesion is reasonably sized and prominent posterior draining veins can also be seen. They are variable intensity on T1WI and high intensity on T2WI. Edema can be noted surrounding the lesions.

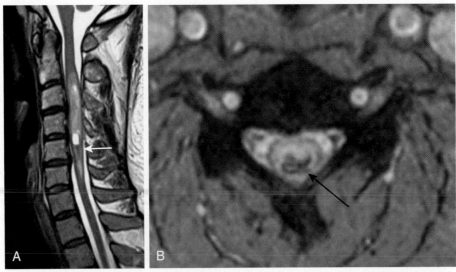

FIGURE 16-16 Ependymoma. **A,** Classic ependymoma in the cervical cord, with "cap sign" (low intensity inferiorly, *arrow*), which is more clearly demonstrated to be hemorrhage *(arrow)* on the gradient echo scan **(B).**

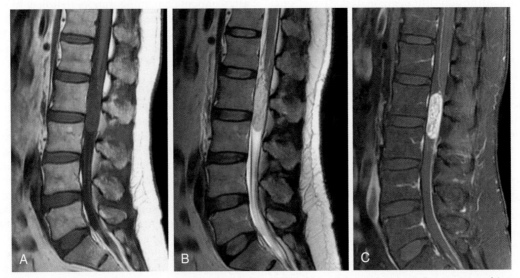

FIGURE 16-17 Myxopapillary ependymoma. **A,** Sagittal T1; **B,** T2; and **C,** postcontrast fat-suppressed images show the intradural extramedullary myxopapillary ependymoma, which is well defined and very bright on T2. The location is classic.

Avid homogeneous enhancement is visualized in the solid portion of the tumor.

Ganglioglioma/Gangliocytoma

The terms ganglioglioma and gangliocytoma are frequently used synonymously. Most important for radiologists is that they look the same, behave the same, and are treated the same in the spinal cord. These are rare tumors (1.1% of all spinal cord neoplasms), most frequent in children and young adults (average age of 12 years), consisting of neoplastic large mature neurons or ganglion cells and neoplastic glial cells. These lesions are slow-growing, relatively benign neoplasms with the majority occurring in the cervical cord and less commonly involving the thoracic cord, conus, or entire spinal cord. Findings in this tumor include involvement consisting of long segments of tumor (commonly extending to over eight vertebral body segments)

associated with scoliosis and bony remodeling, mixed signal on T1WI and high intensity on T2WI, prominent tumoral cysts, patchy enhancement that extends to the pial surface without central enhancement (15% of cases demonstrate no enhancement), lack of edema, hemosiderin, or calcification. Gross resection of the tumor is recommended.

Paragangliomas

The new kid on the block as far as intramedullary masses is the paraganglioma. This is an avidly enhancing well-defined mass of neuroendocrine origin usually seen in adults in the conus and cauda equina region. If you are lucky, you might see a "salt-and-pepper" appearance with the lesion and/or flow voids within the tumor or along the cord surface to help you bring this rare tumor to the forefront of your differential diagnosis.

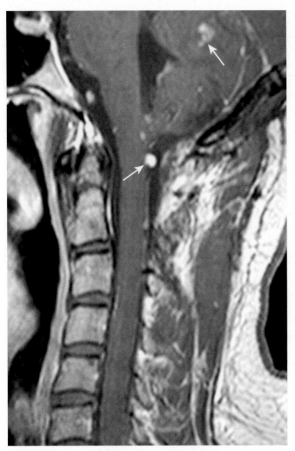

FIGURE 16-18 Hemangioblastoma. The nature of these enhancing lesions in the cerebellum and along the dorsal surface of the cord *(arrows)* on this sagittal postcontrast T1-weighted imaging would be elusive without the known diagnosis of von Hippel–Lindau disease.

Intradural Metastatic Disease

Metastases can deposit on the dura, pia-arachnoid region, and rarely in the cord itself (Fig. 16-19). The incidence of leptomeningeal and intramedullary metastases (up to 3% of patients with metastatic disease) is increasing as patients with cancer live longer. Patients may have nonspecific symptoms (headache, back pain) or focal neurologic symptoms. Intramedullary lesions result from tumor growth along the Virchow-Robin spaces because of CSF spread. Spinal cord metastases may also occur as a result of hematogenous dissemination. Intramedullary metastases enlarge the cord, are associated with cord edema, and enhance. Box 16-7 lists lesions that commonly metastasize to the spinal cord or its coverings.

Intraspinal metastases are usually low intensity on T1WI and high intensity on T2WI (prominent edema surrounding the tumor nodule) with avid homogeneous enhancement of the tumor nodule.

CSF spread of tumor in the leptomeninges is most often diagnosed by enhanced MR, but lumbar puncture in carcinomatous meningitis is abnormal, with positive cytology on serial punctures. MR reveals enhancement of the metastatic nodules, producing an irregular cord margin rather than its usual smooth contour. The entire subarachnoid space can enhance (sugar-coated appearance), and tumor nodules can be seen on nerve roots or in the cauda equina.

Pial metastases reveal linear enhancement on the surface of the cord; however, this can be observed in nonmalignant disease including sarcoidosis and infection. The CSF on T1WI can be more intense than normal and the conspicuity between CSF and cord may be diminished. Cerebral MR is necessary in this situation because leptomeningeal metastases may be the result of CSF seeding from brain parenchymal metastases.

Parameningeal masses are a preferred presentation for leukemia (termed chloromas), which can grow through the intervertebral foramina, a mode of spread also noted in both Hodgkin disease, non-Hodgkin lymphoma, and neuroblastoma. CNS lymphoma rarely occurs as an intramedullary lesion (3%). Leukemic cell spread to the CSF may be invisible on MR and better diagnosed by lumbar puncture.

Extramedullary Intradural Lesions

Meningioma

Meningiomas are well-circumscribed, globular lesions with a female predominance and constitute approximately 25% of spinal canal neoplasms. They are usually located in the thoracic region (approximately 80%) but can occur in any location throughout the spine. They may coexist with schwannomas in patients with neurofibromatosis type 2. The vast majority are extramedullary intradural, but in rare cases they can be both intradural and extradural or purely extradural (higher tendency to be malignant).

These tumors are isointense to slightly hypointense on T1WI and most of the time these typical intradural extramedullary lesions are obvious on nonenhanced images (Fig. 16-20). Meningiomas enhance and may show a dural tail. On T2WI, they are isointense to slightly hyperintense relative to the spinal cord and are usually well demarcated by the brighter CSF. They displace the spinal cord to the opposite side and enlarge the CSF above and below the tumor on the ipsilateral side. Meningiomas are dural based in location, and may compress the cord but do not invade it. Meningiomas may demonstrate calcification and may widen the neural foramen, but far less often than peripheral nerve sheath tumors. Other characteristics that distinguish meningiomas from nerve sheath tumors include: (1) position, with meningiomas more typically posterolaterally and nerve sheath tumors anteriorly located within the canal; (2) a tendency for multiplicity of nerve sheath tumors; and (3) low intensity central regions on postgadolinium T1WI and T2WI can be seen with nerve sheath tumors.

Nerve Sheath Tumors: Schwannoma and Neurofibroma

Two types of benign nerve sheath tumors can be distinguished histopathologically: the schwannoma and the neurofibroma. Schwannomas (also termed neurinoma or neurilemmoma) are encapsulated masses arising from a focal point. They may be associated with hemorrhage, cysts, and fatty degeneration. Neurofibromas consist of diffuse proliferation of Schwann cells and are well circumscribed but not encapsulated. Collagen is a conspicuous element in the neurofibroma as opposed to the schwannoma. Unlike schwannoma, they demonstrate little proclivity toward hemorrhage, vascular change, or fatty degeneration.

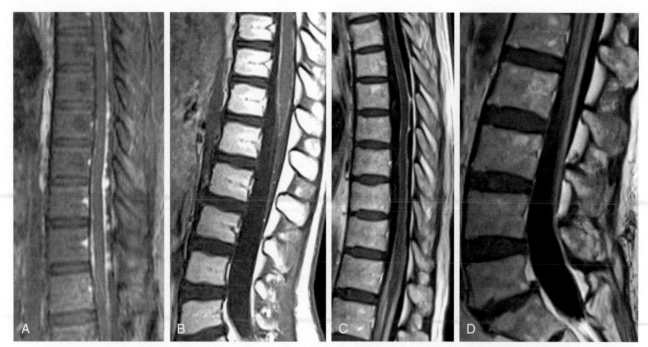

FIGURE 16-19 Leptomeningeal metastases. **A,** Enhanced fat-suppressed sagittal T1-weighted imaging (T1WI) shows diffuse linear and nodular enhancement on the posterior and anterior surface of the spinal cord in this patient with metastatic myxopapillary ependymoma. The appearance is likened to "sugar coating," or as the Germans say, zuckerguss. **B,** On this postcontrast T1WI, leptomeningeal spread along with dropped metastases along the cauda equina nerve roots are shown in this patient with metastatic medulloblastoma. **C** and **D,** Smooth linear enhancement along the surface of the cord and cauda equina on this contrast-enhanced sagittal T1WI in this patient with breast cancer indicates metastatic leptomeningeal involvement. Cancer stinks, no matter how much you zuckerguss it.

In patients with NF1, all spinal nerve root tumors are neurofibromas; however, neurofibromas can occur in patients without NF1. Schwannomas are the most common neurogenic tumors of the spine in all comers. In patients with neurofibromatosis type 2, almost all spinal nerve root tumors are schwannomas or mixed tumors rather than neurofibromas. Schwannomas are seen as solitary lesions (except in neurofibromatosis type 2), whereas neurofibromas are often multiple and occur with neurofibromatosis.

Schwannomas may occur in a purely intradural location (two thirds of cases), be partially intradural and extradural (one sixth), or purely extradural (one sixth). Peripheral nerve sheath tumors can assume a dumbbell shape; however, this configuration is not specific to nerve sheath tumors and may be seen in benign and malignant spinal masses not necessarily of neural origin. As a general rule, benign lesions tend to have well-defined margins and expand/remodel the intervertebral foramina in contrast to malignant lesions with irregular margins and bony destruction.

It is impossible to separate schwannoma from neurofibroma by imaging, and thus they are clumped together. They are usually isointense on T1WI and are hyperintense on T2WI. They also enhance. Areas that are isointense or of high intensity on T1WI and low signal on PDWI and T2WI have been associated with hemorrhage (Fig. 16-21). A low intensity area ("dot") has been seen on postcontrast T1WI and thought to increase specificity for the diagnosis of peripheral nerve sheath tumor. The same is true of the "target sign" (no radiology text would be complete without one!) in which there is low intensity centrally on T2WI

> **BOX 16-7** Tumors that Commonly Metastasize to the Spinal Cord or Leptomeninges
>
> **CHILDREN**
> Lymphoma, leukemia
> Medulloblastoma
> Pineal region tumors (e.g., germinoma, pineoblastoma)
> Choroid plexus tumors
> Ependymoma
> Neuroblastoma
> Retinoblastoma
>
> **ADULTS**
> Lymphoma, leukemia
> Glioblastoma
> Hemangioblastoma
> Melanoma (primary or metastasis)
> Metastases from lung > breast > melanoma > colorectal > renal > gastric carcinomas
> Oligodendroglioma

surrounded by brighter peripheral signal. The substrate of this nonenhancing area includes calcification, old hemorrhage, fibrosis, foam cells, and microcysts. Calcification is rare compared with meningioma.

Other findings that are associated with neurofibromatosis include enlarged neural foramina and vertebral scalloping secondary to dural ectasia, lateral meningocele, or an arachnoid cyst. Lesions having a diffuse multinodular appearance have been termed plexiform neurofibromas, being either schwannomas or neurofibromas, and are

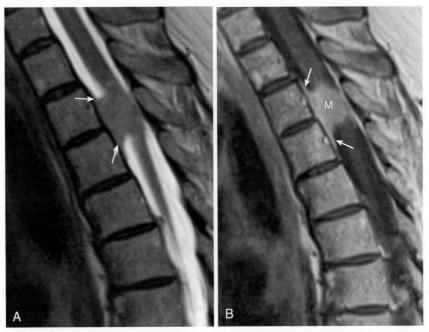

FIGURE 16-20 Meningioma. **A,** There is an intermediate intensity intradural extramedullary mass in the thoracic spinal canal on sagittal T2-weighted imaging (T2WI). Note the dural attachment *(arrows)*. Cord compression is present, and there is faint T2 high signal within the cord above and below the lesion indicating edema. **B,** On contrast-enhanced sagittal T1WI, the mass enhances (M) as does the dural tail *(arrows)*.

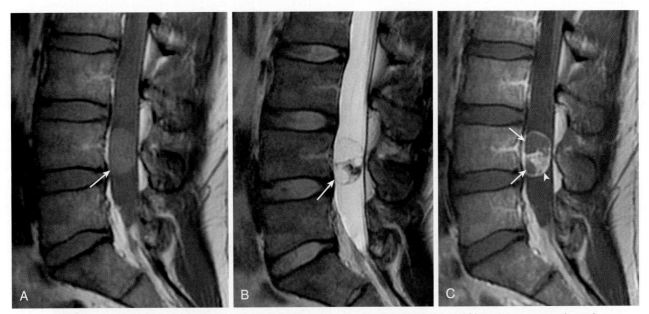

FIGURE 16-21 Schwannoma. **A,** Sagittal T1-weighted imaging (T1WI) shows an ovoid hypointense mass *(arrow)* at the L4 level. **B,** Sagittal T2WI shows that the mass *(arrow)* has heterogeneous components, with areas of low T2 signal indicating hemorrhage. **C,** Following contrast administration, the lesion is shown to have solid *(arrowhead)* and cystic *(arrows)* components, a not uncommon feature of schwannomas.

usually associated with neurofibromatosis. Lateral meningoceles may be associated with wormian bones, malar hypoplasia, developmental delay, short stature, and spontaneous intracranial hypotension.

Hereditary Motor and Sensory Neuropathies

Hereditary motor and sensory neuropathies (HMSN) comprise a group of diseases, in some cases, associated with hypertrophic peripheral and/or cranial nerves. We have

excluded HMSN 2 and 4 because of the lack of imaging findings—we think. These diseases are characterized by concentric proliferation of Schwann cells, interspersed with collagen, in response to multiple episodes of demyelination and remyelination. This process results in an "onion bulb" appearance to the nerve. Patients present with a variety of signs and symptoms depending on the particular disease. On MR, enlargement and, at times, enhancement of the nerve roots can be seen.

Charcot-Marie-Tooth disease (CMT or HMSN 1) is usually an autosomal dominant disease that presents with slowly progressive distal atrophy (common peroneal muscular atrophy) and weakness in conjunction with pes cavus and scoliosis. The posterior columns, optic nerve, and acoustic nerve may be involved. Enlarged peripheral nerve roots can be seen on MRI (Fig. 16-22). There may be minimal or no abnormal enhancement seen.

Dejerine-Sottas disease (HMSN 3) is an autosomal recessive condition with slowly progressive motor and sensory loss and ataxia. Scoliosis and pes cavus are frequent. There are enlarged peripheral and cranial nerves with hypomyelination. Imaging findings included enlarged nerves best demarcated on T2WI with variable enhancement.

Inflammatory Neuropathies

Chronic inflammatory demyelinating polyradiculoneuropathy (CIDP) is an acquired demyelinating condition of peripheral nerves characterized by the slow onset of proximal weakness, paresthesias, and numbness. Cranial and peripheral nerves may be enlarged and can enhance (Fig. 16-23). Imaging findings are often bilateral and symmetric. CIDP is often associated with lesions in the brain that appear similar to those of MS. CIDP is considered to be a chronic version of Guillain-Barré syndrome.

Guillain-Barré syndrome is an autoimmune disorder that can occur following viral or bacterial infection (most commonly *Campylobacter jejuni* infections). Although Guillain-Barré syndrome favors motor neurons, sensory and autonomic nerves can also be affected. Patients typically present with symptoms ranging from ascending weakness to flaccid paralysis. Several subtypes exist, including acute inflammatory demyelinating polyradiculopathy, axonal subtypes, and regional syndromes (including the Miller-Fisher variant in which ataxia and areflexia are the dominant features). In patients with Guillain-Barré syndrome, enlarged and enhancing nerve roots (preferentially ventral nerve roots but dorsal nerve roots can also enhance), usually at the cauda equina level (Fig. 16-24), are seen on MRI.

Additionally, paraneoplastic syndromes have been reported to enlarge cranial and peripheral nerves produced by a microvasculitis associated with inflammatory changes.

Lipoma

Lipomas (Fig. 16-25) are usually seen in the first three decades of life and are most common in the thoracic region but can be seen throughout the spine. In the lumbar region, they are associated with myelodysplasia or tethered cord (see Chapter 8). They are commonly identified in the filum terminale (1.5% to 5% prevalence). Lipomas may be intradural (60% of cases) or extradural (40%). Imaging characteristics include low density on CT or high signal on T1- and T2-weighted images. With fat suppression, they appear dark. These lesions can be extensive and compress the spinal cord. Lipomas could be confused superficially with hemorrhage, but signal intensity is suppressed with fat suppression techniques, whereas hemorrhage is not. Chemical shift artifact can be identified along the frequency encoding axis.

If you observe fat in a sacrococcygeal mass, consider the diagnosis of teratoma. There is a female predominance and a high incidence (as high as 60%) of malignant transformation, especially in male patients. There is also a high association of other anomalies, including anorectal ones. Lesions present as a large expansile mass, with fatty, cystic, and solid components, and may even be hemorrhagic. Elevated levels of α-fetoprotein are associated with the tumor. Teratomas may also occur within the spinal cord.

Extradural Lesions

The vertebral bodies are the most frequent site of metastatic disease to the spine. On plain films and CT, most metastases are osteolytic or a combination of osteolytic and osteoblastic. On MRI, the normal high intensity on T1WI from fatty marrow is replaced by lower signal intensity from tumor cells (Fig. 16-26). Short tau inversion recovery (STIR) imaging shows tumor infiltration well. Later there is destruction of the bony trabeculae, usually without periosteal response. MR is thus more sensitive in most cases at diagnosing metastatic disease than plain films or CT. On conventional T2WI, the tumor-replaced marrow may appear as high

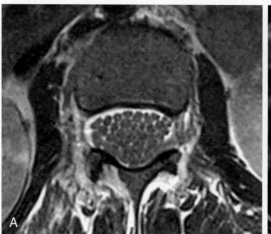
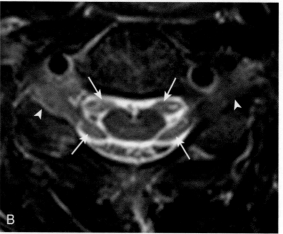

FIGURE 16-22 Charcot-Marie-Tooth disease. **A,** Markedly thickened lumbar nerve roots are present in this patient with Charcot-Marie-Tooth disease. **B,** In this same patient, even the ventral and dorsal nerve roots arising from the cervical cord are massively enlarged *(arrows)* as are the exiting nerve roots within the neural foramina *(arrowheads)*. Holy smokes, those nerves are huge!

signal intensity compared with the normal fatty marrow with lytic lesions, or may appear as lower signal intensity with sclerotic lesions. Metastatic disease is most conspicuous therefore on unenhanced T1WI, and may be difficult to perceive on T2 or fat-suppressed T2 images. Common cancers associated with spinal metastasis include breast, prostate, lung, and kidney. Other lesions include Ewing sarcoma, neuroblastoma, melanoma, lymphoma, leukemia, multiple myeloma, and sarcoma. Pelvic lesions metastasize through the epidural plexus of veins (Batson plexus) to the thoracolumbar bodies. Lymphoma rarely can present as an epidural mass without bone involvement.

Multiplicity of lesions in the vertebral column should raise suspicion for metastatic disease. Oftentimes, the primary tumor is known and metastatic disease is suspected based on patient symptomatology. In these cases, it is important to not only identify metastatic lesions to bone but to also assess for complications of metastatic disease. This includes an assessment for pathologic fracture, extraosseous soft-tissue extension into the perivertebral and intraspinal spaces, cord compression, and cord signal changes.

Osteoporotic Compression Fractures

Osteoporosis is characterized by decreased bone mass. It is associated with compression fractures that are spontaneous or associated with minimal trauma. These are usually located in the midthoracic region and thoracolumbar junction. Osteoporosis is the most common cause of compression fractures and is particularly observed in elderly women. Their prevalence in women older than 50 years has been estimated to be 26%. The pain from an acute fracture usually lasts 4 to 6 weeks and is associated with exquisite pain at the fracture site. The spinal cord may be compressed in these fractures.

It is difficult to distinguish between acute osteoporotic vertebral collapse and acute collapse from metastatic disease (Fig. 16-27). Both may be associated with pain, and compress the spinal cord. Box 16-8 provides CT criteria

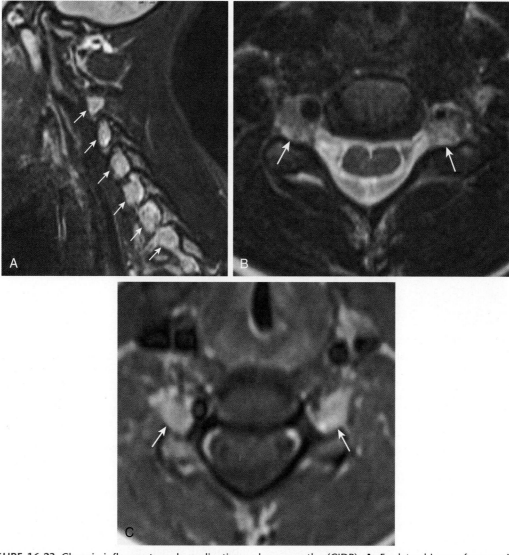

FIGURE 16-23 Chronic inflammatory demyelinating polyneuropathy (CIDP). **A,** Far lateral image from sagittal short tau inversion recovery sequence shows marked enlargement of the cervical nerve roots as they exit the neural foramina *(arrows).* **B,** Axial T2-weighted imaging (T2WI) again shows the marked enlargement of the exiting cervical nerve roots *(arrows),* and **(C)** postcontrast axial T1WI shows these nerve roots to enhance avidly *(arrows)* in this patient with CIDP.

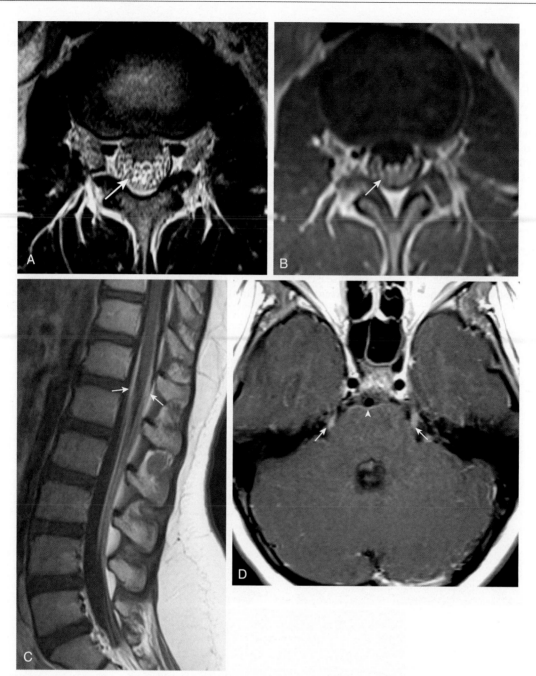

FIGURE 16-24 Guillain-Barré syndrome (GBS). **A,** Axial T2-weighted imaging (T2WI) shows enlargement of the cauda equine nerve roots *(arrow)*. (The dark signal in the ventral aspect of the thecal sac is due to cerebrospinal fluid pulsation artifact—ignore that.) **B,** Axial and **(C)** sagittal postcontrast T1WI show marked enhancement of the cauda equine nerve roots *(arrows)*. **D,** Intracranially, abnormal enhancement of the prepontine segment of the trigeminal nerves *(arrows)* and leptomeninges *(arrowhead)* is also seen in this patient with GBS.

that have been reported to be useful in the separation of benign versus malignant acute compression fracture. The presence of all bone fragments from the original cortex (puzzle sign) is both very constant and quite specific for benign disease. It is interesting that sclerosis, marked comminution of cancellous bone of the vertebral body, Schmorl nodes, or fracture of a pedicle have low specificity for separating benign from malignant disease.

On MR, both nonpathologic and pathologic compression fractures may have low signal on T1WI and high signal on T2WI, and enhance. Both could involve multiple vertebral bodies and be associated with soft-tissue

masses. A paraspinal mass is more likely to be neoplastic; however, soft-tissue hematoma associated with benign compression fracture can make the distinction difficult. Diffuse replacement of the entire vertebral body marrow signal on T1WI strongly suggests malignant disease, whereas compression fractures are more often confined to the endplates. Cortical destruction, abnormal signal throughout vertebral bodies that are not compressed, and involvement of the posterior elements all suggest malignant disease. In the presence of other metastatic lesions in the spine, the collapsed vertebra most likely results from metastatic disease.

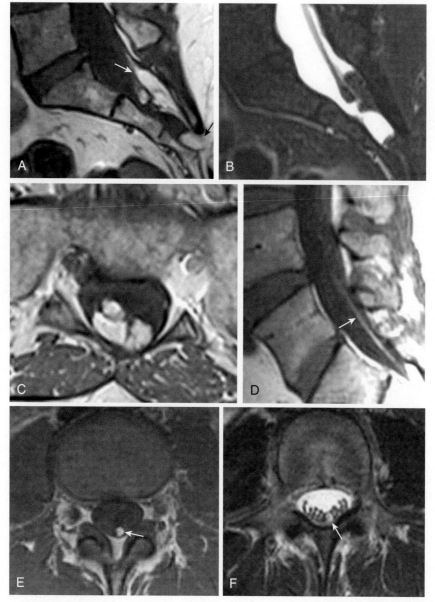

FIGURE 16-25 Fatty intraspinal masses. **A,** Extramedullary lipoma *(white arrow)* is seen on this sagittal T1-weighted image (T1WI) with inferior extent of the lipoma extending through a bony defect in the dorsal sacrum *(black arrow)*. **B,** The lipoma signal suppresses completely on short tau inversion recovery image. **C,** The lipoma is again shown on axial T1WI. **D,** This sagittal T1WI in a different patient shows incidental asymptomatic fatty filum *(arrow)*. Fatty filum in a different patient seen as T1 hyperintensity **(E)** and T2 hyperintensity **(F)**, paralleling the course of the cauda equina *(arrows)*.

Recently, DWI has been applied in an attempt to distinguish benign from malignant fractures; if restricted diffusion is definitively identified, it is likely from hypercellular tumor. Features that might help distinguish nonpathologic and pathologic compression fractures on MRI are summarized in Box 16-9.

In the case of a single collapsed vertebral body in a patient with or without metastatic disease, a repeat examination in 4 to 6 weeks is recommended. In this interval, benign collapse progresses to isointensity on T1WI and T2WI, with decreased enhancement, whereas metastatic disease remains stable or progresses without a change in intensity. Although this waiting may be viewed as a cop-out, many times it is not possible to distinguish between acute, nontraumatic benign and

metastatic vertebral body collapse. Naturally, the patient will have a full workup for occult malignancy, including bone scan to determine whether there are other lesions. If waiting is not an option (and it often isn't), bone biopsy is the only definitive test for distinguishing benign versus malignant disease.

The list of diseases that can produce vertebral compression includes metastases, osteomyelitis, congenitally deficient vertebral endplates, benign and malignant primary vertebral neoplasms, osteomalacia, hyperparathyroidism, hemochromatosis, myeloproliferative disorders, and trauma.

Neural Crest Tumors

Tumors of neural crest origin occur in infancy and arise from the sympathetic plexus. These include

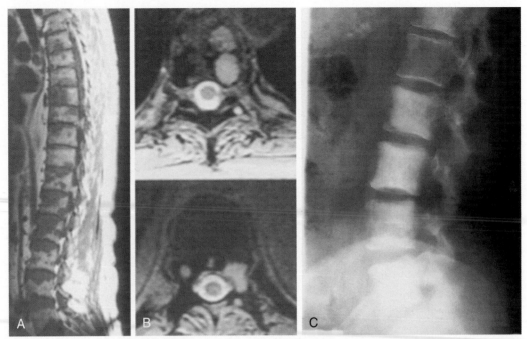

FIGURE 16-26 Metastatic lesions. **A,** Note the low-intensity lesions outlined by residual fat in the vertebral bodies on this sagittal T1-weighted imaging (T1WI) scan in a patient with breast metastases. **B,** Same patient as **A;** axial T2WI scan of one vertebra reveals multiple high-intensity metastases in the vertebral body and pedicles. **C,** Lateral spine radiograph in a different patient with metastatic breast cancer shows extensive osteoblastic metastases.

neuroblastoma (patients <5 years old), ganglioneuroblastoma (5 to 8 years old), and ganglioneuroma (>8 years old) (Fig. 16-28). Calcification occurs in 30% of cases, and hemorrhage has been reported. These paravertebral masses, which usually occur in the thoracic region, can extend through the neural foramina to compress the thecal sac (yet more dumbbells…like us!). They all enhance. Fifty percent of paraspinal neuroblastomas enter the spinal canal.

When evaluating paraspinal masses, extramedullary hematopoiesis should probably cross your mind. This process occurs when normal erythropoiesis in the marrow fails due to disorders in hemoglobin production (like sickle cell disease) and myeloproliferation (like myelofibrosis or chronic myelogenous leukemia), prompting red cell production outside the marrow compartment. Hepatosplenomegaly is more commonly seen, but more rarely, extramedullary hematopoiesis can be seen in the chest as well defined, lobulated, unilateral, or bilateral paraspinal soft tissue masses (see Fig. 16-28).

Hemangioma

Hemangiomas are common benign lesions of the vertebral body and are most often incidental findings. They are composed of fully developed adult blood vessels. More than 50% of solitary hemangiomas occur in the spine and they have been found in approximately 12% of spines at autopsy. Multiple hemangiomas can occur in approximately one third of cases. There is a 2:1 female/male predominance, with most lesions found in the lower thoracic and lumbar vertebrae.

These lesions most often present in the vertebral bodies as round geographic lesions but rarely can be extensive, replacing the entire vertebral body with extension into the pedicles, arches, and spinous processes. The cortical margins are usually distinct. On CT and plain films, thick trabeculae with a striated appearance is seen. On spinal angiography, these lesions are vascular. The vertical striations represent vascular channels interspersed with thickened trabeculae. MR reveals high signal intensity on T1WI and (fast spin echo) T2WI because of their fat and water content (Fig. 16-29).

While benign, vertebral bodies with hemangiomas can have compression fractures or epidural extension and compromise the spinal canal or neural foramen. Curiously, with extraosseous extension, the signal intensity of the exophytic portion of the lesion may be dark on T1WI and high on T2WI. Problems occur when these lesions compress the spinal cord (myelopathy) or when they become hypervascular and produce venous hypertension. Most hemangiomas are asymptomatic; however, pain is associated with growth (especially in women during pregnancy) and compression fracture. Very rarely, they can be aggressive and show extraosseous extension, producing myelopathy from cord compression.

The differential diagnosis is focal fatty marrow replacement. This can be distinguished on standard T2WI by noting decreased intensity of the focal fatty lesion and high signal in the nonfat-containing regions of the hemangioma. If fast spin-echo imaging is used, then the fat and hemangioma are both bright on the T2WI, so that fat saturation (as in STIR imaging) should be used. Not infrequently we see atypical hemangiomas that simulate bone metastases by having low signal on T1WI and brightness on T2WI. Even with these unusual signal intensity combinations, the CT appearance still is classic for hemangiomas and the diagnosis can be made with greater confidence when CT is performed in conjunction with MRI. Otherwise patients may need to go to biopsy ("bloody hell") or undergo follow-up imaging to assess for interval change.

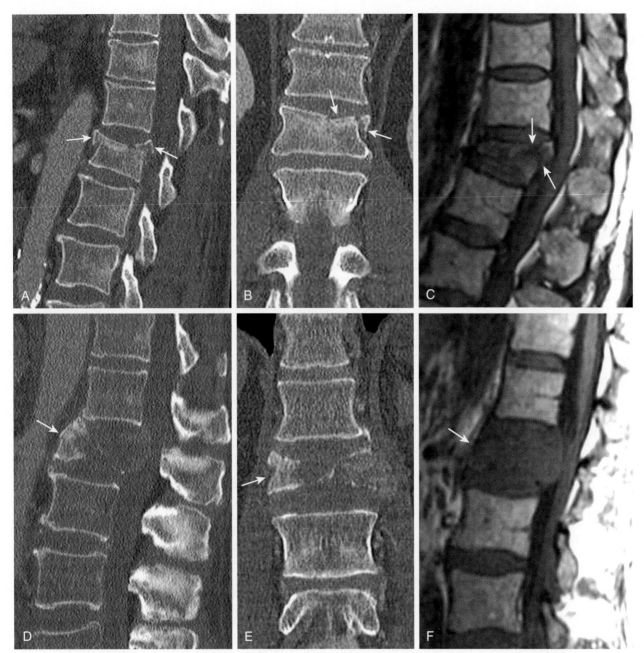

FIGURE 16-27 Osteoporotic compression versus pathologic compression fractures. **A,** Sagittal and **(B)** coronal reconstructions from noncontrast computed tomography (CT) show acute compression deformity with multiple superior endplate fracture fragments that can be pieced together like a puzzle *(arrows)*, a very good indicator of benign fracture. **C,** Sagittal T1-weighted imaging (T1WI) in a different patient with benign fracture also shows compression deformity with fragments that could conceivably be pieced together *(arrows)*. Compare that case to this one in which **(D)** sagittal and **(E)** coronal reconstructions from noncontrast CT show extensive disruption and compression deformity of a single vertebral body *(arrows)*, but the margins of the fracture fragments are not clearly seen and lack the "puzzle" configuration of a benign fracture. **F,** Sagittal T1WI in this same patient shows that the marrow space at this level is completely replaced by soft tissue tumor *(arrow)*, without discernible fracture plane. This is a pathologic compression fracture resulting from metastatic disease.

Chordoma

Chordomas are slow-growing, locally invasive neoplasms derived from remnants of the notochord, with approximately 50% originating in the sacral region and 15% affecting the spine (particularly the cervical region). In the spine, the tumor mass is associated with lytic lesions of the bone, at times with a sclerotic rim and calcification. There is a spectrum of intensity patterns with most lesions demonstrating very high signal on T2WI. These lesions may involve the disk space and multiple vertebral bodies. In the differential diagnosis, think about chordoma in bizarre spinal lesions that resemble large herniated disks, schwannoma, or unusual osteomyelitis (Fig. 16-30).

Chondrosarcoma

Chondrosarcoma may also occur in the spine and is associated with destruction of bone in the body or posterior elements. The hallmark of this lesion is chondroid calcification, which is much better detected on CT or plain film.

Aneurysmal Bone Cyst, Giant Cell Tumor, Osteoblastoma, and Osteoid Osteoma

Benign bony tumors of the spine include aneurysmal bone cyst, giant cell tumor, osteoblastoma, and osteoid osteoma and are best imaged on CT and plain films (by musculoskeletal radiologists, ahem). These lesions have some characteristics that may help distinguish them from one another. Osteoblastoma (>1.5 cm) is larger than osteoid osteoma (<1.5 cm) and is much more common in the spine. Both have a propensity for the posterior elements or transverse process, have a lytic and/or calcific nidus surrounded by a sclerotic rim, and are associated with pain, which in the case of osteoid osteoma is classically described as nocturnal and relievable by aspirin (Fig. 16-31). Giant cell tumors are lytic lesions found commonly in the sacrum and vertebral bodies; tumor margins can be thin or dehiscent and soft-tissue component of the mass may be present. Aneurysmal bone cysts are usually circumscribed multiloculated

lytic lesions, occasionally containing hemorrhage (with fluid levels) and involving the posterior elements, particularly the lamina (Fig. 16-32). Some of these entities coexist within the same lesion or evolve into each other (enough said—read MSK REQUISITES).

Eosinophilic Granuloma

Eosinophilic granuloma (part of the Langerhans cell histiocytosis spectrum) affects children and when seen in the spine predominantly involves the vertebral body. It is a rapidly growing tumor, producing bone loss and vertebral collapse associated with normal vertebral disk spaces (vertebra plana) (Fig. 16-33). Other lesions that can cause vertebra plana are summarized in Box 16-10.

There may be associated soft-tissue extension from the vertebral body posteriorly into the vertebral canal or anteriorly into the prevertebral tissues. MR shows a collapsed vertebral body with heterogeneous intensity on T1WI. A bullet-shaped vertebral body in Hurler (mucopolysaccharidosis I) and Morquio (mucopolysaccharidosis IV) diseases may simulate vertebra plana of eosinophilic granuloma.

Osteochondroma

These lesions usually arise from the posterior elements and commonly involve the thoracic or lumbar vertebrae. They are commonly found in teenagers where they grow and occasionally cause neurologic symptoms. On MR, these lesions are heterogeneous showing high intensity on T2WI (from the cartilaginous portion of the lesion) and low intensity on all pulse sequences from the ossified part of the tumor.

Paget Disease

Paget disease commonly involves the spine and produces enlargement of all the vertebral elements. It usually demonstrates thickened bone cortex with disorganized trabecular pattern (distinguishing it from fibrous dysplasia), and sclerosis (late phase), with low intensity on T1WI and T2WI. An early lytic phase has also been described (with osteoporosis circumscripta when seen in the skull). Involvement is associated with back and neck pain and neurologic dysfunction associated with facet arthropathy, lateral recess syndrome, and stenosis of the spinal canal. It is difficult at times to distinguish from osteoblastic metastatic disease because Paget disease is polyostotic in greater than two thirds of cases. There is the possibility of malignant transformation to osteogenic sarcoma in approximately 10% of cases: the rate of malignancy is 20 times greater in pagetoid bone than in nonpagetoid bones. The appearance of osteogenic sarcoma is variable, depending on whether the lesion is lytic, blastic, or mixed. Osteoid matrix or calcification appears as a signal void on all MR pulse sequences.

Multiple Myeloma

In multiple myeloma, three patterns of marrow involvement have been observed on sagittal MR. Both symptomatic and asymptomatic patients may reveal these patterns early in the course of the disease. The most common type reveals focal lesions with low intensity on T1WI and high intensity on T2WI (Fig. 16-34). Rarely, lesions may be hyperintense on T1WI, probably because of hemorrhage within the myeloma foci. A second pattern consists of a variegated (inhomogeneous)

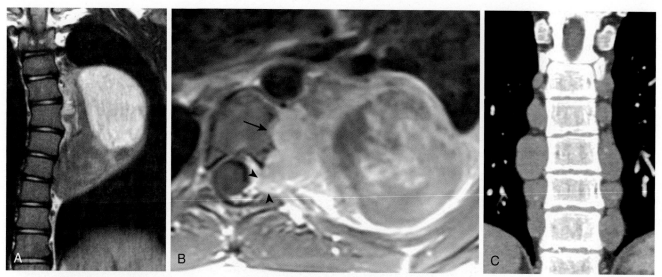

FIGURE 16-28 A, Coronal T2-weighted image (T2WI) on an 8-year-old child shows a bulky mixed intensity left paraspinal mass with associated scoliosis of the adjacent thoracic spine. **B,** Axial postcontrast T1WI shows the same bulky mass scalloping the adjacent thoracic vertebral body *(arrow)* and sneaking its way into the spinal canal (between *arrowheads*) through an expanded neural foramen. This was a ganglioneuroma. **C,** Contrast that with this case, where there are multiple bilateral paraspinal masses. Before you jump into that neural crest tumor differential, you should know this is an adult patient with thalassemia. Yep, this is extramedullary hematopoiesis. Sometimes history helps a lot.

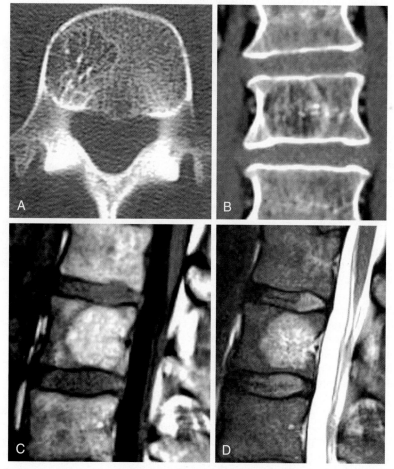

FIGURE 16-29 A, Hemangioma. Unenhanced computed tomography (CT) shows the characteristic speckled appearance due to vertical trabecular striations, with low-density areas and high-density spicules in this hemangioma in the right aspect of this vertebral body. **B,** Coronal CT reconstruction in a different patient shows the vertical trabeculated striations, which accounts for the "corduroy" appearance initially described on plain radiograph. **C,** Sagittal T1-weighted imaging (T1WI) in another patient (we mentioned these were common, right?) shows a high-intensity bony mass in the vertebral body. **D,** The lesion is also bright on T2WI.

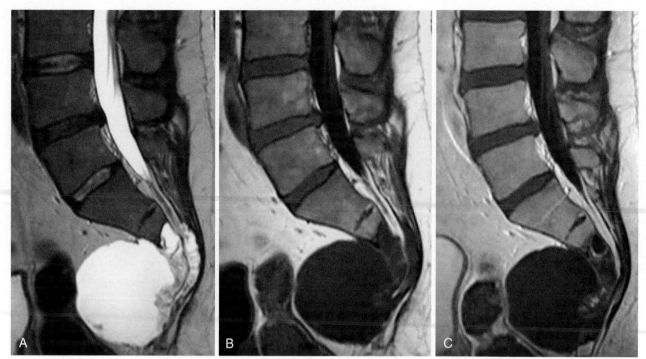

FIGURE 16-30 Sacral chordoma. **A,** Sagittal T2-weighted imaging (T2WI) shows a very hyperintense mass arising from the lower sacral elements and bulging into the presacral space. The T2 hyperintensity is characteristic for chordoma. **B,** Sagittal T1 and **(C)** postcontrast sagittal T1WI show little to no enhancement within the lesion. Histopathology confirmed chordoma.

appearance with tiny foci of hypointensity on T1WI and hyperintensity on T2WI. The third diffuse pattern represents total marrow involvement so that on T1WI the disks are hyperintense to the replaced fatty marrow. The diffuse and variegated appearance on T1WI can be difficult to distinguish from inhomogeneous distribution of fat in older patients. Vertebral compression may occur without the marrow being diffusely abnormal on MR and have a vertebra plana pattern. Vertebral compression fractures occur in 50% to 70% of patients and spinal cord compression in 10% to 15% of these cases. Rarely, multiple myeloma can involve the leptomeninges with enhancement of the nerve roots.

Plain films and CT demonstrate purely lytic lesions or diffuse osteopenia. It is unusual for myeloma to involve the pedicles. Consider myeloma in a patient with diffuse osteopenia and compression fractures (particularly male patients). On CT, destruction of cortical bone is much less common in myeloma than metastatic disease, whereas cancellous vertebral body destruction occurs in both metastatic and myelomatous lesions. Remember that myeloma is one of the diseases that may be difficult to detect with bone scans. A solitary plasmacytoma may be indistinguishable from a lytic metastasis and is a harbinger of multiple myeloma.

Vertebroplasty

This technique has been used to treat pain resulting from both benign and malignant vertebral body collapse. The procedure involves injection of bone cement into the affected vertebral body. Complications result from leakage of cement into perivertebral veins resulting in compression of the spinal cord or nerves or pulmonary embolism. Patients at the highest risk for complications are those

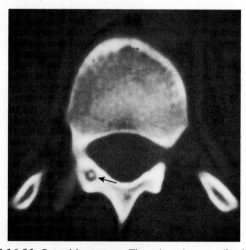

FIGURE 16-31 Osteoid osteoma. There is a circumscribed lytic region in the right lamina on this computed tomography scan with bone windows. In the central portion of the hypodense region *(arrow)* is a punctate area of high density. This is another classic!

with metastatic disease owing to destruction of the posterior cortex.

STIR MR is often used to determine whether an osteoporotic compression fracture is recent because high signal implies bone edema. Those patients who receive vertebroplasty for bright STIR compression fractures get more relief than those whose compression fractures are not bright (presumably more chronic and not causing acute irritation).

Kyphoplasty is a related technique to vertebroplasty where a balloon is blown up in the vertebral body after the insertion of the needle through the pedicle. Cement is injected into the cavity created by the balloon inflation.

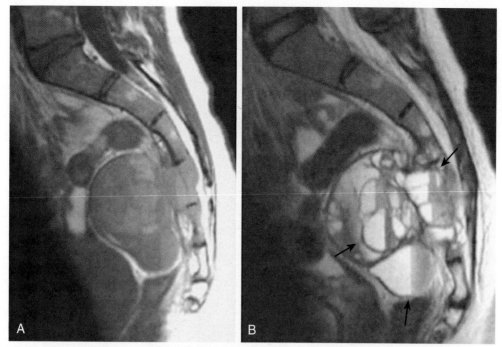

FIGURE 16-32 Aneurysmal bone cyst. **A,** T1-weighted imaging (T1WI) and **(B)** T2WI in the sagittal plane shows blood-fluid levels *(arrows)* in a bone lesion projecting from the lower sacrum into the anterior pelvis. The fluid levels are characteristic of aneurysmal bone cysts.

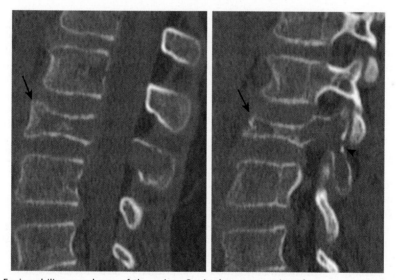

FIGURE 16-33 Eosinophilic granuloma of the spine. Sagittal reconstructions from computed tomography lumbar spine shows loss of height at L3 *(arrows)* with lytic appearance within the medullary compartment including posterior elements *(arrowhead)*. This loss of height, *vertebra plana,* is characteristic of eosinophilic granuloma involving the spine.

Some believe this technique improves height restoration and provides pain relief.

Epidural Lipomatosis

Epidural lipomatosis can occur from a number of different causes including obesity, steroid use (usually after prolonged use of oral steroids), and Cushing syndrome. The fatty tissue is seen most often in the posterior epidural space and can contribute to spinal stenosis and can produce significant cord or cauda equina compression. On MRI, prominent epidural fat is demonstrated to distort the

BOX 16-10   Causes of Vertebral Plana

Osteoporosis
Steroids
Tumors (leukemia, myeloma)
Eosinophilic granuloma
Fracture
Mucopolysaccharidoses

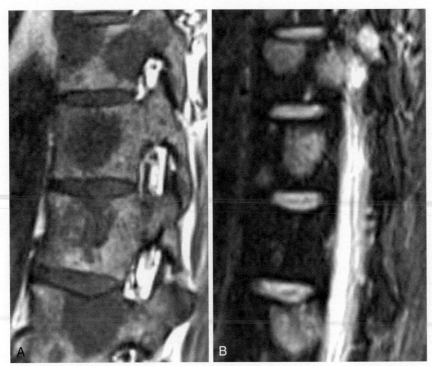

FIGURE 16-34 Multiple myeloma. **A,** Sagittal T1 and **(B)** sagittal short tau inversion recovery (STIR) images through the midthoracic spine show multiple discrete T1 hypointense and STIR hyperintense marrow lesions in this patient with known multiple myeloma.

normal round or oval-shaped configuration of the thecal sac, resulting in crowding and compression of thecal sac contents. Treatment involves weight loss and cessation of steroid use, depending on the cause.

VASCULAR LESIONS

Infarction

Spinal cord ischemia and infarction can occur at any location in the cord but has a propensity for the upper thoracic or thoracolumbar regions particularly at the ventral cord because of the tenuous blood supply. The clinical presentation includes profound impairment of bowel and bladder function, loss of perineal sensation, and moderate impairment of sensory and motor function of the lower extremities. Ischemia to the conus may result from poor collateral supply after occlusion of the dominant blood supply (artery of Adamkiewicz). Rarely, posterior spinal artery infarcts can be seen.

In patients with the acute onset of symptoms referable to the spinal cord, conus medullaris, or cauda equina, and with what appears to be an enlarged cord or conus on MR, consider spinal cord infarction (Fig. 16-35). Infarction can be the result of problems associated with the descending aorta, such as atheroma, aortic surgery, and dissecting aneurysm. Other causes include vertebral occlusion or dissection, arteritis, vascular malformations, pregnancy, hypotension, sickle cell anemia, tuberculosis, meningitis, arachnoiditis, vascular malformation, diabetes, degenerative disease of the spine, and disk herniation with spinal artery injury.

On MR, high signal on T2WI is noted in the cord, most commonly the ventral cord, which is usually enlarged,

and enhancement may or may not be present. Occasionally, vertebral body high signal intensity on T2WI can be identified and is most likely the result of concomitant infarction. Seeing high intensity in a vertebra in association with an acute nontraumatic myelopathy should strongly suggest spinal cord infarction. Careful attention should be paid to the aorta for aortic dissection or aneurysms as a cause. If a technically adequate diffusion-weighted sequence can be performed in the spine, it will be brightly positive like most infarcts and seal the deal.

Vascular Malformations

Spinal vascular malformations are separated into spinal dural arteriovenous fistula (SDAVF), spinal cord arteriovenous malformation (SCAVM), and spinal cord (or perimedullary) arteriovenous fistula (SCAVF) (Fig. 16-36). Table 16-2 summarizes the clinical differences between SDAVF and SCAVM. Treatment of spinal vascular malformations is dependent on many factors, including age, malformation type, and neurologic condition, and consists of embolization, surgery, or a combination of the two. Blood supply to the malformation is very important in determining whether to proceed with embolization or to perform surgery.

When malformations produce increased venous pressure they can produce subarachnoid hemorrhage. If the venous drainage is intracranial (particularly with SCAVF), patients may present with intracranial subarachnoid hemorrhage. The differential diagnosis of subarachnoid hemorrhage and hematomyelia is provided in Box 16-11.

Venous malformations are incredibly unusual in the cord and if they hemorrhage they are reportable.

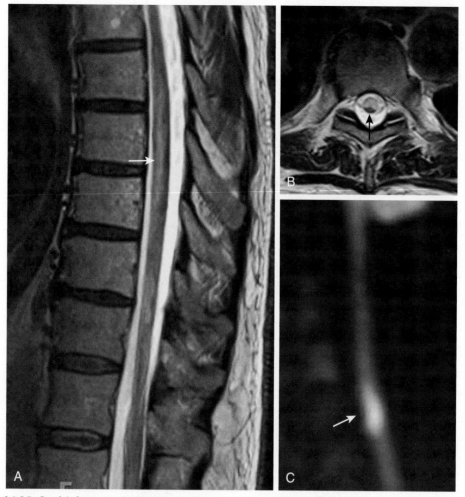

FIGURE 16-35 Cord infarct. **A,** Sagittal T2-weighted imaging (T2WI) shows a long segment abnormal high signal intensity and swelling within the ventral cord at the mid to lower thoracic spine levels *(arrow)*. **B,** Axial T2WI confirms involvement of the ventral cord, with preserved normal signal only within the dorsal aspect of the spinal cord *(arrow)*. While the imaging findings carry a differential diagnosis, the clinical context of abrupt loss of sensation and weakness abruptly following aortic aneurysm repair should make the diagnosis of cord infarct a no-brainer. **C,** Diffusion-weighted imaging from another patient shows high signal at the level of cord infarct *(arrow)* relative to the normal cord.

Spinal Dural Arteriovenous Fistula

SDAVF is the most common spinal vascular malformation having a nidus at or near the nerve root sleeve. This malformation, whose draining veins are most frequently found on the dorsal aspect of the lower thoracic cord or conus medullaris, sometimes has the eponym Foix-Alajouanine syndrome attached to it, to describe the myelopathy secondary to venous hypertension in the cord. In approximately 85% of cases, a single radicular artery with systemic pressure is identified draining directly into spinal pial vein(s). However, there are cases with many arterial feeders originating from either single or multiple levels that may be either unilateral or bilateral. The systemic pressure in the spinal veins initially dilates these vessels with subsequent kinking and poor venous drainage. This results in venous hypertension defined histopathologically by stasis, edema, ischemia, and leading to swelling and subsequent infarction of the spinal cord. The veins may also appear serpentine.

Complaints begin with an insidious onset of lower extremity weakness or sensory changes, associated with nonradiating pain starting in the caudal spinal segments and progressing superiorly. There is a propensity for these to occur in men in their fifth or sixth decade of life. These lesions are difficult to detect, and patients may have persistent neurologic sequelae if the lesions go unrecognized, although partial recovery is possible. This clinical presentation can sometimes be mistaken for degenerative disk disease. CSF protein can be increased mistakenly leading to the thought of a spinal cord tumor.

SDAVF has also been associated with the syndrome of subacute necrotizing myelopathy, which involves progressive myelopathy in combination with necrosis of the spinal cord. Box 16-12 lists diseases that may manifest this syndrome. The bottom line is to consider the diagnosis of SDAVF in *all* patients with progressive myelopathic signs or symptoms and positive MR findings.

The MR findings are key (Fig. 16-37). The spinal cord may be of normal size or enlarged with intramedullary high intensity seen on T2WI. Associated with this are usually (but *not* always) prominent vessels on the posterior aspect of the spinal cord. Look carefully for subtle bumps along the surface of the spinal cord, which is normally

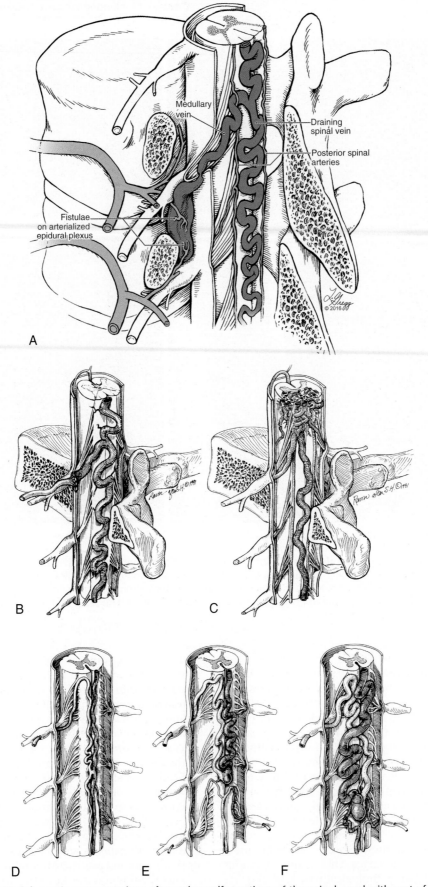

FIGURE 16-36 Schematic representations of vascular malformations of the spinal canal with part of lamina and dura removed. The more common sites of spinal arteriovenous fistulas include the epidural compartment near the nerve root sleeve **(A)**, and in the proximal nerve root sleeve **(B)**. **C,** An intramedullary spinal arteriovenous malformation can have multiple arterial feeders from both the anterior and posterior spinal arteries. The nidus is located within the spinal cord and drains into a dilated venous plexus. **D-F,** Variations of perimedullary arteriovenous fistulas: in all three types the fistulous connection is intradural. These can range from small fistula to large arterial feeder and a massively dilated venous system. Venous aneurysms are often associated with these lesions. (**A** © 2016 Lydia Gregg; **D-F** from Barrow Neurological Institute.)

TABLE 16-2 Summary of the Clinical Differences between Spinal Dural Arteriovenous Fistula and Spinal Cord Arteriovenous Malformation

	SDAVF	SCAVM
Age	Fifth or sixth decade	Third decade
Male/female	>5:1	1:1.1
Onset	Insidious	Acute
Hemorrhage	Very rare	>50%

SCAVM, Spinal cord arteriovenous malformation; *SDAVF,* spinal dural arteriovenous fistula.

BOX 16-11 Spinal Subarachnoid Hemorrhage and Hematomyelia

Trauma
Iatrogenic/postoperative
Coagulation disorders/anticoagulation
Dissections
SCAVM or fistula
Cavernous angioma
Spinal cord tumor
Spinal artery aneurysm
Spinal venous aneurysm
Vasculitis

SCAVM, Spinal cord arteriovenous malformation.

BOX 16-12 Subacute Necrotizing Myelopathy

Acute disseminated encephalomyelitis
Iodochlorhydroxyquin intoxication
Neuromyelitis optica
Herpes zoster infection
Lupus
Mononucleosis
Multiple sclerosis
Mumps
Rubella
Spinal dural AVM
Toxoplasmosis
Tuberculosis
Vacuolar myelopathy in AIDS
Venous hypertension (Foix-Alajouanine syndrome)

AIDS, Acquired immunodeficiency syndrome; *AVM,* arteriovenous malformation.

---

perfectly smooth. (*Okay,* truth be known, you see veins on the back of the cord *all the time,* so it is a judgment call when these are too prominent.) Obviously, these vessels enhance and the spinal cord itself, in addition to the veins, may or may not enhance slightly. Another finding recently recognized is the presence of peripheral hypointensity on T2WI, which in combination with high intensity within the cord should increase the specificity for the diagnosis of venous hypertension. This is thought to represent pial capillaries containing deoxyhemoglobin secondary to venous hypertension. Spinal MR angiography, when performed appropriately, can be used to localize the arteriovenous fistula and even the feeding vessels.

Conventional spinal angiography is suggested in equivocal cases and even in cases where no clear abnormality is seen on MRI but clinical suspicion for high flow vascular malformation remains, rather than going to myelography and CT. Spinal angiography is also necessary even when MR is unequivocal to define all of the feeders to the malformation. Spinal angiography requires careful attention to all possible vascular pathways that may supply the malformation. All the standard vessels including intercostal, vertebral, costocervical, thyrocervical, and subclavian arteries should be injected. The blood supply, however, may arrive through vessels distant from the malformation including the iliac arteries, hypogastric arteries, and sacral arteries.

Usually the anterior spinal artery does not arise at the same level as the SDAVF. If it does then an endovascular approach is contraindicated and surgical treatment is required. If the malformation is supplied by other vessels, embolization with permanent occlusive agents appears to be the procedure of choice.

Occasionally, intracranial dural arteriovenous malformations (fistula) may have intraspinal drainage (<5%). The arterial supply is from meningeal branches of the external carotid artery or the vertebral artery. These usually involve the medulla or cervical spine and may present acutely with hemorrhage, quadriparesis, or medullary dysfunction. High intensity on T2WI in the medulla and/or cervical cord associated with peripheral cord hypointensity should suggest this possible diagnosis.

Spinal Cord Arteriovenous Malformation

SCAVMs are intramedullary lesions fed by branches of the anterior or posterior spinal arteries, into a nidus and then into spinal veins. These are true arteriovenous malformations just like in the brain with similar angioarchitecture. Acute symptoms are caused by intramedullary hemorrhage. Unlike SDAVF, there is no gender predilection and progressive myelopathy is less common. Juvenile AVMs are ridiculously rare (or nonexistent) lesions with potential extensive involvement with intramedullary, extramedullary, and extraspinal components. They are described to occur early in life and tend to have a poor prognosis.

Spinal Cord Arteriovenous Fistula

SCAVF are intradural extramedullary ("perimedullary") arteriovenous fistulas (without an intervening capillary network) located on the pial surface. They involve the anterior or posterior spinal arteries with a single arteriovenous communication. Clinical presentations can vary from progressive myelopathy to hemorrhage including subarachnoid hemorrhage (up to 30%). Venous aneurysm can be seen with this malformation.

Cavernous Angioma

Cavernous angiomas of the spinal cord are uncommon intramedullary vascular lesions. They appear similar to those in the brain, which are common. The age of presentation is 30 to 50 years with a 1:4 male to female ratio. These lesions may be familial. The clinical presentation includes pain, weakness, and paresthesias, which may be progressive or episodic and progressive.

MR is the imaging study of choice, with the malformation appearing round with regions of high signal intensity (methemoglobin) on T1WI and T2WI surrounded by low signal intensity on T2WI (hemosiderin) (Fig. 16-38). There is

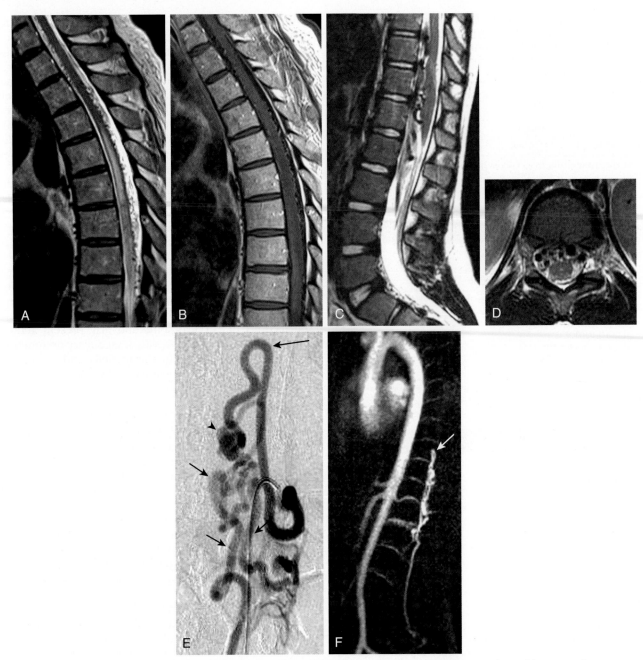

FIGURE 16-37 Arteriovenous malformations. **A,** Sagittal T2-weighted image (T2WI) in a patient with progressive lower extremity weakness shows innumerable flow voids along the surface of the cord, predominantly along the dorsal surface. There is increased T2 signal and expansion of the lower thoracic cord due to venous congestion. **B,** Postcontrast sagittal T1WI shows some enhancement within these prominent flow voids. Spinal angiogram confirmed presence of a dural arteriovenous fistula, which was successfully embolized. **C,** Sagittal and **(D)** axial T2W images show enlarged serpentine flow voids along the ventral aspect of the lower spinal cord without associated cord signal change. Believe it or not, this 6-year-old patient was asymptomatic and the abnormality was picked up incidentally on another imaging study performed for a different reason, prompting this MRI. **E,** Conventional spinal angiogram followed, showing a markedly enlarged and tortuous artery of Adamkiewicz *(large arrow)* with a direct fistulous connection to a dilated venous pouch *(arrowhead)* with multiple enlarged draining veins coursing caudally from the fistula *(small arrows)*. Wowsers! **F,** Follow-up contrast-enhanced spinal MRA after partial embolization of the fistula shows the fistula and artery of Adamkiewicz *(arrow)* to be much smaller in caliber.

minimal mass effect or edema unless there has been recent hemorrhage. These lesions may show minimal contrast enhancement. Myelography may be normal or may reveal atrophy or cord expansion. Calcification has been reported. Angiography is usually negative in these cases, although late venous pooling or abnormal draining veins may be found.

Siderosis

Recurrent hemorrhage from spinal vascular malformations or hemorrhagic tumors such as ependymoma or hemangioblastoma is associated with hemosiderin deposition throughout the leptomeninges as well as intracranially.

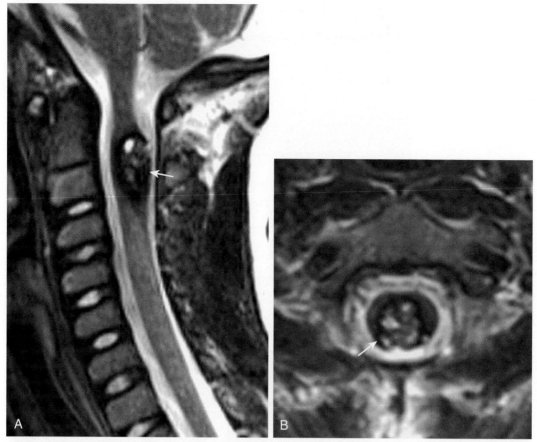

FIGURE 16-38 Cavernoma of cord. **A** and **B,** Just like in the brain, a popcorn-shaped lesion with high signal intensity centrally (extracellular methemoglobin) and dark signal peripherally (hemosiderin) in the cord on T2-weighted imaging is characteristic of a cavernoma *(arrows)*. Note absence of high signal within the cervical cord despite the size of the lesion. Popcorn, anyone?

SPINAL TRAUMA/INJURY

While in many cases the presence of spine injury can be assessed and excluded with good physical examination, in the circumstance of severe pain, distracting injuries, and altered awareness, imaging is required to assess integrity of the spine. In acute spinal injury, CT is the best method for detecting fractures and dislocations of the bony spine and has replaced plain radiograph examination at many trauma centers, at least among adult patients (Fig. 16-39). CT aids in visualizing C1, C2, C6, and C7 vertebrae, which are difficult to visualize in many cases on plain films. The swimmer's view is useful for alignment but is of limited value in detecting C6 and C7 and T1 fractures.

MR helps most with ligamentous injuries and to assess the spinal cord for trauma. Angiography by CT, MR, or conventional techniques can assess the integrity of the carotid and vertebral arteries, which can be traumatized in certain types of spine injuries.

It is very important that thin section axial CT images be reviewed in conjunction with coronal and sagittal reconstructions and areas of suspicion be confirmed in multiple planes. Increases in intervertebral distances are not as evident on axial CT as on plain films; subluxation, dislocation, and abnormal angulation are more difficult to recognize without multiplanar reconstructions. Fractures oriented in the axial plane may be completely missed on axial CT and partial volume averaging may mask or mimic fractures.

MR is the only technique that can reliably reveal intrinsic injury to the cord and ligaments (Fig. 16-40). The fibrous nature of the spinal ligaments results in their normal hypointensity on all MR pulse sequences. They should have well-demarcated margins throughout their course. Tears or partial tears resulting in edema are best visualized on STIR images (and almost as well on fat-suppressed T2WI) as high intensity. The sagittal plane depicts injury to the anterior and posterior longitudinal ligaments, the ligamentum flavum, and the interspinous ligaments. Ligamentous injury is often associated with traumatic disk herniation. High signal within and surrounding facet joints at levels of injury can indicate facet capsular rupture. High T2 signal and ligamentous discontinuity at the tectorial membrane, transverse, and alar ligaments indicate craniocervical ligamentous injury.

Injury to these structures may be present even if no fractures are evident by CT. The presence of prevertebral swelling on radiograph or CT should raise suspicion for soft-tissue injury involving the anterior column. In acute spinal cord injury abnormalities include intramedullary hemorrhage, which may be petechial or diffuse, and swelling, both of which can be appreciated on MR. In severe spinal cord trauma, lacerations and spinal cord transections (also seen in fetal hyperextension, breech delivery, flexion-hyperextension injury in children, and direct crush injuries) can be appreciated by MR. MR can also visualize

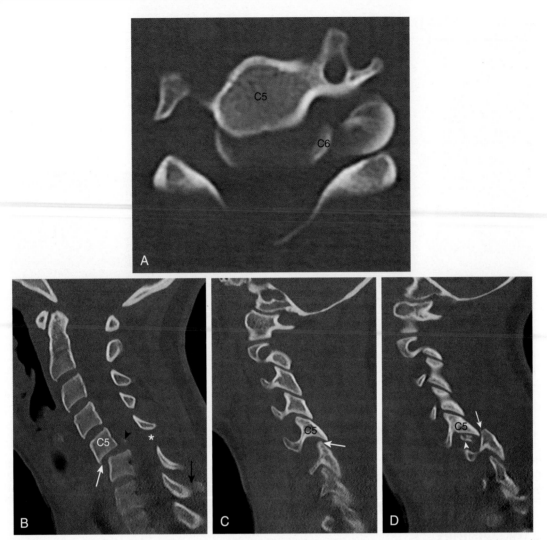

FIGURE 16-39 Cervical spine trauma on computed tomography (CT). **A,** There is something not right on this axial CT image of the cervical spine at the C5-C6 level where both C5 and C6 vertebral bodies are present on the same slice. **B,** Sagittal CT reconstruction helps us understand what is going on, showing traumatic anterolisthesis of C5 on C6 *(white arrow)* and widening of the interspinous interval *(asterisk).* A small chip fragment from C5 got left behind *(arrowhead)* during the migration of C5 anteriorly. Note nondisplaced fracture of the C7 spinous process *(black arrow).* **C,** More laterally on the same reconstructed series, there is abnormal widening of the C5-C6 facet articulation with partial "uncovering" of the facet joint *(arrow),* disrupting the normal "shingled" appearance of the fact joints. **D,** Even worse, on the other side, the inferior articulating facet of C5 has "jumped" over the superior articulating facet of C6 *(arrow),* which is fragmented *(arrowhead).* Three-column injury is *definitely* unstable. A magnetic resonance image is warranted (barring any contraindications) to assess for potential cord and/or ligamentous injuries and/or intraspinal hemorrhages before urgent spine surgery for stabilization.

subluxations and vertebral body fractures, hematomas around ligamentous tears, cord compression, and traumatic disk herniations.

Patients with acute spinal cord injury without radiographic abnormalities and normal MR findings have the best prognosis, whereas patients with hemorrhage and/or low or high signal intensity on T1WI and high signal on T2WI within the cord have the poorest prognosis. Enhancement within the cord has been observed at approximately 2 weeks after injury with a duration of 2 to 3 months. It has been reported to represent early necrosis, absorption, and reorganization of the spinal cord hematoma, and may be an indicator of severe damage. Cases with isointensity on T1WI and high signal on T2WI have some potential for reversible change and improved outcome. The pathologic consequences of significant spinal cord trauma include

(1) atrophy, (2) myelomalacia, (3) posttraumatic syrinx, (4) arachnoid cyst, and (5) arachnoiditis.

The onset of symptoms in a posttraumatic syrinx is extremely variable and not necessarily related to the extent of the lesion, so that MR is appropriate in following up all patients with spinal cord injury, even those who are relatively asymptomatic. MR displays a cystic lesion in the cord that grows over time. Identification is important because this is a treatable cause of deterioration after spinal cord injury. Intrinsic damage to the cord associated with dynamic changes in CSF pressure and arachnoiditis are factors implicated in the development of posttraumatic syrinx.

Plain radiographs can also play an important role in assessing cervical spine instability in both the acute and nonacute settings. Changes in alignment of the spine on

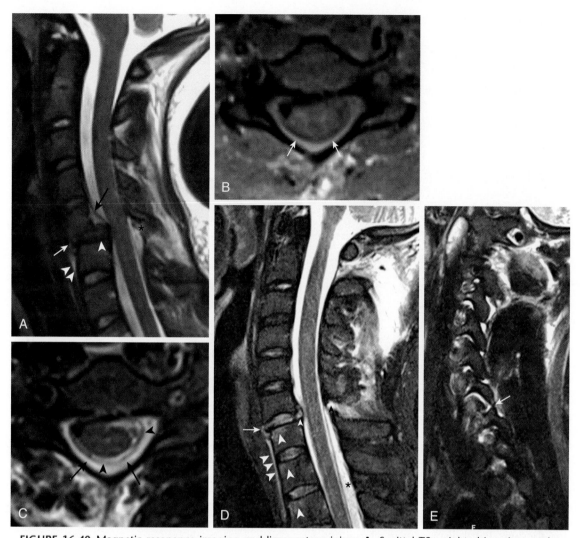

FIGURE 16-40 Magnetic resonance imaging and ligamentous injury. **A,** Sagittal T2-weighted imaging on the same patient as in Fig. 16-39 shows the same traumatic anterolisthesis of C5 on C6 *(white arrow)*. Now we can see the injuries to the soft tissues. There is edema within the prevertebral soft tissues *(double arrowheads)*. There is laxity and disruption of the posterior longitudinal ligament *(black arrow)*. There is elevated T2 signal within the disk, indicating acute disk rupture *(single arrowhead)*. There is disruption of the ligamentum flavum *(asterisk)*, normally seen as dark signal at this level. Note associated widening of the interspinous interval here. The cord looks tight within the canal, but at least the cord signal looks good. **B,** Axial gradient echo, and **(C)** axial T2 images show epidural hematoma surrounding the thecal sac *(arrows)* contributing to the crowding of the cord within the thecal sac. In the cervical spine, the dural margins should approximate the bony canal margins, but in this case, they are lifted off as a result of the epidural hematoma *(arrowheads in **C**)*. **D,** In a different patient with similar injury, sagittal short tau inversion recovery (STIR) shows anterior longitudinal ligament disruption *(arrow)* and associated edema in the prevertebral tissues *(triple arrowheads)*, acute disk rupture *(small arrowhead)*, ligamentum flavum disruption *(black arrowhead)*, and abnormal T2 high signal within the interspinous ligament (at and posterior to the *black arrowhead*), as well as marked edema within the soft tissues of the upper neck. Note the elevation of the dura posteriorly, resulting from epidural fluid collection *(black asterisk)*. Bone contusions along the superior endplates of C7, T1, and T2 *(large single arrowheads)* were not appreciable by computed tomography but are clearly present on STIR. **E,** In this same patient, the inferior articulating facet of C6 is "perched" on the superior articulating facet of C7, and edema around the facet joint *(arrow)* indicates facet capsular rupture. Help!

radiographs obtained in flexed, neutral, and extended positions can indicate ligamentous laxity. These patients can benefit from spine-stabilizing surgery. In the acute setting, these examinations should be performed under the direct supervision of a clinician who knows and has examined the patient before acquiring images. The patient's neck should not be directly manipulated by the technologist or any physician who has not performed a thorough clinical assessment of the patient.

Cervical Nerve Root Avulsions

These are the result of traction injuries on the upper extremities that tear the roots from the spinal cord (Fig. 16-41). The roots are absent on one side and the cord is consequently pulled to the contralateral side. Thin section MR or CT myelographic images can show the nerve root disruption. Pseudomeningoceles are associated with this injury. They result from tears of the arachnoid and dura, which can be identified on MR or CT myelography.

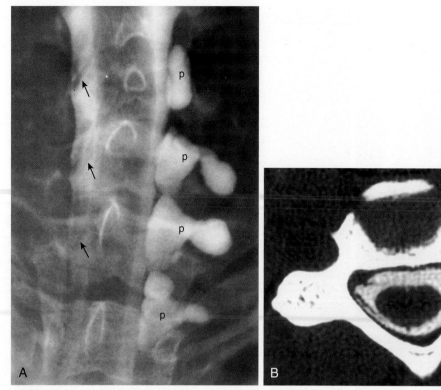

FIGURE 16-41 Cervical root avulsion. **A,** Myelogram illustrates multiple cervical nerve root avulsions manifested by multiple pseudomeningoceles (p) from the torn arachnoid and dura. Ipsilateral nerve roots are not seen, having been torn and retracted. Contralateral nerve roots *(arrows)* can be identified. This is a classic! **B,** Note the absence of anterior and posterior roots on the left side with an incipient pseudomeningocele from the empty root sleeve on this CT myelogram.

After healing, some pseudomeningoceles may not communicate with the subarachnoid space and can present as epidural mass lesions. Focal displacement of epidural fat can confirm the location of the lesion and rule out post-traumatic arachnoid cyst.

Spine Fractures and Dislocations

It is important to appreciate the definition of clinical stability and instability. This turns out to be a controversial topic. Stability has been defined as the capacity of the spine under physiologic loads to limit displacement so as not to compromise or damage the spinal cord and nerve roots. Stability also denotes prevention of deformity or pain secondary to anatomic changes. Conversely, clinical instability may be defined as greater than normal range of motion within a spinal segment so as to assume a risk of neurologic injury. Bones and ligaments are responsible for spinal cord stability. The implication of clinical instability is that surgical intervention is necessary to stabilize the spine and protect the cord. While certain fracture types can be characterized as stable or unstable based on radiographic appearance, not all can, and very often the clinical evaluation also plays a crucial role in determining spine stability in conjunction with imaging findings. Cervical spine injuries that are typically considered to be stable and unstable are summarized in Box 16-13.

The spinal column can be divided into three columns. The anterior column is composed of the anterior

---

BOX 16-13 Stability of Cervical Spine Fractures

TYPICALLY CONSIDERED STABLE

Anterior subluxation
Burst fracture, lower cervical vertebrae
Pillar fracture
Posterior neural arch fracture, atlas
Unilateral facet dislocation
Spinous process fracture
Clay-shoveler fracture
Transverse process fracture
Compression fracture of <25% of vertebral body height
Anterior wedge fracture
Isolated avulsion without ligamentous tear
Type I odontoid fracture
Type III odontoid fracture (usually)
Endplate fracture
Osteophyte fracture (not including corner or teardrop fracture)
Trabecular bone injury

TYPICALLY CONSIDERED UNSTABLE

Atlantooccipital subluxation/dislocation
Bilateral C1-C2 dislocation
Bilateral facet, laminae, or pedicle fractures and/or dislocation
Chance fractures
Extension teardrop fracture
Flexion teardrop fracture
Hangman fracture
Hyperextension fracture-dislocation
Type II odontoid fracture
Jefferson fracture

longitudinal ligament, the anterior annulus, and the anterior portion of the vertebral body. The middle column delineated by the posterior longitudinal ligament, the posterior portion of the annulus, and the posterior aspect of the vertebral body and disk. The posterior column contains the neural arch, facet joints and capsules, ligamentum flavum, and all other posterior spinal ligaments. The middle column is critical as instability occurs when two of the three columns are injured (Fig. 16-42). Compression fractures result in anterior column injury (usually stable), whereas burst fractures affect the anterior and middle columns (unstable). Flexion-distraction injuries affect the middle and posterior columns and fracture-dislocation injuries result in a three column injury.

Hyperflexion is a forward rotation and/or translation of a vertebra with both distraction of the posterior column and compression of the anterior column. Hyperextension injuries involve posterior rotation/translation. Here the anterior and middle columns bear the brunt of the injury (anterior longitudinal ligament, disk, and posterior longitudinal ligament). "Sprains" are minor tears of the supraspinous, interspinous, and facet capsules from flexion-distraction injuries, which can result in anterior subluxation. Anterior subluxation is seen as a hyperkyphotic angulation of the cervical spine with widening of the distance between the spinous processes and interlaminar space when compared with other levels (fanning). Other findings include narrowing of the anterior disk space with widening of its posterior aspect with or without anterior translation of the involved vertebral body, widening of the space between the subluxed vertebral body and the superior articular facet of the lower vertebral body, and subluxation of the facet joints. Flexion-extension radiographs can best demonstrate this injury.

There are a range of facet-related injuries from fractures to facet subluxations to perched facets to facet dislocations (jumped facets) that can be unilateral or bilateral (see Figs. 16-39, 16-40). Rotational facet injury is a generic term used to describe unilateral facet dislocation/subluxation as well as unilateral facet fractures with malalignment. Facet subluxations without fractures occur in only 25% of cases but have a higher rate of cord injury and therefore should be stabilized. It is usually the inferior articulating facet of the vertebral body above that fractures. Facet fractures with vertebral injury lead to rotational instability and require surgical fixation.

Atlantooccipital Dislocation

Atlantooccipital dislocation (otherwise known as craniocervical dissociation) occurs as a result of injury to the stabilizing ligaments at the craniocervical junction (Fig. 16-43). This is customarily considered a fatal injury, although there are many case reports of survival. Severe injuries include pontomedullary junction laceration, contusion or laceration of the inferior medulla and spinal cord, injury to the midbrain, subarachnoid hemorrhage, subdural hemorrhage, and vascular dissection. The incidence is increased in children predominantly because of their small occipital condyles and horizontal plane of the atlantooccipital joints. The high-velocity shearing forces are directed to the face or posterior of the skull. Atlantooccipital dislocation

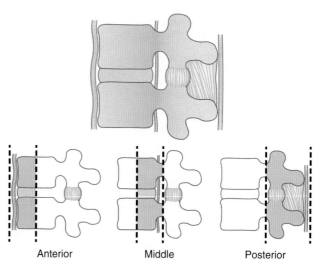

FIGURE 16-42 The three-column classification model of the spine proposed by Denis represented an advance in thinking about spinal injuries over the previous two-column model. The concept of the middle column, composed of the posterior longitudinal ligament, the posterior aspect of the annulus fibrosus, and the posterior portion of the vertebral body and disk were added. (From Garfin SR, Blair B, Eismont FJ, Abitbol JJ. Thoracic and upper lumbar spine injuries. In Browner BD, Jupiter JB, Levine AM, et al, eds. *Skeletal trauma: fracture, dislocations, ligamentous injuries.* Vol 1. 2nd ed. Philadelphia: Saunders; 1998:967.)

Below the figure, labels read: Anterior, Middle, Posterior

can be associated with severe hyperextension, type I odontoid fracture (see Odontoid Fracture subsection), loss of the normal relationship of the occipital condyles to C1, and prevertebral swelling. Longitudinal distraction and anterior or posterior dislocation of the occiput relative to the atlas is observed as a result of tearing of the tectorial membrane and alar ligaments (Fig. 16-44) and is seen as increased distance between the basion and dens (greater than 12 mm on lateral radiograph). CT can detect basion and condylar fractures; however, if there is no associated bony fracture, this severe injury can be altogether missed when reviewing axial CT images alone. Review of multiplanar images is critical, and in this type of injury, coronal and sagittal reformatted images will demonstrate widening between the lateral masses of C1 and the occipital condyles. MR with T2WI and STIR images reveal prevertebral soft-tissue injury, ligamentous tears, epidural hematoma, spinal cord injury, and brain stem injury.

In children, this diagnosis can be problematic because of variable bone ossification in the craniocervical junction, especially in the dens where the basion-dens interval is reported to be unreliable in patients younger than 13 years of age.

Figure 16-45 shows the three types of occipital condyle fractures.

Atlantoaxial Distraction

This injury occurs as a result of severe extension. There is disruption of the articular capsules, alar ligaments, transverse ligament, and tectorial membrane between C1 and C2. Occasionally, there is a type I dens fracture. There is obvious widening of the space between C1 and C2. Look for prevertebral swelling on plain films and CT. On MR, edema can be identified on STIR or T2WI in the prevertebral, interspinous, and nuchal ligaments. Other reported findings may be facet

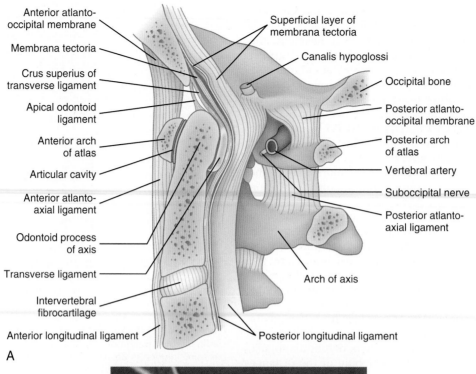

Anterior atlanto-occipital membrane

Membrana tectoria

Crus superius of transverse ligament

Apical odontoid ligament

Anterior arch of atlas

Articular cavity

Anterior atlanto-axial ligament

Odontoid process of axis

Transverse ligament

Intervertebral fibrocartilage

Anterior longitudinal ligament

Superficial layer of membrana tectoria

Canalis hypoglossi

Occipital bone

Posterior atlanto-occipital membrane

Posterior arch of atlas

Vertebral artery

Suboccipital nerve

Posterior atlanto-axial ligament

Arch of axis

Posterior longitudinal ligament

A

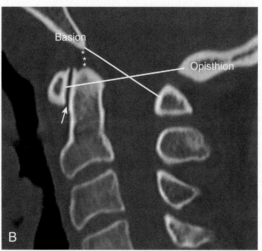

B

FIGURE 16-43 Stabilizing ligaments and useful measurements at the craniocervical junction. **A,** Schematic demonstrates the important stabilizing ligaments at the craniocervical junction. **B,** Useful measurements at the craniocervical junction to keep in mind include the anterior atlantodental interval *(arrow),* which should measure no greater than 3 mm in an adult, and should not change in flexed and extended positions as well as the basion-axial interval *(asterisks),* which should measure less than 12 mm. Powers ratio is calculated as the distance between the basion and spinolaminar line of C1 divided by the distance between the anterior arch of C1 and the opisthion. A ratio greater than 1 implies occipitoatlantal dislocation.

widening, epidural hematoma, and focal increased T2 signal intensity in the spinal cord.

Atlantoaxial Rotation

Atlantoaxial rotation results in torticollis, so that the atlas is rotated and dislocated from the articular processes of the axis. Head rotation does not correct what appears to be an asymmetric widening of the space between the odontoid with regard to the lateral masses of C1, nor does the spinous process of C1 move. CT or plain radiography performed with the head turned to the left, right, and in the neutral position can reveal the lack of correction.

Atlantoaxial Instability

Atlantoaxial dislocation can be caused by trauma with associated fractures and rupture of the transverse ligament, or may be seen in nontraumatic situations associated with transverse ligament laxity or odontoid process malformations (os odontoideum, ossiculum terminale, or agenesis of the odontoid base, apical segment, or the entire base). This injury is a doosie, especially in the nontraumatic setting, because it can be fatal yet patients can be completely asymptomatic. Complete dislocation of the C1-C2 articulation may be another source of injury—this is to be distinguished from rotatory subluxation where the facets are intact. The anterior atlantodental interval can be abnormally widened as a result of ligamentous laxity and

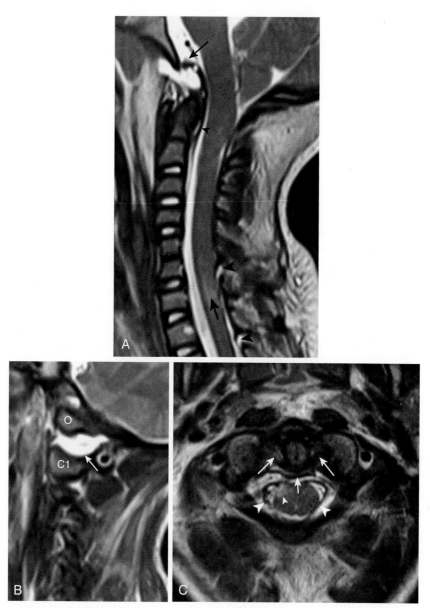

FIGURE 16-44 Craniocervical dissociation. A, Sagittal T2-weighted imaging shows disruption and redundancy of the tectorial membrane *(small arrow).* There is widening of the space between the tip of the clivus and the dens with absence of the apical ligament usually seen connecting the tip of clivus to the dens, also disrupted. The transverse ligament *(small arrowhead)* is elevated off of the posterior aspect of the dens. Despite the severe injury at the craniocervical level, the atlantodental interval is preserved so the C1-C2 relationship is maintained appropriately anteriorly. Multiple foci of ligamentum flavum disruption are present at other cervical levels *(large arrowheads)* and there is T2 high signal indicating edema within the cord at the C7 level *(large arrow).* **B,** Far lateral image from the same series shows abnormal widening and fluid *(arrows)* between the occipital condyle (O) and C1 lateral mass (C1), which was present bilaterally (only one side shown here). Also note that the occipital condyle is ever so slightly anteriorly displaced relative to the lateral mass of C1. **C,** Axial T2 image shows high signal at the attachment of the bilateral alar ligaments at the dens *(large arrows)* and elevation and laxity of the transverse ligament *(small arrow).* There is elevation of the dura within the spinal canal *(large arrowheads)* as a result of epidural hematoma and there is another contusion within the cord *(small arrowhead).* (And you thought *you* were having a bad day….)

can vary in width on patient positioning (Fig. 16-46). Other nontraumatic causes include infections such as tonsillitis and pharyngitis (Grisel syndrome), Down syndrome (trisomy 21), Marfan syndrome, neurofibromatosis, ankylosing spondylitis, rheumatoid arthritis, calcium pyrophosphate deposition disease, and bone tumors. Calcium pyrophosphate deposition disease is another source of thickening of the transverse ligament of the odontoid process (like rheumatoid arthritis), cysts, erosion of the dens, but rarely leads to C1-C2 subluxation. Calcium pyrophosphate

deposition disease is more likely than rheumatoid arthritis to show ligamentous calcification.

Basilar Invagination/Basilar Impression

These are descriptions for congenital or acquired upward migration of the dens above the level of the foramen magnum. The term impression is reserved for cases where the underlying bone is abnormal, whereas invagination implies no intrinsic bone pathology (Fig. 16-47). Congenital cases

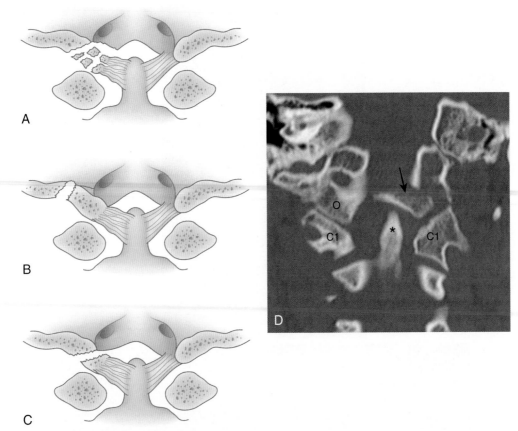

FIGURE 16-45 The classification of Anderson and Montesano describes three basic types of occipital condyle fractures. The first **(A)** is an impaction-type fracture, which is usually the result of an asymmetric axial load to the head; it may be associated with other lateral mass fractures in the upper cervical spine. The next type **(B)** is a basilar skull-type occipital condyle fracture, which may be the result of a distraction force applied through the alar and apical ligament complex. **C,** Type III avulsion-type occipital condyle fracture. **D,** Coronal computed tomography reconstruction shows the normal relationship between the occipital condyle (O) and C1 lateral mass (C1) on the right. The dens of C2 is indicated by *asterisk*. On the left, there is an occipital condyle fracture with displacement and angulation of the occipital condyle fracture fragment *(arrow)*. Disruption of the apical and alar ligaments was demonstrated on subsequent magnetic resonance imaging (not shown). (**B,** Redrawn from Anderson P, Montesano P. Morphology and treatment of occipital condyle fractures. *Spine.* 1988;13:731.)

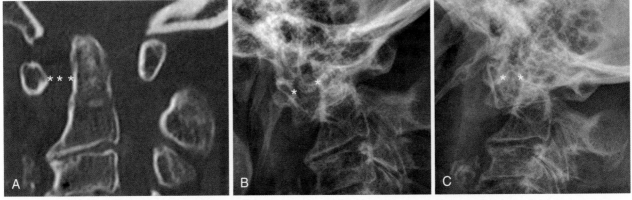

FIGURE 16-46 Atlantoaxial instability. **A,** Sagittal computed tomography image shows abnormal widening of the anterior atlantodental interval (demarcated by *asterisks*). This should be no wider than 3 mm in an adult and is enlarged in this case as a result of degenerative pannus (not shown). **B,** Flexion and **(C)** extension radiographs of the cervical spine show that despite this patient's limited excursion, this interval is wider in the flexed position than the extended position, indicating ligamentous laxity and instability.

can be seen in the setting of osteogenesis imperfecta, Klippel-Feil syndrome, achondroplasia, and Chiari malformations. Pannus formation from rheumatoid arthritis, Paget disease, metastatic disease, and metabolic derangements such as hyperparathyroidism and rickets can also cause vertical migration at the skull base. The displaced dens can impress upon the brain stem, and can cause neurologic compromise, hydrocephalus, syrinx, and can even be fatal. On imaging, if the dens extends greater than 3 mm above a line drawn from the hard palate to the opisthion (Chamberlain line), or greater than 4.5 mm above a line drawn from hard palate to the inferiormost margin of the occipital bone (McGregor line), then basilar invagination is present. Surgical stabilization is typically required in these cases.

Odontoid Fracture

Odontoid fractures have been divided into oblique fractures through the upper portion of the dens (type I), transverse fractures through the junction of the dens with the body of the axis (type II) (unstable), and fractures extending into the cancellous portion of the body of the axis (type III) (Fig. 16-48).

Os Odontoideum

Os odontoideum is a round smooth bony ossicle that is seen superior to the base of the dens. Its position is variable either in the position of the odontoid process (orthotopic) or at the foramen magnum near the base of the occiput (dystopic). On lateral radiographs, there is commonly a smoothly marginated bone ossicle, separated by a zone of radiolucency, above the superior facets of axis (Fig. 16-49). Originally thought to be a congenital or developmental anomaly, it is now thought to be the result of an acquired or posttraumatic cause. There is an association, albeit infrequent, with

Klippel-Feil syndrome, myelodysplasia, Morquio syndrome, Down syndrome, and spondyloepiphyseal dysplasia.

The os is embedded within the transverse atlantoaxial ligament, and is associated with atlantoaxial instability, translating over the axis and at times producing spinal cord compression. This usually requires surgical correction by posterior atlantoaxial fusion. The issue for radiologists is the differentiation of the os from a type I or type II odontoid fracture. Acute fractures do not have smoothly corticated margins as seen with os odontoideum; rather they have sharp radiolucent margins.

Jefferson Fracture

Jefferson fractures are breaks in the ring of the atlas that were classically described as having four sites (junctions of the anterior and posterior arches with the lateral masses, which are where the bone is thinnest), although there are many variations in the number of fractures (Fig. 16-50). They are caused by compressive forces. Tears of the transverse ligament are associated with Jefferson fracture.

Hangman Fracture

Hangman fractures result from hyperextension of the neck and are fracture dislocations of C2. The spectrum of traumatic spondylolisthesis is seen in Figure 16-51. When both pedicles are fractured, there is anterior subluxation of the body of C2 on C3, whereas the posterior ring does not move, being fixed by the inferior articular processes.

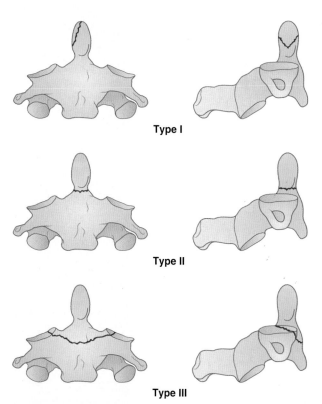

FIGURE 16-48 Types of odontoid fractures. This diagram illustrates the configuration of the three types of odontoid fractures from frontal (left-sided images) and lateral (right-sided images) perspectives. (From Modic M, Masaryk T, Ross J. *Magnetic resonance imaging of the spine.* Chicago: Year Book Medical Publishers; 1988.)

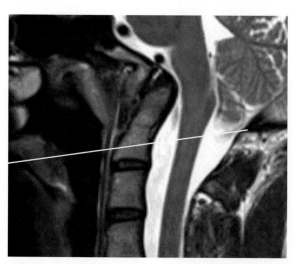

FIGURE 16-47 Basilar invagination. Sagittal T2-weighted imaging scan from a patient with Klippel-Feil syndrome shows superior migration of the dens relative to the McGregor line (*white line* depicted), indicating basilar invagination. The marrow is normal, hence the term invagination, rather than impression. There is mild mass effect on the ventral medulla as a consequence, although (good news!) there is no abnormal signal in the brain stem. Note the flattened appearance of the clivus: the basal angle measured 150 degrees, indicating platybasia.

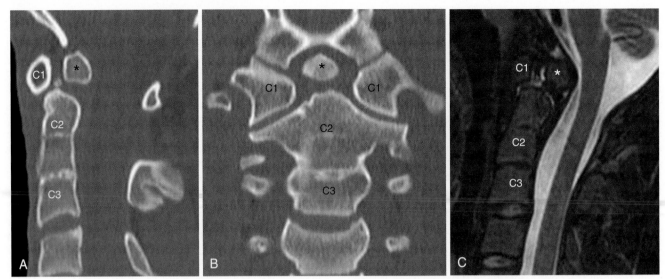

FIGURE 16-49 Os odontoideum. **A,** Sagittal and **(B)** coronal computed tomography (CT) reconstructions along with **(C)** sagittal short tau inversion recovery image show well corticated ossific density *(asterisk)* above the dens of C2, representing the os. There is a calcific density between the anterior arch of C1 and the os on the sagittal CT image, suggesting possible previous trauma. There is incomplete segmentation of the C2 and C3 vertebral bodies.

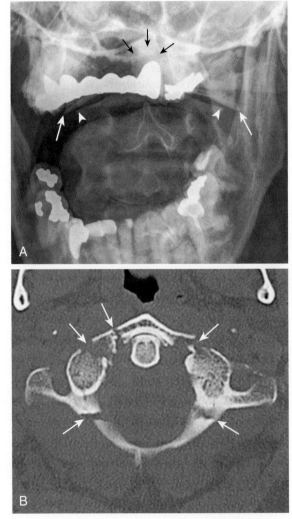

FIGURE 16-50 Jefferson fracture. **A,** Open mouth odontoid view radiograph shows lateral subluxation bilaterally of the lateral masses of C1 *(white arrows)* with regard to the lateral margins of the lateral masses of C2 *(arrowheads).* Normally, these structures should line up with each other. The tip of the dens in indicated by *black arrows.* **B,** Axial computed tomography image from the same patient shows multiple fractures *(arrows)* through the anterior and posterior arches of C1. The anterior atlantodental interval is not widened in this case.

Thoracolumbar Spine Fractures

Numerous classification schemes for categorizing thoracolumbar spine fractures have been proposed to guide clinical management, with the thoracolumbar injury classification and severity score (TLICS) most recently introduced to provide a clearer basis for the need for urgent surgical intervention by including both imaging and clinical parameters in the decision process. This classification scheme takes into account and applies a numerical scoring system (1 to 10) to injury morphology, integrity of the posterior ligament complex, and the patient's clinical symptomatology in determining the need for urgent surgical spine stabilization (score >5).

From the radiologist's perspective, there are several observations that should be recorded in the report that can aid the spine surgeon to make a determination as to whether or not urgent intervention is required (Fig. 16-52). This includes a description of the fracture pattern

(compression, burst, translation/rotation, distraction). Compression fractures are those that involve endplate injuries with or without anterior cortex injuries but with preservation of the posterior vertebral body wall. Burst fractures are comminuted fractures that involve the posterior vertebral body with displacement of superior endplate fragments into the spinal canal and are a result of axial loading injury. Translational injuries result in lateral displacement of one vertebral body with respect to another. Distraction injuries result in widening of bony relationships in vertical plane. For all injury types, percent vertebral body height loss, presence of retropulsion and impact on spinal canal diameter, kyphosis, and presence of intraspinal hemorrhage should be described in the radiology report.

Additionally, presence of posterior ligament complex injury should be assessed, evidenced by facet joint or interspinous distance widening, spinous process avulsion

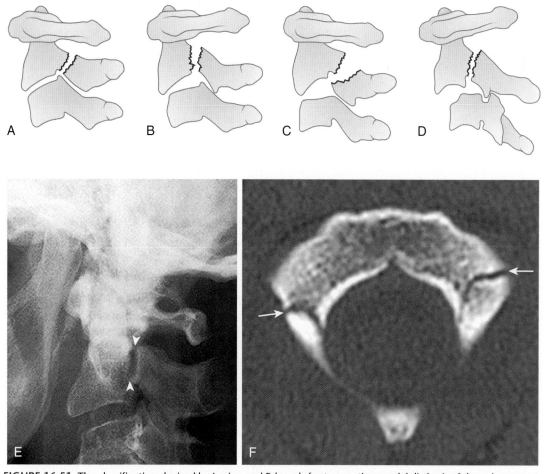

FIGURE 16-51 The classification devised by Levine and Edwards for traumatic spondylolisthesis of the axis accounts for the majority of fractures of this type. **A,** The most common pattern is type I, which is characterized radiographically by a fracture through the neural arch with minimal translation (<3 mm) and minimal angulation. Not pictured in the classification is the later addition of type IA, which has also been called atypical hangman fracture. **B,** Type II fractures have significant angulation (>3 degrees) and translation (>3 mm) and are much more unstable than type I fractures. **C,** Type IIA fractures are identified radiographically by an oblique fracture line with minimal translation but significant angulation. **D,** Type III axial fractures combine bilateral facet dislocation between C2 and C3 with a fracture of the neural arch of the axis. **E,** Lateral radiograph shows acute fracture through the pars interarticularis of C2 *(arrowheads)*. Note slight anterolisthesis of C2 with regard to C3. **F,** Axial computed tomography (CT) image shows a fracture defect extending through the pars interarticularis of C2 bilaterally *(arrows)*. (**A-C,** From Levine AM, Edwards CC. The management of traumatic spondylolisthesis of the axis. *J Bone Joint Surg Am.* 1985;67:217-226; **D,** From Levine AM, Edwards CC. Treatment of injuries in the C1-C2 complex. *Orthop Clin North Am.* 1986;17:42.)

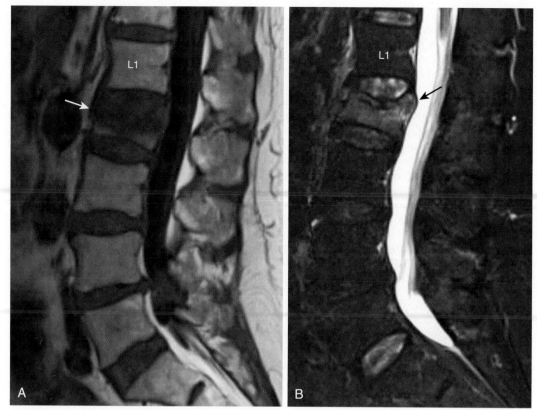

FIGURE 16-52 Acute compression fracture. **A,** Sagittal T1-weighted imaging and **(B)** short tau inversion recovery scans show abnormal signal intensity indicating edema at the L2 vertebral body *(arrows).* Specifically, there is disruption along the superior endplate, resulting in at the most 25% loss of height of L2. There is minimal posterior bulging of the superior L2 endplate into the spinal canal, without frank disruption of the posterior ligament complex and without significant canal compromise. There is no cord signal abnormality and no intraspinal hemorrhage is present. This is all good news, and surgery can be averted.

injury, or vertebral body dislocations on radiograph or CT. On MRI, injury to the supraspinous and interspinous ligaments, ligamentum flavum, facet capsules, disk, and anterior and posterior longitudinal ligaments imply injury to the posterior ligament complex. Increased T2 signal within these structures as well as focal discontinuity of the T2 dark stabilizing ligaments can suggest the diagnosis of posterior ligament complex injury. Presence of intraspinal hemorrhage and cord/nerve root injury should also be reported.

Chance fractures are flexion/distraction injuries that occur at the thoracolumbar junction and typically are seen as a result of lap seat belt injury in a motor vehicle collision or fall from an axial loading injury. The vertebral body including anterior and posterior aspects as well as the posterior elements are fractured. All three columns are involved and there is a high probability of concomitant abdominal injury.

The most common fractures seen in motor vehicle collisions involving the lumbar spine are transverse process avulsion fractures. These are often at multiple levels and bilateral, and occur as a result of psoas muscle contractions leading to avulsions or from extreme rotation or sideways bending from motor vehicle collision impact. These are stable fractures. The rub is that when you see transverse process fractures, you should look for other, potentially unstable fractures elsewhere in the spine. Vertebral compression fractures are next most common.

Insufficiency Fractures of the Sacrum

These pathologic injuries result from the inability of abnormal bone to withstand normal stress and are seen after radiation therapy or secondary to postmenopausal or steroid-induced osteoporosis. In addition to the sacrum, they can occur in the lower extremities, ilium, and pubis. The major finding is linear defect running vertically (parallel to the sacroiliac joints) in the sacral alae, usually at the first through third sacral segments. Radionuclide bone scans reveal increased uptake ("Honda" sign). These are really tricky to pick up on MRI but not impossible, as T1 and T2 hypointense signal reflecting the fracture defect, with or without edema in the adjacent marrow. The underlying marrow disease can make the fractures difficult to appreciate even on CT, if you're not looking for them. Look and you shall see!

Spinal Cord Injury without Radiologic Abnormality

This condition presents as myelopathy after a traumatic event but the imaging of the cord is normal. This is more commonly seen in children with cervical spine injuries. It is presumed that the ligamentous laxity that children have can lead to injuries not apparent on plain film or CT. The cause is usually a hyperextension injury and patients may present with central cord symptoms. MRI may show ligamentous injuries and/or cartilaginous fractures. Treatment is immobilization. And that's a wrap!

Chapter 17

Approach and Pitfalls in Neuroimaging

SEARCH STRATEGY

If you have read the preceding chapters with due diligence, you have exceeded the knowledge base of the authors who have never even read what they have written. But can you read the films? Yes, that's the rub. One can have a huge cache of facts, but applying that genius to reading the studies is a whole different ball of ear wax. For that, we have (ta-dah!) Chapter 17: How to read studies. We hope to present to the reader in this chapter the ultimate Super-Google/Bing/PubMed "Search Engine" for identification and understanding the findings. Our goal in this chapter is to train you to have "good eyes." We always teach our fellows "you see what you know." The preceding chapters have allowed you to know some things. However, the ultimate compliment for the radiologist is that "She is brilliant *and* she has great eyes." Of course such a compliment may transcend the radiology application.

It is time to flex your OCD muscle. Reading neuroradiology requires an organized, thorough review of all images of a study. Many radiologists have picked up parotid masses, posterior triangle lymphadenopathy, cervical syringohydromyelic cavities, nasopharyngeal carcinomas, C1-C2 dislocations, and cervical vascular disease, on sagittal "scout" magnetic resonance (MR) images. How? By having a thorough consistently applied way of looking at the studies. Get in a routine for reading studies. Perhaps create a "macro template" on your cervical spine speech recognition program that forces you to make a comment about the patency of the vertebral arteries or status of the thyroid gland or visualized portions of the brain. We now have the electronic tools for reminding us of our blind spots. So for every miss (and we do not wish those misses on anyone—LOL), add to your template macro a reminder to check that region. To err is human, to get it right is divine. Nothing is more constructive (and instructive) than a missed case. How do you think we became so brilliant?

Another important point to remember is to avoid reading cases in a vacuum. Use the clinical information at hand to help guide your interpretation of the images. This means searching the medical record for clinical indication for the scan, physical examination findings, and overall clinical assessment of this patient. Not only will this help guide your review of the case, it will provide the basis for a more succinct and clinically relevant interpretation of a scan. No clinician wants to read a report that gives a differential diagnosis including congenital, vascular, infectious, inflammatory, demyelinating, traumatic, degenerative, toxic, metabolic, and neoplastic processes. Try to get a sense of what the clinical picture is, so that you can provide a more relevant and clinically useful differential diagnosis.

Do not be reluctant to look at previous scans when they are available. While it might seem tedious at first, recognizing that a lesion has been present on a previous examination or is a new finding is really helpful in your overall assessment. Also, it helps to review the previous report to make sure all findings previously described are accounted for on your scan.

Reading a Brain Computed Tomography Scan

The approach to reading a case varies according to the clinical indication for the study and personal habits. As long as the modus operandi is consistent, thorough, and organized, the exam will be read appropriately. We will provide one technique with a generic example of a commonly scribbled computed tomography (CT) request: "Emergency department (ED) study: rule out bleed." What is the context of the "rule-out bleed" study? Is it in the setting of a stroke, trauma, or worst headache of one's life? An ounce of clinical information is worth a pound of radiologist's tripe. Check the electronic medical record to direct the search. Sometimes the ED doctors are as aphasic on request slips as the Glasgow Coma Scale-3 patient.

Start by taking the central to peripheral approach, looking at each image from first slice to last individually. Your first "gestalt" of a CT is based on an analysis of the ventricles and basal cisterns. Because CT remains the primary screen in the ED for trauma or acute neurologic deficits, the first thing to assess is whether there is mass effect or subarachnoid hemorrhage. Look carefully at the density of the basal cisterns. Is blood present? Is there blood in the interhemispheric fissure, ventricles (beware the occipital horns—they are blowing for you), or sylvian fissure? Are the ventricles shifted from their normal position, and are the cisterns or sulci effaced by mass effect? Assess to what extent structures are shifted, because it may require an emergent intervention to save the patient. Is herniation (e.g., subfalcine, uncal, transtentorial, upward) present, detectable by shift of the lateral ventricles, dilatation of the temporal horn, compression of ambient cisterns, rotation of brain stem, or fourth ventricular displacement? Are all the basal cisterns effaced and the ventricles small, suggestive of diffuse cerebral swelling? Call someone quick! Be careful—subtle sulcal effacement (with even more subtle density changes) may be the only clue to early stroke. Are the ventricles and, in particular, the temporal horns enlarged out of proportion to the degree of sulcal enlargement, suggestive of hydrocephalus, or are they appropriate for the patient's age (quick check at date of birth) and clinical status? A 70-year-old with small ventricles should raise suspicions. Asymmetry of the lateral ventricle size is not uncommon, but it should heighten the search for a unilateral obstructive mass or septation/cyst at the foramen of Monroe. If there is hydrocephalus, is it communicating (all ventricles enlarged) or noncommunicating (isolated ventricular enlargement)?

By the analysis of the ventricles and cisterns, you should have determined whether a surgical emergency is present. Has an aneurysm bled? Does the patient need to be shunted? Is there brain swelling requiring steroids and diuretics or surgical decompression (careful to look in the posterior fossa for cerebellar infarcts that swell, obstruct the fourth ventricle, and lead to acute supratentorial hydrocephalus)? Is there potential herniation from a mass that needs to be removed emergently? Are you going to be up all night doing an angiogram and/or postoperative imaging?

What about the medical emergency? Is the stroke hemorrhagic, negating the opportunity for thrombolysis? Is there bilateral insula or cingulum or temporal lobe edema, suggestive of herpes encephalitis requiring emergent antiviral treatment? Is there added density to the subarachnoid space and mild hydrocephalus, suggestive of meningitis? Antibiotics please. Good eyes, doctor!

Next on the agenda is looking at the periphery of the brain for extraaxial collections. Again, CT is commonly used now in the setting of acute head trauma, and a subdural or epidural hematoma is a major source of morbidity. You may already have a suspicion that the patient might have mass effect on the basis of effacement of sulci, midline shift, buckling of the gray-white matter interface, or compression of ventricles. Search for an extraaxial collection after closed head injury. We routinely do coronal reconstructions in soft-tissue windows for our trauma cases to have a second chance (because we have "blown it" so many times just on the axial scans) at detecting that subtle subdural. We also review all of the thinnest section images from the raw data. A bit of extra "compulsiveness" that can keep you out of "malpracticeness"! The potential pitfalls: Bilateral extracerebral collections that do not cause ventricular shift; the isodense subdural hematoma detected by noticing that the white matter is not approaching the periphery of the brain and there is absence of normal sulcation.

On the peripheral pass through the brain you may detect soft-tissue scalp swelling, which hints to the site of the head trauma. Pay particular attention to the underlying brain as well as to the contrecoup side. If you detect a skull fracture (after your later review of the bone windows), be wary about epidural hematomas over the temporal lobes (middle meningeal artery) or near venous sinuses.

Next, begin to look at the brain parenchyma itself. An initial run-through from central to peripheral will identify any obvious areas of density differences, suggestive of hemorrhage or necrosis. Hypertensive bleeds occur most commonly in deep gray matter structures near the ventricles, so start there. In the setting of trauma, petechial hemorrhages that are rather subtle on CT may be the only evidence of a shearing injury (which will end up so easily detected on susceptibility-weighted MR!). Check the gray-white junction and the splenium for white matter tears. Common trauma sites are the inferior frontal lobes along the gyrus rectus and the anterior temporal lobes from impact along the greater wing of the sphenoid. Check carefully like a drone surveying Tora Bora. Work outward to detect cortical hemorrhages from infarcts or contusions or amyloid angiopathy. Again, you may be directed to look at a particular location coup or contrecoup on the basis of ventricular displacement, scalp findings, or peripheral collections.

For the stroke evaluation, one must look for the most subtle areas of distorted architecture or density differences. Most good clinicians know whether a patient has had a stroke. They do not need a neuroradiologist to tell them that, and hopefully if the patient got in early enough, your telling the neurologist that you do not see anything is the best news yet (that and the intravenous tissue plasminogen activator [IV tPA] is ready to roll at hour one). Our role is to let them know whether the stroke is hemorrhagic; whether there is mass effect, shift, or herniation; and whether it involves more than one third of the middle cerebral artery (MCA) distribution, leading to excessive risk of thrombolytic therapy. If you want to call the stroke, remember the early signs of infarct: clot in the vessel, loss of insular ribbon hyperdensity, loss of distinction between basal ganglia and internal capsule, and blurring of gray-white differentiation in the periphery. Remember your major vessel vascular anatomy to help with the check for strokes and to suggest another diagnosis if the distribution does not fit. Check the watershed areas for strokes caused by hypotension or hypoxia.

Having looked from central to peripheral and then around the periphery of the scan, it is time for the extracerebral abnormalities. Is there a nasopharyngeal or airway mass present? Are the globes normal, and are there orbital masses? Check the scans with bone windows. Are any skull base foramina enlarged or eroded? Are fractures present (orbital floor? lamina papyracea?)? Is sinusitis present accounting for that "worst headache of life"? Neck abscess, pharyngitis, dental abscess, mastoiditis, or otitis media in a search for a source of fever? Temporomandibular joint disease leading to headaches? Always check the scout topogram; you will be surprised how often you will pick up cervical spine injuries, spondylosis, bone metastases, basilar invagination, platybasia, myeloma, an enlarged sella, and other conditions on the basis of the scout view. These findings may not be as readily apparent on the axial images.

Use narrow window settings to view images for subtle stroke density differences and intermediate ones for subdural hematomas. With everything electronic and multiplanar, you can customize your settings to maximize detection of abnormalities based on clinical indications.

Reading a Brain Magnetic Resonance Scan

MR imaging is a lot harder to read than CT, because you are bombarded by information, all of it potentially useful to the analysis of the case, and the plaintiff malpractice lawyer. It is much harder to give a blueprint for how to read MR than CT because of the different pulse sequences, planes, and scan parameters available and coming out each season. CT is to MR as checkers is to chess. Here is a brief summary of one approach.

Use the same routine: Start from central to peripheral. The first image to look at is the sagittal midline image on the T1-weighted imaging (T1WI) scan. If that scan is hard to recognize, then it is likely that a mass displaces the midline, and that mass will be better seen on axial scans. On the midline image, take a look at the corpus callosum, the sella, the clivus, the superior sagittal sinus, vein of Galen, straight sinus, pineal gland, fourth ventricle, cerebellar tonsil position (5 mm or less below the foramen magnum is normal), cervical spinal cord, upper cervical spine, and nasopharynx. Are any of these structures displaced upward or downward, not present at all, or of abnormal signal intensity? From the midline, go to the more peripheral images and make sure you search for extraaxial collections, the vascular flow voids, and abnormal signal intensity (usually high) to suggest hemorrhage. Again, check the temporal tips and the temporal lobes for hemorrhage or sulcal distortion, particularly in the setting of trauma.

Remember that the sagittal image is usually the only one that also gives you a good peek at the cervical spine and neck, because most brain study axial scans are prescribed from the foramen magnum and go up. This is your only chance for that outstanding edge of the film neck call that elevates you in the eyes of the clinicians and separates your read from that of the competing neurologist on his Stark-law-compliant in-office MR scanner. Is posterior triangle lymphadenopathy present? Is a parotid, submandibular gland, lingual, pharyngeal, or laryngeal mass present? Are the orbits normal? That neurologist is only focused on the brain. Crush that! You know ear, nose, and throat (ENT) radiology after Chapters 9 to 14 of *Neuroradiology: The Requisites!*

The next sequence performed is usually a diffusion-weighted imaging (DWI) scan, particularly if a stroke is in question. Read it and read it fast (tPA is waiting!). Not everything that is bright on DWI is an infarct because of the inherent T2 weighting of these studies. Remember the concept of T2 shine-through as another source of DWI trace image brightness, so if you see high signal, refer to your apparent diffusion coefficient (ADC) maps to determine if the bright areas on DWI are from cytotoxic edema (decreased ADC and dark in signal) or from a T2 effect (increased ADC and bright). Remember also that some entities can cause decreased ADC but are not strokes (and this list unfortunately keeps growing), the classics being pyogenic abscesses, hypercellular tumors, (herpes) encephalitis, some demyelinating lesions, certain infectious processes, and the cortex in status epilepticus.

Remember that the evaluation of stroke may not end at a negative DWI scan. In centers with active stroke intervention programs, you may be required to perform a perfusion scan to demonstrate the ischemic penumbra. In some settings, this will lead to medical therapy designed to optimize cerebral blood flow (hypertensive, hypervolemic therapy) or thrombolysis. The brain tissue may not be infarcted (DWI negative) yet, but if you left the patient to their own devices, in 2 hours it may be. If fluid-attenuated inversion recovery (FLAIR) scans are negative and DWI scans are positive, the patient is likely in the less than 6- to 8-hour window where intraarterial thrombolysis may still be an option.

Next comes either your T2-weighted or FLAIR scan. Subtle differences exist between them but often these are best to identify increased water (i.e., vasogenic edema) that most brain lesions elicit. Just as with axial CT, follow the path from central to peripheral, to central, to peripheral again. As you analyze the ventricles and cisterns for displacements, the subdural and epidural spaces for collections, the deep gray matter structures for signal intensity abnormalities, and the peripheral cortex for subtleties of intensity differences and mass effect, recall how easy things seemed during your medical school training.

FLAIR scans will lull you into a state of complacency as the contrast of edema from normal tissue is so exquisite. But *beware!* Posterior fossa signal intensity abnormalities are not as contrasty on FLAIR and you should use T2-weighted imaging (T2WI) to double check this vital real estate. Remember also that as cystic and encephalomalacic areas approach water content in their chronic stage, they will be hypointense on FLAIR scans and may be less conspicuous. If you see something on FLAIR that might be artifactual, check the T2 to see if the finding persists: if it is still there on T2, it is real.

Often there is an enhanced scan to review. By this time, the FLAIR/T2WI/DWI will have been assessed and a mass may be suspected on the basis of morphologic criteria and/or a signal intensity abnormality or its associated vasogenic or cytotoxic edema; however, numerous lesions are apparent only on enhanced scans...especially those tiny metastases, leptomeningeal pathologies (including that isointense meningioma), cranial neuropathies, and gyriform lesions. If you have not performed a pregadolinium axial T1WI scan to compare with, you may have to rely on your sagittal T1WI scan (sometimes reconstructed in axial plane for easy comparison) to make sure what you are seeing is truly enhancement and not fat, blood, high protein, melanin, or hemorrhage.

Next, look at the periphery on your contrast-enhanced series. Is there abnormal enhancement of the meninges, dura, cisterns, exiting nerves, ependyma, or extraaxial fluid collections? Do the arteries enhance? (They should not on non–flow-compensated images because of fast flow, so maybe they are not arteries but veins, or maybe there is slow flow in them.) Finally, look at the areas that normally enhance to determine whether there are nonenhancing abnormalities. Do the adenohypophysis and pituitary stalk enhance uniformly? Do the venous sinuses enhance? Choroid plexi? Area postrema?

If you perform susceptibility-weighted imaging (SWI) (and we highly recommend them in the setting of stroke, trauma, and seizure workups), review these images for hemorrhagic products. Sometimes it can be tricky to distinguish hemorrhage from air and calcification, which are all dark on the SWI sequences. Hopefully reviewing phase maps and looking at the choroid plexus calcification signal intensity will aid you here. If it follows calcification, there you go, maybe not blood. SWI allows a great look at the meninges (hemosiderosis), the subarachnoid space for hemorrhage, gray matter areas of hemorrhagic stroke, gray-white matter hemorrhagic shearing injuries (check splenium), petechial hemorrhages from amyloid angiopathy versus hypertensive crises versus multiple (septic) emboli, not to mention striatocapsular hypertensive foci. SWI is *sweet!*

Return once again to that sagittal sequence and assess the precontrast signal intensity of any lesion you have identified subsequently. Retrace the hidden areas of the spine, neck, and midline structures.

A resident reading this chapter
Thought, "I wonder what the authors are after"
I know how to read 'em
I don't have to heed them
The lawyers could hardly contain their laughter
(As they hung him from the rafter)

Reading Spine Studies

Regardless if you are reading a spine radiograph, myelogram, CT, or MRI of the spine, it is helpful to get an overview of the spine morphology before diving in level by level. To begin, assess spinal alignment (you can start with sagittal and coronal reconstructions on a CT study). Is the expected curvature of the spine maintained or is there straightening, exaggerated curvature, or subluxation present at any particular level? Next assess vertebral body heights and disk spaces, as well as the facet relationships on either side. Is the density/signal of the vertebral marrow within normal limits? Are there endplate degenerative changes present or do they seem more aggressive and destructive? Are the interspinous distances maintained appropriately? Is there appropriate spacing between the clivus and dens of C2? Is there preservation of the normal anterior atlantodental interval? Coronal imaging is very useful in assessing the relationship between the occipital condyles, lateral masses of C1, and the C2 vertebral bodies. Are the prevertebral tissues of normal thickness? Any abnormalities in this sequence of evaluation should raise suspicion for pathology at that particular level and alert you to examine that level in more detail on your axial images.

On MRI, an overall assessment for cord caliber and signal is best obtained on sagittal T2 and short tau inversion recovery (STIR) sequences. If you suspect an abnormality in cord signal, take a look and confirm the abnormality on the axial images, because spine imaging is notorious for artifacts creating pseudolesions.

Proceeding level by level, an assessment of canal and foraminal patency should follow. If you see canal or foraminal stenosis, provide a reason for it. Is there a disk herniation, facet hypertrophy, or intraspinal/foraminal mass or fluid collection to explain the narrowing you are observing? Is there cord or nerve root compression present as a consequence? Remember that cord compressive pathology deserves a call to the clinical team, because urgent decompression might be necessary.

Reading Neck Studies

Nothing seems to elicit more fear in the radiology resident in training than reading the dreaded neck CT or neck MRI. It is true that the anatomy is complex, but the best way to overcome this fear is to have a systematic approach from one scan to the next.

Regardless of whether you are reading a neck CT or MR, an assessment of the following structures is required. Carefully review the aerodigestive tract (nasal cavity, nasopharynx, oral cavity, oropharynx, hypopharynx, and larynx) for asymmetries. Remember our ENT colleagues are much better than we are at identifying superficial/mucosal disease. Our job is to assess the submucosal or deeper extent of a suspected lesion. If you see an asymmetry, make an assessment of the scope of the abnormality, defining the anterior, posterior, superior, inferior, medial, and lateral extent of the abnormality. Are critical structures involved such as vessels, airway, and spinal canal? Then perform the same process for the soft tissues in the neck, using symmetry as your friend.

Posttreatment scans for malignancy are challenging to review, no doubt about it. The more you review, the better you will get at it. In such cases, it is a good idea to have a complete idea of the reason for the scan, the type of treatment the patient has received (surgery, chemotherapy, radiation), and to be aware of any soft-tissue or bony reconstructions performed. This requires a review of the patient's medical record, which can be time-consuming but is necessary. Once you are ready to review the images, it is important to look at your scan side by side with the previous scan, to assess for potential interval changes. Keep in mind that all changes do not necessarily mean tumor is back; there are a host of expected posttreatment changes that can be appreciated on follow-up examinations. Finally, an understanding of patterns of tumor spread for particular tumor types is very helpful. Is perineural tumor spread typical of the tumor in question? If so, take a careful look at the skull base foramina and respective soft tissues for evidence of invasion or denervation injury.

Next, take a look at the cervical lymph nodes and identify any that appear pathologic based on size, number, morphology, necrosis, perinodal inflammation, or calcification. Then evaluate the salivary glands for any asymmetry in terms of size or enhancement. Take a look at the thyroid gland for focal lesions or generalized enlargement. Make sure the major vessels in the neck opacity if the study is performed with contrast. Do not forget to take a peek at the brain, orbits, paranasal sinuses, cervical spine, and lung apices for abnormalities before you conclude your review.

HOW TO INTERPRET AN IMAGE

Knowledge of the implications of signal intensity and density changes in normal and abnormal tissue, together with the morphology, location, and clinical presentation of a lesion, enables accurate diagnosis. At no extra cost we have included tables of useful radiologic gamuts in the online Appendix at ExpertConsult.com. It is readily searchable for your ease of use...and we will keep expanding on it as our own knowledge increases.

Detection of lesions begins with knowledge of normal anatomy and its variants. It is only through reviewing numerous cases day after day that one obtains a mental image of the normal anatomy from which deviations can be readily identified. This is the source of the speed of the readings made by the codgy old professors of neuroradiology—they have burned a CD template into their brain of the pattern recognition for normality. Thus, they can scan an image at light speed and still detect the subtlest abnormalities. The trainees who view the most cases will be served well in this regard—they will have the easiest time detecting lesions.

Let us start with the complicated modality—MR. MR lends itself to image analysis from several perspectives. These are based on intensity, morphology, and location. We believe that you can evaluate a lesion with very little knowledge—how do you think we got this far? Liken yourself to a stock market technical analyst. Note the trends. You really do not have to know that much about the company (pathology) if you just follow the basics.

The first question is the intensity of the lesion on conventional pulse sequences: (1) T1WI, (2) DWI, (3) T2WI, (4) FLAIR, (5) SWI, and (6) postcontrast scans. Remember that the B0 images from a diffusion-weighted scan are a poor man's SWI, performed in just 40 to 50 seconds—use them. Table 17-1 provides useful information concerning the characterization of lesion types with intensity information. Unfortunately, all too often MR lacks specificity, because most lesions are dark on T1WI and bright on T2WI without hemorrhage or restricted diffusion. There is presently no signal intensity pattern hallmark that clearly distinguishes, say, multiple sclerosis (MS) plaques from infarction or tumor. If a lesion decreases in intensity as the T2 weighting increases, you are more likely to be dealing with a lesion that has susceptibility effects. Lesions that are hypointense on pulse sequences emphasizing T2 information often appear significantly hypointense on SWI if magnetic susceptibility (hemorrhage) is present.

The morphology of the lesion is important in its categorization. Several criteria are critical here: mass effect, atrophy, texture, edema pattern, extent of lesion, nature (solid versus cystic), number (single or multiple), distribution (e.g., along vascular supplies, Virchow-Robin spaces, cranial nerves, meninges, white matter tracts), involvement of one or both hemispheres, and enhancement characteristics. The presence or absence of mass effect is usually obvious. Lesions possessing mass effect are usually "active," whereas those that do not may be either old or still very new (within the first hours). Examples of the former include a tumor or new stroke, whereas the latter would include an old stroke or old traumatic injury to the brain. Mass effect can be subtle, with slight effacement of sulci; however, such changes are highly significant with regard to arriving at the correct diagnosis. Careful observation is necessary. The presence of focal atrophy (tissue loss) signifies past insult to the brain. Although the brain parenchyma decreases with age normally, *focal* loss of parenchyma is significant for old strokes, traumatic episodes, postoperative changes, or treated encephalitides. In contrast, *global* atrophic changes exceeding those for age suggest other processes such as steroid use, neurodegenerative disorders, acquired immunodeficiency syndrome (AIDS) infections, or substance abuse (beware those states legalizing marijuana).

Certain lesions have characteristic textures. For instance, oligodendrogliomas have a rather heterogeneous texture, whereas lymphomas are more homogeneous. If edema is associated with a lesion, there is clearly an irritative element involved. The converse of this is not true; that is, if there is no edema, there is nothing harmful. Lesions without bland edema can be virulent: cortical metastases, gliomatosis cerebri, Creutzfeldt-Jakob disease, and human immunodeficiency virus (HIV) infection. The extent of the lesion also gives some clues to the diagnosis. In general, a lesion spanning both hemispheres is most likely tumor, because bland edema does not usually cross the connecting white matter tracts. Dysmyelinating disorders may be the exception to that rule. Generally, lesions that are aggressive are poorly marginated and infiltrative. Again, the converse is not true. Many lesions have cystic components or are themselves cystic yet span the spectrum of aggressiveness. These include colloid cysts, craniopharyngioma, cystic astrocytoma, or necrotic glioblastomas. FLAIR images can distinguish cerebrospinal fluid (CSF) containing structures (hypointense) from more complex cystic lesions, the latter being high signal intensity with contents that have complex constituents including high levels of protein. Multiplicity of lesions changes the radiologic diagnostic gamut (e.g., if multiple suggest metastatic lesions, multiple strokes, multicentric tumor, neurogenic tumors, and MS).

Enhancement is very important. It establishes that the lesion has an abnormal blood-brain barrier. It does not, however, indicate whether a lesion is benign or malignant, nor does it demarcate the border of the lesion. The presence of enhancement is instrumental in increasing our sensitivity to detecting abnormalities, particularly extra-axial and cortical neoplasms. We err on the side of giving contrast because it improves our sensitivity and specificity and enables us to read faster with more conviction. It also is the best way to show necrosis in a lesion seen as central low signal surrounded by enhancing tissue. This

TABLE 17-1   Signal Intensity Characteristics of Tissue on T1-Weighted Images and T2-Weighted Images

High Intensity on T1-Weighted Images	Low Intensity on T1-Weighted Images	High Intensity on T2-Weighted Images (Hyperintense)	Low Intensity on T2-Weighted Images (Hypointense)
High protein	Water (CSF, edema)	Water (CSF, edema)	High protein
Subacute hemorrhage (methemoglobin)	Acute hemorrhage (deoxyhemoglobin)		Acute hemorrhage (deoxyhemoglobin)
Gadolinium	Chronic hemorrhage (hemosiderin)		Chronic hemorrhage (hemosiderin)
Other paramagnetics (manganese, calcium, iron)	Diamagnetic effects (calcification, air)		Early subacute hemorrhage (intracellular methemoglobin)
Blood flow (flow-related enhancement or even-echo rephasing)	Fast flow		Other paramagnetics (melanin, calcium)
Fat	Very viscous protein		Diamagnetics (calcification, air)
Cholesterol	Susceptibility artifact		High concentrations of gadolinium
Melanin	Low protein		Fast flow
	Air		Fat (non–fast spin echo)
	Calcium		Metal
			Susceptibility artifact
			Air
			Calcium

CSF, Cerebrospinal fluid.

TABLE 17-2 Density Characteristics of Lesions on Computed Tomography

Hypodense	Isodense	Hyperdense
Water (CSF, edema)	Intermediate protein	High protein
Fat	Normal brain	Acute hemorrhage
Air		Calcium
Low protein		Metal
		Iodine
		Pantopaque

CSF, Cerebrospinal fluid.

means badness (usually a high-grade astrocytoma, metastasis, abscess, or radiation injury—all of which will ruin your day prepping for the first tee).

Finally, location (just as in real estate) is of critical importance in making the correct diagnosis. Is the lesion intraaxial, extraaxial, or both? Obviously, extraaxial lesions suggest a different (usually more benign, gentile) differential diagnosis from those that are purely intraaxial. Multiplanar images help resolve this question. Certain lesions have a propensity for specific locations; for example, herpes simplex affects the temporal lobe, oligodendroglioma the frontotemporal lobe, and juvenile pilocytic astrocytomas the posterior fossa and suprasellar region. Does the lesion involve the cortex, white matter, or both? This provides the initial diagnostic algorithm. If the lesion is predominantly in the white matter, we might consider MS, whereas if it affects both white and gray matter it may suggest a stroke. In the latter instance, the lesion should follow a vascular distribution (but sometimes lesions do not read our book).

The interpretation of CT overlaps that of MR with regard to location, morphology, presentation, and enhancement features. The exception to this rule is that vessels enhance on CT, whereas, if there is fast flow, they do not enhance on MR. For your limbic pleasure, consider the density characteristics of lesions on CT in Table 17-2.

REVIEW OF BRAIN NEOPLASMS

To repeat for a third time in this book (Murphy's law: that which is repeated most will be the first forgotten), the most fundamental question that neuroradiologists must ask themselves when faced with an intracranial or intraspinal mass is, "Is the lesion intraaxial or extraaxial?" This assumes you have correctly answered the question, "Is there a mass present?" Happily, most extraaxial nonosseous masses are benign and are limited to a few entities (meningiomas, schwannomas, epidermoids, arachnoid cysts) that can usually be parceled out based on enhancement characteristics, diffusion-weighted and FLAIR signal, and morphology. For the purposes of this discussion, intraventricular lesions are considered extraaxial lesions.

Extraaxial Brain Neoplasms

Neuroradiologists ascertain whether a mass is intraaxial or extraaxial by several criteria. Extraaxial lesions tend to push the intraaxial structures rather than infiltrate them. Therefore, one sees buckling of gray and white matter

around the extraaxial mass. Extraaxial masses tend to have flat, broad bases along the skull or spinal canal. On MR one often sees a "cleft" of (1) low-intensity dura being draped around the mass, (2) cortical vessels being displaced inwardly by the mass, and/or (3) CSF trapped around a mass. Occasionally, particularly in the spinal canal, you will see a meniscus of CSF above and/or below the extraaxial mass. Ipsilateral CSF expansion with contralateral compression of the CSF is the hallmark of intradural extraaxial lesions. Of extraaxial masses, meningiomas classically demonstrate dural tails in which enhancement is seen to extend in a triangular manner along the dura, "tailing off" away from the mass.

In contrast to extraaxial lesions, intraaxial lesions tend to infiltrate the white matter and expand the superficial brain tissue. They blur the distinction between white and gray matter. They tend to have tongues of tissue that extend deeply in the white matter and may cross the midline through the white matter tracts of the commissures and corpus callosum.

Often, because of partial volume effects and the way extraaxial lesions may invaginate into the surrounding central nervous system tissue, it may not be possible to determine whether a lesion is intraaxial or extraaxial on the basis of a single plane. This underscores the tremendous advantage of MR and multiplanar multidetector CT reconstructions for evaluating intracranial or intraspinal lesions. The improved soft-tissue discrimination of MR beats CT by helping to differentiate the signal intensity of the extraaxial neoplasm from dura, vessels, gray matter, and white matter.

Having identified an extraaxial lesion, the next step in limiting the differential diagnosis lies in determining whether the lesion is benign or malignant.

Benign Extraaxial Brain Neoplasms

As stated previously, the common benign extraaxial masses of the central nervous system are meningiomas and schwannomas. Less common benign extraaxial lesions include lipomas and dermoids (fat intensity), arachnoid cysts (pure CSF on all sequences), and epidermoids (often bright on DWI and not quite CSF on FLAIR and T1WI). The latter four do not enhance; meningiomas and schwannomas enhance.

What else distinguishes meningiomas and schwannomas? CT may be useful. Generally, on unenhanced examinations, meningiomas are slightly hyperdense as compared with schwannomas, which are isodense or hypodense. The presence of calcification on CT favors a meningioma. Cystic degeneration favors schwannomas, whereas fatty degeneration favors meningioma, although both can occur in either lesion. Hemorrhagic conversion favors a schwannoma. Meningiomas tend to elicit a hyperostotic response in the adjacent bone, whereas schwannomas expand and remodel bone.

On MR, the presence of a dural tail, vascular flow voids within and around the mass, and a broad dural base should lead you to favor meningioma. Obviously, lesions that are located in areas unlikely to have nerves spanning the extraaxial space are more likely to be meningiomas (convexities, along sphenoid wing, falcine). It is only within the spine, around the foramen magnum, ovale, and rotundum, intraorbitally, in the cerebellopontine angle, along the cavernous sinus, and in a suprasellar location that a differential diagnosis of schwannoma and meningioma is debated (okay, so not that uncommon). Widening of a neural foramen, an orbital fissure, or a porus acusticus might favor a schwannoma. Schwannomas are more oblong or dumbbell-shaped, and they are usually brighter on T2WI.

An important caveat: Anytime you say meningioma, also think sarcoid, plasmacytoma, lymphoma, or dural metastasis.

Malignant Extraaxial Brain Neoplasms

The majority of malignant extraaxial masses are due to metastatic bone disease. (Yes, we have to look at the skull too—yawn, yawn.) Although meningiomas may demonstrate some reactivity of the bone in a sclerotic (or, less commonly, lytic) pattern, the bone changes with malignancies tend to be much more destructive, aggressive, and infiltrative. If only the inner table of the skull is involved, favor meningioma; if both tables, suspect metastases. Extension from the bone to the soft tissue in the scalp is more common with metastatic bone disease, although it may occur with meningiomas too. Sometimes patients are referred to the neuroradiologist on the basis of abnormalities on bone scans where multiple lesions are identified. Multiplicity favors metastatic disease.

The other malignant extraaxial masses include the axes of evil: subarachnoid drop metastases, dural metastases, and lymphoma (which can involve the subarachnoid space, epidural space, bone, dura, or parenchyma). These malignant extraaxial lesions have a higher rate of eliciting vasogenic edema in the subjacent brain, but remember because meningiomas are so much more common, they should also be included the differential diagnosis. Lymphoma and sarcoid are good diagnoses to blurt out when you have run out of things to say in conferences, at the boards, or on a blind date.

Intraventricular Neoplasms

The differential diagnosis of intraventricular lesions resides in a combination of the evaluation of the patient's age and the location of the lesion. In the pediatric age group, an enhancing lesion in the region of the trigone of the lateral ventricle is a choroid plexus papilloma until proven otherwise. The presence of calcification and/or hemorrhage within the lesion centered on the glomus of the choroid plexus will cinch the diagnosis. Meningiomas present near the trigone also. If the lesion is in the fourth ventricle or body of the lateral ventricle, consider ependymomas and medulloblastomas.

In an adult, the intraventricular masses to consider include meningiomas, oligodendrogliomas, intraventricular/central neurocytomas, colloid cysts, (sub)ependymomas, subependymal giant cell astrocytomas, and fourth ventricular choroid plexus papillomas. Meningiomas occur near the glomus of the choroid plexus in the lateral ventricles. In the young adult, an enhancing mass at the foramen of Monro may be a subependymal giant cell astrocytoma (check for stigmata of tuberous sclerosis); in older patients, think neurocytoma, glioblastoma, metastases, or lymphoma. If a mass at the foramen of Monro does not enhance, consider a low-grade glioma or colloid cyst if the patient is younger than 30 years of age, and a subependymoma if older than 30 years of age.

Intraventricular neurocytomas can occur anywhere but favor the septum pellucidum, temporal, and frontal horns. Colloid cysts occur at the midline level of the foramina of Monro. Colloid cysts, ependymal cysts, and epidermoids do not enhance; meningiomas do.

Pineal Region Neoplasms

Specific regions of the brain have more limited differential diagnoses than other areas, including the pineal and the suprasellar cistern regions. Pineal region tumors can best be differentiated on the basis of serology and the sex of the patient. The germinoma–germ cell lines of pineal region tumors are very uncommon in female patients. Therefore, if you encounter a pineal region tumor in a female patient, it is usually a pineocytoma, pineoblastoma, or tectal glioma mimicking a pineal region lesion. Just in case, look for the presence of fat and/or cystic areas to ensure that it is not dermoid or teratoma. On the other hand, in a male patient, favor the germ cell series, and if the lesion is hyperdense on unenhanced CT without evidence of fat or cystic areas, favor germinoma. Germinomas occur in a younger age group than pineal cell tumors but the latter, particularly the pineoblastomas, are of lower ADC values.

Suprasellar Neoplasms

In the suprasellar cistern, the differential diagnosis of masses includes vascular lesions such as aneurysms as well as congenital and neoplastic diseases. MR is preferable to CT in evaluating these lesions because it identifies aneurysms nicely with flow voids and/or flow-related artifacts. Start with seeing if you can separate the uninvolved pituitary gland from the suprasellar lesion because suprasellar extension of a pituitary adenoma is by far the most common suprasellar mass in our book…hey, this *is* our book—LOL. If you cannot find the gland, the presence of sellar expansion and the position of the diaphragma sellae (elevated with pituitary lesions and depressed with suprasellar lesions) may indicate the diagnosis of a pituitary adenoma.

A mnemonic for suprasellar masses is SATCHMOE. This mnemonic is an abbreviation for: suprasellar extension of pituitary adenomas and sarcoid, aneurysms and astrocytomas, tuberculum sellae meningiomas and tuberculosis, craniopharyngiomas (Rathke cysts and choristomas), hypothalamic gliomas and histiocytosis X, metastatic lesions and myoblastomas of the posterior pituitary, optic nerve gliomas and optic neuritis, and epidermoid-dermoid-teratomas and Erdheim-Chester disease. As stated previously, a patent aneurysm is easily separated from other lesions on MR owing to flow voids, pulsation artifacts, and signal intensity. The presence of high signal intensity in a suprasellar mass on a T1WI should suggest the diagnosis of a craniopharyngioma, Rathke cyst, hemorrhagic pituitary adenoma, lipoma, teratoma, or clotted (pseudo)aneurysm. The presence of a chemical shift artifact suggestive of fat in the teratoma or dermoid lesion mitigates against a craniopharyngioma or Rathke cyst, where the high signal pitensity is thought to be due to either hemorrhage or hyperproteinaceous secretions. The distinction between optic chiasm gliomas (seen in neurofibromatosis type I) and hypothalamic gliomas is moot because, often, these lesions encompass both regions. The presence of bony reaction along the planum sphenoidale or tuberculum sellae, a dural tail, and the intensity characteristics will suggest the diagnosis of meningioma.

Intraaxial Brain Neoplasms

The majority of intraaxial neoplasms in the brain result from "gliomas" (usually astrocytomas) and metastases. When faced with a *single* tumor in the brain, the numbers suggest that it is nearly a toss-up between primary astrocytomas and single metastases for the most common intraaxial lesion in the adult. MR often shows a well-defined, enhancing, circumscribed mass within the brain, usually with surrounding edema (unless in the cortex) with metastatic disease. The edema associated with a metastasis may be impressive for the size of the enhancing mass. When you identify multiple enhancing lesions in the adult brain, you are most likely dealing with metastases. No enhancement? Probably not metastases. Okay, it could be metastases on steroids, like Alex Rodriguez?

As opposed to metastases, astrocytomas generally tend to be more infiltrative, less well defined, and, for lower-grade tumors, less avidly enhancing. They are less likely to be multiple, although occasionally multifocal glioblastomas and hemangioblastomas may occur.

Arriving at a limited differential diagnosis for gliomas ultimately may be purely mental gymnastics. Often, despite what a radiologist may say, a piece of tissue is required before treatment is instituted. For all the neuroradiologist's ruminations about whether the lesion is a medulloblastoma versus an ependymoma, the neurosurgeon may simply shrug his or her shoulders, hone the scalpel, lop it out, and leave it to the pathologist to sort out.

Adult Intraaxial Neoplasms

The very first thing to do in evaluating an intraaxial mass in the brain is to check the film for the patient's age. The set of tumors seen in children is different from that seen in adults, with only moderate overlap. In an adult, astrocytomas are the most common primary neoplasms. What are the features that suggest a higher-grade astrocytoma? Let us count the ways: (1) increasing age, (2) necrosis, (3) hemorrhage (best seen on susceptibility scans), (4) enhancement, (5) low ADC, (6) elevated relative cerebral blood volume on perfusion-weighted imaging, (7) rapid growth rate, and (8) elevated choline on magnetic resonance spectroscopic imaging.

In adults with an intraparenchymal mass, look for the internal architecture of the lesion. A hyperdense nonhemorrhagic intraaxial tumor on unenhanced CT should suggest lymphoma or a "blastoma" (primitive neuroectodermal tumor). Lymphoma may coat the ependymal surfaces of the brain ("rimphoma") and is one of the tumors that crosses the corpus callosum. Hemorrhage should give you pause. Are you sure it is a tumor and not a hematoma from a vascular lesion, amyloid angiopathy, or hypertension? If there is a solid enhancing nodule nearby, then the gamut of a hemorrhagic tumor should be entertained and would include metastases (e.g., melanoma, choriocarcinoma,

renal cell carcinoma, thyroid) or a high-grade malignancy, usually glioblastoma.

When a calcified cortex involving intraparenchymal mass is encountered, move from astrocytoma to oligodendrogliomas (these enhance in 40% of cases), and the rare intraparenchymal ependymoma. Cystic masses include gangliogliomas and pilocytic astrocytomas. A mass in the posterior fossa in the adult is most likely a metastasis; one with increased vascularity, a mural nodule, and/or cyst formation is usually a hemangioblastoma or pilocytic astrocytoma.

Pediatric Intraaxial Neoplasms

As stated previously, the lesions in children tend to be different from those in adults. Posterior fossa masses predominate in the pediatric age group. The differential diagnosis often reduces to a choice of ependymoma, medulloblastoma (primitive neuroectodermal tumor), and pilocytic astrocytoma. Astrocytomas are usually identified in the cerebellar hemispheres more peripherally and most often are cystic with an associated well-defined, enhancing nodule. In this regard, they simulate the adult hemangioblastoma. Medulloblastomas, choroid plexus papillomas, and ependymomas may grow in the fourth ventricle in children. Medulloblastomas are usually hyperdense on unenhanced CT, and more likely infiltrate and compress the fourth ventricle as they arise from medullary vela. Ependymomas more commonly calcify, usually enlarge the fourth ventricle, and have a propensity for extending out the foramina of Magendie (medially) and Luschka (laterally). These three tumors vary widely in their ADC values (medulloblastomas very low, ependymomas intermediate, and juvenile pilocytic astrocytomas very high). Choroid plexus tumors are more vascular (have more flow voids on MR) and enhance more dramatically than ependymomas. Fourth ventricular choroid plexus papillomas are more often seen in older patients; the lateral ventricles are favored in children.

Brain stem gliomas occur more frequently in children. Usually, they are identified because of the distorted morphology; the density of these tumors (usually fibrillary astrocytomas) may mimic normal tissue. They infrequently enhance, and when they do it is spotty at best. What is in the differential diagnosis? Worldwide, tuberculosis is a common brain stem lesion, distinguished by greater enhancement and associated leptomeningeal disease. Rhombencephalitis from *Listeria monocytogenes* may simulate an infiltrative brain stem mass and is a bugger of a bug to try to grow in culture. Shake some ampicillin at it, though, and maybe it will go away. Lymphomas may occur in the brain stem. Hopefully, they will be hyperdense on CT to help make that diagnosis. Demyelinating (MS, acute disseminated encephalomyelitis, central pontine myelinolysis) and vascular disorders may affect the brain stem; the presentations for these lesions (more abrupt in onset and usually not seen in children) should allow differentiation from an astrocytoma.

In the supratentorial space, gangliogliomas, low-grade astrocytomas, and cortical neuroblastomas predominate in children. These lesions may be identical in appearance, and unless you see cystic areas suggesting gangliogliomas, arriving at a specific diagnosis is very difficult. In children, think ET: PNET (primitive neuroectodermal tumor), DNET (dysembryoplastic neuroepithelial tumors), or the large, aggressive, often necrotic, atypical rhabdoid tumor.

NONNEOPLASTIC BRAIN LESIONS

Infarcts

The previous discussion assumes that a lesion is identified as neoplastic. In the adult, the major differential diagnosis of a lesion that has mass effect but is nonneoplastic is an infarct. As opposed to neoplastic lesions, infarcts follow a vascular distribution, may involve the gray matter preferentially, and generally are wedge shaped and wider at the cortical surface. Infarcts almost always present with an acute ictus and within a week or two lose mass effect and swelling, whereas neoplasms and untreated infections progress. Persistence of mass effect beyond 2 weeks strongly favors a nonischemic cause. The diagnosis of acute infarction is no longer the "stroke of genius" thanks to diffusion-weighted scanning. It is incredibly rare for a new completed infarct not to light up on DWI from lowered ADC.

Difficulty arises when infarcts do not appear to be within an arterial distribution and do not obey the anterior, middle, and posterior cerebral artery territories. In these cases, consider venous infarcts in your differential. Depending on which vein/sinus is occluded, one may see deep gray (internal cerebral vein, straight sinus), temporal lobe (transverse sinus, middle cerebral veins, vein of Labbé), or parasagittal (superior sagittal sinus) lesions. Venous infarcts on the whole are hemorrhagic, and often a thrombosed sinus and/or cortical vein (most likely seen on that sagittal T1-weighted image) are seen in association. Diffusion-weighted scans are usually bright, but there may be an element of DWI reversibility with venous ischemia. If hemorrhagic, it may screw the pooch on the DWI image, making things too confusing for the in-office neurologist.

In young patients with strokes, think about predisposing conditions such as sickle cell disease, dissections (check your vascular "voids"), hypercoagulable states (antiphospholipid antibody syndrome), MELAS (mitochondrial encephalomyopathy, lactic acidosis, and strokelike episodes), and vasculitis.

Infections

Septic emboli may be located in subsegmental distributions of major vessels, usually the MCA. They may induce a large amount of vasogenic edema because of the infection, and they generally arise at the corticomedullary junction, with extension to the gray matter. These lesions are frequently multiple, and the differential diagnosis usually includes metastases. Patients with valve lesions who self-medicate in dark alleys (i.e., shoot up) are the typical hosts.

When septic emboli induce abscesses, the differentiation from neoplasm becomes much more difficult. DWI should help considerably, as the high signal would be rare for a differential diagnosis of metastases and necrotic primary tumors. Let's assume equivocal DWI: What features suggest abscesses? (1) The walls may be thin and regular with a thinner inner border closest to the ventricle, (2) lots of vasogenic edema, (3) on unenhanced T1WI, a hyperintense rim, and (4) associated meningeal enhancement.

The distinction between encephalitis and infarct is generally made on vascular distribution or DWI (although encephalitis may have restricted diffusion, it is not as constant as acute strokes), or given a nurse practitioner taking a good history in the ED, clinical grounds. Encephalitides generally do not respect vascular distributions; however, with herpes virus infection, they may be localized to particular parts of the brain supplied by the MCA (but it spares basal ganglia). Herpes preferentially affects the gray matter and white matter of the temporal lobes and frontal lobes, and may be a unilateral or asymmetric bilateral process. Fever, bilaterality, confusion, and a rapidly progressive course over a few days distinguish herpes from tumors. Offer to perform the lumbar puncture to help, if the clinician looks at you cross-eyed…on the patient, not the clinician.

The other viral encephalitides, including HIV infection and cytomegalovirus infection, may cause focal high signal intensity abnormalities in the white matter that resemble those of ischemic small vessel white matter disease, but are generally more confluent. Once again, the differential diagnosis is generally made on clinical grounds, with the viral infections occurring in a younger population and white matter ischemia in an older population with vascular risk factors. These entities do not usually enhance, whereas MS might (see Demyelinating Disorders subsection in this chapter).

Metabolic Disorders

Metabolic disorders of the brain can cause signal intensity abnormalities in the basal ganglia and deep gray matter structures as well as within the white matter. As opposed to neoplastic, vascular, or infectious causes, the metabolic diseases generally are bilateral and symmetric processes. They may or may not incite much edema or mass effect. Leave these diagnoses to serology, spectroscopy, geneticists, and the fleas—they only deserve four sentences.

Demyelinating Disorders

When one or more lesions are present in the white matter in the brain of an adult, the differential diagnosis may be extensive. Because MS is most often a clinical diagnosis and also the most common demyelinating disorder in adults, the neuroradiologist does not lose face if he or she recommends clinical correlation in the presence of areas of demyelination in the young patient's brain.

Because MS is a polyphasic disease, the presence of areas of abnormal intensity in the white matter that do and do not show enhancement suggests that the lesions are spaced out in time. Therefore, if you identify multiple white matter lesions in the brain, usually without mass effect, some of which enhance and some of which do not enhance, the likelihood of MS is increased. Ask for a history of optic neuritis (check for optic atrophy or bright signal in the nerves to score points with the authors) and you may have your final clue. It should be noted, however, that other polyphasic disorders such as Lyme disease, sarcoidosis, and vasculitides can appear in a similar manner.

At intermediate age groups that overlap the pathologies, you may be asked, "Is it MS or is it ischemic white matter?" Favoring MS would be (1) callososeptal interface and corpus callosum lesions, (2) flame-shaped lesions from the edges of the ventricles, (3) absence of basal ganglionic lacunar disease, (4) enhancing lesions, (5) cerebellar peduncle lesions, (6) lesions in the spinal cord, and (7) lesions in the optic nerves.

The typical differential diagnosis of monophasic white matter lesions includes acute disseminated encephalomyelitis or posttraumatic white matter shearing injuries. Shearing injuries of the white matter almost always have hemorrhage associated with them on susceptibility-weighted images. When hemorrhage is present in white matter lesions, search for a history of trauma. Also think amyloid and microhemorrhages. Usually in the chronic phase, hemosiderin is seen. Typically, posttraumatic injuries, migrainous white matter lesions, progressive multifocal leukoencephalopathy, and dilated Virchow-Robin spaces do not enhance. Acute disseminated encephalomyelitis, on the other hand, may enhance.

As opposed to demyelinating disorders, the dysmyelinating, peroxisomal, and metabolic disorders are spotted zebras and tend to be bilateral, symmetric, and more diffuse (bizarre). They are typically seen in a pediatric or at-risk population. When diffuse white matter disease is associated with macrocephaly, Alexander and Canavan diseases should be suggested. If the white matter abnormality has an occipital predominance and has an enhancing advancing border, adrenal leukodystrophy is the major diagnosis to consider. It is left to serologic and enzymatic analysis to diagnose the most common dysmyelinating disorder, metachromatic leukodystrophy, as well as the lipodystrophies, mucopolysaccharidoses, and other leukodystrophies. Toss a bone to the spectroscopists: Canavan disease is diagnosed by markedly elevated N-acetylaspartate (NAA) peaks.

Hematomas

Let us say you find a hematoma in the brain parenchyma. History sometimes suggests a traumatic hematoma, but make sure the trauma preceded the ictus. Is the bleed the cause of the motor vehicle collision leading to traumatic brain injury (uh-oh, you've got to do an arteriogram!) or did the trauma from the motor vehicle collision cause the bleed (shooo—you're in the clear). Mass effect out of proportion to the amount of hemorrhage, incomplete hemosiderin rings, lack of reduction in mass effect over time, and bizarre intensity progression strongly suggest a tumor. An enhancing nodule would seal the diagnosis. Subarachnoid hemorrhage associated with a hematoma might point to an aneurysm. Enlarged serpentine arteries and veins are seen with arteriovenous malformations, but beware, they may be transiently inapparent because of compression by the hematoma. Make sure the clinicians check the patient's blood pressure to exclude a hypertensive hemorrhage—basal ganglionic and thalamic hemorrhages are common in patients with hypertension. Finally, increased age and multiplicity suggest amyloid angiopathy. Siderosis may

accompany this diagnosis. Tiny dots of parenchymal hemosiderin staining may also be found with amyloid angiopathy, but you must exclude etiologies such as microbleeds from hypertensive crises, radiation-induced telangiectasias, diffuse axonal injuries, and tiny cavernomas.

SPINAL LESIONS

First step in the spine: Is it intradural intramedullary, intradural extramedullary, or is it extradural? Use the intraaxial/extraaxial discussion above to help out (those of you on the electronic version can cut and paste it here).

With regard to intradural intramedullary (intraaxial) lesions of the spine, the presence of *multiplicity* should suggest the diagnosis of a demyelinating disease, sarcoid, hemangioblastomas, neurofibromas, lymphoma, and metastases. Remember to scan the brain for additional lesions depending on the clinical context. Of course, the most common cause of a cord signal abnormality is from spinal stenosis in the cervical spine banging the cord—spondylotic myelopathy.

With a *single* intramedullary lesion, neoplasms should be first on your mind. It has been said in the past that nearly all the neoplasms of the spinal cord enhance, but that's not always the case. Tumor enhancement is generally focal and nodular rather than diffuse streaky as seen in the demyelinating and inflammatory disorders; however, occasionally, unusual spinal lesions such as herpes zoster, acute disseminated encephalomyelitis, collagen vascular disease, or tuberculosis simulate a neoplasm. Cord neoplasms usually span many segments before they present clinically. Transverse myelitis may cross a few segments, but a solitary MS plaque will usually involve just one to two levels.

In patients with high-intensity cords on T2WI, be sure to scrutinize the enhanced sequence for evidence of serpentine veins on the posterior surface of the cord. Because a dural vascular malformation in the proper clinical setting (Foix-Alajouanine syndrome) may simulate a neoplasm with progressive myelopathy, this may be your only chance to spare the patient a cord biopsy. Be vigilant, as these lesions favor the lower thoracic region, may or may not expand the cord, and generally are centrally located. Also with multisegment high T2 signal in the cord, consider infarct in your differential; the clinical context will be most useful in this regard.

Guess what? Only broad generalizations can be made in trying to determine the histology of an intramedullary glioma, but you might as well take the plunge. Better defined, more focal hemorrhagic lesions with hemosiderin caps tend to be ependymomas, whereas diffuse infiltrated masses are typically astrocytomas. Younger age, NF-1 history, and cervical lesions favor astrocytomas; filum involvement, NF-2, and older age favor ependymomas. Hemangioblastomas are vascular and may have flow voids within them.

The characteristics of intradural extramedullary lesions follow the intracranial ones. (Cut and paste again here.) In the spine, we would say the features that favor spinal neurogenic tumors over spinal meningiomas include: (1) cervical and lumbar location, (2) no dural tail, (3) no calcification, (4) traversing from intradural to extradural, (5) cystic components, (6) NF-1, NF-2, and (7) foraminal enlargement not sclerosis.

Extradural lesions typically arise from the perivertebral tissues or disc space before making their way into the spinal canal. Lesions span the gamut of degenerative, infectious, inflammatory, and neoplastic processes. Use your clinical history to guide the differential. From a practical standpoint, it is probably more important to alert the clinical team about extent of spinal involvement and mass effect upon critical intraspinal structures rather than make a precise histopathologic diagnosis.

GRAND FINALE

There are always exceptions to our rules except this rule. Be careful (Tables 17-3 and 17-4). It is a jungle out there, with herds of "zebras" and hungry lawyers stalking in the bushes. Give it your best shot, before the lawyers do. That is the excitement of our profession.

Live, love, learn, and leave a legacy.

TABLE 17-3 Overrated Signs in Neuroradiology

Name of Sign	Presumed Significance	Pitfall
Hyperdense clot (hyperdensity in MCA vessel) on CT	Clot in vessel	Vessels are nearly always brighter than surrounding areas on third-generation or newer CT scanners
Insular ribbon sign (loss of density of cortex in insula) on CT	Sign of MCA infarct	Rarely seen
Murphy's tit (nubbin at apex of aneurysm)	Site of aneurysmal bleed	Too infrequently seen
Chiari I malformation (tonsils below foramen magnum)	Congenital anomaly	Too frequently seen
Partially empty sella (increased CSF in sella)	?	If you do not see it, it may be abnormal (that is how often we see it)
Stalk deviation (stalk pushed off midline)	Adenoma may be on opposite side, pushing stalk over	Seen in 3% of normal patients
Hippocampal-choroidal fissure dilatation	Suggestive of Alzheimer disease	Requires unusual plane of section to see reproducibly, found in department chairpersons
Hyperdense falx or tent (greater than brain density) on CT	Interhemispheric subdural hematoma	Falx and tentorium are naturally dense, often calcify
Dural tail sign	Sign of meningioma	Seen in any lesion involving the dura. The authors are keen observers of this sign.
Periventricular hyperintensity	Has been used to classify multiple sclerosis, age-related "ischemic" lesions in Alzheimer disease and depression	Universally seen. Only nonradiologists would use this "Simple Simon" approach
Bright CSF on FLAIR	Indicative of subarachnoid disease	Basal cisterns often normally bright Anesthesia with use of 100% O_2 turns CSF bright Nonspecific Patients with renal failure after gadolinium will have bright CSF on FLAIR and T1-weighted imaging
Aqueductal flow void accentuation	Indicates normal pressure hydrocephalus	Nonspecific Correlates poorly with shunt success Technique-dependent
CSF dimple-cleft sign	Cortical dysplasia	Still waiting to see first case
Books by other authors	Prettier, more expensive	Have you laughed once reading their books?

CSF, Cerebrospinal fluid; *CT*, computed tomography; *FLAIR*, fluid-attenuated inversion recovery; *MCA*, middle cerebral artery.

TABLE 17-4 Overrated Techniques

Name of Technique	Presumed Benefit	Pitfall
Magnetic resonance angiography		
Cervical occlusion	Evaluation for stenosis or occlusion	Overestimates degree of stenosis; poor for "string sign near occlusions"
Intracranial	Aneurysms, AVMs, stenosis	All overcalled and undercalled depending on sequence, observers, scanners
Gadolinium-enhanced	Origins from aorta; less artifact	Venous contamination Low resolution
Spinal	AVMs of spine	Technically limited; takes too long

Continued

TABLE 17-4 Overrated Techniques—cont'd

Name of Technique	Presumed Benefit	Pitfall
Magnetic resonance venography	Venous or sinus thrombosis	Often equivocal Dependent on plane of imaging and stage of thrombosis Poor for cortical vein evaluation
Diffusion-weighted imaging	Specific for strokes	Also positive with some abscesses, demyelination In other lesions T2 shine-through obscures results Hemorrhage may limit application
Perfusion-weighted imaging	Ischemic penumbra	Does not take into account collateral flow, tandem lesions, bilateral disease No universal postprocessing paradigm Potential benefit is hotly contested
Three-dimensional Fourier transformation gradient echo of spine	Thin sections to assess foraminal disease	Badly overestimates disease Susceptibility artifact, especially postoperatively Unable to evaluate signal abnormality in cord
Computed tomography "stroke windows"	Supposed to show low-density strokes better	Never mind—send to diffusion-weighted imaging (above); irreproducible
CSF pulsation studies	To show effect of tonsils on CSF blood flow at cervicomedullary junction; evaluation for fibromyalgia/chronic fatigue syndrome	No one knows what normal looks like High rate of positive Munchausen
Magnetic resonance spectroscopy	Increased specificity in multiple diseases	Everything except Canavan has low N-acetylaspartate, high choline
Functional magnetic resonance imaging	Localizes eloquent cortex	Movement degrades images Neovascularity of tumor, AVMs shift venous drainage
Postgadolinium magnetization transfer imaging	Shows more "enhancing" lesions in brain	False-positive results Many of these lesions are bright before gadolinium Increases the echo time (TE), decreases the available number of slices Need pregadolinium scans, takes longer

AVM, Arteriovenous malformation; *CSF,* cerebrospinal fluid.

Index

Page numbers followed by *b* indicate boxes; *e,* electronic content; *f,* figures; and *t,* tables.